3rd International Conference on Materials Engineering and Applied Mechanics 2024

Journal of Physics: Conference Series Volume 2808

Taiyuan, China
15-17 March 2024

ISBN: 979-8-3313-0217-7
ISSN: 1742-6588

Printed from e-media with permission by:

Curran Associates, Inc.
57 Morehouse Lane
Red Hook, NY 12571

CURRAN ASSOCIATES INC. proceedings.com

Some format issues inherent in the e-media version may also appear in this print version.

Institute of Physics
Dirac House, Temple Back
Bristol BS1 6BE UK

Phone: 44 1 17 929 7481
Fax: 44 1 17 920 0979

techtracking@iop.org

Additional copies of this publication are available from:

Curran Associates, Inc.
57 Morehouse Lane
Red Hook, NY 12571 USA
Phone: 845-758-0400
Fax: 845-758-2633
Email: curran@proceedings.com
Web: www.proceedings.com

TABLE OF CONTENTS

Preface

Peer Review Statement

Investigation of the interfacial bonding properties of NiAl-based composites 1
Biao Sun, Tong Cao, Bo Qiu, Geng Zhang, Feng Ding

Compressible polybenzoxazine high-strength aerogel with a linear structure for thermal insulation 6
Xin Long, Ji Xue, Jianan Qin, Jubo Tang, Linnan Bi, Ying Lin, Xiongbang Wei, Jiaxuan Liao

Performance analysis on VPTIG repair welding joint of 2219 aluminum alloy 13
Xian Li, Yonglun Song, Zhenyang Lu

Effect of laser power on microstructure and properties of Ti-Al/WC composite coatings 23
Hongwei Liu, Baoquan Dong, Xun Gong, Haojie Chen, Haichao Jiao, Hui Zhao, Haitao Zhao, Li Mei

Modification and characterisation of nickel-based alloy materials for ultra-supercritical power generation 29
Xiaolong Zhang

The analysis and simulation of fracture behavior of fused quartz under dynamic loading 41
Wenbo Duan, Bingyao Zhao, Ping Zhou, Dongming Guo

Research on ballistic protection characteristics of perforated plates made of different steel materials for 12.7 mm penetrator projectile 48
Anshun Shi, Xiaowen Hong, Hongwei Han, Kailang Xia, Xuefei Fan, Chao Qin, Sujie Sun

Effect of ZnO on tribological properties of MoS2 dry film lubrication coatings 60
Tao Duan, Qin Wang, Peng Cheng, Dewei Mi

Ti3AlC2-modified NiCrAlY as a metallic bond coat for thermal barrier coatings: a study on high-temperature oxidation resistance 71
Wenbin He, Junliang Li, Kun Liu

Utilization of waste silty soil as fine aggregate in controlled low-strength material 78
Haomiao Qiu, Senhua Jiang, Lei Shou

Dynamic characteristics of dry-type transmission tower under wind-sand load 86
Fule Li, Binxia Yuan, Jingjing Dai

Mechanical properties of workpieces based on 3D printing 93
Mingxia Zhao

Analysis of hydrodynamic performance of underwater accompanying robot based on CFD 99
Yang Chen, Fanghua Liu, Jiawei Shao, You Miao, Yongxin Cao

The effect of the propellant type size on combustion performance of the multi-nitro ester propellant 111
Tao Wang, Ling Chen, Weidong He

Mode I interlaminar fracture toughness of glare laminates by single cantilever beam 117
Xiaoge Hua, Zheng Tian

Investigation of oxide inclusion films in cast superalloy impeller 130
Lihui Zhang, Jun Chen, Weijie Xing, Wenjin Wu, Fengwei Guo, Qiang Zhang, Changkui Liu

Mass-controlled contact variable stiffness elastic metamaterials for bandgap tuning 135
Ronghao Lyu, Kai Zhang

Study on seismic performance of cold-formed steel frame structures based on software ABAQUS 142
Futing Pan, Junlin Gao, Ying Wang, Tao Wang

Plasticity constitutive theory considering material length parameters 148
Zihan Gong, Gaoxiao Han, Lidan Li, Hao Chen, Wengui Zhang

Buckling characteristics of the multi-crack egg-shaped shell under axial pressure 154
Hao Lv, Yongmei Zhu, Suzhou Zhang

Analysis of the reasons for the fracture of the heat-engine plant turbine vane of the power plant feedwater pump 160
Jun Gao, Liming Xie

Research on the effect of mg doping on the properties of ZnO nanorods 166
Xue Tian, Yingying Yang, Yuanping Sun

Analysis of cracking causes of angle steel in 220 kV transmission tower 172
Pingcheng Fan, Liming Xie, Qingyun Hao, Hao Chen

Effect of acoustic filler on sound insulation properties of millable polyurethane rubber 179
Xiao Han, Zhaoyang Zhang, Ling Zhang, Wei Cheng

Microstructure and properties of Cu-Fe alloy prepared by spray forming 187
Su Huang, Baosen Lin, Zerui Zhang, Jianping Li

Research on multiaxial fatigue life prediction of lifting equipment based on symbol corrected equivalent stress spectrum 192
Li Chen, Keqin Ding

Acceleration of the synthesis of single B-YB2SI2O7 feedstocks through modifying the grain size of SIO2 200
Zelong Wang, Donghui Guo, Xinlei Jia, Shun Wang, Tao Zheng, Runze Jin, Zhiliang Zhou, Ning Zhou, Baosheng Xu

Experimental study on frost resistance of steel fiber reinforced concrete 209
Jiuyang Li, Yuepeng Zhu, Jingwei Luo, Xinmei Fan, Xiaoyu Wang, Li Chen, Fan Li, Zicheng Zhang

Effect of salt concentration on the antibubble lifetime 215
Shuheng Liu, Ji An, Ziqi Liu, Guilin Zhao, Rui Gao

Study on fatigue life of abutment based on multi-sample test method 223
Liangxia Zhang, Yunjing Tian, Mengsun Qiao, Yongning Li, Jinni Luo, Zhifang Hu, Lijuan Zhao

Study on the properties of solidified materials of municipal waste incineration ash cementitious materials 230
Ke Wang, Zhan Weng, Jinghui Zhang, Jiahui Du, Mingfu Li, Tiejun Liu

Study on physicochemical properties of corn starch modified by dielectric barrier discharge (DBD) plasma 239
Ziwen Wang, Jinhai Niu

Experimental study on the high-velocity impact resistance of Vitrimer-based carbon fiber composites 245
Ziyue Lin, Peng Chen

Interfacial morphology and reliability of GaN nanocomposite Sn-Ag-Cu lead-free braze joints 252
Shaoxuan Liu, Min Qu, Hang Li, Yuan Liu, Yan Cui

Effect of process parameters on the temperature field of laser sintering of nylon PA2200 powder 259
Kai Liu

Stability of viscoelastic fluid conveying pipe 264
Yu Shen, Bo Wang

Enhanced photoluminescence efficiency and stability of CsAgCl2 by a NaCl-assisted synthesis 272
Jili Zhang, Sen Li, Peng Yang, Yusheng Zhai, Ruiliang Zhang, Lin Zhang, Long Sun, Zhan Su, Yan Gao, Zhifeng Zhang

Effect of calcium sulfate whiskers on mechanical properties and enhancement mechanism of metakaolin-based polymer mortar 282
Xiong Zheng, Gang Zhao, Ming Jiang, Kuixiang Guo, Yu Cui, Jinsong Wang

Development of a large-area array multi-point synchronous explosive circuit using sheet explosive detonated by a mild detonating fuse 290
Zhiyun Guo, Qiang Lu, Yang Ding, Jin Li, Liangyong Zhang

Biaxial tension test of soft materials 298
Minxing Li, Ming Yang, Huiming Wang, Ronghao Bao

DLP 3D printing of high-resolution, temperature-responsive hydrogel microneedles with rapid customization capability 306
Xinmeng Zhou, Huan Liu, Huanjun Li

Preparation and performance verification of calcium silicate composite electrical fireproof materials 314
Yin Wang, Yu Wang, Dedong Yang, Guanggang Fu, Chong Zheng, Fanfang Kong, Jingkai Nie, Jun Ji

Research on deformation strain field of vehicle body welding based on laser scanning detection 320
Lina Li, Hongchang Sun, Yuanhao Bao

Application evaluation of TORSION DRILLING string 334
Zengbao Zhang, Jianbo Lv, Xiaoming Zhu, Yuxin Liu

Performance analysis of motorized axles 342
Yan Yan, Peng Hang, Liangqiang Cao, Tao Huang, Jingjing Duan, Yahu Wu

Multi-objective optimization of the wave-proof plate layout of the liquid tank vehicle based on NSGA-II and RSM 350
Houjie Xiao, Hongbin Qin, An He

Research on the failure mechanism of non-metallic oil pipeline flange 356
Yu Weng, Dejun Li, Shuxuan Jiang, Shuang Yang, Gang Wu, Zhian Deng

Experimental study on the friction characteristics of the interface between piles and coral reef rock 363
Shoulong Zhang, Dengfeng Sang, Guanghua Feng, Shiding Su

Optimization design of the ducted fan powered high-lift system based on i-sight platform 370
Sen Mao, Gang Li, Chuangxin Zhao, Qiang Guo, Wenliang Feng, Zongming Xie

Analysis of isothermal elastohydrodynamic lubrication on the rough surface of a new type of trident universal joint 379
Yutian Xu, Yan Yang, Guowei Wang, Jiahong Niu

Research on the process of vertical water entry experiment of truncated cone-shaped projectile 386
Zhiyong Zhang

Analysis of circumferential CO2 arc welding welded shield machine hob to investigate the welding residual stress fields 396
Xiaojie Xue, Jianfu Chen, Shuhua Huang, Liang Zhang, Zhaoping Tian

A novel process on recyclable particle materials used for aggregate columns 408
Mingyuan Wang, Yajun Wang, Zuguo Zhang, Huifeng Zheng

Effect of explosion venting effect on explosion response characteristics of box-shaped cabin 419
Yinliang Ma, Chenli Ji

Preparation of butyl bonded poly (styrene-divinylbenzene) microspheres by Grignard reaction 426
Meijuan Jiang, Fuqiang Jin, Hao Zhang, Xiaolu Yi, Xin Zhao

Static response of MEMS double layered simply supported beams containing quasicrystals 432
Yunzhi Huang, Penghui Wang, Yuepeng Liu

Ease-off modification and tooth contact analysis for near net shape face-gear 438
Yinan Han, Miaojiang Ding, Shuaiqi He, Jianjun Yang

Multi-material optimal design of miura origami folds based on genetic algorithm 444
Yahong Jiang, Jinyu Zhou, Yaping Xue, Haotian Hu

Research on engineering selection methods for electromagnetic lines and insulation materials of UHV converter transformers under extreme margin conditions 451
Xinru Yu, Chao Wu, Huanchao Cheng, Chaohua Wang, Yongfeng Zhao, Junyu Deng

Preparation and autocatalytic properties of CNTs/Co3O4/AP Core-shell composite energetic materials 459
Linpeng Li, Shuhong Ba

Effect of boundary layer thickness and dimensionless wall distance y+ on numerical simulation of heat pipes 465
Wenjie Zhao, Jiajia Deng, Dawen Xue, Lin Xu

Study on constitutive model and mechanical performance test simulation of 6082-T6 based on genetic optimization algorithm 473
Suokun Chen, Qinmin Zhang

Influence of pre-tightening force on bearing capacity of composite cylinder 482
Bo Yang

Establishment and prediction of liquid permeability model for metal porous materials 489
Shengcun Lin, Wenxiu Feng

Influence of high strain rate and catalyst concentration on dynamic impact behavior of vitrimers 496
Rui Xiang, Le An

Numerical study on mechanical responses of AHR and BWD packer under a cyclic loading 503
Xin Ma, Bing Shao, Yongliang Zhang, Xiaomei Wang

Strain-rate dependent dynamic constitutive model of tempered glass 508
Lei Li, Bangxian Liu, Erfeng An

Study on the effect of permanent magnet guideway surface defect on the mechanical properties of multi-surface high-temperature superconducting permanent magnet levitation system with halbach-type guideway 517
Jiahao Peng, Xianfeng Zhao, Yulong Guo

Lamb wave measurement method for in-plane elastic constants of balanced symmetric laminates based on machine learning 525
Quan Xie, Yaolu Liu, Jun Fang

Static and dynamic mechanical behavior and constitutive parameters determination of high nitrogen austenitic stainless steel 536
Xiaowen Hong, Anshun Shi, Yuanyuan Ge, Xuefei Fan, Chao Qin, Guoji Yu, Lujie Wang

Analysis of influencing factors on numerical computation of underwater explosion of slender tapered charge 546
Houlin Fang, Cunxu Liu, Qiang Lu, Quanshi Guo, Shiying Tang, Zhiyun Guo, Dezhi Zhang

Mechanical properties analysis of bending-torsion coupled blades with different skin area angles 553
Xiaoyin Jin, Jiang Liu

Impacts of mandrel parameters on forming quality of tube bending with small bending radius 561
Tao Wan, Jun Fang, Wenxuan Shang, Kelu Wang, Shiqiang Lu

Study on the influence of hole shapes on the fluid and mechanical properties of perforated honeycomb sandwich structures 566
Pengfei Xiao, Chunping Zhou, Yunfeng Luo

Simulation and analysis for residual stresses in anode-supported flat-plate solid oxide fuel cells for ships 572
Zhaoqiang Li, Lingling Wang, Haijun Yang, Wendong Ji, Yushan Zhang, Baocheng Lu

Lightweight design of composite sandwich corrugated structures based on variable density method 578
Ce Tian, Zhimin Tian, Xinwei Cao, Shangwei Dong

Research on improving parallel sequence sampling method in ship type optimization 584
Dongqin Li, Lingfeng Lu

Structural optimization and drag reduction research of biomimetic hydrofoils inspired by pufferfish 592
Kairan Wei, Lei Zhao, Dongpo Zhu, Guizhong Tian

Study on thermo-mechanical coupling model in pantograph-catenary system of rail vehicle 601
Hangyan Jiang, Ning Zhou, Yao Cheng, Haifei Wei, Jingke Yan, Weihua Zhang

Study on the structure of the cowl based on optimal water dynamic characteristics 610
Tian Wang, Mingxuan Li, Jia Tang, Yuhan Liu, Wenbo Li

Experimental study on the flow capacity of shale fractures with different degrees of sand filling 616
Ting Chen, Chunlei Yu, Zhigang Sun, Min Zhang, Qiang Sun, Chenwei Liu

Research on creep mechanical properties of flexible screen functional film layer 628
Yuhua Ma, Weiwei Su, Xin Luo, Yubin Hu, Yiming Yang, Geng Zhou, Yachao Xiao

Study on the correlation and calibration of macro and micro parameters of discrete element model for sandstone 638
Jiangchun Hu, Luyang An, Xuewei Huang, Chenxu Wang

Study on the relationship between reflectivity and thickness of radar-absorbing material 648
Song He, Jun Hu, Jinsong Chen, Qiushi Xi

Simulation study on the dynamic response of X80 steel pipelines under explosive loads 653
Yuntian Wang, Min Yang, Danping Hu, Qiujie Wei, Ying Zhang, Yuzhu Guo

Generalized mixed element method of plane problems in cartesian coordinate and its applications 662
Yunze Tian, Guanghui Qing

Split-tensile strength and micro-structure analysis of hybrid fibers coal gangue concrete 673
Jiuyang Li, Jingwei Luo, Xinmei Fan, Yuepeng Zhu, Xiaoyu Wang, Li Chen, Fan Li, Zicheng Zhang

Study on the effect of polypropylene fiber shape and admixture on the mechanical properties of concrete 679
Jiuyang Li, Xinmei Fan, Yuepeng Zhu, Jingwei Luo, Xiaoyu Wang, Jinpeng Guo, Fan Li, Zicheng Zhang

Numerical simulation of impactor velocity under dynamic mesh-based double-sided pressurization 685
Jialiang Zhou, Tong Tang, Feiyin Li, Shaojie Ma, Huifa Shi, Saisai Liang, Yingzhong Zhang

Study on inertial migration behaviours under secondary flow patterns using a fluid-structure interaction approach 692
Xin Tan, Ying Lin

Numerical simulation of single abrasive grain grinding of silicon carbide surface 701
Zhiquan Guo, Luchao Ma

Analysis of the influencing factors of transmission coefficients based on ZnSe/BeTe heterojunction 711
Xingyu Li

Establishment and validation of mixed solid degenerate shell elements 720
Xinlei Jin, Guanghui Qing

Mechanical properties and mechanism analysis of modified red layer filler with slag powder 729
Yuan Gao, Haoying Niu, Bingbing Zhang

Stress analysis and calculation of the inner anchor structure of pressure-type anchor 741
Jian Ma, Qi Zhang, Hanbo Dan, Feng Pan, Gang Xu, Yong Guo, Qing Lü

Author Index

Journal of Physics: Conference Series **2808** (2024) 011001 doi:10.1088/1742-6596/2808/1/011001

Preface

As we gather to commemorate the momentous occasion of the 2024 3rd International Conference on Materials Engineering and Applied Mechanics (ICMEAAE 2024), held from March 15 to 17, 2024 in Changsha, China (hybrid event), it is with immense pride and gratitude that we present this proceedings, a compilation of the ideas, insights, and innovations shared by our esteemed participants.

The ICMEAAE 2024 attracted a diverse and esteemed group of participants. The conference boasted a robust attendance of over 120 scholars, researchers, and professionals from various fields, including a significant contingent of international delegates hailing from diverse countries across the globe. This diversity of perspectives and backgrounds added immeasurable depth and breadth to the discussions and collaborations that took place during the conference.

The conference agenda was designed to cover a wide range of topics within the fields of materials engineering and applied mechanics. It commenced with opening remark from the conference chair, followed by a series of plenary sessions, speeches, and free discussions. One of the highlights of was the series of keynote speeches delivered by leading figures in relevant domains, including Prof. Xiaobao Yang (South China University of Technology, China), Prof. Richard Yang (Western Sydney University, Australia), Prof. Liang Zhang (Xiamen University of Technology, China) and Prof. Shahid Hussain (Jiangsu University, China). These speakers, representing both academia and industry, provided a unique perspective on the current trends, challenges, and future directions of research in their respective areas. Their insights were met with keen interest and lively discussions, fostering a vibrant intellectual atmosphere throughout the conference.

The call for papers generated a robust response from the academic community. The proceedings of ICMEAAE 2024 are a testament to the quality and diversity of the submissions received. These papers cover a range of topics, including but not limited to Materials Science and Engineering, Materials Failure and Protection, Materials Synthesis and Processing Processes, Material Properties, Vibration Science, Experimental Mechanics of Engineering Structure, etc. The authors, representing a diverse array of institutions, have presented their research in a clear and concise manner, making the proceedings a valuable resource for both scholars and practitioners alike.

In conclusion, the 2024 3rd International Conference on Materials Engineering and Applied Mechanics was a resounding success, bringing together experts from around the world to discuss and advance the latest research in materials engineering and applied mechanics. We are grateful to all the participants, speakers, and authors for their contributions to this momentous event, and we look forward to the continued growth and development of these important fields in the coming years.

The Committee of ICMEAAE 2024

Committee Member

Conference Chairman
Assoc. Prof. HUI Kwun Nam, University of Macau, Macau, China

Publication Chair
Prof. Richard (Chunhui) Yang, Western Sydney University, Australia
Prof. Xiaobao Yang, South China University of Technology, China

Technical Program Committee Chair
Assoc. Prof. Alan Lim Teik Cheng, Singapore University of Social Sciences, Singapore

Local Organizing Chair
Prof. Shahid Hussain, Jiangsu University, China

Technical Program Committee
Prof. Zishun Liu, Xi'an Jiaotong University, China
Prof. Enwei Chen, Hefei University of Technology, China
Prof. Donghai Lin, Shanghai Polytechnic University, China
Prof. Ruixiang Bai, Dalian University of Technology, China
Prof. Guiyin Li, Guangdong University of Petrochemical Technology, China
Prof. Asher Yahalom, Ariel University, Israel
Researcher Kun Liang, Chinese Academy of Sciences, China
Assoc. Prof. Jianxun Zhang, Xi'an Jiaotong University, China
Assoc. Prof. Xiaoyan Zhou, China University of Petroleum, China
Assoc. Prof. Yang Jiao, Transport and Energy Arizona State University, America
Assoc. Prof. Muhammad Zubair, Information Technology University, Pakistan
Assoc. Prof. Aleksandra Baron-Wiechec, Guangdong Technion-Israel Institute of Technology, Silesia
Assoc. Prof. Taza, City University of Science and Information Technology, Pakistan
Assoc. Prof. Anuj Kumar, University of Petroleum and Energy Studies, India
Assoc. Prof. Engin Derya Gezer, Karadeniz Technical University, Turkey
Dr. AHMED LEGROURI, International University of Grand-Bassam, Morocco
Dr. Hock Jin Quah, Universiti Sains Malaysia, Malaysia

Organizing Committee
Prof. Xiaopeng Li, Shenzhen University, China
Prof. Jixin Zhu, University of Science and Technology of China, China
Prof. Guiyin Li, Guangdong University of Petrochemical Technology, China
Prof. Li Guo, Hunan University, China
Prof. Liping Tang, Southwest Petroleum University, China
Assoc. Prof. Yabin Yang, Sun Yat-Sen University, China
Assoc. Prof. Haoxi Cong, North China Electric Power University, China
Assoc. Prof. Xiang Zhu, Henan University, China

Journal of Physics: Conference Series **2808** (2024) 011002 doi:10.1088/1742-6596/2808/1/011002

Peer Review Statement

All papers published in this volume have been reviewed through processes administered by the Editors. Reviews were conducted by expert referees to the professional and scientific standards expected of a proceedings journal published by IOP Publishing.

- **Type of peer review:** Single Anonymous
- **Conference submission management system:** Morressier
- **Number of submissions received:** 175
- **Number of submissions sent for review:** 140
- **Number of submissions accepted:** 100
- **Acceptance Rate (Submissions Accepted / Submissions Received × 100):** 57.1
- **Average number of reviews per paper:** 3
- **Total number of reviewers involved:** 18
- **Contact person for queries:**
  Name: Xuexia Ye
  Email: xx.ye@keoaeic.org
  Affiliation: AEIC Academic Exchange Information Centre

Investigation of the interfacial bonding properties of NiAl-based composites

Biao Sun[1*], Tong Cao[1*], Bo Qiu[1], Geng Zhang[1] and Feng Ding[1]

[1]School of Mechatronic Engineering, Xi'an Technological University, Xi'an 710021, China

[*]Corresponding author's e-mail: sunb0102@163.com, tongcao@foxmail.com

Abstract. NiAl-based composites with $Ag_2Mo_2O_7$ addition were prepared by spark plasma sintering technology. The microstructure and interfacial bonding characteristics of NiAl-based composites were explored, and the wear mechanism was revealed at room temperature. The interfacial bonding properties were analyzed at the molecular level based on molecular dynamics. The results suggest that after sintering, $Ag_2Mo_2O_7$ transforms into Ag and MoO_3 phases, forming NiAl/Ag and NiAl/MoO_3 interfaces. Plastic deformation and fatigue wear account for the majority of the wear mechanism in NiAl-based composites at room temperature. The binding energy of the NiAl/Ag interface is about 16% higher than that of the NiAl/MoO_3 interface. Among them, the binding energy of the NiAl/Ag interface is 1.759 kcal/mol·Å^2, and the binding energy of the NiAl/MoO_3 interface is 1.475 kcal/mol·Å^2. The fatigue spalling and microcrack of NiAl matrix composites at room temperature are attributed to the poor interfacial bonding properties between NiAl and MoO_3.

1. Introduction

With the development of high technology, modern industry has put forward higher and higher requirements for the service performance of materials under harsh working conditions. NiAl alloys are considered effective high-temperature matrix materials because of their high hardness, low density, and outstanding oxidation resistance [1]. However, the engineering application of NiAl is hindered by its poor high-temperature tribological performance [2]. Studies have shown that the introduction of trace elements can effectively enhance the microstructure of NiAl intermetallic compounds, to improve their mechanical properties at high temperatures.

In recent years, a large number of scholars have improved the elevated-temperature tribological characteristics of NiAl alloy by adding solid lubricants. Feng et al. [3] prepared NiAl matrix composites with $AgNbO_3$ by hot pressing sintering technology and found that $AgNbO_3$ formed Ag, Nb_2O_5, and NbC during the sintering process, which greatly enhanced the elevated-temperature friction reduction behaviors of NiAl-based composites. Yao et al. [4] prepared NiAl-MoO_3/CuO composites by adding MoO_3 and CuO, studied their elevated-temperature tribological behaviors, and revealed the mechanism of friction reduction and wear resistance. Wang et al. [5] prepared NiAl-TiO_2/Bi_2O_3 composite coatings using plasma spraying technology and found that the combined action of $Bi_4Ti_3O_{12}$ and $NiTiO_3$ generated by high-temperature tribo-chemical reactions improved the high-temperature anti-friction and wear resistance of NiAl alloy. However, these studies did not consider the impact of the interface formed between solid lubricants and NiAl matrix on the composites at room temperature (RT). Therefore, it is essential to further explore the bonding characteristics of the interface between the NiAl matrix and the solid lubricant.

In this paper, NiAl-$Ag_2Mo_2O_7$ composites were synthesized. The microstructure and interfacial bonding characteristics of NiAl-$Ag_2Mo_2O_7$ composites were analyzed at RT, and the wear mechanism at room temperature was revealed. Based on molecular dynamics, the interfacial bonding properties between lubricant and matrix in composites were investigated at the molecular level. This study guides the synthesis and development of NiAl-based solid self-lubricating materials.

2. Experimental and calculation methods

2.1 Preparation of composite materials

NiAl composites with 0wt% $Ag_2Mo_2O_7$ (NA0) and 10wt% $Ag_2Mo_2O_7$ (NA10) were prepared by spark plasma sintering technology. The commercially available NiAl intermetallic compound was used as the matrix, and $Ag_2Mo_2O_7$ was used as the solid lubricant which was synthesized based on previous work [6]. The specific ratios are shown in Table 1. Firstly, $Ag_2Mo_2O_7$ powder was added to NiAl powder and fully ground by a planetary ball mill. Then the mixed powder was loaded into a graphite abrasive tool and sintered by a spark plasma sintering (SPS) furnace. The specific sintering process parameters were following the previous work [6]. Finally, the sintered samples were polished to the mirror.

Table 1. Proportioning of NiAl-$Ag_2Mo_2O_7$ composites.

NiAl based composite	Composition (wt%)	
	NiAl	$Ag_2Mo_2O_7$
NA0	100	0
NA10	90	10

2.2 Characterization and testing of NiAl-based composites

The tribological tests were taken on a ball-on-disk tribometer. The polished NiAl-$Ag_2Mo_2O_7$ composite material was used as a disk, and Si_3N_4 was used as a counterpart ball. The test temperature was room temperature (RT), the normal load was 10 N, the linear velocity was 0.2 m/s, the rotation radius was 5 mm, and the total friction test time was 30 min. The worn surface, microstructure, and element distribution of the samples before and after the experiment were analyzed by scanning electron microscope (SEM) and energy dispersive spectrometer (EDS).

2.3 Calculation of interfacial bonding properties of composite materials

The NiAl/Ag and NiAl/MoO_3 interface models were established by molecular dynamics software. The interfacial bonding properties between NiAl and Ag, NiAl and MoO_3 were analyzed. Firstly, the crystal structure parameters of NiAl, Ag, and MoO_3 were determined by the ICSD database. Then, according to the principle of minimum surface energy, NiAl/Ag and NiAl/MoO_3 interface structures were constructed. The Smart algorithm is used to optimize the energy and geometric structure of the interface model. The regular ensemble (NVT) and Nose-Hoover thermostat were selected to control the simulated ambient temperature, and the temperature was set to 298 K (25°C). Considering the computational efficiency, the maximum iteration step is set to 1000 steps, the convergence accuracy is Fine, and the Universal force field is applied. Finally, the interface binding energy is derived by calculating the energy of the system after equilibrium.

3. Results and discussion

3.1 The phase composition of composite materials

The RXD pattern of NA10 composites is given in Figure 1. The phase of $Ag_2Mo_2O_7$ is hardly observed, while new phases of Ag (No. 04-0783) and MoO_3 (No.05-0508) are found, which is attributed to the in-

Journal of Physics: Conference Series **2808** (2024) 012001 doi:10.1088/1742-6596/2808/1/012001

situ decomposition of $Ag_2Mo_2O_7$ to form Ag and MoO_3 phases during high-temperature sintering. This phenomenon is similar to the previously reported studies [7].

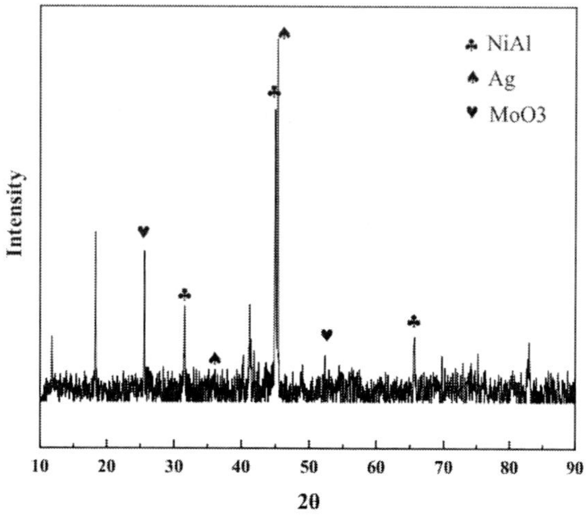

Figure. 1. The XRD diffraction pattern of NA10.

Figure 2 shows the BSE diagram of NA0 and NA10 composites. From the BSE diagram, it can be found that the NA10 composite material with $Ag_2Mo_2O_7$ shows more obvious compactness than the NiAl (NA0) composite material without $Ag_2Mo_2O_7$, and presents different contrast regions. Combined with the XRD spectrum, it is judged that the white area is the Ag-rich zone, the black gray area is the MoO_3 zone, and the dark gray is the NiAl matrix zone.

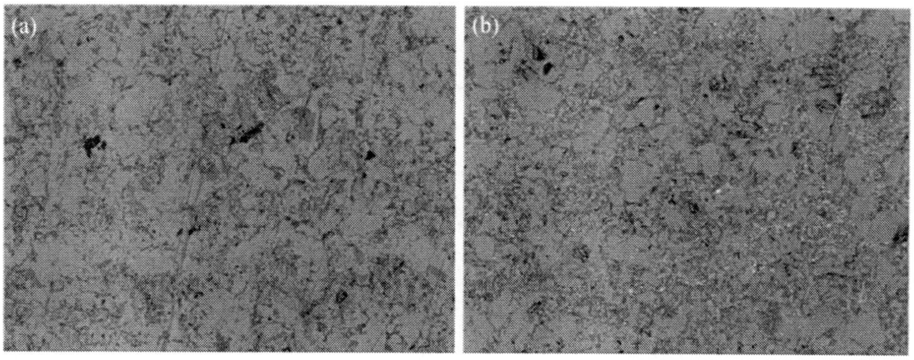

Figure 2. BSE diagram of NiAl-based composites: (a) NA0, (b) NA10.

3.2 Wear morphology of composite materials

Figure 3 presents the SEM graphics of the worn surface of NA10 composites at RT. From Figure 3a, it can be observed that there are numerous peeling pits on the worn surface, and a large amount of microcracks are also found as displayed in enlarged Figure 3b. This indicates that the mechanical characteristic of NiAl- $Ag_2Mo_2O_7$ composites were influenced markedly by the presence of 10wt% $Ag_2Mo_2O_7$. This is due to the formation of NiAl/Ag and NiAl/MoO_3 interfaces with poor compatibility through the sintering process. Thus, the wear mechanism of NiAl- $Ag_2Mo_2O_7$ composites at RT is mainly fatigue wear. It is essential to explore the interface bonding properties of NiAl-$Ag_2Mo_2O_7$ composites.

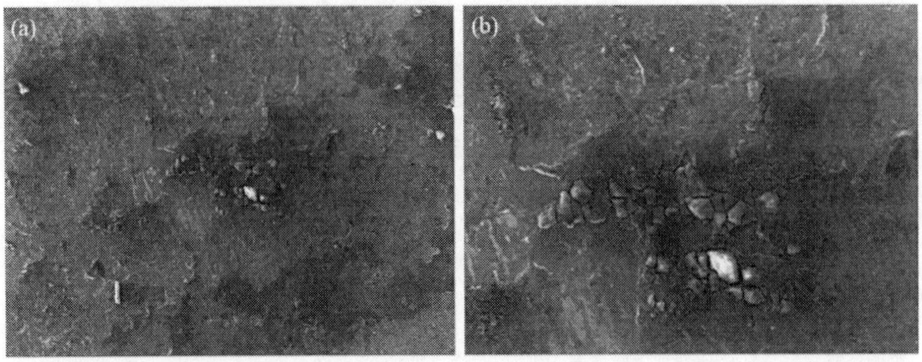

Figure 3. Wear morphology of NA10 (a) and corresponding enlarged image (b).

3.3 Interface bonding performance of composite materials

Figure 4 shows the molecular dynamics models of NiAl/Ag and NiAl/MoO_3 interfaces. The NiAl/Ag model contains 292 atoms, and the NiAl/MoO_3 model contains 288 atoms. The total energy of the system is obtained by molecular dynamics calculation of the interface model. The calculation formula of the interfacial binding energy of the two-layer material is as follows:

$$E_{interface} = \frac{E_{Total} - (E_{up} + E_{down})}{A} \tag{1}$$

where E_{Total} is the total energy of the interfacial structure, E_{up} is the total energy of the upper layer material, E_{down} is the total energy of the lower layer material, and A is the interfacial area.

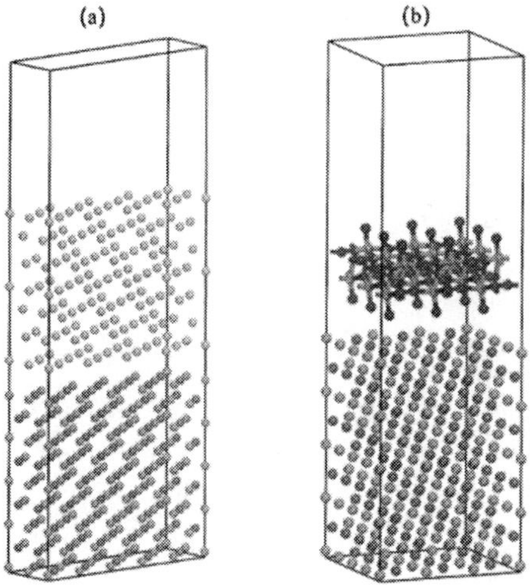

Figure 4. NiAl/Ag (a) and NiAl/MoO_3 (b) interface models.

The interface binding energy is an important index to judge the stability of the interface. The larger the interface binding energy is, the more stable the interface is. Usually, when the composite material is destroyed, the crack or fracture phenomenon occurs at the lowest interface binding energy. Table 1 gives the interface binding energy of NiAl/Ag and NiAl/MoO_3 interface models at RT. It can be found in Table 2 that the NiAl/Ag interface has a higher binding energy than the NiAl/MoO_3 interface, which is about 16% higher. This indicates that NiAl and Ag in the composite material have good bonding properties and are not easy to be damaged. The dangerous interface is mainly the NiAl/MoO_3 interface, which explains why fatigue spalling and microcracks exist on the worn surface of NiAl-$Ag_2Mo_2O_7$ composites at room temperature.

Table 2. Calculation results of interface binding energy.

Interface models	E_{Total} (kcal/mol)	E_{up} (kcal/mol)	E_{down} (kcal/mol)	A (Å^2)	$E_{interface}$ (kcal/mol·Å^2)
NiAl/Ag	245.922	81.329	-40.443	116.518	1.759
NiAl/MoO_3	3198.031	2910.404	86.452	136.315	1.475

4. Conclusions

In this paper, NiAl- $Ag_2Mo_2O_7$ composites were synthesized using the SPS technique. The microstructure and interfacial bonding characteristics of NiAl-$Ag_2Mo_2O_7$ composites were researched at RT. The wear mechanism was illustrated, and the bonding properties of the interface in the composites were explored from the molecular level. The key conclusions are given below:

(1) NiAl/Ag and NiAl/MoO_3 interfaces were formed in NiAl-based composites during the sintering process.

(2) Plastic deformation and fatigue wear account for the majority of the wear mechanism in NiAl-$Ag_2Mo_2O_7$ composites at RT.

(3) The binding energy of the NiAl/Ag interface is about 16% higher than that of the NiAl/MoO_3 interface. The micro-cracks on the worn surface are caused by the poor interfacial bonding among MoO_3 and NiAl.

Acknowledgments

This work is financially supported by the National Natural Science Foundation of China (52305417, 52102075); Natural Science Basic Research Program of Shaanxi Province (2023-JC-YB-347, 2023-JC-QN-0538, 2023-JC-QN-0481); Scientific Research Program Funded by Shaanxi Provincial Education Department (23JK0489); Science and Technology Planning Project of Bei Lin District, Xi'an (GX2316). We are also grateful to the reviewers for their constructive comments.

References

[1] Zhu S, Cheng J, Qiao Z, Yang J. (2019) High-temperature solid-lubricating materials: A review. Tribol Int,133: 206–23.

[2] Sheng LY, Yang F, Xi TF, Guo JT. (2013) Investigation on microstructure and wear behavior of the NiAl–TiC–Al_2O_3 composite fabricated by self-propagation high-temperature synthesis with extrusion. J Alloys Compd, 554: 182–8.

[3] Feng X, Lu C, Jia J, Xue J, Wang Q, Sun Y, et al. (2020) High-temperature tribological behaviors and wear mechanisms of NiAl–NbC–Ag composites formed by the in-situ decomposition of $AgNbO_3$. Tribol Int, 141: 105898.

[4] Yao Q, Jia J, Chen T, Xin H, Shi Y, He N et al. (2020) High-temperature tribological behaviors and wear mechanisms of NiAl-MoO_3/CuO composite coatings. Surf Coat Technol, 395: 125910.

[5] Wang X, Feng X, Lu C, Yi G, Jia J, Li H. (2018) Mechanical and tribological properties of plasma sprayed NiAl composite coatings with addition of nanostructured TiO_2/Bi2O_3. Surf Coat Technol, 349: 157–65.

[6] Sun B, Ding F, Qiu B, Zhao Y, Guo J. (2023) High-temperature tribological properties of NiAl-$Ag_2Mo_2O_7$ composite coatings prepared by spark plasma sintering. Surface and Coatings Technology, 456: 129290.

[7] Liu E, Bai Y, Gao Y, Yi G, Jia J. (2014) Tribological properties of NiAl-based composites containing Ag_3VO_4 nanoparticles at elevated temperatures, Tribol. Int, 80: 25–33.

Compressible polybenzoxazine high-strength aerogel with a linear structure for thermal insulation

Xin Long[1,2], Ji Xue[2], Jianan Qin[2], Jubo Tang[2], Linnan Bi[1,2], Ying Lin[2], Xiongbang Wei[2], Jiaxuan Liao[1,2,*]

[1]School of Materials and Energy, University of Electronic Science and Technology of China, Chengdu, 611731, P. R. China
[2]The Yangtze Delta Region Institute (Quzhou), University of Electronic Science and Technology of China, Quzhou, Zhejiang 324003, P. R. China

[*]Corresponding author's e-mail: jxliao@uestc.edu.cn

Abstract. The vulnerable structures have been the biggest barrier to the progress and use of aerogels, making it impossible to use aerogels alone in practical environments. The reversible sacrificial hydrogen bonding system in polybenzoxazine can greatly affect the strength function, providing a new approach for designing high-strength aerogels. We achieve a long-chain structured benzoxazine monomer as a precursor and a polybenzoxazine aerogel with excellent mechanical performance. The aerogel achieves a high compression strength (18.8 MPa) and a high fracture strain (78.0 %) at a 0.252 g/cm^3 density. Furthermore, the aerogel also exhibits a low room temperature heat conductivity coefficient (0.03343 W/(m·K)) and hydrophobicity (water contact angle of 142.7°), demonstrating its high potential for use in the fields of construction, industry, space flight aircraft, etc.

1. Introduction

With the deepening of global environmental awareness and the gradual aggravation of the energy crisis, how to continuously improve the efficiency of energy use has become a major problem that can not be ignored in the development process, and reducing ineffective heat loss is one of the most critical links [1-3]. Aerogels have a 3-dimensional nanostructured porous network, which simultaneously exhibits zero convection heat transfer, infinite radiation shielding, and long mean free path effects [4]. The resulting thermal conductivity can rival or even be lower than that of stagnant air, making it recognized as the most effective thermal insulation material at the nanoscale. However, despite this, the inherent limitations of the aerogel's own structure have led to many shortcomings in mechanical properties [5]. These factors make aerogels difficult to apply in practical environments, and they can only exhibit their thermal insulation effectiveness when used as a carrier for other materials after pulverization. In this doped manner, the aerogel itself does not act as the main component, resulting in limited thermal insulation effectiveness.

Although researchers have demonstrated the feasibility and rationality of organic polymers as a research system for high-strength aerogels [6], the increase in strength of these aerogels relies too much on density increase. The density is positively correlated with the thermal conductivity of the solid phase, making it difficult to simultaneously optimize the insulation performance and mechanical properties. However, polybenzoxazine can be considered as an exception. The polybenzoxazine structure is rich in N, and O atoms, which can form a sacrificial bond cross-linking network to affect

Journal of Physics: Conference Series **2808** (2024) 012002 doi:10.1088/1742-6596/2808/1/012002

the mechanical properties [7, 8]. Hydrogen bonds, as a reversible sacrificial bond, can preferentially consume energy before the chemical bond breaks, and can also achieve stress redistribution to achieve the purpose of material self-protection [9]. Currently, polybenzoxazine aerogels have high strength, but their deformability is somewhat lacking, and the large number of hydrogen bonds formed leads to hydrophilic drawbacks. Based on this, this study has designed a completely new form of long-chain polybenzoxazine aerogel to further overcome these shortcomings.

2. Experimental

2.1 Materials

Dodecanediamine was supplied by Shanghai Yuan Ye Biotechnology Co., Ltd. Concentrated hydrochloric acid was supplied by Chengdu Colon Chemical Co. LTD. Formaldehyde, p-Cresol, N, N-Dimethylformamide, n-Hexane, and Ethyl alcohol were supplied by Chengdu Cowezhuo Technology Co., LTD. CO_2 was supplied by Chengdu Qiaoyuan Gas Co. LTD.

2.2 Preparation of the polybenzoxazine aerogel

6.0 g of dodecylamine, 3.6 g of formaldehyde, and 6.5 g of p-cresol are reacted in 50 ml of N, N-dimethylformamide, and the mixture is subjected to a hydrothermal reaction at 120°C for 6 hours. The product obtained by washing with n-hexane, followed by drying, is the benzoxazine monomer. Subsequently, 12.0 g of the monomer is dissolved in 80 ml of N, N-dimethylformamide, and after complete dissolution, 4 ml of concentrated hydrochloric acid is slowly added dropwise. Stirring is immediately stopped after the completion of the addition, and gelation occurs within approximately 2 hours. After standing at room temperature for 48 hours, the obtained gel is subjected to impurity removal in 200 ml of N, N-dimethylformamide for 12 hours. The gel is then subjected to solvent exchange in 200 ml of ethanol, with each exchange lasting 12 hours and repeated 3 times. Finally, the alcogel is subjected to supercritical CO_2 drying at 50°C and 12 MPa for 8 hours to obtain the polybenzoxazine aerogel.

2.3. Characterization

The test contents and characterization equipment used in this research are shown in the table below.

Table 1. Test Items and Corresponding Characterization Equipment

Serial Number	Test Item	Characterization Equipment
1	H1-NMR	Ascend Aeon900
2	SEM	GeminiSEM 500
3	BET and BJH	Rise-1010 s Analyzer
4	Porosity	American Mike 9620 Mercury Porosimetry
5	Density	XS204 Density Balance
6	TGA	Setline TGA Thermal Gravimetric Analyzer
7	Compression Stress-strain Curve	SUST CMT-1104 Universal Tester (10 mm/min)
8	Thermal Conductivity	TPS2500 Analyzer
9	XPS	Escalab 250xi Photoelectron Spectrometer
10	Water Contact Angle	Kino SL250
11	FTIR	ThermoFisher i5
12	Mass	ES-E Semi-microbalance

Journal of Physics: Conference Series **2808** (2024) 012002 doi:10.1088/1742-6596/2808/1/012002

3. Results and discussion

Firstly, it is necessary to verify the composition of the synthesized benzoxazine monomer. As shown in Figure 1, the proton peaks in the ranges of 1.26~2.56, 3.70~5.01, and 6.78~7.28 ppm correspond to $-CH_2-$ radicals between the oxazine rings and the peaks of benzyl groups, $-CH_2-$ radicals in the oxazine rings, and the H atoms on the benzene rings, demonstrating the successful preparation of the described benzoxazine monomer.

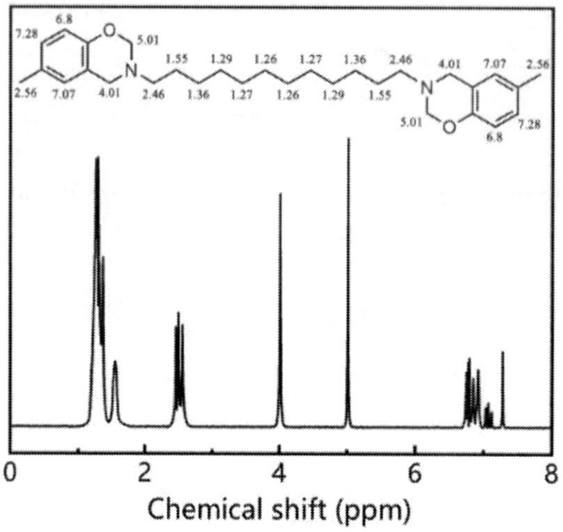

Figure 1. HNMR spectrum and constitutional formula of the benzoxazine monomer.

As shown in Figure 2a, the polybenzoxazine aerogel exhibits excellent compressive mechanical properties, with fracture occurring at a compression strain of 78.0%, corresponding to a high compressive strength of 18.8 MPa. Compared to aerogels with similar density such as SiO_2 aerogels, TiO_2 aerogels, and Al_2O_3 aerogels, the strength performance of the polybenzoxazine aerogel has significant advantages [5]. The long-chain structure between the oxazine rings effectively generates high degrees of freedom of motion, allowing the polybenzoxazine aerogel to exhibit excellent deformability even though it is still composed of nanoparticles (Figures 2b~2d).

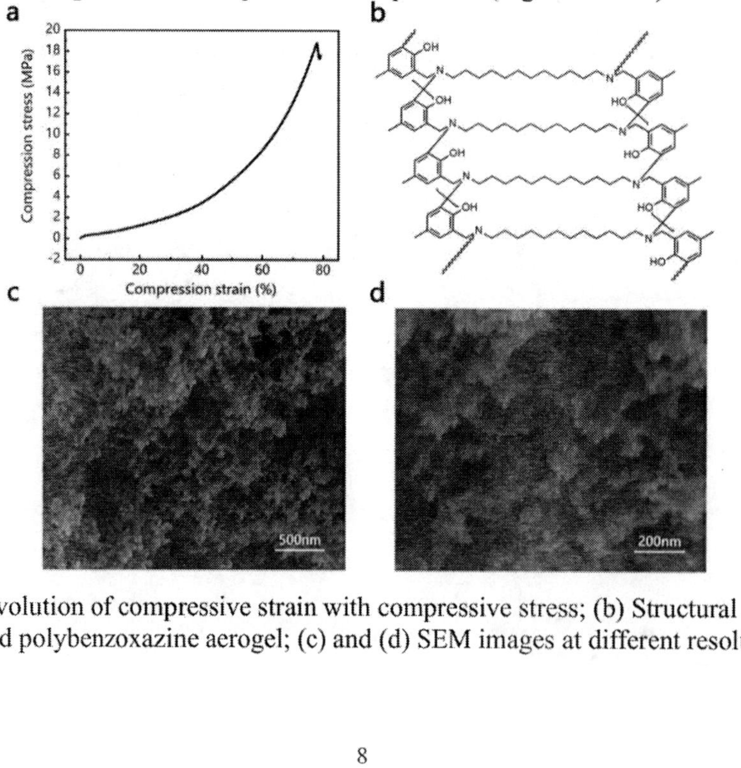

Figure 2. (a) Evolution of compressive strain with compressive stress; (b) Structural formula of the prepared polybenzoxazine aerogel; (c) and (d) SEM images at different resolutions.

Subsequently, we carry out a preliminary verification of the type of hydrogen bonds formed. During the formation of hydrogen bonds, the proton H can covalently polarize the N and O atoms as the acceptors, weakly increasing their binding energy [9]. Therefore, we first use XPS to verify the existence of hydrogen bond structures in the polybenzoxazine aerogel (Figure 3a). The peak binding energies of the C-C, C-N, and C-OH bonds are 284.5, 285.6, and 286.2 eV, respectively (Figure 3b). The peaks of the C-N and C-N···H bonds are 399.6 and 401.7 eV, respectively (Figure 3c), while the peak binding energy of the C-O bond is located at 532.8 eV (Figure 3d), providing preliminary evidence for the formation of a hydrogen bond network.

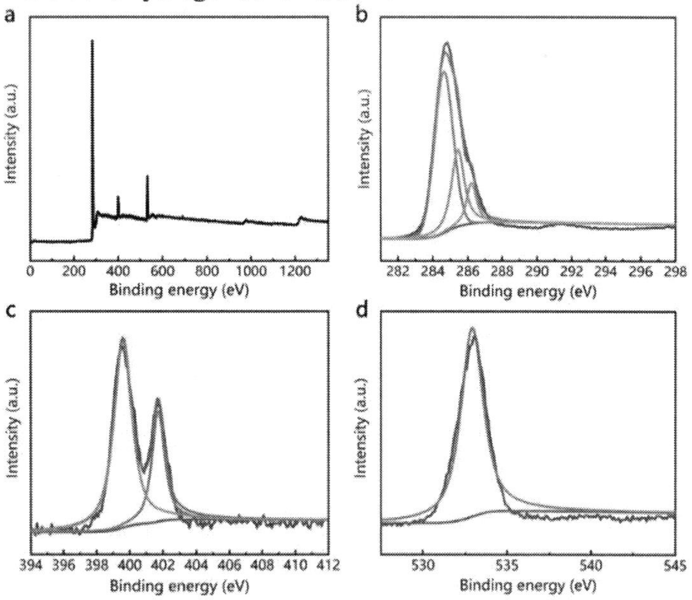

Figure 3. XPS spectrum of the polybenzoxazine aerogel. (a) full spectra; (b)~(d) C_{1s}, N_{1s} and O_{1s} spectra.

Based on this, we use FTIR to quantitatively analyze the formed hydrogen bond system (Figure 4a). First, we transform the 2000~3700 cm^{-1} range into an absorbance curve by using the Beer-Lambert law, and then perform peak fitting by using a Lorentzian Gaussian function, with the results shown in Figures 4b and 4c. In the polybenzoxazine aerogel, there are often -OH···π, -OH···O, -OH···N, and the -O-NH^{+} hydrogen bonds formed by proton transfer equilibrium from Mannich base (in addition to OH···O is the intermolecular hydrogen bond, and the rest are intramolecular hydrogen bonds), with corresponding wavenumbers of approximately 3550, 3400, 3200, and 2830 cm^{-1}, respectively [10]. These different hydrogen bonds form a stable sacrificial bond network, effectively improving the strength function.

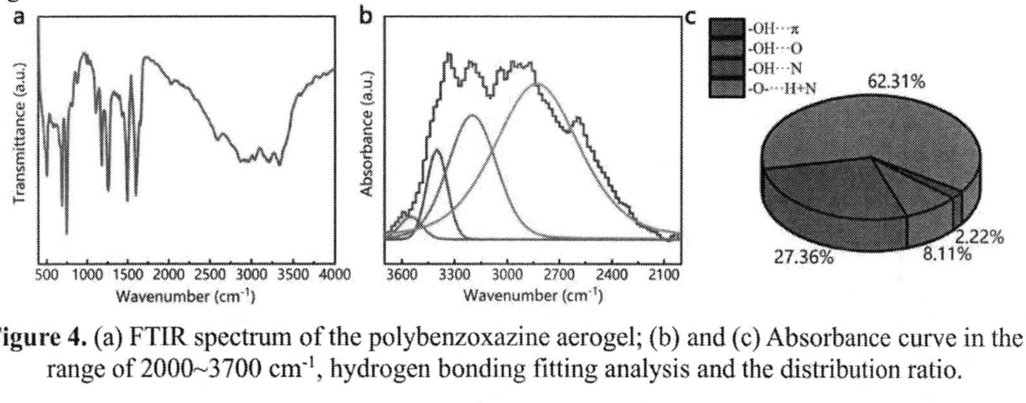

Figure 4. (a) FTIR spectrum of the polybenzoxazine aerogel; (b) and (c) Absorbance curve in the range of 2000~3700 cm^{-1}, hydrogen bonding fitting analysis and the distribution ratio.

As described in Figure 5a, the adsorption isotherm shows a significant increase in the adsorption capacity when the relative pressure P/P_0 is 0~0.05, which is caused by microporous pores. When the P/P_0 is 0.85~1.0, the adsorption capacity suddenly increases without reaching a saturation plateau, showing the coexistence of mesopores and macropores. The analysis of Figure 5b exhibits that the mean pore diameter of the polybenzoxazine aerogel is 39.8 nm (Figure 5b), falling within the mesoporous range of 2~50 nm, effectively suppressing the convective heat transfer. Combined with the low density and high porosity characteristics (Figure 5c), the aerogel has a low thermal conductivity of only 0.03343 W/(m·K). As the temperature continues to increase while maintaining thermal stability (Figure 5d), the increase in thermal conductivity is also relatively slow (Figure 5e). At the same time, a sample with a height of 2.13 cm is placed on a heating stage at 250°C. 15 min before, the top surface temperature gradually increases due to solid-phase heat conduction, until the transmission reaches saturation, and the upper temperature remains basically unchanged. At 30 min, the upper temperature is only 67.1°C, demonstrating the effectiveness of thermal insulation in actual environments (Figure 5f).

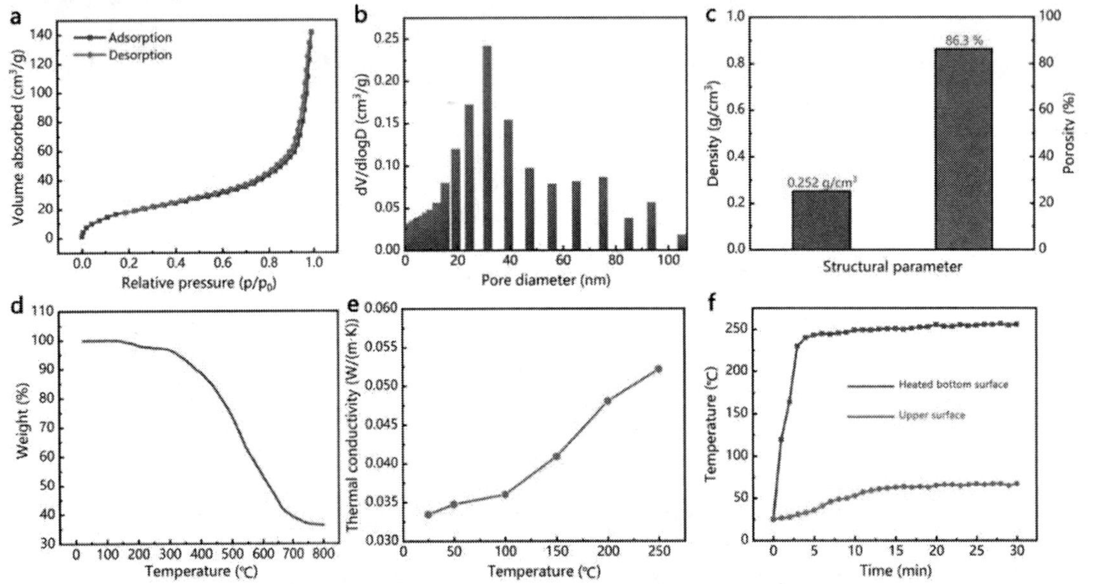

Figure 5. Thermal insulation performance and related structural analysis. (a) BET curve; (b) BJH curve; (c) Density and porosity; (d) Thermogravimetric curve; (e) Evolution of heat conductivity coefficient with temperature; (f) Evolution of the heating base temperature and the top surface temperature over time during the heating process.

When water infiltrates or evaporates from the aerogel, the resulting capillary action can greatly disrupt the three-dimensional connected nanonetwork backbone of aerogels [11]. Therefore, excellent hydrophobic performance is the basic guarantee for the long-term application of aerogels. With the support of long-chain oxazine rings and benzyl groups, the polybenzoxazine aerogel exhibits excellent hydrophobicity, with a high hydrophobic angle of 142.7° (Figure 6a). Even after 10 min, the angle remains stable at 132.1°, indicating good short-term stability (Figure 6b). In the laboratory environment of 25°C and 80% RH, the polybenzoxazine aerogel reaches saturation in about 48 h with only around 0.8 % moisture absorption, and then remains essentially unchanged, demonstrating excellent long-term hydrophobic stability (Figure 6c).

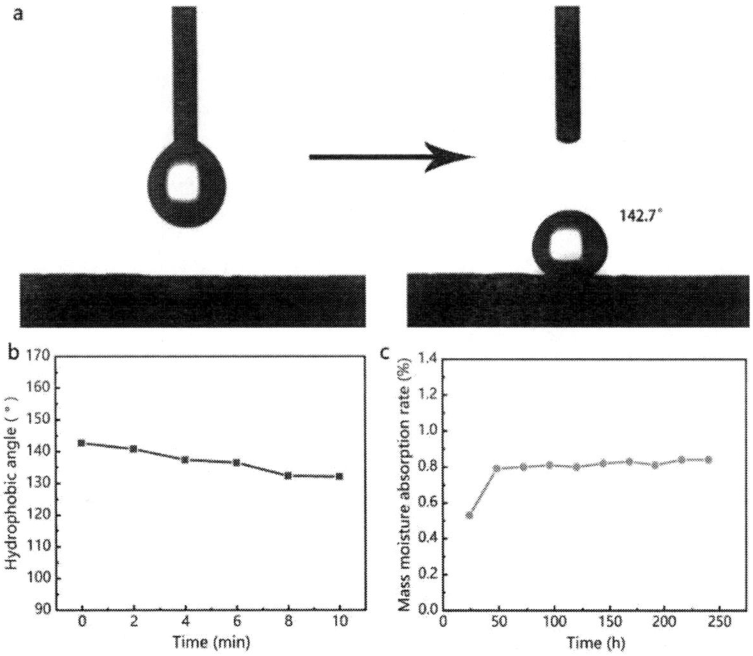

Figure 6. (a) Static contact angle; (b) Evolution of contact angle with time; (c) Change in mass moisture absorption over time.

4. Conclusion

All in all, we achieve a new linear benzoxazine monomer and use it as a precursor to prepare a polybenzoxazine aerogel with excellent mechanical properties, low thermal conductivity, and hydrophobicity. The polybenzoxazine aerogel exhibits a high compression strength of 18.8 MPa and a high fracture strain of 78.0% at a low density of 0.252 g/cm^3. Additionally, it has a 0.03343 W/(m·K) low heat conductivity coefficient and a 142.7° high contact angle. These comprehensive properties not only comprehensively improve the aerogels' mechanical properties in fields related to thermal insulation but also ensure the promising application of polybenzoxazine aerogel as an outstanding thermal insulator.

Acknowledgments

This research was supported by the Municipal Government of Quzhou under Grant (No.2022D015, No.2023D023, 2023D030, No.2022D028, and 2023D002).

References

[1] Li J., Liu Q., He J., Zhang Y., Mu L., Zhu X., Yao Y., Sun C.-L., Qu M. (2024) Aerogel-based solar interface evaporation: Current research progress and future challenges. Desalination, 569: 117068.

[2] Long X., Wu X., Wei X., Yu J., Wang S., Zhou L., Liao J. (2023) SiO2 Nanostructure-Based Aerogels with High Strength and Deformability for Thermal Insulation. ACS Appl. Nano Mater., 6: 14393-14400.

[3] Long X., Yan X., Zhou L., Chen W., Ren S., Qiu Y., Sui L., Wei X., Wang S., Liao J. (2023) Highly Deformable High-Strength SiO(2) Aerogel Designed with an Alternating Structure of Hard Cores and Flexible Chains for Thermal Insulation. ACS Macro Lett., 12: 653-658.

[4] Li J., Ahmad Z., Chen J., Chen H., Tao J. (2024) Fabrication of SiC nanofiber aerogel felt with high-temperature thermal insulation performance. J. Eur. Ceram. Soc., 44: 1923-1931.

[5] Li G., Hu L., Zhang K., Hou S., Fan J. (2023) Synthesis of Al_2O_3-SiO_2 aerogels with low thermal

conductivity and high strength by methyltriethoxysilane as a silica precursor. J. Sol-gel Sci. Techn., 108: 35-46.

[6] Gui Y., Fei Z., Zhao S., Zhang Z., Shao H., Chen J., Li K., Du H., Yang Z. (2023) High-strength and multifunctional honeycomb polyimide aerogel fabricated by a freeze casting-assisted extrusion printing and building block-assembly strategy for sound absorbing metamaterials. Addit. Manuf., 77: 103799.

[7] Liu H., Zhou Z.-y., Li S.-h., Lu B.-a., Zhao H.-w., Liu Q.-q. (2021) The preparation and characterization of high-performance mesoporous carbon from a highly π-conjugated polybenzoxazine precursor. New J. Chem., 45: 8022-8031.

[8] Long X., Tang P., Zhou L., Chen W., Ren S., Qiu Y., Sui L., Wei X., Liao J. (2023) Reversible sacrificial hydrogen bonds enhanced ultra-high strength polybenzoxazine aerogel for thermal insulation. Appl. Surf. Sci., 613: 156004.

[9] Ghaffari M., Masoodi H. R., Bagheri S. (2021) Theoretical insight into the hybridization effect of donor and acceptor atoms on the cooperativity of C-H···N hydrogen bonds. J. Mol. Model., 27: 119.

[10] Kim H.-D., Ishida H. (2002) A Study on Hydrogen-Bonded Network Structure of Polybenzoxazines†. J. Phys. Chem. A, 106: 3271-3280.

[11] Long X., Huang P., Wei X., Yu J., Wang S., Liao J. (2023) Ultra-high strength, highly deformable and superhydrophobic polybenzoxazine@cellulose nanofiber composite aerogel for thermal insulation. Compos. Part A, 175: 107771.

Performance analysis on VPTIG repair welding joint of 2219 aluminum alloy

Xian Li[1], Yonglun Song[2,*] and Zhenyang Lu[2]

[1]Automotive Engineering College Beijing Polytechnic Beijing, China
[2]Beijing University of Technology Beijing, China

*Corresponding author's e-mail: ylsong@bjut.edu.cn

Abstract: When the welded joint of 2219 aluminum alloy has defects, the high-frequency VPTIG is used for welding repair. The mechanism of control is proposed for the repair welding joint, and the process route for satisfying the repair welding joint of 2219-T87 aluminum alloy is formulated to reduce the influence of overall performance on the joint local repair welding position. The research results show that the DC-EN of VPTIG coupled to pulse current at a frequency of 20-25 KHZ is helpful to restrain the metallurgical defects such as pores and refine the grain of the fusion zone, to ensure the comprehensive performance of the repair welding joint.

1. Introduction

2219 aluminum alloy has been widely used in the structure manufacturing of space launch vehicles, mainly using TIG, FSW, etc., to achieve the welding of the initial welding joint. The research on the welding of the initial welding joint is rich, and satisfactory welding performance has been achieved, but there is relatively little research on the repair welding joint in China. For large aerospace aluminum alloy structural parts, in the welding process, due to the weldability of materials, product structure, welding operation, welding environment, and other factors, there will inevitably be holes, slag inclusion, non-penetration, cracks, and other out-of-tolerance defects. For the component with serious defects, if it is used directly, it will be easy to have accidents or not work properly. If it is scrapped directly, it will cause a great waste of manpower, material resources, and financial resources. To ensure the quality of the product, repair welding often becomes a way to repair defects. Based on the barrel principle, the weakest link in welded structural parts is the repair welding joint, which is the key point to evaluate the welding quality of a structural part [2-7].

In this paper, the welding defects of 2219-T87 aluminum alloy were repaired by a welding process in which a 20-25 kHz pulse current was added at DC-EN time of variable polarity TIG. Based on the detection and analysis of microstructure, tensile fracture, and mechanical properties, the optimization method of comprehensive properties of the 2219-T87 aluminum alloy repair welding joint is discussed.

2. Test method

Considering the locality and subtleties of the defect location of 2219 aluminum alloy structural parts, manual repair welding is generally adopted in the repair welding method, using the visibility of the human eye and the flexibility of manual operation to ensure the smooth connection between the repair welding position and the initial welding position. The schematic diagram of the high-frequency VPTIG welding system is shown in Figure 1. The welding heat source is provided by a VPTIG main power supply and a high-frequency coupled pulse power supply. High-purity argon (Ar≥99.999%) is used as the protection gas, and the welding is completed by manual welding. The current waveform output by the welding main power supply and the high-frequency coupled pulse power supply is shown in Figure 2 [1].

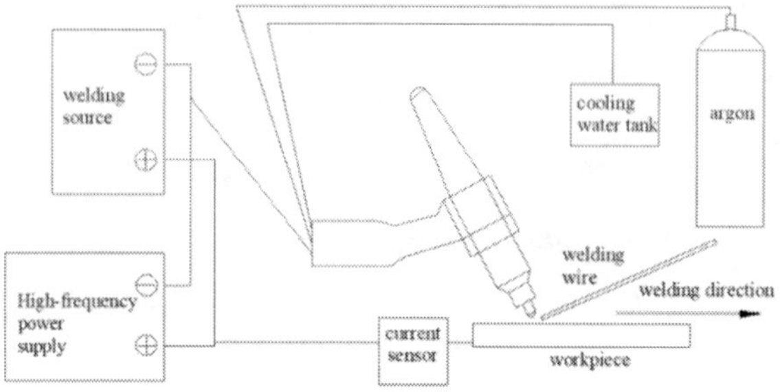

Figure 1. Schematic diagram of welding system.

The repair welding joint for test verification is made of two 300 mm × 150 mm × 5 mm test plates welded by single-sided double-layer high-frequency VPTIG automatic welding. The welding wire grade of 2325 domestic welding wire is selected. The chemical composition of the base metal and the welding wire are shown in Table 1.

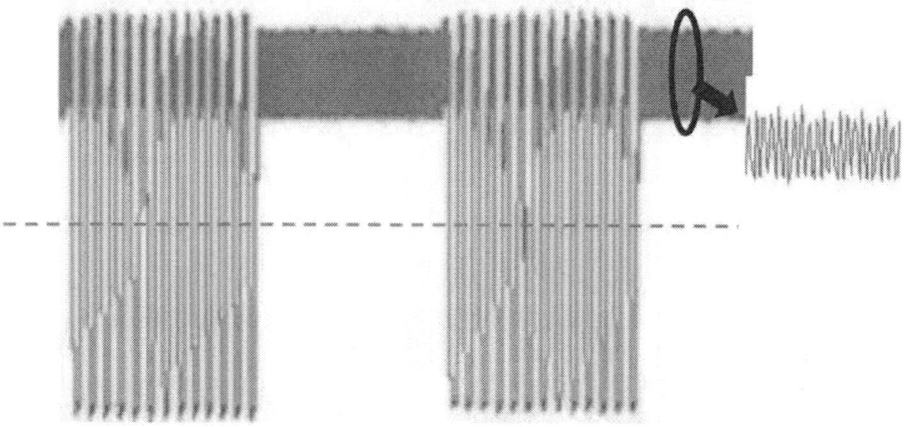

Figure 2. Coupled current waveform.

Table 1. Compositions of 2219 and 2325 (wt. %).

	Cu	Mg	Mn	Fe	Si	Al
2219	5.8-6.8	≤0.02	0.3	≤0.30	≤0.022	Margin
2325	3.6-4.2	1.0-1.5	0.5-0.7	0.06	0.06	Margin

Before starting the repair welding, it is first necessary to remove the defects according to the defect detection position of the joint to be repaired using the mechanical processing method. During the removal process, it is necessary to reduce the scope of removal as much as possible while ensuring that the defects are cleaned. Then the surface of the repaired joint is degreased and deoxidized. After the completion of the repair welding, it is also necessary to repair the surface of the weld, so that it is consistent with the original weld and smooth transition to reduce stress concentration and avoid cracks.

In the process of manual repair welding, the welding process features a two-sided, multi-layer, multi-pass, low-energy, fast, and thin filling layer. By using two-sided multi-layer and multi-pass welding, the heat input of each pass can be reduced by increasing the number of welding strips on both sides, which is conducive to improving the toughness and plasticity of welded joints, and also helps to reduce the stress concentration coefficient and release the internal stress. We move the welding gun at a faster welding speed and reduce wire feed speed, the filling amount of each wire, and the linear energy input, to achieve the goal of reducing the metallurgical damage and local stress of the repaired joint, ensuring the comprehensive performance of the repaired joint. The welding process specifications for repair welding are shown in Table 2, which can be fine-tuned in actual welding. The DC-EN of VPTIG is welded with a pulse current at 23 kHz frequency. Finally, it should be noted that the start and end of each layer and each weld should be staggered to avoid the overlap of welding defects and the concentration of stress.

Table 2. Compositions of 2219 and 2325 (wt. %).

Welding speed (mm/min)	Constant current (A)	Peak value (A)	Fac (Hz)	f (Hz)	Duty cycle (%)	Clear (%)	High frequency peak (A)	high frequency of f (KHz)	Wire feed speed (m/min)
300	195	195	100	4	50	30	100	23	0.5

According to the above welding process characteristics and welding process specifications, the surface topography of repair welding obtained is shown in Figure 3.

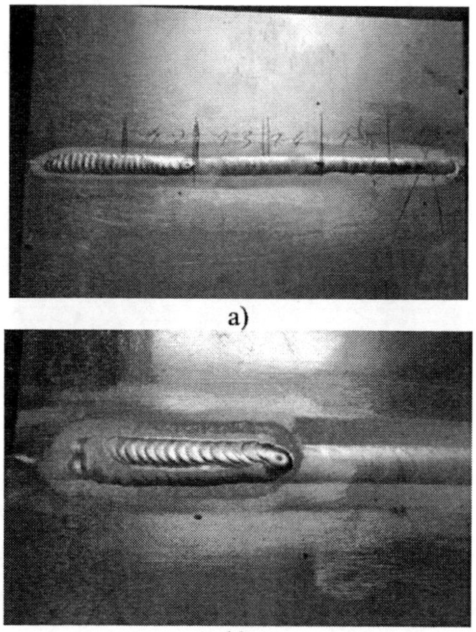

a)

b)

Figure 3. The surface topography of repair welding.

Journal of Physics: Conference Series **2808** (2024) 012003 doi:10.1088/1742-6596/2808/1/012003

3.Test results

3.1 Microporosity detection

In the high-frequency VPTIG welding process, X-ray detection technology was used to detect the porosity of the repair welding seam, and no porosity was found, because X-ray can only detect the porosity with a diameter greater than 0.2 mm. The high-frequency VPTIG process effectively prevented the generation of large porosity. For the detection of micropores, metallographic analysis is needed.

By making metallography and observing the section of the welding joint with a microscope, it can be found that there are almost no micro-pores in the layer of the weld, while some micro-pores can be found between layers, as shown in Figure 4. In the high-frequency VPTIG process, a thin filling layer and high-frequency oscillation are used to promote the porosity in the layer to overflow in time, which effectively inhibits the porosity defects. The micro-pores at the interlayer junction are difficult to clean. The rapid welding process is used in the repair welding process, and the faster welding speed is not conducive to the timely overflow of micro-pores at the interlayer junction. The interlayer junction is bound to be affected by the external environment, which also increases the formation rate of pores. Considering the local nuance of micropores, the performance of the repair welding joint can also accept their presence.

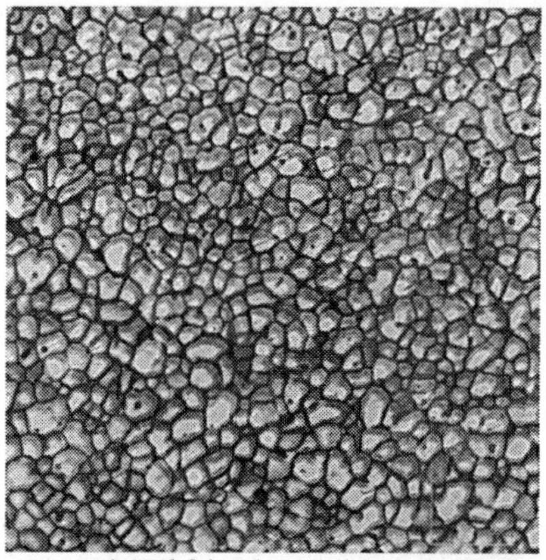

a) Within the weld zone

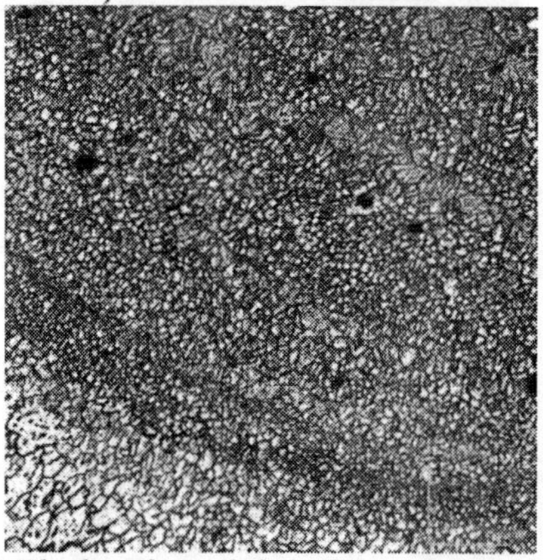

b) The interlayer joint of the weld zone

Figure 4. The metallography of the weld zone for repair welding.

3.2 Microstructure analysis

Due to the use of multi-layer and multi-pass welding, the microstructure of repair welding becomes more complex. Figure 5 shows the microstructure of the weld zone, in which Figure 5 (a) is the microstructure within the weld zone, and Figure 5 (b) is the microstructure of the interlayer joint of the weld zone [8-12].

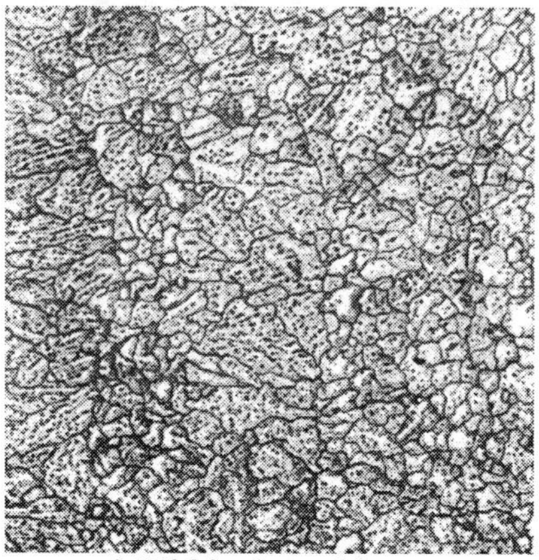

a) Within the weld zone

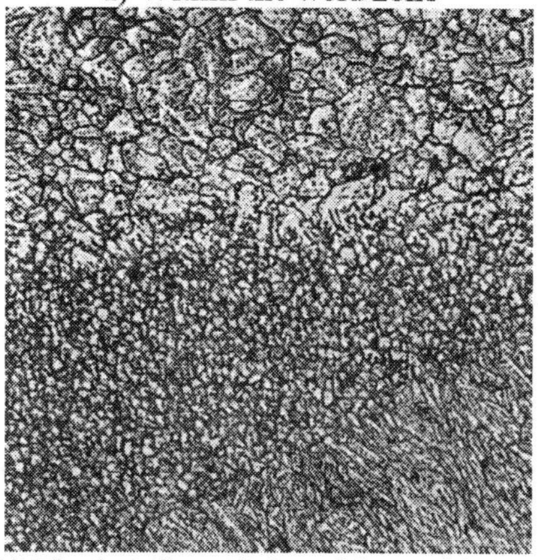

b) The interlayer joint of the weld zone

Figure 5. The microstructure of the weld zone for repair welding.

Comparing the microstructure in the weld zone in Figure 5 (a) with that in Figure 6 before repair welding, it can be found that the grain uniformity becomes relatively poor. At the same time, multiple thermal cycling also increases the grain boundary precipitation relatively after repair welding and presents a network distribution, as shown in Figure 7, which would reduce the joint performance to a certain extent. In addition, in the general repair welding method, due to the high thermal expansion coefficient and solidification shrinkage ability of 2219 aluminum alloy, crystallization cracks may occur in the weld zone. By adopting the multi-layer and multi-pass welding process, the strain rate of the weld metal can be reduced, which is conducive to reducing the tendency of crystallization cracks, and no crystallization cracks are found in the microstructure of the weld zone of the repair welding joint.

Figure 6. The microstructure of the weld zone before repair welding.

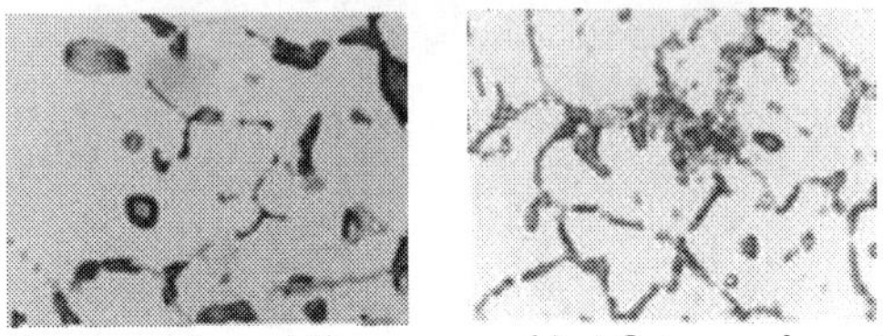

a) Before repair welding b) After repair welding

Figure 7. The contrast before and after repair welding joint for the grain boundary.

It can be seen from the microstructure of the interlayer joint of the weld zone in Figure 5 (b) that fine crystal bands are easily generated at the joint of the weld layer. Although this will also affect the joint performance, the repair welding process adopted will bring more advantages. In multi-layer welding, the high heat input of the rear layer weld will cause the front layer weld to be affected by heat, and the eutectic of the grain boundary with a low melting point on the side of the front layer weld will melt, resulting in liquefaction cracks. However, by using a thin filling layer and low heat input, the rapid welding method can effectively avoid the generation of liquefaction cracks, and no liquefaction cracks are observed at the interlayer joint of the weld area.

The fusion zone of the high-frequency VPTIG welding seam is shown in Figure 8. It can be seen that the structure of the fusion zone is relatively uniform, which effectively refines the grain of the fusion zone. At the same time, almost no fine crystals are produced, and no liquefaction cracks are produced, which is the result of multiple thermal cycles under the high-frequency VPTIG process. The microstructure and properties of the fusion zone largely determine the performance of the welded joint.

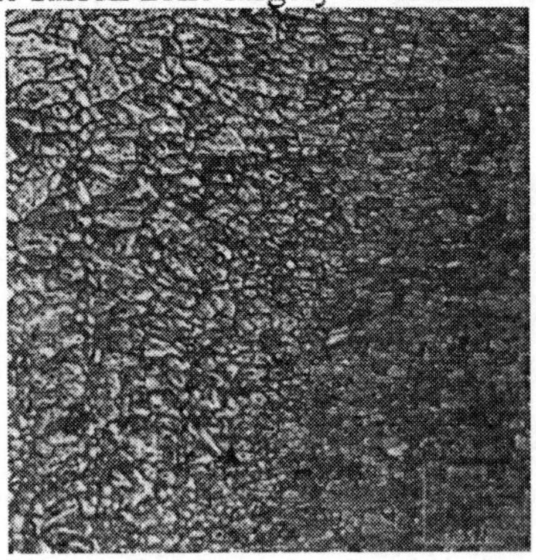

Figure 8. The microstructure of the fusion zone of repair welding.

3.3 Analysis of tensile fracture morphology

Figure 9 shows the fracture morphology of the repair welding joint under scanning electron microscopy. In these two fractures of different multiples, it can be observed that there are a certain number of micro-pores, dimples of uneven size, ductile fracture steps, and local river pattern and cleavage steps with brittle fracture characteristics. Compared with Figure 10, it can be seen that the tensile fracture before repair welding is dominated by a large number of uniform dimples, with ductile fracture steps and a small number of slip planes. After repair welding, the morphology and structure of the fracture are relatively decreased, but some characteristics of the fracture before repair welding are retained, which plays a great role in reducing joint performance [13-14].

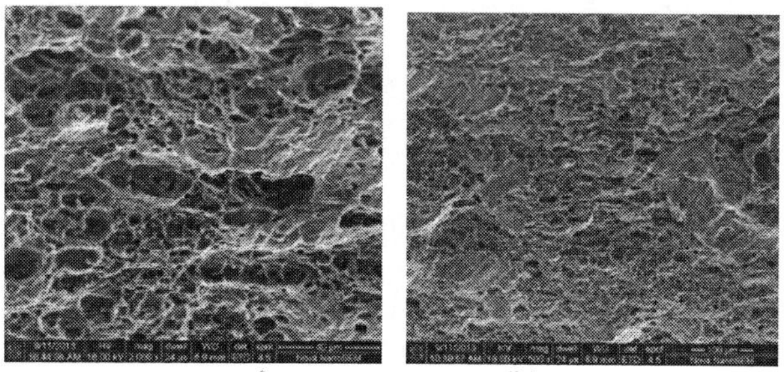

a) (b)

Figure 9. The fracture morphology of repair welding joint.

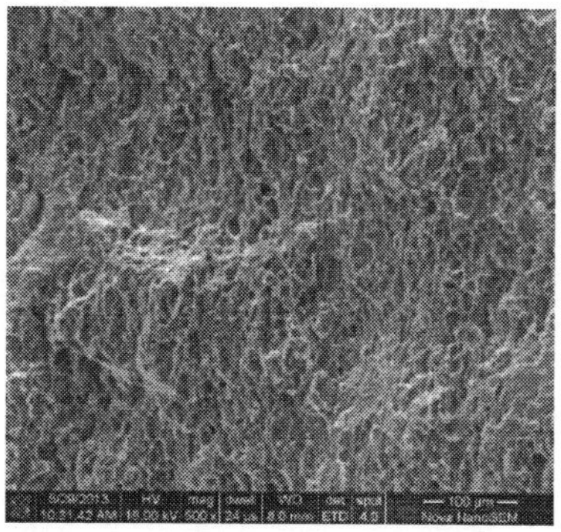

Figure 10. The fracture morphology of the joint before repair welding.

The fracture initiation source of the repair welding joint is shown in Figure 11. It can be seen that no matter how much optimization is made to the welding process, a local brittle slip surface has appeared in the repaired welded fracture. Micro-porosity (whose diameter is less than 0.1 mm) also increases significantly, and porosity exists at the edge of the fracture. With the stretching of the external force, the shape variable of the pore edge perpendicular to the force direction is large, which makes it easy to produce stress concentration and crack initiation. In addition, because the heat-affected zone is subjected to multiple thermal cycles, the softening zone is transferred to this region. Therefore, the fracture first starts from the front edge of the weld, the crack gradually expands to the heat-affected zone side of the base metal, and the heat-affected zone is completely disconnected at the back of the weld.

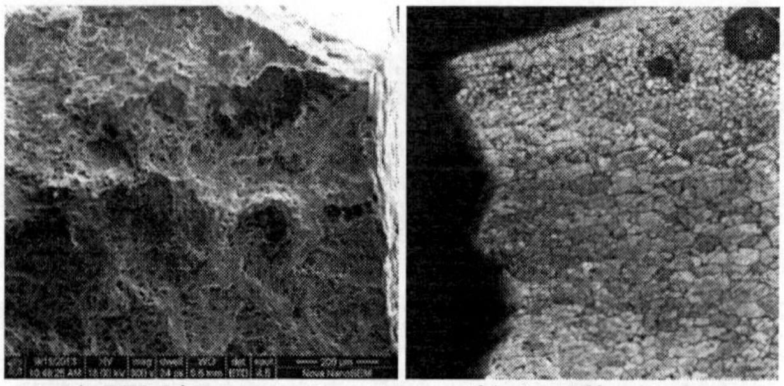

a) SEM image b) Metallography

Figure 11. The fracture initiation source of repair welding joint.

3.4 Analysis of tensile properties

The tensile specimen of the repair welding joint for 2219 aluminum alloy is shown in Figure 12, which also verifies the fracture of the joint. The mechanical properties of the repair welding joint are shown in Table 3. It can be seen that the mechanical properties of the repair welding joint for 2219 aluminum alloy are lower than that of the initial welding joint. In addition, the reduction of tensile strength is greater than the reduction of elongation, which also shows the characteristics of the repair welding process adopted can effectively ensure the plastic properties of the joint. The strength of the repair welding joint is 61.1% of the base material, and the elongation is 40.5% of the base material, which also maintains a high strength coefficient (greater than 0.6) and elongation after rewelding. It also shows again that adopting a reasonable high-frequency VPTIG welding process of 2219 aluminum alloy is the key to ensuring the comprehensive mechanical properties of the welding joint.

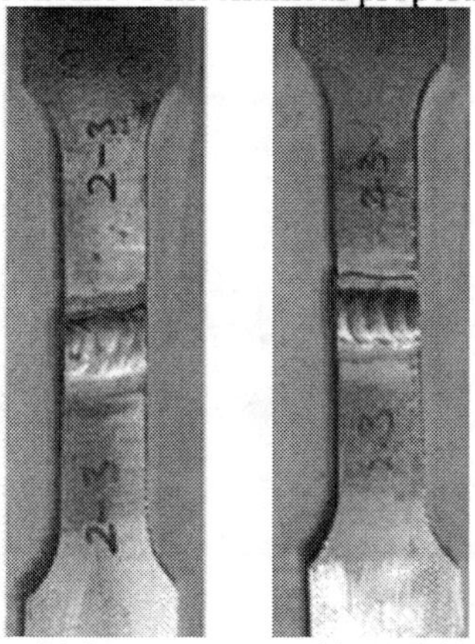

a) Weld front b) Weld back

Figure 12. The tensile specimen of the repair welding joint.

Table 3. The mechanical properties of repair welding joint.

Welding process	Tensile strength σ_b (MPa)	Elongation δ_5 (%)
The repair welding joint	275	4.7
The initial welding joint	305	5.5
The base material	450	11.6

4. Conclusions

(1) Considering the locality and detail of the defect location of 2219 aluminum alloy structural parts, the manual repair welding method is generally adopted in the repair welding process, and the welding process of two-sided, multi-layer, multi-pass, low-energy, fast, and thin filling layer is proposed in the manual repair welding process.

(2) The high-frequency VPTIG (the DC-EN of VPTIG is coupled to the pulse current at the frequency of 20-25 kHz at all times) is used to repair the welding process route, which helps to suppress the porosity defects, but there are still tiny pores at the interlayer junction. At the same time, the grain of the fusion zone can be refined, the liquefaction crack can be inhibited, the strain rate of the weld metal can be reduced, the crystal crack can be avoided, and the metallurgical damage of the repaired welded joint can be reduced.

(3) The reliability of the proposed repair welding process was verified through stoma detection, microstructure analysis, tensile fracture analysis, mechanical property analysis, etc. The strength of the joint reached 61.1% of the base material, and the elongation reached 40.5% of the base material, which was within the acceptable range. However, the mechanical properties of the 2219 aluminum alloy repair welding joint are lower than the initial welding head, and multiple repair welding should be avoided.

References

[1] Li Xian. Study on high-frequency coupling pulse TIG welding technology of 2219 Aluminum alloy [D]. Beijing: Beijing University of Technology, 2014.
[2] Chen Zhiyuan, Zhang Xiaohong, Lin Peng, Fan Yifei, and Huang Xian. Study on microstructure and mechanical properties of TIG welded joint of 2219 aluminum alloy Thick plate [J]. Aerospace Materials Technology, 2023 (1): 64-68.
[3] Han Kuoliang, Wang Huimiao, Liu Hanwei, Wang Yasen, and Song Jianling. Performance Analysis of 2219 Aluminum Alloy TIG Welding Joint under Different Welding Processes [J]. Electric Welding Machine, 2019 (5): 48-51.
[4] Sun Shixuan, Wei Ruigang, Gao Yanjun, Wuhao, Diao Yu, and Zang Mengnan. Study on repair welding performance of 2219 aluminum alloy cross welding joint [J]. Pressure Vessels, 2018 (6): 1-7.
[5] Meng Danyang, Gu Zhonghua, Miao Yongyue, Zhang Dengkui, and Liang Shuangfeng. Effect of Welding Layer Number on Microstructure and Properties of Welded Joints of 2219 Aluminum Alloy [J]. Hot Working Technology, 2020 (19): 44-47.
[6] Yao Xuwei, Cui Guoping, Qiu Yiqing, et al. Defect repair of VPTIG lock bottom weld of 2219 Aluminum alloy and its effect on joint performance [J]. Metal Processing (Hot processing), 2022 (10): 40-44.
[7] Ma Linghang, Li Bo, Zhao Yanguang, et al. Research status of welding defect repair technology for rocket tank [J]. Electric Welding Machine, 2023, 53 (03): 31-45.
[8] Su D., Jin H., and Wu S . Microstructure and mechanical properties of 2219 aluminum alloy VPTIG welds during cyclic thermal treatment [J]. Rare Metals, 2022, 41 (10): 3, 539-3, 545.
[9] Zhandong W., Yue Z., Qiang W., et al. Microstructure-based modeling of the PMZ mechanical properties in 2219-T8 aluminum alloy TIG welding joint [J]. Materials Design, 2022, 223.
[10] Dengkui Z., Aiping W., Yue Z., et al. Microstructural evolution and its effect on mechanical properties in different regions of 2219-C10S aluminum alloy TIG-welded joint [J]. Transactions of Nonferrous Metals Society of China, 2020, 30 (10): 2, 625-2, 638.
[11] Liu Yuning, Zhang Jian, Liu Debo, et al. Effect of TIG repair welding on microstructure and properties of 2219 aluminum alloy friction stir welded joint [J]. Aerospace Materials Technology, 2022, 52 (06): 77-82.
[12] Yang Kang, Tang Yu, Wang Pengju, et al. Effect of Different Atmospheres on microstructure and properties of 2219 Aluminum alloy TIG welded joint [J]. Welding Machine, 2023, 53 (12): 95-100.

[13] Tian Zhijie, Xiong Linyu, Li Ying, Dong Kangying, and Gao Hui. Fracture toughness of VPTIG welded joint of 2219 aluminum alloy [J]. Mechanical Engineering Materials, 2017 (12): 54-58.

[14] Derek Hull, Li Xiaogang, Dong Chaofang, Du Cuiwei, Gao Jin, and Lu Lin (trans.). Fracture Morphology [M]. Science Press, 2009.

Journal of Physics: Conference Series **2808** (2024) 012004 doi:10.1088/1742-6596/2808/1/012004

Effect of laser power on microstructure and properties of Ti-Al/WC composite coatings

Hongwei Liu[1], Baoquan Dong[2], Xun Gong[1], Haojie Chen[1], Haichao Jiao[1], Hui Zhao[1,*], Haitao Zhao[1] and Li Mei[3]

[1]School of Materials Science and Engineering, Shenyang Ligong University, Shenyang, Liaoning, 110519, China

[2]Liaoning Hengtong Metallurgical Equipment Manufacture Co, Ltd, Benxi, Liaoning, 117000, China

[3]Shenyang Jinfeng Special Cutting Tools Co, Ltd, Shenyang, Liaoning, 110015, China

*Corresponding author: Email: zhaohui@sylu.edu.cn

Abstract: To enhance the performance of TC4 titanium alloy, laser cladding technology was used to prepare a Ti-Al/WC composite coating on its surface, and the influence of laser power on the physical phase composition, microstructure, friction and wear properties and hardness of the coating were studied. The studies indicate that the coating consists of Ti_3Al_2, TiAl, Ti_3Al gray matrix organization and TiAl, Ti_2AlC gray-white precipitation organization. When laser power increases, the gray-white precipitated tissue increases significantly, and the precipitated TiC changes from dot-like, flake-like, and short rod-like to dendritic. When the laser power is 1200W, TiC gradually transforms into Ti_2AlC phase, and the shape of the precipitated tissue becomes irregular. In comparison to the substrate, the average hardness of the coatings rises by 1.88-2.27 times, and the maximum average hardness of 635.6HV is achieved at 1000W of laser power. The cladding layer has better wear resistance at laser powers of 1000W and 1200W. The wear mechanisms of the coatings are abrasive and adhesive wear.

1. Introduction

Titanium alloys are characterized by high specific strength, excellent corrosion resistance, and biocompatibility, making them a wide range of high-quality functional materials. In contrast, their adhibition in some fields is limited by poor surface hardness and bad wear resistance. To enhance its wear resistance and hardness, surface treatment is an effective method.

Because of its unique advantages, laser cladding has been more widely used in industry as an advanced surface modification technology. Because of its extensive qualities, including high wear resistance and hardness, ceramic-metal composite coatings have gained a lot of attention in the field of surface modification of titanium alloys. According to related research, intermetallic substances of titanium and aluminum can increase the wettability of the ceramic phase between the substrate and the ceramic phase, and WC not only has a very similar coefficient of thermal expansion to that of titanium alloy, but also has a very good red hardness and excellent wear resistance, and it is considered to be an ideal reinforcing material for the wear-resistant coatings of titanium alloys [1]. Xiao et al. [2] studied the preparation of Ti-Mo-Al-B composite coatings on the surface of TC4 titanium alloy, the average hardness of the coating was 648.69 HV, which was improved by 1.97 times compared to the substrate, and the amount of wear was reduced compared to that of the substrate by 11%. Zhang et al. [3] prepared

Journal of Physics: Conference Series **2808** (2024) 012004 doi:10.1088/1742-6596/2808/1/012004

Ti-Al/WC composite coatings on the surface of titanium alloys, and the study indicated that the addition of WC made the in-situ-generated TiC dispersed in the fused-coating layer. The average hardness of the fused cladding layer was about 1.4 to 1.9 times that of the matrix. Furthermore, researchers have examined the impact of laser process parameters on the forming quality and performance of metal matrix composite coatings reinforced with WC particles. Their findings indicate a notable enhancement in the optimum fused cladding layer's performance [4,5].

In summary, the performance of TC4 titanium alloy can be effectively improved by the creation of ceramic-metal composite coatings on its surface using laser cladding. One essential parameter that significantly affects the preparation of composite coatings is the laser power. As a result, the Ti-Al/WC composite coating described in this study was created by laser cladding the surface of TC4 titanium alloy, and it examined the effects of laser power on the microstructure and characteristics of the cladding layer while maintaining consistency concerning other parameters.

2. Experimental procedure

The substrate utilized for this test was the titanium alloy TC4, and it was cut into blocks of 100 mm × 100 mm × 10 mm using wire cutting. The surface of the substrate was polished with a polishing wheel to remove the oxide skin and ultrasonically cleaned with anhydrous ethanol. The cladding material was selected as 99.9% pure spherical Ti, Al, WC powder, 45-105 um particle size, the composition ratio for the atomic ratio of 1:1 Ti, Al, and 20% mass fraction of WC. Proportional to the powder in the mixing tank in the glove box, the mixer uses the speed of 30 r/min mixing to mix for 6 h and the 120°C vacuum to dry for 6 h. The experiments were carried out using the coaxial powder feeding method, with a spot size of 3 mm, an out-of-focus amount of 17 mm, a scanning speed of 120 mm/min, a lap rate of 40%, and laser powers of 600 W, 800 W, 1000 W, and 1200 W, respectively.

After the fusion cladding was completed, the specimen was cut along the cross-section, and the cut specimen was sanded, polished, and corroded using a corrosion solution. The fused cladding layer's physical phase was examined by XRD, and its microstructure was examined with an inverted metallographic microscope (200TAM). With a loading load of 100 N and a loading duration of 10 s, the hardness of the substrate and the fused cladding layer were assessed using a microhardness tester. Ten points were measured close to the fused cladding layer's center line to determine the average hardness. By avoiding the unmelted TiC particles, the hardness was determined. Friction and wear tests were performed using a reciprocating friction and wear testing apparatus. The load was 10 N, the frequency was 3 Hz, the reciprocating stroke was 5 mm, and the friction pair was GCr15.

3. Results and discuss

3.1. Phase structure study of Ti-Al/WC composite coating

The XRD patterns of coatings with various laser strengths are displayed in Figure 1. The XRD spectra of the coatings showed little variation at 600 W and 800 W laser power, as shown in Figure 1. The fused layer mainly consisted of Ti_3Al_2, TiAl, Ti_3Al and only a small amount of TiC as well as a trace amount of Ti_2AlC was detected. The distinctive diffraction peaks of Ti_3Al_2, TiAl, and Ti_3Al showed an increased intensity at 1000W of laser power, with the maximum intensity of TiC diffraction peaks. The typical diffraction peaks of Ti_2AlC exhibited a considerable enhancement in intensity at 1200W of laser power, while the intensity of Ti_3Al_2, TiAl, Ti_3Al, and TiC characteristic peaks dropped. This indicates the reaction of TiC with intermetallic compounds generated by the reaction of Ti and Al to produce Ti_2AlC. The diffraction peaks of WC were detected at different powers, which indicates the presence of some undissolved WC particles in the coating.

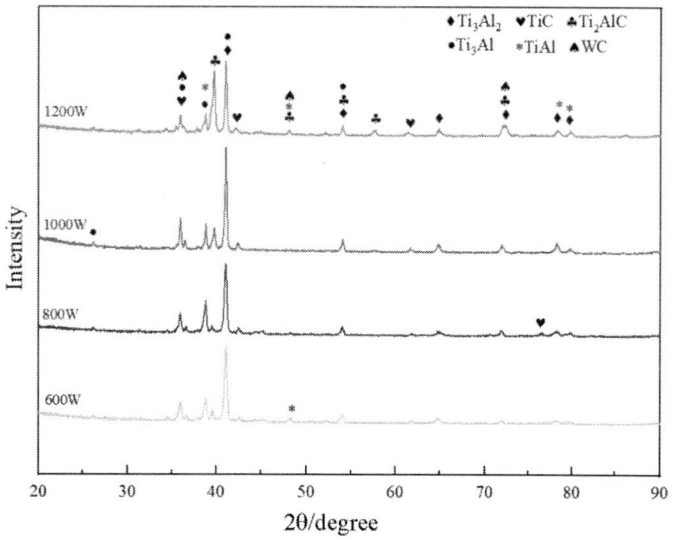

Figure 1. XRD of cladding layers with different laser power.

3.2. Microstructure of Ti-Al/WC composite coating

The microstructure of the coating's center area at various laser powers is displayed in Figure 2. When the XRD data is taken into account, it becomes clear that the coating is made up of gray precipitation tissue with a primary composition of TiAl and Ti_2AlC and gray matrix tissue with a primary composition of Ti_3Al_2, TiAl, and Ti_3Al. As laser power increases, there is a noticeable increase in the gray-white precipitate tissue.

As can be seen in Figure. 2(a), when the laser power is low, the TiC generated in situ is in the form of fine dots, flakes, and some TiC in the form of short rods [6]. When the laser power increases to 800 W, more WC particles decompose the C atoms in the coating will be fully supplied, and the TiC in the alloy will be dendritic at this value [7]. With the further increase of laser power, the dendritic TiC organization increased significantly. A large amount of dendritic TiC formation consumed most of the C atoms in the coating, which led to the subsequent TiCs not being sufficiently supplied with C atoms during the formation process, and thus the flaky TiC organization appeared in Figure. 2(c). When the laser power increased to 1200W, it could be observed that the shape of the precipitated off-white tissue in Figure. 2(d) became irregular. Checking the related literature and analyzing the XRD results, it is known that when the laser power is higher than a certain value, the encapsulation reaction between molten TiAl and solid TiC will consume a large amount of aluminum. Therefore, when the aluminum content in the alloy is low or unevenly distributed, it is possible that the TiC phase will not be completely transformed into the Ti_2AlC phase, which is retained in the solidification process, and the TiC coexists with the Ti_2AlC phase [8] and irregularly shaped precipitation organization appears.

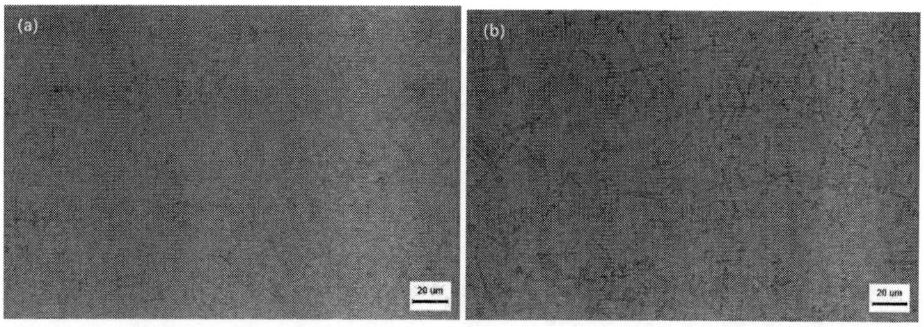

Journal of Physics: Conference Series **2808** (2024) 012004 doi:10.1088/1742-6596/2808/1/012004

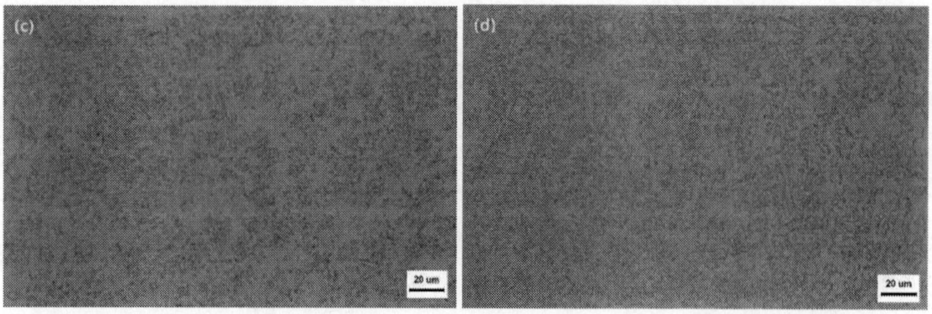

Figure 2. Microstructure of cladding layers with different laser power: (a) 600W (b) 800W (c) 1000W (d) 1200W.

3.3.Microhardness of Ti-Al/WC composite coating

The histogram of the average hardness of the coatings at various laser powers is displayed in Figure 3. The substrate has a hardness of 280 HV, and the fused coating layer with varying laser strengths has a harder surface than the substrate. The hardness of the coating is enhanced most when the power of the laser is 1000W, and the average hardness of the coating is approximately 635.6 HV, 2.27 times that of the substrate; when the laser power is 600 W, the coating is least enhanced in comparison to the substrate; the average hardness of the coating is 526.4 HV, and the degree of improvement is 1.88 times that of the substrate. Compared to 1000 W, the layers became less hard at 1200 W of laser power. This is because an excessively high dilution rate of the cladding layer may adversely affect its performance; that is, the higher the laser power is, the higher the cladding layer's dilution rate is [9].

The gradient curve of the cladding layer's hardness at various laser intensities is displayed in Figure 3 (b). The cladding region's hardness varies with different laser strengths, and it abruptly increases close to the fusing line. The WC particles' breakdown into C and W elements during the laser cladding process and the polarization of these two elements under the stirring effect of the laser beam are the causes of the hardness fluctuations in the cladding area. This uneven distribution of TiC and other reinforcing phases in the cladding layer is the result. More unmelted WC particles are maintained at the fusion line because of the short molten pool existence period and rapid solidification speed, increasing the hardness.

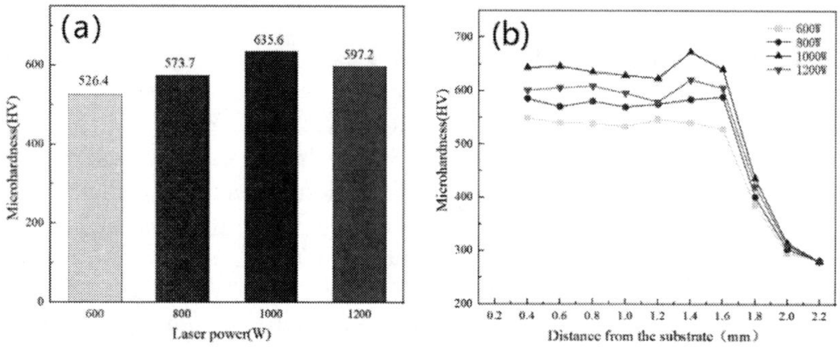

Figure 3. Microhardness of cladding layers with different laser power.

3.4.Frictional wear properties of Ti-Al/WC composite coating

Figure 4 shows the abrasion morphology of the cladding layers after abrasion with different powers. Observation of Figure 4 (a) (b) shows that there is a large distribution of abrasive chips on the surface, accompanied by obvious scratches and stripping pits, which is because a part of the WC particles are not sufficiently dissolved in the case of low laser power, and in the friction and wear process of GCr15 friction vice contact with the WC particles, the WC particles are dislodged or ruptured, and these particles will lead to the broadening of the abrasion marks. At this time, the cladding layers are abrasive wear and slight adhesive wear.

Figure 4 (c) (d) for the laser power of 1000 W, 1200 W under the abrasion morphology, compared with Figureure ab scratches are shallower. Its surface is not formed by the WC particles shedding peeling pits, but it can be observed that the surface of the coating has occurred a significant spalling. This is because the formation process of TiC, Ti_2AlC, the surrounding stress concentration, in the high cyclic load surface is prone to serious delamination, after cutting wear and plastic deformation and shedding, shedding the formation of hard particles as abrasive particles will be cut on the coating surface. When the laser power is high, the wear mechanism of the cladding layers is mainly typical abrasive wear and adhesive wear.

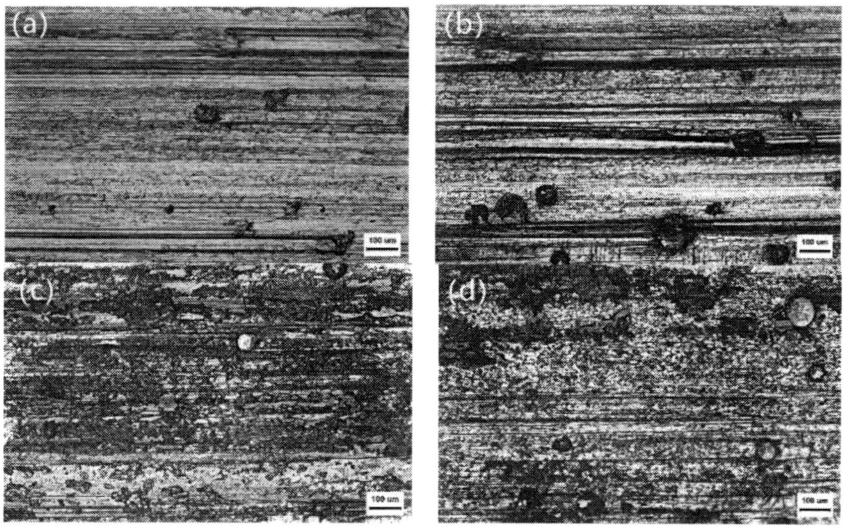

Figure 4. Abrasion morphology of coating with different laser power: (a)600 W (b)800 W (c)1000 W (d)1200 W.

4. Conclusion

(1) The coating is mostly made up of unfused WC, Ti_3Al_2, TiAl, Ti_3Al, TiC, and Ti_2AlC. The TiC content rises and then falls with increasing laser power, changing from dots, flakes, and short rods to dendrites; at 1200 W, the TiC partially transforms into the Ti_2AlC phase, giving rise to the irregular shape of the precipitation organization.

(2) The average hardness of the coating was enhanced by 1.88 to 2.27 times compared to the substrate, and the coating had the highest average hardness when the laser power was 1000W.

(3) When the laser power is 1000 W, the wear resistance of the coatings is best. The wear mechanisms of the coatings are abrasive and adhesive wear.

Acknowledgments

Special thanks for the financial support provided by the Liaoning Provincial Science Program (publish a list of successful candidates: Science and Technology Tackling Special Project) "Energy-saving and High-efficiency Heat Exchange Tubing Complete Set of Equipment" (No. 2022JH1/10400029).

References

[1] Ji S C, Li Z X and Li J L (2018) Titanium surface thick tungsten carbide coating research progress . Journal of titanium industry progress, 35 (6): 20-25.

[2] Xiao H Q, Yin X G and Feng J Y (2023). Microstructure and frictional wear properties of laser cladding Ti-Mo-Al-B composite coating on TC4 titanium alloy. Review the material: 1-12.

[3] Zhang L, Zhao W and Sui X M (2023). Study on Microstructure and properties of Ti-Al/WC coating on Titanium Alloy Surface by laser cladding. The hot working processes, 52(14):108-113.

Journal of Physics: Conference Series **2808** (2024) 012004 doi:10.1088/1742-6596/2808/1/012004

[4] Meng Y,Yang Y and Zhang C (2023). Influence of Bioactive Glass Addition on TC4 Laser Cladding Coatings: Microstructure and Electrochemical Properties. Coatings, 13(9).
[5] Akinlabi T E, Erinosho F M (2017). Characterizing the Effect of Laser Power on Laser Metal Deposited Titanium Alloy and Boron Carbide. Metallurgical and Materials Transactions,48(11).
[6] Li B X (2022). Effect of WC addition on microstructure and properties of titanium laser cladding layer. The China Civil Aviation University.
[7] Zhao X X (2022).Study on microstructure and properties of TiAl coating by laser claddingon TC4 alloy Surface. Guizhou university.
[8] Li F Q, Gao Z Z and Li L Q (2017). Microstructure and properties of composite layer prepared by TC4 surface silk powder synchronous laser cladding. Rare Metal Materials and Engineering,46(01):177-182.
[9] M. E Z ,Farid A Y and Ahmad S E (2021). Effect of Laser Cladding Processing Parameters on Nitinol's Clad Dimensions, Microstructure, and Hardness.Materials Sciences and Applications,12(12):603-613.

Journal of Physics: Conference Series **2808** (2024) 012005 doi:10.1088/1742-6596/2808/1/012005

Modification and characterisation of nickel-based alloy materials for ultra-supercritical power generation

Xiaolong Zhang[1*]

[1]Faculty of Engineering and Physical Sciences, University of Southampton, Southampton, SO17 1BJ, United Kingdom

[*]Corresponding author's e-mail: 2528zxl@gmail.com

Abstract: Ultra-supercritical power generation technology has consistently pioneered energy-efficient utilization techniques, with the development and application of tubing serving as the cornerstone for ultra-supercritical power plant units. Among numerous candidate materials, nickel-based alloys have garnered widespread attention due to their exceptional high-temperature strength and corrosion resistance. This study primarily delineates the modification techniques employed for nickel-based alloy materials utilized in 700°C ultra-supercritical units, encompassing solid-solution strengthening, high-entropy alloys, additive manufacturing, and surface-coating. Additionally, this paper consolidates characterization methodologies for nickel-based alloys, focusing on morphological and crystallinity characterization and mechanical performance evaluations, thereby offering valuable insights and guidance for researchers in this field.

1. Introduction

Coal power generation is an important source of electricity in today's human society, and coal-fired power generation technology will remain the mainstream of global power generation technology in the coming years [1]. Modern ultra-supercritical power plants run at steam temperatures higher than 700°C, in contrast to traditional coal-fired power production systems, at steam pressures in the order of 35 MPa, and at efficiencies approaching or even exceeding 50%. Plant components demand a suitable material with sufficient creep strength and oxidation resistance and corrosion in extreme environments than commonly used materials to run a power plant in such a typical environment. High-temperature alloys are needed for high-temperature elements in the new A-USC power plant, namely superheaters and collector tubes, and the material's fatigue resistance also must be taken into account [2]. As the operation duration under elevated pressure extends and the system size grows, it becomes increasingly challenging to manage the boiler-turbine unit owing to intricate factors like nonlinearity, coupling, and load disturbances. Additionally, supercritical water has unique properties related to heat transfer, including heat transfer deterioration (HTD) as well as enhancement (HTE). While HTD arises, the local wall temperature rises swiftly, raising the possibility of safety issues such as boiler water wall explosions [3]. Therefore, it is critical to identify and develop materials that can meet these stringent conditions [4].

To meet the rigorous operating conditions of ultra-supercritical units, alloy materials with high hardness, superior heat resistance, and robust corrosion resistance are more suitable than single metals. Within the spectrum of alloy materials, nickel-based alloys are particularly favored for ultra-supercritical units due to their commendable mechanical properties, resistance to oxidation under high temperatures, and stress corrosion cracking [5]. Under development and testing, several of the Ni alloys that are suitable for usage at an operating environment of around 700°C and pressures of 310 kg/cm^2 are considered. The high-temperature strength and corrosion resistance of Ni-based heat-resistant alloys surpass those of heat-resistant steels. Elements such as Al and Ti within the matrix facilitate the formation of precipitation-strengthened phases, playing a pivotal role in enhancing creep-fatigue strength. Nevertheless, challenges persist for nickel-based alloys, including insufficient service life and elevated costs.

In this paper, addressing the challenges, we consolidate various modification and manufacturing techniques for nickel-based alloys, encompassing solid-solution strengthening, high-entropy alloys, additive manufacturing, and surface coating. Furthermore, we comprehensively outline the principles of various analytical and testing methods, focusing on morphological and crystallinity characterization and mechanical performance evaluations, and focusing on providing valuable insights and guidance for study in this domain.

2. Modification of nickel-based alloys

2.1 Solid solution strengthening

Solid-solution strengthening refers to the enhancement of strength and hardness in metals after alloying. This is primarily attributed to the distortion of the local microstructure caused by solute elements and the influence of dislocations. For Ni-based alloys used in USC power plants, solid solution strengthening is typically achieved by introducing elements such as tungsten, molybdenum, and cobalt into nickel-chromium alloys to improve their resistance to creep deformation. As far as strength goes, Alloy 617 has already met the 100 MPa at 700°C 100, 000-hour creep-rupture strength criterion [7]. The nickel alloy strengthened using this method exhibits excellent antioxidative and corrosion-resistant properties. It possesses a single-phase austenitic structure, ensuring structural stability with minimal aging tendencies. Furthermore, it demonstrates superior machinability and weldability. The alloy is amenable to plastic deformation, and its heat treatment process is both straightforward and time-efficient. A brief annealing process effectively alleviates internal stresses while allowing for controlled grain size.

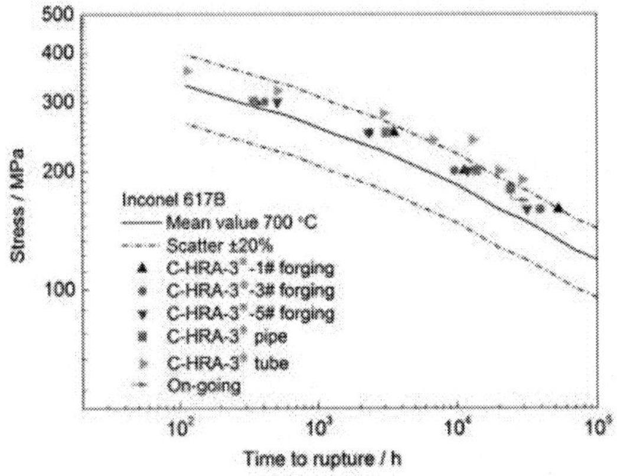

Figure 1. C-HRA-3® alloys' creep life at 700°C [9].

Among the nickel-chromium alloys, Inconel 617, developed by INCO Alloys International, has emerged as a promising material for ultra-supercritical power plants due to its exceptional antioxidative and carburization-resistant properties at elevated temperatures and pressures. Building upon this foundation, Spiegel et al. introduced boron elements into the alloy at concentrations of 0, 30, and 60 ppm. The results indicate that the addition of 30 ppm boron yields the most significant enhancement in the creep resistance of Inconel 617 alloy [8]. While optimizing the boron (B) content, the incorporation of niobium (Nb) and vanadium (V) elements concurrently enhances the high-temperature strength of the material, leading to the development of Inconel 671B alloy. To further enhance the endurance of the

Inconel 617B alloy, Liu et al. refined the molybdenum (Mo) content and introduced tungsten (W) and zirconium (Zr) elements. Through the synergistic effects of W-Mo and B-Zr composite strengthening mechanisms, the performance of the material was optimized, resulting in the C-HRA-3® alloy, which has been employed in various components [9]. Figure 1 illustrates the creep life of the C-HRA-3® alloy at elevated temperatures. The creep life performance of this alloy across various components consistently surpasses the average levels observed in Inconel 617B alloy.

With the incorporation of more precious metal components, the cost of the material escalates. Introducing iron elements into the alloy not only reduces costs but also enhances its machinability. In recent years, certain nickel-iron-chromium alloys have demonstrated a favorable cost-performance ratio and retained excellent mechanical properties. Xu et al. incorporated trace amounts of the rare-earth element yttrium into deformed nickel-iron-based alloys, thereby augmenting the alloy's corrosion resistance [10]. Although significant progress has been made in the research of precipitation-strengthened nickel-based alloys, the stability of the alloy's microstructure diminishes due to the precipitation of η and γ phases during prolonged service at elevated temperatures. Furthermore, when operating above 700°C, such alloys still exhibit insufficient resistance to creep deformation [11].

2.2 High-entropy nickel alloys

Typically, conventional alloys consist of a few primary components with a little addition of some element, which can significantly strengthen the materials' characteristics [12]. Recently, high-entropy materials (HEAs), in which structural strengthening is achieved through the manipulation of added element types, and proportions, with their unique structural characteristics, have sparked a growing interest in environmental science and renewable energy technologies [13]. The title "high entropy" was chosen since these alloys' high configurational entropy of mixing, relative to more conventional alloys, was the main reason why random solid solutions outweigh intermetallic complexes [14]. Yeh et al. introduced the concept of high-entropy alloys and processing parameters [15]. These alloys are composed of a combination of multiple elements (more than five types), and each holds a relatively uniform atomic percentage ranging from 5% to 35%. In this configuration, atoms of each element randomly occupy lattice sites, resulting in a structured crystalline arrangement. Such alloys exhibit a highly ordered microstructure, as opposed to the random distributions observed in alloys with fewer added elements. This characteristic undoubtedly expands their exploration and application potential.

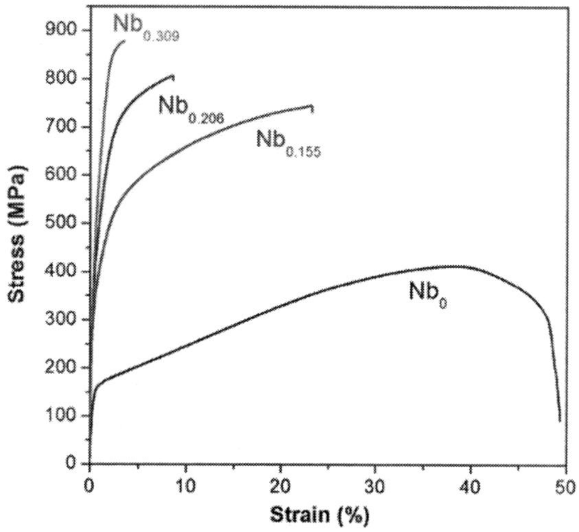

Figure 2. Tensile stress-strain curve of as-cast CoCrFeNiNbx alloy [15].

Within this context, single-phase alloys based on chromium-cobalt-nickel with a face-centered cubic (FCC) crystal structure demonstrate excellent stability, wear resistance, and corrosion resistance, albeit with lower strength. He et al. successfully developed a multiphase FCC HEA utilizing precipitation hardening [16]. Through the addition of Ti and Al elements, precipitation phases with sizes ranging from 20-100 nm were observed, under room temperature, exhibiting a yield strength exceeding 1 GPa. Liu et al. enhanced Nb content, resulting in a cast γ' phase. Based on dislocation observations and estimations, the conclusion was drawn that shear strengthening serves as the primary mechanism [17]. Figure 2 presents stress-strain curves for materials with varying Nb content, while Table 1 enumerates specific values for fracture strength, yield strength, and tensile ductility. Both fracture and yield strengths of the specimens increase with an augmented niobium concentration. Specifically, the fracture strength and yield strength for the Nb_0 alloy are 413 MPa and 147 MPa, respectively. However, for the $Nb_{0.309}$ alloy, these values notably escalate to 879 MPa and 478 MPa. In contrast, the elongation at fracture for the Nb_0 alloy significantly decreases from 48% to 3.5%.

Table 1. Tensile properties of CoCrFeNiNbx alloys.

	$\sigma_{0.2}$	σ_{max}	
Nb_0	147	413	48.0
$Nb_{0.155}$	321	744	23.3
$Nb_{0.206}$	402	807	8.6
$Nb_{0.309}$	478	879	3.5

An approach to further enhance the performance of HEAs is optimizing the solid solution parameters of the material. However, the solid solution parameters of HEAs inherently possess a high degree of randomness. Zhang et al. utilized energy-filtered transmission electron microscopy to examine the short-range order (SRO) characteristics of chromium-cobalt-nickel high-entropy alloys. The degree of nanoscale local order can be directionally altered through thermo-mechanical processing, thereby offering a novel avenue for tailoring the mechanical properties of high-entropy alloys [18]. In addition to FCC, HEAs have also been developed in the body-centered cubic (BCC) field. Despite the commendable strength and performance under the high temperature of HEAs, the distorted lattice structure not only provides enhanced strength but also significantly reduces the material's ductility. Additionally, the substantial incorporation of precious metals elevates the overall cost of this material.

2.3 Additive manufacturing of nickel alloys

Apart from stringent mechanical performance requirements, the design shapes and dimensions of components for ultra-supercritical units pose significant challenges to the material's forming and processing techniques. This necessitates materials with outstanding formability and weldability properties. However, additive manufacturing techniques offer a potential solution to this challenge. Distinct from traditional metal fabrication methods, this approach employs a layer-by-layer deposition process, allowing for the creation of desired material shapes without the need for intricate molds. Furthermore, due to its additive nature, this method minimizes material wastage compared to subtractive and isomorphous manufacturing processes. Murray et al. introduced a defect-resistant, high-strength Co-Ni high-temperature alloy fabricated by using selective laser melting (SLM) and electron beam melting (EBM). As depicted in Figure 3, materials produced by both methods in post-processing exhibit strengths exceeding 1.1 GPa, with tensile elongations surpassing 13% at room temperature [19]. However, the microstructure of nickel alloys fabricated by using metal additive manufacturing techniques often exhibits epitaxial growth characteristics. Additionally, due to the presence of numerous alloying elements, fabricated components frequently manifest micro-segregation and crack defects, resulting in slightly inferior properties compared to their forged counterparts of the same alloy type.

To mitigate crack formation, Zhao et al. introduced zirconium into the nickel-based high-temperature alloy Haynes 230. This incorporation facilitated the formation of continuous interdendritic liquid films, thereby alleviating thermal stresses and effectively inhibiting the initiation of thermal cracking [20].

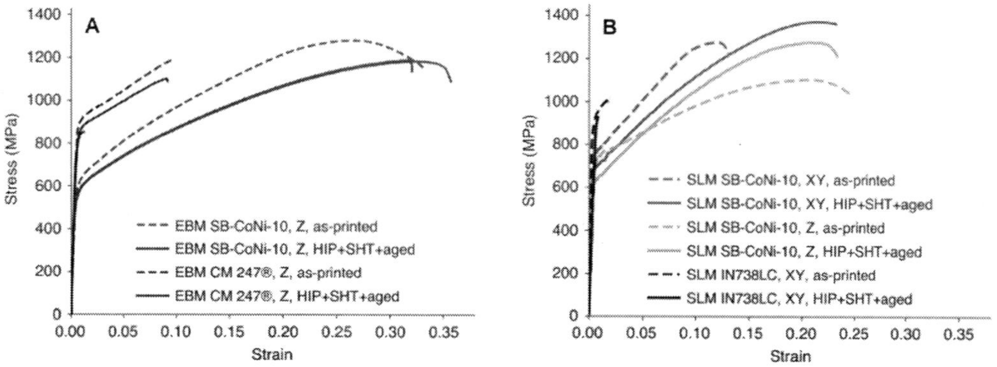

Figure 3. Room temperature stress-strain curve for EBM (A) and SLM (B) SB-CoNi-10 alloys [19].

The stress-strain curves for the Haynes 230 specimens (original type and Zr modified type) are shown in Figure 4, where AP denotes as-printed samples and HT signifies sample post-heat treatment. The specimens modified with Zr exhibit an enhancement in strength by nearly 50%. Additive manufacturing technology holds vast potential for future applications. Currently, there is a need for further optimization of material properties and manufacturing process parameters. Additionally, the commercial viability of the materials and the actual performance of products in service still require validation.

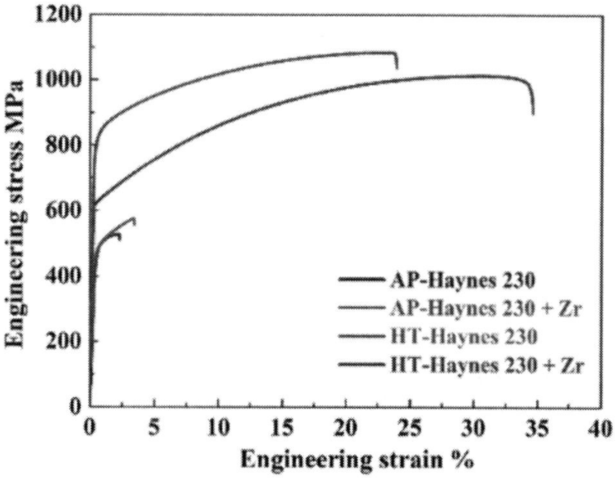

Figure 4. Stress-strain curves of Haynes 230 specimens before and after heat treatment and Zr modification [20].

2.4 Alloy surface-coating

The combination of mechanical properties, high-temperature durability, and oxidation resistance of nickel alloys poses challenges. Surface coating technologies offer new solutions: through methods such as electrochemical deposition, chemical vapor deposition, or thermal spraying, protective coatings can be created on the substrate surface. In the application of ultra-supercritical units, coatings can provide excellent oxidation resistance and corrosion resistance for nickel alloys, significantly reducing the demand for substrate performance and costs. Coating preparation technologies have matured, primarily including the following methods:

(1) Plasma spraying technology: This technique involves melting the coating material using a plasma jet and spraying the material onto the pre-treated surface of the workpiece. This technology can operate at lower temperatures, reducing the risk of thermal stress and deformation on the workpiece. The protective performance of the coating is excellent, but it requires high pre-treatment of the substrate surface, and the cost cannot be ignored.

(2) Electron beam evaporation: This method utilizes electron beam bombardment to evaporate the coating material, depositing it onto the material surface in a high vacuum environment. This technology boasts high material utilization efficiency and a lower risk of thermal damage to the workpiece. However, the coating thickness may not be uniform.

(3) Vacuum arc ion plating: In this process, the cathode material is evaporated during continuous arc discharge and deposited onto the surface of the workpiece to form a coating. Compared to plasma spraying technology, vacuum arc deposition produces coatings with uniform and controllable thickness. However, it suffers from lower production efficiency and the presence of significant macroparticles (MPs) in the coating. To address this issue, Xu et al. introduced a hollow cathode vacuum arc (HCVA) in the arc ion plating chamber, effectively reducing the target voltage (by 29%) and decreasing the size of macroscopic particles in the coating, thereby enhancing deposition efficiency [21]. As illustrated in Figure 5, the HCVA significantly reduces the overall quantity and the macroparticles' average diameter.

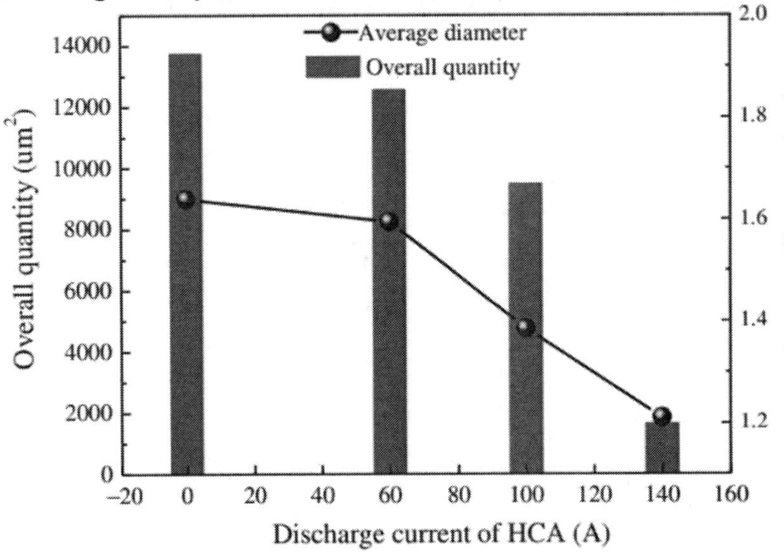

Figure 5. MPs parameters on the AlTiN coating surfaces versus HCVA current [21].

(4) Chemical vapor deposition (CVD): CVD involves the directed transport of chemical reaction gases through gas pipelines, with strict control over the concentration of the reaction atmosphere. This method utilizes gaseous precursor reactants, facilitating chemical reactions between atoms and molecules, resulting in the decomposition of certain components within the gaseous precursor, thereby forming a coating on the substrate surface. Coatings produced by this technique exhibit uniform thickness and excellent adhesion to the substrate.

In addition, depending on the choice of materials, coatings can impart various performance enhancements to the substrate material. Thermal barrier ceramic coating materials typically exhibit low thermal conductivity and high coefficients of thermal expansion. However, its thermal shock resistance, thermal insulation properties, and phase stability need improvement, which can be achieved through doping with rare earth metal oxides, mixtures, and similar approaches [22]. The mismatch in thermal expansion coefficients between ceramic surface layers and alloy materials can affect the bonding strength between them. To address this issue, a metallic bonding layer is introduced, which not only resolves the mismatch but also enhances the oxidation and corrosion resistance of the materials.

Journal of Physics: Conference Series **2808** (2024) 012005 doi:10.1088/1742-6596/2808/1/012005

3. Characterization techniques for nickel-based alloys

3.1 Morphology characterization

Similar to optical microscopes that utilize visible light for observation, the scanning electron microscope (SEM) employs electrons for imaging purposes. Figure 6 illustrates the schematic diagram of an SEM. Electrons, upon interaction and reflection from the material surface, are captured by a detector.

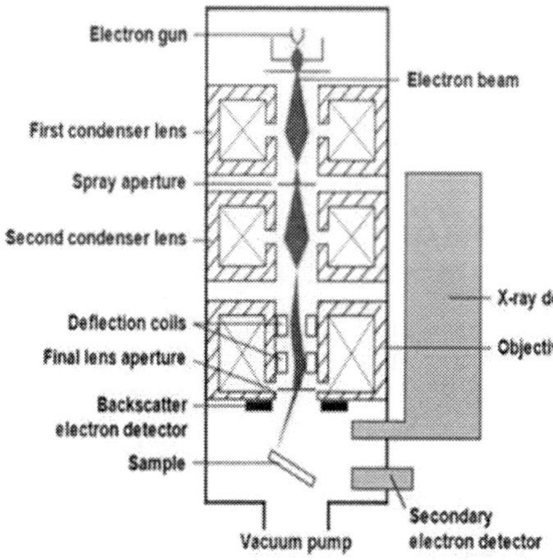

Figure 6. Schematic diagram of SEM structure [23].

The strength of the signal at various positions, after processing, provides a direct representation of the sample's morphology. Due to the significantly shorter wavelength of electrons compared to visible light, the diffraction effects allow SEM to achieve much higher resolution capabilities than optical microscopes, reaching resolutions at the nanometer scale.

In the application of nickel-based alloys, scanning electron microscopy (SEM) is extensively employed for characterizing aspects such as material microstructures, precipitate phases, oxides, and grain morphology. This enables researchers to investigate failure modes, strengthening mechanisms, corrosion behavior, and other properties of the materials. Bhagwat et al. utilized SEM to examine the surface microstructures of Ni-617 oxidized at 650 and 710°C in AUSC steam for varying durations, as depicted in Figure 7. Their findings indicated that the morphology of the oxide layer is significantly influenced by both time and temperature [24]. Furthermore, given that different elements emit characteristic X-rays when excited by an electron beam, SEM coupled with Energy Dispersive Spectroscopy (EDS) enables the characterization of both the types and concentrations of elements within a sample.

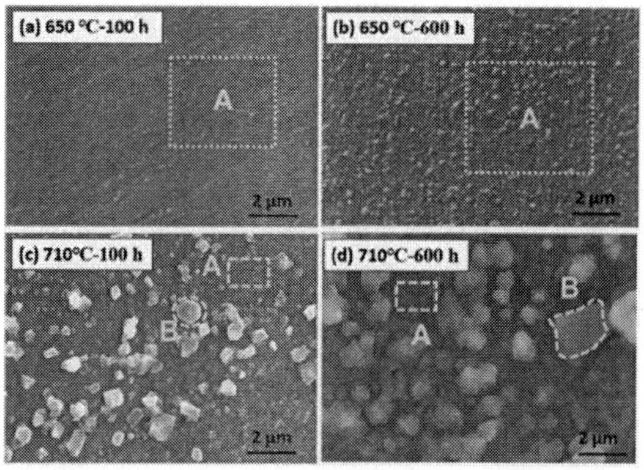

Figure 7. Surface morphology of Alloy 617 after: (a) 650°C-100 h; (b) 650°C-600 h; (c) 710°C-100 h; (d) 710°C-600 h oxidation under the AUSC steam [24].

Wang et al. employed EDS analysis to investigate the hot corrosion behavior of the GH984G alloy. Their findings revealed that a higher chromium content facilitates the formation of a denser and more continuous Cr_2O_3 layer, thereby enhancing corrosion resistance [25]. Nowadays, the advent of the combination offers a powerful tool for scientific research. This combination integrates high-resolution imaging with precise elemental analysis, enabling researchers to gain a comprehensive understanding of a sample's properties and composition. Such techniques have played a pivotal role not only in the characterization of alloy materials but also across various other fields.

Atomic force microscopy (AFM), based on the interaction between a sharp tip and the sample surface, enables high-resolution imaging of surfaces with atomic-scale resolution. The fundamental principle of AFM involves the measurement of forces as the tip is scanned across the surface. By maintaining the forces constant using feedback mechanisms, AFM generates topographic images of the sample surface. Additionally, AFM can provide information about surface properties such as roughness, adhesion, elasticity, and magnetic forces, making it a valuable tool for surface characterization at the nanoscale. In addition to imaging, AFM can be used for nanomanipulation, nanolithography, and nanomechanical measurements. It enables researchers to probe and manipulate individual atoms and molecules, making it an invaluable tool for fundamental studies in nanoscience and nanotechnology.

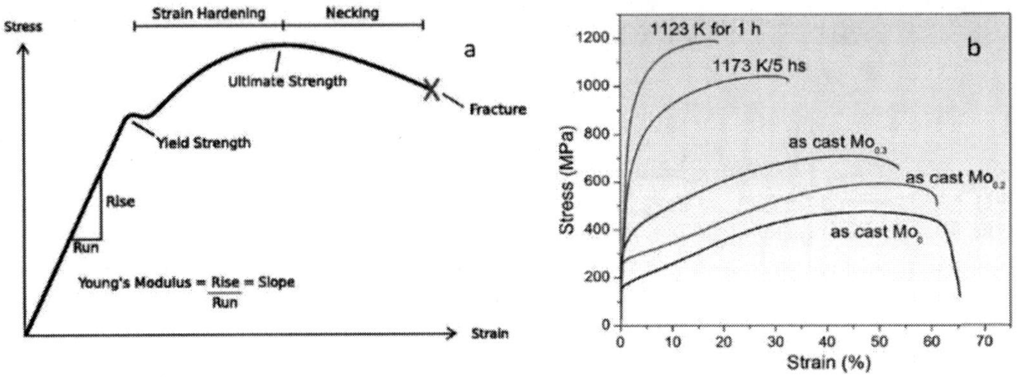

Figure 8. Stress-strain curve in tensile testing (a) and experimental stress-strain curve of the FeCoCrNi alloy (b) [26] [17].

3.2 Mechanical properties characterization

Stress represents the force per unit area applied to a material, while strain denotes the ratio of deformation length to the original length of the material. The change of material stress and strain is termed the stress-strain curve. This curve provides a direct and precise reflection of a material's crucial mechanical properties and its response behavior to external forces. The entire stress-strain curve of a material can be categorized into four distinct stages, as illustrated in Figure 8: (1) Elastic stage, where strain is directly proportional to stress. During this phase, the internal microstructure of the material undergoes no irreversible changes, and upon unloading, the specimen returns to its original dimensions; (2) Yielding stage, characterized by deformation beyond mere elasticity. Only a portion of the deformation in the specimen can be recovered, and changes occur within the material's internal structure; (3) Hardening stage: The specimen undergoes pronounced and uniform plastic deformation. As the strain on the specimen increases, there is a corresponding increase in stress. This phenomenon, where the resistance to plastic deformation rises with increasing deformation, is also known as work hardening or strain hardening; (4) Necking stage: The specimen begins to experience non-uniform plastic deformation, leading to neck formation. The stress decreases until the specimen eventually fractures.

Different stages in the stress-strain relationship signify varying ways in which materials respond to external forces. Analyzing the data at each stage and transitional point provides insights into various mechanical properties of the material. For instance, by tracking the stress values during elastic and plastic deformation, one can determine parameters like yield strength and tensile strength of the material. In characterizing the mechanical properties of nickel alloys, Liu et al. explored the precipitation-strengthening process in the FeCoCrNi alloy [17]. They plotted the stress-strain curves for the FeCoCrNi alloy with varying Mo contents, as well as the FeCoCrNiMo0.3 alloy after aging treatment (refer to Figure 8(b)). By employing the formula for tensile strength calculation $\sigma = F_b/S_o$, where σ denotes the tensile strength (N/mm^2), F_b represents the applied force on the specimen (N), and S_o is the original cross-sectional area of the specimen (mm^2), they determined the ultimate tensile strength of the alloy. The results indicate that post aging at 1, 123 K for 1 hour, the ultimate tensile strength of the specimen increased by approximately 50% compared to its pre-aged state.

3.3 Hardness characterization

Hardness quantifies a material's resistance to penetration by harder substances and reflects its resistance to plastic deformation, serving as a pivotal metric for material performance. The methods employed to measure hardness are rapid, cost-effective, and convenient. Based on the testing techniques and their applicability, three prevalent methods are employed:

1) Brinell Hardness (HB): A specified diameter steel or alloy ball is pressed into the specimen's surface. After maintaining for a predetermined time, the indentation diameter is measured. The Brinell hardness value is determined by dividing the applied load by the spherical area of the indentation.

2) Rockwell Hardness (HR): A diamond cone or a small diameter steel ball is pressed into the material surface under a constant load. Hardness is calculated from the depth of the indentation. The formula is expressed as HR = N-h/S, where N is a constant, contingent upon the indenter type, h represents the indentation depth, and S is the scale unit of the given reference.

3) Vickers Hardness (HV): The Vickers hardness is obtained by dividing the surface area of the indentation by the applied load when a diamond pyramid indenter is used.

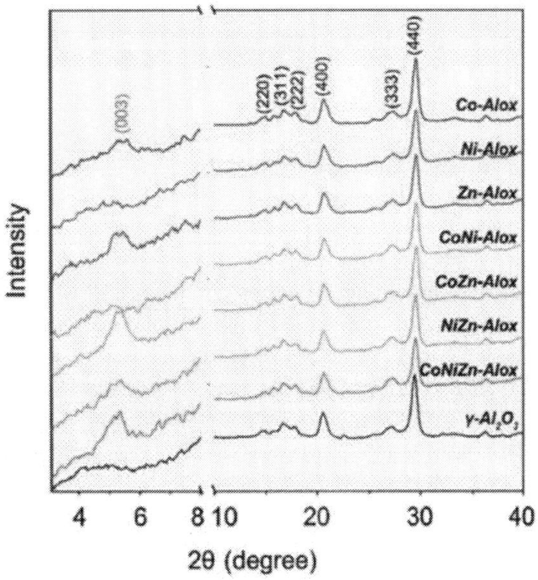

Figure 9. Co, Ni, and Zn XRD pattern and pristine γ-Al_2O_3 at 2θ(3-40°) [28].

3.4 Crystallinity characterization

X-ray diffraction (XRD) is applied for the characterization of crystalline materials by analyzing the diffraction patterns resulting from the interaction of X-rays with the atomic structure of the sample. XRD provides valuable information regarding the crystal structure, phase composition, crystallographic orientation, lattice parameters, and crystallite size of materials.

XRD analysis can be performed in different modes, including powder X-ray diffraction and single-crystal X-ray diffraction (PXRD and SC-XRD). In PXRD, powdered samples are used, and the diffraction patterns provide information about the crystalline phases present in the sample. SC-XRD, on the other hand, is used for determining the atomic arrangement within single crystals, providing detailed structural information.

Li et al. utilized XRD spectra to analyze surface corrosion products of three alloy-coated specimens, namely Al, Al-Si, and Co-Al. They identified NiO as the predominant corrosion product in the Al coating, while Al_2O_3 was found to be the main corrosion product in the Al-Si coating. Similarly, the Co-Al coating primarily produced Al_2O_3 [27].

Gou et al. used XRD to determine the uptake mechanisms of Co, Ni, and Zn on γ-Al_2O_3 at pH 7.5, shown in Figure 9. These multimetallic LDHs appeared to be crystalline, according to XRD, with the type of heavy metal influencing the crystallinity [28].

4. Conclusion

Currently, countries are actively engaged in the research and development of nickel-based alloy tubing for 700°C ultra-supercritical boilers, achieving significant milestones and fostering favorable conditions for advancements in this domain. Nickel-based alloys demonstrate superior high-temperature strength and corrosion resistance. Furthermore, the processing characteristics of these alloys largely meet the stringent requirements for tubing in ultra-supercritical boiler applications. Collectively, these advancements lay a solid foundation for the development of 700°C ultra-supercritical power plants. This review encapsulates the primary methods for modifying nickel-based alloys and commonly employing characterization techniques. The objective is to provide guidance to research professionals interested in nickel-based alloy studies, facilitating simpler but more concise efforts in the development of tubing suited for ultra-supercritical boilers.

References

[1] Somova E. V., Tugov A. N., and Tumanovskii A. G. Overview of Foreign Boiler Designs for Ultra Supercritical (USC) Boilers and Prospects for Development of USC Power Units in Russia. Thermal Engineering, 2021, 68 (6), 417-433. DOI: 10.1134/s0040601521060094.

[2] Kumar A. and Pandey C. Development and Evaluation of Dissimilar Gas Tungsten Arc-Welded Joint of P92 Steel/Inconel 617 Alloy for Advanced Ultra-Supercritical Boiler Applications. Metallurgical and Materials Transactions, 2022, 53 (9), 3, 245-3, 273. DOI: 10.1007/s11661-022-06723-0.

[3] Zhou X., Niu T., Xin Y., Li Y., and Yang D. Experimental and numerical investigation on heat transfer in the vertical upward flow water wall of a 660 MW ultra-supercritical CFB boiler. Applied Thermal Engineering, 2021, 188. DOI: 10.1016/j.applthermaleng.2021.116664.

[4] Cheng C., Peng C., and Zhang T. Fuzzy K-Means Cluster Based Generalized Predictive Control of Ultra Supercritical Power Plant. IEEE Transactions on Industrial Informatics, 2021, 17 (7), 4, 575-4, 583. DOI: 10.1109/tii.2020.3020259.

[5] Song Y., Li Y., Li H., Zhao G., Cai Z., and Sun M. Hot deformation and recrystallization behavior of a new nickel-base superalloy for ultra-supercritical applications. Journal of Materials Research and Technology, 2022, 19, 4, 308-4, 324. DOI: 10.1016/j.jmrt.2022.06.141.

[6] Bhiogade D. S. Ultra supercritical thermal power plant material advancements: A review. Journal of Alloys and Metallurgical Systems, 2023, 3. DOI: 10.1016/j.jalmes.2023.100024.

[7] Ghule B., Sundaresan C., Ningshen S., and Raja V. S. Oxidation behavior of Ni-base superalloy 617 in simulated Advanced Ultra Supercritical (AUSC) steam. Corrosion Science, 2023, 224. DOI: 10.1016/j.corsci.2023.111503.

[8] M. Spiegel, M. B., J. Klöwer, K. H. Tak, H. Tschaffon, and U. Zeislmair. The influence of boron in Alloy 617. Vortragsveranstaltung FVHT, 2011, 34.

[9] Liu Z., Cen Z., He X., and Bao H. Systematical innovation of heat resistant materials used for 630~ 700°C advanced ultra-supercritical (A-USC) fossil fired boilers. Acta Metall Sin 2020, 56 (4), 539-548.
[10] Xu Y. X., Lu J. T., Li W. Y., Yang Z., Dang Y. Y., and Yang X. W. Effect and role of alloyed yttrium on the fireside corrosion behavior of Ni-Fe based alloys for 750°C ultra-supercritical boiler applications. Corrosion Science 2018, 143, 148-156. DOI: 10.1016/j.corsci.2018.08.025.
[11] Viswanathan R., Coleman K., and Rao U. Materials for ultra-supercritical coal-fired power plant boilers. International Journal of Pressure Vessels and Piping 2006, 83 (11-12), 778-783. DOI: 10.1016/j.ijpvp.2006.08.006.
[12] Zhang Yong. High entropy materials. IntechOpen, 2023. DOI: 10.1007/978-981-13-8526-1.
[13] Ma Y., Wang Q., Schweidler S., Botros M., Fu T., Hahn H., Brezesinski T., and Breitung, B. High-entropy energy materials: challenges and new opportunities. Energy & Environmental Science 2021, 14 (5), 2, 883-2, 905. DOI: 10.1039/d1ee00505g.
[14] Li W., Xie D., Li D., Zhang Y., Gao Y., and Liaw P. K. Mechanical behavior of high-entropy alloys. Progress in Materials Science 2021, 118. DOI: 10.1016/j.pmatsci.2021.100777.
[15] J. W. Yeh, S. K. C., S. J. Lin, J. Y. Gan, T. S. Chin, T. T. Shun, et al. Nanostructured high-entropy alloys with multiple principal elements: novel alloy design concepts and outcomes. Adv. Eng. 2004, Mater. 6, 299-303. DOI: 10.1002/adem.200300567 2004.
[16] He J. Y., Wang H., Huang H. L., Xu X. D., Chen M. W., Wu Y., Liu X. J., Nieh T. G., An K., and Lu Z. P. A precipitation-hardened high-entropy alloy with outstanding tensile properties. Acta Materialia 2016, 102, 187-196. DOI: 10.1016/j.actamat.2015.08.076.
[17] Liu W. H., Yang T., and Liu C. T. Precipitation hardening in CoCrFeNi-based high entropy alloys. Materials Chemistry and Physics 2018, 210, 2-11. DOI: 10.1016/j.matchemphys.2017.07.037.
[18] Zhang R., Zhao S., Ding J., Chong Y., Jia T., Ophus C., Asta M., Ritchie R. O., and Minor A. M. Short-range order and its impact on the CrCoNi medium-entropy alloy. Nature 2020, 581 (7, 808), 283-287. DOI: 10.1038/s41586-020-2275-z.
[19] Murray S. P., Pusch K. M., Polonsky A. T., Torbet C. J., Seward G. G. E., Zhou N., Forsik S. A. J., Nandwana P., Kirka M. M., Dehoff R. R., et al. A defect-resistant Co-Ni superalloy for 3D printing. Nat. Commun. 2020, 11 (1), 4, 975, Research Article. DOI: 10.1038/s41467-020-18775-0.
[20] Zhao Y., Ma Z., Yu L., and Liu Y. New alloy design approach to inhibiting hot cracking in laser additive manufactured nickel-based superalloys. Acta Materialia 2023, 247. DOI: 10.1016/j.actamat.2023.118736.
[21] Xu J., Xu M., Wang J., Wang H., Gong C., Yang C., Wang C., Zhang X., Ma T., and Zhao J. Discharge characterization and technological application of an arc ion plating with a hollow cathode vacuum arc. Journal of Adhesion Science and Technology 2022, 37 (4), 649-662. DOI: 10.1080/01694243.2022.2035511.
[22] Belichko D. R., Konstantinova T. E., Maletsky A. V., Volkova G. K., Doroshkevich A. S., Lakusta M. V., Kulik M., Tatarinova A. A., Mardare D., Mita C., et al. Influence of hafnium oxide on the structure and properties of powders and ceramics of the YSZ–HfO2 composition. Ceramics International 2021, 47 (3), 3, 142-3, 148. DOI: 10.1016/j.ceramint.2020.09.151.
[23] Mohammed A. and Abdullah A. Scanning electron microscopy (SEM): A review. In Proceedings of the 2018 International Conference on Hydraulics and Pneumatics – HERVEX, Băile Govora, Romania, 2018, Vol. 2018, pp 7-9.
[24] Ghule B., Sundaresan C., Ningshen S., and Raja V. S. Oxidation behavior of Ni-base superalloy 617 in simulated Advanced Ultra Supercritical (AUSC) steam. Corrosion Science 2023, 224. DOI: 10.1016/j.corsci.2023.111503.

[25] Wang J., Zhuo J., and Yao Q. Investigation of the fireside corrosion of a new wrought Ni-Fe-Cr-based alloy in advanced ultra-supercritical steam boilers. Corrosion Science 2022, 199. DOI: 10.1016/j.corsci.2022.110152.

[26] Breakdown, 2008. Typical Stress vs. Strain diagram for a ductile material [[online] Available]. https://en.wikipedia.org/wiki/Stress%E2%80%93strain_curve. 2008.

[27] Li Yanming, Chi Qingxin, Liu Huan, Tong Wenwei, Zhao Shiwei, and Gao Zhikun. High-temperature behaviors of three coatings on nickel-based superalloy DSM11 [J]. Journal of Aerospace Power, 2021, 36 (7): 1, 499-1, 508. Doi: 10.13224/j.cnki.jasp.20200448.

[28] Gou W., Mo X., Ren C., Wang H., and Li W. Formation of crystalline multimetallic layered double hydroxide precipitates during uptake of Co, Ni, and Zn onto gamma-alumina: Evidence from EXAFS, XRD, and TEM. Chemosphere 2022, 307 (Pt 4), 136, 055. DOI: 10.1016/j.chemosphere.2022.136055.

Journal of Physics: Conference Series **2808** (2024) 012006 doi:10.1088/1742-6596/2808/1/012006

The analysis and simulation of fracture behavior of fused quartz under dynamic loading

Wenbo Duan [1], Bingyao Zhao [1], Ping Zhou [1*] and Dongming Guo [1]

[1]State Key Laboratory of High-performance Precision Manufacturing, Department of Mechanical Engineering, Dalian University of Technology, Dalian 116024, Liaoning, China

[*]Corresponding author's e-mail: pzhou@dlut.edu.cn

Abstract: Fused quartz is widely utilized across various industries due to its superior performance. However, its brittleness often leads to fracture occurrences under variable external loads during processing or usage, making the investigation of its dynamic mechanical properties and fracture mechanisms crucial. In this study, the fracture strength variations of fused quartz under different loading rates were obtained through strength experiments. It was observed that when the velocity is below 2 mm/min, the loading rate has little influence on fracture strength. And when the velocity exceeds 2 mm/min, fracture strength rises with the increase in velocity. According to the fitted fracture strength-strain rate curve, a modified peridynamic failure criterion was developed to simulate the fracture process under different loading speeds. The simulated fracture morphology closely matched experimental results and the error in fracture load values is less than 8%. It validates the accuracy of the simulation. Additionally, it was found that the displacement values at the moment of fracture under different loading rates were relatively similar. It indicated that higher speeds result in greater loading energy, thus leading to the generation of more cracks and fragments. This study provides theoretical support and reference for understanding the fracture behavior of fused quartz.

1. Introduction

Fused quartz, as a typically hard and brittle material, finds wide applications in various fields such as space optics [1] and military industries [2] due to its outstanding optical, thermal, and mechanical properties. However, due to its low fracture toughness, fused quartz often develops internal damage defects during normal processing and usage, leading to component failures. Moreover, the external loads or disturbances it experiences are typically dynamic [3]. Therefore, investigating the fracture mechanisms of fused quartz under dynamic conditions is crucial.

Since the mechanical properties of many materials change under the influence of strain rate, obtaining the dynamic properties of the material is typically necessary before delving into the study of fracture mechanisms [4]. Grady and Lipkin [5] investigated the relationship between fracture strength and strain rate in different rocks and found that the fracture strength of most rock materials follows a power-law relationship with strain rate, with the exponent mostly being 1/3. Peroni et al. [6] conducted experiments by using the Split Hopkinson pressure bar method and Brazilian disk test to investigate the variation of material properties of commercial glass with strain rate. The results showed that the compressive strength and elastic modulus of the glass material were not very sensitive to strain rate, while the tensile strength increased significantly with increasing strain rate. Therefore, it is necessary to explore the dynamic properties of fused quartz before investigating its fracture mechanisms.

With the advancement of computer technology, an increasing number of scholars are using numerical simulation methods to explore brittle fracture processes. Peridynamics, as a non-local model, successfully avoids the issue of singularities around crack tips and has been widely applied in fields such as crack propagation and brittle failure. Jafaraghaei et al. [7] simulated the failure process of glass plates subjected to impact loads at different angles and velocities using peridynamics and analyzed the influence of material porosity on the reaction force and restitution coefficient. Madenci et al. [8] used the PD model to predict the residual strength of laminate composites under impact compression loads, demonstrating that PD can effectively capture the failure modes of compression failure. However, to simulate the failure of brittle materials under dynamic loads, it is also necessary to consider the influence of material dynamic properties.

This study first conducted ball-on-ring strength experiments to test the fracture strength of fused quartz samples under different loading rates and fitted the fracture strength-strain rate curve. Subsequently, this curve was used to adjust the failure criterion of bond-based peridynamics, and simulations of the fracture process in ball-on-ring experiments were performed. Finally, the simulation results were compared with the strength experiments to validate the model's accuracy, and the impact of velocity variation on brittle fracture was analyzed.

2. Experimental procedures and results

The experimental principle of the ball-on-ring strength test is illustrated in Figure 1. The experimental sample is a square fused quartz plate with a side length of $2L$ and a thickness of t. The quartz plate is positioned centrally above a support ring with a circular hole diameter of $2a$, and a spherical indenter with a radius of R applies a load P until the sample fractures.

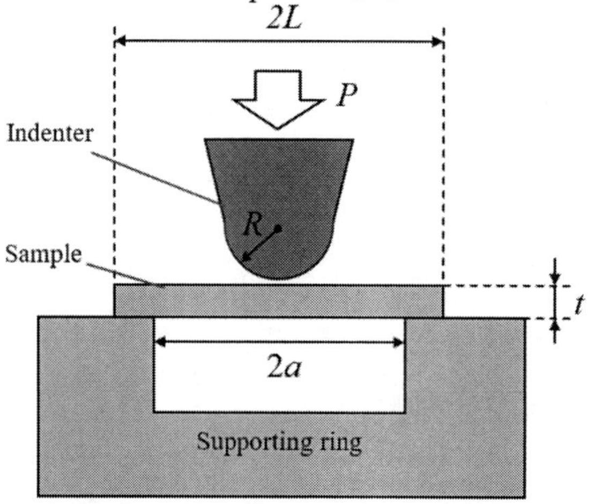

Figure 1. Diagram of the ball-on-ring experiment.

After the indenter contacts the fused quartz, it creates a region of approximately uniform load distribution with a radius of b. Within this region, the maximum radial stress equals the tangential stress, and this maximum stress can be calculated by using Equation (1) proposed by Kirstein and Woolley [9], which was later adopted by Shetty et al. to calculate the maximum principal stress under uniformly distributed loads [10].

$$\sigma_{\max} = \frac{3P(1+\nu)}{4\pi t^2}\left[1+2\ln\frac{a}{b}+\frac{1-\nu}{1+\nu}\left(1-\frac{b^2}{2a^2}\right)\frac{a^2}{L^2}\right] \tag{1}$$

The radius b of the uniformly distributed load region can be calculated based on the relationship between the contact region radius z and the thickness t of the quartz plate, using the formula proposed by With and Wagemans [11].

$$\begin{cases} b = z & z > 1.724t \\ b = (1.6z^2 + t^2)^{1/2} & z < 1.724t \\ b = 0.325t & z \to 0 \end{cases} \tag{2}$$

The value of z can be determined according to the Hertz contact formula, which depends on material properties, Load P, and Radius r of the indenter.

$$z = \left[\frac{3Pr}{4} \left(\frac{1-\mu_1^2}{E_1} + \frac{1-\mu_2^2}{E_2} \right) \right]^{1/3} \tag{3}$$

where E_1 = 201 GPa, μ_1 = 0.26, E_2 = 72 GPa, and μ_2 = 0.17, which refer to the elastic modulus and Poisson's ratio of the fused quartz and the indenter, respectively.

This study conducted ball-on-ring strength experiments on fused quartz glass plates at different loading rates using a mechanical testing machine, testing the fracture strength of quartz glass under different strain rates. The experimental parameters are shown in Table 1. Due to the high variability in the strength experiments, each speed was repeated 10 times, and the maximum load P_{min} at the time of fracture was recorded.

Table 1. Experimental parameters.

Parameter	Value
v (mm/min)	0.2, 1, 2, 5, 10, 20, 50
L (mm)	5
t (mm)	1
a (mm)	4
r (mm)	5

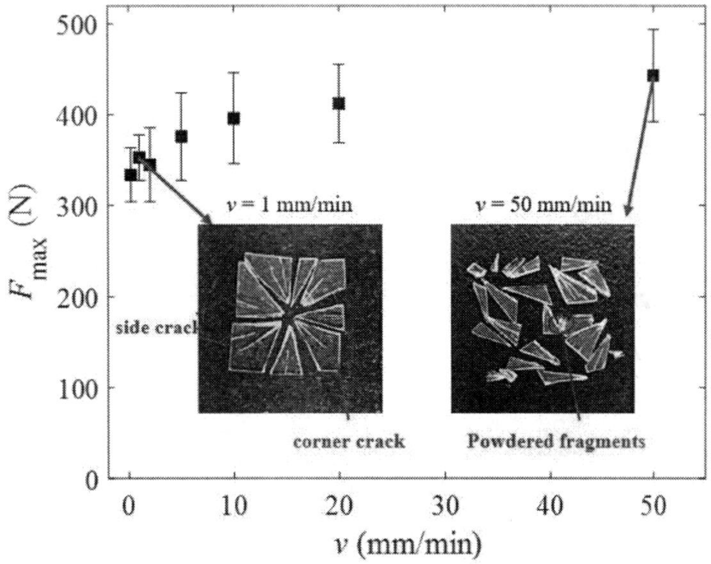

Figure 2. F_{max}-v result and damage morphology images.

Based on the average maximum load for each experimental speed, a plot of the maximum load versus speed variation is shown in Figure 2. From the resulting plot, it can be observed that the maximum load generally increases with the gradual increase in indenter speed. However, between speeds of 0.2 mm/min and 2 mm/min, the variation in maximum load is minimal, remaining around 340 N, with some fluctuations. When the speed reaches 50 mm/min, the average maximum load has already reached 443 N, which is significantly higher than the load at low-speed loading. As the velocity increases, the discreteness of experimental results also increases. The fracture morphology of the quartz plate at a loading speed of 1 mm/min appears radial, with cracks radiating outward from the central region, forming multiple triangular or wedge-shaped fragments. On individual fragments, some cracks have not completely penetrated. These cracks at the corners generally propagate outward from the center, while cracks on the side propagate inward from the outer edge. Additionally, there are a few circular Hertz cracks at the center, resulting in voids at the center of the fragments. At a loading speed of 50 mm/min, the fragmentation results in more numerous and smaller fragments, with some powdery debris, making it difficult to reconstruct the specific fracture morphology.

The strain rate under different loading speeds can be calculated by using Equation (4).

$$\dot{\varepsilon} = \frac{v}{t} \tag{4}$$

According to Equations (1), (2) and (3), the fracture strength of the quartz plate under different loading speeds can be determined. When the strain rate is below 0.03 s^{-1}, the fracture strength is essentially 585 MPa. However, when the strain rate exceeds 0.03 s^{-1}, a power-law fit is applied to the curve of fracture strength versus strain rate, and the fitting expression is as follows:

$$\sigma_f(\dot{\varepsilon}) = p\dot{\varepsilon}^q + l \qquad (\dot{\varepsilon} > 0.03) \tag{5}$$

The fitting results are as follows: p = -930 MPa, q = -0.0476, and l = 1, 684 MPa. The graph of the fitted strain rate-fracture strength curve is shown in Figure 3.

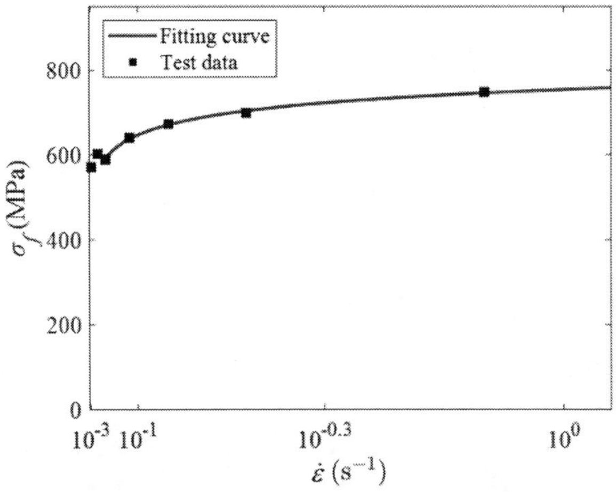

Figure 3. Curve plot of strain rate and fracture strength.

3. Correction and simulation

Based on the observed pattern of strength variation with strain rate, a certain adjustment is made to the failure criterion of bond-based peridynamics materials. This adjustment is then applied to simulate the ball-on-ring experiment and validate it against experimental results. The geometric dimensions of the sample model are identical to those of the experimental sample, with a discrete size of Δx = 0.2 mm. The elastic modulus E = 72 GPa, Poisson's ratio μ = 0.25, and density ρ = 2, 440 kg/m^3 are used. For isotropic brittle materials, the critical elongation ratio s_0 of the bond length is typically considered equal to σ_{crit}/E, where σ_{crit} is the uniaxial tensile strength [12-13]. In this study, the fracture strength σ_f obtained from the ball-on-ring experiment is used. Taking into account the variation of fracture strength under different strain rates, and substituting the bond strain rate for strain rate in Equation (5), the critical elongation ratio s_0 influenced by the bond strain rate in the model is given by:

$$s_0(\dot{s}) = \begin{cases} 0.0078 & \dot{s} \le 0.03 \\ 0.0234 - 0.0129\dot{s}^{-0.0476} & \dot{s} > 0.03 \end{cases} \tag{6}$$

The bond strain rate is calculated in the computation by the change in bond length at each time step.

$$\dot{s} = \frac{s_{t+\Delta t} - s_t}{\Delta t} = \frac{\xi_{t+\Delta t} - \xi_t}{\xi_0 \Delta t} \tag{7}$$

where ξ_0 represents the initial bond length and Δt is the time step.

The established peridynamics model is illustrated in Figure 4, where the blue region represents the free region of the sample, with a bottom diameter identical to the diameter of the ring hole, with a thickness of 8 mm. The green region denotes the boundary constraint area with a thickness of 0.2 mm. According to the experimental scenario, the movement of material points in the negative direction of the z-axis is restricted in this area during computation and all other degrees of freedom remain unconstrained. A C++ program developed in-house is used to compute the model via explicit integration. The study simulated the situation with an indenter velocity of 1 mm/min and 50 mm/min. The results are visualized and analyzed by using MATLAB.

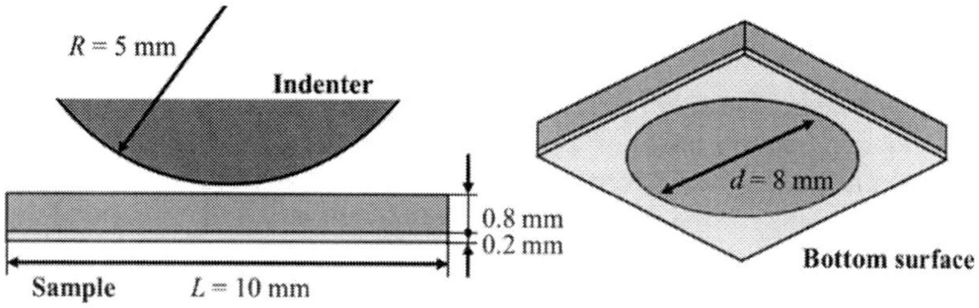

Figure 4. Peridynamics model.

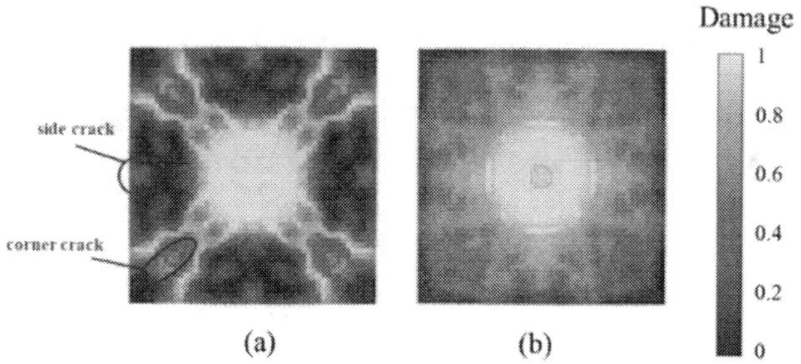

Figure 5. Simulated simulation results.

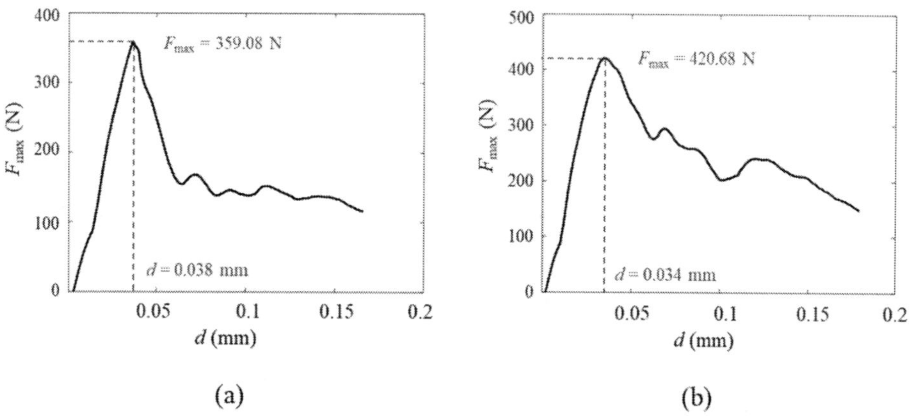

Figure 6. Simulated load-displacement curve.

Figure 5 depicts the damage results obtained from the peridynamics simulation, reflecting the extent of localized damage [14]. When the loading speed is 1 mm/min, several wide cracks radiate from the central region to the outer periphery of the sample, forming multiple large fragments. Within individual fragments, there are also relatively fine cracks that have not completely penetrated. Cracks at the corners extend outward from the center, while those at the edges extend inward from the outer periphery. At a loading speed of 50 mm/min, the central impact damage area enlarges, and the degree of damage is more severe. Surface cracks are more densely distributed and intersect with each other, forming smaller fragments, which aligns with the experimental results. The load-displacement curves in Figure 6 show that the maximum loads under different speeds are 359.08 N and 420.68 N, respectively. The error in fracture load values is less than 8%, validating the accuracy of the simulation model. Moreover, the displacements at the time of fracture are 0.038 mm and 0.034 mm for 1 mm/min and 50 mm/min, respectively, with little difference between them. This is because, under similar displacements, larger loads imply greater energy, requiring more cracks to dissipate this energy, explaining why increasing the loading speed results in more cracks and fragments.

4. Conclusion

This study systematically investigated the variation of fracture strength of fused quartz glass samples under different loading rates, focusing on the influence of strain rate. It was found that when the velocity is below 2 mm/min, the loading rate had little influence on fracture strength. And when the velocity exceeds 2 mm/min, fracture strength rises with the increase in velocity. This phenomenon may be due to an increase in the loading rate, leading to a shorter crack propagation time. In turn, it requires higher stress to overcome the tensile strength of the material and lead to fracture. Based on the fitting curve of fracture strength-strain rate, a modified peridynamic failure criterion was developed to simulate ball-on-ring strength experiments under different loading speeds. Both experimental and simulated results revealed that at lower velocities, the samples exhibited fewer cracks and larger fragment sizes. At higher velocities, cracks became denser, intersecting with each other, resulting in smaller fragments. The simulated results demonstrated good consistency with experimental observations. Additionally, we observed that the maximum loads at different velocities were generally consistent with experimental values, showing an increasing trend with velocity. Moreover, although there was little difference in displacement values at fracture, higher velocities led to greater impact energy on the samples, resulting in more extensive crack propagation and consequently generating more numerous smaller fragments. These findings deepen our understanding of the fracture behavior of fused quartz glass samples and provide valuable insights for evaluating their performance in practical applications.

Acknowledgments

The authors greatly appreciate the National Natural Science Foundation of China (51991372).

References

[1] Li C., Zhang L., Sun L., Yang S., Wu C., Long X., Ding J., and Jiang Z. (2019). A quantitative analysis of the indentation fracture of fused silica. Journal of the American Ceramic Society, 102 (12), 7, 264-7, 277.

[2] Ruggiero A., Iannitti G., Testa G., Limido J., Lacome J. L., Olovsson L., Ferraro M., and Bonora N. (2014). High strain rate fracture behavior of fused silica. In Journal of Physics: Conference Series (Vol. 500, No. 18, p. 182, 036), IOP Publishing.

[3] Chadegani A., Iyer K. A., Mehoke D. S., & Batra R. C. (2015). Hypervelocity impact of a steel microsphere on fused silica sheets. International Journal of Impact Engineering, 80, 116-132.

[4] Cheng J., Yang Z., Wang C., Zhao L., Chen M., Wang J., ... & Xu H. (2022). Effect of scratches on the damage characteristics of fused silica optics under extremely high impact load. International Journal of Mechanical Sciences, 219, 107, 099.

Journal of Physics: Conference Series **2808** (2024) 012006 doi:10.1088/1742-6596/2808/1/012006

[5] Grady D. E. & Lipkin J. (1980). Criteria for impulsive rock fracture. Geophysical Research Letters, 7 (4), 255-258.

[6] Peroni M., Solomos G., Pizzinato V., & Larcher M. (2011). Experimental investigation of high strain-rate behavior of glass. Applied Mechanics and Materials, 82, 63-68.

[7] Jafaraghaei Y., Yu T., & Bui T. Q. (2022). Peridynamics simulation of impact failure in glass plates. Theoretical and Applied Fracture Mechanics, 121, 103, 424.

[8] Madenci E., Yaghoobi A., Barut A., & Phan N. (2021). Peridynamic modeling of compression after impact damage in composite laminates. Journal of Peridynamics and Nonlocal Modeling, 1-21.

[9] Kirstein A. F. & Woolley R. M. (1967). Symmetrical bending of thin circular elastic plates on equally spaced point supports. J. Res. Natl. Bur. Stand. C, 71 (1), 1-10.

[10] Shetty D. K., Rosenfield A. R., McGuire P., Bansal G. K., & Duckworth W. H. (1980). Biaxial flexure tests for ceramics. Am. Ceram. Soc. Bull., (United States), 59 (12).

[11] De With G. & Wagemans H. H. (1989). Ball-on-ring test revisited. Journal of the American Ceramic Society, 72 (8), 1, 538-1, 541.

[12] Wang Y., Han F., & Lubineau G. (2021). Strength-induced aerodynamic modeling and simulation of fractures in brittle materials. Computer Methods in Applied Mechanics and Engineering, 374, 113, 558.

[13] Zhang H. & Qiao P. (2018). A coupled aerodynamic strength and fracture criterion for open-hole failure analysis of plates under tensile load. Engineering Fracture Mechanics, 204, 103-118.

[14] Silling S. A. & Askari E. (2005). A mesh-free method based on the aerodynamic model of solid mechanics. Computers & Structures, 83 (17-18), 1, 526-1, 535.

Research on ballistic protection characteristics of perforated plates made of different steel materials for 12.7 mm penetrator projectile

Anshun Shi[1], Xiaowen Hong[1*], Hongwei Han[1], Kailang Xia[1], Xuefei Fan[1], Chao Qin[1] and Sujie Sun[1]

[1]No. 52 Institute of China Ordnance Industries, Yantai, 264003, China

[*]Corresponding author's e-mail: hongycxw@163.com

Abstract. To compare the ballistic protection characteristics of perforated plates made of different steel materials against 12.7 mm perforating ammunition, the ballistic gun, light curtain velocity measurement system, and high-speed camera were used to study its ballistic performance, and the ballistic protection, perforated plate damage area, core fracture and core deflection characteristics of perforated plates made of different armor steel materials were compared. The results show that the ballistic protection performance of C-2 and H-1 perforated plates is better than that of C-1 and C-3 perforated plates, the protection performance of C-1 and C-3 is equivalent, and the protection coefficient of H-1 can reach or even exceed that of C-2. The hole damage area of an H-1 perforated plate is larger than that of a circular perforated plate, and the number of bullet core fractures is more, but its crack arrest performance is poor. The damage area of a single interconnection hole in perforated plates of different armor steel materials meets the linear relationship, and its linear slope can characterize the degree of hole loss. The damage of a single interconnection hole with hexagonal holes is greater than that of perforated plates with circular holes. At the same time, the projectile deflection angle of perforated plates made of different armor steel materials also conforms to the linear relationship, and the fitting linear slope can measure the degree of core deflection. The higher the hardness of perforated plates with circular holes, the greater the core deflection, while the degree of core deflection of hexagonal holes is greater than that of perforated plates with circular holes.

1. Introduction

The armor material of traditional vehicles mainly uses homogeneous armor steel. Due to its poor resistance to multiple rounds, improving its ballistic protection ability is mainly achieved by increasing the thickness, but this seriously affects the mobility of armored vehicles [1, 2]. To achieve the lightweight and multi-bullet resistance performance of the armor module, researchers have carried out a lot of research on the perforated armor plate. The perforated plate is mainly evenly distributed, with many holes and regular geometric shapes on the homogeneous armor steel. When the projectile penetrates the perforated plate, if the impact point is located near the hole, the projectile body is subjected to the action of asymmetric force, which causes ballistic deflection, projectile head erosion, and core fragmentation [3, 4].

At present, the research on perforated plates is mainly focused on armor steel with different hole shapes and different steel materials and perforated structures. Balos et al. [5] studied the ballistic protection ability of perforated plates made of different types of steel and geometric shapes. The results showed that the perforated plate with optimized structure can make the bullet core break frequently, can be divided into five parts at most, and cannot penetrate the base armor. The mass efficiency of the whole additional armor is 1.76, and the mass efficiency of the perforated plate is 5.91. Mishra et al. [6] studied the influence of the size and shape of regular holes on the ballistic impact performance of high-hardness steel plates and found that the ballistic efficiency of perforated plates is very sensitive to the size of holes, and the protection efficiency of plates with holes close to the diameter of the bullet core is higher. They also found that the presence of pores inhibited the formation of shear bands on perforated plates. Balos et al. [7] evaluated the properties of isothermally quenched ductile iron at different temperatures to determine its applicability as a perforated plate. The study found that 7 mm and 9 mm perforated plates with higher hardness and lower ductility produced by austenitic tempering of isothermally quenched ductile iron at a lower temperature of 275°C can effectively break the bullet core, thus significantly reducing the penetration ability of the bullet core. Garbarz et al. [8] studied the ballistic performance of perforated armor plates made of nano-bainite steel with different thicknesses. According to the procedures specified in STANAG 4569 and AEP-55 Volume 1 specification (format applicable to test plates), the protective performance of nano bainite perforated plates with a hardness of 500 HBW was tested through multiple shots. Based on the analysis of the shooting test results and the macro and micro analysis of perforated plates before and after shooting, the best perforation method was selected, and the optimal perforation geometric size and arrangement under different plate thicknesses were determined. The above research on perforated plates is based on the ballistic performance of perforated plates made of a single material or different heat treatment methods, but the protective performance of perforated plates made of different armor materials is not compared.

In this paper, the ballistic test of a 12.7 mm penetrator projectile penetrating different steel perforated plates will be carried out to obtain the protection performance of different steel perforated plates. Based on the ballistic test results, the remaining penetration depth of different base armor, the perforation damage area of perforated plates, and the number of core fractures will be analyzed. The perforated plate armor steel material with better protection performance will be selected to provide a reference for the material selection of additional armor for the ballistic protection of light armored vehicles.

HNASS (circular) | 30CrNiMoNb (circular) | 22SiMn2TiB (circular) | HNASS (hexagon)

Figure 1. Perforated plates of different armor steel materials.

2. Experimental design

2.1. Test sample and scheme

The perforated plates of different armor steel materials used in the test are shown in Figure 1. The size of the perforated plate is 320 mm × 320 mm × 6 mm, the diameter of the circular hole is 10 mm, the circumscribed diameter of the regular hexagonal opening is 10 mm, and the distance between holes is 15 mm. To compare the bullet resistance of perforated plates made of different armor steel materials, typical 19Cr19Mn2Ni0.88N (HNASS), 30CrNiMoNb, and 22SiMn2TiB armor steel were selected for comparative study. The 32Mn2Si2MoA steel with high hardness and 45 steel with medium and low hardness were used as base target plate materials, respectively. The chemical composition of different steel materials is shown in Table 1.

Table 1. Chemical composition of different steel materials.

Steel grade	Chemical composition (wt. %)										
	C	Si	Mn	P	S	Cr	Ni	Mo	Cu	N	W
22SiMn2TiB	0.19	0.70	1.50	0.015	0.010	—	—	—	—	—	—
30CrNiMoNb	0.27	0.20	0.20	0.015	0.008	0.60	0.60	0.15	—	—	—
HNASS	0.03	—	19.28	—	—	19.32	2.01	0.0001	0.031	0.88	0.005
32Mn2Si2MoA	0.27	0.30	1.40	0.015	0.008	—	—	—	—	—	—
45	0.45	0.17	0.5	—	—	0.25	0.30	—	0.25	—	—

2.2. *Experimental site*

Figure 2 is the schematic diagram of the test layout. The test gun is a 12.7 mm ballistic gun, and a light curtain speed measurement system is arranged in front of the ballistic gun for speed measurement. The bullet used in the test is a 54-Type 12.7 mm armor-piercing incendiary bullet (API) with an average mass of 48.3 g. The bullet is coated with armor. The armor material is F11 copper-clad steel, and the armor weight is 11.3 g. There is a small amount of incendiary agent made of barium nitrate and aluminum-magnesium alloy powder in the head of the projectile, with a mass of 0. 92 g. A lead sleeve with a mass of 6.1 g is embedded between the bullet core and the armor. The core material is T12A tool steel with a mass of about 30.0 g. The distance between the orifice plate and the ballistic muzzle is 100 m. The base target plate is fixed on the target frame. The orifice plate and the base target plate are separated by a support gap and fastened with bolts. The target plate materials are 32Mn2Si2MoA armored steel and 45 # steel, with a size of 500 mm × 500 mm and a thickness of 14 mm and 20 mm, respectively. The distance between the orifice plate and the 45 # steel target plate is 400 mm, and the distance between the orifice plate and the 32Mn2Si2MoA armored steel target plate is 50 mm. During the test, the V_{25} speed of 12.7 mm API met 818^{+7}_{-8} m/s, and the forward penetration test was carried out according to the predetermined target point. At the same time, two Phantom-V12 high-speed digital cameras were used for each test to observe the projectile penetration process.

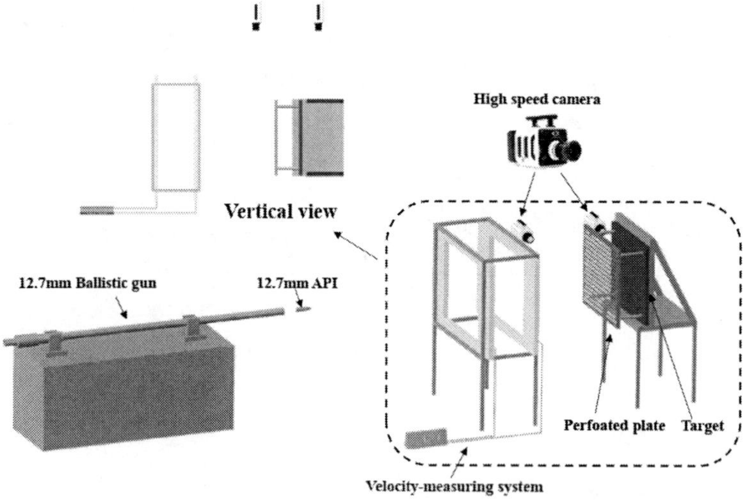

Figure 2. Schematic illustration of experimental layout.

During shooting, one high-speed camera is used to capture the position of the projectile before perforation, and one is used to observe the state of the projectile after passing through the orifice plate.

The shooting rate of the projectile before hitting the target is 50000 frames/s, and the maximum resolution of each frame is 640 (H) × 600 (V). The shooting rate of the projectile after hitting the target is 9000 frames /s, and the maximum resolution of each frame is 640 (H) × 480 (V).

3. Experiment results and analysis

3.1. Ballistic test results of perforated plates made of different armor steel materials

Table 2 shows the ballistic test results of perforated plates made of different armor steel materials. For HNASS, 30CrNiMoNb, and 22SiMn2TiB armor steel materials, the circular perforated plates are represented by C-1, C-2, and C-3, respectively, and HNASS armor steel materials hexagonal perforated plates are represented by H-1. According to the test results, the small penetration depth of the matrix occurred on the perforated plate (C-1-50) installed 50 mm away from the 32Mn2Si2MoA armored steel matrix, but the penetration depth of the perforated plate at the gap of 400 mm away from the 45 # steel matrix (C-1-400) was greater than the 50 mm gap, indicating that the perforated plates of different armored steel materials achieved effective protection against 12.7 mm API under the conditions of 50 mm and 400 mm gap.

The ballistic protection characteristics of perforated plates made of different armor steel materials were evaluated by the protection factor N and the protection efficiency η:

$$N = \frac{T_b \cdot \rho_g}{T_t \cdot \rho_t} \tag{1}$$

where T_b is the penetration depth of a standard homogeneous armored steel semi-infinite target when shooting with a standard projectile (mm); ρ_g is the steel density 7.85 (g/cm^3); T_t is the penetration depth of special armor when fired by the same standard bullet (mm); ρ_t is the special armor density (g/cm^3).

In this paper, the impact of the mass reduction of perforated plates on the protection performance is not considered. Therefore, the ballistic performance index of the perforated plate is evaluated by calculating the protection efficiency:

$$\eta = \frac{T_b - T_{rp}}{T_b} \times 100\% \tag{2}$$

where the T_{rp} is the residual penetration depth of the base armor outside the perforated plate (mm).

Table 2. Ballistic test results of perforated plates made of different armor steel materials

Target designation	Projectile velocity (m/s)	Residual penetration depth (mm)	No. of interconnected holes	Damaged area (mm^2)	Cores fractured	N	η
C-1-50-1	815	3	5	499.22		7.67	86.96
C-1-50-2	819	4	6	481.25		5.75	82.61
C-1-50-3	818	4.5	4	280.08	—	5.11	80.43
C-1-50-4	834	3.5	5	443.75		6.57	84.78
C-1-50-5	820	5	5	425		4.6	78.26
C-2-50-1	820	2	5	606.64		11.5	91.30
C-2-50-2	809	1.5	6	635.16		15.33	93.48
C-2-50-3	828	5	6	539.84	—	4.6	78.26
C-2-50-4	806	3.5	6	505.47		6.57	84.78
C-2-50-5	830	3.5	5	724.22		6.57	84.78
C-3-50-1	818	4	5	722.27		5.75	82.61
C-3-50-2	820	3.5	7	866.02		6.57	84.78
C-3-50-3	827	4	6	645.7	—	5.75	82.61
C-3-50-4	824	4	6	689.84		5.75	82.61
C-3-50-5	816	3.5	6	736.72		6.57	84.78
H-1-50-1	818	3.5	6	735.94	—	6.57	84.78

H-1-50-2	828	3.5	7	771.48		6.57	84.78
H-1-50-3	821	3	8	675.78		7.67	86.96
H-1-50-4	830	3.5	8	928.52		6.57	84.78
H-1-50-5	820	3.5	7	810.94		6.57	84.78
C-1-400-1	810	5.4	6	768.34	2	7.41	86.5
C-1-400-2	802	8.6	5	655.91	2	4.65	78.5
C-1-400-3	804	6.5	4	467.13	5	6.15	83.75
C-2-400-1	823	6.1	7	893.29	3	6.56	84.75
C-2-400-2	823	5.7	7	875.53	1	7.02	85.75
C-2-400-3	826	5.5	5	680.01	4	7.27	86.25
C-3-400-1	811	5.0	4	476.85	3	8.00	87.5
C-3-400-2	820	6.7	6	600.60	2	5.97	83.25
C-3-400-3	819	9.9	4	685.10	1	4.04	75.25
H-1-400-1	820	4.7	7	934.29	8	8.51	88.25
H-1-400-2	818	7.1	6	719.30	1	5.63	82.25
H-1-400-3	828	5.2	10	1290.75	2	7.69	87.00

The comparison of the protection coefficient and protection efficiency of perforated plates made of different armor steel materials is shown in Figure 3.

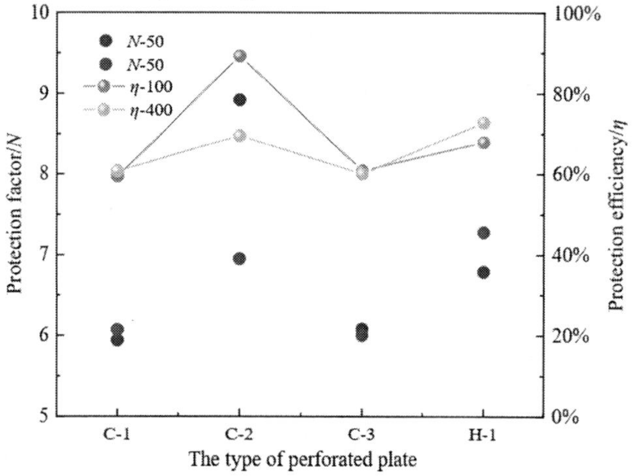

Figure 3. Protection coefficient and protection efficiency of different armor steel perforated plates.

It can be seen from Figure 3 that the protection coefficient of HNASS and 22SiMn2TiB circular perforated plates is lower than that of 30CrNiMoNb circular perforated plates and HNASS hexagonal perforated plates, and the corresponding protection efficiency is the same trend. For 30CrNiMoNb armor steel with the highest initial hardness (534 HBW), its perforated plate has the highest protection coefficient and protection efficiency, while HNASS armor steel with lower initial hardness (368 HBW) has the same protection effect as 22SiMn2TiB armor steel with medium hardness (499 HBW), indicating that HNASS armor steel has more significant impact hardening characteristics. It is worth noting that under the conditions of 50 mm and 400 mm spacing. The protection coefficient of H-1 armor steel perforated plate reaches or even exceeds 30CrNiMoNb high hardness armor steel. In addition, it is found that the protection coefficient and protection efficiency of high-hardness armor steel 30CrNiMoNb are more dominant under the condition of a small gap.

3.2. Analysis of damage area and core fracture characteristics of perforated plates made of different armor steel materials

The damage characteristics of perforated plates made of different armor steel materials are analyzed below. The damage area of perforated plates is mainly considered. The damage of perforated plates is represented by the area of sufficient overlap between the bullet core and the remaining plates [5]. The treatment method for the hole damage area is calculated according to the scheme shown in Figure 4, which is drawn for circular holes and regular hexagonal holes, respectively. The overlapping size of the bullet core and orifice plate edge is *h*, *H* = *D*-*d*, *D* is the diameter of the bullet core, *d* is the perforation diameter of the orifice plate, and the circumscribed diameter of a regular hexagon. The hole damage area is drawn in AUTOCAD 2019 software, and the ruler is set according to the actual size of the orifice plate. Then the AREA command shortcut key is used to quickly calculate the area of the overlapping area, that is, the damage area of the perforated plate.

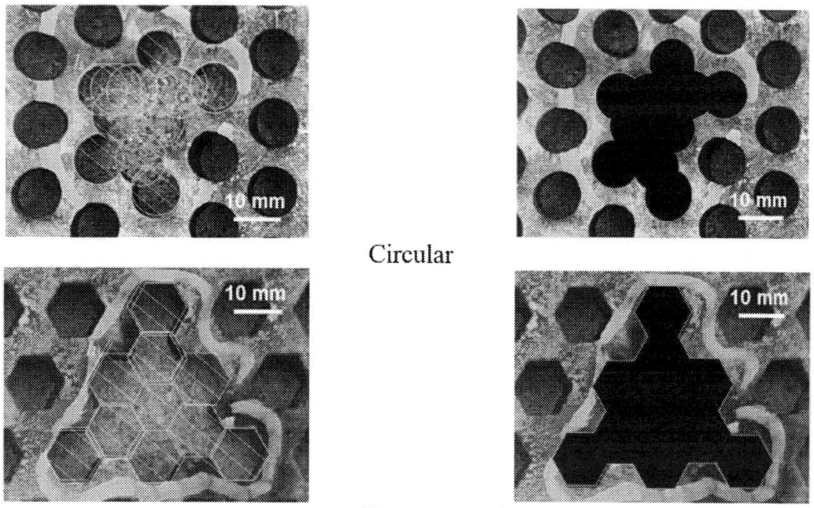

Hexagon

Figure 4. Schematic diagram of hole damage area of an orifice plate.

After the test, the fragment holes of the perforated plate with a 50 mm installation gap are treated to obtain the hole damage areas at different impact points, as shown in Figure 5.

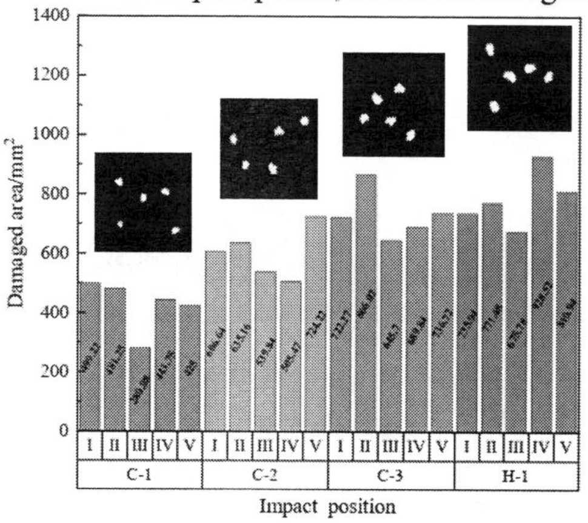

Figure 5. Damage area at each impact point of perforated plates made of different armor steel.

Figure 5 shows that the hole damage area of C-1 armored steel material is the smallest, that of C-3 is the largest, that of C-2 is between them, and that of H-1 is also larger. The hole damage area of the hexagonal perforated plate is larger than that of the circular perforated plate. The damaged area of holes can explain the impact toughness of different armor steel materials, so the impact toughness of C-1 armor steel material is better. At the same time, it was found that the crack arrest performance of the circular perforated plate was better than that of the hexagonal perforated plate.

Due to the change of perforated plate installation clearance and base armor, there is a great difference in the damage state of base armor under the condition of 50 mm and 400 mm installation clearance, as shown in Figure 6.

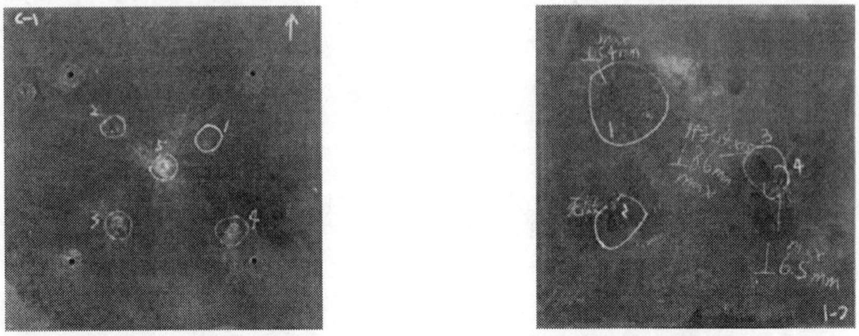

(a) 50 mm installation clearance 32Mn2Si2MoA base (b) 400 mm installation clearance 45# base

Figure 6. Damage state of base armor under different installation clearance conditions.

It can be seen from Figure 6 that the 32Mn2Si2MoA base with higher strength under the condition of 50 mm installation gap forms only one crater, and the bullet pit is not broken, while the 45 # steel base with lower strength under the condition of 400 mm installation gap forms multiple craters, and the bullet pit is broken into multiple. Therefore, the analysis of the bullet pit fracture is mainly carried out for the gap of 400 mm, and the statistics of the damage area of the perforated plate with a 400 mm installation gap and the bullet pit fracture are shown in Figure 7.

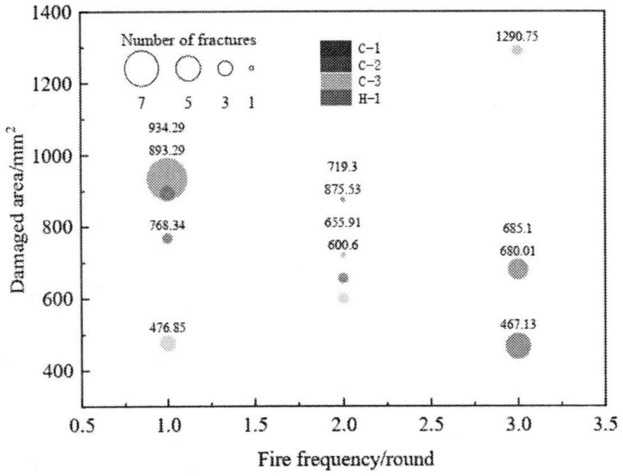

Figure 7. Damage area and core fracture number of perforated plates of different armor steel.

The hole damage area of C-1 armored steel material is smaller than that of C-2 material, the hole damage area of C-3 armored steel material is the smallest, and the hole damage area of H-1 hexagonal perforated plate is still the largest. There is a big difference between the change rule for the damaged area of holes in the C-3 material and the 50 mm installation gap. The reason is that in addition to the different locations of the impact point, when the projectile penetrates the C-3 perforated plate, it passes through the edge of the same damaged area. Therefore, the whole damage area is caused by the tearing of the edge of the perforated plate rather than the impact of the projectile on the complete area. For perforated plates of different armor steel materials with circular holes, there is little difference in the fracture of the bullet cores, which break into 1, 3, and no more than 5. For perforated plate H-1 with regular hexagonal holes, the maximum number of bullet cores can reach 7, indicating that for regular hexagonal holes, although the damage area of the holes is large, its ability to fracture the bullet cores is higher than that of circular holes, that is, the ballistic protection performance is excellent.

The hole damage area of the perforated plate under different gap conditions has great differences due to the different impact points and the number of hole damage pieces, and the data are relatively scattered. Therefore, the damage area of perforated plates with 50 mm and 400 mm gaps is averaged to obtain the change of the damage area of a single interconnection hole of perforated plates with different armor steel materials and opening types, as shown in Figure 8.

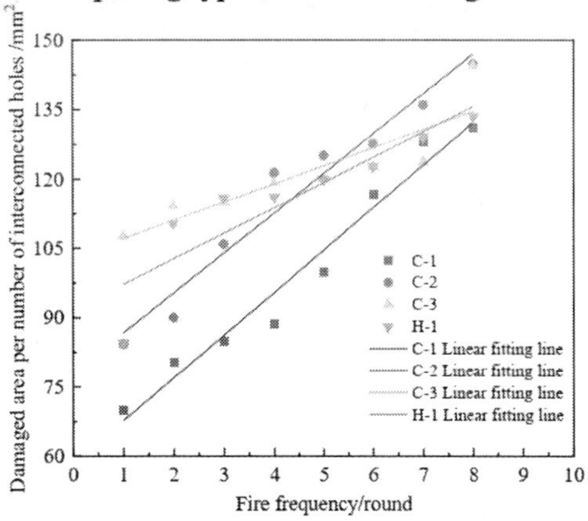

Figure 8. Comparison of damage area of single interconnection hole of perforated plate made of different armor steel materials.

It can be seen from Figure 8 that there is an obvious linear relationship between the damage area of a single interconnection hole of perforated plates made of different armor steel materials: $y = k \cdot x + b$, k represents the slope of the fitting line, and b represents the intercept. For the same opening shape, the damaged area of a single interconnection hole of C-1 is the smallest, the damaged area of a single interconnection hole of C-3 is larger, and the damaged area of C-2 is between the two. Since the projectile passes through the edge of the hole, the fitting straight line only reflects the changing trend of the damaged area of the hole, where k measures the degree of hole loss in the perforated plate, and B can characterize the damaged area of a single interconnection hole. For the H-1 hexagonal opening, the damaged area of a single interconnection hole is larger than that of the circular perforated plate.

3.3. Analysis of the deflection characteristics of the bullet core after perforation

The deflection of the bullet core after perforation is related to the ballistic protection characteristics of the perforated plate. Because a large amount of firelight is generated when the 12.7 mm API penetrates the perforated plate, and the firelight covers the entire field of view of high-speed photography, the desired deflection attitude of the bullet core is not observed by high-speed photography under the conditions of 50 mm gap and 400 mm gap. Therefore, to capture the deflection attitude of the entire bullet core, the gap between the perforated plate and the base armor is increased to 1.5 m. The fracture and deflection process of the bullet core penetrating the perforated plate is shown in Figure 9. To clearly show the deflection state of the main bullet core, the deflection attitude of the bullet core is enlarged, as shown in Figure 10.

Figure 9 shows the state (C-1) of the projectile penetrating between two holes. At t=0~0.74 ms, after the projectile and the perforated plate act, the projectile head is eroded. Then, due to the asymmetric collision of the projectile target, the interior of the bullet core is subjected to asymmetric bending tensile stress. When the bending tensile stress exceeds the tensile strength of the bullet core, the bullet core will undergo brittle fracture. When t=0.74 ms, the projectile broke more, and then the main bullet pit formed in the period of t=0.83~1.29 ms flew out of the firelight coverage area at a faster speed, and the other fragmented bullet pits deviated from the initial trajectory and flew around.

The enlarged view of the pit deflection attitude at t=1.38 ms shows the deflection of the movement attitude of the main pit after fracture, and the final damage state on the base armor is shown in Figure 11. Only the crater of the broken main bullet core is left on the 45 # steel base plate, while the rest of the broken bullet cores are only left with white punctate material, which has not penetrated the base plate.

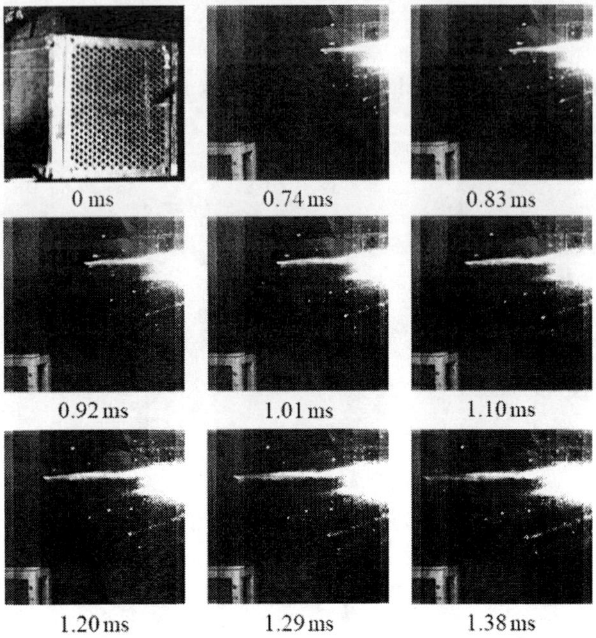

Figure 9. Fracture and deflection process of bullet core after bullet penetration.

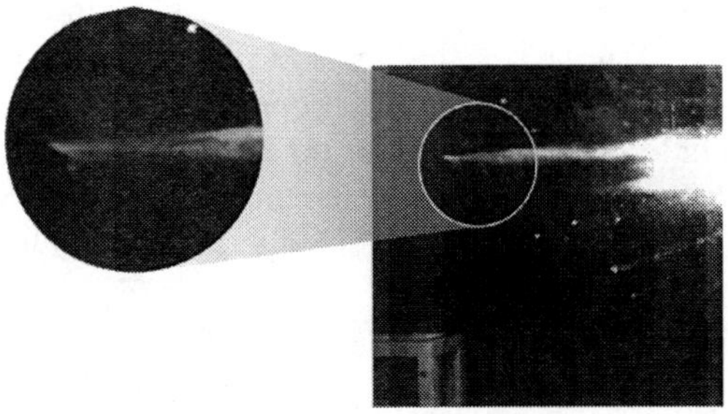

Figure 10. Enlarged view of deflection attitude of bullet pit.

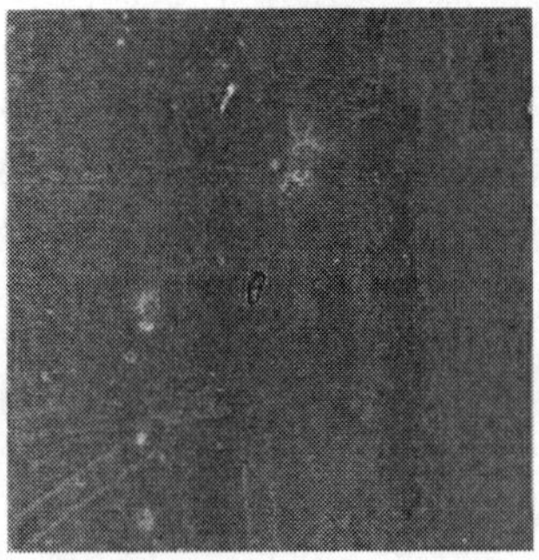

Figure 11. Damage state of base armor.

Journal of Physics: Conference Series **2808** (2024) 012007 doi:10.1088/1742-6596/2808/1/012007

Since the position of the projectile before hitting the target can be calibrated by high-speed photography, the deflection attitude of the bullet core after penetrating the target cannot be determined. Since the number of broken bullet pits with a gap of 50 mm is small, there is only one bullet pit on the base armor. Therefore, according to the crater position on the base plate and the calibration results of high-speed photography in Figure 6 (a), the deflection angle of the bullet pit under the condition of a gap of 50 mm is calculated. The bullet impact position on the base armor before and after hitting the target is shown in Figure 12.

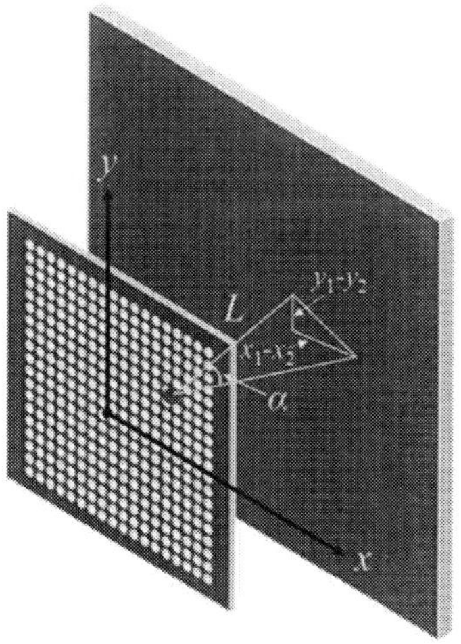

Figure 12. Schematic diagram of impact point position before and after target penetration.

Note that the position of the bullet hitting the perforated plate is (x_1, y_1), and the position of the bullet in the base armor after penetrating the target is (x_2, y_2). The clearance of 50 mm is expressed by L, then the deflection angle of the bullet α, which can be calculated according to trigonometric function:

$$\alpha = \arctan \frac{\sqrt{(x_1 - x_2)^2 + (y_1 - y_2)^2}}{L} \tag{3}$$

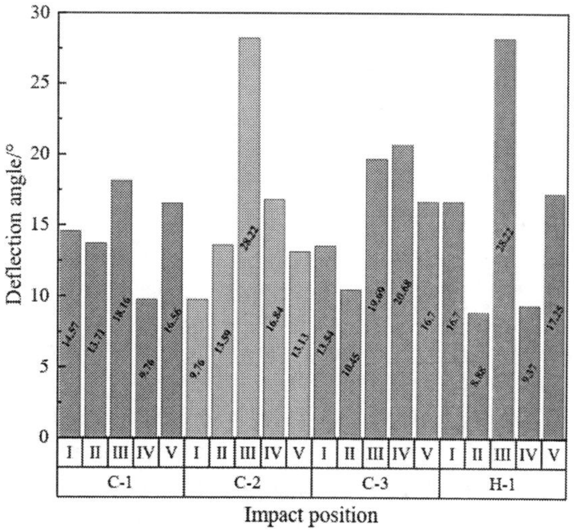

Figure 13. Relationship between projectile deflection angle and impact point of perforated plates made of different armor steel materials.

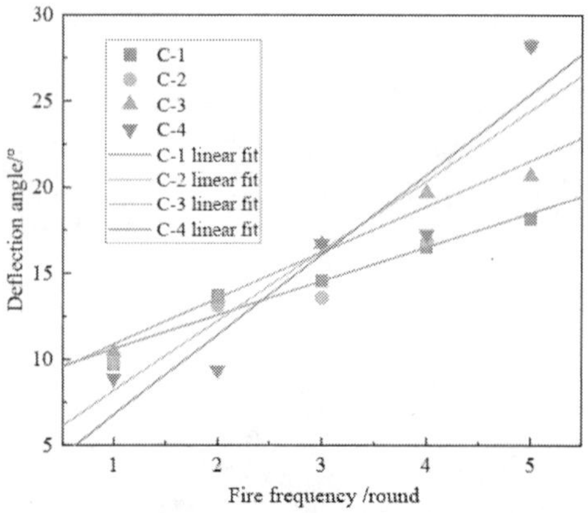

Figure 14. The deflection angle of bullet cores of perforated plates made of different armor steel.

The relationship between projectile deflection angle and impact point position of perforated plates made of different armor steel materials is shown in Figure 13. It can be found that the deflection angles of C-2 and H-1 armor steel perforated plates are large, while the deflection angles of C-1 and C-3 armor steel materials are not different. The calculated projectile deflection angles of perforated plates made of different armored steel materials are sorted from small to large and fitted. It is found that there is also a linear relationship between the projectile deflection angles of perforated plates made of different armored steel materials: $y = k{\cdot}x + b$, see Figure 14. For the same opening shape, the deflection angle of the core of C-1 is the smallest, the deflection angle of C-2 is the largest, and C-3 is between the two, suggesting that the higher the hardness, the greater the deflection of the core. It can be seen from Figure 14 that the fitting straight line reflects the variation trend of the deflection angle of the bullet pit, in which k measures the degree of the deflection of the bullet pit, while the physical meaning of b seems not obvious. For the H-1 regular hexagonal opening, the deflection degree of the bullet core is greater than that of the circular perforated plate. It is worth noting that the location of the impact point has a great impact on the deflection of the bullet core, which is not considered in this paper and is worthy of further study.

4. Conclusion

(1) According to the ballistic test of perforated plates made of different armor steel materials, it is found that the protection coefficient and protection efficiency of C-1 and C-3 perforated plates are lower than those of C-2 and H-1 perforated plates. For the C-2 perforated plate with the highest hardness, its protection coefficient and protection efficiency are the highest. The impact hardening characteristics of C-1 material are more significant. Its protection coefficient and protection efficiency are equivalent to those of C-3 with medium hardness. The protection coefficient of the H-1 perforated plate can reach or even exceed C-2. However, the protection coefficient and protection efficiency of C-2 are more dominant under the condition of a small gap.

(2) By analyzing the damaged area of perforated plates with different installation gaps, it is found that the hole damage area of the H-1 perforated plate is larger than that of C-1, C-2, and C-3 perforated plates. The impact toughness of C-1 armored steel material is better, and the crack arrest performance of circular perforated plate is better than that of hexagonal perforated plate. For perforated plates of different armor steel materials with circular holes, the number of core fractures is less, while the number of core fractures of H-1 perforated plate is more, indicating that the ballistic protection performance of hexagonal holes is better than that of circular perforated plates.

(3) There is an obvious linear relationship between the damage area of a single interconnection hole in perforated plates of different armor steel materials. The slope k of the fitted line measures the degree of hole loss in perforated plates, while the intercept b can characterize the damage area of a single interconnection hole. In the circular opening, the damaged area of a single interconnection hole of C-1

is the smallest, C-3 is larger, and C-2 is between the two, while the damaged area of a single interconnection hole of hexagonal opening H-1 is larger than that of the circular perforated plate.

(4) The projectile deflection angles of perforated plates made of different armor steel materials also meet the linear relationship. In the circular opening, the deflection angle of the bullet core of C-1 is the smallest, that of C-2 is the largest, and that of C-3 is between the two. The fitting straight-line slope k can measure the degree of pit deflection. The deflection degree of the bullet core of the hexagonal hole H-1 is greater than that of the circular hole perforated plate.

References

[1] Killc, Namik, and B. Ekici. "Ballistic resistance of high hardness armor steels against 7.62mm armor piercing ammunition." Materials & Design 44. FEB. (2013):35-48.
[2] A, S. Ryan, et al. "The ballistic performance of an ultra-high hardness armor steel: An experimental investigation." International Journal of Impact Engineering 94 (2016):60-73.
[3] TRASI Y, BEN-MOSHE D, ROSENBERG G. An armor assembly for armored vehicles: EP0209221[P].1987-01-21.
[4] PENG J X, CUI T N, JIN Y X, et al. Research on design and ballistic resistance of perforated armor with oblique holes [J]. Chinese Journal of Applied Mechanics, 2021, 38(3): 893-901.
[5] Balos, Sebastian, et al. "Geometry, mechanical properties and mounting of perforated plates for ballistic application." Materials and Design 31.6 (2010):2916-2924.
[6] Mishra, Bidyapati, et al. "Experimental studies on the effect of size and shape of holes on damage and microstructure of high hardness armor steel plates under ballistic impact." Materials & Design 43. JAN. (2013):17-24.
[7] Janjatovic, Petar, et al. "Ballistic Behavior of Aus tempered Compacted Graphite Iron Perforated Plates." Defense Science Journal 6 (2019).
[8] Garbarz, Bogdan, et al. "Methods to Increase the Protective Effectiveness of Add-on Armor Made of Perforated Ultra-High-Strength Nano bainitic Steel Plates." *Problems of Mechatronics. Armament, Aviation, Safety Engineering* 14.1 (2023): 23-60.

Journal of Physics: Conference Series **2808** (2024) 012008 doi:10.1088/1742-6596/2808/1/012008

Effect of ZnO on tribological properties of MoS_2 dry film lubrication coatings

Tao Duan[1], Qin Wang[2], Peng Cheng[1] and Dewei Mi [3]*

[1]China Productivity Center of Machinery Co., Ltd., Beijing 100044, China
[2]Sichuan Traffic Technician College, Chengdu 610000, China
[3]China Academy of Machinery Science and Technology Group, Beijing 100044, China

*Corresponding author's e-mail: midw@163.com

Abstract: As a key component of the high-speed railway brake system, high-speed railway brake disc bolts are often lubricated with MoS_2 lubrication coating, and the friction performance of the coating is a key factor in ensuring the operation of the equipment. The friction performance of the coating is a key factor to ensure the operation of the equipment. In this paper, the effect of ZnO on the friction performance of MoS_2 dry film lubrication coating and the performance of MoS_2 dry film lubrication coating were studied. SEM and EDS were used to characterize the coating morphology and abrasion marks, SEM and EDS were used to characterize the coating morphology and abrasion marks, and a friction and wear tester was used to study the effect of ZnO addition, particle size, and working temperature on the friction performance of the coating. It was found that the mechanism of ZnO is to harden and support the film-forming structure of the coating to ensure the integrity of the lubricant film and enhance the anti-wear performance of the coating. The optimal amount of ZnO is 5%, the optimal particle size is 30 nm, and the optimal operating temperature is 100°C. The results of this paper provide a theoretical basis for further optimizing the preparation and performance of MoS_2 dry film. The results of this paper provide a theoretical basis for further optimizing the preparation and performance of MoS_2 dry film lubrication coatings.

1. Introduction

As a key component in the train braking system, the stability and reliability of high-speed rail brake disc bolts are crucial to the safe operation of trains. MoS_2 dry film lubrication coating, compared with the traditional lubrication technology using grease, not only improves the lubrication performance, stability, and reliability significantly but also has a better lubrication effect under special environmental conditions [1–4].

The main components of dry film lubrication coatings include solid lubricants, binders, dispersants, and functional fillers [5]. The main components of a dry film lubrication coating include solid lubricants, binders, dispersants, and functional fillers. Therefore, the frictional properties of MoS_2 dry film lubrication coatings are influenced by some factors [6–8]. Functional fillers play an important role in dry film lubrication coatings by enhancing the surface hardness and load-carrying capacity and are an integral part of the coating [9]. Functional fillers are an integral part of the coating. Typically, metals and their oxides are used as fillers to enhance the organizational properties of the coating such as load and wear resistance to meet the needs of different environments. ZnO has excellent physical and chemical properties, such as high melting point, high thermal conductivity, high chemical stability, biocompatibility, etc., which can effectively improve the hardness of the coating [10,11]. ZnO can effectively improve the hardness and wear resistance of coatings [12]. ZnO Relevant studies have shown that the addition of ZnO to diesel oil can increase the viscosity of lubricating oil, significantly improve the tribological properties of pure diesel oil, and reduce the wear of workpieces [13]. The addition of ZnO in diesel oil can significantly improve the tribological properties of pure diesel oil and reduce the wear of the workpiece. The addition of ZnO in the appropriate amount in the automotive engine lubrication system can form a layer of solid lubrication layer, to improve the wear and friction performance of the sliding surface [14]. ZnO is used as a filler in fastener lubrication systems. The addition of ZnO as a filler in the solid lubricant coating of fasteners can form a self-laminating protective film, which improves anti-wear and anti-glue properties [15]. The ZnO is well dispersed in the fastener lubricant to form a monomolecular lubricant film, which reduces adhesive wear and improves friction performance [16]. ZnO has a Mohs hardness of 4.5 and a coefficient of friction of between 0.08 and 0.16. Adding ZnO to dry film lubrication coatings can form a monolayer lubrication film, improve anti-adhesive properties, reduce adhesive wear, and improve friction performance [17]. The friction coefficient is 0.08-0.16. Some scholars have studied the adaptive coatings of ZnO-MoS_2 and ZnO-WS_2 [18], sulfides and $ZnMoO_4$ and $ZnWO_4$ ternary oxides formed during high temperature sliding contact reduced the coefficient of friction values at high temperatures, but the intermediate low-temperature performance and the role of ZnO were not investigated.

This thesis aims to investigate the effect of ZnO on the frictional properties of MoS_2 dry film lubrication coatings. Using MoS_2 as a solid lubricant, $Na_2O{\cdot}nSiO_2$ as the binder, and ZnO as a functional filler, the dry film lubrication coating of MoS_2 was prepared, and friction experiments were carried out by using friction and wear tester to measure the coefficient of friction. SEM and EDS analyses were carried out to investigate the influence laws of ZnO additive amount, particle size, and working temperature on the friction performance of the coating, to provide theoretical and practical bases for designing and optimizing the coating materials.

2. Experimental

2.1 Materials and Preparation

The main materials used in the experiments were: $Na_2O{\cdot}nSiO_2$ (Henan BoRun, modulus 3.32), MoS_2 (Macklin, AR 99%), micron ZnO (Macklin, AR 99%), and nano-ZnO (Macklin, 30 nm, 200 nm, 99.9%). The specific preparation method is as follows: 20 ml of $Na_2O{\cdot}nSiO_2$ is measured, 6.0 g of MoS_2 powder and 0.5 g of ZnO (1000 nm) powder are added to mix thoroughly, and a high-speed homogenizer is used to disperse the mixed solution at 10,000 RPM for 10 minutes. The lubricant continued to be formulated with different additions and different particle sizes of ZnO as described above, and the additive ratios of each component material are shown in Table 1.

The procedure for preparing lubricated specimens is as follows:

(1) Clean specimen

(2) Brush on MoS_2 mixture, 2-3 coats

(3) The lower specimen is heated and cured at 150°C for one hour.

A Pocket-Surfix X-FN coating thickness gauge was used to measure the coating thickness, and it was found that as the amount of solids added to the solution increased, the thickness of the coating showed varying degrees of increase, with the range varying between 30 and 50 μm.

Table 1. Material addition ratio.

Groups	Solution ($Na_2O{\cdot}nSiO_2$) / ml	Lubricants (MoS_2) / g	Additives (ZnO) / g
A1	20	6.0	0.5 (1000 nm)
A2	20	6.0	1.0 (1000 nm)
A3	20	6.0	1.5 (1000 nm)
B1	20	6.0	1.0 (30 nm)
B2	20	6.0	1.0 (200 nm)

2.2 Tribological property tests

In this study, the effect of ZnO on the frictional properties of silicate-based MoS_2 dry film lubrication coatings was evaluated using Rtec MFT-5000. The spherical-disk friction test system consists of a spherical, disk, and loading system as shown in Figure 1. During the test, the spherical disk rotates to produce circular abrasion marks to evaluate the friction performance of the coating by determining the friction force and friction coefficient. The test conditions were set as load 15 N, rotational speed 100 RPM, and duration 600 s. The upper test ball was selected as an E52100 steel ball with a diameter of 6.3 mm, and the lower specimen was selected as a 45 steel disc with a diameter of 70 mm and thickness of 6.35 mm.

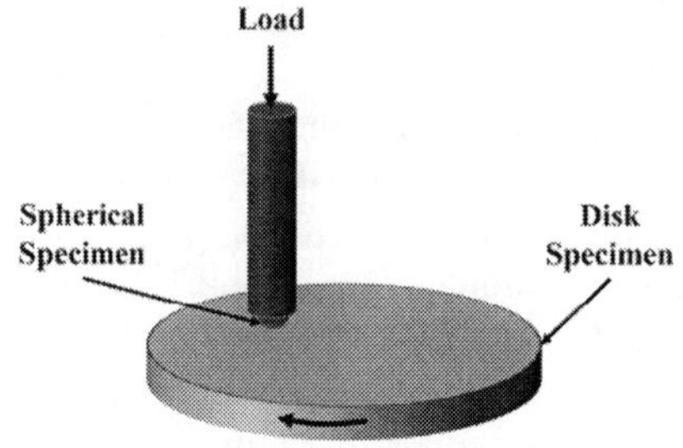

Figure 1. Spherical-disk friction system.

2.3 Coating surface micro-morphological analysis

ZEISS EVO18 tungsten filament scanning electron microscope (SEM) was used to analyze the micro-morphology of the coating surface, and the elemental species and content of the coating constituents were analyzed using an energy spectrometer (EDS). Due to the weak electrical conductivity of the coating, the surface of the coating needs to be carbon sprayed, and the microstructure of the coating was analyzed at a working voltage of 20 keV.

3. Results and discussion

3.1 Surface morphology of coatings with different ZnO additions

Figure 2 demonstrates the microscopic morphology of MoS_2 coatings with different additions of ZnO. It was observed that the solid material (light-colored phase) in the coating could show uniform distribution without agglomeration when the ZnO addition was 2.5% and 5%, respectively, and a better dispersion effect could be achieved; however, when the ZnO addition was increased to 7.5%, some particles appeared to be agglomerated (yellow dashed line selection in the figure), and larger agglomerated particles were formed, which were unable to be completely dispersed. Therefore, the amount of ZnO added should be controlled below 5%.

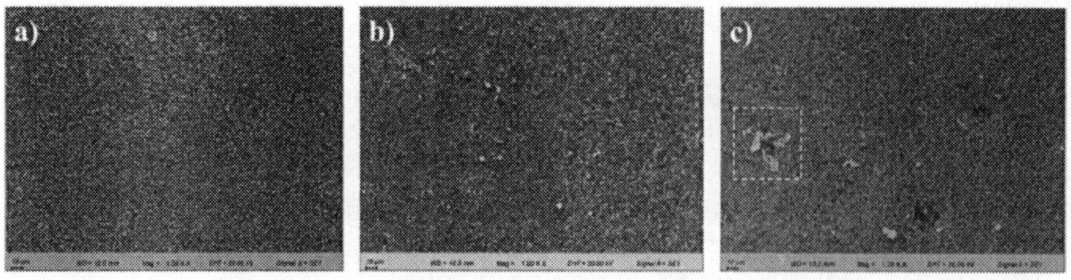

Figure 2. Surface morphology of coatings with different ZnO additions a) 2.5% b) 5% c) 7.5%.

3.2 Surface morphology of coatings with different particle sizes of ZnO additions

The particle size of ZnO nanoparticles is generally in the range of 1-100 nm, which has the characteristics of lightweight, high stability, and high plasticity. The surface effect of ZnO nanoparticles, with the smaller particle size, the number of atoms on the surface of the particles increases rapidly, and

Journal of Physics: Conference Series 2808 (2024) 012008 doi:10.1088/1742-6596/2808/1/012008

the surface free energy and surface area of the ZnO particles are also accelerated.

Figure 3 demonstrates the microscopic morphology of MoS_2 coatings with the addition of ZnO with different particle sizes: a) 30 nm, b) 200 nm, and c) 1000 nm. It is observed that the solid material (light-colored phase) in the coating after the addition of nano-ZnO can show a uniform distribution without the formation of large agglomerated particles, and a better dispersion effect can be achieved; however, when ZnO with a particle size of 1000 nm is added, some of the particles appear to be agglomerated (yellow dashed selection in Figure 3), which cannot be completely dispersed, and the dispersion effect decreases.

Figure 3. Surface morphology of ZnO with different particle sizes a) 30 nm, b) 200 nm, c) 1000 nm.

3.3 Friction test of MoS_2 coatings with different additions of ZnO

Figure 4 shows the friction test results of MoS_2 coatings with different additive levels (0, 2.5%, 5%, and 7.5%) of ZnO, where the MoS_2 additive level is 30%. The experimental results show that the MoS_2 coating with ZnO as an additive has a significant effect in reducing the coefficient of friction. The average coefficient of friction of the coating without ZnO was 0.164 over 600 s. When 2.5% of ZnO was added, the initial coefficient of friction of the coating was about 0.10, which tended to increase as the friction process progressed, with an average coefficient of friction of 0.174 over 600 s, and slight fluctuations occurring when the friction progressed to 350 s. The coefficient of friction of the coating without ZnO was about 0.10. When the amount of ZnO added is 5%, the initial friction coefficient of the coating is about 0.11, the friction coefficient is stable with the friction, the average friction coefficient in the friction process is 0.156, and the whole process is smooth without fluctuation. However, when the amount of ZnO was increased to 7.5%, the initial coefficient of friction of the coating was about 0.10, and the coefficient of friction tended to increase with friction, with an average coefficient of friction of 0.200 in 600 s, and slight fluctuations in the friction process up to 300 s. By comparison, the lowest average coefficient of friction was obtained with 5% ZnO addition, with a decrease of 4.88% compared to the coating without ZnO addition.

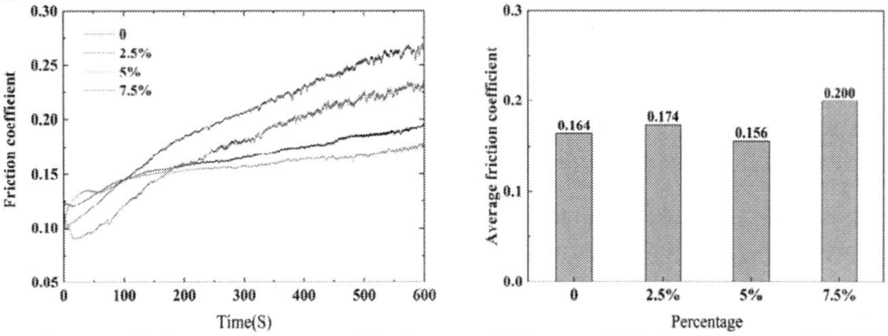

Figure 4. Friction coefficient and average friction coefficients of MoS_2 coatings with different levels of ZnO.

Figure 5 demonstrates the wear trace morphology of MoS_2 coatings with different additions of ZnO. Observation of the wear traces showed that the wear was mainly abrasive, with the formation of plowing and abrasive particles along the friction direction, and no substrate was exposed within the wear marks. When ZnO was not added, the lubrication film formed was overall rough, with abrasive particles generated (white dashed selection in the figure) and a small number of cracks (white arrows in the figure). After adding 2.5% of ZnO, under the action of friction to form a lubricant film surface smooth but cracks (yellow arrows in the figure), analyzing the reason for ZnO (Mohs hardness 4.5) hardness is higher than the MoS_2 (Mohs hardness 1.5), which plays a good support for hardening the role of the coating to ensure that the lubricant film smooth and even, but not enough to ensure that the addition of the lubricant film continuity; increased to 5% of the ZnO, the increase in the amount of ZnO can fully guarantee the continuity of the lubrication film, the coating formed by the lubrication film continuous and smooth surface; however, when the amount of ZnO added to 7.5%, the formation of the lubrication film is continuous and complete, but a small amount of flaking phenomenon (red dotted line in the figure of the constituency) and cracks (red arrows in the figure of the place), the analysis of the reasons for the excessive addition of ZnO led to the dispersion and bonding. The reason for this is that the over-addition of ZnO leads to a decrease in the dispersion and bonding effect, which damages the integrity of the coating. The reason for this is that the excessive addition of ZnO leads to a decrease in the dispersion and bonding effect, which destroys the integrity of the coating.

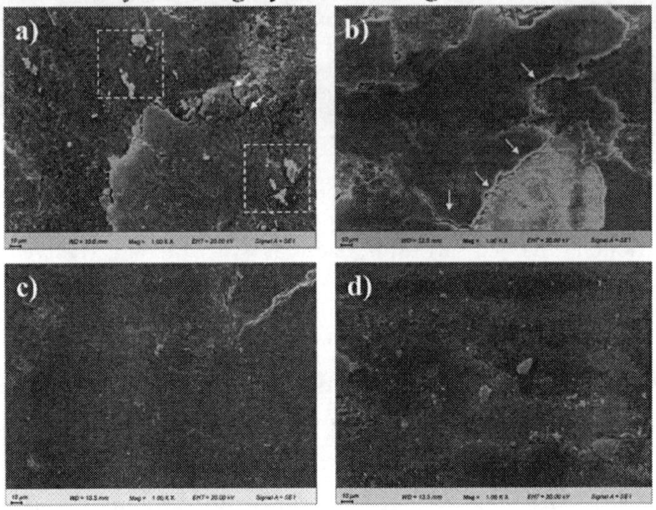

Figure 5. Morphology of abrasive marks with different ZnO additions: a) 0, b) 2.5%, c) 5%, d) 7.5%.

Combined with the friction test results and scanning electron microscopy analysis, it can be found that the ZnO particles have excellent support hardening effects. Due to its small size and strong adsorption capacity, ZnO particles can achieve a good dispersion effect in the coating. During the friction process, these particles can ensure the continuity of the lubricant film and make the surface of the lubricant film harder and smoother, thus significantly reducing the coefficient of friction. However, if the amount of ZnO added is too high, the dispersion effect of the coating may be reduced and the lubricant film becomes brittle and easy to peel off, which leads to an increase in the coefficient of friction. Therefore, the amount of ZnO added to MoS_2 dry film lubrication coatings should be controlled within a reasonable range, with the optimal range being 2.5%-5%.

3.4 Friction test of MoS_2 coatings with different particle size of ZnO

Figure 6 demonstrates the friction test results of MoS_2 coatings with different particle sizes of ZnO, where MoS_2 was added at 30%. The experimental results show that the particle size of ZnO affects the friction coefficient of the coating. The initial friction coefficient of the coating with 1000 nm ZnO was about 0.11, and it was stable during the friction process for 600 s, with an average friction coefficient of 0.156, and the friction process was smooth without fluctuation. The initial friction coefficient of the coating with 30 nm ZnO was about 0.11, and the average friction coefficient was 0.143, which was stable during the 600 s of the friction process, with an upward trend when the process was carried out for about 500 s. For the coating with 200 nm ZnO, the initial friction coefficient was about 0.11, the fluctuation of friction coefficient was small and stable, and the average friction coefficient was 0.148 in 600 s.

Journal of Physics: Conference Series **2808** (2024) 012008 doi:10.1088/1742-6596/2808/1/012008

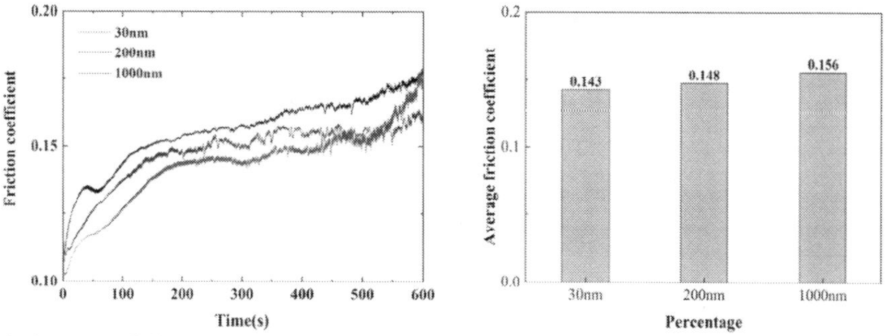

Figure 6. Friction coefficient and average friction coefficients of MoS_2 coatings with different particle sizes of ZnO.

Figure 7 demonstrates the wear trace morphology of MoS_2 coatings with different particle sizes of ZnO added. Observation of the wear traces shows that the wear is mainly dominated by abrasive wear, forming plowing and abrasive particles along the friction direction, with no substrate exposed within the wear marks. After the addition of ZnO with a particle size of 30 nm, a lubrication film with a smooth surface was formed under friction; when ZnO with a particle size of 200 nm was added, the coating formed a continuous lubrication film with a smooth surface and a small amount of flaking (yellow dashed selection in the figure); however, when the particle size of ZnO with a particle size of 1000 nm was added, the formation of the lubrication film was continuous and intact, but there was a larger number of flaking phenomena (red dashed line selection in the figure).

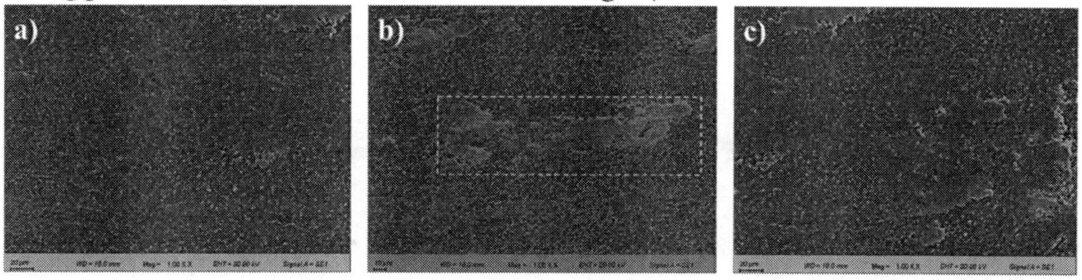

Figure 7. Abrasion mark morphology with different particle sizes of ZnO: a) 30 nm, b) 200 nm, c) 1000 nm.

Combined with the friction test results and scanning electron microscopy analysis, nano ZnO can play a better role in supporting the hardened coating to ensure the continuous integrity of the lubrication film compared with micron ZnO, while smaller size ZnO particles produce less spalling during the friction process. Therefore, the use of nano ZnO in MoS_2 dry film lubrication coating can obtain better friction performance.

3.5 Friction test of coatings at different temperatures

Figure 8 demonstrates the friction test results of MoS_2 coatings at different temperatures with 30% MoS_2 addition and 5% ZnO addition. The experimental results show that the temperature has a significant effect on the coating friction coefficient. At room temperature, the friction coefficient remained around 0.14. When the temperature was increased to 100°C, the friction coefficient decreased significantly with an initial value of about 0.08, which decreased rapidly to 0.04 within 100 s and remained stable with an average friction coefficient of 0.048. When the temperature continued to be increased to 200°C, the initial friction coefficient was about 0.09, which increased slightly after the break-in period and stabilized at about 0.1 after 100 s with an average friction coefficient of 0.105. However, when the temperature rises to 300°C, the coefficient of friction increases significantly, with an initial coefficient of friction of about 0.09, showing a gradually increasing trend and fluctuating after 450 s, with an average coefficient of friction of 0.192 over 600 s.

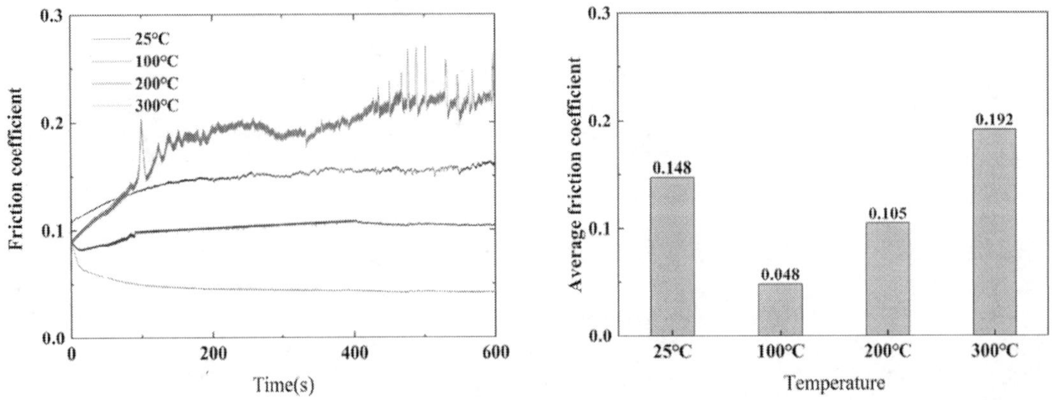

Figure 8. Friction coefficient and average friction coefficient of MoS_2 coatings at different temperatures.

Figure 9 demonstrates the wear trace morphology of MoS_2 coatings at different temperatures. Observation of the wear traces shows that the wear is mainly dominated by abrasive grain wear, with no bare substrate within the wear marks and a few adhesive wear traces, forming a plow cut along the friction direction. At 100°C, a lubrication film with a smooth surface was formed under friction; at 200°C, the coating formed a continuous lubrication film with a smooth surface and a small amount of flaking (yellow dashed selection and yellow arrow in the figure); however, when the temperature was increased to 300°C, the lubrication film formed was continuous and intact with no breakage, but a large number of plowing occurred (red arrows in the figure), and the analysis of the reason was that the temperature increases when the temperature increases, the intermolecular thermal energy increases, destroying crystallization, resulting in the phenomenon of melting and the formation of adhesive wear.

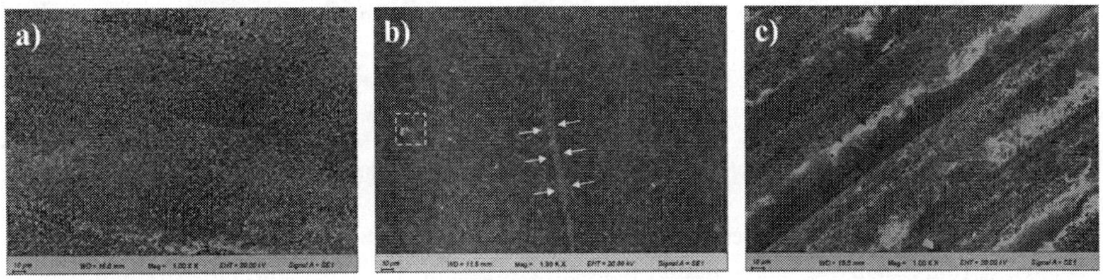

Figure 9. Morphology of abrasion marks at different temperatures: a) 100°C, b) 200°C, c) 300°C.

Combined with the friction test results and scanning electron microscopy analyses, ZnO can still ensure the continuous integrity of the lubrication film within 300°C. However, the working temperature is too high, which will lead to the melting phenomenon of the coating, the lubrication film appears with deep plowing groove marks, and the formation of adhesive wear, resulting in a drastic increase in the coefficient of friction. Therefore, the service temperature of the prepared MoS_2 dry film lubrication coating should be controlled from 25°C to 200°C.

3.6 Elemental analysis before and after friction of membrane layer system

Figure 10 illustrates the distribution of major elements in the coating containing ZnO. The figure shows that the sodium silicate solution cured to form a binder layer covering the entire area. The MoS_2 particles (larger particle size) formed a bond with the surface, while the ZnO particles (smaller particle size) filled the pores between the binder and the MoS_2, which acted as a support for the hardened coating and helped to increase the anti-wear properties of the coating, thus reducing the coefficient of friction of the coating.

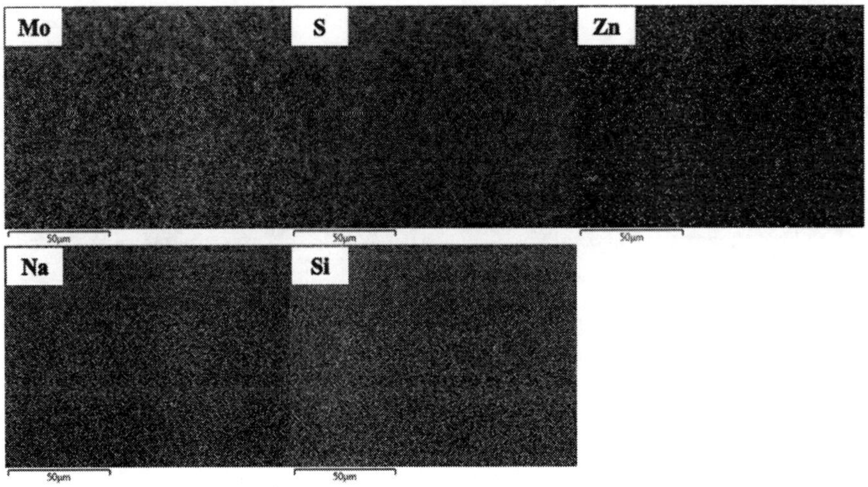

Figure 10. EDS elemental maps of ZnO-containing coatings.

Figure 11 demonstrates the distribution of elements in the ZnO-containing coating after friction. It was observed that the sodium silicate bonding layer was intact after friction, but some of the MoS_2 was detached (white dashed selection in the figure), and the ZnO remained intact (red dashed selection in the figure). In Figure 12, changes in the content of major elements before and after friction are shown. It was observed that Fe element appeared after friction, and the content of Mo element in the coating decreased slightly, indicating that a certain degree of friction loss occurred in the upper specimen during friction, which led to the shedding of MoS_2, and there was no obvious friction loss of ZnO during friction, which was analyzed as the reason that the hardness of ZnO was higher than that of MoS_2, and ZnO played the role of hardening the coating, which could better support the coating, ensure the integrity of the coating, and improve the anti-wear performance of the coating. At the same time, the content of other elements in the coating remains unchanged, indicating that the coating successfully isolates the direct contact between the two friction surfaces and transforms it into the contact between the upper specimen and the coating, which reduces the coefficient of friction and reduces the wear of the lower specimen.

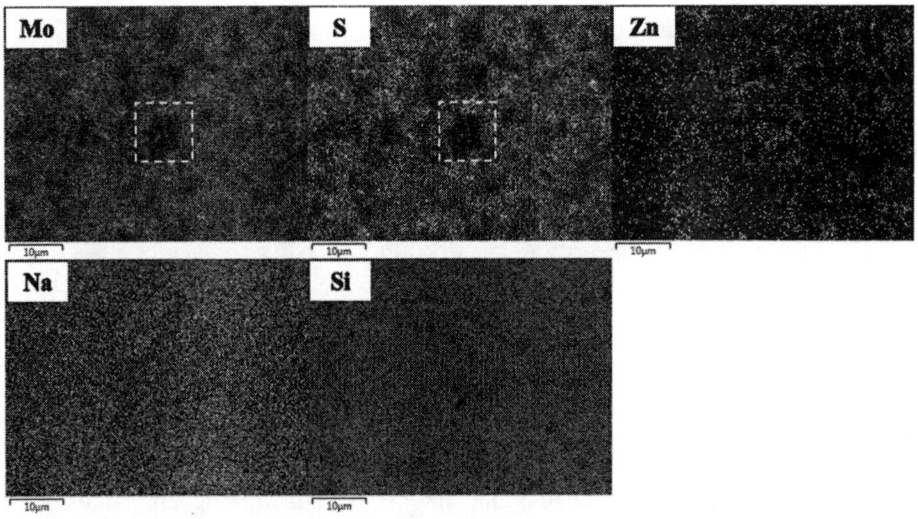

Figure 11. EDS elemental map of abrasion marks of ZnO-containing coatings.

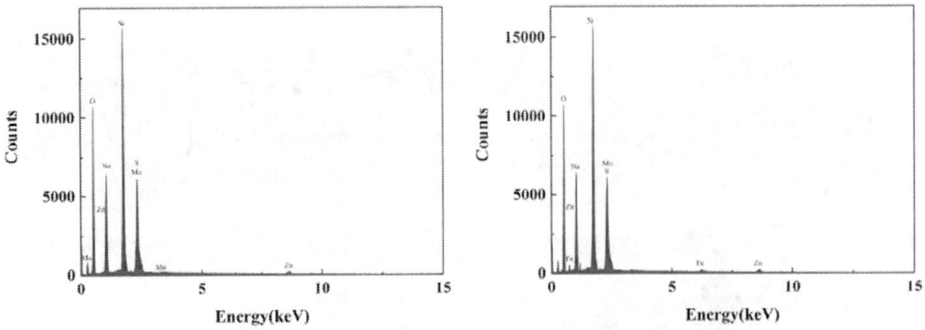

Figure 12. Changes in elemental content before and after friction.

3.7 Analysis of the role of ZnO in MoS_2 dry film systems

Based on the results of the above analyses, the dry film lubrication model structure of MoS_2 containing ZnO in Figure 13 can be constructed. Through this mechanism model, the lubrication action mechanism of the coating can be explored.

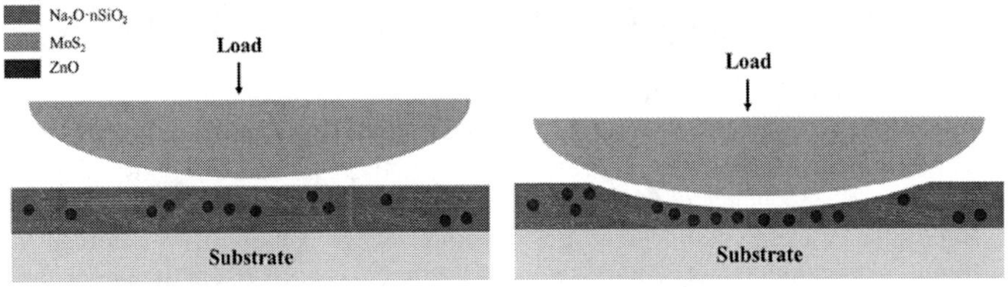

Figure 13. Lubrication mechanism model with added ZnO.

(1) The sodium silicate binder exhibits the capability to permeate throughout the entire coating system, facilitating the formation of a comprehensive bonding layer. MoS_2, serving as the primary lubricating phase, demonstrates the ability to disperse uniformly within the binder system. However, owing to its low hardness, MoS_2 is susceptible to detachment during wear. In contrast, ZnO, characterized by its high hardness, serves as the hardening phase. It functions to support the integrity of the hardened coating while aiding in the formation of the lubricating film. Moreover, ZnO serves to ensure the integrity of the lubricating film even after the detachment of MoS2, thereby enhancing the anti-wear and friction-reduction properties of the coating.

(2) At a ZnO addition of 2.5%, the incorporation of ZnO fails to sufficiently occupy interstitial spaces, thereby limiting its ability to provide additional support for the hardening process. Upon increasing the ZnO addition to 5%, synergistic lubrication between ZnO and MoS2 is achieved, resulting in a reduced coefficient of friction. However, further increasing the ZnO addition to 7.5% diminishes its dispersion effectiveness, consequently leading to an elevation in the coefficient of friction.

(3) When ZnO particles are of considerable size, their dispersion within the system is inadequate, thereby impacting the structural integrity of the dry film matrix, consequently elevating the average coefficient of friction. Conversely, smaller ZnO particles exhibit superior dispersion characteristics, facilitating their role in bolstering the integrity of the hardened coatings and thereby augmenting the formation of a lubricative film.

4. Conclusion

Using MoS_2 as solid lubricant, sodium silicate as binder, and ZnO as functional filler, a dry film lubrication coating of MoS_2 was prepared, and the coefficient of friction was measured using a friction and wear tester, and the effects of ZnO addition, particle size and working temperature on the friction performance of the coating were investigated by SEM and EDS analysis, and the following conclusions were drawn:

(1) The addition of 5% ZnO to a dry film lubricating coating of MoS_2 in a silicate system effectively reduces the coefficient of friction of the coating.

(2) The particle size of the ZnO affects the coating's coefficient of friction. Using ZnO with a particle size greater than 200 nm results in a higher coefficient of friction; lower coefficients of friction can be achieved by using 30 nm ZnO.

(3) The working temperature has a significant effect on the friction properties of the coating, which is suitable for temperatures from 25°C to 200°C, with optimum lubrication achieved at 100°C.

(4) The mechanism of ZnO is to harden and support the film-forming structure of the coating, ensuring the integrity of the lubricant film and improving the anti-wear properties of the coating, thus reducing the coefficient of friction.

Acknowledgments
Funded Project: High Performance Fasteners Project (TC220H063).

References

[1] HAN X, THRUSH S J, ZHANG Z, et al. Tribological characterization of ZnO nanofluids as fastener lubricants[J]. Wear, 2021, 468–469: 203592.
[2] YIN J, YAN H, CAI M, et al. Bonded flake MoS2 solid lubricant coating: An effective protection against fretting wear[J]. Journal of Industrial and Engineering Chemistry, 2023, 117: 450–460.
[3] MUKHTAR S H, GULZAR A, SALEEM S, et al. Advances in Development of Solid Lubricating MoS2 Coatings for Space Applications: A Review of Modeling and Experimental Approaches[J]. Tribology International, 2023: 109194.
[4] KUMAR V P, PANDA D K. Review—Next Generation 2D Material Molybdenum Disulfide (MoS2): Properties, Applications, and Challenges[J]. ECS Journal of Solid State Science and Technology, 2022, 11(3): 033012.
[5] QIN K, FENG Q, LI H, et al. General Situation of Solid Lubricant Coatings[J]. IOP Conference Series: Earth and Environmental Science, 2020, 508(1): 012219.
[6] JOBST A, MERKLEIN M. Applicability of Solid Lubricant Coatings in Cold Rod Extrusion of Stainless Steels[J]. Defect and Diffusion Forum, 2020, 404: 95–100.
[7] KUMAR R, HUSSAINOVA I, RAHMANI R, et al. Solid Lubrication at High-Temperatures—A Review[J]. Materials, 2022, 15(5): 1695.
[8] LI Y, ZHOU Z, HE Y. Solid Lubrication System, and Its Plasma Surface Engineering: A Review[J]. Lubricants, 2023, 11(11): 473.
[9] ZHOU C, LI Z, LIU S, et al. Layered double hydroxides for tribological application: Recent advances and future prospective[J]. Applied Clay Science, 2022, 221: 106466.
[10] LEITNER J, BARTŮNĚK V, SEDMIDUBSKÝ D, et al. Thermodynamic properties of nanostructured ZnO[J]. Applied Materials Today, 2018, 10: 1–11.
[11] BEHROOZIAN A, KACHOEI M, KHATAMIAN M, et al. The effect of ZnO nanoparticle coating on the frictional resistance between orthodontic wires and ceramic brackets[J]. Journal of Dental Research, Dental Clinics, Dental Prospects, 2016, 10(2): 106–111.
[12] SINGH Y, KUMAR SINGH N, SHARMA A, et al. Effect of ZnO nanoparticles concentration as additives to the epoxidized Euphorbia Lathyris oil and their tribological characterization[J]. Fuel, 2021, 285: 119148.
[13] MOUSAVI S B, ZEINALI HERIS S. Experimental investigation of ZnO nanoparticles effects on thermophysical and tribological properties of diesel oil[J]. International Journal of Hydrogen Energy, 2020, 45(43): 23603–23614.
[14] ELAGOUZ A, ALI M K A, XIANJUN H, et al. Frictional performance evaluation of sliding surfaces lubricated by zinc-oxide nano-additives[J]. Surface Engineering, 2020, 36(2): 144–157.
[15] ESSA F A, ZHANG Q, HUANG X, et al. Improved Friction and Wear of M50 Steel Composites

Incorporated with ZnO as a Solid Lubricant with Different Concentrations Under Different Loads [J]. Journal of Materials Engineering and Performance, 2017, 26(10): 4855–4866.
[16]HERNÁNDEZ BATTEZ A, GONZÁLEZ R, VIESCA J L, et al. CuO, ZrO2, and ZnO nanoparticles as anti-wear additives in oil lubricants[J]. Wear, 2008, 265(3–4): 422–428.
[17]CHAI Z, LIU Y, LU X, et al. Influence of crystal structure on friction coefficient of ZnO films prepared by atomic layer deposition[J]. Science China Technological Sciences, 2016, 59(3): 506–512.
[18]PRASAD S V, MCDEVITT N T, ZABINSKI J S. Tribology of tungsten disulfide–nanocrystalline zinc oxide adaptive lubricant films from ambient to 500°C[J]. Wear, 2000, 237(2): 186–196.

Journal of Physics: Conference Series **2808** (2024) 012009 doi:10.1088/1742-6596/2808/1/012009

Ti_3AlC_2-modified NiCrAlY as a metallic bond coat for thermal barrier coatings: a study on high-temperature oxidation resistance

Wenbin He[1,2], Junliang Li[1,2] and Kun Liu[1,2*]

[1]Henan Engineering Technology Research Center for Green Manufacturing and Precision Measurement, Zhengzhou University of Light Industry, Zhengzhou, 450000, PR China

[2]Mechanical and Electrical Engineering Institute, Zhengzhou University of Light Industry, Zhengzhou, 450000, PR China

[*]Corresponding author's e-mail: liukun023@zzuli.edu.cn

Abstract. Thermal barrier coatings (TBCs) are employed in high-temperature gas turbines to reduce the surface temperature of metallic components such as turbine blades. In order to enhance the operational performance of conventional TBCs under high-temperature and high-pressure conditions, this study designs and prepares a novel TBC with YSZ as the ceramic layer and NiCrAlY as the bond coat. Different amounts of Ti_3AlC_2 particles (5wt%, 10wt%, and 15wt%) are added to the bond coat. The microstructure and organization of the coatings are characterized by using scanning electron microscopy (SEM) and energy-dispersive X-ray spectroscopy (EDS). The resistance to high-temperature oxidation of the coatings is evaluated through atmospheric muffle furnace testing. When the Ti_3AlC_2 content is 5wt%, the coating exhibits less weight gain than traditional TBCs after oxidation at 950°C for 80 hours. Compared to coatings with Ti_3AlC_2 content of 10wt% and 15wt%, the coating with 5wt% Ti_3AlC_2 achieves the optimal resistance to high-temperature oxidation.

1. Introduction

The evolution of coating tech is vital in modern industry, spanning corrosion resistance, wear resistance, thermal protection, etc [1, 2]. Thermal barrier coatings (TBCs), crucial for gas turbines in high-temp conditions, endure oxidation that causes the TC-BC interface to bend and wrinkle, and perpendicular stresses lead to coating cracks [3]. MAX phase materials, with a layered structure, exhibit excellent thermal stability and oxidation resistance, blending metal and ceramic properties [4]. At high temps, MAX phases decompose, yielding oxides for diverse applications [5, 6]. Ti_3AlC_2, a standout MAX representative, mimics metals in electrical and thermal conductivity, displaying remarkable resistance to high-temp oxidation [7]. To enhance the high-temperature performance of traditional thermal barrier coatings, Hernández et al. incorporated SiC fibers into the coatings, leading to improved oxidation resistance in isothermal oxidation experiments at 950°C [8]. Chen et al. employed the double glow plasma (DGP) technology to prepare thermal barrier coatings and found enhanced oxidation resistance of the coatings in high-temperature oxidation experiments at 950°C [9]. This experiment uses the supersonic plasma spraying method for an 80 μm NiCrAlY-Ti_3AlC_2 bond coat and a 220 μm 8YSZ ceramic layer. After exposing the coating to 950°C insulation for 80 hours, it is weighed, and its oxidation kinetic curve is analyzed. The internal microstructure and oxidation status of the bond coat after oxidation are observed, aiming to analyze oxidation resistance and mechanisms.

2. Materials and methods

Circular discs (25 mm dia, 5 mm thickness) were cut from high-temp alloy rods as substrates. The substrate was cleaned by using an ultrasonic cleaning device and subjected to sandblasting treatment on the surface. Supersonic plasma spray raw materials: NiCrAlY (Ni 69wt%, Cr 18wt%, Al 5wt%, Y 1wt%), Ti_3AlC_2, and 8YSZ (8wt% Y_2O3-ZrO_2). These materials were agglomerated via a planetary ball mill, and after drying, flowable spray powders were obtained. The coating was prepared by using an ABB six-axis robotic arm for supersonic plasma spraying, and the substrate was sprayed with optimized spraying parameters as detailed in Table 1. Specifically, coatings TBC1, TBC2, TBC3, and TBC4 were applied, with the coating structures outlined in Table 2.

Table 1. Supersonic plasma spraying process parameters.

Powder	Current/A	Voltage /V	Main Gas Flow (Ar)/slpm	Secondary Gas Flow (H_2)/slpm	Delivery Rate/g.min^{-1}	Spraying Distance/mm
TC	380	131	63	22	35	105
BC	420	131	190	8	20	105

Table 2. Four types of thermal barrier coating structures.

Abbreviation	TC (220 μm)	BC (80 μm)
TBC1	YSZ	NiCrAlY- Ti_3AlC_2 (0wt%)
TBC2	YSZ	NiCrAlY- Ti_3AlC_2 (5wt%)
TBC3	YSZ	NiCrAlY- Ti_3AlC_2 (10wt%)
TBC4	YSZ	NiCrAlY- Ti_3AlC_2 (15wt%)

The cleaned and dried coating samples were weighed by using a microbalance with a precision of one part per million, and the initial weight m_0 was recorded. The area S of the thermal barrier coating was measured. The samples were placed in crucibles and then all specimens were placed into a muffle furnace preheated to 950°C. At regular intervals, the samples and crucibles were removed and allowed to cool naturally to room temperature in the air, and the final weight m_1 was recorded. The oxidation weight gain per unit area $\Delta W\%$ during the oxidation process was calculated, and an oxidation weight gain curve was plotted according to Equation (1).

$$\Delta W\% = \frac{(m_1 - m_0)}{S} \times 100\% \tag{1}$$

After the oxidation experiment concluded, the cross-sectional microstructure of the samples was observed by using scanning electron microscopy (SEM), and the internal elemental content of the coating was analyzed by using energy-dispersive X-ray spectroscopy (EDS).

3. Experimental results and analysis

3.1 High-temperature oxidation kinetics

From Figure 1, it is evident that with increasing oxidation time, the oxidation weight gain rate of TBC1 gradually decreased. After 30 hours of oxidation, the weight gain per unit time noticeably decreased. The oxidation weight gain pattern of TBC2 was similar to that of TBC1, but the weight gain remained consistently lower than TBC1. The oxidation weight gain of TBC3 was substantial, with the weight gain within the 20 hours of oxidation surpassing that of TBC1's 80 hours of oxidation. TBC4, after 10 hours of oxidation, showed a substantially higher weight gain than other coatings, and the coating failed after 10 hours of oxidation.

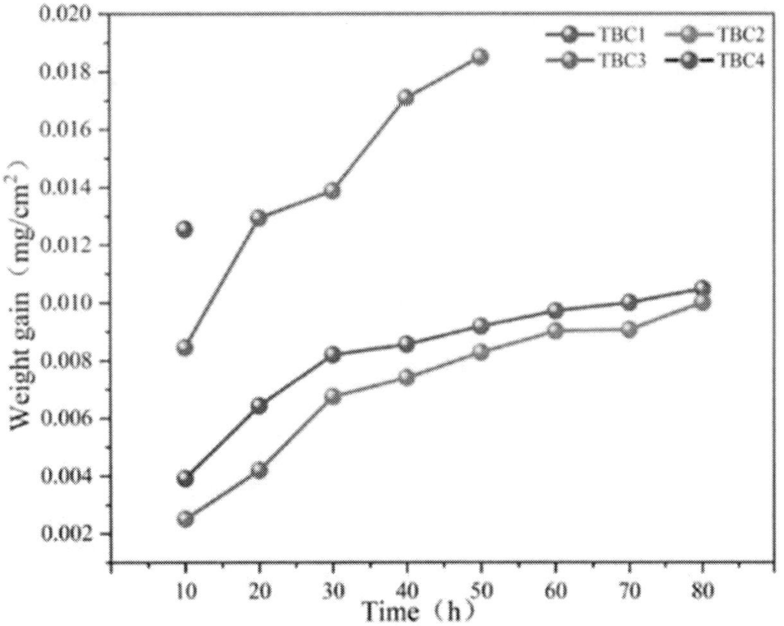

Figure 1. Oxidation weight gain curve of the thermal barrier coating.

3.2 Microstructure of the coating after high temperature

After 10 hours of oxidation, TGO formed at the interface in TBC1, creating a thin, discontinuous oxide layer. Due to the high oxidation reactivity of Al, EDS analysis indicated Al oxidation in the bond coat, gradually forming an Al_2O_3 film. The presence of 5wt% Ti_3AlC_2 facilitated Al oxidation, accelerating Al_2O_3 film formation in TBC2, resulting in a continuous, relatively thick oxide film at the interface. EDS analysis of TBC2 revealed a significant Al_2O_3 presence with trace TiO_2 in the TGO. TBC3 exhibited a TGO layer with a thickness of 1.15 μm, with EDS scans indicating a higher content of Ti oxide than Al. TBC4's coating developed a 5.62 μm thick TGO layer, and EDS scans showed a high Ti content, suggesting that the TGO was mainly composed of TiO_2, as shown in Table 3. The coarse-grained and porous structure of TiO_2 led to the formation of numerous penetrating long cracks at the interface of TBC4, which continued to expand during subsequent oxidation, resulting in the failure of TBC4, as shown in Figure 2.

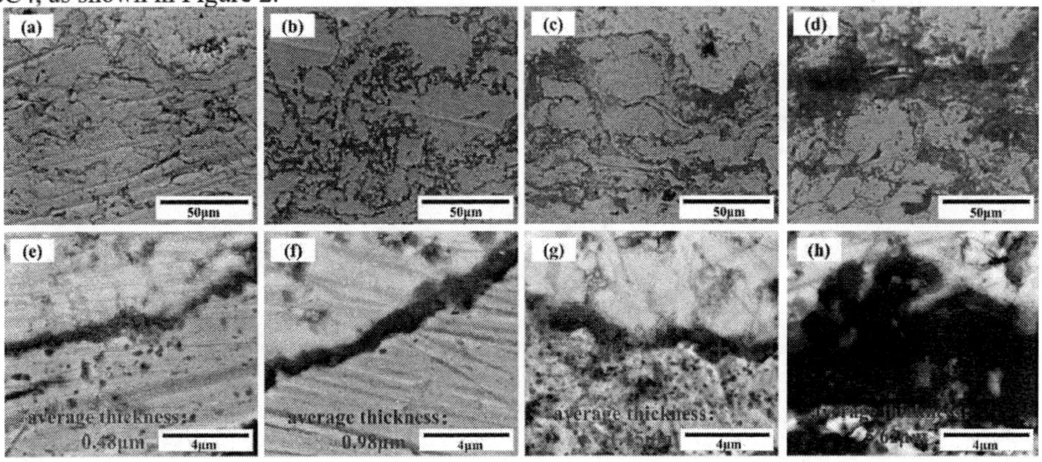

Figure 2. SEM image of the bond coat after 10 hours of oxidation for the thermal barrier coating (a) TBC1 (b) TBC2 (c) TBC3 (d) TBC4. SEM image of the TGO (e) TBC1 (f) TBC2 (g) TBC3 (h) TBC4.

Journal of Physics: Conference Series **2808** (2024) 012009 doi:10.1088/1742-6596/2808/1/012009

Table 3. Elemental point scan table of the bond coat after 10 hours of oxidation using EDS.

Sequence Number	Cr (wt%)	Ni (wt%)	Al (wt%)	Ti (wt%)	O (wt%)
1	51.94	19.34	0.31	0	28.41
2	46.3	15.69	5.55	0.57	31.89
3	36.53	14.01	7.12	11.86	30.48
4	6.12	4.17	7.61	51.94	30.16

After 30 hours of high-temperature oxidation, TGO in the bond coat further thickened. In TBC1, Al underwent oxidation, forming a protective TGO layer. The dense Al_2O_3 film formed, slowing down oxygen ion diffusion into the bond coat. The higher Al mass fraction in TGO of TBC2 indicates further Al oxidation within the bond coat, producing a substantial amount of Al_2O_3 and a small amount of TiO_2. From Figure 1, it is evident that at this point, TBC2's oxidation weight gain was less than that of TBC1, indicating that during the 30-hour oxidation, the oxidation layer composed of a significant amount of Al_2O_3 and trace TiO_2 formed earlier than a pure Al_2O_3 layer. TBC3 had already developed a 2.55 μm thick TGO layer. With a high Ti_3AlC_2 content of 10wt% in TBC3, Table 4 scans revealed that the TGO in TBC3 was primarily composed of TiO_2 at this stage, as shown in Figure 3. TBC4's oxidation failure revealed the risk of excessive TiO_2 in the TGO. At this stage, TBC3 was highly susceptible to failure, posing a significant risk.

After 50 hours of high-temperature oxidation, TBC2's bond coat exhibited lower oxidation than TBC1 and TBC3. From Table 5 scan results and previous studies, it is known that at this point, Cr in TBC1 underwent oxidation [10]. The oxide film composed of Al_2O_3 and Cr oxides effectively slowed down the oxidation rate of the bond coat. In TBC2, the TGO thickness reached 2.23 μm. Although the TGO thickness in TBC2 was consistently greater than that in TBC1, the oxidation weight gain in TBC2 remained lower than that in TBC1. TBC2 contained 5wt% Ti_3AlC_2, and according to EDS scans, a significant proportion of Al occupied TBC2's TGO, indicating that the TGO in TBC2 was composed of a substantial amount of Al_2O_3 and TiO_2. During the 50-hour oxidation process, the oxide layer composed of a significant amount of Al_2O_3 and a small amount of TiO_2 dramatically slowed down the oxidation rate of the bond coat. TBC3 had achieved a TGO thickness of 6.12 μm, thicker than TBC4's TGO before failure, with numerous penetrating cracks appearing at the interface, as shown in Figure 4. This implies that excess Ti_3AlC_2 had an adverse effect on TBC oxidation.

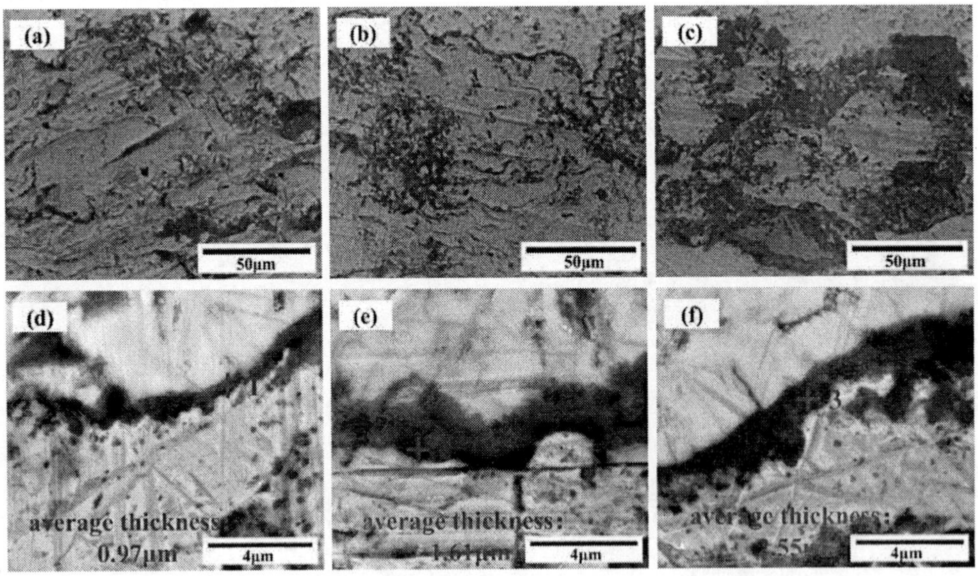

Figure 3. SEM image of the bond coat after 30 hours of oxidation for the thermal barrier coating (a) TBC1 (b) TBC2 (c) TBC3. SEM image of the TGO (d) TBC1 (e) TBC2 (f) TBC3.

Journal of Physics: Conference Series 2808 (2024) 012009 doi:10.1088/1742-6596/2808/1/012009

Table 4. Elemental point scan table of the bond coat after 30 hours of oxidation using EDS.

Sequence Number	Cr (wt%)	Ni (wt%)	Al (wt%)	Ti (wt%)	O (wt%)
1	56.94	17.46	0.78	0	24.82
2	37.56	23.66	8.81	1.07	28.9
3	5.3	4.08	12.91	40.74	36.97

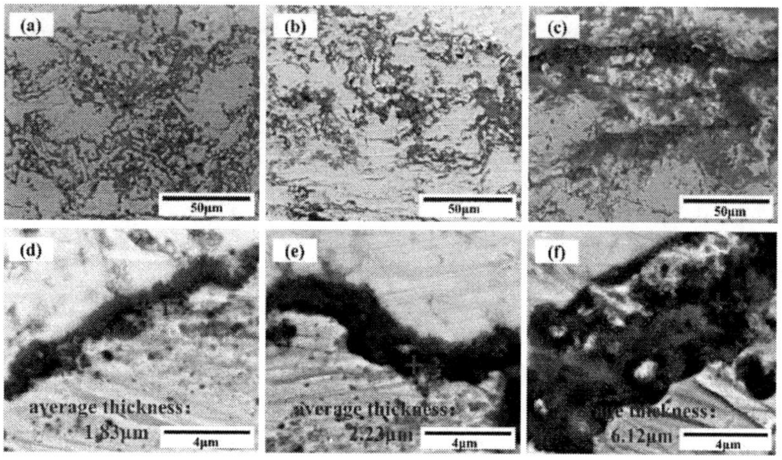

Figure 4. SEM image of the bond coat after 50 hours of oxidation for the thermal barrier coating (a) TBC1 (b) TBC2 (c) TBC3. SEM image of the TGO (d) TBC1 (e) TBC2 (f) TBC3.

Table 5. Elemental point scan table of the bond coat after 50 hours of oxidation using EDS

Sequence Number	Cr (wt%)	Ni (wt%)	Al (wt%)	Ti (wt%)	O (wt%)
1	63.35	3.48	1.19	0	31.98
2	19.32	32.14	15.98	1.4	31.16
3	1.29	2.57	6.76	37.47	51.91

After 80 hours of high-temp oxidation, TBC1's bond coat further oxidized. As per EDS scans and prior research, Cr and Ni in TGO underwent oxidation, forming trace Cr-Ni metal oxides [10]. The combined oxide film with Al_2O_3 slowed oxygen ion diffusion into the bond coat. The oxidation extent of the bond coat in TBC2 was lighter compared to TBC1. According to Table 6, the dominant components in TBC2's TGO were oxides of Al and Ti, forming a protective TGO layer mainly composed of Al_2O_3 and assisted by TiO_2. This composition effectively slowed down the oxygen ingress, resulting in a smaller TGO thickness in TBC2 compared to TBC1 after 80 hours of oxidation. This indicates that the thermal barrier coating containing 5wt% Ti_3AlC_2 exhibited excellent oxidation resistance. TBC2 demonstrated a strong bond at the interface between the bond coat and YSZ layer, without significant cracks or voids, enhancing interlayer bonding strength and extending the coating's lifespan, as shown in Figure 5.

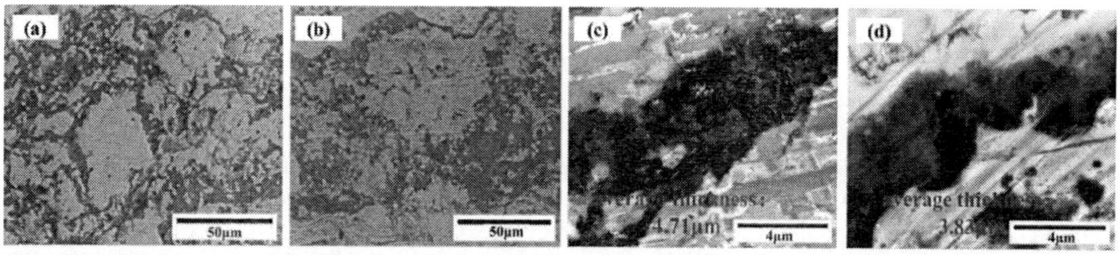

Figure 5. SEM image of the bond coat after 80 hours of oxidation for the thermal barrier coating (a) TBC1 (b) TBC2. SEM image of the TGO (c) TBC1 (d) TBC2.

Journal of Physics: Conference Series **2808** (2024) 012009 doi:10.1088/1742-6596/2808/1/012009

Table 6. Elemental point scan table of the bond coat after 80 hours of oxidation using EDS.

Sequence Number	Cr (wt%)	Ni (wt%)	Al (wt%)	Ti (wt%)	O (wt%)
1	44.77	20.58	7.89	0	26.76
2	24.13	20.35	19.93	1.65	33.94

Based on the aforementioned research results, the mechanism by which Ti_3AlC_2 particles enhance the high-temperature oxidation resistance of coatings can be summarized. Figure 6 depicts an oxidation schematic based on the oxidation weight gain curve of the bonded coatings. After 10 hours of oxidation in TBC1, a continuous Al_2O_3 protective film had not yet formed. However, in TBC2, Ti_3AlC_2 underwent oxidation to generate Al_2O_3 during the initial stages of high-temperature oxidation, as shown in Equation 2. The Al_2O_3 produced from the oxidation of Ti_3AlC_2 supplemented the discontinuous Al_2O_3 film formed from the oxidation of NiCrAlY, forming a continuous and dense Al_2O_3 film at the interface between the ceramic layer and bond coat. Subsequently, Ti_3AlC_2 continued to oxidize to form TiO_2, as indicated in Equation 3. After 10 hours of oxidation, TBC2 had a large amount of Al_2O_3 combined with trace amounts of TiO_2, forming a dense, continuous oxidation film, which slowed down the diffusion of oxygen ions into the bond coat. Because TBC2 formed a continuous and dense oxidation film earlier than TBC1, it exhibited superior high-temperature oxidation resistance compared to TBC1.

$$4Ti_3AlC_2 + (2y + 3)O_2 \rightarrow 4Ti_3C_2O_y + 2Al_2O_3 \quad (2)$$

$$4Ti_3C_2O_y + (20 - 2y)O_2 \rightarrow 12TiO_2 + 8CO_2 \quad (3)$$

Combining Equation 2 and Equation 3, the oxidation reaction equation of Ti_3AlC_2 in the bond coat is represented by Equation 4.

$$4Ti_3AlC_2 + 23O_2 \rightarrow 12TiO_2 + 2Al_2O_3 + 8CO_2 \quad (4)$$

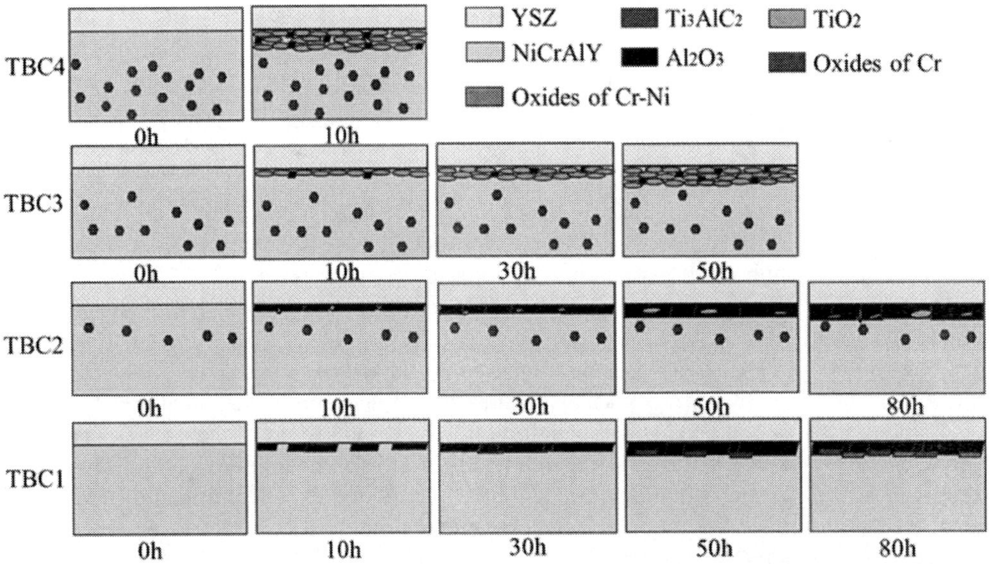

Figure 6. Schematic diagram of the high-temperature oxidation process of four thermal barrier coatings.

In TBC3, the presence of 10wt% Ti_3AlC_2 resulted in substantial oxidation of Ti_3AlC_2, forming an oxide layer composed of a large amount of TiO_2 and a small amount of Al_2O_3. However, the TiO_2 layer had a loose texture, thus failing to impede the diffusion of oxygen into the bond coat, leading to further oxidation of Ti_3AlC_2 and subsequent delamination and failure of the coating. Similarly, in TBC4 containing 15wt% Ti_3AlC_2, even more, TiO_2 was generated after oxidation, resulting in a significant increase in oxidation weight gain and subsequent delamination and failure of the coating. The fundamental reason for the inferior oxidation resistance of TBC3 and TBC4 compared to TBC2 is the rapid formation of an oxide layer consisting of a large amount of loose-textured TiO_2 in a short period.

4. Conclusion

In early high-temperature oxidation, TBC1's Al_2O_3 film formed slowly, while TBC2's Ti_3AlC_2 oxidation produced abundant Al_2O_3 and a small amount of TiO_2, promoting the early formation of a dense TGO layer. This process decelerated the bond coat's oxidation rate. TBC3 and TBC4, with an excess of Ti_3AlC_2, generated plentiful TiO_2, leading to increased mass. The loose nature of TiO_2 accelerated bond coat oxidation, forming a TGO layer mainly composed of TiO_2 and traces of Al_2O_3. Compared to TGO layers with a significant Al_2O_3 and trace TiO_2, this composition demonstrated poorer oxidation resistance in slowing bond coat oxidation. TBC2 with 5wt% Ti_3AlC_2 displayed excellent antioxidation capability.

Acknowledgments

This work was supported by the National Natural Science Foundation of China (Grant No. U2004169) and the Major Science and Technology Projects of Henan Province in 2022 (221100220200).

References

[1] Thakare, JG., Pandey, C., Mahapatra, MM., Mulik, RS. (2021) Thermal Barrier Coatings- A State of the Art Review. Met. Mater. Int., 27:1947-1968.
[2] Gurrappa, I., Rao AS. (2006) Thermal barrier coatings for enhanced efficiency of gas turbine engines. Surf. Coat. Technol., 201:3016-3029.
[3] Liu, JH., Lu, Z., Zhou, YW., Zhang, J., Lyu, GL. (2023) A review of the failure behavior and countermeasures of thermal barrier coatings. J. CERAM. PROCESS. RES., 24:285-307.
[4] Reshetnyak, EN., Kuprin, AS., Prikhna, TA., Bortnitskaya, MA., Belous, VA. (2023) SYNTHESIS, STRUCTURE, AND PROTECTIVE PROPERTIES OF PVD MAX PHASE COATINGS. A REVIEW. PART I. MAX PHASE COATINGS DEPOSITION. PROBL. ATOM. SCI. TECH. pp.111-125.
[5] Tallman, DJ., Anasori, B., Barsoum, MW. (2013) A critical review of the oxidation of Ti_2AlC, Ti_3AlC_2, and Cr_2AlC in air. Mater. Res. Lett., 1:115-125.
[6] Liu, JJ., Yang, J., Yi, GW., Ma, JQ., Yu, Y., Qiao, ZH., Hu, B., Liu, WM. (2020) Effect of high-temperature oxidation on Si_3N_4 containing Ti_3AlC_2. Ceram. Int., 46:14697-14705.
[7] Mao, FZ., Pi, WQ., Li, YQ., Xu, X., Wen, P., Zhou, W. (2022) Improved mechanical properties and toughening mechanism of mullite ceramics reinforced by introducing Ti_3AlC_2 particles. INT. J. APPL. CERAM. TEC., 19:1650-1658.
[8] Hernández, GIV., Albarrán, MAG., de Anda, ER., Montalvo, JDI., Guillén, JCD., Aguilar, JMC. (2023) Study of the Incorporation of SiC Microfibers and Their Effect on Adherence at the Interface of a NiCoCrAlY-7YSZ Thermal Barrier Coating. High Temperature Corrosion of Materials., 101:105-124.
[9] Chen, WL., Liu, M., Zhang, JF. (2016) Impedance Analysis of 7YSZ Thermal Barrier Coatings During High-Temperature Oxidation. J. Therm. Spray Technol., 25:1596-1603.
[10] Cao, Xy., Xiang, Xd. (2011) Research on Inoxidizable Performance of Bond Coat of Thermal Barrier Coating at High Temperature. Dongfang Turbine., pp.71-75.

Journal of Physics: Conference Series **2808** (2024) 012010 doi:10.1088/1742-6596/2808/1/012010

Utilization of waste silty soil as fine aggregate in controlled low-strength material

Haomiao Qiu[1*], Senhua Jiang[1] and Lei Shou[2]

[1]Research Center of Coastal and Urban Geotechnical Engineering, Zhejiang University, Hangzhou, 310058, China

[2]Hangzhou CBD Development Group CO, LTD., Hangzhou, China

[*]Corresponding author's e-mail: 11412030@zju.edu.cn

Abstract. Silty soil excavated in construction projects is usually considered waste material. This work pioneers a high-value utilization of waste soil by exploring the feasibility of repurposing silt soil to prepare controlled low-strength materials (CLSMs). For sustainable purposes, alkali-excited fly ash and waste glass powder are used instead of pure cement, and the necessary additives are introduced to improve the performance of CLSM. Laboratory tests are performed on CLSMs with different mixture ratios to evaluate the flowability, unconfined compressive strength, and setting times. The effects of two critical parameters: the water content (W/G), i.e., the mass ratio of water to solid in CLSM mixtures, and the stabilizer content (S/ML_d), i.e., the mass ratio of stabilizer to dry silty soil are studied. Experimental results indicate that the increase of water content has a favorable effect on flowability, but significantly increases setting time and reduces the compressive strength. Furthermore, when the water content is over 0.4, the initial setting time may exceed 4 hours, limiting the application of CLSMs. The increase of stabilizer content is beneficial to the increase of flowability and strength and markedly accelerates the final setting time. The predictive equations of flowability and compressive strength and 7 days of the silty soil-based CLSM have been developed with W/G and S/ML_d, and the predicted results are compared with the results on the contour map. Based on the prediction model, the recommended intervals of W/G and S/ML_d are given to meet the strength and flowability targets of different applications.

1. Introduction

Recently, the process of industrialization and urbanization has gradually accelerated in China, and the bulk solid waste represented by construction waste, industrial solid waste, and household waste has shown a rapid increase trend, resulting in a series of problems such as resource waste and environmental pollution. Proper disposal and recycling of solid waste has become an unavoidable and important issue in China's economic and social development. In recent years, the Chinese government has been committed to supporting the large-dosage, large-scale, and high-value utilization of bulk solid waste in order to achieve the target of energy conservation and emission reduction [1].

Controlled low-strength material (CLSM) is a casting-type filling material that is formed by specific stirring preparation and mixing with certain fluidity. CLSM has attracted wide attention in various engineering applications due to its comprehensive utilization of solid waste and excellent working performance in recent years [2]. Originally, CLSM is composed of Portland cement, fine aggregate, supplementary binder material, and water. Sand is commonly used as a fine aggregate in CLSMs [2]. For the sake of sustainability, a variety of solid wastes such as industrial by-products [3-4], and household waste [5] are synergistically used to prepare a low-carbon stabilizer, and expansive soil from civil excavation projects is used as fine aggregate [6] in many studies.

Silty soil is the most common type of soil in the excavation of underground space in many coastal cities in China, such as Shanghai, Hangzhou, and Nanjing. However, silty is often considered waste and is difficult to use directly due to its poor engineering properties [7]. Therefore, the utilization of waste silty soil as fine aggregate in CLSMs is studied in this work. It is accepted that the high specific surface area and water absorption of silty soil are unfavorable factors that limit their application in CLSMs. To meet the targeted flowability, more free water is required, which contradicts the requirement of hardened strength [6, 8].

Aiming to prepare a low-carbon silty soil-based CLSM and improve workability, the alkali-excited solid waste stabilizer is selected in this study. To reduce the water requirement of the CLSM system, fly ash (FA) and waste glass powder (GW) are used as binders for their low water absorption and significant lubrication. Certain additives are also introduced into the mixtures in this work to adjust the working performance.

The feasibility of utilization of waste silty soil as fine aggregate in CLSM is demonstrated by experiments. The properties of CLSM are evaluated by flowability, unconfined compressive strength at 7 days, and setting time. Moreover, the effects of water content and stabilizer content on the performance of CLSM are investigated in this work. At last, the reference combinations of water content and stabilizer content are given, depending on the applications.

2. Materials and methods

2.1. Physical properties of the excavated silty soil

In this research, the excavated silty soils are taken from a construction project site in Hangzhou, Zhejiang Province, China, primarily due to its local availability. The liquid limit is 30.1%, the plastic limit is 20.9%, and the water content is 28.9% of the silty soils.

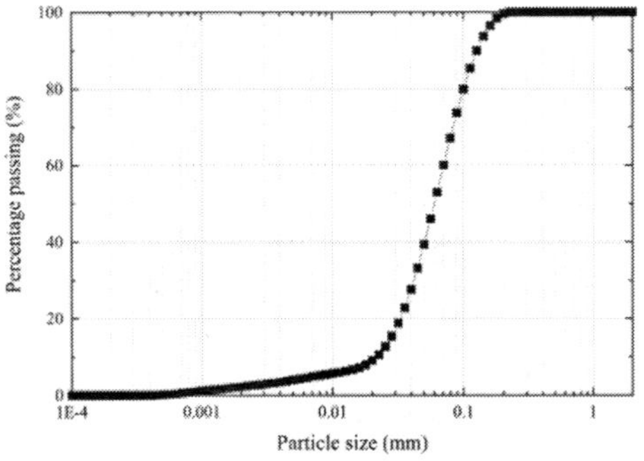

Figure 1. Particle size distribution curve of waste silty soil.

Figure 1 illustrates the particle size distribution of waste silty soil. The sand content is 34%, the silty content is 59%, and the clay content is 7%. According to the silty content and liquid limit, the soil can be classified as low liquid limit silt soil, denoted by ML, and the symbol ML_d stands for dry silt soil in this paper.

2.2. Composition of Stabilizer

The stabilizer in this study consists of three parts: alkaline activator solution, binders, and additives. The alkali activator solution is a sodium silicate solution (denoted by AAS) with a modulus of 1.2 and a solid content of 42%, which is prepared by mixing NaOH and Na_2SiO_3. The mass fraction of AAS is 5% in this work. The binders in the stabilizer are made of Portland cement (denoted by PC), Class F fly ash (denoted by FA), and glass waste (denoted by GW). The particle size distribution curves and the main chemical composition of these binders are shown in Figure 2. The mass fractions of PC, FA, and GW are taken as 30%, 30%, and 20% in this work. Three additives are introduced in the stabilizer, i.e., Calcium oxide (denoted by CO), Sodium hexametaphosphate (denoted by SH), and Nano-silicon dioxide (denoted by NS).

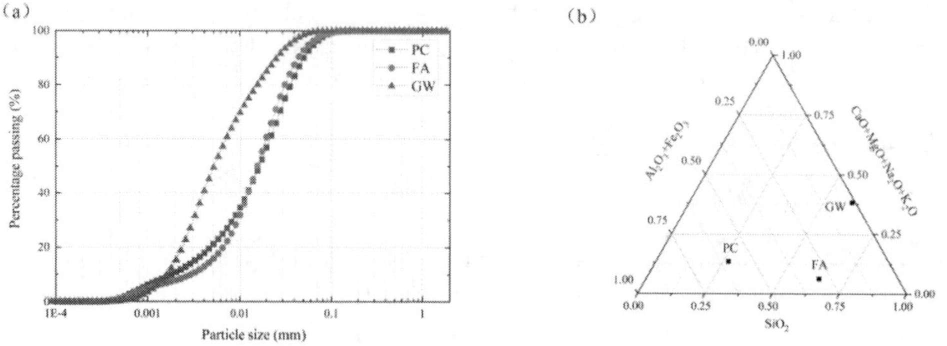

Figure 2. Particle size distribution curves and ternary diagram of the main chemical composition of three binders (PC, FA, GW).

2.3. Experimental tests for CLSM

The notation W/G is the water content, which refers to the mass ratio of water to solid in CLSM mixtures, and the S/ML$_d$ is the stabilizer content, which refers to the mass ratio of stabilizer to dry weight of soil. The W/G is taken as 0.36, 0.38, 0.40, 0.42, and S/MLd is taken as 0.14, 0.16, 0.18, and 0.20 in this work.

The biggest challenge in preparing CLSMs with silty soil as fine aggregate is to solve the problem of flowability. Good flowability distinguishes CLSMs from other backfill materials, which enables the materials to be self-leveling and self-consolidating. The flow tests are performed according to ASTM D6103 [2]. The setting times are estimated by performing the Vicat needle test according to ASTM C403 [9] at a temperature of 20°C. The unconfined compressive strength (UCS) tests at 7 days are performed according to ASTM 4832 [10].

3. Results and discussion

3.1. Flowability of CLSM

Figure 3(a) shows the variation of flowability with W/G. The vertical coordinate represents the flowability *f*, which is divided by 100 in order to minimize the error in the subsequent fitting. It is not difficult to find that the fluidity of CLSM increases significantly with the increase of free water. For instance, the flowability at W/G=0.42 is almost twice as high as that at W/G=0.36, regardless of the stabilizer content. It should be noted that there is a marginal effect on the increase of flowability. That is, the effect of water content on the improvement of flowability is decreasing with increasing free water. Another interesting finding is that the water content requirement is less than silty soil-based CLSM reported by Wan et al. [6]. The difference may be attributed to the lower water requirement and stronger lubricating effect of FA and GW in the stabilizer than PC, and the dispersing effect of AAS and SH. According to the results of the tests, the fitting equations of the relationship between the flowability (*f*/100) and water content (W/G) are obtained with different stabilizer content, as shown in the figure.

To further illustrate the effects of stabilizer content on flowability, the variation of flowability with S/ML$_d$ of the specimen is shown in Figure 3(b). Similarly, the flowability is divided by 100. It can be seen that the flowability of the CLSM increases approximately linearly with the increase of stabilizer content, which reinforces the difference between the stabilizer in this study and the ordinary Portland cement stabilizer. The linear fitting equations for the relationship between flowability (*f*/100) and stabilizer content (S/ML$_d$) are obtained with different water content, as shown in the figure.

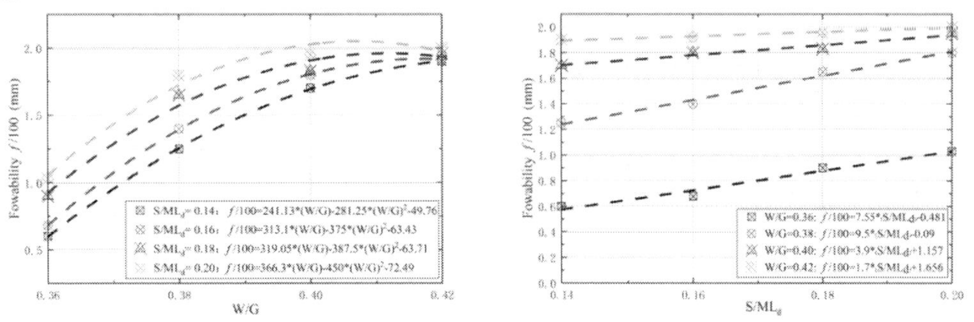

Figure 3. Flowability of the silty soil-based CLSM based on water content and stabilizer content.

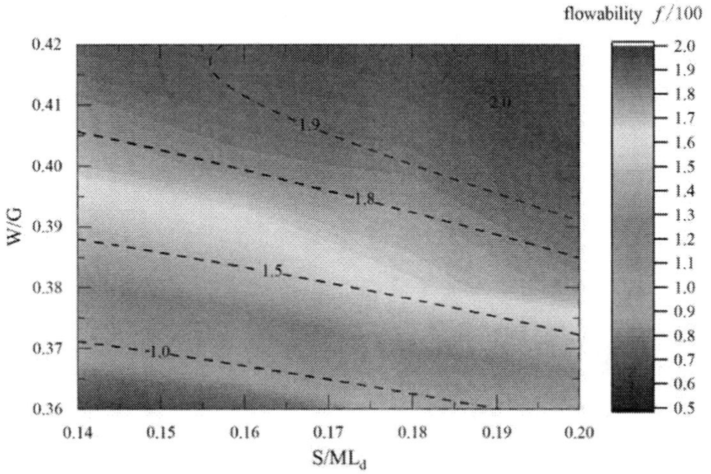

Figure 4. Contour plot of flowability based on stabilizer content and water content.

Figure 4 presents the contour plot of flowability based on stabilizer content and water content, in which the dashed line is the result of surface fitting. According to the results of the tests, the fitting equation is as follows:

$$f/100 = -70.59 + 46.02s + 329.57w + 14.06s^2 - 373.44w^2 - 115.75sw \quad (1)$$

where s is S/ML_d, w is G/W. It can be observed that the predicted results of Equation (1) match the results of the experimental tests well.

3.2. Compressive Strength of CLSM

Figure 5(a) presents the variation of unconfined compressive strength at 7 days with W/G of the specimen. When the water content increases from 0.36 to 0.42, the compressive strength of the specimens shows a linear decrease, and the decrease tends more obvious with the increase in stabilizer content. The linear fitting equations of the relationship between compressive strength c and water content (W/G) are obtained with different stabilizer content, as shown in the figure.

Figure 5(b) shows the variation of unconfined compressive strength with an increase in stabilizer content with different water content of stabilizer. It can be noticed that the increase in stabilizer content has a significant effect on the increase of the strength. For instance, the unconfined compressive strength at S/ML_d=0.20 is almost twice as high as that at S/ML_d=0.14 when W/G=0.36. Similarly, the linear fitting equations of the relationship between compressive strength c and stabilizer content are shown in the figure.

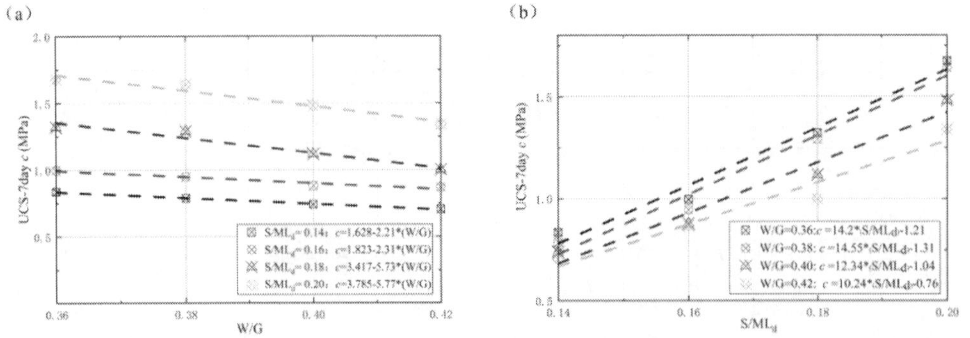

Figure 5. Unconfined compressive strength (7 days) of the silty soil-based CLSM based on water content and stabilizer content.

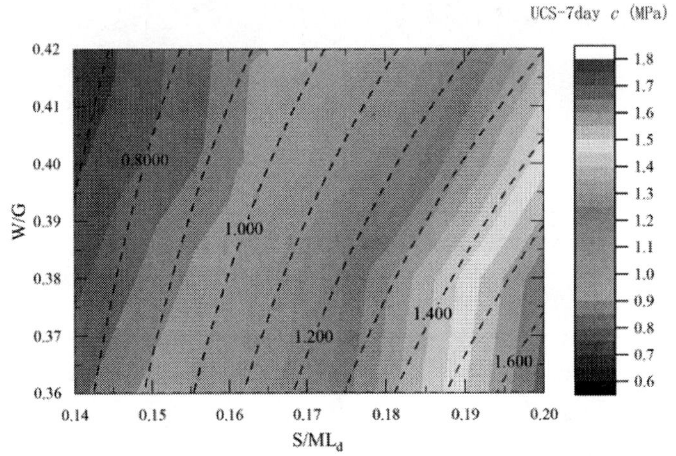

Figure 6. Contour plot of unconfined compressive strength based on stabilizer content and water content.

Figure 6 presents the contour plot of unconfined compressive strength at 7 days based on stabilizer content and water content, in which the dashed line is the result of surface fitting. According to the results of the tests, the fitting equation is as follows:

$$c = 44.014s + 9.22w - 79.127sw - 4.73 \quad (2)$$

where s is S/ML$_d$, w is G/W.

3.3. Setting time of CLSM

Figures 7-8 display the variations of initial and final setting times with water content and stabilizer. As can be seen from the figure, the initial setting time of CLSMs is between 1.8~5 h, and the final setting time is between 7.8~12 h in this work. As the water content increases, the setting times, especially the initial setting time rapidly increase, and it is a good way to strictly limit the water content to less than 0.4 to reach a reasonable initial setting time. It is also important to note that the increase of stabilizer content can also shorten the setting time. However, the increase of stabilizer content has a greater effect on the final setting time than the initial setting time.

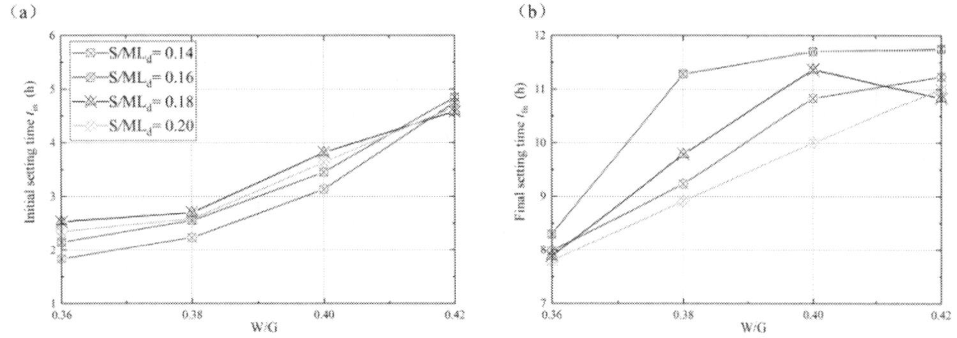

Figure 7. Setting times based on water content with different stabilizer content.

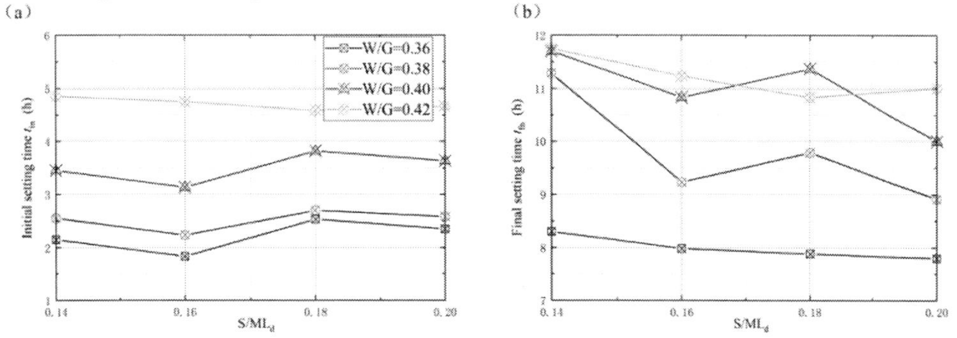

Figure 8. Setting times based on stabilizer content with different water content.

3.4. Predictions and applications

Depending on the needs of different backfilling projects, the flowability can change from rigid to fluid. In this work, the classification system on flowability is according to ACI committee 229R [2], as follows: a) High flowability (f>200 mm, denoted by H); b) Normal flowability (150 to 200 mm, denoted by N); c) Low flowability (f<150 mm, denoted by L). It should be noted that for flowability slump less than 100 mm, the symbol LL is adopted in this work. The compressive strength can be varied from 0.3 to 2.1 MPa, depending on the application. Four intervals can be classified according to the unconfined compressive strength, that is 0.3-0.7 MPa, 0.7-1.0 MPa, 1.0-1.5 MPa, and 1.5-2.1 MPa in this work. It should be noted that when the compressive strength is less than 0.7 MPa, CLSM is often used as a backfill rather than a structural fill.

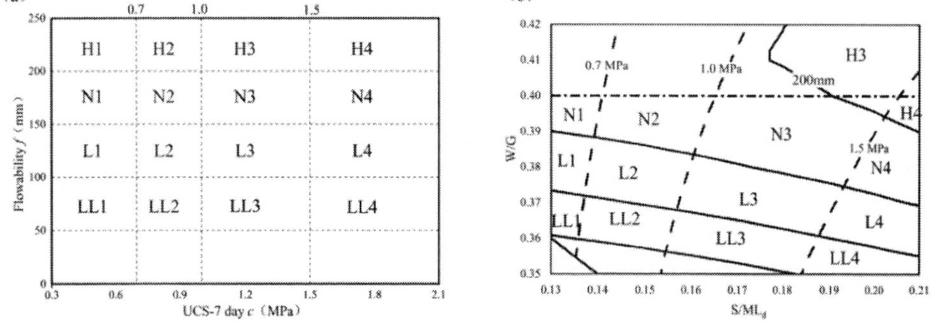

Figure 9. Interval delineation based on flowability and unconfined compressive strength, and prediction of stabilizer content and water content.

Combining the flowability and unconfined compressive strength, 16 different combinations are displayed in Figure 9(a). Based on the above analysis in Sections 3.1-3.3 and Equations (1) and (2), the predicted combinations of stabilizer content and water content corresponding to the 16 different combinations are shown in Figure 9(b). It should be noted that the range below the dashed line W/G = 0.4 is what we need to consider the initial setting time target. It is concluded that the water content and the stabilizer content are two critical parameters that guarantee both the flowability and compressive strength of the silty soil-based CLSMs to meet the engineering requirements, and the different combinations of water content and stabilizer content are suitable for different applications.

4. Conclusions

The present research deals with silty soil excavated in construction projects, and provides a controlled low-strength material using waste silty soil as fine aggregate. Different from cement-based CLSMs, an alkali-activated stabilizer is adopted in this work. In view of the large water demand of silty soil, water demand, and lubricity binders are considered in the design of stabilizer, and the dispersion effect of alkali activators and additives is also considered. Laboratory tests are performed on CLSMs with different commotions of water content and stabilizer content to evaluate the flowability, setting times, and unconfined compressive strength. The limitations and potential applications are also discussed in this work. The main conclusions can be drawn as follows:

- The alkali activator stabilizer in this work has comparative advantages for silty soil-based CLSMs, especially in terms of lowering water demand and improving flowability. Another advantage of this stabilizer is the possibility of utilizing industrial and household solid waste.
- Flowability, setting time, and compressive strength are the most important properties that limit the applications of CLSMs, and the effect of components of CLSMs on these properties needs to be focused on when designing CLSMs.
- Water content (W/G) and stabilizer content (S/ML_d) are two critical parameters to carry out the design of silty soil-based CLSMs. The increase in water content (W/G) is favorable for the increase in flowability but is very detrimental to the compressive strength and setting time. Limiting the water content to less than 0.4 is necessary to obtain a reasonable setting time.
- The increase in stabilizer content (S/ML_d) is beneficial to flowability, setting time, and strength at the same time. However, the cost increases considerably with the increasing stabilizer content. Different from other stabilizers, the stabilizer in this work is beneficial to flowability.
- The applications as well as combinations of water content and stabilizer content are discussed to initially determine the appropriate combinations of different needs.

Acknowledgments

We would like to acknowledge the financial support provided by the Science and Innovation Foundation of Hangzhou CBD Development Group CO, LTD.

References

[1] Notice of The State Council of the People's Republic of China (2021). An action plan for achieving carbon peak before 2030.
[2] American Concrete Institute (2022). Report on Controlled Low-Strength Materials (ACI 229R-13 reapproved 2022).
[3] Lee NK, Kim HK, et al. (2013). Alkal-activated, cementless, controlled low-strength materials utilizing industrial by-products. Constr. Build. Mater., 49, 738.
[4] Singh N.B, & Middendorf B (2020). Geopolymers as an alternative to Portland cement: An overview. Construction & Building Materials, 237
[5] Poorveekan K, Ath KMS, et al. (2021). Investigation of the engineering properties of cementless stabilized earth blocks with alkali-activated eggshell and rice husk ash as a binder. Constr. Build. Mater, 277.

[6] Wan X, Ding J, Jiao, et al. (2023). Preparing controlled low-strength materials using excavated waste soils with polycarboxylate superplasticizer. Environ Earth Sci., 82.
[7] Xiaoying Yang. (2011). Experimental study on engineering properties of Qiantang Alluvial Silts. (Doctoral dissertation, Zhejiang University).
[8] Wang L, Feng W, et al. (2022). Engineering properties of soil-based controlled low-strength materials made from local red mud and silty soil. Constr. Build. Mater, 358: 129453.
[9] ASTM. (2008). Standard Test Method for Time of Setting of Concrete Mixtures by Penetration Resistance. ASTM C403.
[10] ASTM. (2010). Standard test method for preparation and testing of controlled low-strength material (CLSM) test cylinders. ASTM D4832.

Journal of Physics: Conference Series **2808** (2024) 012011 doi:10.1088/1742-6596/2808/1/012011

Dynamic characteristics of dry-type transmission tower under wind-sand load

Fule Li [1], Binxia Yuan [1] and Jingjing Dai [2]

[1]Shanghai University of Electric Power, Shanghai, 200090, China
[2]China Haisum Engineering Co., Ltd, Shanghai, 2000231, China

*Corresponding author's e-mail: yuanbinxia100@163.com

Abstract: The desertification in Northwest China is serious, and the transmission tower has been affected by dust storms. To investigate the harm of wind-sand load on transmission towers, the paper takes the dry-type transmission tower as the research object and uses ANSYS software to study the dynamic characteristics. The wind and sand loads are coupled and superimposed by MATLAB software. The dynamic characteristics of transmission towers under different wind load and sand load conditions are studied. The results show that the horizontal offset and the overall stress of the tower will increase under wind-sand coupling load, which makes the tower stability worse.

1. Introduction

A transmission tower is a kind of high-rise building. Due to the difference in the working environment of the tower, there is great randomness and uncertainty in the load conditions [1]. Generally, the load of the transmission tower can be divided into variable load [2-5], permanent load, and accidental load [6]. Most of the variable loads are loads under natural conditions, such as ice load, sand load, wind load, etc. However, the load under sudden circumstances such as disconnection and earthquake are regarded as accidental loads. Therefore, variable loads, especially wind load and sand load, will cause great harm to the tower in arid areas. As the height of the tower increases, the vibration effect caused by the pulsating-wind will also increase, and the safety problems of the transmission tower will be important. The occurrence of collapse accidents causes serious problems for the stable transportation of the power grid [7].

In particular, there are sandstorms all year round in Northwest China [8-10], and the sandstorm load is harmful to the safety of transmission towers. However, wind and sand loads are usually ignored in the design of transmission towers and lines [11]. In the last ten years, there have been a lot of sandstorms. To decrease the damage caused by wind-sand load on the transmission lines, the dynamic characteristics of transmission towers in sandstorm conditions are very important.

With the increase of sandstorm load intensity, the influence of the wind-sand coupling effect on towers gradually increases, and may even lead to the destruction of transmission towers. In the paper, the single-circuit-dry-type transmission tower is taken as the research object. MATLAB software is used to couple the wind load and sand load. The dynamic characteristics under different wind-sand loads are investigated.

2. Modeling process

2.1 Finite element model of dry-type tower

ANSYS APDL platform is used to model and study the single circuit dry type tower, as shown in Figure 1a. The tower height is 39.5 m, the root length of the tower base is 8.2 m, and the length of the cross arms at the tower head is 11.2 m. The model has a total of 404 nodes, 124 key points, and 1056 units. The tower is constructed of angle steel with Q345 steel as the main tower material and Q235 steel as the auxiliary material.

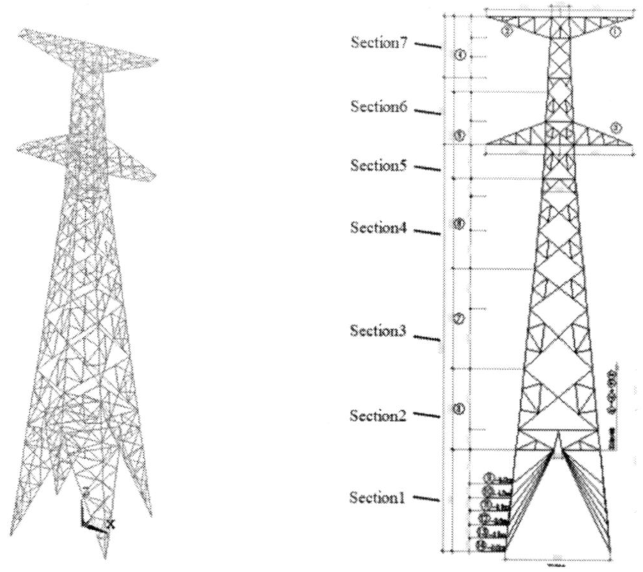

(a) (b)

Figure 1. Finite element model and sectional drawing of tower: (a) Finite element model; (b) Load sectional drawing of tower.

2.2 Static-wind load

Here, the wind angle is 0 °, indicating that the wind is perpendicular to the direction of the tower. The tower is divided into seven segments, and the stress condition of the tower is obtained by calculating the node load force on each section, respectively, as shown in Figure 1b. The calculation formula of the standard value of static-wind load acting on the transmission tower is as follows (1):

$$W_s = \omega_0 \mu_z \mu_s \beta_z A_f \quad (1)$$

$$\omega = v^2/1600 \quad (2)$$

$$\mu_z = 1.379\left(\frac{z}{10}\right)^{0.24} \quad (3)$$

$$\beta_z = 1 + \xi \varepsilon_1 \varepsilon_2 \quad (4)$$

where W_s is the average wind load (KN); ω_0 is the basic wind pressure at 10 m (kN/m^2); v is the basic wind speed; μ_z is the wind pressure variation coefficient at height z; μ_s is the shape coefficient of the transmission tower, which is 3.1; β_z is the wind vibration coefficient at height Z; ξ is the pulsation amplification coefficient; ε_1 is the influence coefficient of wind pressure height change; ε_2 is the influence coefficient of vibration mode; A_f is the area of the transmission tower under wind pressure.

When the wind speed and direction angle are 20 m/s and 0 °, respectively, the results and node load force obtained from different segments are shown in Table 1.

Table 1. Various coefficients, wind load, and node load force of transmission tower.

Height (m)	3.75	10.5	18.2	24.2	28.75	32.5	37.25
ω_0	0.25	0.25	0.25	0.25	0.25	0.25	0.25
μ_z	1.089	1.395	1.592	1.704	1.776	1.829	1.890
μ_s	3.1	3.1	3.1	3.1	3.1	3.1	3.1
ε_1	0.72	0.72	0.63	0.55	0.55	0.55	0.55
ε_2	0.08	0.15	0.4	0.53	0.53	0.74	0.74
$\xi = \omega_0 T^2$	2.31	2.31	2.31	2.31	2.31	2.31	2.31
β_z	1.133	1.249	1.582	1.673	1.673	1.940	1.940
$A_f(m^2)$	4.650	7.260	5.734	4.158	1.175	3.000	2.160
Nodes No.	12	10	12	12	6	10	10
Node load force $F_s(N)$	370	980	930	765	450	824	613

2.3 Pulsating-wind load

The pulsating-wind power spectrum is necessary for applying the random vibration theory to calculate the wind-induced vibration response. The Davenport pulsating wind speed spectra Formula (5) is as follows:

$$S_H(n) = \frac{4KV_0{}^2}{n}\frac{x^2}{(1+x^2)^{\frac{4}{3}}},\ x = \frac{1200n}{V_0}\ n \geq 0 \tag{5}$$

where K is the ground resistance coefficient; n is the frequency.

After simulating the pulsating wind speed based on the Davenport wind speed spectrum, the transient dynamics are carried out. The dynamic wind speed time history curve is obtained, and then the pulsating wind load is calculated.

2.4 Wind-sand coupling load

The density of sand dust (ρ_{TSP}) is used as the main basis to distinguish the classification of sand load. After consulting the relevant literature data analysis [12], the ρ_{TSP}-wind speed relationship curve is fitted by Origin software, and the fitting Formula (6) is as follows:

$$\rho_{TSP} = 2.59118 + 0.12039e^{\frac{v}{3.73584}} \tag{6}$$

where v is the wind speed, in m/s.

According to the pulsating-wind load, the sand-load is obtained through the calculation formula of sand load per volume, as shown in Formula (7):

$$F_s(t) = 4\rho_{TSP}V_s(t)^2A_f \tag{7}$$

where $F_s(t)$ represents the load per volume of sand; $V_s(t)$ represents the approximate volume of each segment; A_f represents the projected area of each segment.

3. Results and discussion

3.1 Dynamic characteristics of tower under static-wind

When the wind load with an average wind speed of 20 m/s is loaded, the displacement nephogram, stress nephogram, and axial nephogram are obtained, as shown in Figure 2. Through the analysis results, it can be found that the dry-type transmission tower has high symmetry. Figure 2a shows the displacement nephogram under the static wind. From the tower foot to the tower top, the model color gradually darkens. It can be found that the displacement offset of the transmission tower increases as the height increases, suggesting that the tower stability gradually deteriorates. The offset of the left cross arm and the right cross arm is equal, which is 8.29 mm. Figure 2b shows the displacement nephogram under the static wind. The stress is distributed on the tower leg. The maximum stress on the tower is 8.16 MPa. Figure 2c is the axial force cloud diagram of the transmission tower. It can be seen that the rods on the leeward side of the transmission tower are subjected to compressive stress, while the rods on the windward side are subjected to tensile stress. At the same time, the node stress on the main tower material decreases with the increase in height. The maximum axial force, tension, and pressure produced are 23.1 kN and -23.4 kN, respectively.

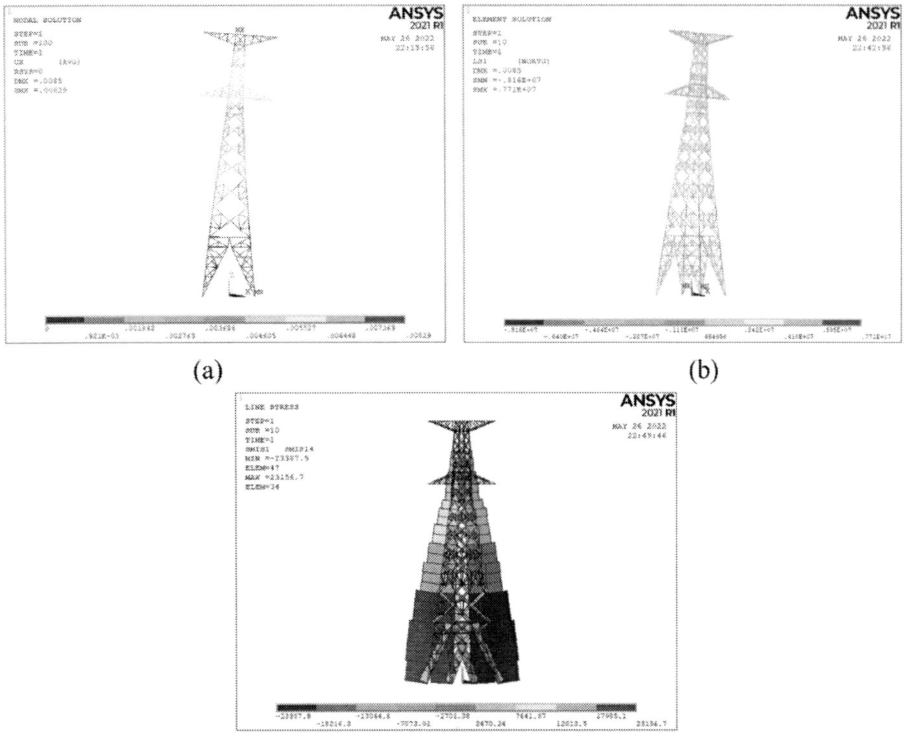

(a) (b)

(c)

Figure 2. Dynamic results under static wind: (a) The displacement nephogram, (b) Stress nephogram, (c) Axial nephogram.

3.2 Dynamic characteristics under pulsating-wind and wind-sand coupling load

The pulsating-wind curve on Davenport wind speed spectra is obtained in Figure 3. Figure 3a is the comparison between the target wind speed power spectrum and the calculated simulation results. The results indicate that the simulation results are distributed on both sides of the target wind speed power spectrum function curve, which is very consistent with the expected results. Figure 3b is a dynamic wind speed diagram, which has 600 s in total. It is sufficient to take any data for 10 seconds during calculation. The data of 345 s~354 s are taken in the paper.

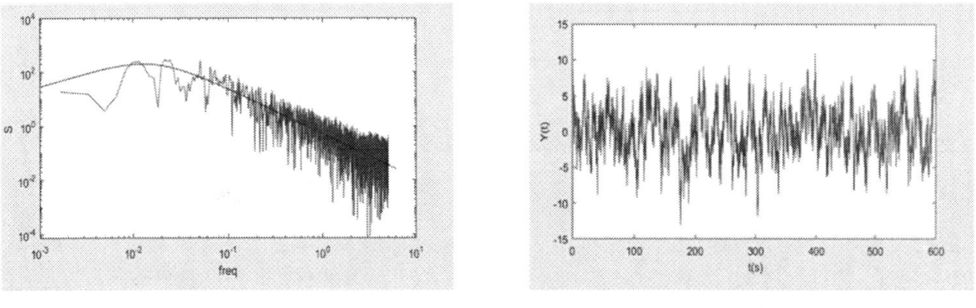

(a) (b)

Figure 3. Pulsating-wind speed time history curve: (a) Comparison of target wind speed power spectrum and simulation results, (b) Dynamic anemometer.

Journal of Physics: Conference Series **2808** (2024) 012011 doi:10.1088/1742-6596/2808/1/012011

Pulsating-wind is substituted into the instantaneous wind load calculation Formula (1). Through calculation, we can get the node load force, and then conduct the transient dynamic analysis of the dry type tower in the pulsatingwind. Figure 4 shows the comparison diagram of wind sand load and wind load displacement. The displacement oscillation trend of the midpoint of each section is similar, and the maximum displacement is taken near the position of time t=1. As the height of the tower becomes higher, the displacement also increases.

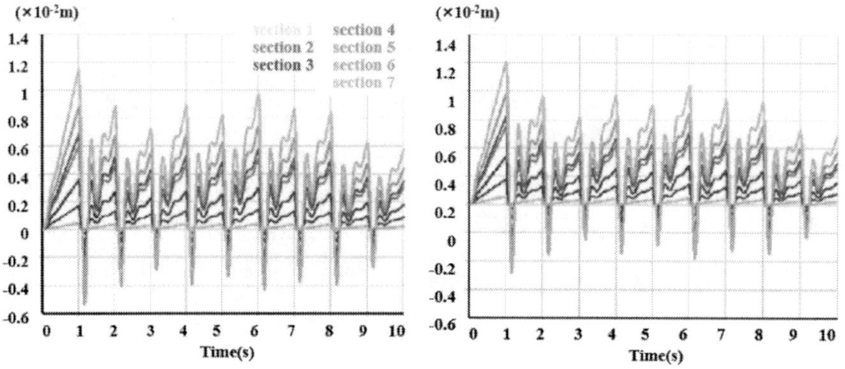

(a) (b)

Figure 4. Comparison diagram of wind-sand load and wind load displacement: (a)Pulsatingwind load, (b) Pulsatingwind-sand load.

According to the maximum displacement obtained under the condition of wind-sand load, the data of t=1 s is extracted. Figure 5 shows the displacement nephogram, stress nephogram, and axial force nephogram. The maximum displacement under pulsating-wind and sand coupling loads are 12.3 mm and 13.7 mm. Compared with the static wind, the increase rates are 48.4% and 65.2%, respectively. The maximum stress under pulsating-wind and wind-sand coupling load is 11.3 MPa and 13.7 MPa, and the increase rates are 38.5% and 67.9%, respectively. The maximum axial tension under pulsating-wind and wind-sand coupling load is 33.9 Kn and 38.5 kN, and the growth rates are 46.7% and 66.7%, respectively. The maximum axial pressure under pulsating-wind and wind-sand coupling load is -34.2 Kn and -39 kN, and the growth rates were 46.1% and 66.7%, respectively. Compared with pulsating-wind, the corresponding maximum displacement, maximum stress, maximum axial tension, and maximum axial pressure under wind-sand coupling load increased by 11.4%, 21.2%, 13.6%, and 14.0%, respectively in Table 2.

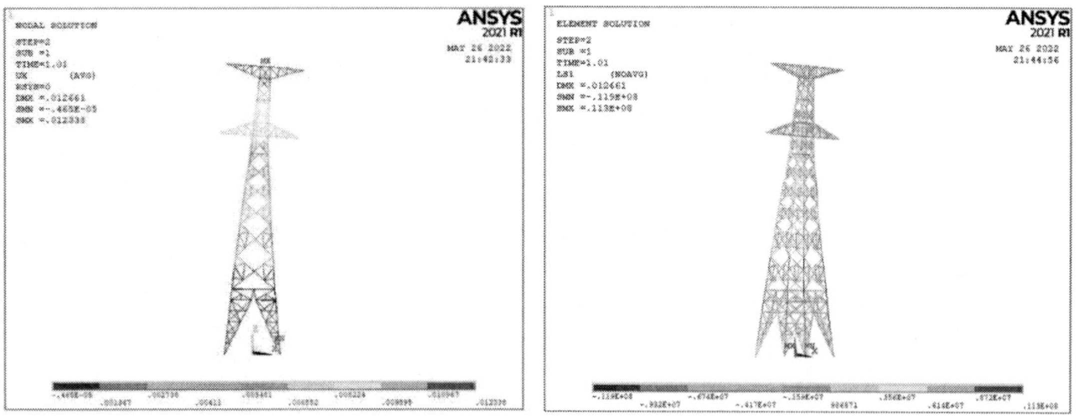

(a) (b)

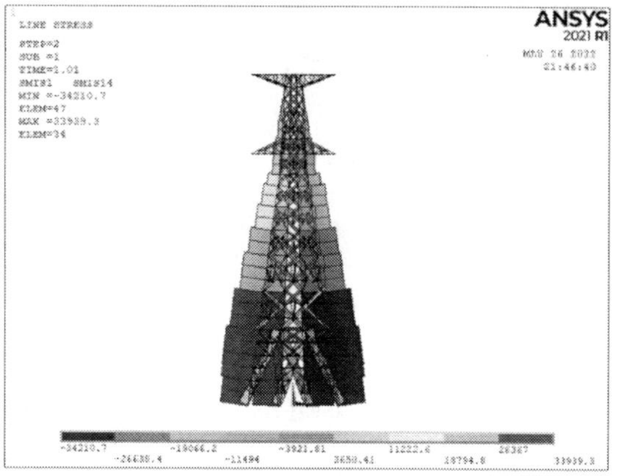

(c)

Figure 5. Displacement nephogram, stress nephogram, and axial force nephogram of transmission tower under the conditon of pulsating-wind-sand load

Table 2. Comparison of maximum values of displacement, stress, and axial force nephogram under different loads.

Load type	Max. displacement (mm)	Max. stress (Mpa)	Max. axial force (kN)
Wind-sand load	13.7	13.7	38.5 -39.0
Pulsating-wind load	12.3	11.3	33.9 -34.2
Static-wind load	8.29	8.16	23.1 -23.4

4. Conclusion

In the paper, a dry-type transmission tower is established by the finite element method. Dynamic characteristics of static wind, pulsating-wind, and wind-sand load are investigated. As the height increases, the displacement offset increases, and its stability gradually deteriorates. Meantime, the stress is mainly distributed in the tower legs, and the node stress on the main material decreases with the increase in height. The bars on the tower's leeward side are subjected to compressive stress, while the bars on the windward side are subjected to tensile stress. Compared with the displacement, stress under pulsating-wind and sand coupling load has increased significantly by >38%. Compared with pulsating-wind load, the corresponding maximum displacement, stress, axial tension, and axial pressure under wind-sand coupling load are increased by 11.4%, 21.2%, 13.6%, and 14.0%, respectively. It can be seen that the wind load is still the main factor affecting the wind-sand load, but the sand load cannot be underestimated. With the continuous improvement of the wind sand level, the destructive power of sandstorms will not continue to rise.

References

[1]LI H N, BAI H F. China Civil Engineering Journal, 2007, 40(2). 39-46.
[2] CHEN J J, Jiang F Q, XUE C X, et al. China Railway Science, 2012, 33(1): 1-5.
[3] WANG Z Q, QI L Z, YANG W G. Guangdong Electric Power, 2016, 29(10): 102-108.
[4] ZHANG Z Q, LI H N, LI S W, et al. Journal of Southeast University (Natural Science Edition), 2018, 48(03). 506-511.
[5] LI H N, REN Y M, BAI H F. Proceedings of the CSEE, 2007, 27(30): 43-48.
[6] Li HN, Bai HF. Progress in Natural Science, 2006, 16(9): 899-911.
[7] WANG Y H, LIU X L, REN W Q, et al. Science Technology and Engineering, 2021, 21(25):

Journal of Physics: Conference Series **2808** (2024) 012011 doi:10.1088/1742-6596/2808/1/012011

10725-10731.
[8] HE R P, ZHANG Y W, GU Z L. China Environmental Science, 2021, 41(08), 3511-3522.
[9] YANG X R, WANG J S, HE Y C, et al. Journal of Desert Research, 2008, (03): 567-571.
[10] WANG X W, HUANG Y X, LIU Z G, et al. Meteorological Science and Technology, 2007, (05), 681-686.
[11] DL/T 5551-2018 load code for the design of overhead transmission line. Beijing: China Planning Press, 2018.
[12] General Administration of quality supervision, inspection and Quarantine of the People's Republic of China GB/T 20480-2006 sandstorm weather grade. Beijing: China Standards Press, 2006.

Journal of Physics: Conference Series **2808** (2024) 012012 doi:10.1088/1742-6596/2808/1/012012

Mechanical properties of workpieces based on 3D printing

Mingxia Zhao[1*]

[1]State Grid Jibei Electric Power Company Limited Skills Training Center Baoding Electric Power VOC.& TECH., Baoding, China

[*]Corresponding author's e-mail: 1850508590@qq.com

Abstract: In view of the quality defects of 3D printed workpieces and insufficient mechanical properties of parts, the optimization of mechanical performance parameters of workpieces based on 3D printing was studied. Taking the FDM molding technology and 3D printing technology of PLA material as the research object, better mechanical properties of the device can be obtained by optimizing the printing temperature, filling rate, and other parameters of the workpiece spline. Experimental results show that for printing temperature, an increase in temperature will increase the ultimate strength of the material, but will also decrease the elastic modulus, so 195°C was finally selected as the processing temperature. For the filling rate and the number of contour turns, the larger the value is, the better the mechanical properties of the processed workpiece will be. The ultimate strength and elastic modulus will increase, but the elongation at break will decrease. Other parameters such as layer thickness, filling angle, etc. have less impact on mechanical properties.

1. Introduction

3D printing is developing rapidly, and many products can be completed through 3D printing technology, making it play an important role in multiple disciplines [1]. 3D printing technology based on the principle of FDM (Fused Deposition Modeling) melts the printing material at high temperatures and extrudes and condenses it through the nozzle to obtain a three-dimensional object [2]. This technology is a rapid prototyping technology that has the advantages of being non-toxic and pollution-free, simple to operate, suitable for making complex hollow parts, and can print a variety of colorful materials. It is the focus of the development of 3D printing technology [3].

Since 3D printed workpieces will have serious defects, especially the mechanical properties that are insufficient to support the mechanical structure, and there are problems such as large roughness, researchers need to study their processing methods. Thrimurthulu et al. [4] studied the relationship between roughness and molding time in 3D printing technology and calculated it based on intelligent algorithms. The results proved that this technology can obtain the best molding direction in 3D molding technology. Masood et al. [5] studied the volume error of FDM printed workpieces. By establishing a mathematical model and optimizing it, they obtained the relationship between the volume error of the workpiece and the forming direction and also obtained a stable workpiece processing method. Liu et al. [6] proved through a large number of experiments that the workpiece is formed by stacking materials in a molten state. The filling method will affect the mechanical properties of the workpiece, and the support problem in the 3D printing process is effectively solved through the generation of filling paths.

3D printing technology has many advantages, but it also has certain shortcomings. 3D printed workpieces based on FDM molding technology have inconsistent mechanical properties in all directions, the connections between layers are not strong, and the performance of the workpiece is poor. Therefore, how to improve the mechanical properties of molded parts has become a problem that needs to be solved. This study optimizes the process parameters of 3D printing, studies the mechanical properties of the workpiece, and explores processing methods for workpieces with higher mechanical properties through analysis.

2. Research on mechanical properties of 3D printed workpieces

2.1 3D printing materials

PLA (Polylactic Acid) is a commonly used 3D printing material that is biodegradable and is a type of polyester material [7]. PLA material has good thermal stability, and the processing temperature in 3D printing is 200±30°C. It can be suitable for processing 3D-printed workpieces through extrusion drawing and other methods. Workpieces processed with PLA have good gloss and transparency, good heat resistance, good biocompatibility, and good performance.

This study uses PLA material suitable for FDM molding technology, with a tensile strength of 52.5 MPa, an elastic modulus of 2000 MPa, and a wire diameter of 1.75±0.03 mm.

2.2 Factors affecting the mechanical properties of 3D printing technology

2.2.1 Printing temperature. For 3D-printed workpieces, different printing temperatures will cause differences in the bonding force between spline layers, and the filaments ejected from the nozzle will also have different bonding effects. Although the printing nozzle is very fast, there will always be a certain temperature gap between the filaments. Especially when the ambient temperature changes and the nozzle cools down quickly, different printing temperatures will affect the bonding force between the filaments. There are obvious changes.

Similarly, the difference in cooling time between layers causes a more obvious temperature difference between the two layers of materials, which in turn leads to a decrease in the bonding effect of the two layers and reduces the mechanical properties of the workpiece.

The bonding force F_{ad} of the spline can be expressed as:

$$F_{ad} = n_i \cdot (F_{ad(1)} + F_{ad(2)} + \cdots + F_{ad(n_a)}) = n_i \cdot \sum_{i=1}^{n_a} F_{ad(i)} \tag{1}$$

in the formula, n_i is the number of filling layers, and $F_{ad}(i)$ is the bonding force of the i-th layer of the corresponding workpiece.

2.2.2 Filling rate. The filling rate mainly affects the density of the material. When the material filling rate reaches 100%, there are no gaps in the spline, and the compact spline has better mechanical strength. When the material filling rate drops to 50%, although the edges and contours of the workpiece are still solid, there are 50% voids in the internal filling part. Through physical analysis, it can be seen that the greater the void rate is, the looser the internal structure of the material will be, resulting in a significant decrease in mechanical strength.

The stress situation of the filled part of the workpiece can be expressed by the formula:

$$F_i = n_i \cdot (\rho \cdot n_a + 1) \cdot F_{a(0)} \tag{2}$$

in the formula, n_i is the number of filling layers, ρ is the filling rate, n_a is the number of wire diameters, and $F_a(0)$ is the stress of the initial layer spline.

The force on the bottom part can be expressed as:

$$F_b = n_b \cdot (n_a + 1) \cdot F_{a(0)} \tag{3}$$

in the formula, n_b is the number of bottom layers.

In the same way, the stress on the top layer can be expressed as:

$$F_t = n_t \cdot (n_a + 1) \cdot F_{a(0)} \tag{4}$$

in the formula, n_t is the number of layers on the top layer.

2.2.3 Number of contour circles. The number of contour turns in 3D printing is mainly the wire diameter of the contour part. The larger the number of contour turns is, the larger the wire diameter of the outer contour of the spline will be, and the filling rate of the contour part always remains 100%, so the size of the number of contour turns will affect the mechanical properties of the external contour of the workpiece.

For splines based on FDM molding technology, the contour force is related to the number of layers n_l and the stress condition F_l of each layer of the contour. Therefore, the contour force can be expressed as:

$$F_s = F_l \cdot n_l = k_l \cdot v \cdot 2 \cdot n_c \cdot n_l \tag{5}$$

in the formula, v is its deformation amount, and k_l is the stiffness of each wire diameter.

2.2.4 Layer thickness. Layer thickness is also an important parameter that affects mechanical properties. For workpieces with fixed dimensions, the larger the spline layer thickness is, the fewer the total number of layers is, which in turn will significantly reduce the time required for printing. Considering the temperature mentioned above, the shorter the printing time is, the higher the mechanical strength of the workpiece will be.

2.2.5 Fill angle. The filling angle is the angle between the spline filling part and the outer contour part of the workpiece. Commonly used filling angles include 0°/90°, 30°/60°, and 45°. The upper and lower layers, in order to ensure the performance of the workpiece, will be filled at different angles. Generally, the wire diameters between the upper and lower layers are perpendicular to each other, but different filling angles will also affect the mechanical properties and mechanical effects of the workpiece.

2.2.6 Filling method. The so-called filling method refers to the path form in the wire during the spline filling process. The most commonly used filling methods are linear and honeycomb. In the case of linear filling, the direction of the wire diameter is a straight line, while in honeycomb filling, the wire diameter will form a hexagonal shape in the filling part, which can enhance the mechanical properties of the structure to a certain extent. When the filling rate of the workpiece reaches 100%, there is no essential difference between the two because it is fully filled.

The final resultant force on the workpiece is:

$$\begin{aligned} F_a &= F_s + F_i + F_b + F_t + F_{ad} \\ &= 2.n_c.n_1.k_1.v + n_i.\sum_{i=1}^{n_{li}} F_{d(0)} + [(n_i + n_b + \rho.n_i).n_a + (n_i + n_b + n_i)].F_{\alpha(0)} \end{aligned} \tag{6}$$

3. Experimental results and data analysis

In order to compare the impact of different process parameters on the 3D printing effect, the six process parameters mentioned above are discussed separately. By adjusting the temperature conditions of 3D printing, the workpiece filling rate, the number of circles of the workpiece outline, and the filling method and filling angle, we analyze and compare its elastic modulus, ultimate strength, and other mechanical performance indicators, and discuss its error range. The material chosen for 3D printing is PLA material.

3.1 Effect of 3D printing temperature on the mechanical properties of the workpiece

For the printing of PLA workpieces, the temperature range is 200±30°C. The temperatures selected here are 190°C, 200°C, and 210°C, and the splines at different temperatures are stretched.

As the temperature changes, the tensile curve of the material becomes significantly different. The increase in temperature causes the stress of the material to increase significantly, and the strain also has an enhanced effect. The mechanical performance indicators of the material at different temperatures are shown in Figure 1.

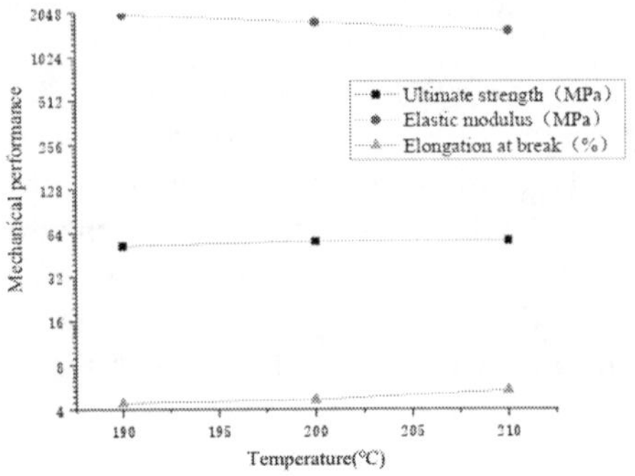

Figure 1. Mechanical performance indicators at different temperatures.

The results in the table show that as the printing temperature gradually increases, the ultimate strength that the spline can withstand gradually increases, and the elongation at the break of the material also increases significantly, proving that the mechanical properties of the material have been significantly enhanced. At the same time, due to the increase in temperature, the elastic modulus has decreased significantly, but the magnitude of the decrease is more obvious than the increase in other parameters. Therefore, it is necessary to select the appropriate printing temperature during 3D printing. After many times It is verified that 195°C is a suitable printing temperature when the mechanical strength of the material is higher.

3.2 Effect of filling rate on mechanical properties of workpiece

Consider the mechanical properties of the workpiece at filling rates of 20%, 40%, 60%, 80%, and 100% respectively. The comparison results are shown in Table 1.

Table 1. Mechanical properties of workpieces under different filling rates.

Fill rate	20%	40%	60%	80%	100%
Ultimate strength US/MPa	19.81±1.35	21.42±1.14	27.19±1.28	34.48±1.63	52.56±1.93
Elastic modulus Es/MPa	848±42	909±44	1138±51	1549±72	1994±102
Elongation at break δ/%	4.309±0.265	5.188±0.280	4.941±0.312	4.670±0.305	4.412±0.241

It can be seen from the results in Table 1 that as the filling rate increases, the ultimate strength of the spline gradually increases, from 19.81 MPa to 52.56 MPa. The strength increases by about 2.65 times, and the effect is significantly improved. As for the elastic modulus, its elastic modulus gradually increased from 848 MPa to 1994 MPa, and the performance increased by about 2.35 times, which also has good effects. In terms of elongation at break, the changing trend is that it first increases and then decreases with the change of filling rate, gradually changing from 4.309% to 4.412%, and the overall fluctuation is not obvious.

From the effect, it can be seen that the higher the filling rate is, the better the mechanical properties of the processed workpiece spline will be. However, the higher the filling rate also means the higher the processing cost, which needs to be weighed according to the actual situation.

3.3 The influence of the number of turns of the workpiece contour on its mechanical properties

The number of contour circles of the workpiece spline was set to 2 to 5 circles respectively, and the mechanical performance parameters under different conditions were compared. The results are shown in Table 2.

Table 2. Effect of different numbers of contour turns on mechanical properties.

Number of contour circles	2	3	4	5
Ultimate strength USs/MPa	52.56±1.93	53.63±1.73	55.32±2.05	55.51±2.88
Elastic modulus Es/MPa	1994±102	2016±85	2040±82	2065±143
Elongation at break δ/%	4.412±0.241	5.650±0.255	4.963±0.312	4.678±0.481

From the results in Table 2, we know that for different contour numbers of turns, the larger the number of turns is, the higher the ultimate strength of the spline is, which gradually rises from 52.56 MPa to 55.51 MPa. The elastic modulus also gradually increased, from 1994 MPa to 2065 MPa, and the elongation at break also gradually increased, from 4.412% to 4.678%.

It can be seen from the experimental results that an increase in the number of contour turns will increase the mechanical properties of the workpiece spline. However, overall, the improvement in mechanical properties by the number of contour turns is small and has a limited impact on performance.

3.4 Effect of workpiece layer thickness on mechanical properties

To compare the workpiece effects of different spline layer thicknesses, the layer thickness of the spline was selected to be 0.15~0.3 mm respectively. The comparison results are shown in Table 3.

Table 3. Effect of workpiece layer thickness on mechanical properties.

Layer thickness/mm	0.15	0.2	0.25	0.3
Ultimate strength USs/MPa	55.29±1.79	52.56±1.93	55.18±2.02	54.08±1.71
Elastic modulus Es/MPa	2041±93	1994±102	2098±107	1979±96
Elongation at break δ/%	4.542±0.315	4.412±0.241	4.668±0.398	4.757±0.442

It can be seen from the results in Table 3 that by changing the layer thickness of the spline, its mechanical properties will change to a certain extent, but there is no obvious trend change overall. Parameters such as ultimate strength and elastic modulus first decreased, then increased, and decreased again. The elongation at break first decreased and then increased. But overall, the changes in performance parameters are not obvious, proving that the thickness of the spline layer has a small impact on the mechanical properties. Considering the influence of spline molding time, you can try to choose a larger layer thickness to make the molding time shorter.

3.5 Effect of fill angle

Different filling angles were compared respectively, and the impact of choosing filling angles of 0°/90°, 30°/60° and 45° on the mechanical properties of the workpiece splines was shown in Table 4.

Table 4. Effect of filling angle on mechanical properties of workpiece.

Fill angle	0°/90°	30°/60°	45°
Ultimate strength USs/MPa	53.74±1.57	54.78±1.66	52.56±1.93
Elastic modulus Es/MPa	1960±135	1961±132	1994±102
Elongation at break δ/%	3.942±0.324	4.158±0.327	4.412±0.241

It can be seen from the results in the table that the different filling angles will cause the mechanical properties of the spline to fluctuate to a certain extent. The ultimate strength gradually decreases, the elastic model gradually increases, and the fracture elongation also gradually increases. But overall, different filling angles have little impact on mechanical properties. Generally, a filling angle of 45° has relatively better mechanical properties.

4. Conclusion

This article studies the selection of mechanical performance parameters of workpieces based on 3D printing. Through comparison of different parameters, it is found that 3D printed workpieces have better mechanical performance parameters. It can be known from the experimental results that for the PLA material in this study, as the printing temperature rises, the ultimate strength increases significantly, while the elastic modulus decreases rapidly, and the elongation at break becomes larger. When the filling rate and the number of contour turns of the material increase, the mechanical properties of the material increase rapidly. However, since the cost of the material also increases accordingly, comprehensive consideration is required. The layer thickness, filling angle, and filling method generally have little impact on the strength of the 3D-printed workpiece of PLA material, so the processing parameters need to be selected according to the actual situation.

References

[1] Peng Zilong, Wu Jinyin, Wang Mengjie, Li Yinan, Lan Hongbo. Electric field driven micro-scale 3D printing mask electrolytic processing microstructure [J]. Transactions of the Chinese Society of Mechanical Engineering, 1-17.

[2] Shen Yuling. The value and application of 3D printing technology in machinery manufacturing automation [J]. Papermaking Equipment and Materials, 2023, 52(12):91-93.

[3] Zhang Jiaming, Ye Shengjia, Yang Xiaoyan, Xin Xianzhen, Zeng Li, Wei Bin. Effect of digital light processing 3D printing parameters on the mechanical properties of guide plate resin [J]. Journal of Dental Materials and Devices, 2023, 32(04): 240-247.

[4] Thrimurthulu K., Pulak M.P., Reddy N.V. Optimum part deposition orientation in fused deposition modeling [J]. International Journal of Machine Tools & Manufacture, 2004(44). 585–594.

[5] Masood S.H., Rattanawong W., Iovenitti P. Part Build Orientations Based on Volumetric Error in Fused Deposition Modelling [J]. Int J Adv Manuf Technol,2000(16):162–168.

[6] Liu Bin, Ruan Feng, Wang Fuzhen. Research on automatic generation algorithm of support structure in melt deposition molding system [J]. Mechanical and Electrical Engineering Technology, 2001(3):23-26.

[7] Xing Zhipeng, Ye Yanqiu. Application of 3D printing technology in the field of automotive machinery manufacturing [J]. Automobile Test Report, 2023, (20): 46-48.

Journal of Physics: Conference Series **2808** (2024) 012013 doi:10.1088/1742-6596/2808/1/012013

Analysis of hydrodynamic performance of underwater accompanying robot based on CFD

Yang Chen [1*], Fanghua Liu[1], Jiawei Shao[1], You Miao[1] and Yongxin Cao[1]

[1]School of Mechanical Engineering, Jiangsu University of Science and Technology, Jiangsu Province, China

*Corresponding author's e-mail: Ryanwork35@163.com

Abstract: We design a novel underwater accompanying robot capable of collaborating with divers for joint operations to improve the efficiency and safety coefficient of underwater operations for divers by studying the impact of speed and drift angles on its hydrodynamic performance under various underwater conditions. Utilizing the computational fluid dynamics (CFD) method, applying the $Realizable\ k-\varepsilon$ turbulence model, and incorporating standard wall functions for near-wall regions, Fluent software is employed to simulate fluid dynamics and solve hydrodynamic parameters for various underwater conditions and postures. Research findings indicate that conditions of horizontal straight navigation and vertical ascent/descent, pressure drag, and viscous drag on the robot exhibit a positive correlation with velocity. Experiencing variations in drift angle, the robot's longitudinal movement is not significantly altered during horizontal maneuvering and remains consistent under the same velocity. Conversely, later resistance distinctly increases with larger drift angles. This study can offer valuable reference and insights for subsequent research on underwater accompanying robots.

1. Introduction

Underwater robots and divers are crucial participants in modern underwater operations of marine engineering. Divers typically work in environments with a water depth of up to 60 meters. Due to factors such as ocean currents, visibility, support vessels, and the equipment used, underwater operations exhibit significant unpredictability, leading to relatively high operational risks [1]. The underwater accompanying robot designed in this study can assist divers in carrying a large number of tools, enabling prolonged collaboration with divers and enhancing their underwater efficiency.

Due to the unique nature of the underwater environment where the underwater accompanying robot operates, complex underwater conditions can introduce significant disturbances to the robot's navigation. It is necessary to analyze the hydrodynamic performance of the robot under various conditions to ensure optimal performance throughout its operation, inducing horizontal straight navigation, horizontal maneuvering, and vertical ascent/descent to understand the resistance characteristics during these movements [2]. It is necessary to construct a reliable hydrodynamic model to obtain accurate hydrodynamic parameters. Commonly used methods are based on semi-theoretical formulas and numerical calculation methods are based on Computational Fluid Dynamics (CFD) technology [3]. Compared to traditional calculation methods, CFD numerical simulation technology has the advantage of obtaining hydrodynamics and flow field details for underwater robots that are closer to reality at a lower cost and in a shorter time frame [4-5]. Currently, there is limited hydrodynamic research on underwater accompanying robots. Therefore, this study employs CFD numerical simulation methods to conduct hydrodynamic simulation analysis on underwater accompanying robots, obtaining relevant hydrodynamic parameters. This provides a basis for propeller configuration selection and structural optimization for the robot.

2. Establishment of hydrodynamic computational model

2.1 Overall structure of underwater accompanying robot

The underwater accompanying robot adopts an open-frame structure with dimensions of 800 mm×800 mm×700 mm, and its overall structural layout is depicted in Figure 1. Its main frame is welded and manufactured from aluminum alloy plates, with the buoyancy module mounted above the frame. There are symmetrically installed oxygen cylinder storage units on both sides of the frame. The middle support plate symmetrically accommodates the installation of steering and propulsion modules, longitudinal propulsion module, buoyancy adjustment device, underwater camera, underwater lighting, pressure-resistant electronic control cabin, gimbal, and ultra-short baseline system. The storage box is mounted below the frame. The underwater accompanying robot designed in this study is equipped with an oxygen storage unit, a storage unit, and a storage box, enabling collaborative operations with divers.

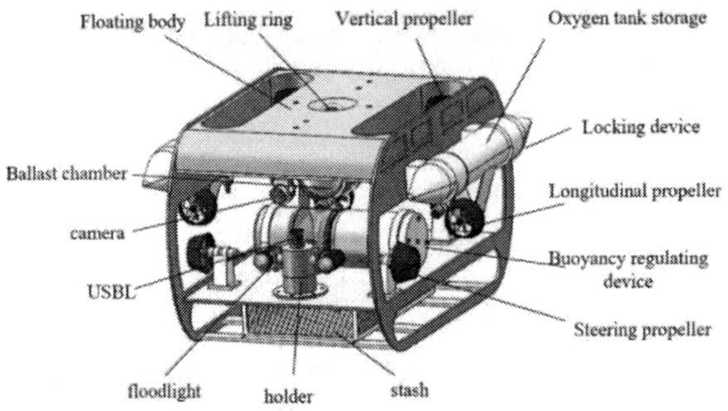

Figure 1. Structure diagram of an underwater accompanying robot.

2.2 Selection and establishment of the coordinate system

This paper employs the Cartesian coordinate system, which includes two coordinate systems, the fixed coordinate system $E-\xi\eta\zeta$ and the moving coordinate system $G-xyz$, as illustrated in Figure 2, both of which determine following the right-hand rule.

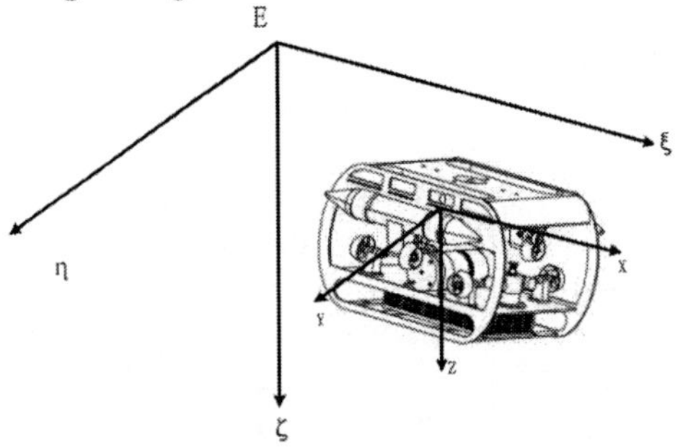

Figure 2. Schematic diagram of the coordinate system of the underwater accompanying robot.

In the moving coordinate system, $E\xi$ plane is parallel to the ground. Placing the underwater accompanying robot into the coordinate system, the analysis of the robot's motion and posture changes is conducted through the coordinate system. The center of gravity (G) of the robot is located at the origin of the moving coordinate system, and the three axes of the coordinate system coincide with the directions of the robot's inertial principal axes. Positive directions of the coordinate axes align with the direction

Journal of Physics: Conference Series **2808** (2024) 012013 doi:10.1088/1742-6596/2808/1/012013

of robot motion, and the definitions of robot motion parameters are provided in Table 1.

Table 1. Six-degree-of-freedom motion parameters of the underwater accompanying robot.

Movement style	Movement expression
Surge	Translation along the X-axis
Sway	Translate along the Y-axis
Heave	Translation along the Z-axis
Roll	Translation along the X-axis
Pitch	Translation along the Y-axis
Yaw	Translation along the Z-axis

The transformation relationship between the fixed coordinate system and the moving coordinate system is given by Equations (1) and (2).

$$\begin{bmatrix} \xi \\ \eta \\ \zeta \end{bmatrix} = \boldsymbol{T} \begin{bmatrix} x \\ y \\ z \end{bmatrix} \tag{1}$$

$$\boldsymbol{T} = \begin{bmatrix} \cos\psi\cos\theta & \cos\psi\sin\theta\sin\varphi - \sin\psi\cos\varphi & \cos\psi\sin\theta\cos\varphi + \sin\psi\sin\varphi \\ \sin\psi\cos\theta & \sin\psi\sin\theta\sin\varphi - \cos\psi\cos\varphi & \sin\psi\sin\theta\cos\varphi - \cos\psi\sin\varphi \\ -\sin\theta & \cos\theta\sin\varphi & \cos\theta\cos\varphi \end{bmatrix} \tag{2}$$

where θ, φ, and ψ represent the pitch angle, roll angle, and yaw angle of the robot, respectively to rotations around the X-axis, Y-axis, and Z-axis.

2.3 Control equations and turbulence models

CFD calculations of the underwater accompanying robot involve solving the motion state of th e water flow field, so water is generally considered an incompressible flow field. This means that in the hydrodynamic calculations and analysis of underwater robots, the compressibility of water is neglected, and the influence of air bubbles is typically not taken into account under normal conditions [6].

According to the principles of physics, it is known that any fluid flow problem must satisfy the laws of conservation of mass, conservation of momentum, and conservation of energy [7].

Thermal exchange is not considered in the hydrodynamic analysis of underwater robots. Therefore, the fundamental governing equations for the water flow field degenerate into the mass conservation equation and the Navier-Stokes equation [8], as shown below.

$$\begin{cases} \dfrac{\partial u_i}{\partial x_i} = 0 \\ \dfrac{\partial u_i}{\partial t} + \dfrac{\partial}{\partial x_j}(u_i u_j) = v\dfrac{\partial}{\partial x_j}(\dfrac{\partial u_i}{\partial u_j} + \dfrac{\partial u_j}{\partial u_i}) - \dfrac{1}{\rho}\dfrac{\partial p}{\partial x_i} \\ \dfrac{\partial \rho}{\partial t} + \dfrac{\partial(\rho u_j)}{\partial x_j} = S_m \end{cases} \tag{3}$$

where ρ represents fluid density; t is time; x_i is the coordinate value component on the axis; u_i is the velocity vector; S_m is the increment of liquid mass per unit time.

The momentum conservation equation (Navier-Stokes) in an inertial reference frame is:

$$\frac{\partial(\rho u_j)}{\partial t}+\frac{\partial(\rho u_i \cdot u_j)}{\partial x_j}=-\frac{\partial p}{\partial t}+\frac{\partial \tau_{ij}}{\partial x_j}+\rho \cdot g_i+F_i \tag{4}$$

where P is the pressure in the flow field; g_i is the component of gravitational acceleration; τ_{ij} is the Reynolds stress; F_i is the external force component along the X-axis.

Due to the inherent difficulty in directly solving the Navier-Stokes equation, various simplified solution methods have been developed in the field of fluid mechanics [9]. Among them, the Reynolds-Averaged Navier-Stokes (RANS) equations have been verified as a suitable approach for engineering solutions.

$$\begin{cases} \frac{\partial(\rho \bar{u}_i)}{\partial x_i}=0 \\ \frac{\partial(\rho \bar{u}_i)}{\partial t}+\frac{\partial(\rho \bar{u}_i \cdot \bar{u}_j)}{\partial \bar{x}_j}=-\frac{\partial p}{\partial x_i}+\frac{\partial}{\partial x_j}(\mu \frac{\partial \bar{u}_i}{\partial x_j}- \\ \qquad \rho \cdot \overline{u_i' u_j'})+\rho \cdot \bar{g}_i+\bar{F}_i \end{cases} \tag{5}$$

$$\begin{cases} \frac{\partial(\rho \bar{u}_i)}{\partial x_i}=0 \\ \frac{\partial(\rho \bar{u}_i)}{\partial t}+\frac{\partial(\rho \bar{u}_i \cdot \bar{u}_j)}{\partial \bar{x}_j}=-\frac{\partial p}{\partial x_i}+\frac{\partial}{\partial x_j}(\mu \frac{\partial \bar{u}_i}{\partial x_j}- \\ \qquad \rho \cdot \overline{u_i' u_j'})+\rho \cdot \bar{g}_i+\bar{F}_i \end{cases} \tag{6}$$

where u_i' is the fluctuating velocity, which is related to turbulent velocity $\rho \cdot \overline{u_i' u_j'}$ corresponding to the Reynolds stresses, reflecting the momentum changes caused by pulsating velocity; μ_t is the turbulent viscosity. Following the eddy viscosity assumption proposed by Boussinesq, a correlation function is established between $\rho \cdot \overline{u_i' u_j'}$ and μ_t.

$$\begin{cases} -\rho \cdot \overline{u_i' u_j'}=\mu_t(\frac{\partial u_i}{\partial x_j}+\frac{\partial u_j}{\partial x_i})-\frac{2}{3}(\rho k+\mu_t \frac{\partial u_i}{\partial x_j})\delta_{ij} \\ \mu_t=\rho C_\mu \frac{k^2}{\varepsilon} \end{cases} \tag{7}$$

where k is turbulent kinetic energy, C_μ is an empirical constant, and ε is the turbulent kinetic energy dissipation rate. This is the context of the standard $k-\varepsilon$ two-equation model.

$$\begin{cases} \frac{\partial(\rho k)}{\partial t}+\frac{\partial(\rho k u_i)}{\partial x_i}=\frac{\partial}{\partial x_j}\left[\left(\mu+\frac{\mu_t}{\sigma_k}\right)\frac{\partial k}{\partial x_j}\right]+G_k+ \\ \quad G_b-\rho\varepsilon-Y_M+S_k \\ \frac{\partial(\rho\varepsilon)}{\partial t}+\frac{\partial(\rho\varepsilon u_i)}{\partial x_i}=\frac{\partial}{\partial x_j}\left[\left(\mu+\frac{\mu_t}{\sigma_\varepsilon}\right)\frac{\partial\varepsilon}{\partial x_j}\right]+C_{1\varepsilon}\frac{\varepsilon}{k} \\ \quad (G_k+G_{3\varepsilon}G_b)-C_{1\varepsilon}\rho\frac{\varepsilon^2}{k}+S_\varepsilon \end{cases} \tag{8}$$

When accompanying the robot to perform underwater operations, due to the interaction between the underwater robot and the fluid, the fluid will be tensioned at high speed, and the standard k-ε model will produce a certain amount of distortion and even negative positive stress when the time-averaged strain rate is particularly large [10], so the modified *Realizable* $k-\varepsilon$ model is used in this paper.

Realizable $k-\varepsilon$ model variable equations are as follows:

$$\begin{cases} \frac{\partial(\rho k)}{\partial t}+\frac{\partial(\rho k u_i)}{\partial x_i}=\frac{\partial}{\partial x_j}\left[\left(\mu+\frac{\mu_t}{\sigma_k}\right)\frac{\partial k}{\partial x_j}\right] \\ \quad +G_k-\rho\varepsilon \\ \frac{\partial(\rho\varepsilon)}{\partial t}+\frac{\partial(\rho\varepsilon u_i)}{\partial x_i}=\frac{\partial}{\partial x_j}\left[\left(\mu+\frac{\mu_t}{\sigma_\varepsilon}\right)\frac{\partial\varepsilon}{\partial x_j}\right]+ \\ \quad \rho C_1 E\varepsilon-\rho C_2\frac{\varepsilon^2}{k+\sqrt{\nu\varepsilon}} \end{cases} \tag{9}$$

2.4 Treatment of near-wall flow field

Due to the low Reynolds number and turbulence development in the nearwall region, special treatment is required for the previously established *Realizable* $k-\varepsilon$ model in the near-wall region.

According to the relevant literature, the near-wall area can be divided into three sub-layers. A viscous sublayer is in the innermost layer, and the flow is similar to a laminar flow. The outermost layer is logarithmic, and the flow is turbulent. The area between the viscous sublayer and the logarithmic layer is the buffer layer [11-12]. Due to the extremely small thickness of the buffer layer, it is often classified as a logarithmic layer in engineering calculations.

There are two ways to deal with multi-level near-wall areas: one is the low Reynolds number $k-\varepsilon$ model method, which analyzes the detailed boundary layer by using a small grid-scale perpendicular to the wall direction; The second is the wall function method, which can save computational resources by applying appropriate conditions to the near-wall region without the need to solve the boundary layer by using the empirical formula.

The computational grid of the two methods is shown in Figure 3, and the standard wall function method applies appropriate conditions to the near-wall region by using an empirical formula, rather than specifically processing the flow in the wall area. In practical calculations, the standard wall function method can be calculated on a coarse mesh, which saves computing resources and time, and it is more suitable for large-scale flow field calculation and engineering applications.

The low Reynolds number $k-\varepsilon$ model method has high requirements for the mesh, and it is necessary to carry out very fine meshing in the near-wall area to solve the laminar flow problem of the viscous bottom layer, which will greatly increase the calculation cost and calculation time. Therefore, this paper introduces the standard wall function method into the model to deal with the multi-level near-wall area.

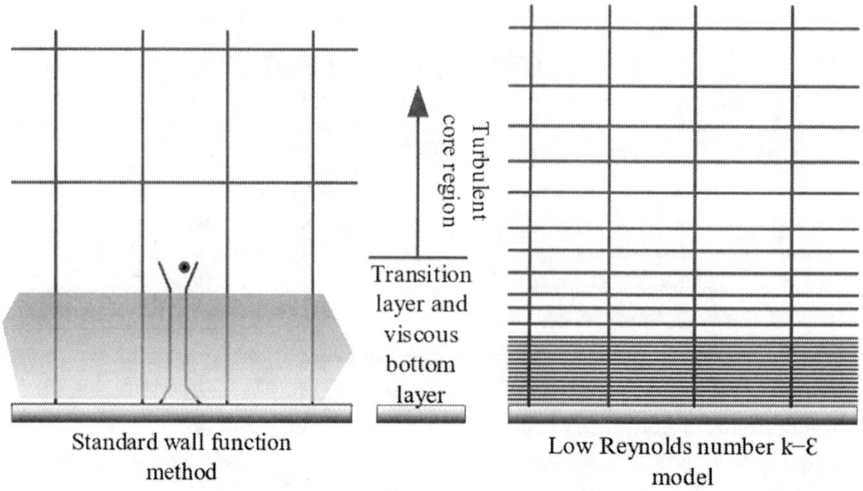

Figure 3. Calculation grids corresponding to the standard wall function method and low Reynolds number A model.

The formula for the standard wall function is:

$$U^{*} = \frac{1}{k} In\left(Ey^{+}\right) \tag{10}$$

$$\begin{cases} \dfrac{\partial(\rho k)}{\partial t} + \dfrac{\partial(\rho k u_i)}{\partial x_i} = \dfrac{\partial}{\partial x_j}\left[\left(\mu + \dfrac{\mu_t}{\sigma_k}\right)\dfrac{\partial k}{\partial x_j}\right] \\ \qquad + G_k - \rho\varepsilon \\ \dfrac{\partial(\rho\varepsilon)}{\partial t} + \dfrac{\partial(\rho\varepsilon u_i)}{\partial x_i} = \dfrac{\partial}{\partial x_j}\left[\left(\mu + \dfrac{\mu_t}{\sigma_\varepsilon}\right)\dfrac{\partial\varepsilon}{\partial x_j}\right] + \\ \qquad \rho C_1 E\varepsilon - \rho C_2 \dfrac{\varepsilon^2}{k + \sqrt{v\varepsilon}} \end{cases} \tag{11}$$

2.5 Computational domain analysis and meshing

In this paper, the ICEMCFD front-end processing is used to analyze the numerical model of the underwater accompanying robot. Due to the complexity of the shape and structure of the underwater accompanying robot, the computer workload is very large and the calculation time is long if it is directly meshed, so the structure of the robot is simplified [13]. The simplified structure is shown in Figure 4.

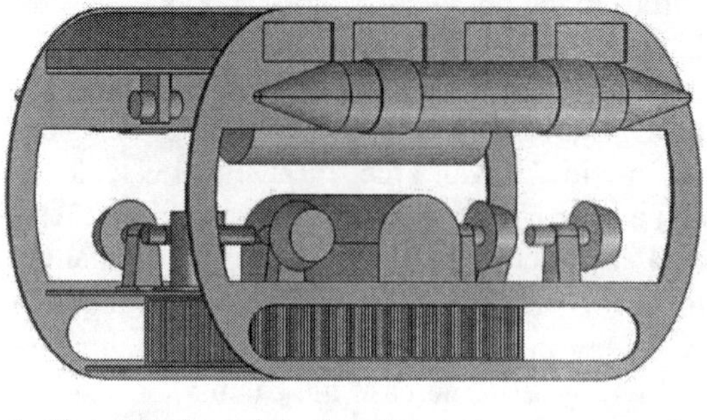

Figure 4. Simplified model of underwater accompanying robot.

The CFD calculation of the hydrodynamics of the underwater accompanying robot is to convert the external circumnavigation problem of the robot into the internal flow by creating a computational domain flow field. Similar to the pool test, the watershed scale and boundary location will have a direct impact on the calculation results. If the scale of the basin is too small, the flow field is not fully developed, which will produce a blocking effect similar to that in the pool test, resulting in inaccurate calculation results. If the basin scale is too large, computing resources will be wasted and computing efficiency will be affected. The overall longitudinal length of the robot is L=1, 100 mm. The whole watershed (Domain 1) is set as a rectangle, so the total length is 9L=9, 900 mm and the accompanying robot is placed in the front half of the watershed. To improve the calculation accuracy of the robot in the near area, a computational domain (Domain 2) is established near the robot, and the length, width, and height are 1.3 times the size of the robot. The distance from the inlet to the front of the robot is 2L=2, 200 mm, the distance from the back end of the robot to the outlet is 6L, and the width and height of the watershed are set to 4L=4, 400 mm.

Meshes can be divided into structured meshes and unstructured meshes according to their structural form. For underwater accompanying robots with many appendages, the geometry with complex shapes and its near watershed (Domain 2) is unstructured, and Domain 1 is structured meshing except for Domain 2. The boundary layer grid delineated in this paper is shown in Figure 5.To accurately capture the flow in the boundary layer, the mesh division closer to the boundary basin of the robot surface is finer, to improve the resolution of the boundary layer mesh. The mesh of the outer basin of the robot boundary adopts a sparse tetrahedral meshing method, which improves the calculation efficiency while ensuring the calculation accuracy, and the total number of meshes after the final sub-regional hybrid meshing is about 6.4 million.

Figure 5. Local diagram of boundary layer grid.

3. Hydrodynamic performance analysis

3.1 Hydrodynamic performance analysis and parameter setting

The underwater accompanying robot designed in this paper has its power system and is mainly in three working conditions: horizontal direct navigation, horizontal navigation change direction, and vertical floating and sinking during underwater operation. It is assumed that the robot is a rigid body with constant mass and volume in water. Since the Fluent solver cannot directly give the resistance characteristics of the robot in the flow field, it needs to be converted according to the empirical formula.

$$F = \frac{1}{2}\rho v^2 L^2 C_d \tag{12}$$

where F is the water resistance, ρ is the fluid density, v is the incoming flow velocity, L is the characteristic length, and C_d is the water resistance coefficient.

In Fluent, the robot is placed in the calculation domain of different flow velocities for fluid simulation analysis, and the specific parameters are set as follows: the turbulence model adopts the improved and adjusted *Realizable* $k-\varepsilon$ model, and the water-liquid in the calculation domain media ignores the influence of self-weight; At the boundary conditions, the flow field inlet (inlet) is set as the velocity inlet (velocity-inlet), the flow field outlet (exit) is the pressure outlet (pressure-outlet), and the wall is symmetric. Using the SIMPLE algorithm, the momentum equation and turbulent flow energy dispersion are both discretized by the second-order welcome style [14].

3.2 Hydrodynamic performance analysis under horizontal direct navigation conditions

In the direct flight simulation of the underwater accompanying robot, the flow velocities were set to 0.1 m/s, 0.3 m/s, 0.5 m/s, 0.8 m/s, 1 m/s, and 1.2 m/s, respectively. After the convergence is calculated, the resistance characteristics of the robot at different direct speeds are obtained. The results are shown in Table 2 below.

Table 2. Resistance characteristics of underwater accompanying robots in direct navigation.

Speed $(m \cdot s^{-1})$	Differential pressure resistance (N)	Sticky resistance (N)	Total resistance (N)	Resistance coefficient (C_d)
0.1	1.73	0.22	1.95	0.610473853
0.3	15.45	1.36	16.81	0.584733075
0.5	43.06	3.19	46.25	0.579167502
0.8	110.88	6.38	117.26	0.573591058
1.0	173.99	8.91	182.90	0.572593168
1.2	246.33	11.21	257.54	0.559905399

As can be seen from Table 2, the differential pressure resistance accounts for the largest proportion of the total resistance, which is caused by the open frame structure of the underwater accompanying robot. Among them, when the underwater robot is accompanied by horizontal direct flight, the differential pressure resistance and viscous pressure resistance are positively correlated with the speed. Because the robot does not involve the change of elevation angle when flying directly, the pitching moment generated is very small and negligible. Because the structure is not completely symmetrical, the underwater robot will be subjected to downward vertical resistance, as shown in Figure 6, and the magnitude of vertical resistance increases with the increase of direct speed. When the direct speed reaches the maximum value of 1.2 m/s, the direct flight resistance of the robot is the largest, which is 257.54 N, and the data obtained by Fluent are imported into the CFD-post software for post-processing. The surface pressure distribution contour and velocity distribution contour of the maximum direct speed of the robot are obtained, as shown in Figure 7 and Figure 8.

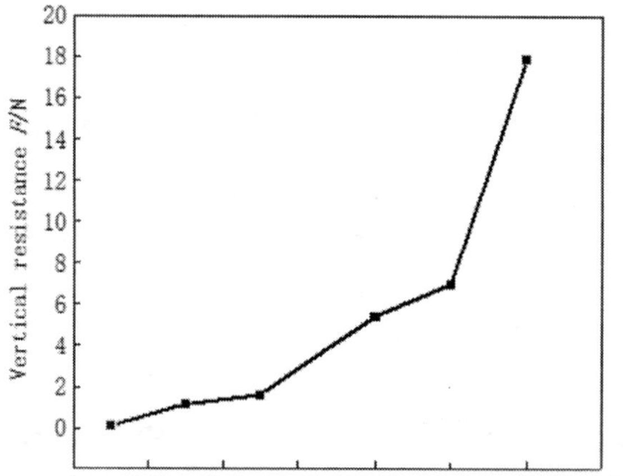

Figure 6. Variation curves of drooping resistance at different direct navigation speeds.

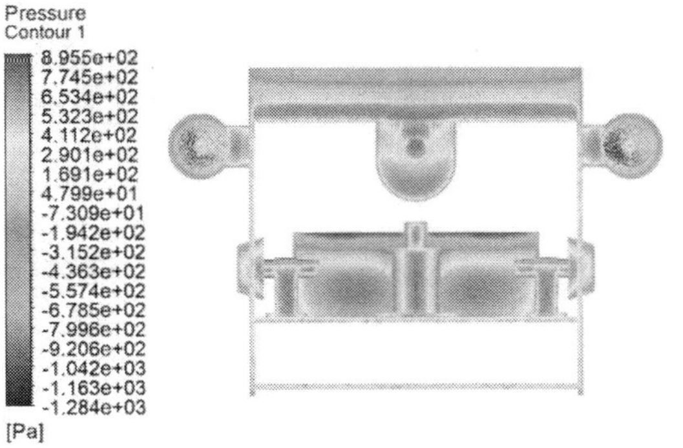

Figure 7. Nephogram of surface pressure distribution at maximum speed.

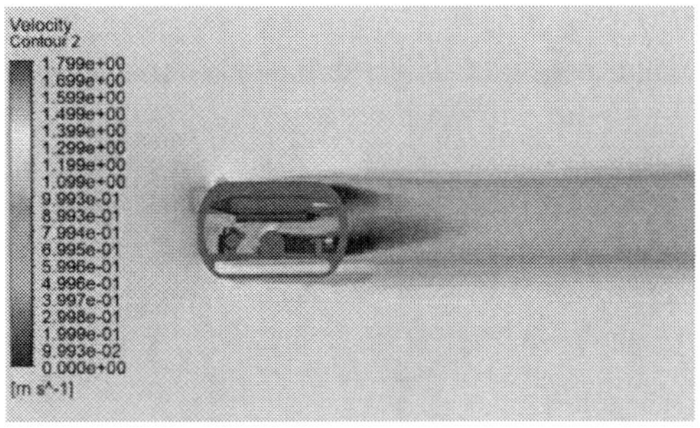

Figure 8. Speed distribution cloud diagram at maximum speed.

It can be seen from Figure 7 that the surface pressure distribution of the underwater accompanying robot is symmetrical at the maximum direct velocity. The pressure on the surface of the water is the largest, with a maximum value of 895.5 Pa, and the pressure in the rest of the parts is relatively small. As can be seen from Figure 8, there is no flow separation and vortex in the surrounding watershed of the robot, and the movement of the robot is relatively stable under the condition of horizontal direct navigation.

3.3 Analysis of hydrodynamic performance under the condition of horizontal navigation under variable direction conditions

Since it is often necessary to change the course according to the diver's actions when the accompanying robot is operating underwater, it is also necessary to analyze the hydrodynamic performance of the robot under the condition of changing direction in horizontal navigation. The default parameters remain unchanged, and the drift angles are set to 2°, 4°, 6°, 8°, and 10° at the flow field velocities of 0.3 m/s, 0.5 m/s, 0.8 m/s, and 1 m/s, respectively. The results obtained are shown in Tables 3 and 4.

Table 3. Lateral force characteristics of underwater accompanying robot under steering at different speeds.

Drift (°)/speed ($m \cdot s^{-1}$)	0.3	0.5	0.8	1.0
2	-3.49	-1.33	2.98	7.31
4	-2.04	0.64	5.91	9.63
6	-0.17	2.54	5.39	14.61
8	0.17	4.11	5.12	12.99
10	1.16	3.67	6.43	17.72

Journal of Physics: Conference Series **2808** (2024) 012013 doi:10.1088/1742-6596/2808/1/012013

Table 4. Longitudinal force characteristics of underwater accompanying robot under steering at different speeds.

Drift (°)/speed (m · s^{-1})	0.3	0.5	0.8	1.0
2	16.31	44.44	114.54	174.65
4	16.04	43.28	112.16	174.03
6	15.15	42.04	109.15	173.11
8	14.48	41.64	105.54	171.77
10	14.06	40.17	108.43	169.86

It can be seen from Table 3 and Table 4 that when the underwater robot accompanies the horizontal navigation direction change motion at a certain speed, with the continuous increase of the drift angle, the lateral resistance of the robot is continuously increased with the positive direction of the robot, which is due to the increase of the drift angle and the corresponding increase of the side area of the robot, resulting in the resistance of the fluid to the larger side area when the robot advances. Therefore, the lateral resistance increases. The longitudinal resistance of the robot decreases with the continuous increase of the drift angle, which is due to the increase of the drift angle. The inclination angle increases, the front of the robot is tilted, and there is a certain streamline so that the longitudinal resistance is slightly reduced. With the continuous increase of speed, the resistance generated by pushing and agitating the surrounding fluid when the robot is advancing is also increasing, and the fluid exerts greater pressure resistance in all directions of the robot, resulting in increased lateral resistance and longitudinal resistance suffered by the robot. With the increase in speed, the difference between the longitudinal resistance and the lateral resistance of the robot at the same speed is more obvious. When the speed is increased to 1 m/s, the longitudinal force can reach 174.65 N, and the lateral resistance of the robot at this speed is 17.72 N.

3.4 Hydrodynamic performance analysis under vertical float and sinking conditions

When the underwater accompanying robot follows the diver for underwater operations, it often needs to dive or float the movement, which requires the hydrodynamic performance analysis of the robot under vertical floating and sinking conditions. According to the actual working conditions of vertical float and subsidence, the top of the watershed is set as the inlet of the flow field, and the bottom of the watershed is set as the outlet of the flow field. The total length of the watershed is 10 liters, and the width and height are both set to 4 liters. In the Fluent simulation environment, the parameters and boundary conditions are the same as before.

In the numerical calculation of vertical float and sink motion, due to the complex structure of the underwater accompanying robot and the fact that the top and bottom are not completely symmetrical, the speed under the vertical float and sink condition should not be too large, and the flow velocities were set to 0.1 m/s, 0.3 m/s, 0.5 m/s, 0.8 m/s, 1 m/s, and 1.2 m/s respectively for simulation calculation. After the convergence is calculated, the resistance characteristics of the robot under different vertical float and sink velocities are obtained. The results are shown in Table 5 below.

Table 5. Vertical floating and sinking resistance characteristics of the underwater accompanying robot.

Speed (m · s^{-1})	Differential pressure resistance (N)	Sticky resistance (N)	Total resistance (N)	Resistance coefficient (C_d)
0.1	5.75	0.12	5.87	1.176117011
0.3	49.59	0.72	50.31	1.120016029
0.5	134.37	1.54	135.91	1.089240633
0.8	341.05	3.28	344.33	1.077971599

Journal of Physics: Conference Series **2808** (2024) 012013 doi:10.1088/1742-6596/2808/1/012013

1.0	350.22	4.98	535.20	1.072330194
1.2	759.96	7.03	766.99	1.067184822

It can be seen from Table 5 that when the robot is vertically floating underwater, the water resistance of the robot increases greatly with the increase of sailing speed, and when the sailing speed reaches a maximum of 1.2 m/s, the water resistance is 766.99 N. Therefore, to reduce energy consumption, the speed of floating and sinking should not be too large. The data obtained was imported into the CFD-post software for post-processing, and the surface pressure distribution contour and velocity distribution contour of the robot's maximum vertical float and sink movement speed were obtained, as shown in Figure 9 and Figure 10.

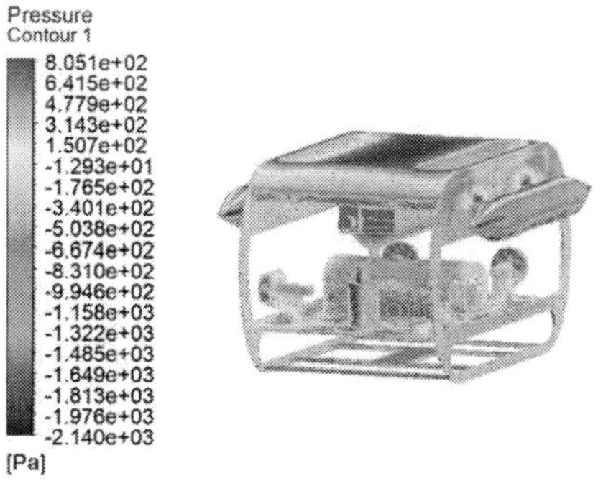

Figure 9. Nephogram of surface pressure distribution at maximum speed.

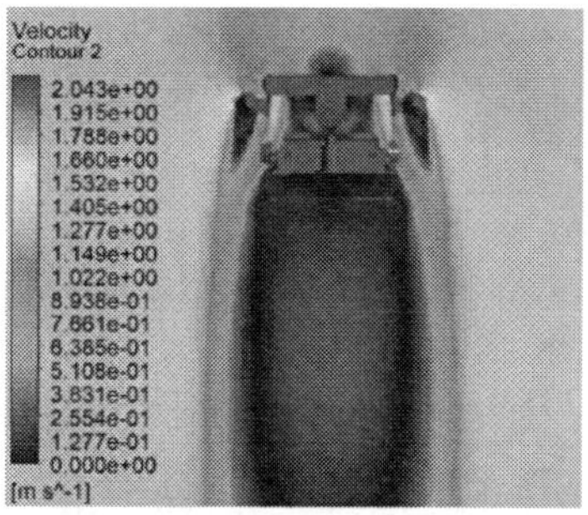

Figure 10. Speed distribution cloud diagram at maximum speed.

As can be seen from Figure 9, the surface pressure distribution of the underwater robot is also symmetrical under the maximum vertical float and sinking velocity, and the frontal surface pressure is the largest, with a maximum value of 805.1 Pa. As can be seen from Figure 10, the partial flow separation phenomenon occurred in the watershed near the tail of the robot, accompanied by vortices, which is caused by the incomplete symmetry of the robot's structure. The pressure difference between the head and tail is formed, which can be further optimized in the follow-up research, so that the robot is more stable when sailing.

4. Conclusion

In this paper, based on the computational fluid dynamics (CFD) method, the fluid motion simulation of the designed underwater accompanying robot is carried out, and the hydrodynamic performance of the underwater operation is studied under three main working conditions: horizontal direct navigation, horizontal oblique navigation, and vertical floating and sinking. The *ealizable* $k-\varepsilon$ turbulence model, which is improved from the standard k-ε model, is used to introduce the standard wall function in the near-wall region processing and divide it into sub-region hybrid meshing. The hydrodynamic performance parameters of the underwater accompanying robot under different working conditions are obtained, and the main conclusions are as follows:

(1) According to the obtained hydrodynamic performance parameters, the maximum water resistance of the robot is 257.54 N when the speed reaches 1.2 m/s under the condition of horizontal direct navigation. Under the condition of horizontal navigation direction, the longitudinal resistance and lateral resistance have a great influence on the robot's movement, the maximum longitudinal resistance of the robot is 174.65 N, and the maximum lateral resistance is 17.72 N. In the case of vertical floating, the maximum water resistance of the robot is 766.99 N.

(2) According to the maximum heading resistance obtained under the three working conditions and the configuration and selection of the thruster, a certain margin needs to be ensured. The longitudinal propulsion system of the designed underwater accompanying robot needs to provide 300 N thrust, the steering propulsion system needs to provide 30 N thrust, and the vertical propulsion system needs to provide 800 N thrust, which can be used as the data basis and theoretical support for model structure optimization and motion control.

References

[1] Zhang X. Risk control in joint operations of ROV and divers [J]. China Petroleum and Chemical Standard and Quality, 2019, 39 (06): 72-73.

[2] Guan Yingjie, Cui Qingjia, Xu Biao, et al. Research on the Influence of Channel Thrusters on the Hydrodynamic Performance of AUV Based on CFD [J]. Control and Information Technology, 2023, (06): 65-71. DOI: 10.13889/j.issn.2096-5427.2023.06.010.

[3] Yan Y., Yu F., & Chen Y. (1986). Calculation of hydrodynamic coefficients and dynamic modeling of open-frame underwater robots [J]. Acta Armamentarii, 2021, 42 (09): 1, 972-1, 986.

[4] Chen T. & Wan D. (2018). CFD analysis of hydrodynamic resistance performance for underwater vehicles [J]. Journal of Hydrodynamics (Ser. A), 33 (02), 143-149.

[5] Yaxin L., Lixiao L., Yu W., et al. Hydrodynamic analysis and motion control of the Coanda-effect jet thruster for underwater robots [J]. Ocean Engineering, 2022, 266 (P4).

[6] Xu Shunyuan, et al. Numerical simulation and hydrodynamics analysis of a tethered underwater robot with control equipment [J]. Journal of Marine Science and Technology, 2021: 1-15.

[7] Huang L. (2012). Study on the mechanism of drag reduction in the spanwise motion of wall turbulence (Doctoral dissertation), Nanjing University of Science and Technology.

[8] Zhi M., Yan G., Sun L., et al. Study on the hydrodynamic characteristics of underwater recovery of airbag-structured navigational bodies [J/OL]. Acta Mechanica Sinica: 1-16 [2023-12-27].

[9] Jiang S., Xia C., Sun P., et al. Structural design and hydrodynamic analysis of a certain underwater salvage robot [J]. Manufacturing Automation, 2020, 42 (05): 61-66.

[10] Qi X., Miao A., & Wan D. (2018). CFD calculation of hydrodynamic derivatives for underwater vehicles [J]. Journal of Hydrodynamics (Ser. A), 33 (03), 297-304.

[11] Luo Yihan, Wu Jiaming, and Zhou Huifeng. Underwater Robot Trajectory Tracking Control Based on CFD Hydrodynamic Parameters [J]. Chinese Journal of Ship Research, 2022, 17 (03): 237-245+272. DOI: 10.19693/j.issn.1673-3185.02739.

[12] Waldemar K., Mirosława K., Tomasz K., et al. Transient hydrodynamic coefficients for a single DOF underwater manipulator of a square cross-section [J]. Ocean Engineering, 2023, 268.

[13] Cheng X., Hu Q., & He Y. (2020). Simulation research on hydrodynamic characteristics of small open-frame underwater robots [J]. Journal of Ship Electronic Engineering, 40 (02), 87-92.

[14] Laxman S. J., Anirban B., Vishwanath N., et al. Hydrodynamic characteristics of an Asian sea bass-inspired underwater body [J]. Applied Ocean Research, 2023, 141.

The effect of the propellant type size on combustion performance of the multi-nitro ester propellant

Tao Wang [1, 2] **, Ling Chen** [1, 2] **and Weidong He** [1, 2, *]

[1]School of Chemistry and Chemical Engineering, Nanjing University of Science and Technology, Nanjing, 210094, China

[2]Key Laboratory of Special Energy Materials (Nanjing University of Science and Technology), Ministry of Education, Nanjing 210094, China

[*]Corresponding author's e-mail: hewedong@njust.edu.cn

Abstract: To study the effect of propellant size on the combustion performance of the propellant, the combustion performance of three propellant types and their different ratio of inner perforation to diameter were studied by using the combustion experiment and data processing method of multi-nitro ester propellant as the research object. The results showed that within the test drug size range, under different initial temperatures and filling density conditions, the larger the ratio of charge length to the diameter of single-perf, 7-perf, and 19-perf propellant is, the larger the burning rate will be under the same pressure at the initial stage of combustion. The larger the length-to-diameter ratio of 7-perf and 19-perf propellants of the same type is, the better the gradual combustion will be, and the less influence there will be on single-perf propellants. The loading density has a certain effect on the constant volume combustion performance of the propellant: the higher the loading density is, the lower the combustion rate will be, which is because the erosion combustion phenomenon of the initial combustion of the propellant decreases with the increase of the loading density.

1. Introduction

The launcher is an indispensable part of the body-controlled weapon system, an important guarantee for the weapon to achieve long-range launch, precision strike and win the war, and is an important strategic material of the country [1]. The combustion performance of propellant in the chamber is of great significance in the study of propellant charge. The combustion condition of propellant in the chamber directly affects not only the muzzle velocity, range, accuracy, and other essential problems that restrict the effectiveness of weapons and equipment, but also the safety and reliability of propellant application.

When the projectile in the gun chamber moves, the volume occupied by the gas and the temperature of the gunpowder gas are constantly changing, which will complicate the relationship between the pressure, the temperature, and the specific volume of the gunpowder gas, so it is necessary to study the law of pressure change under the condition of a constant volume. Under constant-volume conditions, there is no change in volume and no work is done. In addition, the burning property of gunpowder can be easily analysed from the pressure variation rule of this condition [2-5]. Therefore, the study of the gas equation of state at the constant volume is an important part of internal ballistics [6]. The closed vessel of a constant volume commonly used in such experiments is called a closed explosive device. The existing combustion performance analysis is also mainly based on the closed explosive device test. The pressure-time curve measured by the closed explosive device test is used to obtain the relevant parameters of the combustion of the propellant and analyse them.

In order to pursue further development of high-chamber pressure, high-rate of fire, and high-power launching weapons, multi-nitro ester propellant with high energy and excellent mechanical properties has been developed all over the world [7]. As a multi-nitro ester propellant, its main components are diethylene glycol dinitrate (DEGDN), nitroglycerin (NG), and nitrocotton. At present, a lot of work has been done on the synthesis, characterization, and pyrolysis mechanism of nitrate as a plasticizer for propellant and the enhancement of mechanical properties of propellant. In this paper, the effects of the size [8-11], temperature, and loading density on the combustion performance of homogeneous multi-nitro ester propellant were investigated from various angles, so as to further improve the study on the combustion performance of multi-nitro ester propellant.

2. Experimental section

2.1. Test sample

Basic components of multi-nitro ester propellant include NC (60%), NG (28%), and other materials (12%). For the preparation process, absorption tablets and alcohol ketone solution were added first (the ratio of alcohol ketone is 1:1, and the ratio of solution to solute is 0.26). The raw materials are kneaded in the kneading machine for 2.5 h, extruded by the oil press and the mold. The perforation type is 15/1 tube, 9/7 cylinder, and 12/19 lace. They are cut into a certain length after proper drying, and the volatile solvent is removed under certain technological conditions. The perforation type parameters are shown in Table 1. The table shows some deviation in the actual value of the length-diameter ratio of the inner perforation.

Table 1. Shape and size of experimental propellant.

Number	Sample	Type	Shape	Length-Inside diameter ratio	Length /mm	Inside diameter /mm	Outside diameter /mm	Arc thickness /mm
1#	15/1-6.1	15/1	Tubular single perforation	14.3	40.0	2.8	6.5	1.9
2#	15/1-9.2	15/1	Tubular single perforation	21.4	60.0	2.8	6.5	1.9
3#	9/7-1.5	9/7	Cylindrical seven perforations	18.5	11.1	0.6	7.1	1.3
4#	9/7-2.5	9/7	Cylindrical seven perforations	30.7	18.4	0.6	7.1	1.3
5#	12/19-1.2	12/19	Lacy nineteen perforations	22.3	13.4	0.6	12.2	1.5
6#	12/19-1.5	12/19	Lacy nineteen perforations	28.0	16.8	0.6	12.2	1.5

2.2. Constant volume combustion performance test

The experiment was carried out according to GJB770B-2005. The combustion chamber volume of the closed burner was 100.334 ml. Nitrocotton was used as the ignition propellant. The ignition propellant amount was 1.000±0.005 g, the ignition pressure was 10.98 MPa, and the propellant loading density was 0.12 g/cm^3 and 0.20 g/cm^3. The p-t curves of the samples were tested at high temperature (50°C), normal temperature (20°C), and low temperature (-40°C), respectively, and the combustion performance was analysed.

2.3. Experimental data processing methods

The *L-B* and *u-p* curves can be obtained through the following processing processes.

Journal of Physics: Conference Series **2808** (2024) 012014 doi:10.1088/1742-6596/2808/1/012014

- *L-B* curves can be obtained by:

$$L = \frac{1}{p \cdot p_m} \frac{dp}{dt} \tag{1}$$

$$B = p / p_m \tag{2}$$

To analyse the *L-B* curve, it is necessary to adopt the quantitative evaluation method of the gradual combustion of the propellant based on the closed explosive test, and define the gradual combustion factor P_r of the propellant:

$$P_r = \frac{B_m \cdot L_m}{L_{0.1} + L_{0.3}} \tag{3}$$

- *u-p* curves can be obtained by:

$$u = u_1 p^n \tag{4}$$

where p_m is the maximum pressure, MPa; p is the test point pressure, MPa; L is the dynamic activity value of gas, $MPa^{-1} \cdot s^{-1}$; B is the grain width, mm; $L_{0.1}$ and $L_{0.3}$ are the corresponding dynamic activity when the relative pressure is 0.1 and 0.3 respectively; L_m is the dynamic activity corresponding to the combustion splitting point; B_m is the relative pressure corresponding to the combustion splitting point; u is the burning rate, u_1 is the burning rate coefficient, and n is the burning rate pressure index.

3. Analyses of experimental results

In this chapter, the constant volume combustion performance of multi-nitro ester propellant with different propellant sizes was analysed at 50 °C, 20 °C, and -40 °C, and 0.12 and 0.20 $g \cdot cm^{-3}$, respectively.

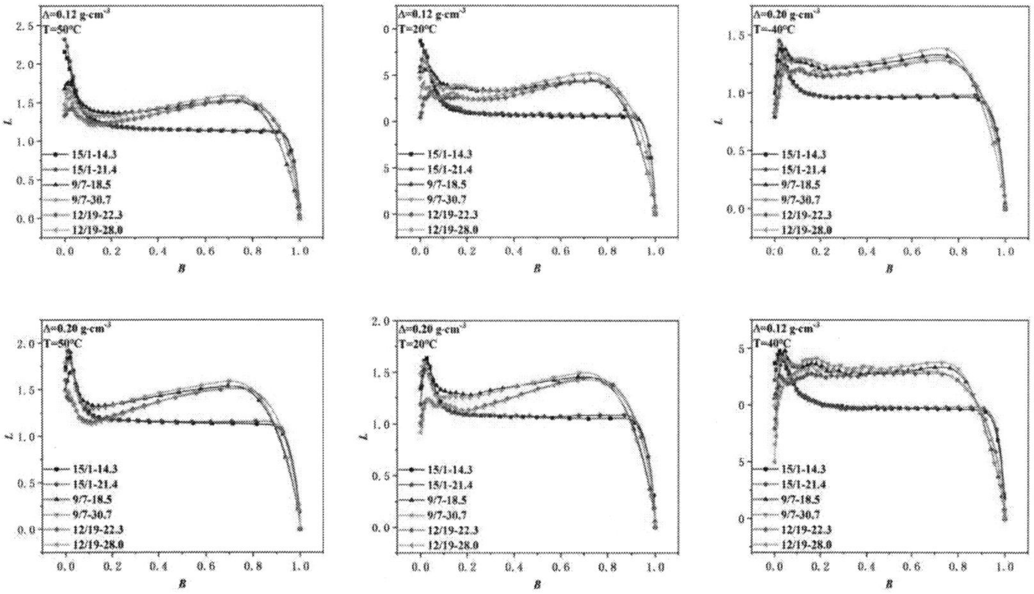

Figure 1. *L-B* curves of multi-nitro ester propellant with different propellant types and sizes under different conditions.

Table 2. Characteristics of gradual combustion in *L-B* curves of multi-nitro ester propellants at different temperatures.

Sample	Temperature /°C	P_{r1}	P_{r2}
9/7-18.5	50	0.3924	0.3982
	20	0.3818	0.3881
	-40	0.3737	0.3800
9/7-30.7	50	0.4139	0.4144
	20	0.4154	0.4134
	-40	0.3754	0.3947
12/19-22.3	50	0.4498	0.4473
	20	0.4209	0.4446
	-40	0.3127	0.3903
12/19-28.0	50	0.4543	0.4591
	20	0.4279	0.4543
	-40	0.3324	0.3964

Note: P_{r1} represents the incremental combustion factor at a loading density of 0.12 g·cm^{-3}, and P_{r2} represents the incremental combustion factor at a loading density of 0.20 g·cm^{-3}.

3.1. L-B curves

As can be seen from the *L-B* curves in Figure 1, the order of initial activity *L* of each curve is single-perf propellant > 7-pref propellant > 19-pref propellant, indicating that 15/1 tubular propellant has a higher burning intensity in the initial stage of combustion, and the *L* value of porous propellant is greater than that of single-perf propellant in the middle stage of combustion. This is because porous propellant has a smaller initial burning surface than tubular propellant. When porous propellant disintegrated at a certain stage of combustion, the burning surface increased rapidly. When *B* is 0.4~0.8, the order of *L* values of each curve is 4#>3#>6#>5#>2#>1#, indicating that in the stable stage of the curve, the order of *L* values is 7-pref propellant > 19-pref propellant > single-perf propellant. The larger the elongate-diameter ratio of 7-pref propellant, 19-pref propellant, and single-perf propellant is, the higher the dynamic activity will be.

According to the P_r values of the progressive combustion factor in Table 2, the larger the ratio of length to diameter of 7-perf and 19-perf propellant of the same propellant type, the better the progressive combustion ability will be.

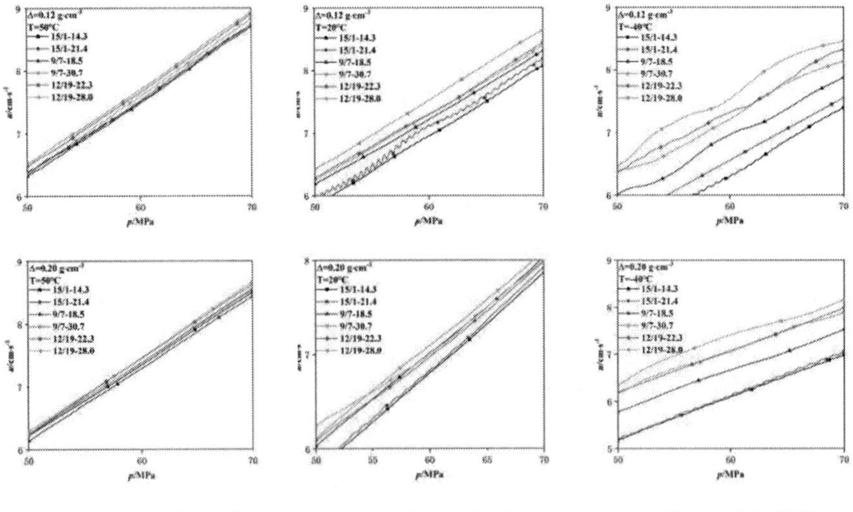

Figure 2. Local magnification of the *u-p* curve of multi-nitro ester propellant with different types and sizes.

3.2. *u-p curve local magnification analysis*

As can be seen from thc *u-p* curve in Figure 2, at 50~70 MPa, the order of burning rate is $2^{\#} > 1^{\#}$, $4^{\#} > 3^{\#}$, and $6^{\#} > 5^{\#}$. At the initial stage of combustion, under the condition of the same propellant type, the larger the slenderness ratio of the propellant is, the faster the burning rate will be. Firstly, the gas generated in the propellant perforation increases the pressure inside the perforation and the pressure difference inside and outside the perforation, thus making the gas flow outward. The larger the length-diameter ratio is, the larger the pressure difference and the higher the burning rate will be; Secondly, the gas generated in the propellant perforation flows outwards, and the erosion of the gas flow on the burning surface increases the burning rate, which is the corrosive combustion phenomenon of the propellant. The corrosive combustion phenomenon of the propellant with a large aspect ratio is more serious, so the burning rate is higher.

3.3. *Comparison of u-p curves at two loading densities*

It can be seen from the *u-p* curve in Figure 3 that in the burning rate rising stage, the higher the loading density, the lower the burning rate of the same type of propellant under the same pressure. It is speculated that the reason is that the higher the loading density and the smaller the free volume of the chamber is, the faster the pressure outside the propellant perforation will increase, which hinders the external flow of gas in the perforation and reduces the erosion combustion phenomenon, resulting in a slower burning rate.

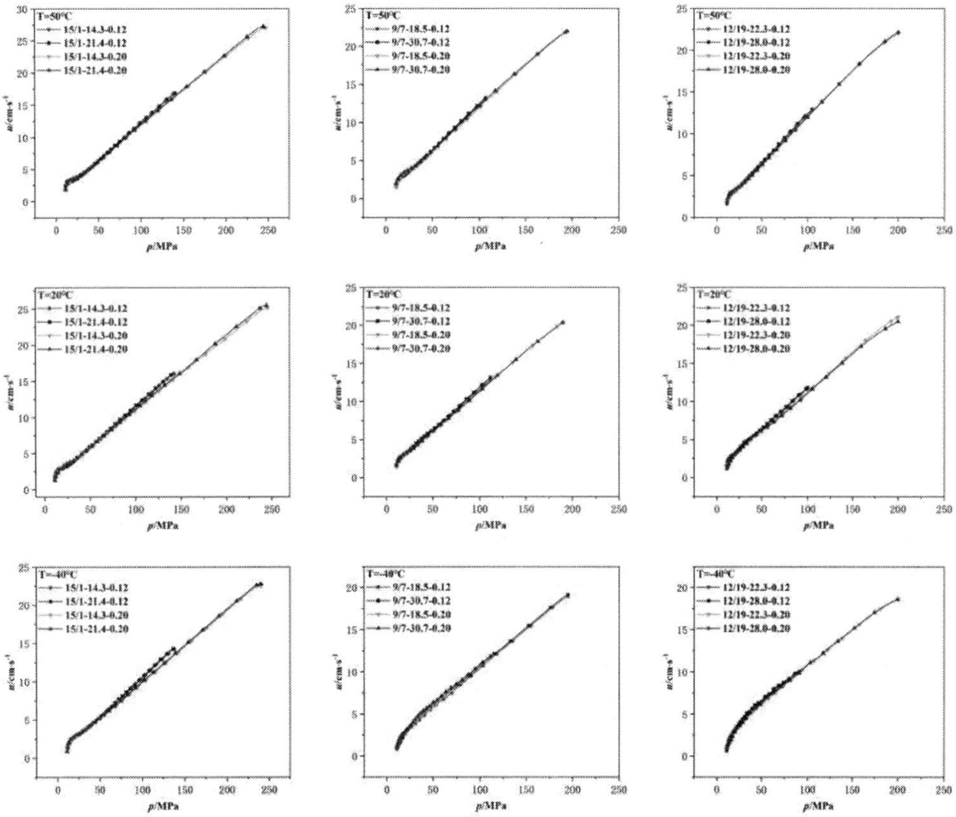

Figure 3. The *u-p* curves of multi-nitro ester propellant with different drug sizes under two loading densities at initial temperatures of 50 °C, 20 °C, and -40 °C, respectively.

4. Conclusion

Under the conditions of temperatures of 50 °C, 20 °C, and -40 °C, and loading densities of 0.12 and 0.20 g·cm^{-3}, the influence of charge type size on combustion performance at constant volume is as follows: In the range of test charge type size, the larger the ratio of charge length-to-diameter of single-perf, 7-perf and 19-perf propellant is, the larger the burning rate under the same pressure at the initial stage of combustion will be. The larger the length-to-diameter ratio of 7-perf and 19-perf propellants of the same type is, the better the gradual combustion will be, and the less influence there will be on single-perf propellants. The loading density has a certain effect on the constant volume combustion performance of the propellant: the higher the loading density is, the lower the combustion rate will be, which is because the erosion combustion phenomenon of the initial combustion of the propellant decreases with the increase in the loading density.

Acknowledgment

This work is supported by the Jiangsu Funding Program for Excellent Postdoctoral Talent and a project funded by the China Postdoctoral Science Foundation (2023TQ0158).

References

[1] Wang Z S, He W D. 2006. Principle and Technology of Gunpowder charge design M. Beijing: Beijing Institute of Technology Press.
[2] Boulkadid M K, Lefebvre M H, and Jeunieaul. 2019. Recovery of solid double base artillery propellants artificially aged by burns J. Journal of High Energy Materials. 38(1): 1-19.
[3] Zhang N, He Z D, and Secretary W. 2019. Calculation and experimental study on combustion performance of variable burning rate propellant J. Solid Rocket Technology. 42(2): 205-209.
[4] Xiao Z G, Ying S J, and Xu F M. 2010. The apparent burning rate model and numerical calculation of the pressure measured in a closed explosive device J. Journal of Nanjing University of Science and Technology. 34(2): 262-265. (in Chinese)
[5] Zhang L H, Xiao Z L, Lu B F, Liu Y P, Ma Z L, He Z D, and He L M. 2008. Analysis of gradual combustion of variable burning rate propellant based on experimental data of closed explosive J. Journal of Explosives and Explosives. 31(3): 67-70.
[6] Jin Z M. 2004. Gun interior Ballistics M. Beijing: Beijing Institute of Technology Press.
[7] Wang Z S. 2006. Introduction to energetic materials M Harbin: Harbin Institute of Technology Press.
[8] Xu Z G, Liang H, Ding Y J. 2020. Calculation of shape function and combustion performance of four-hole cuboid gun propellant J. Chinese Journal of Energetic Materials. 28(6): 491-497.
[9] Yang W T, Clive W, Yan G H. 2020. Calculated combustion behavior of multi-perf disk (MPD) gun propellant in closed bomb J. Chinese Journal of Explosives & Propellants (Huozhayao Xuebao). 44(4): 538-542.
[10] Zhong J H, Tian S M, and Zheng D S. 2009. Research progress of high gradient propellant J. Shanxi Chemical Industry. 29(4): 32-35. (in Chinese)
[11] Wang Q L, Zhao X F, Liu S W, Zhang Y B, Yu H F, Yao Y J, Zhu Y C, and Wei L. 2009. A quantitative evaluation method for gradual combustion of propellant based on Closed explosive test J. Journal of Explosives and Explosives. 32(3): 71-74.

Mode I interlaminar fracture toughness of glare laminates by single cantilever beam

Xiaoge Hua[1*], Zheng Tian [2]

[1]COMAC Shanghai Aircraft Design & Research Institute, Shanghai, 201210, P. R. China

[2]School of Chemistry and Molecular Engineering, East China Normal University, Shanghai, 200241, P. R. China

[*]Corresponding author's e-mail: huaxiaoge@comac.cc

Abstract: GLARE laminates, also known as Glass Fiber Reinforced Aluminum Laminates, have emerged as critical materials for fuselage and wing skin applications due to their outstanding fatigue resistance and good damage tolerance. Among the various modes of failure in GLARE laminates, interfacial delamination plays a significant role in determining their overall performance. However, GLARE laminates suffer complicated failure mechanisms and have a complex multi-interface system. Clearly, it is crucial to accurately determine the feasibility and accuracy of the SCB (Single Cantilever Beam) method for GLARE laminates by combining finite element simulation with experiments. Meanwhile, a detailed investigation was conducted on the influence of fiber laying and the loading rate of GLARE laminates. The findings revealed that the SCB tests resulted in pure Mode I interlaminar fracture failure. Furthermore, it was observed that loading rates range of 1~5 mm/min achieves reasonable and effective Mode I interlaminar fracture toughness. Additionally, finite element simulation results demonstrated the absence of in-plane shear failure within the fiber layer, leading to the production of pure Mode I interlaminar fracture.

1. Introduction

Fiber Metal Laminates (FMLs) represent materials that combine fiber-reinforced composites with metal alloys [1, 2]. GLARE laminates are one of these [3, 4]. These laminates exhibit outstanding impact resistance, fatigue, good residual, blunt notch strength, and convenient manufacturability and repairability, making them widely utilized in critical aircraft components such as the fuselage, wing skin, cabin, floor, and other aspects. Due to their multi-layer structure, GLARE laminates possess notable anisotropic properties and a complex failure mechanism, distinguishing them from traditional single composite laminates [5]. Generally, the evaluation method of resin matrix composites lacks accuracy, and makes the data untrustful [6]. Therefore, the sample structure of the upper and the lower arm is not balanced by DCB. Due to the difference in performance between upper- and lower-arm materials, the unloaded end of GLARE laminates always shifts upward during the loading process, resulting in serious out-of-plane displacement.

Li et al. [7] tested the interfacial fracture toughness obtained by DCB, which was different from the theoretical calculation results and may be related to the loading rate and other parameters. The interlaminar fracture toughness G_{IC} and critical load P_C [8] of crack propagation of laminates were obtained by DCB, and the interfacial fracture performance was accurately evaluated. However, different material properties and structures affected the interlaminar fracture toughness value when the evaluation method of GIC was applied to the FMLs material system. Thus, it is impossible to accurately evaluate G_{IC} of FMLs [9].

In addition, a single cantilever beam was proposed, and displacement-controlled loading was used to promote crack propagation [10, 11]. The investigation of SCB provides a new idea for evaluating the interlaminar fracture properties of FMLs. However, there are no reports or feasibility studies on the application of SCB to the G_{IC} properties of FMLs. Therefore, the test methods, and evaluation criteria for the interface properties of aluminum alloy/glass fiber in GLARE laminates need to be further discussed in order to evaluate the interface properties scientifically and accurately.

For these reasons, the work in this paper primarily focuses on the mechanics principle and failure mode of G_{IC} of GLARE by SCB. Besides, SCB testing was employed to systematically investigate the effects of loading rates, fiber orientation, and the feasibility of utilizing SCB to evaluate the G_{IC} of GLARE. The stress-strain and damage modes of all layers and components under the SCB model were further explored by 3D finite element simulation.

2. Experimental

2.1 Fabrication of specimens

The GLARE was produced utilizing a hand layup technique, incorporating alternating layers of aluminum alloy sheets and S4-glass/epoxy prepregs, as detailed in Table 1. The aluminum alloy sheets were made of 2024-T3, which are 0.3 mm thick per sheet. The S4-glass/epoxy prepregs were 0.125 mm thick per layer.

Table 1. Materials' thickness and layup.

Stacking configuration	Structure	Thickness	
		Experiment	Numerical simulation
Al-[0-0]-Al-[0-0]-Al Al-[90-0]-Al-[0-90]-Al	3/2	1.54±0.02	1.404

Unidirectional and cross-laminated composite materials have been produced and tested for the effects of different layups of GLARE. The production process is shown in Figure 1.

Firstly, the surface treatment of aluminum alloy sheets is carried out for better bonding with fiber-reinforced composites [12, 13]. This typically involved anodizing the sheets at 20 °C, 20 min, and 10 V voltage in a 130 g/L phosphoric acid solution. Then, the sheets are thoroughly rinsed and dried [14, 15]. Secondly, an adhesive was applied to the specimens within 24 hours, after which GLARE-3/2 was laid up with different stacking configurations. A folded aluminum foil is used as a starter defect for 50 mm. Finally, the sheets are bonded in an autoclave. This final step ensures proper bonding between the layers [1].

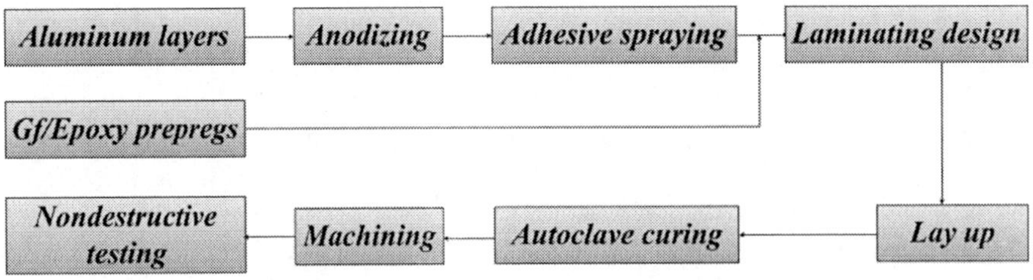

Figure 1. Specimen fabrication process.

The specimens underwent milling to be cut and were affixed with hinges using adhesive. Subsequently, the specimens' structure is illustrated in Figure 2.

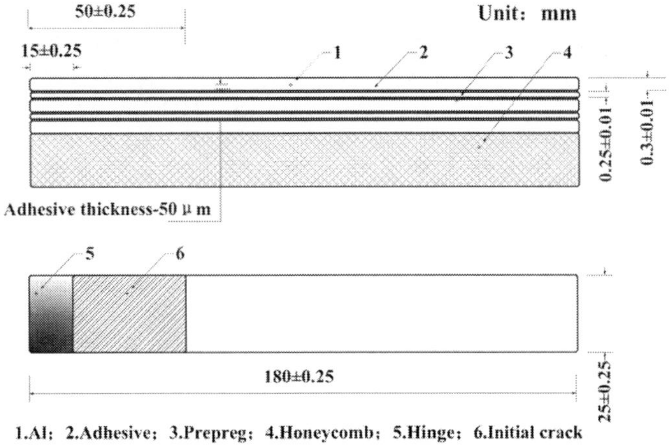

Figure 2. Geometry of GLARE specimens using for SCB, unit: mm.

The Al/Epoxy/Al laminates with a centrosymmetric structure, as shown in Figure 3, were also prepared using a fabrication process similar to that of GLARE. The G_{IC} of Al/Epoxy/Al could be evaluated well using the DCB method since the centrosymmetric structure always exhibited Mode I fracture under DCB loading. Therefore, The DCB and newly developed SCB methods were compared to verify the effectiveness of SCB before the investigations on the interlaminar fracture toughness of GLARE.

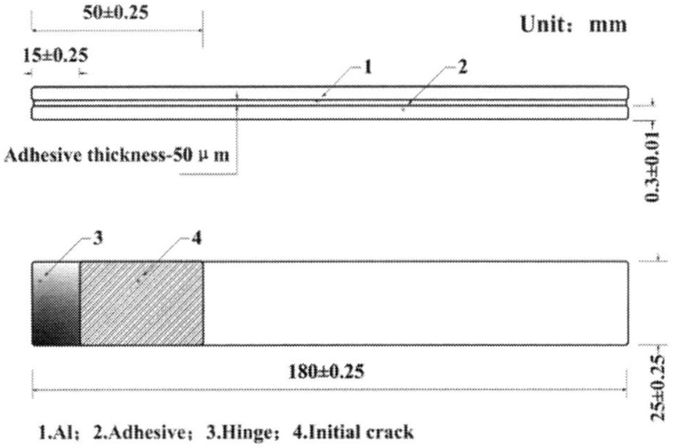

Figure 3. Geometry of Al/Epoxy/Al specimens using for SCB and DCB, unit: mm.

2.2 SCB experiment fixture

The SCB has not yet formed a standard test method for evaluating G_{IC}, and the standardized test fixture cannot be directly purchased. Therefore, the test fixture is designed and processed based on the laminated materials, as shown in Figure 4.

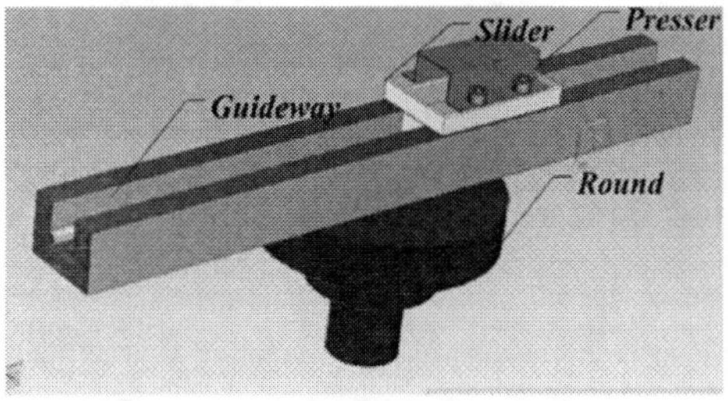

Figure 4. The diagram of the SCB experiment fixture.

The self-designed test fixture of SCB was designed with a guide rail, which had low friction and slid easily, as shown in Figure 5. The clamp is connected by a disc and a stretcher. The guide rail is connected by a screw and a disc. Meanwhile, the sheet is rigidly fixed to the slider by a bolt and moves with the slider. The friction force between the guide rail and the slider is very small compared with the load force of the specimen, which can be neglected basically.

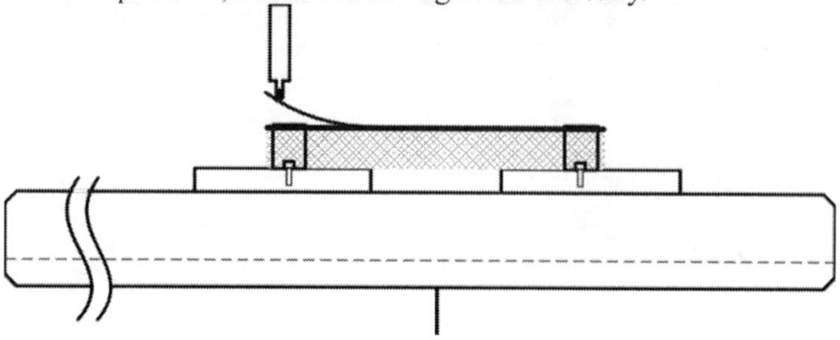

Figure 5. The schematic diagram of the fixture for Mode I interlaminar fracture toughness test by SCB.

2.3 SCB test

SCB tests on GLARE laminates are conducted using a CMT-5105 universal electronic testing machine, as depicted in Figure 6, and employing displacement rates range of 1~5mm/min, respectively. The experimental setup is outlined in Figure 3. Each specimen underwent loading until the crack extended.

The position of the crack tip was recorded, followed by unloading before the subsequent loading cycle. G_{IC} was computed using Equation 1 and Equation 2.

$$G_{IC} = \frac{3P^2ea^2}{2b} \tag{1}$$

$$e = \frac{\sum_{i=1}^{n}(a_i^3 - \bar{a}^3)(c_i - \bar{c})}{\sum_{i=1}^{n}(a_i^3 - \bar{a}^3)^2} \tag{2}$$

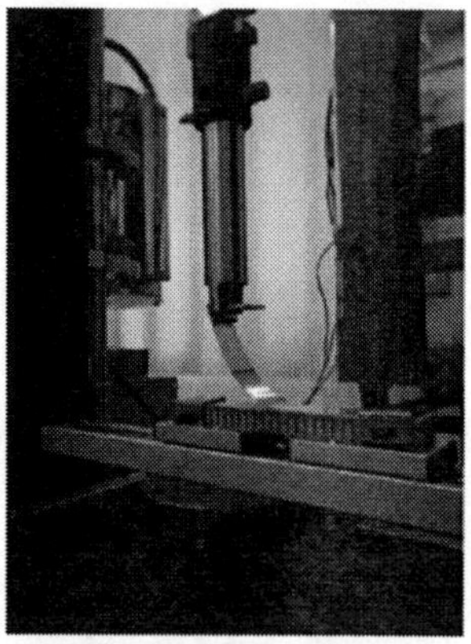

Figure 6. SCB test.

3. Mechanical principle and finite element model

3.1 Mechanical principle of Mode interlaminar fracture

Figure 7 shows the displacement and stress near the crack tip of GLARE. The mechanical principle of SCB is rigidly fixed in the unexpanded region of the crack and the unloaded portion of the crack propagation region, as well as the lower arm of the specimen, according to the setting of the boundary condition of the specimen. Then, the boundary condition is expressed as $u_1 = u_2 = u_3 = 0$, $v_{11} = v_{12} = v_{23} = 0$. The crack propagates stably and produces pure G_{IC} under tensile force. The specimen produces no out-of-plane displacement, and the unexpanded sample remains horizontal under the pure G_{IC} (see Figure 8).

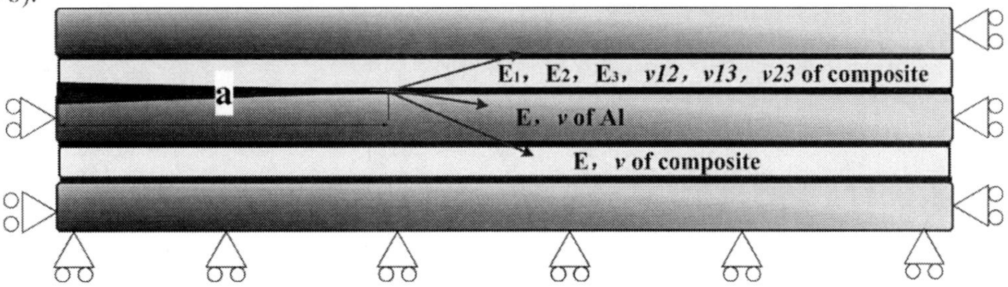

Figure 7. Mechanics principle of SCB.

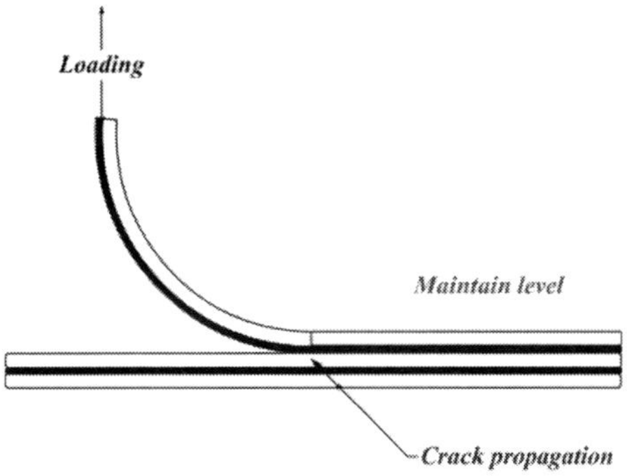

Figure 8. G_{IC} failure mode of SCB.

3.2 Finite element model

Three-dimensional eight-node linear brick solid elements were used to mesh Aluminum alloy layers [16]. Table 2 and Table 3 show the mechanical properties parameters and true stress-strain values of aluminum alloy sheets, respectively.

Table 2. Mechanical properties of 2024-T3 Al.

Material	Elastic Modulus /GPa	Yield Strength /MPa	Tensile Strength /MPa	Poisson's ratio	Fracture strain
2024/T3 Aluminum	72	305	450	0.33	0.19

Table 3. Real stress-strain value of 2024-T3 Aluminum Alloy.

Yield Strength /MPa	True strain /(%)
360	0
366	0.02
375	0.05
387	0.12
400	0.45
415	1.31
425	2.13

438	5.13
447	8.26
455	14.71

Eight-node quadrilateral continuum elements are used to mesh the fiber layers, which are treated as orthotropic linear elastic materials [17]. Among them, the mechanical properties were measured by the tensile test, referring to standard ASTM D3039; Poisson's ratio was obtained by longitudinal and transverse tensile tests of materials; fracture energy was provided by manufacturers (see Table 4).

Table 4. Mechanical properties of GFRP/ epoxy resin composites.

Material	Performance	Direction	Parameter
GFRP/ epoxy resin composites	Elastic Modulus /GPa	E_{11}	55
		E_{22}	9.5
	Ultimate strength /MPa	X_T	1900.95
		X_C	2000
		Y_T	50
		Y_C	150
		S	50
	Poisson's ratio	v_{12}	0.33
		v_{13}	0.45
	Fracture toughness /kJ·m^{-2}	G_f	12.5
		G_m	1

Figure 9 shows the failure criteria of the fiber composite layer. σ_{max} is the failure strength, and ε_i is the strain when the initial failure occurs.

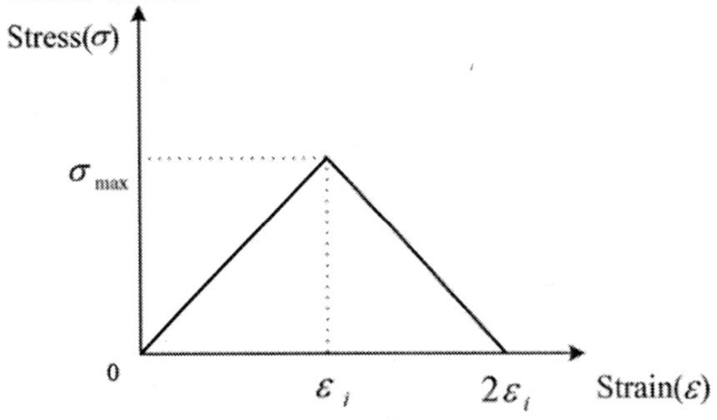

Figure 9. Failure criteria of fiber composite layer.

Additionally, the application of the cohesive zone model [18] proved advantageous in elucidating the processes of delamination inception and progression. Table 5 details the pertinent material characteristics of cohesive layers.

$$\left\{\frac{(\varepsilon_n)}{\varepsilon_n^f}\right\}^2 + \left\{\frac{(\varepsilon_s)}{\varepsilon_s^f}\right\}^2 + \left\{\frac{(\varepsilon_t)}{\varepsilon_t^f}\right\}^2 = 1 \tag{3}$$

$$G_{IC} + (G_{IIC} - G_{IC})\left(\frac{G_{IIC}}{G_T}\right)^{\eta} = G_C \tag{4}$$

where ε_n represents the normal strain, ε_s and ε_t represent shear strains, ε_n^f denotes the allowable normal strain, ε_s^f and ε_t^f denote maximum shear strains, G_{IC} and G_{IIC} are fracture toughness for Mode I and Mode II, and G_T is the shear fracture energy.

Table 5. Mechanical properties of cohesive layers.

Material	Performance	Direction	Parameter
Cohesive layers	Initial normal stiffness /(N/mm)	K_n^0	1.0×10^6
		K_s^0	1.0×10^6
		K_t^0	1.0×10^6
	Interfacial strength /(MPa)	τ_n^0	6.57/5.56/5.37
		τ_s^0	36.62
		τ_t^0	36.62

4. Results and discussion

4.1 Comparison between DCB and SCB using centrosymmetric Al-Epoxy-Al laminates

A simple comparison is made between DCB and newly developed SCB using centrosymmetric Al-Epoxy-Al laminates. The critical load values for crack propagation exhibit significant differences owing to the different loading modes and conditions of the two methods, as shown in Figure 10. Even though consistent fracture toughness is obtained, as shown in Figure 11, the results showed that the average values of G_{IC} differed by only 2.9%.

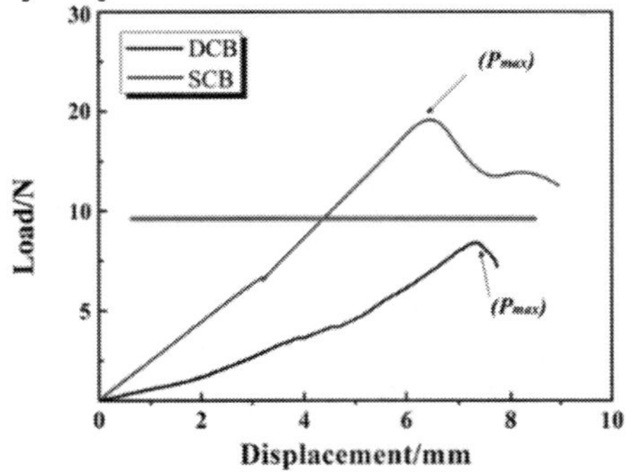

Figure 10. Critical load-displacement curves of Al/Epoxy/Al laminates under DCB and SCB.

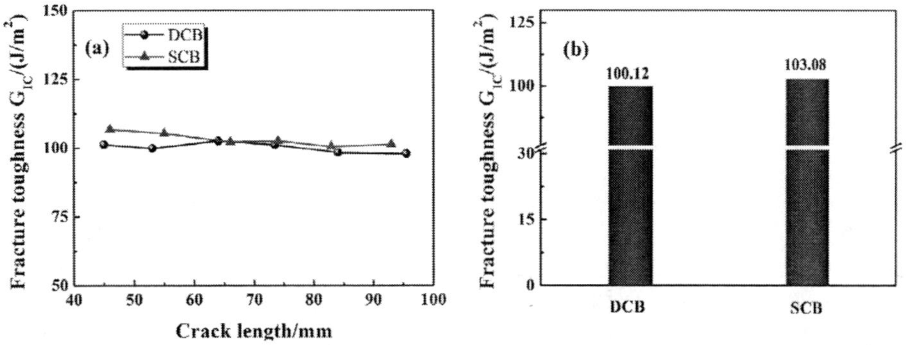

Figure 11. The fracture toughness of Al-Epoxy-Al laminates under DCB and SCB: (a) R curves, (b) average fracture toughness values.

The results above indicated that the newly developed SCB method, together with the fixture, is applicable to evaluate G_{IC}.

4.2 G_{IC} of GLARE by SCB for interface scan

In the case of unidirectional-plies laminates (see Figure 12), a minor number of glass fiber composite fragments persist on the aluminum plate, while the majority of aluminum plates are exposed. This observation suggests that crack propagation in unidirectional laminate primarily occurs along the interface. Consequently, unidirectional-plies laminates predominantly involve adhesion failure. Conversely, for cross-plies laminates (see Figure 13), a significant quantity of glass fiber composites adhere to the aluminum alloy plate. The cohesive failure mode largely characterizes the mixed failure mode of cross-plies laminates.

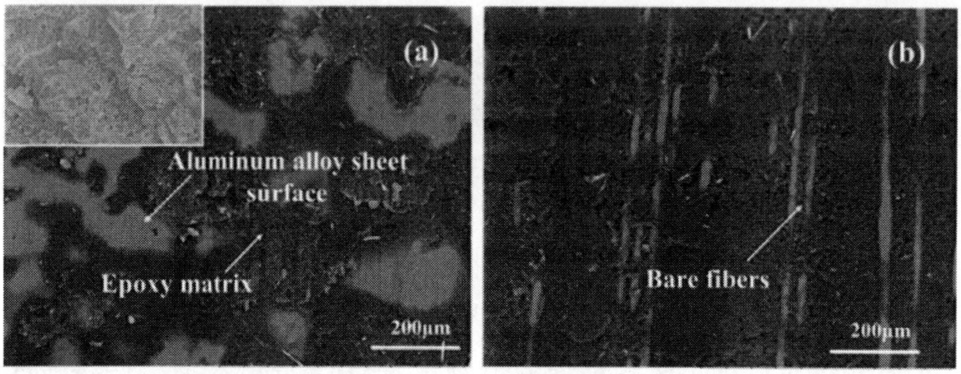

Figure 12. Interface microstructure of unidirectional-plies laminates: (a) upper, (b)lower.

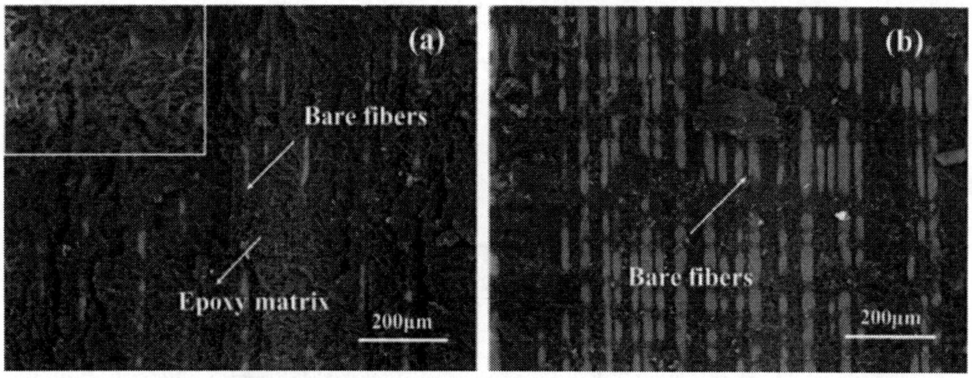

Figure 13. Interface microstructure of cross-plies laminates: (a) upper, (b)lower.

4.3 Effect of fiber orientation on G_{IC} of GLARE laminates

The critical load-displacement curve of the GLARE laminate, as depicted in Figure 14, displays a nearly linear relationship during the initial crack propagation stage. When the maximum load is reached, the crack begins to expand and drops slowly. Meanwhile, there is also a series of small jump-forming extensions with significant load and displacement. The critical loads of crack propagation in unidirectional-plies and cross-plies laminates were 38.46 N and 55.9 N, respectively. Obviously, the critical load of cross-plies laminates was about 1.45 times that of unidirectional-plies laminates. The reason was mainly attributed to the fact that the fiber layering method inside the cross-plies laminates was more resistant to crack propagation. Therefore, the critical load required for crack propagation of cross-plies laminates was larger.

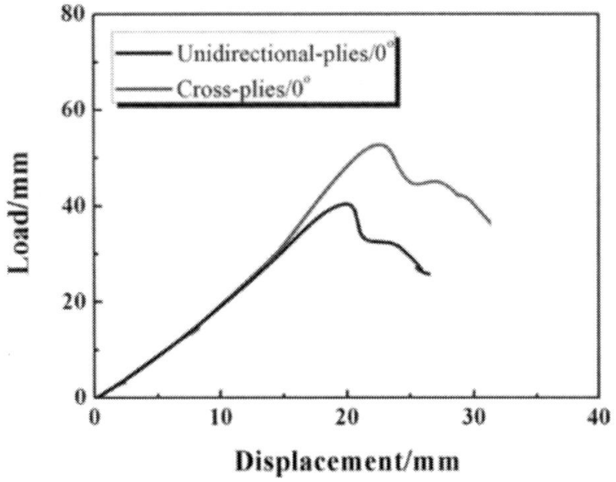

Figure 14. Load-displacement curve.

The relationship curve depicting the connection between flexibility (c) and cubic crack length (a^3) is illustrated in Figure 15. The coefficients required for this analysis were obtained through linear fitting and subsequently applied to Equation 2. The interlaminar fracture toughness values for both GLARE unidirectional and R curves can be observed in Figure 12.

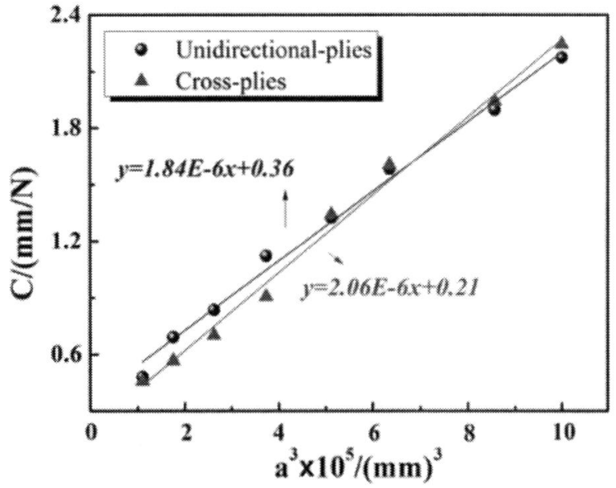

Figure 15. Flexibility fitting curve.

Figure 16 presents the variation of G_{IC}, denoted as Gc, with respect to the R-curve. Specifically, it is observed that the G_{IC} exhibits a slight increase with crack length until it reaches a stable state.

The G_{IC} of unidirectional plies measures 153.04 J/m². As the crack propagates, the G_{IC} gradually stabilizes at 160 J/m². In comparison, the G_{IC} of cross-plies surpasses that of unidirectional-plies. This may be due to the improved crack-bridging capability of cross-plies, which effectively hinders crack propagation.

Furthermore, the R curve demonstrates a smooth and stable trend, indicating smooth crack propagation and well-bonded interfaces within the laminates. These findings collectively support the notion that G_{IC} increases and remains stable with the growth of crack length, exhibiting enhanced resistance to fracture and improved bonding characteristics.

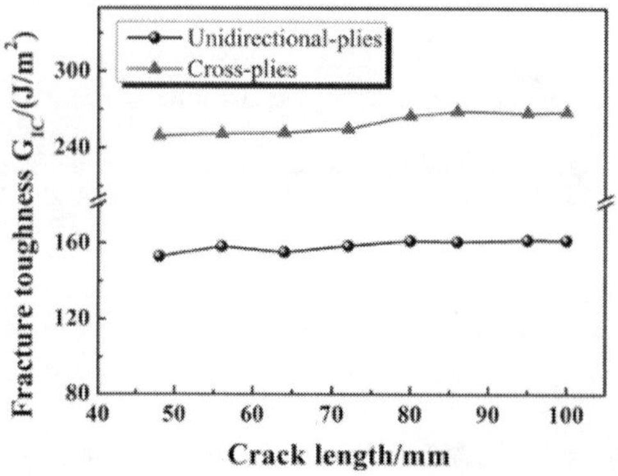

Figure 16. R curves.

4.4. Influence of loading rate on G_{IC} of GLARE

In Figure 17, it is observed that the critical load exhibits an upward trend when the loading rate increases. Figure 17(a) represents unidirectional plies, and Figure 17(b) represents cross-plies.

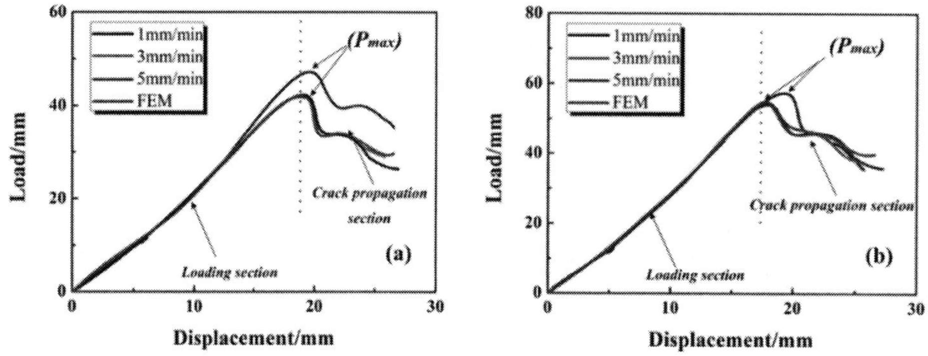

Figure 17. Load-displacement curve.

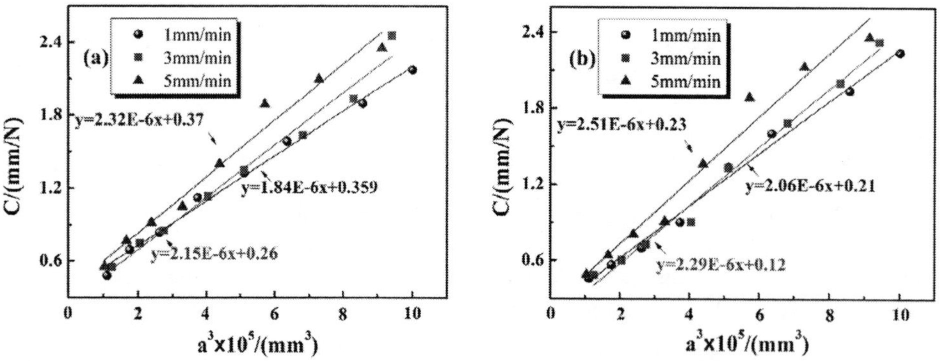

Figure 18. Flexibility fitting curve.

The relationship curve between flexibility (c) and cubic crack length (a^3) at different loading rates was drawn, as shown in Figure 18. Figure 18(a) represents unidirectional plies, and Figure 18(b) represents cross-plies. Based on the above values, the interlaminar fracture toughness values of R curves are plotted in Figure 19. Figure 19(a) represents unidirectional plies, and Figure 19(b) represents cross-plies.

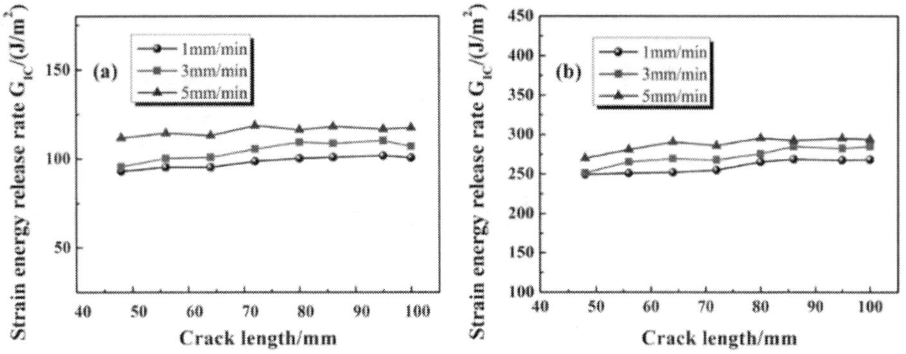

Figure 19. R curves.

As depicted in Figure 19, a marginal increase in the interface was observed with a higher loading rate. Meanwhile, the R curve showed a slight increase trend, but it can basically maintain a stable horizontal expansion. When the crack length reaches 65-70 mm, the fracture toughness remains basically unchanged with the crack propagation. In addition, as the loading rate increases, the fluctuation of the curve becomes larger, indicating that the stability of crack propagation decreases, but the overall variation of the R curve does not change.

It can be seen from Figure 20 that for unidirectional-plies laminates, the average fracture toughness values are 158.83 J/m^2, 170.25 J/m^2, and 187.63J/m^2 at 1, 3, and 5 mm/min, respectively. For the cross-plies laminates, the average fracture toughness values were 259.69 J/m^2, 272.7 J/m^2, and 288.05 J/m^2 at 1, 3, and 5 mm/min. Therefore, the interface fracture toughness value obtained at 5 mm/min was also significant from the viewpoint of improving the experiment efficiency.

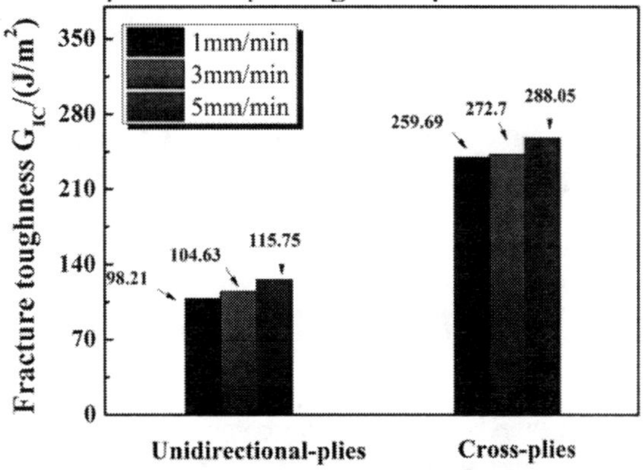

Figure 20. Average interfacial fracture toughness of GLARE laminates.

4.5.Numerical analysis

The ultimate loading moments reveal the normal stress and in-plane shear stress of GLARE, which is evaluated in Figure 21. Notably, the highest normal stress was observed near the pre-crack in the fiber layer. Under maximum loading conditions, the ultimate stresses for the 0°-0° and 0°-90° were determined to be 2051 MPa and 2297 MPa, surpassing the bending strength of the glass/epoxy prepreg (1620.23 MPa). Consequently, the samples experienced Mode I interlayer fracture failure due to tensile stress. Additionally, the maximum in-plane shear stresses in the 0°-0° and 0°-90° laminates were respectively found to be 1.58 MPa and 2.83 MPa. The glass/epoxy prepreg (50.98 MPa) is considerably higher than those of the in-plane shear strength. As a result, no in-plane shear failure was displayed to pure Mode I interlaminar fracture.

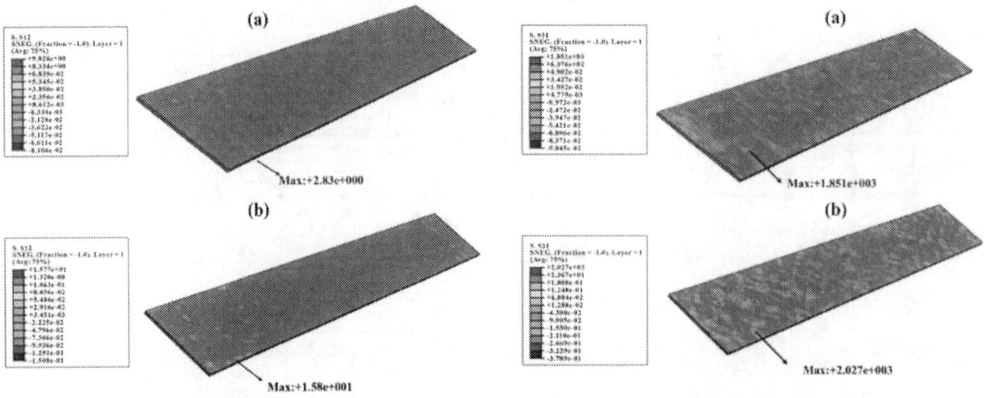

Figure 21. The ultimate loading moments reveal the normal stress distribution and in-plane shear stress distribution patterns exhibited by the prepreg layers.

Figure 22 provides a clear presentation of progressive failure mechanisms observed in different layers surrounding the crack. The stiffness degradation variable, denoted as SDEG = 1 (the red region), indicates complete destruction. SDEG = 0 (the blue region), signifies the absence of lamination. Additionally, 0 < SDEG < 1 suggests partial damage to the cohesion element in this part.

Visualizing the laminate on the *x-y* plane, delamination occurs at the loading point and gradually extends during crack propagation. In terms of the *z* direction, initial delamination failure occurs near the middle aluminum layer. Importantly, this layer exhibits a larger debonding area compared to other layers, which underscores its significance in the overall damage process.

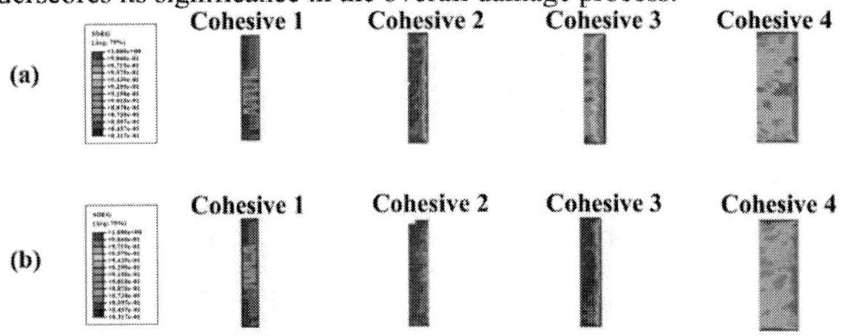

Figure 22. Cohesive layers' predicted damage in the SCB of GLARE.

5. Conclusion

(1) The load-displacement curves exhibited good agreement with the experimental results, suggesting that established models incorporated with suitable parameters were effective in accurately predicting the critical loading conditions. Moreover, the stress-strain nephogram and the damage situation of each layer showed that the pure Mode I interface fracture failure of GLARE laminates occurred based on the self-designed SCB fixture.

(2) 3D finite element models were developed to investigate the interlaminar fracture characteristics of GLARE. The purpose of this research was to establish a refined FEM capable of accurately assessing the G_{IC} of GLARE by SCB.

(3) For the interfacial fracture properties, this research investigated the effect of loading rate for GLARE and indicated that obtained at a 5-mm/min loading rate exhibited a considerable improvement in test efficiency.

References

[1] Böhm M.; Głowacka K. Fatigue life estimation with mean stress effect compensation for

lightweight structures-The case of GLARE 2 composite [J]. Polymers 2020, 12: 251.

[2] Megeri S.; Naik G. N. Numerical studies of the low velocity impact behaviour on hybrid fiber metal laminates [J]. Materials Today 2021, 44: 1860-1864.

[3] Chen Y.; Jin K.; Li H.; Lin Y.; Lu Y.; Hua X.; Yuan Q.; Wang H.; Tao J. Effect of peer ply and cooling rate on the tensile properties of Al/Gf/PP laminate prepared by hot pressing [J]. Journal Of Sandwich Structures & Materials 2020, 22, 2567-2581.

[4] Moriniere F. D., Alderliesten R. C., Sadighi M., Benedictus R. An integrated study on the low-velocity impact response of the GLARE fiber-metal laminate [J]. Composite Structures 2013, 100: 89-103.

[5] Xu Yingmei, Li Huaguan, Yang Yinfei, Hu Yubing, Tao Jie *. Determination of residual stresses in Ti/CFRP laminates after preparation using multiple methods [J]. Composite Structures 2019: 715-723.

[6] Goossens S.; Berghmans F.; Muñoz K.; Jiménez M.; Karachalios E.; Saenz-Castillo D.; Geernaert, T. A global assessment of barely visible impact damage for CFRP sub-components with FBG-based sensors [J]. Composite Structures 2021, 272: 114025.

[7] Li Huaguan, Ding Ying, Li Jianchun, Li Zhaoxuan, Wang Jie, Wu Chao, Xiang Junxian, Tao Jie *. The effect of different interface conditions on the energy absorption characteristics of Al/CFPP hybrid tube [J]. Thin-Walled Structures 2023, 185: 110578.

[8] Cortes P. The fracture properties of a fiber metal laminate based on a self-reinforced thermoplastic composite material [J]. Polymer Composites 2014, 19: 116-122.

[9] Jin Kai, Wang Hao, Tao Jie *, Du Dandan. Mechanical analysis and progressive failure prediction for fibre metal laminates using a 3D constitutive model [J]. Composites Part A: Applied Science and Manufacturing, 2019, 124: 105490.

[10] Lin Yanyan, Li Huaguan, Wang Qilitai, Gong Zhibing, Tao Jie *. Effect of plasma surface treatment of aluminum alloy sheet on the properties of Al/Gf/PP laminates[J]. Applied Surface Science 2020, 507: 145602.

[11] James G., Ratcliffe, James R. R. Sizing a single cantilever beam specimen for characterizing facesheet-core debonding in sandwich structure[J]. Composite Materials, 2011, 45(25): 2669-2684.

[12] Ma K. P., Wu C. W., Tsai Y. T., Hsu Y. C., Chiang C.C. Internal residual strain measurements in carbon fiber-reinforced polymer laminates curing process using embedded tilted fiber bragg grating sensor [J]. Polymers 2020, 12: 1479.

[13] Fiore V, Alagna F, Di Bella G, Valenza A. On the mechanical behavior of BFRP to aluminum AA6086 mixed joints [J]. Composite Part B 2013, 48: 79-87.

[14] Wang Hao, Li Huaguan, Xu Yingmei, Lin Yanyan, Li Hongli, Tao Jie *. Effects of thermal residual stresses on tensile and interlaminar shear behaviors of GLARE laminates [J]. Applied Composite Material 2021, 28: 877-898.

[15] Sun Z Q, Huang M H, Hu G H. Surface treatment of new type aluminum lithium alloy and fatigue crack behaviors of this alloy plate bonded with Ti-6Al-4V alloy strap [J]. Materials and Design 2012, 35: 725-730.

[16] Lin Yanyan, Li Huaguan, Zhang Zhongwei, Tao Jie*. Low-velocity impact resistance of Al/Gf/PP laminates with different interface performance [J]. Polymers 2021, 13(24): 4416.

[17] Xiang Junxian, Li Huaguan, Wang Hao, Li Hongli, Chen Xi, Tao Jie *. Residual stress variation and deformation of shot peened glass fiber reinforced composites/aluminum–lithium laminates under thermal shock and thermal fatigue [J]. Polymer Composites 2021, 42(12): 6523-6533.

[18] Turon A., Camanho P. P., Costa J., et al. Accurate simulation of delamination growth under mixed-mode loading using cohesive elements: definition of interlaminar strengths and elastic stiffness [J]. Composite Structures 2010, 92(8): 1857-1864.

Investigation of oxide inclusion films in cast superalloy impeller

Lihui Zhang [1], Jun Chen [1, 2, 3, 4], Weijie Xing [1], Wenjin Wu [1], Fengwei Guo [1], Qiang Zhang [1], and Changkui Liu [1, 2, 3, 4 *]

[1]AECC Beijing Institute of Aeronautical Materials, Beijing 100095, China
[2]Beijing Key Laboratory of Aeronautical Materials Testing and Evaluation, Beijing 100095, China
[3]Key Laboratory of Science and Technology on Aeronautical Materials Testing and Evaluation, Aero-engine Corporation of China, Beijing 100095, China
[4]Failure Analysis Centre of Aviation Industry Corporation of China, Beijing 100095, China

[*]Corresponding author's e-mail: changkuiliu621@163.com

Abstract: Oxide inclusions are commonly found in casting alloys. During vacuum investment casting of polycrystalline nickel-based superalloy, various forms of oxide inclusions can be observed, including membrane-like inclusions. These oxide inclusion films can significantly reduce the mechanical properties of nickel-based superalloys. A cast superalloy impeller that fractured due to internal oxide inclusion films was analyzed in this article. The objective of this article is to analyze the morphology and composition of oxide inclusion films and investigate their formation mechanisms. This article uses a scanning electron microscope and transmission electron microscope to characterize oxide inclusion film samples. The results show that the oxide inclusions are composed of film-like alumina, magnesium-aluminium spinel, and magnesia particles encapsulated by magnesium-aluminium spinel films. The oxide inclusion films are formed through a reaction between the liquid metal melt and the magnesia crucible. These oxide films are then floating into the nickel matrix through the scouring of the liquid metal melt and become internal inclusions.

1. Introduction

Nickel-based superalloys are important in aviation, aerospace, and weaponry fields due to their excellent high-temperature oxidation resistance, corrosion resistance, and mechanical properties [1]. This alloy makes up over 40% of the high-temperature materials used in aero-engines, including turbine blades, turbine discs, combustion chambers, and other hot-end components [2]. Moreover, aluminum is a key element in nickel-based superalloy used in aerospace engineering, which helps provide excellent fracture resistance and good creep properties at high temperatures. However, Al is also a highly active element in superalloy and significantly contributes to the generation of non-metallic inclusions [3]. Normally these kinds of inclusions cannot be eliminated even with the use of high-purity raw materials and secondary refining processes. Inclusions are typically introduced during the melting process and can be difficult to remove in later processes, such as hot isostatic pressing. These inclusions can significantly reduce the tensile strength and fatigue properties of the superalloy components [4].

Therefore, addressing the formation mechanisms of inclusions is important in the development of superalloys.

An auxiliary power unit (APU) turbine impeller ruptured into two halves during the aircraft's ground preparation, causing the impeller debris to eject out and impacting the sealing structure quickly. This resulted in the rupture of the turbine blade and a series of injurious plastic deformations of APU. The debris of the impeller was scattered on the surface of the aircraft and the ground due to the airflow. The failure analysis revealed that the turbine rupture failure was caused by the presence of large-sized aggregated oxide inclusion films in the center of the turbine impeller. This paper aims to study their formation mechanisms by characterizing the oxide inclusion films comprehensively, and then optimize the casting process of cast superalloy impeller.

2. Materials and methods

Metallographic specimens in this article are prepared by using a wire-electrode cutting machine and then subjected to the standard metallographic preparation system, which involves mounting, grinding, polishing, and etching with a DD6 superalloy corrosive agent. The impeller fracture and metallographic organization were analyzed by using a TESCAN CLARA-GMU scanning electron microscope for microscopic observation and elemental analysis. To further characterize the internal inclusions, the oxide films then were extracted by using the standard focused ion beam (FIB) method and analyzed with FEI Talos F200S transmission electron microscope after thinning.

3. Results and discussion

3.1. Inclusion film's characteristic

The fracture in the main body of the superalloy impeller is rough and light grey. From the macro fracture direction, it appears to originate from a small plane tilted along the centre and expands in all directions, shown in Figure 1 (a). The fracture of the superalloy impeller is mainly overload fracture, and the dimple and cleavage-like features can be observed on the fracture at high magnification, which converges to an inclusions aggregation area at a size of about 7.6 mm × 4.2 mm in the centre of fracture, shown in Figures 1 (b)~(d). Energy disperse spectroscopy analysis shows that the inclusions mainly contain oxides including alumina and magnesia, shown in Table 1. Brittle fracture characteristics with propagation arc and the free surface can be seen in the microscopic morphology of the centre oxide inclusions, shown in Figures 1 (e) and (f).

Figure 1. (a) Morphology of cast superalloy impeller fracture; (b)&(c) SEM observation of dimple and cleavage-like features; (d) SEM observation of fracture origin showing inclusion area; (e)&(f) SEM observation of inclusions showing brittle fracture.

Table 1. Elements composition of oxide inclusion films (weight percent).

	C	O	Mg	Al	Ca	Ti	Cr	Co	Ni	Ta
Area 1	6.37	38.97	13.02	31.97	0.67	1.41	5.09	0.30	2.20	-
Area 2	6.00	42.07	11.14	34.69	0.90	0.95	1.26	0.24	1.87	0.88
Area 3	6.37	38.97	13.02	31.97	0.67	1.41	5.09	0.30	2.20	-
Area 4	-	38.38	16.39	36.99	0.09	0.48	5.36	0.25	1.60	-

Figure 2 shows the elements distribution map of the oxide inclusion films. These inclusions are mainly in the form of a film, but the presence of encapsulated oxide particles can also be observed in some areas. A transmission electron microscope was used to analyze the phase composition of the inclusions, which was lifted out by a focused ion beam. The inclusions were mainly composed of alumina (Al_2O_3) and magnesium-aluminium spinel ($MgAl_2O_4$) phases, while the encapsulated oxide particles were identified as magnesia (MgO) phases. The selected analysis regions, bright-field images of oxide inclusion films, electron diffraction images of each phase, and calibration results are shown in Figure 3.

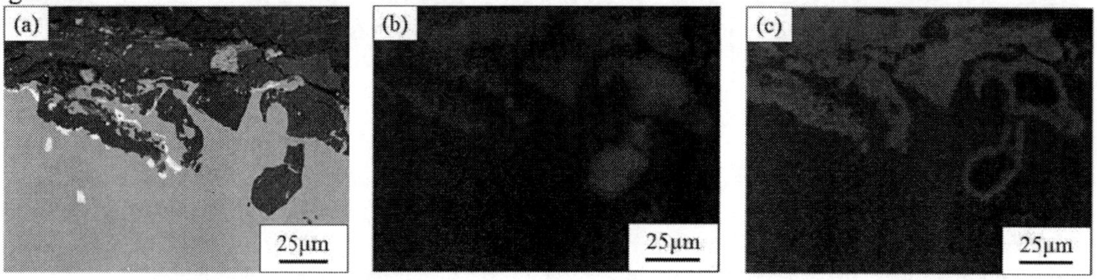

Figure 2. (a) SEM metallography of oxide inclusion films; (b) EDS map-scan of Mg element distribution; (c) EDS map-scan of Al element distribution.

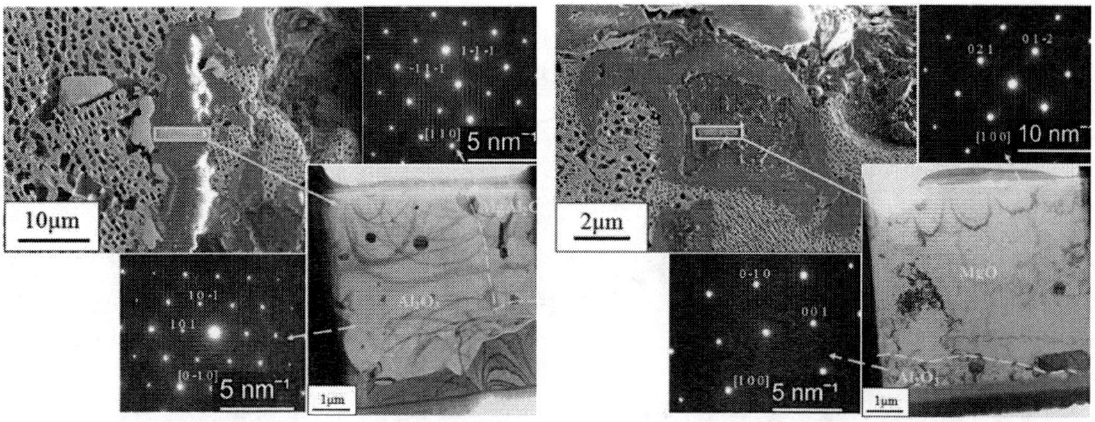

Figure 3. SEM images showing the region selected for FIB lift-out and bright-field images of oxide inclusion films, electron diffraction images, and their calibration results.

3.2. *Inclusion film formation*

From a morphological perspective, oxide inclusions are interdendritic oxide films in the cast superalloy, and a few particles encapsulated with oxide films also exist in the superalloy. From the point of view of elemental distribution, the superalloy impeller matrix does not contain magnesium, so magnesium must come from the magnesia crucible. The crucible used for casting superalloy is typically made of either magnesia or alumina [5~7]. However, the inner surface of this crucible often lacks flatness and contains numerous porosities. As a result, when liquid metal is poured into the crucible, the highly active aluminum atoms within the liquid react with the oxygen in the local crevice or porosity of the crucible, producing alumina that adheres to the surface of the crucible in the form of a film. During the further reaction, magnesium-aluminium spinel is produced as a film-like oxide that thickens and penetrates the loose crevices of the magnesia crucible [8~10]. In the casting process, the liquid metal inside the crucible carries the weakly combined oxide films and magnesia crucible particles into the mould, resulting in the formation of oxide inclusion films in the superalloy impeller centre, shown in Figure 4.

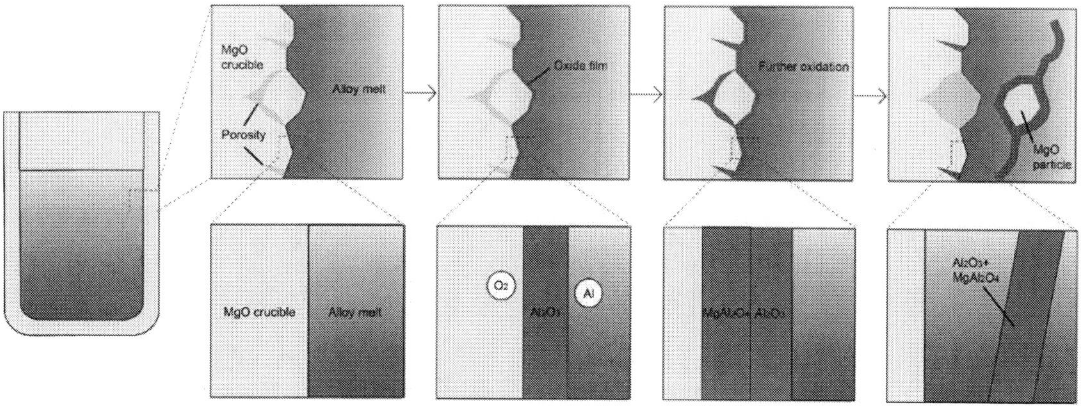

Figure 4. Formation mechanisms of oxide inclusion films.

4. Conclusion

(1) Fractures in cast superalloy impellers are related to the presence of a large area of inclusions at the core.

(2) The inclusions found in the cast superalloy impeller are primarily membrane alumina and magnesium-aluminum spinel films, and a few magnesia particles encapsulated with oxide films can also be seen in some areas.

(3) The internal inclusions of cast superalloy impeller are produced by the oxidation between liquid metal melt and crucible in the casting process. The aluminum inside the matrix first forms a thin film of alumina on the contact surface with the crucible. As the reaction continues, magnesium-aluminium spinel is formed through the reaction of alumina and magnesia crucible. This spinel fills the larger crevices in the crucible. The liquid metal then scours the oxide inclusion films and wraps the magnesia crucible particles, which eventually become internal defects in the liquid metal matrix.

References

[1] Guo J. T. (2010). Materials Science and Engineering for Superalloys. Science Press, Beijing.

[2] Wang L. Z., Li X., Liu L. K., Yang S. F., and Li J. Q. (2010). Control on composition and characteristics of non-metallic inclusions in nickel-base superalloy. J. China Metallurgy, 31 (5): 32-38, 71.

[3] Zhou X. M., Wang D. L., Wang L., and Wang W. Y. (2008). Deformation behavior of non-metallic inclusions nickel-base P/M superalloy. J. Failure Analysis and Prevention, 3 (3): 23-27.

[4] Telesman J., Gabb T. P., Kantzos P. T., Bonacuse P. J., Barrie R. L., and Kantzos C. A. (2021). Effect of a large population of seeded alumina inclusions on crack initiation and small crack fatigue crack growth in Udimet 720 nickel-base disk superalloy. J. International Journal of Fatigue, 142: 105, 953.

[5] Busch J. D., DeBARBADILLO J. J., and Krane M. J. (2013). Flux entrapment and titanium nitride defects in electro slag remelting of INCOLOY alloys 800 and 825. J. Metallurgical and Materials Transactions A., 44 (12): 5, 295-5, 303.

[6] Jeong J., Park D., Shim S., Na H., Bae G., Seo S. J., and Lee J. (2020). Prediction of behavior of alumina inclusion in front of solid-liquid interface in SPFH590 steel. J. Metallurgical and Materials Transactions B., 51 (2): 690-696.

[7] Tabatabaei Y., Coley K. S., Irons G. A., and Sun S. (2017). A multilayer model for alumina inclusion transformation by calcium in the ladle furnace. J. Metallurgical and Materials Transactions B., 49 (1): 375-387.

[8] Fuchs G. E. and Kaplan M. A. (2016). Investigation of oxide bifilms in investment cast superalloy IN100: Part I: Mechanical Properties. J. Metallurgical and Materials Transactions A., 47: 2, 346-2, 361.

[9] Kaplan M. A. and Fuchs G. E. (2016). Investigation of oxide bifilms in investment cast superalloy IN100: Part II: Characterization. J. Metallurgical and Materials Transactions A., 47 (5): 2, 362-2, 375.

[10] Gao X., Zhang L., Qu X., Luan Y., and Chen X. (2020). Investigation on the formation mechanism of non-metallic inclusions in high-aluminum and titanium-alloyed Ni-based superalloy. J. Vacuum, 177: 109, 409.

Journal of Physics: Conference Series **2808** (2024) 012017 doi:10.1088/1742-6596/2808/1/012017

Mass-controlled contact variable stiffness elastic metamaterials for bandgap tuning

Ronghao Lyu[1] and Kai Zhang[1*]

[1]School of Aerospace Engineering, Beijing Institute of Technology, Beijing 100081, China

[*]Corresponding author's e-mail: zhangkai@bit.edu.cn

Abstract. The tunability of elastic metamaterials has emerged as a focal point in recent years within the realm of metamaterial research. However, achieving extensive bandgap tuning at low frequencies and identifying appropriate control mechanisms remain challenging tasks in existing studies. In this research, we propose a novel approach based on mass-controlled contact-variable stiffness locally resonant elastic metamaterials. By altering the mass and orientation of oscillators, the contact state of the variable stiffness structure can be modified within the unit cell, resulting in significant variations in effective stiffness and enabling broad bandgap tuning at low frequencies. Furthermore, the gravitational field control method employed in this study offers simplicity, rapidity, and ease of manipulation. The materials utilized in this research are readily accessible, and the structure's simplicity facilitates ease of fabrication, thus holding promise for practical applications in scenarios such as vibration and noise reduction in ships, vehicles, buildings, and other related fields.

1. Introduction

Metamaterials, artificial periodic materials with novel properties not found in natural materials, have been in high demand in engineering fields such as aerospace and marine industries. Elastic metamaterials, through the design of microstructures within their periodic units, can create bandgaps in certain frequency ranges, thereby enabling control over elastic waves at sub-wavelength scales [1]. Based on the theories of local resonance [2] and Bragg scattering [3], the position and size of these bandgaps are closely related to the stiffness of the microstructures. Therefore, achieving simple and efficient adjustment of the microstructure stiffness in elastic metamaterials has become a focal point in the study of elastic wave control.

In 2014, Wang et al. proposed an elastic wave metamaterial with adjustable capabilities based on beam buckling-induced stiffness variation, allowing for the opening or closing of bandgaps through pre-deformation [4]. In 2018, Amendola et al. designed a prestressed variable stiffness tensile structure, which achieved theoretically arbitrary frequency tunability of elastic waves by adjusting global/local prestress to control stiffness [5]. In 2019, Gao et al. studied the bandgap characteristics of porous structures under uniaxial tension and found that stiffness changes induced by tension led to an increase in the structure's natural frequency [6]. In 2020, Li et al. introduced a magnetically controlled monostable acoustic metamaterial, where an increase in magnetic field reduced the structure's effective stiffness, resulting in a downward shift of the absorption peak frequency [7]. In 2021, Montgomery et al. proposed a magnetically controlled variable stiffness metamaterial with two deformation modes, offering extensive bandgap tuning capabilities [8]. Also in the same year, Zhou et al. demonstrated a

Journal of Physics: Conference Series **2808** (2024) 012017 doi:10.1088/1742-6596/2808/1/012017

one-dimensional adjustable quasi-zero stiffness metamaterial, elucidating the relationship between stiffness and deformation, with bandgaps effectively adjustable by changing the precompression amount, ultimately achieving quasi-zero stiffness and thus super low-frequency bandgaps [9]. In 2023, Gao et al. designed a magnetically controlled tunable ultra-low-frequency local resonance metamaterial, allowing for bandgap adjustment by changing the magnetic field and geometric parameters [10].

In this study, a local resonance-based elastic wave metamaterial is proposed. The main innovations of this research are as follows: (1) A control method based on the variation of effective stiffness within a unit cell induced by structural contact stiffness alteration, which can significantly change the position and range of bandgaps in the metamaterial; (2) Within the designed local resonance unit cell, both the size and orientation of oscillator mass play a significant role in determining the effective stiffness of the unit cell, thus expanding the role of oscillator mass in bandgap control; (3) Utilizing the gravity field as a control mechanism eliminates the need to alter the external configuration of the metamaterial or add additional devices, making the method simple, rapid, and easy to operate. In this paper, the unit cell model of metamaterials is initially proposed. Subsequently, the principles governing the manipulation of bandgaps under the influence of mass are theoretically analyzed. Following this, the dispersion curve characteristics are computed by utilizing simulation software. Finally, validation is conducted through parametric analysis and transmission curve simulations.

2. Metamaterial unit cell model

As shown in Figure 1(a), this paper proposes a two-dimensional metamaterial unit cell model with a variable stiffness structure due to contact. The model consists of three parts: local resonance oscillator made of lead or steel; contact variable stiffness structure (CVSS), consisting of two sections of thick beams and a pair of V-shaped thin beams made by 3D printing with TPU material; unit cell frame made of aluminum alloy. The dimensional parameters of the model can be referred to in Table 1, and the material parameters can be referred to in Table 2. The CVSS can undergo new contact-induced compression, changing the deformation mode from the bending of the thin beams to the compression of the thick beams, resulting in significant stiffness changes, as shown in Figure 1(b). Due to the different bearing areas and deformation modes before and after contact, the effective stiffness of CVSS undergoes a tremendous change of 440 times before and after variable stiffness, as shown in Figure 1(c). The remarkable variable stiffness characteristics of the unit cell make it possible to achieve a wide range of elastic wave bandgap tuning. To this end, two control strategies are proposed based on the effective contact conditions of the variable stiffness structure, namely, the strategy of varying the mass of oscillators and the strategy of varying the orientation of oscillator masses.

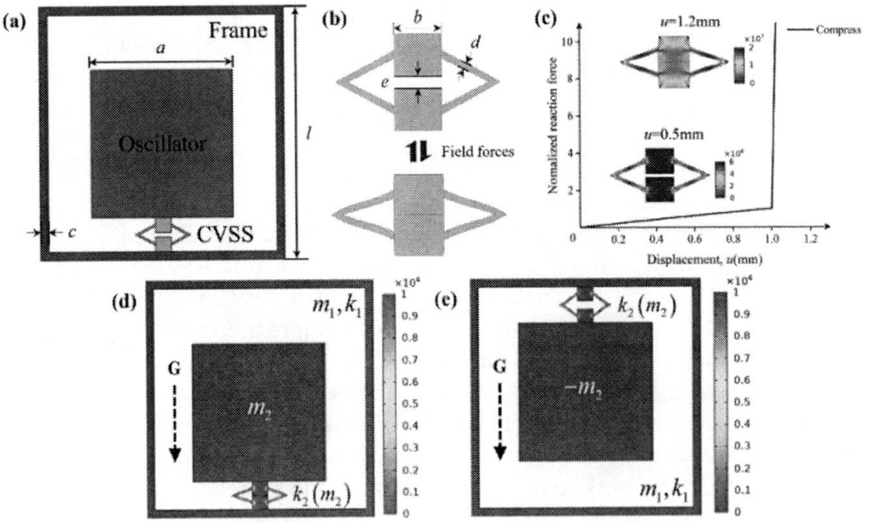

Figure 1. Single-cell metamaterial models, variable stiffness configurations, and two strategies.

Table 1. Unit cell geometric parameters.

l (mm)	a (mm)	b (mm)	c (mm)	d (mm)	e (mm)
60	35	4	2	0.5	1

Table 2. Unit cell material parameters.

Material	Density (kg/m^3)	Elastic Modulus (GPa)	Poisson's Ratio
Lead (Oscillator)	14343	16	0.40
Steel (Oscillator)	7850	180	0.30
TPU (CVSS)	1120	0.05	0.38
Aluminum Alloy (Frame)	2800	70	0.33

Based on the static structural simulation of CVSS under compression load conducted earlier, it is evident that CVSS exhibits two-stage variable stiffness characteristics upon contact. The relationship between stiffness and displacement can be expressed as follows:

$$k_2 = \begin{cases} 5.5\times10^3\left(\text{N/m}^2\right), u \le e \\ 2.5\times10^5\left(\text{N/m}^2\right), u > e \end{cases} \tag{1}$$

The load magnitude when the structure gap e decreases from 1 mm to 0 mm is $F_l = 121$ N/m. Figure 1(d) illustrates the strategy of varying the mass of the oscillator. Case 1: If the oscillator's material property is lead, then the gravitational force acting on the oscillator is $\text{G}_1 = \rho_{\text{A}1}\text{b}^2 g_{const} = 136(\text{N/m}) \ge F_l$. At this point, the relative vertical displacement produced by CVSS is $u_1 = 1.00$ mm $\ge e$, and stress near the contact surface of the coarse beam portion of CVSS is observed in the simulation simultaneously, indicating that CVSS has made contact and undergone stiffness change. Case 2: If the oscillator's material property is steel, then the gravitational force acting on the oscillator is $\text{G}_2 = \rho_{\text{A}2}\text{b}^2 g_{const} = 94(\text{N/m}) < F_l$. In this scenario, the relative displacement produced by CVSS is $u_2 = 0.97$ mm, suggesting that CVSS has not made contact, and significant stiffness variation has not occurred.

Figure 1(e) illustrates the strategy of varying the orientation of oscillator masses: In the upright position, the unit cell's condition is the same as described in the previous scenario. However, in the inverted unit cell state, the gravitational field causes the structure to stretch rather than compress. The relative displacement caused by stretching is $u_3 = 1.11$ mm. Consequently, CVSS does not generate new contact during the loading deformation process. Instead, it can reopen the contact already established in the upright position, achieving the variable stiffness effect.

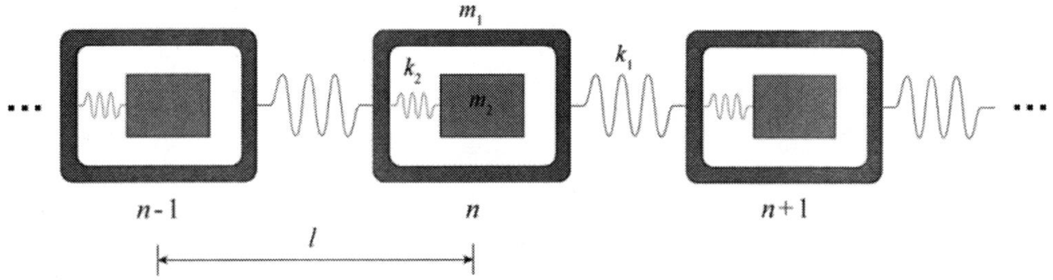

Figure 2. One-dimensional infinite mass-spring chain model.

Journal of Physics: Conference Series **2808** (2024) 012017 doi:10.1088/1742-6596/2808/1/012017

3. Local resonance theory

Elastic metamaterials can generate bandgaps through structural design, which can prevent the propagation of elastic waves within a certain frequency range. The range of the bandgap is highly correlated with the stiffness of the unit cell structure. In this paper, the relationship between the stiffness variation of the unit cell structure and the elastic wave bandgap can be explained by a one-dimensional infinite mass-spring chain model, as shown in Figure 2.

In the figure, m_1 represents the mass of the frame, m_2 denotes the mass of the oscillator, k_1 and k_2 represent the effective stiffness of the frame and the effective stiffness of CVSS, respectively. The model corresponds to the system's equations of motion:

$$\begin{gathered} -m_1\omega_1^2 u_1^j = k_1\left(u_1^{j+1} + u_1^{j-1} - 2u_1^j\right) + 2k_2\left(u_1^j - u_2^j\right) \\ -m_2\omega_2^2 u_2^j = -2k_2\left(u_2^j - u_1^j\right) \end{gathered} \tag{2}$$

where $\omega_1 = \sqrt{k_1/m_1}$, $\omega_2 = \sqrt{k_2/m_2}$.

The Bloch boundary condition is given as follows:

$$u_{n+1} = u_n e^{iql} \tag{3}$$

where q is the wave number; l is the periodic size of the unit.

Furthermore, Euler's equation is also applicable:

$$e^{i\theta} = \cos\theta + i\sin\theta \tag{4}$$

The dispersion equation for this one-dimensional periodic mass-spring system is obtained as:

$$m_1 m_2 \omega^4 - \left[(m_1 + m_2)k_2 + 2m_2 k_1\left(1 - \cos(ql)\right)\right]\omega^2 + 2k_1 k_2\left(1 - \cos(ql)\right) = 0 \tag{5}$$

When $ql = 2a\pi,\ a \in Z$, the minimum value of ω is achieved, at which point $\omega_0 = \sqrt{k_2/m_2}$. Here, ω_0 represents the lower bound frequency of the dispersion curve bandgap within a certain frequency range, also known as the starting frequency. When $ql = (2a-1)\pi,\ a \in Z$, $\omega_E = \omega_0\sqrt{1 + m_2/m_1}$, representing the upper bound frequency of the dispersion curve bandgap within a certain frequency range, also known as the cutoff frequency. Therefore, through structural design to control the stiffness, the position and range of the bandgap can be influenced, thereby achieving tunability of elastic waves.

In previous studies, it has been challenging to achieve significant variations in the bandgap by varying the mass m_2 of the oscillator, typically resulting in changes of only a few orders of magnitude. However, in this paper, from Equation (1), it can be observed that m_2 can influence the stiffness k_2 of CVSS by more than two orders of magnitude through changes in magnitude and direction. Additionally, k_2 significant changes in the position and size of the bandgap can be induced by varying the mass m_2.

4. Simulation of elastic wave bandgap control

4.1 *Band structure simulation*

To obtain the position and range of the designed elastic wave bandgap, this study utilizes multiphysics simulation software to model and simulate the periodic unit cell. The first Brillouin zone of this metamaterial is a square, enabling the study of the band structure along the path of $\Gamma \to X \to M \to \Gamma \to Y \to M$.

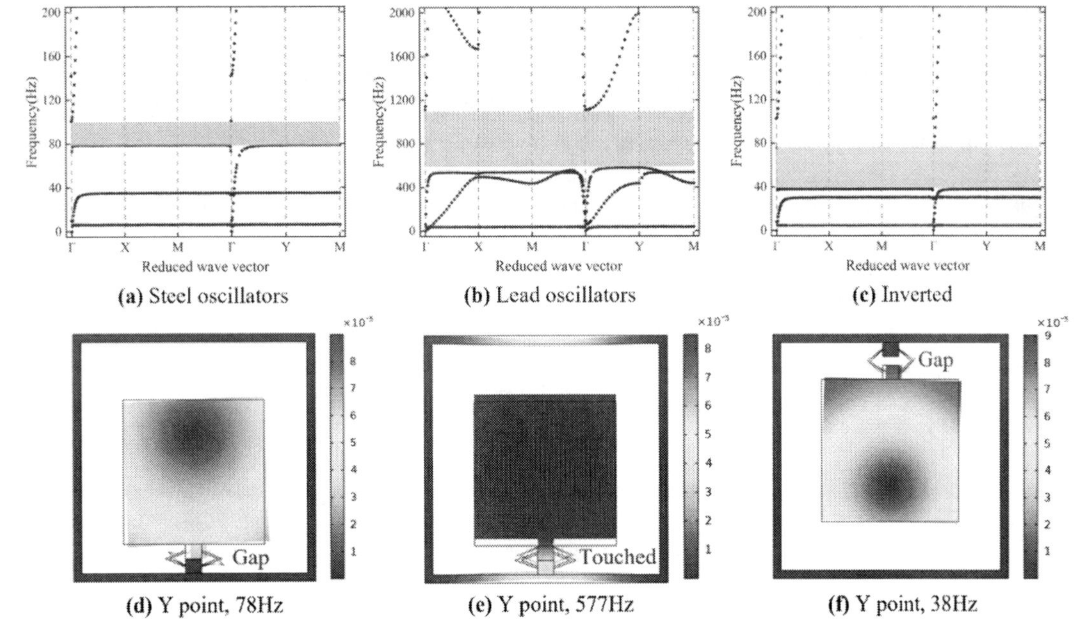

Figure 3. Dispersion curves and resonant modes under two different mass control strategies.

The bandgap range for the steel oscillator is 78 Hz to 101 Hz, falling within the extremely low-frequency range. The bandgap range for the lead oscillator is 577 Hz to 1109 Hz, representing an ultra-wide bandgap in the low-frequency state. The bandgap frequency range in the inverted state is 38 Hz to 75 Hz. The simulation results demonstrate that the stiffness variation induced by the size and orientation of the oscillator mass m_2 affecting the contact variable stiffness structure under different loading conditions can significantly impact the size and range of the bandgap formed by the metamaterial.

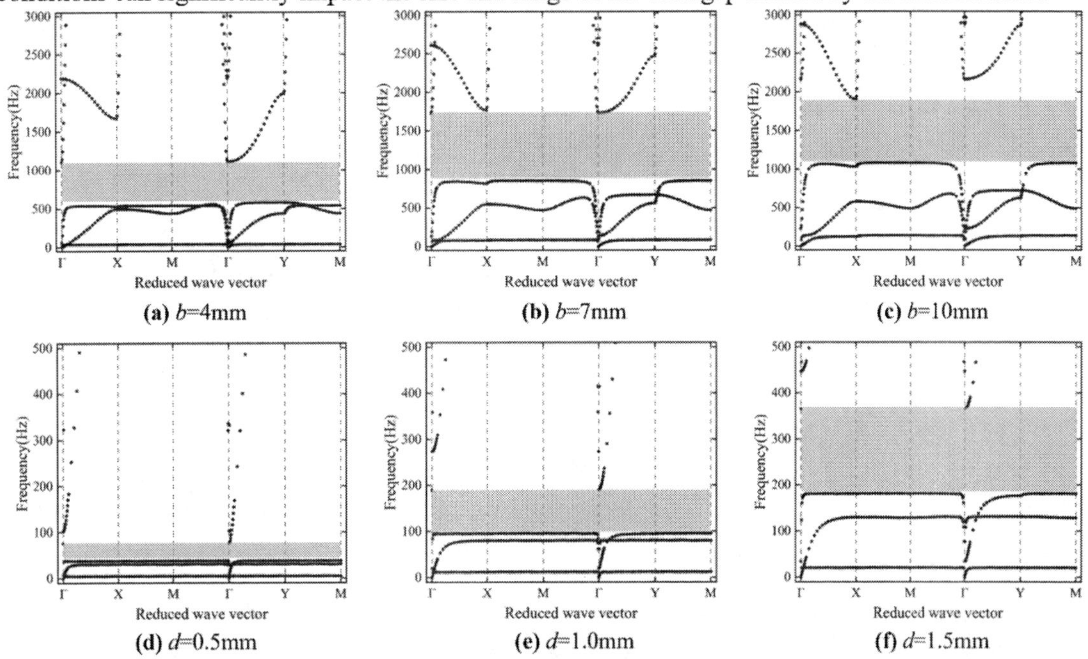

Figure 4. Dispersion curves of the metamaterial unit cell with different size parameters.

Journal of Physics: Conference Series **2808** (2024) 012017 doi:10.1088/1742-6596/2808/1/012017

4.2 Parametric simulation

To investigate the influence of different parameters of the contact variable stiffness beam on the adjustment of the bandgap range, this paper conducts a parametric analysis of the lead oscillator in both upright and inverted states with respect to the contact variable stiffness CVSS. Specifically, parameters b and d are varied as $b = 4,\ 7,\ 10$ mm, $d = 0.5,\ 1.0,\ 1.5$ mm, respectively, and the dispersion curves of the metamaterial are simulated. The results are shown in Figure 4.

In Figure 4, (a-c) plots represent the dispersion curves of the upright unit cell with different thicknesses of the thick beams, with bandgap ranges of 577 Hz-1109 Hz, 848 Hz-1729 Hz, and 1082 Hz-1905 Hz, respectively; (d-e) plots represent the dispersion curves of the inverted unit cell with different thicknesses of the thin beams, with bandgap ranges of 38 Hz-75 Hz, 95 Hz-189 Hz, and 181 Hz-365 Hz, respectively. From the bandgap variations observed in Figure 4, it can be inferred that as the thicknesses b and d of CVSS increase, the band structures shift upwards, consistent with the theoretical expectation of bandgap shift due to increased stiffness. In the case of dispersion curves in the inverted state, they shift upwards with increasing thickness of the thin beams. These results are the same as our theoretical expectations.

4.3 Transmission curve simulation

To obtain the bandgap range under actual elastic wave propagation conditions, this paper constructs a 3×3 array of metamaterial unit cells for frequency domain studies of wave propagation. Specifically, a vertical displacement of 0.1 mm is applied to the center point of the upper or lower boundary of the array structure, and the relative displacement between the center point of the other end boundary and this point is monitored. The displacement transmission curves of the array structure in both upright and inverted states with lead oscillator are obtained, as well as the displacement fields of the array structure at two selected frequencies, as shown in Figure 5.

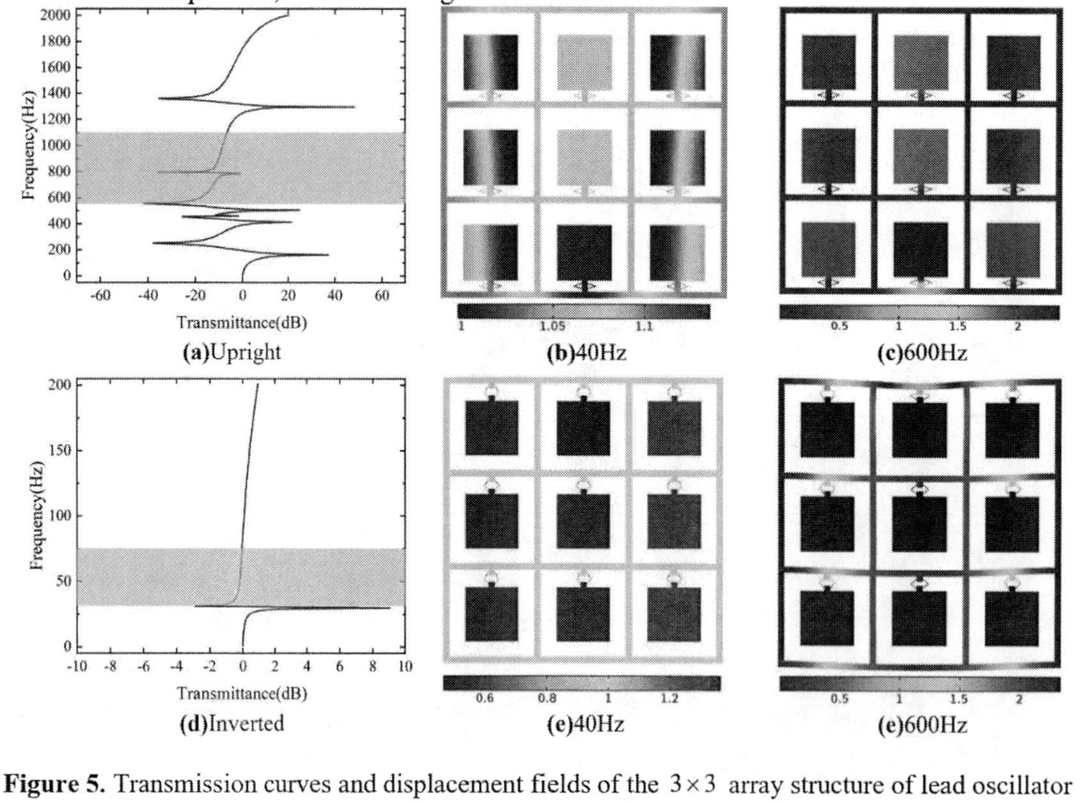

Figure 5. Transmission curves and displacement fields of the 3×3 array structure of lead oscillator unit cells in upright and inverted states.

Comparing Figure 5 with Figure 3, it is evident that the positions and sizes of the bandgaps are highly consistent, thereby validating the correctness of the previous bandgap calculations. Through simulation of elastic wave transmission at two different frequencies, it can be confirmed that control over the transmission of structures at different elastic wave frequencies can be achieved by adjusting the size and orientation of the oscillator's mass. This finding suggests further design possibilities: by assembling oscillators of different masses and orientations, control over the transmission paths of elastic waves at multiple frequencies can be achieved.

5. Conclusion

In conclusion, this study introduces a novel approach for controlling elastic waves utilizing metamaterials through manipulation of contact stiffness variation induced by discrepancies in oscillator mass size and direction. Our findings demonstrate that the designed local resonance unit cell can achieve significant contact stiffness variations, exceeding two orders of magnitude under compressive loads. Incorporating the influence of mass size and direction on stiffness variation into local resonance theory has revealed a dual impact of mass on bandgaps within our study. Furthermore, we have validated the targeted design capabilities of metamaterial bandgap positions and sizes through the combination of various geometric and material parameters. Additionally, assembly techniques enable the realization of wave transmission control for specific frequencies of elastic waves. These results underscore the potential of our method in the design of tunable elastic metamaterials, highlighting its practical applicability across diverse engineering disciplines.

References

[1] Liu, Z., Zhang, X., Mao, Y., Zhu, Y., Yang, Z., Chan, C., Sheng, P. (2000) Locally Resonant Sonic Materials. Science. 289(5485): 1734-1736. doi: 10.1126/science.289.5485.1734.

[2] Huang, H., Sun, C., Huang, G. (2009) On the negative effective mass density in acoustic metamaterials. Int. J. Eng. Sci., 47(4): 610-617. doi: 10.1016/j.ijengsci.2008.12.007.

[3] Kushwaha, M., Halevi, P., Dobrzynski, L., Djafari-Rouhani, B. (1993) Acoustic band structure of periodic elastic composites. Phys. Rev. Lett., 71(13): 2022-2025. doi: 10.1103/PhysRevLett.75.3580.

[4] Wang, P., Casadei, F., Shan, S., Weaver, J. C., Bertoldi, K. (2014) Harnessing Buckling to Design Tunable Locally Resonant Acoustic Metamaterials. Phys. Rev. Lett., 113(1): 014301. doi: 10.1103/PhysRevLett.113.014301.

[5] Amendola, A., Krushynska, A., Daraio, C., Pugno, N. M., Fraternali, F. (2018) Tuning frequency band gaps of tensegrity mass-spring chains with local and global prestress. Int. J. Solids Struct., 155: 47-56. doi: 10.1016/j.ijsolstr.2018.07.002.

[6] Gao, N., Li, J., Bao, R., Chen, W. (2019) Harnessing uniaxial tension to tune Poisson's ratio and wave propagation in soft porous phononic crystals: an experimental study. Soft Matter, 15(14): 2921-2927. doi: 10.1039/c8sm02468e.

[7] Li, X., Xing, T., Zhao, J., Gai, X. (2020) Broadband low-frequency sound absorption using a monostable acoustic metamaterial. J. Acoust. Soc. Am., 147(2): EL113-EL118. doi: 10.1121/10.0000714.

[8] Montgomery, S. M., Wu, S., Kuang, X., Armstrong, D. C., Zemelka, C., Ze, Q., Zhang, R., Zhao, R., Qi, H. J. (2021) Magneto-Mechanical Metamaterials with Widely Tunable Mechanical Properties and Acoustic Bandgaps. Adv. Funct. Mater., 31(3): 2005319. doi: 10.1002/adfm.202005319.

[9] Zhou, J., Pan, H., Cai, C., Xu, D. (2021) Tunable ultralow frequency wave attenuations in one-dimensional quasi-zero-stiffness metamaterial. Int. J. Mech. Mater. Des., 17(2): 285-300. doi: 10.1007/s10999-020-09525-7.

[10] Gao, W., Yang, B., Hong, Y., Guo, K., Sun, P., Sun, J. (2023) Investigation on the tunable low-frequency property of magnetic field induced phononic crystal with Archimedean spiral beams. Mech. Syst. Signal Pr., 185: 109756. doi: 10.1016/j.ymssp.2022.109756.

Journal of Physics: Conference Series **2808** (2024) 012018 doi:10.1088/1742-6596/2808/1/012018

Study on seismic performance of cold-formed steel frame structures based on software ABAQUS

Futing Pan[1*], Junlin Gao[2], Ying Wang[1] and Tao Wang [2]

[1]Key Laboratory of Building Structure of Anhui Higher Education Institutes, Anhui Xinhua University, Hefei, Anhui, 230088, China
[2]Urban Construction College, Anhui Xinhua University, Hefei, Anhui, 230088, China

[*]Corresponding author's e-mail: panfuting@126.com

Abstract. Cold-formed steel has many advantages such as high strength, light weight, good economy, and high construction efficiency, which is conducive to the industrialization and green development of buildings. The frame structure has many advantages, such as a flexible spatial layout, easy handling of facades, and simple construction. This study analyzes the seismic performance of cold-formed steel frame structures from three influencing factors: the floor numbers, the cross-sectional dimensions of cold-formed steel, and the strength of steel. The results indicate that the main factors affecting the seismic performance of cold-formed steel frame structures are the cross-sectional dimension and strength of the steel. The smaller the cross-sectional size of the steel is, the smaller the bending stiffness of the frame is, and the poorer the recoverability of the structure is. Under the same cross-sectional size, when the thickness of the steel is reduced by half, the horizontal bearing capacity of the structure approaches 48%. The higher the strength of the steel is, the faster its bearing capacity decreases and the worse its hysteresis performance is.

1. Introduction

Compared with hot-rolled steel, cold-formed steel has the characteristics of simple, fast, and efficient processing. At the same time, the thickness of the steel plate is small, the structural self weight is light, and the cross-section of the profile is expanded, which can form a large moment of inertia and rotation radius, and has good economic and seismic performance. Cold-formed steel structures belong to prefabricated steel structure buildings, which have many advantages such as a short construction period and high recovery rate. They have their own unique advantages in middle and low rise building construction, and are widely used in structures such as villas, low rise residential buildings, and light steel factories.

At present, the widely used cold-formed steel building structural system is a "box" load-bearing structure composed of composite wall panels, floor slabs, and roof trusses. The horizontal load of the structure is borne by the shear resistant wall, and the vertical load is borne by the columns in the load-bearing wall. The structural form can generally be applied to buildings below three floors. The box-shaped cold-formed steel building structural system is a non-frame load-bearing system with low structural bearing capacity. The cross-sectional types of load-bearing components are too single, which affects the development of this structural system towards multi-story and high-rise buildings. Many scholars have proposed new section combination methods to construct a cold-formed steel frame structure system, with the beams and columns of the combination section as the main load-bearing components, which can better meet the requirements of large openings [1-3]. Composite components are formed by assembling cold-formed thin-walled steel components and connecting steel plates using high-strength bolts. Research has shown that composite components can effectively connect the double limb C-shaped steel, and composite columns have high bearing capacity. At the same time, the bending moment transmission of the connecting nodes is reliable, the overall seismic performance of the frame is good, and all seismic indicators are good. Gao et al. analyzed the seismic performance of cold-formed steel frame buildings, conducted parameter analysis from dynamic amplification factor, maximum displacement angle between elastic-plastic floors, etc., and proposed reasonable design suggestions [4]. Ye et al. conducted experimental research on the mechanical properties of cold-formed steel frame slotted steel plate shear walls and found that the structure has a strong energy dissipation capacity and high energy dissipation safety reserve [5].Some scholars have conducted research on the performance of beams and columns with cold-formed thin-walled steel box sections, and found that their stress performance is good and suitable for engineering applications[6-7].

This study uses the finite element analysis software ABAQUS to establish a numerical simulation model of cold-formed steel frame structures. The analysis is conducted from the aspects of the floor numbers, the cross-sectional dimensions of cold-formed steel, and the strength of steel, providing a reference for the application of cold-formed steel in multi-story and high-rise light steel structures.

2. Finite element numerical simulation analysis

2.1. Structural layout of the framework

Based on the commonly used framework structure layout schemes for civil buildings, this study proposes a horizontal span of 4.8 m and 3 m for the framework structure, and a vertical span of 3 m. The structural layout diagram is shown in Figure 1. Considering the structural layout rules and uniform stress distribution, this article selects a transverse frame corresponding to the 5-axis as the calculation unit to analyze its seismic performance, as shown in Figure 2.

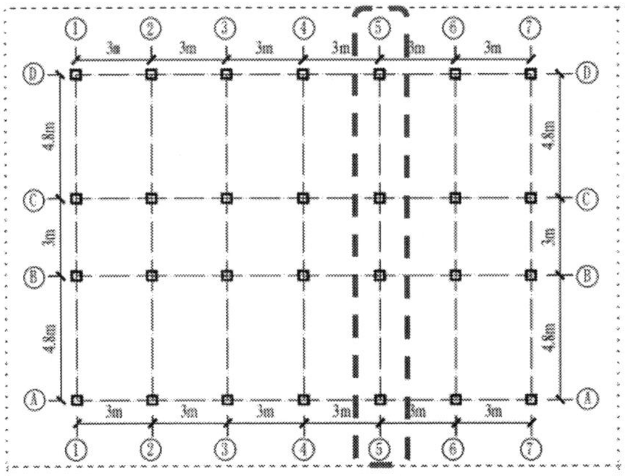

Figure 1. Plan layout of the frame structure.

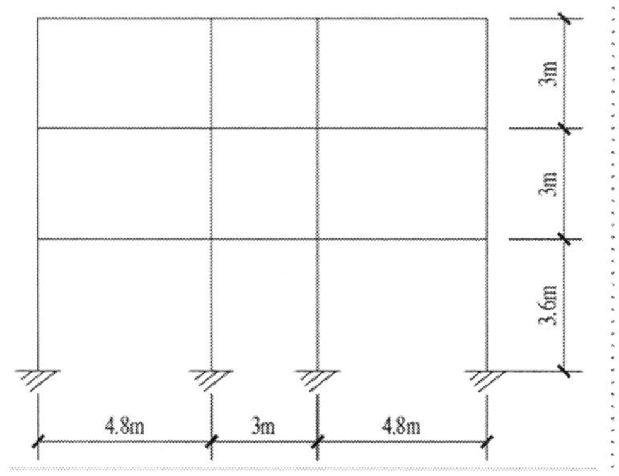

Figure 2. Calculation diagram of a horizontal frame (3 floors).

2.2. Parameter Design

2.2.1. The floor numbers. Due to the susceptibility of cold-formed steel to instability and damage under stress, it can only be used in mid to low rise buildings. The building floors designed in this study include 3 floors, 4 floors, and 5 floors, with a bottom floor height of 3.6 meters and a standard floor height of 3 meters. The calculation diagrams of the frame structure are shown in Figure 2, Figure 3, and Figure 4, respectively.

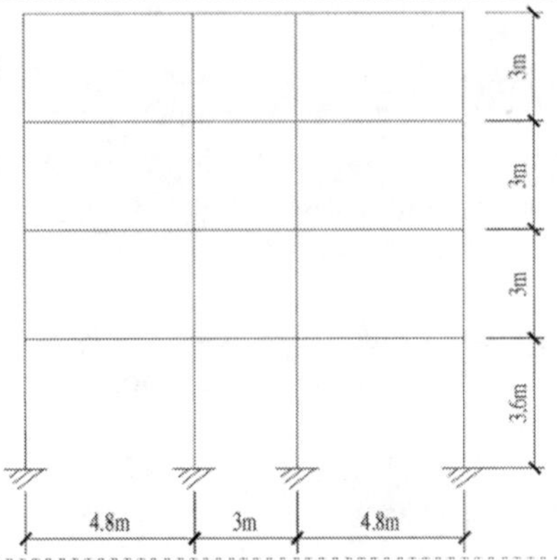

Figure 3. Calculation diagram of a horizontal frame (4 floors).

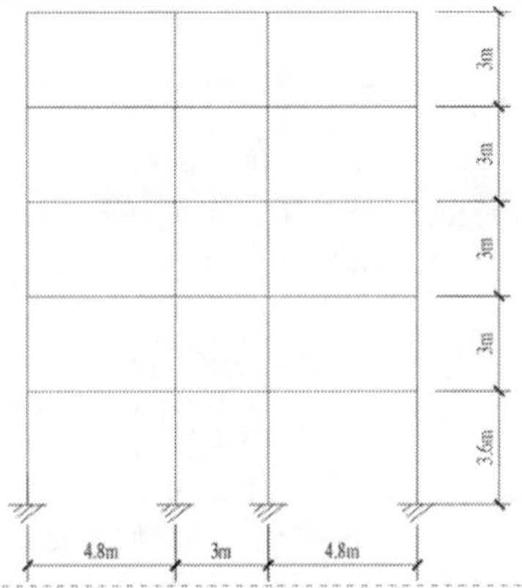

Figure 4. Calculation diagram of a horizontal frame (5 floors).

2.2.2. Section dimensions of cold-formed steel. Most cold-formed steel wall systems use C-shaped steel and U-shaped steel, but the frame structure system designed in this study uses cold-formed thin-walled square steel pipes for beams and columns, with dimensions divided into Size 1 (200 mm × 200 mm × 6 mm), as shown in Figures 5. The other two sizes are Size 2 (200 mm × 200 mm × 3 mm) and Size 3 (140 mm × 140 mm × 6 mm).

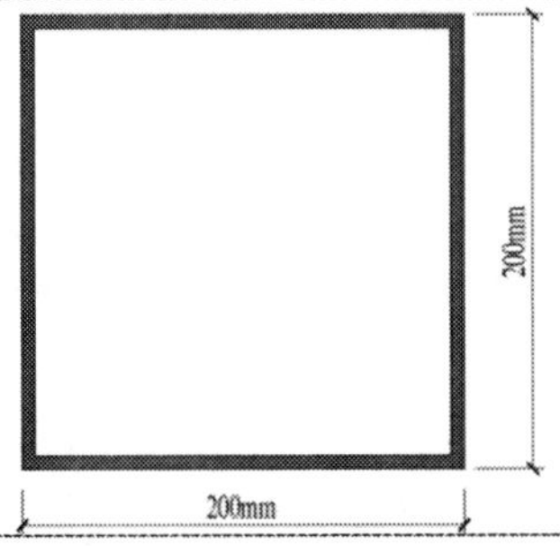

Figure 5. Cold formed thin-walled square steel pipes (Size 1-t = 6 mm).

2.2.3. Strength of cold-formed steel. The strength of steel has a significant impact on the load-bearing performance, and for cold-formed steel that is prone to buckling, whether the strength of high-strength steel can be fully utilized is also an important factor affecting the structural bearing capacity. Therefore, in this study, the strength of cold-formed steel is taken as Q235, Q345, and Q550. The material characteristic parameters of the section steel are shown in Table 1.

Table 1. The material characteristics of steel.

	Elastic modulus (N/mm^2)	Poisson's ratio	Yield strength (N/mm^2)
Q235	2.10×10^5	0.3	235
Q345	2.10×10^5	0.3	345
Q550	2.10×10^5	0.3	550

2.3. Low cycle repeated loading system

Cold-formed steel beams and steel columns are simulated using beam elements. The lower end of the bottom frame column is fixed with constraints, and horizontal displacement is applied at the top edge column node. The displacement controlled loading system is used. Before yielding, it is divided into three levels, $\pm 0.25\Delta y$, $\pm 0.5\Delta y$, $\pm 0.75\Delta y$, $\pm 1\Delta y$, and each level is divided into three cycles of displacement. After yielding, it is divided into five levels, $\pm 2\Delta y$, $\pm 3\Delta y$, and each level is divided into two cycles.

2.4. Numerical simulation analysis results

2.4.1. The failure mode of the structure. Considering the lateral constraint support effect of the floor and roof slabs on the frame beam, the elastic modulus of the beam is relatively large in numerical analysis, so the deformation under horizontal load is concentrated in the frame columns. After repeated low cycle loading, the deformation diagram of the 5-story frame is shown in Figure 6. The stress at the beam end is relatively high, which meets the design concept of strong columns and weak beams in structural design.

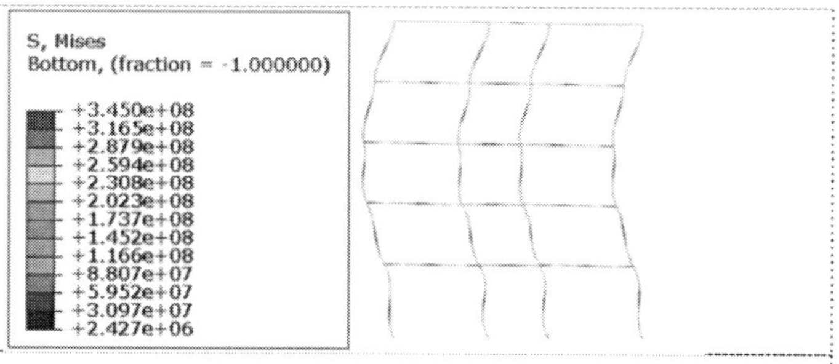

Figure 6. Deformation and stress diagram of a 5-story building.

2.4.2. Hysteresis performance analysis. When the strength of the steel is taken as Q345 and the cross-sectional size of the steel is taken as Size 1, stress analysis was conducted on frame structures with different levels, and the analysis results are shown in Figure 7. Under the action of horizontal low cycle reciprocating loads, the hysteresis loops of the three sets of frame models exhibit a shuttle shape, with a full shape of hysteresis curves and strong plastic deformation ability of the structure. They have good seismic performance and energy dissipation capacity, and there is no significant stiffness degradation in the later stage.

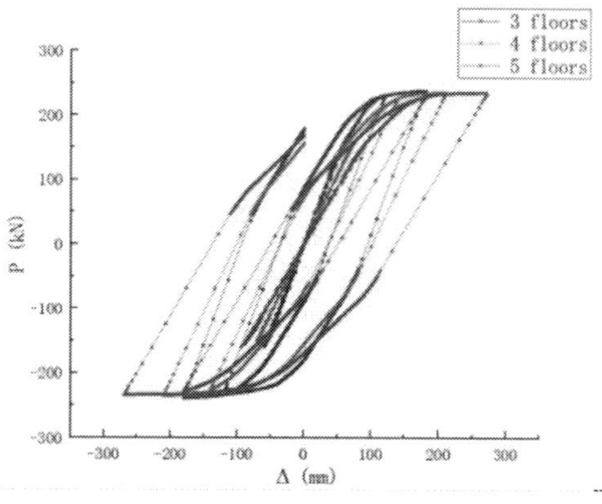

Figure 7. Hysteresis curves for different floors (Q345, Size 1).

When the steel strength is taken as Q345 and the frame structure has 5 floors, seismic performance analysis was conducted on frame structures with different steel section sizes. The analysis results are shown in Figure 8. Under the action of horizontal low cycle reciprocating loads, the hysteresis loops of the three frame models show a shuttle shape. The bearing capacity of the Size 2 and Size 3 frame models is significantly lower than that of the Size 1 frame model, indicating that for cold-formed steel frames, the section size of the steel has a significant impact on the bearing capacity, mainly due to the lateral displacement of the frame structure and the overall instability and local buckling of the steel. The horizontal bearing capacity of the frame structure of Size 2 decreased by 47.96% compared to that of Size 1, while the horizontal bearing capacity of the frame structure of Size 3 decreased by 40.18% compared to that of Size 1.

For a frame structure with 5 floors, the cross-sectional size of the steel section is taken as Size 1, a stress analysis was conducted on frame structures with different steel strengths, and the analysis results are shown in Figure 9. Under the action of horizontal low cycle reciprocating loads, the lower the strength of the steel, the fuller the hysteresis loop, the stronger the plastic deformation capacity of the structure, and the better the seismic performance and energy dissipation capacity. The horizontal bearing capacity of the frame structure using Q345 steel is 102.95% higher than that of Q235, and the horizontal bearing capacity of the frame structure using Q550 steel is 218.59% higher than that of Q235.

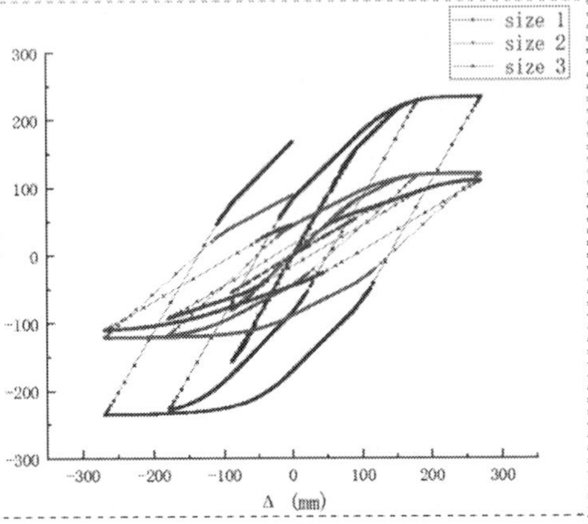

Figure 8. Hysteresis curves of different steel dimensions (Q345, 5 floors).

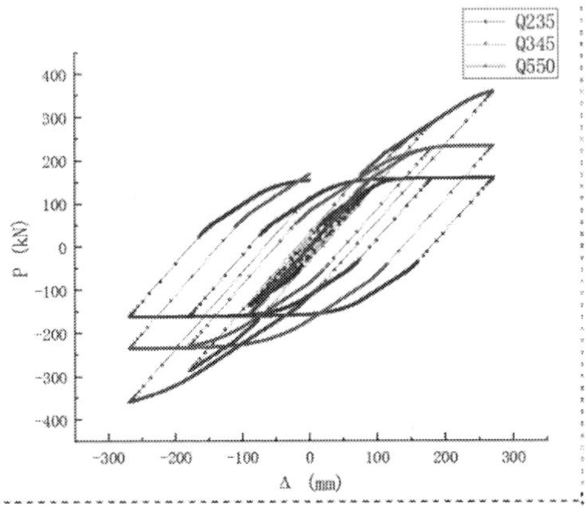

Figure 9. Hysteresis curves of different steel strengths (Size 1, 5 floors).

3. Conclusion

The failure mode of the frame is overall bending instability, mainly concentrated at the beam end and the first floor column base. For ordinary strength cold-formed steel frame structures, the seismic performance is good, the bearing capacity degradation is stable, and it has good energy dissipation capacity.

The main factors affecting the energy dissipation capacity of cold-formed steel frame structures are the strength and cross-sectional dimensions of the steel sections. When the strength of the steel is low and the cross-sectional size is large, the energy dissipation capacity of the frame structure is strong. It can be seen that for cold-formed steel, the high strength of the steel is often not fully utilized due to instability.

In order to fully utilize the strength of steel, it is necessary to strengthen the instability of cold-formed steel components and overall design, in order to achieve the use of light steel structures in mid to high-rise buildings and accelerate the industrialization and green development of the construction industry.

Acknowledgments

The research described in this paper was sponsored by Anhui Xinhua University Campus Quality Engineering Project (2022jy001), Anhui Provincial Quality Engineering Project (2021jyxm0607), the key projects of Key Laboratory of Building Structure of Anhui Higher Education Institutes (Anhui Xinhua University) (KLBSZD202004 and KLBSZD202007), the Natural Science Research Funds for universities of Anhui Province (KJ2020A0795 and KJ2019A0885) and Anhui Province College Student Innovation and Entrepreneurship Project (S202312216037).

References

[1] Chen, M., Huang, J.H., Zhao, G.T. (2016) Research progress of compound section cold-formed thin-wall steel structures. Engineering mechanics., 33: 1–11.
[2] Chen, M., Li, M.Q., Deng, B.F. (2015) A study on seismic behavior of cold-formed thin-wall steel frames., Progress in steel building strucres., 17: 7–12, 51.
[3] Chu, Y.P., Zhong, Y., Luo, P. (2022) Research on seismic performance design method of cold-formed steel buildings based on damage., Industrial Construction., 52: 50–55, 90.
[4] Gao, H.W., Yao, Y., Chu, Y.P., Cheng, D.G., Yang, X.P. (2014) Research progress of compound section cold-formed thin-wall steel structures. Sichuan building science., 40: 196–199, 203.
[5] Ye, L., Wang, Y.H., Shi, Y., Luo, W., Kong, W.B. (2020) Study on the mechanical properties of cold-formed steel framed shear wall with slits. Engineering mechanics., 37: 156–166.
[6] Whittle J., Ramseyer C. (2009) Buckling capacities of axially loaded,cold-formed,built-up C-channels.Thin-Walled Structures., 47: 190—201.
[7] Vijayanand S., Anbarasu M. (2017) Effect of spacers on ultimate strength and behavior of cold-formed steel built-up columns .Procedia Engineering., 173: 1423—1430.

Plasticity constitutive theory considering material length parameters

Zihan Gong [1*], Gaoxiao Han [1], Lidan Li [2], Hao Chen [1] and Wengui Zhang [1]

[1] Lanzhou Jiaotong University, Lanzhou 730070, Gansu, China
[2] Beijing Jiaotong University, Beijing 100044, China

[*]Corresponding author's e-mail: gongzihan369@163.com

Abstract: The traditional continuous medium theory introduces the homogenization assumption that the material remains constant from the macroscopic to the microscopic view, which has been successfully applied to the analysis of the macroscopic mechanical properties. When the dimensions are reduced to the microscopic view, the internal defects of the material start to appear, leading to the inhomogeneity of the material properties, which is, in practice, manifested as a ruler effect. Therefore, it is necessary to introduce the material length parameter into the structural theory to model the mechanical response of new materials. Based on the theory of size effect, many scholars have carried out a large number of studies. The most widely used theories are strain gradient theory and differential nonlocal model, mainly the first strain gradient theory, the second strain gradient theory, and the simplified strain gradient theory. Some scholars define it from the kinetic point of view, but most of these consider the intrinsic relationship of elastic materials. To further investigate the plasticity intrinsic theory, some scholars have proposed the gradient plasticity theory, the nonlocal plasticity theory, and so on. In this paper, based on the previous research results, we briefly summarize the development and outlook of the plasticity eigenstructure theory under the consideration of the length parameter of the material. Then, we derive the plasticity eigenstructure relation equation, the full-volume theoretical model, and the yield criterion corresponding to the Mises material under the consideration of the endowment size of the material from the perspective of the gradient theory of plasticity, and finally, put forward a new plasticity eigenstructure theory-higher-order nonlocal gradient theory. A new plasticity constitutive theory, the higher-order nonlocal gradient theory, is finally proposed, and the defining equations and their constitutive relations are derived in detail. The proposed theory is intended to provide a theoretical basis for analyzing the microdefects in materials.

1. Introduction

The strain gradient theory was first applied to microelectromechanical systems (MEMS), to solve the problem of the size effect which cannot be predicted in the classical theory. This theory was gradually applied to quasi-brittle materials such as rock and soil and concrete, by defining the intrinsic relationship by introducing the strain energy density.

The theory is widely applied to elastic materials, but it is inconvenient in the process of application due to the existence of numerous material length parameters and the relationship between each parameter. Fleck et al. proposed an expression for the strain energy density containing five material parameters, assuming incompressibility of the material in 1997, which was simplified by Lam et al. in 2003. Fleck introduced the decomposition of the Smyshlyaev-Fleck strain gradient tensor in 1993, assuming incompressible material, which was first decomposed into a biased part and a hydrostatic part. The hydrostatic part was decomposed into three parts orthogonal to each other. Zhou Shenjie and Li Anqing decomposed compressible solids into a fully symmetric part and an antisymmetric part, and then the symmetric part was decomposed into two parts with and without traces. The antisymmetric part is decomposed into two parts with symmetric and antisymmetric rotation gradients based on the symmetry of the rotation gradient.

All of the above representations introduce higher-order strain gradients into the constitutive relation, which increases the complexity of the boundary conditions and the constitutive relation. Chen and Wang [4] defined the concept of strain gradient, which is introduced as an internal variable in the plastic hardening relation, thus avoiding the introduction of higher-order stresses into the constitutive relation. Eringen [1] in the 1970s and 1980s proposed the nonlocal theory, also known as the integral theory based on nonlocal strains. Unlike the classical elasticity theory, which assumes that the stress at a point depends only on the strain at that point, the nonlocal theory assumes that the stress at a point should depend on the strains at all points within the elastomer. Based on this assumption, Eringen defined its intrinsic relationship by introducing a weight function, which has various forms. Thus, the nonlocal theory has various forms, the differential form of the nonlocal theory, the integral form of the nonlocal theory [2], and the two-phase nonlocal theory, which takes into account both the local and the nonlocal stresses. Lim et al. (2015) proposed the nonlocal strain gradient theory and the higher-order nonlocal strain gradient theory, which takes into account both nonlocal and strain gradient effects. The stresses in small-scale materials can be explained not only by the nonlocal stress field but also by the pure strain gradient stress field. The effects of the two length scales on the mechanical behaviors can be considered by combining local and nonlocal curvature in the intrinsic relationship.

In recent years, Zhang and Gao et al. proposed a simplified strain gradient theory in 2020, in which only two parameters describing the material scale are included, and the strain energy density is expressed as a function of the strain tensor ε_{ij}, the curvature tensor χ_{ij}, and the symmetric part of the second-order displacement gradient tensor η^{s}_{ijk}.

Another scholar simplified the expression of strain energy density proposed by Eringen in 1999 and proposed a simplified strain gradient containing four parameters. In 2023, Lazar [3] assumed from the kinetic theory that the potential and kinetic energy densities are not only a function of strain and velocity but also a function of their gradients at a fixed point in a certain time, and obtained the elasticity theory of the eigenstructure equations. In 2022, Chen and Wang [4] defined their eigenstructure equations through the strain energy density.

Compared with the development of the elasticity principal theory, the plasticity principal theory considering the length parameter of the material has also been rapidly developed in recent years, and the nonlocal plasticity theory, gradient plasticity theory, etc. have been successively proposed [5]. Sarva Ramini et al. developed the nonlocal plasticity theory in 2019, which is different from the lack of the internal features of the material reflecting the material's fine-scale properties in the traditional plasticity theory finding. The nonlocal plasticity theory delocalizes the material localization variables and indirectly considers the changes and interactions in the internal microstructure of the material scale. Based on the nonlocal plasticity theory, the nonlocalized variables of the material indirectly consider the changes and interactions of the internal microstructure of the material. Rawat et al. [6] proposed a nonlocal plastic damage model for quasi-brittle materials in 2020 and used the method of geometric analysis [9] to avoid the problems of computational accuracy and computational efficiency that would occur during the process of finite element analysis. The application of strain gradient theory in beam bending was proposed by Li et al. [7] in 2022.

Based on the above development of elastic and plastic constitutive theories considering material parameters, this paper will apply the gradient plasticity theory to establish its elastic-plastic constitutive equations and propose a higher-order nonlocal gradient plasticity theory to study the plastic constitutive relations of the theory. To ensure the correctness of the theory, the intrinsic relationships presented in this paper will be applied to specific materials in Abaqus numerical simulations to verify the theory.

2. Theory of gradient plasticity

2.1 Constitutive equation

According to the expression for the body free energy defined by Emilio Martínez-Pañeda and others in metallic crack extension:

$$w(\varepsilon_{ij}^e, \varepsilon_{ij,k}^p) = \frac{1}{2}\varepsilon_{ij}^e C_{ijkl}\varepsilon_{kl}^e + \frac{1}{2}\mu L^2 \varepsilon_{ij,k}^p \varepsilon_{ij,k}^p \quad (1)$$

where C_{ijkl} is the isotropic elastic principal array; μ is the shear modulus; L is the internal length of the material. Introducing the relationship between Stress σ_{ij} and Strain ε_{ij}^e, and higher-order stress η_{ijk}^e and higher-order strain $\varepsilon_{ij,k}^p$, we can obtain:

$$\sigma_{ij} = \frac{\partial w}{\partial \varepsilon_{ij}^e}, \eta_{ijk}^e = \frac{\partial w}{\partial_{ij,k}^p} \quad (2)$$

The expression for the intrinsic relationship can be obtained as:

$$\sigma_{ij} = C_{ijkl}(\varepsilon_{kl} - \varepsilon_{kl}^p)$$
$$\eta_{ijk}^e = \mu L^2 \varepsilon_{ij,k}^p \quad (3)$$

We define the plasticity coefficient as:

$$W^p = \Sigma E^p \quad (4)$$

where Σ is the effective stress; E^p is the effective plastic stroke.

$$E^P = \sqrt{\frac{2}{3}\varepsilon_{ij}^p \varepsilon_{ij}^p + L^2 \varepsilon_{ij,k}^p \varepsilon_{ij,k}^p} \quad (5)$$

Thus, Equation (4) can be expressed as:

$$\delta W^p = \Sigma \delta E^p = q_{ij}^D \delta \varepsilon_{ij}^p + \eta_{ijk}^D \delta \varepsilon_{ij,k}^p \quad (6)$$

The intrinsic relationship can be obtained as:

$$q_{ij}^D = \frac{2}{3}\frac{\Sigma}{E^p}\varepsilon_{ij}^p, \eta_{ijk}^D = \frac{\Sigma}{E^p} L^2 \varepsilon_{ij,k}^p \quad (7)$$

Bringing Equation (7) into Equation (5), we give the effective stress as:

$$\Sigma = \sqrt{\frac{3}{2} q_{ij}^D q_{ij}^D + \frac{1}{L^2}\eta_{ijk}^D \eta_{ijk}^D} \quad (8)$$

2.2Full volume theory of plasticity

Ilyushin, in 1943, proposed an elastoplastic intrinsic theory of hardened materials under small deformations which introduced the following basic assumptions:

(1) Volume changes elastically with no plastic volumetric strain. Therefore, according to the generalized Hooke's law, the volumetric stress is proportional to the volumetric strain:

$$\varepsilon_v = \frac{1-2v}{E}\sigma_v \quad (9)$$

(2) The plastic strain tensor is proportional to the bias stress tensor $\varepsilon_{ij}^p = \eta s_{ij}$, where η is the scaling factor.

(3) There is a one-to-one single-valued functional relationship between the equivalent stress $\overline{\sigma}$ and the equivalent strain $\overline{\varepsilon}$, i.e., $\overline{\sigma} = \overline{\sigma}(\overline{\varepsilon})$ or $\overline{\varepsilon} = \overline{\varepsilon}(\overline{\sigma})$.

According to the above assumptions, for almost incompressible materials, the stress-strain relationship of the full volume theory can be finally expressed as:

$$e_{ij} = \frac{3}{2}\frac{1}{E_s(\overline{\sigma})} s_{ij}, \varepsilon_v = \frac{1-2v}{E}\sigma_v \quad (10)$$

where $E_s(\overline{\sigma})$ is the cut-line modulus of the $\overline{\sigma} \sim \overline{\varepsilon}$-relationship curve, which can be obtained from the uniaxial tensile test [11].

From the above assumptions, the stress-strain relationship of the full volume theory is finally expressed as:

$$\sigma_v = 3K\varepsilon_v, s_{ij} = \frac{2}{3}E_s(\overline{\sigma})e_{ij} \quad (11)$$

where Kdenotes bulk modulus and $K = \frac{E}{3(1-2v)}$.

Bringing in the expression yields the plastic allometric eigenstructure relation under this theory expressed as:

$$\sigma_v = 3K\varepsilon_v, s_{ij} = \frac{2}{3}\left(\frac{\Sigma_p}{\Sigma_p^0}\right)^{\frac{1}{n}} \Sigma_0 \frac{e_{ij}}{\Sigma_p} \tag{12}$$

2.3 Yield criterion

The yield criterion is the discriminant condition to judge whether the material enters into yielding or not. The length parameter of the material is introduced in the gradient plasticity theory to derive the Mises yield criterion of the gradient plasticity theory, and the plasticity isomorphism relationship is derived. The Mises yield criterion is simple and practical, which indicates that a point of an object yields when the distortion energy corresponding to the stress state at that point reaches a certain value. Its expression is given by: $f = \sqrt{3J_2}$.

Aifantis introduced the concept of nonlocalization and proposed a theory of gradient plasticity, which takes into account the influence of strain points in the point domain of the material on the stress state at that point. After a series of refinements [8], the basic framework of the traditional continuum mechanics is retained and a nonlocal amount of equivalent plastic strain is introduced only in the yield function, which can be represented as:

$$\sigma = \sigma_0 + h\langle\varepsilon^p\rangle \tag{13}$$

where $\langle\varepsilon^p\rangle$ represents the nonlocal equivalent plastic strain.

$$\langle\varepsilon^p\rangle = \varepsilon^p + l^2\nabla^2\varepsilon^p \tag{14}$$

Therefore, the Mises yield criterion can be written as:

$$f = \sqrt{3J_2} - \sigma = \sqrt{3J_2} - \sigma_0 - h(\varepsilon^p + l^2\nabla^2\varepsilon^p) \tag{15}$$

Matrix P is introduced based on the relationship between the bias stress and the second invariant:

$$P = \begin{bmatrix} \frac{2}{3} & -\frac{1}{3} & -\frac{1}{3} & 0 & 0 & 0 \\ -\frac{1}{3} & \frac{2}{3} & -\frac{1}{3} & 0 & 0 & 0 \\ -\frac{1}{3} & -\frac{1}{3} & \frac{2}{3} & 0 & 0 & 0 \\ 0 & 0 & 0 & 2 & 0 & 0 \\ 0 & 0 & 0 & 0 & 2 & 0 \\ 0 & 0 & 0 & 0 & 0 & 2 \end{bmatrix} \tag{16}$$

The yield function can be further expressed as:

$$f = \left(\frac{2}{3}\sigma^T[P]\sigma\right)^{\frac{1}{2}} - \sigma_0 - h\varepsilon^p - l^2\nabla^2\varepsilon^p \tag{17}$$

Guidelines for plastic associated flow are:

$$\varepsilon_{ij}^p = \lambda\frac{\partial g}{\partial\sigma_{ij}}$$

where g is the plastic potential function. The plastic intrinsic relationship generally taken as $g = f$ is called the flow criterion associated with the yield function, and then the plastic strain can be expressed as:

$$\varepsilon_{ij}^p = \lambda\frac{\partial f}{\partial\sigma_{ij}} = \lambda\frac{3[P]\sigma}{2\left(\frac{3}{2}\sigma^T[P]\sigma\right)^{\frac{1}{2}}} \tag{18}$$

The elastic-plastic constitutive relation is:

$$\begin{gathered} \tau_{ij} = C_{ijkl}^e\varepsilon_{kl}^e \\ \varepsilon_{ij} = \varepsilon_{ij}^e + \varepsilon_{ij}^p \end{gathered} \tag{19}$$

Bringing in the expression yields:

$$\sigma_{ij} = C_{ijkl}^e\left(\varepsilon_{ij} - \lambda\frac{\partial f}{\partial\sigma_{ij}}\right) \tag{20}$$

3. Two-phase gradient plasticity theory

Based on the two-phase intrinsic model with localized and nonlocalized effects proposed by Eringen in 1978, and the gradient plasticity theory, this paper proposes to establish a new theory, i.e., the two-phase gradient plasticity theory that takes into account the nonlocalized effects and the gradient effects at the same time [10]. Its body-free energy can be expressed as:

$$w\left(\varepsilon_{ij}^{e};\varepsilon_{ij,m}^{p},\varepsilon_{ij,m}^{p\prime},\alpha\right)=\frac{1}{2}\xi_1\varepsilon_{ij}^{e}C_{ijkl}\varepsilon_{ij}^{e}+\frac{l^2}{2}\varepsilon_{ij,m}^{p}C_{ijkl}\int_v\xi_2\alpha(|x-x'|,\alpha)\varepsilon_{ij,m}^{p}\,dV \tag{21}$$

where ξ_1 and ξ_2 denote the local and nonlocal parameters of the model, respectively; α denotes the non-local parameters and α = ea, where e is the internal length parameter of the material and a is the constant associated with the material; l is the strain gradient parameter; $\alpha(|x-x'|,\alpha)$ is the weighting function, which is generally of three forms.

The total stress tensor can be expressed as:

$$t_{ij}=\sigma_{ij}-\sigma_{ij}^{*} \tag{22}$$

where σ_{ij} is the stress tensor and $\sigma_{ij}=\xi_1C_{ijkl}\varepsilon_{ki}$, and σ_{ij}^{*} is higher-order stress tensors.

$$\sigma_{ij}^{*}=\int_v\xi_2\alpha(|x-x',ea|)\varepsilon_{ij,m}^{p}\,dV \tag{23}$$

The differential eigen structure equation can be obtained as:

$$\xi_1C_{ijkl}\varepsilon_{kl}-\xi_2l^2C_{ijkl}\varepsilon_{kl,mm}^{p}=(1-(ea)^2\nabla^2) \tag{24}$$

The above theories consider localized, nonlocalized, and strain gradient theories, which are developments of the strain gradient theory and the nonlocalized gradient theory [12]. A higher-order nonlocal theory that considers strain gradient effects was first proposed by Lin, which achieves a single theory that can characterize both stiffness-hardening and stiffness-softening effects. Therefore, it is also regarded as a combination of the nonlocal theory and the strain gradient theory. The two-phase gradient plasticity theory proposed in this paper contains both the strain gradient parameter and the nonlocal parameter. If the nonlocal parameter is larger than the strain gradient parameter, the material will exhibit stiffening softening; if the nonlocal parameter is smaller than the strain gradient parameter, the material will exhibit stiffening hardening.

4. Conclusions

By introducing the length parameter of the material into the elastic-plastic constitutive relationship, it can effectively deal with the size effect caused by micro-defects. At the same time, considering the influence of nonlocal parameters, it can influence the degree of the mechanical properties of the material under the two theories.

Although the existing nonlocal plasticity theory and gradient plasticity theory gradually tend to mature, there is still the complexity of the study. The theory is not uniform, so the determination of the length of the material parameters does not have a clear physical meaning, not forming a unified theory of determination. In engineering applications, we often use microscopic experiments to calibrate the parameter. For the finite element simulation of the actual problem, mesh delineation of the degree of fineness will seriously affect the calculation of the efficiency and accuracy, complex boundary conditions, and the effect of the material's mechanical properties.

For the finite element simulation of practical problems, the fine degree of mesh division will seriously affect the efficiency and accuracy of the calculation and the establishment of complex boundary conditions, so some areas need to be improved. Based on the above research content, from the theoretical level, this paper derives a new constitutive relationship to solve the problem of size effect caused by material micro-defects. The experimental and numerical simulation of the data support still needs a lot of research.

References

[1] Eringen A. C. Nonlocal polar elastic continua [J]. International Journal of Engineering Science 1972, 10: 1-16.

Journal of Physics: Conference Series **2808** (2024) 012019 doi:10.1088/1742-6596/2808/1/012019

[2] Fernadez Saez J., Zaera R., Loya J. A., and Reddy J. N. Bending of Euler-Bernouli beams using Eringen's integral formulation: A paradox resolved [J]. International Journal of Engineering Science 2016, 99: 107-116.

[3] M. Lazar. Reduced strain gradient elasticity model with two characteristic lengths: fundamentals and application to straight dislocations. Continuum Mech. Thermodyn. (2022) 34:1, 433-1. 454.

[4] Chen S. H. and Wang T. C. A new hardening law for strain gradient plasticity. Acta Material, 2000, 48 (16): 3, 997-4, 005.

[5] Gortsas D. G. and Aggelis D. Polyzos. The strain gradient elasticity via nonlocal considerations. Solid and Structures, 2023, 269: 112-117.

[6] Rawat P., Raghu A., and Rajagopal M. Hossain. Nonlocal plasticity-based damage modeling in quasi-brittle materials using an Isogeometric Approach. Engineering Computations, 10.1108/EC-12-2019-0562.

[7] Anqing Li, Qing Wang, Ming Song, et al. On Strain Gradient Theory and Its Application in Bending of Beam. Computational Physics, 2022, 12 (9): 1, 304

[8] Zbib H. M. and Aifantis E. C. A gradient-dependent model for the Portevin-Le Chatelier effect. Scripta Materialia, 1988, 22: 1, 331-1, 336.

[9] Shunhui Yin, Zhibing Xiao, jingang Liu, et al. Variational Formulations and Isogeometric Analysis of Timoshenko-Ehrenfest Microbeam Using a Reformulated Strain Gradient Elasticity Theory. Crystals, 2022, 12 (6), 752.

[10] Markus Lazar. Reduced strain gradient elasticity model with two characteristic lengths: fundamentals and application to straight dislocations. Continuum Mech. Thermadyn. (2022), 34: 1, 433-1, 454.

[11] Lidiia Nazarenko, Lidiia Nazarenko, and Lidiia Nazarenko. The variational principle in coupled strain-gradient elasticity. Mathematics and Mechanics of Solids, 2022, 27 (10): 2, 256-2, 274.

[12] Ansari R., Oskouie F. M., Roghani M., et al. Nonlinear analysis of laminated FG-GPLRC beams resting on an elastic foundation based on the two-phase stress-driven nonlocal model [J]. Acta Mechanica, 2021, 232 (6): 1-17.

Journal of Physics: Conference Series **2808** (2024) 012020 doi:10.1088/1742-6596/2808/1/012020

Buckling characteristics of the multi-crack egg-shaped shell under axial pressure

Hao Lv [1], Yongmei Zhu [1*] and Suzhou Zhang [1]

[1]Jiangsu University of Science and Technology, Zhenjiang, Jiangsu 212003, China

[*]Corresponding author's e-mail: zymjskjdx@163.com

Abstract. This study focused on the buckling characteristics of egg-shaped shells with single cracks and double cracks under axial pressure. First, the geometric parameters of the egg-shaped shell were designed, and a numerical model of the egg-shaped shell was established. Then, the initial crack was introduced into the equatorial weld of the egg-shaped shell, and the effect of the crack on the buckling characteristics under different wall thicknesses was explored, as well as the influence of crack angle and crack spacing on buckling characteristics under double cracks. The results indicate that there are three buckling modes of cracked egg-shaped shells, and the transition of these three modes is influenced by the length of the crack. When the egg-shaped shell contains double cracks, the impact on the ultimate load is greater when the two cracks are parallel. When the distance between the centers of two cracks exceeds a certain distance, it can be treated as a single-crack situation. The research results have practical reference value for the design of egg-shaped shells subjected to axial loads.

1. Introduction

The pile legs bear the full weight of the offshore platform and are prone to fatigue cracks under the influence of sea winds and ocean currents, which affects its bearing capacity and stability [1]. To improve its axial bearing capacity without changing the integrity of the shell, it can be optimized by changing the shape of the sidewall of the shell [2]. The multi-egg junction structure proposed by Zhang et al. [3] removes the ends with larger curvature without compromising safety, improves space occupancy, and makes it more streamlined and aesthetically pleasing. Zhou et al. [4] studied the axial compression buckling characteristics of multi-egg joint-shaped pile legs by using a segmented modular design method and proved that multi-egg joint-shaped pile legs have smaller mass and better stability.

Nowadays, many rotary shells are manufactured by using nonfilm bulging technology, but during the bulging process of segmented prefabricated parts, crack initiation is prone to occur at the weld seam [5]. Considering the harshness and complexity of offshore working environments, it is particularly important to study the buckling modes of cracked egg-shaped shell structures.

This paper takes the egg-shaped shell formed by nonfilm bulging after welding multiple conical platforms as the research object. A part of the tip and blunt end is cut off to simulate the shape of the end face of the multi-egg intersecting pile leg. An initial crack is introduced at the equatorial weld of the egg-shaped shell, and compared with the non-cracked egg-shaped shell to study the influence of single crack and multiple cracks on the ultimate load and buckling mode of the egg-shaped shell.

2. Numerical modeling of an egg-shaped shell

The egg-shaped shell is a closed rotating shell, and many scholars have studied the structure and mechanical properties of the egg-shaped shell. Among them, Zhang et al. [6] conducted biomimetic research on goose eggs and proposed to derive the contour curve function based on the contour characteristics of goose eggs, and verified that the N-R equation is an ideal pressure shell contour curve function, as shown in Equation (1):

$$\begin{Bmatrix} r = (L^{\frac{2}{n+1}} x^{\frac{2n}{n+1}} - x^2)^{1/2} \\ n = 1.057(L/B)^{2.372} \end{Bmatrix} \tag{1}$$

where n is the shape index of the pressure-resistant shell; L is the long axis of the egg-shaped shell; B is the short axis of the egg-shaped shell; $SI=B/L$ is the egg-shaped coefficient of the egg-shaped shell, which controls the shape of the shell by its numerical value. The size parameters of the egg-shaped shell in this paper are shown in Table 1. The egg-shaped shell is made of Q235 steel, and its mechanical performance parameters are shown in Table 2.

Table 1. Size parameters of the egg-shaped shell.

Parameter	Value
Egg-shaped shell long axis L (mm)	256
Egg-shaped shell short axis B (mm)	178
Egg shape coefficient SI	0.69
Egg-shaped shell thickness t (mm)	0.9

Table 2. Material parameters of Q235 steel.

Elastic modulus E (GPa)	Yield strength σ_s (MPa)	Poisson ratio ν
200	206.8	0.3

The egg-shaped pressure-resistant shell designed in this paper adopts a multi-stage cone welding to form a prefabricated body. If there are multiple segments, the effect of egg shape will be better, but the number of welds will also increase. The uneven stress distribution is prone to occur during bulging, which can cause stress concentration and reduce the safety of the structure. If there are fewer segments, the egg-shaped effect will be reduced. Therefore, an egg-shaped shell structure consisting of six sections of conical table welding combination is selected, and its two-dimensional model is shown in Figure 1. It is divided into six regions from the tip to the blunt end, and the numbers 1-6 in the figure represent one conical table region, respectively. After internal pressure bulging, it can achieve a basic overlap with the ideal egg-shaped shell model curve. We cut off the tip and blunt end by 28 mm each, keeping the upper and lower ends flush for axial compression. The red line indicates the cutting position of the tip and blunt end.

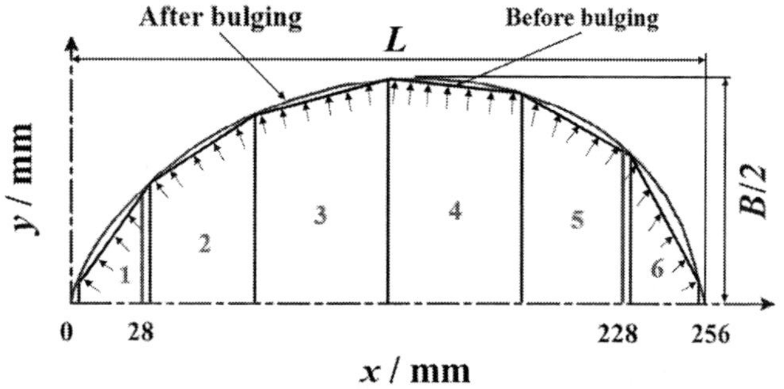

Figure 1. Egg-shaped shell geometry.

Journal of Physics: Conference Series **2808** (2024) 012020 doi:10.1088/1742-6596/2808/1/012020

Based on the dimensions and material parameters of the aforementioned egg-shaped shell, we establish numerical models for crack-free egg-shaped shell and cracked egg-shaped shell, respectively. The crack tip is divided into a three-ring element ring to divide the grid, and pyramid and tetrahedral elements are used for transition away from the crack tip, with a total of 35601 grid elements divided. The lower-end face adopts a fully fixed constraint, while the upper-end face applies a fully coupled constraint to the end face by defining a reference point and a vertical downward axial load to the reference point. The finite element model of a cracked egg-shaped shell is shown in Figure 2.

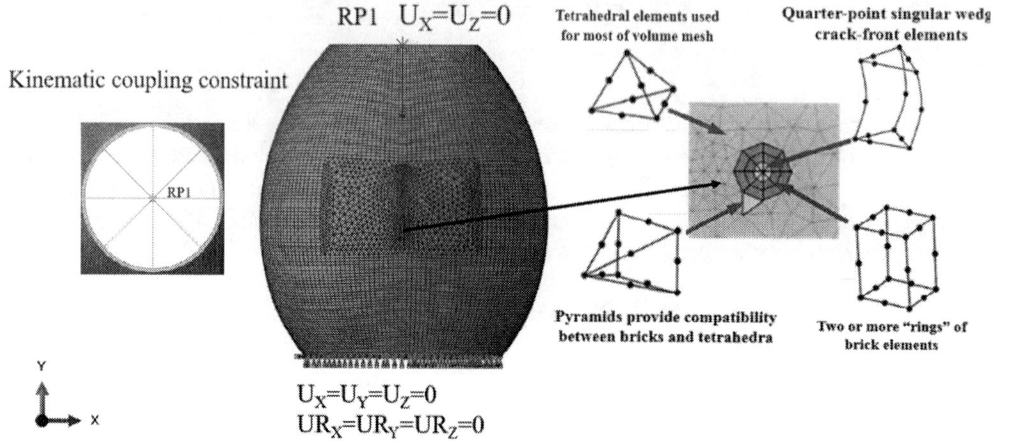

Figure 2. Finite element model of the cracked egg-shaped shell.

3. Numerical results and analysis

We perform eigenvalue buckling calculations on the egg-shaped shell to obtain the buckling shape and ultimate load of the egg-shaped shell, where $\gamma = F_C/F_t$, γ is the residual strength coefficient of the egg-shaped shell, F_c is the buckling load of the egg-shaped shell with cracks, and F_t is the buckling load of the crack-free egg-shaped shell.

3.1. *Influence of cracks on buckling characteristics under different wall thicknesses*

We select three different sets of *t/B* to study the variation of the ultimate load of cracked egg-shaped shells with wall thickness.

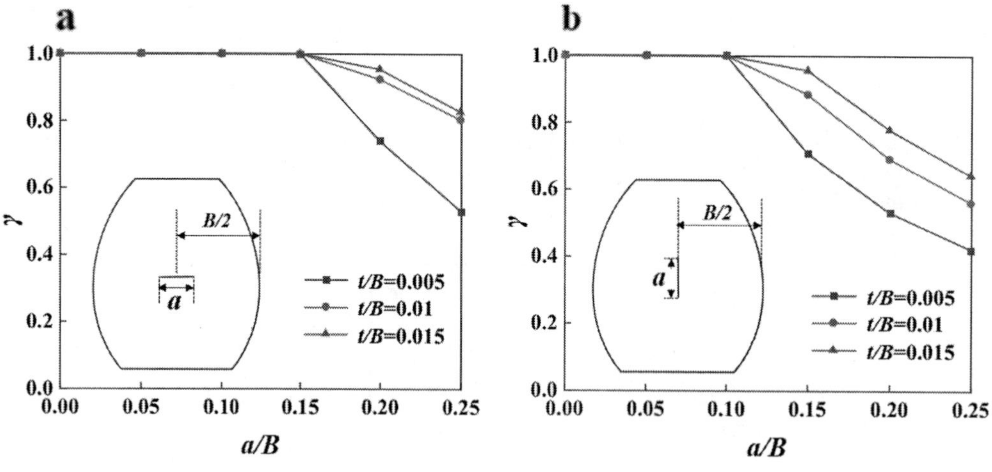

Figure 3. Buckling loads of single-crack eggshells with different wall thicknesses. (a) circumferential crack; (b) longitudinally crack.

Journal of Physics: Conference Series **2808** (2024) 012020 doi:10.1088/1742-6596/2808/1/012020

Figure 3 shows the variation of the ultimate load of the egg-shaped shell with crack length under different wall thicknesses, including a single circumferential ($\alpha = 0°$) and axial ($\alpha = 90°$) crack, where a is the crack length and B is the short axis length of the egg-shaped shell. The results indicate that when the crack length is short, the crack has no significant effect on the ultimate load of the egg-shaped shell, and the residual strength coefficient of the egg-shaped shell is $\gamma \approx 1$. As the crack length gradually increases, the residual strength coefficient of the egg-shaped shell decreases continuously, at which point $\gamma < 1$. For two types of cracks in different directions, both exhibit a thinner egg-shaped shell; as the crack length increases, its residual strength coefficient decays faster. Compared to circumferential cracks, axial cracks have a greater impact on the ultimate load of egg-shaped shells.

To study the transformation of the egg-shaped shells from upper end to local crack buckling with increasing crack length, three different buckling modes were determined:

(1) Upper-end buckling: the buckling mode and load are basically the same as those of the crack-free egg-shaped shell ($0.95 < \gamma < 1$);

(2) Transition: cracks affect the buckling load and buckling mode of the shell, but the buckling position is not localized. The buckling deformation covers the shell between the upper end and the crack. Depending on the thickness of the shell, the buckling load can be as low as 70% of the crack-free egg-shaped shell.

(3) Local crack buckling: the buckling position is localized, buckling deformation occurs only near the crack, and the buckling load can be as low as 40% of the crack-free egg-shaped shell.

The modal deformation diagram of the egg-shaped shell is shown in Figure 4. For an egg-shaped shell without cracks, the buckling position is concentrated at the upper end of the egg-shaped shell, and the buckling mode is similar to that of a cylindrical shell, both of which are wrapped in waves at the buckling position of the shell, as shown in Figure 4 (a). As the crack length increases, the buckling position begins to change from the upper end of the egg-shaped shell towards the vicinity of the crack, and the buckling deformation position exists at the upper end of the crack and egg-shaped shell, as shown in Figure 4 (b). As the crack length further increases, the buckling deformation position only exists near the crack, characterized by an inward depression at the crack tip, as shown in Figure 4 (c).

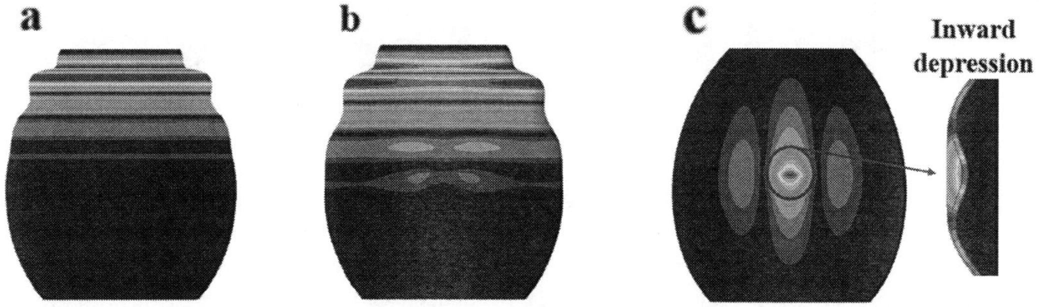

Figure 4. Buckling mode displacement nephogram. (a) upper-end buckling; (b) transition buckling; (c) local crack buckling.

3.2. *Influence of double crack angle on buckling characteristics*

In practical engineering applications, cracks often do not exist alone, and their location and distance have a high degree of randomness. In this section, two equally long cracks are added to the egg-shaped shell to study the changes in the buckling characteristics of the egg-shaped shell under multiple crack conditions. We select a cracked egg-shaped shell model with t/B=0.005 and d/B=0.2 to calculate the effect of double cracks on the buckling load of egg-shaped shells under different crack angles, where β is the angle between two cracks and d is the distance between the centers of the two cracks. Figure 5 shows the ultimate load of an egg-shaped shell with cracks at different angles.

Journal of Physics: Conference Series **2808** (2024) 012020 doi:10.1088/1742-6596/2808/1/012020

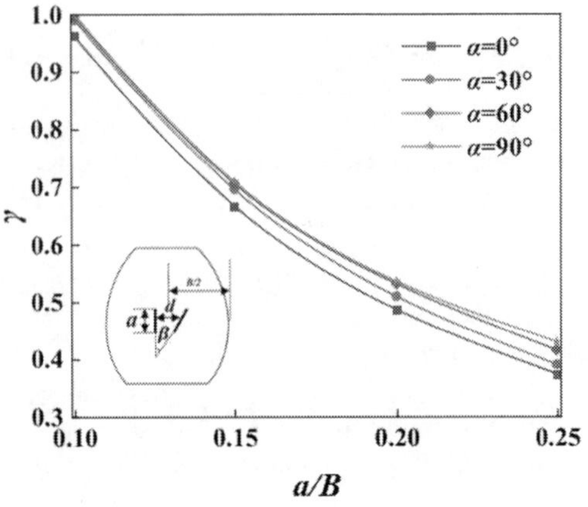

Figure 5. Effect of different angles on buckling load under double crack.

From Figure 5, it can be seen that when the two cracks are in a vertical relationship ($\beta = 90°$), the decrease in ultimate load is the smallest compared to the non-cracked egg-shaped shell. Its buckling mode is shown in Figure 6 (a). As the crack angle decreases, the interaction between the two crack tips will gradually lead to a decrease in the ultimate load of the egg-shaped shell. When both cracks are axial ($\beta = 0°$), the interaction between cracks is most pronounced, and their buckling modes are shown in Figure 6 (d). In summary, when two cracks exist parallel to the axial direction on the egg-shaped shell, their ultimate load decreases the most.

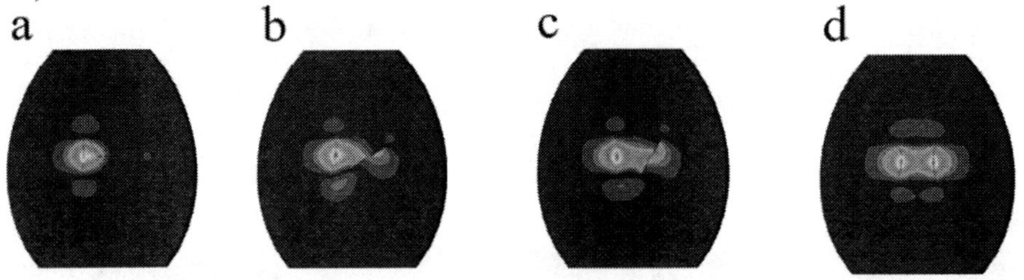

Figure 6. Buckling modes at different crack angles. (a) $\beta = 90°$; (b) $\beta = 60°$; (c) $\beta = 30°$; (d) $\beta = 0°$.

3.3. *Influence of double crack spacing on buckling characteristics*

When there are multiple cracks in the shell, the impact on the ultimate load of the egg-shaped shell is greatest when the crack direction is parallel. Therefore, this section takes the egg-shaped shell with t/B=0.005 and a/B=0.1 as an example to study the influence of crack spacing on its buckling load.

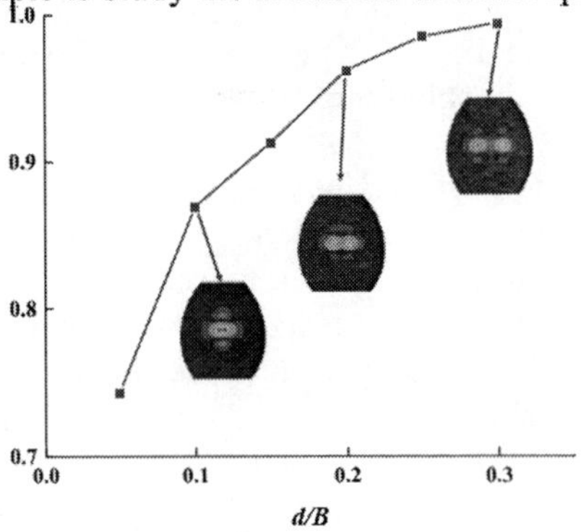

Figure 7. Effect of crack distance on buckling load.

From Figure 7, it can be seen that when there are multiple cracks with different distances in the shell, the buckling load decreases in different proportions, and the closer the two cracks are, the greater the proportion of the decrease is. This is because when the crack spacing is close, the interaction between the cracks is obvious, leading to a sudden decrease in their ultimate load. However, as the distance increases, the interaction between the cracks weakens. When the crack spacing exceeds a certain distance (d/B>0.3), the interaction between the cracks no longer exists. The buckling load is close to that of a single crack when there are double cracks. At this point, the axial compressive ultimate load of the egg-shaped shell can be considered under the condition of a single crack.

4. Conclusion

This paper focuses on the uneven stress distribution at the weld toe of the egg-shaped shell after nonfilm bulging, which leads to stress concentration and crack initiation. The influence of cracks on the buckling shape and ultimate load of the egg-shaped shell is studied, and the following conclusions are finally drawn:

(1) Compared to circumferential cracks, axial cracks have a greater impact on the ultimate bearing capacity of egg-shaped shells containing a single crack.

(2) For egg-shaped shells with cracks, their buckling modes have three states: upper-end buckling, transition buckling, and local crack buckling. When local crack buckling occurs, the maximum decrease in ultimate load occurs, and the length of the crack transformed between each two buckling modes approximately linearly varies with the thickness of the egg-shaped shell.

(3) For an egg-shaped shell with multiple cracks, its buckling characteristics are not only affected by a single crack but also by the interaction between cracks. When both cracks are in the axial direction, the maximum impact on the ultimate load of the egg-shaped shell is greatest, and the farther the cracks are apart, the smaller the interaction between the cracks is. When d/B>0.3, it can be considered as the axial compressive ultimate load of egg-shaped shells under single crack conditions.

Acknowledgments

This paper is financially supported by the National Natural Science Foundation of China [Grant numbers: 52271277], the National College Students ' Innovation and Entrepreneurship Development Program of China [Grant numbers: 202310289008Z]

References

[1] Mahanta, R. (2021). Effect of Deployment of Jack-Up on Piles and Structure of Existing Offshore platform—A Case Study. *Lecture Notes in Civil Engineering*, 134:641-652.
[2] Zhang, J., Wang, F., Wang, F., Zhao, X.L., Tang, W.X., & Chen, F.L. (2021). Buckling properties of bulged barrels under external pressure. *Thin Walled Structures*, 168(3):108226.
[3] Zhang, J., Zuo, X.L., Wang, W.B., & Tang, W.X. (2016). Bionic investigation on multiple intersec ting egg-shaped pressure hulls. *China Mechanical Engineering*, 27(19):2668-2674.
[4] Zhou, T., Tang, W.X., Wang, W.B., & Zhang, J. (2018). Buckling of axially compressed composite egg-shaped shells. *Journal of Ship Mechanics*, 22(02):174-188.
[5] Paolo, L. (2023). Stress intensity factors of elliptical cracks at the weld toe. *Theoretical and Applied Fracture Mechanics*, 128: 104147.
[6] Zhang, J., Wang, M.L., Wang, W.B., Tang, W.X., & Zhu, Y.M. (2017). Investigation on egg-shaped pressure hulls. *Marine Structures*, 52:50-66.

Analysis of the reasons for the fracture of the heat-engine plant turbine vane of the power plant feedwater pump

Jun Gao[1] and Liming Xie[1,*]

[1]Inner Mongolia Electric Power Research Institute Branch, Hohhot, Inner Mongolia, 010010, China

[*]Corresponding author's e-mail:489296683@qq.com

Abstract. The turbine of a 350MW thermal power unit's feedwater pump in the Mengxi region experienced blade fracture and failure during operation. To identify the cause of blade fracture, comprehensive testing and analysis were conducted on the fractured blade using experimental methods such as macroscopic analysis, chemical make-up analysis, mechanical property testing, microstructure testing, and SEM micro area analysis of the fracture surface. The results indicate that the inner shoulder of the first tooth of the blade root is at a 90° right angle and belongs to the weak part of stress concentration. The steam of this unit contains Cl^-, which poses a significant risk to blade materials. Corrosive Cl^- accumulates at the inner arc shoulder of the blade root long and forms tensile stress together with the static and dynamic loads during blade operation, leading to the initiation of SCC microcracks in this area. Under the long-term cyclic excitation stress generated by the high-speed rotation of the rotor, the source of stress corrosion cracks propagates in a fatigued manner, resulting in severe fracture of the blade.

1. Introduction

Vane is an important component for energy conversion in power plant steam turbines. The working conditions of steam turbine vane are harsh, and they undergo rapid rotation in high-pressure steam environments, bearing considerable stress. At the same time, they also need to transmit the torque generated by dynamic steam, and the stress situation is complex. Power plant steam turbines have multi-stage vanes, each with multiple vanes. If the blades of the feedwater pump turbine malfunction or even break, it can cause the feedwater system to shut down, ultimately leading to the shutdown of the unit [1].

On August 13, 2023, the third stage moving blade of the feedwater pump turbine of Unit 2 of a 350MW thermal power plant in the Mengxi region broke during operation. The tile installed on the turbine bearing has been in service for less than a year and is considered to have broken early in service. The feed pump turbine is a TGQ14/6-1 type, single cylinder, dual source, impulse condensing turbine produced by Beijing Electric Power Equipment Factory. The maximum continuous power is 14MW, the exhaust pressure is rated at 4.7-34kPa, and the continuous operating speed range is 3000r/min to 6000r/min. The blade material is 2Cr12NiMo1W1V [2][3].

Based on the operation and on-site maintenance, this article analyzes the causes of blade fracture in the feedwater pump turbine and advises on how to prevent the occurrence of similar failure events and ensure continuous and reliable operation of the unit.

2. Testing and analysis

2.1 Macro-observation and analysis

Perform macroscopic morphology inspection on the fractured vane. The broken vane is the third stage of moving the vane of the pneumatic feedwater pump. There are a total of 108 vanes installed in this stage, with a blade length of 136.58mm and a fungal root type. The fracture site of the broken blade is located at the right angle shoulder of the first tooth of the blade root. The appearance of the blades has not been damaged by unloading. The blade has a flat surface and no obvious plastic deformation. There are obvious fatigue bands visible on the fracture surface [4].

In addition, there is a crack defect on the blade adjacent to the broken blade. The crack is located at the right angle shoulder of the first tooth inner arc of the blade root, with a length of about 20mm. It extends vertically along the blade body from the inner arc shoulder of the blade root, as shown in Figure 1.

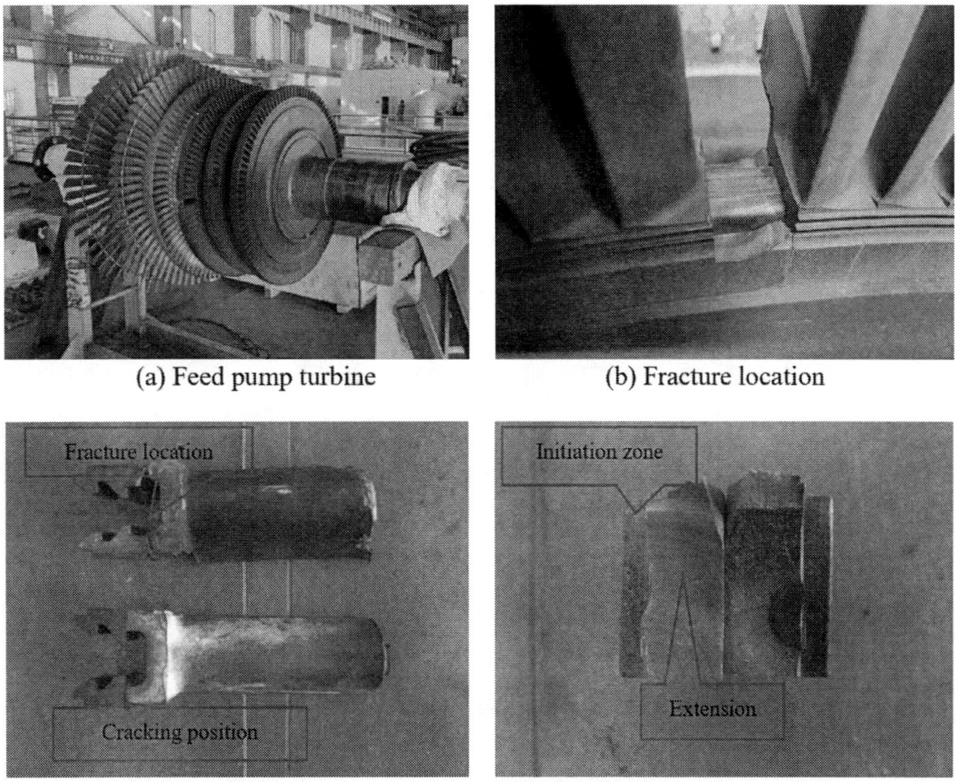

(a) Feed pump turbine (b) Fracture location

(c) The whole (d) Broken blade fracture surface

Figure 1. Macro morphology of fractured vane.

2.2 Detection and analysis of blade fracture morphology

Microscopic inspection and analysis of the fracture surface of fractured blades. There are multiple corrosion pits at the right angle shoulder of the first tooth of the blade root, which continue to the surface of the fracture. The initial fracture zone is formed along the stress concentration area of the corrosion pit due to the action of excited vibration stress. The extended zone can clearly observe fatigue bands with fatigue fracture characteristics, as shown in Figure 2 [5].

Journal of Physics: Conference Series **2808** (2024) 012021 doi:10.1088/1742-6596/2808/1/012021

(a) Corrosion pit

(b) Expansion area (low magnification)

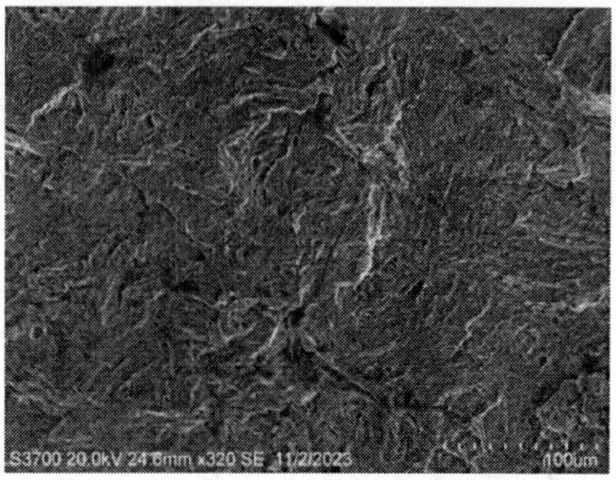

(c) Expansion area (high magnification)

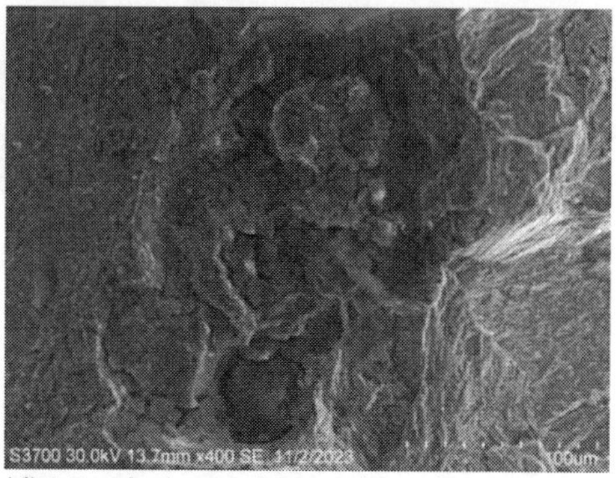

(d) Morphology of corrosion pits

Figure 2. The SEM morphology of the port.

2.3 Metallographic structure detection and analysis

Axio Observer is used to perform microstructure analysis on fractured leaves using an Alm-type metallographic microscope. The metallographic structure of the fractured feedwater pump turbine blade is tempered martensite. No abnormalities were found in organizational status, and microcracks with self-corrosion pit propagation can be seen near the fracture surface. No other abnormalities are observed in the structure. As shown in Figure 3 [6].

(a) Cracks in corrosion pits

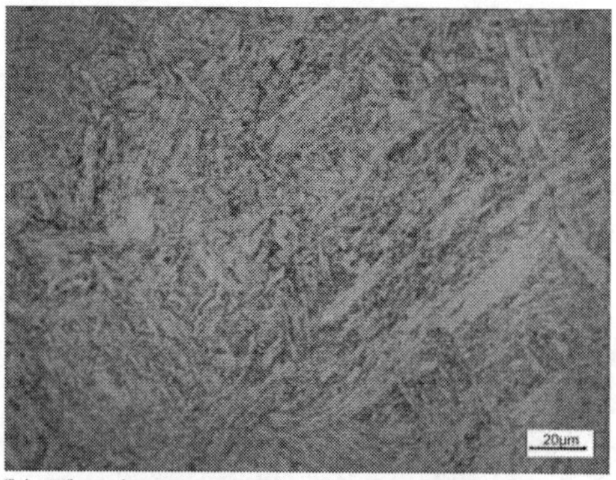

(b) The tissue near the fracture surface

Figure 3. Microstructure of fractured vane.

2.4 Chemical make-up testing

Chemical make-up detection of fractured leaves uses spectroscopic instruments. The chemical make-up analysis of the fractured moving blade was carried out using the SPECZROMAX desktop direct reading spectrometer. According to the test results, it can be concluded that the chemical make-up of the fractured blade is not abnormal. (table 1)

Table 1. The results of chemical make-up testing. %

Element	C	Si	Mn	Cr	Ni	Mo	V	W	Cu	S	P
Detection result	0.20	0.24	0.56	11.26	0.97	1.22	0.21	1.25	0.098	0.001	0.018
Required numerical value	0.20~0.25	≤0.50	0.50~1.00	11.00~12.50	0.50~1.00	0.90~1.25	0.20~0.30	0.90~1.25	≤0.30	≤0.020	≤0.030

2.5 Mechanical property testing and analysis

Mechanical performance testing of fractured blades uses impact testing and a fully automatic Brinell hardness tester. According to the test results, it can be concluded that the mechanical properties of the blades meet the requirements. (table 2)

Table 2. Testing results of mechanical properties.

Project	Brinell hardness/HBW	Impact toughness /J
Detection result	283	68
Standard value	277-311	≥11

2.6 Energy spectrum detection in fracture

Energy dispersive spectroscopy (EDS) analysis was performed on the corrosion pit near the fracture surface, and it was found that there is corrosive Cl^- inside the fracture. The normalized content of Cl^- is 0.58%. The results of the EDS analysis are shown in Figure 4[7].

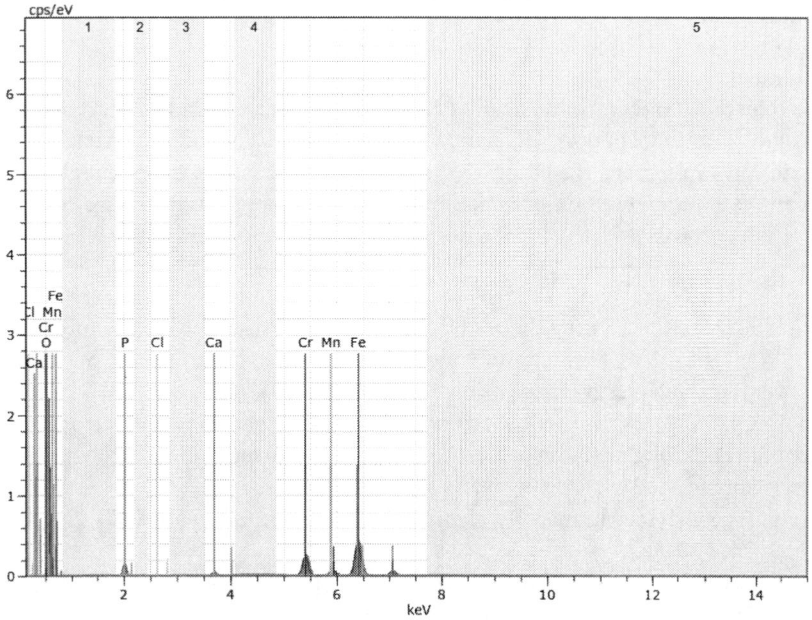

Figure 4. Energy spectrum analysis results.

3. Comprehensive analysis

From analyzing the fracture morphology, the fracture of the blade is caused by a corrosion pit at the right angle shoulder of the first tooth of the blade root.

From the analysis of chemical make-up testing, the chemical make-up and elemental content of the fractured leaves meet the standard requirements.

From a microscopic perspective, the tissue of the fractured blade is a normal tempered martensitic structure, with no abnormalities observed. There are multiple corrosion pits at the right angle shoulder of the inner arc of the first tooth of the blade root, and some of the corrosion pits have initiated cracks.

Analyzing mechanical performance testing results, all mechanical performance data of the fractured blade meet the requirements.

Analyzing energy spectrum detection results from blade fracture surface, excessive CL^- exists in the corrosion pit at the fracture site of the blade, causing martensitic SCC. The crack continues to expand during operation, ultimately leading to the overall fracture of the blade.

From the analysis of the stress state, the vane of the feed pump turbine is subjected to significant tensile stress and cyclic excitation stress for a long time during operation. In addition, the first tooth inner arc shoulder of the blade root is at a 90° right angle, belonging to the stress concentration area.

4. Conclusions and recommendations

Through experimental testing and analysis, the direct reason for the fracture of the vane of the feedwater pump turbine in this group is that the first tooth inner arc shoulder of the blade root is at a 90° right angle, which belongs to the weak part of stress concentration. The steam medium in thermal power units contains a large amount of harmful CL^-, which leads to the SCC of the blades. During the operation of the turbine rotor, it rotates at high speed and experiences strong vibrations, and the source of stress corrosion cracks propagates in a fatigued manner, resulting in a severe fracture of the blade [8].

Suggestion: Firstly, it is necessary to investigate whether other vanes have fractures or cracks and focus on the position of the inner arc shoulder of the first tooth of the blade root. If any problems are found, they should be dealt with in a timely manner. Secondly, the supervision of the quality of steam and water in the unit should be strengthened to avoid excessive accumulation of corrosive Cl^- in the steam and water medium. Furthermore, measures should be taken to minimize the excitation force of the moving vane at all levels as much as possible in order to prevent similar blade fracture failures from occurring again [9][10].

Acknowledgments

I would like to express my gratitude to the experimental team for their efforts and the unit's scientific and technological funding support. (Research on Intelligent Evaluation of Metallographic Structure of Metal Supervision Equipment Based on Deep Learning, 2022-ZC-05; Research on the creep damage mechanism and evaluation method of nickel based high-temperature fastening bolts,2023-12)

References

[1] Zhang F S, Ma Y F, Jiang W H, et al. The optimization for turbine blade forging process parameters based on grey theory [J]. Machinery Design & Manufacture, 2019, (9): 149-151.
[2] XU Lei. Analysis of material failure mode and treatment prevention of steam turbine vane[J]. Power Station System Engineering, 2016, 32(3): 47-50.
[3] DI J, WANG S S, YAN X J, et al. Experimental investigation on effect of surface strengthening process and roughness on water droplet erosion behavior in turbomachinery[J]. Tribology International, 2021, 153: 106647.
[4] Zeman J L. The Deviatoric map – a simple tool in design by analysis[J]. Int. J. Pres. Ves. Piping, 1999, 76, 339-344.
[5] RSE-M. In-service inspection rules for the mechanical components of PWR nuclear islands[S]. Paris: 2020.

[6] Xiaoshan Shi, Wei Yao, Da'an Liu, Kaiwen Xia, Tiewu Tang, Yirui Shi. (2020) Experimental study of the dynamic fracture toughness of anisotropic black shale using notched semi-circular bend specimens, Eng. Fract. Mech. 205 136–151.
[7] Hildebrandt F D, Thomas B W, Ulmer M W. Opportunities for reinforcement learning in stochastic dynamic vehicle routing. Comput Oper Res, 2023, 150: 106071
[8] Mao, L. Embrittlement and recovery of 2Cr12NiMoWV steel high temperature fastening bolt, Mechanical engineering materials, 2021(4): 48-52.
[9] Abe F, Kutsumi H, Haruyama H et al. Corrosion Science[J],2017, 114: 1.
[10] Bugge J, Kjær S, Blum R. Energy [J], 2006, 31(10-11): 1437

Research on the effect of mg doping on the properties of ZnO nanorods

Xue Tian [1], Yingying Yang [1] and Yuanping Sun [1*]

[1]School of Physics and Electronic Information, Yantai University, Yantai, Shandong Province, China

[*]Corresponding author's e-mail: ypsun@ytu.edu.cn

Abstract: Semiconductor materials, as the core of the new generation of information technology, and nano ZnO as the third-generation semiconductor, have high absorption and transparency in the visible light range, as well as a wide bandgap, making it widely applicable in the field of optoelectronic components. This article prepared Mg-doped ZnO nanorods using a low-temperature water bath method and studied the effect of Mg doping on the structure and luminescent properties of the samples using scanning electron microscopy (SEM), X-ray diffraction (XRD), and photoluminescence (PL) analysis. The SEM results indicate that the obtained sample has a double-columnar ZnO nanorod structure. As the Mg doping concentration increases, the size of ZnO nanorods first increases and then decreases. When the Mg doping concentration reaches 2%, the size of ZnO nanorods reaches its maximum, about 7.94 μm. The XRD results indicate that the obtained ZnO sample has a hexagonal wurtzite structure, and there is no additional Mg impurity phase in the sample, indicating that Mg^{2+} has successfully replaced Zn^{2+} into the lattice. The PL spectrum indicates that the obtained sample has a maximum UV emission value of around 395 nm and exhibits green (520 nm, 566 nm) and orange (600 nm) luminescence characteristics. With the increase of Mg doping concentration, the UV emission peak strengthens and shifts red, and the visible emission peak weakens, which improves the photocatalytic performance of ZnO nanorods.

1. Introduction

Semiconductor materials, as the core of the new generation of information technology, and nano ZnO as the third-generation semiconductor, have high absorption and transparency in the visible light range, as well as a wide bandgap, making it have broad application potential in the field of optoelectronic devices. As a wide bandgap semiconductor material, ZnO has a bandgap width of 3.37 eV at room temperature and an exciton binding energy of 60 meV [1], which enables efficient exciton luminescence at room temperature. This characteristic enables ZnO to achieve ultraviolet luminescence at room temperature and plays an important role in manufacturing laser devices. On the other hand, ZnO has good thermal and chemical stability, low or even negative electron affinity potential, high thermal conductivity and carrier mobility, and high breakdown field strength and emission current so that ZnO can be used as a cathode for field emission displays. Nano ZnO is widely used in fields such as photodetectors, solar cells, gas sensors, photodiodes, photocatalytic degradation, and varistors and has broad application prospects.

Le H Q N et al. synthesized Al-doped ZnO and Fe-doped ZnO nanoparticles through a co-precipitation annealing method to create new and safer sunscreen ultraviolet filters. The sunscreen formula using metal doped ZnO nanoparticles shows high SPF values, preventing white marks on the skin, making it very suitable for people of color as a sunscreen [2]. Said et al. used a microwave-assisted synthesis method, calcined at 600°C, and doped Fe or Ni into ZnO nanoparticles. Research has confirmed that ZnO doped with Fe atoms has high saturation magnetization, while ZnO doped with Ni has high coercivity [3]. Doping can regulate the impurity-related luminescent properties of ZnO nanorod materials, enabling them to exhibit excellent luminescent properties in both green and orange light bands. This indicates that under certain experimental conditions, Mg^{2+} can replace the Zn^{2+} lattice in ZnO to form $Zn_xMg_{1-x}O$ alloy within a certain range. Adding Mg^{2+} is beneficial for inhibiting electron-hole recombination and improving its photocatalytic degradation performance. There are many growth methods for $Zn_xMg_{1-x}O$, including chemical precipitation, sol-gel, hydrothermal, low-temperature water bath, magnetron sputtering, etc.

This article successfully prepared Mg^{2+} doped double columnar ZnO nanorods using a low-temperature water bath method. Research has shown that changing the doping concentration of Mg^{2+} can alter the size and luminescent performance of ZnO nanorods. This method has the characteristics of simple experimental equipment, low cost, and environmental friendliness. In addition, the ZnO nanorod structure prepared by the low-temperature water bath method exhibits rich morphology under certain conditions, without or with few clusters, and has high purity and intact crystalline properties. The low-temperature water bath method can uniformly heat the sample and facilitate observation, and there will be no boiling phenomenon in the experiment.

2. Experiment

2.1 Experimental process

In sample preparation, zinc acetate $(CH_3COO)_2Zn\cdot 2H_2O$ and magnesium acetate $(CH_3COO)_2Mg\cdot 4H_2O$ were used as sources of Zn and Mg, potassium hydroxide (KOH) was used as pH buffer, hexadecyl methyl tetramine (CTAB) was used as stabilizer in the reaction, and deionized water (DI) was used as the solvent for the reaction. The drug grades used in the experiment are analytical and have not been further purified.

We weigh 2.19 g of zinc acetate and a certain amount of magnesium acetate into a beaker, add 100 ml of deionized water, and stir electromagnetic stirring for 3 minutes until completely dissolved. Then, 0.56 g of KOH and 0.36 g of CTAB were added to the mixed solution, and the molar percentages of Mg^{2+} in the uniform solution obtained by electromagnetic stirring were 0%, 0.5%, 1%, 2%, and 3%, respectively. We place the prepared solution in a 90°C water bath and react for 24 hours, then naturally cool to room temperature. After 3-5 rounds of ultrasonic cleaning, centrifugation, and filtration, all obtained reaction mixed solutions were placed in a drying oven and dried at room temperature for 24 hours to obtain dry pure ZnO and Mg-doped ZnO powder. Table 1 provides the numbering of the prepared samples and their reaction conditions.

Table 1. Sample numbers and reaction conditions.

Sample number	Temperature/°C	The molar ratio of Mg/Zn	Reaction method	Time/h
D1	90	0	Low-temperature water bath method	24
D2	90	0.005	Low-temperature water bath method	24
D3	90	0.01	Low-temperature water bath method	24
D4	90	0.02	Low-temperature water bath method	24
D5	90	0.03	Low-temperature water bath method	24

2.2 Experimental instruments

The ZnO sample growth equipment is a WB100-2 low-temperature water bath from Qun An Scientific Instruments Co., Ltd. X-ray diffraction (XRD) testing was performed using the SmartLab III X-ray diffractometer from Rigaku, Japan. Scanning electron microscopy (SEM) characterization was

Journal of Physics: Conference Series **2808** (2024) 012022 doi:10.1088/1742-6596/2808/1/012022

performed using a JSM-7900F field emission scanning electron microscope from Japan, and photoluminescence spectroscopy (PL) testing was performed using the FLS980 fluorescence spectrometer from EI, UK.

3. Results and discussion

3.1 Morphological analysis of nanorods (SEM)

Figures 1(a)-(e) show the SEM images of the sample. From the figure, it can be seen that the shape of the sample is a short cylindrical hexagonal prism consisting of two hexagonal prisms. In alkaline environments, due to the different shielding effects of OH^- on each crystal plane, the growth rate of each crystal plane tends to approach. Hence, the samples prepared in alkaline environments are short columnar. We measure the length of ZnO nanorods in the field of view in Figure 1 and take the average value, obtaining the size of the nanorods in the range of 7-8 μm, as shown in Figure 1(f), the particle size diagram of Group D sample. As shown in the figure, the size of zinc oxide nanorods increases with the increase of Mg^{2+} doping concentration, and the change in size first increases to the maximum and then decreases. When the Mg^{2+} concentration reaches 2%, the maximum size of zinc oxide is 7.94 μm. Due to the polarity of ZnO nanorods, when an appropriate amount of Mg^{2+} is added to ZnO nanorods, the growth rate of the polar plane is greater than that of the non-polar plane, so the length of the sample gradually increases. But when the concentration of Mg^{2+} doping is too high, Mg^{2+} cannot be fully incorporated into the sample, and the excess Mg^{2+} will attach to the crystal plane of ZnO nanorods, limiting their directional growth and resulting in a shorter nanorod length.

In hydrothermal reactions, if the solution is strongly alkaline, the solubility of $Zn(OH)_2$ significantly increases, resulting in a large amount of OH^- in the solution. These OH^- affect the interfacial properties of crystal growth, where the (0001) crystal plane begins to grow with Zn^{2+}, while the $(000\bar{1})$ crystal plane begins to grow with O^{2-}, and CTAB can be decomposed into Br^- and CTA^+ in the reaction solution. CTAB can combine the growth unit on the composition plane (0001) to form a hexagonal cylindrical twin nucleus. By adding growth units at the twin interface (0001), the twin grows along the polar c-axis, ultimately forming a special double-column structure.

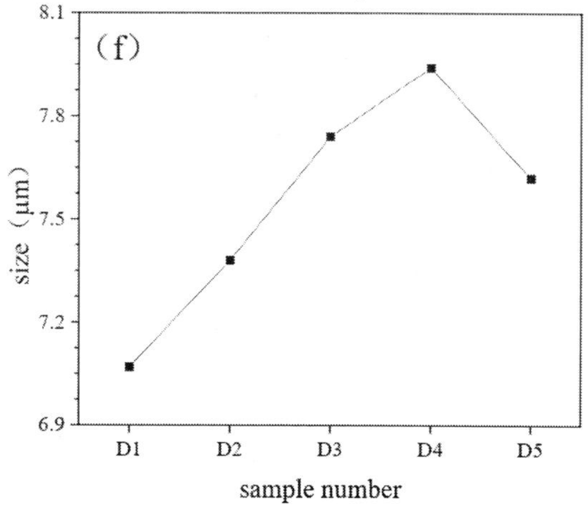

Figure 1. (a)-(e) SEM of group D samples (f) Particle size diagram of group D samples

3.2 XRD analysis

The XRD results of Group D samples are shown in Figure 2(a), and the diffraction pattern of the sample obtained from the experiment is consistent with the standard spectrum of ZnO (JCPDS: 99-0111), indicating that the obtained ZnO samples are all of the wurtzite structure. From the XRD spectrum, it can be observed that with the addition of Mg^{2+}, the positions of various peaks in ZnO undergo varying degrees of shift. As shown in Figure 2 (b), it can be seen that as the Mg^{2+} doping concentration increases, the position of peak (101) in the X-ray diffraction peak of the sample shifts towards the lower angle side. This is because the ionic radii of Mg^{2+} and Zn^{2+} are close, and Mg doping causes a change in the lattice constant of ZnO of about 1%, which may be due to lattice distortion, resulting in tensile stress inside the crystal and causing the diffraction peak to shift towards a lower angle [4].

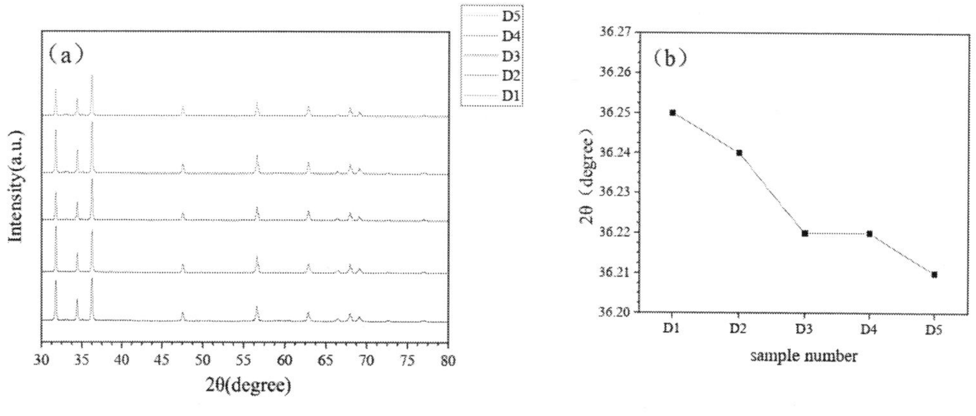

Figure 2. (a) XRD patterns of group D samples (b) 101 Peak position variation chart

3.3 PL spectral analysis of nanorods

We obtain the photoluminescence spectrum of the Group D sample, as shown in Figure 3(a), at an excitation wavelength of 325 nm at room temperature. From the graph, the pure zinc oxide exhibits the maximum UV emission at 395 nm and green (520 nm, 566 nm) and orange (600 nm) emission peaks. As the Mg^{2+} content increases, different intensities of ultraviolet emission peaks (395 nm), green (520 nm, 566 nm), and orange (600 nm) emission peaks can be observed in all doped samples. The peak of visible light (520 nm) comes from the transition of electrons towards the valence band at deep energy levels, which may be related to Vo [5]. The orange emission peak in the sample is mainly due to the optical transition from the bottom of the conduction band to the oxygen gap level. Moreover, moderate Mg^{2+} doping can enhance the visible light emission of ZnO samples, especially the green emission near 520 nm, making it a good material for preparing micro green light-emitting diodes. After normalizing the UV emission peak of the sample, Figure 3(b) was obtained. It can be seen from the figure that doping reduces the relative visible light intensity of the sample, as shown in Figure 3(c), which shows the change in the 520 nm visible light peak after normalization. Moreover, the relative visible light intensity reaches its minimum when the Mg doping concentration reaches 2%. The change in emission intensity indicates that doping Mg ions into zinc oxide can better capture excited electrons, thereby preventing electron-hole pair recombination and effectively prolonging the service life of photo-generated charge carriers, thereby improving the photocatalytic performance of zinc oxide.

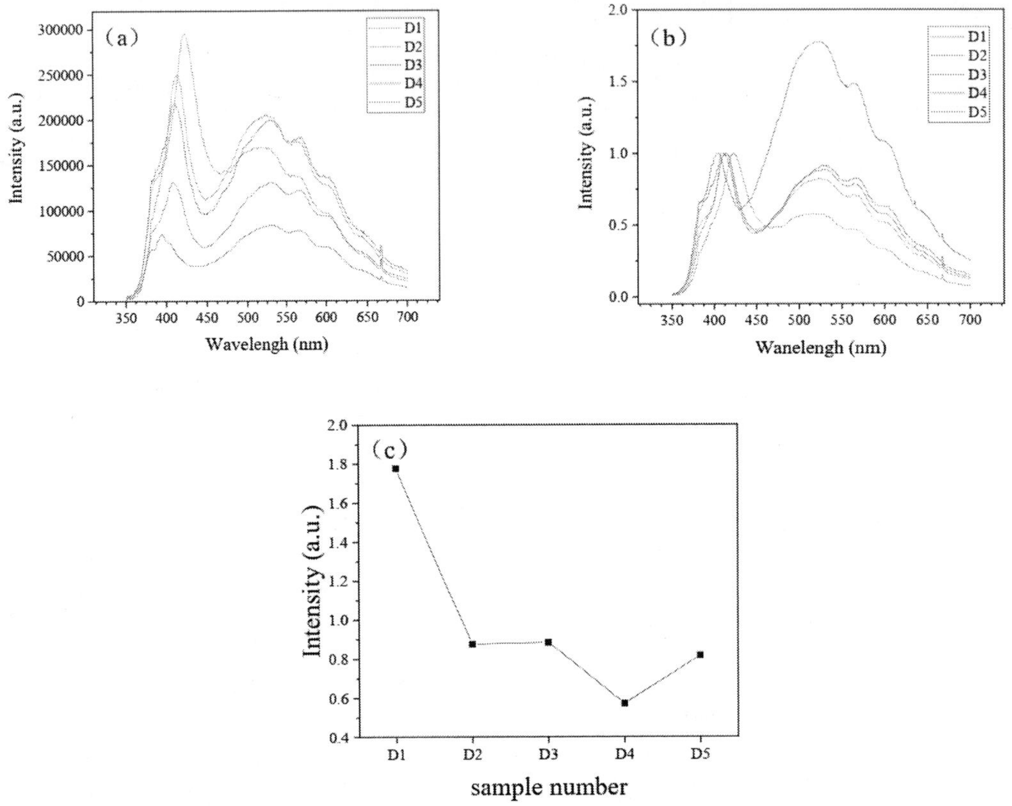

Figure 3. (a) PL diagram of the sample (b) Normalize the UV emission peak (c) Normalized change in visible light peak intensity at 520 nm

Figure 4 shows the changes in the intensity and position of the ultraviolet emission peak. The higher the concentration of magnesium doping results in a stronger UV emission peak that shifts towards red light. When the doping concentration reaches 2%, the peak of UV emission reaches its maximum. Generally, the stronger the UV emission peak intensity is, the better the crystal quality of the sample is. This further indicates that Mg^{2+} doping into nano ZnO is beneficial for improving its crystal quality. Due to the low work function of Mg (3.66 eV) and the 4.45 eV work function of n-type ZnO [6], Mg^{2+} is prone to form Ohmic contact with ZnO, causing the energy band of ZnO to bend downwards. Under the excitation of the incident light, electron-hole pairs are generated in the zinc oxide sample, and electrons accumulate at the interface, participating in interband and defect level recombination, enhancing and redshifting the UV emission peak of ZnO. The peak position of the green emission peak has hardly changed because the doping of Mg^{2+} into the nano ZnO replaces the lattice position of Zn^{2+}, only changing the position of the conduction band bottom without changing the valence band top and deep energy level. This further indicates that the green emission peak is due to the transition of electrons from deep energy levels to valence bands in zinc oxide.

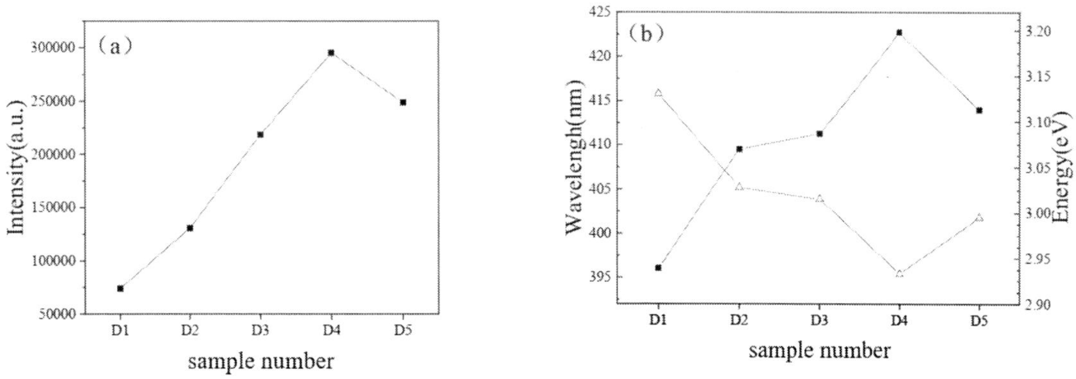

Figure 4. (a) Changes in peak intensity of ultraviolet luminescence (b) Changes in the position and energy of ultraviolet emission peaks

4. Conclusion

A Mg^{2+} doped double columnar ZnO nanorod was successfully prepared using a low-temperature water bath with zinc acetate and magnesium acetate as the zinc and magnesium sources in the reaction, potassium hydroxide as pH buffer, and CTAB as stabilizer. Research has shown that by changing the doping concentration of Mg^{2+}, the size and luminescent performance of ZnO nanorods can be altered. With the increase of Mg^{2+} doping concentration, the XRD diffraction peak of the sample shifts to the left, and the UV luminescence peak shifts to the red. When the Mg^{2+} doping concentration reaches 2%, the particle size and UV emission peak intensity of ZnO nanorods reach their maximum. Different intensities of green (520 nm, 566 nm) and orange (600 nm) emission peaks can be observed, especially the green emission near 520 nm, making it a good material for preparing micro green light-emitting diodes. The doping of Mg^{2+} reduces the relative visible light intensity, effectively captures excited electrons, prevents electron-hole recombination, prolongs the service life of photo-generated carriers, and improves the photocatalytic performance of ZnO nanorods.

Acknowledgments

This work is supported by the National Natural Science Foundation of China (Grant No: 12374031).

References

[1] Oezguer U, Alivov Y I, Liu C, et al. (2005) A comprehensive review of ZnO materials and devices. AIP Publishing, 98: 11-1.
[2] Le H Q N, Ngoc N A T, Nguyen H Q N, et al. (2023)Synthesis of Al-doped and Fe-doped ZnO nanoparticles as UV filters in sunscreens.Vietnam Journal of Chemistry, 61: 52-59.
[3] Said S A, Musa S G G, Mohd N S, et al. (2021) Assessment of Magnetic Properties between Fe and Ni Doped ZnO Nanoparticles Synthesized by Microwave Assisted Synthesis Method. Solid State Phenomena, 5944: 119-124.
[4] Ruziana M, Azri S M A . (2022) Structural and Electrochemical Behaviors of ZnO Structure: Effect of Different Zinc Precursor Molarity. Condensed Matter, 7: 71-71.
[5] Wong K M, Alay-E-Abbas S M, Fang Y, et al. (2013) Spatial distribution of neutral oxygen vacancies on ZnO nanowire surfaces: An investigation combining confocal microscopy and first principles calculations.Journal of Applied Physics, 114.
[6] Fang Y J, Sha J, Wang Z L, et al. (2011) Behind the change of the photoluminescence property of metal-coated ZnO nanowire arrays. Applied Physics Letters, 98.

Journal of Physics: Conference Series **2808** (2024) 012023 doi:10.1088/1742-6596/2808/1/012023

Analysis of cracking causes of angle steel in 220 kV transmission tower

Pingcheng Fan[1, 2, *], Liming Xie[1, 2], Qingyun Hao[3] and Hao Chen[1, 2]

[1] Inner Mongolia Power (Group) Co., Ltd., Inner Mongolia Power Research Institute Branch, Hohhot, 010020, China

[2] Inner Mongolia Key Laboratory of Smart Grid of New-type Power System Technology, Hohhot, 010020, China

[3] Inner Mongolia Power (Group) Co., Ltd., Inner Mongolia Power Communication Company, Hohhot, 010020, China

[*] Corresponding author's e-mail: pingcheng_dky@163.com

Abstract: An experimental analysis was conducted on the cause of cracking in the angle steel of a 220 kV transmission tower using macroscopic inspection, chemical composition analysis, microstructure inspection, and other testing methods. The results show that the angle steel did not have a groove before welding with the filler material at the notch bending position. Severe incomplete penetration defects were formed during the welding process, resulting in a large stress concentration at the defect site. This is the main reason for its corrosion and fracture. It is suggested to focus on checking for cracks during welding construction and strengthening the inspection of the welding quality of the angle steel of the transmission tower during the infrastructure stage for the bending part of the angle steel notch.

1. Introduction

The transmission line tower is a crucial element that supports the high-voltage lines of the power grid. The quality of the main materials used in the tower directly affects its safety, stability, and reliable operation. Under current processing conditions, the angle steel component of the iron tower's main material is manually arc welded. The strength of the weld affects the stability of the entire base of the iron tower after the line operation is hung. Therefore, the quality of the welding has a significant impact on the safety and reliability of the entire iron tower [1-4].

During the inspection process, inspectors from a power supply bureau discovered a serious hidden defect in the main material (inner side) of the right lower arm of the 220 kV transmission line tower angle steel facing the large side. To identify the cause of the cracking of the iron tower angle steel and prevent similar failures from occurring in the future, the author conducted an inspection and analysis.

2. Physical and chemical inspection

2.1 Macro inspection

The macroscopic morphology of the cracked angle steel is observed. The cracking location is located at the fusion line between the angle steel notch bending part and the filler material welding. The cracking originates from the outer arc side of the angle steel bending and extends inward along the fusion line. The fracture surface is flush, without metallic luster, and no obvious plastic deformation is observed.

From the cross-sectional view of the crack, it can be seen that the bending part of the angle steel notch was not beveled when welding with the filler material, resulting in incomplete penetration defects, as shown in Figure 1.

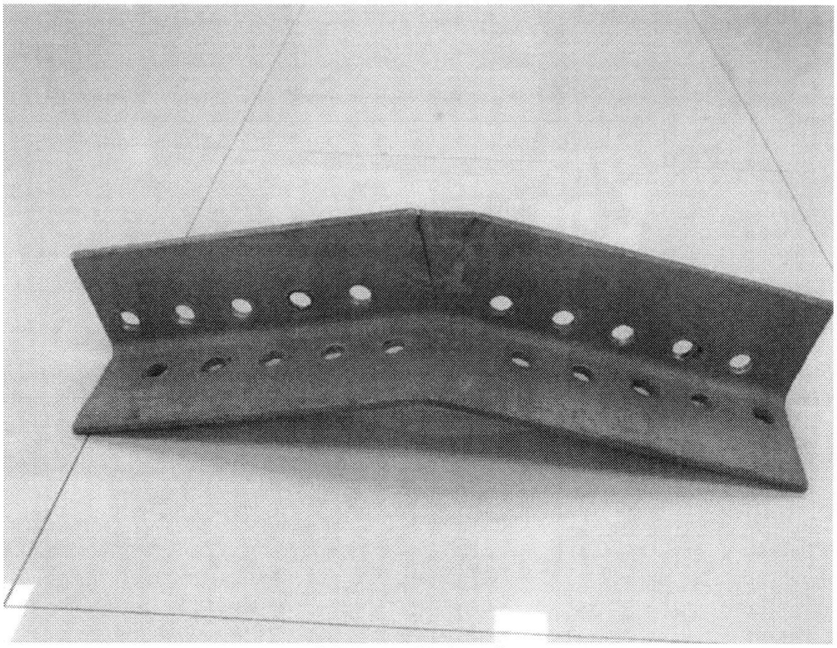

(a) Overall situation

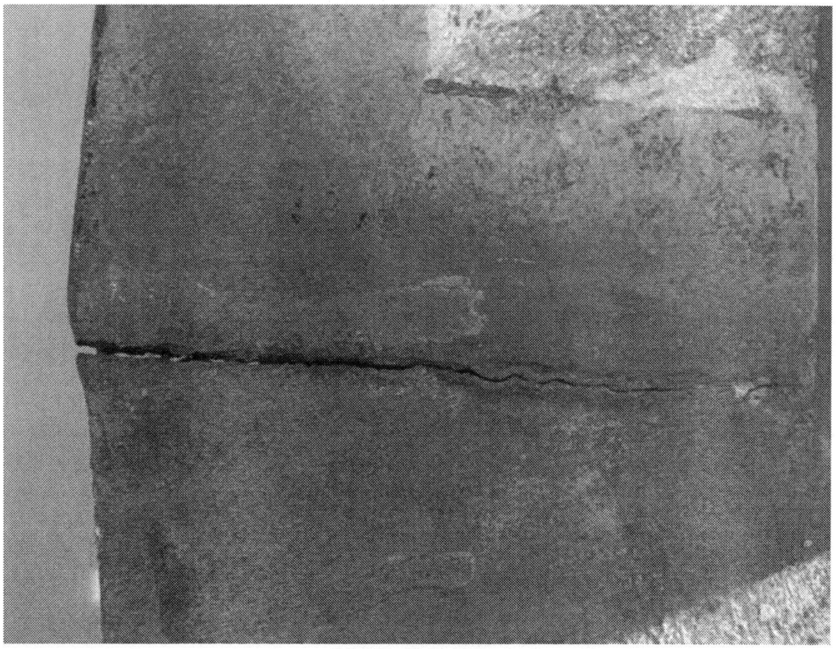

(b) Cracking area

(c) Fracture morphology

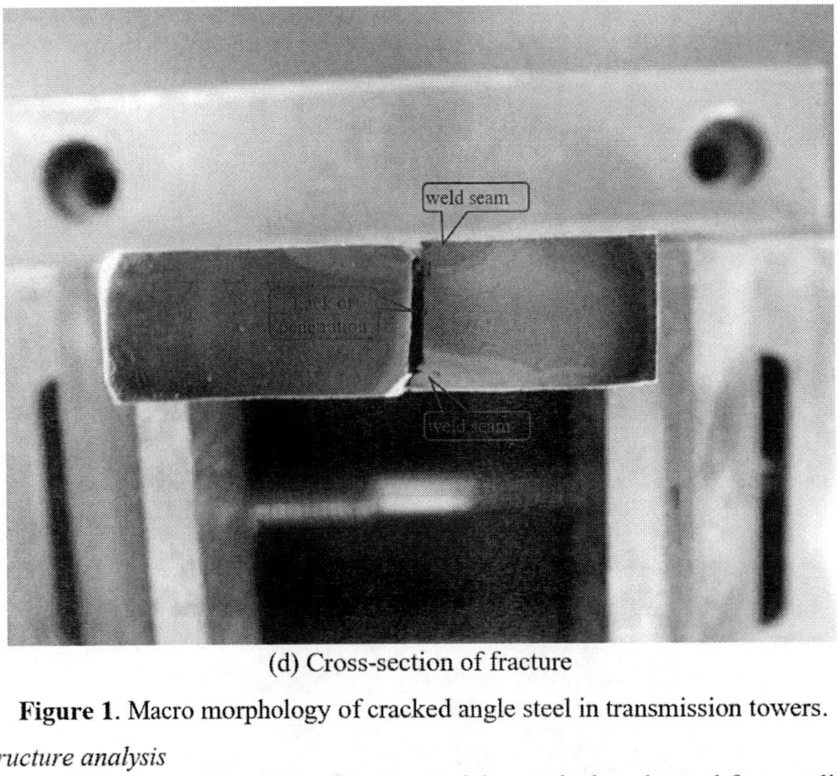

(d) Cross-section of fracture

Figure 1. Macro morphology of cracked angle steel in transmission towers.

2.2 Microstructure analysis

Samples should be taken from the self-cracking area of the cracked angle steel for metallographic and microstructural examination, as shown in Figure 2. The base material of the angle steel has a banded pearlite + ferrite microstructure, while the weld seam has a sorbite microstructure. No serious defects, such as inclusions, cracks, or overheated structures, were found. The cross-section of the crack reveals that the angle steel's inner and outer walls consist of weld seam and heat-affected zone structures, while the middle is composed of the base metal structure. This suggests the presence of a lack of penetration defect in the weld seam [5-7].

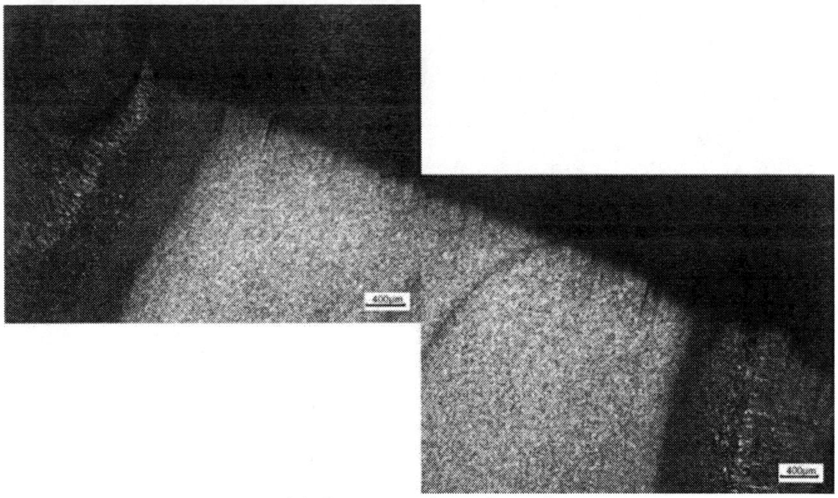

(a) Cross-section of crack.

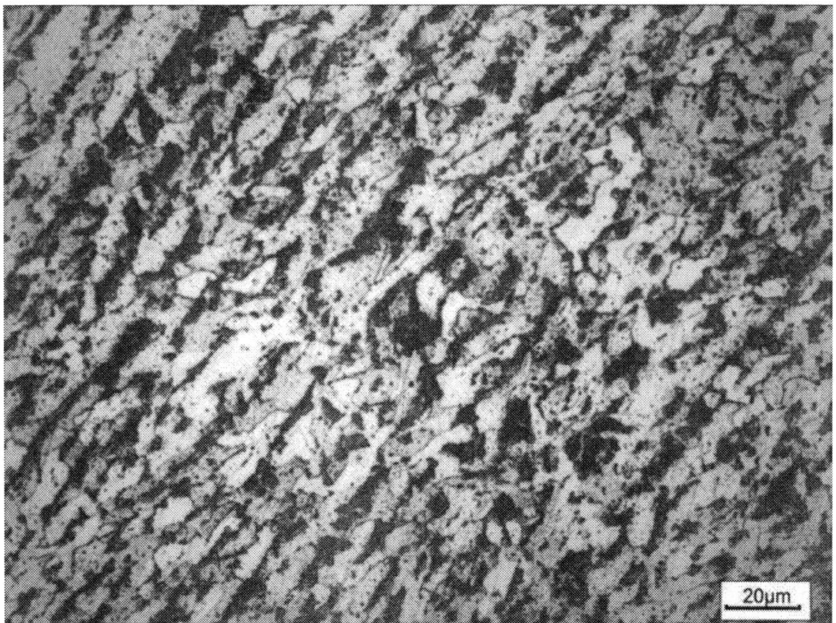

(b) Metallographic structure of the base material

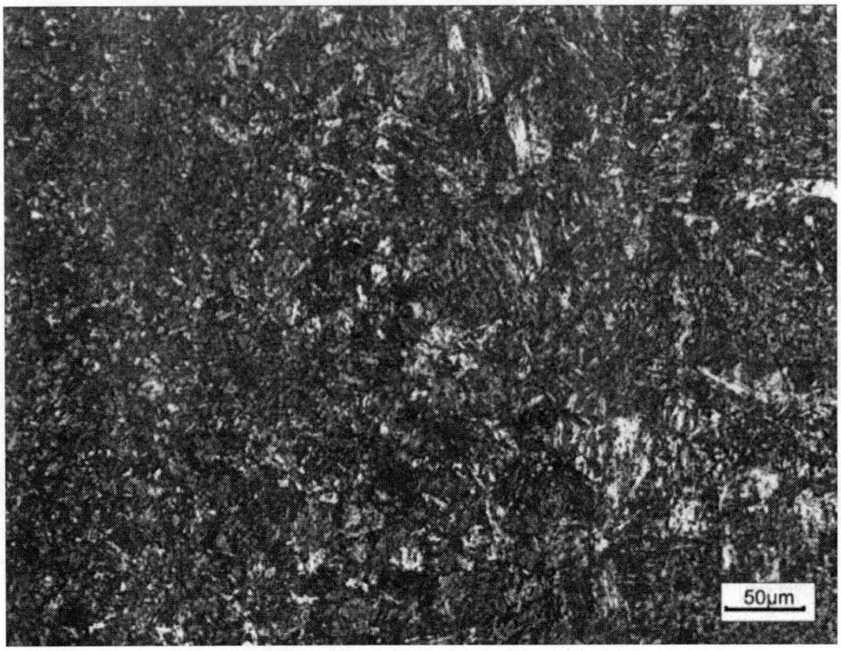

(c) Metallographic structure of weld seam

Figure 2. Metallographic microstructure of angle steel for transmission towers.

2.3 Chemical composition analysis

Chemical composition testing was conducted on the cracked transmission tower angle steel samples, and the test results are shown in Table 1. The results indicate that the chemical composition of the transmission tower angle steel meets the requirements of the standard GB/T 1591-2018 for Q355B steel.

Table 1. Chemical composition testing results of angle steel for transmission towers (wt%).

Detect elements	C	Si	Mn	P	S
Measured value	0.16	0.38	1.36	0.024	0.018
Standard requirement	≤0.24	≤0.55	≤1.60	≤0.035	≤0.035

3. Analysis and discussion

Based on the macroscopic morphology analysis, the cracking occurred at the fusion line between the angle steel notch bending area and the filler material welding. The cracking originated from the outer arc side of the angle steel bending area and extended inward along the fusion line. The fracture surface is flush, without metallic luster, and no obvious plastic deformation was observed. The cross-sectional view of the crack reveals that the bending part of the angle steel notch was not beveled during welding with the filler material, resulting in incomplete penetration defects. Chemical composition analysis indicates that the composition of the cracked transmission tower angle steel meets the requirements of GB/T 1591-2018. No abnormalities were found in the microstructure of the cracked transmission tower angle steel base material and weld seam through metallographic analysis.

4. Conclusions and recommendations

This article analyses the main reasons for the cracking of the angle steel of a 220 kV transmission line tower and draws the following conclusions: the welding process of this angle steel does not comply with the requirements of 6.6 in GB/T 2694-2018 'Technical Conditions for Manufacturing of Transmission Line Towers'. Serious incomplete penetration defects are formed in the groove bending area before welding with the filler material during the welding process. This results in significant stress concentration at the defect location, providing the source of cracking for the angle steel. The crack continues to expand under stress, ultimately leading to cracking [8-11].

To ensure the safe, stable, and reliable operation of transmission towers and to prevent the recurrence of cracking of such tower angle steels, the following suggestions are proposed. Firstly, to ensure welding quality, it is necessary to strictly follow the process and quality management requirements [12]. During processing, the groove must meet welding requirements, and appropriate welding process parameters should be selected. Welding should be carried out according to standard specifications, melting the blunt edges to ensure complete fusion between the welding material and the base metal, and avoiding the phenomenon of incomplete welding between the base metal and the base metal [13-14]. During processing, the groove must meet welding requirements, and appropriate welding process parameters should be selected. Welding should be carried out according to standard specifications, melting the blunt edges to ensure complete fusion between the welding material and the base metal, and avoiding the phenomenon of incomplete welding between the base metal and the base metal [15-16].

Acknowledgments
This work was financially supported by the self-funded technology project of the Inner Mongolia Power Research Institute Branch (Grant No. 2022-ZC-07).

References

[1] National Energy Administration. (2015) Anticorrosivecoating of transmission line steel tower: DL/T 1453-2015. Electric Power Press, Beijing.
[2] National Energy Administration. (2012) Technical Regulations for Design of Tower Structures of Overhead Transmission Lines: DL/T 1453-2015. China Planning Press, Beijing.
[3] Ministry of Housing and Urban Rural Development of the People's Republic of China. (2017) Steel Structure Design Standards: GB 50017-2017. China Architecture & Building Press, Beijing.
[4] Ye D.S. (2020). Analysis and research on the causes of welding cracks in important components of power towe.r China Metal Bulletin, (09), 141-143.
[5] Rao N.P, Knight G.S, Lakshmanan N, Iyer N.R. (2010). Investigation of transmission line tower failures. Engineering Failure Analysis, 17(5), 1127-1141.
[6] Yan Z.L. (2014). Analysis of Factors Influencing the Performance of Q460 Steel. Applied Mechanics and Materials, 599, 7-11.
[7] Piao Z, Ma J, Wang Y, Wang S. (2016) Analysis of Crack Defects in Q420B Angle Steel, Foundry Technology, 37(12):2597-2599.
[8] Liu J. (2015) Experimental Study on Bearing Capacity of Straight Welded Pipe Used in Transmission Towers with Weld Defect. In 2015 International Conference on Sustainable Energy and Environmental Engineering, Bangkok, pp. 108-111.
[9] Le M, Park J, Kim J, Jung G, Lee J. (2018). Nondestructive Testing of Welding Defects in a Tubular-Type Transmission Tower using Magnetic Imaging with Magnetic Sensor Array. Materials Evaluation, 76(6), 853-861.
[10] Wang Q.Z (2023). Optimization of Continuous Cooling Transformation Curve and Arc Welding Process for Q355B Steel. MW Metal Forming, (09), 80-86.
[11] Zeng S, Xu D, Chang J, He F, Sun C, Zhang L. (2019). Development of hot rolled weathering angle steel for transmission towers. Engineering Research Express, 1(2), 025002.
[12] Tian G.S, Qin R, Yang J.J, Li J, Jiang F. (2017). Analysis of the Causes of Cracking in the Angle Steel of the Corner Tower Foot Plate of the 330 kV Line. Power Safety Technology, (09), 36-42.
[13] Wang Z.H, Lu W.X, He X.M, Chen W.Q, Liu G.F, Zhang S. (2017). Analysis of the welding quality status of transmission line angle steel towers. Qinghai Electric Power, (02), 36-40.

[14] Ma Y.H, Na X J. (2012). Research on Welding Process of Low Alloy High Strength Steel High Voltage Iron Tower. Technology Perspective, (06), 110-111.
[15] Chang J.W, Xu D.L, Zhang L, Wang H, Zhang D.Y, Yu C.H. (2011). Steel and its welding technology for transmission towers Hot Working Technology, (15), 148-151.
[16] Wang Z.H, Hua Y.R, Zhou G.P, Jia Q. (2022). Analysis of factors affecting the welding quality of ultra-high voltage transmission towers. Welding Technology, (08), 164-167.

Effect of acoustic filler on sound insulation properties of millable polyurethane rubber

Xiao Han [1*], Zhaoyang Zhang [1], Ling Zhang [1] and Wei Cheng [1]

[1] Luoyang Ship Material Research Institute, Luoyang, 471023, China

[*] Corresponding author's e-mail: 1667714474@qq.com

Abstract: Millable polyurethane (MPU) compound was prepared by mechanical blending with millable polyurethane as matrix material, hollow glass microspheres (HGM), and mica as acoustic fillers. Curing characteristics, mechanical properties, morphology, and damping capacity were characterized by a moving die rheometer, universal testing machine, scanning electron microscope (SEM), and thermomechanical analyzer. The results show that t_{90} has been shortened from 22 minutes and 33 seconds to 19 minutes and 46 seconds, ΔMH increased to 25.82 dN-m, the modulus increased, the maximum loss factor decreased, and the damping temperature range domain remained unchanged. The mechanical properties were better, the tensile strength was 20.5 MPa, the morphology was more complete, and the sound insulation was higher when HGM/mica content was 5/5 phr.

1. Introduction

Equipment operation and impacted vibration of external wind and water flow generate noise in the military and civil industrial fields. Noise pollution has become a common problem and brought many troubles to people's lives and work. Therefore, researching and developing sound insulation materials was provided with significance. Among many damping acoustic materials, rubber material has been widely used at home and abroad because of its excellent noise reduction and vibration damping. Besides structure, installation condition, frequency of the sound wave, and incident direction, the properties of a rubber material such as chemical composition, primary structure, secondary structure, and damping capacity had a great influence on sound insulation performance. Most of the acoustic energy entered the material and was converted into internal energy and dissipated when the characteristic impedance of rubber material matches with that of the propagation medium [1-4].

To further improve the sound insulation performance of rubber materials, inorganic fillers were often used to modify polymer materials in which polymer materials were continuous phase. Inorganic fillers were dispersed phase, which was uniformly distributed in the continuous phase by blending method [5]. Inorganic fillers possess a strong interfacial binding effect, which could effectively improve the damping capacity of polymer materials, and the preparation method was simple and easy to operate.

Millable polyurethane (MPU) was composed of polyol (OH-R-OH), diisocyanate (NCO-R-NCO), and a chain extender. The type of MPU depended on the type of macromolecular long-chain polyol. It could be processed by general rubber processing machinery. This work investigated the effects of adding mica powder and hollow glass microspheres (HGM) to MPU blended rubber on the mechanical properties, damping capacity, and sound insulation properties of MPU vulcanized rubber [8-11].

2. Experimental section

2.1. Materials and instruments

The materials include MPU1055 (Polyether-type, Mooney viscosity 40), Mica powder (325 mesh), and hollow glass microspheres k46 from American 3M Company. Other compounding agents and raw materials are commonly used in commercial industrial products.

2.2. Instruments

The instruments include a rolling kneader (LRMR-S-150), a Banbury mixer (X(S)3-25-100), a 63T flat vulcanizer (XLB-Y630), a rotor-less vulcanizer (M-2000A), a dynamic thermo-mechanical analyzer (SATA861[e]), an analytical balance (ES-520A), an acoustic standing wavepipe (MFT-120), and a field emission scanning electron microscope (QUANTA 650 FEG).

2.3. Preparation of rubber compound

The formula was as follows: MPU 100 phr (weight parts), stearic acid 0.3, N330 20, N550 10, DOP 5, DCP 1, and HVA-2 1. The dosage of mica powder and HGM is shown in Table 1.

Table 1. The dosage of HGM and mica powder.

Property	Sample			
	1#	2#	3#	4#
HGM (phr)	0	5	10	15
Mica (phr)	0	5	10	15

We weighed the raw materials according to the basic formula, adjusted the nip width of the rolling mill to 2 mm, and added MPU for plasticizing 5 times. We added a compounding agent except for a vulcanization system into the internal mixer in batches, mixed uniformly, and took it out. Then, we adjusted the nip width of the rolling mill to 1 mm and mixed the rubber compound for about 5 minutes. Next, we added curing agents according to different curing systems, mixed them for about 5 minutes, took them out, and stored them for 16 hours after mixing evenly. The samples were numbered 1#, 2#, 3#, and 4# according to the different mica powder and HGM added.

2.4. Characterizations

The curing temperature was 160°C. Curing characteristics were tested by M-2000A to obtain the curing parameters of compounds with different dosages of acoustic fillers.

Mechanical property was performed on tensile test machine according to GB/T 528-2008. Stretch mode was a unidirectional stretch and the tensile rate was 500 mm/min.

SEM was used to observe the microstructure of the rubber section after spray-gold.

Sound insulation performance was tested by using a standing wave pipe and transfer function method. The testing standard is CB/T 3674-1995. The sound insulation property is tested at a frequency range of 1 to 10 kHz. The temperature range is 15 ± 5°C. Five samples per group are tested and the average of the results are taken.

The dynamic mechanical properties of the samples were tested by DMA according to GJB 981-1990. The testing mode was the shear mode. The heating rate was 2 °C/min, the vibration frequency was 10 Hz, and the temperature range was -100°C-100°C.

3. Results and discussion

3.1. Curing characteristics

The curing characteristics were tested by M-2000A to obtain the curing parameters of compounds with different dosages of acoustic fillers, and the results are shown in Table 2.

Table 2. Curing characteristics of MPU with different acoustic fillers.

Property	Sample			
	1#	2#	3#	4#
Temperature (°C)			160	
MH (dN·m)	16.33	19.66	22.45	25.82
ML (dN·m)	1.12	1.15	1.26	1.27
MH-ML (dN·m)	15.21	18.51	21.19	24.55
t_{10} (m/s)	1:52	1:28	1:14	1:52
t_{90} (m/s)	22:33	21:22	20.24	19:46

The torque disparity of MPU rubber and vulcanization rate increased, and the scorch time t_{10} and optimum cure time t_{90} were shortened with the increase of acoustic filler dosage. This indicated that acoustic fillers were beneficial for improving crosslinking density and shortening crosslinking rate.

Metal compounds such as aluminum oxide, zirconia, and sodium silicate on the surface of HGM and mica exhibited strong interfacial binding effects under temperature which could react with MPU to form partial hydrogen bonds or subvalent bonds during the vulcanization process. The number of binding points increased, which led to an accelerated vulcanization rate and an increase in torque difference.

3.2. Physical and mechanical properties

The physical and mechanical properties of MPU vulcanizate with different amounts of acoustic filler are shown in Table 3. The hardness and density of MPU vulcanized rubber increased and the tensile strength first slightly increased and then decreased as the amount of acoustic filler increased.

Table 3. Physical and mechanical properties of MPU vulcanizate with different amounts of acoustic filler.

Property	Sample			
	1#	2#	3#	4#
Shore A hardness	65	70	72	76
Tensile strength (MPa)	17.2	20.4	20.5	15.1
Elongation at break (%)	434	456	454	404
Modulus at 50% (MPa)	3.84	4.66	4.81	4.51
Modulus at 100% (MPa)	4.82	6.78	7.22	6.44
Modulus at 300% (MPa)	9.46	13.21	14.07	10.44
Tension set (%)	2.5	2.5	2.5	1.8
Density ($g \cdot cm^{-3}$)	1.189	1.185	1.223	1.241

Acoustic filler reacted with isocyanate groups, alcohol groups, and other functional groups on the surface of the MPU to form secondary bonds and chemical bonds during the vulcanization process. The increase in the number of molecular entanglements led to an increase in the hardness and density of MPU vulcanizate with the increase of acoustic filler dosage. The secondary bonds between the molecules and atoms of the filler played the first role when stretching occurred, so the modulus at 50% increased with the increase of acoustic filler dosage.

Both HGM and mica belonged to inorganic fillers, while MPU belonged to polymer materials. MPU was the continuous phase and HGM and mica were the dispersed phase when HGM and mica were used to modify MPU. It could form a dispersed phase uniformly distributed in the continuous phase and the mechanical properties continued to increase when the filler reached a certain amount in the blend system.

Subsequently, the aggregation phenomenon of filler led to stress concentration, uneven distribution of stress in the molecular chain network, and a decrease in tensile strength and elongation at break as the dosage further increased.

3.3. Microscopic analysis of surface morphology

To further investigate the reasons for the performance changes of MPU vulcanizate under different acoustic filler dosages, scanning electron microscope analysis was carried out on MPU vulcanizate after spray-gold. Surface morphology is shown in Figure 1.

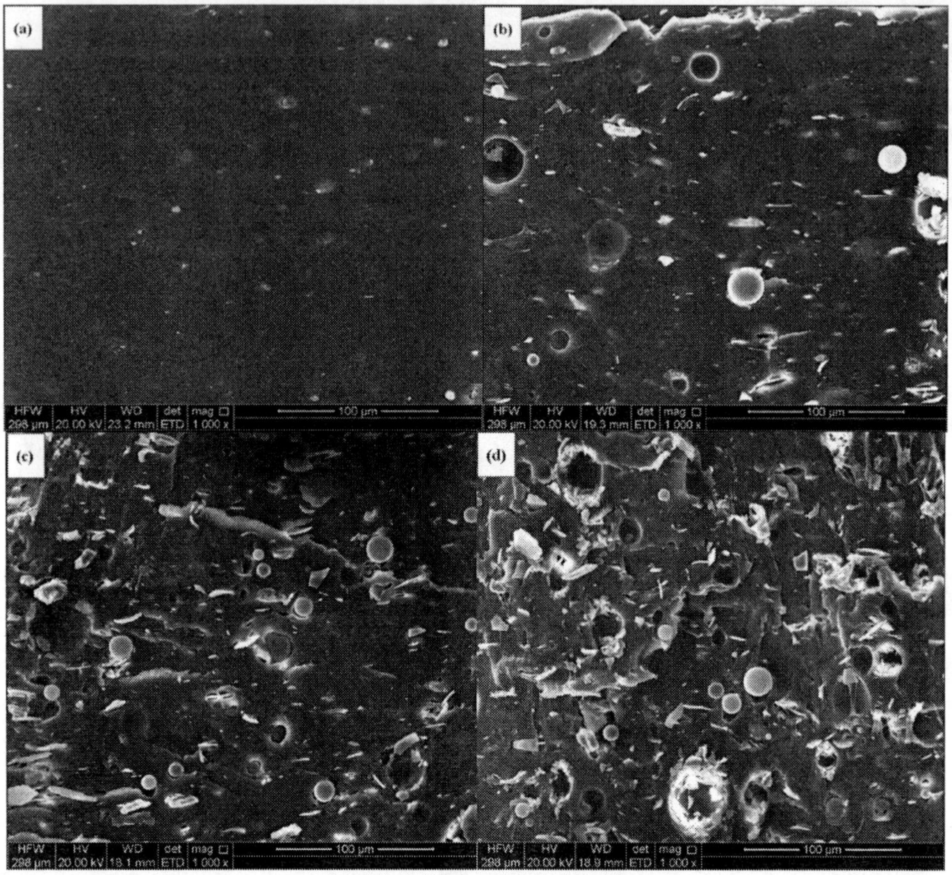

(a) 1#; (b) 2#; (c) 3#; (d) 4#

Figure 1. SEM morphology of MPU vulcanizate with different amounts of acoustic fillers.

The spherical and sheet-like structures of the sample cross-section gradually increased as the amount of HGM and mica increased. As shown in Figure 1 (b), the spherical HGM and sheet-like mica were evenly distributed and structurally intact alternately when the amount of HGM and mica was 5 phr. Acoustic fillers formed a large number of secondary bonds with the rubber, which resulted in higher tensile strength. As shown in Figure 1 (c), the dispersion uniformity of acoustic fillers in MPU was reduced and the free movement of rubber molecular chains was limited to a certain extent, but the acoustic filler structure was intact when the dosage of HGM and mica was 10 phr. As shown in Figure 1 (d), when the amount of HGM and mica was 15 phr, the spherical HGM and sheet-like mica exhibited an irregular distribution and the sheet-like mica structure underwent agglomeration. Some spherical HGMs broke or detached from their original positions and remained in a free state in MPU vulcanizate. The agglomerated sheet-like mica and broken and displaced HGM were prone to become stress concentration points, which resulted in lower tensile strength and a decrease in elongation at break when vulcanizate was subjected to external forces [12].

3.4. Acoustic performance

In practical applications, acoustic materials were affected by external factors such as temperature, pressure, and frequency. The acoustic testing environment used in this experiment was at a temperature of 15±5°C. The pressure points were set from low to high as pressure points A, B, C, and D. The frequency range was from 1 to 10 kHz. The acoustic performance of MPU samples with different acoustic filler dosages under pressure point A is shown in Figure 2.

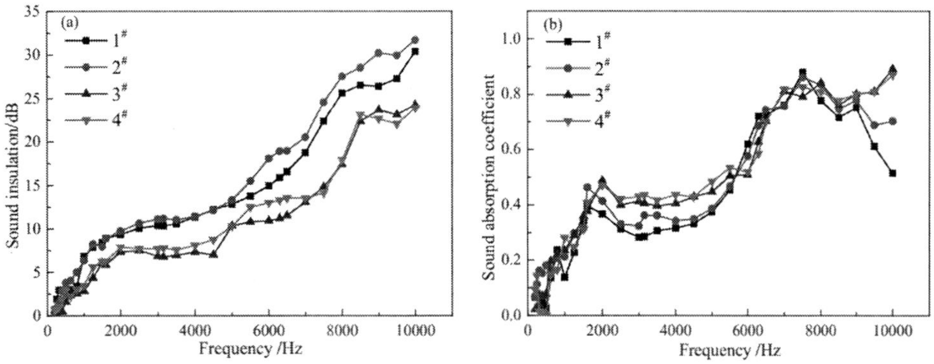

Figure 2. The acoustic performance of MPU samples with different acoustic filler dosages.

Figure 2 (a) shows the sound transmission loss-frequency curve of acoustic fillers with different dosages, and the sound transmission loss of MPU increased as the frequency increased. The sound transmission loss of MPU first increased and then decreased as the amount of HGM and mica increased, which was maximum when the amount of HGM and mica was 5 phr each. As shown in Figure 3, sonic waves would reflect and dissipate some energy when they reached the surface of spherical and sheet-like fillers after adding mica and HGM. The angle formed between spherical and sheet-like fillers and the surface of the MPU was also randomly due to the random distribution of fillers in the material. Therefore, the propagation distance of the sonic wave in the MPU increased, thereby increasing the consumption of acoustic energy. In addition, the sheet-like mica also changed some compression deformation into shear deformation, which increased sound transmission loss. As the dosage further increased, the decrease in dispersion uniformity of acoustic fillers in the system led to agglomeration and limited the free volume of rubber molecular chains, which resulted in a decrease in sound transmission loss.

Figure 2 (b) shows the sound absorption coefficient-frequency curve of acoustic fillers with different dosages. The sound absorption coefficient of MPU increased with the increase of HGM and mica dosage. When sonic waves passed through the hollow structure of HGM, they generated a large amount of scatter and forced vibration, acting as secondary sources to radiate acoustic energy into the surrounding medium, which led to further weakening of sound energy. At the same time, mica with a sheet-like structure increased the probability of sonic waves entering the HGM cavity by reflecting sonic waves, which increased the sound absorption coefficient.

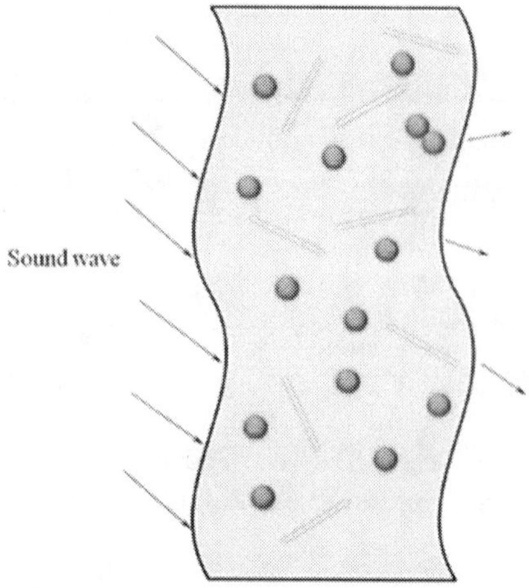

Figure 3. Schematic diagram of sonic wave propagation.

As shown in Figure 4, the influence of hydrostatic pressure on the acoustic performance of MPU was studied with 2# as the object.

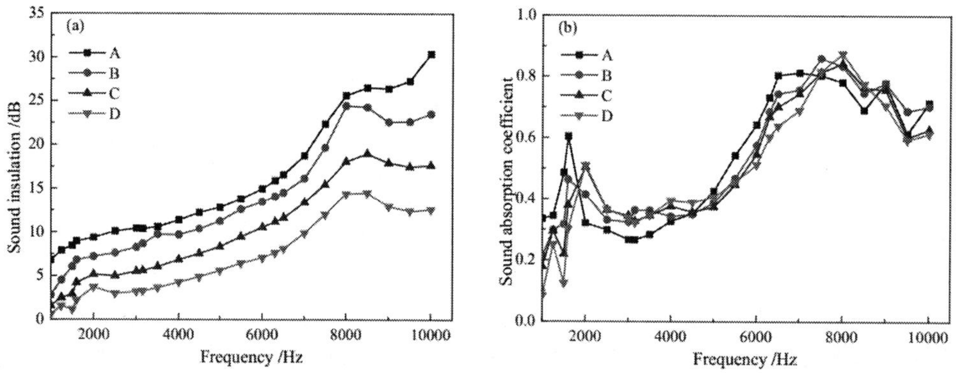

Figure 4. Influence of hydraulic pressure on the acoustic performance of MPU.

Figure 4 (a) shows the sound transmission loss-frequency curve of 2# under different hydrostatic pressure. As the frequency increased, the sound transmission loss of MPU showed an upward trend and tended to flatten out in the high-frequency range overtaking 8 kHz. Due to the squeezing effect of hydrostatic pressure on the acoustic cavity, the degree of deformation of the acoustic cavity led to a decrease in sound insulation properties as the hydrostatic pressure increased.

Figure 4 (b) shows the sound absorption coefficient-frequency curve of 2# under different hydrostatic pressure. Two absorption peaks appeared as the frequency increased. The first absorption peak is in the 2 kHz region. The second absorption peak was in the 8 kHz range and both absorption peaks shifted towards higher frequencies as the hydrostatic pressure increased. The higher frequency increased the movement speed of the rubber molecular chains and the rate of energy dissipation.

3.5. Dynamic mechanical properties

To study the chain segment motion and modulus change of vulcanizate with different filler dosages, the storage modulus (G′) and loss modulus (G″) are shown in Figure 5. The peak value (G") of MPU vulcanizate increased after adding mica and HMG. Both G 'and G″ increased as the amount of acoustic filler increased when the temperature was higher than the glass transition temperature. Acoustic fillers generated more secondary bonds with MPU molecular chains, which limited the movement of molecular chains and led to an increase in modulus.

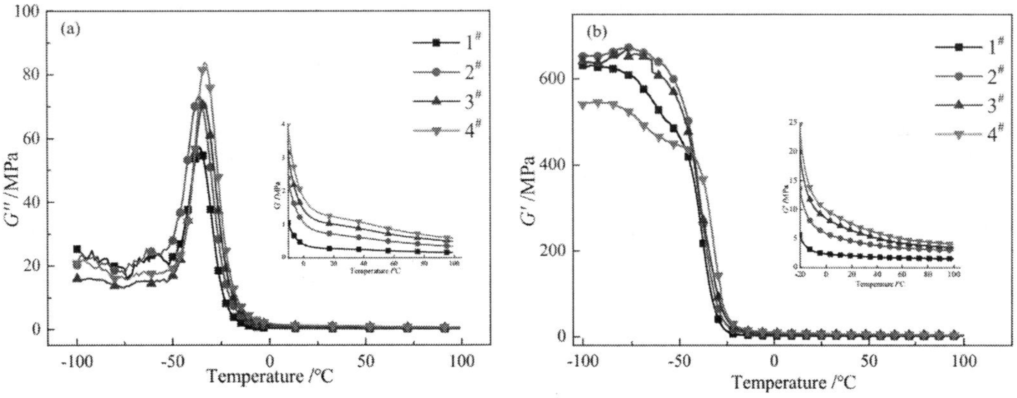

Figure 5. Storage modulus and loss modulus of MPU vulcanizate with different amounts of acoustic fillers.

To further study the viscoelasticity of vulcanizate with different acoustic fillers, this work plotted the loss factor-temperature curve under shear stress, as shown in Figure 6.

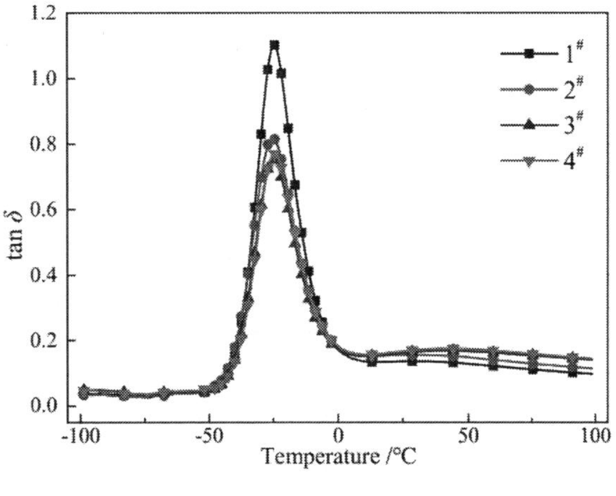

Figure 6. Loss factors of vulcanizate with different amounts of acoustic filler.

The peak value of the loss factor was significantly reduced after adding mica and HGM. Mica and HGM had a larger specific surface area. MPU molecular chains were inserted between the sheet structure and the spherical structure, increasing the contact area and generating more secondary bonds for cross-linking under the action of shear force and temperature. The relative motion and sliding difficulty between rubber molecular chains increased. The damping temperature range remains unchanged.

4. Conclusions

a) The vulcanization rate of the MPU rubber compound accelerates and the value of ΔM_H increases as the amount of acoustic filler increases.

b) The mechanical properties were better when the dosage of HGM/mica was 5/5 phr and 10/10 phr. HGM and mica were evenly distributed in MPU and had a more complete morphology at this dosage.

c) When the amount of HGM/mica was 5/5 phr, the sound transmission loss of MPU vulcanizate was higher. The sound transmission loss decreased as the hydrostatic pressure increased. The sound transmission loss increased as the testing frequency increased.

d) The modulus of MPU vulcanizate increased, the loss factor decreased, and the damping temperature range remained unchanged with the increase in HGM/mica dosage.

References

[1] Wen X. X., Peng Z. T., and Peng L. (2021). Study on Properties of Low-Frequency Sound-absorbing Butyl Rubber Composite. *Chinese Journal of Colloid & Polymer* **3**, 168-170.

[2] Zhang M. X. and Liu P. Z. (2019). Study on sound absorption performance of NBR/high styrene rubber underwater acoustic materials. *Noise and Vibration Control* **39**, 249-253.

[3] Zhang C. and Wang W. F. (2021). Study on the influence of the molecular structure of solution polymerized styrene-butadiene rubber on its acoustic properties. *Material Development and Application* **36**, 34-38.

[4] Gao N. and Hou H. (2017). Low-frequency acoustic properties of a honeycomb-silicone rubber acoustic metamaterial. *Modern Physics Letters B* **31**, 51-62.

[5] Zhu W. W., Xu W., and Pan B. K. (2021). Study on Preparation and Acoustic Performance of Superfine Barium Sulfate/Butyl Rubber Composite. *Chinese Journal of Colloid & Polymer* **4**, 64-67.

[6] Dong J. and Tian P. (2020). Review of underwater sound absorption materials. *IOP Conference Series Earth and Environmental Science* **508**, 5-17.

[7] Li X., Xing S., and Yang J. (2019). Mechanical and Underwater Acoustic Properties of the Polyurethane Elastomers with Different Hard Segment Contents. *IOP Conference Series: Materials Science and Engineering* **490**, 19-32.

[8] Yan Y., Li X. X., and Tian Y. M. (2017). Effect of polyols on the morphology and properties of polyurethane microcellular elastomer based on terephthalic acid. *Applied Chemistry* **34**, 1, 110-1, 116.

[9] Shi K., Jin G., and Liu R. (2019). Underwater sound absorption performance of acoustic metamaterials with multilayered locally resonant scatterers. *Results in Physics* **12**, 95-112.

[10] Sun X. and Liang W. (2016). Cellular structure control and sound absorption of polyolefin microlayer sheets. *Composites Part B: Engineering* **87**, 21-26.

[11] Fu W., Wang L., and Huang J. N. (2022). Preparation and Properties of Polyethylene Glycol Functionalized Graphene Oxide/Polyurethane Composites. *Polymer Materials Science & Engineering* **9**, 101-109.

[12] Hu S., Cai J., and Fu Q. (2019). Preparation and properties of organic montmorillonite/NBR-PVC sound insulation composites. *Materials Science and Technology* **27**, 7-11.

Microstructure and properties of Cu-Fe alloy prepared by spray forming

Su Huang[1], Baosen Lin[1], Zerui Zhang[1] and Jianping Li[1,*]

[1]State Key Laboratory of Rolling and Automation, Northeastern University, Shenyang, Liaoning, 110819, China

[*]Corresponding author's email: ljp@mail.neu.edu.cn

Abstract. To prevent or reduce the issue of macroscopic segregation in metastable immiscible Cu-Fe alloys during solidification, the alloys were produced using spray forming technology and subsequently cold rolled. The Cu10Fe alloy microstructure characteristics were investigated, and its mechanical properties were evaluated and compared with those of cast alloys. The results show that the alloy produced by spray formed has a well-dispersed Fe phase and the grain size is smaller, which effectively suppresses macroscopic segregation. The mechanical properties of spray-formed Cu-Fe alloys are higher than those of cast alloys, and the tensile strength and elongation of the alloys after cold rolling are 584 MPa and 2.2% respectively. The study provides valuable insights into the optimization of spray deposition techniques for producing high quality Cu-Fe alloys.

1. Introduction

Cu-Fe alloy is a novel structural functional material, which possesses high electrical conductivity and high strength, good magnetic and electromagnetic shielding properties. The alloy is inexpensive and abundant in raw materials, and it has good prospects for application in the fields of aerospace, automotive, and electronic communications [1,2]. However, Cu is incompatible with Fe, has a positive enthalpy of mixing, and is a typical immiscible alloy [3]. When Cu-Fe alloys are produced by traditional melting methods, there is an immiscible separation phenomenon between liquid Cu and Fe, which often leads to severe segregation of the microstructure during solidification, limiting their application. Therefore, Cu-Fe alloy research has focused on producing alloys with uniform composition and fine, uniform distribution of Fe particles, which is technically challenging.

Currently, researchers are using alloying, vacuum cast, and double melt mixed casting [4-6] to prepare Cu-Fe alloys are characterized by microstructure uniformity and fine grain size. The studies above have demonstrated that Fe particles produced through conventional melting methods are larger in size or coarser dendritic, with an average size of more than ten micrometers. This limits the further development and industrial application of the alloy by affecting its subsequent processing, properties, and product quality.

Spray forming technology, as a rapid solidification process, is an important part of modern rolling technology, equipment, product development, and innovation platforms. It is a near-final, semi-solid processing technology that combines the processes of molten metal atomization and deposition forming to produce preformed billets directly from liquid metal. Rapid solidification prevents or reduces the segregation of Cu-Fe alloys during solidification, resulting in a homogeneous microstructure and fine grain size. However, there are fewer studies and reports on Cu-Fe alloys produced through spray forming.

Journal of Physics: Conference Series **2808** (2024) 012025 doi:10.1088/1742-6596/2808/1/012025

Therefore, based on the exploration and research of new materials, this study uses the pilot experimental machine for spray forming and then rolling to prepare Cu-Fe alloys and investigate their microstructure and properties. As an effective research method for advanced materials development, this study explores the preparation technique for resolving macroscopic segregation of high-density differential two-phase metals such as Cu-Fe.

2. Materials & methods

A schematic outline of the spray forming equipment and process is presented in Figure 1. For this experiment, Cu-Fe slabs were prepared using pure Fe and electrolytic Cu with spray forming equipment that was independently developed. Spray forming technology involves melting electrolytic Cu and pure Fe in an electromagnetic induction heating furnace crucible and insulating it for a certain period of time. The molten metal is then atomized under an argon pressure of 0.12 MPa and a spray height of 280 mm at 1500°C. The reciprocating movement of the water-cooled receiving plate constantly deposits atomized droplets, ultimately forming Cu-10Fe alloy slabs with near-final density. The deposited samples were milled on both sides and cold rolled in multiple passes, and rolling was terminated when rolled to 0.2 mm thickness.

Optical and scanning electron microscopy were used to characterize the alloy's microstructure. Further, tensile specimens were prepared, and tensile experiments were also performed on the samples after cold rolling to assess the tensile properties of the alloys.

Figure 1. Schematic outline of spray-forming devices

3. Results & discussion

3.1 Deposited characterization

The Cu10Fe alloy ingot's microstructure comprises a light-colored Cu and a dark-colored Fe, as displayed in Figures 2(a) and (b). Fe phases are mainly distributed as dendrites on the Cu matrix. The Fe dendrites are fully developed and form complete dendrites with uneven sizes ranging from 50-150 μm. This phenomenon can be attributed to the natural cooling of the ingot within the mold, where the cooling rate is slow and the temperature is difficult to diffuse. This induces the atoms in the liquid and solid phases to have a sufficiently long time to diffuse fully. As the Fe phase grows, a concentration gradient forms gradually at the interface between the solid and liquid phases. Furthermore, a smaller temperature gradient is formed, producing a greater compositional supercooling that encourages the Fe phase to grow in a dendritic. The dendrites of Fe are well-developed due to the release of latent heat during melting at the crystallization front. Figures 2(c) and (d) display the Cu10Fe alloy microstructure after deposition. The figure illustrates a relatively uniform distribution of Fe particles and fine-grained.

The Fe phase is primarily present in the shape of rods or spherical particles on the Cu matrix, which have an average grain size of 5.5 μm. Spray forming is a process in which high-temperature metal is cooled by high-pressure atomized gas. This results in a faster cooling rate than natural cooling, which shortens the time of elemental diffusion. As a result, macro segregation is significantly reduced and grain refinement is achieved.

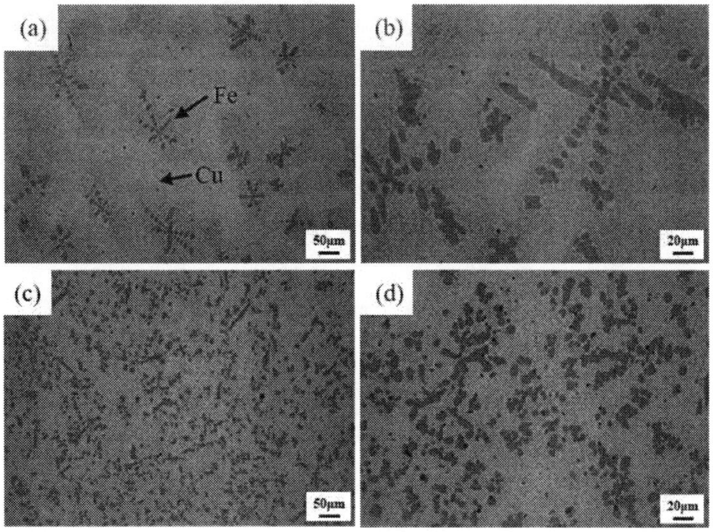

Figure 2. OM microstructure of samples: (a) and (b) cast; (c) and (d) deposited.

3.2 Microstructure characteristics of alloys after cold rolled

The microstructure of Cu10Fe alloy after cold rolling deformation is depicted in Figure 3. The figure illustrates that, under rolling stress, Fe grains elongate into bands in the direction of cold rolling after undergoing large rolling deformation. However, the Fe phase is not completely deformed and its deformation is inhomogeneous. It is rolled into a band and spindle distribution in the Cu matrix, which is a soft-phase Cu and hard Fe two-phase coordinated deformation of the results. During cold rolling processing, the Cu phase yield stress and Young's modulus are relatively low, and the Cu phase preferentially undergoes plastic deformation, while the relatively hard Fe phase remains elastic. When the flow stress of the Cu matrix increases due to work-hardening from cold rolling, the Fe phase performs plastic deformation and evolves into a band shape to achieve coordinated deformation. Further observation showed that dendritic Fe basically disappears in the cast alloy after rolling, and Fe phase distribution is more uniform in the deposited state after rolling than in the cast state. However, there are still fine spherical Fe particles in the deposited state that have not been deformed. These partially spherical particles present in the matrix keep the particles spherical because the stress in the direction of rolling is smaller than that in the dendritic direction, resulting in the particles maintaining their spherical shape.

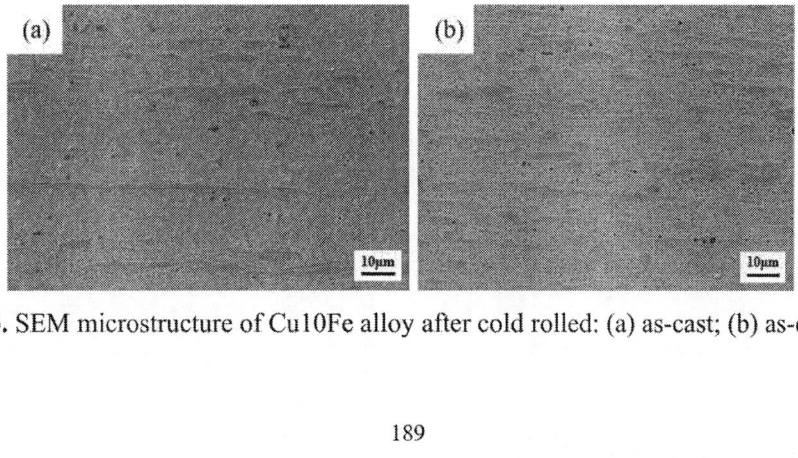

Figure 3. SEM microstructure of Cu10Fe alloy after cold rolled: (a) as-cast; (b) as-deposited.

3.3 Mechanical property

Figure 4 displays the tensile curves of the cast and deposited samples after cold rolling. The plot confirms that the tensile strength of the deposited samples after cold rolling deformation is better than that of the cast ones. Specifically, the tensile strength of the cold-rolled deposited samples reached 584 MPa with 2.2% elongation, whereas the cast samples had only 542 MPa tensile strength with 3.9% elongation. In this work, high-strength samples were achieved by accumulating numerous dislocations due to cold rolling deformation as well as the elongation of Fe phases into bands, grain breaking, and refinement after rolling. The spray-deposited sample has finer and more uniformly distributed grains compared to the as-cast state, resulting in a more pronounced grain refinement enhancement in the Cu matrix.

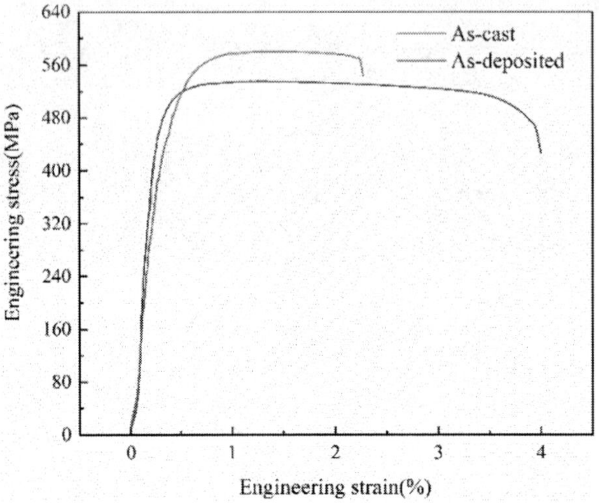

Figure 4. Tensile curves of the Cu10Fe alloy after cold rolled.

4. Conclusions

(1) The Cu10Fe alloy was prepared using spray forming technology, with the Fe phase distributed in the form of rods or spheres. The grains are finer compared to the Fe dendritic grains found in casting samples. Compared to the Fe dendrites in the as-cast samples, the grains are of a finer size and are distributed more diffusely and evenly within the Cu matrix.

(2) During cold rolling deformation, coordinated deformation occurs between the Fe and Cu phases, with the Fe phase being elongated and evolving into bands and spindles.

(3) The tensile strength of the cold rolled spray-formed Cu10Fe alloy was higher than the cast sample at 584 MPa with an elongation of 2.2%.

Acknowledgments

This paper is one of the phase results of the National Natural Science Foundation of China (51274063).

References

[1] Funkenbusch, P.D., Courtney, T.H. (1981) Microstructural strengthening in cold worked in situ Cu–14.8 Vol. % Fe composites. Scr. Mater.,15: 1349.

[2] Sarkar, S., Srivastava, C., Chattopadhyay, K. (2018) Development of a new class of high strength copper alloy using immiscibility route in Cu-Fe-Si system: Evolution of hierarchical multi-scale microstructure. Mater. Sci. Eng., A, 723: 38-47.

[3] Nakagawa, Y. (1958) Liquid immiscibility in copper-iron and copper-cobalt systems in the supercooled state, Acta Mater., 6: 704-711.

[4] Yuan, D.W., Xiao, X.P., Luo, X., Wang, H., Han, B.J., Liu, B.X., Yang, B. (2022) Effect of multistage thermomechanical treatment on Fe phase evolution and properties of Cu-6.5Fe-

0.3Mg alloy. Mater. Char., 185: 111707.
[5] Jeong, Y.B., Jo, H.R, Kim, J.T., Hong, S.H., Kim, K.B. (2019) A study on the micro-evolution of mechanical property and microstructures in (Cu-30Fe)-2X alloys with the addition of minor alloying elements. J. Alloy. Compd., 786: 341-345.
[6] Wang, M., Jiang, Y.B., Li, Z., Xiao, X., Gong, S., Qiu, W.T., Lei, Q. (2021) Microstructure evolution and deformation behaviour of Cu-10 wt% Fe alloy during cold rolling. Mater. Sci. Eng., A, 801: 140379.

Research on multiaxial fatigue life prediction of lifting equipment based on symbol corrected equivalent stress spectrum

Li Chen [1,2,*], **Keqin Ding** [1,2]

[1]China Special Equipment Inspection and Research Institute, Beijing, 100029, China
[2]Technology Innovation Center of Structural Health Monitoring for State Market Regulation, Beijing, 100029, China

[*]Corresponding author's e-mail: ChenLi111528@163.com

Abstract. During the cyclic operation of lifting equipment, critical structural fatigue damage and fracture failure often occur, causing huge economic losses and even leading to major accidents. In practical work, lifting equipment not only bears tensile and compressive loads but also significant shear loads. Under its actual working conditions, it causes multiaxial fatigue damage. This article studies the multiaxial load spectrum obtained by deploying fiber Bragg gratings strain rosette in high-risk and prone-to-cracking areas of lifting equipment and using a multiaxial stress equivalent method based on symbol correction, the multiaxial complex stress spectrum is equivalent to a modified uniaxial stress spectrum. Combined with the developed dual parameter four peak valley rain flow fast counting algorithm program, the cumulative damage caused is calculated using the Miner linear cumulative fatigue damage model. Finally, a prediction is made for the multiaxial load fatigue life of high-risk and prone-to-cracking areas in lifting equipment. The results show that the multiaxial fatigue life prediction method for high-risk and prone-to-cracking areas of lifting equipment based on symbol corrected Von Mises equivalent stress spectrum in this article is more accurate than the prediction results obtained from classical Von Mises equivalent stress spectrum and uniaxial stress spectrum calculations, which to some extent solves the problem of difficult engineering application of current multiaxial fatigue theory.

1. Introduction

With the intensification of international market competition, lifting equipment has been used at full capacity for a long time and operates at an excessive frequency. At the same time, driven by economic interests, exceeding the service life of lifting equipment also occurs from time to time. In recent years, critical structural fatigue damage and fracture failure often occur in lifting equipment during periodic cyclic operations, causing huge economic losses and even leading to major accidents. Therefore, the analysis of structural fatigue damage and prediction of the remaining life of lifting equipment is of great significance for ensuring its service safety and reliability. Wang et al. [1] described the optimal parameter probability distribution model for the load stress history of a gantry crane and predicted the fatigue life of the gantry crane. Qu et al. [2] studied the feasibility of using damage mechanics combined with stress spectrum to evaluate the early life of bridge cranes. Wu et al. [3] studied lifting equipment and used ABAQUS and Fe safe joint simulation methods to analyze the fatigue life of trapezoidal frames, pull rods, and the entire crane. Qi et al. [4] used neural networks to predict the load characteristics of cranes and used Miner's linear damage accumulation theory and linear elastic fracture mechanics method to predict the fatigue life of key parts of lifting equipment structures.

The current research mainly evaluates the fatigue life of key components or critical parts of lifting equipment based on software simulation or unidirectional stress testing. Due to the fact that lifting equipment not only bears tensile and compressive loads but also significant shear loads in actual work, its actual working condition is under the action of multiaxial loads. However, there are no relevant reports on the fatigue damage analysis and life prediction of lifting equipment under multiaxial loads based on real-time monitoring data. This article is based on strain rosette monitoring data in high-risk and prone-to-cracking areas of lifting equipment. A multiaxial stress equivalent method based on symbol correction is adopted to equivalent the multiaxial complex stress spectrum to a modified uniaxial stress spectrum. Combined with the developed dual parameter four peak valley rain flow fast counting algorithm program, the Miner linear cumulative fatigue damage model is used to calculate the cumulative damage caused. Finally, the multiaxial load fatigue life of high-risk and prone-to-cracking areas of lifting equipment is predicted.

2. Materials and methods

2.1. Prediction method for high cycle fatigue life

The stress on the lifting equipment structure is within the elastic range. The general analytical framework is shown in **Figure 1**.

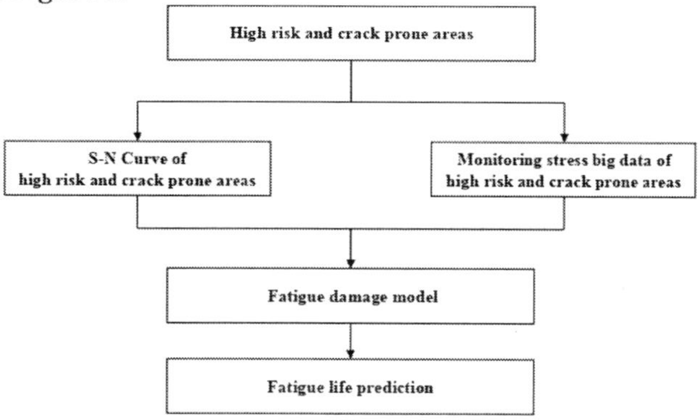

Figure 1. Prediction method for high cycle fatigue life.

As shown in **Figure 1**, the stress spectrum of high-risk and crack-prone areas can be obtained through structural health monitoring systems, and the S-N curve can be achieved through standard specimen fatigue tests.

As shown in **Figure 2**, the standard S-N curve can be expressed as:

$$N\sigma^m = C \tag{1}$$

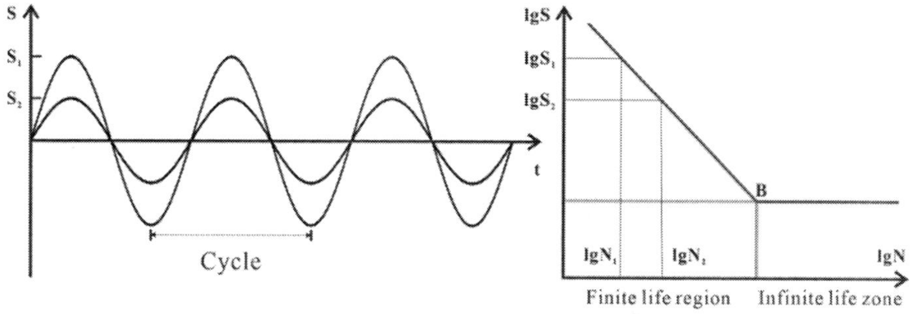

Figure 2. S-N curve.

To obtain the S-N curve of high-risk and crack-prone areas for lifting equipment, comprehensive consideration should be given to the effects of various other factors such as stress mode and surface quality and a correction coefficient $K_{\sigma D}$ should be introduced. The formula of the S-N curve of high-risk and crack-prone areas is:

$$N(K_{\sigma D}\sigma)^m = C \tag{2}$$

The logarithmic form of Formula (2):

$$\lg(N) = \lg(C) - m\lg(K_{\sigma D}\sigma) \tag{3}$$

2.2. Equivalent stress calculation method based on symbol correction

As shown in Formula (4), the total equivalent stress of Von Mises is greater than or equal to zero. In the equivalent process, the compressive stress represented by the negative sign will experience a loss of the negative sign [5]. The definition of equivalent stress based on symbol correction used in this article is as follows:

$$\begin{aligned}\sigma_{eq} &= \frac{1}{\sqrt{2}}\sqrt{\left(\sigma_x-\sigma_y\right)^2+\left(\sigma_y-\sigma_z\right)^2+(\sigma_z-\sigma_x)^2+6\left(\tau_{xy}^2+\tau_{yz}^2+\tau_{zx}^2\right)} \\ &= \frac{1}{\sqrt{2}}\sqrt{(\sigma_1-\sigma_2)^2+(\sigma_2-\sigma_3)^2+(\sigma_3-\sigma_1)^2}\end{aligned} \tag{4}$$

where σ_{eq} is the equivalent stress, and σ_x, σ_y, σ_z, τ_{xy}, τ_{yz}, and τ_{zx} are the stress tensor components, respectively. σ_1, σ_2, and σ_3 are the first, second, and third principal stresses, respectively.

For the common plane stress situation of lifting equipment, the Formula (4) is illustrated as follows:

$$\sigma_{eq} = \sqrt{(\sigma_1)^2+(\sigma_2)^2-\sigma_1\sigma_2} \tag{5}$$

For the 0-45-90° strain rosette, if its monitoring strains for strain are $\varepsilon_{0°}$, $\varepsilon_{45°}$, and $\varepsilon_{90°}$, then its principal strain can be expressed as:

$$\left.\begin{matrix}\varepsilon_1\\ \varepsilon_2\end{matrix}\right\} = \frac{\varepsilon_{0°}+\varepsilon_{90°}}{2} \pm \sqrt{\left(\frac{\varepsilon_{0°}-\varepsilon_{90°}}{2}\right)^2+\left(\frac{\varepsilon_{0°}+\varepsilon_{90°}-2\varepsilon_{45°}}{2}\right)^2} \tag{6}$$

Based on constitutive formulas:

$$\begin{aligned}\sigma_1 &= \frac{E}{1-v^2}(\varepsilon_1+v\varepsilon_2) \\ \sigma_2 &= \frac{E}{1-v^2}(\varepsilon_2+v\varepsilon_1)\end{aligned} \tag{7}$$

Substituting Formula (7) into Formula (4) yields

$$\sigma_{eq} = \frac{E}{1-v^2}\sqrt{(1-v+v^2)(\varepsilon_1^2+\varepsilon_2^2)+(4v-1-v^2)\varepsilon_1\varepsilon_2} \tag{8}$$

Equivalent stress σ_{eq} is an energy concept whose value is always positive and often results in the loss of symbols. To this end, a symbol correction formula is introduced:

$$\sigma_{eq}^{sm}=\mathrm{sign}\big((\sigma_1)^2\mathrm{sign}(\sigma_1)+(\sigma_2)^2\mathrm{sign}(\sigma_2)+\sigma_1\sigma_2\mathrm{sign}(\sigma_1)\mathrm{sign}(\sigma_2)\big) \tag{9}$$

The real (signed) stress time history is obtained:

$$\sigma_{eq}^{true} = \sigma_{eq}^{sm}\sigma_{eq} \tag{10}$$

2.3. Linear fatigue cumulative damage models

Assuming that the loading sequence is composed of a total of l different stress amplitudes such as $\sigma_1, \sigma_2, \cdots, \sigma_l$, with corresponding fatigue lives of $N_{1,f}, N_{2,f}, \cdots N_{l,f}$, and corresponding loading times of $n_1, n_2, \cdots, n_l$, then the damage caused by each stress amplitude is [6]:

$$D_i = n_i / N_{i,f} \tag{11}$$

When Formula (12) is met, it is judged as a material failure:

$$D = \sum_{i=1}^{l} D_i = \sum_{i=1}^{l} n_i / N_{i,f} \geq 1.0 \tag{12}$$

2.4. Stress monitoring data analysis method

Figure 3 shows the improved rain flow counting method (RFM) [7]. By truncating the docking process from the maximum peak, multiple rain flow counts will only need one to complete the final cycle count.

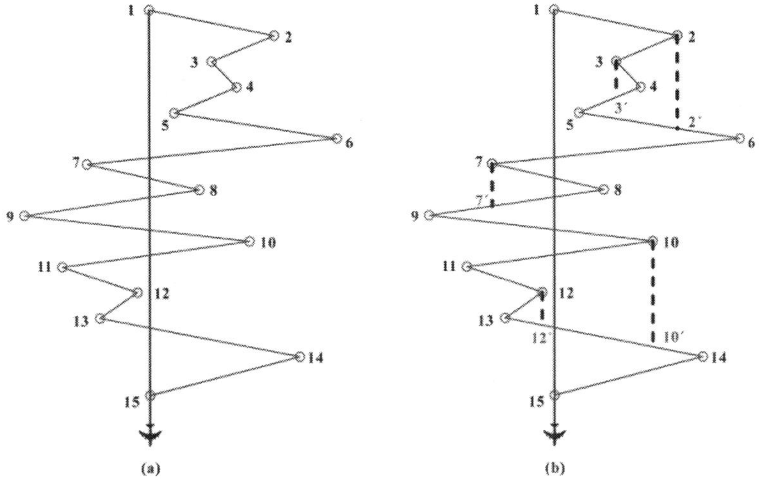

Figure 3. The improved RFM.

The monitoring data and the result data that needs to be saved are defined as independent data variables, avoiding multiple changes in data location and effectively improving algorithm efficiency.

3. Results & discussion

3.1. The distribution of monitoring points for lifting equipment

The distribution of monitoring points of lifting equipment based on FBG is shown in **Figure 4**. Based on monitoring data analysis, A2 is the most dangerous high-risk area, prone to cracking.

Figure 4. FBG distribution of monitoring points of lifting equipment.

3.2. Stress state monitoring data

Figure 5 shows the stress monitoring data of high-risk crack-prone zone A2. Figure 5 shows the 0-degree directional monitoring data, 45-degree directional monitoring data, 90 degree directional monitoring data, first principal stress monitoring data, Von Mises monitoring data and Symbol correction Von Mises monitoring data from left to right and from top to bottom, respectively.

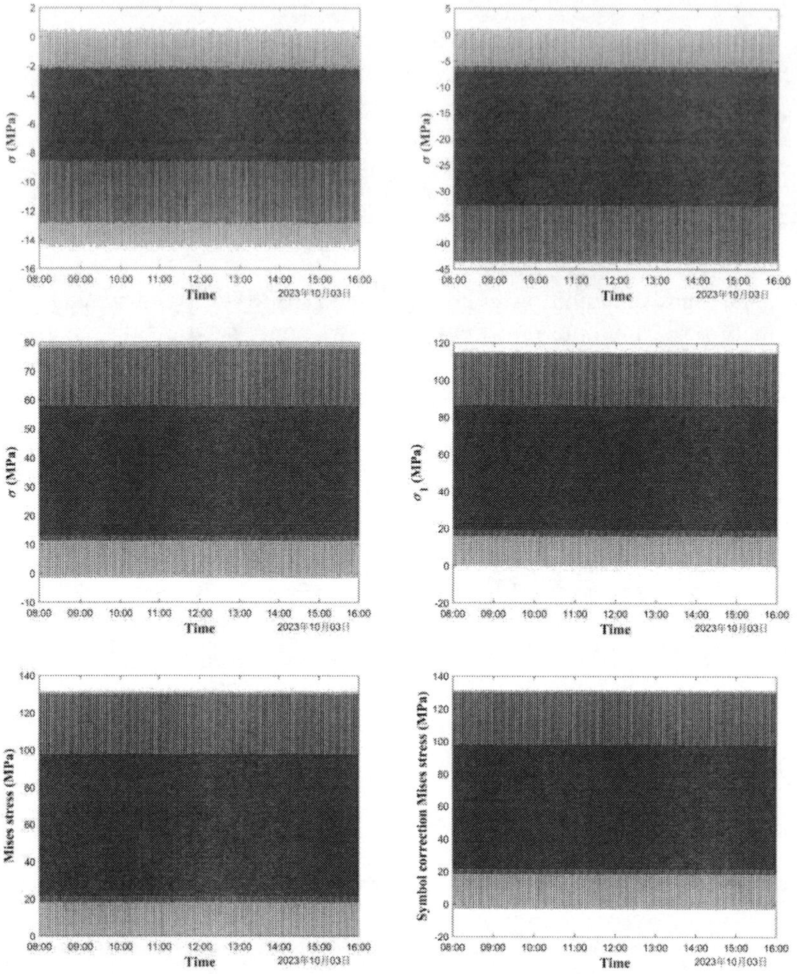

Figure 5. Stress monitoring data of high-risk crack-prone zone A2.

3.3. Results

Table 1. Based on symbol correction Von Mises equivalent stress spectrum and damage.

Grade	Stress amplitude (Mpa)	Frequency (/year)	Accumulated frequency	Damage
1	5	13908	17367	6.93e-13
2	15	2739	3459	1.41e-08
3	25	0	720	0
4	35	0	720	0
5	45	240	720	1.28e-4
6	55	0	480	0
7	65	240	480	0.0061
8	75	240	240	0.0275

Journal of Physics: Conference Series **2808** (2024) 012026 doi:10.1088/1742-6596/2808/1/012026

According to the results in Table 1, the fatigue damage situation and fatigue life of measuring point A2 of the lifting equipment can be further obtained as follows:

$$\text{damage}= 0.0337, \text{Fatiguelife}=\frac{1}{0.0337} = 29(\text{year}) \tag{13}$$

Table 2. Based on Von Mises equivalent stress spectrum and damage.

Grade	Stress amplitude (Mpa)	Frequency (/year)	Accumulated frequency	Damage
1	4.5	14187	17787	2.34e-13
2	13.5	2880	3600	4.91-09
3	22.5	0	720	0
4	31.5	0	720	0
5	40.5	240	720	4.23e-05
6	49.5	0	480	0
7	58.5	240	480	0.0020
8	67.5	240	240	0.0091

According to the results in Table 2, the fatigue damage situation and fatigue life of measuring point A2 of the lifting equipment can be further obtained as follows:

$$\text{damage}= 0.0111, \text{Fatiguelife}=\frac{1}{0.0111} = 89(\text{year}) \tag{14}$$

Table 3. Based on the first principal stress spectrum and damage.

Grade	Stress amplitude (Mpa)	Frequency (/year)	Accumulated frequency	Damage
1	4	14737	17617	7.04e-14
2	12	2160	2880	1.07e-09
3	20	0	720	0
4	28	0	720	0
5	36	240	720	1.23e-05
6	44	240	480	0.0001
7	52	0	240	0
8	60	240	240	0.0026

According to the results in Table 3, the fatigue damage situation and fatigue life of measuring point A2 of the lifting equipment can be further obtained as follows:

$$\text{damage}= 0.0027, \text{Fatiguelife}=\frac{1}{0.0027} = 364(\text{year}) \tag{15}$$

Table 4. Based on 90-degree directional stress spectrum and damage.

Grade	Stress amplitude (Mpa)	Frequency (/year)	Accumulated frequency	Damage
1	2.55	13566	17166	5.71e-16
2	7.65	2880	3600	1.25e-11
3	12.75	0	720	0
4	17.85	0	720	0
5	22.95	240	720	1.08e-07
6	28.05	0	480	0
7	33.15	240	480	5.15e-06
8	38.25	240	240	2.32e-05

According to the results in Table 4, the fatigue damage situation and fatigue life of measuring point A2 of the lifting equipment can be further obtained as follows:

$$\text{damage}= 2.844\text{e-}5,\ \text{Fatiguelife}=\frac{1}{2.844\text{e-}5} = 35162(\text{year}) \tag{16}$$

3.4. Discussion

According to the calculation results from Tables 1 to 4, it can be seen that the life predictions based on symbol corrected Von Mises equivalent stress spectrum, Von Mises equivalent stress spectrum, first principal stress spectrum, and 90-degree directional stress spectrum are 29, 89, 364, and 35162 years, respectively. The design life of lifting equipment is generally 20-30 years, indicating that the multiaxial fatigue life prediction results based on symbol corrected Von Mises equivalent stress spectrum calculation are more accurate and safer. The results obtained by using the Von Mises equivalent stress spectrum and the first principal stress spectrum are relatively risky. The results calculated using a 90-degree stress spectrum (uniaxial stress) deviate significantly and are not suitable for use.

4. Conclusion

This article studies the installation of fiber Bragg gratings in high-risk and prone-to-cracking areas of lifting equipment to obtain their multiaxial load spectra. Using a symbol corrected multiaxial stress equivalent method, the multiaxial complex stress spectrum is equivalent to a modified uniaxial stress spectrum. Combined with the developed dual parameter four peak valley rain flow fast counting algorithm program, the Miner linear cumulative fatigue damage model is used to calculate the cumulative damage caused. Finally, based on the cumulative damage degree D, the fatigue life of high-risk and prone-to-cracking areas of lifting equipment under multiaxial loads is predicted.

According to the fatigue life analysis results of high-risk and prone-to-cracking areas in lifting equipment, the multiaxial fatigue life prediction results calculated based on the symbol corrected Von Mises equivalent stress spectrum are more accurate and safer. The results obtained by using the Von Mises equivalent stress spectrum and the first principal stress spectrum are relatively risky. The results calculated using a 90-degree stress spectrum (uniaxial stress) deviate significantly and are not suitable for use.

The multiaxial fatigue life prediction method based on symbol corrected Von Mises equivalent stress spectrum in high-risk and prone-to-cracking areas of lifting equipment adopted in this article has to some extent solved the problem of difficult engineering application of current multiaxial fatigue theory.

Acknowledgments

This research was supported by the National Key Technologies Research & Development program through 2018YFF0214700.

Reference

[1] Wang, J., H. Qing, Q. Bian. (2021) Stress spectrum and main girder fatigue life estimate for gantry crane. Engineering Journal of Wuhan University, 54(03): 263-268.

[2] Qu, X., X. Wen, X. Zhang. (2021) Assessing and analysis of the fatigue life of the bridge-type crane based on the damage mechanics. Journal of Safety and Environment, 21(03): 1012-1016.

[3] Wu, F., Z. Duan, X. Yang, P. Qi. (2019) Fatigue Life Analysis of Quayside Container Crane's Based on ABAQUS and Fe-Safe. Journal of Lanzhou Jiaotong University, 38(02): 80-85+92.

[4] Qi, Q., C. Li, Q. Dong, Y. Chen, H. Xu. (2023) Prediction of load spectrum for crane life cycle and structural optimal design based on fatigue life. Chinese Journal of Engineering Design, 30(03): 380-389.

[5] Bao, M., D. Shang, H. Chen. (2012) New Fatigue Life Prediction Method Under Random Multiaxial Loading. Journal of Mechanical Strength, 34(05): 737-743.

[6] Mei, J., P. Dong. (2016) A new path-dependent fatigue damage model for non-proportional multi-axial loading. International Journal of Fatigue, 90(sep.): 210-221.

[7] Huang, W. (2017) The frequency domain estimate of fatigue damage of combined load effects based on the rain-flow counting. Marine Structures, 52: 34-49.

Acceleration of the synthesis of single B-YB₂SI₂O₇ feedstocks through modifying the grain size of SIO₂

Zelong Wang[1], Donghui Guo[1], Xinlei Jia[1], Shun Wang[1], Tao Zheng[1], Runze Jin[1], Zhiliang Zhou[1], Ning Zhou[1, 2], and Baosheng Xu[1, 2, *]

[1]Institute of Advanced Structure Technology, Beijing Institute of Technology, Beijing 100081, China

[2]Beijing Key Laboratory of Lightweight Multi-functional Composite Materials and Structures, Beijing Institute of Technology, Beijing 100081, China

[*]Corresponding author's e-mail: xubsh@bit.edu.cn

Abstract: The purpose of this study is to create single-phase β-$Yb_2Si_2O_7$ feedstocks. Firstly, the single-phase β $Yb_2Si_2O_7$ feedstocks were synthesized by using a micro-nano compound method. By adjusting the grain size of SiO_2, the synthesis of β-$Yb_2Si_2O_7$ was accelerated, and its mechanism was further investigated. Based on this, the appearance and phase structure of β-$Yb_2Si_2O_7$ were studied, and the physical properties of β-$Yb_2Si_2O_7$ were characterized. The results reveal that in the solid-phase synthesis of β-$Yb_2Si_2O_7$, nano SiO_2 exhibits greater advantages compared to micron SiO_2 due to its nano-adsorption capacity and heightened reactivity when reacting with Yb_2O_3. This enhanced reactivity accelerates the production of β-$Yb_2Si_2O_7$. When the molar ratio of Yb_2O_3 and SiO_2 is 1:2.3, the single phase β-$Yb_2Si_2O_7$ feedstocks were effectively manufactured by sintering at 1, 500°C for one hour. The β-$Yb_2Si_2O_7$ feedstocks have a particle size distribution with d_{10}, d_{50}, and d_{90} values of 20.56 μm, 33.85 μm, and 52.65 μm, respectively. The β-$Yb_2Si_2O_7$ feedstocks have apparent and tap densities of 1.66 g/cm^3 and 2.05 g/cm^3, respectively. In summary, single-phase β-$Yb_2Si_2O_7$ feedstocks can be the preferred materials for EBCs.

1. Introduction

The ceramic matrix composites represented by silicon carbide fiber-reinforced silicon carbide (SiC_f/SiC) exhibit characteristics such as low density, mechanical properties, and excellent oxidation resistance [1]. Because of its lightweight and temperature resistance, it has emerged as a potential candidate material for hot-end components of aero engines.

The service environment of aero-engines is highly complex and severe, rendering them susceptible to the corrosive effects of molten calcium magnesium aluminum silicate (CMAS) in the atmosphere and water vapor in the combustion chamber, ultimately leading to a decrease in the mechanical properties and service reliability of structural components [2]. Thus, it is imperative to apply an environmental barrier coating (EBC) on the surface of components to prevent or mitigate the corrosion and damage of ceramic matrix composite high-temperature components in the engine gas environment.

Rare-earth silicate is internationally acknowledged as the preferred material for EBCs, which is expected to have a long service life in the gas environment at temperatures up to 1, 300°C or higher [3]. The family of rare-earth silicate materials is huge and the system is rich. To explore the optimal material, numerous scholars have conducted extensive research on the intrinsic characteristics of different rare-

Journal of Physics: Conference Series **2808** (2024) 012027 doi:10.1088/1742-6596/2808/1/012027

earth silicate materials [4]. The results indicate that rare-earth disilicates (REDS), represented by $Yb_2Si_2O_7$ (YbDS), exhibit a lower thermal expansion coefficient compared to rare-earth monosilicates (REMS) and better compatibility with SiCf/SiC substrates [5]. Simultaneously, most REDS experience a polycrystalline transformation phenomenon at high temperatures. However, $Yb_2Si_2O_7$ maintains a stable single phase during the temperature rise process, exhibiting exceptional high-temperature stability.

There are various methods for the synthesis of $Yb_2Si_2O_7$ powders, with the sol-gel method and high-temperature solid phase synthesis method being the most widely employed. Liu et al. prepared $Yb_2Si_2O_7$ powder through the sol-gel process, followed by calcination at 1, 500°C for ten hours [6]. However, they discovered that the synthesized powder was not a single phase but contained a small amount of Yb_2SiO_5 (YbMS) second phase. Zhao et al. employed Yb $(NO_3)_3 \cdot 6H_2O$ and ethyl orthosilicate as raw materials to prepare $Yb_2Si_2O_7$ powder through high-temperature heat treatment of aged gel using the sol-gel method for 60 hours [7]. Li et al. synthesized $Yb_2Si_2O_7$ powder through a high-temperature solid phase synthesis method at 1, 500°C for six hours of calcination, but a small amount of Yb_2SiO_5 second phase was still present [8]. Ueno et al. synthesized $Yb_2Si_2O_7$ powder through a high-temperature solid-phase reaction method, with a molar mixing ratio of Yb_2O_3 and SiO_2 set at 1:2. The raw material powder was uniformly mixed, dried, and calcined in argon at 1, 500°C for 12 hours.

The sol-gel method has numerous process parameters, resulting in a long experimental period and significant time consumption. Conversely, the solid-state reaction method has emerged as the preferred synthesis process for most powder materials due to its simple preparation process and suitability for mass production. Numerous studies have demonstrated the presence of Yb_2SiO_5 second phase in $Yb_2Si_2O_7$ powders synthesized through solid-state synthesis. Additionally, as shown in Figure 1, the Yb_2O_3-SiO_2 binary thermodynamic phase diagram reveals that the preparation conditions for single-phase $Yb_2Si_2O_7$ are exceedingly stringent. In summary, there are limited reports on the high efficiency and advanced preparation technology of single-phase ytterbium disilicate.

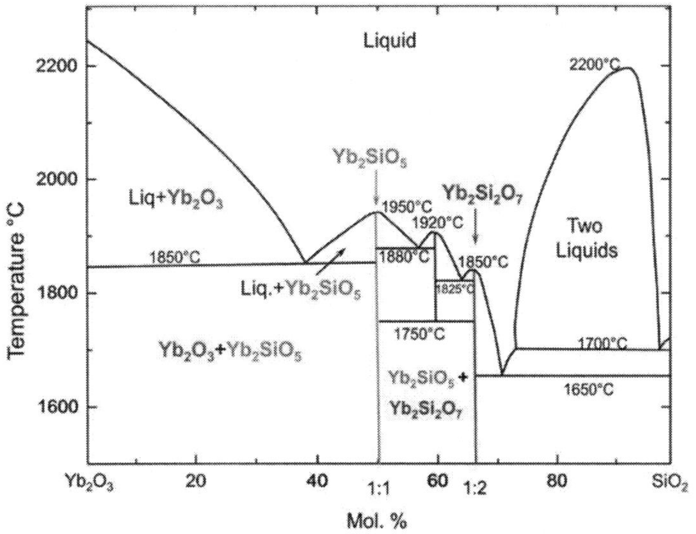

Figure 1. Yb_2O_3-SiO_2 pseudo-binary phase diagram.

In this work, the single phase β-$Yb_2Si_2O_7$ feedstocks were successfully synthesized by using the raw material micro-nano composite method. The consequence of particle size of SiO_2 on the synthesis of single-phase β-$Yb_2Si_2O_7$ feedstocks was investigated. Furthermore, we conducted a comprehensive analysis of the reaction mechanism, feed morphology, and phase structure of the resulting β-$Yb_2Si_2O_7$ feedstocks.

Journal of Physics: Conference Series **2808** (2024) 012027 doi:10.1088/1742-6596/2808/1/012027

2. Experiment

2.1. Raw materials

Micron Yb_2O_3 powder (99.9%, Beijing InnoChem Science and Technology Co., Ltd.), micron SiO_2 powder (99.9%, Shanghai Aladdin Biochemical Technology Co., Ltd.), and nano SiO_2 powder (99.9%, Beijing InnoChem Science and Technology Co., Ltd.) were used as original materials, while deionized water served as an auxiliary material.

2.2. Synthesis of $Yb_2Si_2O_7$ feedstocks

Two systems were designed by using the micro-nano composite method: Yb_2O_3-SiO_2 (Si/Yb) and Yb_2O_3-nano SiO_2 (NSi/Yb). Based on this, the single phase β-$Yb_2Si_2O_7$ feedstocks were successfully synthesized by using the solid-phase synthesis method. The preparation process encompasses three main elements: ball milling, spray drying, and solid-phase synthesis.

Firstly, Yb_2O_3 and SiO_2 were dispersed in deionized water at a molar ratio of 1:2.3, with the addition of 5.0wt% PVA as an adhesive agent. The resulting mixture was then ball-milled for 24 hours to produce a slurry. Successively, the slurry was exposed to spray granulation in a spray drying apparatus (LPG-5, Jiangsu Xianfeng Drying Engineering Co., Ltd., Changzhou, China). The granulated powder was then sintered in a high-temperature muffle furnace at a heating rate of 5 °C/min.

2.3. Characterization

The crystal phases of the calcined powders were identified by an X-ray diffractometer using a scanning rate of 10 °/min and a scanning range of 10° to 70°. The phase content of the calcined powders was quantitatively analyzed by MDI Jade 6, and the results of the phase content are shown in Figure 7.

Scanning electron microscopy (SEM) and energy-dispersive spectroscopy (EDS) were employed to examine the microstructure of two raw materials (micron SiO_2), agglomerated powders, and single-phase $Yb_2Si_2O_7$ materials. Transmission electron microscopy (TEM) was employed to determine the grain size of nano SiO_2 and single-phase $Yb_2Si_2O_7$ materials. Additionally, the Zeta potential analysis instrument was used to measure the potential values of the raw materials. The particle size of β-Yb2Si2O7 feedstocks was characterized by a laser particle size analyzer (BT-9300H). Tap density tester (ZS-102) and Hall flowmeter were used to characterize the tap density and fluidity of β-Yb2Si2O7 feedstocks.

3. Results and discussion

The micro-morphology of the three raw materials was characterized, and the results are shown in Figure 2. The particle size of micro Yb_2O_3 and micro SiO_2 is approximately 10 μm, as shown in Figures 2 (a) and (b). The particle size of nano SiO_2 is around 20 nm, which is significantly different from the previous two, as shown in Figure 2 (c). Studying the particle size and morphology of raw materials is beneficial for the analysis of subsequent experimental results.

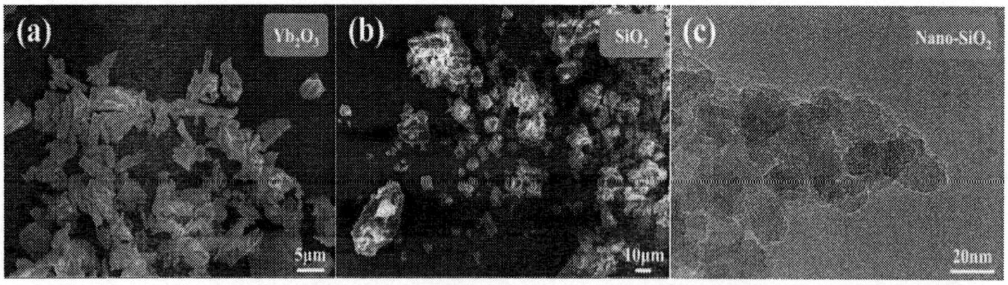

Figure 2. Morphology characterization of three raw materials: (a) SEM of Yb_2O_3; (b) SEM of SiO_2; (c) TEM of nano SiO_2.

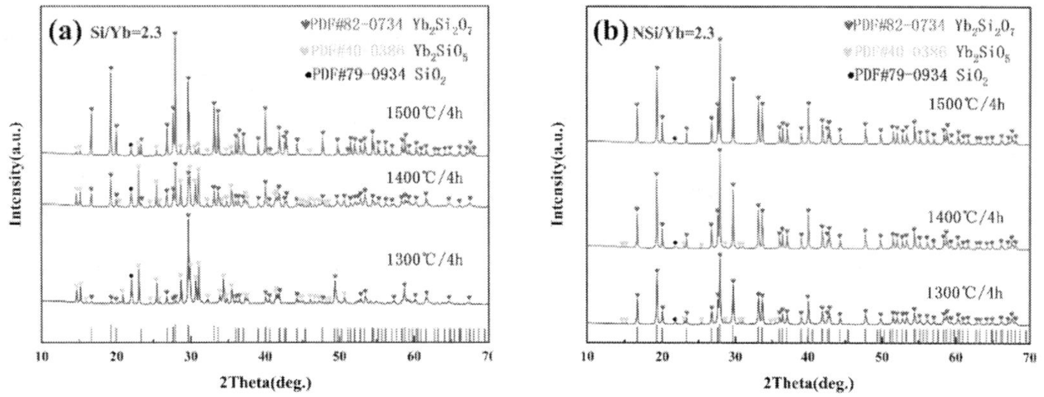

Figure 3. XRD patterns of the two Yb_2O_3-SiO_2 (Si/Yb and NSi/Yb) systems at different sintering temperatures.

The XRD patterns of the two Yb_2O_3-SiO_2 systems (Si/Yb and NSi/Yb) are presented in Figure 3, which displays the sintering results after being heated at different temperatures for four hours. During the sintering process at 1, 300°C, as shown in Figures 3 (a) and (b), it is evident that Yb_2O_3 and SiO_2 in the two systems engage in a reaction to generate a significant amount of metastable X_2-Yb_2SiO_5 (PDF#40-0386) and a small quantity of β-$Yb_2Si_2O_7$ (PDF#82-0734). Based on the above analysis, we can deduce the reaction between Yb_2O_3 and SiO_2 at 1, 300°C and the formation process of β-$Yb_2Si_2O_7$ products as follows:

$$Yb_2O_3 + SiO_2 \rightarrow X_2\text{-}Yb_2SiO_5 \quad (1)$$

$$X_2\text{-}Yb_2SiO_5 + SiO_2 \rightarrow \beta\text{-}Yb_2Si_2O_7 \quad (2)$$

Between 1, 300°C and 1, 500°C, the amount of X_2-Yb_2SiO_5 in Figure 3 (a) decreases gradually as the temperature rises, but it is still present at 1, 500°C. In contrast, the content of X_2-Yb_2SiO_5 in Figure 3 (b) decreases significantly with the increase of calcination temperature and completely transforms into β-$Yb_2Si_2O_7$ at 1, 500°C.

Based on this, we can conclude that β-$Yb_2Si_2O_7$ is synthesized under identical calcination conditions, while nano SiO_2 possesses additional advantages in the reaction process. The explanation behind this phenomenon lies in the fact that nano-sized SiO_2, compared to micron-sized SiO_2, possesses a larger specific surface area, increased surface energy, enhanced activity, as well as stronger adsorption and catalytic properties. These characteristics confer an advantage in its reaction with Yb_2O_3.

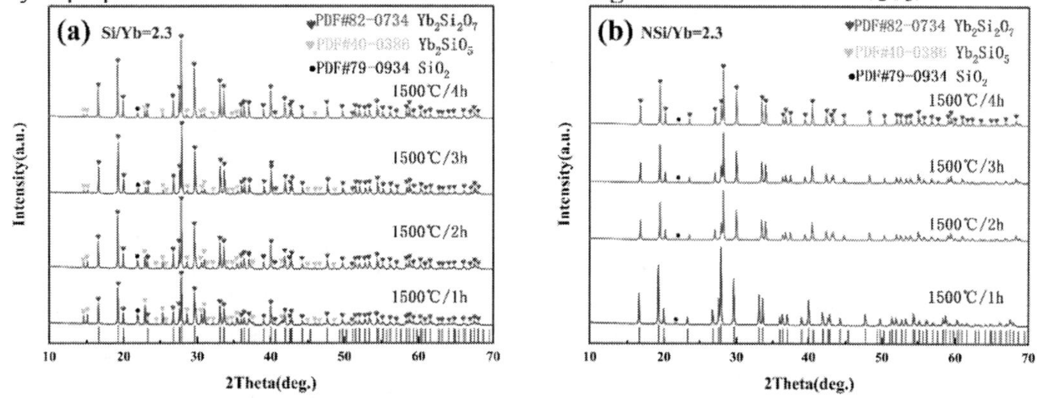

Figure 4. XRD patterns of the two Yb_2O_3-SiO_2 (Si/Yb and NSi/Yb) systems after sintering at 1, 500°C for different hours.

Based on the above results, two Yb_2O_3-SiO_2 systems (Si/Yb and NSi/Yb) were sintered at 1, 500°C for different lengths of time. The XRD patterns of these systems are presented in Figure 4. It can be observed that all peaks exhibit a second phase compared to Figure 4 (a). The Yb_2O_3-SiO_2 (NSi/Yb) system in Figure 4 (b) has completely reacted into a single phase β-$Yb_2Si_2O_7$ after being sintered at 1, 500°C for one hour. This result indicates that nano SiO_2 exhibits a higher reaction preference than micron SiO_2 in solid-phase reactions.

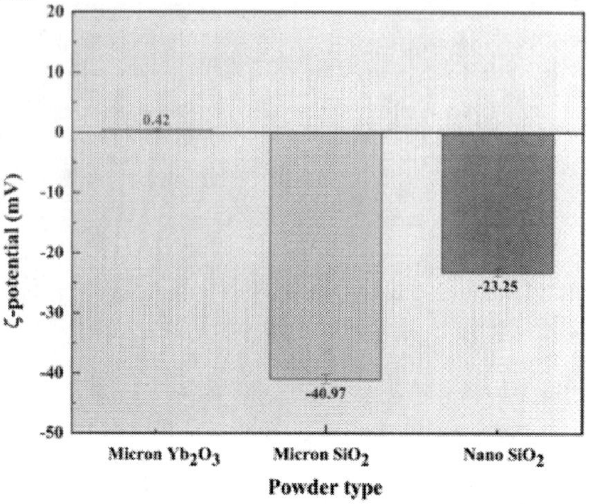

Figure 5. ζ-potential of three raw materials: micron Yb_2O_3, micron SiO_2, and nano SiO_2.

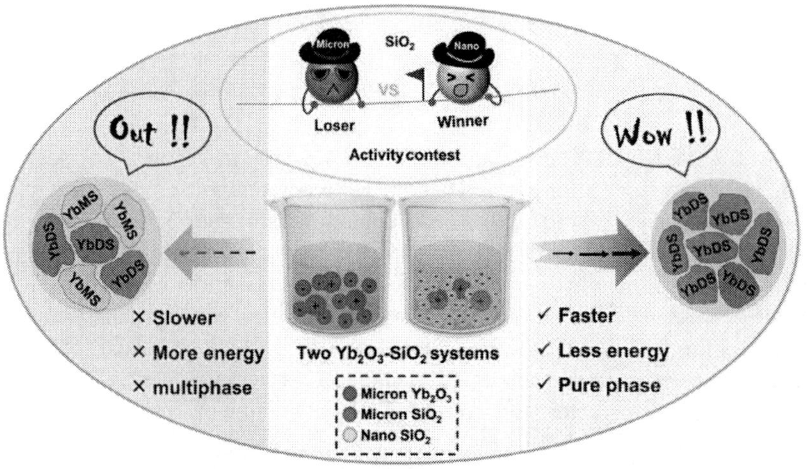

Figure 6. Comparison of reaction advantages of SiO_2 with different grain sizes.

To further investigate the mechanism behind the reaction priority of nano SiO_2, we conducted ζ-potential measurements on three different raw materials: micron Yb_2O_3, micron SiO_2, and nano SiO_2. ζ-potential is a numerical form of the surface charge of a suspended particle, providing insights into its electrostatic behavior [10]. Typically, particles with opposite charges are routinely attracted to each other due to static electricity. Based on the ζ-potential values presented in Figure 5, the charge magnitudes of micron Yb_2O_3, micron SiO_2, and nano SiO_2 are +0.42 mV, -40.9 mV, and -23.25 mV, respectively. Micron Yb_2O_3 is close to electroneutrality, suggesting that electrostatic attraction has minimal influence in the Yb_2O_3-SiO_2 system. The higher the absolute value of ζ-potential is, the more stable the system is. The charge magnitude in micron SiO_2 is approximately twice that of nano SiO_2. This implies that micron SiO_2 possesses significant inertia, but nano SiO_2 exhibits a higher activity. Drawing from the previous results, the Yb_2O_3-SiO_2 (NSi/Yb) system possesses additional advantages in the high-temperature solid-state reaction synthesis of β-$Yb_2Si_2O_7$. In Figure 6, the contrast of reactivity of silicon oxide with different particle sizes is shown. The Yb_2O_3-SiO_2 (NSi/Yb) system can produce single phase β-$Yb_2Si_2O_7$ more quickly and efficiently than the Yb_2O_3-SiO_2 (Si/Yb) system, with a lower required reaction activation energy.

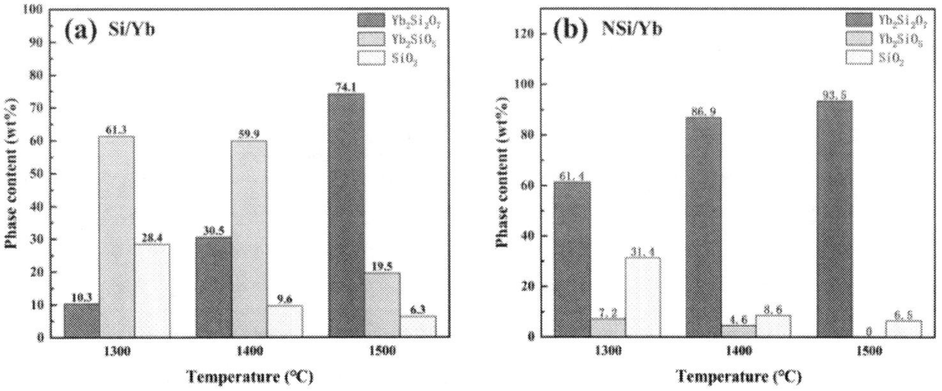

Figure 7. Phase contents of two Yb_2O_3-SiO_2 systems calcined at high temperature for 4 h: (a) Si/Yb; (b) NSi/Yb.

The phase composition of two Yb_2O_3-SiO_2 (Si/Yb and NSi/Yb) systems was calculated after being calcined at different temperatures for four hours. The content of the $Yb_2Si_2O_7$, Yb_2SiO_5, and SiO_2 phases are presented in Figure 7. As seen in Figure 7, the content of Yb_2SiO_5 and SiO_2 decreased significantly with increasing temperature, suggesting a reaction between Yb_2SiO_5 and SiO_2, which further validates the reaction (2). From 1, 400°C to 1, 500°C, the content of Yb_2SiO_5 and SiO_2 decreased rapidly, indicating that an increase in temperature was beneficial for promoting the reaction (2).

Compared to Figure 7 (b), 19.5wt% of the second phase Yb_2SiO_5 remains in Figure 6 (a) after being calcined at 1, 500°C for four hours. Therefore, the micro-nano composite system of Yb_2O_3-SiO_2 (NSi/Yb) has more evident advantages in the synthesis of single-phase $Yb_2Si_2O_7$. It has been reported that during the spraying process of $Yb_2Si_2O_7$ feed, the loss of silicon source results in the formation of a second phase (Yb_2SiO_5) [11]. The introduction of an appropriate amount of SiO_2 into the system can compensate for the loss of silicon source and reduce the presence of the second phase.

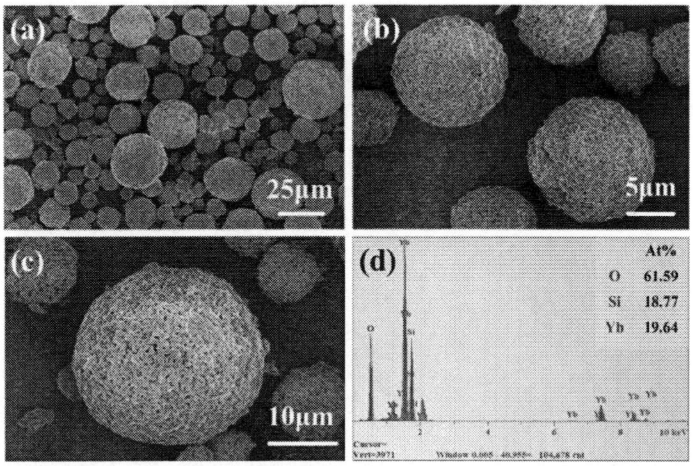

Figure 8. SEM images of β-$Yb_2Si_2O_7$ feedstocks: (a-c) surface topography; (d)EDS.

The SEM results of $Yb_2Si_2O_7$ feedstocks prepared by using the Yb_2O_3-SiO_2 (NSi/Yb) micro-nano composite are presented in Figure 8. The surface morphology of the $Yb_2Si_2O_7$ feedstocks is illustrated in Figures 8 (a)-(c). The prepared feedstocks exhibit a uniform and spherical morphology with a smooth surface, which facilitates an improved feedstock flow and is suitable for plasma spraying applications [12]. The EDS results of the $Yb_2Si_2O_7$ feedstocks are presented in Figure 8 (d). The content of Yb, Si, and O elements is 19.64%, 18.77%, and 61.59%, respectively, which is close to the standard ratio of 2:2:7. This provides strong evidence that the $Yb_2Si_2O_7$ feedstocks were successfully prepared through the micro-nano combination of Yb_2O_3-SiO_2 (NSi/Yb).

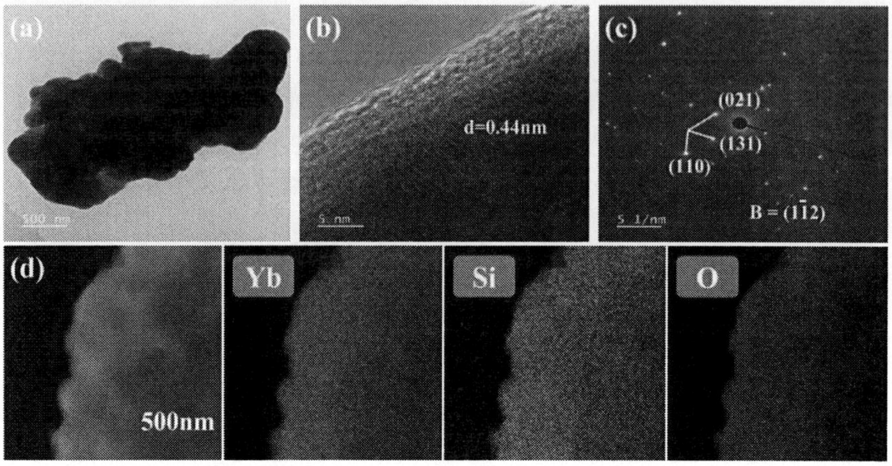

Figure 9. TEM analysis of the β-$Yb_2Si_2O_7$ feedstocks: (a) TEM; (b) HRTEM; (C) SAED; (d) Mapping.

Figure 9 displays the TEM analysis of β-$Yb_2Si_2O_7$ feedstocks. Although the feedstocks exhibit clear agglomeration in Figure 9 (a), their particle composition remains uniform. Figures 9 (b)-(c) present the HRTEM and SAED patterns of β-$Yb_2Si_2O_7$. Clear lattice orientation and white spots are observable, which indicates that the β-$Yb_2Si_2O_7$ are effectively generated. Figure 9 (d) shows the distribution of Yb, Si, and O elements, and it is clear that the three elements are evenly distributed.

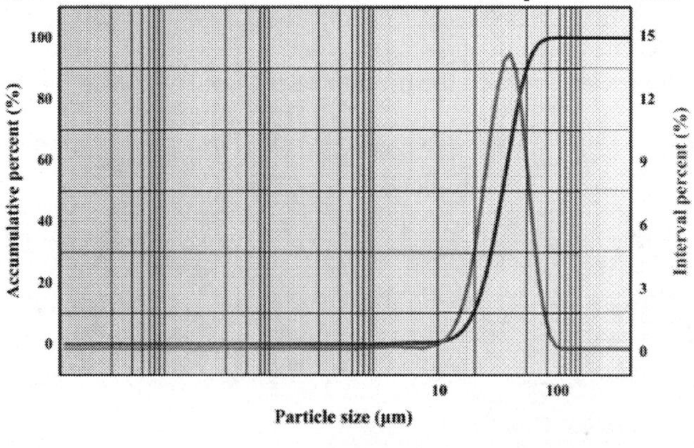

Figure 10. Particle size distribution of the β-$Yb_2Si_2O_7$ feedstocks.

Table 1. Physical properties of β-$Yb_2Si_2O_7$ feedstocks.

d_{10} (μm)	d_{50} (μm)	d_{90} (μm)	Apparent density (g/cm^3)	Tap density (g/cm^3)	Flowability (s/50 g)
20.56	33.85	85.65	1.66	2.05	54.7

The particle size distribution of the β-$Yb_2Si_2O_7$ feedstocks is displayed in Figure 10. By analyzing the data, we can obtain the significant parameters: d_{10}, d_{50}, and d_{90}. The d_{50} value signifies that 50% of the particles have a size equal to or less than this value, while d_{10} and d_{90} carry similar meanings. The d_{50} of the prepared β-$Yb_2Si_2O_7$ feedstocks is 33.85 μm, making it suitable for atmospheric plasma spraying. Additionally, the physical properties of the β-$Yb_2Si_2O_7$ feedstocks were characterized and summarized in Table 1. The apparent density and tap density are 1.66 g/cm^3 and 2.05 g/cm^3, individually, while the feeding flowability reached 54.7 s/50 g. In conclusion, the above physical properties directly suggest that β-$Yb_2Si_2O_7$ feedstocks are appropriate for atmospheric plasma spraying of EBCs.

4. Conclusions

In this work, single-phase β-$Yb_2Si_2O_7$ feedstocks for atmospheric plasma spraying were prepared successfully for the first time by using the micron-nano composite method. The consequence of the particle size SiO_2 on the generation of β-$Yb_2Si_2O_7$ was thoroughly investigated. The mechanism of accelerating the generation of β-$Yb_2Si_2O_7$ through the use of nano SiO_2 was investigated. Additionally, the physical properties of these β-$Yb_2Si_2O_7$ were estimated. The significant conclusions are as follows:

(1) The β-$Yb_2Si_2O_7$ feedstocks were prepared by setting the molar ratio of Yb_2O_3 to nano SiO_2 as 1:2.3 and sintering at 1, 500°C for one hour.

(2) Compared to micron SiO_2, nano SiO_2 displays exceptional reactivity, leveraging its nano-adsorption effect to offer a reaction advantage in solid-phase synthesis. This advantage enables the efficient and accelerated preparation of single-phase β-$Yb_2Si_2O_7$ feedstocks.

(3) The synthesized β-$Yb_2Si_2O_7$ feedstocks possess a spherical shape with a uniform and smooth surface. The feedstocks have a d_{50} value of 33.85 μm. Additionally, the apparent density and tap density of β-$Yb_2Si_2O_7$ feedstocks were 1.66 g/cm^3 and 2.05 g/cm^3, respectively.

Acknowledgments

This work is supported by the National Science and Technology Major Project (2017-VI-0020-0093) and the National Natural Science Foundation of China (Nos.12090031 and 11602125).

References

[1] Z. Tian, L. Zheng, W. Hu, et al. Tunable properties of $(Ho_xY_{1-x})_2SiO_5$ as damage self-monitoring environmental/thermal barrier coating candidates, Scientific Reports, 9 (2019), 415.

[2] W. Song, S. Yang, M. Fukumoto, et al. Impact interaction of in-flight high-energy molten volcanic ash droplets with jet engines, Acta Materialia, 171 (2019), 119-131.

[3] D. Guo, J. Liu, J. Shi, et al. Synthesis and characterization of single-phase X2-Lu_2SiO_5 nanostructured feedstocks for advanced plasma-sprayed environmental barrier coatings, Surface and Coatings Technology, 476 (2024), 130, 230.

[4] L. Dong, M. J. Liu, X. F. Zhang, et al. Pressure infiltration of molten aluminum for densification of environmental barrier coatings, Journal of Advanced Ceramics, 11 (2022), 145-157.

[5] P. Chen, P. Xiao, X. Tang, et al. Corrosion behavior and failure mechanism of SiC whisker and c-$AlPO_4$ particle-modified novel tri-layer $Yb_2Si_2O_7$/mullite/SiC coating in burner rig tests, Journal of Advanced Ceramics, 11 (2022), 1, 901-1, 917.

[6] J. Liu, L. Zhang, Q. Liu, et al. Calcium-magnesium-aluminosilicate corrosion behaviors of rare-earth disilicates at 1, 400°C, Journal of the European Ceramic Society, 33 (2013), 3, 419-3, 428.

[7] C. Zhao, F. Wang, and Y. Sun. Synthesis and characterization of β-$Yb_2Si_2O_7$ powders, Ceramics International, 39 (2013), 5, 805-5, 811.

[8] B. Li, X. Zhang, Z. Zhang, et al. Preparation and characterization of a novel nanostructured $Yb_2Si_2O_7$ feedstock used for plasma-sprayed environmental barrier coatings, Ceramics International, 49 (2023), 10, 897-10, 905.

[9] F. Mao, Y. Niu, X. Zhong, et al. Oxidation behaviors and mechanisms of Yb_2O_3-doped silicon coatings fabricated by vacuum plasma spray, Ceramics International, 47 (2021), 19, 906-19, 913.

[10] P. Zhang, Y. Zhao, J. Huang, et al. Enhanced mechanical and wear properties of Ni-W-SiC composite coatings by the synergistic influence of micro-nano SiC mixture, Surface and Coatings Technology, 467 (2023), 129, 678.
[11] H. Wang, J. Zhang, L. Sun, et al. Microstructure and phase composition evolution of dual-phase ytterbium silicate coatings plasma sprayed from stoichiometric $Yb_2Si_2O_7$ feedstock powder, Surface and Coatings Technology, 437 (2022), 128, 373.
[12] F. Zhou, C. Deng, Y. Wang, et al. Characterization of multi-scale synergistic toughened nanostructured YSZ thermal barrier coatings: From feedstocks to coatings, Journal of the European Ceramic Society, 40 (2020), 1, 443-1, 452.

Journal of Physics: Conference Series **2808** (2024) 012028 doi:10.1088/1742-6596/2808/1/012028

Experimental study on frost resistance of steel fiber reinforced concrete

Jiuyang Li [1], Yuepeng Zhu [1, *], Jingwei Luo [1], Xinmei Fan [1], Xiaoyu Wang [1], Li Chen [1], Fan Li [2] and Zicheng Zhang [3]

[1]School of Civil Engineering, Changchun Institute of Technology, Changchun, 130012, China;
[2]School of Traffic Engineering, Wuhan Vocational College of Communications, Wuhan, 430065, China;
[3]China Fifth Construction Co., LTD., ChinaConstucion First Group, Shenzhen, 518000, China

[*]Corresponding author's e-mail: zyp13865950257@163.com

Abstract: To study the application of steel fiber reinforced concrete in cold areas, a rapid freeze-thaw test was carried out for C30 grade steel fiber reinforced concrete with a steel fiber content of 0.4%, 0.8%, 1.2%, and 1.6%, respectively. The results show that in the case of increasing freezing and thawing times, the failure pattern of the specimen is aggravated obviously. However, with the addition of steel fiber, the mass loss rate of the specimen is gradually reduced, and the initial tangent modulus and flexural properties are improved. The freezing resistance of the specimens has been improved continuously. Steel fiber can effectively resist the durability damage of concrete after the freeze-thaw cycle.

1. Introduction

Concrete is an indispensable material in construction projects and is widely used in various projects because of its robustness and durability [1]. To meet the ever-increasing demand for performance and safety, steel fiber reinforced concrete has been widely used and applied because of its better mechanical properties than ordinary concrete [2]. However, at present, many scholars focus on the mechanical properties of steel fiber-reinforced concrete. In cold areas or places with large temperature fluctuations, Freeze-thaw damage may be the main reason for the failure of concrete structures [3-5]. Therefore, it is particularly important to study the durability of steel fiber-reinforced concrete after repeated freezing and thawing.

In this experiment, beam specimens were prepared by adding steel fibers of different proportions (0, 0.4%, 0.8%, 1.2%, 1.6%), and tested by rapid freeze-thaw method to study the failure forms, mass loss rate, and initial tangent modulus and bending toughness changes of steel-fiber concrete before and after freeze-thaw cycles. The test results were analyzed to understand the effect of steel fiber incorporation on the durability of concrete.

2. Test preparation

2.1 Design of test materials and mix ratio

The cement used in the test was P.O 42.5 cement. The coarse aggregate size of 4.5-16 mm ordinary

Journal of Physics: Conference Series **2808** (2024) 012028 doi:10.1088/1742-6596/2808/1/012028

gravel. Fine aggregate is natural river sand with a particle size of 2.61. Milling steel fiber is selected, and the water is ordinary tap water.

The SF is referred to as steel fiber below, and the SF concrete mix ratio is calculated according to the C30 grade and according to the "Ordinary Concrete Mix ratio Design Regulations" [6]. The benchmark fit is shown in Table 1. The SF was added to the sample by volume ratio of 0.4%, 0.8%, 1.2% and 1.6%, respectively. The samples were numbered 100mm×100mm×400mm and named SF04, SF08, SF12 and SF16 respectively.

Table 1. C30 concrete mix ratio.

Water /kg·m^{-3}	Cement /kg·m^{-3}	Cobble /kg·m^{-3}	Sand /kg·m^{-3}	Water-cement ratio	Sand ratio /%
184	307	1155	804	0.62	41%

2.2 Freeze-thaw test scheme

The freeze-thaw test of steel fiber concrete was carried out by using the fast freezing method in the Standard of Test Methods for Long-term Performance and Durability of Ordinary Concrete [7], and the TDR-16 fast freezing and thawing testing machine was used. The content of steel fiber 0.4%, 0.8%, 1.2%, and 1.6% were divided into 4 groups, and ordinary concrete (PC) was set as the control group.

To ensure the uniform dispersion of steel fibers, cement, sand, and stone are added to the mixer first, and then the steel fibers are added to the mixer in the way of mixing and throwing. After mixing is completed, they are poured into the mold, stood for 24 h, and put into the curing room. The curing specimen 24d is soaked in (20±2) water and taken out to the freeze-thaw testing machine for the freeze-thaw cycle after 4d. During the cycle, the temperature of the specimens was (-18±2)°C and (5±2)°C, respectively, and the mass and initial tangent modulus of the specimens were tested after 50 freeze-thawing sessions.

3. Test results and analysis

3.1 Failure pattern

Figure 1 shows the schematic diagram of surface damage of PC, SF04, SF08, SF12, and SF16 after 200 freeze-thawing sessions. As can be seen from the figure, the mortar of the outer layer of the PC fell off, and the coarse aggregate was exposed, causing the most serious freeze-thaw damage. The surface protection layer of SF12 and SF16 is relatively complete. According to the above experimental phenomena, SF significantly improves the frost resistance of concrete, and the surface sample damage slowly decreases with the increase of SF content.

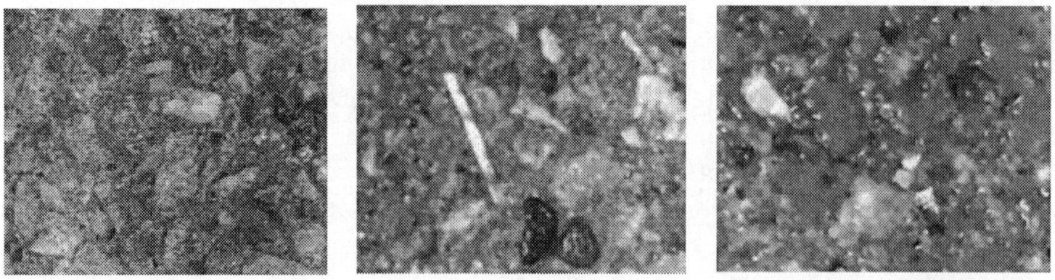

(a) PC (b) SF04 (c) SF08

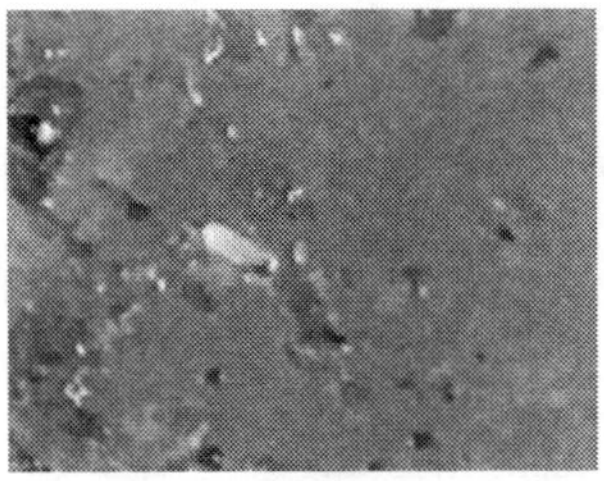

(d) SF12

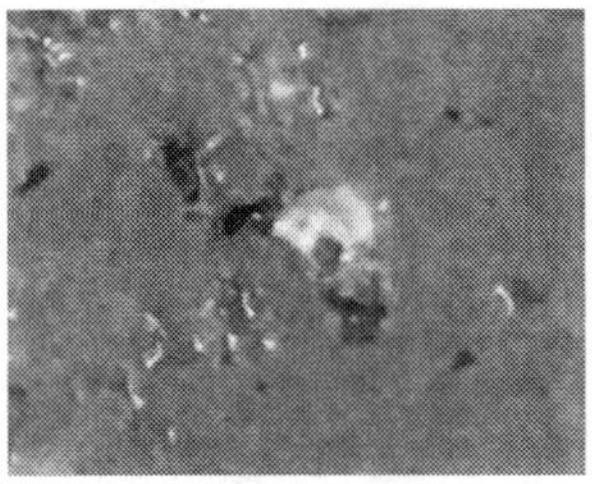

(e) SF16

Figure 1. The morphology of each specimen was frozen and thawed 150 times.

3.2 Quality loss rate

The mass damage rate is an important index to judge the freezing resistance of the specimen. With the augment in the cycle of freezing and thawing of concrete, the surface mortar continues to fall off, and the specimen reaches the failure state when the mass loss rate reaches more than 5%.

According to the test, the quality of the specimen was measured and the data were sorted out. The specific results are shown in Table 2 and Figure 2.

Table 2. Quality loss rate of specimens before and after freezing and thawing.

Frequency of freeze-thaw	Mass loss rate (%)				
	PC	SF04	SF08	SF12	SF16
50	1.11	0.92	0.72	0.52	0.58
100	2.59	1.87	1.58	1.06	0.91
150	4.23	3.09	2.68	2.18	1.81
200	6.28	4.13	3.47	3.03	2.88

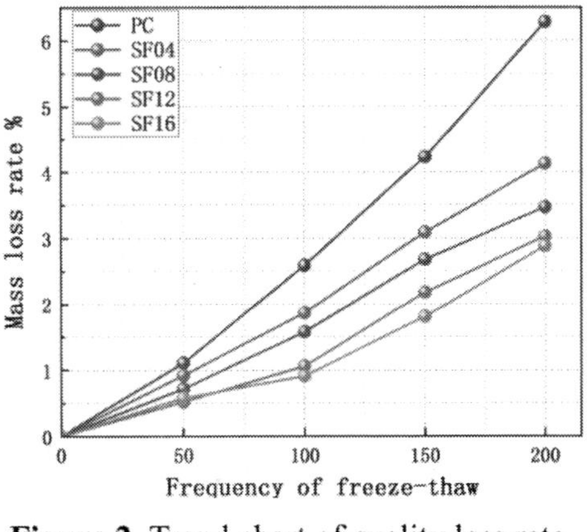

Figure 2. Trend chart of quality loss rate.

According to the figure and table, the order of freeze-thaw resistance loss of concrete specimens after 200 cycles of freezing and thawing is SF16<SF12<SF08<SF04<PC. The mass loss rate of PC specimens exceeds 5%, reaching the failure condition. The mass loss rate of other types of SF specimens is below 5%, indicating the good frost damage resistance performance of SF concrete. The addition of steel fiber has good protective effects on the quality loss of concrete under the cycle of freezing and thawing conditions. The main reason is that under freeze-thaw conditions, the main damage to concrete is surface mortar peeling. With the increase of freeze-thaw cycles, the surface peeling situation becomes more obvious. After adding steel fibers, effectively reducing the crack width of concrete specimens, thereby suppressing the occurrence of mortar peeling on the surface of specimens.

3.3 Initial tangent modulus

The initial tangent modulus of concrete is an important index to test the internal structure change of concrete under a freezing and thawing environment. When the relative initial tangent modulus loss rate of the specimen after the freeze-thaw cycle reaches below 60%, the specimen reaches the failure state.

After the initial tangent modulus of the specimen was determined according to the test, Table 3 was sorted out and Figure 3 was drawn.

Table 3. Initial tangent modulus under freezing and thawing cycles.

Frequency of freeze-thaw	Relative initial tangent modulus (%)				
	PC	SF04	SF08	SF12	SF16
0	100	100	100	100	100
50	86.1	91.5	95.5	93.5	97.4
100	81.5	87.9	90.7	91.8	94.3
150	74.3	81.1	84.8	88.5	90.4
200	66	75.3	80.2	85.4	86.9

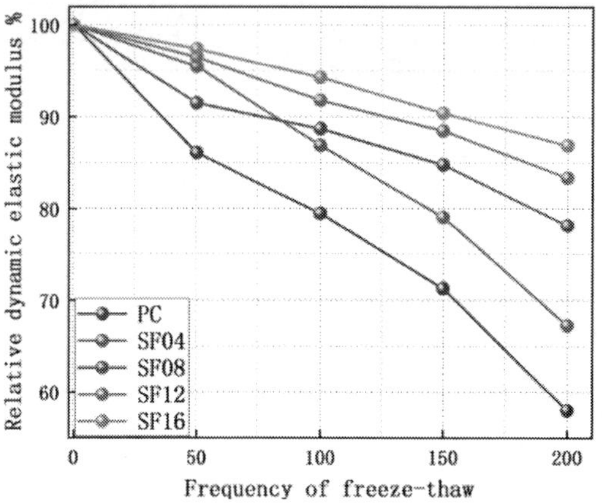

Figure 3. Trend chart of relative initial tangent modulus rate.

It can be seen from the chart that after 200 cycles of freezing and thawing, the relative initial tangent modulus of PC is lower than 60% of the original specimen, reaching the failure condition, while the loss of initial tangent modulus of steel fiber-reinforced concrete specimen is significantly inhibited. This is because the initial tangent modulus is determined by the ultrasonic probe transmitting sound waves and measuring the passing rate of sound waves between aggregate and mortar inside the specimen. With the increase of freeze-thaw times, cracks and porosity in the specimen gradually increase, and the sonic pass rate gradually decreases, resulting in a continuous decline in the initial tangent modulus of the specimen. However, the incorporation of SF can effectively inhibit cycle freezing and thawing failure of the specimen, reduce the occurrence of internal holes and cracks, and thus improve the initial tangent modulus of the specimen. The experiment shows that the incorporation of SF can improve the freezing resistance of concrete.

3.4 Folding strength

The flexural strength of concrete is one of the important indicators for determining the quality of concrete. In this experiment, the flexural strength of concrete before and after the cycle of freezing and thawing was tested. The concrete specimens before the cycle of freezing and thawing and after 200 cycles of freezing and thawing were placed on a 100 KN microcomputer-controlled electro-hydraulic servo testing machine for bending toughness testing. The loading rate was 0.01 KN/s, and a three-point loading method was used as shown in Figure 4, with a support spacing of 300 mm.

Concentrated load acts on the midpoint of the beam specimen.

Figure 4. Loading diagram of bending toughness.

According to the measured data of the test, the calculation was carried out. Table 4 was obtained after sorting, and the histogram was drawn as shown in Figure 5.

Table 4. Flexural strength before and after freeze-thaw.

Frequency of freeze-thaw	Flexural strength				
	PC	SF04	SF08	SF12	SF16
0	6.2	6.51	7.13	7.63	8.16
200	3.61	4.07	4.72	5.15	5.69

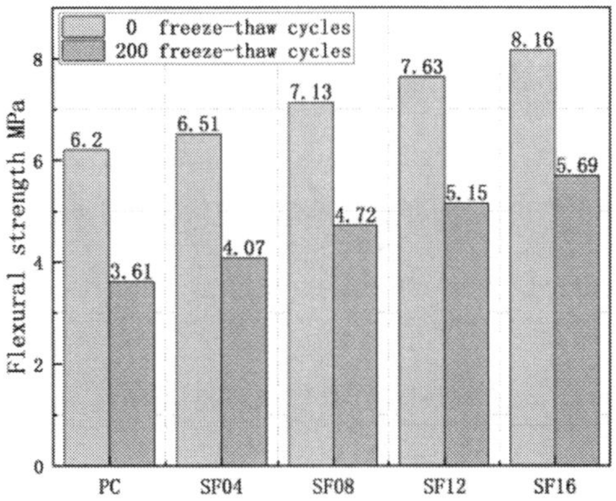

Figure 5. Flexural strength before and after freeze-thaw.

It can be seen from the chart that after 200 cycles of freezing and thawing, the bending strength of PC decreased by 42%, and the bending strength loss was the most serious. Moreover, with the incorporation of SF, the bending loss gradually decreased, and the bending damage of SF16 was the lowest of 30%, indicating that with the incorporation of SF, the freezing resistance of concrete specimens gradually increased, which may be because the resistance to freezing damage of concrete specimens was gradually enhanced with the incorporation of SF. The steel fiber network structure with staggered distribution is formed inside the concrete, and its bonding ability to the concrete is gradually enhanced, which inhibits the internal cracks of the concrete during the cycle of freezing and thawing. Moreover, the rough appearance of the concrete after freeze-thaw also provides friction for the steel fiber during the tension process, thus enhancing the freeze resistance of the concrete as a whole.

4. Conclusion

(1) With the increase of SF content, the freezing erosion resistance of concrete is gradually enhanced.

After 200 cycles of freezing and thawing, the outer surface of ordinary concrete is seriously eroded. SF can effectively inhibit the falling off of the protective layer of concrete and has a good inhibition effect on concrete damage.

(2) After 200 cycles of freezing and thawing, the mass loss rate and initial tangent modulus of ordinary concrete both reached the damage limit, while the specimens with steel fibers effectively inhibited the generation of cracks and holes, effectively resisted the freeze-thaw damage of concrete, and significantly transformed the cycle of freezing and thawing resistance of concrete.

(3) With the increase of SF content, not only is the bending strength of concrete improved, but also the formed network structure can effectively delay the flexural loss rate after the cycle of freezing and thawing. After 200 cycles of freezing and thawing, the bending loss rate of ordinary concrete is 42%, while the bending loss rate of steel fiber concrete with 16% content is reduced to 30%, and the SF has a significant effect on the bending strength loss after the cycle of freezing and thawing.

References

[1] LI Jiuyang, Luo Jingwei, Fan Xinmei et al. Visual analysis of research hotspot and Frontier trend of steel fiber reinforced concrete [J]. Journal of Liaoning Normal University (Natural Science Edition),2023,25(02):55-63.
[2] LI Jiuyang, Fan Xinmei, Luo Jingwei et al. Effect of Steel fiber type and content on Mechanical properties of concrete [J]. Journal of Xinxiang University,2023,40(09):49-52.
[3] Niu Ditao, Jiang Lei, Bai Min. Experimental study on frost resistance of steel fiber reinforced concrete [J]. Civil, Architectural and Environmental Engineering,2012,34(04):80-84+98.
[4] Gao Danying, Xie Xiaopeng. Properties of steel fiber reinforced concrete under freezing and thawing action [J]. Industrial Building,2006(10):65-68.
[5] Li Zhengwei, Jin Jinlin, Lu Jinchi et al. Experimental study on mechanical properties of steel fiber reinforced concrete under freezing and thawing conditions [J]. Sichuan Building Science Research,2019,45(06):74-77
[6] JGJ 55-2011, Code for Design of Common Concrete Mix Ratio [S]. Beijing: China Architecture and Construction Press, 2011.
[7] GB/T 50082-2009, Standard for test methods for long-term performance and durability of ordinary concrete [S].

Effect of salt concentration on the antibubble lifetime

Shuheng Liu[1], Ji An[1, *], Ziqi Liu[1], Guilin Zhao[1] and Rui Gao[1]

[1]Merchant Marine College, Shanghai Maritime University, Shanghai, 201306, China

[*]Corresponding author's e-mail: anji@shmtu.edu.cn

Abstract. Antibubble lifetime has attracted increasing attention since it limits the practical applications of antibubbles. This paper presents an experimental study on the effect of salt concentration on the antibubble lifetime. With a uniform salt solution by fully stirring, the antibubble lifetime is observed to increase first but then decrease with the salt concentration. The lifetime extension is not significant. However, in a nonuniform salt solution without stirring, the generated heavy antibubbles suspended near the container's bottom have a much longer lifetime, about an order of magnitude higher. Along with the free dissolution of salt, a blurred liquid layer emerges near the container's bottom due to the formation of worm-like micelles. These micelles endow the liquid viscoelastic properties and slow down the air drainage in the film, leading to a significantly longer lifetime. This study provides a feasible method to investigate the air drainage in the film because of the slow drainage speed. Moreover, the long lifetime makes the following experimental processes of antibubbles more convenient.

1. Introduction

Antibubble has been regarded as a promising candidate for microfluid manipulation because of its unique structure: a thin air film separating the inner and outer liquids [1]. This structure can protect the inner liquid from any outer contamination. On the other hand, compared with the double emulsion particles (a liquid film separates the innermost and outer liquids), air is much easier to exhaust from the system due to the larger density difference. However, the practical applications of antibubbles are limited by the short and broadly distributed lifetime [2-8].

Since the first report of antibubbles in 1932, the lifetime has attracted much attention [9]. Hughes and Hughes reported that the antibubbles could maintain stability for at least one minute, although they immediately burst on being touched. In 1974, Stong demonstrated that an antibubble located in a spinning tube could be stable as long as the spinning speed exceeded a threshold value [10]. According to Dorbolo et al, the antibubble lifetime is governed by the air drainage in the film. The gravity of the inner globule drives air from the bottom to the top of the antibubble. As a result, the bottom air film thins and finally collapses as the thickness reaches a threshold value, of about 100 nm [2]. Liquid properties play the dominant role in the antibubble lifetime. Increasing the bulk viscosity and the surface modulus was observed to elongate the antibubble lifetime [11]. Moreover, the standard deviation of the lifetime decreased and the antibubble stability during the first second of their life was enhanced. Kim and Vogel studied the effect of micellar structures, contamination, and external pressure on the antibubble lifetime [3, 4]. They found that antibubble lifetime decreases with increasing micellar structures, contamination, and external pressure. In 2012, Scheid et al. upgraded the air drainage model where the air-liquid interface was regarded as slippery [12]. The antibubble lifetime and the threshold air film thickness for the collapse were found to increase with the surface shear viscosity. Meanwhile, the upgraded model indicates that the relation between the antibubble radius and the lifetime also depends on surface viscosity: (i) for small surface viscosity, the lifetime decreases with the radius; (ii) for large surface viscosity, the lifetime increases with the radius. With a well-designed apparatus, Vitry et al. eliminated the disturbance of dust particles on the antibubble lifetime and verified the relation between the antibubble lifetime and the radius [4]. Meanwhile, the air drainage model was finally completed through the addition of the Marangoni effect. Despite the air drainage, air dissolution plays an important role in the lifetime of antibubbles [13]. Air in the thin film dissolves into the liquid, which leads to the thinning of the film and accelerates its collapse. On the other hand, antibubbles of extremely short or long lifetimes have been observed in specific liquids. By impacting a higher-viscosity drop onto a lower-viscosity silicon oil pool, Beilharz et al. observed the formation of an antibubble whose lifetime is about 0.1 s [14]. Long-lived antibubbles with a lifetime of at least one hour were generated by coating the air-liquid interfaces with hydrophobic particles [15]. Without any change in liquid properties, antibubbles can also be stabilized through mechanical methods, like vertical vibrations [16]. In this way, the antibubble lifetime was not only elongated but also controllable by adjusting the vibration intensity. In this paper, we study the effect of salt concentration on the antibubble lifetime, which has been widely investigated in the stability of bubble or foam stability [17]. In the fully stirred uniform solution, the effect of salt concentration on the antibubble lifetime is finite. The unexpected experimental results show that the lifetime of the heavy antibubbles generated in a nonuniform salt solution extends by more than one order of magnitude. Elongation of the antibubble lifetime is attributed to the formation of worm-like micelles which renders the liquid-air interfaces viscoelastic and hinders the air drainage in the film.

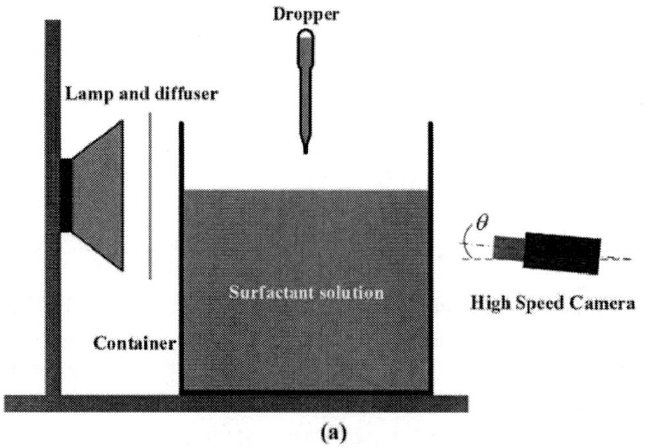

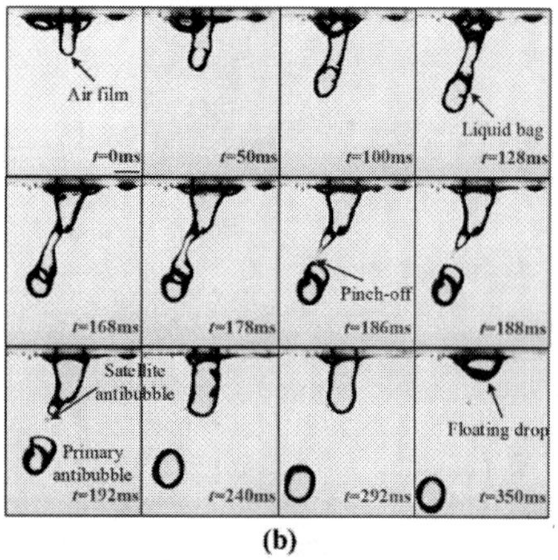

Figure 1. (a) Schematics of the experimental setup. The high-speed camera is slightly tilted to capture the dynamics on the liquid surface. (b) Detailed dynamics of the antibubble formation with a jet impact on the liquid bath. The black strike indicates the surrounding air film.

2. Experimental setup

Figure 1(a) shows the schematics of the experimental setup. To minimize the disturbances caused by the flow field, a large acrylic container with dimensions of 25 cm in length, width, and height is used and the liquid depth is 20 cm. On the other hand, the newly generated antibubbles tend to wander towards the container walls after rising up to the liquid surface [20]. The antibubble collapses immediately as it touches the container wall, resulting in large deviations in the recorded lifetime. This large liquid surface can inhibit the premature collapse of antibubbles. A commercial detergent FAIRY® containing both anionic and nonionic surfactants is used to prepare the surfactant solution. The weight ratio between the detergent and water is 0.12%. The surface tension and density of the detergent solution are measured as σ=30 mN/m and ρ=1000 kg/m^3, respectively. A dropper is used to produce antibubbles by impacting a jet onto the liquid surface. This method has been widely utilized in previous studies [21]. The whole process is captured by a high-speed camera (Phantom V2012, Vision Research, USA). The background light is a light-emitting diode (LED) lamp together with a diffuser. To capture the dynamics on the liquid surface, the camera is slightly tilted with an angle θ<2°. As shown in Figure 1(b), the impinging jet penetrates downwards along with the surrounding air film (t=0 ms). Due to total reflection, the air film appears as a black stripe whose width does not indicate the real thickness of the air film. During the penetration, the kinetic energy of the liquid is partly converted into the surface tension energy and partly dissipated by the liquid viscosity, leading to the deceleration of the injected liquid. The injected liquid converges into a liquid bag at the bottom (t=128 ms) which finally pinches off, generating a primary antibubble (t=186 ms). Subsequently, the residual liquid jet pinches into satellite antibubbles (t=192 ms) or retracts into a floating droplet on the liquid surface (t=350 ms). The key to antibubble generation is the stability of the bottom air film of the liquid bag during penetration [20].

3. Lifetime of antibubble in uniform of NaCl

To investigate the effect of salt concentration c on the antibubble lifetime, sodium chloride (NaCl) is added to the detergent solution. After fully stirring, all the salt dissolves. Fifteen minutes later, we start to produce antibubbles and record the lifetime. The antibubble lifetime is measured in the movies captured by the high-speed camera. For each concentration, at least fifty antibubbles are recorded. Figure 2(a) shows the relation between the antibubble lifetime τ and diameter D. The lifetime disperses widely and varies randomly with the antibubble diameter, which is similar to previous studies [2-4]. Figure 2(b) shows the variation of the average antibubble lifetime τ_0 with the NaCl concentration. Despite the large error bar, the average lifetime tends to increase at first with the salt concentration but then decreases as the salt concentration exceeds a critical value. Moreover, the recorded longest lifetime for each concentration follows the same trend. In the present condition, the critical concentration of NaCl is about 0.3 mol/L, where the average lifetime doubles.

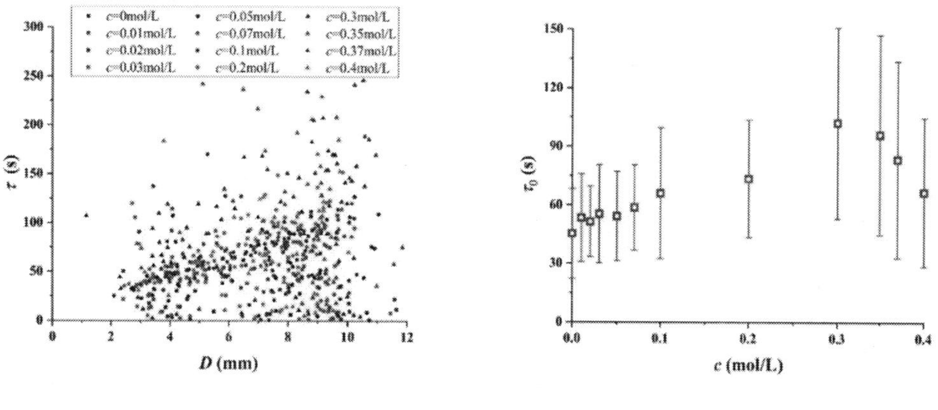

(a) (b)

Figure 2. Effect of NaCl concentration on the antibubble lifetime: (a) the relation between the antibubble lifetime and diameter; (b) the average lifetime varies with the salt concentration.

A simple model can be built to calculate the antibubble lifetime. As shown in Figure 3, the airflow can be simply regarded as a Poiseuille flow between two parallel plates with the length $l = \pi D/2$ and the pressure difference $\Delta p = \rho_l g D$, where ρ_l and g indicate the liquid density and gravity acceleration. Then we have the simplified one-dimension Navier-Stokes equation:

$$-\frac{\partial p}{\partial x}+\mu_a\frac{\partial^2 v}{\partial z^2}=0 \tag{1}$$

where μ_a and v are the air viscosity and flow speed, respectively. Integrating Equation (1) with the boundary conditions $v(0)=0$ and $v(h)=0$, the mean speed can be expressed as:

$$\bar{v}=\frac{\rho_l g h^2}{6\pi\mu_a}\sim h^2 \tag{2}$$

This classical Reynolds thinning of the film has a lifetime $\tau_r\sim D\mu_a/\rho_l g h_c^2$, with h_c being the critical thickness for the collapse of the air film [2]. In this way, the calculated antibubble lifetime is about 1 h, which is significantly larger than the experimental results, as shown in Figure 2. Benoit et al. attributed the shorter experimental lifetime to the slipping boundary [4]. The boundary slip can be weakened by changing the liquid properties, like the liquid viscosity and surface modulus. Increasing the viscosity and surface modulus has been reported to elongate the antibubble lifetime [11].

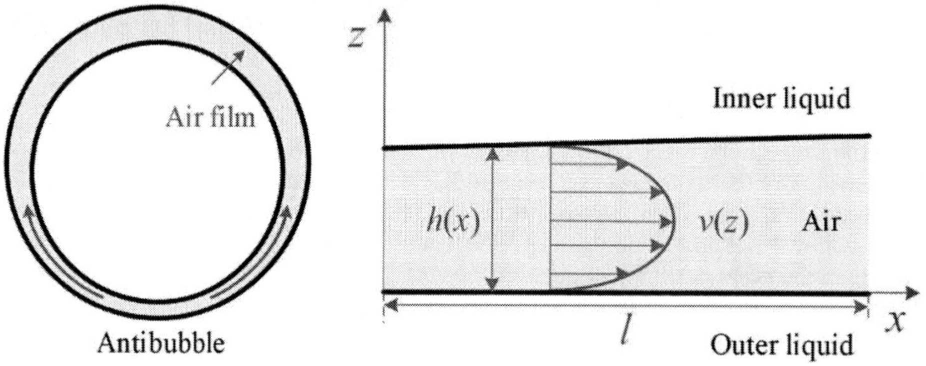

Figure 3. Sketch of the air drainage in the film, where h and v indicate the air film thickness and airflow speed, respectively.

Adding salt leads to an increase in surface viscosity [17], which benefits the stability of antibubbles. Thus, the antibubble lifetime first increases with the salt concentration. However, it starts to fall with the further increase of the salt concentration. Meanwhile, the solution becomes blurred, as shown in Figure 4, especially when the salt concentration reaches c=0.4 mol/L. A similar phenomenon has been reported by Kim that the solution transparency significantly decreases after a week [3]. They attributed this decrease to the formation of wormlike and spherical micellar structures due to the presence of impurities. The presence of saline ions may accelerate the formation of wormlike or threadlike micelles. This may contribute to the decrease of the antibubble lifetime at higher salt concentrations.

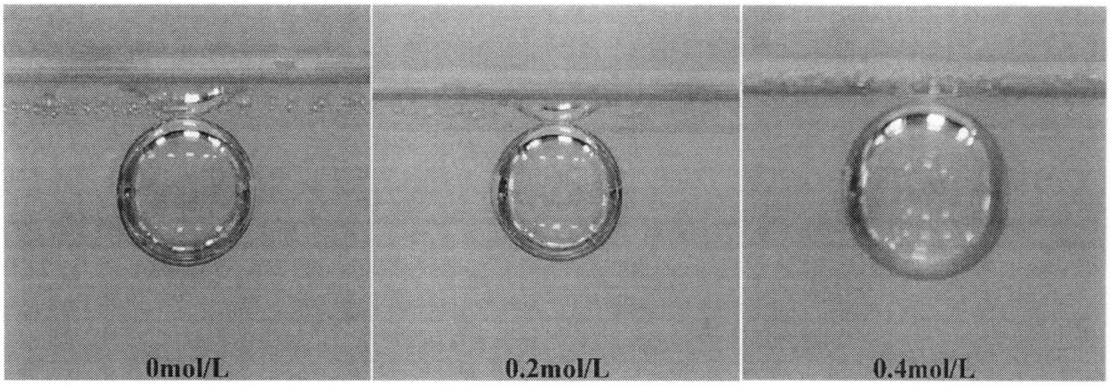

Figure 4. The detergent solution becomes blurred with the increase of the salt concentration.

4. Lifetime of heavy antibubbles

Based on previous studies, the antibubble lifetime is sensitive to external disturbances. As the experiments are conducted in an open environment, contamination on the liquid surface may lead to the immature collapse of antibubbles. To avoid this contamination, heavy antibubbles are produced by adding extra salt into the injecting liquid. The ideal condition is to generate suspended antibubbles so that the buoyancy of the air film is balanced by the density difference between the inner globule and the surrounding liquid. As the air film thickness is uncontrollable, the dissolution or release of air through the liquid-air interfaces makes it difficult to maintain the antibubble suspended. The antibubble may rise to the liquid surface and exist for a time before the collapse, or fall down to the bottom wall of the container where it collapses immediately.

To generate long-time suspended antibubbles, a density gradient is produced by adding salt into the detergent solution without stirring. The free dissolution of salt generates a density gradient that the liquid density decreases from the bottom to the liquid surface. During the salt dissolution, many tiny bubbles are released and rise upward to the liquid surface. These bubbles come from the air entrapped in the salt particles. We start to produce antibubbles as the rising bubbles vanish and the salt totally dissolves. Without stirring, there are still some tiny bubbles attached to the bottom wall, and occasionally one bubble detaches and rises to the liquid surface. Moreover, we observe a blurred layer at the container's bottom whose thickness increases with the amount of the added salt, as shown in Figure 5. The equivalent concentration c_e is defined as the salt mass equal to that added in the uniform solution. The thickness of the blurred layer increases with the amount of salt.

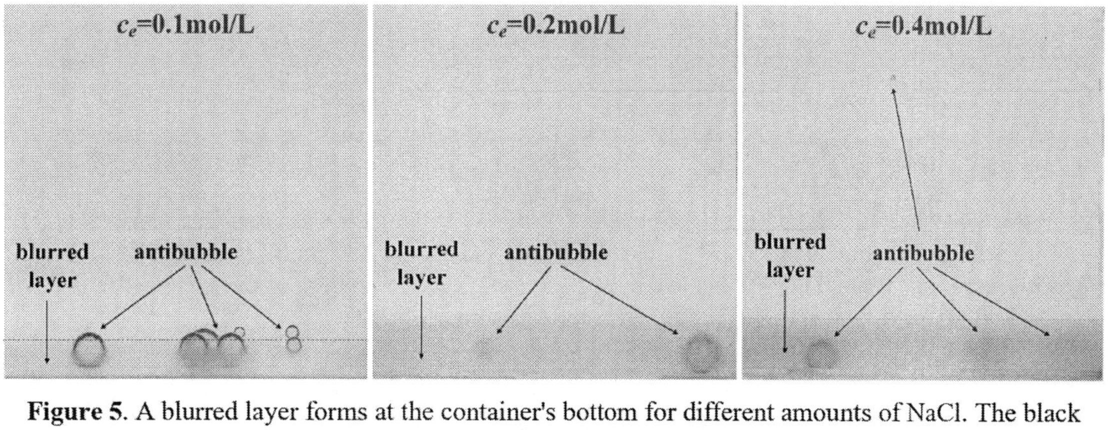

Figure 5. A blurred layer forms at the container's bottom for different amounts of NaCl. The black dashed lines indicate the upper boundary of the blurred layer. The heavy antibubbles are generated with the liquid at the container's bottom. The smaller antibubbles suspend at a higher position.

Antibubbles are generated by injecting the heavier liquid that is absorbed below the surface. In this way, the generated antibubbles of different air film thicknesses can be suspended at different heights. Figure 6 shows the variation of the heavy antibubble lifetime with diameter where d indicates the depth of the absorbed liquid. The dashed lines indicate the average lifetime of antibubbles. The antibubbles generated with the liquid at higher positions (*d*=1 cm and 5 cm) have an average time that is even smaller than those produced in the uniform liquid. This is caused by the larger gravity of the inner globule which accelerates the air drainage in the film. Antibubbles tend to suspend at a position higher than the injected liquid because of the buoyancy of the air film. With the same injected liquid, the smaller antibubbles are observed to suspend at higher positions, as shown in Figure 5. The antibubbles generated with the liquid at *d*=1 cm all rise up to the liquid surface. For *d*=5 cm, some antibubbles can suspend under the liquid surface. However, for *d*=10 cm, all the antibubbles suspend near the container's bottom. The part or the whole of the antibubble is immersed in the blurred layer, as shown in Figure 5. Under this condition, the generated antibubbles have a much longer lifetime than those in the uniform solution by an order of magnitude, as shown in Figure 6.

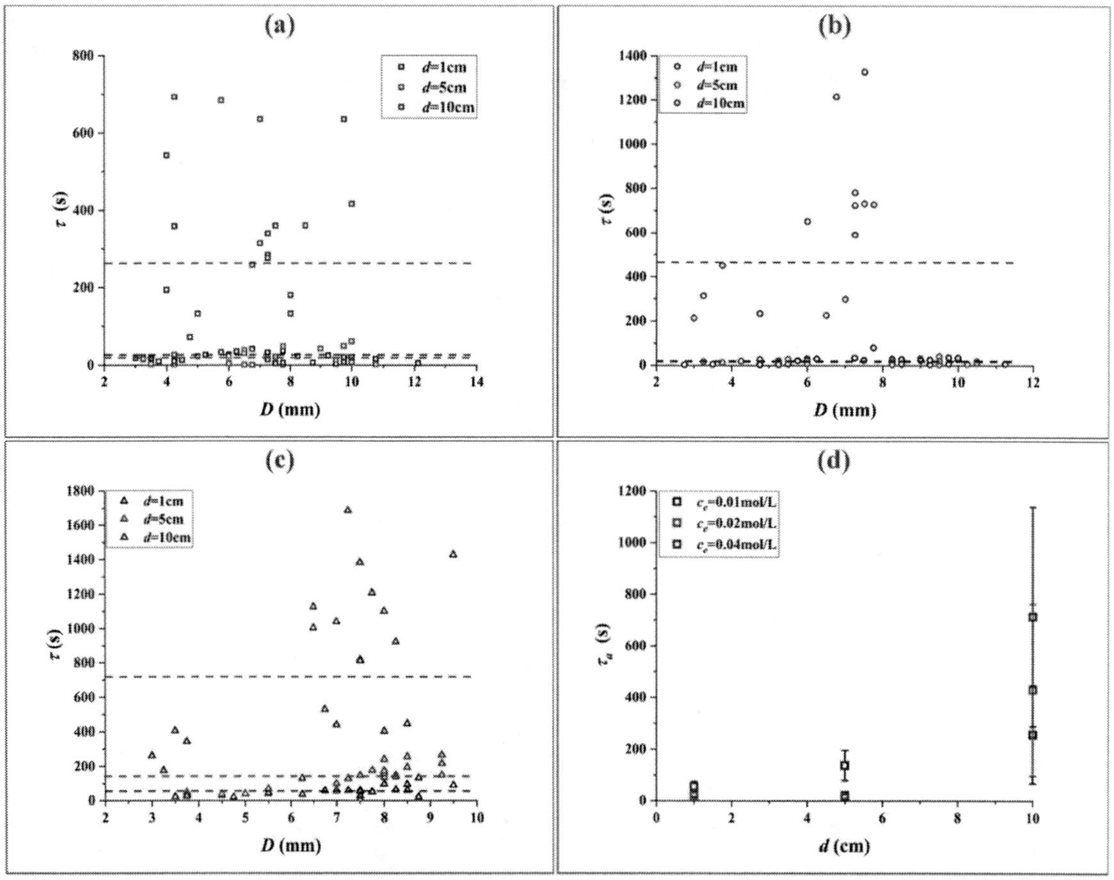

Figure 6. Variation of the heavy antibubble lifetime with diameter. Different amounts of salt are added and the equivalent concentrates are (a) 0.01 mol/L, (b) 0.02 mol/L, and (c) 0.04 mol/L, respectively. The dashed lines indicate the average lifetime. (d) Variation of the average lifetime with the liquid position.

5. Discussion

The significant increase in the heavy antibubble lifetime is unexpected since the previous studies indicate that the antibubble lifetime decreases with the outer pressure, even if there exists a critical pressure above which the antibubble collapses immediately [3]. However, in the present condition, the heavy antibubbles suspending near the container's bottom wall have the longest lifetime. Moreover, the long-lived antibubbles are located in the blurred layer where worm-like micelles form. Based on the previous study, the presence of worm-like micelles should damage the antibubble lifetime [3]. The variation of the antibubble lifetime in the uniform solution also verifies this point since the antibubble lifetime decreases in the blurred solution.

The hypothesis is that the micelles formed in the blurred layer are different from those in the uniform solution due to the ultra-high local salt concentration. In general, surfactants in dilute solution usually form spherical micelles. However, in the presence of salt, the small micellar aggregates undergo enormous growth in one dimension, forming worm-like micelles or even precipitation [22]. Moreover, the micelle length increases significantly at higher concentrations of salt [22]. In the present condition, the salt concentration at the bottom should be saturated which contributes to the formation of ultra-long worm-like micelles. The ultra-long worm-like micelles have high surface activity and exhibit viscoelastic behavior analogous to the flexible polymer solution, which weakens the boundary slip on the air-liquid interfaces. As a result, the air drainage in the film slows down and the antibubble lifetime elongates [4].

Figure 7(a) shows the variation of the average antibubble lifetime with time T at c_e=0.4 mol/L. Here T=0 indicates the salt is added to the solution. It takes about 2 hours for the salt to totally dissolve. Moreover, all the antibubbles are generated with the liquid near the container's bottom and suspend near the bottom wall, as shown in Figure 5. For each case, at least twenty antibubbles are recorded. The long-lived antibubbles can be generated for a much longer time compared with the previous study [3]. For each case, the recorded maximum lifetime τ_m also decreases with time. The maximum lifetime recorded can be maintained longer than 10 min for more than one week, even when the blurred layer becomes invisible.

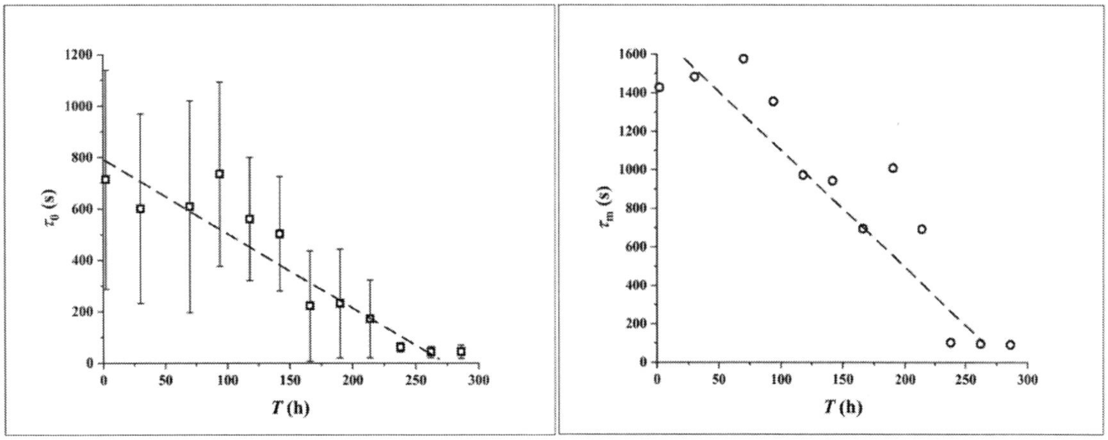

(a) (b)

Figure 7. (a) Variation of the average antibubble lifetime with time. (b) Variation of the maximum lifetime recorded for each time. The black dashed lines are just for guidance.

6. Conclusions

This paper presents an unexpected phenomenon that the heavy antibubbles suspending near the container's bottom have a much longer lifetime than the common antibubbles. This significant increase in the antibubble lifetime is attributed to the formation of the ultra-long micelles which renders the air-liquid interfaces viscoelastic properties and weakens the air drainage in the film. Compared with previous studies, the elongation of the antibubble lifetime needs long worm-like micelles that form at high salt concentrations. We suspect this study should stimulate further research on the micelle structure and the corresponding effect on the liquid properties. Moreover, this long lifetime makes the study on antibubbles, or the applications of antibubbles more convenient.

References

[1] J. E. Silpe and D. W. Mcgrail, (2013) Magnetic antibubbles: Formation and control of magnetic macroemulsions for fluid transport applications. J. Appl. Phys. 113,17: 17B304.
[2] S. Dorbolo, E. Reyssat, N. Vandewalle, and D. Quere (2005) Aging of an antibubble. Europhys. Lett. 69, 6: 966-970.
[3] P. G. Kim and J. Vogel (2006) Antibubbles: Factors that affect their stability. Colloids Surf. A: Physicochem. Eng. Aspects 289, 1: 237-244.
[4] J. Jiang, A.T. Poortinga, Y. Liao, T. Kamperman (2023) High-Throughput Fabrication of Size-Controlled Pickering Emulsions, Colloidosomes, and Air-Coated Particles via Clog-Free Jetting of Suspensions, Advanced Materials, 35: 2208894.
[5] Y. Vitry, S. Dorbolo, J. Vermant, and B. Scheid (2019) Controlling the lifetime of antibubbles. Adv. Colloid Interface Sci. 270: 73-86.
[6] W. Wang, G. Zhao, Z. Liu, F. Lin, P. Wang, J. An (2024) Antibubble formation through single drop impact: Effect of density difference, Physics of Fluids, 36: 032004.
[7] W. Wang, F. Lin, X. Wei, J. Zou, S. Dorbolo (2022) Arbitrary extension of the antibubble lifetime, Physical Review Fluids, 7: L061601.
[8] R. Zia, A. Nazir, A. T. Poortinga, and C. F. van Nostrum (2022) Advances in antibubble formation and potential applications. Adv. Colloid Interface Sci. 305: 102688.
[9] W. Hughes and A. R. Hughes (1932) Liquid drops on the same liquid surface. Nature 129, 59.
[10] C. L. Stong. (1974) Curious bubbles in which a gas encloses a liquid instead of the other way around [J]. The Amateur Scientist, 230(6): 122-128.
[11] S. Dorbolo, D. Terwagne, R. Delhalle, J. Dujardin, N. Huet, N. Vandewalle, and N. Denkov (2010) Antibubble lifetime: Influence of the bulk viscosity and of the surface modulus of the mixture. Colloids Surf. A 365: 43.
[12] B. Scheid, S. Dorbolo, L. R. Arriaga, and E. Rio (2012) Antibubble dynamics: The drainage of an air film with viscous interfaces. Phys. Rev. Lett. 109, 26: 264502.
[13] B. Scheid, J. Zawala, and S. Dorbolo (2014) Gas dissolution in antibubble dynamics. Soft Matter 10, 36: 7096-7102.
[14] D. Beilharz, A. Guyon, E. Q. Li, M.-J. Thoraval, and S. T, Thoroddsen (2015) Antibubbles and fine cylindrical sheets of air. J. Fluid Mech. 779: 87-115.
[15] A. T. Poortinga (2011) Long-lived antibubbles: Stable antibubbles through pickering stabilization. Langmuir, 27, 6: 2138-2141.
[16] W. Wang, F. Lin, X. Wei, J. Zou, and S. Dorbolo (2022) Arbitrary extension of the antibubble lifetime. Phys. Rev. Fluids 7: 061601.
[17] O. Obisesan, R. Ahmed, and M. Amani (2021) The effect of salt on the stability of aqueous foams. Energies, 14, 2: 279.
[18] A. Somoza, A. Soto, J. Pang, and K. K. Mohanty (2023) Experimental evaluation of foams stabilized by ionic liquids for enhanced oil recovery. J.CO_2 Util. 72: 102507.
[19] S. Dorbolo, H. Caps, and N. Vandewalle (2003) Fluid instabilities in the birth and death of antibubbles. New J. Phys. 5, 1: 161.
[20] Z. Liu, J. An, W. Chen (2023) Effect of drainage on antibubble collapse, Physics of Fluids, 35: 123108.
[21] Y. Song, L. Zhang, E.N. Wang (2020) Criteria for antibubble formation from drop pairs impinging on a free surface, Physical Review Fluids, 5:123601.
[22] A. S. Rafique, S. Khodaparast, A. S. Poulos, W. N. Sharratt (2020) Micellar structure and transformations in sodium alkylbenzenesulfonate (NaLAS) aqueous solutions: effects of concentration, temperature, and salt [J]. Soft Matter, 16(33): 7835-7844.

doi:10.1088/1742-6596/2808/1/012030

Study on fatigue life of abutment based on multi-sample test method

Liangxia Zhang[1*], Yunjing Tian[1], Mengsun Qiao[2], Yongning Li[1], Jinni Luo[1], Zhifang Hu[1] and Lijuan Zhao[1]

[1]Ningxia Institute of Science and Technology, Shizuishan, 753000, China
[2]Shizuishan Second People's Hospital, Shizuishan, 753000, China

*Corresponding author's e-mail: zlx0524@126.com

Abstract: The key to increasing the service life of dental implants is to improve the stability of dental implants. In the dental implant system, the abutment is the key structure connecting the upper crown and the lower implant and achieving anti-rotation and anti-loosening between them. During daily chewing, the abutment-implant connection will often be twisted by force, and the implant operation requires torque. To ensure its stable use, the abutment-implant connection should have sufficient anti-torsion and anti-deformation ability. The connection structure has an important influence on the overall stability of the implant system.
The research object of this paper is a group of common and representative commercial implant systems. The fatigue life of the abutment is studied by experiments. According to the implant dynamic fatigue test standard (BS EN ISO 14801:2016), the worst conditions of implant fatigue loading were determined, and fatigue tests were carried out on the implant system with this type of connected structure. Through the test results, the maximum cyclic load that the implant can bear in the case of fatigue fracture is determined, and whether the product can be used safely under this load force is verified. The main conclusions of this study are as follows: the abutment can withstand more than 5 million cycles of loading under 240 N load without being damaged. In the absence of special conditions, patients can use the implant product for a long time without replacing it, avoiding the risk of multiple operations.

1. Introduction

After the dental implant system is implanted in the human body, it is generally not removed, and it can replace the natural teeth to perform the chewing function. Once a fatigue fracture occurs, clinical surgery is needed to remove it, which brings great trouble to patients. In the process of implant service, long-term chewing may lead to system instability, and the abutment and screws may loosen or break. Therefore, it is hoped that the fatigue performance of the abutment can be studied through this experiment, the fatigue life of the abutment can be determined, and the safe use range of the implant can be determined to solve the clinical difficulties of implant implantation for patients. The dental implant fatigue test standard ISO14801 has stipulated the fatigue test program, including the amplitude, frequency, and waveform of the load, as well as the stress mode and environment of the implant [1]-[4]. Therefore, under the guidance of fatigue test standards, this experiment was conducted on the commercial implant abutment. The model number of the commercial implant abutment was 404525, that is, the abutment had an outer diameter of 4.0 mm, a gingival penetration length of 4.5 mm, and an angle of 25°. The abutment structure is shown in Figure 1, and Figure 2 shows the connection diagram

Journal of Physics: Conference Series **2808** (2024) 012030 doi:10.1088/1742-6596/2808/1/012030

of each component of a dental implant. According to the test results, the fatigue life of the 404525 abutment was determined.

Figure 1. Outline structure diagram of abutment.

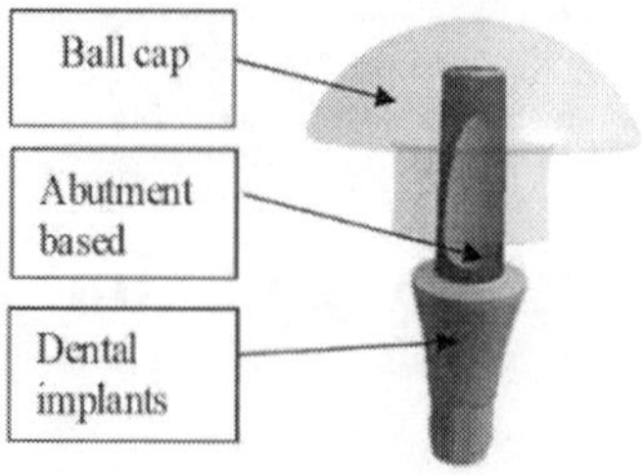

Figure 2. Connection diagram of dental implant components.

2. Fatigue performance test

2.1 Test equipment

The test device required for this test is the Carem-3000 electromagnetic fatigue mechanics testing machine produced by Tianjin Kell Measurement and Control Co., Ltd. (as shown in Figure 3), which is widely used in dynamic and static mechanical fatigue tests of various materials. The fixture material is stainless steel with an elastic modulus of 200 GPa, and the embedded material is epoxy resin with an elastic modulus of 5.86 GPa. The fixture is shown in Figure 4.

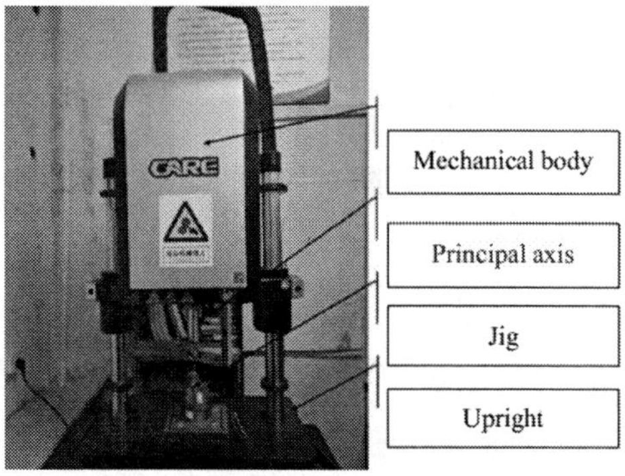

Figure 3. Endurance testing.

Figure 4. Fixture.

2.2 Test scheme

This test was conducted according to YY/T 0531-2018 "Dynamic Fatigue Test of Dental Implants in Bone" and ISO 14801: 2007 Standard for Dentistry-Implants-Dynamic Loading Test for Endosseous Dental Implants. The implant and the abutment are connected with Morse taper to achieve good sealing and avoid penetration. Both are self-locking by taper to ensure a solid connection. The fatigue test of the abutment-implant system was carried out, and the assembly torque of the abutment and dental implant was 32 N/cm. We explore the force corresponding to the static load failure value and the dynamic cycle reaching 5 million times. First, the static load test is carried out to find the maximum load and the dynamic test cycles between the nominal peak of static load 0.8, 0.6, 0.5, 0.4, 0.2, and 10% of the nominal peak, and finally, we find the corresponding force when the cycle reaches 5 million times.

The conditions for stopping the experiment were as follows: 1. Obvious plastic deformation or direct fracture of dental implants; 2. The number of fatigue cycles reaching 5 million times.

1. Static test research on abutment

The test samples were produced by Weigao Jiekang Biomaterials Co., Ltd., and the test conditions were set at 25°C in an air medium.

The purpose of the test is to study the maximum static load range of the sample.

Test implementation process: Among the 12 groups of samples used in this test, 2 groups were randomly selected for maximum static load test. After the sample of the abutment implant system was placed on the test machine and clamped, the static load was applied slowly at a speed of 0.05 mm/s until the implant system produced obvious plastic deformation or fracture, and the fatigue testing machine automatically stopped the test.

Table 1 shows the maximum static load of the two groups of samples. Through the static load test, the average maximum static load is 1, 193 N.

Table 1. Static test results.

Sample number	Maximum load (N)
1	1, 303
2	1, 083
Mean value	1, 193

Figure 5 shows the displacement-load curve of the specimen subjected to static load. When the implant system is bearing load, tensile stress occurs on one side and compressive stress occurs on the other side. Therefore, a comprehensive analysis of the displacement-load curve of the implant system should be combined with compressive stress-strain and tensile stress-strain curves of the material. The failure process can be divided into three processes: elastic deformation stage, yield stage, and fracture stage. Segment a-b on the curve in the figure is the stage of elastic deformation of the implant system. In this stage, the displacement and load are approximately linear, which accords with Hooke's law of materials. Segment b-c of the curve is the yield stage of the implant, during which the implant system mainly deforms plastic, and the load and displacement are no longer linear. When the curve reaches Point c, the implant system breaks, and the load value at this time is the static load failure value. The c-d stage is the relationship between the displacement amount and the load after the fracture of the implant system, indicating that the ability of the system to resist deformation drops sharply after the fracture occurs, and the load value continues to decrease with the increase of the displacement amount.

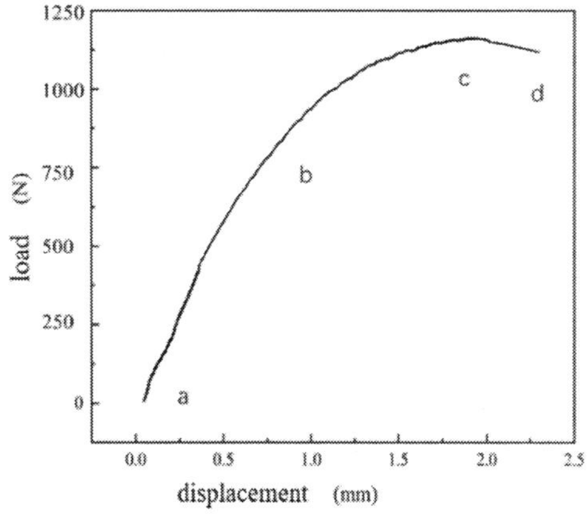

Figure 5. Displacement-load curves.

2. Dynamic experimental study of the base station

Through the static test study in the above section, the maximum static load of 1, 193 N is obtained. According to the test scheme, the maximum static load of 1, 193 N was multiplied by the static load coefficient according to Formula (1), and the numerical rounding was performed to obtain the dynamic test load value, as shown in Table 2.

$$F = F_1 \approx F_{max} \times \mu \tag{1}$$

where F denotes dynamic test loading load (N); F_1 denotes load after rounding (N); F_{max} denotes

Journal of Physics: Conference Series **2808** (2024) 012030 doi:10.1088/1742-6596/2808/1/012030

average maximum load (N); μ denotes static load coefficient.

Table 2. Dynamic test load.

Average maximum load (N)	Static load coefficient	Calculated load (N)	Load after rounding (N)
	0.8	954.4	960
	0.6	715.8	720
1, 193	0.5	596.5	600
	0.4	477.2	480
	0.2	238.6	240

Two groups of samples were selected under each load, and a total of 10 samples were divided into 5 groups. Table 3 shows the dynamic test results of Samples 1 to 10.

From the load-cycle values in Table 3, the curve shown in Figure 6 is fitted. Since the fracture of the implant system is caused by fatigue fracture, the classical fatigue strength analysis model S-N curve can be used to analyze the data obtained from the experiment [5]-[10]. The S-N curve is one of the basic fatigue characteristic curves of structural materials, which not only shows the relationship between the working stress and the fatigue life of the structure under damage but also can be used to show the fatigue resistance of the structure. Under the action of sinusoidal alternating load, unstable variable stress with a certain regularity is generated inside the implant system, so Formula (2) is used for fitting analysis.

$$y = a \times x^{b} \tag{2}$$

where y is the maximum load value of sinusoidal load; x is the fatigue life of the implant system; a and b are the parameters.

According to the test data in Figure 5, the fatigue curve of the implant system was fitted with the number of cycles before fatigue failure as the horizontal coordinate and the maximum absolute value of the corresponding dynamic load as the vertical coordinate. The results are shown in Figure 6 and the fitting fatigue curve equation is shown in Formula (3). The fatigue life of the implant system increased continuously. When the load reached 240 N, the fatigue life of the sample reached 5×106 times, and no fracture occurred.

$$y = (3900.99744 \pm 1158.80755) \times x^{(-0.17013 \pm 0.02932)} \tag{3}$$

Table 3. Dynamic test results.

Sample number	Load (N)	Cycle number	Failure mode
1	960	3, 186	The implant was cracked.
2	960	10, 824	The implant was cracked.
3	720	11, 907	The implant was cracked.
4	720	99, 510	The implant was cracked.
5	600	17, 890	The implant was cracked.
6	600	82, 745	The implant was cracked.
7	480	146, 202	The implant was cracked.
8	480	178, 592	The implant was cracked.
9	240	5, 000, 000	No apparent failure
10	240	5, 000, 000	No apparent failure

Figure 7- Figure 14 shows the variation of maximum and minimum dynamic loads with the number of cycles under loads 96-960 N, 72-720 N, 60-600 N, and 48-480 N respectively. Figure 15- Figure 16 shows the variation of the maximum and minimum load without fracture with the number of cycles under load 24-240 N.

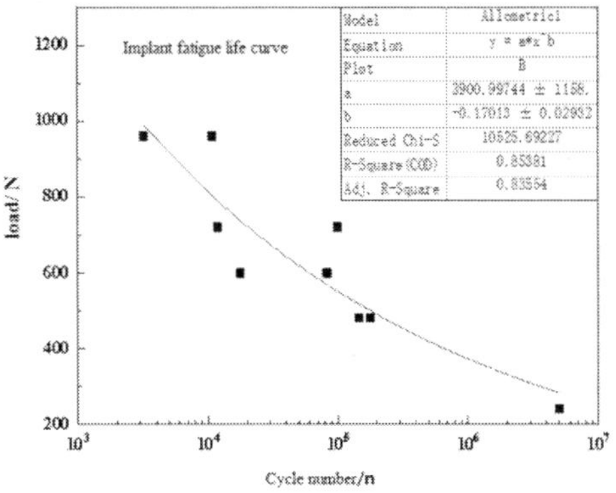

Figure 6. The Load-cycle graph.

It can be seen from Figures 7-14 that the load in the initial stage of the cycle test has a rising process because the load generated by the fatigue testing machine is a servo from a lower value to the target load. When the load value reaches the target load, the relationship between the maximum load and the minimum load value with the number of cycles is almost a straight line, which indicates that the sinusoidal alternating load changes smoothly during the test and meets the test requirements. At the end of the test, the load suddenly decreased, because the sample suddenly broke at this time. There was a sudden change in the load value, and the number of cycles corresponding to the mutation was the fatigue life of the implant system. As shown in Figures 15-16, when the sample was loaded 24-240 N, no sudden decrease of the above load occurred, no mutation occurred, and the sample did not break. When the cycle reached 5 million times, the test was stopped.

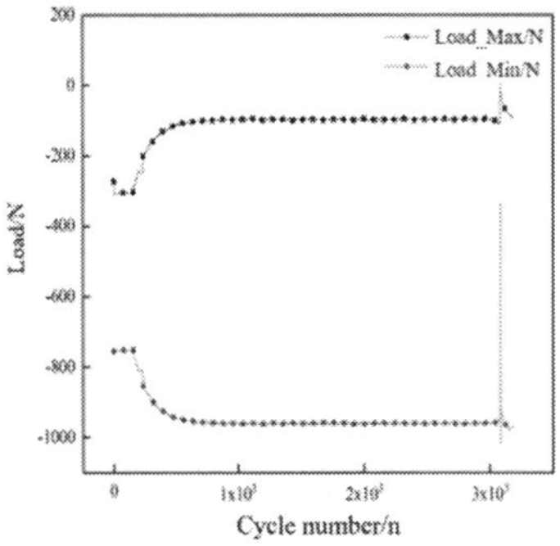

Figure 7. The 60-600 N load cycle of sample 1.

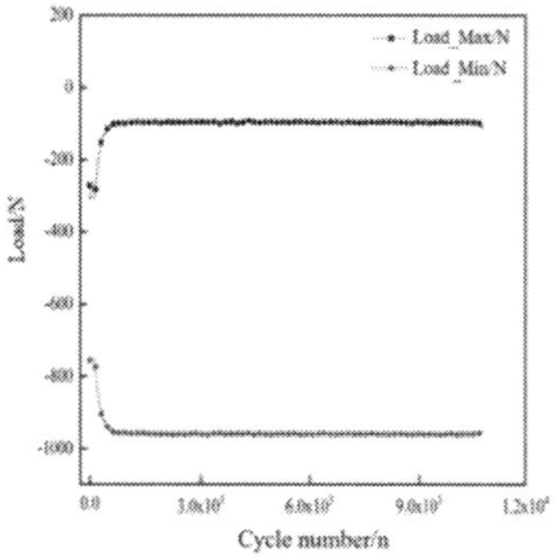

Figure 8. The 96-960 N load cycle of sample 2.

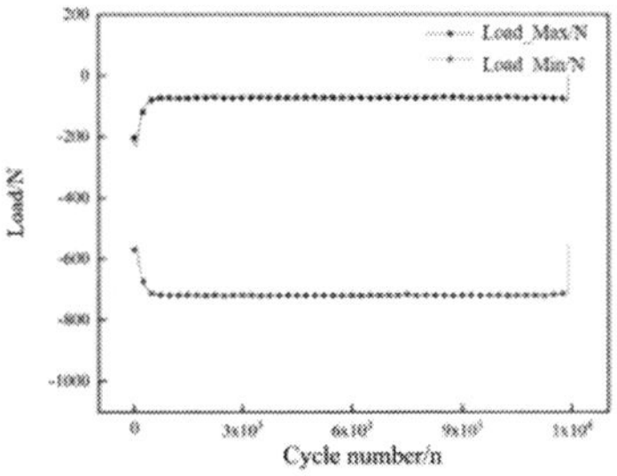

Figure 9. The 72-720 N load cycle of sample 3.

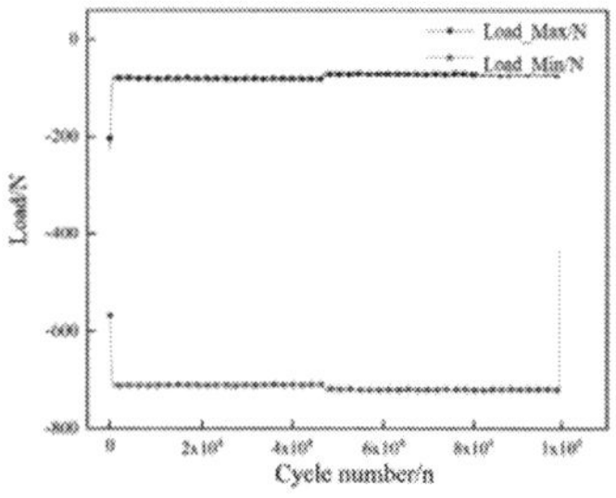

Figure 10. The 72-720 N load cycle of sample 4.

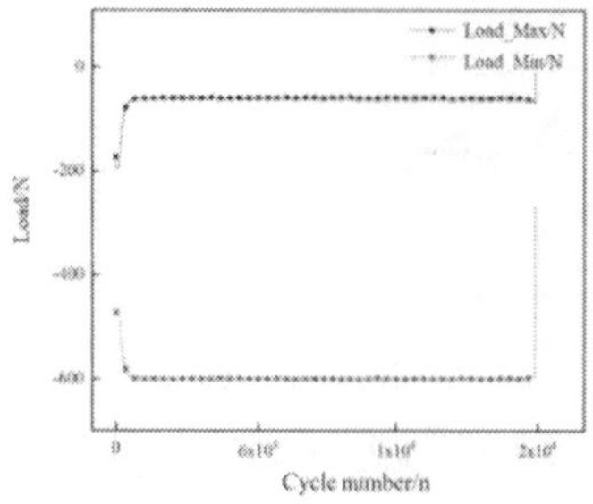

Figure 11. The 60-600 N load cycle of sample 5.

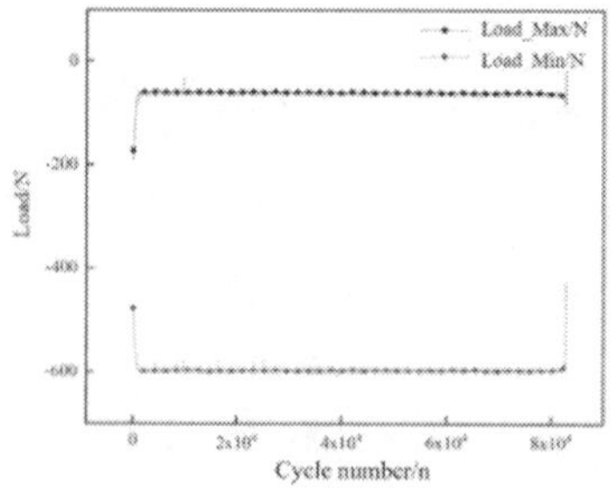

Figure 12. The 60-600 N load cycle of sample 6.

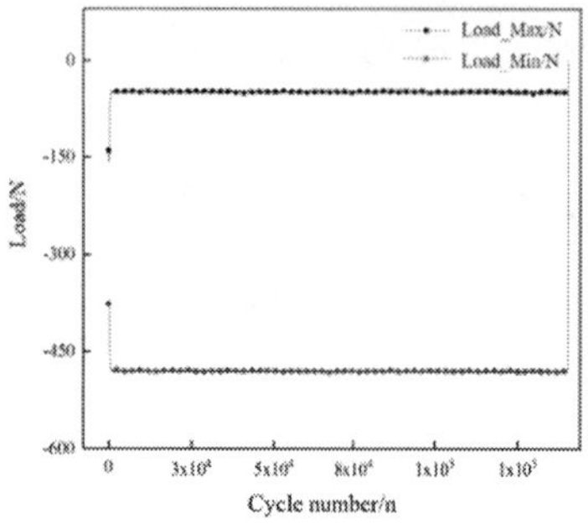

Figure 13. The 48-480 N load cycle of sample 7.

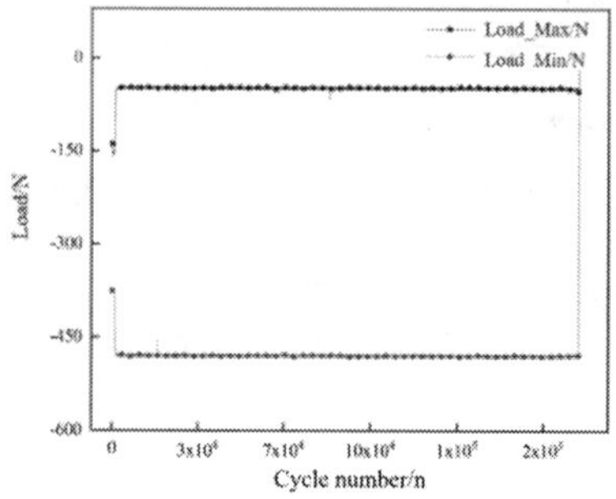

Figure 14. The 48-480 N load cycle of sample 8.

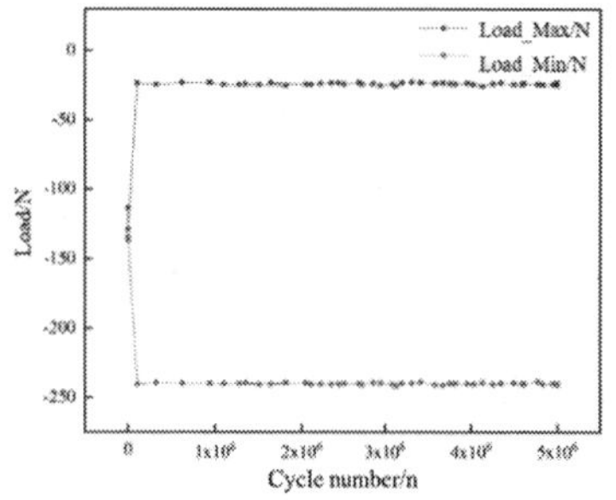

Figure 15. The 24-240 N load cycle of sample 9.

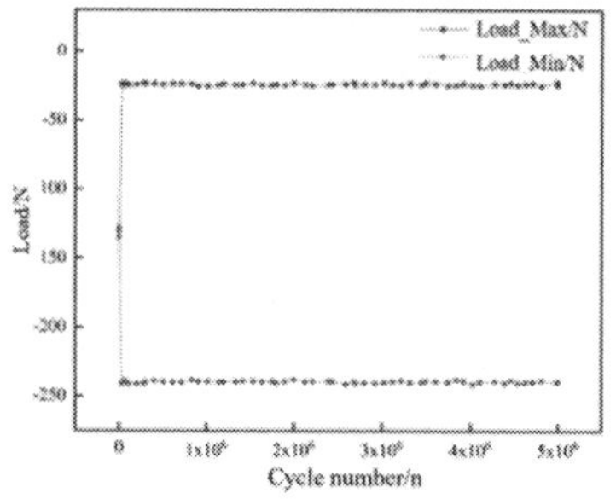

Figure 16. The 24-240 N load cycle of sample 10.

3. Conclusion

Under the guidance of the standard of dynamic fatigue test of dental implants, two groups of tests, static test, and dynamic fatigue test, were carried out. First, two groups of 12 samples were randomly selected to carry out a static load test, and the average maximum static load value of 1, 193 N was obtained to reduce the error. Then, 80%, 60%, 50%, 40%, and 20% of the maximum static load value were used as the peak value of sinusoidal alternating load for dynamic fatigue test, and two samples were selected in each group to record the change of fatigue properties of samples.

Through this test, it can be concluded that: (1) The fracture characteristics and rules of the dental implant system were found when the sample was damaged by dynamic load in the range of 40%-80% of the maximum static load value; (2) This series of products can withstand 5 million cycles without failure under 240 N cycle load. It is safe and reliable for patients to use implants within the range of these conditions.

Acknowledgments

This work was financially supported by the Study and Optimization of Fatigue Life of Dental Implant

Personalized Abutment (NGY2020092), Mechanical Engineering Materials and Thermal Processing – University Curriculum Ideological and Political Demonstration Project (NXSZ20220216), and Exploration and Research on the Construction of Teaching Resources of Mechanical Engineering Control Basis – Based on the Construction of First-class Major of Mechanical Engineering (bjg2023101).

References

[1] Sun Fei. Research on implant stability based on friction contact [D]. Northeastern University, 2017.
[2] Wang Shitang. Finite Element analysis and fatigue life study of implant-abutment connection system [D]. Northeastern University, 2016.
[3] Tang Chunbo, Han Ke, and Zhou Yanmin. Design and study of implant-abutment joint structure [J]. Journal of Oral and Maxillofacial Prosthodontics, 2002, 3 (001): 61-63 (in Chinese).
[4] Zhang Haoliang. Biomechanical analysis and animal experimental study of dental implant connection system based on reverse engineering. Northeastern University, 2014.
[5] Kuangta Yao, Hung Chan, et al. Mechanical performance of conical implant-abutment connections under different cyclic loading conditions [J]. Journal of the Mechanical Behavior of Biomedical Materials, 2018.
[6] Wang Yuqi, Tian Kebin, and Wu Shilian. Research status of the influence of angle base biomechanics [J]. Medical Biomechanics, 2020, 35 (3): 391-396 (in Chinese).
[7] Teja Guda, Thomas A., Ross Lisa A., and Lang Harry R. Probabilistic analysis of preload in the abutment screw of a dental implant complex [J]. The Journal of Prosthetic Dentistry, 2008.
[8] Cibirka R. M., Nelson S. K., Lang B. R., et al. Examination of the implant-abutment interface after fatigue testing [J]. Journal of Prosthetic Dentistry, 2001, 85 (3): 268-275.
[9] Liu X., Sun C., and Hong Y. Effects of stress ratio on high-cycle and very-high-cycle fatigue behavior of a Ti–6Al–4 V alloy [J]. Acta Metallurgica Sinica, 2015, 622: 228-235.
[10] Gratton D. G., Aquilino S. A., and Stanford C. M. Micromotion and dynamic fatigue properties of the dental implant-abutment interface [J]. Journal of Prosthetic Dentistry, 2001, 85 (1): 47-52.

Study on the properties of solidified materials of municipal waste incineration ash cementitious materials

Ke Wang[1, *], Zhan Weng[1], Jinghui Zhang[1], Jiahui Du[1], Mingfu Li[1] and Tiejun Liu[1]

[1]China Nuclear Power Engineering, Beijing, China

[*]Corresponding author's email: wktracy@163.com

Abstract. MSWI FA needs to be treated effectively to prevent harmful effects on the environment and human beings. The use of different ratios of cementitious materials for the solidification of fly ash provides innovative ideas for the environmentally friendly treatment of MSWI FA. In this study, some raw materials were used to solidify fly ash, and it was revealed that the incorporation of appropriate concentrations of fly ash (20 and 30%) could improve the compressive strength, fluidity, pore structure, and rheological properties of the solidified materials, and reduce the setting time, the bleeding rate, and inhibit the phenomenon of drying shrinkage of the solidified materials. However, when the content of MSWI FA exceeded 40%, the performance of the solidified material decreased significantly due to the poor volcanic ash activity and the high water absorption capacity of MSWI FA.

1. Introduction

In China, the amount of municipal solid waste generated and stored has increased dramatically. The municipal solid waste in China increases by about 8-10% every year, seriously endangering the ecological environment of China [1]. Incineration has many advantages such as small size and light weight of the final product, fast processing speed, heat recycling, etc., which is a critical approach to dealing with city garbage. During the incineration process, toxic metals are centralized in MSWI FA [2], which is highly toxic and hazardous, and its content accounts for about 5% of the original municipal waste, which poses a tremendous strain on human health and the environment. China classifies MSWI FA as hazardous waste and stipulates that MSWI FA must be treated as harmless.

GBFS is a waste produced by rapid cooling during pig iron smelting, which has excellent activity after grinding and can be stimulated by chemicals such as alkalis and sulfates. MSWI FA has inferior cementitious activity and excessive incorporation may negatively affect the hydration reaction [3]. However, MSWI FA has a good complex excitation effect on GBFS when used with FGDG. FGDG can be a substitute for cement to solidify MSWI FA [4]. Therefore, the use of this method not only adds the strength of the solidified material and declines the disposal fee but also avoids the energy consumption and CO_2 emission of cement [5].

In this study, municipal fly ash was solidified by different cementitious materials, and the influence of municipal fly ash on the performance of solidified materials, such as pressure resistance, fluidity, rectangular properties, setting time, shrinkage, and pore structure was investigated.

2. Experimental preparation

2.1 Raw materials
The fly ash, tailings, slag, and desulfurization gypsum were obtained from Hebei Province, which should be dried and crushed before testing.

2.2 Formulation of samples
The formulations of the solidified materials are shown in Table 1, the concentration of solids in the different samples was 80 wt.% and the mass ratio of cementitious material to tailing sand (C/S) was 1:4.

Table 1. Formulation of GBFS-MSWI FA solidified material (g).

Sample	TS	Cementitious Materials			Water
		GBFS	MSWI FA	FGDG	
M0	2000	450	0	50	625
M10		400	50		
M20		350	100		
M30		300	150		
M40		250	200		
M50		200	250		
M70		100	350		
M90		0	450		

3. Performance tests for solidified materials

3.1 Fluidity
The fluidity of the mortar was conducted according to the national GB/T 2419-2005 for the test, and the diameter of the mortar in two mutually perpendicular directions is the fluidity of the mortar.

3.2 Mechanical performance test
The mechanical strength test was done as per GB/T17671-1999, and the results were the average of three experiments, accurate to two decimal places.

3.3 Setting time test
The setting time test of the samples was tested as per JGJ/T70-2009. The resistance to penetration was measured after 2 hours of curing, then 0.5 hours until 0.3 MPa was reached, and then every 15 minutes until it reached 0.7 MPa. The time taken to reach these two values is the time of the initial and final setting of the mortar, respectively.

3.4 Rate of bleeding
The rate of bleeding test was carried out according to the JC/T 2153-2012. The newly prepared mortar is poured into the measuring cylinder and the total mass M1 is called. After 30 minutes, the water secretion was removed from the surface of the mortar every 5 minutes and the total weight was measured until the value did not change anymore and the total weight of water secretion (M_2) was recorded. The formula for calculating the rate of bleeding (R_b) is shown below:

$$Rb = 1 - M_2 / M_1$$

3.5 Shrinkage test
A shrinkage test is performed according to JGJ/T70-2009. The samples were cured for different days and the length was measured in the marked direction.

3.6 Rheological properties test
The main steps of testing the rheological properties of cured mortar are: before each measurement, the mortar sample is fully mixed, and then the mixing blade is inserted into the specimen [6].

4. Results of the experiment

4.1 Characterization of materials
After detection, the XRF results of different materials are listed in Table 2, with XRD spectra presented in Figure 1.

Table 2. XRF results of the raw materials.

Raw materials	Ca	Cl	Si	Mg	S	Na_2	Al	Fe	K	LOI
TS	16.36	0.04	50.62	8.02	1.32	1.25	7.48	11.54	2.00	3.6
DGFG	50.96	0.26	3.85	1.47	40.37	0.07	1.37	0.55	0.18	10.7
GBFS	40.86	0.03	29.27	9.75	1.48	0.32	14.76	1.46	0.46	0.8
MSWI FA	39.85	22.45	4.32	3.93	10.28	5.99	1.59	2.03	5.96	2.8

4.1.1 Tail sand. From Table 2, the base compositions of TS are Fe, Si, Ca, Mg, and Al, while the main phases of tailings are SiO_2 and a small amount of $CaCO_3$ (Figure 1(a)).

4.1.2 GBFS. According to Table 2, the main compositions of GBFS are Mg, Si, Al, and Ca. Figure 1(b) shows that the GBFS is mainly amorphous, characterized by a distinct broadband without any peaks and only a small amount of $CaCO_3$.

4.1.3 FGDG. Ca and S make up the main chemical composition of FGDG, and $CaSO_4 \cdot 2H_2O$ is the main mineral phase, as shown in Figure 1(c).

4.1.4 Fly ash. The base components of fly ash are Ca, Cl, S, Na, and K, and chloride and calcium salts are the main mineral phases.

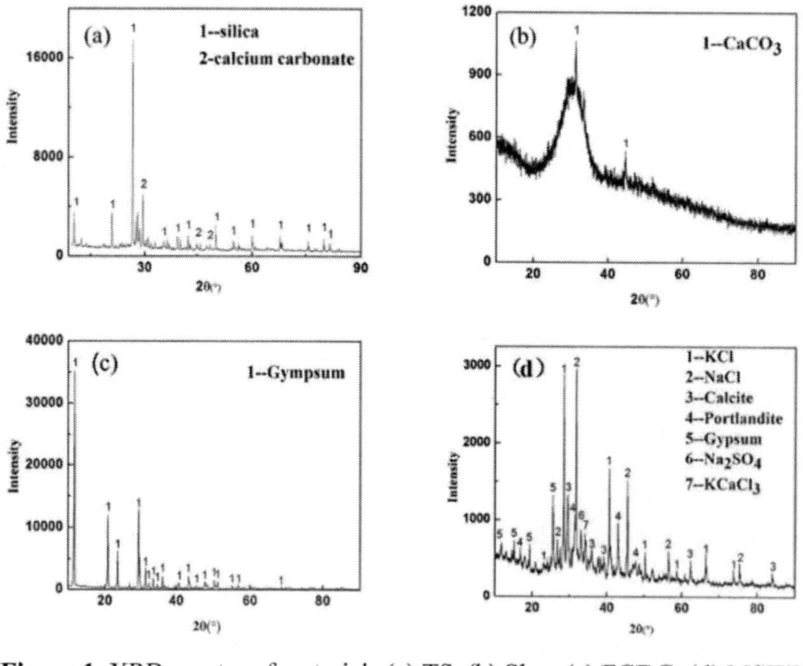

Figure 1. XRD spectra of materials (a) TS, (b) Slag, (c) FGDG, (d) MSWI FA.

4.2 Solidified materials performance

4.2.1 Mechanical properties and fluidity. The impact of the fly ash dosage was studied and the consequences were shown in Figure 2. As shown in the figure, the fluidity of sample M0 was 265 mm, and with the doping of MSWI FA, the fluidity of Samples M10-M30 were 269 mm, 274 mm, and 284 mm, and promoted by 1.51%, 3.40%, and 7.17% from Sample M0, respectively. The small amount of municipal ash can improve the fluidity of the solidified material. However, when the dosage of MSWI FA is further increased, the fluidity decreases, and the fluidity of Samples M40-M90 is reduced by 7.39%, 13.73%, 31.34%, and 45.77%. Compared with that of Sample M30, the MSWI FA is highly water absorption, which increases the water requirement of the cementitious system in large quantities, and the cementitious material forms a highly viscous mud-like slurry rapidly, which leads to the fluidity of the samples significantly reduced.

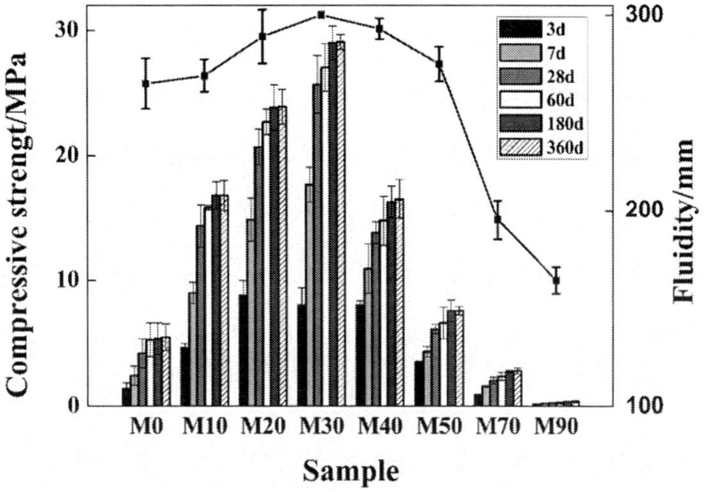

Figure 2. Results of mechanical properties and fluidity of samples.

The compression strength of the solidified material first increased and then decreased with the blending of MSWI FA. For M0, the compression strength was 5.66 MPa after curing for 360 days. The compression strength of bulks increased was the blending of municipal fly ash, and then reached the highest value of 30.18 MPa for M30. Yet, with the further blending of municipal fly ash, the compression strength of the bulks decreased drastically, and the endurance strength of Sample M70/90 was lower than 2.9 MPa after curing 360 d. When the samples had no fly ash, the hydration reaction could not be fully performed, resulting in a low compression strength of the solidified materials. However, with the blending of municipal fly ash, the slag was gradually activated with MSWI FA, resulting in an improvement of hydration reaction, which improved the compression strength of solidified materials [7]. When the fly ash dosage is reiteratively increased, the low gelling ability of fly ash causes the content of active things in the gelling materials to become smaller, the chemical reaction becomes weak, the production of hydration products decreases greatly, and the compression strength decreases.

4.2.2 Setting time. The detection results of different setting times of the samples are shown in Figure 3. We can see that the first setting time of M0 was 11.8 hours. With the increase in the amount of MSWI FA, the setting time of samples accordingly decreased. For Sample M50, the initial setting time was 7.7 hours, and the final setting time was 9.1 hours. The decreasing trend of Samples M70 and M90 is obvious, and the M90 is 2.1 hours and 2.3 hours.

For M0, due to no MSWI FA, the chemical reaction of the gelling bulk was weak. When added MSWI FA to the solidified material, the slag was wrecked by alkaline substances, and the $[Ca_2Al(OH)_6 \cdot 2H_2O]^+$ from GBFS hydrated with the SO_4^{2-} from FGDG and Cl^- from MSWI FA, which shorten the reaction time. However, MSWI FA further increases, with the high absorption of water of MSWI FA resulting in rapid adhesion of mortars, forming a highly viscous slurry and greatly decreasing the first setting times and last setting times [8].

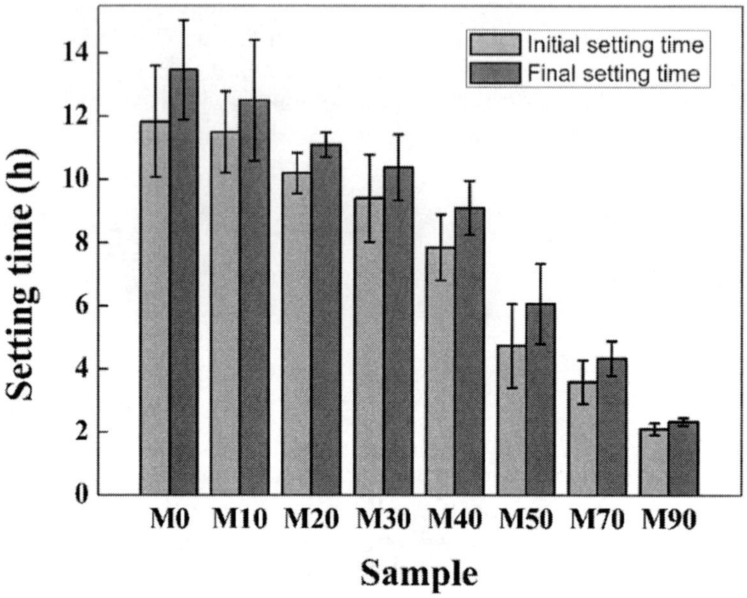

Figure 3. Effect of the MSWI FA content on the setting times of mortars.

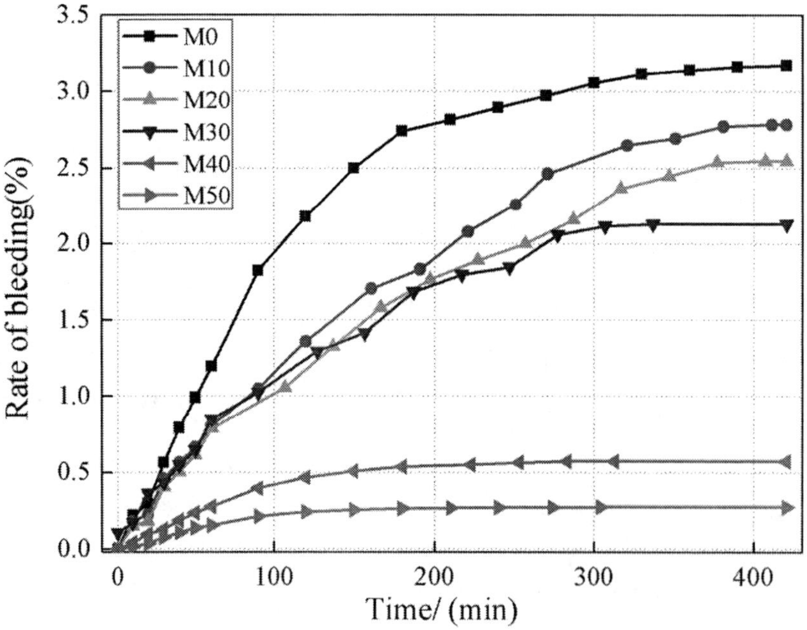

Figure 4. Bleeding rate of different solidified materials.

4.2.3 Rate of bleeding. Bleeding is unavoidable during the preparation of the solidified material, but an excessive rate of bleeding leads to the defective structure and texture of the solidified material. Figure 4 shows the municipal fly ash on the bleeding rate of solidified material.

As observed, we can see that the bleeding rate of M0, M10, M20, and M30 was reduced gradually, and the stable state time was shortened from 425 minutes to 367 minutes, which indicated that the adhesive cohesion and homogenization of the mortar were enhanced after the incorporation of MSWI, and the separateness of the cementing fraction from the water was reduced, which in turn lowered the bleeding rate. For both M40/50 samples, the level of hydration is quite subdued because only a small amount of hydration products was produced by a handful of active ingredients, resulting in a loosely structured mortar. Because of the absorption of the municipal fly ash, the bulks were viscous. Nevertheless, due to the MSWI FA's very high absorption, the mortar was thick and the free water was not lost. This correlates with a bleeding rate of under 0.1%, and the time required to equilibrate drops dramatically to 255 and 151 minutes for the M40 and M50 samples, respectively.

4.2.4 Rheological properties. The rheological properties of samples M0, M10, M20, M30, M40, and M50 are shown in Figure 5, and the shear stress of the mortars improves with growing municipal fly ash. GBFS leads to a weaker reaction and fewer products, thus leading to a slacker structure. Consequently, the mortar exhibited less impedance to the rotor. For M10, M20, M30, and M40, the alkalinity of the mortar is increased by the addition of municipal fly ash, which leading a progressive depolymerization of the surface of the GBFS, dramatically reactivating the hidden activity. As a result, the inner structure of the mortar gets denser and the viscosity is increased. For M50, the aggressive absorption of MSWI FA contributes to the viscosity of the mortar, which greatly increases the shear stress.

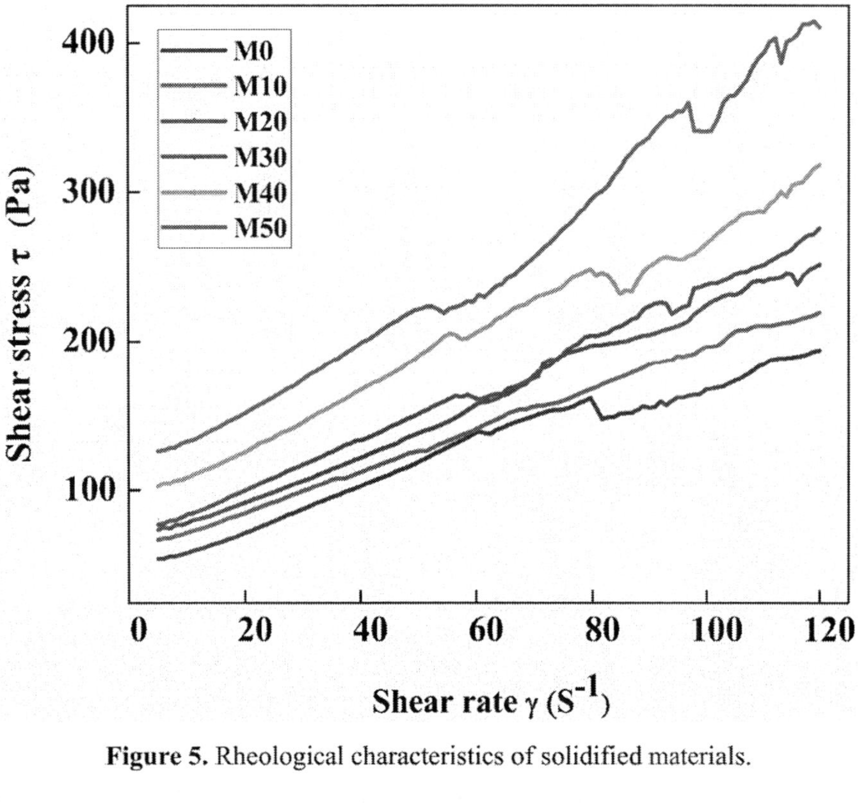

Figure 5. Rheological characteristics of solidified materials.

According to Figure 5, the rheological curves of each sample are nearly a linear line, so it can be determined that the types of fluid of the solidified material are Bingham fluids [9]. The simulated rheological parameters of different samples are presented in Table 3.

Table 3. Results of the simulation of rheological parameters

Sample	η_{pl}	τ_0	R^2
M0	1.22	56.29	0.98
M10	1.41	58.93	1.00
M20	1.74	60.33	1.00
M30	1.80	65.28	1.00
M40	1.92	95.24	0.99
M50	2.69	95.68	0.98

4.2.5 Pore structure. This experiment investigates the municipal fly ash dosing on the pore structure of hardened pastes with different cementitious materials P0, P10, P20, P30, P40, and P50. Figure 6 shows the pore volume distribution curves (Figure 6(a)) and total pore volume curves (Figure 6(b)) of different MSWI FA-slag-based cementitious materials at 28 d of hydration.

As can be seen from Figure 6(a), the pore capacity distribution curves of the pastes showed different patterns in different particle size ranges: in the range of 0-10 nm, the pore capacity distributions of different samples were P 50 > P 40 > P 0 > P 10 > P 20 > P 30 in descending order; in the range of 10-50 nm, the pore capacity distributions of different cementitious material samples were P 0 > P 50 > P 40 > P 20 > P 10 > P 30; in the range of particle sizes larger than 50 nm, the pore volume distribution is P 0 > P 50 > P 10 > P 40 > P 20 > P 30 in descending order, and the results show that the municipal fly ash has different influences on the pore volume distribution for different particle sizes of the cementitious materials. The P30 sample, on the other hand, had the least pore volume of all the samples.

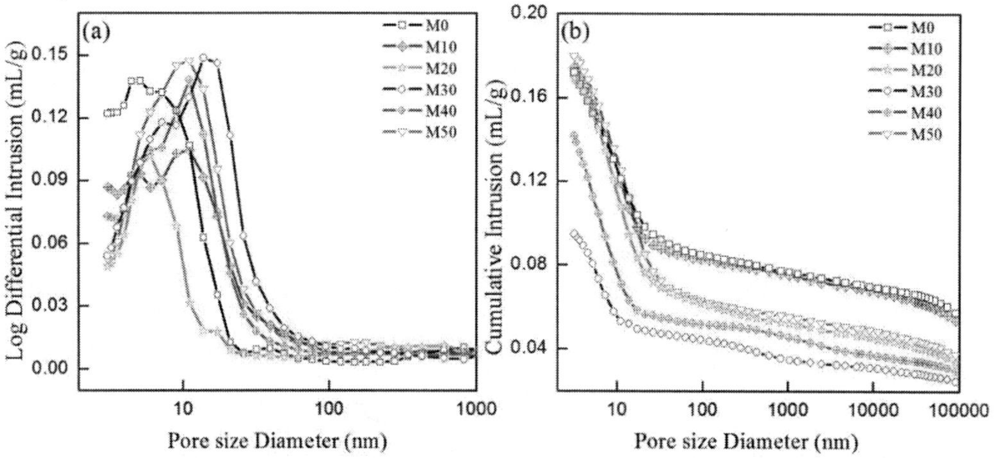

Figure 6. Municipal fly ash dosing on the pore structure of pastes with different cementitious material.

We can see from Figure 6(b) that after 28 d of hydration, a decreasing trend and then an increasing trend with the addition of municipal fly ash appear, and the total pore volume in descending order was P 0 > P 10 > P 50 > P 20 > P 40 > P 30. With a little fly ash doping, the porosity of the paste gradually decreased. When the appropriate amount of MSWI FA was doped to 30%, the total pore capacity of the hardened paste reached the lowest, which may be due to the doping of MSWI FA facilitating the production of hydration products. The pore space in the hardened paste was maximized to fill the pore space, and the pore capacity was reduced substantially. Along with the increase of municipal ash, the evaporation of water adsorbed by MSWI FA led to an increase in the pore content of the hardened paste, a loss in the internal structural compactness, and a decrease in the working performance of the solidified material.

From the above results, municipal ash has a big effect on the structure of the hardened slurry, which is mainly due to two reasons: 1) The inert substances (SiO_2 and $CaCO_3$, etc.) in the MSWI FA are embedded in the paste internal crevices, which play the role of physical filler; 2) The OH^- released from the MSWI FA can greatly make GBFS promote the reaction, and the generated products make the structure denser. In the actual hydration process, physical filling and hydration product filling often exist simultaneously.

Some scholars divided the pores in the hardened pastes into four categories according to the pore size and the state of water presence in the pastes: (1) < 20 nm are harmless pores, which only affect the shrinkage and creep of the pastes; (2) 20 ~ 50 nm are less harmful pores (water, medium surface tension); (3) 50 ~ 200 nm are harmful pores (volumetric water); (4) > 200 nm are multi-hazardous coarse pores. Figure 7 shows the distribution of pore structure in the hardened pastes of MSWI FA-slag-based cementitious materials in the range of particle sizes statistically according to the pore classification method [10].

As can be seen from Figure 7, for the pores with particle size between 20 - 50 nm and 50 – 200 nm, the effect of MSWI FA dosing on the percentage of the two kinds of pores is not significant. For the pores > 200 nm, the percentage of pores in this size decreased to different degrees with the municipal ash, with the lowest percentage of pores in this size for sample M30, which was 33%. The percentage of pores with particle size < 20 nm also increased with the increase of MSWI FA doping, and reached the highest at 30% of MSWI FA doping, which was 48%. With the incorporation of MSWI FA, pores in the hardened paste tend to decrease in particle size, the harmful coarse pores, and the proportion of harmless pores increases, which is thus reflected in the macroscopic strength of the corresponding solidified material [11].

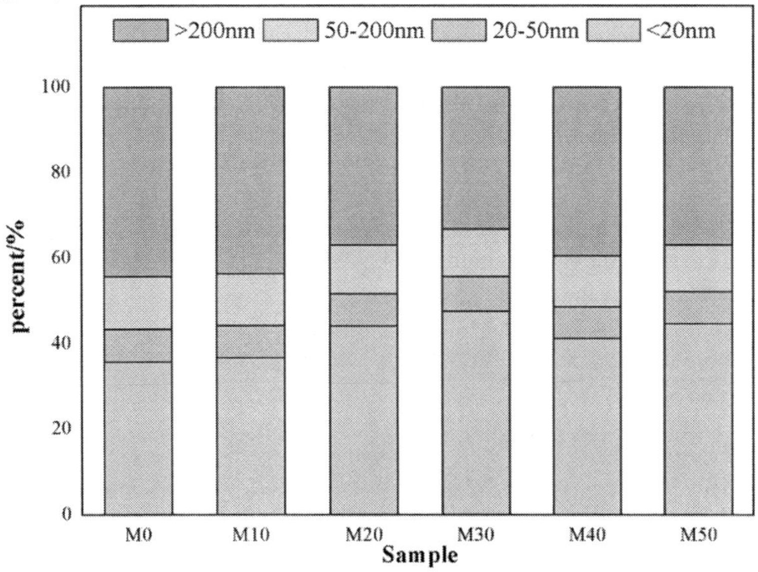

Figure 7. Effect of MSWI FA dosage on the pores of different sizes in hardened pastes.

5. Conclusion

With the addition of MSWI FA doping, the strength and fluidity of the solidified material were the trend of increasing and then going down. When the MSWI FA doping was 30%, the solidified material had the highest compressive strength and showed excellent fluidity. Meanwhile, the sample had the shortest setting time, and the urinary rate tended to be balanced to reach 2.1% after being left to stand for about 420 minutes. The rate of bleeding tended to equilibrate after about 420 minutes and reached 2.1%. The incorporation of MSWI FA led to a rise in the shear stress of the solidified materials at the same shear rate. The shear stresses of Sample M0 and Sample M50 finally reached 193.77 MPa and 410.28 MPa. The fluid types of different solidified materials were all Bingham fluids and with the incorporation of MSWI FA. From the pictures of the pore structure test, the MSWI FA content has a big effect on the structure of the hardened paste. With a moderate amount of MSWI FA incorporated up to 30%, the percentage of multi-hazardous coarse pores and hazardous pores reached the lowest, and the percentage of non-hazardous pores reached the highest.

References

[1] Aixiang W, Yong W, Hongjiang W. Status and Prospects of the Paste Backfill Technology [J]. Metal Mine, 2016.
[2] Bie R, Chen P, Song X, et al. Characteristics of municipal solid waste incineration fly ash with cement solidification treatment [J]. Journal of the Energy Institute, 2016. DOI: 10.1016/j.joei.2015.04.006.
[3] Jin Z, Sun W, Zhao T, et al. Chloride binding in concrete exposed to corrosive solutions [J]. Journal of the Chinese Ceramic Society, 2009.
[4] Igor C, Maria H, Carmen T, et al. Synthesis and characterization of a binder cement replacement based on alkali activation of fly ash waste [J]. Process Safety and Environmental Protection, 2018, 119:23-35. DOI: 10.1016/j.psep.2018.07.011.
[5] Richardson I G. The calcium silicate hydrates [J]. Cement & Concrete Research, 2008, 38(2):137-158.
[6] Mellado A, Borrachero M V, Soriano L, et al. Immobilization of Zn (II) in Portland cement pastes [J]. Journal of Thermal Analysis & Calorimetry, 2013, 112(3):1377-1389.
[7] Tommaseo C E, Kersten M. Aqueous solubility diagrams for cementitious waste stabilization systems. 3. mechanism of zinc immobilization by calcium silicate hydrate [J]. Environmental Science & Technology, 2002, 36(13):2919-2925.
[8] Gougar D M L, Scheetz B E, Roy D M. Ettringite and C-S-H Portland cement phases for waste ion immobilization: A review [J]. Waste Management, 1996, 16(4):295-303.
[9] Vespa M, Dähn R, Grolimund D, et al. Spectroscopic Investigation of Ni Speciation in Hardened Cement Paste [J]. Environmental Science & Technology, 2006(40):2275-2282.
[10] Moore A, Taylor H. Crystal Structure of Ettringite [J]. Nature, 1968, 218(5146): 1048-1049.
[11] Chrysochoou M, Dermatas D. Evaluation of ettringite and hydrocalumite formation for heavy metal immobilization: Literature review and experimental study [J]. Journal of Hazardous Materials, 2006, 136(1):20-33.

Journal of Physics: Conference Series **2808** (2024) 012032 doi:10.1088/1742-6596/2808/1/012032

Study on physicochemical properties of corn starch modified by dielectric barrier discharge (DBD) plasma

Ziwen Wang[1*], Jinhai Niu[2]

[1]Criminal Investigation Police University of China, Shenyang, Liaoning, China
[2]Minzu University, Dalian, Dalian, Liaoning, China

[*]Corresponding author's e-mail: wangzwvip@163.com

Abstract. Starch, as a cheap and renewable biological material, has been widely used in the chemical industry, oil drilling, and other fields. However, there are still some shortcomings in natural starch, which often needs to be modified by some methods. The development and application of modified starch have attracted much attention. The use of low-temperature plasma methods to modify starch has gradually attracted widespread attention because there is no chemical impurity residue. DBD is a type of discharge that produces low temperature plasma. At normal temperature and pressure, plasma can be generated in a DBD reactor which uses air as discharge gas and adopts various electrode structures such as wire-column and plate-plate. The high-energy electrons, reactive nitrogen, and reactive oxygen species generated in the discharge can break chemical bonds when they act on the surface of biological materials, and then change the surface topography and physical and chemical properties of the materials.

1. Introduction

Starch is a natural polymer polymer, which is widely found in plants such as corn, wheat, rice, potato, cassava, etc., and is the main source of carbohydrates for human beings. In addition to being a food source, starch is also a common chemical raw material. Starch as a natural biopolymer material has been widely used in food, pharmaceutical, oil drilling, and other industries. The physical and functional properties of natural starch are poor, such as insolubility, enzymatic resistance, poor machinability, poor solubility, poor shear, and thermal stability, which cannot meet the needs of industrial production. Starch can meet the functional requirements of industrial production after modification, with broad market prospects. Usually, starch modification mainly uses chemical modification, biological modification, physical modification, and other methods. Chemical modification for acid, alkali modification, and oxidation modification. The use of gene modification methods, or enzyme modification methods, is the biological modification of starch. Physical modification has gradually attracted widespread attention because there is no chemical impurity residue. Physical modification of starch includes high pressure, high temperature, ultrasonic, irradiation, and low temperature plasma technology. High temperature and high pressure modification in physical modification have higher energy consumption, and there are more uncontrollable factors in ultrasonic modification. Plasma technology has advantages such as low energy consumption, multiple discharge methods, obvious modification effect, and no introduction of new chemical impurities. In recent years, researchers in various countries have carried out the modification of starch by plasma technology. Among many plasma technologies, DBD plasma technology has carried out a lot of research work on the inhibition of pathogenic bacteria on medical instruments and fruit surfaces in the preparation of optoelectronic film materials and deposition of medical film materials [1].

Plasma technology is widely used in the sustainable reuse of agriculture and the treatment of industrial wastewater [2]. There are a large number of active particles, free radicals, ozone, and other active substances in the low-temperature plasma generated by air or oxygen, which has a strong oxidation performance [3]. The common active substances in plasma are active oxygen, such as O_3, O_2^- and O, and also includes OH^-, H_2O^+, OH, $H_2O_2^-$

2. Materials and methods

2.1. Materials
Corn starch belongs to Shanghai Aladdin Biochemical Technology Co., LTD. It is a reagent grade sample.

2.2. Instrument and equipment
The XRD model is XRD-6000, the SEM model is S-4800, and the FTIR model is Nicolet iS10.

2.3. Experimental method
As shown in Figure 1, corn starch was modified by using a wire-barrel DBD plasma reactor. The quartz tube has an outer diameter of 12 mm and an inner diameter of 10 mm. The high voltage electrode is in the center of the adaptation tube, with a diameter of 1 mm, the ground electrode is outside the quartz tube, and the length of the discharge area is 40 mm. The reactor was filled with corn starch and air was controlled by a mass flow controller to enter the plasma reactor at a flow rate of 300 ml/min. After 10 to 20 minutes of air, it is important to make sure that there is no excess component in the reactor. At room temperature and pressure, a voltage is applied to a high voltage electrode to generate plasma in the reactor. The voltage was 14 kV and 16 kV respectively, and the discharge time at each discharge voltage was 30 s, 2, 5, 10, and 20 min respectively, with a frequency of 6 kHz.

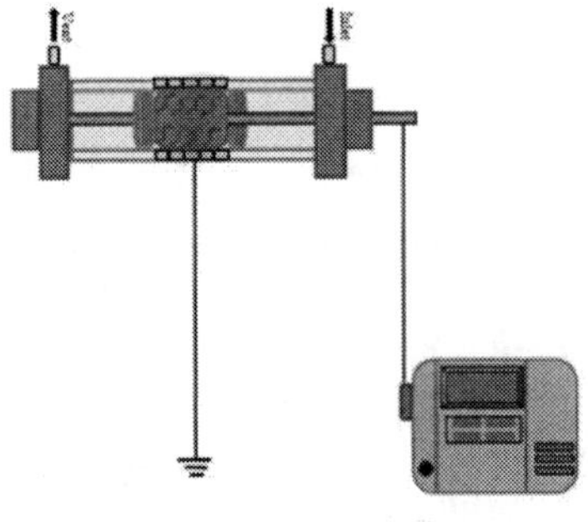

Figure 1. Schematic diagram of the experimental setup for starch modification by DBD plasma.

3. Results and discussion

3.1. XRD
Starch is a semi-crystalline compound as a polymeric biomass material. According to the different X-ray diffraction peak positions, starch can be divided into four crystalline starch types: A, B, C, and V [4]. Figure 2 shows the XRD pattern of corn starch after modification by air DBD plasma with different applied voltages and modification times. It can be seen that the diffraction characteristic peaks of corn starch appear at 10.2°, 15.5°, 17.1°, 18.0°, 19.7° and 23.0°, which belong to typical type A starch [5]. The double helix molecular structure in type A starch is composed of monoclinic crystal cells, so its

structure is more compact than that of other crystal types. When air plasma is used to modify corn starch, the diffraction peak intensity of corn starch decreases after 20 minutes of discharge at the same discharge voltage. However, no new diffraction peak appears in the XRD pattern of starch after plasma modification, so it can be inferred that no new crystal phase is produced in the discharge modification process, and part of the crystallization may change to an amorphous structure. When Ge et al. used low-temperature plasma to treat red bean starch, they also found that the crystallization mode of modified starch did not change, but the crystallinity of starch usually decreased [6]. The pH value of corn starch and cassava starch decreased, and the water-binding capacity and swelling capacity increased [7]. Du et al. modified cassava starch with plasma and found that the solubility of the starch was significantly increased. They believe that the increase in solubility is due to the plasma action that causes the starch to depolymerize or form smaller starch fragments [8]. The modification of starch with plasma found that the solubility of the starch was significantly increased [9].

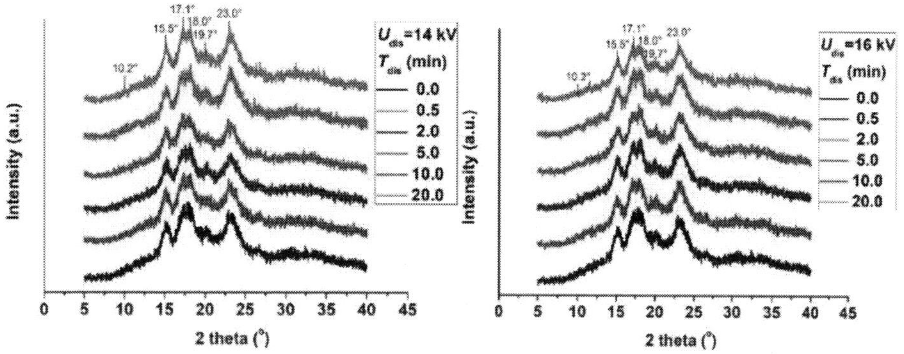

Figure 2. XRD patterns of corn starch modified by air plasma at 14 kV and 16 kV with different discharge time.

3.2. SEM

The surface of the starch sample was pretreated by gold spraying by an ion sputtering instrument. The gold spraying time was 20 s, and the surface topography of the modified starch sample was analyzed under the conditions of scanning voltage 5 kV and current 5 μA. As shown in Figure 3(a), the unmodified corn starch was irregular spherical particles with smooth surface and particle diameter distribution of about 5-20 μm. Air DBD discharge plasma was used to modify corn starch. Figure 3(b) shows the shape of corn starch particles modified by air plasma. It can be seen from the figure that when the discharge was 30 s, a small number of pits and holes appeared in the corn starch particles. As a commonly used physical modification method, plasma modified starch can cause changes in the surface morphology and physical and chemical properties of starch. After modification, the adhesion stress, frostbite stability, water absorption capacity and thermal stability of starch are significantly enhanced, and cracks and holes are formed at the same time. The natural surface of starch is different according to the type. The surface of cereal starch (such as rice, corn, sorghum, wheat, and pseudo-grain crops such as tartary buckwheat and quinoa) has a natural microporous structure. The particle size of potato starch (potato, cassava, sweet potato, taro, etc.) is larger and the structure is loose. The bean starch such as peas can be divided into smooth starch and wrinkled starch without obvious cracks and holes. Starch particles with pores on the surface are prone to generate more and larger pores after plasma treatment. If the starch particles are small, they tend to aggregate. The energetic active substances in the plasma can enter the starch particles through the microporous structure, thus degrading the starch chains and producing more pore structures from the outside to the inside. It was found that the higher the amylose content is, the more filamentous particles there are, and the smaller the particles are, the less obvious plasma etching is. Both waxy corn starch and ordinary corn starch have holes, but waxy corn has more holes than ordinary corn, which is called etching. The morphology of starch was changed by plasma etching [10]. Chen et al. analyzed the crystallinity of corn starch modified by plasma and found that the modified starch had an amorphous

Journal of Physics: Conference Series **2808** (2024) 012032 doi:10.1088/1742-6596/2808/1/012032

structure. Meanwhile, SEM was used to analyze the morphologic characteristics of modified corn starch and found that plasma had an etching effect on starch [11].

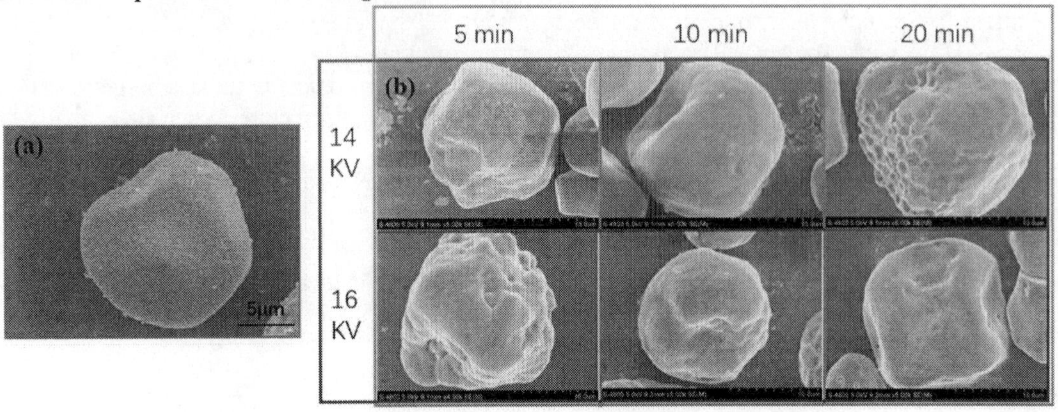

Figure 3. Morphology of corn starch modified by air plasma.

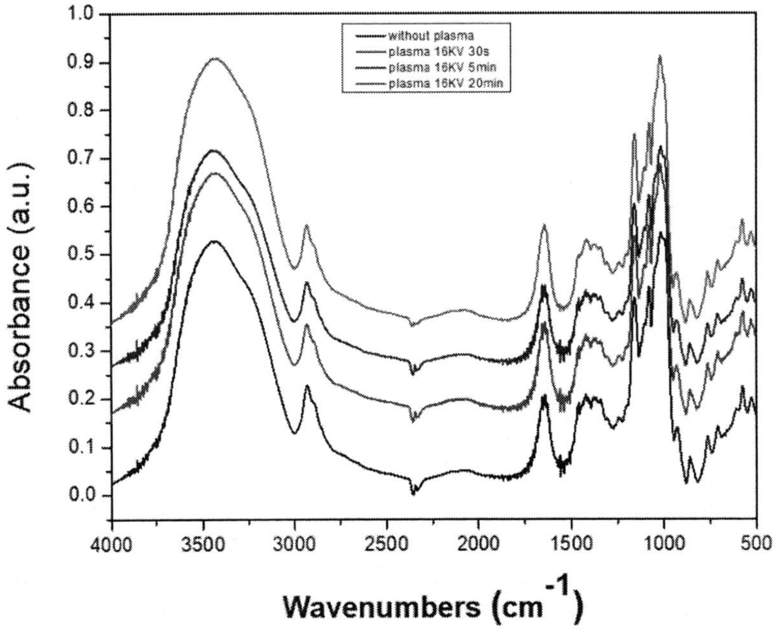

Figure 4. FTIR spectra of corn starch modification process by air plasma at a discharge voltage of 16 kV.

3.3. FTIR

As shown in Figure 4, the surface of corn starch has a wide and strong infrared absorption band, usually in the range of 3600-3100 cm^{-1}, mainly from the tensile vibration of O-H of different structures in starch molecules. The contraction vibration absorption peaks of two distinct C-H bonds at 2930 cm^{-1} and 2895 cm^{-1} mainly belong to CH_2 and CH. The absorption peak of 1645 and 1244 cm^{-1} is O-H bending vibration, and the absorption peak of 1165 cm^{-1} is C-O-C tensile vibration [12]. The results show that the absorption peak of 1157 cm^{-1} is C-O, 1105 cm^{-1} is C-C, and 982 cm^{-1} is C-OH. The effects of these three groups on type A starch in FTIR are greater than those of type C and type B starch samples. The obvious characteristic peak of type A starch appears at 1022 cm^{-1}, 3600-3000 cm^{-1}, and 1200-800 cm^{-1} are also significantly higher than those of type B and type C starch, indicating that there is stronger hydrogen bond interaction in type A starch. The density of hydrogen bonds of type B starch is the lowest, but the content of water molecules between double helices of type B starch is higher than that of type A starch [4]. In the process of plasma modification of natural starch, the active substance interacts with starch molecules to change various properties of starch, such as cross-linking, depolymerization, introduction of functional groups, changes in amylose and amylopectin content, and plasma etching. Through these different mechanisms, plasma can cause surface modification of biodegradable polymers. Zhang et al. After modification of cassava starch by plasma, it can be seen from FTIR that the C-O-C asymmetric tensile bond, C-OH bending bond, C-O-C asymmetric tensile bond, C-H bending bond and O-H stretching bond, and the chemical changes of corresponding functional groups can be predicted by the change of peak strength of the absorption peak. It can be inferred from the peak value of FTIR spectra that starch molecules undergo cross-linking reactions during plasma treatment, resulting in gelatinization and rheological changes [13]. Plasma treatment reduced the relative ratio of amylose to amylopectin.

4. Conclusion

Corn starch was modified by air plasma, and the diffraction peak intensity of corn starch decreased after discharge for 20 min. However, no new diffraction peak appeared in the XRD pattern of corn starch after plasma modification, which indicated that no new crystal phase was generated during the discharge modification process, and part of the crystallization may be transformed into an amorphous structure. SEM analysis showed that the surface of corn starch modified by DBD plasma appeared different degrees of dents and holes, and the dents became denser with the increase of discharge voltage and discharge time.

References

[1] Sheng, X, W. Yan, W, J. Zhao, L, L. et al. (2023) Cold plasma-222 nm UV: A new cold sterilizing method for food contact surfaces. J. Food Control, 152: 109870.

[2] Kooshki, S. Pareek, P. Mentheour, R. et al. (2023) Efficient treatment of bio-contaminated wastewater using plasma technology for its reuse in sustainable agriculture. J. Environmental Technology & Innovation, 32: 103287.

[3] Kopuk, B. Gunes, R. Palabiyik, I. (2022) Cold plasma modification of food macromolecules and effects on related products. J. Food Chemistry, 382: 132356.

[4] Liang, Y, S. Wang, F. Ma, R, R. et al. (2024) Structural properties of the intra- and interhelical cavities of V6-type crystalline starches. J. Carbohydrate Polymers, 330: 121835.

[5] Wang, R. He, Z, S. Cao, Y, J. et al. (2023) Impact of crystalline structure on the digestibility of amylopectin-based starch-lipid complexes. J. International Journal of Biological Macromolecules, 242(4): 125191-125191.

[6] Gupta, R, K. Guha, P. Srivastav, P, P. et al. (2023) Effect of high voltage dielectric barrier discharge (DBD) atmospheric cold plasma treatment on physicochemical and functional properties of taro (Colocasia esculenta) starch. J. International Journal of Biological Macromolecules, 253(2): 126772-126772.

[7] Chauhan, M. Kalaivendan, R, G. Eazhumalai, G. et al. (2023) Atmospheric pressure pinto-plate cold plasma effect on physicochemical, functional, pasting, thermal, and structural characteristics of proso-millet starch. J. Food Research International, 173(2): 113444-113444.

[8] Du, Z, X. Li, X, X. Zao, X, Y. et al. (2024) Multi-scale structural disruption induced by radio frequency air cold plasma accelerates enzymatic hydrolysis/hydroxypropylation of tapioca starch. J. International Journal of Biological Macromolecules, 260(2): 129572.

[9] Gebremical, G, G. Tappi, S. Laurita, R. et al. (2023) Effects of plasma activated water (PAW) on rheological, thermal, hydration and pasting properties of normal maize, waxy maize and potato starches. J. Food Hydrocolloids, 144:109006.

[10] Khachatryan J, M. Khachatryan, G. et al. (2023) Comparison of Physicochemical Properties of

Silver and Gold Nanocomposites Based on Potato Starch in Distilled and Cold Plasma-Treated Water. J. International Journal of Molecular Sciences, 24(3): 2200-2200.
[11] Chen, C, S. Tong, F. Sun, R, H. et al. (2023) Plasma Effects on Properties and Structure of Corn Starch: Characterization and Analysis. J. Foods, 12(21): 2304-8158.
[12] Qi, M, s. Song, J. Jiang, Y, J. et al. (2024) Understanding the degradation mechanisms of cyanide and starch in cassava flour during extrusion processing. J. Innovative Food Science and Emerging Technologies, 91: 103548.
[13] Zang, K, X. Zhao, Z, H. Zhao, M. et al. (2022) Low-pressure plasma modification of the rheological properties of tapioca starch. J. Food Hydrocolloids, 125:107380.

Journal of Physics: Conference Series **2808** (2024) 012033 doi:10.1088/1742-6596/2808/1/012033

Experimental study on the high-velocity impact resistance of Vitrimer-based carbon fiber composites

Ziyue Lin[1,*] and Peng Chen[1]

[1]School of Aerospace Engineering, Xi'an Jiaotong University, Xi'an, 710049, P. R. China

[*]Corresponding author's e-mail: zy_linee@163.com

Abstract: Vitrimer-based composites combine the excellent performance of thermoset materials with the recyclability of thermoplastic materials. In this paper, based on ballistic impact experiments, the high-velocity impact resistance of CFRP before and after matrix modification and the failure mechanism of Vitrimer-based composites are investigated. The better impact resistance of Vitrimer-based composites is demonstrated in terms of ballistic limit, energy absorption, and failure modes. The failure mechanism of Vitrimer-based composites, which is different from that of conventional Epoxy-based composites, is found by CT scanning and verified in the strain signals.

1. Introduction

With the rapid development of aerospace, automotive, and other industrial fields, carbon fiber-reinforced polymer composites have become increasingly popular, due to their excellent specific modulus and specific strength [1-3]. Consequently, the demand and consumption have also increased dramatically. In view of the fact that thermosetting CFRP composites such as epoxy resins are often used as the material of choice in these fields, the issue of recycling and reusing waste and end-of-life parts generated during the manufacturing process poses a serious challenge in the context of an economically and environmentally friendly era [4-6]. To address this problem, the adaptive thermosetting resin with dynamic covalent bonding (Vitrimer) [7] proposed by Leibler's team [8] makes the recycling self-healing of composites possible. The reversible exchange of dynamic bonds can effectively dissipate energy, and the static mechanical properties of conventional epoxy resins are well preserved, while the energy absorption effect of epoxy-based vitrified resins at high strain rates has been verified [9]. However, the impact resistance of Vitrimer-based carbon fiber composites has yet to be investigated, which will determine the prospects of its application in aerospace.

In recent years, many scholars have conducted research on matrix modification to improve the impact resistance of composites [10-13]. The reversible exchange mechanism of dynamic bonds can effectively dissipate energy, which not only retains the excellent static mechanical properties of traditional epoxy resins but also verifies the energy absorption effect of Vitrimer under high strain rates [14]. This paper conducts ballistic impact tests on two types of composite materials before and after matrix modification, aiming to explore the performance and underlying mechanisms of Vitrimer-based composites under impact conditions.

Journal of Physics: Conference Series **2808** (2024) 012033 doi:10.1088/1742-6596/2808/1/012033

2. Materials and methods

2.1. Material design and preparation

Composite laminates made of two different matrix components are selected for this experimental design, and the fiber cloth used is T300 plain weave carbon fiber cloth provided by the manufacturer. One of the resins is a traditional epoxy resin, which is prepared from epoxy monomer (DGEBA) and glutaric anhydride (GA) under the action of curing accelerator 2,4,6-triphenyl (referred to as DMP-30). For the Vitrimer resin, the preparation method of Leibler's team is followed, which involves replacing DMP-30 with $Zn(acac)_2$ as a catalyst, adding DGEBA and GA in a proportional and sufficient mixture, and then curing for 12 h at 140°C. Resin and carbon fiber scrim at 40% and 60% mass ratio are made into composite laminates by hot press tank molding process with alternating layers at 0°/45° and then cut into specimens measuring 150 mm×150 mm×4 mm by using a high-pressure water cutting machine. The projectile used in the test is a cylindrical flat-head projectile made of heat-treated alloy steel.

2.2. Ballistic impact test program

The high-speed gas cannon test system is shown in Figure 1, which is composed of an 80 mm caliber first-class light gas cannon, high-speed cameras, speed measurement system, and data acquisition system. Before the test, the projectile is placed in the cartridge tray, and the speed of bullet ejection is controlled by controlling the change of gas pressure in the gas chamber. The rapid pressure valve is opened after the gas pressure in the gun chamber stabilizes to reach a predetermined value, and the high-pressure gas expands to instantly push the cartridge tray and the bullets to glide in the gun tube. A high-speed camera is placed in front, back, and on the side of the laminate to accurately record the attitude and speed changes during the impact of the projectile, as well as the real-time damage of the composite laminate.

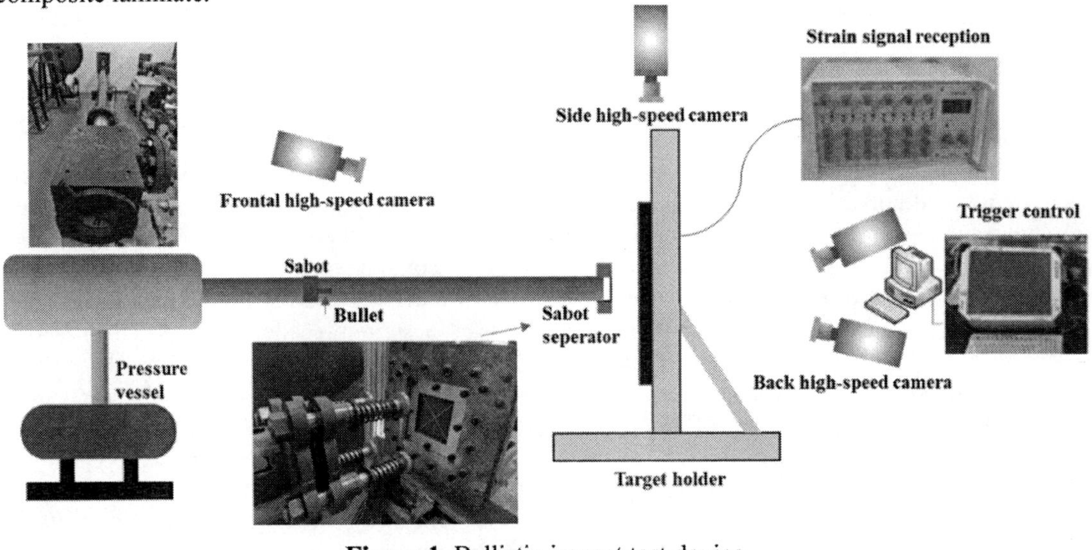

Figure 1. Ballistic impact test device.

3. Results & discussion

3.1. Ballistic limits and energy absorption

In this paper, five tests with different impact velocities are conducted on each of the composite laminates of the two materials, including multiple penetrations and one rebound data. The specific test results are shown in Table 1. The impact resistance is evaluated by comparing the ballistic limit and energy absorption effect of the flat-head projectile penetrating the two composite plates. The initial

Journal of Physics: Conference Series **2808** (2024) 012033 doi:10.1088/1742-6596/2808/1/012033

and residual velocities before and after impact are plotted in a coordinate system, and the ballistic limit fitting is realized by the following equation [15].

$$V_r = (V_i^p - V_{bl}^p)^{\frac{1}{p}} \quad (1)$$

where V_r is the residual velocity, V_i is the initial velocity, V_{bl} is the ballistic limit velocity, p is the fitting parameter, for simplicity, the value of p is taken as 2. The ballistic limit curve obtained by fitting is shown in Figure 2, and the intersection point of the curve and the horizontal axis is the ballistic limit.

Based on the analysis of the experimental results obtained from the curves, the ballistic limit velocities of the Vitrimer-based composite and Epoxy-based composite are determined to be 97.9 m/s and 106.1 m/s, respectively. When comparing the two, the ballistic limit velocity of the Vitrimer-based composite is found to be 8.7% higher than that of the Epoxy-based composite. Furthermore, the kinetic energy of both materials at their critical penetration velocities is calculated. The result shows that the energy absorption capacity of the Vitrimer-based composite at the point of initial penetration is improved by 17.5% compared to the unmodified composite. To provide a more intuitive representation of the energy absorption characteristics of both materials during impact, this study calculates the energy absorbed by the composite plates as the difference in kinetic energy before and after the bullet impacts the target plates. The data is plotted with the initial kinetic energy on the horizontal axis and the energy absorption on the vertical axis to create a trend chart illustrating the relationship between energy absorption and impact energy. It is evident from Figure 3 that the Vitrimer-based composite exhibits superior overall energy absorption compared to the Epoxy-based composite, demonstrating its significant advantage in terms of impact resistance.

Table 1. Summary of ballistic impact test results.

Label	Initial Velocity (m/s)	Residual Velocity (m/s)	Penetration
V-1	150.3	108.4	Yes
V-2	134.6	84.6	Yes
V-3	116.6	51.7	Yes
V-4	111.1	29.5	Yes
V-5	93.4	0	No
E-1	153.6	113.1	Yes
E-2	138	92.6	Yes
E-3	116.6	56.3	Yes
E-4	104.3	43.1	Yes
E-5	74.9	0	No

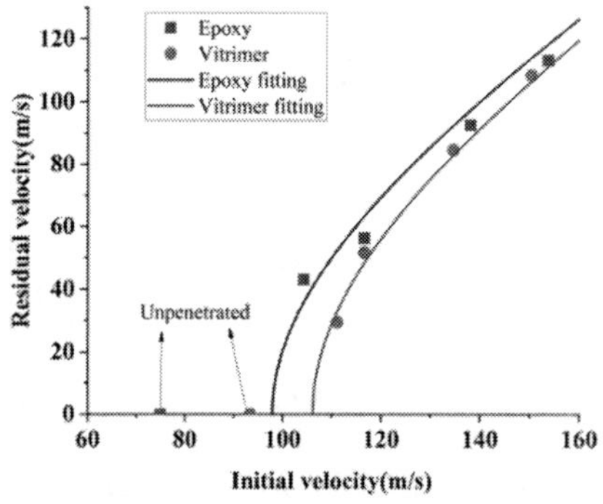

Figure 2. Ballistic limit curves.

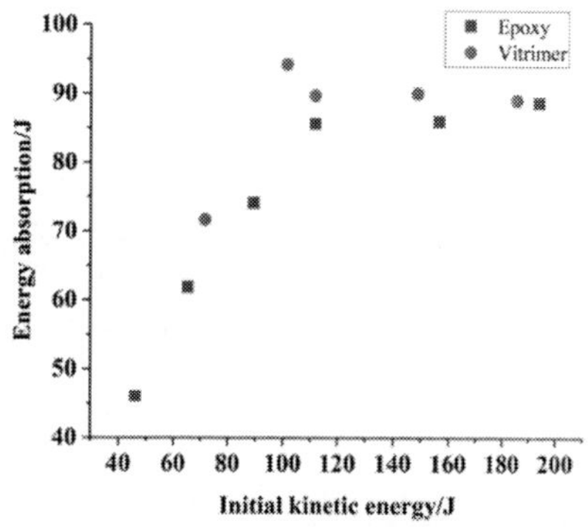

Figure 3. Energy absorption at different initial energies.

Journal of Physics: Conference Series **2808** (2024) 012033 doi:10.1088/1742-6596/2808/1/012033

3.2. Failure mechanism

The high-speed camera shows the damage process of the bullet on the Vitrimer-based composite under the speed of 150.3 m/s. As shown in Figure 4, after the bullet contacts the target plate, it first compresses the target plate to make its center concave, and then circular damage begins to appear on the back side of the plate. When the projectile enters the target plate as a whole, the back side of the plate flies out of the shear formation of the punch block. Until the projectile completely penetrates the target plate, the front and back surfaces of the laminate are sprayed with varying degrees of matrix powder. Additionally, some fiber strips may detach from the back of the laminate. Observing the macroscopic damage morphology after the impact shown in Figure 5, a punch hole of the same size as the projectile body is evident on the projectile face. The fracture in the damaged area appears flat, indicating that the front fiber is primarily experiencing shear fracture. On the other hand, the back surface of the projectile shows obvious traces of fibers being pulled off, with irregular fiber fractures. Additionally, there is obvious fiber delamination on the side, suggesting that the region near the back surface of the projectile is mainly damaged by fiber tensile fracture and matrix cracking.

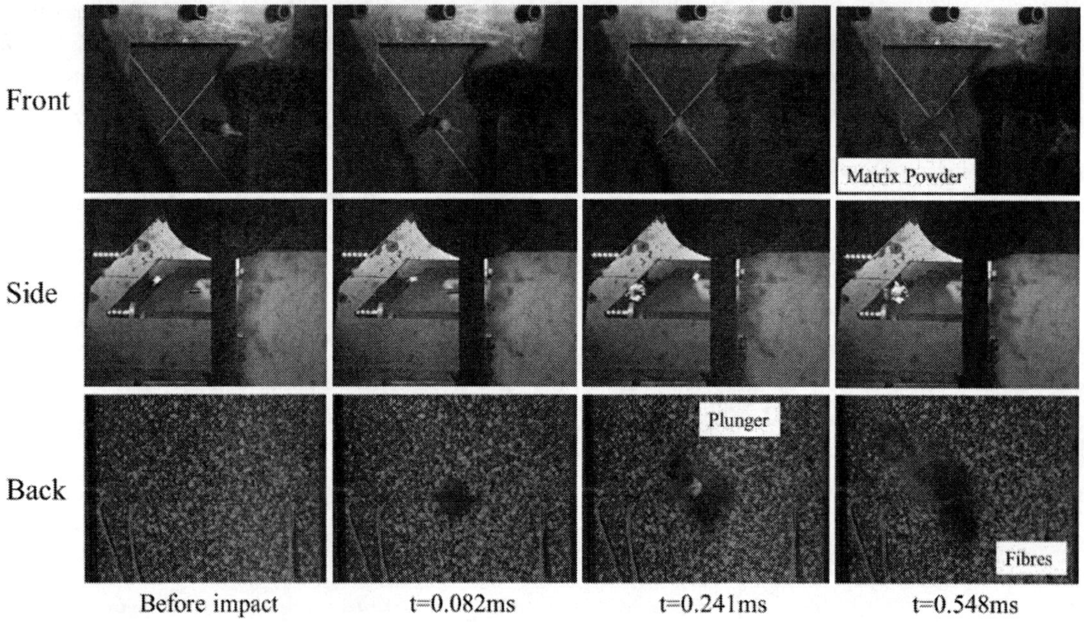

Figure 4. Pictures from the high-speed videos showing the impact process of Specimen V-1 at different moments.

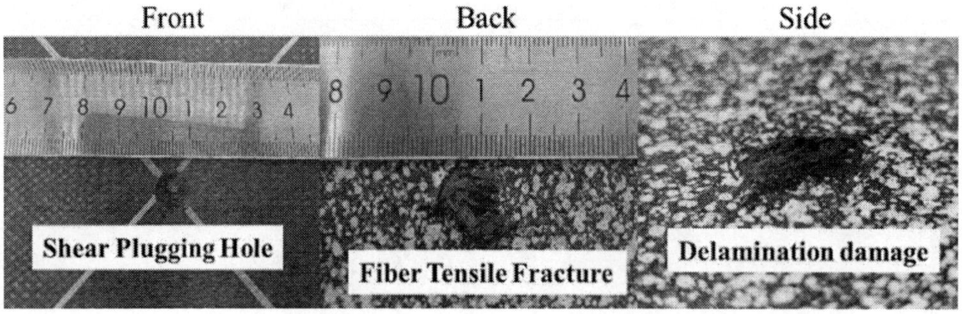

Figure 5. Macroscopic damage morphology of the Specimen V-1.

To further reveal the failure mechanism of Vitrimer-based composite panels under impact, two laminate samples after impact at the same velocity (150 m/s) are selected in this study, and the damage morphology of the center section along the thickness direction is analyzed in detail by using high-precision 3D CT scanning technology. As shown in Figure 6, the damaged cross-section of the Vitrimer-based composite exhibits a unique combination of rectangular and trapezoidal region features. The rectangular region corresponds to the shear damage zone during impact, while the trapezoidal region represents the tensile damage zone. This type of failure is originally found more often in composites impacted by spherical bullets [16]. This finding suggests that the modification treatment of Vitrimer effectively suppresses the shear impaction mechanism of the composite when subjected to impact. In contrast, the damaged cross-section of the Epoxy-based composites after impact presents a more regular rectangular morphology of the incision, showing that the material is almost completely penetrated by the projectile due to the shear action during the impact process, which further highlights the advantages of the Vitrimer-based composites in terms of impact resistance.

After a thorough analysis, we have summarized the failure mechanism of Vitrimer-based carbon fiber composite materials under flat-nosed projectile impact. Specifically, when a projectile impacts the front surface of the laminate at a specific velocity, the laminate undergoes compression, and stress waves propagate in the form of compressive waves along its thickness direction. Due to the large contact area between the flat-ended projectile and the laminate, and the fast impact velocity, the front part of the laminate forms a shear punch plug hole during the impact process. The velocity of the projectile gradually decreases due to the blockage of the composite plate, resulting in insufficient force to sustain further shear impingement effects. As the compression wave propagates to the back of the laminate and is reflected to form a tensile wave, when the tensile wave is reflected to meet with the front end of the slug, the tensile stress on the fibers at the back of the laminate reaches its strength limit and the fibers undergo a tensile fracture. Consequently, in the later stages of the impact process, distinct tensile damage zones are observed in the CT images.

This failure mechanism reveals the complex damage evolution process of Vitrimer-based carbon fiber composites under the impact of a flat-head projectile. This offers a crucial theoretical basis for enhancing our comprehension of the impact resistance of composites.

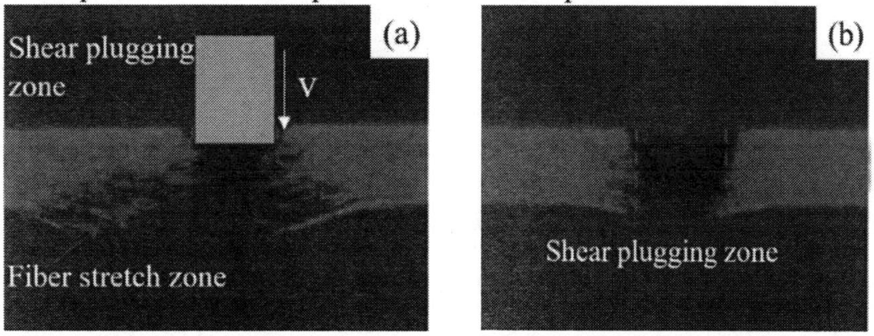

Figure 6. Pictures of the cross-sections under CT scanning (a) CFR-Vitrimer (b) CFR-Epoxy.

3.3. Strain signal verification

Before the test, we attach strain gauges to the back of the plate in the horizontal direction 40 mm away from the center in order to obtain the change curve of the x-direction strain signal on the back of the laminate with time. As shown in Figure 7, we find that the strain signals show an alternating tensile and compressive form throughout the impact process, which reflects the vibration process of the plywood under the action of the stress wave. In particular, we notice that the strain signal reaches its amplitude position in about 60-80 μs of time. By comparing the strain amplitudes of the two materials, we find that the Vitrimer-based composite exhibits a larger tensile strain amplitude upon impact, with a 27.5% increase in maximum tensile strain compared to the Epoxy-based composite. This result suggests that the fiber tensile fracture effect is more pronounced on the back of the laminate when Vitrimer-based composites are subjected to impact. Additionally, fiber tensile fracture

Journal of Physics: Conference Series **2808** (2024) 012033 doi:10.1088/1742-6596/2808/1/012033

is the most efficient failure mode for energy absorption in composite impacts [17, 18]. Therefore, we can prove that Vitrimer-based composites show more excellent performance in terms of high-velocity impact resistance.

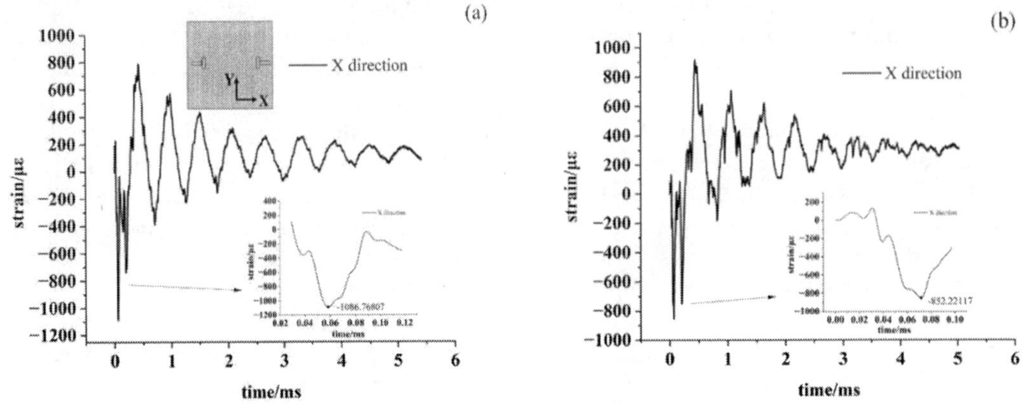

Figure 7. Strain-time curves measured for the Specimen V-1 (a) and the Specimen E-1 (b).

4. Conclusions

In this paper, the comparison of the high-velocity impact resistance of the two composites before and after matrix modification is investigated, and the failure mechanism of Vitrimer-based composites is analyzed. With the help of a ballistic impact test system and CT scanning technology, it is observed that the failure modes of Vitrimer-based composites are mainly dominated by shear plugging and fiber tensile fracture under the impact of a flat-head bullet; in contrast, Epoxy-based composites almost fail in pure shear plugging mode. After matrix modification, the ballistic limit of Vitrimer-based composites is increased by 8.7% compared with that before modification, and its ballistic limit energy absorption capacity is increased by 17.5%. The findings in this paper might provide help in understanding and improving the impact resistance of composites.

Acknowledgments

The authors gratefully acknowledge the financial support of NSFC (11902246).

References

[1] Frhaan, W. K. M., Abu Bakar, B. H., Hilal, N., et al. (2021) CFRP for strengthening and repairing reinforced concrete: a review. Innovative Infrastructure Solutions, 62: 49.

[2] Sayam, A., Rahman, A. N. M. M., Rahman, M. S., et al. (2022) A review on carbon fiber-reinforced hierarchical composites: mechanical performance, manufacturing process, structural applications and allied challenges. Carbon Letters, 325: 1173-1205.

[3] Zhu, W., Fu, H., Li, F., et al. (2022) Optimization of CFRP drilling process: a review. The International Journal of Advanced Manufacturing Technology, 1235-6: 1403-1432.

[4] Liu, T., Zhao, B., Zhang, J. (2020) Recent development of repairable, malleable and recyclable thermosetting polymers through dynamic transesterification. Polymer, 194: 122392.

[5] Ahrens, A., Bonde, A., Sun, H., et al. (2023) Catalytic disconnection of C-O bonds in epoxy resins and composites. Nature, 6177962: 730-737.

[6] Shi, Q., Jin, C., Chen, Z., et al. (2023) On the Welding of Vitrimers: Chemistry, Mechanics and Applications. Advanced Functional Materials, 3336: 2300288.

[7] Van Zee, N. J., Nicolaÿ, R. (2020) Vitrimers: Permanently crosslinked polymers with dynamic network topology. Progress in Polymer Science, 104: 101233.

[8] Montarnal, D., Capelot, M., Tournilhac, F., et al. (2011) Silica-Like Malleable Materials from Permanent Organic Networks. Science, 3346058: 965-968.

[9] An, L., Xiang, R., Jin, C. (2023) Dynamic behavior of vitrimers under high-strain rate loadings. Polymer, 282: 126149.
[10] Vieille, B., Casado, V. M., Bouvet, C. (2013) About the impact behavior of woven-ply carbon fiber-reinforced thermoplastic- and thermosetting-composites: A comparative study. Composite Structures, 101: 9-21.
[11] de Ruijter, C., van der Zwaag, S., Stolze, R., et al. (2009) Liquid crystalline matrix polymers for aramid ballistic composites. Polymer Composites, 314: 612-619.
[12] Khodadadi, A., Liaghat, G., Bahramian, A. R., et al. (2019) High-velocity impact behavior of Kevlar/rubber and Kevlar/epoxy composites: A comparative study. Composite Structures, 216: 159-167.
[13] Tang, F., Dong, C., Yang, Z., et al. (2022) Protective performance and dynamic behavior of composite body armor with shear stiffening gel as buffer material under ballistic impact. Composites Science and Technology, 218: 109190.
[14] Yu, B., Karthikeyan, K., Deshpande, V. S., et al. (2017) Perforation resistance of CFRP beams to quasi-static and ballistic loading: The role of matrix strength. International Journal of Impact Engineering, 108: 389-401.
[15] Ipson, T. W., Recht R. F. (1975) Ballistic-penetration resistance and its measurement. Composite Structures, 15(7): 249-257.
[16] López-Puente, J., Zaera, R., Navarro, C. (2007) An analytical model for high velocity impacts on thin CFRPs woven laminated plates. International Journal of Solids and Structures, 449: 2837-2851.
[17] Naik, N. K., Shrirao, P., Reddy, B. C. K. (2006) Ballistic impact behavior of woven fabric composites: Formulation. International Journal of Impact Engineering, 329: 1521-1552.
[18] Dogan, U., Sunbuloglu, E. (2023) A novel simplified ballistic impact theory for woven composites International Journal of Impact Engineering, 177: 104530.

Interfacial morphology and reliability of GaN nanocomposite Sn-Ag-Cu lead-free braze joints

Shaoxuan Liu[1*], Min Qu[1], Hang Li[1], Yuan Liu[1] and Yan Cui[1]

[1]School of Mechanical and Materials Engineering, North China University of Technology (NCUT), Beijing, 100044, China

[*]Corresponding author's e-mail: liushaoxuan1999@126.com

Abstracts. GaN nanoparticles, in varying mass fractions (0 wt%, 0.20 wt%, 0.40 wt%, 0.60 wt%, and 0.80 wt%), were integrated into a Sn-3.0Ag-0.5Cu lead-free solder using mechanical mixing followed by reflow soldering. The solders underwent isothermal aging at 170°C for different durations to assess the influence of GaN content on the melting characteristics, wettability, and growth behavior of the intermetallic compound (IMC) in the Sn-3.0Ag-0.5Cu solder. Findings indicate that GaN incorporation does not markedly alter the melting point of SAC305. However, a moderate GaN addition enhances the wettability of Sn-3.0Ag-0.5Cu, with the optimal effect observed at 0.40 wt% GaN, where the wettability is improved by 61.5% relative to the base solder. Furthermore, GaN addition serves to curtail the IMC layer's growth, which, while it thickens with prolonged isothermal aging, does so at a rate governed by diffusion. At 0.40 wt% GaN, the diffusion coefficient is determined to be 1.453×10^{-15} m^2/s, underscoring the superior performance of the novel SAC305-0.4GaN composite solder.

1. Introduction

Lead solder, traditionally recognized as a hazardous substance, has been utilized extensively over the years owing to its superior properties. However, recent awareness regarding environmental and human health implications has underscored that even minuscule quantities of lead can significantly impact human health [1]. Consequently, industries, including electronics, have progressively shifted from lead to lead-free solder, especially in electronic packaging applications. This transition aligns with the industry's push towards miniaturization and high integration, a trend that has gained momentum in the 5G era where dependable electronic packaging is crucial for stable data transmission. This encompasses technologies like system integration packaging and 3D packaging [2]. In this evolving landscape, Sn-based solder materials such as Sn-Ag-Cu, Sn-Cu, Sn-Ag [3], Sn-Zn [4], and Sn-Bi [5] have gained prominence. Notably, the SAC system is emerging as a favored alternative to Sn-Pb solder, attributed to its comparatively superior performance.

While Sn-Ag-Cu solder material is a promising alternative to traditional Sn-Pb solder, it does exhibit certain limitations. For instance, during the soldering process, an intermetallic compound (IMC) layer forms at the interface between the solder and the base material [6]. This IMC layer, possessing different mechanical properties from the solder alloy, can lead to stress concentrations at the solder joint interface, potentially causing cracks that often initiate at the junction between the brazing alloy and the IMC layer [7]. This is a critical concern, as an increase in the IMC layer's thickness correlates with a decrease in the joint's reliability [8]. Moreover, Sn-Ag-Cu solder faces challenges such as a high melting point, inferior wettability, and elevated surface tension, which necessitate resolution [9]. A promising strategy to address these issues involves integrating a reinforcing phase into the solder material, with metal compounds and intermetallic compounds being commonly used. Among potential candidates, gallium nitride (GaN) stands out due to its excellent electrical, mechanical, and thermal properties [10]. GaN, characterized by its high hardness, superior thermal conductivity, and robust corrosion resistance, can mitigate IMC layer growth and shows great promise in electronic packaging applications. This study explores the preparation of GaN-reinforced, lead-free solder material using a mechanical mixing method. By varying the content of GaN nanoparticles, the research aims to enhance the mechanical and electrical properties and overall reliability of the solder, ultimately identifying the composition that delivers the best comprehensive performance.

2. Experiment

We used Sn-3.0Ag-0.5Cu (SAC305) brazing material with a particle size of 25-45 μm and GaN with an average particle size of 50 nm.

2.1 Preparation of composite brazing material

Firstly, we obtained different weights of brazing material and GaN by proportioning and weighing and poured the weighed SAC305 brazing material and GaN into the ball milling jar of the planetary ball mill, with zirconium oxide ball as the ball milling medium. The ball ratio of the material was 5:1, the ball milling time was 60 min, and the rotational speed was 120 r/min. After the ball milling was finished, the mixed powder was blended with the flux paste according to the ratio of 1:8. The mixture was stirred manually for 30 min, and stirred until the brazing material was mixed with the flux paste at the ratio of 1:8. After mixing uniformly, new composite lead-free brazing solder pastes with different content of SAC305-xGaN (x= 0 wt%, 0.20 wt%, 0.40 wt%, 0.60 wt%, 0.80 wt%) were produced.

Experiments using 15 mm * 15 mm * 2 mm polished copper substrate with 99.99% purity. Different copper substrates were coated with solder paste of different compositions with a diameter of 6-7 mm. 30 mm * 5 mm * 2 mm Cu bars were used for shear experiments. The brazing material after screen printing to the same position, a series of solder joints and samples were obtained by reflow soldering on an F4N nitrogen reflow soldering machine. The obtained solder joints were put into a vacuum drying oven and subjected to isothermal aging at 170°C for 0, 24, 72, 168, and 360 h.

2.2 Test methods

In this study, a Differential Thermal Scanning Calorimeter is used to determine the melting point of composite brazing material. The powder after ball milling was taken 20-30 mg into the small crucible of the differential thermal scanning calorimeter for DSC measurement, the heating rate as well as the cooling rate of the test was 10 °C/min and the maximum temperature was set to 300°C. The DSC curves were finally obtained.

Figure 1. Wetting simulation schematic.

The wetting angle was determined through observations made with an optical microscope, and the corresponding simulation outcomes are depicted in Figure 1, where α represents the wetting angle.

Additionally, the IMC layer's examination was conducted using a Scanning Electron Microscope following an etching process, enabling the acquisition and subsequent fitting of the relevant data.

3. Results and discussion

3.1 Melting characteristics

The melting point of a brazing alloy is one of the important indexes for evaluating the solderability of this brazing material. The DSC curves of Sn-3.0Ag-0.5Cu-xGaN are shown in Figure 2. It is known from the graphs that the addition of nano GaN particles does not have much effect on the melting performance of the brazing materials in terms of the various temperature points. This is because for a multi-component alloy such as SAC305, the addition of a small amount of GaN is not sufficient to significantly change the lattice structure of the alloy at the macroscopic level and produce enough lattice distortion to change its melting temperature. At the same time, GaN itself has a much higher melting point than SAC305 and very good thermal stability, making it less likely to chemically react or decompose during the melting process of SAC305 alloy, so that its effect on the melting temperature of the alloy is limited.

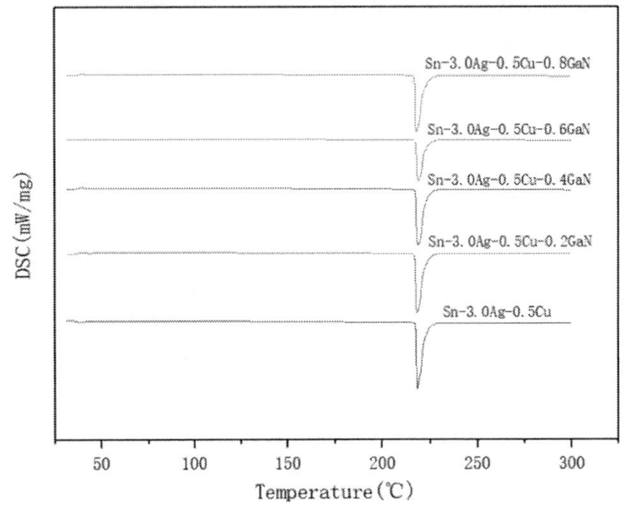

Figure 2. DSC curves of composite brazing materials with different contents.

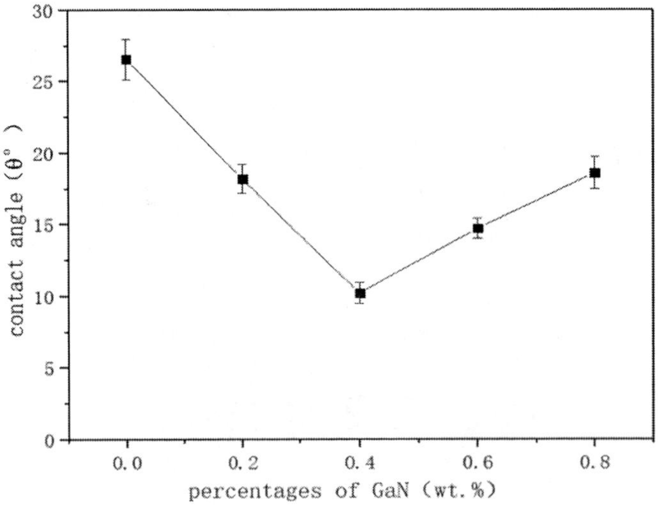

Figure 3. Wetting angle variation of Sn-3.0Ag-0.5Cu-xGaN composite brazing materials.

3.2 Wettability

Figure 3 shows the variation of composite brazing material wetting angle with GaN content. When the GaN content was 0 wt%, the wetting angle was 26.55°. When different contents of GaN were added, the wetting angle changed. The wetting angle reached a minimum value of 10.2° when the GaN content was 0.40 wt%, which was 61.5% lower compared with SAC305. With the increase of GaN content, the wetting angle showed a decreasing and then increasing trend. The results indicate that the addition of GaN can improve the wettability of the composite brazing material.

In order to gain a deeper understanding of the reasons for the change in the wetting angle, we can analyse it in conjunction with Young's equation and the Gibbs adsorption equation, where Equation 1 is Young's equation and Equation 2 is the Gibbs adsorption equation:

$$cos\theta = \frac{\gamma_{sg}-\gamma_{sl}}{\gamma_{lg}} \tag{1}$$

where γ_{sg} represents the surface tension between solid and gas; γ_{sl} represents the surface tension between liquid and solid; γ_{lg} represents the surface tension between liquid and gas.

$$\Gamma = -\frac{1}{\mathrm{RT}}\left(\frac{\partial\gamma}{\partial lnC}\right)_T \tag{2}$$

where Γ denotes the adsorption amount per unit area of GaN particles on the surface of SAC305 alloy; R is the ideal gas constant with a value of 8.314 J/(mol·k); T denotes the absolute temperature; γ denotes the surface tension of SAC305 alloy, which is the surface tension under the influence of GaN particles adsorbed onto the surface of SAC305 alloy in this experiment; C denotes the GaN particle concentration in SAC305 alloy; $\partial\gamma/\partial lnC$ is the partial derivative of the surface tension with respect to the logarithm of the GaN particle concentration C, which indicates the rate of change of the surface tension with the logarithm of the GaN particle concentration at constant temperature, and this derivative quantifies the change of the surface tension of SAC305 alloy when the concentration of GaN particles increases.

The incorporation of GaN particles plays a surfactant-like role within the solder matrix, significantly impacting the solder's surface properties. By forming an adsorption layer on the liquid solder's surface, GaN reduces the surface tension, as this layer diminishes the interaction between the solder's molecules and the surrounding air, thereby lowering the liquid-gas surface tension (γ_{lg}). Concurrently, GaN particles' dispersion alters the solder's microstructure, refining the grains and creating a distinctive microstructure at the interface. This change enhances the contact points between solid and liquid phases, consequently reducing the solid-liquid interfacial tension (γ_{sl}). Additionally, according to Gibbs' adsorption isotherm, an increased GaN concentration heightens the adsorption (Γ) at the solder's surface, which further decreases γ_{lg}. Collectively, these modifications contribute to a reduction in the wetting angle, enhancing the solder's wettability.

The minimum wetting angle at 0.40 wt% GaN content is due to the fact that, at 0.4% GaN content, the optimum equilibrium point of chemical compatibility and interfacial reaction between the solder and the base material is reached, which means that at this ratio and content, the GaN particles promote the most effective chemical bonding and physical adsorption, thus minimising the γ_{sl} and achieving the minimum wetting angle. At the same time, the GaN particles at this content achieve the most uniform distribution in the solder, avoiding the aggregation phenomenon, and optimising the microstructure and surface roughness of the solder at the same time. This uniform distribution has been optimised for the microstructure to help improve the contact efficiency between the solder and the substrate, thus reducing γ_{sl} and achieving the minimum wetting angle.

With the further increase of GaN content, the wetting angle rises slowly. This is because with the increase of GaN content, GaN particles agglomerate in the solder, and particle aggregation will lead to an inhomogeneous surface of the solder, which will affect the effective contact area and interfacial reaction between the solder and the substrate and increase the solid-liquid surface tension (γ_{sl}), leading to an increase in the wetting angle. Meanwhile, according to the Gibbs adsorption equation, with the increase of GaN content, the surface will reach an adsorption saturation point, which means that further increase of GaN will not lead to more GaN molecules adsorbed to the surface. At this stage, $\partial\gamma/\partial lnC$ may be close to 0 or become positive, which indicates that the surface tension no longer decreases with the increase of GaN content but increases because of the particle agglomeration. The surface tension increases, and the wetting angle increases.

3.3 IMC growth behaviour

Figure 4 shows the IMC layer thickness versus GaN content of composite brazing material after aging at 170°C for 72 hours.

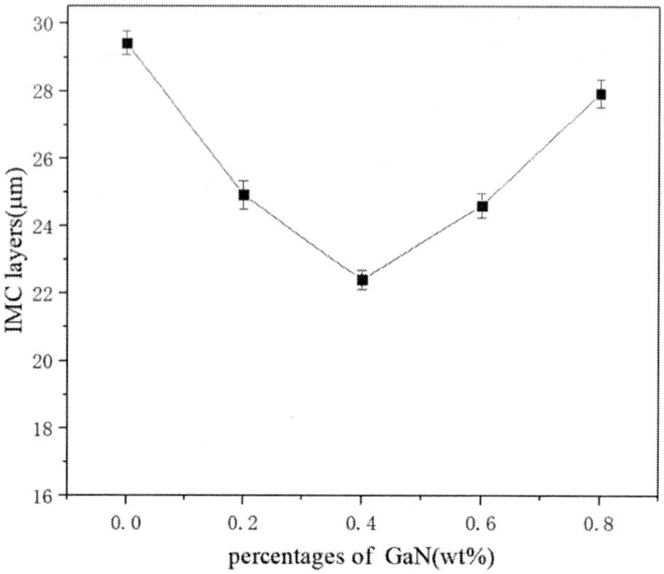

Figure 4. Thickness of IMC layer with different GaN content after 72 h at 170°C

Figure 4 illustrates a distinctive trend in the IMC layer's thickness following the incorporation of nano-GaN: an initial decrease followed by an increase. At 0 wt% GaN content, the IMC layer measures 29.37 μm thick. This thickness diminishes progressively with the introduction of GaN, hitting a nadir of 22.49 μm at 0.40 wt% GaN content, marking a 30.5% reduction compared to the SAC305 solder. This decrease is attributed to the uniform dispersion of GaN particles within the solder matrix, which act as diffusion barriers, impeding the mutual diffusion of copper and tin. As atoms navigate around these GaN particles, the IMC's growth rate slows, consequently thinning the IMC layer. However, as GaN content escalates to 0.80 wt%, the IMC layer's thickness slightly increases to 27.99 μm but remains less than that of SAC305. This phenomenon is linked to the nano-scale effects and Van der Waals forces among GaN nanoparticles, promoting agglomeration. Such agglomerations create large, distinct regions within the solder, altering the physical and chemical uniformity and affecting Cu and Sn atoms' diffusion behavior. Around these agglomerated areas, diffusion channels or pathways form, expediting atom migration to the solder joint interface, thus slightly increasing the IMC layer's thickness. Despite these variations, the overall trend post-GaN addition shows a decrease followed by a slight increase in IMC layer thickness, yet consistently staying below that of SAC305. Specifically, at a given isothermal aging time, the Sn-3.0Ag-0.5Cu-0.4GaN composite solder exhibits a thinner IMC layer at the solder joint interface.

The influence of aging time on the IMC layer's thickness is quantifiable, with the relationship between this thickness at the solder joint interface and the aging time being articulated through Equation 3. This correlation is particularly relevant during the isothermal aging of the solder joint, a phase where the diffusion of the base material's elements predominantly dictates the IMC layer's expansion [11].

$$d - d_0 = \sqrt{Dt} \quad (3)$$

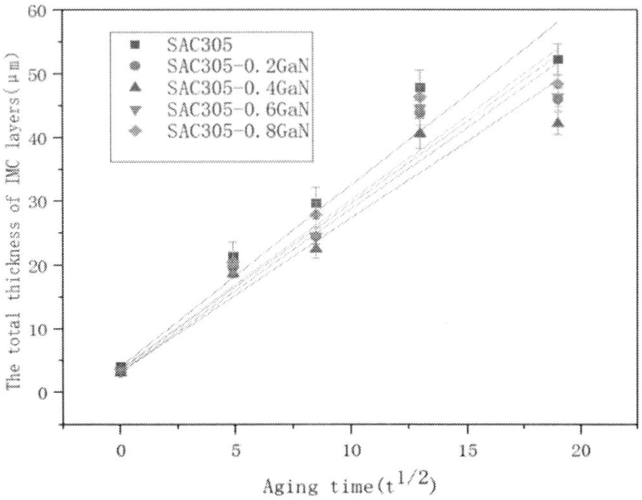

Figure 5. IMC layer thickness of SAC305-xGaN solder joints at different aging times

In this context, 't' denotes the isothermal aging time, 'd' represents the IMC layer's thickness at time 't', and 'd_0' indicates the IMC layer's initial thickness at the onset of isothermal aging. The symbol 'D' stands for the diffusion coefficient of the IMC layer.

Utilizing the aforementioned equation, this experiment establishes a relationship between the IMC layer's thickness and the isothermal aging time. The fitting results are illustrated in Figure 5, where a linear correlation between the IMC thickness and the square root of the isothermal aging time is observed, signifying that diffusion governs the IMC layer's growth. The diffusion coefficient 'D' is determined by calculating the slope of the fitted line. Additionally, the diffusion coefficients for composite solder materials with varying GaN content are summarized in Table 1.

Table 1. Diffusion coefficients of SAC305-xGaN brazing materials with different GaN contents.

GaN content (wt%)	0	0.2	0.4	0.6	0.8
Diffusion coefficient/ 10^{-15} $m^2 \cdot s^{-1}$	2.5	1.82	1.453	1.71	2.28

When the GaN content was 0 wt%, the diffusion coefficient of the composite brazing materials was 2.5×10^{-15} m^2/s, and the diffusion coefficients of the composite brazing materials were changed after adding different contents of GaN. When the GaN content is 0.40 wt%, the diffusion coefficient of the composite brazing material reaches the minimum value of 1.453×10^{-15} m^2/s, which is 41.88% lower than that of SAC305; with the continued addition of GaN content, the diffusion coefficient increases. When the GaN content is 0.80 wt%, the diffusion coefficient of the composite brazing material is 2.28×10^{-15} m^2/s, which is still smaller than that of the SAC305 brazing material, indicating that the addition of enhancement-phase GaN leads to a decrease in the rate of atomic diffusion, which effectively inhibits IMC growth. The addition of the reinforcing phase GaN can effectively reduce the diffusion coefficient, but the inhibitory effect is weakened as the content of the reinforcing phase reaches a certain value and then continues to be added.

4. Conclusion

The addition of GaN did not significantly change the melting point, but it could effectively improve the

SAC305 brazing material wettability. At the optimal amount of its addition of 0.40 wt%, the wetting angle decreased by 61.5% relative to that of SAC305 braze material. Thereafter, with the continued increase in the content of GaN, the wetting angle rose, and the wetting angle was still smaller than that of SAC305 brazing material. The addition of GaN inhibited the growth of IMC at the interface of SAC305 brazing material, in which the inhibition effect was most obvious when the GaN content was 0.40 wt%. Thereafter, with the continued increase in the GaN content of the reinforcing phase, the thickness of the IMC increased, the inhibition effect was weakened, and its thickness was still smaller than the thickness of the IMC layer of SAC305 brazing material. With the increase of aging time, the IMC layer thickness gradually increased. The thickening of the IMC layer was confirmed to be controlled by diffusion by Equation. After adding GaN particles, the diffusion coefficient decreased, and the diffusion coefficient of the composite brazing material reached a minimum value of 1.453×10^{-15} m^2/s at a GaN content of 0.40 wt%, which is 41.88% lower compared with that of SAC305. As the GaN particles continue to increase, the diffusion coefficient of the composite brazing material slowly increases, but it is still lower than that of the SAC305 brazing material. In summary, the new composite brazing material SAC305-0.4GaN has better performance.

References

[1] Xu, Sha, et al. "Effect of carbon nanotubes and their dispersion on electroless Ni–P under bump metallization for lead-free solder interconnection." Journal of Materials Science: Materials in Electronics 25 (2014): 2682-2691.

[2] Shen J, Chan Y C. Research advances in nano-composite solders [J]. Microelectronics Reliability, 2009, 49(3): 223-234.

[3] Ashutosh Sharma, Ashok K Srivastava, and Byungmin Ahn. Microstructure, Wetting, and Tensile Behaviors of Sn-Ag Alloy Reinforced with Copper-Coated Carbon Nanofibers Produced by the Melting and Casting Route [J]. Metallurgical and Materials Transactions, 2019: 1-11.

[4] Manasijević Dragan et al. Microstructure, Melting Behavior and Thermal Conductivity of the Sn–Zn Alloys [J]. Thermochimica Acta, 2021: 178978-.

[5] Sun Lingyao et al. A New Low-Temperature Solder Assembly Technique to Replace Eutectic Sn-Bi Solder Assembly [J]. Micromachines, 2022, 13(6): 867-867.

[6] Zaimi N S M, Mohd Salleh M A A, Al Bakri A M M, et al. Influence of Non-Metallic Particles Addition on Wettability, Intermetallic Compound Formation, and Microhardness of Sn-0.7 Cu Lead-Free Solder Paste [J]. Solid State Phenomena, 2018, 280: 169-174.

[7] Hua L, Chan Y C, Wu Y P, et al. The determination of hexavalent chromium (Cr6+) in electronic and electrical components and products to comply with RoHS regulations [J]. Journal of Hazardous Materials, 2009, 163(2-3): 1360-1368.

[8] Yang D, Chan Y C, Tu K N. The time-dependent melting failure in flip chip lead-free solder interconnects under current stressing [J]. Applied Physics Letters, 2008, 93(4).

[9] Puttlitz K J, Stalter K A. Handbook of lead-free solder technology for microelectronic assemblies [M]. CRC Press, 2004.

[10] Antoniammal P, Arivuoli D. Size and shape dependence on melting temperature of gallium nitride nanoparticles [J]. Journal of Nanomaterials, 2012, 2012: 8-8.

[11] Shen J, Zhao M, He P, et al. Growth behaviors of intermetallic compounds at Sn–3Ag–0.5 Cu/Cu interface during isothermal and non-isothermal aging [J]. Journal of alloys and compounds, 2013, 574: 451-458.

Journal of Physics: Conference Series **2808** (2024) 012035 doi:10.1088/1742-6596/2808/1/012035

Effect of process parameters on the temperature field of laser sintering of nylon PA2200 powder

Kai Liu[1*]

[1]School of Mechanical and Electrical Engineering, Wuhan Polytechnic, Wuhan, Hubei, 430074, China

[*]Corresponding author's e-mail: liukaizxm@163.com

Abstract: Utilizing the ANSYS finite element analysis software, a three-dimensional heat transfer finite element model for the SLS forming process of nylon PA2200 powder was constructed. The APDL parametric language was employed to control the movement of the Gaussian heat source, enabling numerical simulations of the temperature field during the single-layer, single-pass sintering process. Through numerical simulation analysis, the varying pattern of the melting width was derived. Based on the optimal overlap rate required for satisfactory SLS forming quality, the optimal parameter combination for nylon PA2200 powder sintering was determined to be a laser power of 40 W, a scanning speed of 5000 mm/s, a layer thickness of 0.12 mm, and a preheating temperature of 165°C. These findings provide valuable insights for optimizing the SLS process and enhancing the quality of the final sintering parts.

1. Introduction

Selective Laser Sintering (SLS) technology, with its versatility in formable materials, high material utilization, ease in fabricating intricate structural parts, and the absence of support structures has emerged as a leading 3D printing technology. Its widespread applications in engineering fields are evidence of its popularity [1]. However, the process parameters of SLS play a crucial role in determining the quality and mechanical properties of the final parts. The intricate interplay of these parameters further complicates the quest for the optimal combination. Relying solely on experimentation to obtain this optimal combination is labor-intensive, inefficient, and costly, often making it challenging to pinpoint the exact set of parameters that yield the best results. Consequently, numerous scholars turn to simulation and model prediction methods to streamline this process and identify the optimal combination of process parameters more effectively.

Meng et al. used ANSYS software to simulate the changes in temperature and stress fields in the selective laser sintering and forming process, optimized the scanning speed and scanning path, and improved the problem of numerical oscillations of the temperature field [2]. Monteiro et al. investigated the effect of different preheating times and preheating procedures on the crystallization, densification, and properties of sintering nonstoichiometric ceramics [3]. Yao et al. characterized the three-dimensional characteristics of the pore space of the powder-layered layer and investigated the distribution of the pore space by characterizing the relative volume of the pore space [4]. Mohammad et al. conducted a simulation analysis of the SLS powder-forming process by using finite element software, establishing a mathematical model that connects SLS forming process parameters with the warping deformation of the formed parts, and analyzing the influencing factors of such deformation [5]. Manshoori et al. utilized finite element software to establish a process scanning algorithm model, analyzing the warping issues and processing time reduction challenges of SLS-formed parts. The model was then compared to traditional scanning algorithms, demonstrating its advantages [6]. Fang used ANSYS to carry out finite element numerical simulations to analyze the changing rule of the temperature field and the changing trend of stress and strain to achieve the purpose of improving the quality of molding [7]. WANG et al. used the finite element analysis software ANSYS to simulate the transient temperature field of single-layer sintering and multilayer sintering at different ambient temperatures and investigated the variation of energy accumulation during the sintering process by analyzing the distribution of the temperature field [8].

The SLS process involves the heating of powder materials using a laser heat source. During the entire sintering process, there is a transfer of laser energy. However, the thermophysical properties of the materials vary with temperature, which in turn affects the transmission of laser energy. This has a significant impact on whether the material can be successfully formed, its forming accuracy, and its mechanical properties. Therefore, analyzing the distribution of the temperature field is crucial for studying the SLS sintering process.

2. Simulation experiment

2.1 Determination of thermophysical parameters of test materials

Selective laser sintering simulation is a nonlinear transient analysis process in which thermophysical parameters such as specific heat capacity and heat transfer coefficient change with temperature. By referring to engineering materials manuals, the thermophysical properties of nylon PA2200 powder material were obtained as shown in Table 1. The melting point of PA2200 powder material is 176°C, and it is necessary to preheat the powder before sintering.

Table 1. Thermophysical parameters of PA2200.

T/°C	C/ J/(kg·°C)	K/W/ (m·°C)
20	1.8	0.25
100	1.9	0.28
180	1.92	0.3
220	1.93	0.27
400	1.96	0.25

2.2 Modelling of heat sources

During the SLS forming process, the laser heat source was input into the powder bed in the form of laser power energy density. The intensity distribution of the irradiation area of the CO_2 laser selected in this study exhibited a Gaussian distribution characteristic. In the simulation, the heat source model [9] was established as a volumetric heat source model according to Equation (1). The heat source was then cyclically loaded by using APDL programming to conduct numerical simulations of the temperature field during the sintering process.

$$q = \frac{2(1-R)\ P\gamma}{\pi\omega^2} \bullet e^{\left(-2\times\frac{(x^2+(y-v\times t)^2}{R^2}\right)} - (\gamma z) \quad (1)$$

where q is the laser energy density; P is the power of the laser heat source; $x^2+(y-v\times t)^2$ is the distance from the particle surface to the center of the laser spot; π is the caliber of the laser heat source; R is the transmission of laser energy in powder beds (%); γ is the absorbance in the z-direction.

2.3 Establishment of solid model and meshing

The established three-dimensional simulation model was a cube with dimensions of 1 mm*1 mm*1 mm. Considering the spot diameter of 0.5 mm, scanning distance of 0.3 mm, and powder particle size of 50 μm, the model grid size was divided into 25 μm, which met the requirements of the simulation model in this paper. The grid division of the model is illustrated in Figure 1.

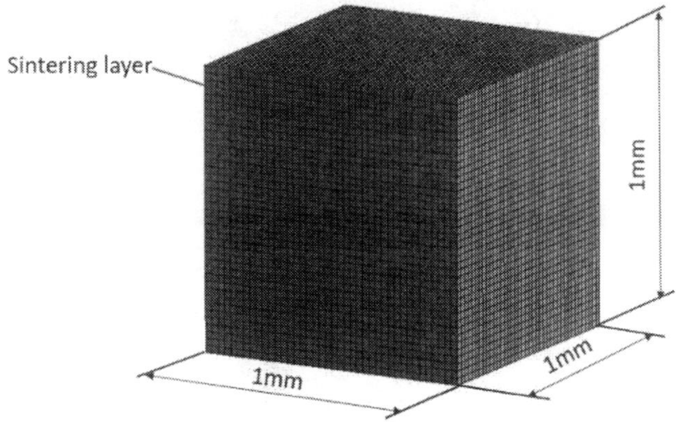

Figure 1. Schematic diagram of model meshing.

2.4 Simulation of experimental program design

The SLS forming process involved numerous parameters, and their impacts varied depending on the structural characteristics of the product. To minimize the extensive computational requirements for model solving, this paper selected four key factors that significantly affected forming quality based on existing research: preheating temperature, laser power, scanning speed, and layer thickness. These factors served as the variables for the orthogonal test. For each test variable, three experimental levels were chosen to conduct a "four-factor, three-level" orthogonal test. The table of test factor levels is presented in Table 2.

Table 2. Comparison of the numerical results and correlations.

Test No.	A (Laser Power)/W	B (Scanning Speed)/mm·s	C (Preheating Temperature)/°C	D (Layer Thickness)/mm
1#	1 (30)	1 (3000)	1 (165)	1 (0.1)
2#	1 (30)	2 (4000)	2 (168)	2 (0.12)
3#	1 (30)	3 (5000)	3 (170)	3 (0.14)
4#	2 (40)	1 (3000)	2 (168)	3 (0.14)
5#	2 (40)	2 (4000)	3 (170)	1 (0.1)
6#	2 (40)	3 (5000)	1 (165)	2 (0.12)
7#	3 (50)	1 (3000)	3 (170)	2 (0.12)
8#	3 (50)	2 (4000)	1 (165)	3 (0.14)
9#	3 (50)	3 (5000)	2 (168)	1 (0.1)

Note: 1#, 2#, and 3# indicate levels, and the data in parentheses are the values of the factors.

3. Results and discussions

Empirical values for the forming parameters of nylon PA2200 powder material were as follows: powder density of 930 kg/m^3, modulus of elasticity of 1646 MPa, elongation at break of 18%, and durometer of 75. This means that during the sintering process when the powder temperature in the powder bed exceeded 176°C, the nylon PA2200 material could fully melt into a liquid state. At this point, the powder particles could fully fuse, resulting in a high-density formed part and optimal laser sintering effect. Using the factors and level parameters in Table 2, a numerical simulation analysis was conducted according to the orthogonal test scheme. This analysis revealed the variation pattern of the melting width of the sintering powder layer under different combinations of process parameters, as

Journal of Physics: Conference Series **2808** (2024) 012035 doi:10.1088/1742-6596/2808/1/012035

shown in Figure 2 (1-9).

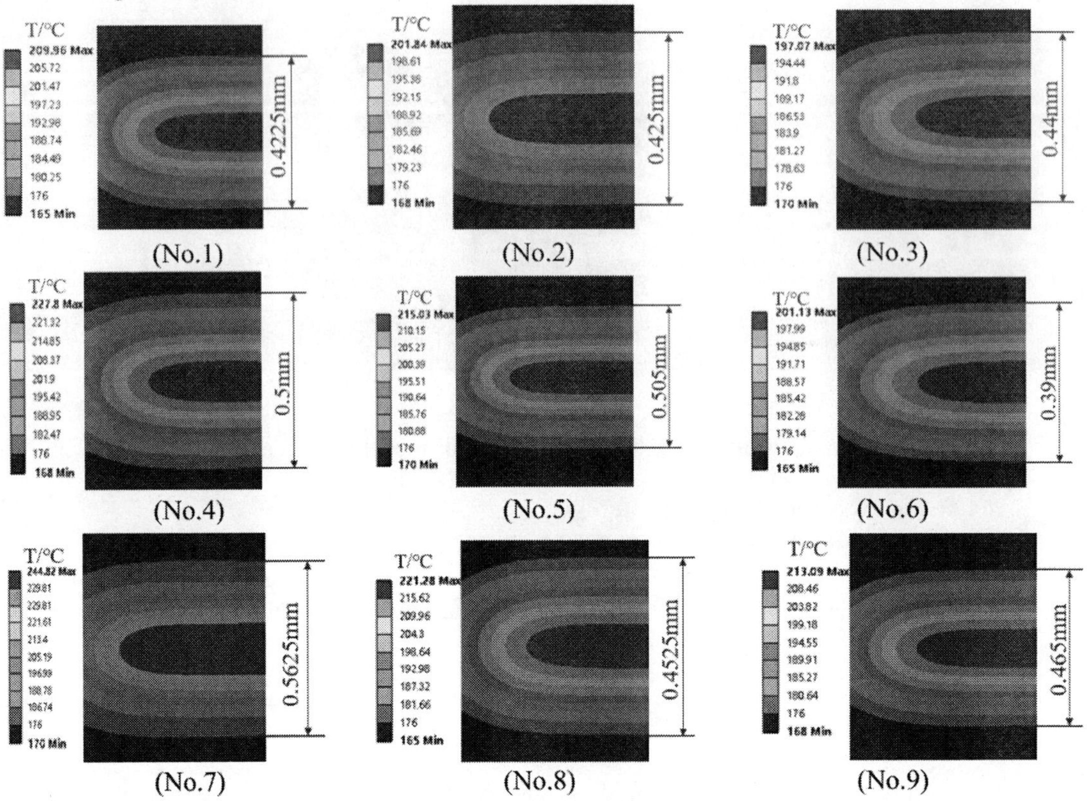

Figure 2. Values of melting width in temperature field for different process parameter groups.

The simulation analysis of single-layer, single-pass sintering determines the melting width under the then-current process parameters, serving as a fundamental prerequisite for optimizing the process parameters and providing a basis for three-dimensional sintering. In the SLS forming process, the overlap ratio is a crucial parameter indicator. It is closely related to the firmness of the adjacent sintering passes, the rationality of the selection of scanning spacing, the uniformity of the sintering surface, and the efficiency of the forming process.

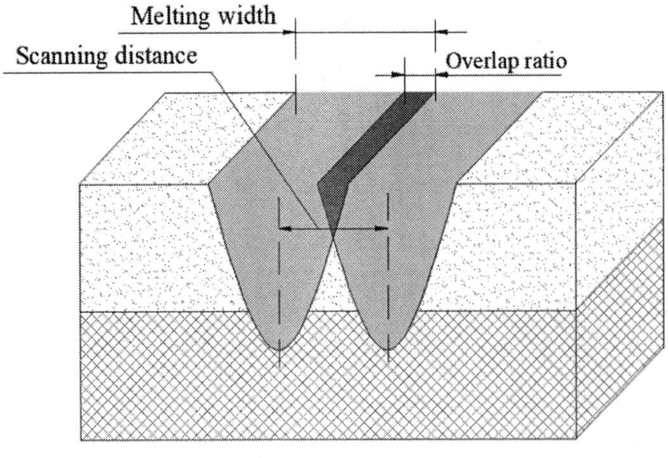

Figure 3. Overlap ratio.

The overlap ratio is defined as the width of the overlapping portion of adjacent sintering channels as a percentage of the melting width (W).

$$\eta = \frac{D}{W} \times 100\% \tag{2}$$

where η is the overlap rate, D is the width of the overlapping section between adjacent sintering tracks, and W is the melting width.

According to the literature [10], the optimal overlap ratio in the rapid prototyping process ranges from 20% to 35%. Based on existing research, the optimal overlap ratio is 25% with an optimal scanning spacing of 0.3 mm. According to Equation (2), when the melting width is 0.39 mm, the overlap ratio is 23%, which is close to the optimal overlap ratio value. Therefore, the optimal combination of process parameters is determined to be a laser power of 40 W, a scanning speed of 5000 mm/s, a layer thickness of 0.12 mm, and a preheating temperature of 165°C.

4. Conclusion

This paper focused on the nylon PA2200 material and simulated the temperature field variations in selective laser sintering by using ANSYS software. By comparing different melting width values, the optimal combination of process parameters was obtained, providing a theoretical basis for the study of nylon powder-forming process parameters. This also serves as a guiding reference for selecting experimental process parameters in selective laser sintering.

Acknowledgments

This work is supported by the school-level subject of Wuhan Polytechnic(2022YK024).

Reference

[1] POWELL D, RENNIE A E W, GEEKIE L, et al. Understanding powder degradation in metal additive manufacturing to allow the upcycling of recycled powders [J]. Journal of Cleaner Production, 2020, 268:122077.
[2] MENG J. Research on key technologies of numerical simulation of selective laser sintering of polymers [D]. Shanghai: Shanghai Jiao Tong University, 2020.
[3] MONTEIRO R H, RODRIGUES PAIS ALVES M F, STRECKER K, et al. Effect of preheating and isothermal holding time on the crystallization, densification, and properties of a sintered lithium silicate glass ceramic [J]. Ceramics International, 2022,48(4): 5590-5600.
[4] YAO D, LIU X, WANG J, et al. Numerical insights on the spreading of practical 316 L stainless steel powder in SLM additive manufacturing [J]. Powder Technology, 2021,390: 197-208.
[5] Ahmadi A, Movahhedy M R, Akbari J. Optimization of process parameters for reducing warpage in selected laser sintering of polymer parts [J]. Additive Manufacturing, 2017, (18): 285-294.
[6] Manshoori Y A, Movahhedy M R, Khodaygan S. An efficient scanning algorithm for improving accuracy based on minimizing part warping in selected laser sintering process [J]. Virtual & Physical Prototyping, 2019, 14(1)59-78.
[7] FANG Z. Thermo-mechanical coupling research on compression molding of straw powder [D]. Beijing: Beijing Forestry University, 2020.
[8] WANG C F, LING Z C, GUAN Z M, et al. Numerical simulation and experimental study of energy accumulation in selective laser sintering of nylon 6 powder [J]. Modern Manufacturing Engineering,2021(8):20-25.
[9] Kundakcolu E, Lazoglu I, Poyraz Z, et al. Thermal and molten pool model in selective laser melting process of Inconel 625 [J]. The International Journal of Advanced Manufacturing Technology, 2018, 95(2): 1-8.
[10] CAI L L, DING H L, MENG J, et al. Powder heating simulation of polymer selective laser sintering [J]. Mold Technology, 2021(5):50-54.

Stability of viscoelastic fluid conveying pipe

Yu Shen [1] **and Bo Wang** [1, *]

[1]School of Mechanical Engineering, Shanghai Institute of Technology, Shanghai, 201418, China

*Corresponding author's e-mail: b.wang@live.com

Abstract: In this paper, the dynamic stability of a pulsating fluid transport viscoelastic pipe under transverse vibration in the supported case is investigated, and a new variable-length beam element taking into account the beam rotation effect is proposed, which incorporates Kelvin-type viscoelastic intrinsic relations to describe the material properties of the beam-type pipe. Stability conditions are given for the subharmonic and combined resonance cases. Numerical models were used to analyze the stability of viscoelastic fluid pipes under the influence of different factors.

1. Introduction

Pipelines are widely used in aviation, civil engineering, marine, and other major projects. The flow pipeline comprises multiple parts, such as support constraints, pipes, fluids, etc. The physical and chemical properties of these components will affect the stability of the entire flow pipeline system, causing problems in engineering applications. Establishing a reliable mathematical model that considers multiple factors is an important means of solving this engineering problem. This is also a method commonly used by scholars to study the vibration problems of flow pipelines. Garziera et al. [1] took a cylindrical flow pipeline supported at both ends as the research object, analyzed its nonlinear dynamic characteristics based on Donnell's nonlinear thin plate theory, and calculated the natural frequency and mode of the fluid-solid coupled flow pipeline. Pellicano and Amabili [2] further studied the dynamic stability under the action of axial static load and dynamic load through mathematical methods. Sinir et al. [3] combined the fractional calculus method to study the nonlinear vibration characteristics of a flow pipeline with zero stiffness supported at both ends under the action of constant flow velocity fluid and obtained the analytical solution of the natural frequency. Chernik et al. [4] used the integral operator to study the vibration characteristics of a Kelvin-Voigt-type flow pipeline supported at both ends and calculated and analyzed the impact of fluid flow velocity on the stability of the pipeline. The results show that divergent instability occurs in the pipeline system as the flow rate gradually increases. Sofiyev et al. [5] combined the classical beam theory to study the vibration problem of the viscoelastic model and the Winkler foundation model flow pipeline[6] under the action of intermittent airflow. In short, mathematical methods have been proven to be effective methods for studying flow pipelines and are widely used in engineering. In this research, we mathematically study the stability of viscoelastic fluid delivery pipes[7,8]. The fluid in the pipe will inevitably vibrate the system when it undergoes axial movement. There are usually two models for different systems and materials: the second-order beam model and the fourth-order beam model[9]. At the same time, combining multiscale methods[9] with partial differential equations of transverse vibration[10,11] can further expand the application of mathematical models[12] in fluid pipeline systems[13,14]. A fluid-structure coupling model was further established to study the interaction between fluid and pipeline, taking into account the effects of factors such as flow velocity, pressure, and temperature on system stability.

2. Governing equation

Based on the classic fourth-order beam model, velocity-dependent damping terms and non-uniform terms caused by axial velocity and prestress are introduced to analyze the behavior of variable-length beams under axial motion under specific conditions, combined with Newton's second law. The governing equation can be obtained:

$$\begin{aligned}&\frac{\partial^2 w}{\partial t^2}+2\beta^{\frac{1}{2}}c\frac{\partial^2 w}{\partial x\partial t}+\left[c^2-\bar{P}+\beta^{\frac{1}{2}}\dot{c}(1-x)\right]\frac{\partial^2 w}{\partial x^2}+\frac{\partial^4 w}{\partial x^4}+\alpha_1\left\{\frac{\partial^3 w}{\partial t^3}\right.\\&+c\left(\beta^{-\frac{1}{2}}+2\beta^{\frac{1}{2}}\right)\frac{\partial^3 w}{\partial x\partial t^2}+2\beta^{\frac{1}{2}}\dot{c}\frac{\partial^2 w}{\partial x\partial t}+\left[3c^2+\bar{P}-\beta^{\frac{1}{2}}\dot{c}(1-x)\right]\frac{\partial^3 w}{\partial x^2\partial t}\\&\left.+\left[3c\dot{c}-\beta^{\frac{1}{2}}\ddot{c}(1-x)\right]\frac{\partial^2 w}{\partial x^2}+\beta^{-\frac{1}{2}}\left[c^3+c\bar{P}-\beta^{\frac{1}{2}}c\dot{c}(1-x)\right]\frac{\partial^3 w}{\partial x^3}\right\}\\&+\alpha_2\left(\frac{\partial^5 w}{\partial x^4\partial t}+c\beta^{-\frac{1}{2}}\frac{\partial^5 w}{\partial x^5}\right)=0\end{aligned}\tag{1}$$

Among them, α_1 and α_2 represent the viscosity coefficient, which is used to calculate the viscoelastic effect of the system. Dimensionless parameters are:

$$\begin{aligned}&\alpha_1=\frac{\eta}{E_1+E_2}\sqrt{\frac{E_1E_2I}{(E_1+E_2)(m_p+m_f)}},\alpha_2=\frac{\eta}{E_2}\sqrt{\frac{E_1E_2I}{(E_1+E_2)(m_p+m_f)}}\\&x\leftrightarrow\frac{x}{l},w\leftrightarrow\frac{w}{l},t\leftrightarrow\frac{t}{l^2}\sqrt{\frac{E_1E_2I}{(E_1+E_2)(m_p+m_f)}},\beta=\frac{m_f}{m_p+m_f},\bar{P}\leftrightarrow\frac{(E_1+E_2)l^2}{E_1E_2I}\bar{P},\end{aligned}\tag{2}$$

Then, Equation (1) can be transformed into a dimensionless form:

$$\begin{aligned}&\ddot{w}+2b_0c_0\dot{w}'+c_0^2w''+w''''+\epsilon\{\alpha_2(\dot{w}''''+b_1c_0w''''')+2b_0\sin(t\Omega)c_1\dot{w}'\\&+\left[b_0(1-x)\Omega\cos(t\Omega)+2\sin(t\Omega)c_0\right]c_1w''\\&+\alpha_1\left[\dddot{w}+(2b\theta c_0+b_1c_0)\ddot{w}'+(c_0^2+2b\theta b_1c_0^2)\dot{w}''+b_1c_0^3w'''\right]\}=0\end{aligned}\tag{3}$$

$$c(t)=c_0+c_1\sin\omega t\tag{4}$$

3. Method of multiple scales (MMS)

Multi-timescale approaches allow us to distinguish and study physical processes occurring in the system at different rates. We introduce the different time scales T_0 and scale T_1. The former is a fast time scale, and the latter is a slow time scale. We analyze the dependence of variable v on parameter ε through perturbation expansion:

$$\frac{\partial}{\partial t}=\frac{\partial}{\partial T_0}+\varepsilon\frac{\partial}{\partial T_1}+O(\varepsilon^2),\quad\frac{\partial^2}{\partial t^2}=\frac{\partial^2}{\partial T_0^2}+2\varepsilon\frac{\partial^2}{\partial T_0\partial T_1}+O(\varepsilon^2)\tag{5}$$

$$v(x,T_0,T_1;\varepsilon)=v_0(x,T_0,T_1)+\varepsilon v_1(x,T_0,T_1)+O(\varepsilon^2)\tag{6}$$

We set the coefficient of each power of ε to 0 to obtain the equations of ε^0 (zero-order) and ε^1 (first-order), which are used to describe the system without disturbance and the first-order effect introduced by the disturbance respectively:

$$\varepsilon^0:\ Mv_{0,T_0T_0}+Gv_{0,T_0}+Kv_0=0\tag{7}$$

$$v_0\big|_0^1=0,\quad v_{0},_x\big|_0^1=0 \tag{8}$$

ε^1:

$$\begin{aligned}&2w_0^{(0,1,1)}(x,t,\varepsilon t)+w_1^{(0,2,0)}(x,t,\varepsilon t)+\alpha_1 w_0^{(0,3,0)}(x,t,\varepsilon t)+2b_0c_0w_0^{(1,1,0)}(x,t,\varepsilon t)+\\&2b_0c_1\sin(\omega t)w_0^{(1,1,0)}(x,t,\varepsilon t)+2b_0c_0w_1^{(1,1,0)}(x,t,\varepsilon t)+(2b_0c_0+b_1c_0)\alpha_1 w_0^{(1,2,0)}(x,t,\varepsilon t)+\\&(b_0(1-x)\omega\cos(\omega t)+2c_0\sin(\omega t))c_1w_0^{(2,0,0)}(x,t,\varepsilon t)+c_0^2w_1^{(2,0,0)}(x,t,\varepsilon t)+\\&(c_0^2+2b_0b_1c_0^2)\alpha_1 w_0^{(2,1,0)}(x,t,\varepsilon t)+b_1c_0^3\alpha_1 w_0^{(3,0,0)}(x,t,\varepsilon t)+w_1^{(4,0,0)}(x,t,\varepsilon t)+\\&\alpha_2 w_0^{(4,1,0)}(x,t,\varepsilon t)+b_1c_0\alpha_2 w_0^{(5,0,0)}(x,t,\varepsilon t)\end{aligned} \tag{9}$$

$$v_1\big|_0^1=0,\quad v_{1},_x\big|_0^1=0 \tag{10}$$

Equation (11) gives the definitions of mass M, gyroscopic G, and linear stiffness operator K:

$$M=1,\quad G=2\beta^{\frac{1}{2}}c_0\frac{\partial}{\partial x},\quad K=\left(c_0^2-\bar{P}\right)\frac{\partial^2}{\partial x^2}+\frac{\partial^4}{\partial x^4} \tag{11}$$

Then, the assumed solution to Equation (6) can be given by:

$$v_0(x,T_0,T_1)=\sum_{n=1}^{\infty}A_n(T_1)\phi_n(x)e^{i\omega_n T_0}+cc \tag{12}$$

where cc represents the complex conjugate of all previous terms on the right side of Equation (12); $A_n(T_1)$ is a time-dependent amplitude function, indicating that the amplitude of each mode changes with the slow time scale T_1; $\phi_n(x)$ represents the spatial shape function of the nth vibration mode. Then, the displacement of the beam changes with position under the state:

$$\begin{aligned}\phi_n(x)=&e^{i\beta_{1n}x}-\frac{(\beta_{4n}^2-\beta_{1n}^2)(e^{i\beta_{3n}}-e^{i\beta_{1n}})}{(\beta_{4n}^2-\beta_{2n}^2)(e^{i\beta_{3n}}-e^{i\beta_{2n}})}e^{i\beta_{2n}x}-\frac{(\beta_{4n}^2-\beta_{1n}^2)(e^{i\beta_{2n}}-e^{i\beta_{1n}})}{(\beta_{4n}^2-\beta_{3n}^2)(e^{i\beta_{2n}}-e^{i\beta_{3n}})}e^{i\beta_{3n}x}\\&-\left[1-\frac{(\beta_{4n}^2-\beta_{1n}^2)(e^{i\beta_{3n}}-e^{i\beta_{1n}})}{(\beta_{4n}^2-\beta_{2n}^2)(e^{i\beta_{3n}}-e^{i\beta_{2n}})}-\frac{(\beta_{4n}^2-\beta_{1n}^2)(e^{i\beta_{2n}}-e^{i\beta_{1n}})}{(\beta_{4n}^2-\beta_{3n}^2)(e^{i\beta_{2n}}-e^{i\beta_{3n}})}\right]e^{i\beta_{4n}x}\end{aligned} \tag{13}$$

The modal function $\phi_n(x)$ should satisfy:

$$\phi_n''''+\left(c_0^2-\bar{P}\right)\phi_n''+2i\beta^{\frac{1}{2}}\omega_n c_0\phi_n'-\omega_n^2\phi_n=0 \tag{14}$$

Substituting into Equation (12), Equations (13) and (14) can be transformed into:

$$\phi_n(x)=A_{1n}e^{i\beta_{1n}x}+A_{2n}e^{i\beta_{2n}x}+A_{3n}e^{i\beta_{3n}x}+A_{4n}e^{i\beta_{4n}x} \tag{15}$$

$$\beta_{jn}^4-\left(c_0^2-\bar{P}\right)\beta_{jn}^2-2c_0\beta^{\frac{1}{2}}\omega_n\beta_{jn}-\omega_n^2=0 \tag{16}$$

where j=1,2,3,4.

σ is a detuning parameter that is introduced to qualify the derivation of fluid pulsation frequency ω from $\omega_n+\omega_m$, so ω can be described by:

$$\omega=\omega_m+\omega_n+\varepsilon\sigma \tag{17}$$

Taking into account the possible resonance, Equation (12) can be written as:

$$v_0(x,T_0,T_1)=A_n(T_1)\phi_n(x)e^{i\omega_n T_0}+A_m(T_1)\phi_m(x)e^{i\omega_m T_0}+cc \tag{18}$$

Substituting Equation (18) and (17) into Equation (9) leads to:

$$
\begin{aligned}
&Mv_{1,tt}+Gv_{1,t}+Kv_1 \\
&=-e^{i\omega_m t}\left\{\begin{array}{l}
i\alpha_1\omega_m^3A_m\phi_m-2i\omega_m\phi_m\dfrac{dA_m}{dT}+2b_0c_0\alpha_1\omega_m^2A_m\dfrac{d\phi_m}{dx}+b_1c_0\alpha_1\omega_m^2A_m\dfrac{d\phi_m}{dx}-2b_0c_0\dfrac{dA_m}{dT}\dfrac{d\phi_m}{dx}+b_0e^{iT\sigma}c_1\omega_n\overline{A_n}\dfrac{d\overline{\phi_n}}{dx}- \\
ic_0^2\alpha_1\omega_m^2A_m\dfrac{d^2\phi_m}{dx^2}-2ib_0b_1c_0^2\alpha_1\omega_mA_m\dfrac{d^2\phi_m}{dx^2}-\dfrac{1}{2}b_0e^{iT\sigma}\varepsilon\sigma c_1\overline{A_n}\dfrac{d^2\overline{\phi_n}}{dx^2}+\dfrac{1}{2}b_0e^{iT\sigma}x\varepsilon\sigma c_1\overline{A_n}\dfrac{d^2\overline{\phi_n}}{dx^2}+ie^{iT\sigma}c_0c_1\overline{A_n}\dfrac{d^2\overline{\phi_n}}{dx^2}- \\
\dfrac{1}{2}b_0e^{iT\sigma}c_1\omega_m\overline{A_n}\dfrac{d^2\overline{\phi_n}}{dx^2}+\dfrac{1}{2}b_0e^{iT\sigma}xc_1\omega_m\overline{A_n}\dfrac{d^2\overline{\phi_n}}{dx^2}-\dfrac{1}{2}b_0e^{iT\sigma}c_1\omega_n\overline{A_n}\dfrac{d^2\overline{\phi_n}}{dx^2}+\dfrac{1}{2}b_0e^{iT\sigma}xc_1\omega_n\overline{A_n}\dfrac{d^2\overline{\phi_n}}{dx^2}-b_1c_0^3\alpha_1A_m\varphi_m^3- \\
i\alpha_2\omega_mA_m\phi_m^4-b_1c_0\alpha_2A_m\phi_m^5
\end{array}\right\} \\
&-e^{i\omega_n t}\left\{\begin{array}{l}
i\alpha_1\omega_n^3A_n\phi_n-2i\omega_n\phi_n\dfrac{dA_n}{dT}+2b_0c_0\alpha_1\omega_n^2A_n\dfrac{d\phi_n}{dx}+b_1c_0\alpha_1\omega_n^2A_n\dfrac{d\phi_n}{dx}-2b_0c_0\dfrac{dA_n}{dT}\dfrac{d\phi_n}{dx}+b_0e^{iT\sigma}c_1\omega_m\overline{A_m}\dfrac{d\overline{\phi_m}}{dx}- \\
ic_0^2\alpha_1\omega_nA_n\dfrac{d^2\phi_n}{dx^2}-2ib_0b_1c_0^2\alpha_1\omega_nA_n\dfrac{d^2\phi_n}{dx^2}-\dfrac{1}{2}b_0e^{iT\sigma}\varepsilon\sigma c_1\overline{A_m}\dfrac{d^2\overline{\phi_m}}{dx^2}+\dfrac{1}{2}b_0e^{iT\sigma}x\varepsilon\sigma c_1\overline{A_m}\dfrac{d^2\overline{\phi_m}}{dx^2}+ie^{iT\sigma}c_0c_1\overline{A_m}\dfrac{d^2\overline{\phi_m}}{dx^2}- \\
\dfrac{1}{2}b_0e^{iT\sigma}c_1\omega_m\overline{A_m}\dfrac{d^2\overline{\phi_m}}{dx^2}+\dfrac{1}{2}b_0e^{iT\sigma}xc_1\omega_n\overline{A_m}\dfrac{d^2\overline{\phi_m}}{dx^2}-\dfrac{1}{2}b_0e^{iT\sigma}c_1\omega_n\overline{A_m}\dfrac{d^2\overline{\phi_m}}{dx^2}+\dfrac{1}{2}b_0e^{iT\sigma}xc_1\omega_n\overline{A_m}\dfrac{d^2\overline{\phi_m}}{dx^2}-b_1c_0^3\alpha_1A_n\phi_n^3- \\
i\alpha_2\omega_nA_n\phi_n^4-b_1c_0\alpha_2A_n\phi_n^5
\end{array}\right\} \\
&+cc+\mathrm{NST}
\end{aligned} \tag{19}
$$

where A_n and A_m represent the changes in pipe vibration amplitude at slow time scales T_1. Moreover, the following orthogonal relationships need to be satisfied to make the equation solvable:

$$\dot{A}_r+\left(\alpha_1\vartheta_r+\alpha_2\xi_r\right)A_r+e^{i\sigma T_1}c_1\kappa_{rs}\overline{A}_s=0 \quad \left(r=m,n;\ s=n,m\right) \tag{20}$$

where,

$$
\begin{aligned}
\vartheta_r=&\frac{-i\omega_r^3\int_0^1\phi_r\overline{\phi}_r\mathrm{d}x-c_0\omega_r^2\left(\beta^{-\frac{1}{2}}+2\beta^{\frac{1}{2}}\right)\int_0^1\phi_r'\overline{\phi}_r\mathrm{d}x}{2\left(i\omega_r\int_0^1\phi_r\overline{\phi}_r\mathrm{d}x+c_0\beta^{\frac{1}{2}}\int_0^1\phi_r'\overline{\phi}_r\mathrm{d}x\right)} \\
&+\frac{i\omega_r\int_0^1\left(3c_0^2-\overline{P}\right)\phi_r''\overline{\phi}_r\mathrm{d}x+c_0\beta^{-\frac{1}{2}}\left(c_0^2-\overline{P}\right)\int_0^1\phi_r'''\overline{\phi}_r\mathrm{d}x}{2\left(i\omega_r\int_0^1\phi_r\overline{\phi}_r\mathrm{d}x+c_0\beta^{\frac{1}{2}}\int_0^1\phi_r'\overline{\phi}_r\mathrm{d}x\right)} \quad \left(r=m,n\right)
\end{aligned} \tag{21}
$$

$$\xi_r=\frac{i\omega_r\int_0^1\phi_r^{(4)}\overline{\phi}_r\mathrm{d}x+c_0\beta^{-\frac{1}{2}}\int_0^1\phi_r^{(5)}\overline{\phi}_r\mathrm{d}x}{2\left(i\omega_r\int_0^1\phi_r\overline{\phi}_r\mathrm{d}x+c_0\beta^{\frac{1}{2}}\int_0^1\phi_r'\overline{\phi}_r\mathrm{d}x\right)} \quad \left(r=m,n\right) \tag{22}$$

where ξr is a positive real number; ϑr is negative real number. And they are opposite numbers to each other.

$$\kappa_{rs}=\frac{-\beta^{\frac{1}{2}}\omega_s\int_0^1\overline{\phi_s'}\overline{\phi}_r\mathrm{d}x-ic_0\int_0^1\overline{\phi_s''}\overline{\phi}_r\mathrm{d}x+\beta^{\frac{1}{2}}\frac{\Omega}{2}\int_0^1(1-x)\overline{\phi_s''}\overline{\phi}_r\mathrm{d}x}{2\left(i\omega_r\int_0^1\phi_r\overline{\phi}_r\mathrm{d}x+c_0\beta^{\frac{1}{2}}\int_0^1\phi_r'\overline{\phi}_r\mathrm{d}x\right)} \quad \left(r=m,n;\ s=n,m\right) \tag{23}$$

We need to introduce some new variables to help us better analyze the stability of the system:

$$A_n=B_ne^{0.5i\sigma T_1},A_m=B_me^{0.5i\sigma T_1} \tag{24}$$

Substituting Equation (23) into Equation (19) yields the ordinary differential equations for B_n and B_m:

$$\begin{aligned} &\dot{B}_n + \left(\alpha_1\vartheta_n + \alpha_2\xi_n + \frac{1}{2}\mathrm{i}\sigma\right)B_n + c_1\kappa_{nm}\bar{B}_m = 0 \\ &\dot{B}_m + \left(\alpha_1\vartheta_m + \alpha_2\xi_m + \frac{1}{2}\mathrm{i}\sigma\right)B_m + c_1\kappa_{mn}\bar{B}_n = 0 \end{aligned} \tag{25}$$

The solution of Equation (19) can be expressed as:

$$B_n = b_n e^{\lambda T_1}, B_m = b_m e^{\bar{\lambda} T_1} \tag{26}$$

where b_m and b_n are amplifications of the slow-scale modulations, substituting Equation (25) into Equation (24) yields the linear ordinary differential equations of b_m and b_n with respect to T_1, and we let $m=n$:

$$\begin{aligned} &\lambda^2 + \left(\alpha_1\vartheta_n + \alpha_2\xi_n + \frac{1}{2}\mathrm{i}\sigma\right)\left(\alpha_1\bar{\vartheta}_n + \alpha_2\bar{\xi}_n - \frac{1}{2}\mathrm{i}\sigma\right) \\ &+\lambda\left[\left(\vartheta_n + \bar{\vartheta}_n\right)\alpha_1 + \left(\xi_n + \bar{\xi}_n\right)\alpha_2\right] - c_1^2\kappa_{nn}\bar{\kappa}_{nn} = 0 \end{aligned} \tag{27}$$

$$\frac{\sigma^2}{4} + \left(\alpha_1\vartheta_n + \alpha_2\xi_n\right)\left(\alpha_1\bar{\vartheta}_n + \alpha_2\bar{\xi}_n\right) - c_1^2\kappa_{nn}\bar{\kappa}_{nn} > 0 \tag{28}$$

4. Fluid solid coupling lateral vibration characteristic function

The system's total kinetic energy includes pipeline kinetic energy and fluid kinetic energy, which can be expressed as:

$$T_P = \int_0^l \frac{1}{2} m_P \dot{y}^2 \,\mathrm{d}x \tag{29}$$

Considering the compressibility of the fluid inside the pipeline and the variability of the fluid linear density inside the pipeline, the fluid linear density after pipeline deformation can be expressed as:

$$m_F = \rho_F A_F \tag{30}$$

where ρ_F is the fluid mass density after the pipe is deformed due to pressure change, and A_F is the flow area before the pipe is deformed. The kinetic energy of the fluid can be expressed as:

$$T_F = \int_0^l \frac{1}{2} m_F \left[V^2 + \left(y^{\Omega} + Vy'\right)^2\right]\mathrm{d}x \tag{31}$$

where V is the average flow velocity of the fluid. Then, the total kinetic energy of the system can be expressed as:

$$T = T_P + T_F = \int_0^l \frac{1}{2} m_P y^2 dx + \int_0^l \frac{1}{2} m_F \left[V^2 + \left(y + Vy'\right)^2\right] \tag{32}$$

Similarly, by calculating the pipe and fluid potential energy separately, we can calculate the total potential energy of the system. Pipe potential energy is mainly generated when the pipe bends and changes in temperature. The potential energy generated by bending can be expressed as:

$$V_{P1} = \int_0^l \frac{1}{2} EI\left(y''\right)^2 \mathrm{d}x \tag{33}$$

The potential energy generated by temperature changes can be expressed as:

$$\begin{aligned} V_{P2} &= \int_0^l \frac{A_F}{2E}[E\theta - \gamma(\Delta T)]^2 \mathrm{d}x \\ &= \int_0^l \left[\frac{1}{2} EA_F\theta^2 + \frac{A_F\gamma(\Delta T)^2}{2E} - A_F\gamma(\Delta T)\theta \right] \mathrm{d}x \\ &= \int_0^l \left[\frac{1}{8}(1-2\mu)EA_F\left(y'\right)^4 + \frac{1}{2}EA_F\alpha^2\Delta T^2 - \frac{1}{2}(1-2\mu)EA_F\alpha\Delta T\left(y'\right)^2 \right] \mathrm{d}x \end{aligned} \tag{34}$$

If higher-order quantities are omitted, it can be simplified to:

$$V_{P2} = -\int_0^l \frac{1}{2}(1-2\mu)EA_F\alpha\Delta T\left(y'\right)^2 \mathrm{d}x \tag{35}$$

The total potential energy of the pipe can be expressed as:

$$\begin{aligned} V_P &= V_{P1} + V_{P2} \\ &= \int_0^l \frac{1}{2}EI\left(y''\right)^2 dx - \int_0^l \frac{1}{2}(1-2\mu)EA_F\alpha\Delta T\left(y'\right)^2 \mathrm{d}x \end{aligned} \tag{36}$$

The potential energy of a fluid is the pressure potential energy generated by fluid pressure:

$$V_F = -\int_0^l \frac{1}{2}A_F p(1-2\mu)\left(y'\right)^2 \mathrm{d}x \tag{37}$$

Then, the system's total potential energy can be obtained:

$$\begin{aligned} V &= V_P + V_F \\ &= \int_0^l \frac{1}{2}EI\left(y''\right)^2 \mathrm{d}x - \int_0^l \frac{1}{2}EA_F\alpha\Delta T(1-2\mu)\left(y'\right)^2 \mathrm{d}x - \int_0^l \frac{1}{2}A_F p(1-2\mu)\left(y'\right)^2 \mathrm{d}x \end{aligned} \tag{38}$$

We substitute Equation (31) and Equation (37) into Hamilton's principle:

$$\begin{aligned} \delta\int_{t_1}^{t_2} \Bigg\{ & \int_0^l \frac{1}{2}m_P(\dot{y})^2 \mathrm{d}x + \int_0^l \frac{1}{2}m_F\left[V^2 + \left(\dot{y} + Vy'\right)^2\right] \mathrm{d}x - \int_0^L \frac{1}{2}EI\left(y''\right)^2 \mathrm{d}x \\ & + \int_0^l \frac{1}{2}A_F p(1-2\mu)\left(y'\right)^2 \mathrm{d}x + \int_0^l \frac{1}{2}EA_F\alpha\Delta T(1-2\mu)\left(y'\right)^2 \mathrm{d}x \Bigg\} \mathrm{d}t = 0 \end{aligned} \tag{39}$$

5. Numerical results

5.1 MMS model verification and the effects of viscosity and mass on system stability

We apply mathematical models to specific cases for analysis and calculation. As shown in **Figure 1**. We treat the actual flow pipeline as a simply supported model with $\rho = 0.4$, $\gamma_0 = 2$, and $\kappa = 1$. The simulated curve separates the plane $\sigma - \gamma_1$ into two parts. The upper part of the curve represents the unstable region of the system, while the lower part represents the stable region of the system. By calculating the plotted curves, we compared the stability of the systems at different viscosities.As shown in **Figure 2** and **Figure 3**, as the viscosity increases, the curve boundary gradually moves upward, and the unstable region shrinks. This means the higher the viscosity coefficient is, the more stable the system is.

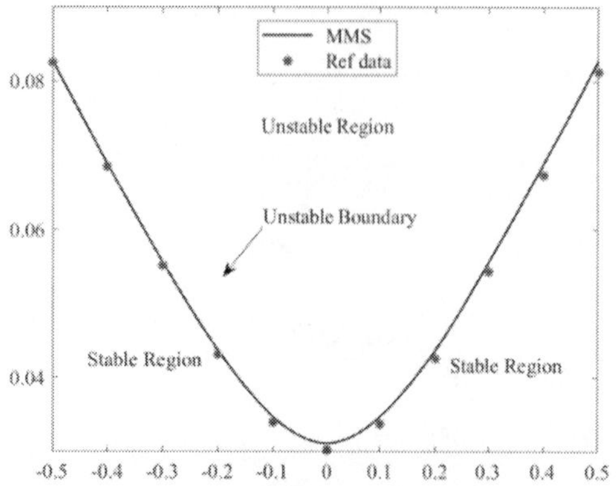

Figure 1. Combination resonance.

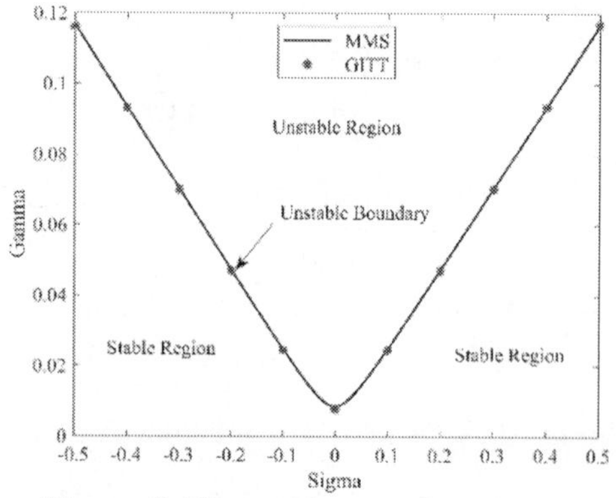

Figure 2. First subharmonic resonance.

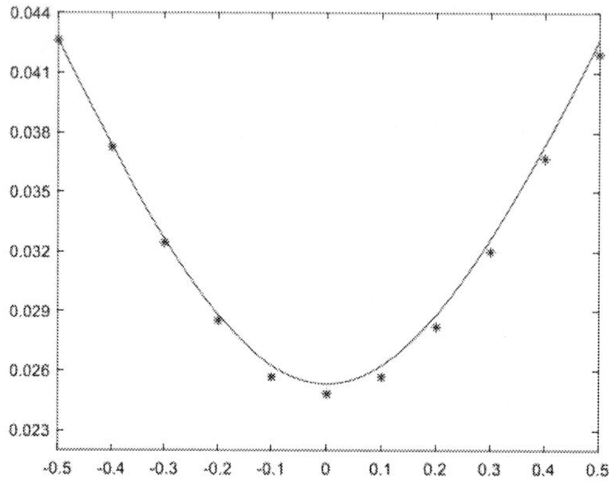

Figure 3. Second subharmonic resonance.

6. Conclusions

The stability of viscoelastic fluid pipelines was analyzed through theoretical analysis and mathematical modeling. The main conclusions are as follows:

(1) If the axial velocity of the pipe fluid fluctuates at a frequency approaching twice the sum or subharmonic resonance of any combination of resonances, the system will become unstable. For different viscosity coefficients, introducing the detuning parameter can help us quantify the deviation value, thereby quantitatively calculating the stability boundary of the system on the amplitude plane.

(2) As the viscosity coefficient increases, the unstable region displayed by the amplitude plane decreases rapidly. This calculation shows that the viscosity coefficient contributes to a large extent to keeping the system stable.

References

[1] Amabili M, Garziera R. Vibrations of circular cylindrical shells with nonuniform constraints, elastic bed and added mass. Part II: Shells containing or immersed in axial flow [J]. Journal of Fluids and Structures, 2002, 16(1): 31-51.
[2] Pellicano F, Amabili M. Dynamic instability and chaos of empty and fluid-filled circular cylindrical shells under periodic axial loads [J]. Journal of Sound and Vibration. 2006, 293(1): 227-252.
[3] Sinir B G, Dönmez D. The analysis of nonlinear vibrations of a pipe conveying an ideal fluid [J]. European Journal of Mechanics-B/Fluids, 2015, 52(8): 38-44.
[4] Chernik V. Analysis of dynamics and controllability of a system described by the equation of a pipeline with a moving fluid [J]. IFAC-Papers Online, 2018, 51(2): 156-161.
[5] Sofiyev A H. About an approach to the determination of the critical time of viscoelastic functionally

graded cylindrical shells [J]. Composites Part B: Engineering, 2019, 156(1): 156-165.
[6] Li, Q., Liu, W., Lu, K., et al. Nonlinear parametric vibration of a fluid-conveying pipe flexibly restrained at the ends. Acta Mechanical, 2019, 33(3): 327–346.
[7] Wang, B., Chen, L.Q. Asymptotic stability analysis with numerical confirmation of an axially accelerating beam constituted by the standard linear solid viscoelastic model. Journal of Sound and Vibration, 2009, 328(4–5): 456–466.
[8] Wang, B. Asymptotic analysis on weakly forced vibration of an axially motion viscoelastic beam constituted by the standard linear solid model. Applied Mathematics and Mechanics (English Edition), 2012, 33(6): 817–828.
[9] An C., Su J. Dynamic behavior of axially functionally graded pipes conveying fluid. Mathematical Problems in Engineering, 2017, 2017: 6789634.
[10] Wang, L.H., Hu, Z.D., Zhong, Z., et al. Hamiltonian dynamic analysis of an axially translating beam featuring time-variant velocity. Acta Mechanical, 2008, 206(3–4): 149–161.
[11] Wang, L.H., Hu, Z.D., Zhong, Z., et al. Dynamic analysis of an axially translating viscoelastic beam with an arbitrarily varying length. Acta Mechanical, 2010, 214(3–4): 225–244.
[12] Zhou, X.Q., Yu, D.Y., Shao, X.Y., et al. Research and applications of viscoelastic vibration damping materials: a review. Composite Structures, 2016, 136: 460–480.
[13] Chen, L.Q., Yang, X.D. Stability in parametric resonance of axially moving viscoelastic beams with time-dependent speed. Journal of Sound and Vibration, 2005, 284(3–5): 879–891.
[14] Chen, L.Q., Yang, X.D. Vibration and stability of an axially moving viscoelastic beam with hybrid supports. European Journal of Mechanics - A/Solids, 2006, 25(6): 996–1008.

Journal of Physics: Conference Series **2808** (2024) 012037 doi:10.1088/1742-6596/2808/1/012037

Enhanced photoluminescence efficiency and stability of $CsAgCl_2$ by a NaCl-assisted synthesis

Jili Zhang [1], Sen Li [*1,2], Peng Yang [1], Yusheng Zhai [1], Ruiliang Zhang [1], Lin Zhang [3], Long Sun [3], Zhan Su [4], Yan Gao [4] and Zhifeng Zhang [*1]

[1] School of electronic information, Zhengzhou University of Light Industry, Zhengzhou, Henan, 450001, China
[2] Henan Key Laboratory of Magnetoelectric Information Functional Materials, Zhengzhou University of Light Industry, Zhengzhou, Henan, 450001, China
[3] White Pigeon Abrasive Tools Co., Ltd, Zhengzhou, Henan, 450000, China
[4] Henan Academy of Sciences Institute of Applied Physics Co., Ltd, Zhengzhou, Henan, 450000, China.

*Corresponding author's e-mail: 2019065@zzuli.edu.cn (S. Li), 2009041@zzuli.edu.cn (Z. Zhang)

Abstract: Here, a NaCl (x = Na/Ag = 0.5) assisted strategy was used to synthesize pure phase $CsAgCl_2$ micro-crystals (MCs) with enhanced photoluminescence (PL) efficiency and stability. Adding a certain amount of NaCl during the preparation could result in high purity of final $CsAgCl_2$ products and higher photoluminescence quantum yield (PLQY), better heat, UV light irradiation, and moisture stability. The $CsAgCl_2$ MCs synthesized in this study displayed a wide PL spectrum spanning from 350 to 750 nm, featuring a peak emission at 544 nm and achieving a remarkable peak PL quantum yield (PLQY) of 73%. The extensive PL spectrum and a substantial Stokes shift of 260 nm provide evidence for the emission mechanism attributed to self-trapped excitons (STEs) within the $CsAgCl_2$ MCs.The dual STE emission originated from two emitting triplet excited states and was observed in the low-temperature PL spectra. A thermal-assisted energy transfer happened with the temperature increase (10~300K), contributing to the efficient STE emission of $CsAgCl_2$ MCs. In conclusion, the pristine $CsAgCl_2$ MCs were employed as a sole phosphor material for encapsulating a 270 nm UV chip for white light-emitting diode (WLED) applications. Despite exhibiting a half-lifetime of 35 minutes under a continuous driving current of 10 mA, the WLED demonstrated notable stability in CCT, CRI, and CIE color coordinates. The resulting device emitted warm white light, characterized by Commission Internationale de l'Éclairage (CIE) color coordinates of (0.48, 0.45), a low correlated color temperature (CCT) of 2734 K, and a high color rendering index (CRI) of 81.9. These results demonstrated the unique advantages of $CsAgCl_2$ MCs in high-performance single-phosphor WLEDs.

1. Introduction

In recent years, notable advancements have been achieved in the realm of lead halide perovskite solar cells, particularly those denoted as $APbX_3$ (where A represents Cs, CH_3NH_3, or NH_2CHNH_2, and X signifies Cl^-, Br^-, or I^-), in the conversion of solar energy into electrical energy[1-3]. The excellent material properties, such as the long carrier diffusion length, low defect density, and easy and low-cost preparation, contribute to the wide concern of the scientific community. Additionally, lead halide

Journal of Physics: Conference Series **2808** (2024) 012037 doi:10.1088/1742-6596/2808/1/012037

perovskites have demonstrated suitability for integration into various other optoelectronic devices, such as light-emitting diodes (LEDs)[4-6], detectors[7], and lasers[8], among others. These limitations restrict the industrialization of these devices. Consequently, pursuing lead-free and stable materials with optoelectronic properties on par with lead-based halide perovskites becomes imperative.

This work presents the synthesis of $CsAgCl_2$ MCs using a NaCl-assisted anti-solvent method. The $CsAgCl_2$ powder prepared in this study exhibited a pure orthorhombic phase with a high PLQY of 73% and a large Stokes shift of 260 nm. Adding NaCl (x = Na/Ag = 0.5) during the preparation resulted in better heat, UV light irradiation, and moisture stability in final $CsAgCl_2$ MCs. When employed as a standalone phosphor, the developed WLEDs showcase a low CCT of 2734 K and a high CRI of 81.9. The CIE color coordinates measure (0.48, 0.45), resulting in warm white illumination. Notably, when the WLEDs are operated in continuous current mode, they demonstrate remarkable stability in CCT, CRI, and CIE color coordinates, alongside a lifespan of 35 min under ambient air conditions.

2. Materials and methods

2.1. Materials

All chemicals employed in this study, namely cesium chloride (CsCl, 99.9%), silver (I) chloride (AgCl, 99.9%), and sodium chloride (NaCl, 99%), were sourced from Innochem. Additionally, dimethyl sulfoxide (DMSO, 99.8%), hexane (98.0%), and ethanol (99.5%) were obtained from Alfa Aesar. Before their use, no further purification was conducted on any of the chemicals above.

2.2. Synthesis of $CsAgCl_2$

In the experiment, $CsAgCl_2$ MCs were synthesized using a NaCl-assisted anti-solvent method. Figure 1 illustrates the sequential synthetic procedures. At the outset, a colorless transparent solution was obtained by stirring vigorously for 1 hour at 60°C a mixture comprising CsCl (0.1 mmol, 0.0327 g), AgCl (0.1 mmol, 0.0282 g), and varied amounts of NaCl (x = Na/Ag = 0, 0.25, 0.5, 0.75) in 2 mL of DMSO. Note that adding NaCl could largely facilitate the dissolving of the AgCl in DMSO. Without the addition of NaCl, a long time of more than 5 hours was needed to prepare the precursor solution. The precursor solution was gradually introduced into 4 mL of ethanol after vigorous stirring at room temperature. As the precursor solution was added, white crystals began forming in the anti-solvent and grew larger over time. Following the decantation of the supernatant solution, the sediments were gathered, subjected to two hexane washes, and dried at 80°C for 1 hour.

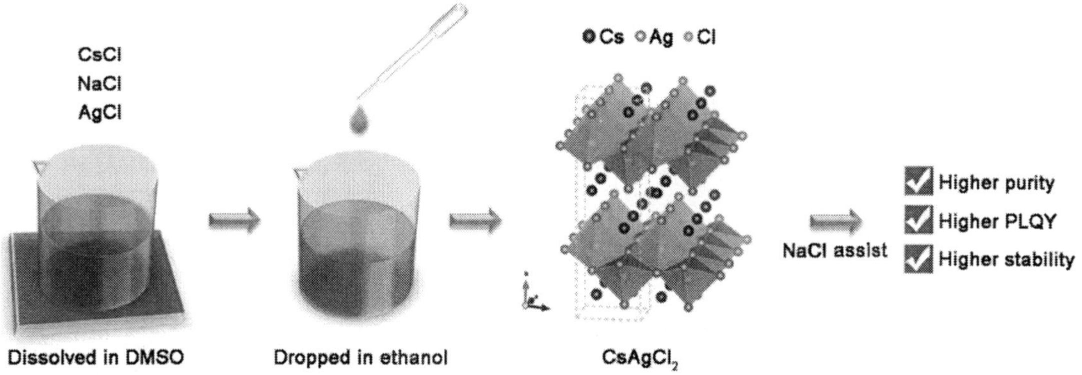

Figure 1. Schematic synthesis procedures of the $CsAgCl_2$ MCs.

2.3. Fabrication of WLED

At first, $CsAgCl_2$ MCs were homogeneously dispersed in hexane through ultrasonic agitation. Following this stage, the mixture was applied onto the surface of a UV LED chip emitting at 270 nm. Ultimately, through evacuation to remove hexane, $CsAgCl_2$ microcrystals were deposited with a uniform and dense distribution. Next, organic silicone (627 *SYLGARD* 184 *Silicone Elastomer*) from Dow Corning was added to the $CsAgCl_2$ MCs. The mixture was then dried at 100°C for 1 hour in air to solidify the organic silicone.

3. Results & discussion

3.1. Structure and morphology

The product structures underwent testing through XRD analysis. Figure 2a presents a comparison of the observed XRD pattern and local magnification. It can be seen that, for the $x = 0$ sample, without the assistance of NaCl, aside from the low peaks that indexed to the standard orthorhombic phase $CsAgCl_2$ (PDF: 76-2238), very strong diffraction peaks (marked with *) existed, which originated from the impurity of AgCl (PDF:31-1238) in the final product. Then, with the addition of 25% NaCl ($x = 0.25$), the intensity of the expected diffraction peaks of $CsAgCl_2$ increased enormously, which indicated the production of $CsAgCl_2$ in a large proportion. Next, after increasing the amount of NaCl to 50%, the diffraction peaks of $CsAgCl_2$ rose continuously, and the peaks of AgCl disappeared.

At last, for $x = 0.75$, the intensity of the diffraction peaks of $CsAgCl_2$ declined again, and the relative intensity of AgCl enlarged. The variation in the intensity of the diffraction peaks indicated the change in the crystallinity and relative content of $CsAgCl_2$ and AgCl in the final products. It was shown that adding NaCl promoted the growth of $CsAgCl_2$ while also preventing the formation of AgCl impurity. The corresponding local magnification of the XRD showed no obvious shift with the additional amount of NaCl, which indicated that the Na^+ ions had not been incorporated into the lattice of $CsAgCl_2$.

Then, SEM measurements were performed to reveal the effect of NaCl on the morphology of the $CsAgCl_2$ MCs, and the pictures are shown in Figure 2b-e. For $x = 0$, the MCs with a diameter of about 2-5 μm were synthesized, but the surfaces seemed not so smooth, indicating there were many defects in the MCs, at least at the surface of the MCs. Note that some MCs had confined plan facets, and others had spherical shapes (Figure 2b). In Figures 2c and 2d, it is observed that for x values of 0.25 and 0.5, there was a notable increase in the size of the confined plan facets MCs, alongside a significant decrease in the number of spherical shape MCs, particularly evident in the $x = 0.5$ sample. In addition, the surfaces of the MCs became smoother than those in Figure 2b, which indicated better crystallinity of the MCs. For $x = 0.75$, as shown in Figure 2e, the MCs were broken into irregular pieces with a large distribution range, which demonstrated too much NaCl would harm the crystallization of $CsAgCl_2$ MCs. Next, the EDS mapping characterization was used to distinguish the element composition of the MCs with different shapes. As shown in Figure 2f-k, for $x = 0.25$, the confined plan facets MCs mainly consisted of Cs, Ag, and Cl elements, and the spherical MCs just consisted of Ag and Cl elements, which indicated that the formed might be $CsAgCl_2$ MCs, and the latter might be of AgCl impurity, which was consisted with the XRD results in Figure 2a. As expected, a very weak and inhomogeneous distribution of Na signals was detected, demonstrating that Na^+ ions had not been incorporated into the lattice of $CsAgCl_2$. Further, for $x = 0.75$, as shown in Figure 2l to 2q, unregularly areas consisting of Ag and Cl increased obviously, confirming the increase of AgCl impurity under this NaCl amount.

Next, the EDS measurements were used to verify the chemical composition of the MCs further, as shown in Figure 2r. With $x = 0$, the sample revealed the presence of Cl, Ag, and Cs, as quantified by the EDS spectrum, showing an atom ratio (At%) of 50.16:41.53: 8.32. This outcome suggests a notably elevated concentration of Ag and Cl relative to the standard stoichiometric ratio of $CsAgCl_2$, consistent with the XRD result in Figure 2a. For $x = 0.25$, the amount of Ag and Cl declined obviously, demonstrating the reduction of AgCl in the product. For $x = 0.5$, the quantified atom ratio (At%) of 47.51:27.99:24.43 was approximate to the standard stoichiometric ratio of the $CsAgCl_2$. For $x = 0.75$, the amount of Ag and Cl increased again. Note that, although for a very large amount of addition in the precursor solution, the Na showed rather a low ratio in different products. HRTEM measured the local magnification of the MCs for $x = 0.25$. Figures 2s and 2t show that the photographs showed clear and well-resolved lattice stripes. The crystal plane spacing measured 2.76 and 2.40 Å for spherical and confined plane MCs, respectively, corresponding to the D spacing between the (200) crystal plane of

AgCl and the (101) crystal plane of $CsAgCl_2$. Together with the XRD results in Figure 2a, we believed that Na had not been incorporated into the $CsAgCl_2$ lattice, but it promoted the growth of $CsAgCl_2$ MCs.

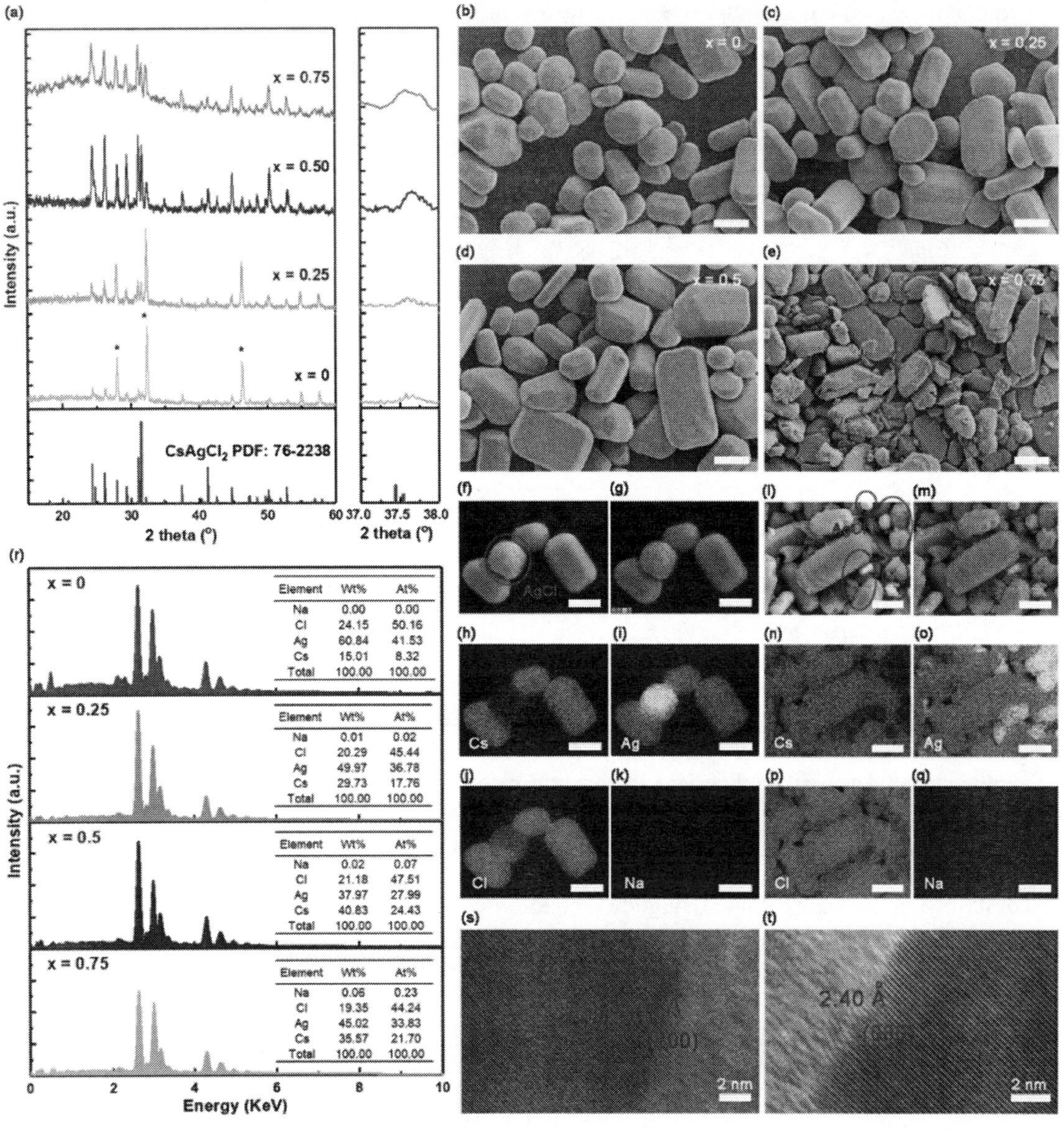

Figure 2. (a) XRD (left) and local magnification XRD (right) of samples with varying NaCl amounts, (b-e) TEM images of samples with different NaCl amounts, with inset scale bars measuring 2.5 μm, (f-k) EDS mapping pictures of the $x = 0.5$ sample, the inset scale bars are 2.5 μm, (l-q) EDS mapping pictures of the $x = 0.75$ sample, the inset scale bars are 2.5 μm, (s, t) HRTEM images show local magnifications of the MCs for $x = 0.25$ sample.

3.2. NaCl regulates $CsAgCl_2$ electron and band structure.

Employing first-principles DFT calculations, we extended our examination to explore the electronic properties of $CsAgCl_2$, specifically emphasizing its energy band structure and DOS. Figures 3a and 3b show that $CsAgCl_2$ MCs possess a direct band gap of 2.34 eV, extending from the valence band edge at the Brillouin zone's point T to the conduction band edge at point Γ. The CBM of $CsAgCl_2$ primarily consists of Cs 3*d* and Cl 2*p* orbitals, contrasting with the VBM primarily constituted by Ag 3*d* and Cl

2*p* hybrid orbitals. The *Tauc* diagram (Figure 3c) for x = Na/Ag = 0.5 reveals an indirect band gap of 4.38 eV, as derived from the data in Figure 7b. The transition from a band gap of 2.34 eV to 4.38 eV signifies a shift like the band gap of $CsAgCl_2$ MCs from direct to indirect due to trace NaCl doping. This alteration in band gap is accompanied by corresponding changes in the material's energy level structure and optical properties.

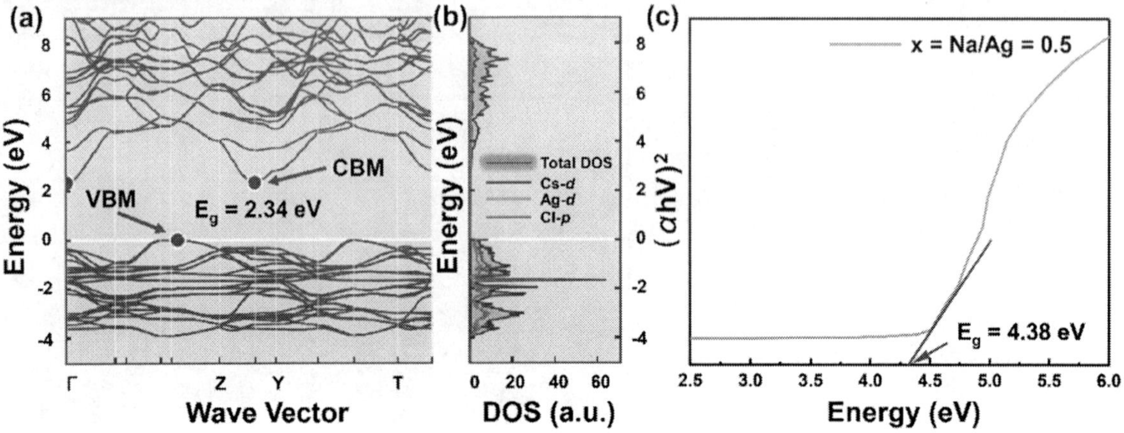

Figure 3. (a) The electronic band structures computed via density functional theory (DFT); (b) Projected density of state (DOS) profiles for the $CsAgCl_2$ MCs; (c) The *Tauc* diagram corresponding to x = Na/Ag = 0.5.

3.3. Optical characteristics

Moreover, a thorough examination of the samples' optical attributes was undertaken. Notably, Figure 4a shows that all samples presented a white appearance under natural lighting conditions. Notably, upon exposure to a 254 nm UV lamp, the samples emitted a vivid warm-white luminescence, particularly prominent for samples with x = 0.25 and 0.5. Furthermore, Figure 4b displays the PL spectra of the samples, showing a significant peak at 544 nm with a considerable FWHM of around 125 nm. Moreover, the PL intensity increased from 0 to 0.5 and then decreased at 0.75.

The samples exhibited absolute PLQY values of 47%, 58%, 73%, and 41% for x = 0, 0.25, 0.5, and 0.75, respectively. Then, enhancement in PLQY may have originated from better purity and crystallinity, as described in previous results. Using the second-order decay function to fit the PL decay spectra, the average PL lifetime of 37, 43, 56, and 40 were obtained for x = 0, 0.25, 0.5, 0.75 (Figure 7a, Supporting information), respectively. The prolongation of the lifetime observed at x = 0.25 and 0.5 could be attributed to a decline in the presence of charge-carrier trapping states. This guarantees effective utilization of excited carriers, leading to high PLQY and prolonged PL lifetime. Although adding a certain amount of NaCl enhanced the light-emitting of the products, all samples showed similar absorption spectra with the absorption onset at about 284 nm (Figure 7b, Supporting information). The difference between the PL peak and the absorption onset revealed the large Stokes shift of 260 nm. However, the PLE spectra of the MCs with different NaCl amounts in Figure 7c (Supporting information) presented a similar peak position of about 270 nm, demonstrating that adding NaCl had not influenced the emission origin of $CsAgCl_2$. The substantial Stokes shift and broad FWHM observed indicate that the PL mechanism of $CsAgCl_2$ cannot be solely ascribed to indirect band emission. Instead, it suggests exciton STE or structural reorganization occurring in the excited state.

To comprehend the optical recombination mechanisms of the $CsAgCl_2$ MCs, we conducted temperature-dependent PL measurements across the temperature spectrum of 10 to 300 K, as shown in Figure 4c. Within the lower temperature range, specifically 10 to 160 K, as shown in Figure 4d, an extra PL peak centered at 455 nm (referred to as peak 1) was observed alongside the emission peak at 580 nm (referred to as peak 2), which closely corresponds to the PL peak position at room temperature. With increasing temperature, the emission intensity of peak one gradually decreases until it disappears completely, while peak two steadily increases. The dependence of dual emissions intensity on temperature suggested excited state energy transfer between two excited states. As the temperature rises, exciton STEs acquire sufficient thermal energy to surpass the energy barrier separating the two emitting triply excited states. In the higher temperature range (from 200 to 300 K), a thermal burst ensues from the thermal dissociation of excitons, resulting in a decline in the intensity of the photoluminescence peak and a shift towards shorter wavelengths. Figure 4e shows the variation in peak position and peak area of peak 1 and peak 2 across the temperature range from 10 to 300 K. It seems that the peak position changed slightly with the temperature. Note that the increase of the area of peak 2 was equal to the decrease of peak 1, which may indicate an energy transfer from peak 1 state to peak 2 state with the increase of temperature. Consequently, Figure 4f depicts the hypothesized photo-physical processes occurring in both rooms and at low temperatures.

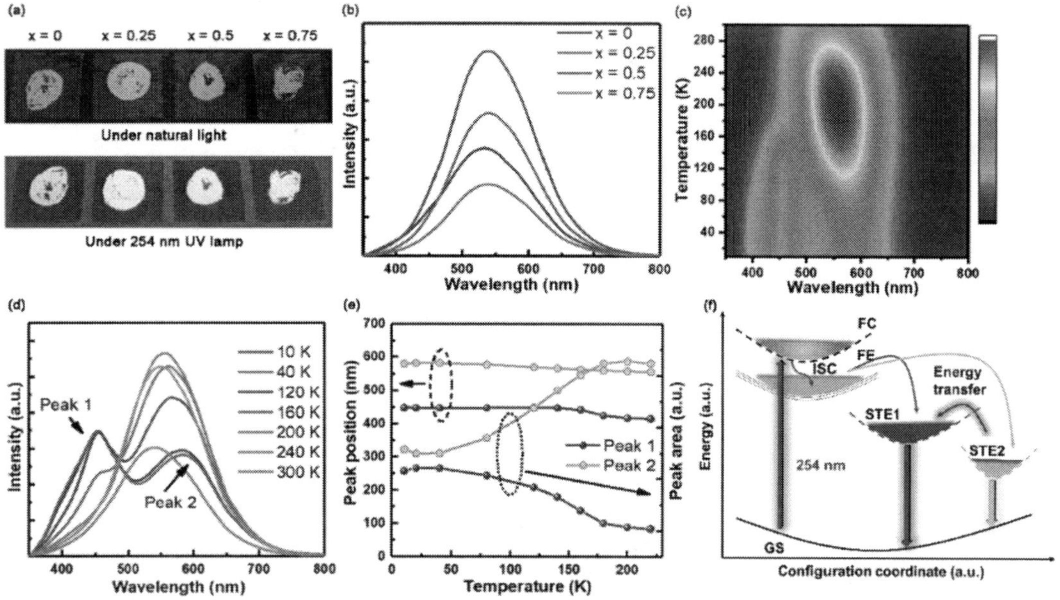

Figure 4. (a) Images of $CsAgCl_2$ samples taken under natural light (top) and 254 nm UV lamp (bottom), synthesized with varying amounts of NaCl, (b) PL spectra of $CsAgCl_2$ samples prepared with different NaCl concentrations, (c) Temperature-dependent PL mapping spectra of $CsAgCl_2$ excited at 266 nm, (d) Representative curves illustrating the temperature-dependent PL spectra of $CsAgCl_2$ samples as a function of temperature, (e) Variations in peak positions and peak areas of $CsAgCl_2$ MCs at different temperatures, and (f) Hypothetical photo-physical processes of $CsAgCl_2$ MCs.

3.4. Stability

The synthesized $CsAgCl_2$ MCs underwent an assessment of their environmental stability, which entailed investigating how heat, UV light exposure, moisture, and storage conditions affected their PL properties. The temperature's impact on PL intensity was analyzed across 30-100°C. Normalized PL spectra of $x = 0$ and $x = 0.5$ were plotted against temperatures ranging from 30 to 100°C, as shown in Figures 5a and b. PL intensity diminishes with increasing temperature. Specifically, at a high temperature of 100°C, approximately 52% and 60% emission decay was observed for $x = 0$ and $x = 0.5$, respectively. This phenomenon is likely attributed to the emergence of additional non-radiative recombination centers at high temperatures. Then, under continuous heating at 100°C, continuous emission dropping happened. In the first 10 h, fast decays of about 10% and 20% happened for $x = 0$ and 0.5, respectively. In the following time, the decay rates became much lower, and total PL intensity decays of about 20% and 30% happened in 90 hours for $x = 0$ and 0.5, respectively. The samples before and after the test shown as insets in Figure 5c agreed well with the results. Afterward, we opted to assess the stability of the products under UV light irradiation by employing a portable UV lamp emitting light at a wavelength of 256 nm, with an intensity of 20 mW/cm^2 serving as the irradiation source. As shown in Figure 5d, rather rapid declines happened under continuous irradiation, with about 40% and 42% of the initial PL intensity retained after 13h for $x = 0$ and 0.5, respectively. To evaluate moisture stability (Figure 5e), the samples were enclosed within a sealed container maintained at a RH of 75%. The evolution of PL over time was monitored. The PL intensity decayed rapidly by about 20% in the initial 10 h, and about 83% and 60% remnant of the initial intensity were left after a long term of 70 hours for $x = 0$ and 0.5, respectively. To assess storage stability, samples corresponding to $x = 0$ and 0.5 were exposed to the atmospheric environment, and XRD measurements were conducted at specific time intervals. As shown in Figure 5f, no noticeable changes happened for all of the samples, demonstrating perfect storage stability. In all, with the assistance of NaCl, the $CsAgCl_2$ MCs exhibited obvious enhancement in heat, UV irradiation, and humidity stability.

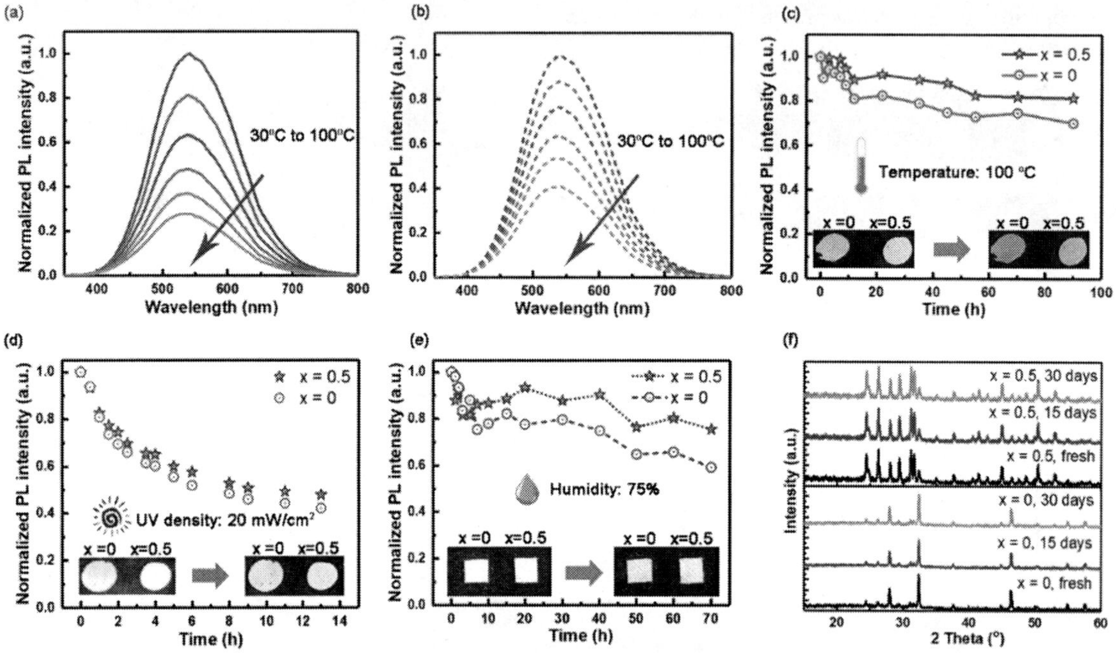

Figure 5. (a) Normalized PL spectra of $CsAgCl_2$ MCs excited by 270 nm, prepared with (a) $x = 0$ and (b) $x = 0.5$, recorded at temperatures ranging from 30 to 100°C, (c) Under continuous heating at 100°C, the normalized PL intensity of samples for $x = 0$ and $x = 0.5$ was measured. Insets showing the samples before (left) and after (right) testing are provided. (d) Samples were exposed to continuous UV light radiation (254 nm, 20 mW/cm^2), and their normalized PL intensity was measured. Insets show the samples before (left) and after (right) testing for $x = 0$ and 0.5, (e) Under RH of 75%, the normalized PL intensity of samples was assessed. Insets show the samples for $x = 0$ and 0.5 before (left) and after (right) testing, as well as (f) XRD spectra of samples evaluated after various storage durations.

3.5. Application

To explore the potential applications of NaCl-assisted synthesized $CsAgCl_2$ MCs as a single down-conversion phosphor, we integrated them onto a commercially available 270 nm UV LED chip to assemble a WLED. Operating the WLED at a current of 10 mA resulted in a bright, warm-white light with excellent color rendering, as shown in Figure 6a. The emission spectra of the operational WLED, shown in Figure 6b, exhibited characteristic peaks spanning from 450 to 800 nm, consistent with the PL spectra in Figure 4b. In addition, the WLED's CIE color coordinates, placed within the warm-white light region, were shown in Figure 6c, with the CCT measured at 2734 K and the CRI at 81.9. Figure 6d demonstrates the emission spectra of the device under various driving currents. As the driving current increased from 1 to 30 mA under UV excitation, a linear enhancement in emission intensity was noted, showing no indication of premature saturation. To assess the operational stability of the WLED, emission spectra were examined at various operating durations while maintaining a driving current of 10 mA. Figure 6e shows a gradual decline in emission intensity over time, resulting in a 50% decay after 35 min, corresponding to a half-lifetime (T_{50}) of 35 min. Nonetheless, minimal changes in CCT, CRI, and CIE color coordinates were observed across different operating durations, as evidenced by Figure 6f and Figure 7d (Supporting Information), underscoring the unique advantages of the single phosphor WLED.

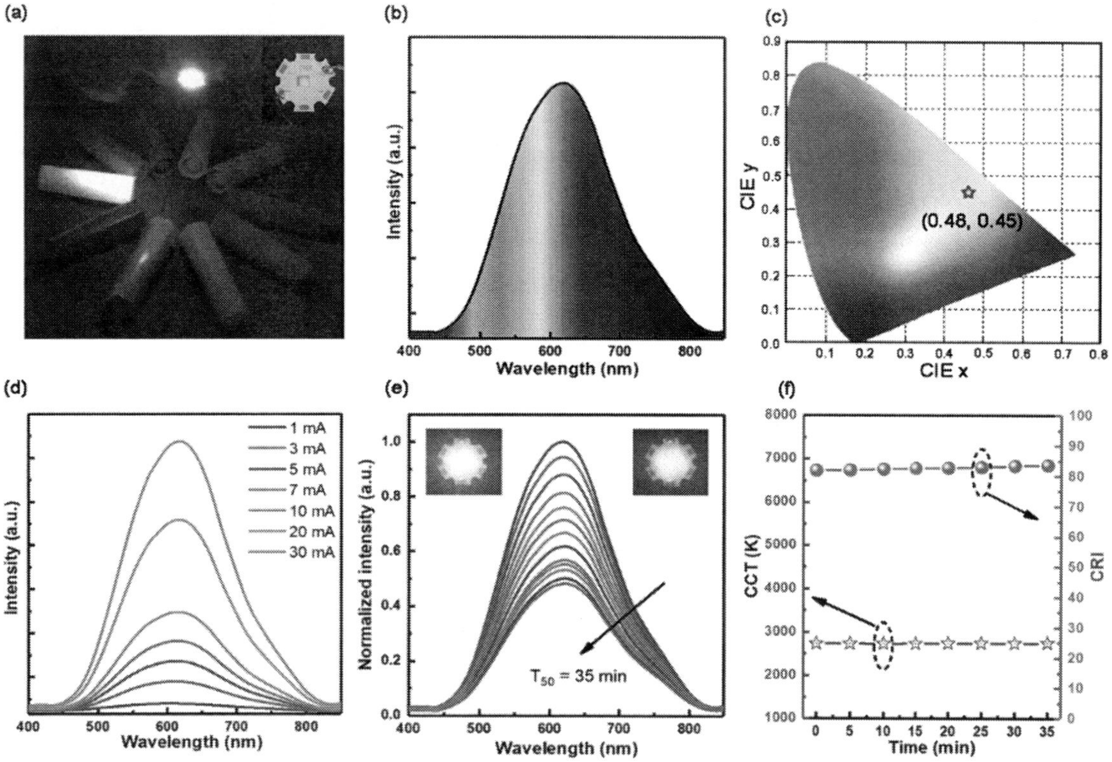

Figure 6. (a) Visual representations and (b) characteristic emission spectra of the WLED constructed with $CsAgCl_2$ MCs operating at 30 mA, (c) CIE color coordinates of the WLED are visualized and depicted on the CIE1931 color space chromaticity chart, (d) Emission spectra of the WLED were observed across a range of driving currents from 1 to 30 mA, (e) Emission intensity of the WLED was normalized and monitored over a 35-minute duration. A comparison of the device before (left) and after (right) testing is shown in the inset. (f) CCT and CRI of the WLED fluctuate as operating times vary.

4. Conclusions

In summary, the critical role of the NaCl additive in the synthesis of $CsAgCl_2$ MCs has been demonstrated. The assistance of NaCl resulted in high purity of final $CsAgCl_2$ products, higher PLQYs, better heat, UV light irradiation, and moisture stability. A dual STE emission to one emission evolution was observed in the temperature-dependent PL, which demonstrated a thermal-assisted energy transfer contributed to the efficient STE emission of $CsAgCl_2$. The benefits above facilitate the utilization of $CsAgCl_2$ MCs as a sole phosphor in WLED applications. Ultimately, the fabricated WLED produced a warm-white light featuring CIE color coordinates of (0.48, 0.45), a CCT of 2734K, and a CRI of 81.9. Remarkably, it sustained a T_{50} of 35 min under a continuous driving current of 10 mA. The WLED demonstrated exceptional stability in terms of CCT, CRI, and CIE color coordinates.

5. Supporting information

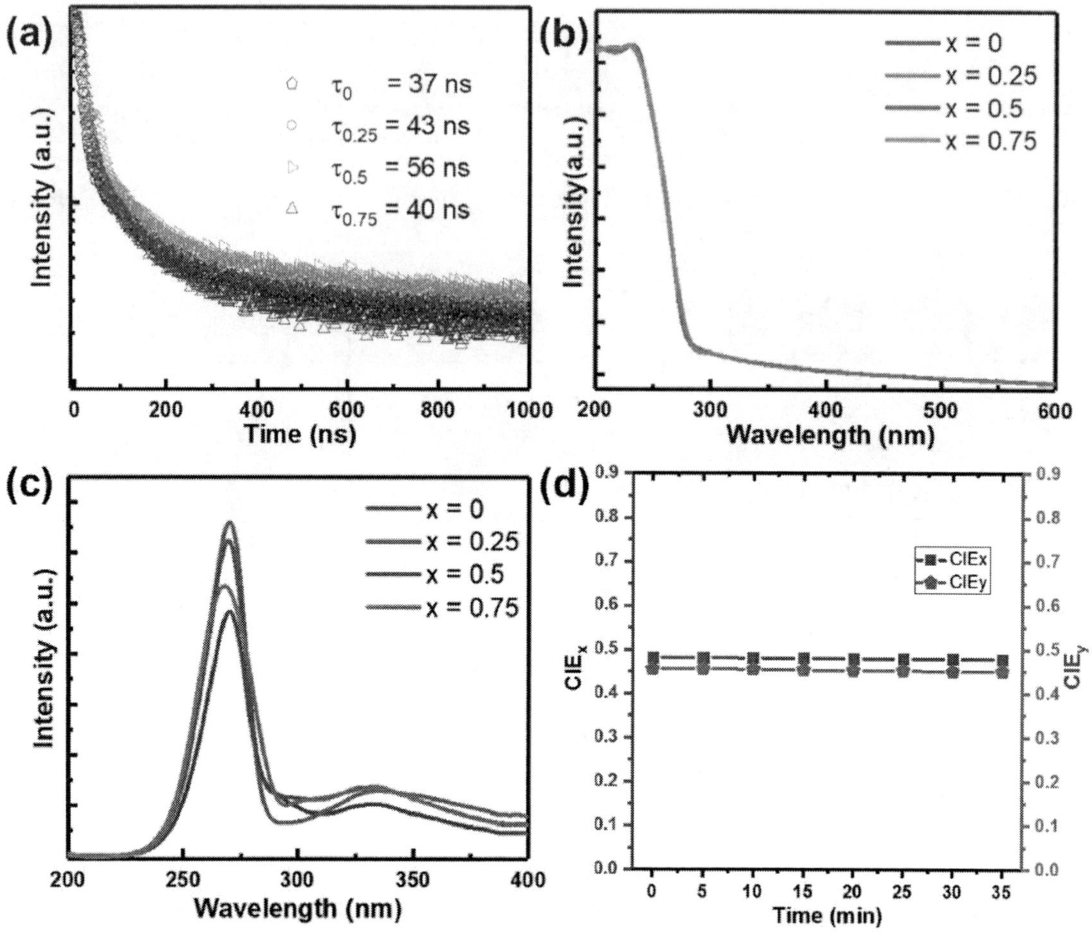

Figure 7. (a) PL decay curves of samples with different NaCl amounts, (b) Normalized absorption, PLE spectra of $CsAgCl_2$ MCs prepared with different NaCl amounts, (c) PLE spectra of $CsAgCl_2$ MCs prepared with different NaCl amount, (d) CIE color coordinates of the WLED were recorded at different intervals during operation.

Acknowledgments
The research described in this paper received support from several funding sources, including the Key Research Projects of Higher Education Institutions in Henan Province (Grant No. 23B416001), the Henan Major Science and Technology Project (Grant No. 221100230100), the Key Research and Development Special Projects in Henan Province (Grant No. 231111222800), the Department of Science and Technology of Henan Province (Grant No. 242102211064), and the Natural Science Foundation of Henan Province (Grant No. 242300420294).

References

[1] Xing G., Mathews N., Sun S., Lim S.S., Lam Y.M., Grätzel M., Mhaisalkar S., Sum TC (2013)Long-range balanced electron-and hole-transport lengths in organic-inorganic $CH_3NH_3PbI_3$. *Science*, 342: 344-347.

[2] Tan H., Jain A., Voznyy O., Lan X., García de Arquer F.P., Fan J.Z., Quintero-Bermudez R., Yuan M., Zhang B., Zhao Y. (2017)Efficient and stable solution-processed planar perovskite solar cells via contact passivation. *Science*, 355: 722-726.

[3] Zhao Y., Ma F., Qu Z., Yu S., Shen T., Deng H.X., Chu X., Peng X., Yuan Y., Zhang X., You J. (2022)Inactive $(PbI_2)_2RbCl$ stabilizes perovskite films for efficient solar cells. *Science*, 377: 531-534.

[4] Chu S., Zhang Y., Xiao P., Chen W., Tang R., Shao Y., Chen T., Zhang X., Liu F., Xiao Z. (2022)Large-area and efficient sky-blue perovskite light-emitting diodes via Blade-coating. *Advance Materials*, 34: 2108939.

[5] Matsushima, T., Bencheikh, F., Komino, T., Leyden, M. R., Sandanayaka, A. S., Qin, C., Adachi, C. (2019)High performance from extraordinarily thick organic light-emitting diodes. *Nature*, 572: 502-506.

[6] Abdi-Jalebi, M., Andaji-Garmaroudi, Z., Cacovich, S., Stavrakas, C., Philippe, B., Richter, J. M., Stranks, S. D. (2020)Author Correction: Maximizing and stabilizing luminescence from halide perovskites with potassium passivation. *Nature*, 581: 1-1.

[7] Liu T., Shi W., Tang W., Liu Z., Schroeder B.C., Fenwick O., Fuchter M.J. (2022)High Responsivity Circular Polarized Light Detectors based on Quasi Two-Dimensional Chiral Perovskite Films. *ACS Nano*, 16: 2682-2689.

[8] Wang C., Dai G., Wang J., Cui M., Yang Y., Yang S., Qin C., Chang S., Wu K., Liu Y. (2022)Low-threshold blue quasi-2D perovskite laser through domain distribution control. *Nano Letters,* 22: 1338-1344.

Journal of Physics: Conference Series **2808** (2024) 012038 doi:10.1088/1742-6596/2808/1/012038

Effect of calcium sulfate whiskers on mechanical properties and enhancement mechanism of metakaolin-based polymer mortar

Xiong Zheng[1], Gang Zhao[2], Ming Jiang[2], Kuixiang Guo[2], Yu Cui[2] and Jinsong Wang[1, *]

[1]College of Civil Engineering, University of South China, Hengyang, Hunan, China
[2]Shanghai Urban Construction Design Research Institute (Group) Co., LTD, Shanghai, China

[*]Corresponding author's e-mail: xhwjs@163.com

Abstract. In consideration of the requirements of the repair mortar environment and in conjunction with the characteristics of the $CaSO_4$ whisker, the addition method of the $CaSO_4$ whisker was employed to enhance metakaolin-base polymer mortar, aiming to achieve reinforcement and toughening effects. Through experimentation, the influence of varying mass percentages (0.5%, 1%, 1.5%, 2%, 2.5%, and 3%) of calcium sulfate whisker on metakaolin-based polymer mortar was studied, focusing on rational energy utilization and mechanical properties. Optimal performance and mechanical strength were determined by proposing an appropriate content of calcium sulfate whisker and liquid-solid ratio for metakaolin-based polymer mortar, while microscopic experiments were conducted to analyze the strengthening mechanism behind polymer mortar with calcium sulfate whisker. The experimental findings show that with an increase in the calcium sulfate whisker content, the setting time of metakaolin-based polymer mortar initially rises, then declines, reaching the maximum setting time at 1% content. Additionally, at a liquid-solid ratio combined with a content level of 1.5% for calcium sulfate whisker, both bending and compressive strength are optimized for metakaolin-based polymer mortar. Calcium sulfate whiskers can alleviate the stress concentration at the tip of cracks within the geopolymer mortar, hindering crack development, and the bridging effect of the whiskers and the interface of the geopolymer mortar matrix in the high territories realizes the overall strengthening and toughening.

1. Introduction

In recent decades, the demand for cement has significantly increased due to advancements in transportation and construction. Nevertheless, cement production consumes a significant quantity of natural resources and accounts for 5-7% of global CO_2 emissions. Therefore, finding alternative cementing materials for the construction industry has become an urgent issue [1]. Polymer-based geopolymers offer excellent properties including high-temperature stability [2], freeze resistance [3], carbonization resistance [4], and acid resistance [5]. Geopolymer production emits less carbon dioxide compared to ordinary Portland cement, making it a sustainable and promising green alternative with exceptional performance capabilities [6]. Consequently, both domestic and international scholars have utilized common raw materials such as high soil, slag, fly ash, silica fume, and red mud as additives to develop composite cementitious materials that effectively consume large quantities of solid waste.

Research results demonstrate that incorporating slag, fly ash, silica ash, or their combinations into

Journal of Physics: Conference Series **2808** (2024) 012038 doi:10.1088/1742-6596/2808/1/012038

phosphogypsum-based composites can significantly enhance mechanical properties and water resistance. Additionally, utilizing phosphogypsum-red mud composites can immobilize heavy metal elements present in red mud while reducing environmental harm caused by the composite material. Pillay et al. [7] studied MK-modified concrete and it revealed that MK enhances mechanical strength and durability of concrete structures. Al-Akhras carried out [8] an investigation on substituting MK for cement and examined different substitution rates' effects on sulfate resistance durability in concrete; and found that under different curing environments, concrete's sulfate resistance increased with the increase of MK substitution rates.

$CaSO_4$ whisker is a sub-nanofiber material derived from processing industrial by-product gypsum, and it boasts a diverse array of applications. Calcium sulfate whisker exhibits properties such as high-temperature resistance, resistance to acids and alkalis, and an appropriate aspect ratio of length to diameter. In the realm of concrete, certain scholars have examined the impact of calcium sulfate whiskers on the evolution of concrete's mechanical properties. Their research reveals that incorporating $CaSO_4$ whisker notably enhances the fundamental mechanical characteristics of concrete, particularly in its early stages, and increasingly stabilizes as it ages. An optimal amount of $CaSO_4$ whisker is essential for enhancing the mechanical properties of concrete, both low and high contents are not conducive. The recommended content of the $CaSO_4$ whisker is 5 kg/m^3. The enhancement of mechanical properties of concrete by $CaSO_4$ whisker occurs through a microscopic mechanism where the cement slurry is enveloped by the whisker, leading to the formation of a spatial skeleton structure.

Additionally, some whisker particles infiltrate the internal pores of the concrete. Several researchers incorporated calcium sulfate whisker into cement mortar, and findings revealed that the addition of 5 wt% $CaSO_4$ whisker led to a 28.3% increase in bending strength and an 8.5% increase in compressive strength of the mortar specimens. This improvement was attributed to the ability of the $CaSO_4$ whisker to effectively retard the development of micro-cracks and suppress the propagation of existing micro-cracks. Simultaneously, the non-threatening porosity of cement mortar rose from 9.33% to 10.62%, while the detrimental porosity decreased from 2.08% to 1.90%. As a result, whiskers can enhance the pore size distribution of the resultant cement mortar [9].

In summary, it is found that calcium sulfate whisker is mostly used for modification, but there are few studies on alkali excited composite gel systems modified by calcium sulfate whisker, and the related studies on the effect of $CaSO_4$ whisker on the properties of alkali excited composite gel materials are also lacking. Therefore, in this paper, silica fume and high territory are used as the precursors of the composite gelling system. Through the mechanical properties and dry shrinkage test, the influence and mechanism of calcium sulfate whisk at different dosages on the alkali-excited composite gelling materials are explored, and corresponding references are provided for the practical application and promotion of alkali-excited composite gelling materials.

2. Materials and methods

2.1 Experimental material

The high territory is produced from Shanxi Chaopai Calcined kaolin Co., LTD., the average particle size is 10 μm, the 28d activity index is 110, and the fineness is 800 mesh. Silica fume from Henan Yixiang New material, SiO_2 content is 98.1%, 28d activity index is 105. Sodium silicate solution (liquid-finished sodium silicate solution, NaOH, and deionized water were used to adjust the modulus to 1.4 and the concentration to 52.3%) was used as the alkali activator in the laboratory. The sodium silicate was produced and purchased from Jiaxing Yourui Material Chemical Co., LTD., Zhejiang Province. The modulus was 2.3 and the concentration was 43.75%. The mass fraction ratio of SiO_2 is 29.99%, and the mass fraction ratio of Na_2O is 13.75. NaOH comes from industrial-grade flaky sodium hydroxide produced by Xinjiang Zhongtai Chemical Co., LTD. (purity is 99%). The sand used for the preparation of mortar is the standard sand produced by Xiamen ASio Standard Sand Co., LTD. Figure 1 is the particle size analysis diagram of raw materials.

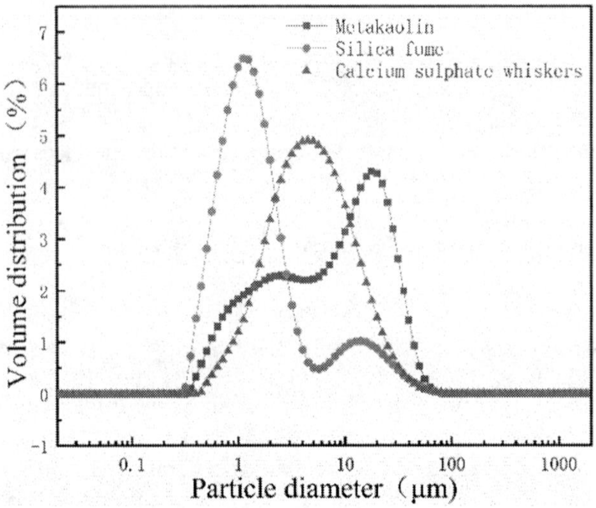

Figure 1. Particle size analysis plots of biotite territory, silica fume, and calcium sulfate whiskers.

2.2 Experimental scheme

The content of the $CaSO_4$ whisker is mixed into the geopolymer mortar according to the proportions of 0%, 0.5%, 1%, 1.5%, 2%, 2.5%, and 3% of the gelling material. The liquid-solid ratio is proposed to be 0.28, 0.30, and 0.32 according to previous experience. The addition method is adopted to add calcium sulfate whisker into the polymer mortar of the metakaolin base. The study examined the fluidity, setting time, dry shrinkage, bending resistance, and compressive strength of mortar made with a high-territory base, and explored how the addition of whiskers influenced the physical and mechanical properties of the mortar. In terms of physical properties, the influences of whisker on slurry fluidity and setting time of metakaolin-based mortar are mainly studied.

3. Results and analysis

3.1 Effect of calcium sulfate whisker on the fluidity of geopolymer

In this study, a series of slightly alkaline earth-based mortars were meticulously prepared according to specific water-to-solid ratios (0.28, 0.30, 0.32), fixed ash-to-sand ratio (0.14), and varying amounts of calcium sulfate whiskers (0%, 0.5%, 1%, 1.5%, 2%, 2.5%, 3%). The objective was to explore how these variables affect the fluidity of the mortar. The experimental results are displayed in Figure 2.

From the data in Figure 2, it is observed that under a fixed water-to-solid ratio, the spread diameter of the mortar initially increases and then decreases with the increase in calcium sulfate whiskers content, indicating that the fluidity first improves and then declines, with the optimum fluidity observed at a 1.5% addition. The fluidity continuously increases when the whisker content increases from 0% to 1.5%, with an average increase of 2.5%; however, when the content increases from 1.5% to 3%, the fluidity continuously decreases, with an average reduction of 1.5%. These results suggest that under a certain water-to-solid ratio, when the calcium sulfate whiskers content is less than 1.5%, the change in mortar fluidity is minimal, without significant deterioration in fluidity. This is primarily due to the ultrafine needle-like structure and higher elastic modulus of the calcium sulfate whiskers, which allow for an even distribution within the mortar, reducing the likelihood of entanglement and clumping. The smooth surface of the $CaSO_4$ whiskers also reduces the internal flow and shear resistance of the mortar matrix to some extent, thereby minimizing fluidity loss. However, when the whisker content is excessively high, the increased surface area and distribution density per unit volume require more slightly alkaline earth-based mortar to coat the whiskers, leading to reduced continuity of the mortar and, consequently, decreased fluidity.

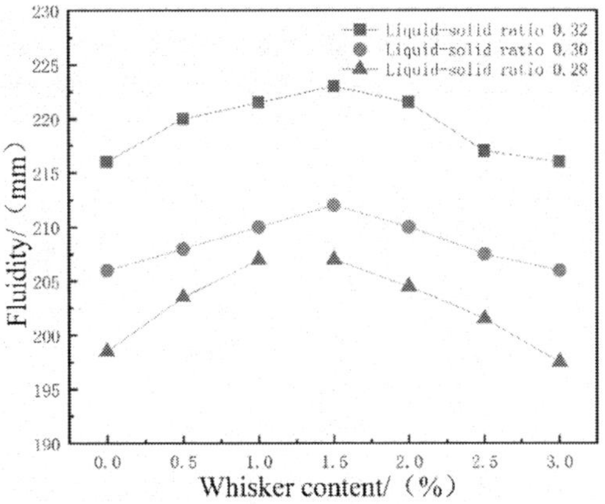

Figure 2. Effect of calcium sulfate whisker on polymer fluidity of metakaolin base.

3.2 Influence of calcium sulfate whisker on setting time of base mortar with high territory

Under the condition of maintaining standard water usage, calcium sulfate whiskers were added in different proportions of 0%, 0.5%, 1%, 1.5%, 2%, 2.5%, and 3% to measure the initial and final setting times in slightly alkaline earth-based mortars, aiming to explore the specific impact of $CaSO_4$ whiskers on the mortar's setting time. According to the data in Figure 3, the addition of CaSO4 whiskers effectively extends the initial and final setting times of the slightly alkaline earth-based mortar, exhibiting a notable retardation effect. However, this retardation effect disappears when the $CaSO_4$ whiskers content exceeds a certain threshold, instead leading to a reduction in the geopolymer mortar's setting time. Specifically, when the content of $CaSO_4$ whiskers is 1.0% or less, it can effectively delay the setting time of the mortar. Yet, once the content exceeds 1%, the setting time is shortened.

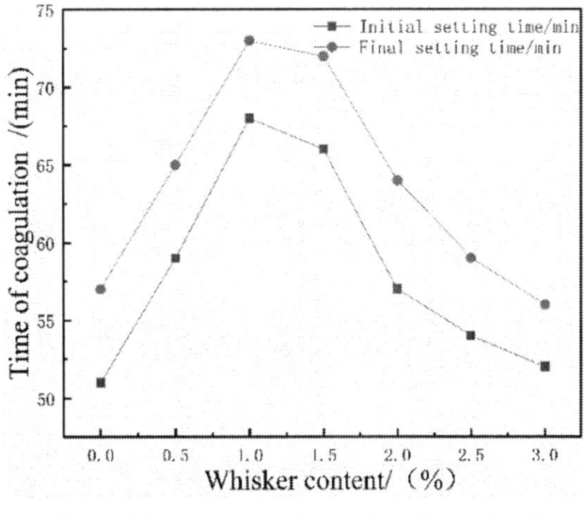

Figure 3. Effect of calcium sulfate whisker on condensation time of polymers at high territorial bases.

This is mainly because the calcium sulfate whisker will partially dissolve in the base mortar slurry on the high side, which can delay the setting time [10]. However, the dissolution of calcium sulfate whisker requires water consumption, and it has a hetero-tropic distribution in the geo-polymer slurry. The geo-polymer wrapping on the whisker increases the consistency of the geo-polymer slurry. When the dosage of $CaSO_4$ whisker is small, it can play a role in retarding coagulation. However, if the dosage is too high, exceeding the critical value, the consistency of the geopolymer slurry is too high, resulting in a shorter descending distance of the test rod and a shorter coagulation time [11].

Journal of Physics: Conference Series **2808** (2024) 012038 doi:10.1088/1742-6596/2808/1/012038

3.3 Influence of calcium sulfate whisker on mechanical properties

To thoroughly investigate the impact of the water-to-solid ratio and the amount of calcium sulfate whiskers added on the performance of slightly alkaline earth-based composite mortars, a series of triplex mold specimens were prepared with different water-to-solid ratios (0.28, 0.30, 0.32) and varying proportions of calcium sulfate whiskers (0%, 0.5%, 1%, 1.5%, 2%, 2.5%, 3%) relative to the total mass of the cementitious materials, along with fixed ash to sand ratio (0.14). After fabrication, the specimens were placed in a curing chamber for standard curing and then extracted at predetermined times (3, 7, 28 days) for comprehensive mechanical testing according to the "Strength Test Method for Geopolymer Mortar". The experimental findings indicated that for water-to-solid ratios of 0.28, 0.30, and 0.32, the compressive and flexural strengths of the slightly alkaline earth-based mortar initially increased with the addition of calcium sulfate whiskers, but subsequently decreased. Notably, at a water-to-solid ratio of 0.32, the increase in strength was relatively lower due to the higher ratio, hence this condition was not a focal point of discussion. Figure 4 shows the effect of calcium sulfate whisker dosing on the mechanical properties of mortar at different liquid-solid ratios. For water-to-solid ratios of 0.30 and 0.28, a more significant increase in strength was observed. Further analysis of flowability and setting time revealed that at a water-to-solid ratio of 0.28, the most substantial increase in mortar strength was achieved with minimal impact on flowability and normal setting time. From these analyses, it is concluded that a water-to-solid ratio of 0.28 is most conducive to the formation and development of the strength of geopolymer mortar.

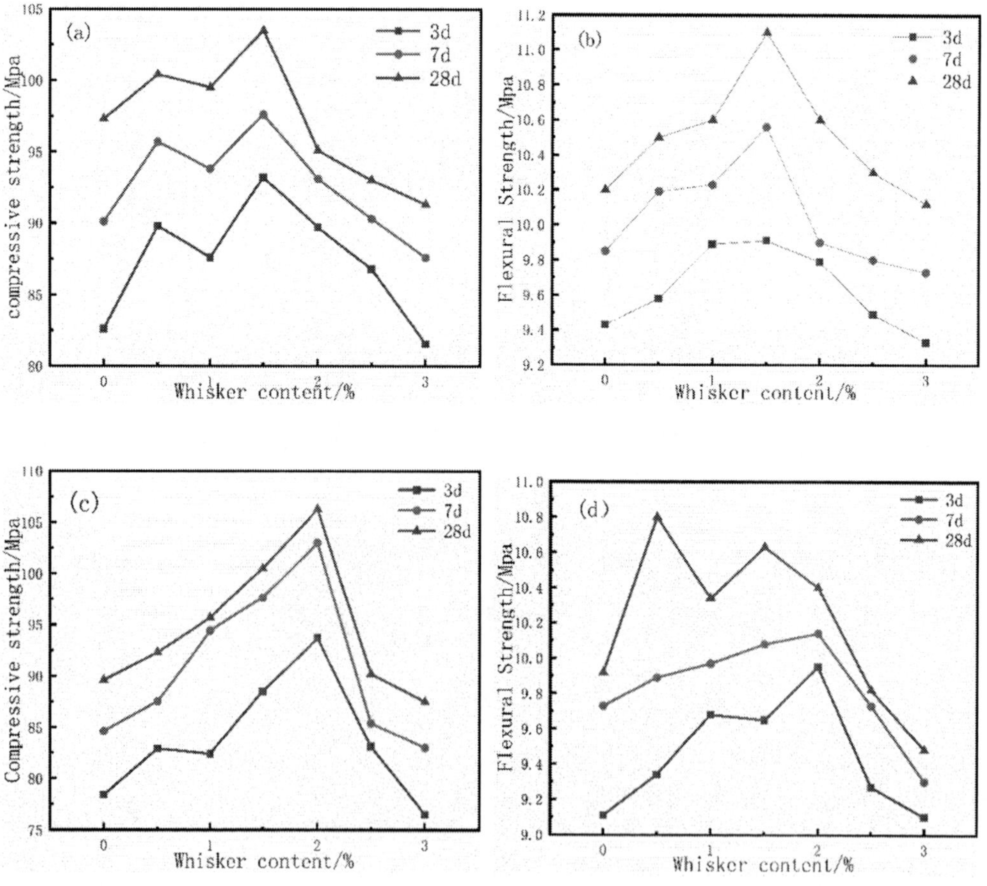

Journal of Physics: Conference Series **2808** (2024) 012038 doi:10.1088/1742-6596/2808/1/012038

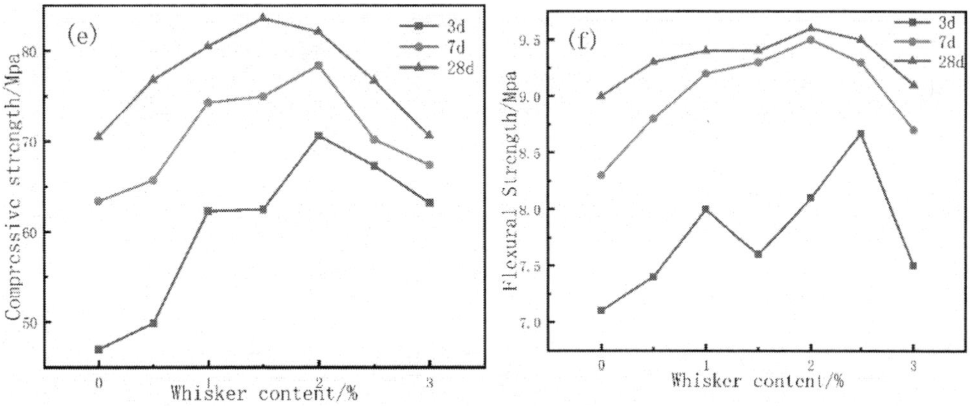

Figure 4. (a) Compressive strength when the liquid-solid ratio is 0.28; (b) Flexural strength when the liquid-solid ratio is 0.28; (c) Compressive strength when the liquid-solid ratio is 0.30; (d) Flexural strength when the liquid-solid ratio is 0.30; (e) Compressive strength when the liquid-solid ratio is 0.32; (f) Flexural strength when the liquid-solid ratio is 0.32.

3.4 Influence of calcium sulfate whisker on dry shrinkage performance

This study investigates the impact of incorporating varying proportions of calcium sulfate whiskers into slightly alkaline earth-based composites on their drying shrinkage rates. The experimental outcomes, depicted in Figure 5, unequivocally demonstrate that the addition of calcium sulfate whiskers significantly mitigates the shrinkage of slightly alkaline earth-based mortar. Notably, an optimum suppression effect is observed when the addition rate of calcium sulfate whiskers is at 1.0%, at which point the shrinkage rate of the slightly alkaline earth-based composites is reduced by 13%. However, it is important to highlight that beyond a 1.0% addition rate, the shrinkage rate of the mortar progressively increases with the increment of calcium sulfate whiskers. Once the addition surpasses 1.5%, the shrinkage rate not only exceeds that of the control group without calcium sulfate whiskers but also exacerbates the overall shrinkage phenomenon of the mortar.

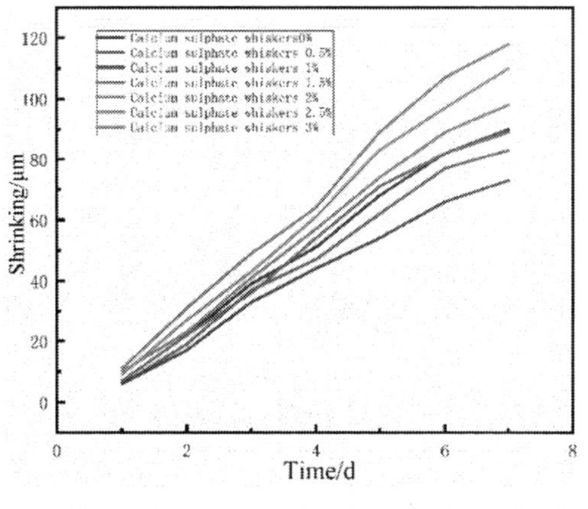

Figure 5. Effect of calcium sulfate whisker on shrinkage value of polymers at metakaolin base.

3.5 SEM analyze

Figure 6 shows the scanning electron microscope (SEM) images taken after 28 days of curing reveal a good connection between the calcium sulfate whisker particles and the hydration products. However, these hydration products predominantly accumulate on the surface of the particles, with the edges of most metakaolin (MK) particles remaining distinctly visible. This indicates that when the calcium sulfate whiskers content is low, the pH of the pore solution is insufficient to fully dissolve the MK particles, thereby affecting the compressive strength of the sample. As the content of calcium sulfate whiskers increases, the number of hydration products significantly rises, forming a more compact network gel structure, resulting in a denser microstructure of the cementitious material. In this context, no obvious spherical particles were observed. This is due to the increase in pH within the pore solution, which promotes the dissolution of particles and leads to the formation of more hydration products like C-(A)-S-H gel to fill the voids. However, with a further increase in the content of calcium sulfate whiskers, unreacted $CaSO_4$ particles begin to be observed.

The SEM images demonstrate that when the content of calcium sulfate whiskers is excessively high, the whiskers may form agglomerates or not dissolve completely, compromising the internal compactness of the paste and adversely affecting the workability and mechanical properties of the paste.

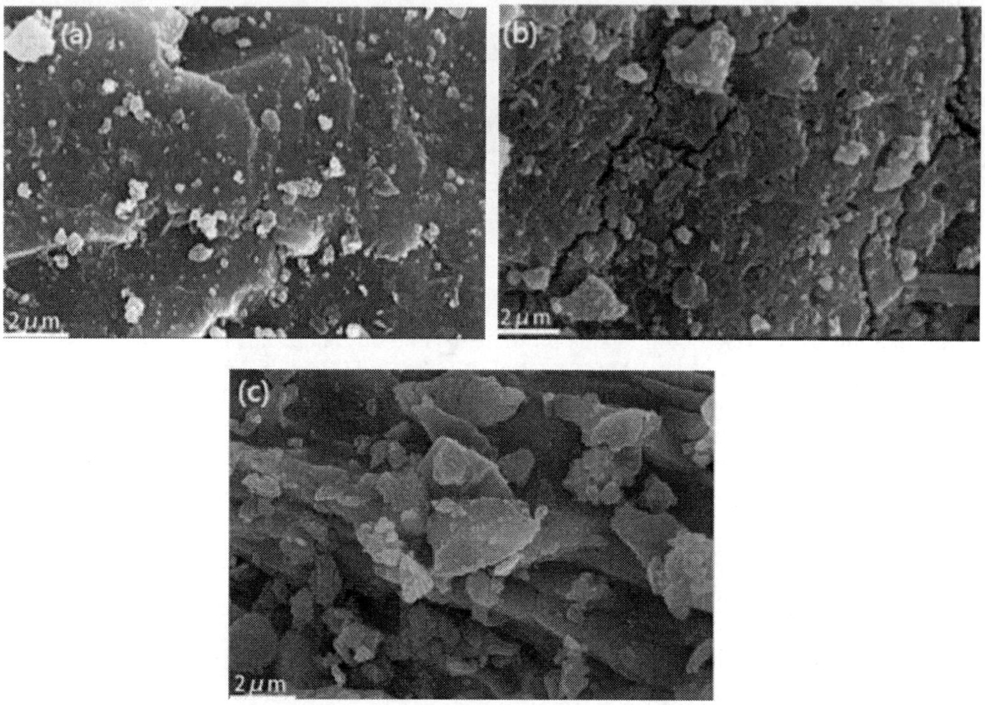

Figure 6. (a) SEM of calcium sulfate whiskers at 0% doping; (b) SEM of calcium sulfate whiskers at 1.5% doping; (c) SEM of calcium sulfate whiskers at 3% doping.

4. Conclusion

(1) Incorporating calcium sulfate whiskers into slightly alkaline earth-based composite mortars can significantly enhance their mechanical properties. However, this enhancement is limited: as the content of $CaSO_4$ whiskers increases, the bending and compressive strength of mortar increases first and then decreases. Experiments show that when the addition of calcium sulfate whiskers reaches 1.5% of the total amount of cementitious materials, both the mechanical and shrinkage properties of the composite mortar are optimized. Compared to the control group, the 28-day compressive and flexural strengths are increased by 5% and 9% respectively, while the shrinkage rate is reduced by 13%.

(2) Calcium sulfate whisker can not only improve the mechanical properties and shrinkage properties of metakaolin-base composite mortar, but also improve the construction performance of mortar, the setting time can be extended by up to 33%, and the fluidity can be increased by up to 5%.

(3) SEM analysis indicates that calcium sulfate whisker possesses a filler effect, enabling it to fill the cracks and pores present in the high-territory-based composite mortar. This contributes to an enhancement in the mortar's density, thereby leading to improvements in its mechanical properties and overall performance

Acknowledgments
This paper is one of the phased achievements of the National Natural Science Foundation Project (42177074).

Technology and Application Research of Urban Water Supply, Drainage and Gas Pipe Network Testing, Repair and Operation and Maintenance (JK2023201A).

References

[1] Singh K. Experimental study on metakaolin and baggashe ash based geopolymer concrete [J]. Materials Today: Proceedings, 2021, Vol.37(Part 2): 3289-3295.

[2] Lemougna P N, Adediran A, Yliniemi J, et al. Thermal stability of one-part metakaolin geopolymer composites containing high volume of spodumene tailings and glass wool [J]. Cement & Concrete Composites, 2020, Vol.114(No.0): 103792.

[3] Yuan Y, Zhao R, Li R, et al. Frost resistance of fiber-reinforced blended slag and Class F fly ash-based geopolymer concrete under the coupling effect of freeze-thaw cycling and axial compressive loading [J]. Construction & Building Materials, 2020, Vol.250(No.0): 118831.

[4] Albidah A, Alqarni A S, Abbas H, et al. Behavior of Metakaolin-Based geopolymer concrete at ambient and elevated temperatures [J]. Construction and Building Materials, 2022, 317.

[5] Luo Z, Zhang B, Zou J, et al. Sulfate erosion resistance of slag-fly ash based geopolymer stabilized soft soil under semi-immersion condition [J]. Case Studies in Construction Materials, 2022, 17.

[6] Dai S, Wang H, An S, et al. Mechanical Properties and Microstructural Characterization of Metakaolin Geopolymers Based on Orthogonal Tests [J]. Materials, 2022, Vol.15(No.8): 2957.

[7] Pillay D L, Olalusi O B, Kiliswa M W, et al. Engineering performance of metakaolin based concrete [J]. Cleaner Engineering and Technology, 2022, Vol.6: 100383.

[8] Al-Akhras N M. Durability of metakaolin concrete to sulfate attack [J]. Cement and Concrete Research, 2007, Vol.36(No.9): 1727-1734.

[9] Wan L, Pan R, Xu J. Mechanical Properties and Microstructure of CaSO4 Whisker Reinforced Cement Mortar (Article) [J]. Journal Wuhan University of Technology, Materials Science Edition, 2019, Vol.34(No.5): 1170-1176.

[10] Zhang C, Cao M. Mechanical properties of calcium carbonate whisker-reinforced high-strength cement mortar(Article) [J]. Journal of Southeast University (English Edition), 2012, Vol.28(No.3): 331-336.

[11] Acuña-Gutie´Rrez I O, Escobedo-Bocardo J C, Almanza-Robles J M, et al. Development of LiCl-containing calcium aluminate cement for bone repair and remodeling applications [J]. Materials Science and Engineering: C, 2017, Vol.70(Part 1): 357-363.

Journal of Physics: Conference Series **2808** (2024) 012039 doi:10.1088/1742-6596/2808/1/012039

Development of a large-area array multi-point synchronous explosive circuit using sheet explosive detonated by a mild detonating fuse

Zhiyun Guo[1], Qiang Lu[1,*], Yang Ding[1], Jin Li[1] and Liangyong Zhang[1]

[1]National Key Laboratory of Intense Pulsed Radiation Simulation and Effect, Northwest Institute of Nuclear Technology, Xi'an, Shaanxi, China

[*]Corresponding author's e-mail: luqiang@nint.ac.cn

Abstract- To meet the requirements of synchronous explosion loading of the large-area array in special experiments, a large-area array multi-point synchronous explosive circuit based on the thin sheet explosive detonated by a mild detonating fuse is designed. The explosive circuit is composed of a detonator, a mild detonating fuse, and a sheet explosive booster. In this study, the synchronization of initiation is analyzed theoretically, and the reliability test and synchronization test of the large-area array multi-point explosive circuit are carried out. The results show that the main factors affecting the initiation synchronicity include the length of the mild detonating fuse and the deviation of the detonating speed of the mild detonating fuse. The comprehensive initiation rate of the multi-point synchronous explosive circuit is 99.5%. The asynchrony ranges for the two-stage, three-stage, and four-stage synchronous explosive circuits are 0.32 μs, 0.37 μs, and 0.50 μs, respectively. This study provides a simpler and more efficient solution for achieving synchronous explosion loading of the large-area array.

1. Introduction

The multi-point synchronous explosive circuit is a detonation system that can realize single-point input and multi-point synchronous output. Based on the carriers involved, the explosive circuits can be categorized into rigid synchronous and flexible synchronous circuits [1-2]. The rigid synchronous explosive circuit utilizes a rigid substrate as the carrier, enabling multi-point synchronous output through the engraving of multiple grooves on the substrate and the subsequent generation of detonation waves following charging and explosion. The flexible synchronous explosive circuit consists of a mild detonating fuse and a beam-splitting transfer mechanism, achieving multi-point synchronous output via serial and parallel connections between multiple mild detonating fuses [3].

Shen et al. adopted a typical plate structure to design an eight-point ring synchronous explosive circuit for the warhead of a shaped charge [1]. The maximum deviation of synchronization output of a cylindrical charge with a detonation diameter of 50 mm was 210 ns. Wen et al. designed a rigid surface synchronous explosive circuit with "I"-shaped network channels connected and a square lattice structure as the output end [4]. The results showed that the two-stage explosive circuit with 16 output points can synchronously detonate cylindrical charges with a diameter of 80 mm, and the detonation synchronicity can be controlled within 120 ns. However, experimental verification has not been conducted. Xu et al. implemented a layout strategy where each vertex of the equilateral triangle served as a detonation point. They devised a 27-point synchronous explosive circuit using this equilateral triangle as the fundamental unit [5]. Experimental results demonstrated that the output synchronization time for 27 points, spanning approximately 180 mm of detonation transmission line, was less than 0.6 μs. Shi et al. designed a rigid multi-point synchronous explosive circuit with "one input and eight outputs" for shaped charges [6]. The results showed that the synchronization error of eight-point output with a detonation diameter of 75 mm was less than 100 ns. Wu et al. designed a flexible multi-point synchronous detonation device with "one input and eight outputs" for the detonation requirements of the warhead by using a cluster plate to constrain 8 silver detonating fuses at the input end [7]. The detonating fuse transmission length was approximately 200 mm, and the synchronization range for the 8-point output was as low as 0.6 μs. To meet the requirement of synchronous detonation on a curved surface, Hu et al. devised a flexible and rigid multi-point synchronous explosive circuit with "one input and 12 outputs" [8]. The results demonstrated that the synchronization standard deviation of 12-point detonations, using a detonation cable length of 450 mm, was 0.6 μs. Guo et al. achieved the miniaturization of flexible synchronous explosive circuits by optimizing the input, output, and conversion joints [9]. The designed "one input and two outputs" and "one input and four outputs" multi-point synchronous explosive circuits have synchronization ranges of 144 ns and 280 ns, respectively. Zhang et al. proposed a method to reduce the size of the bundle converter by directly threading the mild detonating fuse through the converter [10]. The prototype of the "one input and two outputs" and "one input and four outputs" flexible line synchronous explosive circuits has synchronization deviations of 34 ns and 73 ns, respectively, for the shock wave output. Zhang et al. developed a flexible linear six-point synchronous detonation device for achieving linear multi-point synchronous detonation and multi-explosion-shaped projectile warheads [11]. The results suggested that the detonation time deviation of this synchronous detonation device was less than 1.9 μs.

In summary, the current research on multi-point synchronous explosive circuits is mostly aimed at using detonation methods for penetrating the weapon-shaped charge to enhance the penetration-hitting effect. The detonation points are typically limited to single digits and mainly employed for synchronously detonating cylindrical charges with a diameter not exceeding 200 mm. However, there is currently no publicly available literature reporting on large-scale synchronous explosive circuits with more than 100 detonation points.

In order to meet the requirements of large-scale synchronous explosion loading in a specific experiment, in this study, a comprehensive multi-point synchronous explosive circuit based on PETN-based thin sheet explosive with mild detonating fuses is designed. Additionally, the detonation reliability verification experiment, initiation synchronicity theoretical analysis, and experimental tests are conducted.

2. Design of multi-point synchronous explosive circuit

2.1. Design principle

Figure 1 illustrates a two-stage synchronous explosive circuit. The first-stage mild detonating fuse is initiated by a detonator, while the second-stage mild detonating fuse is initiated by a thin sheet explosive booster.

In theory, one detonator can simultaneously detonate multiple Φ1 mm lead-covered mild detonating fuses. Each mild detonating fuse can initiate one thin sheet explosive booster, and each booster can initiate multiple mild detonating fuses. Through several stages of successive amplification, multiple-point initiation can be achieved. The prerequisite for achieving synchronous detonation is the uniformity of detonation time from the detonator to each detonation point, which implies that the length of the flexible detonation cable at each point should be identical. By changing the length of the mild detonating fuse and the shaping of the thin sheet explosive, explosive loading with a large array, specific spatial configuration, and specific timing can be achieved efficiently.

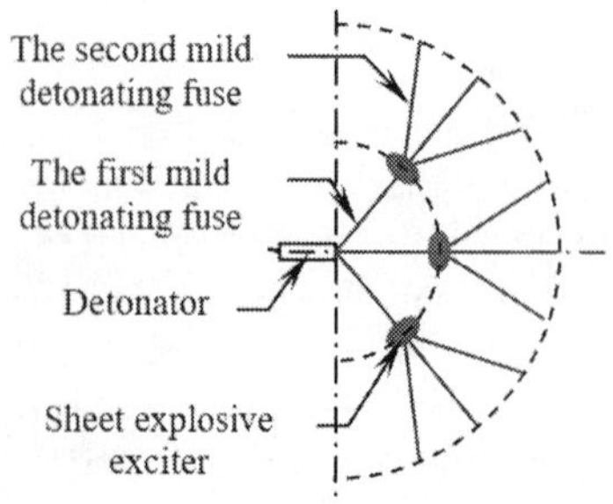

Figure 1. Schematic diagram of the multi-stage explosive circuit.

2.2. Structural design of the thin sheet explosive booster

In the multi-level synchronous explosive circuit illustrated in Figure 1, the thin sheet explosive booster, which functions with a "one input, multiple outputs" capability, plays a pivotal role in bridging various components and ultimately determines the success of the entire explosive circuit. For ease of installation, a "one input, nine outputs" thin sheet explosive booster has been designed. The schematic diagram of the design principle is shown in Figure 2. The input end consists of one mild detonating fuse, which initiates a thin sheet explosive with a diameter of 4 mm and a thickness of 0.4 mm at the center. The output end consists of nine mild detonating fuses, each with its center located on a concentric circle with a diameter of 3 mm, all initiated by the thin sheet explosive. Each output end is at an equal distance from the input end, ensuring the time synchronicity of the thin sheet explosive booster.

The self-developed thin sheet explosive uses PETN as the main explosive. Comprising 90% PETN, it boasts a thickness of 0.4 mm and achieves a detonation velocity of 7.44±0.03 km/s. It can be reliably initiated by lead-covered mild detonating fuses with a charge weight of 0.2 g/m and a diameter of 1 mm.

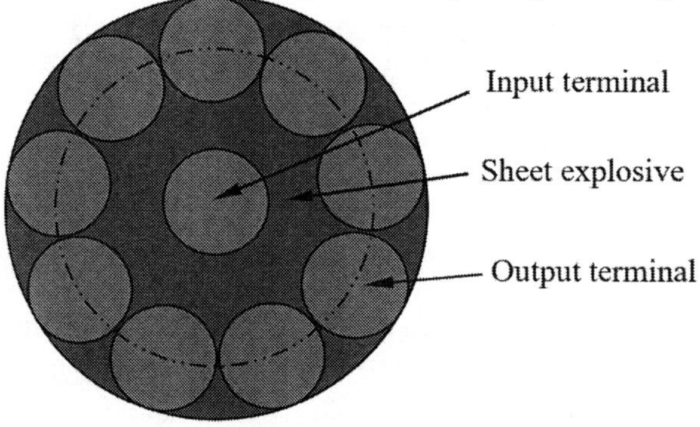

Figure 2. Schematic diagram of the thin sheet explosive booster.

Figure 2 illustrates that the success of the thin sheet explosive booster depends on two factors: firstly, the ability of the mild detonating fuse to initiate the thin sheet explosive (the input end), and secondly, the ability of the thin sheet explosive to initiate the mild detonating fuse (the output end). The former factor has been confirmed during the thin sheet explosive development phase. The subsequent focus will be directed towards conducting verification experiments for the latter aspect.

The layout of the verification experiment is shown in Figure 3, with thin sheet explosives attached to an aluminum plate. The explosives in the upper and lower rows have thicknesses of 0.50 mm and 0.33 mm, respectively, both initiated by Φ1 mm lead-covered mild detonating fuses with a charge weight of 0.3 g/m. The initiated mild detonating fuses are pressed directly against the surface of the explosive with organic glass plates. Eight Φ1 mm mild detonating fuses are arranged, with four of them having a charge weight of 0.2 g/m and the remaining four having a charge weight of 0.3 g/m. The detonation condition of the mild detonating fuse can be ascertained by observing the explosion marks on the surface of the aluminum plate. The blast marks indicate that all eight mild detonating fuses are initiated, demonstrating the feasibility of the designed thin-sheet explosive booster. Figure 4 shows a physical representation of the designed "one input, nine outputs" thin sheet explosive booster.

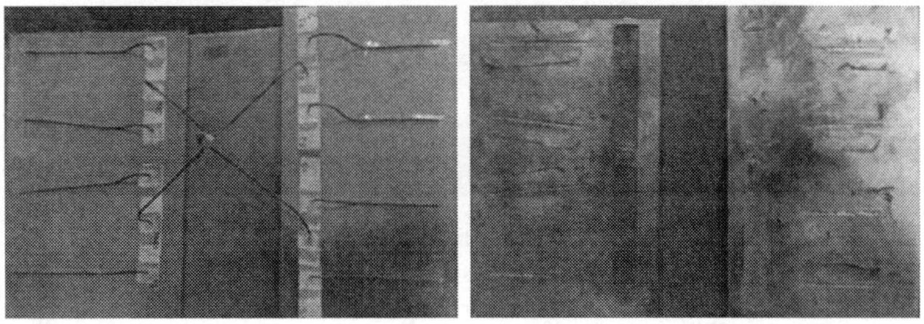

Figure 3. Verification experiment of the thin sheet explosive initiating the mild detonating fuse.

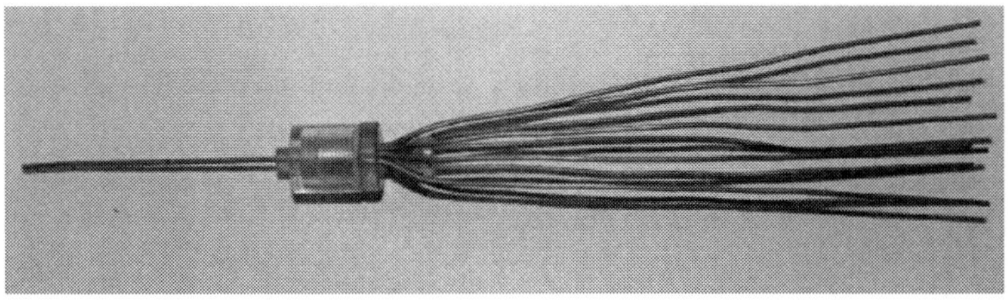

Figure 4. Physical representation of the thin sheet explosive booster.

3. Verification experiments on initiation reliability

To achieve synchronous detonation in a large array, more than 100 initiation points are arranged. The mild detonating fuses at each level may interact with each other, potentially leading to the failure of some initiation points to detonate. Therefore, three rounds of verification experiments on detonation reliability are conducted, progressively increasing in difficulty. In these experiments, the thin sheet explosives are all entire pieces attached to the surface of aluminum plates, being initiated synchronously by the mild detonating fuses arranged, as shown in Figures 5 to 7.

Figure 5 shows the experimental setup for the "one input, nine outputs" two-stage explosive circuit, with a total of nine initiation points (1 to 9). The post-detonation blast scar indicates that all 9 points are initiated, and the initiation rate is 100%.

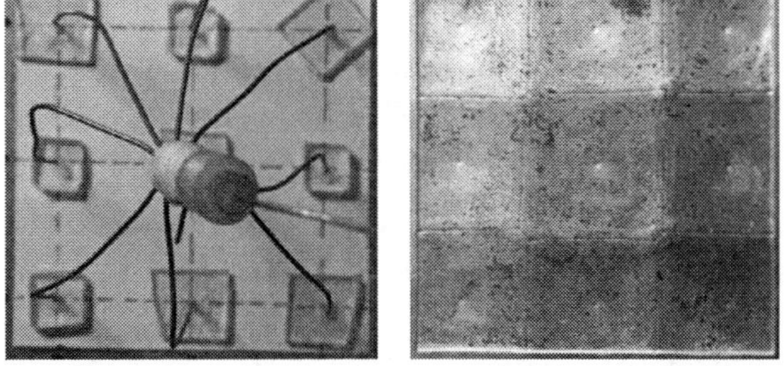

Figure 5. Experiment on the two-stage explosive circuit.

Figure 6 shows the experimental setup for the "one input, 25 outputs" three-stage explosive circuit, with a total of 25 initiation points (from 1 to 4, then to 25). The post-detonation blast scar shows that all 25 points are initiated, and the initiation rate is 100%.

Journal of Physics: Conference Series **2808** (2024) 012039 doi:10.1088/1742-6596/2808/1/012039

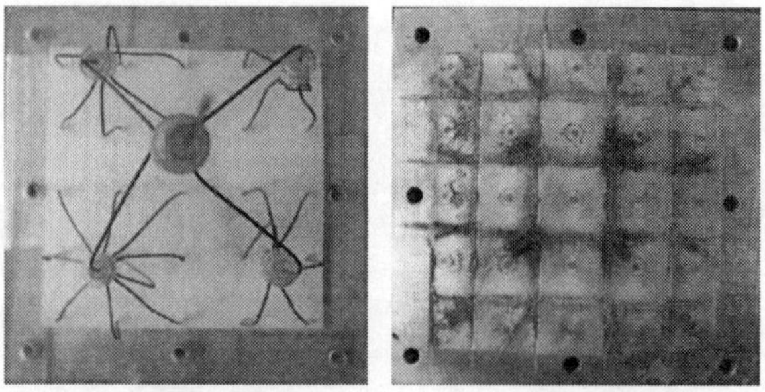

Figure 6. Experiment on the three-stage explosive circuit.

Figure 7 shows the experimental setup for the "one input, 144 outputs" four-stage explosive circuit, with a total of 144 initiation points (from 1 to 4, then to 16, and then to 144). The blast scar after detonation illustrates that one point is not initiated (as shown in the red dot on the right), and 143 points are successfully initiated, with an initiation rate higher than 99.3%.

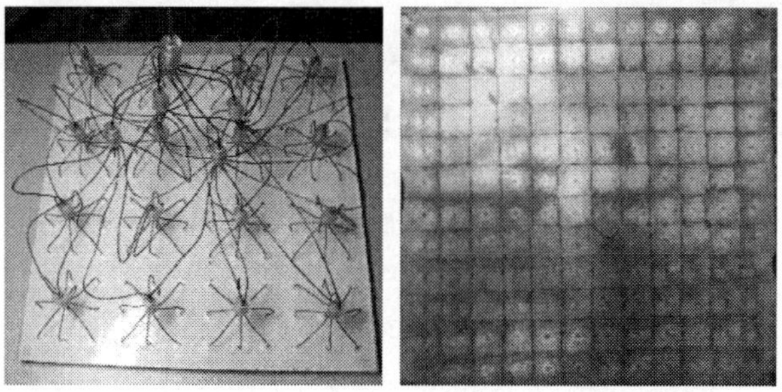

Figure 7. Experiment on the four-stage explosive circuit.

4. Analysis and testing of initiation synchronicity

4.1. Theoretical analysis

In a multi-point synchronous explosive circuit, as the loading configuration becomes more complex and the number of initiation points increases, it necessitates additional levels of the explosive circuit and longer propagation length of mild detonating fuses, and results in a more noticeable lack of synchronicity in initiation. The main factors affecting initiation synchronicity include deviations in mild detonating fuse lengths, instability in the detonation velocity of mild detonating fuses, thin sheet explosive boosters, and deviations in the spacing between initiation points. When laying a mild detonating fuse, its large bending radius and arc bending enable the disregard of any inconsistencies in detonation velocity resulting from the bending [12].

For a multi-stage explosive circuit, the propagation time of detonation from the initiation end to the terminal of initiation is:

$$t = (n-1)\frac{r}{D_s} + \sum_{i=1}^{n}\frac{l_i}{D_r} \tag{1}$$

According to Equation (1), the detonation time asynchrony range of each detonation point is [13-14]:

$$\Delta t = (n-1)\left(\frac{\Delta r}{D_s} + \frac{r\Delta D_s}{D_s^2}\right) + (n-1)\frac{\Delta l}{D_r} + \frac{L\Delta D_r}{D_r^2} \tag{2}$$

where n is the number of stages in the explosive circuit; r represents the propagation distance from the input end to the output end of the thin sheet explosive booster; $\triangle r$ denotes the deviation in propagation distance; l_i stands for the length of the mild detonating fuse in the i-th stage; $\triangle l$ is the deviation in length of the mild detonating fuse; L is the sum of the lengths of the mild detonating fuses in each stage; D_r denotes the detonation velocity of the mild detonating fuse; $\triangle D_r$ represents the deviation in detonation velocity of the mild detonating fuse; D_s is the detonation velocity of the thin sheet explosive; $\triangle D_s$ stands for the deviation in detonation velocity of the thin sheet explosive. By substituting the above parameters into Equation (2), the maximum asynchrony initiation time for each initiation point in the initiation network can be calculated.

r, $\triangle r$, and $\triangle l$ are set to 3 mm, 0.3 mm, and 1 mm, respectively. For the two-stage, three-stage, and four-stage explosive circuits, the values of L are 120 mm, 240 mm, and 400 mm, respectively. D_s, $\triangle D_s$, D_r, and $\triangle D_r$ are set to 7.44 km/s, 0.06 km/s, 6.70 km/s, and 0.06 km/s, respectively. Upon calculation, the maximum asynchrony ranges in initiation time for each initiation point in the two-stage, three-stage, and four-stage synchronous explosive circuits are 0.35 μs, 0.71 μs, and 1.11 μs, respectively.

Equation (2) shows that the total length of the mild detonating fuses (L) and the deviation in detonation velocity of the mild detonating fuses (ΔD_r) are key factors affecting the synchronicity of the explosive circuit. Among them, L is determined by objective factors such as the shape and size of the explosive loading, so choosing mild detonating fuses with better detonation velocity consistency is the main measure to improve initiation synchronicity.

4.2. Experimental testing

When measuring synchronicity, all factors are taken into consideration. The assessment involves measuring the maximum time differences between all initiation points, which represent the maximum asynchrony in initiation time. This measurement is crucial in evaluating the synchronicity performance of the explosive circuit. For the three experiments shown in Figures 5 to 7, the simple and convenient blasting mark method is used to calculate the initiation synchronicity. The blast mark method is derived from the explosive detonation velocity measurement method based on the detonation fuse, and it has been widely used. The schematic diagram of the principle is shown in Figure 8.

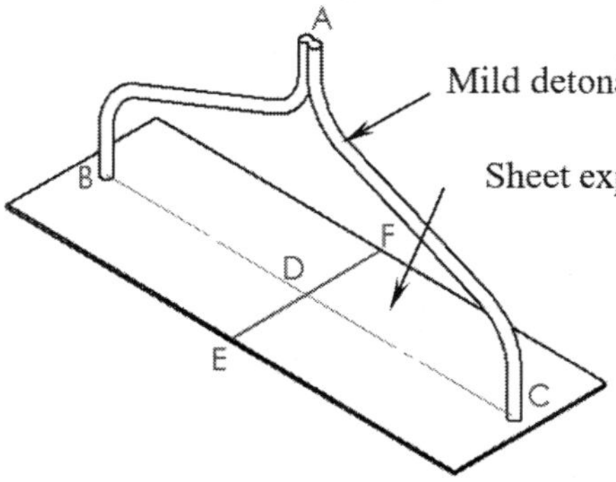

Figure 8. Principle schematic diagram of the blast mark comparison method.

The two mild detonating fuses are aligned at Point A, with their other ends directly pressed against the thin sheet explosive, which is attached to the effect aluminum plate. After the mild detonating fuses are simultaneously initiated, they reach Points B and C through different paths, initiating the adjacent thin sheet explosives. The shock waves from both ends meet at Line EF, forming a distinct mark on the aluminum plate. Moreover, Points B and C also leave circular marks. The initiation time difference between Points B and C can be expressed as:

$$\Delta t_{BC} = (L_{BD} - L_{CD}) / D_s \tag{3}$$

In Equation (3), Ds is the detonation velocity of the thin sheet explosive, and L_{BD} and L_{CD} are the distances between Points B and D, and Points C and D, respectively.

For the experiments shown in Figures 5 to 7, when the entire thin sheet explosive is initiated by the multi-point synchronous explosive circuit, circular marks will form at the initiation points on the effect plate. The initiation will generate stress waves that spread outward on the effect plate. When these stress waves intersect, stress will accumulate, forming a straight line on the effect plate. If the initiation times of the initiation points are perfectly synchronized, square grid marks will form on the effect plate. Due to the deviation in the initiation times of each initiation point, the marks of each square grid on the effect plate are different. By measuring the maximum deviation in distance Δl_{max} between each pair of initiation points, the maximum asynchrony in initiation time Δt_{max} for each initiation point in the initiation network can be obtained as:

$$\Delta t_{max} = \Delta l_{max} / D_s \tag{4}$$

Using the above method, the maximum asynchrony initiation time for the explosive circuit shown in Figures 5 to 7 can be calculated to be approximately 0.32 μs, 0.37 μs, and 0.50 μs respectively, all of which are smaller than the theoretical analysis results.

Table 1 summarizes the results of the three experiments shown in Figures 5 to 7.

Table 1. Summary of the Three Experiments.

Experiment No.	Number of Initiation Points Deployed	Number of Successful Initiations	Initiation Rate	Maximum Asynchrony in Initiation Time (μs)
1	9	9	100%	0.32
2	25	25	100%	0.37
3	144	143	99.3%	0.50
Total	199	198	99.5%	

Table 1 shows that the initiation rate of the three multi-point synchronous explosive circuit experiments exceeds 99%, with a total of 199 initiation points deployed and 198 successful initiations. Only one point fails to initiate, resulting in a comprehensive initiation rate of 99.5%. This indicates that the large array multi-point synchronous explosive circuit based on the sheet explosive detonated by the mild detonating fuse has high reliability. As the number of initiation levels and initiation points increases, the initiation time asynchrony also escalates. For the four-stage synchronous explosive circuit with 144 initiation points, the asynchrony range is 0.50 μs. Compared to other related research results, the synchronous explosive circuit designed in this paper has better synchronicity.

5. Conclusions

In the present work, a large-area array multi-point synchronous explosive circuit based on the sheet explosive detonated by the mild detonating fuse is proposed. In this study, verification experiments on initiation reliability, theoretical analysis, and experimental testing on initiation synchronicity are performed. The main conclusions are as follows:

(1) The thin sheet explosive can initiate mild detonating fuses with charge weights of 0.2 g/m and 0.3 g/m, and the thin sheet explosive booster developed based on this principle plays a crucial role in the explosive circuit.

(2) The initiation rates of the two-stage, three-stage, and four-stage synchronous explosive circuits all exceed 99%, with a comprehensive initiation rate of 99.5%, indicating that the explosive circuit has high reliability.

(3) Theoretical analysis results indicate that the length of mild detonating fuse propagation and the deviation in detonation velocity of the mild detonating fuse are key factors affecting initiation synchronicity. Choosing mild detonating fuses with better detonation velocity consistency is the main measure to improve initiation synchronicity.

(4) The test results show that the maximum asynchrony ranges in initiation time for the two-stage, three-stage, and four-stage synchronous explosive circuits are 0.32 μs, 0.37 μs, and 0.50 μs, respectively. These values are all lower than the theoretical analysis results.

(5) By changing the length of the mild detonating fuse and the shaping of the thin sheet explosive, explosive loading with a large array, specific spatial configuration, and specific timing can be achieved efficiently.

Journal of Physics: Conference Series **2808** (2024) 012039 doi:10.1088/1742-6596/2808/1/012039

References

[1] Shen, H.M., Li, W.B., Wang, X.M., et al. (2016) Design and synchronization analysis of structural parameter of a multi-point explosive circuit. Chinese Journal of Energetic Materials, 24(3): 238-243.
[2] Zheng, Y., Wang, X.M., Huang, Y.S., *et al.* (2008) Design and experimental investigation on multi-point synchronous explosive logic circuit. Initiators & Pyrotechnics, 1: 1-4.
[3] Ning, Z., Chen, K., Jian, H.T., *et al.* (2024) Review on Multi-point Initiation Technology and its Applications. CHINESE JOURNAL OF ENERGETIC MATERIALS, 32: 102-115.
[4] Wen, Y.Q., Jiao, Q.J., Li, G.X., *et al.* (2001) Structural design and time analysis of synchronous explosive circuit in a rigid surface. Acta Armamentarii, 1: 45-47.
[5] Xu, B.Y., Li, G.F., Yuan, D.P., *et al.* (2010) Analysis on the output performance of plane multi-point simultaneous detonation transfer circuit. Initiators & Pyrotechnics, 6: 5-9.
[6] Shi, C.C., Wang, B.G., Chen, T.F., et al. (2014) Design of multi-point synchronous initiation circuit for shaped charge. Initiators & Pyrotechnics, 6: 1-4.
[7] Wu, Y.P., Wei, H.N., Xue, R.H., et al. (2022) Design and experimental study of flexible multipoint synchronous initiation device. Initiators & Pyrotechnics, 3: 14-17.
[8] Hu, H.Q., Pei, M.J., Zhang, D.Z., et al. (2008) Design on multi-point simultaneous initiating circuit for a curved surface. Initiators & Pyrotechnics, 6: 5-8.
[9] Guo, H.W., Wen, Y.Q. (2010) Study on modularization design of flexible synchronous explosive circuit. Initiators & Pyrotechnics, 4: 5-9.
[10] Zhang, Z.W., Li, X.G., Wen, Y.Q., et al. (2014) Design and tests of a flexible linear synchronous explosive circuit. Chinese Journal of Energetic Materials, 22: 401-405.
[11] Zhang, X.H., Zhao, J.Q., Shi, J. (2021) Research on the technology of simultaneously detonating the MEFP warhead. Journal of Projectiles, Rockets, Missiles and Guidance, 41: 88-91.
[12] Wang, T., Liu, T., Jiang, L., et al. (2022) Research on Influencing Factors of Detonation Velocity Consistency of Small Diameter Flexible Detonating Cord. INITIATORS & PYROTECHNICS, 4: 27-30.
[13] Liang, A.D., Sun, X.Y., Li, G.J. (2019) Estimation of detonation output synchronization deviation on the linear synchronous explosion circuit. Journal of Ordnance Equipment Engineering, 40: 237-240.
[14] Bai, Y.W., Zhang, R., Li, Z. (2009) Design on mild multi-point synchronous explosive circuit. Chinese Journal of Energetic Materials, 17: 225-228.

Biaxial tension test of soft materials

Minxing Li[1], Ming Yang[2], Huiming Wang[1, 3] and Ronghao Bao[1, 3, 4*]

[1]Department of Engineering Mechanics, Zhejiang University, Hangzhou, Zhejiang, 310027, China
[2]Hangzhou Intellisense Technology Co., Ltd, Hangzhou, Zhejiang, 311107, China
[3]Key Laboratory of Soft Machines and Smart Devices of Zhejiang Province, Hangzhou, Zhejiang, 310027, China
[4]Center for Soft Machines and Smart Devices, Huanjiang Laboratory, Zhuji 311816, China

[*]Corresponding author's e-mail: brh@zju.edu.cn

Abstract. A specific loading machine is designed and fabricated to test the constitutive properties of soft materials by biaxial tension in this study. The finite element model under ABAQUS is launched to simulate the biaxial tension of a squared specimen, and it is verified by a theoretical model. Using the finite elements model, the effects of clamping conditions of the squared samples, including the clamping width, clamping depth, and clamping numbers, are systematically studied, which helps improve future machine design. The comparison of numerical results with experimental results is conducted by equibiaxial test and pure shear test. Finally, a novel testing procedure is suggested with the combination of numerical simulation and practical experiments to enhance the testing result.

1. Introduction

Due to their inherent flexibility, capacity for large deformation, and quick response to external stimuli like mechanical, electrical, and magnetic fields, soft materials are now widely utilized to fabricate flexible electronic devices, such as flexible sensors, actuators, and screens [1-2]. In the forthcoming intelligence era, soft materials are anticipated to play more important roles in various areas, such as medicine, healthcare, human-computer interaction, and energy and environment [3-4].

To fully realize the aforementioned functional designs in practical applications, it is very crucial to understand the basic mechanical properties of such materials [5]. However, unlike traditional hard materials, soft materials exhibit hyperelasticity and undergo significant deformation in most applications, where the relation between strain and stress is complicated. In the realms of scientific research and industrial design, hyperelastic constitutive models are frequently employed to characterize the mechanical properties of soft materials, specifically applied in the analysis of electromechanical systems in the design of flexible electronic devices. Additionally, they are instrumental in analyzing the deformation modes of devices under external loads in operational environments.

Uniaxial tensile experiments alone are insufficient to adequately determine the hyperelastic constitutive model of soft materials. More kinds of tests should be supplemented simultaneously to acquire the proper constitutive model, such as the uniaxial compressive test, equibiaxial tension test, pure shear test, and so on [6]. Currently, the equibiaxial tension test on planar specimens (such as cruciform specimens, squared specimens, and circular specimens) is the most feasible way to be

Journal of Physics: Conference Series **2808** (2024) 012040 doi:10.1088/1742-6596/2808/1/012040

conducted in mechanical laboratories, owing to its simplicity of tensile machine. For cruciform specimens, the efficient testing area (denoted as a nearly uniform distributed stress or strain state in this paper) is very limited during the test, even the geometry of the specimens has been optimized by numerous researchers [7-8], which will consequently affect the accuracy of final material parameters. Regarding to cruciform specimen, it can maximally achieve a uniformly distributed equibiaxial strain region of no more than 30% [9] due to the size restriction of the testing machine, which is still inadequate to fully capture the mechanical properties. Avanzini et al. [10] conducted biaxial tests using natural rubber with various geometric shapes and demonstrated that squared specimens could attain a wider efficient testing area in comparison with cruciform specimens, which means the more accurate constitutive model parameters. However, limited work has been done in the mechanical testing on squared specimens [11], so there remains a deficiency of commercial apparatus available for researchers to conduct such experiments in laboratories.

Regarding the biaxial testing on squared specimens, the working frames and clamping system are very crucial to the result, which should enable the clamps to move freely along the perpendicular sides to the load direction, resulting in the free expansion of specimens to achieve a wide efficient testing area [12]. The finite element method (FEM) is utilized to investigate the optimal procedures for conducting biaxial tensile tests on squared specimens. Fujikawa et al. [13] proposed to introduce the appropriate notches near the fixture area to obtain accurate nominal stress-strain relation through experiment. However, the effective cross-sectional area to calculate the nominal stress is not correctly suggested in their work, leading to a significant deviation from theoretical expectations for conventional specimens. Luo et al. [14] concentrated on the comparative analysis of two clamping methods: two-corner fixed stretching and single-corner fixed stretching by FEM, showing that the latter could improve the overall stress distribution effectively. However, the displacement boundary conditions in their model were imposed on the distributed lines rather than the clamps, neglecting the effects of the clamping system, which is very crucial to such experiments.

In this work, customized testing equipment is designed and fabricated to apply biaxial tension test on squared planar soft material specimens or standard dog bone specimens to acquire an accurate constitutive model with various tension conditions. A finite element model is built to simulate the biaxial tension test as well, which is verified by the theoretical model. Furthermore, the effects of the clamping system on the testing result are systematically investigated by the numerical model. The experimental results and numerical analysis both validate that the proposed experimental approach could yield highly accurate testing results. Finally, a calibrating procedure is suggested for practical experiments with our testing apparatus.

2. Experimental testing methods and corresponding finite element models

2.1. Experimental test methods

2.1.1 Biaxial tensile machine. To achieve better testing results, a biaxial tensile machine on soft materials, as shown in Figure 1, is developed in our laboratory at Zhejiang University with the cooperation of Hangzhou IntelliSense technology company (Model: ZY-BiFSM). Squared specimens (or standard dog-bone specimens) can be attached to the platform of the machine by four rows of clamps as illustrated in Figure 1(a), where the specimens can be stretched through the travel mechanism driven by four independent servo motors as shown in Figure 1(b), with an arbitrary tension ratio between the two loading directions. Owing to the specific design, the clamps attached to the four guide rails can be freely moved during the test, resulting in sufficient deformation of the whole specimen. As can be seen from Figure 1(a), even grid lines were marked on the surface of the specimen before the test. A wide effective deformation area could be obtained within our test since the grids progressed uniformly during the test. A series of load cells with appropriate loading capacity is installed between the clamps and sliding blocks on the guide rails to measure the loading force by our own developed data acquisition system as demonstrated in Figure 1(c). The mechanism is driven by the control algorithm embedded in

Journal of Physics: Conference Series **2808** (2024) 012040 doi:10.1088/1742-6596/2808/1/012040

the corresponding software, and all the loading and deformation data are recorded for further data analysis.

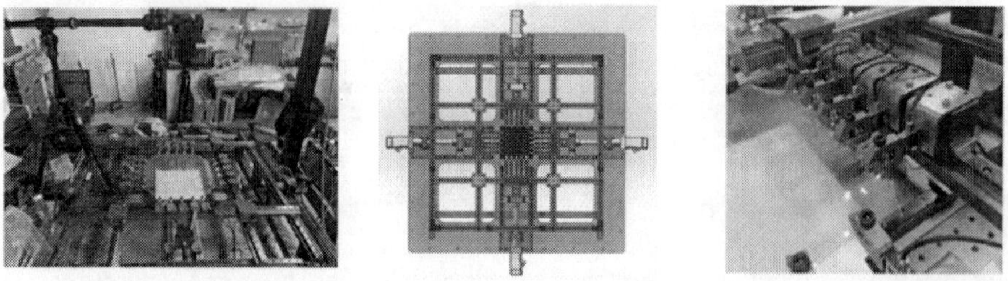

(a) (b) (c)

Figure 1. (a) Experimental setup on squared specimens; (b) Schematic diagram of the main structure of the biaxial tensile machine; (c) Details of the load cells and fixture clamping system.

2.1.2 *Experimental procedure and experimental data processing.* As mentioned above, when the grids were marked uniformly on the surface of the specimen, the grids around the central region were found to move outward evenly during the tension, which is sufficient to assume that the central part remained uniform strain state during the test. To acquire the accurate stretch ratio, four dark points are marked on the central area (30 mm × 30 mm) of the specimen (220 mm × 220 mm), as shown in Figure 2. A digital camera with high resolution is installed on the overhead center of the specimen to capture the traces of marked points. The distortion of the image is calibrated before the test. The digital images are recorded and utilized to calculate the stretch ratio at each frame. Then the nominal strain can be obtained according to the stretch ratio. Regarding the nominal stress, it can be calculated as $S = \sum_i F_i / (L_0 \times d_0)$, where L_0 represents the original length of the sample (220 mm), and d_0 denotes the initial thickness of the sample (1.0 mm) before the test, respectively. Here the summation means the overall of the loads along each side, and the load is averaged within four sides.

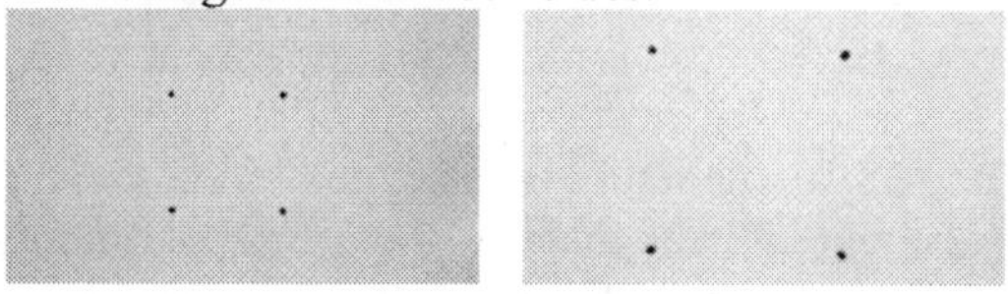

(a) (b)

Figure 2. (a) Image of the four dark points before the test; (b) Image of the four dark points when the recording time is 300 s, taken from the digital camera.

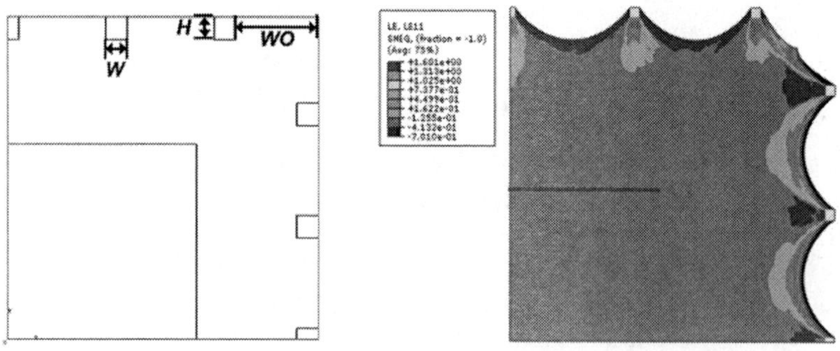

(a) (b)

Figure 3. (a) Schematics of finite element model in ABAQUS with a quarter of specimen. (b) Contour plot of strain distribution along y-direction by FEM.

2.2. Finite element model

Based on the experimental setup, a quarter of the specimen is modeled using ABAQUS for further parameter analysis as depicted in Figure 3(a), owing to the symmetry of the test. Appropriate symmetric boundary conditions are applied. The shell element, S4R, is adopted in the finite element model with sufficient mesh elements. The displacements of the clamps are applied as boundary conditions while the reaction forces are taken as the applied loads. The primary key parameters include W (the clamp width), H (the clamping depth), WO (the distance between the corner edge and the neighboring clamp), and N (the number of clamps at each side), respectively. With our primary analysis, the length (or width) L of the FEM is selected as 90 mm (which is a little different from the testing specimen), and the thickness is taken as 0.5 mm. N is in the range of (4,8), W is in the range of (7-17 mm), H is in the range of (4-10 mm), and WO is in the range of (10-28 mm), respectively. Since the specimen is fixed by the clamps tightly enough, the clamping areas are modeled as rigid bodies for simplicity. A classic Ogden model with 3rd order is adopted in FEM, with strain energy density function (SEDF) as

$$W(\lambda_1,\lambda_2,\lambda_3)=\sum_{i=1}^{3}\frac{\mu_i}{\alpha_i}\left(\lambda_1^{\alpha_i}+\lambda_2^{\alpha_i}+\lambda_3^{\alpha_i}-3\right) \tag{1}$$

where the parameters are [15]:

$$\mu_1=0.618\text{MPa},\alpha_1=1.3,\mu_2=1.18\times10^{-3}\text{MPa},\alpha_2=5.03,\mu_3=-9.81\times10^{-3}\text{MPa},\alpha_3=2.0\,.$$

Under equibiaxial loading conditions, the theoretical formula of nominal stress can be expressed as [13]:

$$S(\lambda)=\sum_{i=1}^{3}\mu_i\left(\lambda^{\alpha_i-1}-\lambda^{-2\alpha_i-1}\right) \tag{2}$$

where λ is the stretching ratio of either side during the test.

Within the FEM simulation, the stretch ratio and nominal stress are calculated as follows:

$$\lambda(t)=\frac{\bar{u}(t)+a_0}{a_0},\quad S(t)=RF_{total}(t)/(L_0\times d_0) \tag{3}$$

where a_0 is the original size of the observation area with nearly uniform deformation, taken as 40 mm; $\bar{u}(t)$ is the average displacement along the edge of the observation area (red line as shown in Figure 3b) at each frame. $RF_{total}(t)$ is the sum of the reaction forces on each side. The original length L_0 and thickness d_0 in the finite element model before tension are 90 mm and 0.5 mm, respectively. With the FEM result, the relation between nominal stress and nominal strain (stretch ratio minus 1) can be obtained and will be compared with the theoretical one presented by Equation (2). Systematic analysis of the parameters on the results will be further conducted.

3. Results and discussion

3.1. Finite element result analysis

3.1.1. Effects of key parameters. Figure 4(a-d) illustrates the influences of WO, W, H, and N on the nominal stress-nominal strain relation obtained by Equation (3), respectively. All the results show that the FEM results fit very well with the theoretical one when the nominal strain is less than 1.4, demonstrating the validity of the FEM model of the equibiaxial test on squared specimens. It can also be seen from the figures that WO and H exhibit relatively limited effects on the nominal stress-nominal strain curve, while W and N have notable influences when the nominal strain is greater than 1.5. It can be inferred that the nominal stress-strain curve is mainly related to W and N. An increase (decrease) in N within the studying range elevates (lowers) the nominal stress-nominal strain curve, and then W needs to be reduced within the studying range to align well with the theoretical one. Regarding experiments, W and N need to be optimized for good results, which corresponds to an optimal specimen size for test preparation.

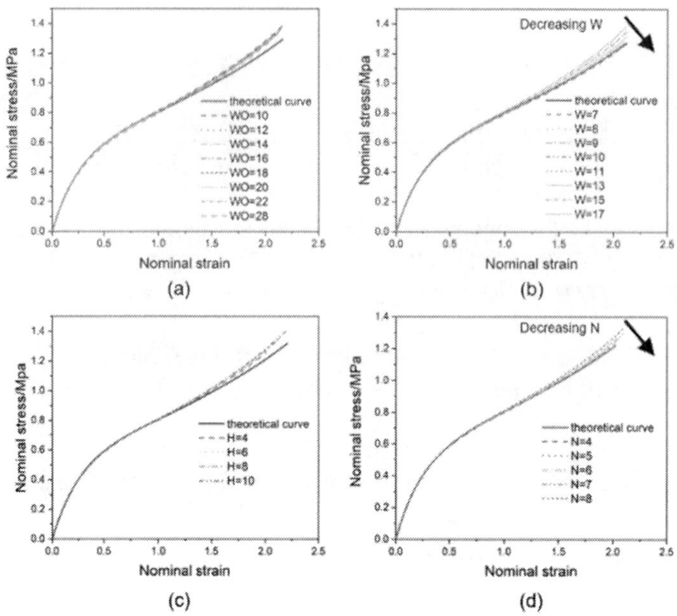

Figure 4. The influence of geometric parameters on the nominal stress-strain curve, *WO*(a), *W*(b), *H*(c), and *N*(d).

3.1.2. Experimental methods based on finite element results. Based on the discussion presented in 3.1.1, when the clamping number N in the experiment is fixed as 5, W will emerge as the primary factor for the result. The constitutive model of soft material is not available before the test. Therefore, the optimal W (corresponding to specimen size) could not be rightly determined by FEM. To achieve a better constitutive model by experiment, it is significant to select an appropriate specimen size considering that W is almost fixed when the testing apparatus is finished. Therefore, we propose a novel experimental testing procedure with the cooperation of FEM simulation, which is schematically illustrated in Figure 5. When the theoretical curves in 3.1.1 were replaced by those obtained through corresponding experiments, the optimal specimen size can be gradually improved to make the FEM one fit well with the experimental one, so that the constitutive model obtained by the experiment can be used by another simulation model for further analysis.

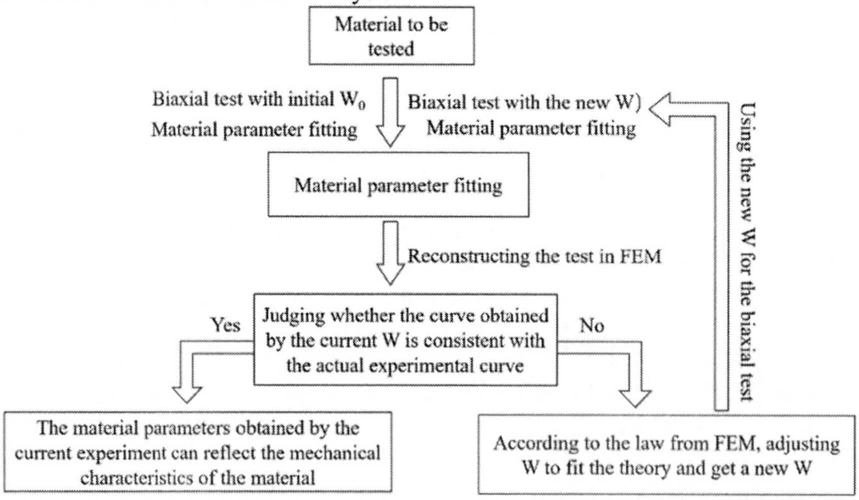

Figure 5. Proposed testing procedure with the aid of FEM.

3.2. Experimental results

3.2.1. Equibiaxial tensile results. Renowned for their remarkable flexibility and adaptability, silica gels are leading materials in contemporary scientific research across diverse fields. Particularly in flexible electronics, silica gels stand out as a pivotal material, offering unmatched versatility in crafting flexible electronic devices. Its distinctive properties render it ideal for a wide array of applications, spanning from wearable sensors to stretchable circuits. Hence, this study opts for silica gel as a prime example of functional soft materials for conducting mechanical property testing experiments. Squared specimens made of FC2211 silica gel (size 220 mm × 220 mm × 1 mm) were prepared for the equibiaxial test by our testing machine presented in 2.1.1. N, W, H, and WO were set as 5 mm, 8 mm, 8 mm, and 30 mm, respectively. The maximal tensile was 180 mm, performed evenly within 300 s, resulting in a stretch ratio of 1.818. Meanwhile, a series of uniaxial tensile tests on the same materials with standard dog-bone specimens were conducted both on an Instron universal testing machine and our testing machine. The results show good agreement of these two apparatus. By combining the equibiaxial and uniaxial experimental data, the following SEDF with 3rd reduced polynomial can be obtained with curve fitting ($R^2 = 0.99589$)

$$W(\lambda_1,\lambda_2,\lambda_3)=\sum_{i=1}^{3}C_{i0}\left(\lambda_1^{\,2}+\lambda_2^{\,2}+\lambda_3^{\,2}-3\right)^i \tag{4}$$

where C_{10}=0.18942, C_{20}=-0.026151, C_{30}=0.0066057, respectively. The fitted curves are plotted in Figure 6(a) in solid lines. A new finite element model in ABAQUS was then built with the constitutive parameters identified by the above experiments as well as the identical clamping conditions with the practical experimental test. Figure 6(b) plots the curves between nominal stain and nominal stress, for FEM and experiment results, respectively. It can be affirmed that the finite element model agrees very well with the experimental one. Therefore, it can be further utilized to improve the experimental test result with the procedure presented in 3.1.2.

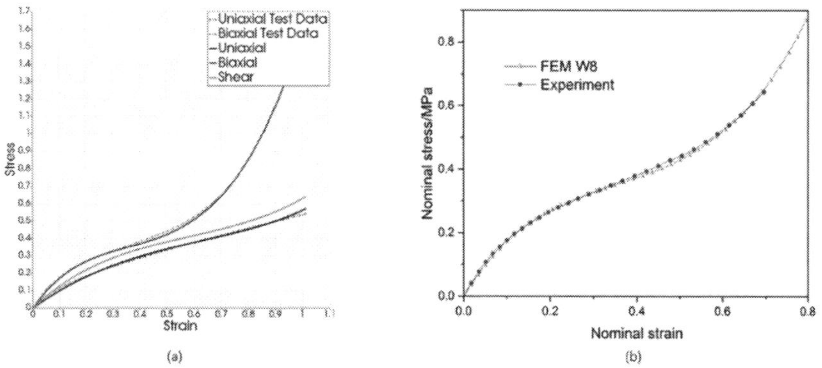

Figure 6. (a) Curve fitting by uniaxial and biaxial experimental data, (b) Comparison of FEM and experimental results under equibiaxial tension.

3.2.2. Validation by pure shear tests. To further validate the finite element model, a pure shear test [13] was taken by the simulation and experiment simultaneously, with identical test conditions. The specimens were fixed on two opponent sides without any motion, while the other two opponent sides were stretched gradually by a given displacement. The stretch ratio curves were obtained for both directions. Figure 7 depicts the changing of forces along two directions (x represents the tensile direction, and y represents the fixed direction) with the stretch ratio, showing the validity of constitutive parameter identification and the finite element model as well.

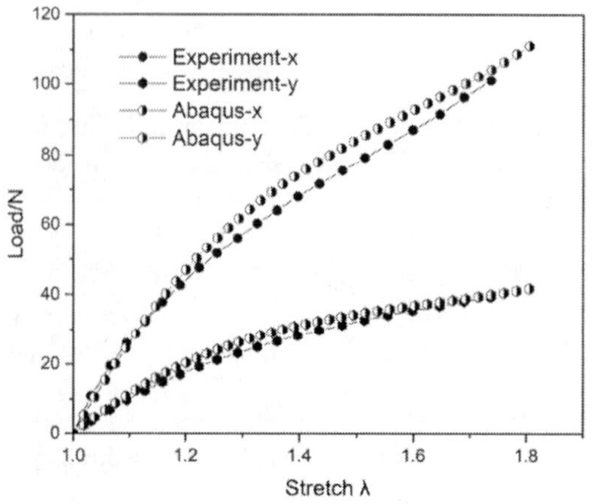

Figure 7. Comparison of finite element results and experimental results of pure shear test.

4. Conclusions

Biaxial tension tests on soft materials are conducted in this paper with FEM and our own developed experimental machine. Based on the FEM, the effects of the clamping system on the testing result are systematically explored, and a novel testing procedure to combine the FEM with the experimental results is presented. The following conclusions are reached:

1) The clamping width and clamp number are the key parameters to affect the testing results.
2) The finite element model can predict the constitutive model well with the theoretical one.
3) The finite element model agrees well with experiment results, and our testing machine can produce a wide efficient deformation area.
4) A novel testing procedure to combine the finite element model and experimental testing is presented, which can be utilized to determine an optimal specimen size during practical tests with specified materials.

Acknowledgments

This work is financially supported by the National Natural Science Foundation of China (12072315), funded by the specialized research projects of Huanjiang Laboratory, Zhuji, Zhejiang Province as well.

References

[1] Pilz da Cunha, M.; Ambergen, S.; Debije, M. G.; Homburg, E. F. G. A.; den Toonder, J. M. J.; Schenning, A. P. H. J. (2020) A soft transporter robot fueled by light. Adv. Sci., 7 (5), 1902842.
[2] Ren, Z.; Hu, W.; Dong, X.; Sitti, M. (2019) Multi-functional soft-bodied jellyfish-like swimming. Nat. Commun. 10 (1), 2703.
[3] Kight, A.; Pirozzi, I.; Liang, X.; McElhinney, D.B.; Han, A.K.; Dual, S.A.; Cutkosky, M. (2023) Decoupling transmission and transduction for improved durability of highly stretchable, soft strain sensing: applications in human health monitoring. Sensors, 23, 1955.
[4] Yin, R.; Wang, D.; Zhao, S.; Lou, Z.; Shen, G., (2021) Wearable sensors-enabled human-machine interaction systems: from design to application. Adv. Funct. Mater. 31 (11), 2008936.
[5] Collins, I.; Hossain, M.; Dettmer, W.; Masters, I., (2021) Flexible membrane structures for wave energy harvesting: a review of the developments, materials, and computational modeling approaches. Renewable Sustainable Energy Rev. 151, 111478.
[6] Sasso, M.; Palmieri, G.; Chiappini, G.; Amodio, D., (2008) Characterization of hyperelastic rubber-like materials by biaxial and uniaxial stretching tests based on optical methods. Polym. Test. 27 (8), 995-1004.

[7] Palacios-Pineda, L.M.; Perales-Martínez, I.A.; Moreno-Guerra, M.R.; Elías-Zúñiga, A. (2017) An Optimum Specimen Geometry for Equibiaxial Experimental Tests of Reinforced Magnetorheological Elastomers with Iron Micro- and Nanoparticles. Nanomaterials, 7, 254.
[8] Putra, K. B.; Tian, X.; Plott, J.; Shih, A., (2020) Biaxial test and hyperelastic material models of silicone elastomer fabricated by extrusion-based additive manufacturing for wearable biomedical devices. J. Mech. Behav. Biomed. Mater. 107, 103733.
[9] Keerthiwansa, R.; Javorik, J.; Kledrowetz, J.; Nekoksa, P., (2018) Elastomer testing: The risk of using only uniaxial data for fitting the Mooney-Rivlin hyperelastic-material model. Mater. Tehnol, 52, 3-8.
[10] Avanzini, A.; Battini, D., (2016) Integrated experimental and numerical comparison of different approaches for planar biaxial testing of a hyperelastic material. Adv. Mater. Sci. Eng, 6014129.
[11] Esmaeili, A.; George, D.; Masters, I.; Hossain, M., (2023) Biaxial experimental characterizations of soft polymers: A review. Polym. Test. 128, 108246.
[12] Jiang, M.; Sridhar, R. L.; Robbins, A. B.; Freed, A. D.; Moreno, M. R., (2021) A versatile biaxial testing platform for soft tissues. J. Mech. Behav. Biomed. Mater. 114, 104144.
[13] Fujikawa, M.; Maeda, N.; Yamabe, J.; Kodama, Y.; Koishi, M., (2014) Determining stress-strain in rubber with the in-plane biaxial tensile tester. Exp. Mech. 54 (9), 1639-1649.
[14] Luo, H.; Zhu, Y.; Zhao, H.; Ma, L.; Zhang, J. (2023) Simulation analysis of equibiaxial tension tests for rubber-like materials. Polymers, 15, 3561.
[15]Holzapfel, G. A., (2002) Nonlinear solid mechanics: A continuum approach for engineering science. 37 (4), 489-490.

Journal of Physics: Conference Series **2808** (2024) 012041 doi:10.1088/1742-6596/2808/1/012041

DLP 3D printing of high-resolution, temperature-responsive hydrogel microneedles with rapid customization capability

Xinmeng Zhou[1], Huan Liu[1] and Huanjun Li[1,*]

[1]Key Laboratory of Medical Molecule Science and Pharmaceutics Engineering of Ministry of Industry and Information Technology, School of Chemistry and Chemical Engineering, Beijing Institute of Technology, Beijing 100081, China

[*]Corresponding author's e-mail: lihj@bit.edu.cn

Abstract: Hydrogel microneedles (MNs) are emerging as versatile transdermal drug delivery platforms for the treatment of various diseases. However, their conventional preparation method is plagued by a complex process, time-consuming nature, and low resolution. To address these challenges, we present a novel method for preparing temperature-responsive hydrogel MNs using a Digital Light Processing (DLP) strategy. The printed NIPAM-AA-AM hydrogel MNs can be rapidly customized (within 1 hour) into various shapes and heights with high resolution. The mechanical tests and H&E staining validate the ability of the MNs to penetrate the skin. In addition, the temperature-responsive drug delivery of the MNs was demonstrated through experiments and COSMOL simulations. Notably, these biocompatible microneedles showed non-toxicity in NIH-3T3 cells and low hemolysis rates, thereby indicating promising prospects for safe in vivo utilization. This study offers a simple and rapid alternative for the fabrication of smart responsive microneedles, opening up new avenues in personalized and efficient transdermal drug delivery strategies.

1. Introduction

Transdermal drug delivery, after more than three decades of development, ranks third among drug delivery methods and exhibits potential for further expansion. However, the skin's stratum corneum presents a challenge, particularly for macromolecular drugs aiming to penetrate the body through transdermal routes. Microneedles (MNs), resembling miniature needle-like structures, provide a promising solution by piercing the skin's stratum corneum, thereby overcoming the limitations of traditional transdermal drug delivery, which relies on passive diffusion [1~3]. With an increasing demand across various applications and advancements in MN preparation technology, the main types of MNs include solid, encapsulated, dissolved, hollow, and hydrogel variants [4]. Among these, hydrogel MNs exhibit biocompatibility, biodegradability, electrical conductivity, mechanical properties, stress responsiveness, and dissolution characteristics, contributing to their growing popularity [5~8]. Meanwhile, the landscape of polymer MN preparation methods is evolving rapidly. Hydrogel MNs are predominantly prepared through techniques such as microplastic molding, three-dimensional stretching lithography, droplet blowing, and electrical pulling, each typically requiring preparation durations from one to three days [9~11].

In recent years, 3D printing technology has witnessed widespread adoption across various engineering domains, including medical devices, aerospace structures, energy equipment, and soft robotics. Printing materials, primarily polymers, and metals, are becoming increasingly abundant [12].

Journal of Physics: Conference Series **2808** (2024) 012041 doi:10.1088/1742-6596/2808/1/012041

Compared to metals, polymers offer advantages such as lighter weight, enhanced corrosion resistance, and greater malleability. The rapid and precise manufacturing capabilities of 3D printing hold immense promise for MN preparation, enabling the fabrication of intricate geometric structures [13~14]. To address the challenges of time-consuming and costly microneedle fabrication, we utilized Digital Light Processing (DLP) printing technology. DLP, characterized by its ultra-fast speed, high precision, and cost-effectiveness, emerges as the preferred choice for hydrogel MN fabrication compared to other 3D printing technologies such as Stereolithography (SLA), Two-Photon Polymerization (2PP), and Fused Deposition Modeling (FDM) [15~16]. In this study, we achieved the preparation of a batch of MNs within merely one hour, with MN tips measuring approximately 20 μm and possessing sufficient mechanical strength for skin insertion. Additionally, the incorporation of NIPAM conferred temperature-sensitive properties to the MNs, offering a new avenue for the fabrication of stimulus-responsive hydrogel MNs.

2. Experimental

2.1 Materials

N-isopropyl acrylamide (NIPAM), acrylic acid (AA), acrylamide (AM) Ethanol (≥99%), and ethyl (2, 4, 6-trimethylbenzoyl) phenylphosphinate (TPO-L) were purchased from Macklin, Shanghai. $Al(OH)_3$ nanoparticle sols (AH NPs) (33 wt%, particle size of 2-5 nm) was purchased from Suzhou Nadi Electronics Co., Ltd. Salicylic acid (SA), Sodium salicylate (SS), 2-Hydroxy-4'-(2-hydroxyethoxy)-2-methylpropiophenone (IR2959), montmorillonite (MMT), and iphenyl (2, 4, 6-trimethylbenzoyl) phosphine oxide (TPO) were acquired from Shanghai Aladdin Biochemical Technology Co., Ltd.

2.2 Preparation of NIPAM-AA-AM MNs

Preparation of MNs printing ink: the formulation of the MNs printing ink involved dissolving NIPAM, AA, and AM in a molar ratio of 2.5:1:1 into 4 g of deionized water, under continuous stirring until complete transparency was achieved. Subsequently, the photoinitiator and crosslinking agent were incorporated into the homogeneous mixture, following which the stirring was maintained for an additional 10 min to ensure thorough dispersion. Print prescriptions are presented in Table 1.

3D modeling of MNs: we use TinkerCAD.com to design a 3D model of the MN array, with each MN measuring 1.2 mm in height, 0.4 mm in bottom diameter, and an infinitely small (0) diameter for the needle tip. The needle spacing was set at 0.8 mm, and MN patches were arranged in a 10×10 matrix, with a substrate thickness of 0.3 mm.

Printing of MNs: to proceed with the actual printing process, the MNs model was converted into a DLP-compatible format using Photonwork software. Once formatted, it was imported into the DLP 3D printer. Following this, the printing ink was carefully introduced into the printer's liquid reservoir. Leveraging the speed and precision of the DLP printing technology, the entire MN array could be fabricated within a single hour.

2.3 Characterisation of NIPAM-AA-AM hydrogel MNs

The morphology of the MNs was examined by using industrial microscopy, while the breaking force of the hydrogel MNs was evaluated by using a universal machine. MN insertion was visualized through H&E staining. The biocompatibility experiments were conducted with NIH-3T3 cells, and the cell viability was assessed by using the CCK-8 assay. The hemolysis rate of MNs was determined by measuring using fresh mouse blood.

2.4 In vitro drug release of NIPAM-AA-AM MNs

In vitro drug release experiments were conducted by using NIPAM-AA-AM MNs loaded with approximately 3.75 mg of drug per MN. SS or SA-loaded MNs were acquired by adding SS or SA uniformly to the ink before printing. A single NIPAM-AA-AM MN was immersed in 8 mL of PBS solution, and aliquots (3 mL) of the release medium were collected at 1, 2, 3, 5, 7, 9, and 24 hours. To

Journal of Physics: Conference Series **2808** (2024) 012041 doi:10.1088/1742-6596/2808/1/012041

maintain a constant volume, 3 mL of fresh PBS solution was replenished in the glass vial. The absorbance was measured by using a UV spectrophotometer, and the concentration was determined by referencing a standard absorbance curve.

2.5 COMSOL simulation of NIPAM-AA-AM MNs in vivo drug release experiments

COMSOL was employed to simulate the in vivo drug release from NIPAM-AA-AM MNs, with separate models created for both the MNs and skin. The drug flow direction was set vertically downward, while flux within the skin remained continuous.

3. Results and discussion

3.1 Preparation and morphological observation of NIPAM-AA-AM MNs

Printing hydrogel MNs using DLP technology offers obvious advantages over traditional mold-based preparation methods. Typically, the conventional mold method involves multiple steps, including mold preparation, casting, polymerization, drying, and demolding, culminating in a production timeline ranging from 24 to 72 hours for a single batch. In contrast, the DLP printing methodology streamlines the process to a single-step polymerization event, thereby drastically reducing the manufacturing time to merely one hour, significantly impacting both time and cost efficiency. Figure 1 illustrates the streamlined workflow: A meticulously designed MN model is imported into slicing software where it is assigned precise printing parameters—each layer receives an exposure time of 20 seconds, the bottom of the exposure time: of 20 seconds, the layer thickness during printing is set at 0.02 mm, with a 2-second interval between layer exposures to prevent excessive curing. Upon setting these parameters, the sliced dlp. file is transmitted to the DLP printer, which utilizes a layer-by-layer additive manufacturing technique to precisely construct the MNs. This method ensures both rapid prototyping and intricate control over the MN geometry, further highlighting the superiority of DLP printing in the realm of hydrogel MN fabrication.

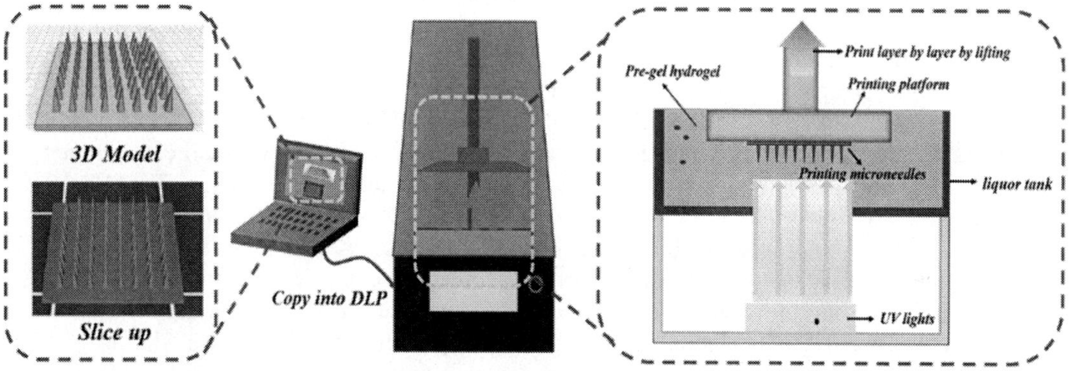

Figure 1. Schematic diagram of DLP printing NIPAM-AA-AM MNs.

To achieve optimal needle sharpness and fidelity in the printed MNs, various printing formulations were screened, as detailed in Table 1. It was observed that maintaining a monomer concentration of 8 mol/L and utilizing TPO-L as the photoinitiator along with AH NPs as the crosslinking agent, while setting the printing time at 20-30 seconds per layer, consistently resulted in the production of stably printed MNs with well-defined needle features. These specific conditions facilitated the achievement of fine and intricate needle shapes, thereby enhancing the overall performance and functionality of the printed hydrogel MNs.

Journal of Physics: Conference Series **2808** (2024) 012041 doi:10.1088/1742-6596/2808/1/012041

Table 1. Different print prescriptions and MNs' photographs.

Monomer concentration (mol/L)	Photoinitiator (0.1 g)	Crosslinking agent (1.0 g)	Print time per layer (s)	Photograph
4	TPO	AH NPs	120	
6	TPO	AH NPs	120	
8	TPO	AH NPs	120	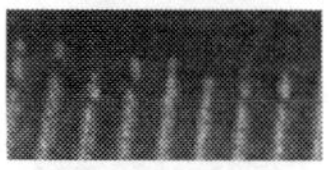
8	IR2959	AH NPs	120	
8	TPO-L	AH NPs	120	
8	TPO-L	AH NPs	60	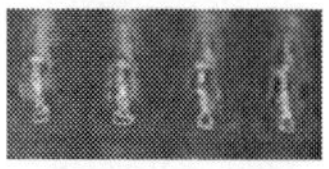
8	TPO-L	AH NPs	30	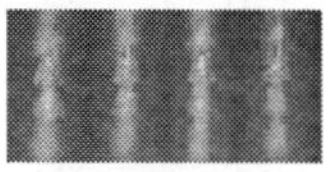
8	TPO-L	AH NPs	20	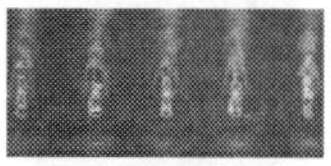
8	TPO-L	AH NPs	10	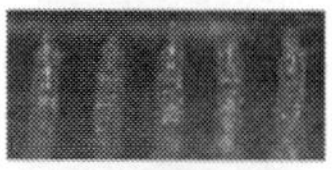
8	TPO-L	AH NPs	5	
8	TPO-L	Bis	30	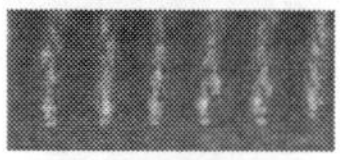
8	TPO-L	MMT	30	

The NIPAM-AA-AM MNs fabricated using DLP printing technology (depicted in Figure 2) demonstrate consistent manufacturability and versatility in accommodating a broad array of shapes and heights concurrently. Comparative analysis between the designed models and the as-printed MNs reveals a marginal decrease in the final MN height. As evidenced by microscopic photographs, the produced MNs exhibit a height of approximately 900 μm, a tip diameter of 20 μm, a base diameter of 300 μm, and an interspacing of 800 μm, while 3D modeling MNs show a theoretical height of 1200 μm. This disparity signifies a minor shrinkage phenomenon during the DLP printing process, which is a common occurrence attributed to factors such as material curing dynamics and the layer-by-layer additive manufacturing technique. Nonetheless, the close correspondence between the modeled and actual MN geometries underscores the precision and accuracy of the DLP printing technology in producing microneedle arrays with tailored dimensions.

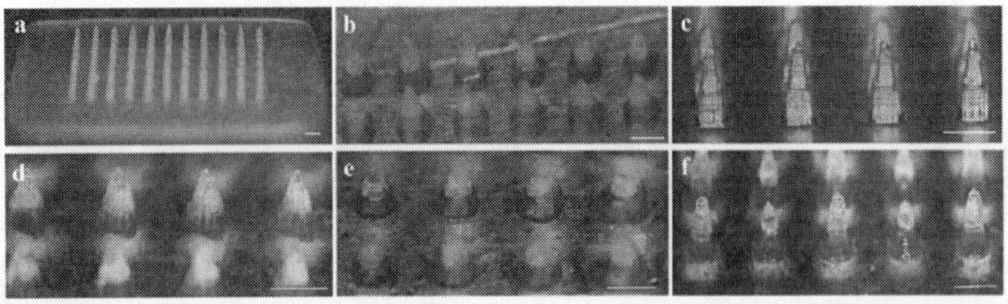

Figure 2. Microscope photographs of NIPAM-AA-AM MNs: (a) A full view of one MNs patch, (b)(c) enlarged view from (a), (d) trigonal-conical, and (e) conical MNs, (f) MNs with different heights on one patch (Scale: 500 μm).

3.2 Evaluation of the effect of MNs insertion

The insertion force exerted by MNs is crucial, as it must satisfy the necessary puncture requirements while ensuring a comfortable and pain-free experience for the user. As illustrated in Figure 3a, the noticeable peak (corresponding to a force of 12.5 N/100 needles) on the force-displacement curve denotes the momentary force applied upon MN fracture. Generally, MNs with a strength ranging from 0.1 N to 0.3 N exhibit the ideal balance, enabling successful skin penetration without inflicting undue harm. In Figure 3b, the results of H&E staining vividly illustrate the penetration depth of MNs into fresh porcine skin, estimated to be around 100 μm. This penetration depth is selected to preclude contact with pain-sensing nerves and thereby minimize any potential discomfort. This careful optimization of MN strength and insertion depth underscores the meticulous design considerations aimed at maximizing patient compliance and comfort in MN-based transdermal applications.

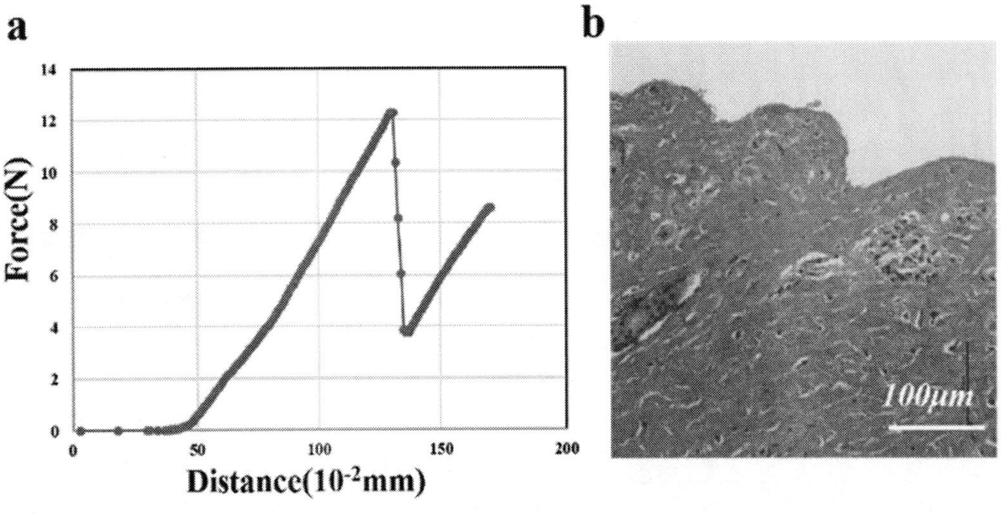

Figure 3. (a) Displacement-fracture force curve with the point of abrupt change in the curve being the point of fracture, and (b) photographs of H&E staining experiments of the NIPAM-AA-AM MNs punctured into pig skin.

3.3 In vitro drug release of NIPAM-AA-AM MNs

The incorporation of NIPAM renders MNs with temperature-sensitive properties. As depicted in Figure 4a, MNs loaded with either SS or SA exhibit a temperature-dependent differential release profile when subjected to temperatures of 30°C, 40°C, and 50°C. An inverse relationship is observed between the release rate and temperature increase; however, below the Lower Critical Solution Temperature (LCST), the NIPAM-AA-AM hydrogel matrix does not undergo volume contraction. Instead, the swelling behavior of the hydrogel enhances the osmotic pressure gradient between the saturated drug within the hydrogel network and the surrounding environment, thereby promoting a faster drug release. Figure 4b illustrates an overall release curve displaying a noticeable "sawtooth" pattern when MNs are intermittently placed in pure water at 25°C and 60°C, underscoring their temperature sensitivity. The intermittent changes in temperature lead to variations in the hydrogel's swelling state, affecting the release kinetics accordingly. Furthermore, Figure 4c presents the cumulative drug release percentages from NIPAM-AA-AM MNs after 1, 2, and 3 weeks of storage at room temperature. At the end of the first week, approximately 23.4% of the drug is released, escalating to 27.1% by the third week. This progression indicates that despite the gradual release over time, a significant portion of the drug retained within the MN patches maintains its activity even at room temperature conditions, highlighting the suitability of these temperature-sensitive MNs for sustained and controlled drug delivery applications.

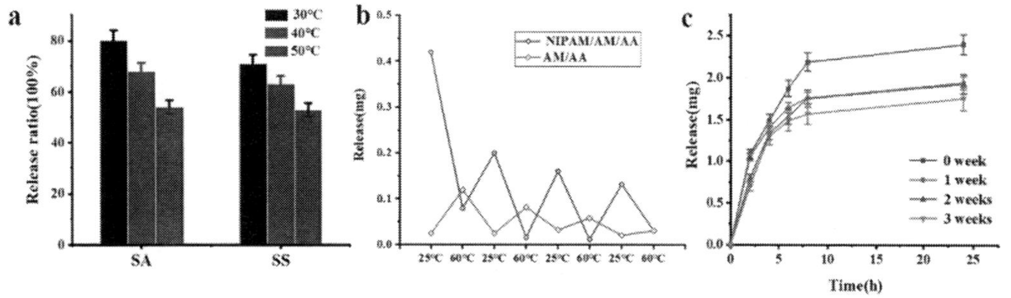

Figure 4. In vitro drug release from NIPAM-AA-AM MNs (a) drug release at 30, 40, and 50 °C (b) alternating release at 25°C and 60°C (10-minute exchange), and (c) graphical representation of drug release over 1 to 3 weeks.

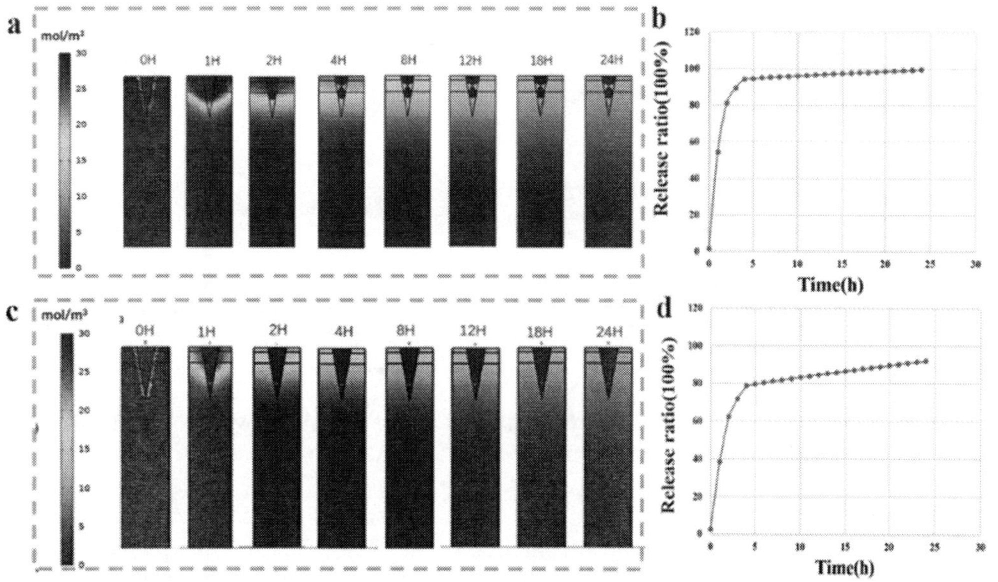

Figure 5. Penetration of the drug in the skin section at 0, 1, 2, 4, 8, 12, 18, and 24 hours and drug release profiles of four-pronged conical MNs (a, b) and conical MNs (c, d).

3.4 Analysis of COMSOL simulated NIPAM-AA-AM MNs in vivo drug release experiment

Skin tissue samples were collected at multiple time intervals (0, 1, 2, 4, 8, 12, 18, and 24 hours) to systematically evaluate the distribution of the administered drug, as depicted in Figure 5a. The MNs demonstrated an initial burst release phase within the first four hours, subsequently transitioning into a steady-state release pattern. In Figure 5b, the cumulative drug release kinetics were modeled by using the computational capabilities of COMSOL software, which revealed a rapid drug release profile, reaching nearly 90% release within four hours, and culminating in full permeation throughout the skin layers within 20 hours post-administration. Moreover, a comparative analysis of drug release rates between conical MNs and four-pronged cone MNs at the 24-hour mark was undertaken. Results indicated that the conical MNs delivered approximately 80% of the payload, marginally lower than the four-pronged counterparts. This discrepancy can be attributed to the potentially less invasive nature of the conical design, leading to a somewhat reduced interaction with the underlying muscle tissue, as evidenced in Figures 5c and 5d. This finding underscores the significance of MN geometry in modulating drug release kinetics and minimizing tissue trauma, thereby informing future design optimization strategies for MN-based transdermal drug delivery systems.

3.5 Biocompatibility of NIPAM-AA-AM MNs

The safety of MNs was also assessed, as illustrated in Figures 6a and 6b, focusing on the cytotoxicity and hemocompatibility of the hydrogel MNs. The viability of NIH-3T3 cells cultivated in the presence of 200 to 1000 μg/mL of the extracted dry gel material for a period of 24 hours was found to be near 100%, exceeding the minimum acceptable threshold of 70% stipulated by the national standard GBT 16886.5-2017. Additionally, the hemolytic potential of the MNs was scrutinized by incubating rat erythrocytes with varying concentrations of 2, 6, and 10 mg/mL of the dry gel extract for 60 minutes. The resulting hemolysis rates were measured at 0.61%, 1.22%, and 0.47%, respectively, all of which fell significantly below the upper limit of 5% as defined in the national standard YY/T.1651.1-2019. These findings collectively attest to the excellent biocompatibility and hemocompatibility of the hydrogel MNs, validating their safety for potential biomedical applications.

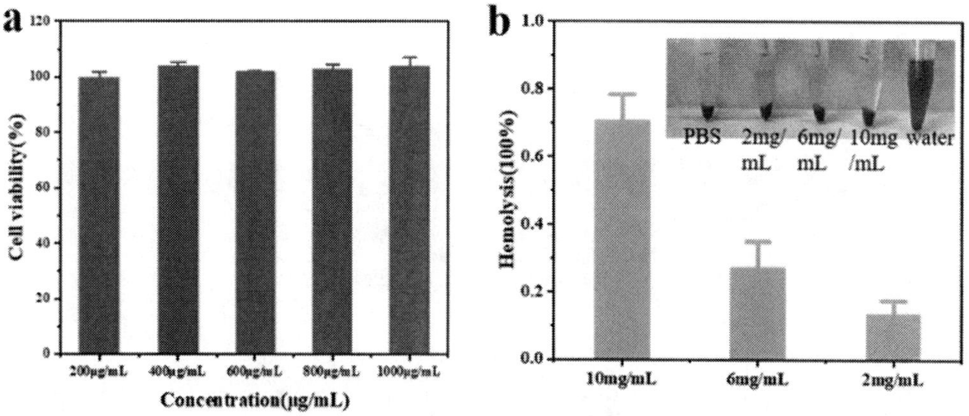

Figure 6. (a) The viability of NIH-3T3 cells cultured in NIPAM-AA-AM MNs' extracts for 24 h; (b) Hemolysis rate of NIPAM-AA-AM MNs' extracts.

4. Conclusion

In this study, we present the successful preparation of NIPAM-AA-AM hydrogel MNs by using DLP printing, with the entire process completed within one hour. Our findings highlight the significant impact of monomer concentration, choice of photoinitiator and crosslinker, and printing duration on MNs' printing efficacy. Fracture force tests and H&E staining experiments demonstrate the MNs' mechanical strength for skin insertion. The introduction of NIPAM enables enhanced drug release at low temperatures, while temperature-sensitive properties are validated through temperature alternation experiments. COMSOL simulation results align with actual observations. Importantly, the material exhibits non-toxicity and safety for human skin application. Although preliminary, these findings offer a convenient, rapid, and cost-effective method for hydrogel MN preparation, opening new possibilities in transdermal drug delivery applications.

References

[1] Morteza A, Sahar S, Bradley J N and Metin S (2018) Recent Advances in Wearable Transdermal Delivery Systems. Adv. Mater., 30: 1704530. [CrossRef]

[2] Rezvan J, Pooyan M, Cynthia K Y, Tarun A, Raffaele V, Sun W, Tapas K M, Franklin R T and Paolo A N (2020) Engineered Microneedle Patches for Controlled Release of Active Compounds: Recent Advances in Release Profile Tuning. Adv. Ther., 3: 2000171. [CrossRef]

[3] Liu H, Zhou X M, Nail A, Yu H, Yu Z L, Sun Y, Wang K, Bao N B, Meng D C, Zhu L R and Li H J (2024) Multi-Material 3D Printed Eutectogel Microneedle Patches Integrated with Fast Customization and Tunable Drug Delivery. J. Control. Release, 368: 115-130. [CrossRef]

[4] Lin L, Wang Y Q, Cai M, Jiang X, Hu Y, Dong Z, Yin D, Liu Y, Yang S, Liu Z, Zhang J, Xu Y, Guo C and Chang L (2022) Multimicrochannel Microneedle Microporation Platform for Enhanced Intracellular Drug Delivery. Adv. Funct. Mater., 32: 2109187. [CrossRef]

[5] Dul M, Stefanidou M, Porta P, Henri M, Levin Y, Kochba E, Wong F S, Dayan C, Coulman S and Birchall J C (2017) Hydrodynamic Gene Delivery in Human Skin Using a Hollow Microneedle Device. J. Control. Release, 265: 120-131. [CrossRef]

[6] Vrdoljak A, Mcgrath M G, Carey J B, Draper S J, Hill A V S and Moore A C (2012) Coated Microneedle Arrays for Transcutaneous Delivery of Live Virus Vaccines. J. Control. Release. 159: 34-42. [CrossRef]

[7] Salvador E S, Tuan-Mahmood T M, Mcelnay J C, Mccarthy H O, Monney K, Woolfson D and Donnelly F (2015) Potential of Hydrogel-Forming and Dissolving Microneedles for Use in Paediatric Populations. Int. J. Pharm. 489: 158-169. [CrossRef]

[8] Aung N N, Ngawhirunpat T, Rojanarata T, Patrojanasophon P, Pamornpathomkul B and Opanasopit P (2020) Fabrication, Characterization, and Comparison of α-Arbutin Loaded Dissolving and Hydrogel Forming Microneedles. Int. J. Pharm., 586: 119508. [CrossRef]

[9] Donnelly R F, Singh T R, Garland M J, Migalska K, Majithiya R, Mccrudden C M, Kole P L, Mahmood T M, Mccarthy H O and Woolfson A D (2012) Hydrogel-Forming Microneedle Arrays for Enhanced Transdermal Drug Delivery. Adv. Funct. Mater., 22: 4879-4890. [CrossRef]

[10] Huang S H, Liu H L, Huang S S, Fu T L, Xue W and Guo R (2020) Dextran Methacrylate Hydrogel Microneedles Loaded with Doxorubicin and Trametinib for Continuous Transdermal Administration of Melanoma. Carbohydr. Polym., 246: 116650. [CrossRef]

[11] Sabri A H B, Anjani Q K, Utomo E, Ripolin A and Donnelly R F (2022) Development and Characterization of a Dry Reservoir-Hydrogel-Forming Microneedles Composite for Minimally Invasive Delivery of Cefazolin. Int. J. Pharm., 617: 121593. [CrossRef]

[12] Chen G, Xua Y, Philip C L K and Kang L (2020) Pharmaceutical Applications of 3D Printing. Addit. Manuf., 34: 101209. [CrossRef]

[13] Lindsay B, Jacob Q, Hossein D, Mohamadmahdi S, Fariba A, Ali F, Laleh A, Sidi B, Adnan M, Pooria M and Ali T (2021) 3D-Printed Hydrogel-Filled Microneedle Arrays. Adv. Healthc. Mater., 10: 2001922. [CrossRef]

[14] Sophia N E, Wan T S, Jasim U, Li H, Manuel J, Andrew R, Dennis D, James F C and Adrian P (2121) A Novel 3D Printed Hollow Microneedle Microelectromechanical System for Controlled, Personalized Transdermal Drug Delivery. Addit. Manuf., 38: 101815. [CrossRef]

[15] Merima S, Amina T, Kenan M, Lamija H, Jasmina H, Ahmet C, Derzija B, Ognjenka R, Maida C, Almir De and Edina V (2021) 3D Printing—A "Touch-Button" Approach to Manufacture Microneedles for Transdermal Drug Delivery. Pharmaceutics, 13: 924. [CrossRef]

[16] Usanee D, Emma M, Aaron R J H, Eneko L and Ryan F D (2022) The Role of 3D Printing Technology in Microengineering of Microneedles. Small, 8: 2106392. [CrossRef]

Journal of Physics: Conference Series **2808** (2024) 012042 doi:10.1088/1742-6596/2808/1/012042

Preparation and performance verification of calcium silicate composite electrical fireproof materials

Yin Wang[1], Yu Wang[1], Dedong Yang[1], Guanggang Fu[1], Chong Zheng[1], Fanfang Kong[1], Jingkai Nie[2] and Jun Ji[2,*]

[1]Wenzhou Power Supply Company, State Grid Zhejiang Electric Power Co., Ltd., Wenzhou, 325000, China
[2]State Grid Smart Grid Research Institute Co., Ltd., Beijing, 102209, China

[*]Corresponding author's e-mail: jijun2006@163.com

Abstract. The wall bushings of the converter transformer are often sealed with rock wool for fire prevention, but the fire resistance is not enough for hydrocarbon fire for 1 h, which leads to frequent fires in the converter station. In this paper, calcium silicate-bound vermiculite lightweight materials with a density of less than 900 kg/m^3 were prepared by static hydrothermal autoclaving reaction with calcium silicate-rich powder as raw material, expanded vermiculite as aggregate, and fiber reinforced. The effect of material ratio on bulk density, strength, and high temperature resistance was discussed. The results show that the micropore structure developed in the binding phase is mainly composed of Tobe mullite and xonotlite, and the interlamellar pores between Tobe mullite and vermiculite contributed to the high porosity of the material. At the same time, the combination phase and vermiculite interpenetrating combination structure, so that the material has a high strength. Low shrinkage and low thermal conductivity characteristics of the composite under electric fire.

1. Introduction

In recent years, the importance of power resources in China's economic development has become increasingly prominent, and the rapid development of social and economic also led to a sharp increase in the demand for electricity [1]. In order to meet the increasing power demand, most of the power equipment, such as the transformer in the converter station, is overloaded, which leads to the fire accident of oil-filled equipment. At present, rock wool is often used to seal the wall bushing of the converter transformer, and there is a short board with insufficient fire resistance.

Tobermorite ($5CaO{\cdot}6SiO_2{\cdot}5H_2O$) displays the dehydration of crystal and constitutional water at 300°C and 724°C, respectively, accompanied by the production of wollastonite and high-temperature lime. The related products have a high service temperature of up to 650°C [2]. After the dehydration, hard xonotlite ($6CaO{\cdot}6SiO_2{\cdot}H_2O$) is transformed into wollastonite, and the highest usage temperature of the products is up to 1050°C [3]. The theoretical dehydration capacity of xonotlite is 12.82%; in contrast, the capacity of wollastonite is only 2.52% [4]. A higher dehydration capacity indicates more heat absorption but is associated with larger high-temperature shrinkage. Expanded vermiculite is a type of heat-resisting lightweight aggregates [5]. The application of expanded vermiculite to a calcium carbonate autoclave system can effectively suppress the high-temperature shrinkage of calcium carbonate materials [6-8].

This study selected expanded vermiculite as the lightweight aggregate to prepare a calcium silicate–vermiculite composite using the static method. The effects of vermiculite addition and molding pressure

Journal of Physics: Conference Series **2808** (2024) 012042 doi:10.1088/1742-6596/2808/1/012042

on both bulk density and strength of the material were investigated. The pore structure characteristics of the material were analyzed through microscopic observation and pore tests. The prepared material had a low high-temperature shrinkage rate and thermal conductivity, with potential application in the field of electric fire prevention.

2. Test process

2.1 Test materials

This study prepared expanded vermiculite with a particle size of 0.3~1 mm using Xinjiang Weili vermiculite with thermal expansion at a temperature of 400°C~600°C. The loose bulk density of the prepared vermiculite was measured to be 180 kg/m^3. Calcium hydroxide [$Ca(OH)_2$] powder with a particle size less than 20 μm was used as the calcium-based raw material. The $Ca(OH)_2$ content in the powder was found to be 88.44%, and the SiO_2 content was 7.08%; the other impurities accounted for 4.47%. Silica sand with a particle size of 400 mesh was used as the silica-based raw material. The content of SiO_2 in the sand was found to be 94.33%. Figure 1 shows the particle morphology of the silica sand. The cellulose fiber sheets of 5 × 5 mm^2 were used in this study as the reinforced fibers, which were then dispersed to form individual fiber filaments, as shown in Figure 2.

Figure 1. Morphological images of natural powdered quartz.

(a) (b)

Figure 2. Morphological images of cellulose fibers (a) Fiber sheets and (b) Fiber monofilaments.

2.2 Preparation and analysis of samples

2.2.1 Mixing of raw materials and molding. In this study, the Ca/Si molar ratios in Ca- and Si-based raw materials were set to 1.0. The additional amount of cellulose fibers was 4.5% of the total mass of Ca- and Si-based raw materials. Additionally, the amounts of expanded vermiculite added were 20%, 30%, and 40% of the total solid raw materials, respectively. The water-to-solid ratio was adjusted following the amount of vermiculite added, resulting in values of 1.15, 1.25, and 1.50, respectively, as shown in Table 1.

Table 1. Parameters and serial numbers of the prepared samples.

Addition amount of vermiculite (%)	Molding pressure (MPa) 1.0	2.0	3.0	4.0	Water-to-solid ratio	Ca/Si molar ratio
20	201#	202#	203#	204#	1.15	1.0
30	301#	302#	303#	304#	1.25	1.0
40	401#	402#	403#	404#	1.50	1.0

The mixture slurry of raw materials was then poured into a mold with a dimension of 210 × 210 × 70 mm^3 and compacted under different pressures of 1.0, 2.0, 3.0, and 4.0 MPa. The mold was held under constant pressure for 5 min before releasing the pressure, and the pressed pellets with a thickness of 20 mm were obtained after demolding. The samples were labeled based on the vermiculite content and the compaction pressure (Table 1). For example, the No. 204# sample showed the addition of vermiculite at a rate of 20%, and the compaction pressure was set at 4.0 MPa.

To investigate how compression change, water drainage, and compaction pressure are interrelated during the molding process, we conducted a compressive test on a material mixture slurry containing 40% vermiculite. The thicknesses of the pressed pellets and the amounts of discharged water were measured under different pressures ranging from 0 to 4.5 MPa.

2.2.2 Sample composition, microstructure, and performance test. The flexural strength of the sample was measured using a three-point bending method with a span of 60 mm. For the compressive strength test, the compression area of the sample was 20 × 20 mm^2. For the tensile strength test, a screw with a diameter of $\Phi = 3.8$ mm was anchored into the test sample and left with a protruding end of 20 mm. The uniform load was applied vertically to the head of the screw until it became loose, and the maximum load P at this point was recorded. The tensile strength was then calculated by dividing the maximum load P by the cross-sectional area A of the screw. The samples were calcined in a muffle furnace at 1050°C for 3 h to investigate the changes at high temperatures. For each sample, both masses and lengths before and after calcination were measured to calculate the mass loss rate and the linear shrinkage in the length direction. The thermal conductivity was tested by commissioning the National Refractory Quality Supervision and Inspection Center using the water flow plate method according to the Test Method for Thermal Conductivity of Refractories (YB/T 4130-2005). The thermal conductivities of the samples were measured at different heat transfer surface temperatures of 200°C, 500°C, and 1000°C.

3. Test results

3.1 Volume density and strengths of different samples

Figure 3 displays the test results of the bulk density, flexural strength, compressive strength, and tensile strength of the samples after first autoclaving and drying. With the same additional amount of vermiculite, the bulk density of the material increased linearly with the molding pressure. Under the same molding pressure, a larger additional amount of vermiculite indicated lower bulk density, as shown in the left part of Figure 3(a). The No. 401# sample showed the lowest bulk density of 538 kg/m^3, while the No. 204# sample exhibited the highest bulk density of 859 kg/m^3. The tensile strength decreased with the increasing addition of expanded vermiculite. When the additional amount of vermiculite was 40%, the effect of compaction pressure on tensile strength became more significant, as shown in the right part of Figure 3(a). The tensile strengths of No. 401# and No. 204# samples were 20.0 and 43.4 MPa, respectively.

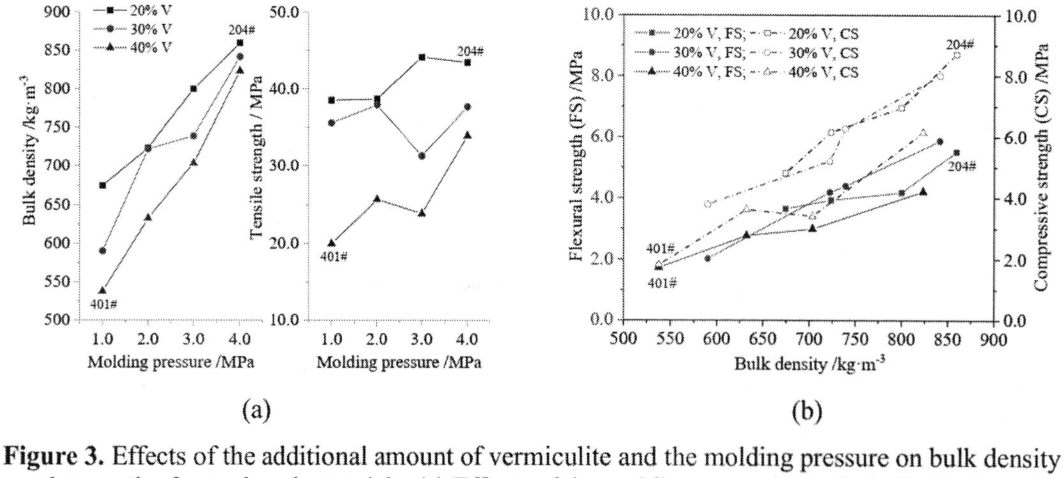

(a) (b)

Figure 3. Effects of the additional amount of vermiculite and the molding pressure on bulk density and strength of autoclaved materials: (a) Effects of the molding pressure on the bulk density and tensile strength; and (b) Relationship between bulk density and strength.

Journal of Physics: Conference Series **2808** (2024) 012042 doi:10.1088/1742-6596/2808/1/012042

Figure 3(b) reflects the relationship among the bulk density, flexural strength, and compressive strength. From the perspective of flexural strength, the strength of the sample with a vermiculite addition of 30% varied more significantly with bulk density. In terms of compressive strength, the strength curves of the samples with 30% and 20% vermiculite addition were quite close, and the strength values were higher than those of the samples with 40% vermiculite addition and the same bulk density. Based on the comprehensive test results of flexural strength and compressive strength, it was concluded that samples with a vermiculite addition of 30% had better strength performance.

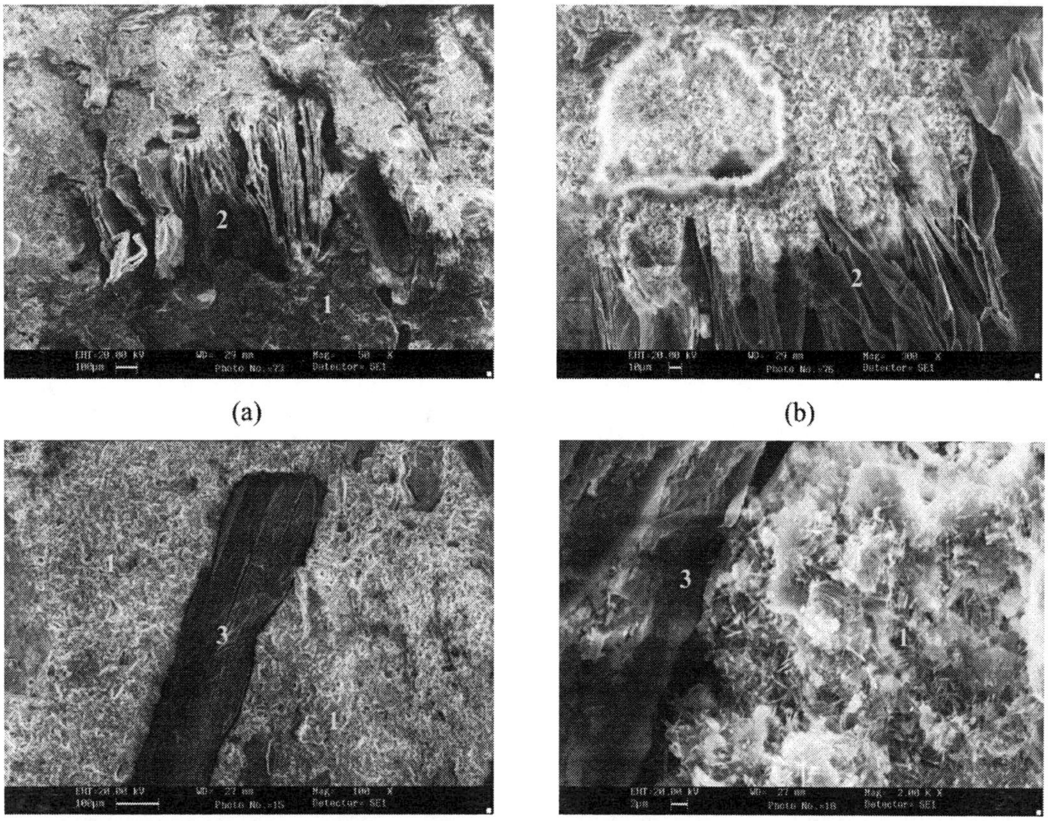

(a) (b) (c) (d)

Figure 4. SEM images of the fractures of the samples after adding expanded and unexpanded vermiculites (a) After the addition of expanded vermiculite particles, (b) Interspersion of the binder phase into the expanded vermiculite flakes, (c) After the addition of unexpanded vermiculite particle, (d) Junction surface between unexpanded vermiculite and high-temperature hydrated calcium carbonate (1 - High-temperature hydrated calcium silicate mineral; 2 - Expanded vermiculite particle; 3 - Unexpanded vermiculite particle.).

3.2 Microscopic structures

Figure 4(a) displays the microstructures of the samples under a scanning electron microscope (SEM). It was observed that the expanded vermiculite flakes were open, and the particles were distributed in a matrix composed of tobermorite and xonotlite. On magnifying the interface between the expanded vermiculite particles and the matrix, it was noted that high-temperature hydration calcium silicate penetrated into the expanded vermiculite flakes, thereby forming an interlaced binding structure, as shown in Figure 4(b). High-temperature hydration calcium silicate existed in the form of flake-like and fibrous crystals, corresponding to tobermorite and hard whiting, respectively. Well-developed internal fine pores were observed in the binder phase, which mainly corresponded to the pores below 0.3 μm in the mercury intrusion porosimetry test curve [9-10]. In the expanded vermiculite particles, large fusiform pores and slightly thinner pores among flakes were detected, which corresponded to the pores above 5 μm and between 0.3 and 5 μm in the mercury intrusion porosimetry curve, respectively. For the samples with a large additional amount of expanded vermiculite, a more dense distribution of vermiculite particles was observed. Also, some vermiculite particles were bent and deformed at a greater molding pressure of more than 3.0 MPa. Figure 4(c) displays the microstructure of the sample with the addition of unexpanded vermiculite under the same autoclaving condition for comparison. It was observed that the vermiculite particles were wrapped by high-temperature hydration calcium silicate. After amplification, the gaps between the vermiculite particles and high-temperature hydration calcium silicate were noted, as shown in Figure 4(d). Therefore, when using expanded vermiculite, the aggregate particles and the matrix were well bonded. According to the related studies, adding unexpanded vermiculite to calcium silicate materials could enhance the toughness of composite materials but contributed nothing to fireproofing [11].

3.3 High-temperature weight loss ratio, linear shrinkage, and thermal conductivity of the prepared samples

The samples showed no significant deformation but only darker surfaces, after the calcination at 1050°C for 3 h. Figure 5 displays the variations in both the weight loss rate and linear shrinkage rate of each sample after calcination under the molding pressure. In contrast, when the additional amount of vermiculite was fixed, both high-temperature weight loss rate and linear shrinkage varied slightly with the molding pressure. The samples with 30% and 40% vermiculites showed similar weight loss rates and linear shrinkage rates, which were slightly higher than the values of the samples with the addition amount of 20%. The thermal conductivity of the No. 302# sample was also measured at high temperatures. At a hot side temperature of 200°C, 500°C, and 1000°C, the thermal conductivity was 0.177, 0.204, and 0.209 W/(m·K), respectively.

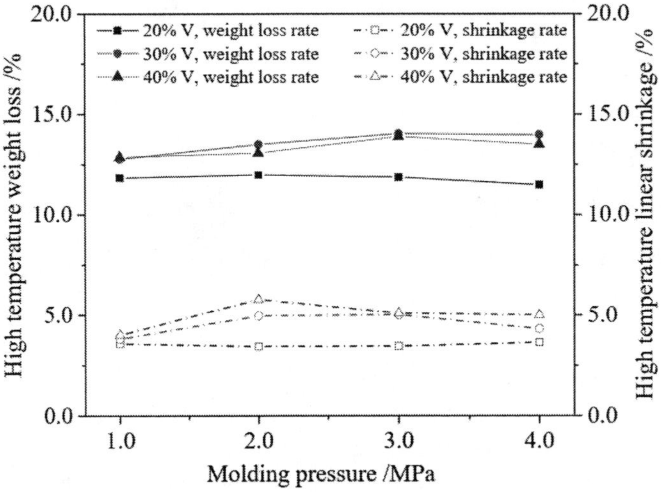

Figure 5. Measured weight loss rates and linear shrinkage rates of the samples after calcination at 1050°C for 3 h.

4. Conclusions

(1) In this study, samples were prepared with different addition amounts of expanded vermiculite (20%, 30%, and 40%) and a Ca/Si ratio of 1.0 using calcium hydroxide as calcium raw material and quartz powder as silicon raw material; the molding pressure ranged from 1.0 to 4.0 MPa. Next, high-temperature hydration calcium silicate and vermiculite lightweight materials with a volume density of 538–859 kg/m^3 were prepared after autoclaving at 210°C and 2.0 MPa for 6 h in saturated water vapor. The results showed that high vermiculite addition and small molding pressure were beneficial for

preparing low-density materials.

(2) The prepared material could withstand a high temperature up to 1050°C without strong shrinkage deformation and had relatively low thermal conductivity at high temperatures. For the sample with 30% vermiculite addition at a molding pressure of 2.0 MPa, the thermal conductivity was measured as 0.177, 0.204, and 0.209 W/(m·K) at hot side temperatures of 200°C, 500°C, and 1000°C.

(3) High addition of vermiculite and low molding pressure is favorable for the preparation of low bulk density materials, which have the requirements of light weight and fire resistance for power equipment.

Acknowledgments
The work was supported by the Science and Technology Project of State Grid Zhejiang Electric Power Company Ltd., (5211WZ220006).

References

[1] Xing F (2000) Study and Application of the Xonotlite Based Calcium Silicate Material. J. Journal of Shenzhen University, (6): 14-15.
[2] Zhao Q Y, Wang Z Z, Xu X and et al. (2020) Research status of crystal structure, thermal behavior and synthesis of tobermulite. J. Journal of the Chinese Ceramic Society, 48 (10): 1536-1551.
[3] Tobermorite Mineral Data (2023) http://www.webmineral.com/data/Tobermorite.shtml.
[4] Chen Y D, Zhang H L, Bao S B E, et al. (2023) Preparation of hydromagnesite-vermiculite fire barrier and investigation of fireproof performance. J. Non-metallic Mines, 46 (1): 6-10.
[5] Shi R (2023) Preparation of basic magnesium carbonate-vermiculite lightweight composites and performance investigation. D. A master's degree from University of Science and Technology Beijing.
[6] Zheng F (2011) Preparation of the fireproof plates with calcium silicate-vermiculite compound steel structure and performance investigation. D. A master's degree from University of Science and Technology Beijing.
[7] Akao M, Fukuda Y, Yamazaki A, et al. (2006) Preparation and characterization of calcium silicate/vermiculite composite. J. Journal of Hard Tissue Biology, 15 (3): 101-104.
[8] Tobermorite Mineral Data. http://www.webmineral.com/data/Tobermorite.shtml, 2023-7-7.
[9] Liu S D, Li Z L, Tong T T, et al. (2022) Molecular dynamics simulation on effect of humidity on mechanical properties of calcium silicate hydrate. J. Bulletin of the Chinese Ceramic Society, 41 (5): 1554-1561.
[10] Liu X, Feng P, Shen X Y, et al. (2021) Advances in the understanding of cement hydrate: calcium silicate hydrate (C-S-H). J. Materials Reports, 35 (9): 9157-9167.
[11] Wang J, Xu W, Li H. (2015) Review on effect of different conditions on the structure and performance of the synthesized calcium silicate hydrated. J. Bulletin of the Chinese Ceramic Society, 34 (5): 1296-1301.

Journal of Physics: Conference Series **2808** (2024) 012043 doi:10.1088/1742-6596/2808/1/012043

Research on deformation strain field of vehicle body welding based on laser scanning detection

Lina Li[1], Hongchang Sun[2*] and Yuanhao Bao[3]

[1]Department of Automobile Engineering, Transportation Technical College, Tianjin, China

[2]School of Mechanical Engineering, Tianjin University of Technology and Education, Tianjin, China

[3]School of Mechanical Engineering, Guichi District Vocational School, Chizhou, Anhui, China

[*]Corresponding author' s e-mail: sunhongchang@tute.edu.cn

Abstract: At present, the traditional welding temperature field and stress prediction mainly rely on the experience curve or empirical formula based on experiment and statistics, which cannot comprehensively predict and analyze the impact of welding on the overall structural mechanical characteristics and objectively evaluate the welding quality. In this paper, based on simulated welding software, the temperature field stress field and residual stress deformation after welding are simulated by the finite element method, and a set of feasible 3D welding temperature field calculation methods, stress field, and deformation dynamic simulation analysis methods are established.

1. Introduction

In the modern manufacturing industry, welding is an indispensable technology, which has been playing a pivotal role in the field of material processing. The welding stress and deformation caused by the welding process will affect not only the manufacturing process of the welded structure but also the service performance of the welded structure [1]. Therefore, accurate simulation and analysis of the welding process, mastering the temperature field, stress field, and deformation law are of vital significance for improving the welding quality and optimizing the welding process, so the stress and deformation characteristics of the weld should be paid attention during the design and construction [2]. To improve production efficiency, shorten the construction period, and reduce the cost, we are supposed to develop a method that can effectively control the welding deformation [3]. At present, the influence of mechanical constraints on welding deformation and residual stress has been deeply studied [4]. In practical applications, to reduce the heat input of the near-weld and avoid excessive thermal stress caused by local heating, it usually needs specific constraints to ensure its safety, reliability, and stability [5]. Fanous used the welding numerical simulation technology to analyze the welding process under three boundary conditions [6] and compared the results with the actual results to find an optimal boundary condition setting method, to achieve the purpose of controlling the residual stress of the weld [7][8].

M. Adak et al. 's theory and experiment were combined to prove that different boundary conditions would have a great influence on the lateral residual stress of the weld from thermal elastic-plastic, plastic theory, and experiment [9][10]. Schenk et al. carried out welding deformation experiments of T-welds with a thickness of 6 mm under two rigid fixed states and analyzed the mechanical properties of welds under different temperatures, different preheating states, and different unloading states [11]. Through numerical simulation, Zhang et al. studied the influence of different clamping methods on welding deformation and residual stress of thin plates, obtaining obvious reduction and control results. Japanese scholar Ueda [12] first proposed the concept of inherent strain in the 1970s [13]. The inherent strain of welding problems consists of phase change strain, thermal strain, and plastic strain [14], which usually exists in the weld and its vicinity. Li D [15] and Luo Qingguo [16] combined the Galerkin [17] method in the space domain and the central difference method in the time domain to conduct a numerical simulation of stress change of a fixed circular plate under repeated action of high [18] and low temperature, and compared it with test data to verify the correctness of this method [19][20]. Based [21] on thermal elastic-plastic theory, Professor Chen et al. [22] from Shanghai Jiao Tong University developed a set of finite element numerical simulation programs, which is suitable for welding samples and compared it with the theoretical calculation results, verifying that the developed program can better predict the deformation and residual stress of welding seams. Wang Yang and Luo Yu respectively used ABAQUS, ANSYS, and JWRIAN software to carry out thermal elastic-plastic and plastic finite element calculation of the three kinds of software, and compared them with the actual measurement results [23]. They found the differences in the analysis and calculation of welding mechanics of the three kinds of software [24]. Wu et al. used Marc finite element calculation software to conduct a numerical simulation of the welding of the mining robot arm and compared it with the test results, indicating that the method can not only ensure the accuracy of the calculation results but also improve the calculation efficiency [25]. Professor Zhou et al. from Jiangsu University of Science [26] and Technology [27] used the thermal elastic-plastic finite element method to conduct a numerical simulation of the welding process of an LNG single tank composed of 304 stainless steel and proved that the method can effectively predict deformation during welding process of a single tank by comparing with test data.

Although scholars have carried out a large number of simulations, most of them have not introduced real deformation detection. To obtain better and more accurate deformation data, this paper will use laser scanning technology for deformation detection, and combine finite-free simulation analysis to carry out welding deformation prediction research.

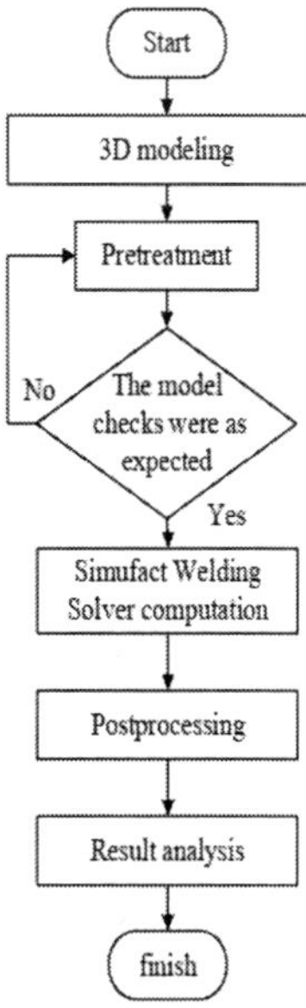

Figure 1. Simu-fact Welding Finite element analysis flow chart.

2. Finite element analysis of welding deformation

2.1 Finite element analysis

This paper intends to approximate the field function (such as displacement, stress, etc.) in each unitary grid, and use the function interpolation method to represent the field function in the grid by the numerical value of each node in the grid so that the known field function (including its derivative) in each node in the grid becomes a new unknown quantity (the number of degrees of freedom). The continuous infinite degree of freedom problem is transformed into a discrete finite degree of freedom problem.

The main research object of this paper is the welding of plates, and its specific finite element analysis process is shown in Figure 1.

2.2 Welding temperature field analysis

To predict the welding temperature field effectively, it is necessary to study the nonlinear dynamic behavior of the welding thermal process and establish a mathematical model to simulate and analyze its dynamic characteristics. In the welding process, the governing equations of heat transfer follow a series of strict regulations.

$$c\rho\frac{\partial T}{\partial t}=\frac{\partial}{\partial x}(\lambda\frac{\partial T}{\partial x})+\frac{\partial}{\partial y}(\lambda\frac{\partial T}{\partial y})+\frac{\partial}{\partial z}(\lambda\frac{\partial T}{\partial z})+Q \tag{1}$$

where c is the specific heat capacity of the material, in *J*/(kg · K); ρ is the density of the material, in kg/m³; *T* is the temperature field distribution function; *t* is the heat transfer time, the unit is s; λ is the thermal conductivity, the unit is W/(m · K); *Q* is the heat source intensity in the solution domain; These parameters *c, p,* λ all vary with temperatures.

2.3 Welding stress field analysis

During the welding process, with the change of temperature, the instantaneous stress will also change; After the welder is cooled to the initial temperature, the residual stress is the residual stress generated during the welding process. The instantaneous welding stress can be caused by the thermal expansion of the material itself, or by the amount of thermal deformation in the welding process or improper welding process. According to the different directions of welding stress in the welded parts, it can be divided into three directions: longitudinal, transverse, and thick stress [28].

The relationship between thermal elastic-plastic stress and strain increment is as follows:

$$d\{\sigma\}=[D]d\{\varepsilon\}-\{C\}dT \tag{2}$$

where $\{C\}$ - vector $[D]$ related material performance, elastic-plastic matrix, the elastic zone $[D]$ = $[D^e]$, in the plastic zone $[D] = [D]_{ep} = [D]^e - [D]_p$, $[De]$-elastic matrix, $[D]_{ep}$– elastoplastic matrix $\{\alpha\}$ - linear expansion coefficient vector. The incremental expression of the finite element equilibrium formula is shown in Formula (3).

$$[K]^e\Delta\{\delta\}=\Delta\{R^e\} \tag{3}$$

where $[K]^e$ - Element stiffness matrix$\{\delta\}$. The displacement variable caused by stress loading, $\{R^e\}$ is the equivalent node force vector.

$$\Delta\{R\}^e=\iint_e[B]^T\{C\}\Delta Tdxdy \tag{4}$$

where $[B]$—Node distribution geometric matrix.

The process of thermal elastoplastic stress finite element analysis is as follows: according to the calculation result of the temperature field, the pre-calculated temperature increment is gradually added to each element. Then, by solving Formula (3), the displacement increment $\Delta\delta$ for each node is obtained, and the stress-strain increment for each element is calculated by Formula (4). Finally, the calculation results of the transient stress field and residual stress field of the whole welding can be obtained based on these data [29].

2.4 Heat source modeling

According to the welding temperature field theory, a new welding heat source model, the double elliptic heat source model, is established and numerically simulated. At present, heat source models are widely used in the welding field, including the double ellipsoid heat source model and the Gaussian heat source model. Given these two commonly used heat source models, there are many improvements, and a new heat source model - a double ellipsoid heat source model is proposed. Considering the non-uniformity of heat source distribution in the welding process, the double ellipsoid heat source model has higher accuracy in reflecting the actual welding conditions. The Gaussian heat source model is a new heat source model developed based on the double circular heat source model, which has better calculation accuracy. Considering many factors, the double ellipsoidal heat source model is chosen in this paper, and its energy density distribution function presents two forms: (5) and (6). On this basis, a mathematical model is established to calculate the shape and temperature field change of double wire submerged arc welding pool, and verified by experiments. A three-dimensional model is presented in Figure 2, which shows the energy density distribution of a double ellipsoidal heat source [30].

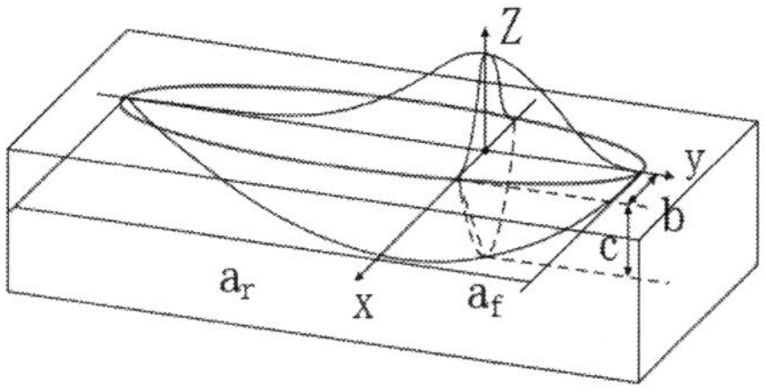

Figure 2. Energy density distribution model of double ellipsoid heat source.

The energy density distribution function of the ellipsoidal heat source in the first half is:

$$q_1(x,y,z)=\frac{6\sqrt{3}f_f\eta UI}{a_f bc\pi\sqrt{\pi}}\exp\{-3[(\frac{x}{a_f})^2]+(\frac{y}{b})^2+(\frac{z+v(\tau-t)^2}{c})^2\} \tag{5}$$

The energy density distribution function of the ellipsoidal heat source in the latter half is:

$$q_2(x,y,z)=\frac{6\sqrt{3}f_r\eta UI}{a_r bc\pi\sqrt{\pi}}\exp\{-3[(\frac{x}{a_r})^2]+(\frac{y}{b})^2+(\frac{z+v(\tau-t)^2}{c})^2\} \tag{6}$$

where, *af, ar, and bc* are the shape parameters of the heat source; η is regarded as the efficiency of welding heat sources; *U* is the arc voltage; *I* is the welding current; *V* is the welding speed; τ is the lag time factor of the power supply; *ff* and *fr* are the distribution coefficients of heat flux density.

The parameters of the double ellipsoidal heat source model adopted in this paper are shown in Table 1:

Table 1. Parameters of the double ellipsoidal heat source model.

Argument	Numerical value
Front axle length a_f (mm)	6
Rear axle length a_r (mm)	10
Width b (mm)	8
Depth d (mm)	8

2.5 Software construction

- Build the initial welding model using SolidWorks software.
  - Hyper Mesh is used to divide and process medium-term finite element mesh.
  - Welding is used as the final data processing platform and signal acquisition software to achieve a seamless connection between data processing and signal acquisition.

3. Finite element simulation results analysis and accuracy verification

3.1 Welding deformation simulation analysis

The front roof panel of the engineering vehicle is a key component of the vehicle body, with the material of 6060 aluminum alloy. The simulation platform is configured with a 6342ES processor, 128 GB memory, and a 5 T hard disk. The simulation was set at 20°C, lasting 7559 s, including 2053 s of welding and 5506 s of cooling. The welding parameters were set to 24 V and 210 A with a welding speed of 8 mm/s. The resulting deformation cloud diagram is shown in Figure 3 by Simu-fact-Welding. The peak deformation near the weld seam and heat-affected zone reached 2.57 mm, attributed to localized expansion and shrinkage of the welding material due to heat.

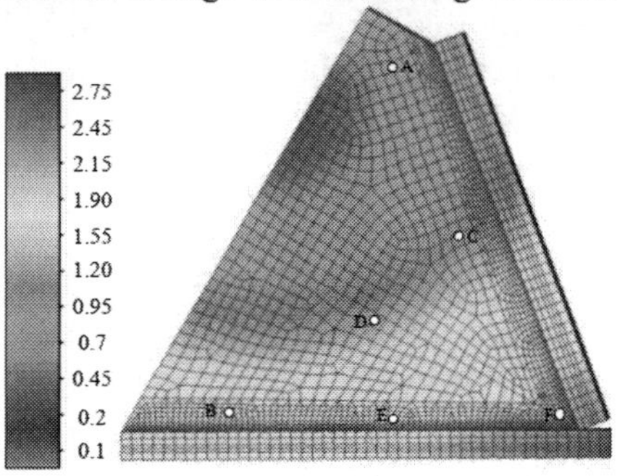

Figure 3. Deformation from finite element analysis.

3.2 Verification of welding deformation simulation accuracy

Considering that simulations are based on ideal conditions, real-world results may differ. Therefore, to verify the finite element model's accuracy, a welding robot experiment was set up with identical parameters. Special attention was given to aluminum's properties, using reverse polarity direct current and cathode atomization to effectively remove surface oxidation. ER5356 aluminum alloy wire was used for MIG welding with 99.99% pure argon as the shielding gas.

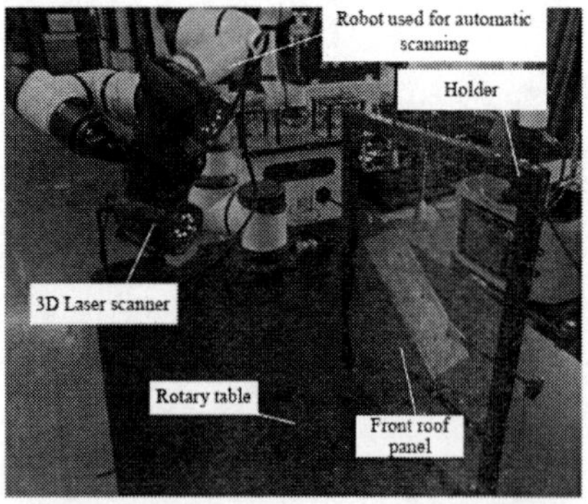

Figure 4. Laser scanning detection.

The welded front roof panel will be cooled for 1.5 hours cooling, measurements were taken using a 3D laser scanner which is widely used in automobiles, aerospace, and military enterprises, with 22 crossed beams and 1 red laser, scanning speed of 1300000 measurements per second, 420*380 mm^2 scanning area as well as a scanning accuracy of 0.044 mm. Figure 4 shows the use of a 3D laser scanner to scan the front roof panel. The 3D point cloud data, obtained from scanning before and after welding, can be imported into the supporting software Polyworks to obtain the measurement results of welding deformation. Polyworks uses scanning data before welding deformation as a reference and compares the point cloud data after welding deformation with the point cloud data before welding deformation to obtain the welding deformation amount.

To ensure the accuracy of the verification results, this article welded 5 identical front roofs and conducted measurement experiments using a 3D laser scanner. The 3D point cloud data measured by the 3D laser scanner is shown in Figure 5. Compared with the 3D point cloud data obtained from the welding deformation simulation in Figure 3, it can be seen that the deformation amount obtained from the experiment has a consistent trend with the deformation amount obtained from the simulation. At the same position in Figure 3 and Figure 5, A~E (6 points) were taken. The measured values of welding deformation, finite element simulation values, and errors are shown in Figure 6. It can be seen that the welding deformation simulation results of the six points fit well with the experimental results, and the relative error is controlled within 10%. Therefore, the finite element simulation method in this article can accurately reflect the actual welding deformation situation. The related deformation and stress analysis based on this are also scientifically reliable.

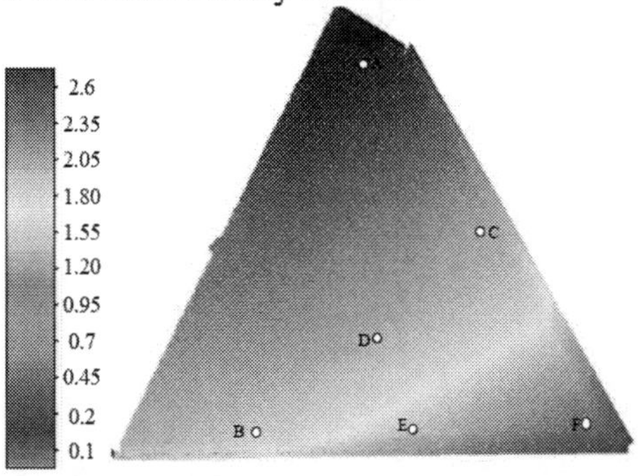

Figure 5. 3D point cloud data measured by the 3D laser scanner.

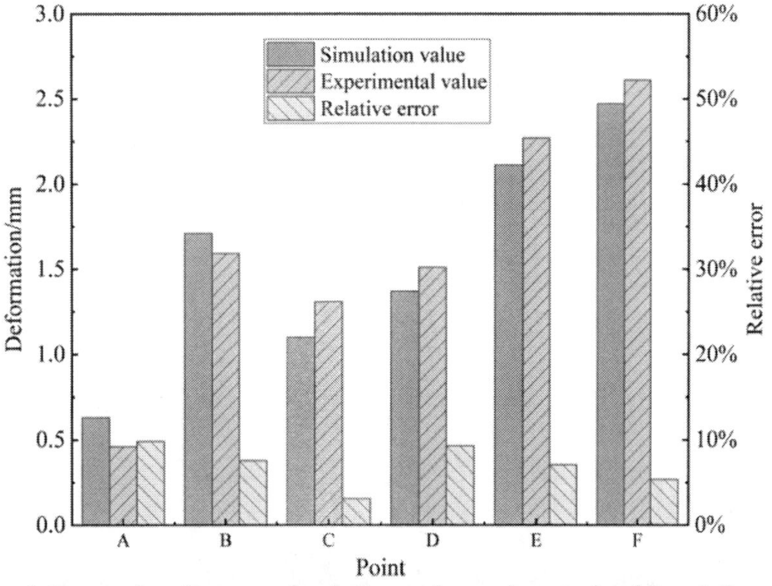

Figure 6. Comparison between simulation and experiment of welding deformation.

4. Influence of different groove angles on residual stress after welding

The finite element model is established, which is divided into horizontal plate and vertical plate. The dimensions of the horizontal plate are 250*150*30 mm, and the dimensions of the vertical plate are 250*150*15 mm, and the material is 6060. To facilitate calculation, the initial temperature of the welding temperature field is assumed to be 20°C. To study the influence of welding voltage on the temperature field and strain field, the welding voltage is changed to observe the normal welding parameters, as shown in Table 2, welding parameters. Therefore, the welding material is 6060 and the maximum temperature is 659°C, so the maximum temperature remains constant.

Table 2. Table type styles.

Welding speed/(mm/s)	Ambient temperature /°C	Welding voltage/V	Welding current/A	Heat source efficiency
7	20	20	200	0.75

4.1 Mesh division of different groove Angle models

The finite element model of 40°, 50°, and 60° is established using transition mesh, in which the number of joints in the weld area is more, while the number of joints away from the weld area is gradually reduced. The welded models of these three grooves contain a total of 7696, 7437, and 7474 nodes, and 5364, 5076, and 4968 units. In addition, to better simulate the actual welding process, the finite element meshing of each pass is also performed according to the sequence in Figure 7.

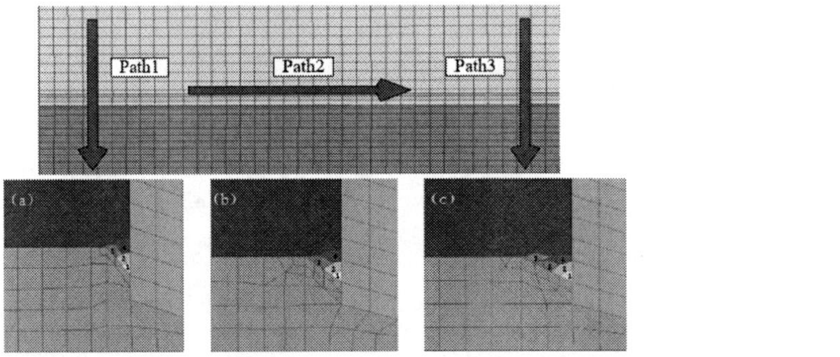

(a) 40°; (b) 50°; (c) 60°

Figure 7. Grid division of conventional bevel.

4.2 Welding residual stress distribution for different groove angle models

To quantitatively analyze the influence of different groove angles on the distribution of residual stress, the distribution of transverse residual stress and longitudinal residual stress of welded joints along Paths 1, 2, and 3 under different groove angles is studied in Figure 8 and Figure 9.

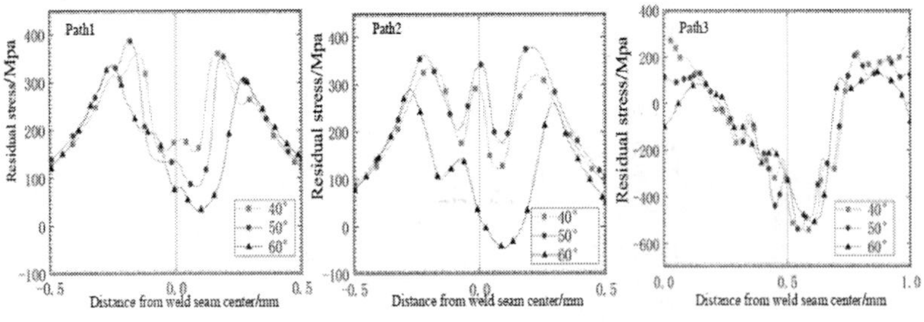

Figure 8. Lateral residual stress along different paths.

Journal of Physics: Conference Series **2808** (2024) 012043 doi:10.1088/1742-6596/2808/1/012043

The lateral residual stress of three different groove angles along the path is distributed in Figures 8 (a), (b), and (c). When the groove Angle is 40°, 50°, and 60°, the peak stress is concentrated near the heat-affected zone, and the peak lateral residual stress is 359, 390, and 341 MPa.

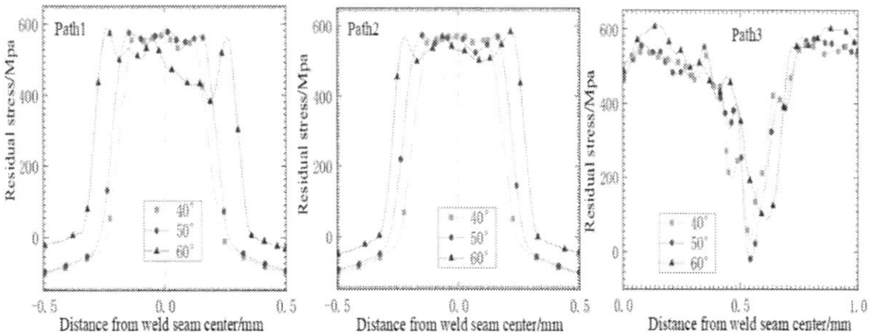

Figure 9. Longitudinal residual stress along different paths.

Figure 9 shows the distribution of longitudinal residual stress in three directions for welded joints with three different groove angles, from the 40°, 50°, and 60° groove angles, while the distance between the position of peak stress and the weld center line will also become farther, and the range of longitudinal residual stress distribution will also become larger.

4.3 Distribution of welding residual deformation in different groove angle models

Due to the similarity of Paths 1 and 3, the distribution of longitudinal deformation of welded joints with different groove angles only on Paths 1 and 2. The distances from weld seam center of Path1 and Path2 are shown in Figure10. For Path 1, the welded joint with a 60° groove angle has the largest deformation on Path 1, reaching a peak value of 7 mm. The welded joint with a 50° groove angle is second, and the peak deformation is 6.3 mm; The welded joint with a 40° groove angle has the smallest deformation, with a peak value of only 5.9 mm. Similar to Path 2, a smaller groove angle can reduce the deformation of welded joints and improve the welding quality.

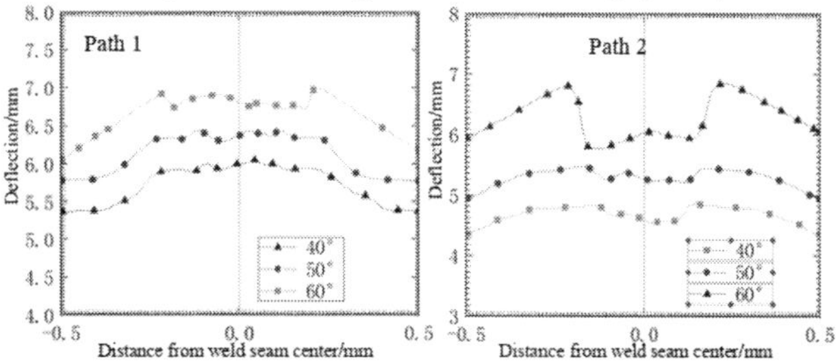

Figure 10. Calculation results of welding residual deformation.

5. Welding simulation verification of the front top deck of the armored vehicle

As the temperature of the weld increases, the grains in the area around the solder joint will grow and form new grains. As the welding part gradually cools, the temperature gradually decreases, resulting in solidification and contraction of the material in the heat-affected zone and the metal melting zone of the welding part, resulting in deformation. Therefore, the internal structure and mechanical properties of the welding parts will also change, and eventually cause different degrees of plastic deformation of the welding parts. With the welding part completely cooled to room temperature, due to the influence of plastic yield, a certain degree of deformation will inevitably occur during the welding process.

According to the arc welding process, the welding groove angle is 40°, the welding current is 220 A, the welding voltage is 24 V, and the welding speed is 8 mm/s. The deck model has four welds, each of which has four welding processes, which is a typical multi-layer and multi-pass welding model, as shown in Figure 11. The specific welding process is shown in Table 3.

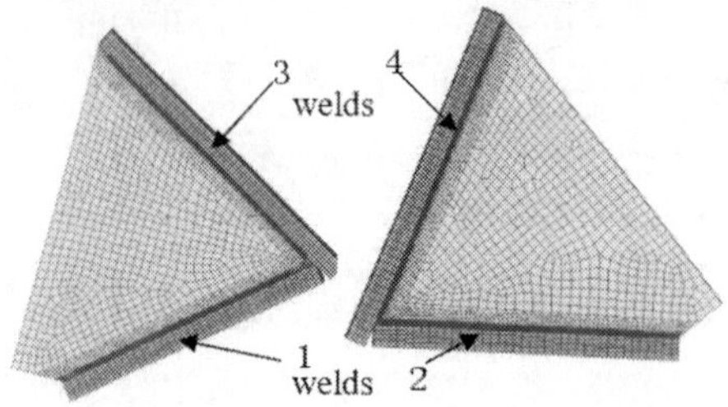

Figure 11. Schematic diagram of welding sequence of the deck model.

The welding parameters are changed to observe the normal welding parameters, as shown in Table 4. The welding material is 6060 and the maximum temperature is 659°C, and the maximum temperature does not change. According to Figure 12 (a), it is known that when the voltage is 24 V, the current is 220 A, and the speed is 8 mm/s. The deformation during welding will gradually increase with the increase of welding time, and when it is about 300 seconds, the deformation will start to decline from the first peak. After falling to about 540 seconds, it began to gradually increase until the deformation after the final welding was maintained at 2.54 mm. According to Figure 12 (b), the residual stress left after welding was between 0.022 Mpa and 114.62 Mpa.

Table 3. Parameters of multi-layer and multi-pass welding.

Weld number	Number of weld layers	Weld Length	Initial time	End time	Welding speed
1	1	498.95	0	62	8
	2	494.533	63	124	8
	3	490.186	125	185	8
	4	493.033	186	247	8
2	1	499.217	248	309	8
	2	494.67	310	371	8
	3	491.293	372	432	8
	4	493.038	433	494	8
3	1	543.838	495	562	8
	2	544.038	563	630	8
	3	540.938	631	698	8
	4	545.685	699	766	8
4	1	545.685	766	834	8
	2	543.685	835	902	8
	3	541.905	903	970	8
	4	542.251	971	1037	8

Table 4. Welding parameter.

Welding speed /(mm/s)	Ambient temperature/°C	Welding voltage /V	Welding current /A	Heat source efficiency
7-9	20	22-26	200-240	0.8

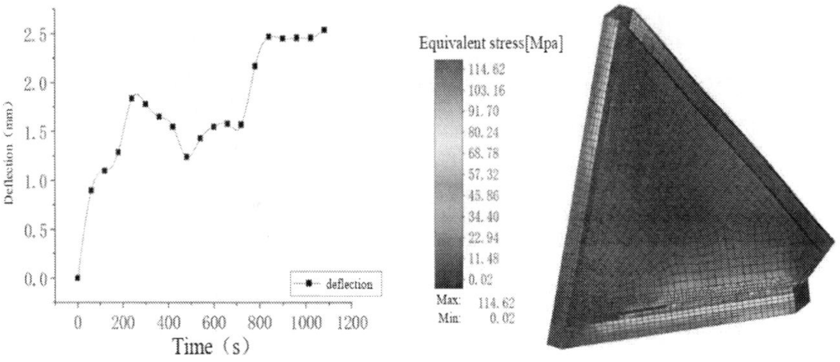

(a) Deflection (b) Residual stress

Figure 12. Deformation and residual stress diagram@24 V-220A-8 mm/s.

5.1 Analysis of residual stress of deformation when welding voltage changes

First, the deformation is shown in Figure 13(a) when the welding voltage is reduced to 22 V. Secondly, according to Figure 13(a), when the voltage is 22 V, the current is 220 A, and the speed is 8 mm/s, the deformation during welding gradually increases with the increase of welding time. It continues to increase until the welding is complete, and there is no decreasing trend in the middle of the welding. Finally, the deformation after welding remains at 3.78 mm. Through Figure 13(b), the residual stress left after welding is between 0.032 Mpa and 114.71 Mpa. By comparing with Figure 12, it can be seen that when the welding voltage is reduced, the increase in deformation during welding also becomes larger, but the change in residual stress is not significant.

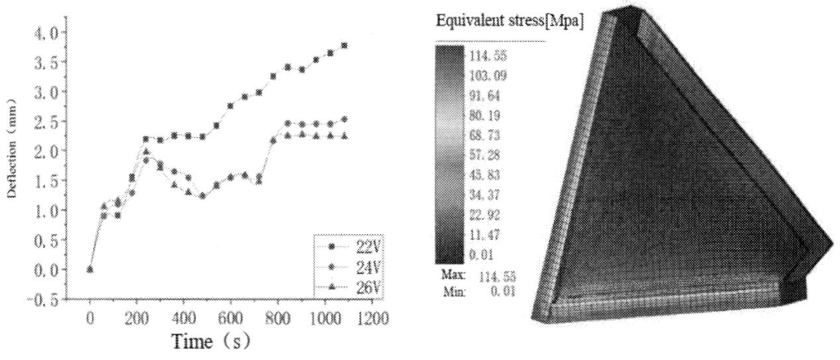

(a) Deflection (b) Residual stress

Figure 13. Deformation and residual stress diagram@22V-220A-8 mm/s.

5.2 Influence of welding current variation on residual stress of deformation

By reducing the welding current to 200 A, the deformation during welding is shown in Figure 14(a). Secondly, when the voltage is 26 V, the current is 200 A, and the speed is 8 mm/s, the deformation during welding gradually increases with the increase of welding time. However, at around 300 seconds, there is a small fluctuation, and afterward, until the welding is complete, there is no obvious decreasing trend. Finally, the deformation after welding remains at 5.12 mm. Through Figure 14(b), the residual stress left after welding is between 0.03 Mpa and 114.71 Mpa. By comparing with Figure 12, it can be seen that as the welding current decreases, the increase in deformation during welding also becomes larger, but the change in residual stress is not significant.

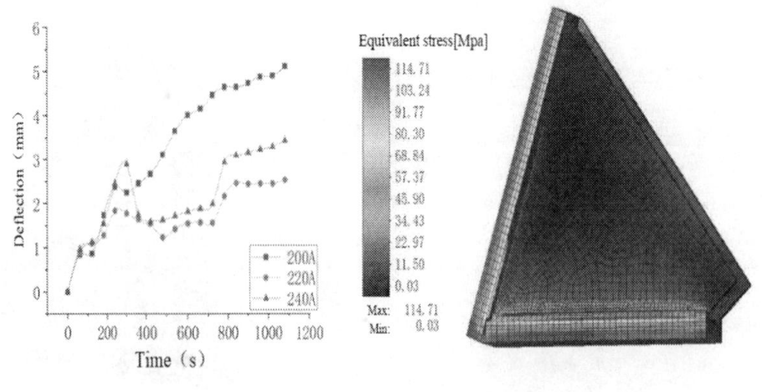

(a) Deflection (b) Residual stress

Figure 14. Deformation and residual stress diagram@26 V-200 A-8 mm/s..

5.3 Influence of welding speed on residual stress of deformation

As the welding speed increases, the energy is reduced, leading to a decrease in both the penetration depth and width of the weld. This is because the amount of wire metal deposited per unit length of weld is inversely proportional to the welding speed, resulting in a decrease in the height of the weld bead. Additionally, the relationship between the weld width and the square root of the welding speed is inversely proportional.Firstly, the deformation during welding is shown in Fig.15(a) by reducing the welding speed to 7mm/s. Secondly, from Fig.15 (a), it can also be observed that the deformation during welding increases gradually with increasing welding time when the voltage is 24V, the current is 220A, and the speed is 7mm/s. As time goes on, it starts to decrease at the first peak and gradually increases at about 600 seconds until the welding is completed. The deformation after welding is maintained at 3.04mm from Fig.15(b).

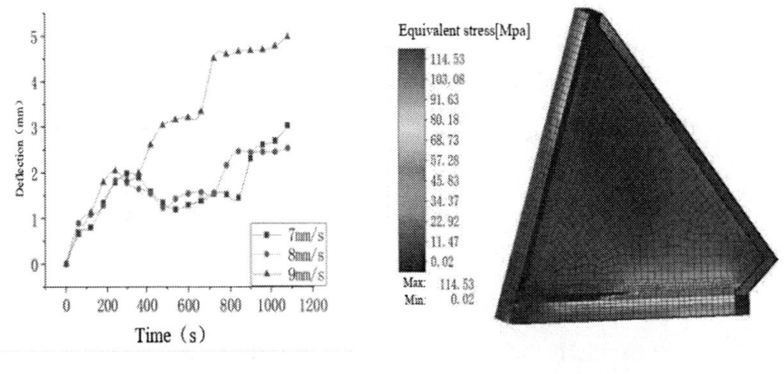

(a) Deflection (b) Residual stress

Figure 15. Deformation and residual stress diagram@24 V-220A-7 mm/s.

By reducing the welding speed to 7 mm/s, the deformation during welding is shown in Figure 15(a). Secondly, according to Figure 15(a), when the voltage is 24 V, the current is 220 A, and the speed is 7 mm/s, the deformation during welding gradually increases with the increase of welding time. However, at around 300 seconds, the first peak starts to decrease, and at around 600 seconds, it starts to gradually rise until the welding is complete. Finally, the deformation after welding remains at 3.04 mm. Through Figure 15(b), the residual stress left after welding is between 0.02 Mpa and 114.53 Mpa. By comparing with Figure 12, it can be seen that as the welding speed decreases, the deformation during welding increases by approximately 0.5 mm compared to that under normal parameters, but the residual stress slightly decreases.

6. Conclusion

Welding requires the establishment of welding models to understand welding materials and parameters before welding can be carried out. This article uses two models, using the method of corner connection, to study the changes in welding groove angle and residual stress, as well as the changes in different welding parameters and residual stress and deformation. Through observation, the following conclusions are obtained:

1) The accuracy of the deformation obtained by finite element simulation was verified by a three-dimensional laser scanner. The relative error between the simulation results and the experimental results was controlled within 10%.

2) The changes in welding groove angle residual stress and residual deformation are analyzed through L-shaped plates. Mainly through three groove angles to observe, it is found that when the welding groove angle is 50°, the transverse residual stress is the largest, and when it is 60°, the transverse residual stress is the smallest, but the longitudinal residual stress is not much different. When analyzing its residual deformation, it is found that when the welding groove angle is 60°, the residual deformation is the largest, and when it is 40°, the residual deformation is the smallest.

3) The changes in welding parameters and deformation are analyzed through the front top deck of the armored vehicle. Mainly through three directions of welding voltage, welding current, and welding speed to observe. The range of welding parameters is shown in Table 4. It is found that when the welding voltage is smaller, the deformation is larger, and the residual stress is larger when the welding voltage is larger. When the welding current is smaller, the deformation is larger, and the residual stress is larger when the welding current is larger. When the welding speed is faster, the deformation will also increase, and the residual stress will decrease with the increase of welding speed.

Acknowledgement

This paper is supported by Tianjin Education Commission Research Program Project 2022ZD034, China.

References

[1] Hou, Z. (2006). Prediction and control of welding deformation of sheet structures. PhD dissertation, Huazhong University of Science and Technology. DOI: 10.7666/d. d009604.

[2] Zeng, J. (2005). The research for prediction of welding deformation of stiffened panels (Master's thesis, Harbin Engineering University). DOI: 10.7666/d. y670799.

[3] Wang, Y., Tang, X. (2012). A review of numerical analysis on welding stress and deformation of aluminum alloy. Welding Technology, (01), 1-4. DOI: 10.13846/j.cnki.cn12-1070/tg.2012.01.002.

[4] Zhao, L., Zhang, K. (2012). Numerical simulation of the welding deformation based on a new inherent strain method. Engineering Mechanics, (08), 33-38. DOI: 10.6052/j.issn.1000-4750.2010.11.0861.

[5] Guo Nan, Yu Yongjian, Yin Xianqing, & Yang Fang. (2019). Prediction of buckling deformation in thin plate TIG surfacing. Journal of Welding, 40(10), 6. DOI: 10.12073/j.hjxb.2019400272.

[6] Lu, A., Shi, Q., Zhao, H., Wu, A. (2001). Three-dimensional finite element numerical simulation of temperature field and stress field in thick plate welding process. China Mechanical Engineering, (02).

[7] Wu, Z., Wang, F., An, G., Ding, P. (2015). Analysis of welding residual stress of Structure by thermal elastic-plastic finite element method. Heat Treatment of Metals, (13), 201-204+207. DOI: 10.14158/j.cnki.1001-3814.2015.13.061.

[8] Yang Wen, Shi Yongji, Wang Yuanqing, & Shi Gang. (2007). Three-dimensional finite element analysis of residual stress in structural steel welding. Journal of Jilin University: Engineering Edition, 37(2), 6. DOI: 10.3969/j.issn.1671-5497.2007.02.020.

[9] Chen, C., Wang, J., Yang, H. (1984). Solution of thermal processing stress using thermo-elastic-plastic finite element method. Shanghai Mechanics, (04), 39-46. DOI: CNKI: SUN: SHLX.0.1984-04-004.

[10] Chen Zhanglan & Xiong Yunfeng. (2010). Numerical prediction of welding deformation of hull plate columns based on inherent strain. Journal of Jimei University: Natural Science Edition, 15(3), 5. DOI: 10.3969/j.issn.1007-7405.2010.03.011.

[11] Wang Wei & Wan Lujun. (2010). A brief discussion on the causes and prevention of welding deformation and stress. Science Communication (7), 2. DOI: CNKI: SUN: KJCB.0.2010-07-083.

[12] Ueda, Y. & Fukuda, K. (1983). Simplified measuring method of three-dimensional welding residual stresses. JOURNAL OF THE JAPAN WELDING SOCIETY, 52(2), 110-117. DOI:10.2207/qjjws1943.52.110.

[13] Ueda, Y., Yuan, M. G., Mochizuki, M., Umezawa, S., & Enomoto, K. (1993). Experimental verification of a method for prediction of welding residual stresses in T joints using inherent strains 4th report: Method for prediction using a source of residual stress. DOI: 10.1080/09507119309548506

[14] You Min, Zheng Xiaoling, & Yu Haizhou. (2003). Discussion on the formation mechanism of welding residual stress. Journal of Welding, 24(2), 5. DOI: CNKI: SUN: HJXB.0.2003-02-013.

[15] Li, D. (2003). Numerical simulation study on welding stress and deformation. (Doctoral dissertation, Wuhan University of Technology). DOI: 10.7666/d.y496681.

[16] Ma Xiangping, & Luo Qingguo. (2002). Study on finite element method for transient temperature field solution. Journal of Academy of Armored Force Engineering (2), 5. DOI: 10.3969/j.issn.1672-1497.2002.02.005.

[17] Khurram, A. (2013). Welding Residual Stress Investigation in Pipe-Plate Joint of Austenitic Stainless Steel & Low Carbon Steel. (Doctoral dissertation, Harbin Engineering University). DOI:10.7666/d. D595095.

[18] Shi, W. J. (2006). Some studies on the finite element method and its applications. (Doctoral dissertation, Guilin University of Electronic Technology). DOI:10.7666/d. D561889.

[19] Langer, W. (2002). An overview of in-stream mining of sand and gravel resources. DOI:10.3133/ofr02153.

[20] Kromm, A., Mee, V. V. D., Kannengiesser, T., & Kalfsbeek, B. (2015). Properties and weldability of modified low transformation temperature filler wires. Welding in the World, 59(3), 413-425. DOI:10.1007/s40194-014-0215-y.

[21] Gannon, L., Liu, Y., Pegg, N., & Smith, M. (2010). Effect of welding sequence on residual stress and distortion in flat-bar stiffened plates. Marine Structures, 23(3), 385-404. DOI: 10.1016/j.marstruc.2010.05.002.

[22] Fallahi, A., Jafarpur, K., & Nami, M. R. (2011). Analysis of welding conditions based on induced thermal irreversibility in welded structures: cases of welding sequences and preheating treatment. Scientia Iranica, 18(3), 398-406. DOI: 10.1016/j.scient.2011.05.030.

[23] A, Z. C., A, Z. C, & B, R. A. S. (2015). Influence of welding sequence on welding deformation and residual stress of a stiffened plate structure - science direct. Ocean Engineering, 106, 271-280. DOI: 10.1016/j.oceaneng.2015.07.013.

[24] Zhang, H. W., Wang, H., & Wang, J. B. (2007). Parametric variational principle-based elastic–plastic analysis of materials with polygonal and Voronoi cell finite element methods. Finite Elements in Analysis & Design, 43(3), 206-217. DOI: 10.1016/j.finel.2006.09.001.

[25] Ma Zheshu, Yao Shouguang, & Ming Xiao. (2003). Microscale Heat Transfer and its research progress. Natural Journal, 25(2), 4. DOI: 10.3969/j.issn.0253-9608.2003.02.003.

Journal of Physics: Conference Series **2808** (2024) 012043 doi:10.1088/1742-6596/2808/1/012043

[26] Liu Zhenbo, Zhao Junjing, Li Yang, Li Yinmao, Yi Fan, Chen Yue, & Luhua Ying. (2017). Research on deformation control technology for multi-layer thick plate welding structure products. Welding Technology (06), 81-85. DOI: 10.13846/j.cnki.cn12-1070/tg.2017.06.023.

[27] Qian Weichang. (1979). Research on the generalized variational principles in elastic theory and its application in finite element calculations. Mechanics in Practice (01), 18-26. DOI: CNKI: SUN: LXYS.0.1979-01-004.

[28] Li, Donglin. (2003). Numerical Simulation Study on Welding Stress and Deformation. (Dissertation, Wuhan University of Technology). DOI:10.7666/d. y496681.

[29] Liao Guofeng, & Lu Dandan. (2015). Research on welding process optimization of loader structural parts. Equipment Manufacturing Technology (8), 3. DOI: CNKI: SUN: GXJX.0.2015-08-082.

[30] Chen, Chu, Wang Jianhua, & Yang Hongqing. (1983). Finite Element Analysis and Calculation of Nonlinear Welding Heat Conduction." Welding Journal, vol. 03, 34-43. DOI: CNKI: SUN: HJXB.0.1983-03-003.

Journal of Physics: Conference Series **2808** (2024) 012044 doi:10.1088/1742-6596/2808/1/012044

Application evaluation of TORSION DRILLING string

Zengbao Zhang[1, 2*], Jianbo Lv[1, 2], Xiaoming Zhu[1, 2] and Yuxin Liu[1, 2]

[1]Sinopec Xinjiang Xinchun Petroleum Development Co., Ltd., Dongying, 257000, Shandong, China

[2]Drilling technology research institute of Sinopec Shengli Petroleum Engineering Co., Ltd., Dongying, 257000, Shandong, China

*Corresponding author email: zhangzb.slyt@sinopec.com

Abstract: The drill string torsion system, represented by Slider and Rocket, has been widely used around the world. It provides a solution with a high performance-price ratio to solve the problem of supporting pressure in sliding drilling. In recent years, the torsional system of drill string has been applied to a certain extent in China and has achieved good results in reducing resistance and increasing speed. Based on the analysis of the system composition, system functions, technical advantages, and application of the TORSION DRILLING string TORSION system, this paper summarizes the results of the field application of the system, some problems that need to be solved urgently and the improvement measures to be taken are analyzed. The results show that: (1) the application statistics of more than ten wells show that the torsional system of drill string can reduce the friction by more than 70% and increase the directional penetration rate by 20-300% when the supporting pressure is serious; (2) the angle torsion control is superior to torque torsion control in practicability, tool face placement and control, but the angle control depends on the availability of rotary table or top drive motor encoder, which brings great inconvenience to the popularization and application; (3) the current drill string torsion system suffers from a lack of automation, relying heavily on the expertise of field personnel, which limits its effectiveness. In response to these shortcomings, there is an urgent need to develop automatic torsion control software to enhance the system's intelligence and efficiency. It is concluded that the TORSION DRILLING string TORSION system can significantly reduce directional friction, decrease the number and time of swinging tool face, increase the length and time of single directional DRILLING, and greatly increase the directional rop, to meet the requirements of torsion directional process.

1. Research purpose of drill column torsion system

Sliding drilling is still the most commonly used directional well, horizontal well trajectory control. In the course of sliding direction, the drilling tool does not rotate, and the friction between the drilling tool and the wellbore is intermittent static friction. It is characterized by the challenge of tool face placement/control, the difficulty of drilling tool lowering, and susceptibility to pump hold-ups, which seriously affects drilling efficiency and increases the risk of differential pressure sticking. Current solutions include the use of rotary steering, the addition of downhole vibration equipment (such as hydraulic oscillators), or the addition of lubricants. Owing to the risk of equipment falling down and the high cost of operation, economical and effective solutions are urgently needed. The drill string torsion system is a kind of auxiliary system developed in recent years that is applied to the conventional sliding direction. The drill string is divided into three sections from the surface to the bottom of the well, namely, the surface rotation control zone, the static friction zone, and the reverse torque zone (see Figure 1). The torsional control system accurately controls the two-way torsional amplitude of the drill string, maximizes the rotation control area, reduces the static friction area, and does not affect the reverse torque area, improving the efficiency of orientation [1-4].

Figure 1. The distribution of drill string torsional directional motion.

Some typical TORSION DRILLING systems at home and abroad are the Slider system of Schlumberger company, Rocket System of CANRIG company, intelligent ground orientation system of Polaris, Pipe Rock System of Chuanqing Drilling Engineering Co., Ltd. of PetroChina, and TORSION DRILLING system of DRILLING technology research institute of Shengli Petroleum Engineering Co., Ltd. of Sinopec. The Slider system has been applied to more than 2,000 wells in the Middle East and US shale gas blocs, the Rocket system serves more than 1,000 wells worldwide, and the Polaris intelligent ground orientation system was successfully applied in North America in 2016. Pipe rock began to be used in Sichuan and Chongqing in 2017, and by the end of 2019, it had been used in more than 100 wells. Torsion DRILLING string control system has been used since July 2019. By June 2020, 12 wells had been extended in Shengli, Pengzhou, and other areas, and good results had been achieved in drag reduction and acceleration [5-7].

2. TORSION DRILLING Introduction of drill string torsion system

2.1. Study on the vibration mechanism of the drill column

$$\begin{aligned} J_r \ddot{\theta}_r + c(\dot{\theta}_r - \theta_b) + k(\theta_r - \theta_b) + c_r \dot{\theta}_r &= T_m \\ J_b \ddot{\theta}_b + c(\dot{\theta}_b - \theta_r) + k(\theta_b - \theta_r) + c_b \dot{\theta}_b &= -T_{fb}(x) - T_{ab} \end{aligned} \tag{1}$$

The kinetic equation of drill column is:

The simplified equation for the torsion dynamics of the drill column is:

$$\frac{d^2\theta_r}{dt^2} = -\frac{c}{J_r}\left(\frac{d\theta_r}{dt} - \frac{d\theta_b}{dt}\right) - \frac{k}{J_r}(\theta_r - \theta_b) - \frac{c}{J_r}\frac{d\theta_r}{dt} + \frac{1}{J_r}T_m$$
$$\frac{d^2\theta_b}{dt^2} = -\frac{c}{J_b}\left(\frac{d\theta_b}{dt} - \frac{d\theta_r}{dt}\right) - \frac{k}{J_b}(\theta_b - \theta_r) - \frac{c_b}{J_b}\frac{d\theta_b}{dt} + \frac{1}{J_b}T_{fb} \tag{2}$$

2.1.1.Establishment of the mathematical model of the drill column system. In order to study the sticky slip vibration between the bit and rock, this section is based on the classical drilling torsion pendulum model modeling the drilling column system. However, given the intricate nature of the actual drilling system, encompassing factors such as structural design, shaft conditions, and external influences on comprehensive movement, it proves challenging to directly articulate through mathematical models. For research convenience, this paper adopts certain simplified and idealized assumptions, outlined as follows:

(1) The influence of drilling fluid on the drilling column is simplified to the viscous damping effect, and the friction between the rock layer and the drill bit is treated centrally and described by the nonlinear friction model. (2) Only the sticky slip vibration of the drill column is studied here, excluding the longitudinal beating and transverse vibration. (3) The top drive device, drill tool combination, drill bit, and so on are simplified into centralized mass blocks, and the drill pipe is simplified into centralized mass and connected by springs. The spring has torsional stiffness.

After the study, the kinematic differential equation of the whole drill column system is established as follows:

$$\begin{cases} J_r\ddot{\theta}_r = T_m - c_r\dot{\theta}_r - k_1(\theta_r - \theta_1) - c_1(\dot{\theta}_r - \dot{\theta}_1) \\ J_1\ddot{\theta}_1 = -k_1(\theta_1 - \theta_r) - c_1(\dot{\theta}_1 - \dot{\theta}_r) - k_2(\theta_1 - \theta_2) - c_2(\dot{\theta}_1 - \dot{\theta}_2) \\ J_2\ddot{\theta}_2 = -k_2(\theta_2 - \theta_1) - c_2(\dot{\theta}_2 - \dot{\theta}_1) - k_3(\theta_2 - \theta_3) - c_3(\dot{\theta}_2 - \dot{\theta}_3) \\ J_3\ddot{\theta}_3 = -k_3(\theta_3 - \theta_2) - c_3(\dot{\theta}_3 - \dot{\theta}_2) - k_4(\theta_3 - \theta_4) - c_4(\dot{\theta}_3 - \dot{\theta}_4) \\ \vdots \\ J_n\ddot{\theta}_n = -k_n(\theta_n - \theta_{n-1}) - c_n(\dot{\theta}_n - \dot{\theta}_{n-1}) - k_{np}(\theta_n - \theta_p) - c_{np}(\dot{\theta}_n - \dot{\theta}_p) \\ J_p\ddot{\theta}_p = -k_{np}(\theta_p - \theta_n) - c_{np}(\dot{\theta}_p - \dot{\theta}_n) - k_{pb}(\theta_p - \theta_b) - c_{pb}(\dot{\theta}_p - \dot{\theta}_b) \\ J_b\ddot{\theta}_b = -k_{pb}(\theta_b - \theta_p) - c_{pb}(\dot{\theta}_b - \dot{\theta}_p) - c_b\dot{\theta}_b - T_{fb} \end{cases} \tag{3}$$

The differential equation can be converted to:

$$\begin{cases} \ddot{\theta}_r = -\frac{k_1}{J_r}(\theta_r - \theta_1) - \frac{c_1}{J_r}(\dot{\theta}_r - \dot{\theta}_1) + \frac{T_m}{J_r} - \frac{c_r}{J_r}\dot{\theta}_r \\ \ddot{\theta}_1 = -\frac{k_1}{J_1}(\theta_1 - \theta_r) - \frac{c_1}{J_1}(\dot{\theta}_1 - \dot{\theta}_r) - \frac{k_2}{J_1}(\theta_1 - \theta_2) - \frac{c_2}{J_1}(\dot{\theta}_1 - \dot{\theta}_2) \\ \ddot{\theta}_2 = -\frac{k_2}{J_2}(\theta_2 - \theta_1) - \frac{c_2}{J_2}(\dot{\theta}_2 - \dot{\theta}_1) - \frac{k_3}{J_2}(\theta_2 - \theta_3) - \frac{c_3}{J_2}(\dot{\theta}_2 - \dot{\theta}_3) \\ \ddot{\theta}_3 = -\frac{k_3}{J_3}(\theta_3 - \theta_2) - \frac{c_3}{J_3}(\dot{\theta}_3 - \dot{\theta}_2) - \frac{k_4}{J_3}(\theta_3 - \theta_4) - \frac{c_4}{J_3}(\dot{\theta}_3 - \dot{\theta}_4) \\ \vdots \\ \ddot{\theta}_n = -\frac{k_n}{J_n}(\theta_n - \theta_{n-1}) - \frac{c_n}{J_n}(\dot{\theta}_n - \dot{\theta}_{n-1}) - \frac{k_{np}}{J_n}(\theta_n - \theta_p) - \frac{c_{np}}{J_n}(\dot{\theta}_n - \dot{\theta}_p) \\ \ddot{\theta}_p = -\frac{k_{np}}{J_p}(\theta_p - \theta_n) - \frac{c_{np}}{J_p}(\dot{\theta}_p - \dot{\theta}_n) - \frac{k_{pb}}{J_p}(\theta_p - \theta_b) - \frac{c_{pb}}{J_p}(\dot{\theta}_p - \dot{\theta}_b) \\ \ddot{\theta}_b = -\frac{k_{pb}}{J_b}(\theta_b - \theta_p) - \frac{c_{pb}}{J_b}(\dot{\theta}_b - \dot{\theta}_p) - \frac{c_b}{J_b}\dot{\theta}_b - \frac{T_{fb}}{J_b} \end{cases} \tag{4}$$

2.2. System composition

2.2.1. System hardware. TORSION DRILLING system hardware, including the main control box, TORSION communication box, and data acquisition unit, is displayed in Figure 2. The torsion control box is installed in the driller's room for the driller's directional engineering operation. The torsion communication box is installed in the VFD room. The data acquisition unit mainly collects the motor speed and angle signals, brake, fan, and oil pump status signals. The torsional system of drill string can be operated independently in parallel with the top drive (or rotary table) electric control system. The directional end can switch back to the top drive (or turntable) of the original electronic control system, and the two systems can switch between convenience.

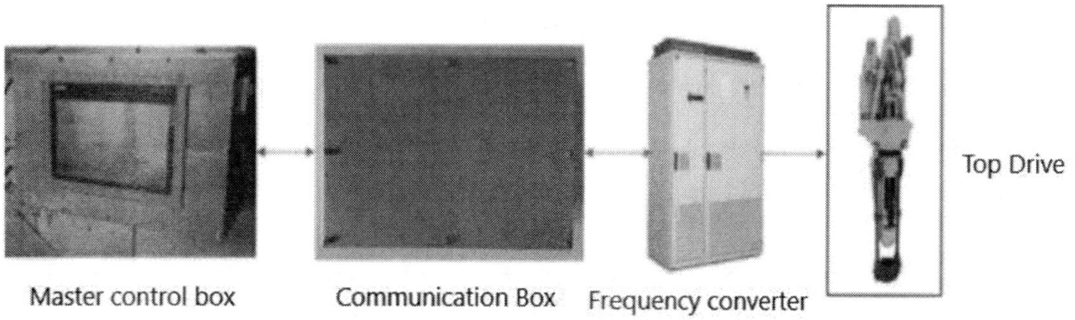

Figure 2. TORSION DRILLIN system hardware schematic.

2.2.2. System software. TORSION DRILLING system software includes touch screen operation software and PLC control software, respectively, installed in the main control box on the touch screen and PLC. The touchscreen operation mainly includes four interfaces: compound drilling, angle torsion, torque torsion, and parameter setting.

2.3. System functions

(1) To solve the supporting pressure and speed up the directional section, through the bidirectional torsion of the drill string, the length of the static friction zone is reduced to the greatest extent, the supporting pressure is reduced, and the bit weight transmission is improved. The speed of the directional section is increased by 20-300%.

(2) Quick and automatic adjustment of tool face: the tool face can be adjusted in real time by the function of torsion and impact, and the tool face can be controlled automatically if the data of the tool face is introduced into the system.

(3) quick tool face placement: tool face placement during torsion to minimize the static time of the drill tool, twist and lower the drill tool to improve the directional efficiency.

2.4 Technical advantages

(1) A TORSION DRILLING system can be applied to all types of electric rotary tables or top drives. The TORSION system runs independently in parallel with the original electric control system of the rotary table/Top Drive, and the switching between the two systems is convenient.

(2) The system has many functions, such as angle torsion, torque torsion, angle impact, torque impact, etc.

(3) It is a surface control system that has no downhole components and is not affected by high temperature, high pressure, and formation instability.

(4) Electronic control system, high reliability, low maintenance and use cost, cost-effective.

3. Field application and effect analysis of torsion control system

Since July 2019, TORSION DRILLING control systems have been installed in the Yellow River DRILLING, Bohai Sea DRILLING, and so on. There are four sets of electric rotary table torsion control systems, and 12 wells have been popularized and applied.

3.1. Analysis of key wells

3.1.1. Y13-X40#. This well is a straight-increasing-stabilizing-lowering four-section casing structure. The designed depth is 2719.02 m, the maximum deviation is 33.73 o, the azimuth is reduced from 220.86 o (1342.72 m) to 183.18 O (1615.14 m), and the deviation is reduced from 33.73 o (1777.47 m) to 5.48 O (2719.02 m). During the actual drilling, due to the long downhole section, the downhole point gradually deepens, the friction gradually increases, and the supporting pressure is serious. The drill string torsion system was used from 2380m to 2670m seven times. The torsion angle was 500-900 °, the torsion speed was 12-22 rpm, and the torsion directional footage was 50.54 m. The friction is reduced from 160 KN-320 KN to 20-30 kn, the directional velocity is increased from 1.5 m/h to 4 m/h, the effect of drag reduction and the speed increase is better, and finally, the downhole deviation is 5.6 °, which meets the design requirement.

3.1.2 P3-4#. Well P3-4 is a natural gas development well. It is designed for a three-open well structure and a three-dimensional trajectory horizontal well. The application interval is in the lower part of the second open hole. The length of the open hole interval is 2521m (well depth 3128-5649m), and the hole size is large (241.3 mm), featuring high drilling fluid density (1.97 g/cm^3) and high circulating pressure (28 Mpa). This makes it challenging to improve drilling efficiency due to high friction, severe supporting pressure, and sticking. In order to prevent the drilling tools from sticking, the drilling tool is lifted up every 20 minutes, and the top drive drill string torsion system is used five times. The drilling tool length is 5649-5969 m, the friction of sliding drilling is 180-200 kn, and the directional drilling time is 32-45 min/m. The drilling tool is lifted up every 20 minutes, and the top drive drill string torsion system is used five times. The friction is reduced from 180-200 kn to 40-60 kn, the drilling time is reduced to 15-25 min/m, and the time interval of lifting the drilling tool is extended to 40 min.

Journal of Physics: Conference Series **2808** (2024) 012044 doi:10.1088/1742-6596/2808/1/012044

3.1.3. Y13-X19#. The designed depth of the well is 2740.47 m, the maximum deviation is 34.86 °, and the azimuth is increased from 155.31 ° (2028.31 m) to 188.13 ° (2346.82 m). In the 2121.80-2304.21 m interval, the drill string torsion system is used, the cumulative torsion directional footage is 75.90 m, the torsion angle is 400-900%, and the torsion speed is 8-15 rpm. After using the system, the azimuth increased from 160 to 184, the friction decreased from 160-200 kn to 20-40 kn, and the orientation velocity increased from 0.5-0.8 m/h to 3-6 m/h. The designed second open drilling period is 13.58 d, and the actual drilling period is 9.5 d. The drilling fluid lubrication material is not used during the course of torsion orientation, which saves the cost of drilling fluid, and the effect of increasing drilling speed is remarkable.

3.2. Effect evaluation

(1) With 11 wells used in the oil field, their depths are generally between 2700-3700m, and the directional friction can be up to 300KN because of high deviation, late torsion direction, improper adjustment of drilling fluid performance, etc. The phenomenon of supporting pressure is serious, which causes difficulty in orientation and seriously affects the speed-up of drilling. After applying the TORSION DRILLING string TORSION system, by optimizing the TORSION parameters, the friction can be reduced to less than 40kN, the directional aging can be increased by 20-300%, and the effect of drag reduction and speed increase is noticeable.

(2) Comparing angle torsion and torque torsion, angle torsion is better than torque torsion in terms of convenience of operation, control, and adjustment of the tool face.

(3) The application of a top-drive version torsion control system in deep directional wells in marine formations has achieved certain results. Still, the application interval is less, and the time is shorter. The torsion technology and parameter optimization have great differences compared with Shengli shallow wells, and the deep well torsion directional technology is not mature yet. Hence, it needs to be applied and perfected continuously.

4. Existing problems and improvement measures

4.1. Existing problems

(1) To improve the compatibility of the drill string torsion system, whether it is the top drive version or the electric rotary table version, the installation and application of the system should match with the original Top Drive/electric rotary table electric control system and mechanical unit. After research, the current torsion control system and the original electrical control system can be completely independent of parallel operation, thus achieving a real-time switch between the systems. The main problem is that the angle-related torsional motion, such as angle torsion and impact, all depend on the installation of the encoder of the top drive/rotary motor, basically not available; the old top-drive encoder damage probability is larger, some do not have the encoder, only the new top-drive encoder can be used normally. Encoder damage does not affect normal rotary drilling, so the general well team maintenance is not affected. In the event of damage to the top drive or rotary motor encoder in a drill string torsion system, the purchase and installation of a new encoder are necessitated. Additionally, if the top drive or rotary motor lacks an encoder altogether, the process requires onsite mapping for installation sizing and equipment processing. Moreover, in more severe cases, the mapping and installation can only be carried out during dry well team movements. Thus, from site inspection, mapping, and installation to normal usage, it may require waiting for 1-2 wells' time, significantly hampering promotion efficiency.

(2) With the increase of well depth and friction, the reverse torque increases, which increases the risk of drill string trip. In the application of the drill string torsion system, the reverse torque sometimes reaches 15000-18000n when the well depth is more than 5000m, or the friction is more than 300kN. The torque of the top drive and back tongs used by the driller is 24000n. M, with a reverse torque of 18,000 n. M, indicating that the risk of unloading is very high.

(3) The automation level of drill string torsion control needs to be improved. There are two main

problems: first, the setting of ground drill string parameters such as torsion angle or torque value, torsion speed, and torsion holding time depends on the experience of site engineers; second, there is no automatic adjustment of tool face.

4.2. Improvement measures

(1) The characteristic parameters of the motor are collected, and the angle is calculated indirectly through the algorithm. Specifically, by collecting motor speed, torque, current, voltage, characteristic curve, acceleration and deceleration time, and other parameters, a special software algorithm is used to calculate angle parameters instead of relying on the installation of motor encoders to further improve the compatibility of the drill string torsion system and improve efficiency.

(2) It can be used with a hydraulic oscillator to reduce the risk of unbuckling. When the depth of the well exceeds 6000m or the directional friction is higher than 300KN, it is recommended to use a hydraulic oscillator. The torsional system of the drill string controls the bidirectional torsional of the upper drill string to reduce friction, and the hydraulic oscillator reduces the friction of the lower drill string. Thus, the reverse torque can be reduced, and the risk of buckle unfastening can be reduced.

(3) Real-time acquisition of engineering parameters and directional data facilitates establishing a drill string torsion control closed-loop model. By introducing engineering parameters such as pump pressure and suspension weight into the torsion system, it can automatically judge whether the supporting pressure occurs according to the change of pump pressure and suspension weight and automatically correct the torsion parameters in real time. The MWD data, such as deviation, azimuth, and tool face, are collected in real time and compared with the design parameters of wellbore trajectory to realize the automatic correction of tool face.

5. Conclusion

(1) The drill string torsion system can significantly reduce friction, reduce supporting pressure, and improve the slide guiding efficiency. Meanwhile, the system has high reliability, low risk, low cost of use and maintenance, and a high performance-price ratio, so the market application prospect is good.

(2) TORSION DRILLING string TORSION system has achieved good results in the field application. However, the system compatibility still needs to be improved, primarily since the angle TORSION control relies heavily on the encoder. It is imperative to improve the torsion system installation and use of convenience in order to promote large-scale applications.

(3) The TORSION DRILLING string system has a low degree of automation at present. The setting of TORSION parameters and the adjustment of the tool face depend on the personal experience of site engineers, and the use effect is greatly subject to human factors. Therefore, it is urgent to explore a new generation of torsion control software that can automatically set torsion parameters and adjust tool faces on the basis of expanding application scale and perfecting construction technology.

References

[1] Yi Xianzhong, Li Zhipeng, Zhou Yuanhua, etc. Research progress on releasing down-hole friction by torsional pendulum of surface drill string [J], drilling and production technology, 2014,37(4): 73-77.

[2] Li Zhipeng, Yi Xianzhong, Tao Ruidong, etc. Research on a new control method of directional sliding drilling. Oil drilling technology [J], 2014,42(4): 59-63.

[3] Li Zhipeng, Xu Jingguo, Jiao Tao, etc. Research on directional sliding drilling control system and control method, petroleum industry computer [J], 2017,25(3): 6 -10.

[4] Xu Wen, Liu Xinli, Ma Rui, etc. Steering control technology for top-driven sliding drilling, petroleum machinery [J], 2013,41(3): 27 -30.

[5] Roberto H. Tello Kragjcek, et al. Successful Application of New Sliding Technology for Horizontal Drilling in Saudi Arabia, Saudi Aramco Journal of Technology Fall 2011.

[6] Colin Gillan, Scott Boone, et al. Applying Computer Based Precision Drill Pipe Rotation and Oscillation to Automate Slide Drilling Steering Control, SPE 148192, November 2011.

[7] Han Lie Xiang, Zhang Ji Chuan, Liu Wei. Pipe ROCK drill string torsion pendulum system application evaluation [J], drilling engineering, 2018,38(11): 73-78.

Performance analysis of motorized axles

Yan Yan[1,*], Peng Hang[1], Liangqiang Cao[1], Tao Huang[1], Jingjing Duan[1] and Yahu Wu[1]

[1]Department of Vehicle Technology, Air Force Early Warning Academy, Wuhan, Hubei, 430000, China

[*]Corresponding author's e-mail: 284528454@qq.com

Abstract. As the electric wheel requires the integration of the hub motor and brake within the hub, the integration scheme and structural parameters of each component in the electric wheel play a crucial role in the coordination with the electric wheel and its running performance. The finite element analysis is conducted on the axle of the electric wheel to judge whether the structural parameters meet the performance requirements, providing a basis for the axle design of the electric wheel. This will have a decisive role in the safety performance and life of the car, effectively improving the working stability of the electric wheel.

1. Introduction

As an important branch of a new generation of electric vehicles, electric wheel vehicles use a plurality of independently controlled hub motors to drive the wheels respectively, which can not only realize the perfect integration of the car's steering portability and the driver's road sense but also make the safety and flexibility of the car coordinated and unified, with good application prospects and technical development potential. At present, the research on electric wheel vehicles mainly focuses on electric wheel drive technology, and there are few studies specifically for the performance analysis of electric wheel axles.

Lv and Zhou [1] analyzed the common faults of the electric wheel steering system and put forward a targeted solution. Zhao and Zhang [2] used finite elements to analyze the steering knuckle in a large car. Wang [3] analyzed the front axle of the heavy-duty truck and confirmed its performance requirements under different working conditions. Zhang [4] designed the structure and layout of the power components in the new electric wheel, and the performance analysis of its dynamic performance was carried out.

During the driving of the vehicle, the in-wheel motor bears the impact and continuous vibration of the road surface transmitted by the wheel, and the axle will also change its normal working performance due to force deformation. In this paper, starting from the axle of the electric wheel, the finite element analysis software ABUQUS is used to analyze the static strength of the axle and the modal analysis of the axle under three typical working conditions: crossing uneven road surface, emergency braking and side skidding, to confirm its performance requirements.

2. Finite element analysis of motorized axles

The three-dimensional model of the electric wheel structure is shown in Figure 1, and the internal structure of the in-wheel motor is shown in Figure 2.

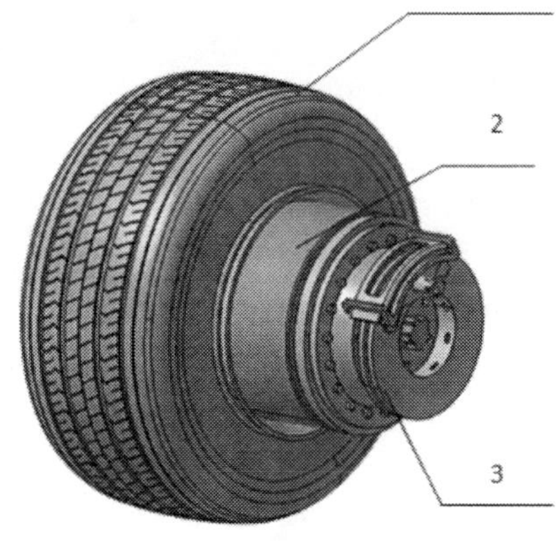

Figure 1. Electric wheel assembly diagram.
1-tire; 2-hub motor, including an external rotor, inverter, stator, etc.; 3-brake.

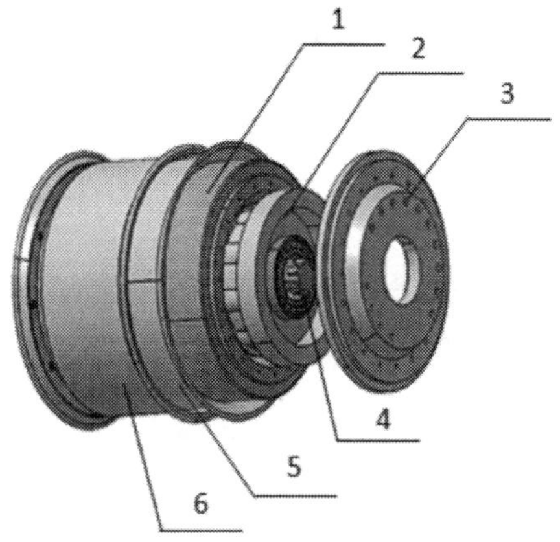

Figure 2. Internal structure diagram of in-wheel motor.
1-stator; 2-inverter; 3-pressure plate;4-motor axle;5-external rotor;6-hub.

2.1. Static strength analysis under typical operating conditions

When performing a strength analysis of the axle, process structures such as keyways and other rounds can be ignored. Process structures such as keyways and other rounds can be ignored when performing strength analysis of axles [5]. The built 3D model is imported into ABUQUS software, and the axle is meshed by using the hexahedral division method to obtain the axle mesh model as shown in Figure 3, in which the number of elements is 22823 and the number of nodes is 25441.

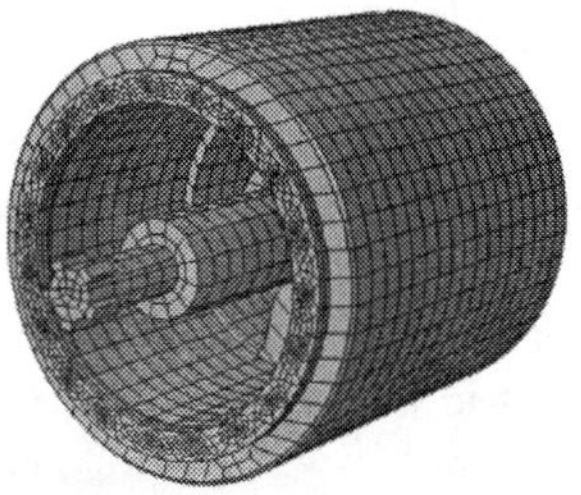

Figure 3. Grid model of wheel axle.

The force analysis of the electric wheel is carried out, the load boundary conditions of the wheel axle under three typical working conditions are calculated, and the calculated load parameter values of the whole vehicle required are shown in Table 1, and Figure 4 is the three-dimensional model of the wheel axle.

Table 1. Calculated load parameter values for the whole vehicle.

Static load on the front axle (N)	Dynamic load factor	Pavement adhesion coefficient	Wheel roll radius (m)	Mass distribution factor
73500	2	0.8	0.568	1.4

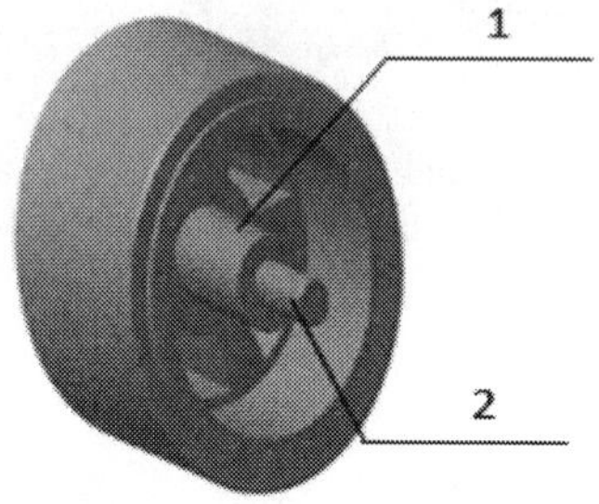

Figure 4. Three-dimensional model of the wheel axle.
1-fix the hub motor;
2- provide the positioning surface of the steering knuckle and the motor in the axial direction.

(1) Case 1: Over uneven road conditions

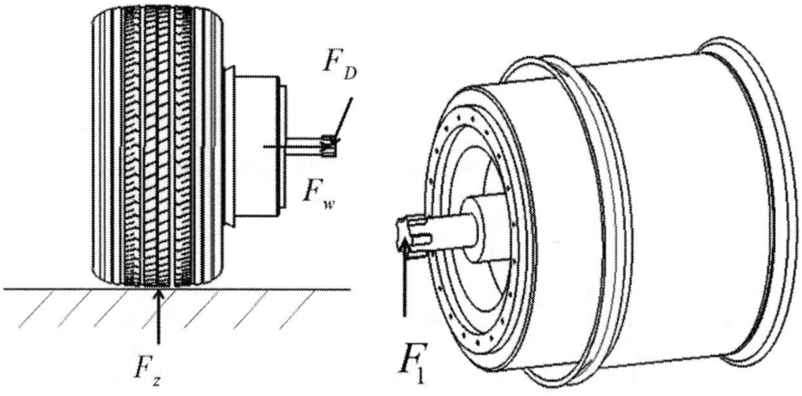

Figure 5. Crossing uneven road conditions

When the car crosses uneven road conditions, the electric wheel is subjected to three forces, namely the vertical reaction force F_Z of the ground, the sprung mass F_D, and the lateral reaction force of the wheel hub F_W. Currently, the vibration of the car is transmitted through the wheels to the axles, which are subjected to the impact vertical load of the ground, and the dynamic load coefficient is the largest currently. The stress of the electric wheel and axle is shown in Figure 5. Bringing in the data to calculate, it can be obtained that the impact vertical load of the ground F_1 = 73500 N. The stress

Journal of Physics: Conference Series **2808** (2024) 012045 doi:10.1088/1742-6596/2808/1/012045

contours and deformation diagrams of the axle under the condition of crossing the uneven road surface are obtained in ABAQUS, as shown in Figure 6.

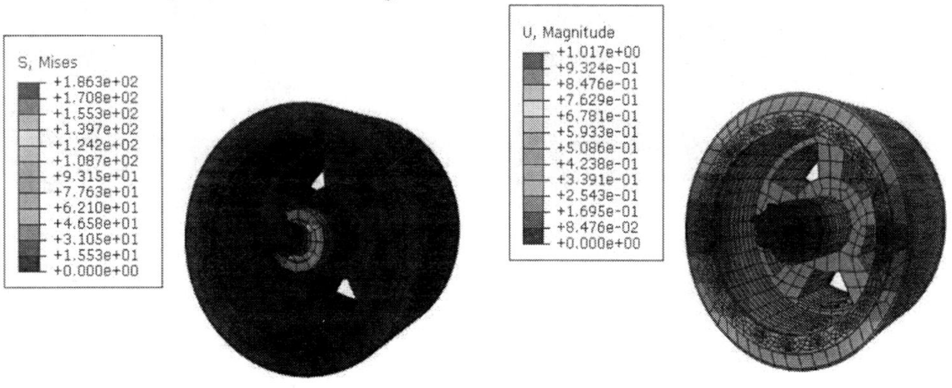

Stress contour diagram Deformation diagram

Figure 6. Working conditions for crossing uneven roads.

It can be seen from Figure 6 that when the automobile passes through the uneven road surface, the junction of Axle Segment 2 and Axle Segment 1 is the main stress part, the maximum stress of the wheel axle at this moment is 186.3 MPa, and the maximum deformation is 1.017 mm. It can be seen from the figure that at the junction of shaft Section 1 and shaft Segment 2, the stress change gradient is large, and the stress concentration phenomenon is serious. The material used for the axle is 45 steel, and its yield strength is 355 MPa, which is much greater than the maximum stress of the axle and meets the strength requirements.

(2) Case 2: Emergency braking condition

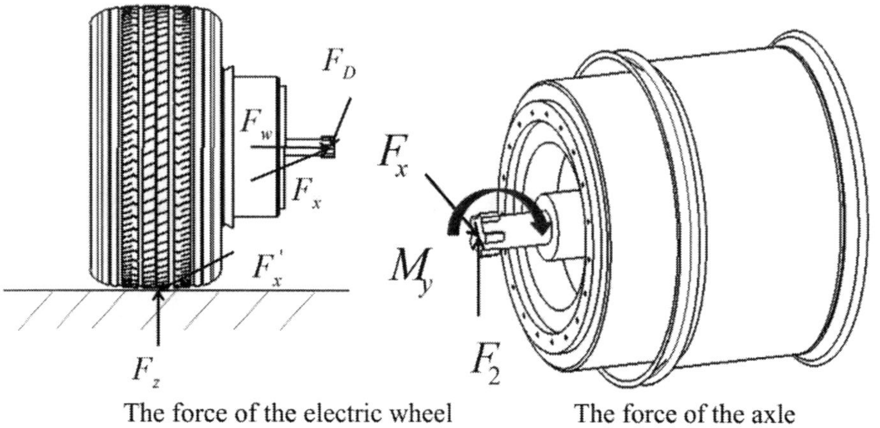

The force of the electric wheel The force of the axle

Figure 7. Emergency braking condition.

When the car is under emergency braking conditions, the electric wheel is subjected to the vertical reaction force F_Z, the longitudinal reaction force F'_X, the sprung mass F_D, the lateral reaction force FW of the wheel hub, and the longitudinal inertia force F'_X on the ground. The axle is subjected to the vertical load, longitudinal load, and a longitudinal load moment on the ground, and the force of the electric wheel and the axle is shown in Figure 7. Bringing in the data to calculate, it can be obtained, subject to the vertical load F_2 = 56344.5 N, longitudinal load F_X = 41160 N, longitudinal load moment of the ground M_y = 23378.9 Nm. The stress contours and deformation diagrams of the axle under emergency braking conditions are obtained in ABAQUS, as shown in Figure 8.

Journal of Physics: Conference Series **2808** (2024) 012045 doi:10.1088/1742-6596/2808/1/012045

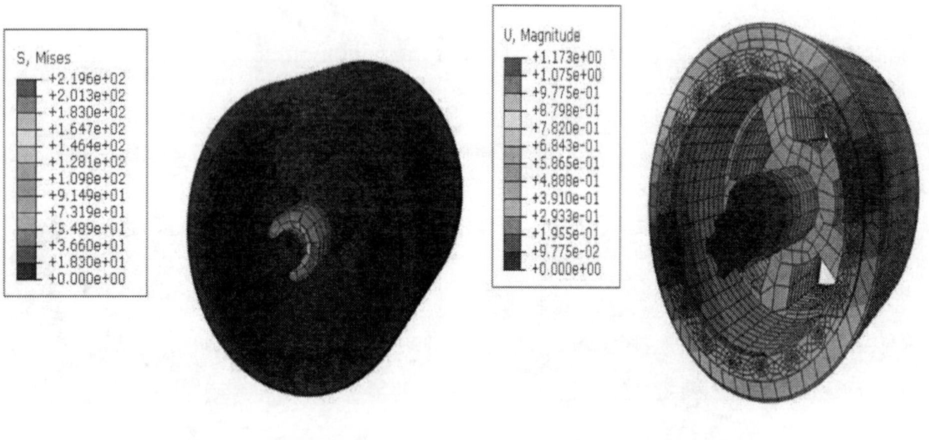

Stress contour diagram Deformation diagram

Figure 8. Emergency braking conditions.

It can be seen from Figure 8 that when the automobile is in emergency braking, the tail part of the wheel Axle 2 and the front end of the wheel Axle 1 are the main stress parts, the maximum stress of the wheel axle at this moment is 219.6 MPa, and the maximum deformation is 1.173 mm. It can be seen from the figure that at the front end of the shaft Section 1, the stress change gradient is large, the stress concentration phenomenon is serious, and the maximum stress is increased compared with the working condition of crossing the uneven road surface, but it still meets the strength requirements of the material.

(3) Case 3: Steering sideslip condition

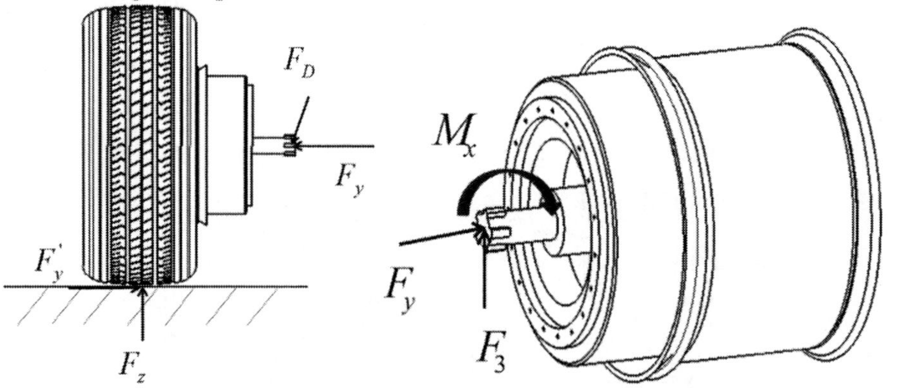

The force of the electric wheel The force of the axle

Figure 9. Steering sideslip condition.

When the car is in the left side-slip condition, the left and right front wheels are subjected to lateral forces of different magnitudes, and the lateral force of the left front wheel is greater than that of the right wheel. The electric wheel is subjected to the vertical reaction force F_Z, the lateral reaction force F'_y, the sprung mass F_D, the lateral force F_y of the hub, the vertical load of the axle, the lateral load, and a lateral load moment of the axle, and the force of the electric wheel and the axle is shown in Figure 9. Bringing in the data to calculate, it can be obtained that the vertical load $F_3 = 68250$ N, lateral load $F_y = 65520$ N, and side load momentary subjected to the ground $M_X = 23378.9$ Nm. The stress contours and deformation diagrams of the axle under the left sideslip condition were obtained in ABAQUS, as shown in Figure 10.

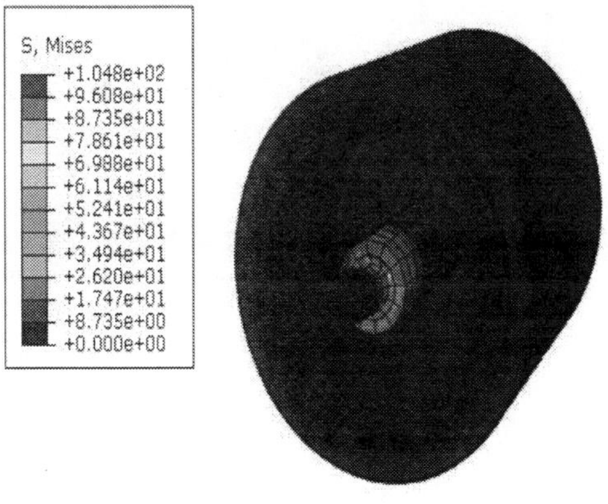

Stress contour diagram

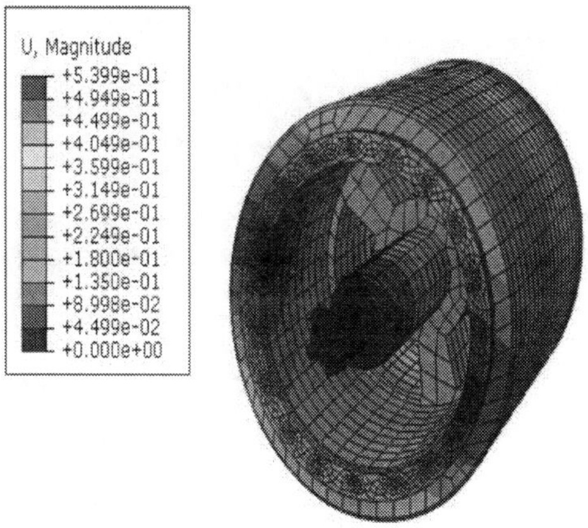

Deformation diagram

Figure 10. Steering sideslip condition.

As can be seen from Figure 10, when the automobile is steering sideslipping, the tail of Axle 2 and the front end of Axle 1 are the main stress parts, the maximum stress of the axle at this moment is 104.8 MPa, and the maximum deformation is 0.54 mm. As can be seen from the figure, at the maximum stress, the stress change gradient is large, and the stress concentration phenomenon is serious, which also meets the strength requirements of the material.

Therefore, through the finite element analysis of the axle under three working conditions, it was determined that the axle structure met the requirements.

2.2. Modal analysis

When an automobile vibrates, the axle is subject to vibrations transmitted from the ground, and to prevent it from resonating with the car and affecting the structure, it is necessary to perform a modal analysis of the motorized axle [6].

In modal analysis, the first 6 orders are rigid body modes and are not considered in the analysis. The natural frequencies and mode shapes of the first 6 orders from the 7th order onwards are usually selected for analysis. The mode shape diagram of the first 6th order was obtained by using ABAQUS software, as shown in Figure 11. Table 2 shows the first six natural frequencies of the axle[7].

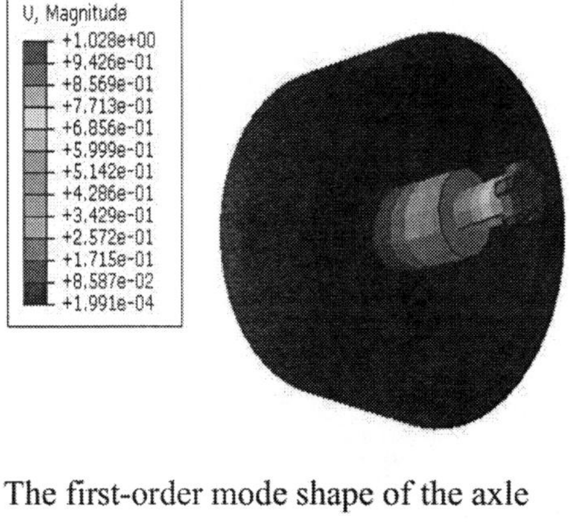

(1) The first-order mode shape of the axle

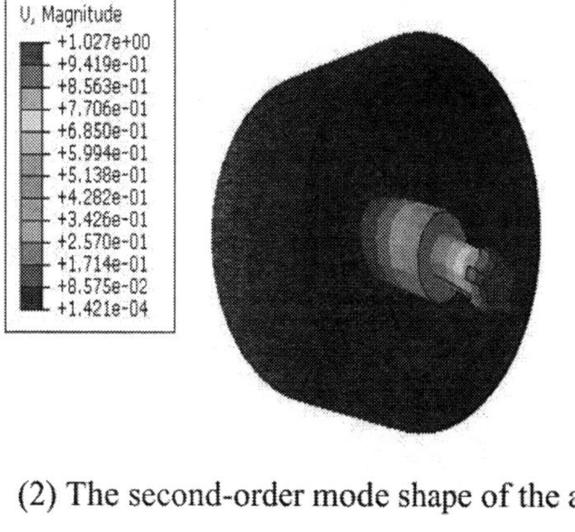

(2) The second-order mode shape of the axle

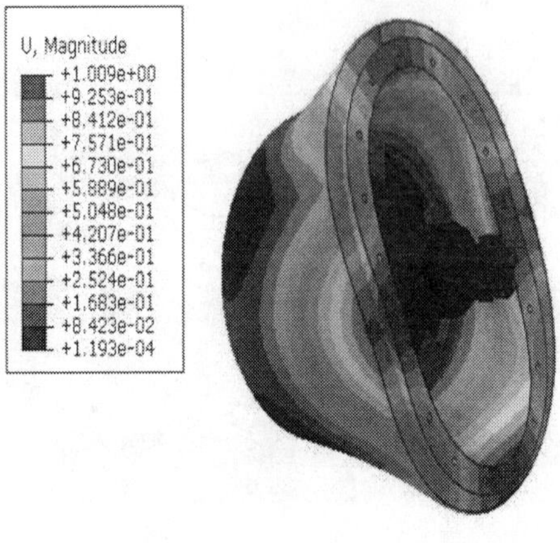

(3) The third-order mode shape of the axle

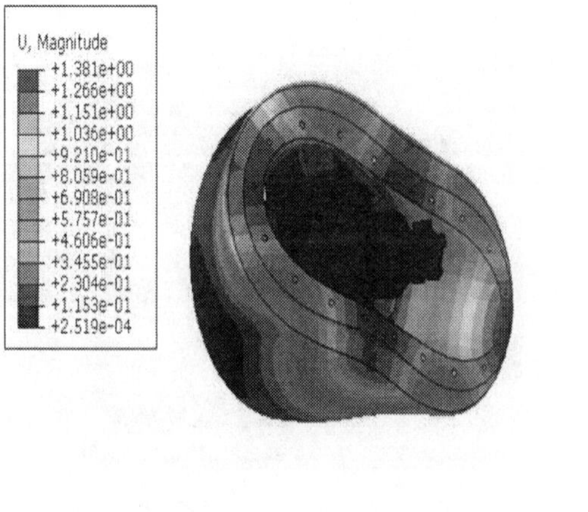

(4) The fourth-order mode shape of the axle

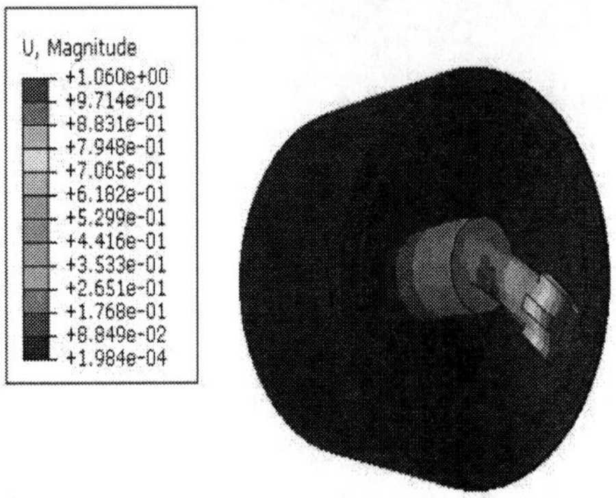

(5) The fifth-order mode shape of the axle

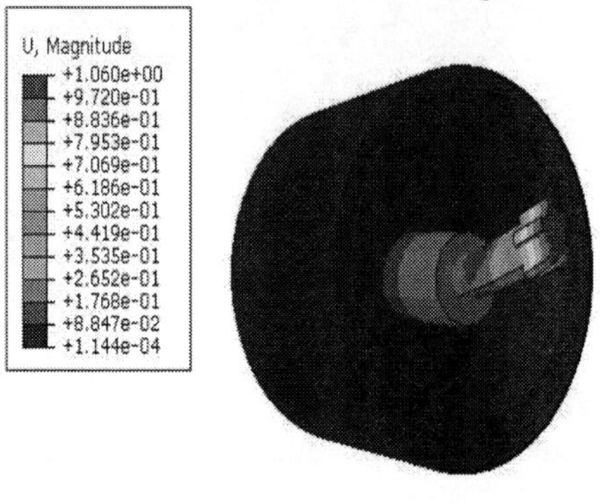

(6) The sixth-order mode shape of the axle

Figure 11. Axle stress and displacement diagram.

As can be seen from Figure 11, in the first-order mode shape, the front end of the axle is bent upward; In the second-order mode shape, the front end of the axle is bent downward; In the third- and fourth-order mode shapes, the overall bending of the axle is not large; In the fifth-order mode shape, the front end of the axle is torsioned to the lower left; In the sixth-order mode shape, the axle front end is twisted upwards to the right[8].

Table 2. The first six natural frequencies of the axle.

Modal order	Natural frequency (Hz)
First order	512.06
Second order	513.25
Third order	1102.0
Fourth order	1109.0
Fifth Order	1373.5
Sixth order	1383.1

As can be seen from Table 2, the first 6th-order natural frequencies of the axle vary from 512.06 Hz to 1383.1 Hz. In the process of movement, the frequency of the car changes in the range of 0.5 Hz ~ 20 Hz, in which the vibration frequency of the automobile body is 0.5 Hz ~ 2 Hz, and the vibration frequency of the wheel is 15 Hz ~ 20 Hz, and these frequency ranges are far from the frequency range of the wheel axle, so there is no resonance; In addition, the natural frequency of a human varies in the range of 0.5 Hz ~ 20 Hz, and it does not cause resonance. Therefore, the axle will not resonate with the car, nor with the person sitting in the car, so the axle structure meets the performance requirements.

Conclusion

In this paper, ABAQUS is used to analyze the key components in the electric wheel, and the static strength analysis of the electric wheel axle under three typical working conditions (over uneven road conditions, emergency braking conditions, and steering sideslip conditions) is carried out by using the finite element method. The values are 219.6 MPa and 1.173 mm, respectively, and the results meet the strength requirements of the axle material. The key components in the electric wheel meet the design requirements, which can effectively improve the working stability of the electric wheel[9].

References

[1] Lv Juhua, Zhou Yuhua. Analysis and treatment of common faults in the steering system of large electric wheel vehicles [J]. Light equipment. 2021(10)
[2] ZHAO Jiayun, ZHANG Lei. Finite Element Analysis Based on Steering Knuckle of a Hyper Works Cast Automobile [J]. Light Vehicle Technology. 2011(3): 23-26.
[3] WANG Zhengxiang. Static and dynamic strength analysis and experimental study of the front axle of heavy-duty truck [D]. Hefei: Anhui University of Science and Technology, 2016.
[4] ZHANG Ruijun. Structural design and performance analysis of new electric wheels [D]. Zhenjiang: Jiangsu University, 2016.
[5] Li Zhiqing. Optimal design of dynamic stability of high-speed axis based on MATLAB genetic algorithm [D]. Coal Mining Machinery, 2012, 33(4).
[6] LI Menghua. Automotive Research Report [EB/OL]. Automotive Industry Research Network, 2018. www.auto-stats.org.cn.
[7] Matsumoto, Masahide, Abe.Structural Optimization for Configuration Design of the motorcycle body. Proceedings of the 6th International Pacific Conference On Automotive Engineering, 2016, 8: 1399-1406.
[8] Ultra Light Steel Auto Suspension Final Report. Lotus Engineering Services, Inc. 2013.
[9] Matsumoto, Masahide, Abe.Structural Optimization for Configuration Design of the motorcycle body. Proceedings of the 6th International Pacific Conference On Automotive Engineering, 2009, 12: 1399-1406.

Multi-objective optimization of the wave-proof plate layout of the liquid tank vehicle based on NSGA-II and RSM

Houjie Xiao[1], Hongbin Qin[1*] and An He[1]

[1]Wuhan University of Technology, Wuhan, Hubei, China

*Corresponding author's e-mail:whutqhb@163.com

Abstract: Under the braking condition of the tanker, the liquid in the tank will have an impact on the tank body due to shaking. The arrangement spacing of the wave-proof plate has a great influence on the wave-proof effect. In order to study the optimal spacing of the wave-proof plate, a liquid sloshing model was established based on FLUENT, and the stress peak of the front head and the wave-proof plate 1 was determined as the optimization goal. The response model of the front head and the wave-proof plate 1 on the space of the three wave-proof plates was obtained through the response surface experiment, and the multi-objective optimization mathematical model of the wave-proof plate spacing was established based on RSM. The NSGA-II algorithm was used to solve the model, and the optimal combination of parameters of the spacing of the wave-proof plate was obtained. After optimization, the peak stresses of the front head and the first wave board are reduced by 8.6% and 5.2%.

1. Introduction

Liquid sloshing is a common problem in the transport of liquid cargo, especially in closed containers such as tankers. He et al. [1] analyzed the trajectory of tank liquid centroid motion with different cross-sectional shapes according to the calculation principle of homogeneous plane centroid. Hermann et al. [2] described the nonlinear equation of motion of the liquid in a rectangular container by modal method. Xu et al. [3] calculated the response of the air pressure in the on-board LNG cylinder over time under different working conditions through the fluid-structure interaction method. Duan et al. [4] experimentally concluded that the wave shield changes the natural frequency of the liquid and reduces the peak value of the impact caused by the liquid shaking. Kim [5] used a neural network and genetic algorithm to analyze the wave-proof plate of the rectangular container under translational load, and obtained the optimal installation height of the wave-proof plate. Li et al. [6] proposed an inverted V-shaped wave protection plate, and found that installing three V-shaped wave protection plates with an angle of 150° had the best wave-proof effect. In this paper, for a certain type of tanker, FLUENT software is used to obtain the optimal parameter combination by multi-objective optimization with the help of NAGA-II genetic algorithm for the response of the space of the wave plate to the impact force generated by the liquid shaking.

2. Tank truck design model and its parameters

In this paper, the tank assembly of a semi-trailer tanker is the research object, and its structure is shown in Figure 1. The length of the tank is 9600 mm, its cross-section is circular, the diameter is 1600 mm, and the volume is 17.5 m^3. In the tank, the material of the wave-proof plate, the cylinder and the two heads are Q235B, and the material properties are shown in Table 1.

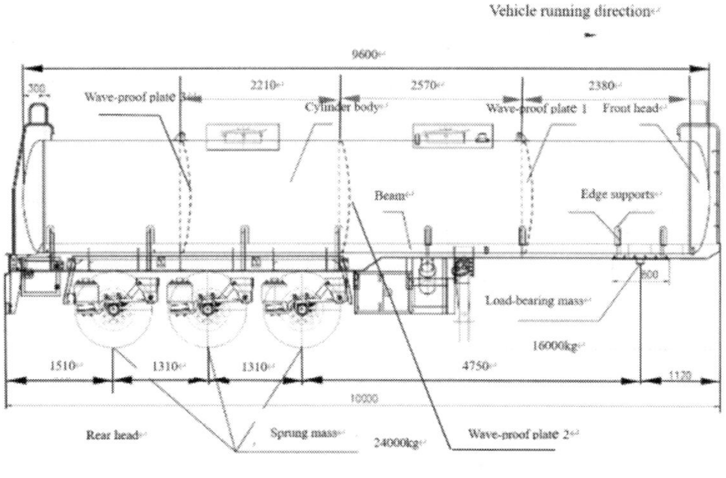

Figure 1. Tank structure diagram of a semi-trailer tanker.

Table 1. The main material performance parameters of the tank body of the tanker.

	Yield strength (Mpa)	Allowable stress (Mpa)	Density (kg/m^3)	Poisson's ratio	Modulus of elasticity (N/mm^2)
Q235B	235	131.6	7850	0.26	2.1e+05

3. Mathematical model for optimization of the wave-proof plate layout in tank

3.1. Tank optimization objective determination

Under braking conditions, the tank car in 1-3*s* decelerates at 6 m/s^2 acceleration. In the VOF model, the first phase is set as the gas phase (air) with a density of 1.225 kg/m^3, and the second phase is set as the liquid phase (sulfuric acid) with a density of 1830 kg/m^3. The analysis results of FLUENT in the liquid tank car are shown in Figure 2. The results show that the peak stress of the front head is the largest, and among the three boards, the peak stress of the first board is obviously greater than that of the other two boards. Therefore, the peak stress of the front head and the first board is selected as the optimization index in the subsequent optimization.

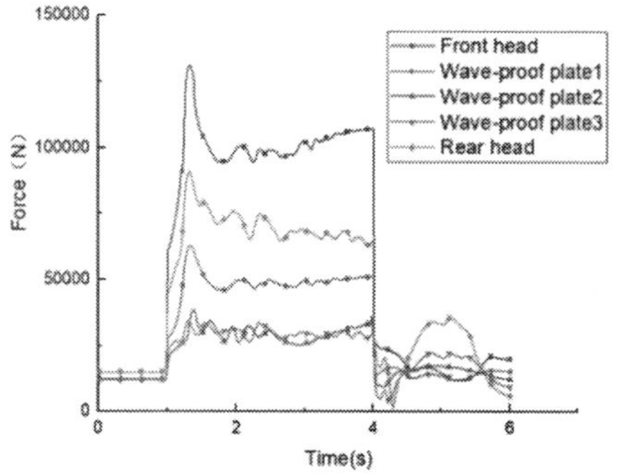

Figure 2. The results of FLUENT analysis of internal fluid in tank.

3.2. The establishment of the response surface model of the front head and the wave-proof plate 1

The Response surface methodology (RSM) is a classic optimization method. The optimization steps are as follows. In the first stage, a series of experimental designs are used to obtain the results of output variables at different levels of design variables. In the second stage, regression modeling is carried out on the basis of sufficient experimental results in the first stage, and the nonlinear relationship between the design variable and the output variable is approximated by a determined mathematical relation. In the above analysis, the response surface is established by using the previous head and the peak force of the wave-proof plate 1 as the output variable.

3.2.1. Design of response surfaces experiment. The distance between the three wave guards and the front head is selected as the design variable, and the sum of the force peaks of the head and the force peaks of the three wave plates is taken as the response value. The specific levels of the three design variables are shown in Table 2. The distance between the wave-proof plate 1 and the front head is named L_1, the distance between the wave-proof plate 2 and the front head is named L_2, the distance between the wave-proof plate 3 and the front head is named L_3, the stress peak of the front head is named F_1, the stress peak of the first wave guard is named F_2, and the force peak of the rear head is named F_3.

Table 2. Design of response surfaces experiment.

Factor	level		
	Lower limit value (*mm*)	Median value (*mm*)	Upper limit value (*mm*)
L_1	1250	2250	3250
L_2	3500	4500	5500
L_3	5750	6750	7750

According to the variable factor level designed in Table 2, the response surface experiment with Box-Behnken as the experimental design method was completed, with a total of 17 experimental points, and the internal relationship between the three design variables and the stress peak of the front head was explored, and the experimental results are shown in Table 3.

Table 3. Response surface experiments and results.

Experimental sites	L_1 (*mm*)	L_2 (*mm*)	L_3 (*mm*)	F_1 (*N*)	F_2 (*N*)
1	2250	3500	7750	137781	76746
2	2250	4500	6750	129782	61765
3	3250	4500	7750	139564	58251
4	3250	4500	5750	133497	41347
5	2250	4500	6750	129782	61765
6	1250	3500	6750	134606	82944
7	3250	5500	6750	142378	46585
8	2250	5500	7750	140386	52742
9	1250	4500	7750	130568	57529
10	3250	3500	6750	149846	35706
11	2250	5500	5750	125863	71245
12	1250	4500	5750	119375	85603
13	2250	4500	6750	129782	61765
14	2250	4500	6750	129782	61765
15	2250	3500	5750	120484	69347
16	1250	5500	6750	128347	81424
17	2250	4500	6750	129782	61765

3.2.2 The construction of the response surface model of the front head and the wave-proof plate 1. In order to establish the relationship between the stress peak of the front head and the three design variables, the quadratic polynomial with the cross term was used as the response surface model to replace the liquid-solid coupling model, which can be expressed as

$$y = \beta_0 + \sum_{i=1}^{3} \beta_i x_i + \sum_{i=1}^{3} \beta_{ii} x_{ii} + \sum_{1 \le i < j \le 3} \beta_{ij} x_i x_j + \varepsilon \tag{1}$$

where y is the response, x_i and x_j (i, j=1, 2, 3) are design variables, β_0, β_i, β_{ii} and β_{ij} are the coefficients to be determined, and ε is the error term. According to Table 3, the response surface models of the front head and the wave-roof plate 1, stress peak can be obtained by multiple linear regression analysis. The specific expression is:

$$\begin{aligned} F_1 &= 1.298 \times 10^5 + 6548.63L_1 - 717.87L_2 + 6135.00 \\ & L_3 - 302.25L_1L_2 - 1281.50L_1L_3 - 693.50L_2L_3 + 431 \\ & 7.37L_1^2 + 4694.87L_2^2 - 3348.37.37L_3^2 \end{aligned} \tag{2}$$

$$\begin{aligned} F_2 &= 61765 - 19951.37L_1 + 31.63L_2 + 1090.75L_3 + 30 \\ & 99.75L_1L_2 + 2744.50L_1L_3 - 3225.50L_2L_3 - 2843.88L_1^2 \\ & +2743.62L_2^2 + 6261.38L_3^2 \end{aligned} \tag{3}$$

To confirm the precision and dependability of the model, the statistical index complex correlation coefficient R^2 and the modified complex correlation coefficient R^2_{adj} were used to test, and the complex correlation coefficient and modified complex correlation coefficient of the response value are shown in Table 4.

Table 4. Error analysis of regression model.

Response value	R^2	R^2_{adj}
F_1	0.9063	0.8659
F_2	0.9856	0.9692

3.3. The establishment of the optimization model of the front head and the wave-proof plate 1

According to the relevant national standards [7], the volume between the wave-proof plates shall not be greater than 7.5 m^3, and according to the relevant parameters of the tanker truck, the distance between the two adjacent wave-proof plates and the front and rear heads and the adjacent wave-proof plates shall not be greater than 3750 mm. According to the design requirements, the distance between the three wave-proof plates and the front head is the design variable, the objective function is the peak force of the front head and the wave-proof plate 1, and the optimization model of multi-objective genetic algorithm for the spacing of the wave-proof plate is established.

$$\begin{cases} Min \quad Y(X) = \{y_1(X); y_2(X)\} \\ s.t. \qquad X = (L_1, L_2, L_3) \\ \qquad 1250 < L_1 < 3250 \\ \qquad 3500 \le L_2 \le 4980 \\ \qquad 5750 \le L_3 \le 7230 \end{cases} \tag{4}$$

4. Solving of wave-proof plate space optimization model based on NSGA-II

4.1. Pareto optimal solution set of the wave-proof plate space optimization model

Compared with many traditional multi-objective optimization methods, such as weighting, constraint and linear programming, the NSGA-II algorithm has the advantages of fast running speed and good solution set convergence. After the model is solved, the obtained set of non-inferior solutions is shown in Table 5.

Table 5. Multi-objective optimization of Pareto non-inferior solution set.

Serial number	L_1 (mm)	L_2 (mm)	L_3 (mm)	F_1 (N)	F_2 (N)
1	3199	4156	5290	126158	45872
2	2765	4442	5325	119737	59509
3	2738	3878	5482	123471	55872
4	2789	4044	5277	120222	57598
5	3104	4553	5277	123897	51543

4.2. Optimization result analysis

Considering that the impact force on the front head is the largest, the impact force on the front head should be reduced as much as possible when the design parameters are selected to ensure that the force on the first wave board is reduced. Therefore, Solution No. 2 is selected as the optimal solution. The FLUENT simulation results of the optimized model and the original model are shown in Table 6, where F_3 is the peak stress of the wave board 2, F_4 is the peak stress of the wave board 3, and F_5 is the peak stress of the rear head.

Table 6. Multi-objective optimization of Pareto non-inferior solution set.

	F_1 (N)	F_2 (N)	F_3 (N)	F_4 (N)	F_5 (N)
Optimization model	119737	59509	42374	36502	97518
Original model	131024	62806	38755	34666	91126

The results show that compared with before optimization, the peak force of the front head and the wave-proof plate 1 is reduced by 8.6% and 5.2%, respectively, and the stress of the head and the three wave-proof plates is more balanced and reasonable, indicating that the reasonable layout of the wave-proof plate can significantly reduce the impact force of the liquid on the tank and ensure the safety of the liquid tanker transportation.

5. Conclusion

(1) The shaking of the liquid tanker under braking conditions was simulated by FLUNET, and the results showed that the peak impact force of the front head was the largest, followed by the rear head. Among the three wave-proof plates, the peak impact force of wave-proof plate 1 is the largest, and the wave-proof plate is the second. Therefore, the force peaks of the front head and the wave-proof plate 1 are determined as two optimization targets.

(2) The response surface experiments on the spacing of the three wave-proof plates under the force of the front head and the wave-proof plate 1 were designed, and the response surface model of the peak force of the front head and the wave-proof plate 1 was obtained through multiple linear regression analysis. The constraints for the optimal design were determined, and a mathematical model of multi-objective optimization of wave-proof plate spacing was established.

(3) The NSGA-II algorithm was used to solve the optimization model of the spacing of the wave-proof plates, and the distances between the three optimized wave-proof plates and the front head were 2765 *mm*, 4442 *mm* and 5325 *mm*, respectively. The results show that the peak force of the front and rear head and the three wave-proof plates are 119737 *N*, 59509 *N*, 42374 *N*, 36502 *N*, and 97518 *N*, respectively. Compared with the optimization before, the peak forces of the front head and the wave-proof plate 1 are reduced by 8.6% and 5.2% respectively, and the stress of the head and the three wave-proof plates is more balanced and reasonable. It shows that the reasonable layout of the wave-proof plate can significantly reduce the impact force of the liquid on the tank and ensure the safety of tanker transportation.

References

[1] He L.Y, Li G.J, Zhou Y. (2022) The characteristics of the center of mass of the liquid in the tank with different cross section shapes. Mechanics in Engineering., 44(02):1109-1116.

[2] Hermann M, Timokha A. (2008) Modal modelling of the nonlinear resonant fluid sloshing in a rectangular tank II: secondary resonance. MATHEMATICAL MODELS & METHODS IN APPLIED SCIENCES, 18(11):1845-1867.

[3] Xu H.Y, Yang N.X, Liang L, et al. (2021) Comparative study on numerical analysis methods of liquid sloshing in vehicle-mounted LNG cylinders. Pressure Vessel Technology., 2021, 38(11):44-52.

[4] Duan L.L, Yang Z.R, Zhang Y.M, et al. (2023) Experimental study of the influence of baffle on liquid sloshing in a tank model. Mechanics in Engineering., 45(05):1109-1116.

[5] Kim H, Lee Y. (2008) Optimization design technique for reduction of sloshing by evolutionary methods. JOURNAL of MECHANICAL SCIENCE AND TECHNOLOGY., 22(1):25-33.

[6] Li B, Ding Y, Zhao Y.Q, et al. (2022) Liquid-solid Coupling Analysis and Optimization Design of Tank Vehicle. Journal of Mechanical Engineering., 58(22):321-333.

[7] GB 18564.1-2019. (2019) Road transport of liquid dangerous goods - Tank vehicles - Part 1: Technical requirements for metal atmospheric tanks.

Journal of Physics: Conference Series **2808** (2024) 012047 doi:10.1088/1742-6596/2808/1/012047

Research on the failure mechanism of non-metallic oil pipeline flange

Yu Weng[1*], Dejun Li[1], Shuxuan Jiang[1], Shuang Yang[1], Gang Wu[1] and Zhian Deng[2]

[1]College of Petroleum Engineering, Xi'an Shiyou University, Xi'an, Shaanxi, 710065, China
[2]Quanzhou Vocational and Technical University, Quanzhou, Fujian, 362000, China

[*]Corresponding author's e-mail: wengyu009@hotmail.com

Abstract. Fiberglass flanges are gradually being applied in oil field mixed transportation pipelines due to their advantages such as corrosion resistance and easy forming. However, at the same time, the anisotropy and local discontinuity are present in glass fiber components. Starting from the discovery of cracking, leakage, and other issues in the fiberglass flange of a certain oilfield sewage export pipeline, a combination of experimental research and numerical analysis is used to study the on-site fiberglass flange joint. Considering the internal pressure, bending moment, bolt load, and other loads in actual operating conditions, mechanical performance tests are conducted on the flange substrate, and the mechanical properties and damage conditions are obtained. A finite element model including pipeline gasket flange is constructed, and the flange damage under preloading, operation, external bending moment, and other operating conditions is analyzed based on the nonlinear Hashin failure criterion and the Tserpes sudden stiffness degradation model. It is found that bolt holes are more prone to large-scale failure damage due to stress concentration. As the external bending moment increases, it will simultaneously reduce the gasket compressive stress and cause flange surface failure and damage. According to research, it is recommended that the pre-tightening force of bolts during pipeline operation should not exceed 400000 N, and the bending moment generated by factors such as wind load and soil load should not exceed 200 N·m.

1. Introduction

The research on leakage of bolted flanges at home and abroad mainly focuses on the leakage between steel flanges, with a focus on pipeline vibration under pressure and temperature, fatigue of steel structures, gasket creep, gasket material itself, and bolt torque [1]. Research on composite materials often focuses on the stress and thermal analysis of bolted connections or laminated plates. Therefore, the leakage mechanism of fiberglass pipe flanges is not yet mature, and there is relatively little experimental research, as well as a lack of validation of numerical models based on relevant experimental data. This article takes the 4-hole fiberglass bolt flange and steel flange joint in a low-pressure pipeline system as the research object [2]. Based on sealing failure and fiberglass flange strength failure, experimental research is conducted to obtain the physical parameters of the composite material. Through numerical research [3], the failure characteristics of fiberglass flanges in different working conditions such as pre-tightening, operation, and pipeline bending moment are studied, providing a reference for actual construction.

However, as a composite material and with weaker stiffness than steel structures, fiberglass pipe flanges are prone to damage when subjected to large service loads [4], resulting in damage and leakage failure at the connection components used in water injection and sewage collection production operations [5]. In addition, composite materials exhibit significant anisotropy, with interlaminar normal strength significantly lower than tangential strength, making the factors leading to the failure of composite structures uncertain [6]. Therefore, this study focuses on multiple cases of cracking and leakage of fiberglass flanges in sewage export pipelines in a certain oilfield and conducts an in-depth exploration of the failure phenomenon of fiberglass flanges [7].

2. Research method

2.1. Material testing methods

This study analyzes the flange connection of low-pressure fiberglass pipes with a diameter of D=50 mm. The size structure of fiberglass flanges is mostly made based on the size of steel structural flanges [8]. According to HG/T20592-2009 Steel Pipe Flange (PN Series) and the data provided by the manufacturer, the specific dimensions of the model are shown in Table 1.

According to HG/T20592-2009 Steel Pipe Flanges (PN Series), HG/T20613-2009 Fasteners for Steel Pipe Flanges (PN Series), and the "Practical Manual for Pressure Vessel Materials: Carbon Steel and Alloy Steel", the selected flange material is WCA, which is consistent with the pipe material. The alloy structural steel 35CrMo is selected for bolts and nuts, and the relevant material parameters are shown in Table 2. For the material parameters of the fiberglass pipe flange section, the mechanical test was conducted on a single-layer fiberglass reinforced epoxy resin board with a fiber content of 50% [9].

Table 1. Steel Flange Dimensions.

Nominal Diameter	Outside Diameter *D/mm*	Bolt Hole Center Circle Diameter *K/mm*	Bolt Hole Diameter *L/mm*
DN50	165	125	18
Nominal Diameter	Number of Bolt Holes *n/ individual*	Bolt *Th*	Flange Thickness *T/mm*
DN50	4	M16	18

Table 2. Material Performance Parameters.

Name	Elastic Modulus/ GPa	Poisson's Ratio	Yield Strength/ MPa
Flange and Connecting Pipe	212	0.288	205
Bolts and Nuts	214	0.286	735

2.2. Numerical analysis methods

This study uses ABAQUS software for numerical analysis, taking 1/2 of the circumference direction to establish a joint model, which includes bolts, nuts, fiberglass pipe flanges, steel flanges, and flange gaskets. The above geometric modeling is meshed. Hexahedral solid element C3D8R is used for the overall model, the GK3D8N element is used for the gasket to simulate contact performance, and the eight-node three-dimensional bonding element COH3D8 is used to simulate the cohesive bonding layer. Figure 1 shows the finite element model after meshing [10].

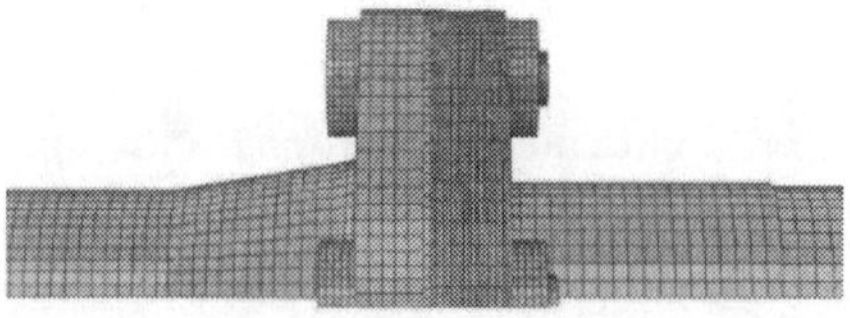

Figure 1. Finite element model.

In the process of flange joint analysis, the bolt load is set on a tangent plane in the middle of the bolt, and displacement constraints are added at the flange connection: displacement constraints are set on the upper connection of the fiberglass steel flange, complete fixed constraints are applied to the end face of the lower connection of the steel flange, and the end face of the upper connection of the fiberglass flange is free, as shown in Figure 2.

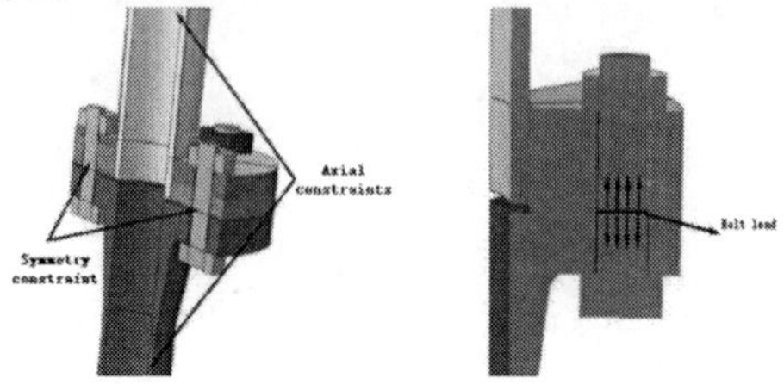

Figure 2. Boundary.

By setting a larger bolt load under pre-tightening conditions and continuously increasing the load, it is determined whether it will affect the integrity of the connection and cause strength damage to the fiberglass flange [11]. The calculation formula is taken from the national standard GB150-2011 fixed pressure vessel.

Five different bending moments of 200-1000 N·m are applied to the end face of the connecting pipe on the fiberglass flange as external conditions, and two different gaskets, chloroprene rubber flange gasket, and flexible graphite gasket, are studied under bolt loads of 20000-300 N and 30000-40000 N.

The Hashin [12] failure criterion can accurately determine various damage failure modes, and later scholars have revised and improved the original Hashin criterion. Therefore, this article chooses the three-dimensional Hashin failure criterion to predict the failure initiation of laminated plates.

Based on the application scenarios of different reduction schemes, this article ultimately selects the material parameter degradation method proposed by the Tserpes model to reduce the stiffness of the damaged area [13], corresponding to different failure damage modes of the three-dimensional Hashin progressive damage criterion. After comprehensive consideration, a combination of the three-dimensional Hashin asymptotic failure criterion and the Tserpes sudden stiffness degradation model is selected for the analysis of fiberglass flanges. A UMAT progressive damage subroutine is written by using the FORTRAN language, taking into account the cohesion model, laying a solid foundation for the development of finite element simulation.

3. Research and analysis

3.1 The effect of bolt load on the strength of flange joints under pretensioning conditions

Based on the three-dimensional Hashin failure damage criterion, when SDV≥1, it can be considered that complete strength failure damage occurs. By applying bolt loads of 40000 N, 50000 N, 60000 N, 70000 N, and 80000 N to the bolts, the strength damage cloud maps of fiberglass flanges under different bolt loads are obtained, as shown in Figure 3 to Figure 6. It can be seen that as the bolt load increases, the most common strength failure damage modes in the radial strength-dominated fiberglass pipe flange connection are matrix tensile failure and matrix compression failure.

Journal of Physics: Conference Series **2808** (2024) 012047 doi:10.1088/1742-6596/2808/1/012047

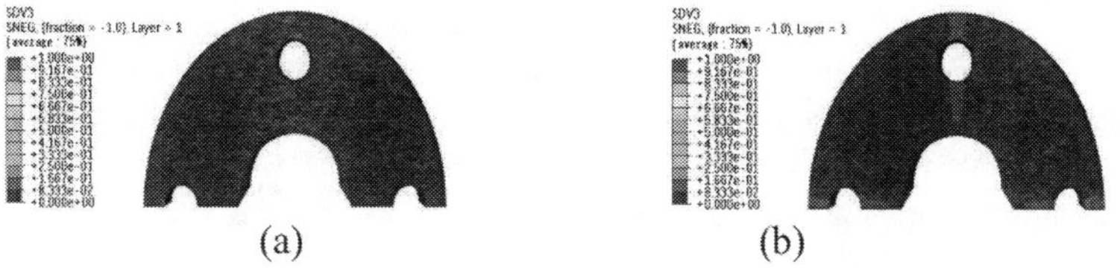

(a) (b)

Figure 3. Tensile failure diagram of chloroprene rubber fiberglass flange matrix at 50000 N (a) and 80000 N (b) bolt load.

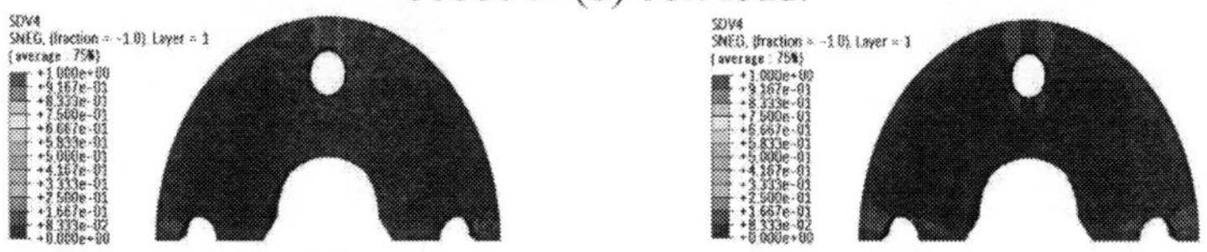

(a) (b)

Figure 4. Compression failure diagram of chloroprene rubber fiberglass flange matrix at 60000 N (a) and 80000 N (b) bolt load.

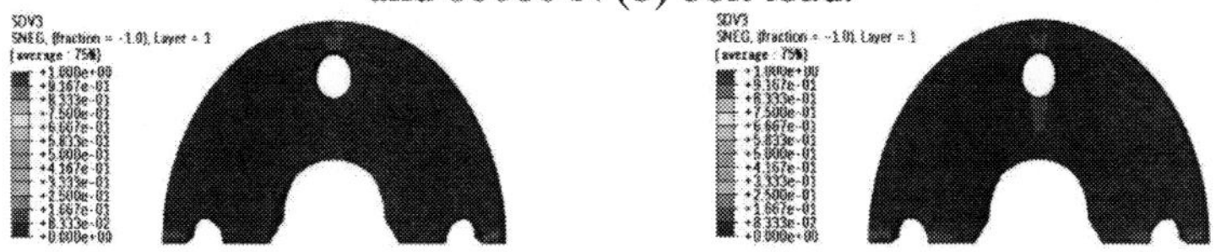

(a) (b)

Figure 5. Tensile failure diagram of flexible graphite fiberglass flange matrix at 60000 N (a) and 80000 N (b) bolt load.

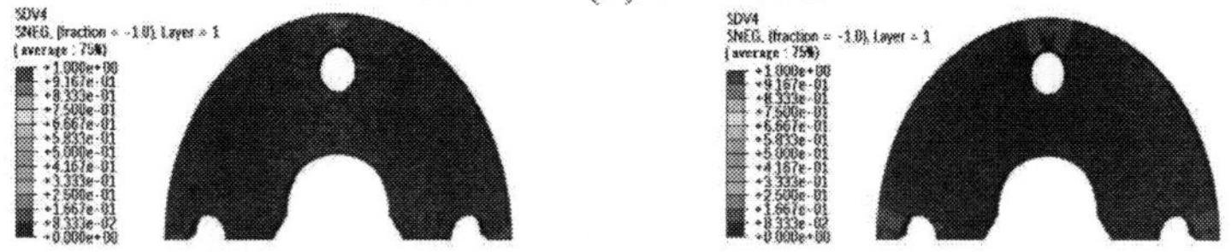

(a) (b)

Figure 6. Compression failure diagram of flexible graphite fiberglass flange matrix at 60000 N (a) and 80000 N (b) bolt load.

By comparing the strength damage cloud maps of fiberglass flanges under different types of gaskets, it is found that under the same bolt load, the deformation degree of chloroprene rubber is greater. The trend of the tensile and compressive failure area of fiberglass flanges under different bolt loads using chloroprene rubber and flexible graphite gaskets under pre-tightening conditions is shown in Figure 7. For fiberglass flanges with predominantly latitudinal strength, when loaded to a maximum of 80000 N, subtle matrix tensile failure occurs only on the surface layer of the flange under bolt load.

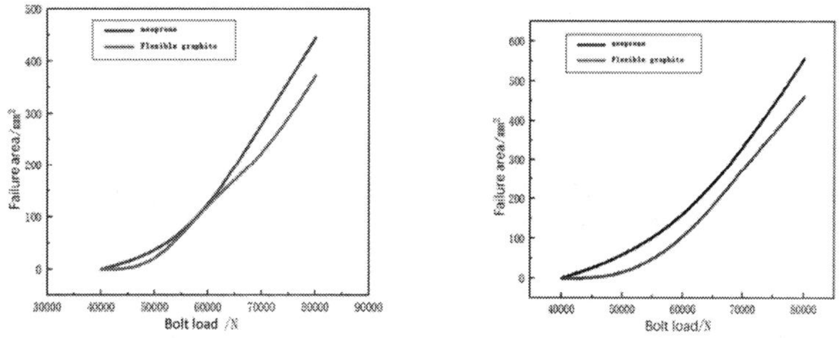

(a) (b)

Figure 7. Failure and damage trend of fiberglass flange under pre-tightening condition.

3.2. Failure analysis of flange joints under operating conditions

After being subjected to the axial force caused by internal pressure in the pipeline, matrix tensile and compressive failure damage is more likely to occur on fiberglass flanges with predominantly meridional strength compared to pre-tightening conditions. In the flange gasket connection system using chloroprene rubber, matrix tensile failure occurs under a bolt load of 40000 N, while matrix compressive failure occurs under a bolt load of 50000 N. In the flange connection system where flexible graphite is used as a flange gasket, the tensile failure of the matrix occurs under a bolt load of 50000 N, while the compressive failure of the matrix remains unchanged compared to the pre-tightening condition, but the failure range has increased.

The trends of the tensile and compressive failure area of fiberglass flanges under different bolt loads using chloroprene rubber and flexible graphite gaskets under operating conditions are shown in Figure 8.

For fiberglass flanges with predominantly latitudinal strength, the axial force generated by internal pressure in the pipeline has a relatively small impact on the fiberglass flange joint under operating conditions. Only in flange joints where chloroprene rubber is used as a flange gasket, does matrix tensile failure occur on the bolt load bearing surface, which still only occurs at the maximum bolt load of 80000 N, but the failure range has increased. For the bolted flange connection system using flexible graphite as a fastening gasket, the structural safety is still maintained and there has been no failure.

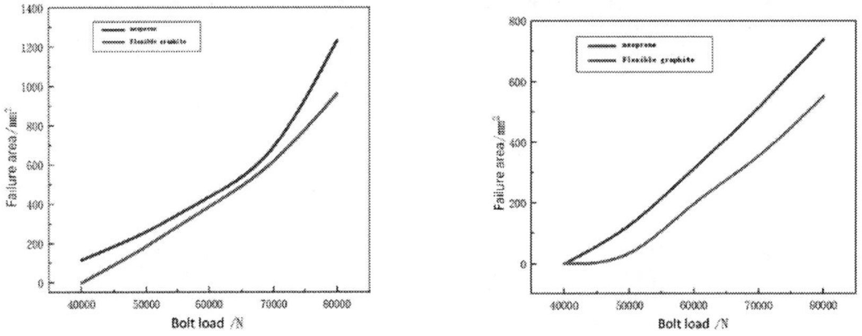

(a) (b)

Figure 8. Trend of failure and loss of fiberglass flanges under operating conditions.

3.3. Numerical analysis of flange joint failure under external bending moment

As the external bending moment applied to the pipeline joint continues to increase, it may have an impact on the strength of the fiberglass flange. By applying a bending moment to the end face of the connecting pipe on the fiberglass flange and increasing it, the following failure cloud maps of the fiberglass flange structure can be obtained based on the three-dimensional Hashin criterion and stiffness degradation (SEDG). The following are examples of the failure damage cloud maps of the chloroprene rubber fiberglass flange joint.

The failure damage cloud diagram of the chloroprene rubber gasket fiberglass flange joint under a maximum bending moment of 1000 N·m is shown in Figure 9 and Figure 10.

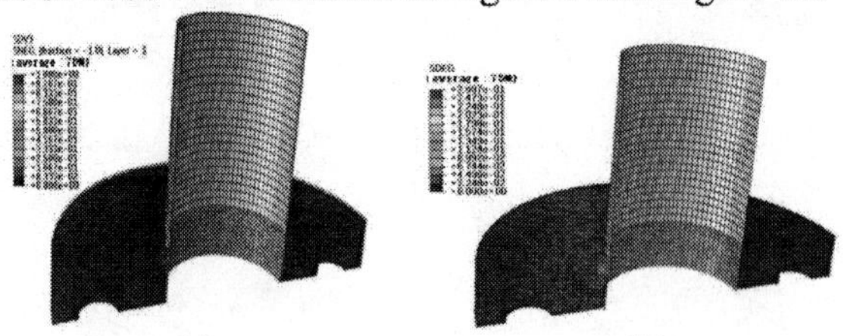

(a) (b)

Figure 9. Matrix tensile (a) and Delamination (b) failure damage diagram of fiberglass flange under bolt load of 20000 N.

Journal of Physics: Conference Series **2808** (2024) 012047 doi:10.1088/1742-6596/2808/1/012047

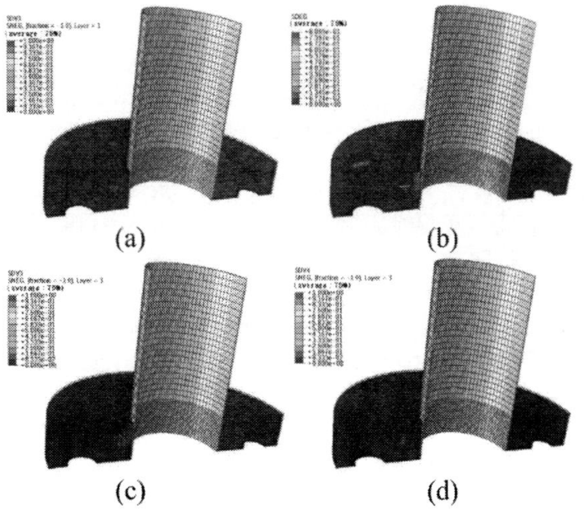

Figure 10. Fiber tensile (a), Delamination (b), Matrix tensile (c) and Matrix compression (d) failure damage diagram of fiberglass flange under 30000 N bolt load.

For the study of strength failure caused by external bending moments in fiberglass reinforced plastic (FRP) mainly with latitudinal strength, the failure damage cloud map of the chloroprene rubber FRP flange joint is also taken as an example. For example, the failure damage cloud map of the chloroprene rubber gasket FRP flange joint under a maximum bending moment of 1000 N·m is shown in Figure 11.

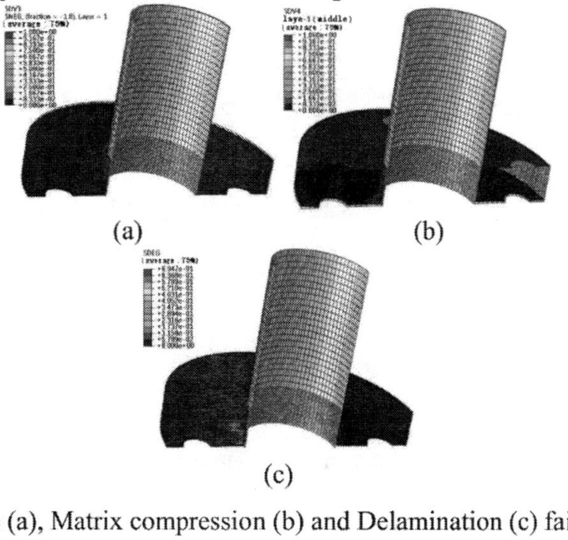

Figure 11. Matrix tensile (a), Matrix compression (b) and Delamination (c) failure damage diagram of fiberglass flange under bolt load of 20000 N.

4. Conclusion

Taking the fiberglass flange connection of a certain oilfield sewage export pipeline as the research object, considering material anisotropy and using the determination of mechanical performance parameters of fiberglass epoxy resin board as the numerical simulation analysis basis, the failure problem of fiberglass flange and steel flange connection under longitudinal and latitudinal strength, bolt load, internal pressure, and external bending moment is discussed. The strength failure damage of fiberglass flange is numerically studied and judged. The conclusions of this study are as follows:

(1) Among the load conditions involved, the strength of the flange itself and the external bending moment are the two main factors leading to the failure of the flange joint. Although bolt load, pipeline

internal pressure, and gasket type have a certain impact on the failure of the joint system, their impact is relatively weak.

(2) With the increase of bolt load, stress concentration occurs at the bolt hole and matrix failure damage occurs. The degree of tensile and compressive failure damage to the substrate on the flange surface is inversely proportional to the strength in the weft direction. Therefore, it is advisable to choose a substrate with higher latitudinal strength when using it to withstand greater bolt loads.

(3) The failure and damage mode exhibited under the bending moment and bolt load is mainly caused by the inward tensile failure of the matrix generated from the flange neck, that is, the longitudinal strain generated under the external bending moment has serious damage to the adhesive performance between the flange and the pipe wall. Therefore, in design, the focus should be on strengthening the flange neck and improving the strength of the adhesive curing agent.

This work was financially supported by the Natural Science Basic Research Program of Shaanxi (Program No. 2021JM-402), and the Research Fund of Key Laboratory of High-Performance Manufacturing for Aero Engine (Northwestern Polytechnical University), Ministry of Industry and Information Technology (Grant No. HPM-2022-03).

References

[1] Liu L, Wang X, Wu Z, et al. Effect of fiber architecture on tension–tension fatigue behavior of bolted basalt composite joints [J]. Engineering Structures, 2023.

[2] Duc T Q, Nhut P, Matsumoto Y. Strengthening of Multi-bolted Connections in Pultruded GFRP by Glass Fiber Sheets: An Experimental Study [J]. Applied Composite Materials, 2023, 30:833 - 856.

[3] Zhang B, Jin X, Yu T, et al. Sealing characteristics of aramid fiber reinforced rubber gasket with scratch depth and width defects [J]. Polymer Composites, 2023.

[4] Behera S, Mohapatra D R, Mondal S. Free Vibration Analysis of FRP Sandwich Composite Plate and Quantification of Uncertainties in its Vibration Response [J]. International Journal of Structural Stability and Dynamics, 2023, 23(03).

[5] Rino Nelson N, Siva Prasad N, Sekhar A S. Stress distribution and performance of threaded bolts in gasket flange joints [J]. Materials Science & Engineering Technology, 2021(9):52.

[6] Wu C L, Kan J C, Wang Q H, et al. FEM analysis of the modular prefabricated steel-concrete composite beam-column internal joint under reciprocating action [J]. Steel & Composite Structures: An International Journal, 2021(1):41.

[7] Signetti S, Bosia F, Ryu S, et al. A combined experimental/numerical study on the scaling of impact strength and toughness in composite laminates for ballistic applications [J]. 2021.

[8] Reed M. Baker Hughes Offers Composite Flexible Pipe for Onshore Pipelines [J]. Pipeline & Gas Journal, 2021(3):248.

[9] Zhou Y X, Chen B, Li J Y. Application and Evaluation of Non-Metallic Pipeline in Lamadian Oilfield [J]. Advanced Materials Research, 2013, 694-697:521-525.

[10] Huan-Zhi C, Shao-Jun L, Jun-Chao Z, et al. Research on the Problem of Sealing Heat Exchanger Pipe Flange and Pipe Plate [J]. Guangzhou Chemical Industry, 2014.

[11] Zhou W, Wang H. Study on sealing performance of bolted flange structure after creep based on ABAQUS [J]. IOP Conference Series Earth and Environmental Science, 2021, 687(1):012150.

[12] Hashin Z. Fatigue Failure Criteria for Unidirectional Fiber Composites [J]. Journal of Applied Mechanics, 1981.

[13] Nguyen-Hoang M, Becker W. Efficient Stress and Failure Analysis of Bolted Joints for Composite Laminates [C]// Deutscher Luft-und Raumfahrtkongress 2021.

Experimental study on the friction characteristics of the interface between piles and coral reef rock

Shoulong Zhang[1,2,3*], Dengfeng Sang[1,2,3], Guanghua Feng[4] and Shiding Su[1,3]

[1]Harbour Engineering Research Institute Co. Ltd., Guangzhou Guangdong, 510230, China
[2]South China University of Technology, Guangzhou Guangdong, 510640, China
[3]Key Laboratory of Environment and Safety Technology of Transportation Infrastructure Engineering, Guangzhou, 510230, China
[4]Fourth Navigation Bureau First Engineering Company Limited, Guangzhou, 510420, China

[*]Corresponding author's e-mail: zhanghaoyi2hokkaido@gmail.com

Abstract: The study aims to conduct coral reef rock-pile wall interfacial shear tests using a modified pile-soil interfacial ring shear instrument to gain insight into the friction characteristics between the two. The tests examined the interfacial friction behavior between the pile and the coral reef rock by simulating the engineering environment. Changing the different pressure conditions systematically recorded the change pattern of the friction characteristics. The test results show that there are significant differences in the friction characteristics between piles and coral reef rock under different environmental conditions, which provides a valuable reference for the design and construction of pile foundations in actual engineering.

1. Introduction

The shear characteristics of calcareous sand could significantly influence the safety and stability of marine structures. The extensive examination of bored piles has resulted in the discovery of numerous significant findings regarding their bearing capacity. Zhang et al. [1] explored the bearing characteristics of bored piles in coral reef environments via laboratory tests, field experiments, and numerical simulations. Their study illuminates key factors such as wall shapes, normal stress, shear force, interface failure mechanisms, and the impact of combined load effects. Feng et al. [2] presented an innovative sediment shear system with multifunctional capabilities tailored for studying shear performance. Employing this apparatus, steel plates with diverse surface roughness underwent varying normal stress levels, shear displacement rates, and temperature conditions. Examination of test data revealed insights into the relationship between different factors and the shear strength of the steel-sand interface. Yu et al. [3] examined the influence of particle corner breakage on pile load performance in coral sand, analyzing both macro-scale and particle-scale phenomena through indoor pile load model testing and coupled simulations using the Discrete Element Method and Finite Difference Method (DEM–FDM). Liu et al. [4] discussed the lateral response of pile-soil interaction in coral and silica sand foundations through comparative numerical investigations. Utilizing the Hypoplastic constitutive model integrated into Abaqus software, the study simulates the stress-strain behavior of coral sand and silica sand with identical particle gradation. Li et al. [5] investigated the shear failure characteristics of coral reef limestone-concrete interfaces through large direct shear tests. The study explored five types of interfaces:

planar, sloped, arc-shaped, stepped, and triangular. Additionally, it analyzed the shear failure characteristics and variations in shear strength across three interface connection types: smooth, mud-filled, and cemented. Liu et al. [6] performed a constant normal stiffness direct shear test on the interface between a pile and a sample of coral reef calcarenite. Their focus was on observing the friction mechanism at the interface and the resulting shear dilation behaviors. The findings indicate that shear failure primarily takes place within the coral reef calcarenite. This is attributed to the diffusion and accumulation of cement paste within the highly porous structure of the coral reef calcarenite during construction. Coop [7] argued that the compressive deformation of calcareous sand is mainly plastic deformation, and the compression of calcareous sand is mainly controlled by particle fragmentation when the pressure exceeds a certain limit. Kaggwa [8] and others studied two calcareous sands from the northwest coast of Australia by using a cyclic triaxial test and comparing it with the dynamic triaxial test of quartz sand. The results of the tests found that the mechanical properties of quartz sand and calcareous sand did not differ much at low-stress levels. Li [9] simulated the wave load by the multifunctional shear instrument developed by Dalian University of Technology, carried out the vertical one-twist cyclic shear test on calcareous sand, and compared and analyzed the dynamic properties of calcareous sand and quartz sand, and the deformation characteristics of the two were similar under the wave load. However, the characteristics of the pore pressure were different from those of quartz sand, and the cyclic effect of the pore pressure in calcareous sand was significantly strengthened with the increasing of the angle of the direction of the initial principal stress, while quartz sand does not have such a law.

With the gradual implementation of the “One Belt, One Road” initiative, the coral reef geology project construction demand is growing. Especially in the “Maritime Silk Road Economic Belt” along the typical countries, all have coral reef distribution. There are already a lot of projects with construction demand, and the future construction demand will be even greater. As we all know, whether the pile foundation can play its vertical bearing role depends largely on the shear friction between the pile body and the soil body on the pile side, so the friction characteristics of the pile-soil interface are the key to studying the bearing capacity and settlement of pile foundation. The ring shear test carried out in this paper is mainly used to simulate the change of pile-soil interface friction characteristics during the pile driving process, and to analyze the difference between the coral reef geology and the conventional geology, which can not only be used to accumulate experience for the subsequent study of this geology, but also provide theoretical support for the subsequent bearing capacity test and analysis of the pile sinking construction.

Figure 1. Blocky coral reef rock obtained from field drilling. (a) Fully weathered coral reef rock block (b) Strongly weathered coral reef rock.

2. Test specimen preparation

The main components of coral reef rocks are calcium carbonate and other insoluble carbonates. Converting all of these equivalently into an equivalent calcium carbonate content, the content can be more than 90%. By conventional geomechanical standards, when this content exceeds 50%, its mechanical properties have changed more significantly. To a certain extent, the calcium carbonate content determines the difference in engineering properties between coral reef rock and other

Journal of Physics: Conference Series **2808** (2024) 012048 doi:10.1088/1742-6596/2808/1/012048

conventional soils. As can be seen from the composition of the coral reef rock obtained in the field, the physical characteristics of the fully weathered coral reef rock are described as highly soft, mainly consisting of a mixture of clayey minerals weathered from the parent rock coral reef rock and gravel-sized coral reef clasts with dimensions roughly ranging from 3 to 60 mm, as shown in Figure 1(a). The physical characteristics of strongly weathered coral reef rock are described as incredibly soft to very soft strongly weathered coral reef rock, with the parent rock weathered to a mixture of clayey minerals with gravel-size or rounded-stone-size coral reef clasts, with sizes ranging mainly between 5 and 70 mm, as shown in Figure 1(b).

Figure 2. Field drilling sampling diagram.

Due to the brittle and friable nature of coral reefs, the impact method of drilling and sampling will damage the integrity of coral reef rocks. Therefore, fewer specimens can be collected and processed to meet the requirements of indoor tests. The specimens that can be tested indoors are mainly coral reef debris and incompletely weathered massive coral reef rock. Most of the drilled rock and soil samples obtained from the borehole investigations in this project were in the form of clasts, as shown in Figure 2. The test soil samples were obtained through survey boring operations at the location of the test piles for the static load tests. The test soil samples consisted of a portion of the reef rock ground gravelly soil and coral reef debris. Three sets of samples were then prepared by crushing and grinding them to a gravelly texture and passing them through a sifter to retain a portion of the particles with a particle size of less than 2 millimeters.

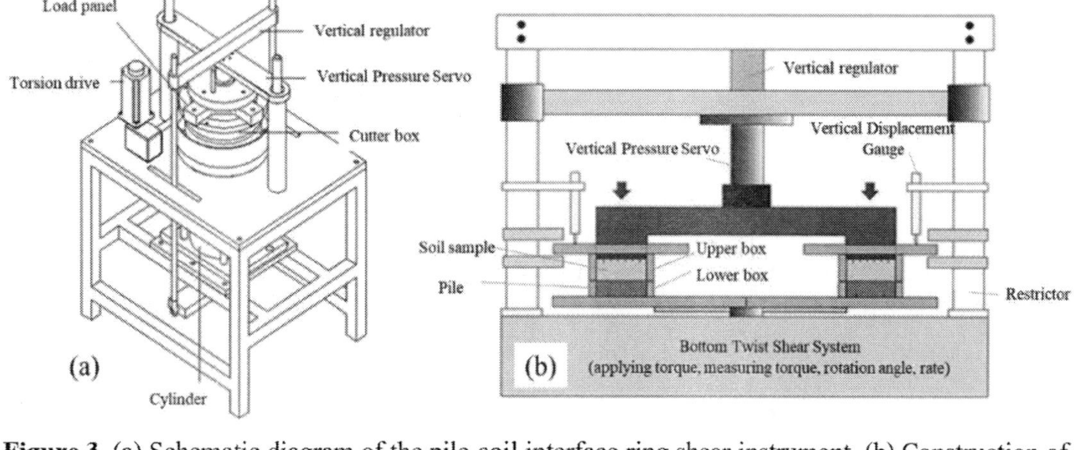

Figure 3. (a) Schematic diagram of the pile-soil interface ring shear instrument, (b) Construction of the ring shear instrument construction.

3. Ring shear test equipment

The ring shear instrument used in the test was upgraded by upgrading the Bishop ring shear instrument, which is widely used at home and abroad. This equipment can carry out large deformation shear at a fixed interface under continuous displacement conditions to accurately assess the pile-soil interface shear characteristics after piling, and the detailed structure of the ring shear instrument is shown in Figure 3.

The lower box soil body is replaced by the pile rigid interface (steel plate), and the lower box is manipulated to carry out ring shear, so that the pile-soil interface unit shear curve and parameters can be obtained. The two separated upper and lower annular specimen boxes are used, with dimensions of 30 cm outer diameter, 20 cm inner diameter, and a total height of 10 cm for the two shear boxes. An observation window (Plexiglas plate) is set on the box's outer side for observing the soil body's shear disturbance layer.

During the test, the normal stress is directly applied to the upper part of the annular soil sample by the force plate, and the upper surface of the sample is fixed by the cutter on the force plate. The lower shear box rotates to apply torque to the specimen, and its damaged surface is the interface between the upper and lower boxes. During the circumferential shear process, the soil sample and the pile interface of the lower box move relative to each other in shear under the torque. The normal load, rotational moments, and angular displacements can be obtained directly from the test, and the shear strength and shear deformation can be converted from the moments and angular displacements. Given that the width of the specimen in the test is narrower compared to the radius of the soil ring, it can be inferred that the shear stress is evenly distributed along the shear surface. The ring shear instrument can accurately reproduce the residual shear stress state of the soil unit after large (theoretically infinite) deformations in the field during the test.

4. Analysis of test results

The prepared specimens were placed in the shear box, and the vertical stresses of 50 kPa, 100 kPa, and 200 kPa were applied to the shear box, and the saturated drainage shear test was conducted at a shear rate of 0.8 mm/min, respectively. The shear method was selected as a single shear, the whole shear process was carried out through the clockwise rotation of the lower shear box, and the soil body was sheared clockwise and positively. The shear stress-shear displacement curves of three groups of specimens under 50 kPa, 100 kPa, and 200 kPa vertical stresses are shown in Figures 4, 5, and 6.

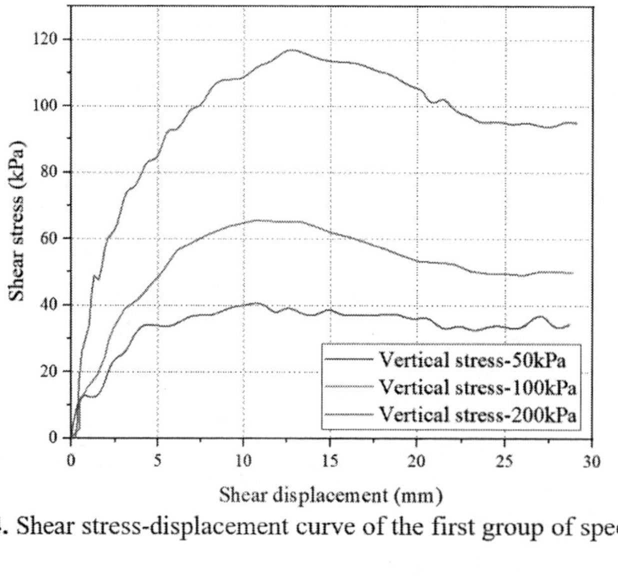

Figure 4. Shear stress-displacement curve of the first group of specimens.

Journal of Physics: Conference Series **2808** (2024) 012048 doi:10.1088/1742-6596/2808/1/012048

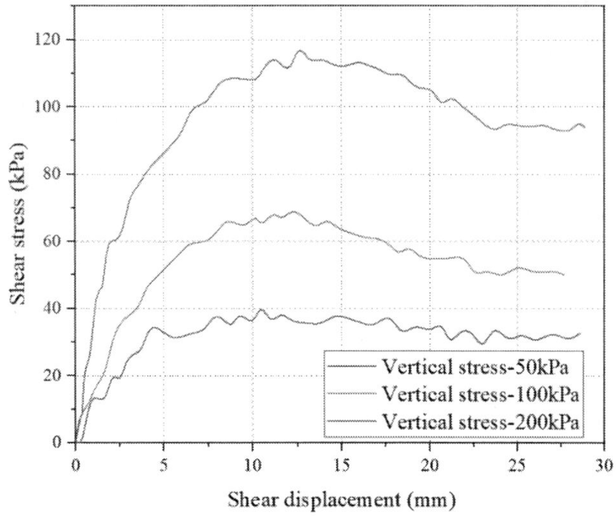

Figure 5. Shear stress-displacement curve of the second group of specimens.

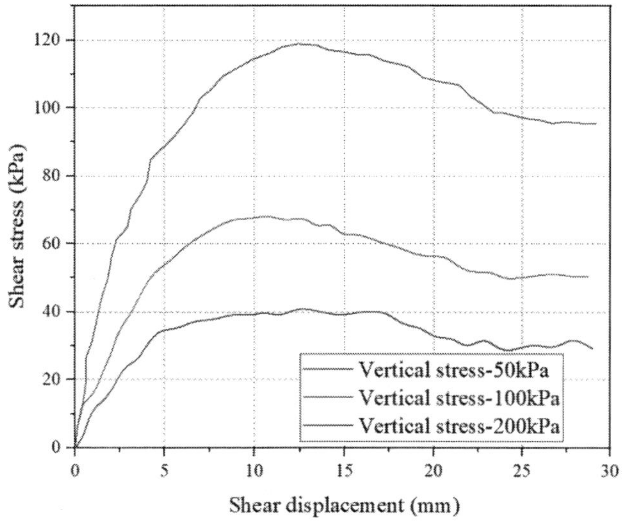

Figure 6. Shear stress-displacement curve of the third group of specimens.

From the shear stress-displacement change curves of the three groups of specimens in the shear process in the figure, it can be seen that when the shear box starts to rotate clockwise, the shear stress suffered by the specimen increases with the increase of the shear displacement. From the rapid growth at the beginning to the subsequent gradual moderation, the shear stress reaches the peak at a certain displacement value. Among them, under the action of 100 kPa and 200 kPa vertical stress, when the shear displacement is increased again, the shear stress is gradually reduced and finally fluctuates at a stable value. This stable value is the residual shear stress of the specimen. By comparing the difference between the peak value and the residual value of the shear stress of the three groups of specimens, it can be seen that the specimen shear strength decreases less under the action of 50 kPa vertical pressure. This indicates that the fragmentation of coral reef rock particles is not obvious under low-pressure conditions, while particle fragmentation becomes more and more serious when the pressure increases, resulting in the change of specimen shear properties. The specific peak shear stress and residual values of the three groups of specimens are shown in Table 1.

Journal of Physics: Conference Series **2808** (2024) 012048 doi:10.1088/1742-6596/2808/1/012048

Table 1. Ring shear test results of three groups of specimens.

No. of specimen	Normal stress (kPa)	Peak shear stress (Pa)	Residual Shear stresses (kPa)	Q345 steel-soil interface friction angle	
				Peak value	Residual
1	50	39.6	35.0	38.4°	35.0°
	100	65.0	50.0	33.0°	26.6°
	200	116.1	95.0	30.1°	25.4°
2	50	41.0	40.2	39.4°	31.0°
	100	67.9	50.0	34.2°	26.6°
	200	115.9	95.2	30.1°	25.5°
3	50	43.0	34.8	40.7°	34.8°
	100	66.7	48.6	33.7°	25.9°
	200	117.1	94.2	30.3°	25.2°

The steel (Q345) - soil interface friction angle δ_f is calculated from Equation (1), τ_f denotes the shear stress on the shear surface of the steel-soil specimen, and σ_n denotes the vertical stress applied to the steel-soil shear surface in the test.

$$\tau_f = \sigma_n \tan\delta_f \tag{1}$$

As can be seen from the test results, the peak friction angle at the steel-soil interface of the reef rock fragments (calcareous sand dominated) was 30.1°~40.7°, and the residual friction angle was 25.2°~35.0°. The changes in the shear properties of the rock-soil samples made from coral reef rock throughout the shear process were mainly caused by the crushing of soil particles. The crushing of calcareous sand particles was mainly due to the fragmentation of $CaCO_3$ or other insoluble carbonates during shear. The degree of crushing gradually increases with the increase of vertical stress. Since the ring shear test simulates the shear friction between the pile and the soil around the pile, the vertical stress applied by the shear box is equivalent to the horizontal effective stress applied to the soil around the pile. The driven construction method during pile formation caused structural damage to the soil on the pile side, particle fragmentation, and reduction of the horizontal effective stress, resulting in a reduction of the pile side frictional resistance, which is much smaller than that of silty sand. Based on the test results, it can also be obtained from Equation (2) for calculating the ultimate lateral friction resistance per unit pile side.

$$q_s = \sigma'_n \tan\delta_f = K\sigma'_v \tan\delta_f \tag{2}$$

where σ'_n is the horizontal effective stress acting on the pile, σ'_v is the vertical effective stress of the soil, K is the static earth pressure coefficient, and δ_f is the friction angle at the pile-soil interface.

The results of the ring shear test showed that the gravel soil prepared from coral reef rock was prone to particle fragmentation during shear. The normal stiffness of the contact interface decreases sharply during shear, but the normal deformation does not increase, which causes the normal stress to decrease and the interface friction to decrease. The peak friction angle of the test soil samples and Q345 steel decayed to a residual friction angle, and the interaction of the pile-soil interface contact was presented as a residual stress state.

The soil around the pile was in a high-stress state during pile driving, which led to the crushing of soil particles and changed the contact friction characteristics between the pile and soil. The main reason for this phenomenon is that compared with the rounded particles of quartz sand, the irregularly shaped angular calcareous sand particles have fewer contact points and less contact area with neighboring particles. Under stress, the stress is higher, and the low strength makes the contact surface more prone to breakage. The crushing of soil particles causes the filling of the intergranular space of soil particles, directly affecting the volume change characteristics of the soil, thus affecting the play of soil strength.

5. Conclusion

Through the geotechnical test and high compression pressure ring shear test, the interface friction characteristics between the pile and coral reef rock were studied, and the following conclusions were mainly obtained:

(1) Coral reef rock is rich in calcium carbonate, which can be divided into fully weathered, strongly weathered, and moderately weathered coral reef rock according to the number of standard penetration strikes and the degree of integrity of the core after weathering, in which the weathered components that become gravelly should belong to lose to slightly dense, compressive sandy soil.

(2) From the results of the ring shear test, the peak friction angle of the interface between Q345 steel and coral reef rock is 30.1°~40.7°, and the residual friction angle is 25.2°~35.0°.

(3) According to the analysis results of the ring shear test, the gravel soil composed of a weathered layer of coral reef rock is prone to particle crushing under high-stress piling conditions. The crushing of particles not only affects the play of soil strength but also reduces the friction angle of the pile-soil interface significantly, which causes the reduction of horizontal effective stress on the pile side, thus affecting the play of bearing capacity of subsequent steel pipe piles.

References

[1] Zhang, Y., Liu, W., Liu, Y., Ding, J., Chen, P., Luo, H., & Yao, X. (2023) Study on Bearing Characteristics of Bored Piles in Coral Reef. Geotechnical and Geological Engineering, 1-21.

[2] Feng, W. Q., Zhang, T. Y., Li, Y. X., & Xu, D. S. (2024) Experimental study of interface performance between calcareous sand and steel pile with different surface roughness and temperatures. Ocean Engineering, 292, 116469.

[3] Peng, Y., Yin, Z. Y., & Ding, X. (2023) Micromechanical analysis of the particle corner breakage effect on pile load performance in coral sand. Acta Geotechnica, 18(12), 6353-6370.

[4] Liu, H., Wang, C., Ding, X., & Zhang, Y. (2023) Comparative numerical analysis of the response of laterally loaded pile in coral and silica sands. Acta Geotechnica, 1-21.

[5] Li, D. J., Shi, C., Ruan, H. N., Li, B. Y., Li, W. Y., & Yao, X. C. (2022) Study on shear behavior of coral reef limestone - concrete interface. Marine Georesources & Geotechnology, 40(4), 438-447.

[6] Liu, H., Zhu, C., Wang, R., Cui, X., & Wang, T. (2021) Characterization of the interface between concrete pile and coral reef calcarenite using constant normal stiffness direct shear test. Bulletin of Engineering Geology and the Environment, 80, 1757-1765.

[7] Coop M R. (1990) The mechanics of uncemented carbonate sands. Geotechnique, 40(4):607-626.

[8] Kaggwa W S. (1990) Influence of Stress Path on the Elastic Moduli of Calcareous Sand. In: Seminar on Computing in Geomechanics: Seminar Papers. Adelaide, S. Aust.: Australian Geomechanics Society, 32-41.

[9] Li, J. G., Wang, R., Yu, H. Z., He, Y., & Xu, C. S. (2005). Experimental research on effect of initial principal stress orientation on dynamic properties of calcareous sand. Yantu Lixue (Rock Soil Mech.), 26(5), 723-727.

Optimization design of the ducted fan powered high-lift system based on i-sight platform

Sen Mao[1*], Gang Li[1], Chuangxin Zhao[1], Qiang Guo[1], Wenliang Feng[1] and Zongming Xie[1]

[1]Chengdu Aircraft Industry (Group) Co., LTD., Chengdu, Sichuan, China

[*]Corresponding author's e-mail: icaus_2023@126.com

Abstract. The vertical/short take-off and landing (V/STOL) unmanned aircraft vehicle (UAV) equipped with a ducted fan has a broad application prospect, and the design of the ducted fan-powered high-lift system has become one of the key factors in the design and development. The i-Sight platform is used to optimize the design of the 2D ducted fan-powered high-lift system in the paper. First, the parametric modeling of the main control parameters is carried out, and then the i-Sight platform is built to carry out optimization calculations. Finally, three optimal design points are found for detailed aerodynamic characteristics analysis, and then the optimal scheme is selected. The i-Sight platform is proven to be accurate and efficient. By moving the ducted fan horizontally forward from the initial position to 160 mm, a larger lift coefficient can be obtained, while moving the ducted fan vertically from the initial position to 300 mm, a larger lift resistance ratio can be obtained.

1. Introduction

UAVs have been widely used and developed rapidly in both military and civilian fields in recent years. The high-speed V/STOL UAV equipped with ducted fan can be used for the urgent needs of express logistics in rural and remote areas, the rapid deployment of materials under military field conditions, and the transportation of future cities, etc., with broad application prospects [1-2], as shown in Figure 1. However, the position of the ducted fan is constantly rotating and changing relative to the wing and fuselage in the transitional flight stage of this kind of aircraft, and the downwash flow of the ducted fan interacts with the wing and fuselage to form a complex unsteady flow field. Aerodynamic interference during takeoff, landing, and transition flight has brought many adverse effects on flight control, flight performance, and control stability of aircraft during flight [3]. Therefore, the design of the ducted fan-powered high-lift system has become one of the key factors in the design and development of these aircraft.

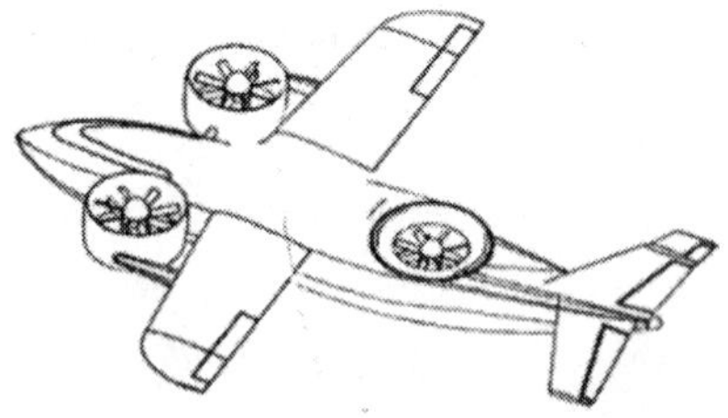

Figure 1. The high-speed V/STOL UAV is equipped with a ducted fan.

Figure 2 shows the schematic diagram of the ducted fan-powered high-lift system. The high-pressure and high-speed flows from the ducted fan with a large range. When the center axis of the ducted fan is like the wing, one part of the flow along the upper surface of the wing flap increases the energy of the boundary layer and delays the separation [4]. Some flows also flow along the lower surface of the wing through the slit and jet to the upper surface of the flap, thereby improving the surrounding flow field, increasing the amount of the wing loop, and improving the wing lift. It should also be noted that the flow deflects through the flaps, increasing the flap lift while also increasing the bow torque. By adjusting the space position of the ducted fan, the high-speed flow above and below the wing can be controlled, and combined with the flap slot size and deflection angle, the overall lifting effect of the powered high-lift system can be affected [5-6].

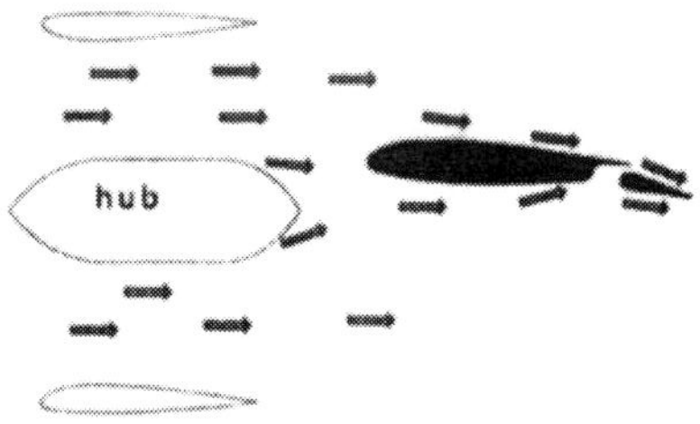

Figure 2. Schematic diagram of the ducted fan-powered high-lift system.

The optimization design of the 2D ducted fan-powered high-lift system is carried out in this paper First, the parametric modeling of the main control parameters was carried out, and then the i-Sight optimization platform was built to integrate all the processes of the parameterization of graphic key design variables and Fluent numerical simulation calculation. Then the three optimal design points were found through cyclic calculation. At last, the aerodynamic characteristics are analyzed in detail and the optimal solution is determined.

2. i-Sight optimization platform

The i-Sight optimization platform will be used in this paper to integrate all processes of aerodynamic numerical simulation to obtain the desired optimization results. The i-Sight software was originally developed by engineers of General Motors during their participation in the research and development of aircraft turbofan engines. As a result, the burden of repeating complicated design steps is effectively reduced and an engine design optimization scheme with excellent performance is obtained [7].

MIGA (Multi-Island Genetic Algorithm) is chosen as the algorithm for optimization design in this paper for multi-objective optimization. MIGA (multi-island Genetic Algorithm) has a stronger global search ability and higher computational efficiency than traditional genetic algorithms which are widely used today. The new population in the traditional genetic algorithm becomes the current population in the next iteration of evolution. The key to a multi-island genetic algorithm is to expand from a single population to multiple populations (or become multiple islands), not only to carry out mutation, selection, and crossover operations within the population but also to periodically carry out targeted migration operations between populations (between multiple islands). Thus the optimal solution can be found quickly and efficiently.

3. Model parameterization

First of all, the optimization objective is determined. During the cyclic calculation and solution of the optimization platform, the maximum lift coefficient of the powered high-lift system cannot be obtained. Considering that the lift coefficient and angle of attack have a wide range of linear regions before stall, the maximum lift coefficient of the configuration must be large if the lift coefficient of the powered high-lift system is relatively large when the angle of attack is 8°. Therefore, the goal of designing the optimal 2D ducted fan-powered high-lift system is defined as finding a certain configuration. The lift-drag ratio L/D at 8° angle of attack and the lift coefficient CL at this time are both maximum, so the multi-objective optimization process is determined.

CATIA software is used to construct the geometric shape in this paper. CATIA torque is first used to draw the 2D cross-section of the ducted fan-powered high-lift system, and the parametric design function of CATIA is applied to the main control parameters of the system so that the shape of the powered high-lift system can be automatically changed with the change of the optimization assignment during the optimization process. To facilitate subsequent grid division [8], the design variable to be optimized is the main control parameter of the 2D ducted fan-powered high-lift system. These parameters are defined here, and the parametric modeling function is used in CATIA software to correlate the main control parameters to the design table through the relationship, and then the external array document is generated from the design table. The subsequent optimization process only needs to change the shape of the high lift system through the numerical modification of the array document and Fluent numerical simulation calculation. As shown in Figure 3 for the position of the fixed wing in space, the horizontal translation distance of the ducted fan is defined as X _ fan, and the vertical translation distance of the ducted fan is defined as Y _ fan. Then, a reference coordinate system is established along the chord line of the careful shape of the reference base machine, and the position 0 of 0.8125c on the chord length of the basic wing is taken as the reference point. It is used to determine the flap's spatial position. The horizontal direction to the right along the string line is X positive direction, the vertical chord length line down is Y positive direction, the distance between the flap and the reference point O in the X direction is X _ flap, the distance between the flap and the reference point O in the Y direction is Y _ flap, and the angle between the flap string and the reference basic wing string is O_TF, that is, the flap deflection angle.

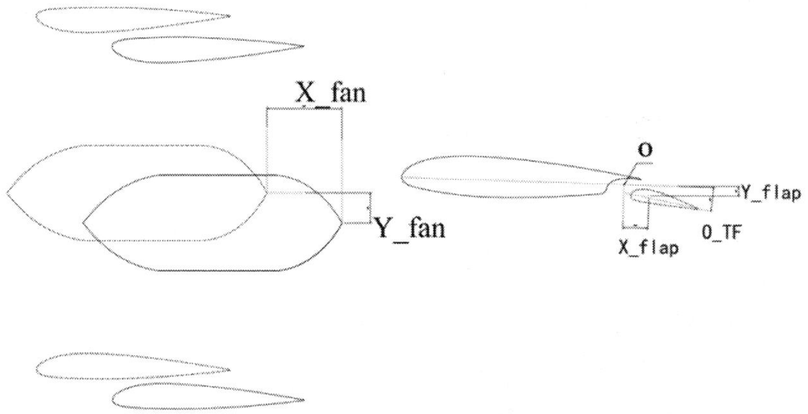

Figure 3. The parameter definition of the 2D ducted fan-powered high-lift system.

Some relevant factors are considered, such as the influence of the horizontal forward and forward movement of the ducted fan on the aircraft's center of gravity, whether the ducted fan interferes with the leading edge of the fixed wing when the ducted fan is in vertical take-off and landing mode, the Fuller flap configuration slot parameters of the general airfoil of NACA23012 and so on. These above parameters are referred to determine the optimized range, as shown in Table 1.

Table 1. Optimization parameters and objectives.

Optimization parameters	Optimization objectives
X _ fan ∈ [-180 0] mm	Maximum: (α=8°)
Y _ fan ∈ [-150 300] mm	Cl
X _ flap ∈ [40 60] mm	L/D
Y _ flap ∈ [18 23] mm	
O _ TF ∈ [8° 14°]	

Journal of Physics: Conference Series 2808 (2024) 012049 doi:10.1088/1742-6596/2808/1/012049

4. Construction of the optimized platform

The flow of numerical simulation optimization using the i-Sight platform is usually shown in Figure 4. Its main idea is to use the Sim-code program integration component to integrate steps such as contour generation modeling, pre-processing grid division, simulation numerical calculation, and result reading operation, and then let the optimization task component change the parameter input in combination with specific algorithms. That is, the specific design variable value, in which the external program used by each part is driven by the script file corresponding to each program [9].

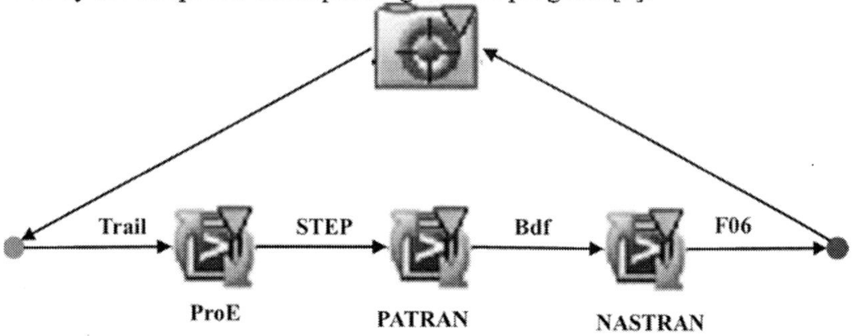

Figure 4. Integrated optimization process.

The integrated platform of the 2D ducted fan-powered high-lift system is optimized in this paper, as shown in Figure 5. The sim-code component is used to invoke CATIA, Pointwise, and Fluent three external programs, and corresponding input and output files in turn, and corresponding data processing and calculation are done. The Optimization process dynamics and results can be viewed in the optimization component.

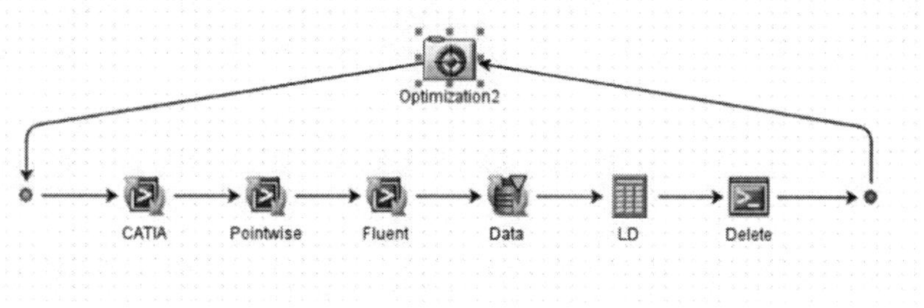

Figure 5. Construction of the optimized platform.

Optimization task module: Multi-island genetic algorithm is selected as the optimization component, and the number of islands (population number) is set to 8, the number of offspring islands to 5, and the genetic algebra to 8 generations. The total number of optimized lift configurations is 320.

5. Optimization result analysis

The design distribution after cyclic iterative optimization is shown in Figure 6, which contains 320 design points of the configuration of the ducted fan lift system. The x-coordinate represents the lift-drag ratio of the configuration at an 8° angle of attack, and the y-coordinate represents the lift coefficient of the configuration at a 6° angle of attack. According to the requirements of the optimization goal, the more the design point is in the upper right corner, its performance is excellent, that is, the greater the lift-drag ratio at an 8° angle of attack, the better the performance is. The lift coefficient at an 8° angle of attack also increases. The curve connected by the points at the top right corner is often called the "Pareto front". The set of design points contained in this curve is the optimal solution set, and the optimal solution is found from this optimal set by combining the analysis of other characteristics. Three design points were selected from the figure, corresponding to the configurations of the three power lift devices, and were further compared as analysis objects to select the best configuration. The three points were defined as configurations A, B, and C.

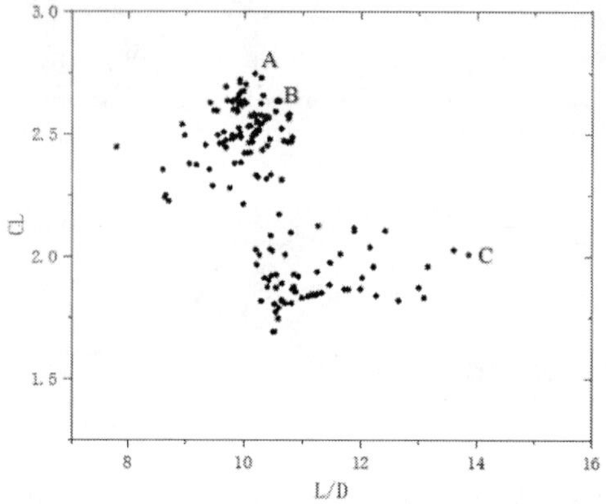

Figure 6. Configuration optimization design point distribution of the ducted fan-powered high-lift system.

Figure 7 shows the relationship between the lift coefficient of the ducted fan-powered high-lift system at an 8° angle of attack and the five optimization variables. The distribution of design points is somewhat random, some have some obvious laws, and the figure can be used for certain trend analyses.

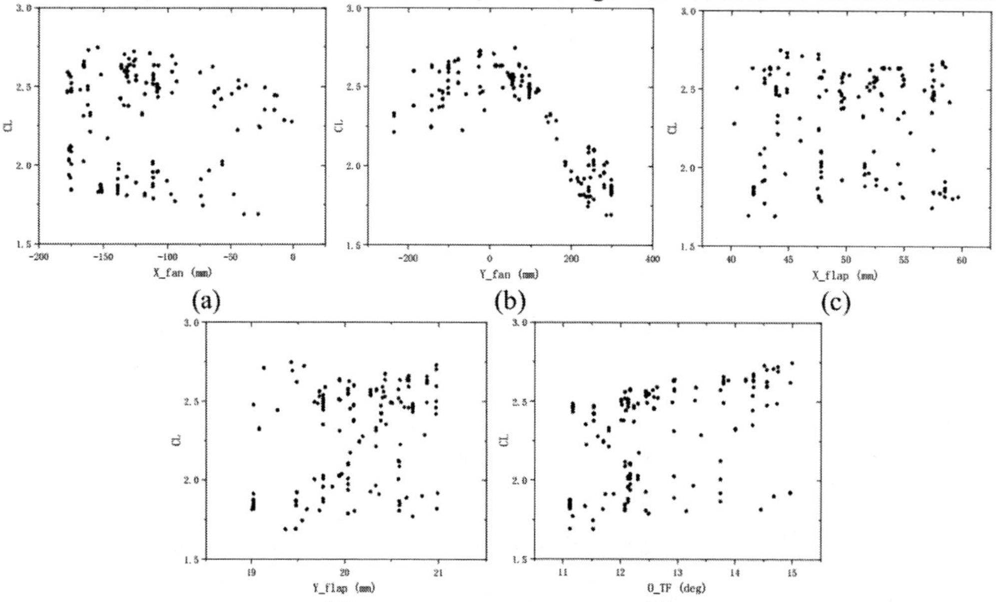

Figure 7. Relationship of Lift coefficient and optimization variable.

As can be seen from Figure 7, the lift coefficient of the powered high-lift system has no obvious relationship with the X _ fan, X _ flap, and Y _ flap. When the displacement of the fan (Y _ fan) in the vertical direction exceeds a certain value, the lift coefficient decreases significantly. When the flap deflection angle O_TF increases, the lift coefficient will also increase. Figure 8 shows the relationship between the lift-drag ratio and five optimization variables of the ducted fan-powered high-lift system at an 8° angle of attack.

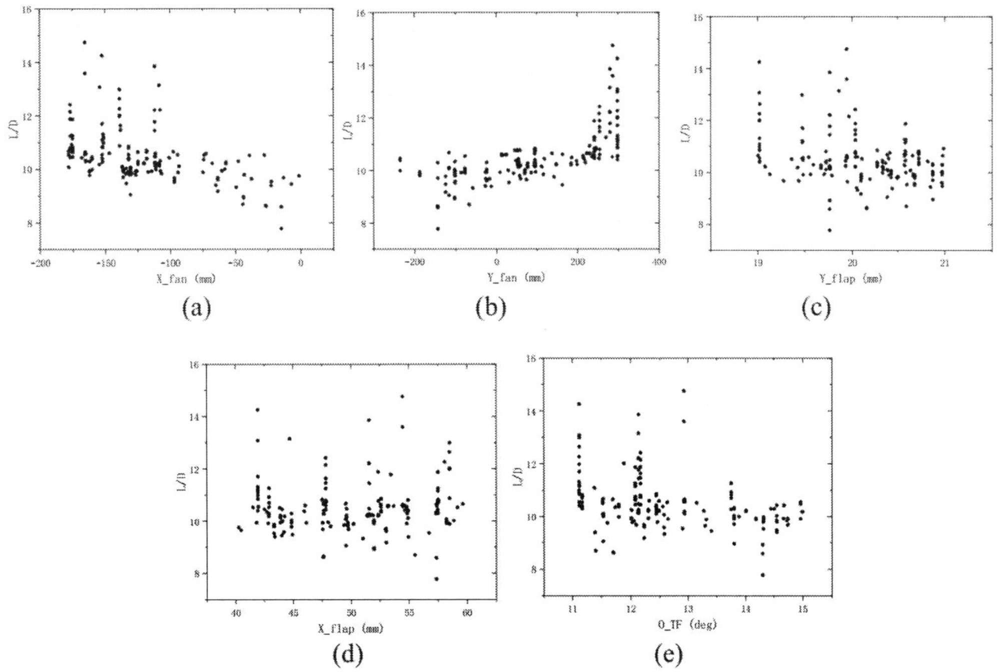

Figure 8. Relationship of lift-drag ratio and optimization variable.

As can be seen from Figure 8, the lift-drag ratio of the ducted fan power lift system has no obvious relationship with the X _ flap in the direction of relative chord length. However, when the Y _ flap and flap deflection angle O_TF in the direction of vertical chord length is small, a relatively large lift-drag ratio can be obtained. The maximum lift-drag ratio can be obtained when the duct fan is moved vertically up to around 300 mm. In addition, the space position of the duct fan relative to the wing has an obvious influence on the lift-drag ratio. When the X _ fan of the culvert relative to the wing is larger, the lift-drag ratio is also larger. When the upward translation of the fan (Y _ fan) in the vertical direction exceeds a certain value, the lift-drag ratio increases significantly. Combined with Figure 7 above, to obtain a larger lift coefficient, the ducted fan should be moved horizontally forward, and the lift-drag ratio will also increase. However, the influence of the ducted fan's forward movement on the center of gravity shift should also be considered. In the vertical direction, to obtain a larger lift-drag ratio, it is necessary to shift the ducted fan vertically upward by a larger distance, which contradicts the larger lift coefficient when the ducted fan is close to the wing, which needs comprehensive consideration.

The three configurations of the 2D ducted fan-powered high-lift system and the position diagram of the prototype are shown in Figure 9. The vertical direction of the ducted fan hub center in configuration A is closer to the wing, and the position of the ducted fan in configurations A, B, and C is like that of the wing, while the flap deflection angle of configuration A and B is larger, the gap is smaller, and the lift coefficient is larger. C configuration ducted fan has a relatively large upward migration in the vertical direction, so the lift-drag ratio at an 8° angle of attack is greater than that of other configurations, which also conforms to the rule analysis above. The lift-drag ratio of the three configurations to be selected is higher than that of the original configuration, the lift coefficient of the two configurations A and B is higher than that of the original configuration, and the lift coefficient of the C configuration is lower than that of the original configuration despite the large lift to drag ratio. The slot parameter values of the three configurations are close to the range of slot parameter selection in the aircraft design manual.

Journal of Physics: Conference Series **2808** (2024) 012049 doi:10.1088/1742-6596/2808/1/012049

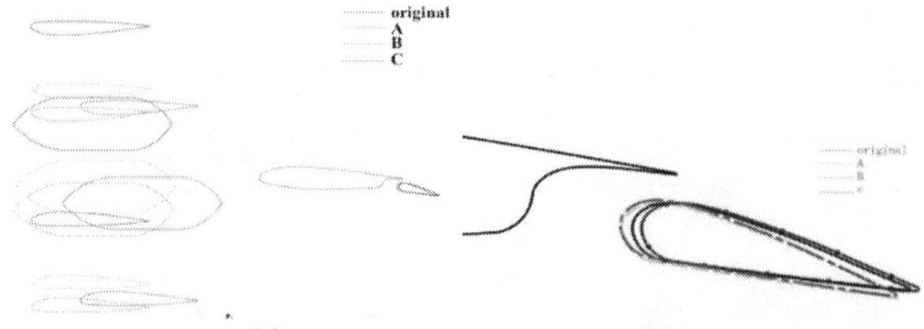

(a) (b)

Figure 9. Comparison between the three configurations and the original configuration.

CFD is used to analyze the aerodynamic characteristics of the four configurations. Figure 10 shows the aerodynamic characteristics curves of the four configurations. In (a), the lift coefficient of the two configurations A and B is greater than that of configuration C and the original configuration at each angle of attack, and the stalling angle of attack of the four configurations except for configuration C is about 12°. This may be because the configuration C ducted fan is higher relative to the vertical direction of the wing than other configurations, and the high ducted fan has more high-speed airflow washing over the upper surface of the wing at the high angle of attack, delaying airflow separation and increasing the stalling angle of attack. In (b), configuration C has a larger lift-drag ratio than other configurations when the lift coefficient is the same, followed by configuration B. This also verifies the trend analysis of the above optimization results from the CFD simulation process of the full angle of attack, that is, when the vertical force wing of the ducted fan is larger, the lift-drag ratio becomes significantly larger. In (c), the bow moment of configuration A and B is larger than the original configuration, while the upward moment of configuration C is smaller, which is related to the deflection angle. The deflection angle of A and B is larger, while the deflection angle of C is the smallest. The increase of deflection angle increases the curvature of the wing, and the lift coefficient also increases, but the upward moment is also increased. In (d), In configuration A and B, a larger suction peak is generated on the upper surface of the wing near the leading edge, resulting in a larger corresponding lift on the main wing, and the flap part pressure coefficient of the two is not significantly increased compared with the original configuration, while the flap pressure coefficient distribution is significantly smaller in configuration C, which also results in a smaller overall wing lift in configuration C.

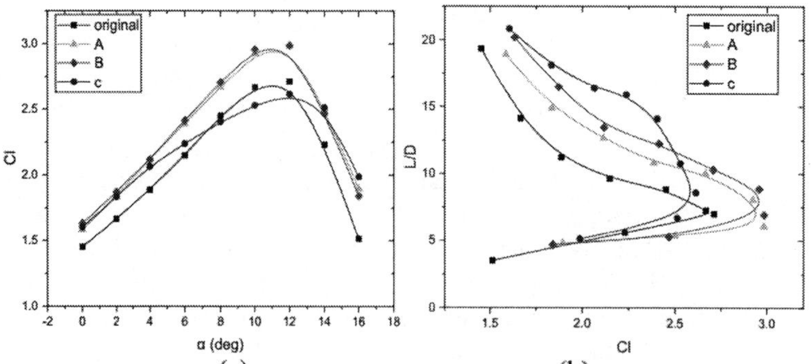

(a) (b)

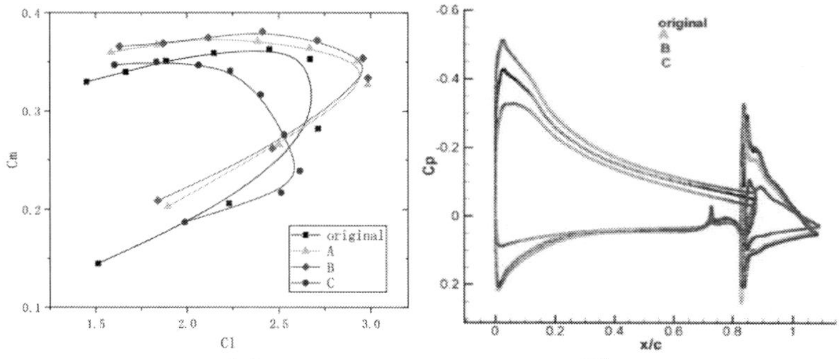

(c) (d)

Figure 10. Comparison of aerodynamic characteristics of four configurations.

Figure 11 shows the flow diagram and pressure cloud diagram of A, B, and C and the original four configurations. It can be seen from the diagram that the leading edge of the lower surface of the wings of A, B, and the original configuration has an obvious high-pressure region, while the upper surface of the wings has a wide range of negative pressure regions, among which the range of A and B is more obvious, indicating that a large suction peak is generated. However, the high-pressure region on the lower surface of the C configuration wing is smaller, which corresponds to the pressure coefficient distribution diagram in Figure 10 (d). At an 8° angle of attack, there is almost no significant separation flow around the four flap configurations (except in the cavity at the rear edge of the main wing). The slot parameters are different, and a separation bubble is generated at the trailing edge of the main wing, but good adhesion flow is still maintained around the flap.

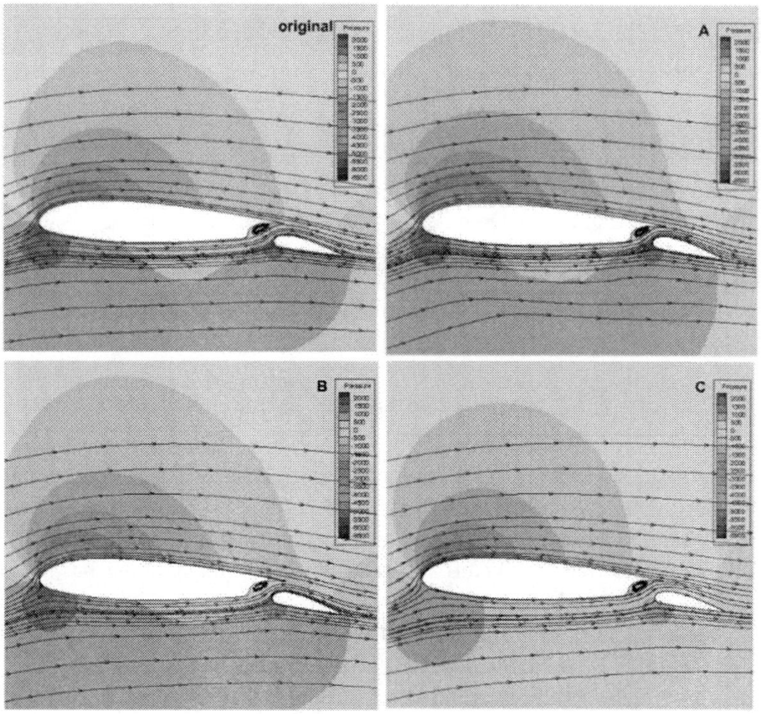

Figure 11. Pressure distributions and flow diagrams of the four configuration.

As a conclusion, configuration B has a higher lift-drag ratio at an 8° angle of attack when the lift coefficient is like configuration A but much larger than the original configuration and configuration C. Although the upward torque coefficient is slightly larger than configuration A, the upward torque generated by the ducted fan is sufficient to reduce this adverse effect. Therefore, the shape of the 2D ducted fan high lift increase system represented by configuration B is chosen as the 2D best take-off configuration.

6. Conclusion

i-Sight platform is used to optimize the design of the ducted fan high lift increase system in this paper. Firstly, the optimization objective, the optimization variables, and their scope are determined, and the construction process of the optimization platform is described in detail. After optimization, the three configurations were compared in detail from multiple perspectives, such as the shape of the selected configuration, its lift coefficient comparison, lift-drag ratio comparison, and pressure coefficient distribution comparison. Finally, an optimal take-off configuration of the ducted fan-powered high-lift system was selected from the three configurations. Three conclusions are made:

- Compared with linear flutter velocity, nonlinear flutter velocity is improved, and the flutter problem does not occur within the 15% safety margin of the maximum flight velocity of each design
- The i-Sight platform adopted in this paper is accurate and efficient in dealing with the optimal design of the ducted fan high lift system. By comparing the lift-drag coefficient and pressure distribution cloud map, the shape of the 2D ducted fan high lift system represented by configuration B at an 8° angle of attack is obtained as the 2D optimal take-off configuration, which will guide the subsequent aerodynamic layout optimization design of the whole aircraft.
- The disadvantage of this method is that the i-Sight optimization design process is carried out on a 2D level. In the process of drawing and dividing the grid, it is found that the position of the ducted fan needs to move freely with the optimized assignment, which will cause uncertainty in the direction of the flow field of the ducted fan's pressure increased surface, which will have a certain impact on the quality of the automatically generated grid. In the future, it is planned to optimize the meshing method of ducted fans to improve the optimization efficiency of i-Sight.

References

[1] Yang Lei, Ye Zhengyin. Unsteady aerodynamic forces during the transition phase of tilting bypass [J]. Journal of Aerodynamics, 2015, 30 (1): 15-16.
[2] Putsep K, Nerep T, Tiismus H. Development and testing of vertical take-off and landing aerial vehicle with tandem electric ducted fan motor[J]. Proceedings of the Estonian Academy of Sciences, 2023(2):72.
[3] Gao Y, Xu Y. The Overall Design of Variable Diameter Ducted Fan in the Aircraft[J]. Aerospace, 2022.
[4] Si Zong, Dong Jianhong. Research on the application of external blowing power increases Technology in large transport aircraft [J]. Aeronautical Science and Technology, 2006 (02): 33-38.
[5] Englar R, Blaylock G, Gaeta R, et al. Recent experimental development of circulation control airfoils and pneumatic powered-lift systems [CV/48th AIAA Aerospace Sciences Meeting Including the New Horizons Forum and Aerospace Exposition. 2010:345.
[6] Yi Ji, Yanbin Li, Xiaochen He. Wind load analysis of radar antenna based on response surface model[J]. Engineering Mechanics, 2019.
[7] Dada Zhou, Guojin Chen, Youping Gong. Research on Multidisciplinary Optimization Method based on i-Sight [J]. Mechanical and Electrical Engineering, 2009, 26 (12): 78-81.
[8] Su Kaijing. Design and Implementation of CATIA 3D Parametric System [D]. Huazhong University of Science and Technology, 2016.
[9] Xiaopeng Wang, Zhenghong Gao. Aerodynamic optimization of airfoil based on Genetic algorithm [J]. Journal of Aerodynamics, 2000.9, 18 (3): 324-329.

Journal of Physics: Conference Series **2808** (2024) 012050 doi:10.1088/1742-6596/2808/1/012050

Analysis of isothermal elastohydrodynamic lubrication on the rough surface of a new type of trident universal joint

Yutian Xu[1*], Yan Yang[1], Guowei Wang[1] and Jiahong Niu[1]

[1]Weifang Engineering Vocational College, Qingzhou, Shandong, 262500, China

[*]Corresponding author's e-mail: quiettree8506@163.com

Abstract. Taking the roughness into consideration, this chapter mainly studies the isothermal elastohydrodynamic lubrication between the slider and the balls in the new three-pronged universal joint and gives the relevant equations under isothermal conditions. The Reynolds equation is the primary tool for iteration, and Matlab is used for programming and plotting. In the simulation experiment, the roughness is given as 0. Then the amplitude and frequency of the roughness are varied to observe the oil film thickness and pressure respectively. With the increase of surface fluctuation, the film thickness in the center of the contact range becomes thicker, the pressure changes more and more sharply, and the peak value of secondary pressure is basically unchanged. As the frequency of the sinusoidal roughness increases, the film thickness in the center of the contact area becomes thinner as the frequency of roughness enhances, the fluctuations become more and more frequent, the pressure fluctuates around the original curve more and more often, and the amplitude becomes larger and larger.

1. Introduction

This article mainly studies the elastohydrodynamic lubrication characteristics of point contact of new universal joints that have appeared in recent years. This new three-prong universal joint was invented by Professor Chang Degong. The author of this article, Xu Yutian, once studied and used it in his laboratory. This new type of three-pronged universal joint has better transmission performance. Its lubricity properties require further study.

In recent years, scholars at home and abroad have conducted many similar lubrication studies. For example, when Wu et al. [1] discovered a central protrusion near the contact center, the pressure experienced a fluctuating process of rapid increase, sudden decrease, and further increase; the thinnest film thickness appears at the contact center, and a depression is generated in front of the contact center. Wang et al. [2] found that the thinnest film thickness generates on the left side or necking area of the single concave point contact elastohydrodynamic lubrication problem when studying the lubrication characteristics. The radius and depth of the pit significantly affect the overall oil film thickness. Mohamed [3] found that the influence of friction is more pronounced when the oil film is thinner. Quiñonez [4] discovered that lateral vibration increases the thinnest point contact of oil film thickness elastohydrodynamic lubrication. Kumar et al. [5] found that the pressure generated in the contact area is highest in serrated bearings with transverse roughness and lowest in sinusoidal bearings with longitudinal roughness. Abd [6] found that increasing the final suction speed ultimately leads to rapid accumulation of film thickness, increasing the film thickness jump at the moment of motion initiation. Liu [7] found that as the size of local defects increases, the equivalent contact stiffness of the defect area decreases, while the contact load of non-defect areas increases. Krupka et al. [8] found that in the transition zone between the elastic viscous and elastic equiviscous modes near the elastohydrodynamic lubrication region, the smooth circular contact operation and the significant deviation between the liquid film thickness of the two modes and the predicted model depend on the inlet temperature. The author of this article recently studied point contact elastohydrodynamic lubrication [9], but did not study frictional properties. This paper focuses on the factors related to friction and studies the friction surface.

2. New three-prong universal joint slider ball lubrication model

Balls are added between the slider and the three-fork groove of the new three-prong universal joint, which changes sliding to rolling, thereby reducing friction. Lubricating oil is added to the slideway. The balls roll and the slider slides. The relationship between the balls and the slider can be regarded as point contact [10], and the simplified model is shown in Figure 1.

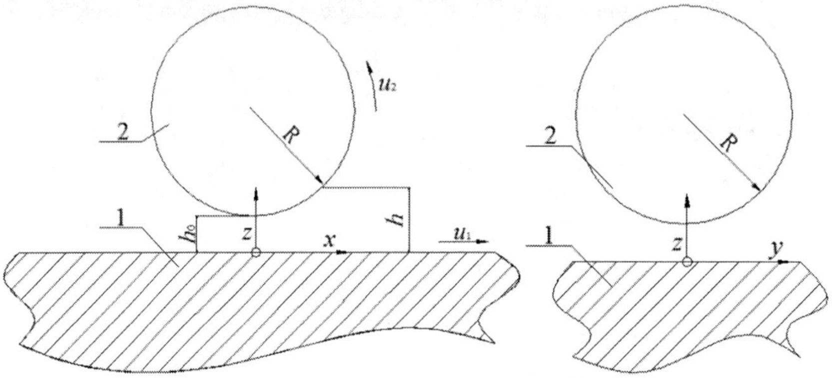

1. Slider 2. Steel ball

Figure 1. A simplified model of slideway lubrication for new structural drawing of tripod universal joint.

Among them, the ball is a sphere. The x direction is located on the surface of the slider and points to the axis of the trifurcated groove. The y direction is located on the surface of the slider and is perpendicular to the axis of the trifurcated groove. The z-direction is perpendicular to the surface of the slider and points to the center of the ball. The slider moves along the trifurcated groove which is the x-axis.

3. Elastohydrodynamic lubrication equation

This chapter mainly studies point contact lubrication, and gives the two-dimensional Reynolds equation as follows:

$$\frac{\partial}{\partial x}\left(\frac{\rho h^3}{\eta}\frac{\partial p}{\partial x}\right)+\frac{\partial}{\partial y}\left(\frac{\rho h^3}{\eta}\frac{\partial p}{\partial y}\right)=12u_s\frac{\partial}{\partial x}(\rho h) \tag{1}$$

where $u_s=|u_1-u_2|$. u_1 is the slider speed, u_2 is the ball speed, and u_s is the relative speed of u_1 and u_2.

The film thickness equation is shown as follows:

$$h(x,y)=h_0+\frac{x^2+y^2}{2R}+v(x,y)+r(x,y) \tag{2}$$

where v (x, y) is the elastic deformation, as shown in Equation (3), and r(x,y) is the roughness deformation. If the roughness is not considered, there will be no such term, and the sinusoidal roughness is as follows:

$$v(x,y)=\frac{2}{\pi E}\iint_{\Omega}\frac{p(s,t)}{\sqrt{(x-s)^2+(y-t)^2}}\mathrm{d}s\mathrm{d}t \tag{3}$$

$$r(x,y)=\mathrm{DA}\sin(2\pi x) \tag{4}$$

Considering that the slider only moves in the X direction, Equation (4) only uses the sine function in the X direction of the slider. v (x, y) in Equation (3) is the pressure function, including the comprehensive deformation of the ball and slider. The expression of E is shown in Equation (5)

$$\frac{1}{E}=\frac{1}{2}\left(\frac{1-v_1^2}{E_1}+\frac{1-v_2^2}{E_2}\right) \tag{5}$$

where the values of E_1 and E_2 are 210 GPa, n_1 and n_2 are both 0.3, and E is calculated to be approximately 230 GPa.

The viscous pressure equation adopts the Roelands model, as shown in Equation (6).

$$\eta=\eta_0\exp\left\{(\ln\eta_0+9.67)[-1+\left(1+\frac{p}{p_0}\right)^z]\right\} \tag{6}$$

where $z=\alpha/[5.1\times10^{-9}(\ln\eta_0+9.67)]$.

The dense pressure equation is shown in Equation (7).

$$\rho=\rho_0\left(1+\frac{0.6p}{1+1.7p}\right) \tag{7}$$

where the pressure boundaries are $p(x_0, t)=0$, $p(xe, t)=0$, $p\geq0$, and the unit is GPa.

The load equation is shown as follows:

$$w=\int_{x_0}^{x_e}\int_{y_0}^{y_e}p(x,y)dxdy \tag{8}$$

The discrete form of the Reynolds equation is shown as follows:

$$\frac{\varepsilon_{i-0.5,j}\mathrm{P}_{i-1,j}+\varepsilon_{i+0.5,j}\mathrm{P}_{i+1,j}+\varepsilon_{i,j-0.5}\mathrm{P}_{i,j-1}+\varepsilon_{i,j+0.5}\mathrm{P}_{i,j+1}-\varepsilon_0\mathrm{P}_{ij}}{\Delta\mathrm{X}^2}=\frac{\rho_{ij}^*\mathrm{H}_{ij}-\rho_{i-1,j}^*\mathrm{H}_{i-1,j}}{\Delta\mathrm{X}} \tag{9}$$

4. Initial conditions

The lubricating oil should be a P100 model [10]. The specific parameters are shown in Table 1.

Table 1. Lubricating oil of P100 parameters.

h_0 (Pa · s)	α (Pa^{-1})	ρ_0 (kg/m^3)
0.214	2×10^{-8}	870

Calculation is performed using the multigrid method. The number of nodes is N=65×65. The starting and end coordinates in the X direction are X_0 equal to -2.5 and XE is equal to 1.5 in the dimensional unification process. The starting and end coordinates in the Y direction are Y_0 equal to -2.0 in the dimensional unification process. 2.0 and YE are equal to 2.0; R is 0.002 m, and Us is 1.0 m/s.

5. Isothermal elastohydrodynamic lubrication simulation results without considering roughness

The above equation is discretized using the difference method, and the Reynolds difference equation is used for iteration. This method is simple to program, but the accuracy is not as good as the multigrid method. The accuracy is set to 10-4, and Matlab software is used for simulation drawing. In the following figure, the X direction refers to the plane where Y is 0, and the Y direction refers to the plane where X is 0. The roughness amplitude in this section is 0. Figure 2 displays the oil film thickness curve, and Figure 3 displays the oil film pressure curve:

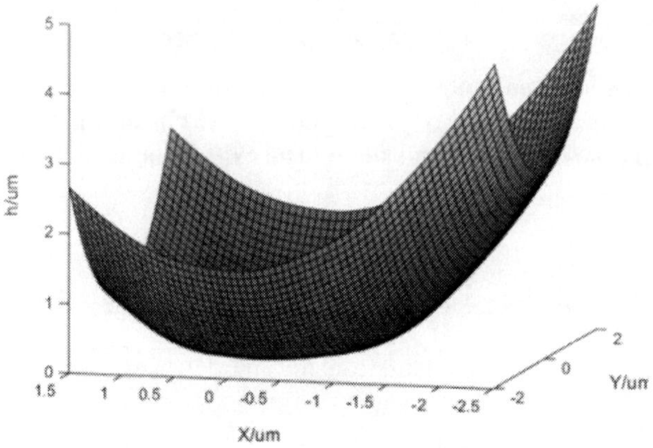

Figure 2. EHL (Elastohydrodynamic Lubrication) film thickness curve.

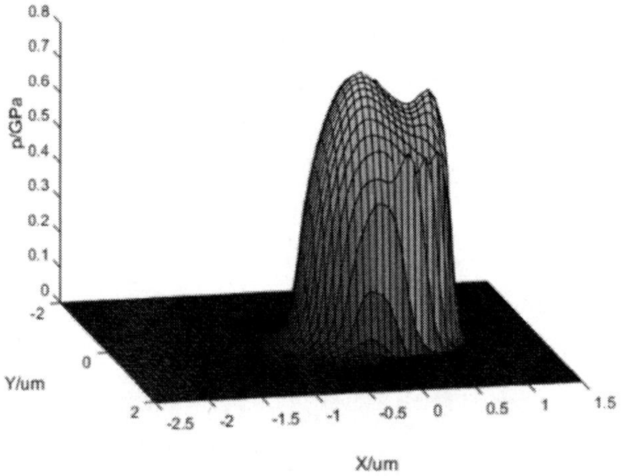

Figure 3. EHL film pressure curve.

As shown in Figure 2, the film thickness presents a state of being thick around all sides and thin in the middle. The film thickness in the X-direction contact area remains almost constant. At the end of the contact area, there is a slight decrease in film thickness, and upon leaving the contact area, the film thickness increases. The film thickness exhibits symmetrical distribution. Upon entering the contact area, the film thickness significantly increases, and the contact area is relatively stable. When leaving the contact area, the film thickness decreases noticeably and then rises rapidly. The thinnest oil film thickness is close to the end of the touch range.

As shown in Figure 3, the touch area presents the shape of a quadratic function surface. At the end of the touch area, a secondary pressure peak is generated, and the pressure outside the touch area is 0. Figure 3 shows that X The pressure curve in the directional contact area is approximately a quadratic curve. When leaving the contact area, a secondary pressure peak is generated. The pressure is symmetrically distributed, and the overall pressure curve is approximately a quadratic curve. The strongest pressure is at the touch range center.

6. Effect of sinusoidal roughness amplitude on isothermal elastohydrodynamic lubrication

When considering roughness, sinusoidal roughness is used to investigate. Considering the amplitude of roughness, the amplitudes are set to 0.01 mm, 0.02 mm, and 0.03 mm respectively, with a frequency of 1 Hz to observe the changes in film thickness and pressure.

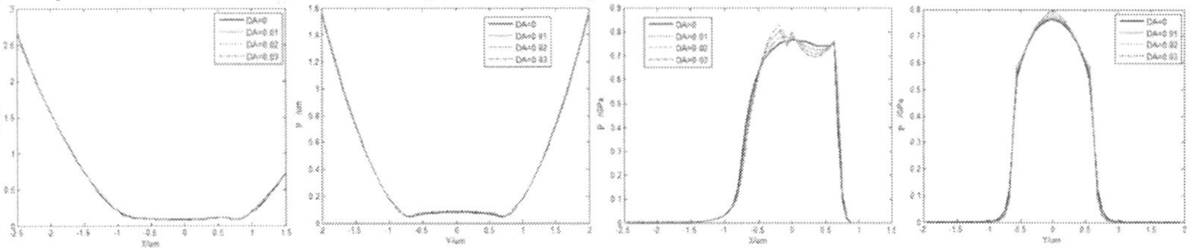

(a) (b) (c) (d)

Figure 4. The curve of film thickness and pressure with sinusoidal roughness amplitude.

From Figures 4(a) and 4(b), as the roughness amplitude becomes larger, the film thickness fluctuates slightly, whether in the X direction or the Y direction. The film thickness in the touch center area becomes thicker as the amplitude of roughness increases, and the edge of the contact area remains almost unchanged.

From Figure 4(c), it can be clearly seen that as the roughness amplitude increases, As the amplitude of roughness increases, the pressure at the front of the contact center increases. At the rear of the contact center, the pressure becomes smaller and smaller, but the secondary pressure peak value is almost unchanged; from Figure 4(d), it can be seen that the pressure in the touch area is symmetrically distributed, and as the roughness increases, the pressure in the contact center area gets bigger and bigger, and the curvature is bigger.

According to Equation (3) and lubrication theory, the oil film is formed on the surface between the slider and the ball. The surface is uneven due to different roughness amplitudes, so the oil film will also form along the uneven surface. The coordinate system is located in the rough. At the location where the degree is 0, the current film thickness is equivalent to the original film thickness plus or minus the roughness amplitude, and there is a Reynolds equation relationship between film thickness and pressure, which causes the pressure to change, and the pressure will change the initial film thickness. Thickness, therefore, the roughness amplitude mainly affects the film thickness.

7. Effect of sinusoidal roughness fluctuation frequency on isothermal elastohydrodynamic lubrication

Now the frequencies are set to 1 Hz, 2 Hz, 3 Hz, and 4 Hz, and the amplitude is set to 0.02 mm, to observe the changes in film thickness and pressure. The roughness, film thickness, and pressure changes are shown in Figures 5(a), 5(b), 5(c) and 5(d) respectively.

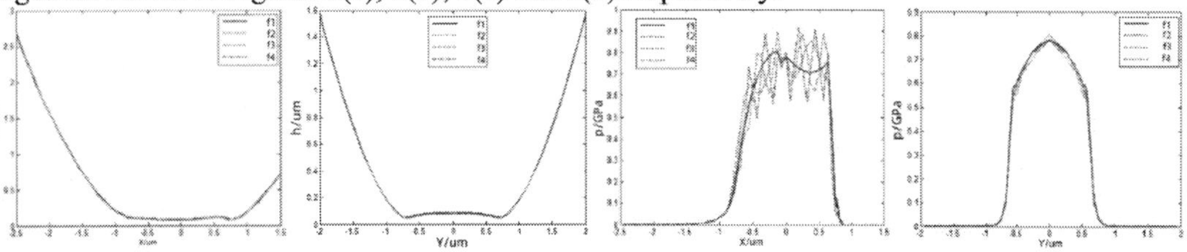

(a) (b) (c) (d)

Figure 5. The curve of film thickness and pressure with sinusoidal roughness frequency.

As shown in Figures 5(a) and 5(b), it can be observed that in the increases when the frequency is 3 Hz and 4 Hz, and the roughness frequency increases faster, the film thickness fluctuates more frequently, and the film thickness in the Y direction is symmetrically distributed. As the frequency upward, the film thickness value reduces and the film thickness becomes more uneven.

From Figure 5(c), it can be clearly seen that as the roughness frequency increases, the pressure in the X direction fluctuates more and more times centered on the original curve, and the amplitude becomes larger and larger. The secondary pressure peak becomes the last fluctuation. From the observation in Figure 5(d), it can be seen that the pressure in the touch area is symmetrically distributed, and as the roughness frequency increases, the pressure in the touch area fluctuates back and forth. The maximum pressure in the center area does not violently change, but the curvature becomes larger and gradually becomes a bulge.

According to Equation (3) and lubrication theory, it can be seen that the surface fluctuates at various speeds due to different roughness frequencies. Oil films will also form on surfaces with various fluctuation frequencies. The film thickness fluctuates faster as the roughness frequency increases. The film thickness and the frequency of pressure fluctuation are changed according to the Reynolds equation. The pressure will feedback the initial film thickness, thus affecting the numerical value of the film thickness, and then changing the pressure amplitude. Therefore, the roughness frequency affects the lubrication characteristics by changing the fluctuation frequency of the film thickness.

8. Conclusion

Taking the roughness into consideration, this chapter mainly studies the isothermal Elastohydrodynamic lubrication between the slider and the balls in the new three-pronged universal joint. According to the movement law of the slider, only the X-direction roughness is considered. The main conclusions are as follows:

(1) First, the Reynolds equation under isothermal conditions, the film thickness equation considering roughness, the viscous pressure, and dense pressure equations and their discrete forms are given. The Reynolds equation is mainly used for iteration, and Matlab is used for programming and drawing. The results show the situation where the roughness is 0, and then the amplitude and frequency of the roughness are changed respectively to observe the lubrication characteristics.

(2) It was found that in isothermal elastomeric flow, as the amplitude of sinusoidal roughness increases, the film thickness in the center of the touch area becomes thicker as the amplitude of roughness increases, the pressure fluctuation becomes larger and larger, and the secondary pressure peak value is almost unchanged. As the frequency of sinusoidal roughness increases, the film thickness in the center of the touch area becomes thinner as the frequency of roughness increases, and the fluctuations become more and more frequent. The number of pressure fluctuations centered on the original curve becomes more and more frequent, and the amplitude becomes larger and larger.

(3) During the processing process, the roughness must be controlled, and under certain cost conditions, the roughness amplitude and frequency should be reduced as much as possible.

References

[1] Wu, H., Wang, X., Hu, R., (2023) Lubricating Properties on Central Bump under Elastohydrodynamic Lubrication Point Contacts, Lubrication Engineering, 48(06), 150-155.

[2] Wang, X., Hu, R., Cheng, X., (2020) Numerical Analysis on Single Dent in EHL Point Contacts, Journal of Ningbo University of Technology, 32(04), 24-30.

[3] Mohamed Abd Alsamieh., (2022) Ultrathin film formation under combined effect of surface roughness and surface force for elastohydrodynamic lubrication of point contact problems, Journal Industrial Lubrication and Tribology, 74(6),597-608.

[4] A, Félix Quiñonez. (2018) Numerical analysis of transverse reciprocating velocity effects in EHL point contacts EHL, Tribology International, 126, 1-8.

[5] Kumar, R., Gopaliya, M, K., Khan, H., (2023) Analysis of thrust pad bearing under mixed-EHL regime considering combined effect of deterministic and stochastic surface roughness, Journal of the Brazilian Society of Mechanical Sciences and Engineering, 45(11).

[6] Abd, A, M., (2023) Study the behaviour of transient elastohydrodynamic lubrication contact at motion start-up in the presence of surface roughness, Industrial Lubrication and Tribology, 75(7), 799-816.
[7] Liu, K., (2023) Study on mechanical characteristic of angular contact ball bearing with local defects under elastohydrodynamic lubrication, Lubrication Science, 36(1), 9-21.
[8] Krupka, J., Dockal, K., Sedlacek, T., (2023) Viscoelastic Response of Elastohydrodynamically Lubricated Compliant Contacts below Glass-Transition Temperature, Polymers, 15(11).
[9] Xu, Y., Yao, C., Yang, Y., (2023) Analysis of Point Contact Isothermal Elastohydrodynamic Lubrication Characteristics of a New Type of Trigeminal Universal joint for Automobile, In: Journal of Physics: Conference Series, 2419.
[10] Zhou, Y., (2017) Research on Friction and Lubrication of Double Joint Universal joints for Automobiles Qingdao University of Science and Technology.

Journal of Physics: Conference Series **2808** (2024) 012051 doi:10.1088/1742-6596/2808/1/012051

Research on the process of vertical water entry experiment of truncated cone-shaped projectile

Zhiyong Zhang[1*]

[1]Jiangsu Automation Research Institute of CSIC, Lianyungang, 222061, China

[*]Corresponding author's e-mail: zhangzhi900720@163.com

Abstract. In order to study the influence of the diameter and velocity of the truncated head on the cavitation and motion characteristics of the projectile when it enters the water vertically, based on the high-speed camera method, the contrast experiment of the different diameter of the truncated head is carried out, and the influence law of the diameter and velocity of the head on the cavitation and motion velocity is obtained. The experimental results show that the velocity change law of the projectile in the deep closed state is basically the same, and the velocity change law in the process of turning the surface closed state begins to change; With the increase of the initial velocity of the projectile, the earlier the surface closure occurs, the earlier the separation of the cavitation from the tail cavity is, and the later the tail cavity contacts with the liquid surface; In the stage of surface closure, the general velocity rule of the projectile is: first increasing then decreasing, then increasing after smooth transition. When the velocity is high, the law of velocity changes to continuous decline.

1. Introduction

The low-velocity water entry process of the projectile is a complex two-phase coupling process, and the water entry impact will produce a series of transient physical problems. In the process of crossing the free liquid surface, air will be rolled into the water to form cavitation. The formation, expansion, closure, and collapse of the cavitation will affect the motion characteristics and hydrodynamic properties of the water entry voyage, in which the structure and shape of the projectile head have an important impact on the vacuole and ballistic characteristics.

In recent years, scholars at home and abroad have studied the cavitation problem when the projectile enters the water at low velocity from different perspectives. Bergmann et al. [1] experimentally and numerically investigated the development of cavitation when disks entered the water with a Froude number of less than 200. Truscott et al. [2] summarized a large number of analyses carried out by foreign scholars on the water entry problem. Scolan [3] investigated the water-entry process of three-dimensional projectiles and analyzed the changing law of physical quantities on the surface of wetted projectiles. Huang et al. [4] used a high-speed camera to record the evolution of the cavitation morphology when a cone-head projectile entered the water at a low velocity. Tang et al. [5] experimentally obtained the effect of the chamfered angle size of the oblique head projectile on the water entry cavitation. Song et al. [6] utilized high-speed camera technology to record the vertical entry process of five types of asymmetric-head projectiles with the head tangent angle ranging from 5° to 25° under the same entry condition and comparatively analyzed their cavitation morphology and velocity changes. Lv et al. [7] used the overlapping mesh technique and introduced the Schnerr and Sauer cavitation model to consider the cavitation phenomenon during the entry process of the spinner, and numerically simulated the entry process of the spinner with different head sizes from low to high speeds. Zheng et al. [8] extracted the contours of the water-entry cavitation from the high-speed camera images and quantitatively analyzed the morphology and characteristics of the water-entry cavitation.

Comprehensive existing literature, the influence of vertical water entry cavitation characteristics and kinematic parameters of the truncated cone head type have not been reported. In this paper, we first analyze the process of vertical water entry with 4.9 m/s initial velocity and different head types of projectiles and verify that the truncated cone projectile has better drag reduction performance. Then the vertical water entry process of the truncated cone projectile with different head diameters and velocities is researched, and the evolution of the cavitation morphology is analyzed, as well as the velocity change law of the projectile in the deep closure and the surface closure.

2. Schematic diagram of the experimental setup

The experimental device is shown in Figure 1, including a glass water tank, a high-speed camera, a computer, two light sources, an electromagnet, three brackets, a coordinate paper, and a protective layer. The water tank is made of 15 mm thick ordinary glass adhesive, and the protective layer is set to cover the bottom of the water tank. In the experiment, the 1300 W advertising light source is used to illuminate the front and back, the high-speed camera is used to shoot the evolution of the water-entry cavitation at a shooting rate of 4100 FPS, and the experimental water is used as tap water.

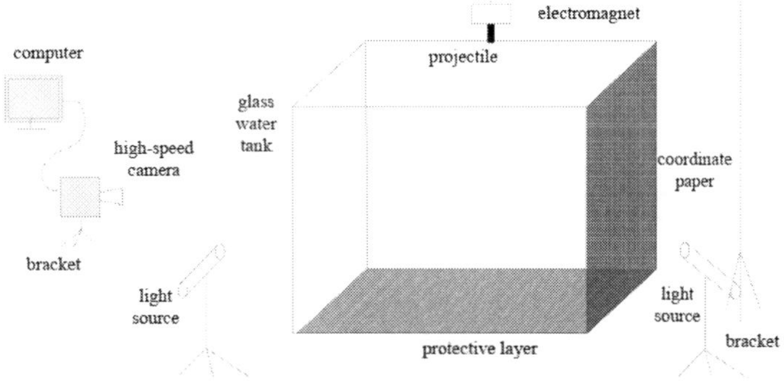

Figure 1. Experimental setup for water entry.

3. Verification of drag reduction characteristics

From the study of Wang et al. [9], when the L/D ratio is about 8, the truncated cone head or the round head projectile has less resistance than the flat head, and the pointed head projectile enters with a 2.8 m/s initial velocity. In this section, in order to verify the universality of this law in the vertical case of low-velocity water entry, the projectile with slightly different aspect ratios is used for vertical water entry experiments, and the dimensions are shown in Figure 2.

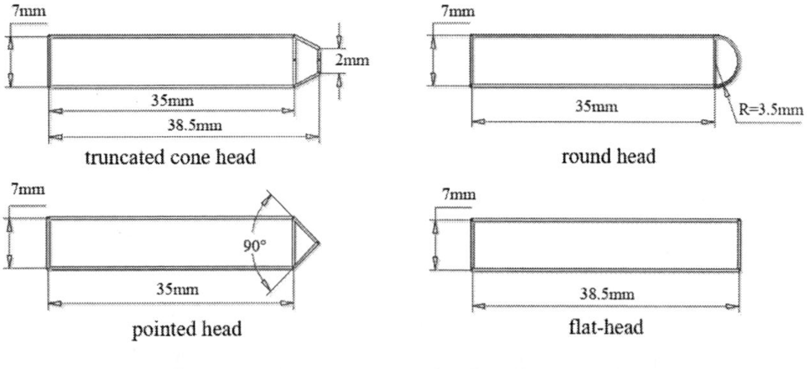

Figure 2. Projectile size for four head types.

Journal of Physics: Conference Series **2808** (2024) 012051 doi:10.1088/1742-6596/2808/1/012051

As shown in Figure 3, the attitudes of each projectile after water entry are quite different with the initial velocity being 4.9 m/s. Except for the flat-head projectile, surface closure occurs in the other three types of projectiles. The difference with the literature is that in the literature, with the initial velocity being 2.8 m/s, each projectile has produced deep closure phenomena after entering the water.

From Figure 4, within the same time period, the truncated cone head projectile and round head projectile descend deeper, which means the truncated cone projectile and round head projectile have less resistance, consistent with the literature. There has been more research about the analysis of the round head projectile's water-entry process at home and abroad. This paper focuses on the analysis of the truncated cone head projectile's water-entry process.

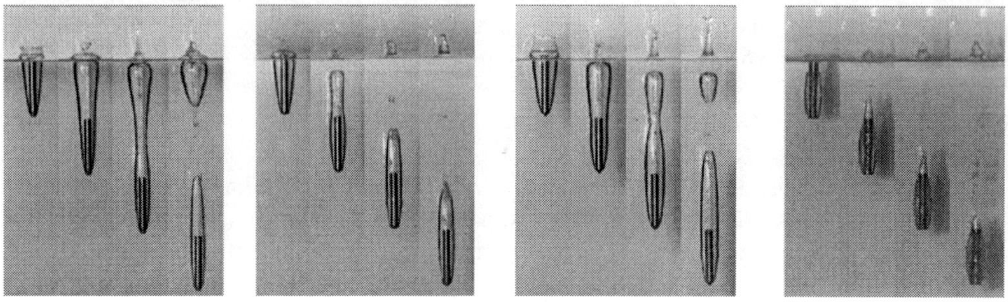

(a) Truncated cone head (b) Round head (c) Pointed head (d) Flat-head

Figure 3. The phase diagrams of different head types of projectiles.

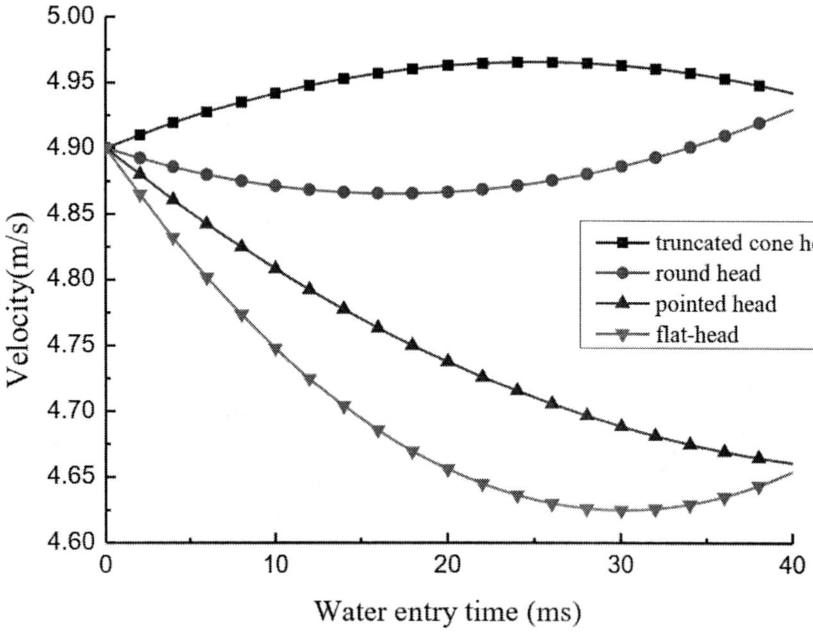

Figure 4. Variation in water entry velocity for each head type of projectile.

4. Truncated cone head projectile model

As shown in Figure 5, the projectile model is composed of a truncated cone head and a cylinder. The length of the cylindrical part is 35 mm, and the diameter is 7 mm. The angle between the cylindrical part and the truncated cone part is 60 deg. The truncated cone head diameter is 2 mm, 3 mm, and 4 mm respectively. The projectile material is No. 45 steel.

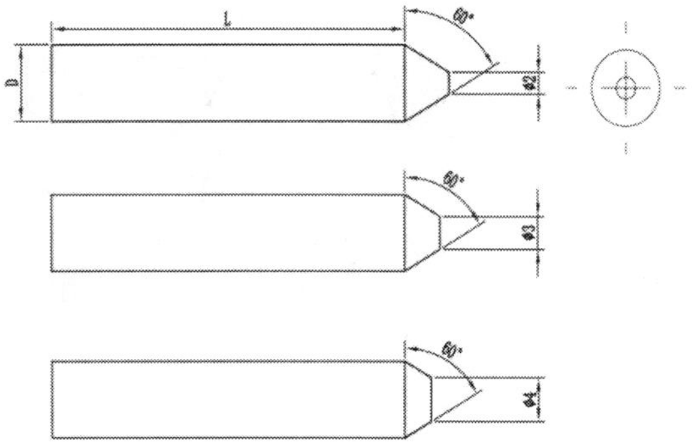

Figure 5. The projectile model.

5. Experimental results and analysis

To study the effect of the initial velocity and head diameter of the projectile on the water-entry cavitation and its motion characteristics under the low-velocity vertical water entry state, the above-established experimental test system is used to carry out experimental research on the projectile's low-velocity vertical water entry. This paper takes the moment when the head of the projectile touches the water surface as the moment t = 0.

5.1. The effect of initial velocity magnitude

This chapter focuses on the deep closure and surface closure and their transition state after the truncated cone-head type projectile enters the water. Different drop heights for Model A are adopted to get different initial velocities. The deep closed moment of each phase diagram is shown in Figure 6, and the velocity and acceleration changes are shown in Figure 7 and Figure 8. The surface closed moment of each phase diagram is shown in Figure 9, and the velocity and acceleration changes are shown in Figure 10 and Figure 11.

From Figure 6, with the initial velocity being 4 m/s, 4.12 m/s, and 4.25 m/s, the cavitation deep closure time is 40.4 ms. However, the deep closure time changes to 39.6 ms when the initial velocity is 4.31 m/s. As verified by several repetitive experimental data, when the initial velocity changes from 3.3 m/s to 4.25 m/s, the deep closure time is 40.4 ms, and the length of the cavitation is positively correlated with the initial velocity. However, when the velocity is 4.31 m/s, the trend is different. The length becomes shorter, and there is a tendency to close the cavitation at the liquid surface. The closure mode is gradually transitioned from deep closure to surface closure. Due to the high velocity, the cavitation at the liquid surface gradually gathers and a downward jet is produced, which produces an effect on the cavitation wall. Additionally, at a higher velocity, the impact is larger and the kinetic energy gained by the aqueous medium is larger, resulting in a shorter and thicker follower cavitation.

From Figures 7 and 8, at a lower initial velocity, the entry velocity increases faster at first, then levels off, and finally increases again. At a higher initial velocity, the entry velocity shows a tendency to decay at first, and the higher the initial velocity is, the more pronounced the decay tendency is. When the initial velocity is 4.31 m/s, the velocity decreases abruptly and then increases gradually, and begins to show a decaying trend again after the deep closure. The reason is that at a lower initial velocity, there is less upward impact force. The direction of the combined force is downward, making the velocity increase. The lower the initial velocity is, the faster the velocity increases. At a higher initial velocity, the impact force is higher. The direction of the combined force is upward, making the velocity decrease. The higher the initial velocity is, the faster the velocity decreases. As the projectile enters the water, the resistance is less than the force of gravity and the velocity slowly increases. When the entry velocity is 4.31 m/s, the follower cavitation is significantly larger than that at lower velocity, the resistance is larger, and the velocity produces attenuation. While at a lower velocity, the gravity of the projectile and the follower cavitation is larger than the entry resistance, and the velocity increases.

From above, the velocity change law of the projectile in the deep closed state is basically the same, and the velocity change law in the process of turning the surface closed state begins to change.

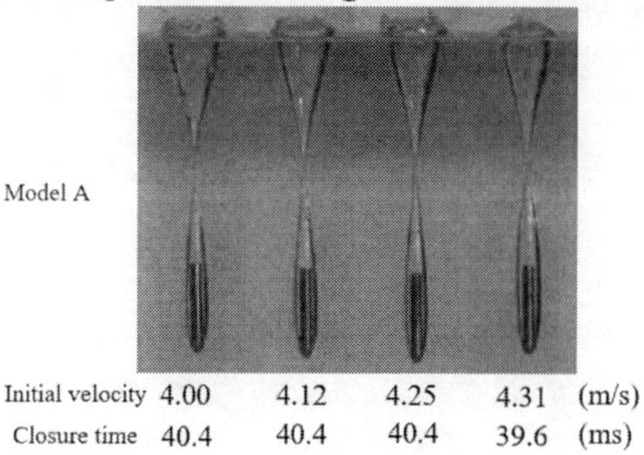

Figure 6. The phase diagram of Model A at the deep closed moment.

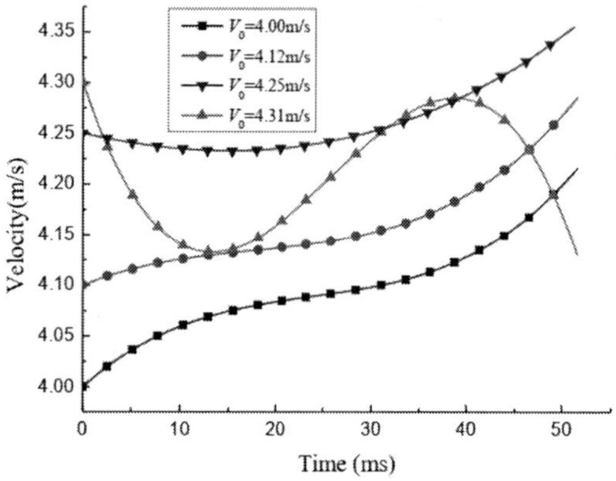

Figure 7. Velocity variations of Model A with different initial velocities.

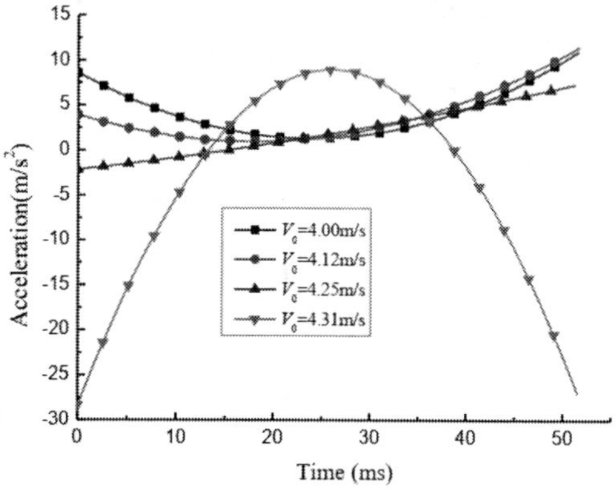

Figure 8. Acceleration variations of Model A with different initial velocities.

From Figure 9, when Model A enters the water with the initial velocity being 4.50 m/s and 4.66 m/s, the cavitation all happens surface closure, and the attitudes of the cavitation are more similar, only the upward jets are different. After the projectile enters the water, the cavitation gradually expands, surrounded by splashes. At about 20 ms, under the action of differential pressure, the water splashes converge to the center and form a closed space. The closure of the cavitation occurs. At this time, due to the impact of the water splash in all directions, a downward jet is produced. The jet is very fast and impacts the wall of the bubble.

From Figures 10 and 11, with the initial velocity being 4.50 m/s and 4.66 m/s, the velocity change law is similar, which first increases, then decreases, and then increases again after a smooth transition. During the cavitation formation phase, the velocity increases due to the dominant role of gravity, consistent with the deep closure phase. During the cavitation expansion phase, the velocity decreases due to high energy dissipation. During the surface closure stage, the cavitation is accompanied by the projectile. At this time, there is an approximate convergence between the projectile resistance and gravity so that the projectile velocity changes slowly. After the separation of the follower cavitation and the tail cavitation, the follower cavitation becomes small. At this time, gravity plays a dominant role and the velocity increases. However, if the initial velocity is too fast, the closure form is close to the class surface closure, and the resistance increases dramatically after entering the water. The velocity is a continuous downward trend as shown in Figure 15(a).

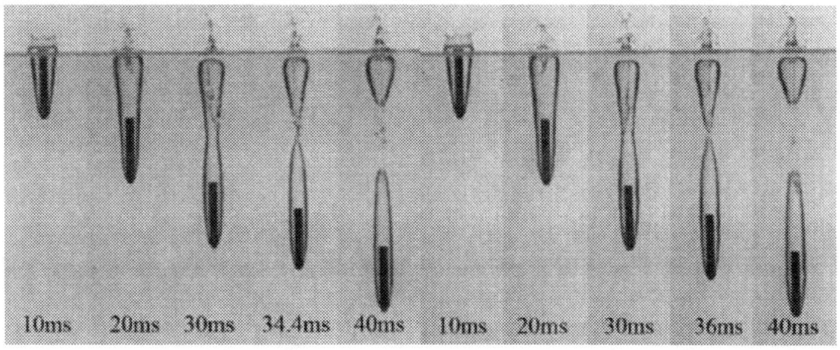

(a) 4.50 m/s initial velocity (b) 4.50 m/s initial velocity

Figure 9. Morphological changes of Model A with different initial velocities.

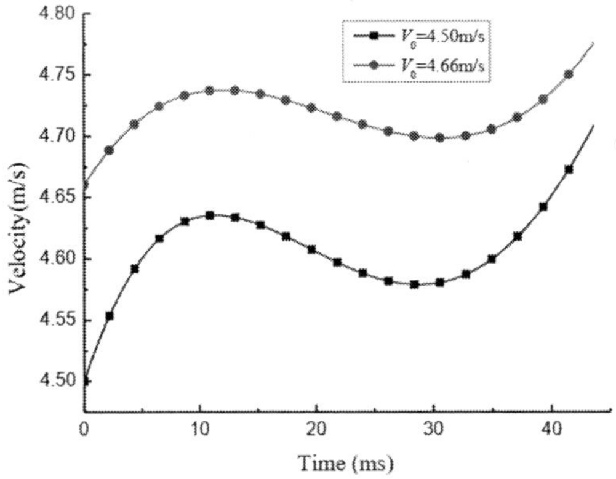

Figure 10. Velocity variations of Model A with different initial velocities.

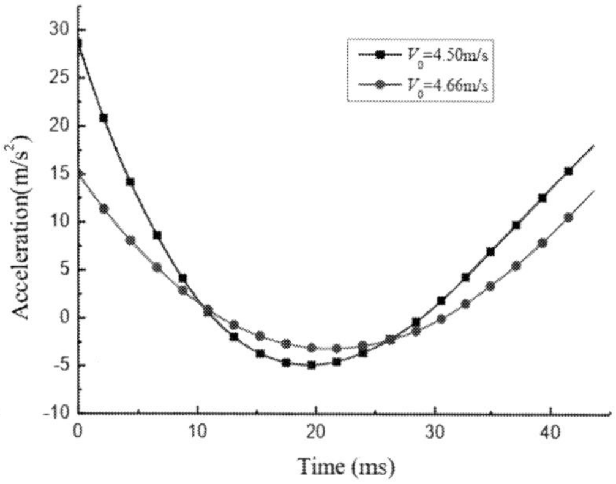

Figure 11. Acceleration variations of Model A with different initial velocities.

5.2. The effect of projectile head diameter size

As shown in Figures 12 and 13, the deep closure times for Models A, B, and C are all 40.4 ms with an initial velocity of 4.0 m/s, and the velocity changes for each model are all upward trends. But Model A has the fastest increase, Model B is the next fastest, and Model C is the slowest. The head plane of each model touches the water surface directly at the very beginning, which is the main factor of drag. Model C has the largest plane area, so Model C is the slowest. The velocity change law with Model C's initial velocity being 4 m/s is the same as that with Model A's initial velocity being 4.31 m/s. Because the resistance of Model C is greater than that of Model A, Model C shows such a pattern of velocity change earlier.

Figures 15, 16, and 17 show the motion in the water, the change in velocity, and the change in acceleration with the initial velocity being 5.45 m/s. From the figures, the velocity change law of each model is basically the same, first decreasing and then increasing, with Model A having the fastest velocity decay, Model B having the next slowest, and Model C having the slowest. Their closure modes are all surface closure, and the formation order is Model C, B, and A. This is due to the fact that from Figure 16, Model C is the fastest after entering the water, and from the above, it is clear that Model C changes from deep closure to surface closure earlier. Similarly, as shown in Figure 15, the isolated cavitation of Model C collapses fastest, meaning it changes from surface closure to a class surface closure earlier.

Due to the limited number of camera frames, sudden changes in velocity and acceleration could not be captured. However, from Figure 17, the acceleration before entering the water is gravitational acceleration in the downward direction. After entering the water, the velocity changes abruptly to the opposite direction. Between 30 ms and 40 ms, the velocity decay levels off and the projectile begins to

Journal of Physics: Conference Series **2808** (2024) 012051 doi:10.1088/1742-6596/2808/1/012051

accelerate downward. This is due to the fact that the upward impact force gradually disappears and the projectile begins to gradually accelerate downward under the action of buoyancy and gravity.

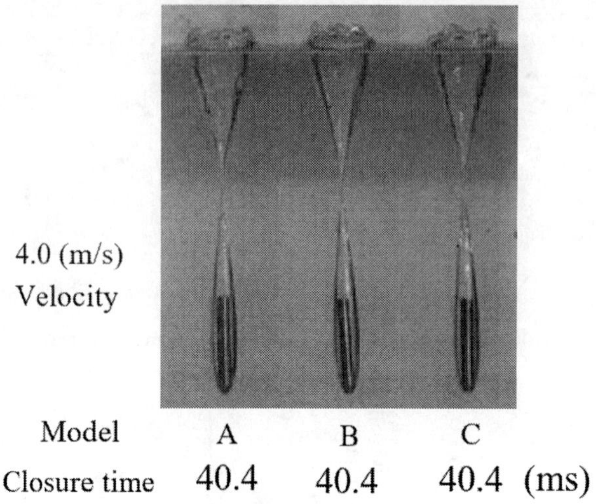

Figure 12. The phase diagram of Model A, B, and C at the deep closed moment.

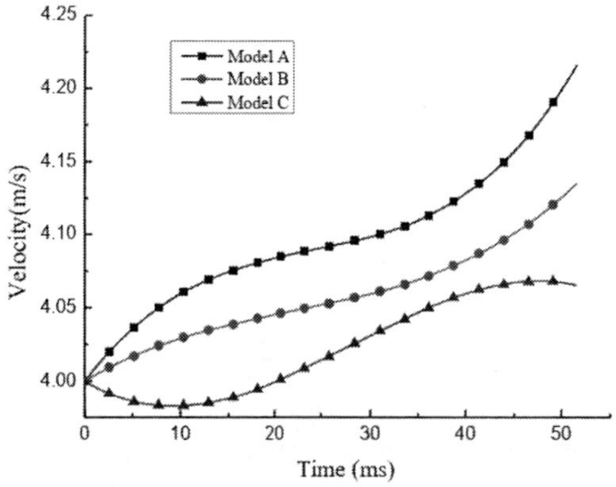

Figure 13. Velocity variations of different models with an initial velocity of 4.0 m/s.

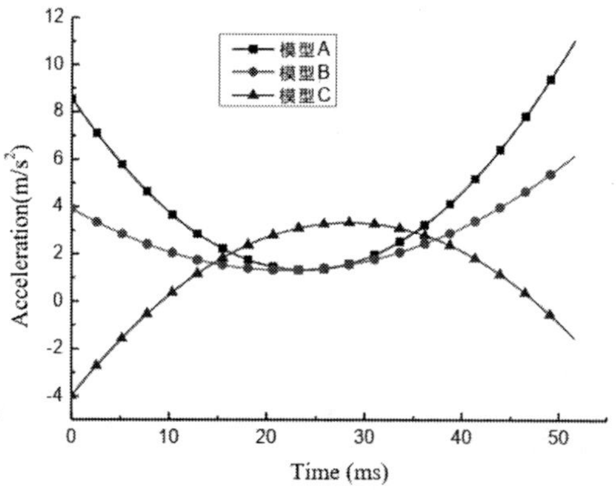

Figure 14. Acceleration variations of different models with an initial velocity of 4.0 m/s.

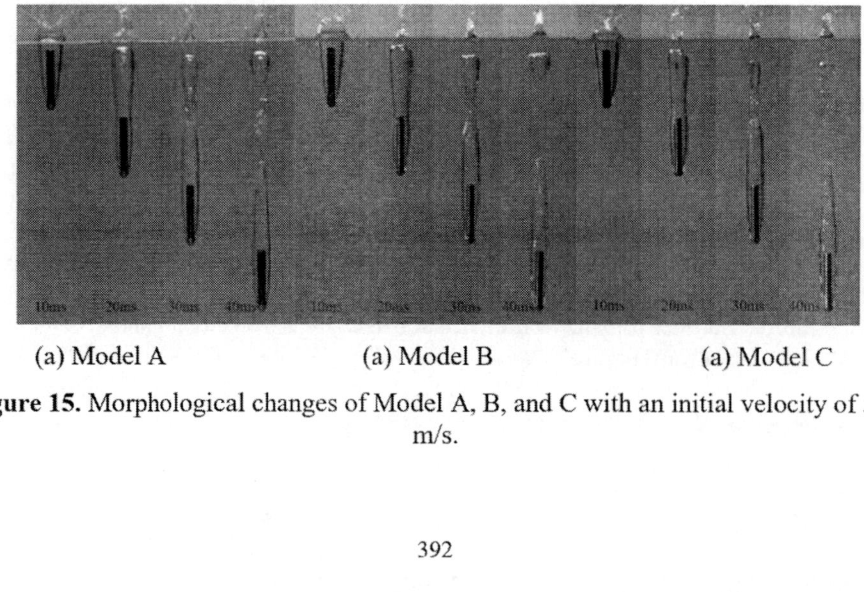

(a) Model A (a) Model B (a) Model C

Figure 15. Morphological changes of Model A, B, and C with an initial velocity of 5.45 m/s.

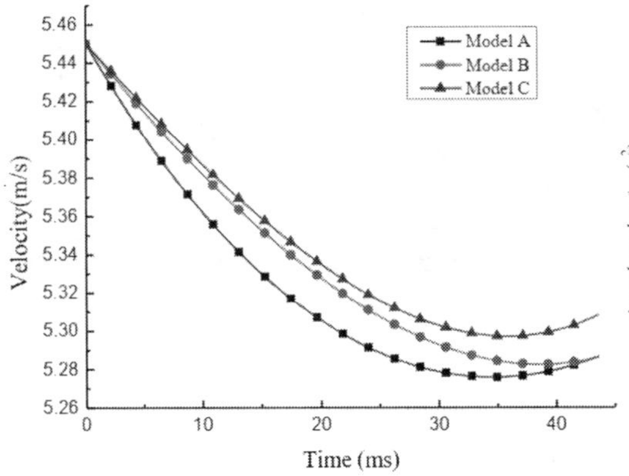

Figure 16. Velocity variations of different models with an initial velocity of 5.45 m/s.

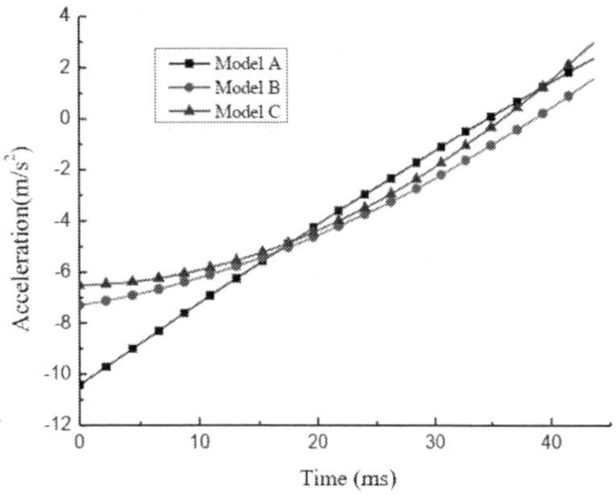

Figure 17. Acceleration variations of different models with an initial velocity of 5.45 m/s.

5.3. Fluctuation phenomena at the wall surface of the cavitation

From the literature [9], the wall fluctuation phenomenon occurs when only deep closure occurs, while from section 4.2, the wall fluctuation occurs when surface closure occurs under certain conditions. To investigate their similarities and differences, this section addresses the phenomenon of wall fluctuations during the surface closure of Model C. In an inertial coordinate system, each typical moment is selected for analysis after the separation of the follower cavitation from the liquid surface cavitation with an initial velocity of 5.45 m/s. Each moment is spaced 0.8 ms apart, as shown in Figure 18. The horizontal line represents the same depth and marks the wave valley. It is proved that the fluctuation will not propagate in the longitudinal direction.

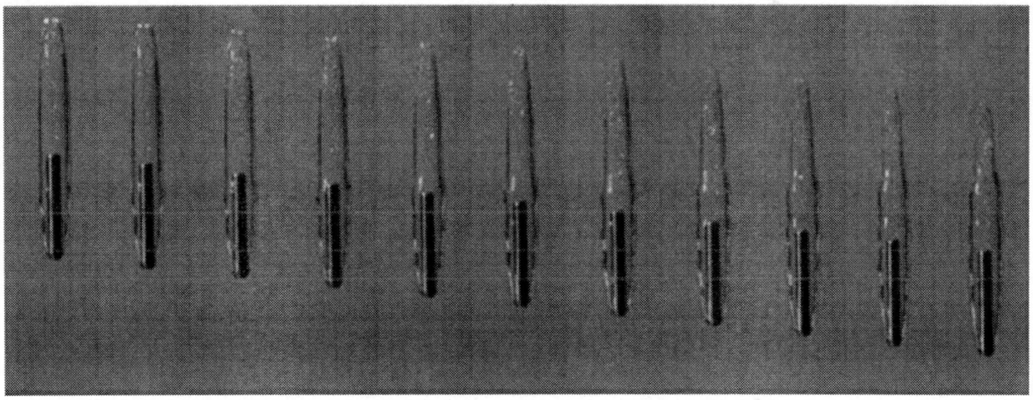

Figure 18. Wall fluctuation change graph of Model C with an initial velocity of 5.45 m/s.

A coordinate system of the projectile is established by placing its bottom in the same plane. From Figures 19(a) to 19(e), the fluctuation goes from trough to peak to trough again, consistent with fluctuations from Figures 19(e) to 19(i). The change varies periodically, with time being 3.2 ms. The fluctuation period in the deep closed state is mentioned as 2.4 ms in the previous literature [8]. In contrast to the previous literature's phenomenon, the longitudinal direction of the follower cavitation gradually shrinks, which means its length becomes shorter. From the surface closure phenomenon in Figure 19, the longitudinal direction of the follower cavitation is gradually lengthening, which exactly explains the reason why the wall fluctuation cycle time becomes longer.

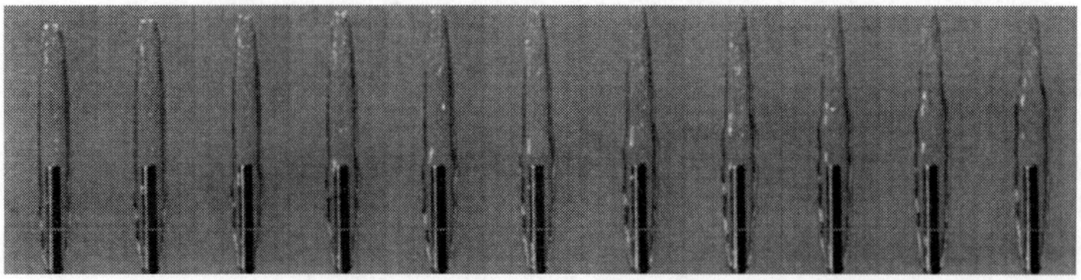

(a) (b) (c) (d) (e) (f) (g) (h) (i) (j) (k)

Figure 19. Wall fluctuation change graph of Model C with an initial velocity of 5.45 m/s in the projectile coordinate system.

6. Conclusion

In this chapter, an experimental study is carried out on the problem of low-velocity vertical water entry of projectiles with different diameters of the truncated cone head type, and the following conclusions are drawn under the experimental conditions:

1. The deep closure time remains essentially constant when only deep closure occurs, while the deep closure time begins to decrease after surface closure occurs. The velocity change law of the projectile in the deep closed state is basically the same, and the velocity change law in the process of turning the surface closed state begins to change.
2. With the increase of the initial velocity of the projectile, the earlier the surface closure occurs, the earlier the separation of the cavitation from the tail cavity is, and the later the tail cavity contacts the liquid surface.
3. In the stage of surface closure, the general velocity rule of the projectile is: first increasing then decreasing, and then increasing after a smooth transition. When the velocity is high, the law of velocity changes to continuous decline.
4. The effect of head diameter on the entry velocity is obvious. When the initial velocity is 4 m/s, the change in velocity is basically an upward trend. The smaller the diameter is, the faster the velocity increases. When the initial velocity is 5.45 m/s, the change in velocity is a decreasing trend. The smaller the diameter is, the faster the velocity decays.
5. When surface closure occurs, the wall fluctuation phenomenon occurs after the separation of the tail cavitation and the follower cavitation. The larger the diameter is, the more pronounced the wall fluctuation phenomenon is.

References

[1] Bergmann, R., etc. (2009) Controlled impact of disk on a water surface: Cavity dynamics [J]. Journal of Fluid Mechanics. 633: 1-34.

[2] Truscott, T. Epps, B. Belden, J. (2013) Water entry of projectiles [J]. Annual Review of Fluid Mechanics, 46: 355-378.

[3] Scolan, Y. (2015) Oblique water entry of a three-dimensional body [J]. International Journal of Naval Architecture and Ocean Engineering, 6: 1197-1208.

[4] Huang, Z., Wang, R., Chen, Z., etc. (2018) Experimental study of cavity characteristic induced by vertical water entry impact of a projectile with a 90° cone-shaped head at different velocities [J]. Explosion and Shock Waves, 38(06): 1189-1199.

[5] Tang, C,. Huang, Z., Chen, Z., etc. (2020) Experimental study of the low-speed vertical water entry process of oblique head projectile [J]. ACTA ARMAMENTARII, 41(S01): 5.

[6] Song, L., Yu, H., Jie, B., etc. (2021) Experimental study of low-speed vertical water entry with moving object of asymmetric head type [J]. Journal of Ordnance Equipment Engineering, 42(03): 35-39, 44.

[7] Lv, H., Xu, L., Wang, Z., (2022) Numerical research on flow field characteristics of axisymmetric bodies with different head shapes during initial water entry [J]. Journal of Ordnance Equipment Engineering, 43(12): 34-42.

[8] Zheng, W., Duan, J., Sun, T., (2023) Study on the morphometric and characteristic measurement method of water-entry cavity based on the image edge detection technology [J]. Journal of China University of Metrology, 34(2): 163-171.

[9] Wang, R., Huang, Z., Guo, Z., etc. (2017) Experimental Study of Low-Speed Vertical Water Entry with Different Head Shape Projectiles [J]. Journal of Ordnance Equipment Engineering, 38(11): 45-50.

Journal of Physics: Conference Series **2808** (2024) 012052 doi:10.1088/1742-6596/2808/1/012052

Analysis of circumferential CO_2 arc welding welded shield machine hob to investigate the welding residual stress fields

Xiaojie Xue[1], Jianfu Chen[2,*], Shuhua Huang[2], Liang Zhang[2] and Zhaoping Tian[2]

[1]Guangzhou Expressway Corporation Limited, Guangzhou, 510000, China
[2]China Railway 14th Bureau Group Shield Tunneling Corporation Limited, Nanjing, 211800, China

[*]Corresponding author's email: njit450220439@163.com

Abstract: This paper investigates the distribution and formation rules of welding residual stress during the process of shield machine hob welding using CO_2 arc welding. A corresponding three-dimensional digital model is created using the FEA software to research the distribution of welding residual stress during the single pass wire filling welding. The numerical model of a shield machine hob is verified by the welding experiments. This study utilizes numerical computational methods to describe the causes and changes of welding residual stress and strain during the welding process. According to the calculations, the results indicate that compressive axial residual stress is observed on the outer surface of the hob body near the weld line. Conversely, tensile axial residual stress is evident on the inner surface in contrast. Additionally, both the outer and inner surfaces of the hob body exhibit tensile hoop residual stress. Near the weld line, the upper surface of the hob ring shows tensile transversal and longitudinal residual stresses. The main reason for this phenomenon is that different plastic strains are generated because of the thermal contracting deformation at the various surfaces and internal stress self-balance. Moreover, the internal residual stress distribution of fillet weld is investigated, which is obviously different from the external of fillet weld.

1. Introduction

CO_2 arc welding is a common metal welding method that utilizes carbon dioxide gas as a shielding agent. In this welding process, an electrode (typically made of steel) forms an arc with the workpiece, while CO_2 gas is introduced around the welding area to protect the molten pool from contamination by impurities such as oxygen and nitrogen in the atmosphere. CO_2 arc welding is a simple, economical, efficient, and widely used method for metal welding, playing a significant role in industrial production, especially in thin-walled structure applications. Circumferential CO_2 arc welding (including butt welding and fillet welding) is commonly used in pipe systems (pipe to pipe and tube to tubesheet, etc.).

Residual welding stress is an internal stress generated during the welding process. Welding is the process of joining two or more metal components together using heat or pressure. During welding, the welded area is subjected to heating and cooling, causing temperature fluctuations and heat transfer, resulting in uneven contraction and deformation of the metal materials. Among these, tensile residual stress is a typical type of welding residual stress, referring to the internal stress that remains in a material even after it has been subjected to tensile forces and returned to its original shape. This type of stress typically occurs during welding processes. During welding, the material undergoes localized heating followed by rapid cooling, leading to non-uniform expansion and contraction. As a result, the material near the weld area may undergo plastic deformation, while the surrounding areas may remain relatively unaffected. This non-uniform deformation results in the generation of residual stresses, which can negatively impact the performance and stability of welded components, such as causing cracks, deformation, and reduced fatigue life [1]. Therefore, predicting the residual strains and stresses generated during welding is of paramount importance for optimizing design, assessing structural performance, and preventing potential problems.

But it is difficult to accurately predict the welding residual stress and strain, because of the complexity of the whole welding process. The FEA method is an effective tool for predicting thermal-structural behavior in circumferential arc welding structures. Furthermore, this method can also be used for predicting the rules of welding residual stress and strain. In the past two decades, many researchers have been dedicated to establishing numerical models for arc welding, using finite element analysis to study the welding residual stresses and strains generated during circumferential arc welding. The establishment of these models aims to gain a deeper understanding of the material behavior during welding and how to optimize welding parameters to reduce residual deformation and improve the quality of weld joints. This approach provides essential theoretical foundations and engineering guidance for designing safer and more reliable welded structures.

Kyong-Ho Chang employed finite element analysis to predict the distribution pattern of residual welding stresses in circumferentially welded steel pipes. This research aims to gain deeper insights into stress behaviors during welding and provides a further understanding of welding structure performance, offering valuable information for enhancing welding quality and designing more reliable welded structures. In the end, the study confirmed a close alignment between the distribution of residual stresses within the welded pipes and the results simulated by the finite element method.

Deng presented a 3D FE model for predicting the temperature and residual stress fields in welded stainless steel pipes and SUS304 girth-welded pipes [2]. The primary objective of this model is to gain insights into the distribution of temperature and stress during the welding process, with particular emphasis on the stress distribution near the start and end locations of the weld. To validate the accuracy and reliability of this model, the author conducted a comprehensive series of experiments. Through comparison and analysis of the experimental results, the effectiveness of the proposed numerical model was confirmed. This study provides important theoretical support and experimental evidence for further improving welding processes and enhancing the quality of welded structures.

In the welded structure components, plate butt welding, pipe butt welding, and tube-to-tubesheet welding are important welding components in engineering projects. The welding of the hob body and the hob ring of the shield machine hob can be compared to the welding of the tube-to-tubesheet. The distributions and magnitudes of the tube-to-tubesheet are significantly different from plate and pipe butt welding under the same welding parameters. There have been a number of investigations for plate and pipe butt welding, and then only regarding the generation of welding residual stress in welded structures. However, there is limited research on tube-to-tubesheet welding, and even if there is some research, the focus is on researching the distribution of welding residual stress, and no explanation is given.

In this study, firstly, an advanced finite element analysis software was employed to construct a sequentially coupled three-dimensional thermo-elastic-plastic numerical model. This model aimed to accurately predict the numerical values and distribution of welding residual stress generated during circumferential CO_2 arc welding fillet welding of shield machine hob. Subsequently, in order to validate the accuracy and reliability of the established numerical model, a series of comprehensive corresponding welding experiments were conducted. These experiments were designed to simulate real welding processes and evaluate the predictive capabilities of the model by comparing the results with those of the numerical model. Ultimately, the good consistency between the experimental results and the numerical model validated the accuracy of the current finite element model, providing a solid foundation for further research. Then, the formation rules of welding residual stress throughout the circumferential CO_2 arc welding fillet welded shield machine hob process were analyzed by the proposed 3D FE model in this paper. Finally, a reasonable explanation was given for the residual stress distribution of the hob body and hob ring at a 90° location.

Journal of Physics: Conference Series **2808** (2024) 012052 doi:10.1088/1742-6596/2808/1/012052

2. Finite element simulation

Figure 1 shows a full 3D thermal-elastic-plastic numerical model developed using finite element analysis software. This model is employed to predict the distribution of the residual stress field and temperature field during the circumferential CO_2 arc welding fillet welding process. The hob body measures 50 mm in outside diameter, 50 mm in length, and 3 mm in thickness. The hob ring measures 150 mm in diameter and 10 mm in thickness.

Due to the high temperature gradient, stress gradient, and flux gradient present in the weld zone, fusion zone, and heat-affected zone, a denser mesh is used within a 10-millimeter area on both sides of the weld line to ensure accuracy during grid partitioning. Along the weld direction of the hob body, the element size remains at 1 mm. In the thickness direction of the hob body, there are three elements of 1 mm each. In the axial direction of the hob body, a minimum element size of 1 mm is used within the anticipated FZ and HAZ area. In this study, the accuracy of the mesh size was evaluated. The element size in the current model is appropriate, ensuring accuracy while speeding up calculation time.

Figure 1. The 3D FE model.

Figure 2. The experimental model.

The 0Cr18Ni8 alloy, also known as 304 stainless steel, is a common stainless steel alloy primarily composed of iron, chromium, and nickel. This alloy exhibits excellent corrosion resistance, mechanical properties, and machinability, making it widely used in the manufacturing of various equipment and components. In the study, 0Cr18Ni8 is used not only as the base metal but also as the filler metal. Figure 3 shows the temperature-dependent characteristics of the material in the experiment.

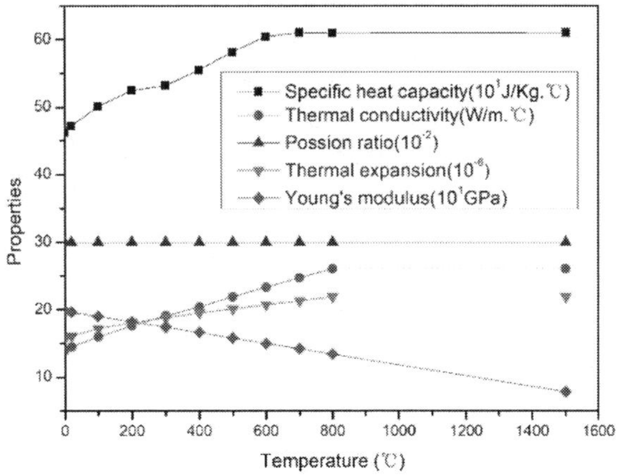

Figure 3. The temperature-dependent characteristics of the material in the experiment.

2.1. Thermal analysis approach

During the thermal analysis of circumferential CO_2 arc welding of the shield machine hob, precise numerical simulation methods were employed, taking into account various boundary conditions such as welding speed, current density, arc length, ambient temperature, etc. By considering these comprehensive factors, it is possible to simulate the temperature distribution during the welding process more accurately. Equation (1) represents the governing equation for transient heat transfer analysis.

$$\rho c\frac{\partial T}{\partial t}(x,y,z,t)=\nabla q(x,y,z,t)+Q(x,y,z,t) \tag{1}$$

where ρ represents density. (g/mm^3); T represents current temperature (°C); t represents time (s); c represents specific heat capacity (J/(g°C)); Coordinates in the reference system are represented by x,y,z (mm); ∇ represents spatial gradient operator; q represents the vector of heat flux (W/mm^2); Q represents the rate of internal heat generation (W/mm^3); Equation (2) provides the constitutive equation for Fourier heat flux for non-linear isotropic media.

$$q=-k\nabla T \tag{2}$$

where k represents thermal conductivity, and it varies with temperature.

The double ellipsoidal heat source model is a mathematical model used to describe the heat transfer behavior of a heat source in welding or other thermal processes. In this model, the shape of the heat source is assumed to be two intersecting ellipses, with each ellipse representing the thermal boundary and temperature distribution of the heat source. By adjusting the size, position, and orientation of the ellipses, it is possible to effectively simulate the heat transfer behavior of heat sources of different shapes and sizes during the welding process. The double ellipsoidal heat source model is commonly employed in numerical simulations and engineering analyses to predict parameters such as temperature field distribution, heat-affected zone, and residual stresses during welding processes, thereby guiding optimization and control of welding procedures. This model can be employed to represent the circumferential CO_2 arc welding process. The model has the features of Gaussian distribution in circumferential welding, enabling more accurate predictions of temperature distribution and heat transfer on the workpiece surface caused by the heat source. Equations (3) and (4) represent the dual-ellipsoidal heat source model established by Goldak et al.

$$q_f(x,y,z)=\frac{6\sqrt{3}f_fQ}{abc_f\pi\sqrt{\pi}}\exp\{-3(\frac{x^2}{a^2}+\frac{y^2}{b^2}+\frac{z^2}{c_f^{\ 2}})\} \tag{3}$$

$$q_r(x,y,z) = \frac{6\sqrt{3} f_r Q}{abc_r \pi\sqrt{\pi}} \exp\{-3(\frac{x^2}{a^2} + \frac{y^2}{b^2} + \frac{z^2}{c_r^2})\} \tag{4}$$

where x, y, and z represent the local coordinates of the heat source. f_f and f_r represent the parameters of sedimentation in the front and rear parts of the heat source, respectively, and $f_f + f_r = 2$. Q represents the total power of the heat input, expressed by Equation (5).

$$Q = \eta UI \tag{5}$$

where η represents arc efficiency, U represents voltage, and I represents current, respectively. The arc efficiency is set to 80%. a, b, c_f, c_r are used to define the geometry characteristics of heat sources. Tables 1 and 2 respectively display the welding parameters and the thermal source parameters.

Table 1. Welding process parameters.

Parameter	Welding current	Welding voltage	Welding speed	Arc efficiency
Value	220A	13V	3mm/s	80%

Table 2. Numerical value of heat source parameters.

Parameter	Width of heat source a	Depth of heat source b	Length of front ellipsoidal c_f	Length of rear ellipsoidal c_r
Value	10	3	5	15

During the CO_2 arc welding process, thermal convection and radiation are considered. It is well known that thermal convection plays an important role at lower temperatures and radiation is dominant at higher temperatures near welded zones [3]. Therefore, Equation (6) represents the final thermal conductivity coefficient, taking into account the influence of temperature on the thermal conductivity coefficients of convection and radiation [4].

$$\begin{cases} h_{total} = 0.0668T \ \ (W/m^2) \ \ 0<T<500°C \\ h_{total} = 0.231T\text{-}82.1 \ \ (W/m^2) \ \ T \geq 500°C \end{cases} \tag{6}$$

2.2. Structural analysis approach

The above-mentioned heat transfer analysis method involves a comprehensive thermal analysis by meticulously calculating the temperature history of nodes. Throughout this process, the system records the variation of temperature over time for each time increment. Subsequently, these temperature histories are utilized as input for thermal loading to compute the thermal stress and strain of the welded structure at different time points. This method offers a comprehensive understanding of temperature distribution changes during the welding process, enabling a more accurate assessment of the thermal response and performance of the welded structure.

In the welding process, various factors, such as heat and force, lead to the generation of multiple types of strains. These strains are not only diverse but also exhibit unique characteristics and behaviors. Firstly, thermal strain arises from the heat generated during the welding process, causing dimensional changes in the welded material as temperatures fluctuate. Secondly, elastic strain refers to the reversible deformation of a material under stress, meaning it can return to its original shape upon removal of the load. In contrast, plastic strain results in permanent deformation of the material, even after the external load is removed. Additionally, creep strain involves gradual deformation of the material over prolonged periods of stress, typically occurring under high temperatures and high stress conditions. Lastly, phase transformation strain involves changes in the material's phase, such as solid-to-liquid or liquid-to-solid transitions, leading to significant alterations in volume and shape. By considering these different types of strains, a more comprehensive understanding of material behavior during the welding process can be achieved. However, in the welding process, the high temperature residence time is very short, so the creep effect is not significant and can be neglected [5]. In addition, in low-carbon steel, phase transformation can be neglected because it has minimal impact on welding residual stresses and strains [6]. Therefore, in this study, Equation (7) can represent the total strain during the welding process.

$$\varepsilon^{total} = \varepsilon^{t} + \varepsilon^{e} + \varepsilon^{p} \tag{7}$$

where ε^{total} represents the total strain, ε^{t} represents thermal strain, ε^{e} represents elastic strain, and ε^{p} represents plastic strain.

During the thermal and structural analysis process, we extensively apply the "Full Newton-Raphson" iterative solution technique to ensure accurate results. This technique involves multiple iterations, considering the material's thermal expansion characteristics, elastic deformation, and plastic behavior, comprehensively simulating the structure's response in complex temperature environments. When calculating thermal strain, we precisely consider the temperature's effect on the material's thermal expansion coefficient. In computing elastic strain, we incorporate temperature-dependent Young's modulus and Poisson's ratio, comprehensively reflecting the material's elastic properties. Additionally, we utilize temperature-dependent yield strength and the von Mises yield criterion to determine the material's plastic strain, providing a more comprehensive description of the structure's behavior under temperature and stress. Through these meticulous and comprehensive calculation methods, we can more accurately predict the performance and behavior of structures under actual working conditions. The structural boundary conditions are constrained around the hob ring.

3. Verification

We conducted a series of circumferential CO_2 arc welding experiments on a shield machine hob. The design of these experiments not only took into account the geometric dimensions and welding parameters shown in Figure 2, but also thoroughly considered various situations and influencing factors that may arise during actual operations. Through these experiments, we were able to gain a comprehensive understanding of key parameters in the actual welding process. We compared the experimental results with the predictions of the finite element model to assess the accuracy and reliability of the model. The center hole drilling strain gauge method is a commonly used experimental technique for measuring residual stress. When a hole is drilled into a material, the surrounding material undergoes deformation, which is influenced by the residual stresses within the material. By measuring the strain around the hole, it is possible to infer the residual stress at the hole location. In order to obtain experimental data on residual stress, the center hole drilling strain gauge method is employed to measure residual stress (including axial stress, hoop stress, transverse stress, and longitudinal stress) at various specified locations. These locations include Points 1-3 on the outer surface of the hob body, located 180° from the weld start position and at distances of 10, 20, and 30 mm from the weld line. Additionally, Points 4-6 are measured on the upper surface of the hob ring, also at a section 180° from the weld start position and at distances of 10, 20, and 30 mm from the weld line.

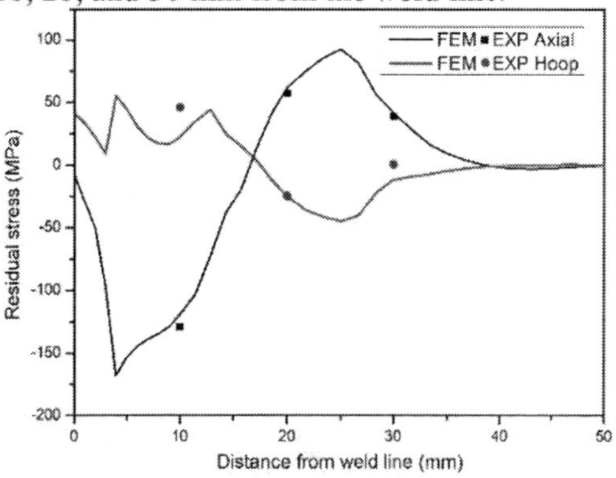

Figure 4. Hoop and axial residual stress at 180° on the outer surface of the hob body.

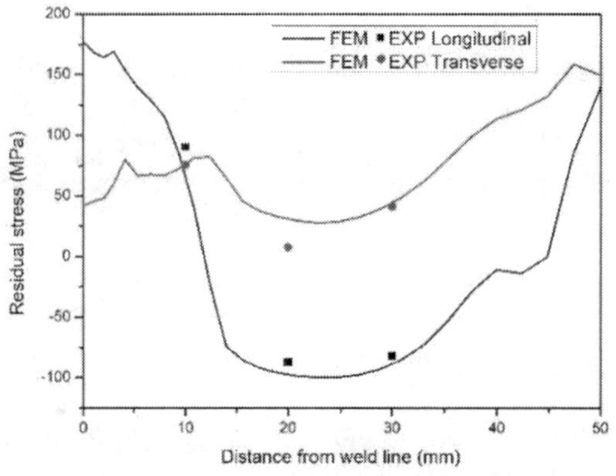

Figure 5. Transverse and longitudinal residual stress on hob ring upper surface at a section 180°.

Figure 4 shows the predicted and measured hoop residual stress and axial residual stress at 180° on the outer surface of the hob body. The data of transverse and longitudinal residual stress on the hob ring upper surface at a section 180° are shown in Figure 5.

From Figure 4 and Figure 5, there is inconsistency in individual data between measured data and predicted data. The primary reason may be attributed to measurement errors and insufficiently accurate material property data during welding, particularly at high temperatures. Nevertheless, the predictions from the Finite Element Method (FEM) align well with the majority of measured data. Therefore, the finite element model proposed in this paper has been validated and proven to reliably predict the residual stresses induced by circumferential CO_2 arc welding on both the hob body and hob ring.

4. Results and discussion

4.1. Characteristics of residual stress on hob body outer/inner surface

In this section, the characteristics of axial and hoop residual stress on hob body outer/inner surface at Sections 0°, 90°, 180°, and 270°are described and discussed. Figure 6 and Figure 7 display the distribution of residual stress.

Figure 6 shows that the axial stress distribution on the outer surface of the hob body is different from that on its inner surface. On the outer surface of the hob body, we can observe the presence of compressive axial residual stresses near the weld line. Especially near the toe of the weld (3 millimeters), these compressive axial residual stresses reach their highest point. The further away it is from the toe of the weld, the smaller the compressive stress is, which diminishes to zero at about 18 mm. Beyond 18 mm the stress reverses from compressive to tensile, increases first and then decreases, and finally decreases to zero.

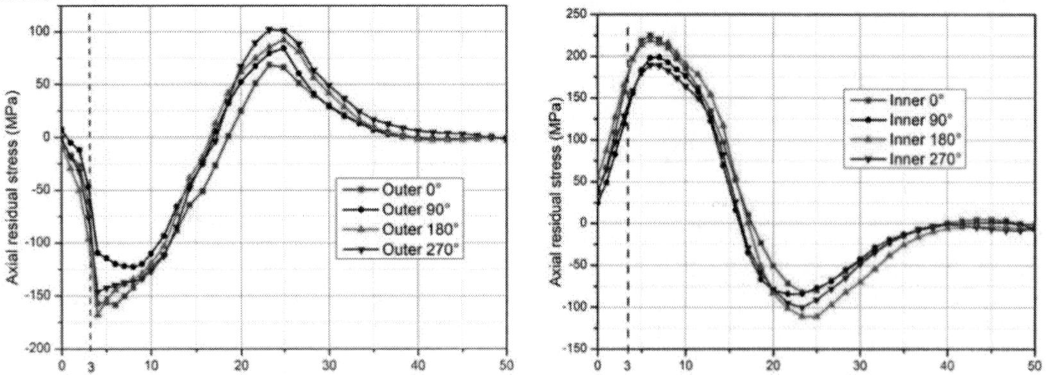

Figure 6. The axial residual stresses on the outer and inner surfaces of the hob body.

Journal of Physics: Conference Series **2808** (2024) 012052 doi:10.1088/1742-6596/2808/1/012052

When inspecting the inner surface of the shield machine hob body, it can be clearly observed that there are significant high tensile axial residual stresses near the weld line. This phenomenon may be attributed to the thermal stresses generated during the welding process and changes in material structure. Additionally, there is a maximum tensile stress of approximately 6 millimeters. With increasing distance from the weld line, the residual stress along the axial direction gradually decreases and tends towards zero. Reversing to compressive stress is also observed. Finally, the stresses approach to the zero.

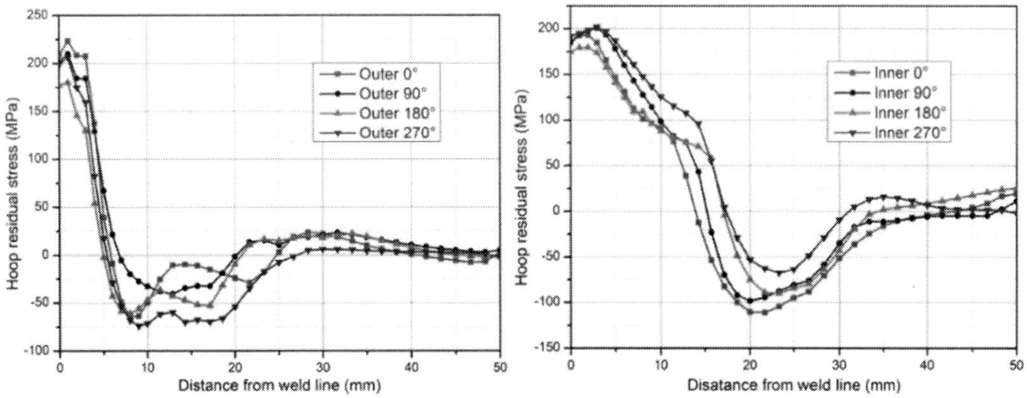

Figure 7. The hoop residual stresses on the outer and inner surfaces of the hob body.

From Figure 7, the distribution of hoop residual stress is roughly the same for the hob body's outer surface and inner surface. The trend of hoop residual stress is consistent with previous research findings, indicating a certain level of consistency. High tensile residual stresses are formed near the weld line. As the distance increases, these residual stresses gradually decrease until approaching zero, and eventually transition to compressive stresses, which finally converge to a zero value. The difference is that the highest compressive stress on the hob body's outer surface is about 10 mm away from the weld line, and that on the hob body's inner surface is about 20 mm.

To investigate the reason for the residual stress characteristics of welded hob body and hob ring, the behavior of plastic strain during the welding is discussed. The difference in stress on the hob body's outer and inner surface is attributed to different plastic strains during the welding. Figure 8 shows plastic stresses at 13, 14, and 16 s as the arc arrives at the 90°location.

For the hob body, the welding deformation is mainly caused by the thermal expansion and contraction in the radial direction because of the same thermal cycle of the inner and outer surfaces. During the welding process, when the arc reaches the 90° position, the hob body undergoes complex strain variations. Initially, there is axial tensile plastic strain inside, likely due to material deformation and temporary stresses induced by welding heat. However, as the arc center moves to the 90° position, this tensile strain gradually transitions to compression. Interestingly, the compression plastic strain on the inner surface of the hob body is more significant compared to the outer surface. This may be because the inner surface is more directly affected by the welding heat, leading to more pronounced plastic deformation internally. Simultaneously, circumferential compression plastic strain occurs on both inner and outer surfaces, indicating the lateral stress on the material during welding. More notably, the internal constraint strength of the inner surface of the hob body is slightly higher than the outer surface. This may suggest that post-welding, the inner surface is more constrained by surrounding structures, resulting in higher internal stresses. At a distance of 10 millimeters from the welding line, circumferential plastic strain on both inner and outer surfaces tends to stabilize. This might indicate that within this distance range, the strain variations induced by welding have reached some equilibrium state, demonstrating the relative stability of the hob body material in this area.

Journal of Physics: Conference Series **2808** (2024) 012052 doi:10.1088/1742-6596/2808/1/012052

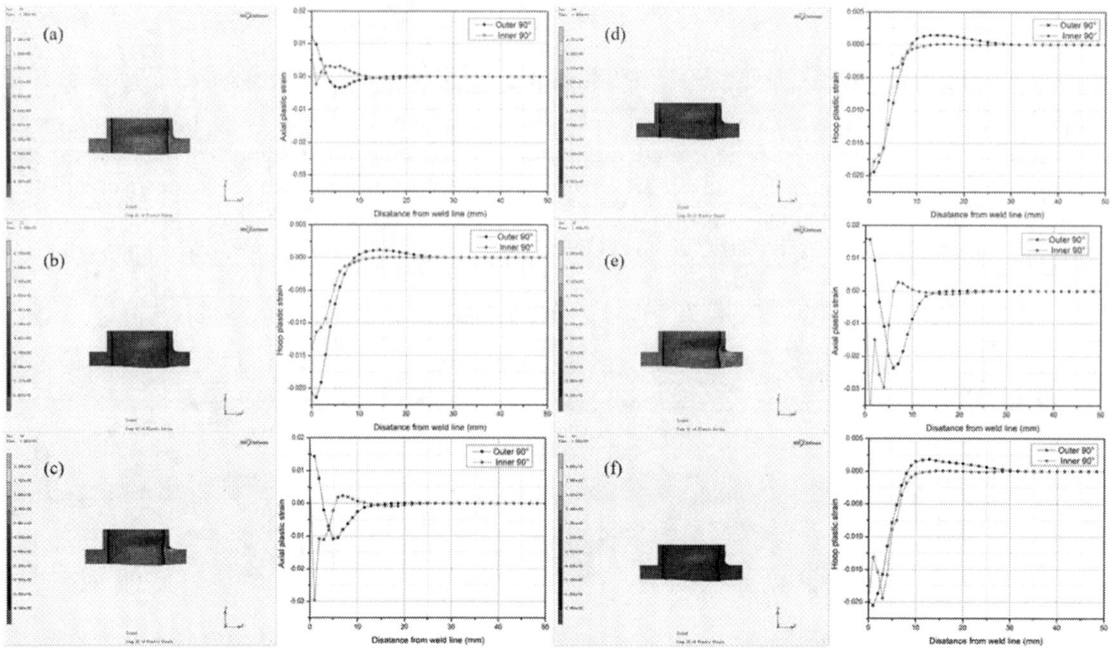

Figure 8. Evolution process of axial and hoop plastic strain of hob body (Factor=6): (a) Axial plastic strain of hob body at 13 s (Factor=6), (b) Hoop plastic strain of hob body at 13 s (Factor=6), (c) Axial plastic strain of hob body at 14 s (Factor=6), (d) Hoop plastic strain of hob body at 14 s (Factor=6), (e) Axial plastic strain of hob body at 16 s (Factor=6), (f) Hoop plastic strain of hob body at 16 s.

From Figure 8, it can be seen that the axial plastic strain at the inner surface of the hob body increases gradually from 13 s to 16 s. However, the plastic strain at the outer surface of the hob body increases slowly. It is due to the fact that although the thermal cycle at the inner and outer surface of the hob body is almost the same, the direction of thermal deformation is different. When the center of the arc arrives at the 90°location, thermal expansion deformation is inward. Therefore, after welding is completed, the significant compression plastic strain on the inner surface not only triggers residual tensile stresses, both axial and circumferential, within the hob body, but also influences the outer surface, resulting in corresponding residual compressive stresses. This dual stress state reflects the complex deformation and stress distribution of the material during the welding process. During the cooling process, as internal heat dissipates, the material gradually returns to its original shape. However, residual stresses leave traces in the material, persistently impacting its performance and stability.

4.2. Characteristics of residual stress on plate upper surface

Figure 9 shows the distribution of transversal and longitudinal residual stress on the hob ring's upper surface. It can be observed that, unlike the characteristics of plate butt welding, significant high transverse tensile residual stresses are formed at the end of the hob ring. This difference may stem from the unique geometry of the hob ring and the stress transfer paths during the welding process. In this scenario, thermal stresses and strains generated during welding may lead to the accumulation of residual stresses at the end of the hob ring, resulting in a distinctive distribution of transverse tensile residual stresses. Understanding this residual stress distribution is crucial for predicting and managing the performance of welded components. The longitudinal tensile residual stress is observed near the weld line, and then the high compressive residual stress is formed at 20-30 mm. There is longitudinal stress mutation at the 0° location. The transversal tensile residual stress always exits on the upper of the hob ring, which is mainly due to the radial shrinkage of the circumferential weld.

Journal of Physics: Conference Series **2808** (2024) 012052 doi:10.1088/1742-6596/2808/1/012052

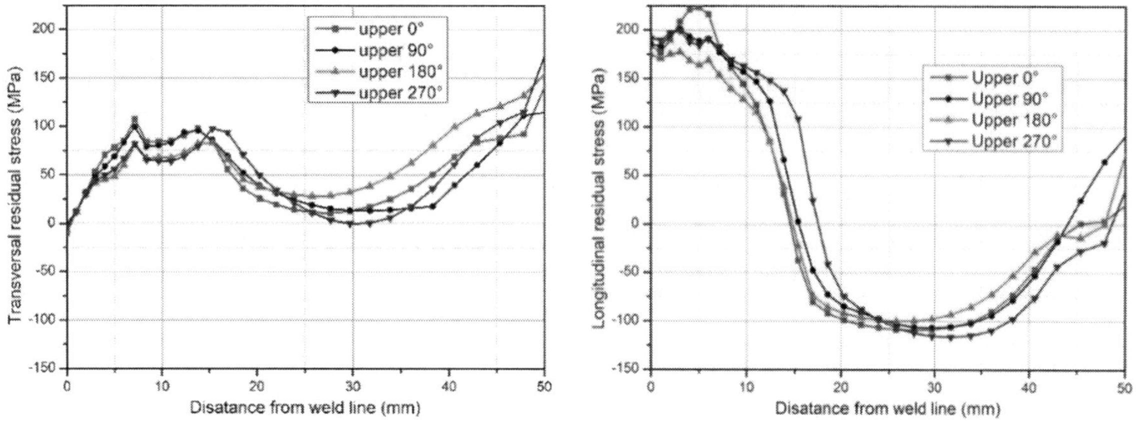

Figure 9. Transversal and longitudinal residual stress on hob ring upper surface.

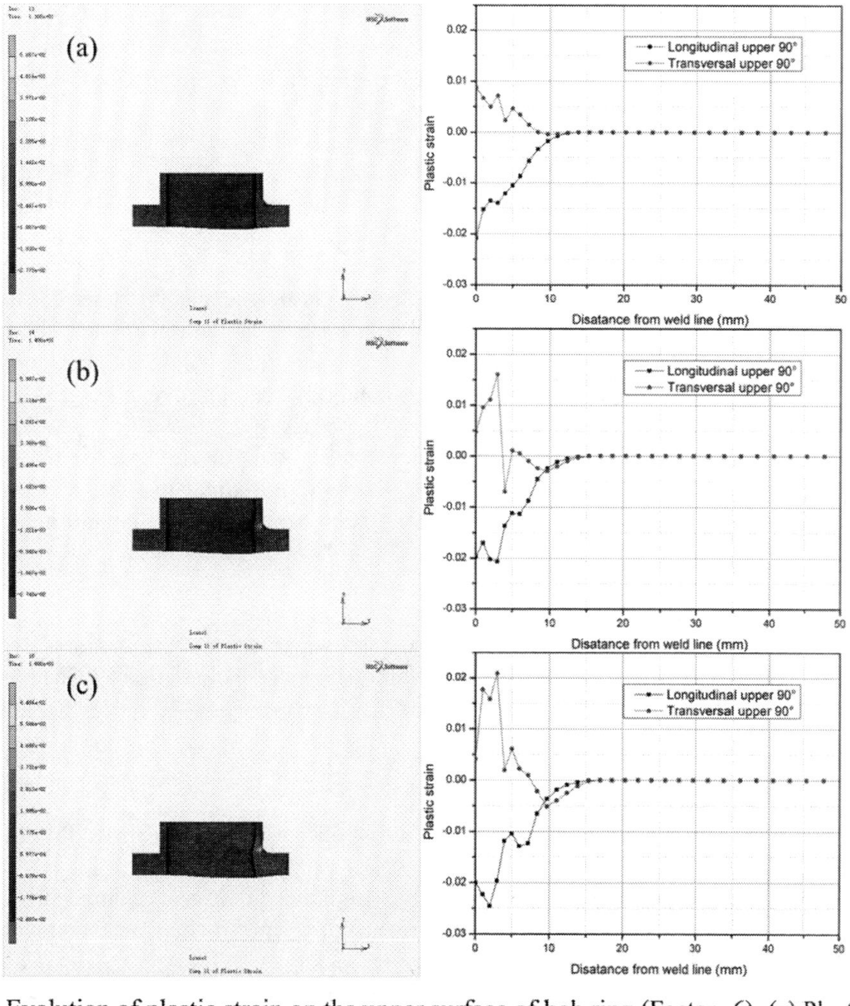

Figure 10. Evolution of plastic strain on the upper surface of hob ring (Factor=6): (a) Plastic strain of hob ring at 13 s (Factor=6), (b) Plastic strain of hob ring at 14 s (Factor=6), (c) Plastic strain of hob ring at 16 s (Factor=6).

The transversal and longitudinal plastic strain distribution and evolution on the upper surface of the hob ring at 90° are shown in Figure 10. The welding arc approaches the point at 90° from 13 s to 14 s, and the center of the arc arrives at the 90° location. It can be seen that with the approaching of the arc, transversal tensile plastic strain near the weld line increases, and longitudinal compressive plastic strain increases, too. During the welding, the transversal plastic strain shows tensile, and longitudinal shows compressive. It is well known that the compressive plastic strain during welding will appear as tensile stress after cooling, so tensile residual stresses are generated near the weld line after the cooling stage.

5. Conclusions

In this study, the FE analysis based on the computational methodology is carried out to predict the magnitude and distribution of weld-induced residual stress and plastic strain in circumferentially CO_2 arc welding fillet welded shield machine hob. Finite Element Analysis (FEA) is a powerful numerical simulation method, and successfully implemented in this study, with its results showing significant consistency with experimental data. By employing the principle of plastic strain, we gain a deep understanding of the distribution characteristics and formation process of welding residual stresses in the hob body and hob ring at the 90° location. Based on the results presented in this study, we can draw the following important conclusions.

(1) According to the analysis, in the vicinity of the welding line, tensile residual stresses are observed on the inner surface of the hob body, while compressive residual stresses are formed on the outer surface along the axial direction. It is noteworthy that both the outer and inner surfaces of the hob body exhibit characteristics of only tensile hoop residual stresses.

(2) For the hob ring, the tensile transversal residual stresses are produced on the upper surface from the weld line to the hob ring edge. However, as the distance from the weld seam increases, we can observe changes in the residual welding stresses. Initially, these stresses exhibit characteristics of residual longitudinal tensile stress, but as the distance increases, they gradually transform into residual longitudinal compressive stress.

(3) A thorough investigation of the plastic strain distribution at the 90° location reveals that the compressive axial and hoop plastic strains on the inner surface of the hob body significantly exceed those on the outer surface. Consequently, thermal contraction deformation induces compressive axial and hoop residual stresses on the outer surface of the hob body, leading to tensile residual stresses on its inner surface.

(4) Because of thermal contracting on the upper surface of the hob ring, tensile transversal residual stresses are generated after welded cooling. Compressive longitudinal residual stresses occur at the middle of the hob ring because of contracting plastic strain near the weld line and internal stress self-balance.

Acknowledgments

The authors would like to sincerely thank the financial support from the National Natural Science Foundation of China (Grant No. 51771119), the Natural Science Foundation of Shanghai (Grant No. 17ZR1419600), and the Opening Project of Jiangsu Key Laboratory of Advanced Structural Materials and Application Technology (ASMA201610).

References

[1] Y. Ueda, H. Murakawa, N. Ma, Welding Deformation and Residual Stress Prevention, Elsevier Science, 2012.
[2] D. Deng, H. Murakawa, Numerical simulation of temperature field and residual stress in multi-pass welds in stainless steel pipe and comparison with experimental measurements, J. Comput. Mater. Sci. (2006) 269-277.
[3] I.F.Z. Fanous, M.Y.A. Younan, A.S. Wifi, Study of the effect of boundary conditions on residual stresses in welding using element birth and element movement techniques. ASME J. Pressure Vessel Technol. 125 (2003) 432-439.

[4] C. Wu, J.W. Kim, Analysis of welding residual stress formation behavior during circumferential TIG welding of a pipe, Thin-Walled Structures, 132 (2018) 421-430.
[5] D. Deng, Experimental and numerical investigations of welding distortion induced by CO_2 gas arc welding in thin-plate bead-on joints, Mater. Des. 52 (2013) 720-729.
[6] D. Deng, FEM prediction of welding residual stress and distortion in carbon steel considering phase transformation effects, Mater. Des. 30 (2) (2009) 359-366.

Journal of Physics: Conference Series **2808** (2024) 012053 doi:10.1088/1742-6596/2808/1/012053

A novel process on recyclable particle materials used for aggregate columns

Mingyuan Wang[1,2], Yajun Wang[3*], Zuguo Zhang[1], and Huifeng Zheng[2]

[1]Zhejiang Huadong Geotechnical Investigation & Design Institute Co., LTD, Hangzhou, Zhejiang, 310030, China
[2]Huadong Engineering Corporation Limited, Hangzhou, Zhejiang, 311122, China
[3]School of Marine Engineering Equipment, Zhejiang Ocean University, Zhoushan, Zhejiang, 316002, China

[*]Corresponding author's e-mail: wangyj@zjou.edu.cn

Abstract. The study focused on the creative development of aggregate column materials and technologies. The seawater (SW) and alkaline agents (AA) were invited to create the recyclable particle materials (RPMs) that were used for the construction of the aggregate columns in the marine environment. The building solid waste (BSW) was utilized to produce the fine powder the chemical activity of which was restored with the help of AA and marine processes. The created products in the study were the RPMs that were also cured in the marine environment. The mechanical and chemical properties were investigated with comprehensive approaches including UCT (uniaxial compression tests), EDS (Energy Dispersive Spectrometer), SEM (Scanning Electron Microscopy), and XRD (X-ray Diffraction). The investigation on the mechanical and chemical properties of the RPMs was carried out at three ages inclusive of 8-day, 49-day, and 583-day. The results from UCTs reported that the UCS (uniaxial compression strength) of RPMs reached 7.29 MPa at 583-day age even in the marine environment. EDS analysis detected the predominant element of RPMs and it was guaranteed that Al, Ca, and Si developed continuously with the age accumulation. Based on XRD and SEM exploration, the principal hydration products and compounds of RPMs used in the aggregate columns included C-S-H, C-A-H, C-A-S-H, N-A-S-H, and $CaCO_3$. Particularly, the salt complexes generated in RPMs filled compactly the microscopic pores of RPMs, which resulted in its densified framework.

1. Introduction

The application of sand or gravel compaction piles is nowadays an accepted choice for soft basement improvement guidelines. However, the exhausting consumption of fossil minerals hindered the general application of sand or gravel compaction piles. The recyclable think was borrowed from the materials sciences and the mechanical performance in engineering cases showed the precise advantages [1-3]. The diverse sources of the particle materials spread the application of the aggregate columns in the soil foundation cases [4]. The aggregate columns using particle materials produced with recycled resources also showed diverse functions in engineering applications. Meanwhile, the design schemes had a direct impact on the aggregate columns' behaviors [5]. The analytical and numerical technologies were also suggested to understand the performance of the aggregate columns in the interfacial environment [6]. The creation of the aggregate columns' construction was also the focus of the geotechnical engineering the conventional method of which was improved by the help of hydraulic jack preloading effects [7]. The study on the static-dynamical characteristics of the aggregate columns was another focus that ascertained the rational application [8-10]. Due to the promising and sustainable peculiarity, the recyclable materials derived from demolition waste have shown the possibility of manufacturing environmentally friendly cement with alkaline activation [11]. Meantime, with the help of the proper modification of the cement compositions, for instance, with the addition of limestone powder and fly ash, the alkaline polymer was able to achieve enhanced compressive strength in the marine environment [12].

The particle materials in the study were produced with alkaline agents (AA) including NaOH and $Na_2O{\cdot}nSiO_2$ and seawater (SW). Particularly, the particle materials were cured in the seawater. Hence, the marine process was the main approach in the study to develop the recyclable particle materials (RPMs) for the aggregate columns technology.

The study on recyclable materials that can be applied to the particle materials in aggregate columns shows the promising capacity and novelty in the infrastructures. Particularly, the marine process for the recyclable particle materials helps consolidate the soft strata with the compressive and draining effects and reduces the exploitation of fossil minerals and freshwater resources. The comprehensive methods were conducted with resort to the uniaxial compression strength (UCS), Energy Dispersive Spectrometer (EDS), Scanning Electron Microscopy (SEM), and X-ray Diffraction (XRD). The chemical and mechanical advantages of RPMs were systematically researched in the study.

2. Conceptions on recyclable particle materials in aggregate columns

The conventional materials for aggregate columns were natural minerals including limestone, basalt, and sandstone the excavation of which caused serious environmental disasters (Figure 1).

Figure 1. Quarry in Zhejiang, China.

Furthermore, the freshwater resource in coastal cities of China was being devastated due to industrial waste pollution and ever-lasting heavy exploitation. Particularly, the increasing demand from infrastructures resulted in the insufficiency of freshwater resources.

The creation of recyclable particle materials that are produced with building solid waste (BSW) and seawater helps eradicate the malady of conventional aggregate columns in infrastructure applications. The solid waste derived from the old constructions was collected and re-grounded into fine powder. The alkaline agents were invited in the mixing of the re-ground into fine powder with seawater. The recyclable paste was moulded with the designed moulds (Figure 2) and the particle materials were prepared (Figure 3).

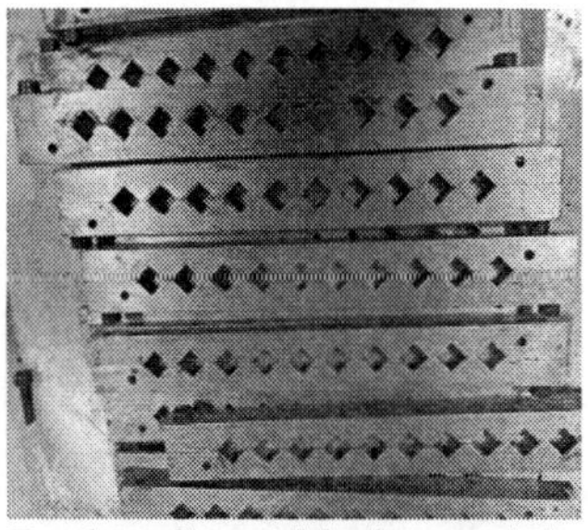

Figure 2. Designed moulds for particle materials.

Figure 3. Particle materials.

The configuration of the particle materials produced in the study is the cubic one the dimension of which is 1 mm × 1 mm × 1 mm. The particle materials produced with solid waste and seawater will be assembled into aggregate columns in the strata of the infrastructures such as embankments that can be reinforced by the particle materials (Figure 4).

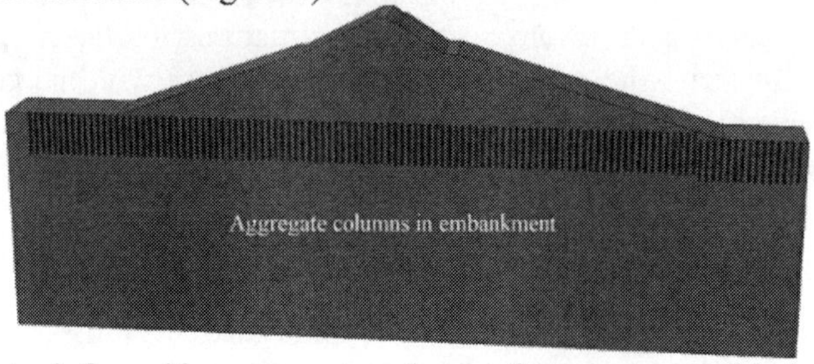

Figure 4. Embankment reinforced by aggregate columns that were assembled with particle materials.

3. Process on recyclable particle materials in aggregate columns

The marine process and alkaline mixture were created in the study to produce the recyclable particle materials. Moreover, the solid waste derived from the old constructions was collected and re-grounded into fine powder (Figure 5). The fine powder was defined as recyclable particle material from paste (RPP).

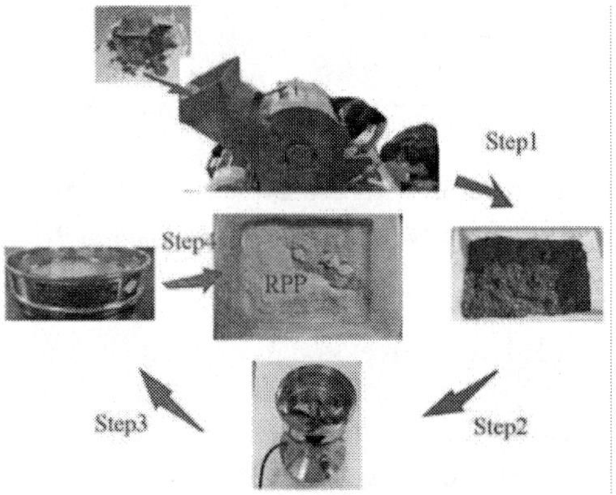

Figure 5. Flowchart of fine powder production.

The Zhejiang native seawater was used for the cement paste production coupled with NaOH and $Na_2O{\cdot}nSiO_2$. The alkaline agents and the cation and anion in seawater activated the cementitious capacity of the fine powder. Especially, no Portland cement was added to the recyclable particle materials in this paper. The study prepared the recyclable particle material with the marine alkaline cement paste and cured them in the Zhejiang native seawater. The oldest age of the recyclable particle material in the study reached 583 days. The temperature and relative humidity of the curing conditions were 20±2°C and 80±5% which was exactly consistent with the native marine environment. The orthogonal methodology was invited to optimize the mass proportion of the recyclable particle material and the goal composition included BSW: AA: SW=1:0.06:0.48.

4. Performance on recyclable particle materials

The mechanical behaviour was explored with the uniaxial compression tests (UCTs). The 5 kN/s loading velocity was adopted uniformly in the study. The expectation of the uniaxial compression strength (UCS) reached 5.26 MPa at 8-day age (Figure 6).

Figure 6. Mechanical behaviour of recyclable particle materials at 8-day age.

Meantime, the UCS level of the recyclable particle materials produced with $Na_2O{\cdot}nSiO_2$ was higher than that produced with NaOH. The UCS maximum of the recyclable particle materials produced with $Na_2O{\cdot}nSiO_2$ attained 7.2 MPa at 49-day age when cured in the Zhejiang native seawater, which indicated that the recyclable particle materials used in aggregated columns were the promising ones even in the stern marine environment (Figure 7).

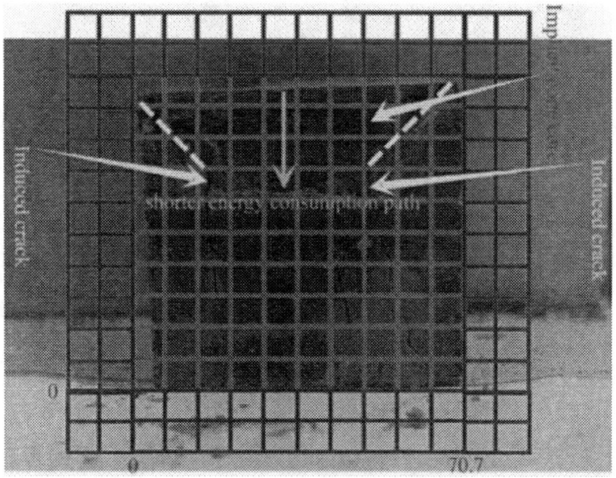

Figure 7. Mechanical behaviour of recyclable particle materials at 49-day age.

With the marine process, the UCS of the recyclable particle materials produced with $Na_2O{\cdot}nSiO_2$ was 7.29 MPa at 583-day age (Figure 8). Therefore, it can be deduced that the UCS of the recyclable particle materials achieved an everlasting rise with age accumulation. Furthermore, the marine environment where the aggregate columns always worked did not cut down the UCS of the recyclable particle materials. The mechanical behaviour showed the superb applicability of the recyclable particle materials created in the study.

Figure 8. Mechanical behaviour of recyclable particle materials at 583-day age.

The EDS analysis (EDAX Inc., Mahwah, NJ, USA) was also invited in the study to investigate the chemical behaviors of the recyclable particle materials. The results from EDS reported that C, O, Na, Mg, Al, Si, S, Cl, K, Ca, and Fe were the principal elements in the hydration products of the recyclable particle materials.

In terms of damaged zones in the recyclable particle materials, the weight percentages wt%s of Na and Cl at 8-day age from NaOH process were higher than that from $Na_2O{\cdot}nSiO_2$ process. By contrast, the weight percentage wt% of Si from non-damaged zones at 8-day age from the NaOH process was lower than that from the $Na_2O{\cdot}nSiO_2$ process (Figure 9).

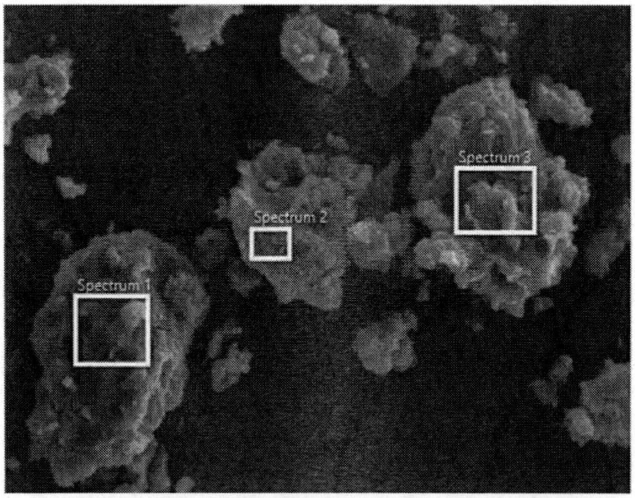

Figure 9. EDS pattern of recyclable particle materials at 8-day age.

In terms of damaged zones in the recyclable particle materials, the weight percentage wt% of C at 49-day age from the $Na_2O{\cdot}nSiO_2$ process was lower than that from the NaOH process. On the contrary, the weight percentages wt%s of Si and Ca at 49-day age from the $Na_2O{\cdot}nSiO_2$ process were higher than that from the NaOH process (Figure 10).

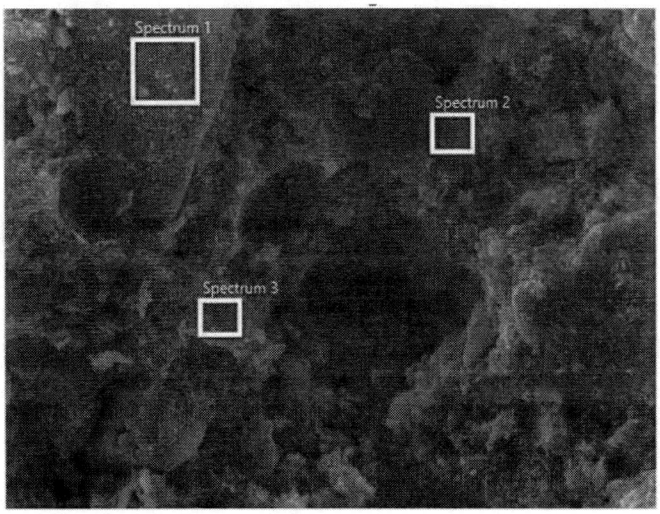

Figure 10. EDS pattern of recyclable particle materials at 49-day age.

The element compositions of the recyclable particle materials at 583-day age reported a precise difference from that at 8-day and 49-day ages. In particular, the recyclable particle materials produced with the $Na_2O{\cdot}nSiO_2$ process showed the exact integrity of the chemical structure owing to the continuous generation of Al, Si, and Ca. At the same time, SO_4^{2-} and Cl^- were assimilated promisingly with the parental structures of the recyclable particle materials at 583-day age (Figure 11).

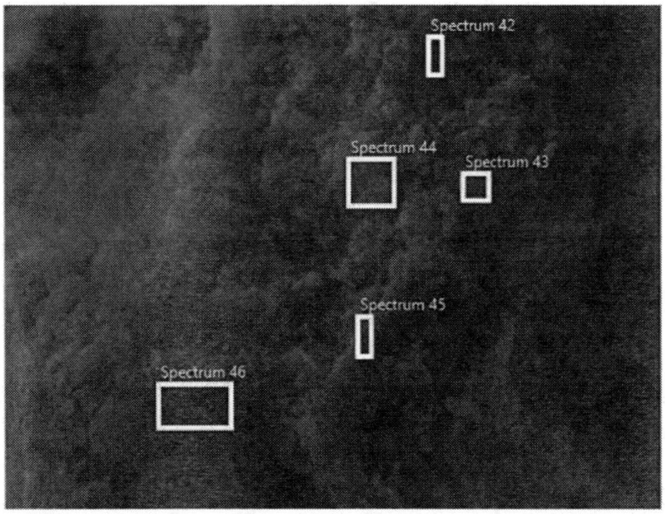

Figure 11. EDS pattern of recyclable particle materials at 583-day age.

X-ray diffraction (XRD) in the study was another approach to the investigation of the chemical behaviours of recyclable particle materials (Figures 12, 13, and 14).

The recyclable particle materials produced with the $Na_2O{\cdot}nSiO_2$ process showed higher main peaks in the XRD patterns. Meantime, their main hydration products were C-S-H and C-A-S-H. Furthermore, there lived more complicated hydration product compositions in the non-damaged zones inclusive of C-S-H, C-A-S-H, and N-A-S-H that generated the spatial framework of the recyclable particle materials (Figure 12).

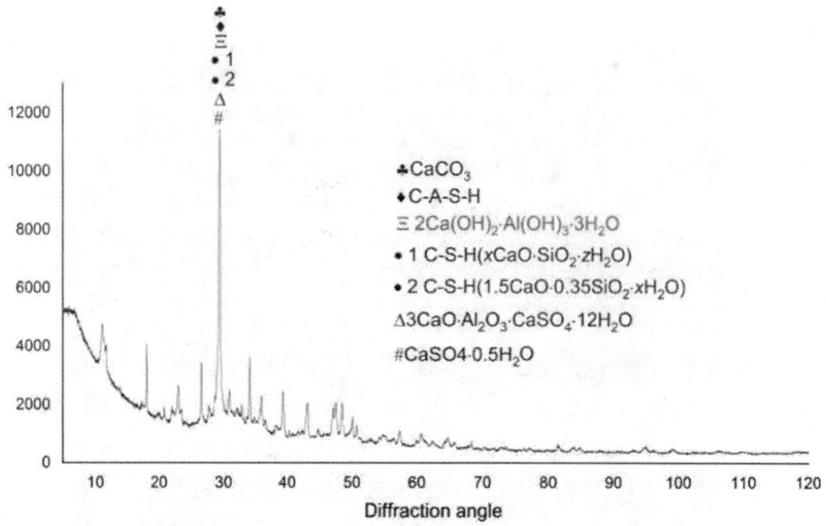

Figure 12. XRD pattern of recyclable particle materials at 8-day age.

The typical Friedel' salt $3CaO·Al_2O_3·CaCl_2·10H_2O$ was certainly generated in the recyclable particle materials due to the marine process (Figure 13). Moreover, Friedel's salt $3CaO·Al_2O_3·CaCl_2·10H_2O$ existed in the recyclable particle materials at 8-day, 49-day, and 583-day ages. The differences in Friedel's salt $3CaO·Al_2O_3·CaCl_2·10H_2O$ included the location, intensity, and connection with the parental structures. The non-damaged zones from the $Na_2O·nSiO_2$ process reported higher main peaks than those from NaOH. Meantime, the framework of the recyclable particle materials produced with the $Na_2O·nSiO_2$ process was more compacted and continuous than that produced with the NaOH process (Figure 13).

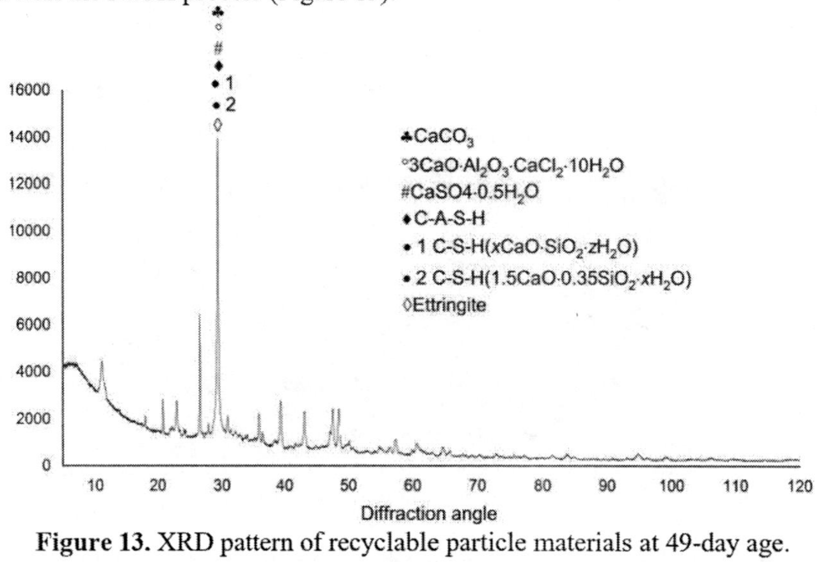

Figure 13. XRD pattern of recyclable particle materials at 49-day age.

The recyclable particle materials form both non-damaged and damaged zones from the $Na_2O·nSiO_2$ process generated at 583-day age the complicated hydration products structures including C-A-H, C-S-H, C-A-S-H, and N-A-S-H, etc (Figure 14). Furthermore, the recyclable particle materials produced with the $Na_2O·nSiO_2$ process achieved a 583-day age more compacted framework. Especially, the density of $Na_2O·nSiO_2$ also affected the framework of the recyclable particle materials.

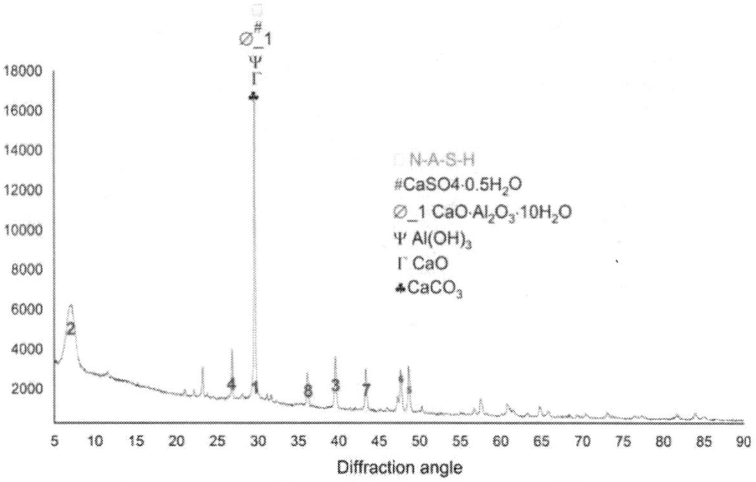

Figure 14. XRD pattern of recyclable particle materials at 583-day age.

The study adopted SEM (Scanning Electron Microscopy) to detect the microscopic structures of the recyclable particle materials at all ages, by which the working behaviours of the aggregate columns can be explored thoroughly. The images from Figures 15, 16, and 17 indicated that the microscopic structures of the recyclable particle materials were compacted continuously with age accumulation. Hence, it was deduced that the recyclable particle materials with AA and marine process played successfully the compressive and draining functions in the aggregate columns that were applied even in the marine environment. The recyclable particle materials application resulted in the exact consolidation in the soft strata by the superb mechanical and chemical performance.

Figure 15 showed that the recyclable particle materials generated Na^{+} and Ca^{2+} alkaline complexes even at 8-day age which guaranteed the compacted framework. The microscopic structure of the recyclable particle materials at 8-day age was the lamellar one.

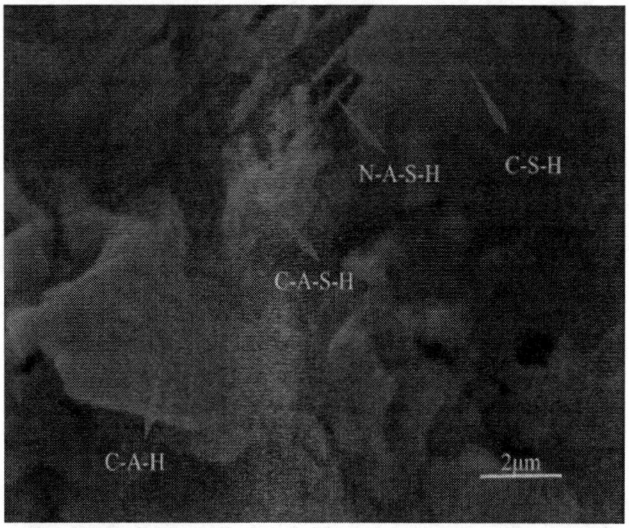

Figure 15. SEM image of recyclable particle materials at 8-day age.

The microscopic structure of the recyclable particle materials at 49-day age developed into the fibre-cluster one. The hydration products including C-S-H and C-A-S-H grew up continuously and hardened intensively, which densified the spatial structures' of fibre-clusters (Figure 16).

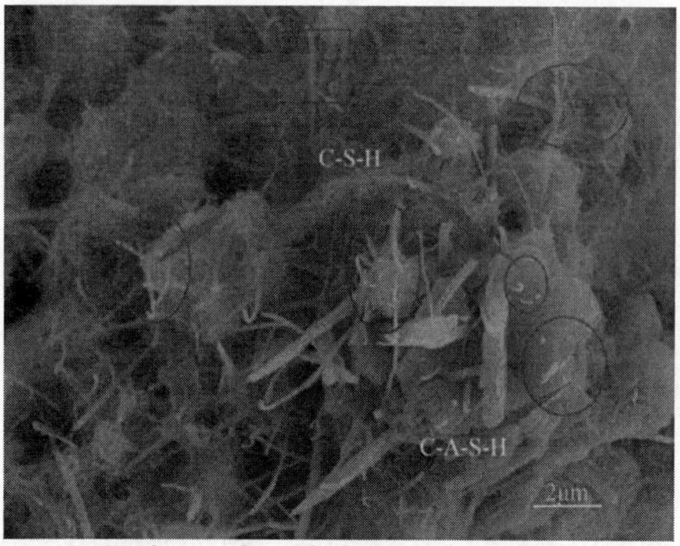

Figure 16. SEM image of recyclable particle materials at 49-day age.

The microscopic structures of the recyclable particle materials at 583-day ages showed the consolidated framework which was formed by the grown hydration products and compounds including C-S-H, C-A-S-H, and $CaCO_3$. The microscopic pores in the recyclable particle materials were filled compactly by the salt complex the predominant composition of which was Friedel's one (Figure 17). Moreover, the SEM images indicated that the size of the crystals' mixture that was composed of C-A-H, C-S-H, C-A-S-H, N-A-S-H, and $CaCO_3$ rose certainly with the age accumulation even in the marine environment (Figures 15, 16, and 17).

Figure 17. SEM image of recyclable particle materials at 583-day age.

5. Discussions

The marine process coupled with AA demonstrated several advantages in the case of the aggregate column including freshwater saving, independence from fossil minerals, direct applicability, lower energy consumption, lower carbon emission, and continuous mechanical capacity. BSW utilization, at the same time, freed the cultivated land from pollution. AA process helped realize the efficient recycling of BSW without heavy carbon emissions. The marine process exhibited the promising application of SW in infrastructure cases, by which the freshwater resource was protected.

6. Conclusions

The recyclable particle materials used for the aggregate columns helped the marine environment consolidate the soft strata with continuous mechanical performance. AA process developed in the study restored the chemical activity of BSW and facilitated the direct application of BSW with lower carbon emission. Particularly, the chemical activity of BSW reported the exact effects of the marine preparation and curing process. The creative cognition on the marine process was that the complicated ions environment from SW accommodated the sufficient active elements resource including Al^{3+}, Si^{4+}, Na^{+}, and Ca^{2+}. Furthermore, AA adopted in the study helped accelerate the bond breach reaction at the early ages (Figures 15 and 16), by which the active elements from BSW were freed from the fine powder. The active elements from SW then accumulated with those from BSW. The reaction in the study was defined as the marine de-polymerization. AA re-collected, consequently, the active elements in the young RPMs and generated the new crystal complexes (Figures 11, 14, and 17) in the grown RPMs. The reaction in the study was defined as the marine re-polymerization. Therefore, the marine de-polymerization in the young RPMs helped loosen the elements and the marine re-polymerization achieved the elements' active reaction. The whole procedure was accomplished during the chemical activity restoring the fine powder. The macroscopic behaviours and microscopic structures of the recyclable particle materials showed promising workability.

Acknowledgements

This study was supported by the National Natural Science Foundation of China, grant numbers 51879236 and 51109118 and Zhejiang Provincial Natural Science Foundation of China, grant number LHY21E090003.

References

[1] You, S.K., Lee, J., Gabr, M.A. (2016) Experimental evaluation of recycled aggregate porous concrete piles for soft ground improvement. Mar. Georesour. Geotec., 34(8): 712–720.
[2] Wang, Y.J., Zhang, C.H., Wang, J.T., Xu, Y.J., Jin, F., Wang, Y.B., Yan, Q., Liu, T., Gan, X.Q., Xiong, Z. (2019) Experimental study on foci development in mortar using seawater and sand. Materials, 12(11):1–24.
[3] Sheng, Z.G., Wang, Y.J., Huang, D. (2022) A promising mortar produced with seawater and sea sand. Materials, 15(17): 1–20.
[4] Adamu, B.N. (2023) Comparative analysis on the effectiveness of crushed aggregate and scoria gravel as columnar reinforcement material under raft foundation. Results in Engineering, 19: 101254.
[5] Zhou, Y., Wang, Y.Q. (2022) Experimental study on the treatment effect of vibro-flotation gravel piles for saturated sand foundation in coastal areas. Arab. J. Sci. Eng., 47(4): 4833–4848.
[6] Ter-Martirosyan, A.Z., Ter-Martirosyan, Z.G., Sidorov, V.V. (2016) Interaction of compaction piles with surrounding soil with due regard for pile diameter expansion. Soil. Mech. Found Eng., 53(3): 166–173.
[7] Comina, C., Mandrone, G., Arato, A., Chicco, J., Duò, E., Vacha, D. (2021) Preliminary analysis of an innovative soil improving system by sand/gravel injections-Geotechnical and geophysical characterization of a first test site. Eng. Geol., 293:106278.
[8] Lee, K.H., Kim, S.K., Yea, G.G. (2014) Assessment of dynamic characteristics for compaction piles with geophysical logging in soft ground. Mar. Georesour. Geotec., 32(1): 62–75.
[9] Jongpradist, P., Krairan, K., Jamsawang, P., Chen, X.B. (2022) Geotechnical Engineering Properties of Cement Fly Ash Gravel Mixtures for Application as Column-Supported Highway and Railway Embankments. Materials, 15:3972.
[10] Sáez. E., Pardo, G.S., Ledezma, C. (2015) Seismic response of a pile-supported excavation on Santiago gravel. Soil. Dyn. Earthq. Eng., 76:2–12.

[11] García-Díaz, A., Delgado-Plana, P., Bueno-Rodríguez, S., Eliche-Quesada, D. (2024) Investigation of waste clay brick (chamotte) addition and activator modulus in the properties of alkaline activation cements based on construction and demolition waste. J. Build. Eng., 84: 108568.
[12] Rashad, A.M., Ezzat, M., ElNagar, A.M., El-Nashar, M.H. (2023) Valorization of limestone powder as an additive for fly ash geopolymer cement under the effect of the simulated tidal zone and seawater attack. Constr. Build. Mater., 369: 130616.

Effect of explosion venting effect on explosion response characteristics of box-shaped cabin

Yinliang Ma[1*] and Chenli Ji[1]

[1]Chinese Flight Test Establishment, Xi'an, Shaanxi, China

*Corresponding author's e-mail: 2225309761@qq.com

Abstract: In this paper, based on dynamics software ANSYS/LS-DYNA, a numerical model is established to simulate the dynamic response of a box-shaped cabin, and the validity of the numerical model is confirmed by comparing its results with existing literature or experimental data. Using the validated model, the study delves into dimensionless damage number of box-shaped cabin structures in relation to implosion responses, taking into account the explosion venting effect. Numerical analysis is performed to determine the relationship between dimensionless damage number D_{in} and deflection thickness ratio δ/H of the cabin's plates. The study then considers the influence of the explosion venting effect on the implosion response of box-shaped cabins and proposes dimensionless damage number D_{in}', which can predict structure damage of constraint explosion.

1. Introduction

An explosion in a cabin is a typical constraint explosion. Due to the closed environment, propagation characteristics of explosion loads are often more complicated than free field explosion, and the damage to cabin structure is more serious. For the problem of internal explosion, two methods of experiment and numerical simulation are currently used for research. Theoretical prediction methods are still in the development stage and feature less data. Hou et al. [1] conducted experiments on typical cabin shrinkage models to analyze load distribution characteristics and destruction modes. At the same time, in response to the structural response under impact load, scholars at home and abroad are actively exploring the use of dimensionless damage to characterize the damage deformation of the structure. Huang et al. [2] studied the impact of the protective cabin opening on the explosion effect. Chen et al. [3] simplified temporal and spatial aspects of cabin explosion to establish a rapid forecasting method for internal explosion loads. Shi et al. [4] proposed a predictive algorithm for cabin structure destruction modes based on velocity judgment. For the solid branch board under unconstrained explosion, Nurick and Martin [5] introduced a dimensionless damage number $\phi_q=\frac{I}{2H^2(bl\rho\sigma_0)^{1/2}}$ for unconfined explosions on solid plates, building on the work of Johnson [6]($D_n=\frac{I^2}{\rho\sigma_0 H^2}$). Jacob [7] considered the locality of loads on the basis of ϕ_q, and introduced a local load coefficient (1+ln$A/A0$) to get a dimensionless damage number $\phi_{ql}=\frac{I(1+ln(A/A_0))}{2H^2(bl\rho_s\sigma)^{1/2}}=\frac{I(1+ln(bl/\pi R_0^2))}{2H^2(bl\rho_s\sigma)^{1/2}}$. We can see that ϕ_q and ϕ_{ql} include impulse, but the input of structure under the internal explosion effect is almost impossible to obtain, so ϕ_q and ϕ_{ql} are not suitable for internal explosions. Yao [8] used the Kirchhoff-Love thin plate theory to propose a

Journal of Physics: Conference Series **2808** (2024) 012054 doi:10.1088/1742-6596/2808/1/012054

dimensionless damage number $D_{in} = \frac{E_e}{\sigma_0} \frac{V}{L^3} \frac{L}{H} = \frac{Q}{\sigma_0 L^2 H}$ (In the formula, $Q = E_e V$ is the total energy generated by explosions). Dimensionless damage number D_{in} considers the effects of internal explosion load strength, material strength, and structural size but does not take into account the impact of explosion venting effect on the total energy of the explosion, so it has certain limitations in solving the problem of the internal explosion of the actual ship's cabin. The explosive load can be leaked from opening, and total energy Q is not all acting on the structure of the cabin.

In response to a simple box-shaped cabin structure, this article uses a single variable method to investigate the influence of explosion venting effects on internal explosion response and proposes a dimensionless damage number D'_{in} considering the explosion venting effect. The research of this paper is based on the dynamics software ANSYS/LS-DYNA.

2. Materials and methods

2.1. Geometry and meshing

The box-shaped cabin structure is used as a simplified model of the actual ship's cabin structure, as shown in Figure 1. The side of the box-shaped cabin is 600 mm, the thickness of the plate is 4 mm, and the material is Q235 steel. In order to simulate an armor-piercing missile penetrating the hull into the room, the center of the box is opened. In order to ensure that the load of various wall surfaces of the box is uniform, TNT explosives are cubes and are arranged in the center of the box, so that the edge length of the explosives is parallel to the box side length. In order to consider the influence of the neighborhood's border constraint on the explosion cabin, boundary plates are set at the boundary of the box. The boundary plates' width is 1/5. In view of the symmetry of load and structure, only a 1/4 finite element model is established to save computing resources. The cabin structure is simulated by Shell 163 unit, and the Lagrange unit algorithm is selected. The grid size of the box-shaped cabin is 8 mm. In order to fully consider the coupling of structure and air, an air domain of 900 mm×900 mm× 900 mm is established, as shown in Figure 2. Air adopts a solid 164 physical unit based on a multi-material Ale algorithm for discrete, with a grid size of 8 mm and local encryption processing performed in a central area. No reflected boundary conditions are set on the outside of the air to simulate infinite airspace. Explosives directly use initialization keywords (*initial_ volume_ fraction_ geometry) to be filled in air. The interaction between structure and air is defined by the flow solid coupling keyword (*Constrained_ Lagrange_ In_ Solid).

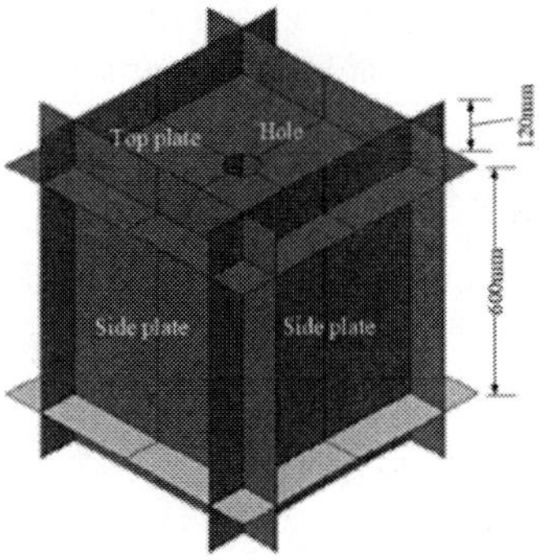

Figure 1. Geometric model of box-shaped cabin.

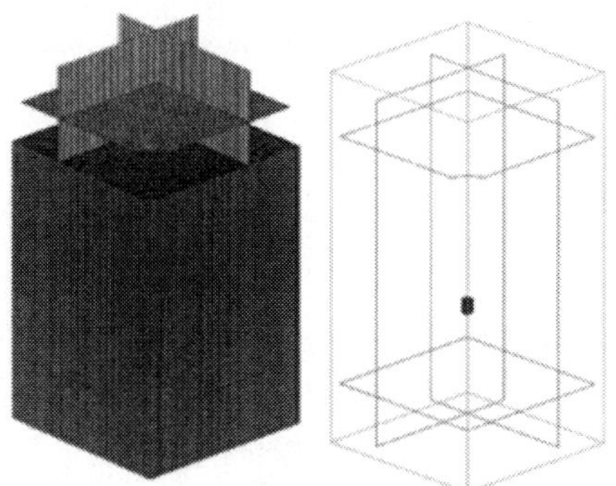

Figure 2. 1/4 FE model of box-shaped cabin.

2.2. Material model

The Johnson-Cook material model describes the dynamic behavior of Q235 steel. The material model takes into account the impact of strain, effect of strain rate, and temperature, which can more accurately simulate the mechanical behavior of Q235 steel under explosive load. Q235 steel's Johnson-COOK material model-specific parameters [9] are shown in Table 1.

Table 1. Johnson-Cook material model parameters used for Q235 steel.

ρ (kg/m³)	*E* (GPa)	*G* (GPa)	*v*	*A* (MPa)	*B* (MPa)	*n*	*c*	$\dot{\varepsilon}_0$ (s^{-1})	*m*
7830	210	80.8	0.3	370	438	0.60	0.01	1.00	0.669

TNT explosive is defined by the keyword *Mat_ High_ Explosive_ Burn. JWL status equation is used to describe the relationship between pressure of explosive product and relative volume and internal energy, and the specific expression is Formula (3):

$$P=A(1-\frac{\omega}{R_1 V})e^{-R_1 V}+B(1-\frac{\omega}{R_2 V})e^{-R_2 V})+\frac{\omega E}{V} \tag{1}$$

In the formula, P is pressure; E is the internal energy of the unit volume explosive; V is current relative volume; A, B, R_1, and R_2 are *JWL* status equation parameters [10]. Assuming that air is an unsigned ideal gas, we use the keyword *MAT_NULL and gas state equation *EOS_LINEAR_POLYNOMAL to describe it. The linear polynomial state equation is as follows:

$$P=C_0+C_1\mu+C_2\mu^2+C_3\mu^3+(C_4+C_5\mu+C_6\mu^2)E \tag{2}$$

In the formula, P is pressure, μ is relative volume, and E is internal energy per unit volume of material. When considering air as an ideal gas, $\gamma=C_p/C_v$ is the specific heat capacity of the ideal gas, $C_0=C_1=C_2=C_3=C_6=0, C_4=C_5=\gamma-1$, γ takes 1.4, and the density of air is 1.29 kg/m³.

2.3. Numerical model validation

The explosion experiments of steel box structures published [9] are used as verification objects to verify the effectiveness of the simulation method. Three cubic boxes [9] are 300 mm, 450 mm, and 600 mm, respectively, and the thicknesses of plates are 2 mm, 3 mm, and 4 mm. Each cabin has three different explosive conditions, with a total of 9 working conditions. Figure 3 compares typical deformation characteristics of the box in different conditions. The numerical simulation results indicate that the plate surface of the box undergoes substantial plastic deformation under the effect of internal explosion, resulting in significant stress and flexing boundary plate. There are wrinkles that deform near the 1/2 position of boxes, which is relatively high in accordance with experimental results [9]. Figure 4 gives experimental data and numerical data of the deformed deflection of the side plate. The scattered points are near the perfect fitting line, and the average error of deformed deflection is 8.7%. It can be considered that the numerical simulation method used in this article has good forecast accuracy and can be used for follow-up research work.

Figure 3. Typical deformation characteristics of box cabin subjected to internal blast loading.

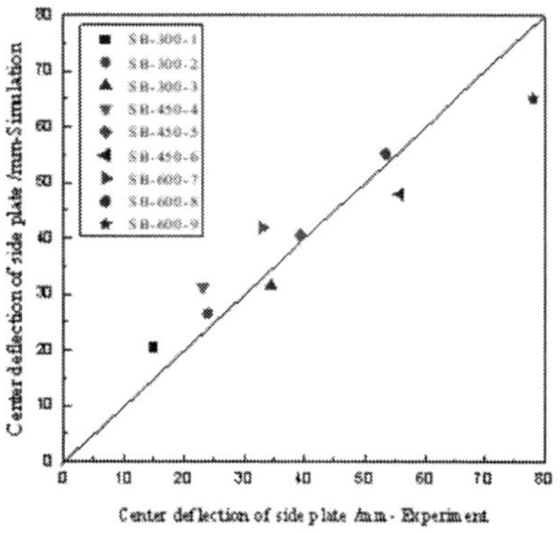

Figure 4. Deflection of the side plate of the box cabin.

Journal of Physics: Conference Series **2808** (2024) 012054 doi:10.1088/1742-6596/2808/1/012054

3. Results & discussion

3.1. The relationship between dimensionless damage number D_{in} and deflection thickness ratio δ/H.

The work conditions designed in this section only take into account the strength of the internal explosion load. TNT explosive equivalent is used as a variable. A total of 12 types of loading conditions are designed. Under these 12 types of working conditions, the destructive mode of a box-shaped cabin is large plastic deformation. Table 2 shows specific computational conditions and numerical results. Dimensionless damage numbers D_{in} ($D_{in} = \frac{E_e}{\sigma_0} \frac{V}{L^3} \frac{L}{H} = \frac{Q}{\sigma_0 L^2 H}$) corresponding to each working condition are listed in the table.

Table 2. Computational conditions and numerical results.

Serial Number	TNT Side Length (mm)	TNT Equivalent(g)	Dimensionless damage number D_{in}	Side plate deflection(mm)	Deflection Thickness Ratio δ/H
1	10	1.6	0.011	9.6	2.40
2	20	13.0	0.090	15.6	3.90
3	30	44.0	0.304	24.9	6.23
4	40	104.3	0.721	37.6	9.40
5	45	148.5	1.026	43.3	10.83
6	50	187.5	1.408	49.6	12.40
7	55	271.2	1.874	53.3	13.33
8	60	352.1	2.432	59.1	14.78
9	65	447.6	3.093	64.7	16.18
10	70	559.1	3.863	72.3	18.08
11	75	687.7	4.751	77.4	19.35
12	80	834.6	5.766	85.3	21.33

According to the simulation data in Table 3, D_{in} is horizontal coordinate, δ/H as vertical coordinates, draws the scatter figure as shown in Figure 5. When the explosion load is gradually increased, the deformation of the structure can be divided into two stages: bending deformation and membrane stretching deformation [11]. At $D_{in} > 1$, data scatter dot is linear distribution, indicating membrane stretching deformation; when $D_{in} < 1$, data scatter dot is non-linear distribution, and deformation of the structure is bending deformation. Fitting the data at $D_{in} > 1$, we can get a functional relationship shown in the formula (3):

$$\delta/H = 39.984 D_{in}^{0.3357}, \quad D_{in} < 1$$
$$\delta/H = 2.161 D_{in} + 9.246, \quad D_{in} > 1 \qquad (3)$$

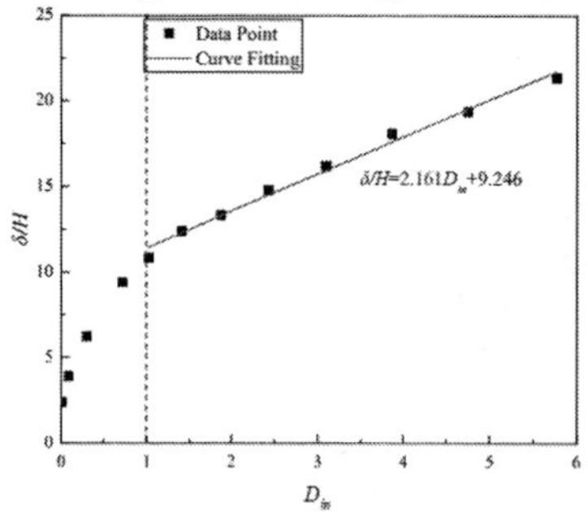

Figure 5. The relationship between dimensionless damage number D_{in} and deflection thickness ratio δ/H.

3.2. Dimensionless damage number D_{in}' considering explosive venting effect.

The following work is mainly concentrated on the case of membrane stretching deformation ($D_{in} > 1$). Under the internal explosion effect, shock waves inside the cabin can leak from cracks or openings. The size of explosive venting can affect the dissipation rate of shock wave energy and energy absorption of the structure. In fact, we think that the explosion venting effect is equivalent to the reduction of explosion energy, and the reduced total energy is Q. Taking the opening size of a box-shaped cabin as a variable, a single variable method is used to study the influence of explosive venting effect on the structure response of the cabin.

It is proposed to apply a dimensionless number $K_2 = \frac{S_{Hole}}{S}$ to indicate an explosive venting effect, where S_{Hole} is the area of the venting hole, and S is the area of the plate where the venting hole is located. The explosive venting is a round hole. The size of the aperture changes from 0-600mm. The most extreme case is to remove all the plates on one side of the cabin, so the range of K_2's changes is 0~1. The internal blast load with three TNT equivalent is 187.5 g TNT, 352.1 g TNT, and 559.1 g TNT. See Table 3 for the specific computational conditions and numerical results.

Table 3. Computational conditions and numerical results.

Serial Number	Venting Hole Diameter(mm)	Dimensionless Venting Hole Area K_2	187.5 g TNT Side Plate Deflection (mm)	187.5 g TNT δ/H	352.1 g TNT Side Plate Deflection (mm)	352.1 g TNT δ/H	559.1 g TNT Side Plate Deflection (mm)	559.1 g TNT δ/H
0	0	0	49.6	12.40	59.1	14.78	72.3	18.10
1	60	0.008	54.7	13.68	68.2	17.05	77.5	19.38
2	100	0.022	53.4	13.43	61.2	15.30	72.3	18.08
3	180	0.071	53.5	13.38	65.5	16.38	74.7	18.68
4	240	0.126	52.6	13.15	63.4	15.85	72.5	18.13
5	300	0.196	52.8	13.20	65.5	16.38	75.5	18.88
6	360	0.283	51.2	12.80	63.1	15.78	73.4	18.35
7	420	0.385	49.5	12.38	60.1	15.03	70.9	17.73
8	480	0.503	48.3	12.08	60.2	15.05	71.5	17.88
9	540	0.636	45.6	11.40	56.2	14.05	68.3	17.08
10	600	0.785	45.0	11.25	54.5	13.63	67.9	16.98
11	NO	1	40.3	10.08	45.9	11.48	61.2	15.30

*Note: "NO" means that all side plate areas (square, 600 mm × 600 mm) are regarded as opening.

Based on the simulation data presented in Table 3, dimensionless number K_2 serves as the horizontal axis, while the deflection thickness ratio of the side plate is plotted on the vertical axis. The resulting scatter plot is illustrated in Figure 6. Scattering points at three loads exhibit a linear distribution, indicating a linear correlation between K_2 and δ/H when the TNT Equivalent is fixed. That is, the total energy of the explosion after reduction shows a linear relationship with K_2. Consequently, we propose the following mathematical expression (6) for dimensionless damage number D_{in}' considering the explosion venting effect.

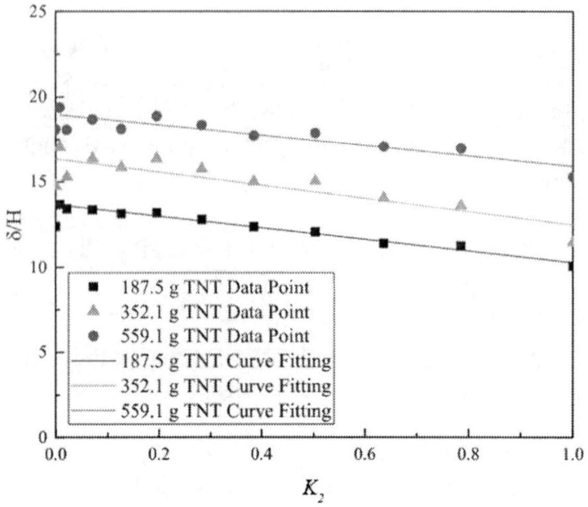

Figure 6. The deflection thickness ratio δ/H of the side plate of the box-shaped cabin is changed by the dimensionless number K_2.

$$D_{in}' = \frac{Q'}{\sigma_0 L^2 H} = \frac{E_e}{\sigma_0}\frac{V}{V_s}\frac{L}{H}(x_1 K_2 + y_1) = \frac{E_e K_1 L(x_1 K_2 + y_1)}{\sigma_0 H} \tag{4}$$

Bringing formula (4) into (3) yields:

$$\delta/H = 2.161\frac{E_e K_1 L}{\sigma_0 H}(x_1 K_2 + y_1) + 9.246 \tag{5}$$

In formula, $\frac{V}{V_s} = K_1 = a^3/L^3$, represents the strength of the implosion effect, a is the side length of cubic TNT, L is the side length of cubic box-shaped cabin, $K_2 = \frac{S_{Hole}}{S}$. When $K_2=0$, the venting effect is not considered, $D_{in}' = D_{in}$, so $y_1=1$. x_1 is a variable affected by K_1, and the relationship between x_1 and K_1 is obtained according to data fitting as follows:

$$x_1 = -|-734.85K_1 + 1.5133| \tag{6}$$

4. Conclusions

Based on dynamic software ANSYS/LS-DYNA, this article has established a box-shaped cabin dynamic response numerical model under internal explosion. It uses existing literature to verify the rationality and effectiveness of the numerical model. Through analysis of numerical calculation, the relationship is constructed between dimensionless damage number D_{in} in this article and deflection thickness ratio δ/H. On this basis, considering the explosion venting effect on explosion response characteristics, dimensionless damage number D_{in}' is proposed. The dimensionless damage number obtained in this article can be modified by experiments to predict damage to internal explosion problems.

References

[1] H. L. HOU, X.ZHU, W.LI, et al. (2010) Experimental studies on characteristics of blast loading when exploded inside ship cabin [J]. Journal of Ship Mechanics, 14(8): 901-907. DOI: 10.3969/j.issn.1007-7294.2010.08.011.

[2] K. Huang, S. F. Liang, L. Z. Wang, Q. R. Wang. (2021) The impact of venting on TNT explosion load in the protective cabin. Protective Engineering, 43(01): 16-21.

[3] CHEN P Y, HOU H L, JIN J, et al. (2020) Simplified calculation model for explosion loading in ship cabin [J]. Ship Science and Technology, 42(17): 22-29. DOI: 10.3404/j.issn.1672-7649.2020.09.005.

[4] R. Shi, X. B. Li, H. J. Wang. (2021) Implosion Damage Effect of Typical Cabin Structure. Marine

Technology, 49(03): 17-22+68.
[5] G.N. Nurick, J.B. Martin. (1989) Deformation of thin plates subjected to impulsive loading—a review Part II: Experimental studies. International Journal of Impact Engineering, 8(2): 171-186.
[6] W., Johnson, Author, Stephen, L., Rice, Reviewer. (1974) Impact Strength of Materials. Journal of Applied Mechanics, 41(2): 331.
[7] N. Jacob, S. Chung Kim Yuen, G.N. Nurick, D. Bonorchis, S.A. Desai, D. Tait. (2004) Scaling aspects of quadrangular plates subjected to localized blast loads—experiments and predictions. International Journal of Impact Engineering, 30(8-9): 1179-1208.
[8] Yao S J. (2016) Investigations on the scaling method and damage features of box-shaped structures under internal blast loading [D]. Changsha: National University of Defense Technology.
[9] S.J. Yao, D. Zhang, F.Y. Lu, X.G. Chen, P.D. Zhao. (2017) A combined experimental and numerical investigation on the scaling laws for steel box structures subjected to internal blast loading. International Journal of Impact Engineering, 102: 36-46.
[10] G.C. Chen, Y.S. Cheng, P. Zhang, J. Liu, C.H. Chen, S.P. Cai. (2020) Design and modeling of auxetic double arrowhead honeycomb core sandwich panels for performance improvement under air blast loading. Journal of Sandwich Structures & Materials, 23(8): 3574-3605.
[11] C. Liu, S.L. Zhang. (2021) Influence of Vertical Surroundings on the Dynamic Response of Fixed Square Plate Subjected to Blast Loading. Journal of Ship Mechanics, 25(10): 1390-1403.

Preparation of butyl bonded poly (styrene-divinylbenzene) microspheres by Grignard reaction

Meijuan Jiang[1]*, Fuqiang Jin[2]*, Hao Zhang[1], Xiaolu Yi[2] and Xin Zhao[1]

[1]Shandong Institute of Non-metallic Materials, Jinan, Shandong, 250031, China
[2]Energy Research Institute of Qilu University of Technology (Shandong Academy of Sciences), Jinan, Shandong, 250014, China

*Corresponding author's e-mail: 15866780735@163.com (Jiang), fuqiangjin@126.com (Jin)

Abstract. The alkylated poly(styrene-divinylbenzene) (PSD) microspheres are attractive chromatographic column packings for high-performance liquid chromatography. In this paper, the butyl bonded PSD microspheres were prepared by the reaction of brominated PSD microspheres with n-butylmagnesium bromide. Fourier transform infrared (FTIR) spectroscopy and solid-state ^{13}C nuclear magnetic resonance (NMR) spectroscopy were used to analyze the polymers. FTIR spectra analysis could establish the occurrence of bromination of PSD microspheres and the reaction of brominated PSD microspheres with n-butylmagnesium bromide. And ^{13}C NMR results illustrated that the Grignard reaction had the largest conversion at the 1-bromobutane adding amount of 0.05 mol according to the analysis of the peak height ratios. The prepared microspheres are expected to serve as chromatographic packing.

1. Introduction

The poly (styrene-divinylbenzene) (PSD) copolymers are widely used as polymeric sorbents since they have a hydrophobic surface, which overcomes many of the limitations of bonded silica, particularly those related to limited pH stability or the presence of silanol groups. They have been chemically modified with hydroxymethyl, acetyl, alkyl, benzoyl, o-carboxy benzoyl, and trimethyl ammonium groups to improve their performance [1, 2]. Among them, the monodisperse alkylated PSD microspheres are attractive packing materials for high-performance liquid chromatography due to their good chemical stability over a wide pH range and high column efficiency [3, 4]. Typically, their synthesis needs two steps, i.e. the preparation of monodisperse PSD microspheres and the modification of PSD with alkyl. Friedel-Crafts reaction is the most common method to modify the polymer beads [5, 6], but the structures of alkyl groups are not unitary. Indeed, the isomerization of primary halides to secondary or tertiary halides occurs in the presence of aluminum chloride under the Friedel-Crafs reaction conditions [7].

Grignard reagents have been frequently used in organic synthesis technology. They have a high reactivity and good stereoselectivity, i.e., the isomerization almost does not occur. Therefore, it is favorite and convenient to utilize Grignard reagents not only in solution-phase reactions but also in polymerization reactions and surface grafting technology [8]. In this paper, the introduction of butyl to the PSD microspheres, to prepare butyl bonded poly(styrene-divinylbenzene) microspheres (BuPSD), was carried out by the reaction of brominated PSD (BrPSD) microspheres with n-butylmagnesium bromide.

2. Materials and methods

2.1. Materials

Styrene (Fuchen (Tianjin) Chemical Reagent Co., Ltd) was used after vacuum distillation. 2, 2-Azobisisobutyronitrile (AIBN, Shanghai NO.4 Reagent & H.V. Chemical Co., Ltd) and benzoylperoxide (BPO, Tianjin Yongda Chemical Reagent Co., Ltd) were recrystallized using ethanol and chloroform-methanol respectively. Divinylbenzene (containing ca. 45% of ethylvinylbenzene, Shenzhen Aituo Chemical Co., Ltd), diethyl ether and 1-Bromobutane were respectively distilled.

2.2. Preparation

The core polystyrene microspheres were prepared by dispersion polymerization according to the procedure we used before [9]. The preparation conditions: Styrene 10.00 g; Ethanol 36.0 ml; Water 4.0 ml; Polyvinylpyrrolidone 1.00 g; AIBN 0.10 g; polymerization at 70°C for 24 h.

Seed swelling polymerization method was used to prepare PSD microspheres [9, 10]. 3.40 g dibutyl phthalate, 1.50 g polystyrene microspheres, and 300 ml of 0.25 wt.% sodium dodecyl sulfate water solution were added in a 500 ml flask. Then the mixture was stirred slowly for 24 h and 15.00 g styrene and 3.00 g divinylbenzene were added. Then 60 ml 5 wt.% poly (vinyl alcohol) water solution was added and the mixture was stirred for 24 h at 35°C and 12 h at 70°C. The resulting samples were washed with water and ethanol, dried in a vacuum, and refluxed for 24 h with chloroform. The samples were dried in a vacuum.

Bromination of PSD microspheres: In a four-mouth flask fitted with a reflux condenser, a mechanical stirrer, a constant pressure funnel, and a thermometer, 20.00 g PSD microspheres and 100 ml 1, 2-dichloroethane were mixed and stirred for 4 h. 1.00 g P_2O_5 was added and stirred for 1 h at 40°C, after which 0.10 g aluminum powder and 1.00 g aluminum chloride were added. Then 28.70 g bromine was added dropwise at 25°C. The reaction was carried out at 35°C for 10 h. The obtained samples were washed with 100 ml aqueous 10 wt.% sodium hydrogen sulfite, with hot water to neutral pH, and dried at 80°C for 24 h. Thus, the BrPSD microsphere sample (containing 0.00468 mol Br element per gram of the microsphere sample, calculated by gravimetric analysis) was obtained.

Table 1. Preparation conditions of BuPSD microsphere samples.

Samples	BrPSD		1-Bromobutane		Magnesium turnings	
	in g	in mol Br [a]	in ml	in mol	in g	in mol
BuPSD1	5.34	0.025	1.35	0.0125	0.34	0.0139
BuPSD2	5.34	0.025	2.7	0.025	0.67	0.0275
BuPSD3	5.34	0.025	5.4	0.05	1.34	0.055
BuPSD4	5.34	0.025	10.8	0.10	2.68	0.11

[a] mol of Br element contained in the added BrPSD sample.

The Grignard reaction of the brominated PSD microspheres with n-butylmagnesium bromide: 15 ml anhydrous diethyl ether, 0.04 g iodine, and a certain amount of magnesium metal turnings were added in a four-mouth flask fitted with a reflux condenser, a mechanical stirrer, a constant pressure funnel, and a thermometer. 15 ml of anhydrous diethyl ether and a certain amount of anhydrous 1-bromobutane were mixed and added to the constant pressure funnel. Then the mixed solution was increased to 25°C to initiate the reaction, after which 25 ml anhydrous diethyl ether was added with stirring. The residual mixed solution of 1-bromobutane and diethyl ether was added dropwise. Ten minutes later, 5.34 g brominated PSD microspheres were added and the reactants were stirred for 12 h at 33°C. The resulting microspheres were washed with 100 ml aqueous 10 wt.% hydrochloric acid,

Journal of Physics: Conference Series **2808** (2024) 012055 doi:10.1088/1742-6596/2808/1/012055

and then washed 2-4 times with ethanol and hot water, followed by drying at 80°C for 24 h. Four BuPSD microsphere samples with different amounts of 1-bromobutane and magnesium turnings were synthesized and shown in Table 1.

2.3. Characterization

Fourier transform infrared (FTIR) spectroscopy was used by a Bruker FTIR Tensor 27 Spectrometer. Wave number: 600~2000 cm^{-1}; resolution: 2 cm^{-1}. Solid-state ^{13}C nuclear magnetic resonance (NMR) spectra were measured with a Varian Infinityplus 300 spectrometer (magnetic field of 7.1 T). The experiments employed a Varian Chemagnetic 4 mm resonance probe. The cross-polarization contact time was 1.0 ms. And the recycle delay was 5.0 s. The spinning rate was 10.0 kHz. The chemical shift was based on the resonance of the methyl group of hexamethylbenzene as an external reference.

3. Results & discussion

Figure 1 shows the FTIR spectra of PSD, BrPSD, and BuPSD microsphere samples with different adding amounts of 1-bromobutane, respectively. The spectrum of PSD (Figure 1a) showed three peaks at 1600, 1493, and 1450 cm^{-1}, belonging to C=C bonds of aromatic rings. The bands at 699 and 754 cm^{-1} were associated with the mono-substituted benzene ring. The bands at 2920 and 2850 cm^{-1} belonged to CH_2 stretching vibration, and 2870 cm^{-1} could be attributed to CH stretch vibration in the backbone of PSD, respectively. For BrPSD (Figure 1b), the spectrum showed a new absorption band at 818 cm^{-1} attributed to the para-substituted benzene ring, and the bands of the mono-substituted one (699 and 754 cm^{-1}) disappeared. The band at 611 cm^{-1} could be ascribed to C–Br stretching vibration. The peaks at 1484, 1450, and 1406 cm^{-1} belonged to the benzene skeleton vibration, and the bands 1074 and 1009 cm^{-1} belonged to the bending vibration of =C–H in the benzene ring. For BuPSD samples, the intensity of the band at 611 cm^{-1} was reduced in Figure 1c, and this band (611 cm^{-1}) disappeared completely in Figures 1d-f, indicating a high yield of transformation in this step. The band at 2920 cm^{-1} gradually split into two peaks for BuPSD, likely due to the emergence of butyl in the samples. These results could establish the occurrence of the proposed transformation in each step.

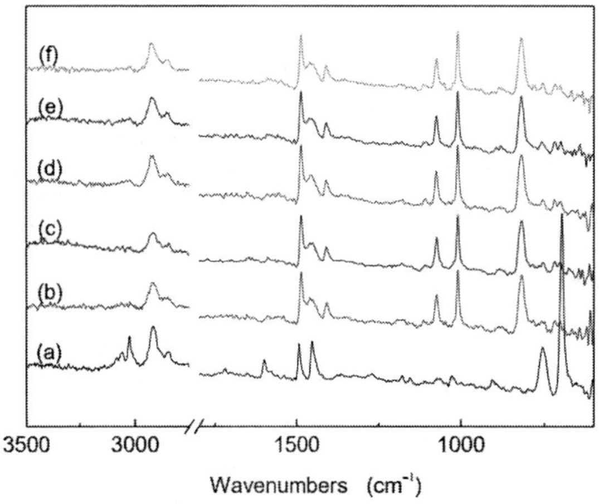

Figure 1. FTIR spectra of PSD (a), BrPSD (b), and BuPSD microsphere samples with 1-bromobutane adding amounts of 0.0125 mol (c), 0.025 mol (d), 0.05 mol (e) and 0.10 mol (f).

Figure 2 shows the ^{13}C NMR spectra of PSD, BrPSD, and BuPSD microsphere samples with different amounts of 1-bromobutane, respectively. For the PSD sample (Figure 2a), the peaks at 146.38 and 128.25 ppm were attributed to C1 and C2–6 in the benzene ring, respectively. The peaks at

Journal of Physics: Conference Series **2808** (2024) 012055 doi:10.1088/1742-6596/2808/1/012055

41.28, 30.54, and 15.96 ppm were ascribed to C7, C8, and $-CH_3$ in the backbone of PSD, respectively. For the BrPSD sample (Figure 2b), the spectrum showed an additional peak at 122.29 ppm, attributed to C4' (C–Br), which indicated that the bromination had taken place. For the four BuPSD samples (Figures 2c-2f), the peak of C1 shifted to a higher field, likely due to the para-position changes of the benzene ring. The intensity of the peaks of C8–11 and C12 increased to some degree, indicating the occurrence of Grignard substitution.

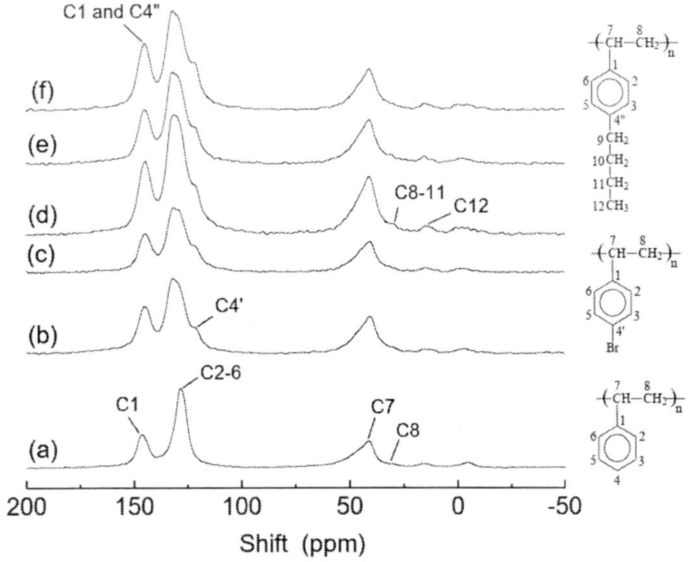

Figure 2. ^{13}C NMR spectra of PSD (a), BrPSD (b), and BuPSD microsphere samples with 1-bromobutane adding amounts of 0.0125 mol (c), 0.025 mol (d), 0.05 mol (e) and 0.10 mol (f).

Additionally for all the BuPSD samples, the intensity of the peak at 122.29 ppm had no significant change, suggesting that the samples still contained much residual Br element, which was not shown in the FTIR spectra mentioned above. Considering that ^{13}C NMR spectroscopy is a non-destructive technique for investigating the microenvironment of materials in solid state and FTIR is a surface analytical technique, we could attribute the residual Br element, shown in the ^{13}C NMR spectra of BuPSD samples, to the unreacted one in the particle interior.

Theoretically, the groups of C9–11 and C12 would emerge, and the amounts of C2–6 and C7 had no significant changes after the Grignard substitution. Thus the ^{13}C NMR peak height ratio of C8–11/C2–6, C8–11/C7, C12/C2–6, and C12/C7 could suggest the reaction conversion. Figure 3 shows the peak height ratio of the BuPSD samples with different amounts of 1-bromobutane. As can be seen, the peak height ratio of C8–11/C2–6, C8–11/C7, C12/C2–6, and C12/C7 all increased with increasing the amount of 1-bromobutane from 0 mol to 0.05 mol and decreased at 0.10 mol. These results indicated that the Grignard reaction had the largest conversion at the 1-bromobutane adding amount of 0.05 mol, and this conclusion can also be drawn from the product yield (Figure 4) of the BuPSD samples with different adding amounts of 1-bromobutane. The product yield was minimum and the corresponding conversion rate of the BuPSD samples was maximum when the 1-bromobutane adding amount was 0.05 mol. It was likely because the excessive concentration of alkyl halide led to a significant side reaction, i.e., the coupling reaction between butyl magnesium bromide and 1-bromobutane. These microspheres are expected to serve as chromatographic packing.

Journal of Physics: Conference Series **2808** (2024) 012055 doi:10.1088/1742-6596/2808/1/012055

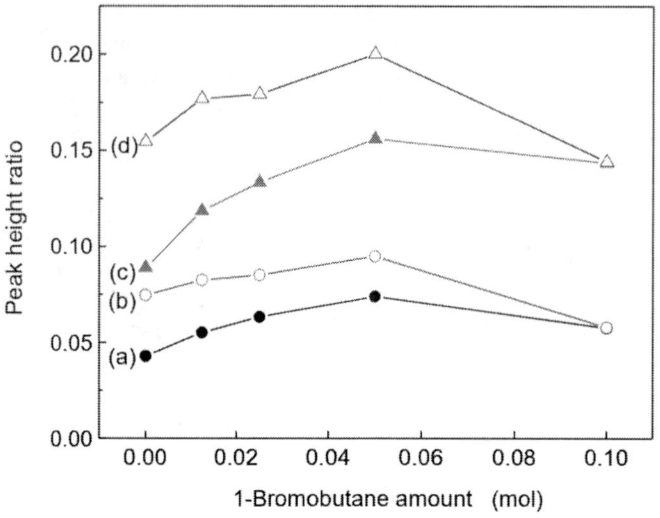

Figure 3. ^{13}C NMR peak height ratio of the BuPSD samples with different adding amounts of 1-bromobutane. (a) C12/C2–6, (b) C8–11/C2–6, (c) C12/C7, (d) C8–11/C7.

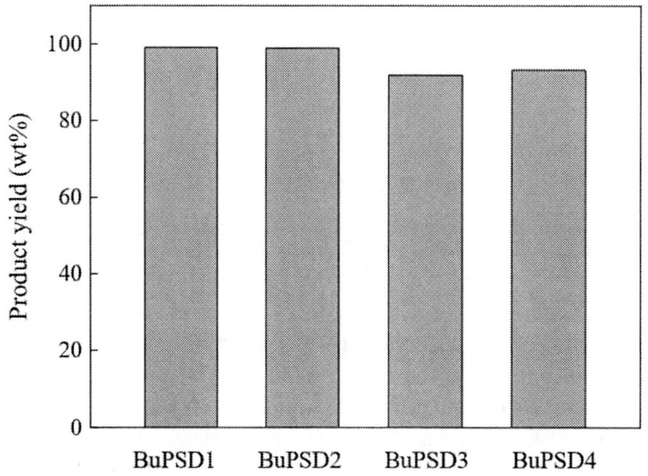

Figure 4. The product yield of the BuPSD samples with different added amounts of 1-bromobutane.

4. Conclusions

The butyl bonded PSD microspheres were prepared by the reaction of brominated PSD microspheres with n-butylmagnesium bromide. FTIR spectra and ^{13}C NMR analysis results indicated the occurrence of the reaction between brominated PSD microspheres and n-butylmagnesium bromide. These microspheres are expected to serve as chromatographic packings for high-performance liquid chromatography. The research on the application of these microspheres will be carried out in the future.

References

[1] Zheng, Y., Wang, Q., et al. (2019) Experimental study on mass transport mechanism in poly (styrene-co-divinylbenzene) microspheres with hierarchical pore structure. Microchemical Journal, 139: 183-192.
[2] Ying, A., Li, M., et al. (2023) Sandwich like poly (ionic liquid)s functionalized microspheres: Efficient interfacial catalysts for preparation of biodiesel. Chemical Engineering Journal, 473: 145361.
[3] Li, C., Che, W., et al. (2023) Hypercrosslinked microporous polystyrene: from synthesis to properties to applications. Materials Today Chemistry, 29: 101392
[4] Liu, X., Wang, Y., Cong, H., et al. (2021) A review of the design of packing materials for ion chromatography. Journal of Chromatography A, 1653: 462313
[5] Tang, R., Zhou, J., et al. (2022) Facile "one-pot" preparation of phosphonate functional polythiophene based microsphere via Friedel-Crafts reaction for selective enrichment of phosphopeptides from milk. Analytica Chimica Acta, 1190: 339268.
[6] Zhou, S., Lin, Q., Ren, Z., et al. (2023) Synthesis of phenyl-based hyper-crosslinked porous organic polymers via Friedel-Crafts reaction for efficient organic dye adsorption. Microporous and Mesoporous Materials, 362: 112765.
[7] Masoumi, H., Ghaemi, A., et al. (2021) Exploiting the performance of hyper-cross-linked polystyrene for removal of multi-component heavy metal ions from wastewaters. Journal of Environmental Chemical Engineering, 9(4): 105724.
[8] Franzén, R.G. (2000) Utilization of grignard reagents in solid-phase synthesis: A Review of the literature. Tetrahedron, 56: 685–691.
[9] Jin, F., Li, Y., et al. (2011) Poly (styrene-divinylbenzene) microspheres prepared by seed swelling polymerization in the presence of tung oil methyl esters. Mater. Sci. Forum, 675-677: 341-344.
[10] Meng, X., Qiu, D. (2020) Fabrication of monodisperse asymmetric polystyrene particles by crosslinking regulation in seeded emulsion polymerization. Polymer, 203: 122799.

Journal of Physics: Conference Series **2808** (2024) 012056 doi:10.1088/1742-6596/2808/1/012056

Static response of MEMS double layered simply supported beams containing quasicrystals

Yunzhi Huang[1*], Penghui Wang[1] and Yuepeng Liu[1]

[1]Beijing Institute of Aerospace Testing Technology, Beijing, 100074, China

[*]Corresponding author's e-mail: huangyz199410@163.com

Abstract: The static response of Micro-electromechanical system (MEMS) double-layer simply-supported beams made of quasicrystalline (QC) and crystal (C) is studied considering the influence of nonlocal scale effects. In light of the theory of nonlocal elasticity for QCs, the closed responses of phonon, phason, and electric fields are derived using the state vector formula and propagator matrix method. Calculation results indicate that theoretical responses are consistent with the simulation ones via the finite element. Among the three micro-distributed forces, the magnitude of their effects is in order of electric field, Casimir, and van der Waals micro-forces. The rising van der Waals and Casimir micro-forces elevates the displacement with the stress of beams. The displacement decreases slowly when nonlocal scale parameters grow. The phonon displacement of microbeams with C/QC stacking order is larger than the corresponding value of QC/C stacking order. This paper's results support the design and evaluation of MEMS structures containing QCs.

1. Introduction

Since the beginning of this century, significant advancements have been made in the preparation and analysis of micro-nano quasicrystalline (QC) alloys [1-5]. Research has revealed that micro-nano QCs exhibit numerous mechanical properties that surpass those of conventional crystal alloys [2]. Laboratory synthesis of micro-nano QCs has enabled their application in electronic devices (actuators, piezomagnetic sensors, as well as other semiconductor apparatuses [6-9]. Nevertheless, there is currently limited research on the theory regarding QC laminated beams and their static response. Previous studies by researchers [10] have focused on control formulas with fundamental responses for one-dimensional as well as two-dimensional nonlocal QCs that serve as a base for investigating microbeams of QCs. Theoretical means for the state vector formula can be utilized to analyze multi-layer constructions for various fields with operating conditions. By applying appropriate functions for displacement with stress, Sheng et al. [11-16] derived three-dimensional theoretical responses for shells and plates with different thicknesses as well as structural boundaries. Gao et al. [17] investigated the vibration of simply-supported piezoelectric plates in free conditions via the propagator matrix approach as well as the three-dimensional displacement approach.

In recent years, researchers have conducted investigations into the static response characteristics of Micro-electromechanical system (MEMS) QC laminated microbeams. Scholars have employed the state vector method to solve MEMS multi-layer QC plates [18] and obtained numerical results for boundary value problems. Here we utilize this method [19] to address the static response of MEMS QC beams while considering nonlocal effects. In nonlocal elasticity theory, material-specific nonlocal scale parameters are incorporated into the constitutive equation as a means to introduce additional material characteristics. Subsequently, a model for a QC laminated beam is created under the theory of nonlocal elastic QCs, and the state equations governing these laminated microbeams are processed through the state vector formula. Finally, through finite element simulation and analytical examples, we reveal both the static response and electromechanical characteristics exhibited by QC laminated beams. The influence of scale parameters for nonlocal effect, Casimir micro-force, van der Waals micro-force, electrostatic field micro-force, and stacking orders on quantities of phonon, phason, and electric fields in laminated beams are discussed.

2. Solving process

In the space coordinate system (x, y, z), consider establishing an elastic double-layer QC microbeam model with unit width b, length L, as well as thickness h_{total} (see Figure 1) based on nonlocal elasticity theory, where only contains variables of x and z. The upper and lower layers of the double-layer microbeam are respectively a QC layer (thickness: h_{QC}) as well as a crystal one (thickness: h_C). The QC atoms on the top surface exhibit quasi-periodic rotational arrangement in the z-direction. Distances between the surface of the QC beam and the substrate are set to the primary gap, and an external voltage load is applied to the surface of the beam. Setting variable u_x, u_z, w_z, D_x, D_z, σ_{xx}, σ_{zz}, σ_{xz}, H_{zz}, H_{xz}, as well as ϕ is unconnected with y in the coordinate system.

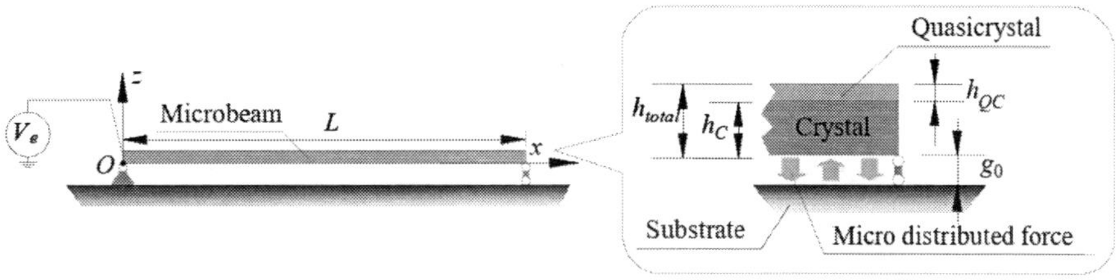

Figure 1. Diagram of piezoelectric QC microbeam with double-layer, where the base point of the coordinate system (x, y, z) is located at the lower left corner of the beam.

The state vector method has simple steps and is computationally efficient [11]. This method is utilized in the following to deal with the equation of QC beams. By choosing u_x, u_z, w_z, σ_{zz}, σ_{xz}, H_{zz}, ϕ, as well as D_z to form the primary state vector (X), choosing H_{xz}, σ_{xx}, as well as D_x forms the additional state vector (Y). We combine basic equations of QC [10] to obtain:

$$\partial \mathbf{A} / \partial z = \mathbf{XA}, \qquad \mathbf{B} = \mathbf{YA} \tag{1}$$

where **X** and **Y** are operator matrices. In addition, by setting the superscript T as transpose, $\mathbf{A} = [\sigma_{zz}\ H_{zz}\ u_x\ D_z\ u_z\ w_z\ \sigma_{xz}\ \phi]^T$ as well as $\mathbf{B} = [\sigma_{xx}\ D_x\ H_{xz}]^T$ in equation (1).

The surface of the beam is subjected to a load along z-direction, and the two ends of the beam are set as simply supported boundaries, where the conditions for this boundary are [9]:

$$\sigma_{xx} = 0,\ \ D_x = 0,\ \ u_z = 0,\ \ w_z = 0\ \ (x = 0 \text{ or } L) \tag{2}$$

Substituting Equation (2) into Equation (1) yields

$$d\bar{\mathbf{A}}/dz = \mathbf{M}\bar{\mathbf{A}}, \qquad \bar{\mathbf{B}} = \mathbf{N}\bar{\mathbf{A}} \tag{3}$$

where $\bar{\mathbf{A}} = [\bar{\sigma}_{zzm}\ \ \bar{H}_{zzm}\ \ \bar{u}_{xm}\ \ \bar{D}_{zm}\ \ \bar{u}_{zm}\ \ \bar{w}_{zm}\ \ \bar{\sigma}_{xzm}\ \ \bar{\phi}_m]^T$ and $\bar{\mathbf{B}} = [\bar{\sigma}_{xxm}\ \ \bar{D}_{xm}\ \ \bar{H}_{xzm}]^T$ are state vectors, and **M** and **N** are also operator matrices.

The solution of the primary formula in the j-th layer in Equation (3) becomes:

$$\bar{\mathbf{A}}_j(z) = \mathbf{C}_j \bar{\mathbf{A}}_j(0) \tag{4}$$

where $\mathbf{C}_j(z)=\exp[\mathbf{E}_j z]=\mathbf{I}+\mathbf{E}_j z+0.5\mathbf{E}_j^2 z^2+\ldots$, and $\mathbf{I}$ is the identity matrix. This paper assumes that the state vector is continuous on two interfaces between layers with $\mathbf{D}$ matric and can obtain:

$$\overline{\mathbf{A}}_p(h_p)=\mathbf{D}\overline{\mathbf{A}}_1(0) \tag{5}$$

To acquire the unknown quantities in $\mathbf{A}(z)$ of different thickness, this paper assumes that z is $z_j \leq z \leq z_{j+1}$ for the j-th layer and solves using Equations (4) and (5).

3. Numerical examples

We analyze the static response of microbeams composed of QC layers as well as layers of crystalline silicon (C) under external voltage influence [1, 10], that is, the electrical and mechanical response characteristics of the structure under electrical or force loads. Firstly, set the size of the microbeam to $L \times h_{total} = 400 \times 10^{-6}$m $\times$ 2.5 $\times 10^{-6}$m, g_0 = 1.5 $\times 10^{-6}$m. Unless otherwise specified, the stacking order of the double-layer microbeams in this example is QC/C. The thickness ratio for two layers of the microbeam is h_{QC}/h_C=1/4. This section uses material data from one-dimensional orthogonal QCs. The silicon layer is a crystal. The equation for the crystal layer can be gained by determining the phason field coefficients of the relevant equations in the QC layer to small values. We set boundary conditions that are only related to the phonon field [9, 10]. In the following content, we will discuss in detail the effects of external forces, nonlocal scale parameters, and double-layer stacking order on the physical quantities of microbeams.

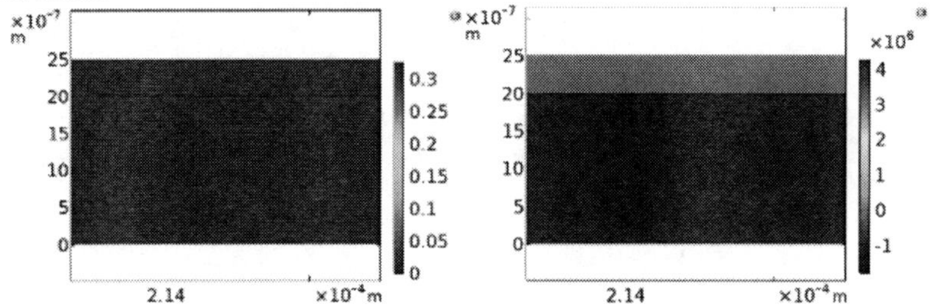

Figure 2. The finite element results of the model: (a) $\bar{u}_z$; (b) $\bar{H}_{zz}$.

The effectiveness and accuracy of the current model are validated by calculating the response of the double-layer beam with the zero-for-scale parameter. The outcomes of the present model agree with solutions from the classical theory of elasticity [13]. This section first compares the analytical results with the literature results [13] and finite element results and then uses an analytical model with phason fields to analyze the characteristics of QC microbeams. Microbeam loading with sinusoidal distributed potential $\phi = 1 \times 10^{-2}$ V at the surface of microbeam with $x = 0.5\,L$ ($z = h_{total} = 2.5 \times 10^{-6}$m). The boundary is simply supported at both ends of the microbeam. This article uses the second method to solve displacement in 10-30 steps. For the convenience of comparison, the following numerical results have been dimensionless, and a horizontal line has been added above the symbol for dimensionless numerical results. Figure 2 reveals the vertical displacement and phason stress cloud map of the model via the finite element method in Comsol software, where the total number of elements in the model reaches 45218. Figure 2 (b) shows the phenomenon of stress stratification, which is consistent with analytical results. Analytical results also meet the boundary conditions, and the trend of physical quantity changes in the analytical solution is consistent with the results in references [10, 13]. In addition, the above results are consistent with the qualitative comparison of stress and displacement curves in relevant experiments [10, 13], indicating the effectiveness as well as the correctness of the results.

A sinusoidally distributed potential is located at the surface of the microbeam ($z = h_{total}$ = 10 μm). Figure 3 shows the physical quantity results under different distributed forces (consisting of electrostatic field micro-force f_{ele}, van der Waals micro-force f_{vdW}, as well as Casimir micro-force f_{Cas}) when an external voltage is applied. The calculation of the model also considers different nonlocal scale

Journal of Physics: Conference Series **2808** (2024) 012056 doi:10.1088/1742-6596/2808/1/012056

parameters. This figure illustrates the size effect of nonlocal elasticity and the effect of micro-distributed forces on physical quantities of the elastic field and electric field of microbeams with a stacking order of QC/C.

Figure 3 (a) shows the $\bar{u}_z$ changes under different forces. The value $\bar{u}_z$ under the f_{ele} effect accompanied by f_{Cas} is smaller than the displacement results under the f_{ele} effect accompanied by f_{vdW}, and greater than the result considering only f_{ele}. This indicates that f_{vdW} and f_{Cas} promote attraction between the surfaces of the microbeam as well as the substrate. In Figures 3 (b) and (c), only the sum of two response values $\bar{w}_z$ and $\bar{H}_{xz}$ of f_{ele} is considered to be greater than the response values under f_{ele} with f_{vdW} and f_{Cas}. This indicates that f_{ele} plays a crucial role in the phason field among these micro-forces, and f_{Cas} has a more prominent impact on the field of phason than f_{vdW}. The displacement of the phason field $\bar{w}_z$ reflects the misalignment of QC atoms at the microscopic level, suggesting that the internal microstructure of QCs changes under the influence of external forces. The size effect of nonlocal elasticity and the effect of micro-distributed forces on horizontal component $\bar{D}_x$ is shown in Figure 3 (d), where the influence of f_{vdW} on $\bar{D}_x$ is slightly greater than that of f_{Cas}.

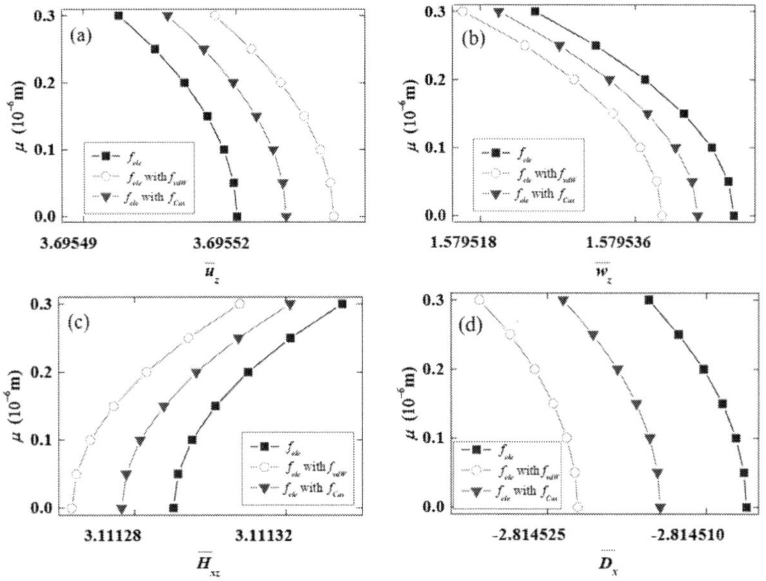

Figure 3. Dimensionless quantities for the phason, phonon, as well as electric fields of the microbeam: (a) $\bar{u}_z$ at bottom surfaces; (b) $\bar{w}_z$ at top surfaces; (c) $\bar{H}_{xz}$ at top surfaces; (d) $\bar{D}_x$ at bottom surfaces of microbeam.

Figure 4 shows the nondimensional displacement maps of microbeams along the z-direction of QC/C with different parameters of nonlocal scale under external voltage. This figure shows the action of parameters μ on the phonon and phason displacement along the z-direction of the microbeam. The scale range of nonlocal parameters refers to the relevant settings in piezoelectric QCs and crystal materials [9, 10]. When μ increases, $\bar{u}_z$ [Figure 4 (a)] and $\bar{w}_z$ [Figure 4 (b)] slowly decrease. In response to QC layered structures, the influence of nonlocal effects and micro-distributed forces cannot be ignored.

Figure 5 shows the nondimensional displacement maps of different parameters along the z-direction of C/QC microbeams under external voltage. This graph reflects the impact of parameters μ on the displacement $\bar{u}_z$ [Figure 5 (a)] and $\bar{w}_z$ [Figure 5 (b)] of microbeams stacked in C/QC order. When μ is increasing, the absolute value of $\bar{w}_z$ and $\bar{u}_x$ decreases, consistent with the trend in Figure 4. Compared with the stacking order of QC/C in Figure 4, the phonon field displacement of the microbeam with the

C/QC stacking order in Figure 5 is larger, and the phason field displacement is also different. This result corresponds to the corresponding conclusion in [3], providing a reference for the design of laminated beams with QC coatings.

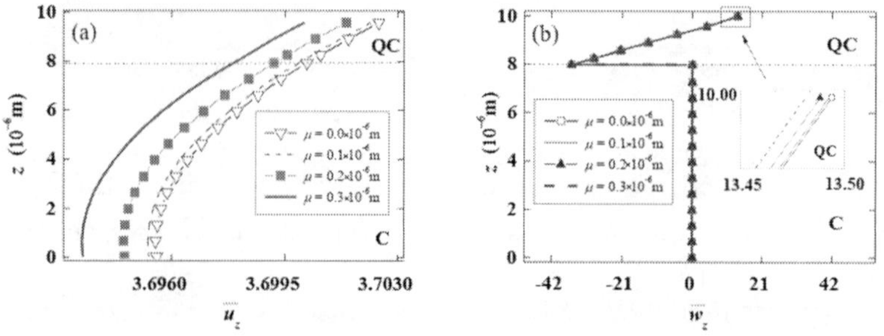

Figure 4. Dimensionless displacement diagram along the thickness direction of the microbeam: (a) $\overline{u}_z$ and (b) $\overline{w}_z$ in the microbeam, where an enlarged view of $\overline{w}_z$ on the peak surface is unveiled at the small image. The layers of quasicrystal and the Silicon are respectively represented by QC and C, and the boundary between layers is denoted through the black dashed line.

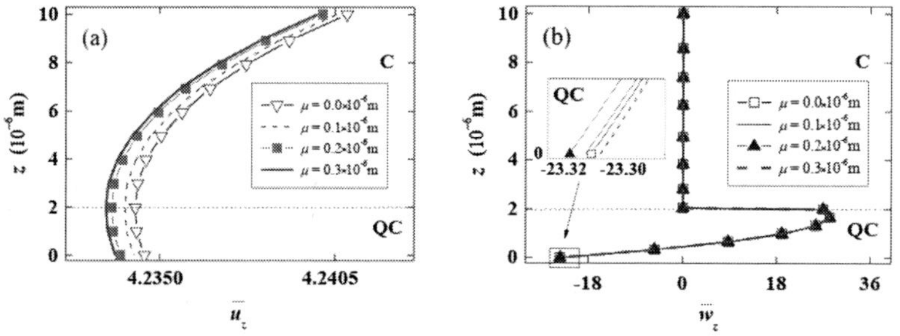

Figure 5. Dimensionless displacement diagram along the thickness direction of the microbeam: (a) $\overline{u}_z$ and (b) $\overline{w}_z$ in the microbeam, where an enlarged view of $\overline{w}_z$ on the surface is illustrated in the thumbnail image.

4. Conclusion

This article investigates the static response of MEMS QC double-layer microbeams. Micro-distributed forces (electrostatic field, Casimir, and van der Waals micro-forces) are present between the beam and the substrates. We have established an elastic deformation model for laminated microbeams composed of QCs and monocrystalline silicon under external voltage and solved the control equations of the model using the secant method.

The numerical results indicate that the results of the established model correspond to the literature. The effectiveness of the theoretical solution was verified by checking the theoretical results and corresponding finite element ones in Comsol software. Nonlinear micro-distributed forces such as electrostatic field, Casimir, and van der Waals micro-forces affect the mechanical behavior of QC laminated microbeams. The van der Waals and Casimir micro-forces hoist the phonon displacement and stress but decrease the phason displacement as well as stress. The presence of nonlocal scale parameters reduces the displacement in the van der Waals micro-force, Casimir micro-force, phonon field, and phason field. The QC laminated microbeam model created here provides a theoretical reference for the design and optimization of other MEMS QC structures.

Acknowledgments
Authors wish to acknowledge assistance and encouragement from colleagues.

Journal of Physics: Conference Series **2808** (2024) 012056 doi:10.1088/1742-6596/2808/1/012056

References

[1] Wang, S., Zhang, S., Xu, Y. (2023) Hot Deformation Behaviour and Deformation Microstructure of an Al-Zn-Ce Damping Alloy. Mater. Trans., 1(9): 64.

[2] Gurusamy, P., Muthuraman, V., Lokeshkumar, J., Hari Krishna Raj, S. (2022) Fabrication and mechanical behaviour of Al-6061/Al2O3 metal matrix composite fabricated through powder metallurgy technique. Int. J. Veh. Struct. Syst. 14.1.11.

[3] Inoue, A., Takeuchi, A. (2004) Recent progress in bulk glassy, nanoquasicrystalline and nanocrystalline alloys. Mater. Sci. Eng. A, 375(7): 16-30.

[4] Ustinov, A.I., Movchan, B.A., Polishchuk, S.S. (2004) Formation of nanoquasicrystalline Al-Cu-Fe coatings at electron beam physical vapour deposition. Scripta Mater., 50(4): 533-537.

[5] Huang, Z., Xing, Y., Gao, Y. (2023) Effective inertia coefficients prediction and cell size effects in thickness direction of periodic composite plates. Int. J. Struct. Stab. Dy., 10.1142.

[6] Bacariza, M.C., Graa, I., Lopes, J.M., (2018) Henriques, C. Enhanced activity of CO2 hydrogenation to CH4 over Ni based zeolites through the optimization of the Si/Al ratio. Micropor. Mesopor. Mat., 267: 9-19.

[7] Sun, Z., Li, X., Li, F. (2022) Study on damage evolution beneath spherical indenter through experiments and mechanism-based strain gradient crystal plastic simulations. Eng. Fract. Mech., 259:108147-.

[8] Wang, Z.F., Zhao, W.M., Qin, C.L., Cui, Y., Fan, S.L., Jia, J.Q. (2012) Direct preparation of nano-quasicrystals via a water-cooled wedge-shaped copper mould. J. Nanomater., (1687-4110): 208-212.

[9] Pedrazzini, S., Galano, M., Audebert, F., Siegkas, P. (2019) High strain rate behaviour of nanoquasicrystalline Al93Fe3Cr2Ti2 alloy and composites. Mater. Sci. Eng. A, 764: 138201.

[10] Zhang, L.L., Zhang, Y.M., Gao, Y. (2014) General solutions of plane elasticity of one-dimensional orthorhombic quasicrystals with piezoelectric effect. Phys. Lett. A, 378(37): 2768-2776.

[11] Waksmanski, N., Pan, E. (2017) Nonlocal analytical solutions for multilayered one-dimensional quasicrystal nanoplates. J. Vib. Acoust., 139: 021006.

[12] Sheng, H.Y., Fan, J.R. (1997) The laminated thick-walled cylinder in a state of non-plain strain. Chin. J. Appl. Mech., 14(2): 64-71.

[13] Hsu, C.W., Hwu, C. (2023) Green's function for thick laminated composite plates with coupled stretching-bending and transverse shear deformation. Compos. Struct., 10: 1016.

[14] Peng, L.X., Chen, S.Y., Wei, D.Y., Chen, W., Zhang, Y.S. (2022) Static and free vibration analysis of stiffened FGM plate on elastic foundation based on physical neutral surface and MK method. Compos. Struct., 290: 115482.

[15] Wang, Y., Liu, X., Li, Z., Feng, Z., Pan, C. Zhang. J., Xu, J. (2023) Nitsche-based isogeometric approach for free vibration analysis of laminated plate with multiple stiffeners and cutouts. Int. J. Mech. Sci., 244: 108041.

[16] Zhang, S.Q., Zhao, Y.F., Wang, X., Chen, M., Schmidt, R. (2022) Static and dynamic analysis of functionally graded magneto-electro-elastic plates and shells. Compos. Struct., 281: 114950.1-114950.14.

[17] Gao, J.X., Shen, Y.P., Wang, Z.K. (1998) Three-dimensional exact analysis for free vibration of simply supported piezoelectric composite laminates. Acta Mech. Sinica., 14(2): 168-177.

[18] Pan, E., Heyliger, P.R. (2002) Free vibrations of simply supported and multilayered magneto-electro-elastic plates. J. Sound Vib., 252(3): 429-442.

[19] Huang, Y.Z., Li, Y., Yang, L.Z., Gao, Y. (2019) Static response of functionally graded multilayered one-dimensional hexagonal piezoelectric quasicrystal plates using the state vector approach. J. Zhejiang Univ. A., 20(2): 133-147.

Ease-off modification and tooth contact analysis for near net shape face-gear

Yinan Han [1], Miaojiang Ding [1], Shuaiqi He [1] and Jianjun Yang [2, 3*]

[1]Zhejiang Jiecang Linear Motion Technology Co., Ltd., Shaoxing City, Zhejiang Province, 312500, China
[2]School of Mechatronics Engineering, Henan University of Science & Technology, Luoyang City, Henan Province, 471000, China
[3]Longmen Laboratory, Luoyang City, Henan Province, 471000, China

*Corresponding author's E-mail: hohooyang@126.com

Abstract: An ease-off modification method for face gear by near-net shape forming was proposed to improve transmission performance. Firstly, based on the involute cylindrical gear surface, the completely conjugated tooth surface of the face gear was obtained, and the ease-off surface was constructed using second-order and fourth-order polynomials. Then, the tooth contact analysis for the modified flank is carried out. According to the loaded tooth contact analysis (TCA), the transmission error (TE) and contact path were consistent with the predesigned parameters. The noise experiments confirmed that the meshing performance of the ease-off modification tooth surface is better than that of the ordinary design method, which provides a reference for the topological modification design method for the die tooth surface of face-gear.

1. Introduction

The face-gear drive is widely used in automobiles, helicopters, and industry fields with the advantages of simple structure and low sensitivity to axial installation errors [1-2]. More extensive attention has been paid to the flank modification technology of face gear to improve transmission performance [3-5]. Due to the structure characteristics of large pitch angles, applying the near-net shape-forming method makes large-quantity batch production possible. Rather than the conventional cutting or grinding process, the method of near-net shape forming [6] has the obvious advantage of better utilization of raw material, higher productivity, and more convenient gear flank topology modification.

To improve and control the transmission performance, the active design method using ease-off topology modification for gear surface is proposed by Shih et al. [7]. Tsay et al. [8] presented a profile modification method to improve the contact performance of die-pressing face-gear drives. This paper presents the active topology modification using ease-off differential tooth surface to improve TE and decrease meshing noise.

2. Establishment of target tooth surface

2.1. Conjugate tooth surface establishment

Based on the conjugate principle of gear surface, the face-gear flank is enveloped by an involute helix cylindrical one according to the given transmission ratio. The three-dimensional model and generation relationship between the involute cylindrical helix gear and face one is illustrated in Figure 1 and Figure 2. The coordinate systems $S_n(O_n{:}X_n,Y_n,Z_n)$ and $S_m(O_m{:}X_m,Y_m,Z_m)$ are fixed with two gears, respectively. $S_1(O_1{:}X_1,Y_1,Z_1)$ and $S_2(O_2{:}X_2,Y_2,Z_2)$ are rotated around axes Z_n and Z_m, respectively. φ_1 and φ_2 are the corresponding rotating angles in the meshing process. So, the transmission ratio is $i_{21}=\varphi_1/\varphi_2$.

Figure 1. Three-dimensional model for face-gear

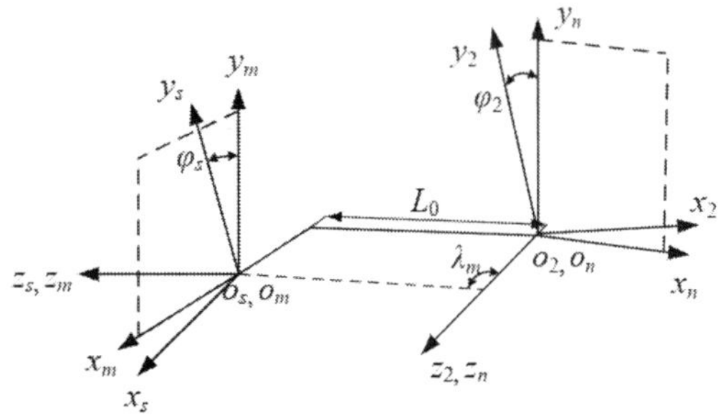

Figure 2. Generation relationship of face-gear

The helix gear surface $\boldsymbol{r}_1$ and the corresponding normal vector $\boldsymbol{n}_1$ are represented as:

$$\boldsymbol{r}_1(\varepsilon_1,\theta_1)=\begin{bmatrix} r_b\left(\cos(\theta_1+\eta_1\pm\varepsilon_1)+\theta_1\sin(\theta_1+\eta_1\pm\varepsilon_1)\right) \\ \pm r_b\left(\sin(\theta_1+\eta_1\pm\varepsilon_1)-\theta_1\cos(\theta_1+\eta_1\pm\varepsilon_1)\right) \\ p\varepsilon_1+L_0 \\ 1 \end{bmatrix} \tag{1}$$

$$\boldsymbol{n}_1(\varepsilon_1,\theta_1)=\frac{\dfrac{\partial \boldsymbol{r}_1}{\partial\theta_1}\times\dfrac{\partial \boldsymbol{r}_1}{\partial\varepsilon_1}}{\left|\dfrac{\partial \boldsymbol{r}_1}{\partial\theta_1}\times\dfrac{\partial \boldsymbol{r}_1}{\partial\varepsilon_1}\right|}=\begin{bmatrix} \pm\cos\beta_b\sin(\theta_1+\eta_1\pm\varepsilon_1) \\ -\cos\beta_b\cos(\theta_1+\eta_1\pm\varepsilon_1) \\ \sin\beta_b \end{bmatrix} \tag{2}$$

Here, symbols± means the left and right flank, respectively. α_1 means the pressure angle of the transverse profile. θ_1 and ε_1 represent the unfolding angle and rotation angle. L_0 means the distance between the origin of the involute curve and that of the fixed coordinate system. β_b is the base circle spiral angle. η_1 is tooth slot half angle, which can be written as:

$$\eta_1=\pi/2N_1-\mathrm{inv}(\alpha_1) \tag{3}$$

N_1 means the tooth number. p is a spiral parameter, which can be calculated as:

$$p=2\pi r_b/\tan\beta_b \tag{4}$$

r_b is the base circle radius. The surface equation $\boldsymbol{r}_0$ of conjugate gear surface Σ_0 can be determined by Equations (5) and (6), namely,

$$f(\varepsilon_1,\theta_1,\varphi_1)=\boldsymbol{n}_1\cdot\boldsymbol{v}_1^{(12)}=0 \tag{5}$$

$$\boldsymbol{r}_0(\varepsilon_1,\theta_1,\varphi_1)=\boldsymbol{M}_{21}\cdot\boldsymbol{r}_1(\varepsilon_1,\theta_1) \tag{6}$$

where $f(\varepsilon_1, \theta_1, \varphi_1)=0$ is the meshing condition. Matrix $\boldsymbol{M}_{21}$ is the transformation relationship from S_1 to S_2. $\boldsymbol{v}_1^{(12)}$ is the relative motion speed at the meshing point.

2.2. Construction of ease-off tooth surface

To obtain the modification of conjugate surface Σ_0, an ease-off surface is superposed on Σ_0 to construct the topological modification of the gear surface. Figure 3 shows the topology diagram of the ease-off surface, and the surface function with two parameters (u, v) can be expressed as:

$$\delta(u,v)=a_0+a_1u+a_2v+a_3u^2+a_4v^2+a_5uv+a_6u^4+a_7v^4 \tag{7}$$

Here, a_0 is a constant term equal to zero if the tooth surface clearance is eliminated. The coefficients a_1, a_3, a_6 and a_2, a_4, a_7 are used to control the shape of the U and V coordinate direction, respectively, and a_5 is used to control the torsion degree of the surface.

Taking the conjugate tooth surface Σ0 as the reference plane, the modification values along the corresponding normal vector direction of surface Σ0 can be calculated according to Equation (8), which is also called ease-off surface. The deviation value Z_d can be expressed as follows:

$$Z_d = \delta(u,v) \tag{8}$$

Take u and v as plane coordinates and Z_d as vertical ones, the modified surface Σ_2 is derived by superimposing the ease-off values on the conjugate tooth surface, which can be expressed as:

$$\boldsymbol{r}_2 = \boldsymbol{r}_0 + \delta(u,v)\boldsymbol{n}_0 \tag{9}$$

Here, $\boldsymbol{n}_0$ is the unit normal vector of surface Σ_0.

3. TCA for ease-off surface

3.1. Ease-off surface

According to the gear generation process, angle φ_{1i} corresponds to ($u_{i,j}$, $v_{i,j}$) values at meshing point i. So, a series of deviation values Z_{dj}^{i} can be obtained based on Equation (9). These deviation values represent an instantaneous modification contact line S, called the ease-off curve in this paper. Here, i is the sequence of the differential ease-off curve, and j represents the sequence of points on each curve. The tangent direction of the curve represents that of the long axis of the contact pattern, and the corresponding direction angle η is illustrated in Figure 3, which can be obtained by traversing the rotation angle φ_1 and expressed as:

$$S_i = S(\varphi_{1i}),\quad \varphi_{1i} = \varphi_1\left[u_{(i,j)}, v_{(i,j)}\right] \tag{10}$$

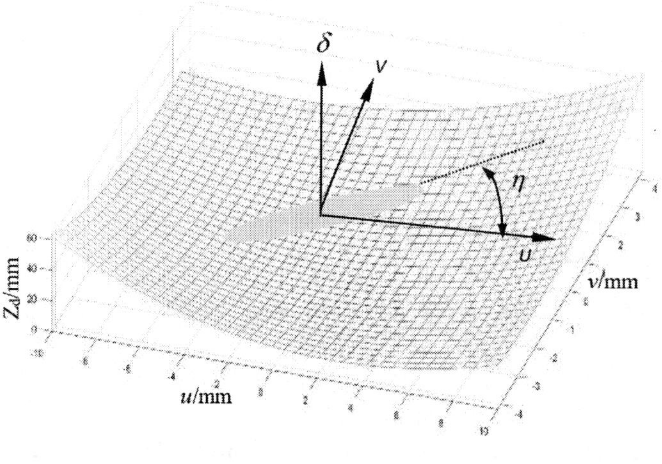

Figure 3. Ease-off surface

3.2. Contact path and TE

The contact point (CP) of a series of ease-off curves is the lowest position of this curve during the meshing process of the gear tooth. The contact path composed of a series of CPs is shown in waterfall diagram (Figure 4), which reflects the CPs on the tooth surface and the direction of the contact path.

The minimum values $Z_{d\min}^{i}$ among Z_{dj}^{i} represent the TE with the helical gear rotation angle φ_{1i}. By converting the deviation values into the angle ones, the gear TE $\Delta\varphi_{2i}$ can be expressed as:

$$\Delta\varphi_{2i} = \frac{Z_{d\min}^{i} \times 10^6}{\boldsymbol{n}_0 \cdot (\boldsymbol{c} \times \boldsymbol{r}_0)} \tag{11}$$

where $\boldsymbol{c}$ is the unit vector in the direction of the axis of the face-gear. $\boldsymbol{r}_0$ and $\boldsymbol{t}$ represent the position and velocity vector at the CP.

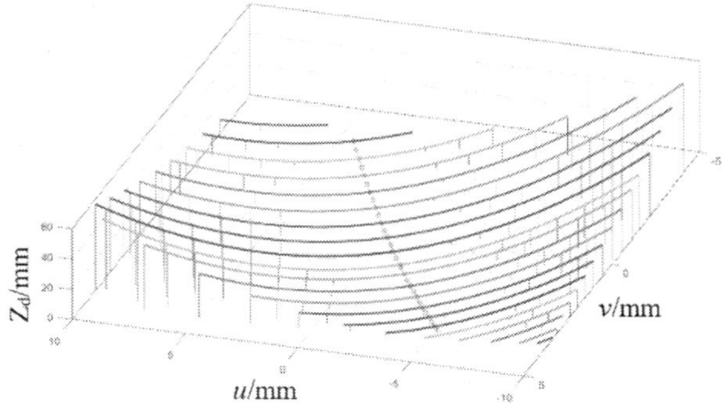

Figure 4. Ease-off curve waterfall

4. Active design of tooth surface topological

An ease-off tooth surface can be constructed by employing the active design for TE along the contact path and ease-off curve. Taking the conjugate surface as a base plane, a straight line *S* is derived by projecting the contact curve on this plane (shown in Figure 5). Point M represents the selected calculation point on the face-gear surface. Points A, B, C, and D are specific points along the contact path, and points E and F are along the ease-off curve.

Based on design requirements, corresponding to angles φ_1, a predetermined second-order and fourth-order polynomial are superposed upon the contact curve S and contact path.

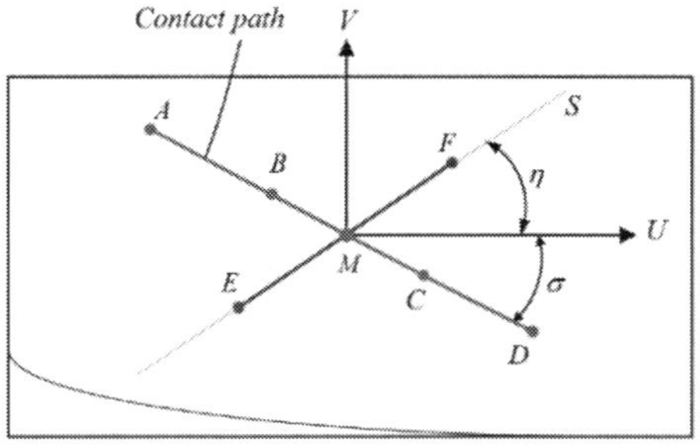

Figure 5. Ease-off calculation points

According to Equations (7), (8), and (9), the deviation values at points A, B, C, D, E, and F can be determined by Equation (12).

$$\begin{bmatrix} z_{d1} \\ z_{d2} \\ z_{d3} \\ z_{d4} \\ z_{d5} \\ z_{d6} \\ z_{d7} \end{bmatrix} = \begin{bmatrix} u_1 & v_1 & u_1^2 & v_1^2 & u_1v_1 & u_1^4 & v_1^4 \\ u_2 & v_2 & u_2^2 & v_2^2 & u_2v_2 & u_2^4 & v_2^4 \\ u_3 & v_3 & u_3^2 & v_3^2 & u_3v_3 & u_3^4 & v_3^4 \\ u_4 & v_4 & u_4^2 & v_4^2 & u_4v_4 & u_4^4 & v_4^4 \\ u_5 & v_5 & u_5^2 & v_5^2 & u_5v_5 & u_5^4 & v_5^4 \\ u_6 & v_6 & u_6^2 & v_6^2 & u_6v_6 & u_6^4 & v_6^4 \\ u_7 & v_7 & u_7^2 & v_7^2 & u_7v_7 & u_7^4 & v_7^4 \end{bmatrix} \begin{bmatrix} a_1 \\ a_2 \\ a_3 \\ a_4 \\ a_5 \\ a_6 \\ a_7 \end{bmatrix} \tag{12}$$

Equation (12) is a linear equation system for solving coefficients $a_1, a_2, a_3, a_4, a_5, a_6$ and a_7. The modified tooth surface can be constructed according to the second and fourth-order polynomials.

5. Case analysis

To confirm the effectiveness of the active design method for topological modification of the gear surface, a pair of plastic face gear used in water flosser products was taken as an example. The

Journal of Physics: Conference Series **2808** (2024) 012057 doi:10.1088/1742-6596/2808/1/012057

working load range of the motor is 4.12 N·mm-13.45 N·mm. The basic geometric parameters are shown in Table 1.

Table 1. Gears and material parameters.

<table>
<tr><th>Parameters</th><th>Cylindrical gear</th><th>Face-gear</th></tr>
<tr><td>Tooth number</td><td>10</td><td>40</td></tr>
<tr><td>Module/mm</td><td colspan="2">0.5</td></tr>
<tr><td>Pressure angle/(°)</td><td colspan="2">20</td></tr>
<tr><td>Tooth width/mm</td><td>5</td><td>3.09</td></tr>
<tr><td>Modification coefficient</td><td>+0.25</td><td>-0.25</td></tr>
<tr><td>Pitch diameter/mm</td><td>5</td><td>20</td></tr>
<tr><td>Outside radius/mm</td><td></td><td>13.09</td></tr>
<tr><td>Inside radius/mm</td><td></td><td>10</td></tr>
<tr><td>Density/(g·cm⁻³)</td><td colspan="2">1.42</td></tr>
<tr><td>Elastic Modulus/(MPa)</td><td colspan="2">2900</td></tr>
<tr><td>Poisson's ratio</td><td colspan="2">0.37</td></tr>
</table>

According to the preset second-order and fourth-order polynomial ease-off curves, the coefficients in Equation (12) are solved, as shown in Table 2. A pair of modified plastic face-gear is injected and assembled, as shown in Figure 6.

Table 2. Modification coefficient.

a_1	a_2	a_3	a_4	a_5	a_6	a_7
-0.001	-0.002	0.002	0.0005	0.001	0.02	0.09

Figure 6. Plastic face gear used in water flosser

In general, loaded TE is considered a key excitation of noise or vibration of the transmission system. Figure 7 shows the loaded TE curve under loads of 5, 10, and 20 N.mm. The TE fluctuation amplitude of 10 N.mm is the smallest, and the amplitude will increase with the load increase. According to the variation trend of load TE in Figure 7, it is clear that the face-gear transmission is relatively stable within the load range of the product motor.

According to the enterprise standard of water flosser products made by JIMOK and BAIR companies, the noise value is no more than 65 dB. Based on the design noise value analysis results, the average noise value for the face-gear surface without modification is 65.8 dB, a little more than that of foreign enterprises. After ease-off surface medication, the value is decreased to 61.6 dB, which is 4.2 dB lower than the existing products. So, the modified face-gear transmission designed in this paper can effectively reduce the noise of this product.

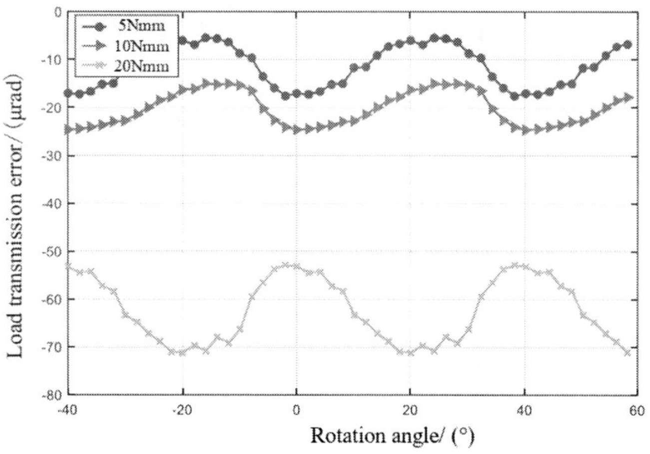

Figure 7. Load TE

6. Conclusion

1) The ease-off differential tooth surface containing meshing information is constructed using a cylindrical gear to develop the conjugate outfacing tooth surface. The polynomial surface equation controls the tooth surface topology shape.

2) Taking the injection mold as an example, the surface modification was designed for a product. The feasibility and correctness of the topological modification design of the face-gear die are verified. The noise testing experiment results show that the noise of the gearbox containing the modified surface gear is reduced, and the water spraying performance meets the demand of the product.

Acknowledgments

Financial assistance was provided by the National Natural Science Foundation (52275054) and the Frontier Exploration Project of Longmen Laboratory (LMQYTSKT023).

References

[1] Litvin F. L. Gear Geometry and Applied Theory [M]. Englewood-Cliffs, NJ: Prentice-Hall, Inc., 2004: 627-696.
[2] Litvin F. L., Fuentes A., Zanzi C., et al. Face gear drive with spur involute pinion: geometry, generation by a worm, stress analysis [J]. COMPUT METHOD APPL M, 2002, 191(23-26): 2783-2813.
[3] Zhou, Y. S., Wu, Y. H., Wang, L. M., et al. A new closed-form calculation of envelope surface for modeling face-gears [J]. MECH MACH THEORY, 2019, 137, 211-226.
[4] Li, Z. M. Q., Fan, J. H., Zhu, R. P. Construction of Tooth Modeling Solutions of Face Gear Drives with an Involute Pinion [J], IJST-T MECH ENG, 2018, 42(1): 35-39
[5] Tetsuo Inoue, Syuhei K.. Characteristics and effectiveness of the face gear with transmission-error-controlled curve considering the path of contact [J], J ADV MECH DES SYST, 2023, 17(2): 1-13.
[6] J. Deng, C. Jiang, X.Z. Deng et al. Numerical simulation and experimental study on hot rolling forming of spur face gears [J], ADV MECH ENG, 2023 15: 9
[7] Shih Y. P., Fong Z. H. Flank Modification Methodology for Face-hobbing Hypoid Gears Based on Ease-off Topography [J]. J MECH DESIGN, 2007, 129(12): 1294-1302.
[8] Tsay M. F., Fong Z. H. Novel profile modification methodology for moulded face-gear drives [J]. P I MECH ENG C-J MEC, 2007, 221(6): 715-725.

Multi-material optimal design of miura origami folds based on genetic algorithm

Yahong Jiang [1], Jinyu Zhou [2*], Yaping Xue [1] and Haotian Hu [3]

[1]School of Mechanical Engineering, Jiangsu University of Technology, Changzhou, Jiangsu, 213001, China
[2]School of Mechanical and Electrical Engineering, Jinling Institute of Technology, Nanjing, Jiangsu, 211199, China
[3]Key Laboratory of Advanced Ship Materials and Mechanics, Harbin Engineering University, Harbin, Heilongjiang, 150001, China

[*]Corresponding author's e-mail: jyjh767@163.com

Abstract. Based on the theoretical model of Miura origami, this paper uses Python and Abaqus software to optimize the design of multi-material positions at the crease of Miura origami by using a genetic algorithm. Firstly, the parametric modeling of Miura origami is carried out; then linear cell meshing is performed at the origami crease, and polylactic acid (PLA) and thermoplastic polyurethane (TPU) are selected as the optimized materials; finally, the multi-material position distribution is optimized by the genetic algorithm according to the target value, to search for the optimal distribution of the materials. Therefore, the angular displacement of the Miura origami model under the thermal drive meets the desired target value. This method solves the problem of finding the optimal multi-material distribution design under the target deformation and provides important ideas and references for the design of space expandable mechanisms simultaneously.

1. Introduction

Miura origami is a type of four-fold origami proposed by Miura, a Japanese astrophysicist [1], where each node has four creases, three mountain folds, and one valley fold, or three valley folds and one mountain fold [2]. The current research is divided into two main categories, the first research direction is the problem of the thickness of origami, which can be divided into zero-thickness and non-zero-thickness origami. For zero-thickness origami, Cai et al. [3] used a polyimide film to design an origami membrane unit to simulate the wrinkling of the membrane and used the effective modulus of elasticity of the material to simulate the crease of the membrane to study the spatial unfolding process of the Miura origami structure. For non-zero-thickness origami models, Tachil [4] proposed a tapered-panel method for rigid foldable origami structures with thickness information while maintaining the ideal origami surface dynamics behavior; Chen et al. [5] proposed a non-zero-thickness spatial origami theory and designed a single-degree-of-freedom kinematically movable origami mechanism by adjusting the folds and panel thicknesses to the thicknesses of the origami structure design with the currently existed method of mechanism analysis, which makes the folding motion of the origami structure easier to control and stabilize.

Another type of research direction is materials for origami. Metamaterials are new programmable man-made composites, mainly determined by the topological and deformation properties of the internal artificial structures, whose mechanical functions can be changed on demand [6, 7], and which can be shape-shifted driven by specific conditions such as heat, electricity, magnetism, and light. Silverberg et al. [8] designed origami geometrical properties through the use of programmable metamaterial structures and self-assembling robots; Kotikian et al. [9] constructed thermo-responsive liquid-crystal elastomer metamaterials through 3D printing, which thermally actuated large deformations of the material to achieve the transformation from 2D to 3D structures or between different 3D topologies. Liu et al. [10] achieved topological bistability of the motion of origami structures through the design of spatial topological motion of origami metamaterials.

In this paper, the mathematical model of Miura origami is used as the basis of design, the crease of the non-zero thickness Miura origami structure is the object of study, the coefficient of thermal expansion of the selected metamaterials, the minimum mean square error (MSE) between the unfolded angular displacements of the origami design and the angular displacements of the simulation model is the target value, and the model-independent variable parameter design is carried out by using the joint invocation of Python and Abaqus software. Through the linear cell division at the crease, the grid cells are digitally filled with dual materials, 0 for active (PLA) materials and 1 for passive (TPU) materials, the initial material random distribution is carried out, and by the preset target value, the genetic algorithm is used to carry out iterative optimization, to obtain the optimal position of the material distribution. Through the validation of the resultant accuracy of the simulated displacements and theoretical displacements obtained, the method can provide a certain theoretical basis for the unfolding mechanism to realize the structure-material design.

2. Space geometric modeling and model design

2.1. Miura origami unit space geometry model

The Miura origami cell [1, 11, 12] consists of four parallelograms with side lengths a and b and an acute angle α subtended by the lengths of the two sides, as shown in Figure 1 (a). The geometric model of origami space is shown in Figure 1 (b). From the geometric conditions:

$$h = a\sin\alpha\sin\lambda \tag{1}$$

$$w = 2b\frac{\cos\lambda\tan\alpha}{\sqrt{1+\cos^2\lambda\tan^2\alpha}} \tag{2}$$

$$\sin\frac{\beta_1}{2} = \frac{\cos\lambda}{\cos\alpha\sqrt{1+\cos^2\lambda\tan^2\alpha}} \tag{3}$$

$$\beta_2 = \pi - 2\arctan\frac{1}{\cos\alpha\tan\frac{\beta_1}{2}} \tag{4}$$

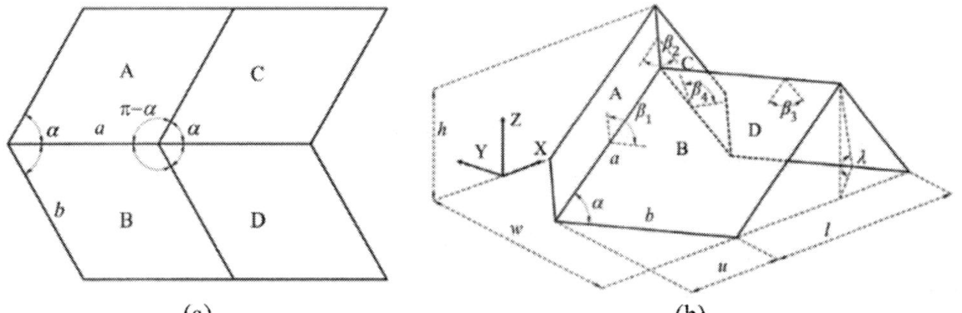

Figure 1. Origami geometry. (a) Miura origami unit (b) Geometric model of space.

2.2. Parametric modeling of miura origami

In this paper, the Python platform is used for the parametric modeling of Miura origami based on the geometrical model of Miura origami, and according to the actual needs of the design, two independent variables are created, namely, the crease width variable z and the model thickness t. By setting different values for the six independent parameters, the Miura origami model is established according to Table 1, as shown in Figure 2.

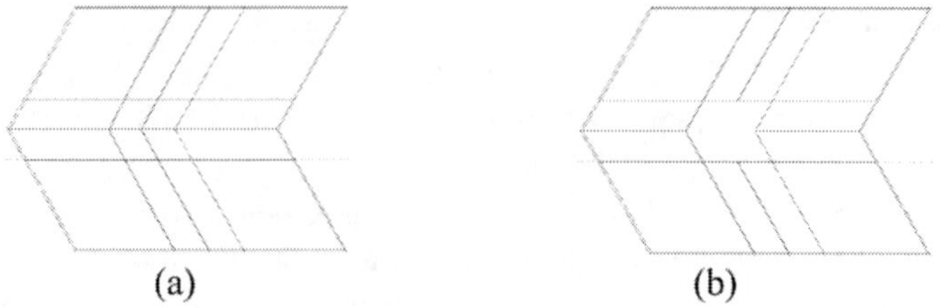

Figure 2. Parametric modeling models. (a) Miura origami unit (b) Miura origami optimisation unit.

Table 1. Miura origami parameter variables.

Parametric variable	Length a /mm	Length b /mm	Width of crease z /mm	Thicknesses t /mm	Plane angle α /°	Two-sided angle λ /°
Value	10	10	2.5	1	60	0

2.3. Parametric modeling of miura origami

According to the finite element model established, the meshing cells are divided into crease regions and non-crease regions according to the design needs, and the C3D8R linear reduced solid cells are used for simulation analysis. The crease region is deformed in the process of imitating the real process, the regional mesh division is the most dense, and the cell size is about 0.25 mm; the non-crease region is almost not deformed, only the rigid body movement occurs; to improve the efficiency of the calculation, the regional mesh is sparse compared to the crease region, the cell size of 0.4 mm~1 mm; as for the thickness direction of the division of the four-layer cell, the cell size is 0.25 mm. The results are shown in Figure 3, with a total of 24, 000 cells and 31, 000 nodes, including 9, 600 cells and 13, 000 nodes at the crease. According to the symmetry of the Miura origami unit and the crease, the cross-section of the crease is selected to simulate the large deformation of non-zero-thickness origami in the process of thermal drive.

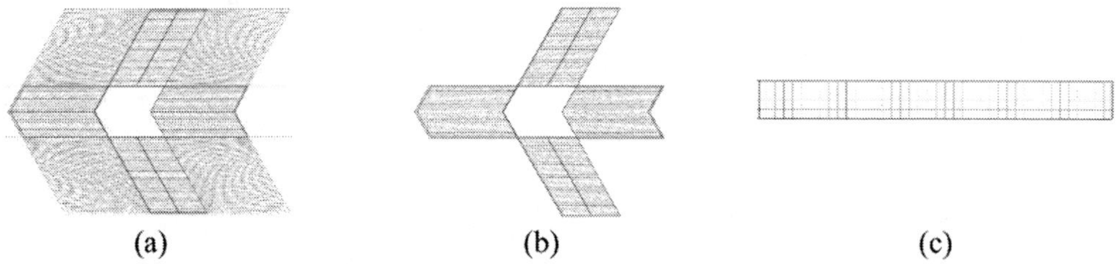

Figure 3. Model meshing. (a) Miura origami gridded (b) Crease gridding (c) Gridded cross-section.

3. Optimization model design and results

3.1. Model optimization build

3.1.1. Mathematical modeling for model design optimization. The mathematical relationship for the crease optimization model of Miura origami is given by:

$$\begin{cases} X=[\theta_i,\varphi_i](i=1,2,\cdots,N) \\ \min f(x)=\sum_{i=1}^{N}\sqrt{[\theta_i-\varphi_i]^2} \\ s.t.\begin{cases} G_1(X)=0.174\le\theta_i\le1.570 \\ G_2(X)=0.174\le\theta_i\le1.570 \\ G_3(X)=|\theta_i-\varphi_i|\le0.085 \end{cases} \end{cases} \tag{5}$$

where the origami model crease at the variable X; $G_i(X)$ for the value of the angular displacement judgment function, constraints on the angle of the angular displacement occur as well as the target angular displacement and the simulation of the difference between the two; the smaller the target function $f(X)$ is, the more to meet the requirements of the optimal design.

3.2. Genetic algorithm optimization model solving

3.2.1. Code. The two selected materials were randomly distributed to each unit and the numbers were coded as individual genomes for the genetic algorithm. Parameters were represented by binary coding, specifying the material with a number, 0 for the active material PLA and 1 for the passive material TPU, as shown in Figure 4.

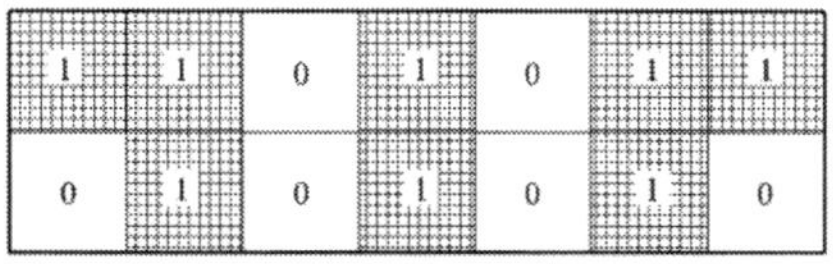

Figure 4. Genome encoding.

3.2.2. Adaptation evaluation. Based on the whole population, individuals were evaluated from the first iteration, and the simulation and analysis of each iteration were completed in Abaqus finite element software, where the data of the angular displacements were extracted and compared with the target angular displacements, and the fitness of the individuals was calculated based on the fitness function [13], which was

$$F(X)=1/f(X)=1/\left(\frac{1}{N}\sum_{i}^{N}\sqrt{(\theta_i-\varphi_i)^2}\right) \tag{6}$$

where $F(X)$ is the fitness function, θ_i is the target angular displacement that occurs when the model is thermally driven, φ_i is the simulated angular displacement that occurs when the model is thermally driven, and the fitness function is the maximization problem.

3.3. Comparison of crease optimization results
According to Equation (8), crease 1 and crease 3 are selected as the optimization targets, and the target angular displacements are 1.919 rad and 1.606 rad, respectively. The symmetry of the Miura origami unit and crease is selected, half of the cross-sectional areas of crease 1 and crease 3 are optimized, and the target angular displacements are half of the total target values, i.e. the angular displacements of crease 1 and crease 3 are 0.959 rad and 0.802 rad, respectively. Fixed in the middle position for 0~1 s, the external temperature changes from 25°C to 125°C in 1 s to optimize the position of the material and achieve the target angular displacement. The parameters of the genetic algorithm are shown in Table 2, and according to the results of 2500 iterations of optimization, the simulated angular

displacements of crease 1 and crease 3 are 0.944 rad and 0.796 rad, respectively. The optimal distribution of the material position is shown in Figure 5 (a) and Figure 5 (b), and the simulation results are shown in Figure 6.

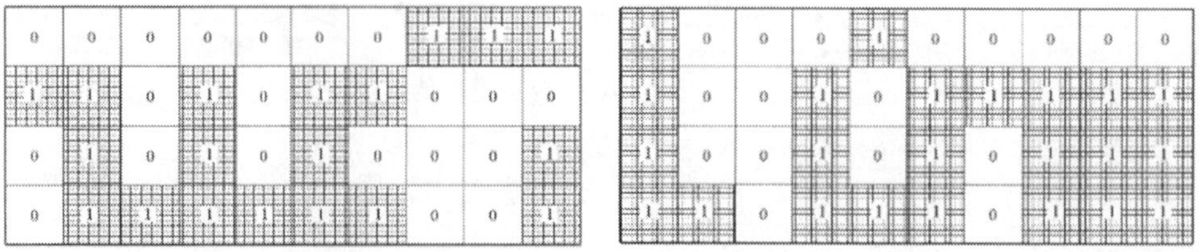

(a) (b)

Figure 5. Crease optimization results. (a) Optimal materials distribution for crease 1(b) Optimal materials distribution for crease 2.

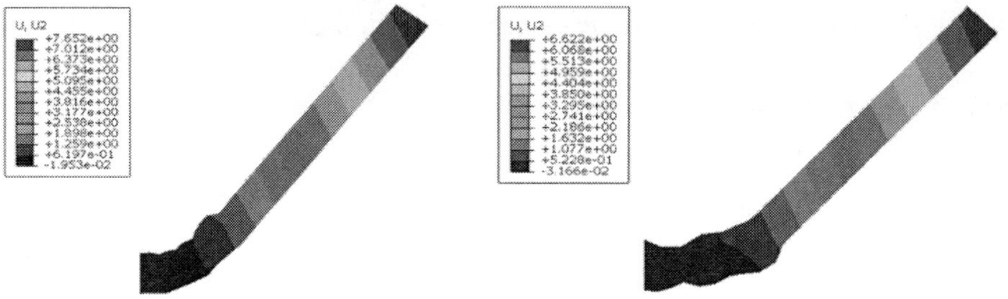

(a) (b)

Figure 6. Displacement of the optimal solution. (a) Crease 1 (b) Crease 2.

Table 2. Parameters of genetic algorithm.

Parametric	Population size in generation	Number of generations r	Crossover probability	Mutation probability
Value	50	50	0.8	0.1

The deformation of the model in Abaqus finite element is constrained to be in space XY, the frames at the six-time points of 0.0 s, 0.5 s, 0.7 s, 0.8 s, 0.9 s, and 1.0 s are extracted, and the representative simulated angular displacements are output and compared with the target angular displacements, as shown in Figure 7. The errors of the target and simulated angular displacements of crease 1 and crease 2 are 0.0064 rad and 0.0143 rad, respectively, which are in line with the error range.

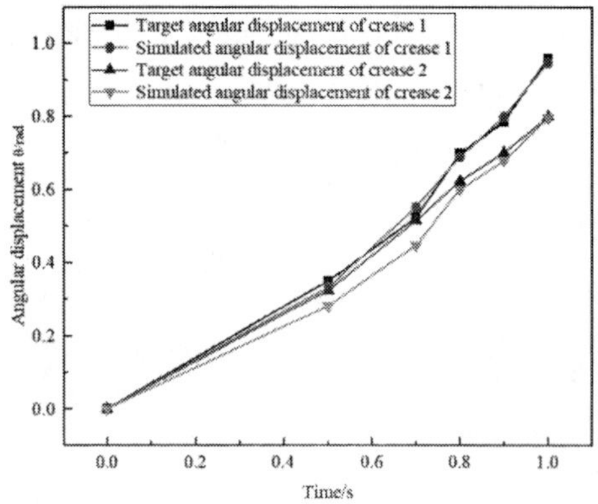

Figure 7. Angular displacement curve at creases.

3.4. Miura Origami Unit Deformation Results

The results of the crease-optimized material distribution were verified by performing a Miura origami cell, and the results are shown in Figure 8.

(a) (b) (c)

Figure 8. Unfolded diagram of miura origami unit. (a) Initial state (b) Unfolding process (c) Show results.

4. Conclusion

In this paper, firstly, parametric modeling is carried out for Miura origami to achieve the complex construction of the model; at the same time, the design is simplified at the crease and the mesh linear division is carried out to provide the basis for model optimization. Secondly, based on the genetic algorithm, bi-material optimization is carried out at the crease of Miura origami with non-zero thickness, and the optimal solution of material distribution is obtained, so that the optimized model undergoes angular displacements of 0.944 rad and 0.796 rad of the crease as the temperature rises from 25°C to 125°C in 0~1 s, which verifies the feasibility of the optimal design of structure-material of the spatial unfolding mechanism. Finally, the dual-material-based optimal design of origami deformation provides a reliable basis for the controllable deformation design of multi-material structures.

Acknowledgments

This work was financially supported by the National Natural Science Foundation of China (52075232), the Natural Science Foundation of Jiangsu Province (BK20201112), and the Graduate Practice and Innovation Project of Jiangsu Province (SJCX22_1471).

References

[1] Koryo M. Method of packaging and deployment of large membranes in space[J]. The Institute of Space and Astronautical Science report, 1985, 618: 1-9.
[2] Cai J, Wang L M, Feng J. Research on unfoldable form and performance based on origami elements[M]. Beijing: Science Press, 2020.
[3] Cai J, Ren Z, Ding Y, et al. Deployment simulation of foldable origami membrane structures[J]. Aerospace Science and Technology, 2017, 67: 343-353.
[4] Tachi T. Rigid-foldable thick origami[J]. Origami, 2011, 5: 253-264.
[5] Chen Y, Peng R, You Z. Origami of thick panels[J]. Science, 2015, 349(6246): 396-400.
[6] Chen Y. Large deformation metamaterials based on mechanism motion[J]. Journal of Mechanical Engineering, 2020, 56(19): 2-13.
[7] Fang H, Wu H, Liu Z, et al. Progress in the dynamics of origami structures and origami metamaterials[J]. Journal of Mechanics, 2022, 54(01):1-38.
[8] Silverberg J L, Evans A A, McLeod L, et al. Using origami design principles to fold reprogrammable mechanical metamaterials [J]. Science, 2014, 345(6197): 647-650.
[9] Kotikian A, Truby R L, Boley J W, et al. 3D printing of liquid crystal elastomeric actuators with spatially programmed nematic order[J]. Advanced materials, 2018, 30(10): 1706164.
[10] Liu B, Silverberg J L, Evans A A, et al. Topological kinematics of origami metamaterials[J]. Nature Physics, 2018, 14(8): 811-815.

[11] Chang B, Xu X, Liang D, Zhang H. Geometric design and motion analysis of Miura folding mechanism for thick plates[J]. Chinese Space Science and Technology, 2022, 42(04): 146-157.
[12] Ye H, Wang Zi, Wang X. Analysis and optimal design of quasi-static compressive mechanical properties of origami superstructures [J]. Journal of Beijing Institute of Technology, 2022, 48(02): 105-113.
[13] Peng X, Liu G, Wang J, et al. Controllable deformation design for 4D-printed active composite structure: optimization, simulation, and experimental verification [J]. Composites Science and Technology, 2023, 243: 110265.

Journal of Physics: Conference Series **2808** (2024) 012059 doi:10.1088/1742-6596/2808/1/012059

Research on engineering selection methods for electromagnetic lines and insulation materials of UHV converter transformers under extreme margin conditions

Xinru Yu[1,*], Chao Wu[1], Huanchao Cheng[1], Chaohua Wang[2], Yongfeng Zhao[2] and Junyu Deng[1]

[1]High Voltage Laboratory, China Electric Power Research Institute, Beijing, 100192, China

[2]State Grid Henan Electric Power Company, Zhengzhou, 450018, China

[*]Corresponding author's e-mail: yuxinru@epri.sgcc.com.cn

Abstract. As the transmission capacity of the ± 800 kV ultra-high voltage project is increased to 10 GW and the ± 1100 kV project capacity is increased to 12 GW, the requirements for the flow level of various equipment have been improved. Due to the limited overall design size, the design margin of the converter transformer is approaching its limit. Due to the weight of a single converter transformer reaching hundreds of tons and the huge amount of material used, the physical property testing for a single electromagnetic wire and a single insulating cardboard in the existing IEC standards can no longer meet the requirements for engineering use. This study focuses on the main materials of converter transformers and researches material selection methods for the operating conditions of ultra-high voltage equipment. The selection scheme for electromagnetic wires, insulating cardboard, and insulating screws of ultra-high voltage converter transformers has been determined. The results of this study can be used to select converter transformer materials in ultra-high voltage engineering and improve the reliability of power equipment operation.

1. Introduction

During the manufacturing phase of the converter transformer in the ± 800 kV ultra-high voltage transmission project, several equipment failures occurred due to defects in raw materials. The main reason for this is that as the transmission capacity of the ± 800 kV ultra-high voltage project is increased to 10 GW and the ± 1100 kV project capacity is increased to 12 GW, the requirements for the flow level of various equipment have been improved. Due to the limited overall design size, the design margin of the converter transformer is approaching its limit. The main raw materials for converter transformers include electromagnetic wires, insulating cardboard, silicon steel sheets, transformer oil, etc. According to statistics, insulating cardboard and electromagnetic wire are the two materials with the highest defect rate. Due to the weight of a single converter transformer reaching hundreds of tons and the huge amount of material used, the physical property testing for a single electromagnetic wire and a single insulating cardboard in the existing IEC standards can no longer meet the requirements for engineering use. It is necessary to combine the actual operating conditions of the converter transformer, starting from its material characteristics, to research the selection method of insulation cardboard and electromagnetic wire for ultra-high voltage converter transformers.

Journal of Physics: Conference Series **2808** (2024) 012059 doi:10.1088/1742-6596/2808/1/012059

2. Characteristics and performance selection method of electromagnetic wire for uhv converter transformers

2.1. The effect of multiple physical fields on the properties of copper electromagnetic wire materials

2.1.1. Variable temperature short-term annealing and microhardness evolution. During the service process of the electromagnetic wire of the converter transformer, it may face short-circuit currents of up to tens of thousands of amperes, which can cause instantaneous temperature rise, local short-term overheating, and other problems. Setting up a short-term annealing experiment with variable temperature, the initial microhardness of semi-hard copper is HV 87.5, as shown in Figure 1[1]-[3]. After annealing at 100°C for 1 hour, the microhardness of the sample remains almost unchanged. After annealing at 200°C for 1 hour, its microhardness slightly decreased. After annealing at 400°C for 1 hour, the microhardness of semi-hard copper significantly decreased. Considering that the initial state of semi-hard copper contains high-density dislocations, and there is no significant change in the grain size distribution in the semi-hard copper sample after annealing at 400°C for 1 hour, the decrease in microhardness is due to the evolution of dislocation structure[4][5].

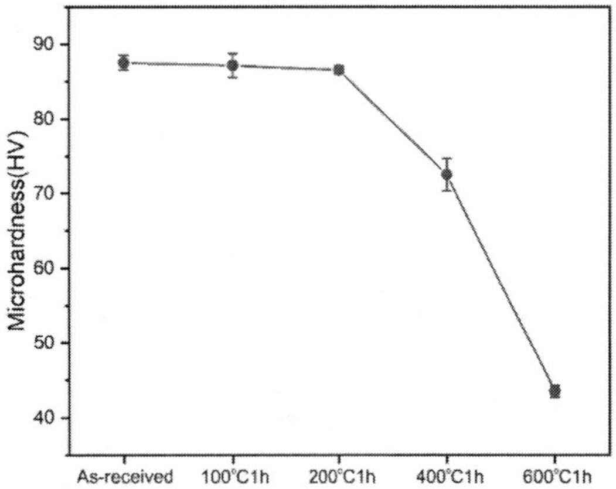

Figure 1. Microhardness test results of semi-hard copper samples subjected to variable temperature short-term annealing.

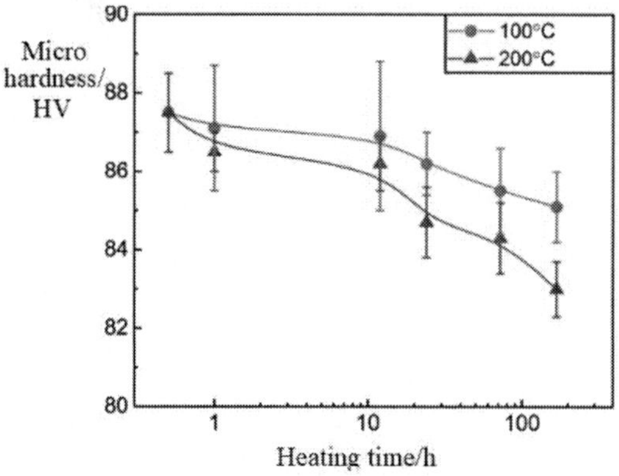

Figure 2. Microhardness test results of semi-hard copper samples annealed at constant temperature for a long time.

*2.1.2. Constant temperature long-term annealing and microhardness evolution.*Under the service condition of semi-hard copper, the stable operating temperature is about 100-120°C. The short-term peak temperature obtained from remote infrared temperature measurement is about 180°C. The microhardness results of semi-hard copper after long-term annealing at 100°C and 200°C are shown in Figure 2. The higher the annealing temperature and the longer the annealing time, the lower the microhardness of semi-hard copper. As shown by the blue line in the figure, the hardness of semi-hard copper after heating at 100°C for 24 hours is 86.2 HV. Extending the heating time to 72 and 168 hours, the hardness of the samples is 85.5 HV and 85.1 HV[6]-[9]. The hardness of the samples heated at 200°C for 24, 72, and 168 hours is 84.7 HV, 84.3 HV, and 83.0 HV, respectively. After annealing at 200°C for 168 hours, the hardness of the semi-hard copper conductor decreased by 5%, although seven days was only 0.04% of the component's design life. Even at 100°C, the hardness loss of the transposed wire still reaches 2.7%. The results of the constant temperature long-term heating experiment have shown that even under normal working conditions, the mechanical properties of the transposed wire deteriorate significantly over time, even reaching several percentage points within one week[10]-[12].

2.2. The influence of multiple physical fields on the force/electrical properties of the electromagnetic field of the converter

During the converter transformer winding operation, a current of several thousand amperes under normal operating conditions will produce a significant Joule heating effect, inevitably leading to a temperature rise in semi-hard copper. If there is a sudden short circuit situation, the external short circuit current is tens of times higher than the normal operating current, which can inject a large amount of heat in a very short time, causing the temperature of semi-hard copper to rise quickly to a higher level[13]. The strength of semi-hard copper at different temperatures will affect its resistance to short-circuit impact, ultimately affecting the service safety of windings and converter transformers. This study conducted in-situ thermo mechanical coupling deformation experiments on semi-hard copper under transmission electron microscopy and quantitatively studied the deformation strength and behavior of semi-hard copper under a temperature rise environment. The testing method used for high-temperature tensile deformation of macroblock semi-hard copper is uniaxial tensile deformation, with quasi-static tensile tests conducted at room temperature, 100, 150, 200, 300, and 400°C, as shown in Figure 3.

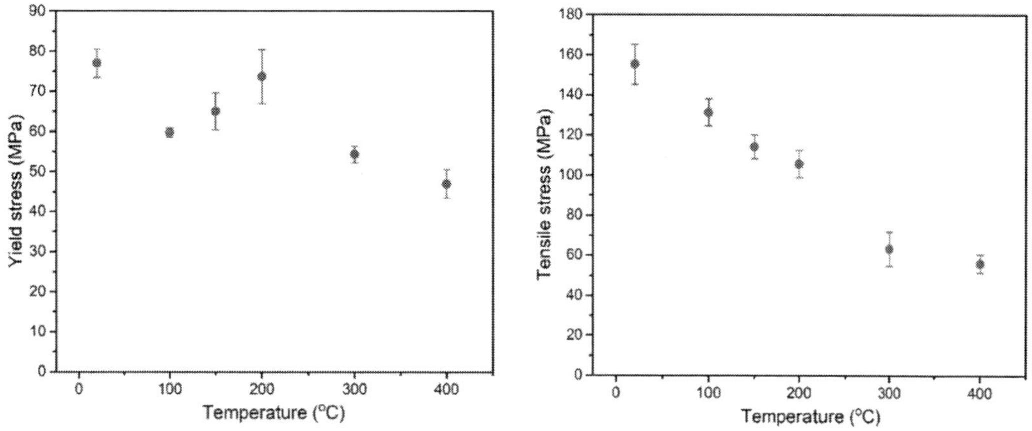

Figure 3. Macro thermo mechanical coupling tensile temperature strength relationship of semi-hard copper samples.

2.3. Selection method for copper materials suitable for electromagnetic wires of ultra-high voltage converter transformers

Based on the experimental results, a selection method for copper materials suitable for electromagnetic wires of ultra-high voltage converter transformers was proposed, as shown in Table 1.

Table 1. Selection scheme for electromagnetic wires of ultra-high voltage converter transformers.

Parameters		Technical requirement
Mechanical performance	Prescribed non-proportional elongation strength Rp 0.2	The measurement results should meet the ordering parameters proposed by the user, and the decrease in Rp 0.2 of all semi-hard wires between the measured values at room temperature (20°C) and high temperature (140°C) should be less than 20%.
	Bond strength	The bonding strength at room temperature should not be less than 10 N/mm^2. The bonding strength at 95°C should not be less than 8 N/mm^2. The bonding strength at 110°C should not be less than 5 N/mm^2.
	Overall bending strength	The measured bending rigidity "EI" value shall not be less than 80% of the calculated value.

Journal of Physics: Conference Series **2808** (2024) 012059 doi:10.1088/1742-6596/2808/1/012059

	Elongation	Elongation greater than 15%.
Electrical performance	Resistivity	The electrical resistivity at 20°C shall not exceed 0.01720 Å x mm^2/m, and the resistance imbalance rate shall not exceed 0.5%.
	DC voltage resistance	The adjacent enameled flat wires of the transposed conductor should withstand DC 1000 V for no less than 5 seconds without breakdown.
	Paint film breakdown voltage	At room temperature, at least four out of five samples should have a breakdown voltage greater than or equal to 4 kV, and the other sample should have a breakdown voltage greater than or equal to 3 kV. At a high temperature of 120°C, at least four out of five samples should not experience breakdown at voltages less than or equal to 3 kV, and the other sample should not experience breakdown at voltages less than 2 kV.
	Wire path	Measure each enameled flat wire in the transposed wire with a multimeter, and there should be no open circuit phenomenon.

3. Characteristics and performance selection methods of insulation paperboard and insulation screw materials for uhv converter transformers

3.1. Surface discharge performance of insulating paperboard

3.1.1. Experimental platform for discharge characteristics of insulated paperboard under composite electric field. To study the discharge characteristics and flashover mechanism of the oil-paper interface under composite voltage, an experimental platform, as shown in Figure 4, was constructed. The platform consists of a voltage regulator, a test transformer, an experimental chamber, and a measurement system. The UDC terminal is a negative polarity DC high-voltage generator, which is connected to the sample through a current limiting resistor R2. UAC is an AC high-voltage generator that adjusts the AC side voltage regulator and DC side voltage output to change the AC/DC ratio applied to both ends of the sample, which is connected to the test sample through resistor R1 and isolated capacitor C1. The flashover voltage is measured by the parallel connection of resistance and capacitance-voltage dividers at both ends of the sample.

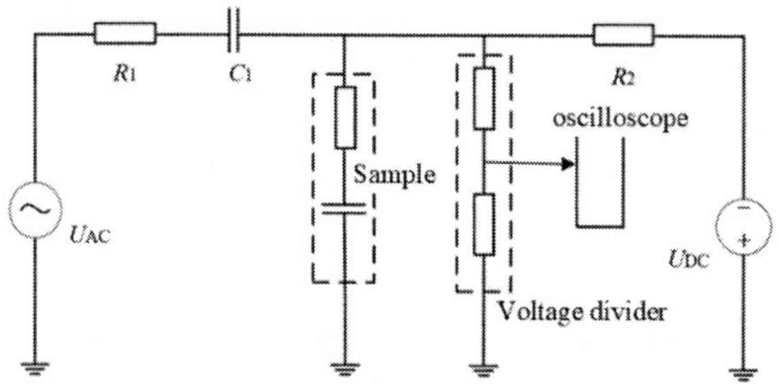

Figure 4. Experimental circuit diagram.

3.1.2. Research on surface discharge characteristics of insulated paperboard. As shown in Figure 5, comparing the mean values of AC voltage components under different tangential normal ratios, it can be seen that when the normal component is small, the change in DC ratio has a smaller impact on flashover, and the peak value of composite voltage increases linearly with the increase of DC ratio; When the tangential normal component is 1:3, that is, when the proportion of the normal component is large, the impact of DC voltage on flashover increases. As the proportion of DC increases, the amplitude of AC voltage peak decrease increases, and the composite voltage peak curve slows down. The electric field in the weak insulation section of the converter transformer coincides with this characteristic, which has a large normal and DC component, which is also the objective reason for the easy occurrence of faults in the converter insulation.

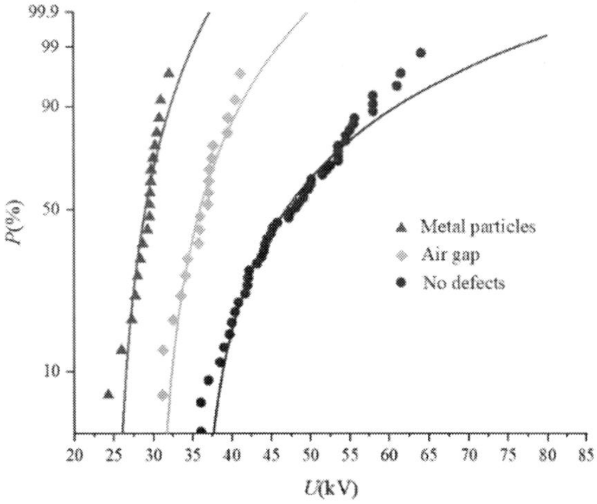

Figure 5. Flashover voltage under different surface impurities on.

Table 2. Three parameters Weibull distribution parameters of flashover field strength along the surface of insulating paperboard under different inclusions.

Webull distribution parameters	Metal particles	Air gap	No defects
E_0	24.31	31.77	36.24
α	8.30	6.06	15.75
β	1.80	1.73	1.56

Sharp protrusions on the edges of metal debris can induce charges of opposite polarity, increasing the electrical voltage around the metal debris and causing local breakdown, neutralizing the tip charges, as shown in Table 2. As the voltage enters the next half cycle, partial discharge will occur again under the influence of the external electric field and residual charge at the tip, and the above discharge process will be repeated in the next cycle. The larger the diameter of the debris, the more sharp protrusions and accumulated charges there are. The sample is more prone to partial discharge at low background voltage, but this effect tends to saturate as the diameter of the inclusion metal increases. When impurities are distributed on the insulation surface, the distorted electric field on the surface makes it easy for the insulation oil to ionize and generate charge carriers, further causing a flashover of the entire insulation surface.

3.2. Mechanical and electrical properties of insulating screw materials

3.2.1. Surface flashover characteristics of insulated screws. In the study, the influence of screw length on the surface flashover voltage of insulated screws was experimentally investigated. As shown in Figure 6, with the increase in screw length, the flashover voltage of insulated screws under AC and DC voltages also increased, and the relationship was approximately linear. Compared to AC voltage, insulated screws have a higher flashover voltage under DC voltage. The tensile strength of epoxy screws is the highest, while the tensile strength of bakelite screws is the lowest, as shown in Figure 7.

Journal of Physics: Conference Series **2808** (2024) 012059 doi:10.1088/1742-6596/2808/1/012059

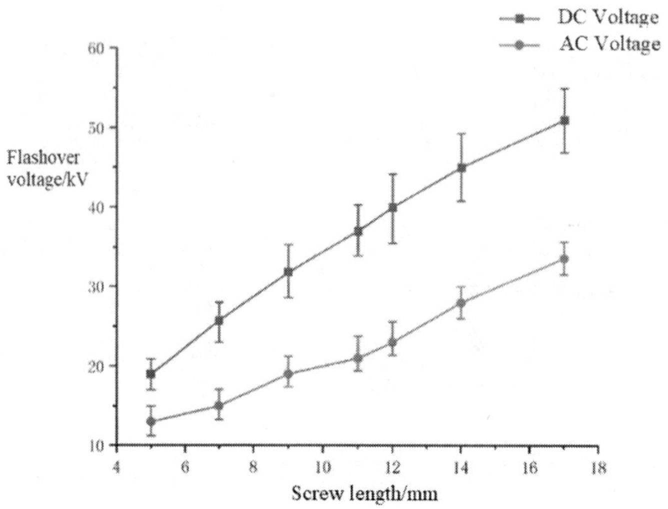

Figure 6. The relationship between surface flashover voltage and length of insulated screws under different voltage types.

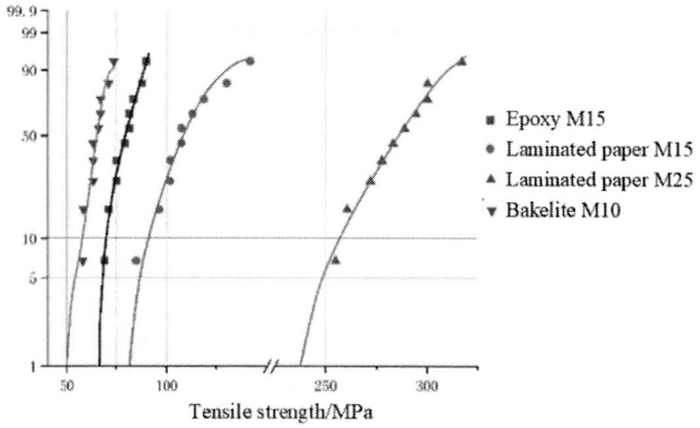

Figure 7. Statistical distribution of tensile strength of different insulated screws.

3.3. Mechanical and electrical properties of insulating screw materials

Selection method for insulating paperboard.Randomly sample insulating cardboard samples at a percentage of 1% to 10%, using a sampling method that ensures that each part of the population has the same chance of being selected. The number of acceptance criteria is chosen based on the Oc curve. The starting voltage of partial discharge is measured under AC voltage. The measured result Ui is divided by the thickness of the specimen to obtain the initial partial discharge field strength E of the specimen, and the modified value is compared with the selection threshold. If E is greater than the selection threshold, the sample is qualified. Otherwise, the sample is unqualified. Table 3 shows the selection scheme for surface flashover of insulating cardboard.

3.3.1. The number of non-conformities m is counted for all samples. If m is lower than the acceptance judgment number, the entire batch of products is qualified and can be put to engineering use. If m is higher than the acceptance judgment number, the entire batch of products is unqualified and cannot be put to engineering use.

Journal of Physics: Conference Series **2808** (2024) 012059 doi:10.1088/1742-6596/2808/1/012059

Table 3. Selection scheme for surface flashover of insulating paperboard.

Detection object	Testing items	Sample requirements	Test method	Testing requirements
Insulated paperboard	Surface adhesion of high-density particles	Paperboard size 30 mm × 100 mm, with a moisture content of less than 1%, fully immersed in oil	AC surface flashover test, selecting a finger-shaped electrode with a back electrode	The flashover field strength along the surface is higher than 12.08 kV/mm

3.3.2. Selection method for insulated screw. The sample used for the experiment needs to be cut into a 10 mm long sample, and a 2 mm deep hole should be drilled on one surface of the screw. The whole diameter of the M10 screw is 4 mm, the M15 screw is 6 mm, and the M25 screw is 10 mm. Each sample is cut into five specimens, and the flashover voltage of the five specimens is obtained. The average value is taken as the surface flashover voltage U of the sample, and the flashover electric field E is calculated. *Table 4 shows the selection scheme for insulation screw surface flashover.* After the testing of the entire batch of samples is completed, the magnitude of the flashover electric field E and the selection threshold of each sample are compared. If the flashover electric field is greater than the selection threshold, the sample is qualified. If the flashover electric field is greater than the selection threshold of excellent products, the sample is excellent; otherwise, it is unqualified. The size of the number of unqualified samples m is compared with the acceptance judgment number. If m is lower than the acceptance judgment number, the entire batch of products is qualified and can be put to engineering use. If m is higher than the acceptance judgment number, the entire batch of products is unqualified and cannot be put to engineering use.

Table 4. Selection scheme for insulation screw surface flashover.

Detection object	Testing items	Sample requirements	Test method	Testing requirements
Insulated screw	Insulation screw surface withstands strength	Screw length 10 mm, fully dried with a moisture content below 1%, paper screw fully immersed in oil	DC surface flashover test, step pressure method, electrode selection 5-10 electrodes used	Epoxy M15: Surface flashover field strength higher than 4.87 kV/mm Laminated paper M15: Surface flashover field strength higher than 3.37 kV/mm Laminated paper M25: Surface flashover field strength higher than 2.51 kV/mm Bakelite M10: The flashover field strength along the surface is higher than 4.07 kV/mm

The mechanical properties of four commonly used insulation screws for rheological properties were measured: laminated paper M15, laminated paper M25, epoxy M15, and Bakelite M10 screws. Among the four types of screws, epoxy screws have the highest tensile strength, while rubber screws have the lowest tensile strength. From the statistical distribution parameters, the position parameter of the epoxy M15 screw is the highest, which corresponds to the highest tensile strength, and its dispersibility is relatively high. The position parameters of the rubber wood M10 screw are the lowest, and the tensile strength is the lowest, but its dispersibility is relatively small, indicating that the product quality is relatively uniform.

4. Conclusion

This study focuses on the main materials of converter transformers and researches material selection methods for the operating conditions of ultra-high voltage equipment. The main conclusions are as follows: the electromagnetic wires of converter transformers were analyzed under external stress conditions under operating conditions, and the microstructure and mechanical evolution properties of the materials were studied under temperature rise, electrodynamic force, and short-circuit impact load characteristics. Aiming at the discharge characteristics of insulation materials in converter transformers under different voltage forms and electric field components, an AC/DC composite voltage experimental platform for insulation materials was established, and a three-parameter Weibull distribution model was proposed to describe the initiation law of partial discharge in insulation materials. A selection scheme for the quality of insulation materials was established based on their discharge characteristics under different voltage forms and electric field components. By using the more scientific OC curve sampling method compared to percentage sampling, the selection plan for insulating paperboard and insulating screw has been determined, which can be used for material selection of converter transformers in ultra-high voltage engineering.

Acknowledgments

This work was financially supported by the SGCC technology project "Research on main material selection method and reliability improvement technology of UHV converter main equipment under limit "GYB17202000119.

References

[1] Meng Y, Ju X, Yang X. The measurement of the dislocation density using TEM[J]. Materials Characterization, 2021, 175: 111065.
[2] Zaefferer S. Interacting with current [J]. Nature Materials, 2021, 20 (4): 448-449.
[3] Zhang Yanbing, Zheng Jin, Wang Dehua, et al. Development of ultra-high voltage DC converter transformers [J] High Voltage Technology, 2010, 36 (001): 255-264.
[4] Liu Z. Ultra-high voltage AC/DC grids[M]: Academic Press, 2014.
[5] Zhou H, Qiu W, Sun K, et al. Ultra-high voltage AC/DC power transmission[M]: Springer, 2018.
[6] Butrymowicz DB, Manning JR, Read ME. Diffusion in Copper and Copper Alloys. Part I. Volume and Surface Self-Diffusion in Copper[J]. Journal Of Physical and Chemical Reference Data, 1973, 2 (3): 643-656.
[7] Tian L, Yang Y-Q, Meyer T, et al. Environmental transmission electron microscopy study of hydrogen charging effect on a Cu-Zr metallic glass[J]. Materials Research Letters, 2020, 8 (12): 439-445.
[8] Csikor Ferenc F, Motz C, Weygand D, et al. Dislocation Avalanches, Strain Bursts, and the Problem of Plastic Forming at the Micrometer Scale[J]. Science, 2007, 318 (5848): 251-254.
[9] Wang Z-J, Li Q-J, Shan Z-W, et al. Sample size effects on the large strain burst in submicron aluminum pillars[J]. Applied Physics Letters, 2012, 100 (7): 071906.
[10] Lu L, Shen Y, Chen X, et al. Ultrahigh strength and high electrical conductivity in copper[J]. Science, 2004, 304 (5669): 422-426.
[11] Qi Bo, Wei Zhen, Li Chengrong, et al. Analysis of the Influence of AC and DC Voltage Ratio on the Internal Air Gap Discharge Characteristics of Oil Paper Insulation [J]. Chinese Journal of Electrical Engineering, 2015, 35 (01): 247-254.
[12] Qi Bo, Wei Zhen, Li Chengrong, et al. The development process of internal air gap discharge in oil paper insulation under the action of AC and DC superimposed voltage [J]. Chinese Journal of Electrical Engineering, 2014, 34 (36): 6554-6561.
[13] Wu Xutao, Li Xiuguang, Li Junhao, etc. Research on Partial Discharge Characteristics of Oil Paper Insulation Along Surface under AC/DC Composite Voltage [J] High voltage electrical appliances, 2016, 52 (04): 128-131+138.

Preparation and autocatalytic properties of CNTs/Co_3O_4/AP Core-shell composite energetic materials

Linpeng Li [1] and Shuhong Ba [1*]

[1]School of Equipment Engineering, Shenyang University of Technology, Shenyang, Liaoning, China

[*]Corresponding author's email: 2319976189@qq.com

Abstract: To change the thermal decomposition property of AP, band structure/state density of nickel oxide, tricobalt tetraoxide carbon nanotubes, and graphite were emulated by using Materials Studio. The results indicate that the catalytic performance of the indirect gap semiconductor Co_3O_4 and the indirect gap semiconductor CNTs was excellent, and the band gaps were 0.076 eV and 0.168 eV, respectively. CNTs/Co_3O_4/AP were made by using the solvent-nonsolvent method. The construction of CNTs/Co_3O_4/AP composites was analyzed by using XRD and TG-DSC according to the tests. The samples were tested for thermal decomposition performance by using DSC, and the results indicated that the high-temperature temperature of AP significantly decreased by 155.58°C. The above results make clear that the CNTs/Co_3O_4/AP has good autocatalytic performance.

1. Introduction

AP is widely used in military fields because of its strong oxidative properties. The addition of catalysts to AP affects the various properties of AP in applications, such as thermal properties and mechanical properties. Therefore, comprehensive consideration should be considered when selecting catalysts.

Transition metal materials are widely used in various catalytic fields, such as photocatalysis and electrocatalysis, due to their superior performance, including transition metal powders (Zn, Cu, and Fe) and transition metal oxide (NiO Co_3O_4). Carbon materials have good physicochemical properties, for example, high specific surface area and high flexibility, which have become a research hotspot in various fields. Carbon materials can be used as desensitizing agents in highly energetic explosives and in their composite with highly energetic explosives, prepared as a new type of explosive. It can be seen from the above study that, the composite transition metal materials have a better catalytic effect than the single transition metal materials, and the synergistic effect between the two materials after the composite will be developed to increase the catalytic active sites and enhance the thermal decomposition performance of AP. Therefore, this article chooses transition metal oxides and carbon materials to be prepared into binary composite materials. Based on the binary materials, a solvent nonsolvent method is used to prepare core-shell composite energy materials.

This article uses Materials Studio software to conduct in-depth research on the band structure and density of state characteristics of transition metal oxides (including NiO[1]and Co_3O_4) and carbon materials (such as CNTs and Cn). By comparing and analyzing these different types of materials, the performance differences in catalyzing AP thermal decomposition were explored. Based on theoretical predictions, we selected transition metal oxides and carbon materials with excellent performance and prepared CNTs/Co_3O_4/AP core-shell composite energetic materials by mixing them with AP.

Journal of Physics: Conference Series **2808** (2024) 012060 doi:10.1088/1742-6596/2808/1/012060

Subsequently, we conducted a series of detailed characterizations of this composite material to comprehensively evaluate its performance.

2. Simulation section

2.1 Performance simulation of transition metal oxide

Two transition metal oxides, NiO and Co_3O_4, were selected for the performance simulation of band structure and density of states, and one transition metal oxide was selected for comparison with experimental data. The transition metal atoms in the transition metal oxides are non-full-band electron systems, and in addition, due to the specificity of the molecular structure of Co_3O_4, the three transition metal oxides were numerically simulated by using the DMOL3 module in Materials Studio [2].

2.1.1 Simulation of NiO properties. According to the lattice parameters of NiO (ICSD: 61544) in the inorganic chemistry crystal library [4], a=b=c=4.1764 Å, the space group is Fm-3m (225) [3], and it is modeled. Its structure is optimized by using the GGA-PBE pseudopotential and the Brillouin zone of 4×4×4 in the DMOL3 module. The optimized cellular structure is adopted with the value of the Brillouin zone of 6×6×6, band structure and state density are calculated, and the results have been shown in Figure 1 and Figure 2.

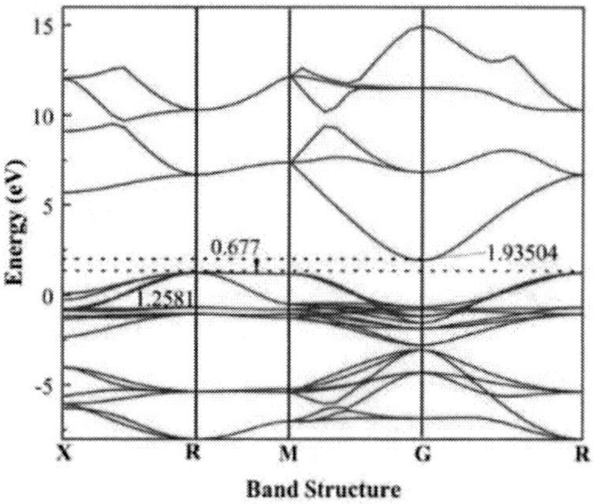

Figure 1. Band structure of NiO.

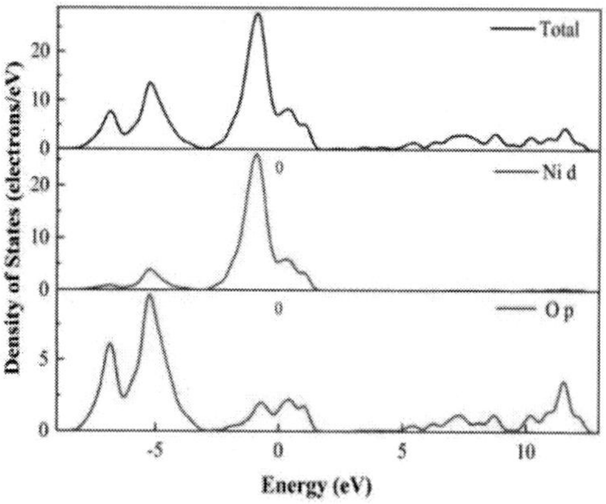

Figure 2. The density of states of NiO.

By the band structure from Figure 1, NiO is an indirect bandgap transition metal oxide semiconductor with a bandwidth of 0.677 eV. The top of the valence band is located at the R point of 1.258 eV, and the bottom of the conduction band is located at the G point of 1.935 eV. From the density of states in Figure 2, the Ni atom d orbitals occupy the valence band low-energy region and contribute to the valence band top, indicating that the Ni atoms have strong interactions with the O atoms, which have a relative locality. Therefore, the density of states to the Fermi level is not 0. The energies around -1.5 eV are mainly contributed by the d orbitals of Ni atoms, and the conduction bands in the higher energy positions above 3 eV were contributed by the p orbitals of O atoms.

2.1.2 Simulation of Co_3O_4 properties. The Co_3O_4 crystal was modeled by using the Co_3O_4 (ICSD: 36256) lattice parameter a=b=c=8.065 Å and the space group Fd-3m (227) according to the inorganic chemical crystal library . Meanwhile, because the spinel structure of Co_3O_4 Co^{2+} and Co^{3+} coexists in the crystal, the splitting of d orbitals makes it antimagnetic below 40 K and becomes paramagnetic at higher temperatures. Two magnetic moments need to be added to the Co atoms in the Co_3O_4 crystals for the performance of the properties and the parameter calculations by using the DMOL3 module. Therefore, the magnetic moments of +5 and -5 were added to the Co atoms in the constructed Co_3O_4 crystal cell, and then it was optimized by using Brillouin zone 2×2×2 with a convergence step of 0.01 Ha. The Brillouin zone of the optimized Co_3O_4 cell structure is set to 4×4×4, and the band structure and density of states were calculated for it. The results are shown in Figure 3 and Figure 4.

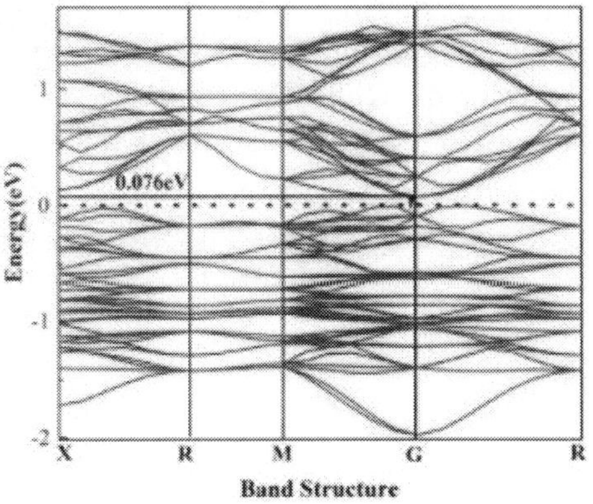

Figure 3. Band structure of Co_3O_4.

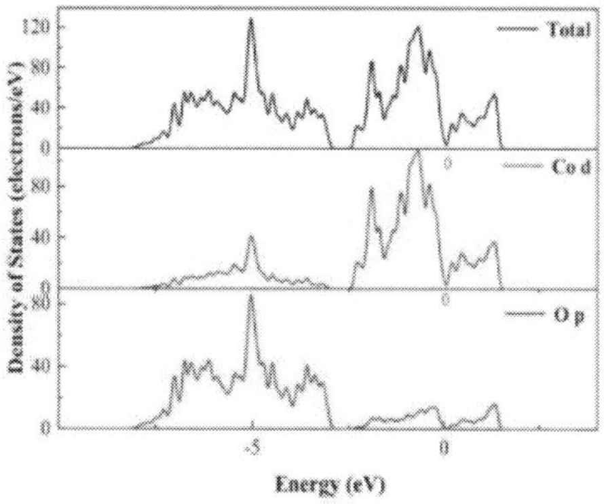

Figure 4. The density of states of Co_3O_4.

According to Figure 3, Co_3O_4 exhibits an indirect transition metal oxide semiconductor with a bandgap width of 0.076 eV after introducing a magnetic moment. The lowest point of the conduction band is located at the G point, while the highest point of the valence band is close to the G point. Further observation of the density of states analysis in Figure 4 reveals that Co atoms exhibit strong localization characteristics and form significant interactions with O atoms. Therefore, on both sides of the Fermi level, the valence band and conduction band are both contributed by the d-electron state of the Co atom. In the lower energy range, the d-orbitals of cobalt atoms dominate the top structure of the valence band, while in the higher energy range, the p-orbitals of O atoms contribute significantly to the valence band. Especially at -5 eV energy, a hybrid peak is formed between the d orbitals of Co atoms and the p orbitals of O atoms, indicating that the interaction between Co and O atoms is stronger than that in NiO.

The band structures and density of states of NiO and Co_3O_4 were simulated by the DMOL3 module in Materials Studio. The data is shown in Table 1 by comparing the band structures of two transition metal oxides in perfect crystalline form and combining them with the catalytic effect of different transition metal oxides on AP [5].

Table 1. Transition metal oxide band gap and its catalytic effect on AP.

Types of transition metal oxides	Add content (%)	HTD (°C)	Band gap (eV)	Reference
NiO	2	93	0.677	[9]
Co_3O_4	2	203	0.076	[10]

As can be seen from Table 1, NiO is an indirect band gap transition metal oxide semiconductor with a bandwidth of 0.677 eV, and Co_3O_4 is an indirect band gap transition metal oxide semiconductor with a bandwidth of 0.076 eV. The d orbitals of the transition metal oxides with a stronger localization and narrower bandwidth are better in the catalysis of AP. The transition metal oxides with narrower energy bands have better catalytic effect, and the transition metal oxides with narrower energy bands have stronger orbital localization of metal atoms, larger work function, and lower difficulty of valence band hole jumping to the Fermi energy level, which is conducive to the catalytic thermal decomposition of AP. Therefore, Co_3O_4 was chosen as the transition metal oxide material.

2.2 Carbon material performance simulation

In this paper, two types of materials, CNTs and Cn, were selected for the simulation study of band structure and density of states. Due to a large number in the atoms in CNTs, the CASTEP module in Materials Studio is slow and too demanding on computer performance. Therefore, the DFTB+ module was chosen to simulate the properties of carbon materials.

2.2.1.CNTs performance simulation. In the Materials Studio software, serrated single-walled CNTs (m=0 and n=10) were established , the lattice constant and space angle in the a and b directions were fixed, and the structure optimization was carried out in the c direction. The empirical parameter of Koster was defined as CHNO interaction during the structure optimization, and the k points were set to 1×1×10. The band structure and density of states were calculated, and the results are shown in Figure 5 and Figure 6.

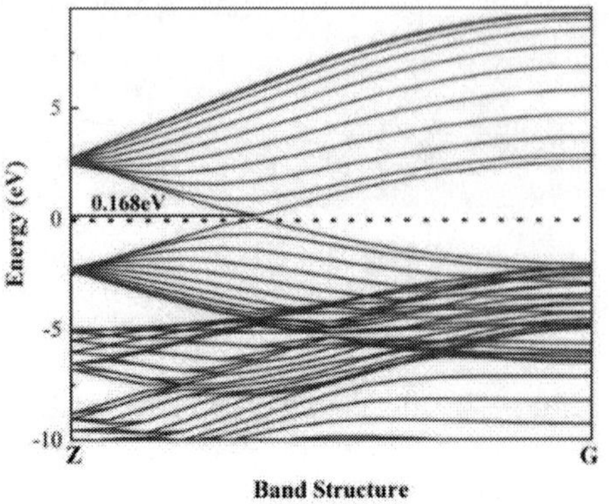

Figure 5. Band structure of CNTs.

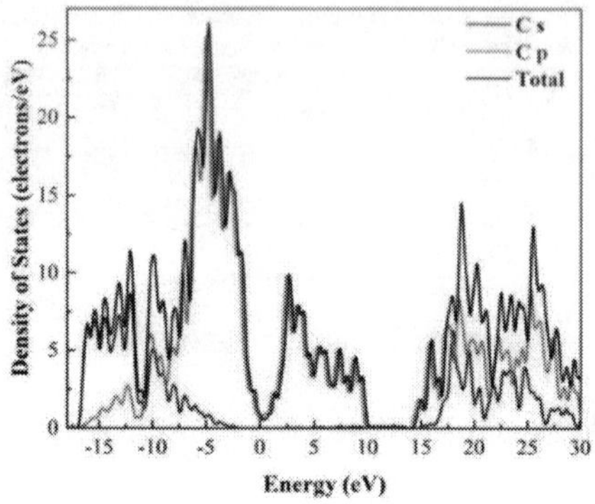

Figure 6. The density of states of CTNs.

According to Figure 5, CNTs are a direct bandgap semiconductor with a bandwidth of 0.168 eV. According to Figure 6 of the density of states, the conduction band and valence band near the Fermi level are composed of p orbitals of c atoms, and the valence band at the high-energy position of the -13 eV valence band is composed of s orbitals of c atoms.

In addition, the s and p orbitals of the c atom are superimposed on the p orbitals in the conduction bands of -10 eV, 18 eV, and 25 eV, with multiple spikes appearing in the superposition.

By simulating the band structure and density of states of CNTs and Cn by the DFTB+module in Materials Studio, combined with the study , the catalytic results of different carbon materials on AP are shown in Table 2.

Table 2. The effect of different materials on AP.

Content and type of carbon material added	LTD (°C)	HTD (°C)	ΔT (°C)
-	314	433	-
2% CNTs	297	374	17/59
4% CNTs	-	370	-/63
2% Cn	332	415	-18/18
4% Cn	338	402	-24/31

Through the performance simulation of CNTs and Cn, it is found that the band of Cn presents the conductor properties, and CNTs are direct band gap semiconductors with a bandwidth of 0.168 eV. As can be seen in Table 2, the catalytic effect of the semiconducting nature of CNTs is stronger than that of Cn, therefore CNTs were chosen as the carbon material.

3. Experiment part

3.1 Preparation of CNTs/Co_3O_4/AP core-shell composite energetic materials

The solvent-nonsolvent method was used to deposit AP on the 25% CNTs and 75% Co_3O_4 composite. 30 ml of pure water was heated in a 40°C water bath, to which AP was added, in a magnetic stirrer at 900 rpm to fully dissolve. The $C_4H_8O_2$ was cooled to 5°C and placed in an ice water bath with a magnetic stirrer at 900 rpm. The AP solution was dripped uniformly into the $C_4H_8O_2$ at a rate of 3 milliliters per minute, and the AP was completely precipitated by a magnetic stirrer for 20 min after the end of the titration. We filtered and dried to obtain the CNTs/Co_3O_4/AP core-shell composite energetic materials.

3.2 Performance test

(1) XRD test

X-ray diffraction analysis was performed on the CNTs/Co_3O_4/AP core-shell composite energetic materials samples to determine whether AP was deposited onto the surface of the 25% CNTs and 75% Co_3O_4 composite. As shown in Figure 7.

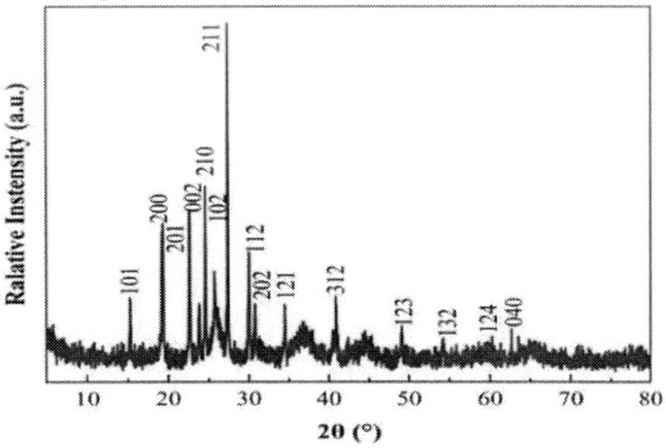

Figure 7. XRD patterns of CNTs/Co_3O_4/AP core-shell composite energetic materials

(2) TG-DSC test

The thermal decomposition performance of CNTs/Co_3O_4/AP core-shell composite energetic material samples was tested under an air atmosphere with a heating range of 0~550°C and a heating rate of 10 °C/min. The DSC curve is in Figure 8.

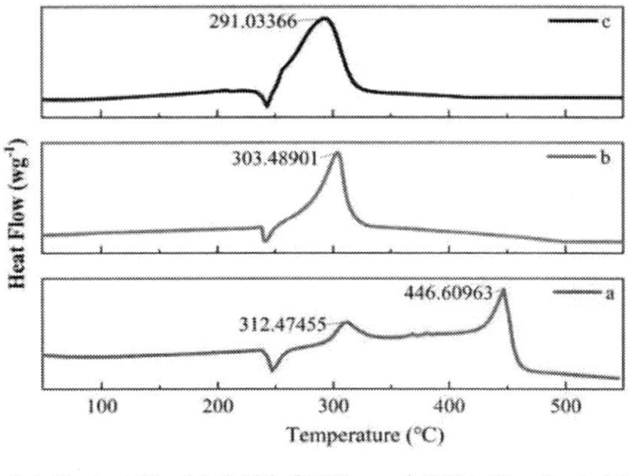

Figure 8. DSC curve: (a) Pure AP; (b) 25% CNTs and 75% Co_3O_4 + AP; (c) CNTs/Co_3O_4/AP.

According to the DSC curve, the CNTs/Co_3O_4/AP core-shell composite energy material reduced the peak temperature by 155.58°C and then 12.46°C on the peak temperature of AP catalyzed by 25% CNTs and 75% Co_3O_4 composites sharply.

4. Conclusion

(1) NiO is an indirect bandgap transition metal oxide material with a bandwidth of 0.677 eV; Co_3O_4 is a transition metal oxide. Transition metal oxide materials with narrower bands exhibit strong localization of d-orbitals of metal atoms, resulting in larger escape work and reduced difficulty in

valence band hole hopping to the Fermi level, which is beneficial for catalyzing the thermal decomposition of AP.

(2) CNTs/Co_3O_4/AP energetic materials were prepared by depositing AP on 3# composites using the solvent-nonsolvent method; the AP temperature decomposition peak temperature was reduced by 155.58°C, the catalytic effect is good.

References

[1] Jebraeil S. M. and Eslami A. Hydrothermal synthesis of graphene oxide/NiO and its influence on the thermal decomposition of ammonium perchlorate [J], 2022.

[2] Mani G., Jos J., Nair P. R., et al. Investigation of kinetic parameters for ammonium perchlorate thermal decomposition in the presence of CN/CuO by TG-MS analysis and kinetic compensation correction [J]. Journal of Solid-State Chemistry, 2021, 301: 122, 301.

[3] Probert M. Electronic Structure: Basic Theory and Practical Methods, by Richard M. Martin. Scope: graduate-level textbook. Level: theoretical materials scientists/ condensed matter physicists/ computational chemists [J], 2011.

[4] Wen T., Feng Y., Bi Y., et al. 3D porous nano Co_3O_4/C composite catalyst for the thermal decomposition of ammonium perchlorate [J]. Materials Today Communications, 2022, 33: 104, 294.

[5] Wang Y., Zhu J., Yang X., et al. Preparation of NiO nanoparticles and their catalytic activity in the thermal decomposition of ammonium perchlorate [J]. Thermochimica Acta, 2005, 437 (1-2): 106-109.

Effect of boundary layer thickness and dimensionless wall distance y+ on numerical simulation of heat pipes

Wenjie Zhao [1]*, Jiajia Deng [2], Dawen Xue [1] and Lin Xu [3]

[1]School of Naval Architecture and Marine Engineering, Zhejiang Ocean University, Zhoushan 316022, China
[2]Guangxi Electrical Polytechnic Institute, Nanning 530007, China
[3]School of Petrochemical Engineering and Environment, Zhejiang Ocean University, Zhoushan 316022, China

*Corresponding author's e-mail: 20101001021@alu.cqu.edu.cn

Abstract: To consider the sway of boundary layer thickness and dimensionless wall distance y+ on the numerical simulation results of the thermo tube, taking the straight thermo tube as an example, using the continuity equation, momentum equation, energy equation, and turbulence model equation as the controlling equations, and adopting the finite volume method for the spatial discretization, the numerical simulation study of the heat pipe is carried out by controlling the changes of boundary layer thickness and dimensionless wall distance y+. Numerical simulation studies show that the boundary layer thickness and y+ have a large effect on the numerical simulation of heat pipes. Reasonably setting these two parameters within a certain range can improve the accuracy of heat pipe numerical calculations, and the results of the study can provide a reference for the security analysis of thermo tube reaction vessels.

1. Introduction

A thermo tube is a two-pass heating appliance using evaporation and condensation of working fluid. To explore the sway of the confine layer thickness and the dimensionless wall distance y+ on the numerical simulation results of heat tubes, taking the straight heat pipe as an example, using the continuity equation, momentum equation, energy equation, and turbulence model equation as the controlling equations, and adopting the finite volume method for the spatial discretization, the numerical simulation study of the heat pipe is carried out by controlling the changes of boundary layer thickness and dimensionless wall distance y+. The calculation results show that the boundary layer thickness and y + have a great influence on the figure simulation fruits of the thermo tube. Reasonably setting these two parameters within a certain range can improve the exactitude of the numerical computation of the thermo tube, and the results of the study can provide a reference for the security analysis of thermo tube reaction vessels.

The heat pipe is generally divided into evaporation portion, thermal isolation portion, and condensation portion. There is a capillary tube core on the inner wall of the container with the same shape as the inner wall, and the liquid-phase workpiece fills the whole tube core. When the workpiece material is in the evaporation section of the heat evaporation, steam with the latent heat of steam is conveyed to the other end of the condensation and releases latent heat of steam, and then condensate comes back to the evaporation section in the thin pipe tube under the action of the pumping force to complete a cycle. In this way, heat energy is effectively transferred from one end to the other. The thermo tube characteristics are as follows: (1) A thermo tube heat exchanger is a new type of high-efficiency conduct heat unit, whose conduct heat behavior predominantly depends on the phase change process of the work material (including liquid metal), which generates a large amount of latent heat of evaporation during the heat transfer process. However, most of the work material (or liquid metal) in the heat pipe has a high latent heat of evaporation, so a large amount of heat energy is delivered without consuming a large amount of water vapor; (2) Isothermal wall temperature field distribution of heat pipe is mainly influenced by the wall temperature distribution, the temperature difference in the phase transition process, and the range of temperature intervening the wall and the perpendicular cliffside, and other factors. Since the temperature distinction intervenes with the flow of moisture in the saturated state, the phase change process is very small, and the wall-to-wall of the tube is very thin, an extremely high isothermal surface can be achieved at a very low heat flow density. The isothermal performance of the heat pipe augments as the identical conductivity for the heat of the heat pipe increases. Its equivalent thermal conductivity is tens or even hundreds of times higher than that of the same material; (3) Using the phase-separated steam and condensation space inside the thermo tube, tall calorie current density is injected in the vaporization section, low heat flow density is injected in the condensation section, and low heat flow density is injected in the cooling section. This conversion ratio can be controlled to a large extent; (4) In the heat pipe filled with a certain amount of inert gas, we can adjust its heat release area, to regulate the amount of heat transmigration process of heat transfer and achieve the evaporation segment of the heat source of the constant temperature control; (5) It is characterized by thermal diodes and thermal switches.

The gas phase flow characteristics of heat pipes have great significance for the performance and security dissection of thermo tube reactors, so scholars domestically and internationally have conducted extensive research. Xu et al. [1] studied and analyzed a tubular heat pipe with a capillary suction core, obtaining that the more the bends happen, for the same amount of heat transfer, the greater the temperature difference intervenes the thermo tube at either end. Mahboobe and Amir [2] used a three-dimensional numerical model to assess the effect of bending on a water-copper cylindrical heat tube, assessing the effect of one 90° bend and two 90° bends on the thermo tube function, and identifying that the bending affected the steam pressure drop to some extent. However, this change was not significant compared to the working pressure of the heat pipe. Shen et al. [3] used Fluent software to establish a physical model corresponding to the heat pipe cooling structure of laptop computers under natural convection conditions, changed the bending angle of the flexible heat pipe, and deduced that under the conditions of constant conductivity for the heat of the flexible thermo tube and the conductivity for heat with the angle change, the model heat dissipation capacity is increased with the reduction of the bending angle of the heat pipe.

Jin et al. [4] analyzed the effect of the variation of gravity thermo tube parameters on its heat transfer characteristics by numerical simulation method. The results show that the VOF model can capture the evaporation condensation phenomenon inside the gravity heat pipe. When the liquid injection rate increases rapidly and the heating power increases, the resistance decreases; in a certain temperature interval, the thermal resistance decreases with increasing pipe diameter. Shen et al. [5] wrote a UDF program for evaporative condensation phase transition based on the VOF model in FLUENT software. On this basis, the two-phase flow and heat-conducting properties in parallel heat pipe bundles under different power conditions are studied. It is found that at high heating power, the mutually excited oscillatory flow of the work mass is more intense and the heat is transported over a longer distance.

Shen et al. [6] used experimental means to investigate the impression of the vapor mass drift rate on the performance of the coagulum heat transmission. It is found that the gas-liquid two-phase current speed increases with the improvement of the inlet flow rate, increasing the shear stress during the air and hydraulic interfaces and the Reynolds number. Simultaneously, the condensation heat pass-on factor of the whole tube is increased. Ma et al. [7] used computational fluid dynamics procedures. A dimension heat conduction conductivity mold about a single module of a thermo tube chill-down pile was smoothly calculated, and the effects of the air gap thickness, air gap thermal conductivity, radiative heat transfer, and surface roughness on the temperature field of the reactor core under different power densities were calculated, and an approximation of the thermal conductivity correction for the effect of the radiative heat transfer and surface roughness factors was proposed, which was verified with the existing experimental results. Gao et al. [8] creatively proposed a new spring-vane hybrid heat exchanger tube.

Scholars have also carried out relevant studies on customized wall functions. Fang et al. [9] evaluated and illustrated the effectiveness and occasions of using the additional terms by simulating the neutral atmospheric boundary layer with a large aerodynamic roughness length at a downsizing ratio of 1:300 and the winding of a TTU low building model inside the neutral atmospheric boundary layer at a downsizing ratio of 1:50. Fang et al. [10] simulated the constant bypass of a TTU model and compared it with wind tunnel tests and field measurements, and re-examined the effect of blockage rate on the surface wind pressure coefficient. Luo et al. [11] summarized the theoretical research results of this problem at home and abroad in recent years according to four aspects: inflow restrictive requirements, turbulence mold arguments, additive cause term model, and perpendicular cliffside function model.

Although the heat evacuation characteristics and the parameters affecting the heat evacuation performance of the thermo tube have been widely studied, the factors affecting the gas-phase simulation of the heat pipe have rarely been addressed, and thus the mechanism of the effect inside the heat pipe is still not well understood.

By using numerical simulation methods, the effects of boundary layer thickness and dimensionless wall pitch y+ on the gas-phase simulation results of heat pipes are investigated, and the effects of these two factors on the pressure drop of gas-phase flow in heat pipes are elucidated. These research results furnish a momentous guideline for the rapid numerical simulation of similar heat pipes, help to deepen the understanding of the gas doping flow behavior inside the thermo tube, and provide useful guidance for the devise and optimization of the thermo tube.

2. Numerical model

2.1. Mathematical-physical model

The gas flow in the adiabatic thermo tube components is numerically simulated and the results are compared. The basic assumptions of the model are as follows:

① The gas-liquid two-phase flow in the heat pipe is assumed to be a steady flow.

② On this basis, the parameters of the thermo tube are assumed to be constant.

The controlling equations include the continuity equation, momentum equation, energy equation, and turbulence model equation.

$$\frac{\partial \rho}{\partial t}+\frac{\partial}{\partial x_i}(\rho u_i)=0 \tag{1}$$

$$\frac{\partial u_i}{\partial t}+u_j\frac{\partial u_i}{\partial x_j}-\frac{1}{\rho}\frac{\partial}{\partial x_j}[(\mu+\mu_t)\frac{\partial u_i}{\partial x_j}]=-\frac{1}{\rho}\frac{\partial p}{\partial x_i}+g \tag{2}$$

$$\frac{\partial}{\partial t}\left(\rho\left(e+\frac{v^2}{2}\right)\right)+\nabla\left(\rho v\left(h+\frac{v^2}{2}\right)\right)=\nabla\left(K_{eff}\nabla T-\sum_j h_j\vec{J_j}+\bar{\bar{\tau}}_{eff}\cdot\vec{V}\right)+S_h \tag{3}$$

$$\rho\frac{dk}{dt}=\frac{\partial}{\partial x_i}[(\mu+\frac{\mu}{\sigma_k})\frac{\partial k}{\partial x_i}]+G_k+G_b-\rho\varepsilon-Y_M \tag{4}$$

$$\rho\frac{d\varepsilon}{dt}=\frac{\partial}{\partial x_i}[(\mu+\frac{\mu}{\sigma_k})\frac{\partial \varepsilon}{\partial x_i}]+C_{1\varepsilon}\frac{\varepsilon}{k}(G_k+C_{3\varepsilon}G_{b)}-C_{3\varepsilon}\rho\frac{\varepsilon^2}{k} \tag{5}$$

$$\mu_t=\rho C_\mu\frac{k^2}{\varepsilon} \tag{6}$$

The liquid working medium, when heated at the evaporator end, forms a vapor and moves toward the condenser rear end of the thermo tube. When a gaseous working medium is formed at long last of the vaporizer of the thermo tube and moves along the heat pipe to the end of the condenser, it is cooled, condenses to a liquid, and then returns to the evaporator end to begin the cycle again. Counter Current Flow (CCF) is formed during the flow of the gaseous mass from the evaporator to the condenser and the mass of the liquid stream flowing from the condenser to the evaporator. Due to the difference in velocity of the counter-current flow, the effect on the gas-liquid two-phase well-up is large, and the pressure loss of the air and hydraulic two-phase flow increases.

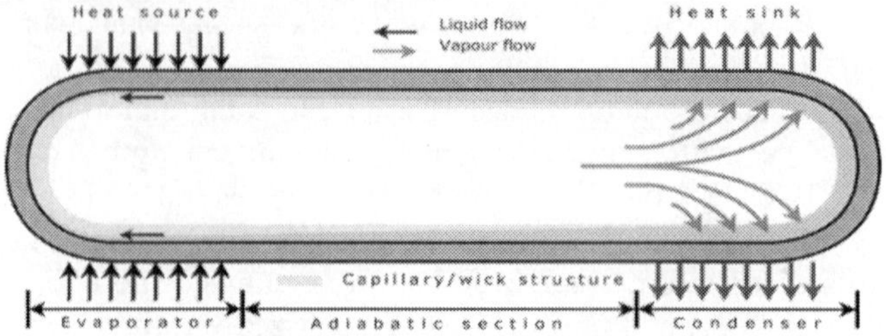

Figure 1. Physical model.

Table 1. Simulated heat pipe parameters.

Project	Tube length	Caliber	Temperature	Mass flow	Water vapor density	Dynamic viscosity	Reynolds number
Parameters	0.6 m	0.008 m	373.15 K	3.01×10^{-4} kg/s	0.6 kg/m^3	1.23×10^{-5}	3, 912.3

2.2.Model setting

Figure 2 shows a schematic of the thermo tube modeling. To visualize the structural schematic of the heat pipe, the modeling is shown by using curved segments. Due to the regular structure of the pipe, a hexahedral mesh is used and the model boundary conditions are set to set the speed entrance in the blast-hole portion. The pressure egress in the outlet section is set, and the surface between the inlet section and the outlet section is set as a solid wall.

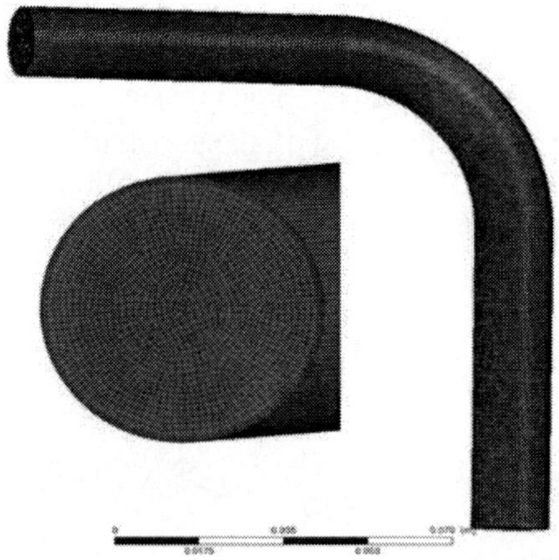

Figure 2. Schematic diagram of heat pipe modeling.

2.3.Model validation

Taking the gas-phase flow experimental device of the straight pipe as a benchmark, the straight pipe model is constructed, and different boundary layer thicknesses and dimensionless distances y+ are used to imitate the gaseous phase stream pressure drop. The applicability of different boundary conditions to the simulation of gas-phase flow pressure cut down in the thermo tube is evaluated.

The following figure shows the gas-phase flow pressure drop values of the straight pipe obtained under different boundary layer thicknesses, and the parameters are adopted from the heat pipe parameters in Table 1.

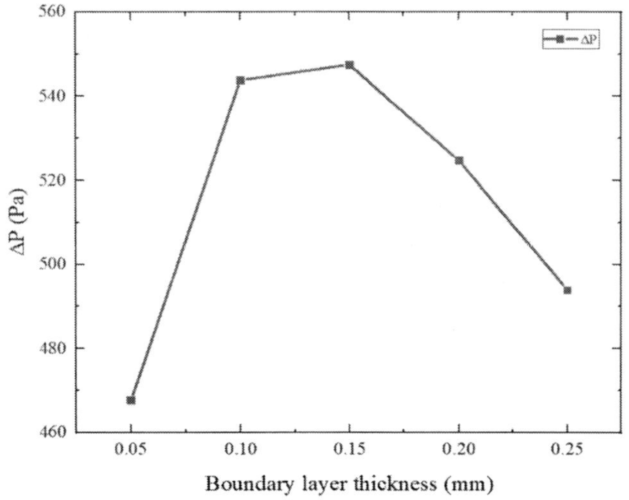

Figure 3. Effect of boundary layer thickness on pressure drop in straight pipe.

The outcome in Figure 3 shows that the gas-phase pressure drop in the heat pipe tends to increase and then decrease as the gauge of the bounding layer enhancement.

For the dimensionless distance y+, the custom wall function factor is adjusted by using the custom wall function process. The basic idea of customizing the wall function can be found in the wall roughness model. The effect of the rough wall surface making the flow field pressure loss increase is realized by reducing the velocity at the first mesh layer of the wall surface,as shown in Figure 4. Therefore, based on the standard wall function, a coefficient less than 1 is added to simulate the shear pressure loss and interface wave pressure loss caused by the countercurrent of the atmosphere and fluid in the heat pipe. In the calculation, it is found that the adjustment coefficient has a big mark under the pressure loss value of the heat pipe, and after many comparisons and constant adjustments, a reasonable coefficient is finally obtained to investigate the gas phase flow features of the thermo tube.

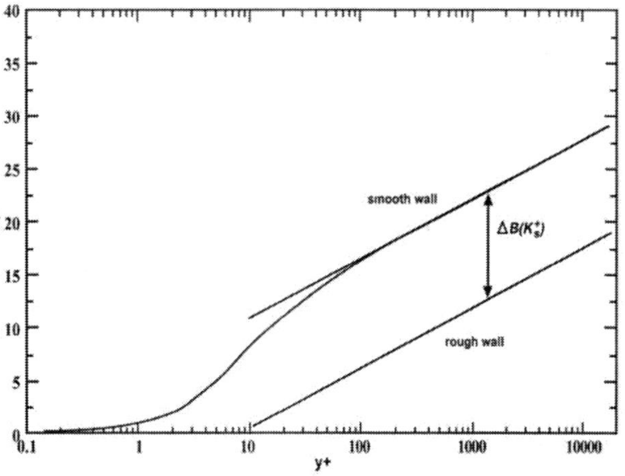

Figure 4. Schematic diagram of wall roughness reducing wall velocity.

Using the thermo tube parameters in Table 1 and adjusting the customized wall function coefficients to simulate the model, the following figure represents the pressure drop values of gas-phase flow under different customized wall function coefficients.

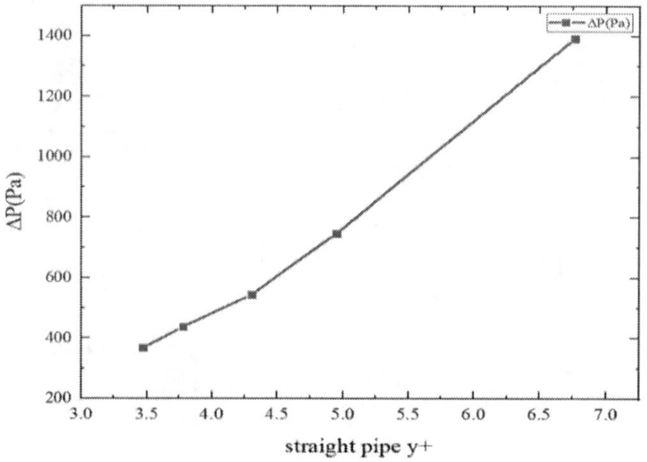

Figure 5. Effect of straight tube y+ change on pressure drop in gas phase flow.

The outcome is represented in Figure 5. We use the heat pipe parameters in Table 1 and adjust the customized wall function coefficients to unfold the numerical calculations for the heat pipe. Calculations show that the gas doping pressure lower in the heat pipe increases as the dimensionless wall distance y+ increases.

2.4. Analysis and discussion

From Figure 6 and 7 to 8, with the gradual rise in the gauge of the border layer, the change in the gas flow rate is not obvious. However, as the gauge of the border layer rises, the gas-phase flow pressure drop increases from 467.62 Pa to 547.51 Pa and then decreases from 547.51 Pa to 493.93 Pa.

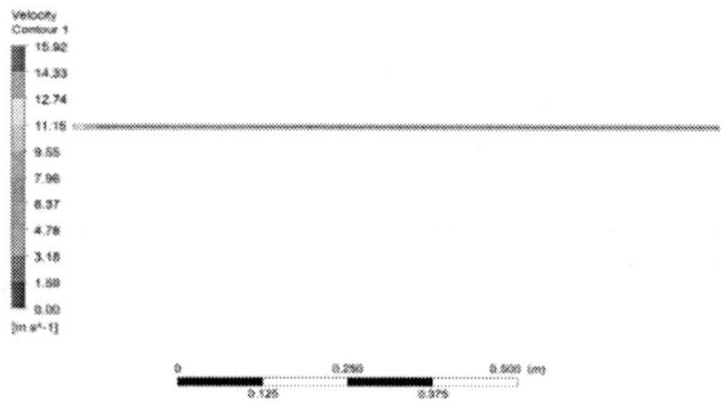

Figure 6. Velocity distribution of gas-phase flow with a boundary layer thickness of 0.05 mm.

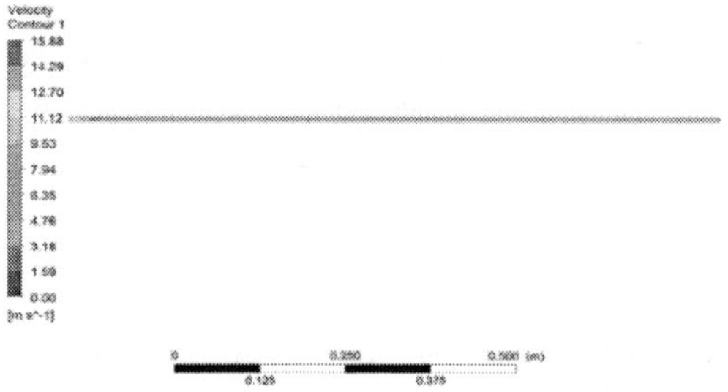

Figure 7. Velocity distribution of gas-phase flow with a boundary layer thickness of 0.15 mm.

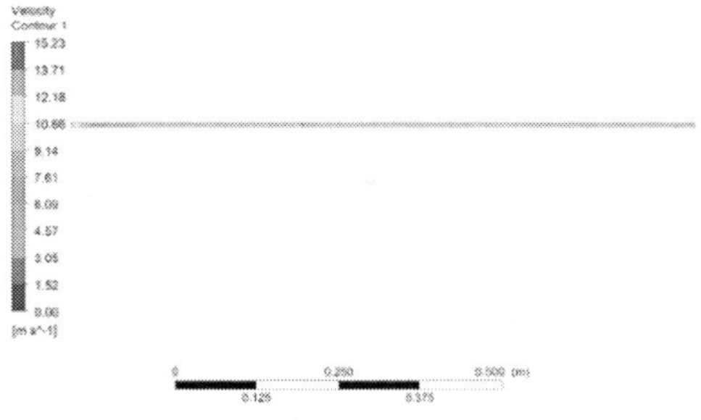

Figure 8. Velocity distribution of gas-phase flow with a boundary layer thickness of 0.25 mm.

From Figure 9 and 10 to 11, the gas discharge velocity rises significantly when the value of y+ gradually increases, the maximum velocity increases from about 15.3 m/s to about 18.2 m/s, and the gas-phase flow pressure drop increases from 367.85 Pa to 1, 392.49 Pa.

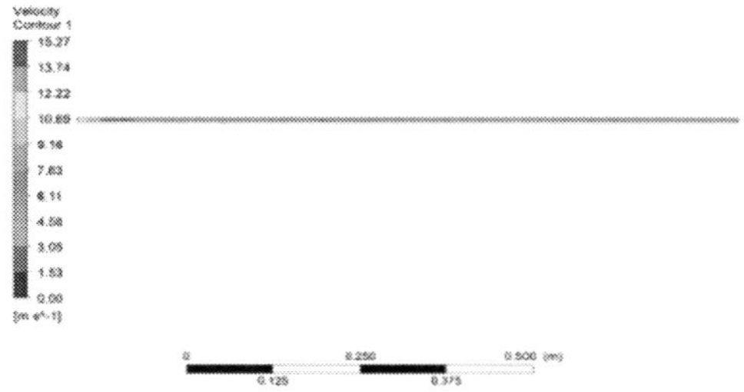

Figure 9. Velocity distribution of gas-phase flow for values of y+3.49.

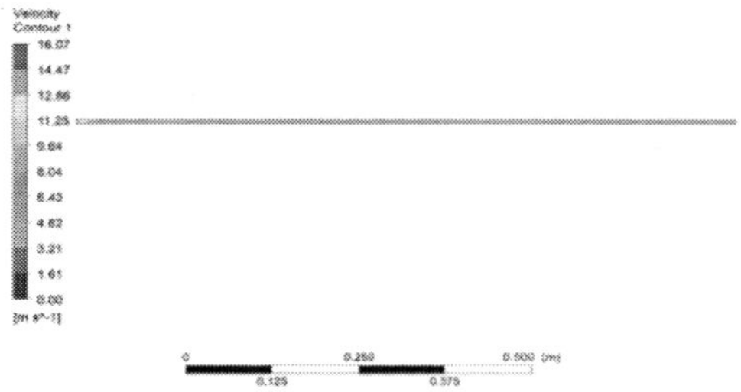

Figure 10. Velocity distribution of gas-phase flow for values of y+4.24.

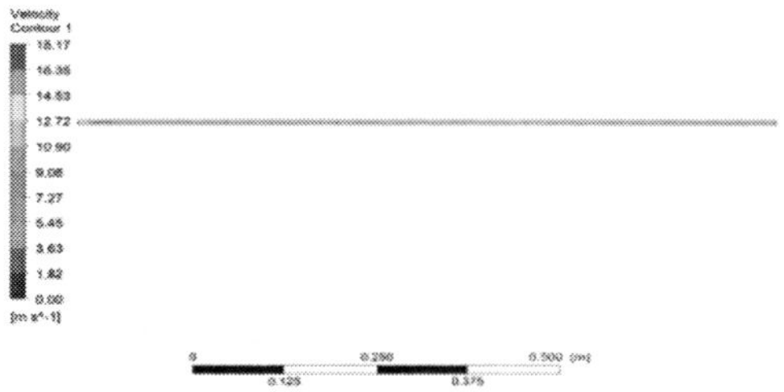

Figure 11. Velocity distribution of gas-phase flow for values of y+6.76.

In summary, by varying the boundary layer thickness and dimensionless wall distance y+, the velocity gradient between the gas-phase fluids in the tube is altered, which results in a change in the maximum gas-phase discharge velocity, correspondingly impacting the gas-phase flow pressure drop for a certain number of other parameters.

3. Conclusions

To go deeply into the law of influence of various elements on the characteristics of gas-phase flow in a thermo tube, the gas-phase flow pressure diminishes gradually with the increase of boundary layer thickness, mainly because of the increase of border layer gauge and the increasing frictional resistance between airflow and wall. However, when the border layer gauge increases to a certain value, the value of the gas-phase flow pressure drop begins to decrease. This is because as the boundary layer thickness increases, the velocity gradient becomes smaller and the frictional resistance gradually decreases. At this stage, the decreasing effect of the frictional resistance begins to exceed the increasing effect of the inertial resistance, which leads to a decrease in the gas-phase stream pressure.

The gas-phase stream pressure gradually increases as the dimensionless wall distance y+ value increases. In the smaller range of y+ values, the gas flow velocity near the wall is low and the effect of inertial drag on the pressure drop is relatively small. However, as the y+ value increases, the wall gas flow rate increases and the inertial resistance begins to significantly affect the pressure drop, increasing overall flow pressure loss.

References

[1] H. Y. Xu, Z. Li, T. J. Shen, et al. Experimental study on heat transfer properties of bent heat tubes [C]. 2009 Annual Conference of Shanghai Refrigeration Association, Shanghai, China, 2009.
[2] Mahboobe M., Amir F. The Effects of Bending on Heat Pipes [C]. ASME 2021 Heat Transfer Summer Conference Collocated with the ASME. 2021 15th International Conference on Energy Sustainability, 2021.
[3] Shen Di, Li Xunfeng, Zhou Jingzhi, et al. Numerical simulation study of laptop heat dissi pation process based on flexible heat pipe [J/OL]. Agricultural Equipment and Vehicle Engineering, 1-7 [2023-12-12]. http://kns.cnki.net/kcms/detail/37.1433.TH.20231031.1628.002.html.
[4] Jin Liqiang, Zhu Jianjun, Wang Yi, et al. Numerical study on the influence parameters of the thermal characteristics of the gravity heat pipe [J]. Fan Art, 2023, 65 (05): 52-57. DOI: 10.16492/j.fjjs.2023. 05.0009.
[5] Shen Chao, Wang Zhuxuan, Liu Yujuan, et al. Numerical simulation study of flow and heat transfer in a parallel flow heat tube [J]. Thermal Science and Technology, 2023, 22 (04): 358-364. DOI: 10.13738/j.issn.1671-8097.020201.
[6] Shen S., Liu R., Yang Y., et al. Condensation character of a stratified flow inside a horizontal tube [J]. Desalination and Water Treatment, 2011, 33 (1-3): 218-223.
[7] Ma Yugao, Liu Minyun, Zhou Dianzhuo, et al. Numerical simulation of heat transfer in heat pipe cooled reactor [C]. 2019: 7. DOI: 10.26914/c.cnkihy. 2019. 062605.
[8] Haibo G., Yuqiu Z., Yongqi L., et al. Numerical simulation of heat transfer performance of different heat exchange tubes in heat recovery steam generator [J]. Process Safety and Environmental Protection, 2022, 163, 506-512.
[9] Fang Pingzhi, Gu Ming, and Han Zhihui. Research on the method of solving the wall function problem in the numerical simulation of atmospheric boundary layer [J]. Vibration and Shock, 2015, 34 (02): 85-90.
[10] Fang Pingzhi, Gu Ming, and Luan Guihan. Numerical simulation of the effect of blocking rate on the surface air pressure coefficient [J]. Journal of Building Science and Engineering, 2013, 30 (03): 101-106.
[11] Luo Kevin, Yang Yi, and Xie Zhuangning. Comparative study of the atmospheric boundary layer in equilibrium states based on the k- ε model [J]. Engineering Mechanics, 2018, 35 (02): 21-29.

Journal of Physics: Conference Series **2808** (2024) 012062 doi:10.1088/1742-6596/2808/1/012062

Study on constitutive model and mechanical performance test simulation of 6082-T6 based on genetic optimization algorithm

Suokun Chen[1], Qinmin Zhang[2*]

[1]China aluminum International Yunnan Aluminum Application Engineering Co. ltd, Qujing, 655000, China

[2]Faculty of Civil Engineering and Mechanics, Kunming University of Science and Technology, Kunming ,650500, China

[*]Corresponding author's e-mail address: kustzqm@163.com

Abstract: Through the tensile test of 6082-T6 aluminum alloy sample, the mechanical properties parameters of the material (specified non-proportional tensile strength, tensile strength, elastic modulus E, elongation at break, proportional limit, Poisson's ratio and strain hardening coefficient N) and the calculation method of strain hardening coefficient N of the constitutive model are analyzed and studied. A method of compiling a genetic algorithm (Ga) based on the Matlab platform to solve the hardening index of aluminum alloy materials is proposed. The results show that 6082-T6 aluminum alloy has good tensile strength and ductility, and the mechanical parameters are less than 3% compared with the national standard. The optimized algorithm is more stable than the traditional method in solving the hardening index n=21. The real stress-strain curve verifies the correctness of the obtained constitutive model, which can be used to predict the mechanical properties of aluminum alloy materials.

1. Introduction

Aluminum is the most abundant metal in the earth's crust, which has the characteristics of lightweight, atmospheric corrosion resistance, low-temperature resistance, bright and beautiful color, easy molding, and wide strength range (elastic limit from 10 MPa to 500 MPa), and is widely used in aviation, manufacturing, and construction. Since 1940, other western countries, such as the United States and Europe, have begun to apply aluminum alloys to the construction industry [1-4]; With the development of the aluminum alloy industry and the promulgation of GB/T 50429-2007 Code for Design of Aluminum Alloy Structures [12], the application of domestic aluminum alloys in building structures has been greatly promoted. Domestic scholars have researched the structural mechanical properties of aluminum alloy [5-11]. 6082 aluminum alloy is an alloy, which is a relatively new nameplate number in 6 series aluminum alloys. Its specified non-proportional extension strength [14] (also called specified plastic extension strength [13], specified non-proportional extension stress or nominal yield strength [12]) and tensile strength are 8% and 17% higher than those of 6061-T6 [12]. On the premise of not losing the extrusion performance of the profile, its corrosion resistance and weldability are superior to other aluminum alloys.

Understanding the mechanical properties and constitutive model of aluminum alloy materials is the key to its popularization and application. Scholars at home and abroad have researched the constitutive

Journal of Physics: Conference Series **2808** (2024) 012062 doi:10.1088/1742-6596/2808/1/012062

relation of aluminum alloy [5-7, 15-17]. Ramberg – Osgood model [17] is a classic model of aluminum alloy, which is constrained by elastic modules, specified non-proportional extension strength, and hardening index. GB/T50429—2007 Code for Design of Aluminum Alloy Joints [12] gives the standard values of elastic modulus and specified non-linear tensile strength but does not give the standard value of hardening index n. The traditional calculation methods of n are the two-point method [18] and the least square fitting method [17]. To ensure the safety of aluminum alloy structure, accurate constitutive relation is extremely important for the mechanical properties analysis of components and materials. Therefore, in this paper, quasi-static tensile test and simulation test are carried out on 6082-T6 aluminum plate sample, and the constitutive relation and mechanical properties are simulated, studied, and analyzed, and a method of solving the exponent in model hardening by genetic algorithm is proposed. Compared with traditional calculation methods, the constitutive relation of materials is obtained, which provides a reference for material engineering applications.

2. Experiment

According to the specification [12], the 6082-T6 proportional sample is designed. The sample size is shown in Figure 1, and the material element composition is shown in Table 1.

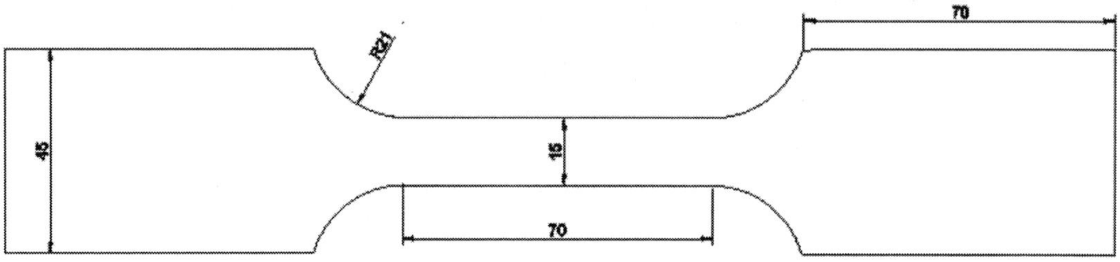

Figure 1. Sample size (unit: mm).

The tensile test instrument is a WDW-200 electronic universal material testing machine. The experimental temperature was room temperature, and the tensile deformation was measured by an extensometer with the same specification as the sample gauge. The longitudinal strain rate was controlled at about $0.7 \times 10^{-4} S^{-1}$, and the micro-strain was measured by a resistance strain gauge.

Table 1. Element composition of 6082 aluminum alloy.

Si.	Fe	Cu	Mn	Mg	Cr	Zn	Ti	Al
0.7~1.3	0.50	0.10	0.4~1.0	0.6~1.2	0.25	0.20	0.10	Bal

3. Experimental results

3.1 Statistical analysis of mechanical parameters

For the same material, its residual stress, extrusion degree, and thickness are the main reasons affecting its mechanical properties. The non-proportional extension strength limit is specified for the longitudinal tensile of 6082-T6 profiles at room temperature under different thicknesses (when the wall thickness t ≤ 5 mm, $f_{0.2} \geq 250$ MPa). The statistical results of elastic modulus E, tensile strength f_u, elongation after fracture e_f, Poisson's ratio ν, $f_{0.1}$ (stress at a plastic elongation of 0.1%), $f_{0.2}$ (stress at a plastic elongation of 0.2%), $f_{0.2}/f_{0.1}$ are shown in Table 2. The measured elastic modulus E, Poisson's ratio ν, elongation at break e_f, and tensile strength f_u of all samples are shown in Figure 2.

Table 2. Statistical values of mechanical properties parameters.

	E/GPa	$f_{0.1}/MPa$	$f_{0.2}/MPa$	f_u/MPa	$e_f/\%$	ν	$f_{0.2}/f_{0.1}$
Average value μ	71.54	290.04	295	328.99	13.8	0.31	1.017
Standard deviation σ	0.013	17.88	18.82	3.178	1.3	0.034	0.005
Variable coefficientσ/μ	0.017	0.061	0.063	0.009	9.6	0.108	0.005
$\mu - 1.645\sigma$	69.522	260.62	264.03	323.77	11.6	0.26	1.008

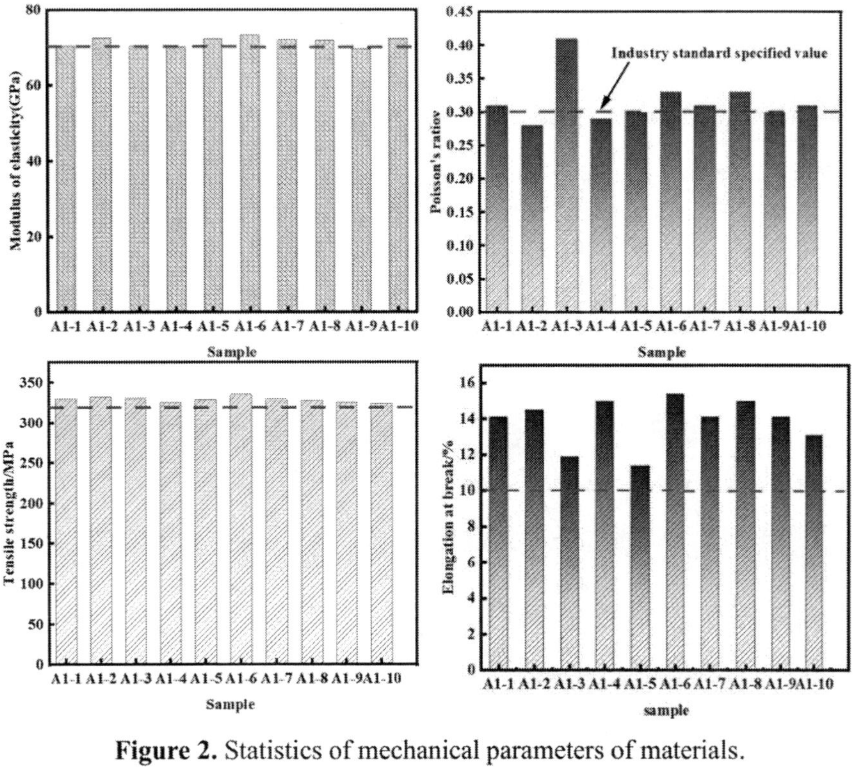

Figure 2. Statistics of mechanical parameters of materials.

According to the 'Code for Design of Aluminum Alloy Structures' [12], the specified non-proportional elongation stress standard value $f_{0.2} \geq 260$ MPa, elastic modulus E = 70 GPa, tensile strength $f_u \geq 310$ MPa, and elongation $e_f \geq 10\%$ of 6082-T6 aluminum alloy under all wall thickness conditions are specified according to the 'Code for Design of Aluminum Alloy Structures' [12]. Observing Fig. 2 and Table 2, we get the following conclusions : (1) The measured value of the elastic modulus E of the 10 6082-T6 aluminum alloy experimental samples involved in this paper is about 70 GPa, with an average value of 70.54 GPa, and the standard value error is +2%; (2) Except that the Poisson's ratio of sample A_{1-3} is greater than 0.4, the Poisson's ratio of other samples is maintained at about 0.3, and the error with the standard value is +3%. (3) The measured range of tensile strength is [329.27, 335.5 MPa]. The 0.005 quantile of the probability distribution function of non-proportional elongation strength $f_{0.2}$ is 264.03 MPa, and the error with the standard value is 1.5 %. (4) The elongation at the break of the samples is greater than 10 %, and the minimum elongation at the break of the sample is 11.9 %.

3.2 Optimization of constitutive model

The Ramber-Osgood power hardening constitutive model [18] is often used to predict the constitutive relationship of materials, and its prediction is close to the actual situation of the constitutive relationship of materials. This model is widely used to describe the constitutive relationship of building materials, passing through a fixed point $(\varepsilon_0 + f_{\varepsilon_0}/E, f_{\varepsilon_0})$. Its general form is:

$$\varepsilon = \frac{\sigma}{E} + \varepsilon_0 (\frac{\sigma}{f_{\varepsilon_0}})^n \tag{1}$$

In Equation 1, E is the elastic modulus; f_{ε_0} is the inertial elastic limit; ε_0 is the plastic deformation corresponding to f_{ε_0}; n represents the degree of strain hardening of the material (the n value range of non-heat-treated aluminum alloy is about 8~15, and the n value of general heat-treated aluminum alloy is greater than 20) [12].

Taking 6061-T6 as an example to reveal the effect of hardening exponent n on the constitutive relationship, the mechanical parameters of 6061-T6 are $E = 68.9\ GPa$, $\varepsilon_0 = 0.002$, $f_{\varepsilon_0} = f_{0.2} = 240\ MPa$, as shown in Figure 3. All stress-strain curves pass through the point P (0.0054, 240 MPa). When n = 1, the model is a straight line through P point; when n > 1, the model is a continuous curve through P point, and the curve change trend is affected by n. As n increases, the stress on the left side of point P and the slope of the linear elastic section of the model gradually increase, and the stress on the right side of point P decreases and tends to be gentle. When n tends to infinity, the model is a bilinear ideal model.

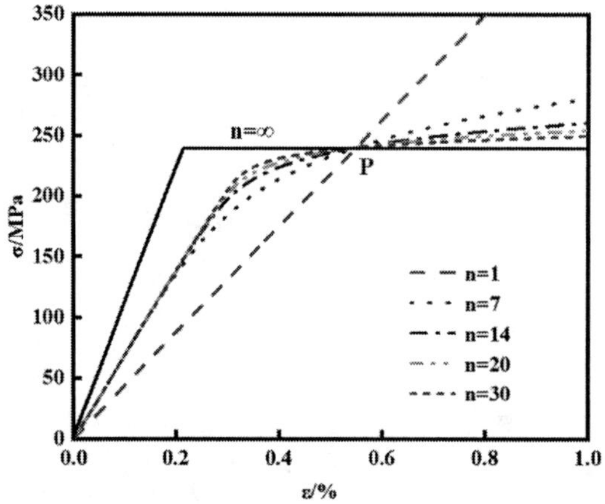

Figure 3. Influence of hardening exponent on constitutive model.

3.2.1 Traditional method to solve hardening exponent

3.2.1.1 Two-point method. By substituting the elastic limit f_{ε_0} and the corresponding plastic strain ε_0, another stress f_x and the corresponding plastic strain ε_x into Equation 1, the coefficient n can be expressed as:

$$n = \frac{ln(\varepsilon_x/\varepsilon_0)}{ln(f_x/f_{\varepsilon_0})} \tag{2}$$

There are many ways to select the reference points $(f_{\varepsilon_0}, \varepsilon_0)$,$(f_x, \varepsilon_x)$ [18]. If f_{ε_0} adopts $f_{0.2}$, then the R-O model is expressed as:

$$\varepsilon = \frac{\sigma}{E} + 0.002(\frac{\sigma}{f_{0.2}})^n \tag{3}$$

Equation 3 is a common constitutive model of aluminum alloy. Substituting $f_{0.1}$ and plastic strain 0.1% into Equation 3, a two-point method for studying the hardening exponent of aluminum alloy microstrain is obtained. The expression is:

$$n_L = \frac{ln(2)}{ln(f_{0.2}/f_{0.1})} \tag{4}$$

The form of Equation 4 is very simple, and the value of n_L is determined by the stress ratio $f_{0.2}/f_{0.1}$. The stress ratio range [1.008, 1.023] was obtained by tensile test, and the range of strain hardening index n [30.381, 77.368]. Since the strain hardening exponent n solved by the two-point method is determined by the stress ratio $f_{0.2}/f_{0.1}$, when the stress ratio decreases, the change rate of n becomes larger. The mechanical properties of materials are affected by the application environment and processing technology. Assuming that $f_{0.1}$ is constant, if $f_{0.2}$ changes slightly, the hardening index will change greatly. This method calculates the hardening index, which is discrete and unstable.

3.2.1.2 Least squares method. The constitutive model is a power function for nonlinear fitting problems. Firstly, the logarithm of the two sides of the equal sign is taken, and then the coordinate is transformed into a linear relationship. Finally, the curve is fitted by MATLAB programming the least square method. That is, after taking logarithms on both sides of Equation 3, let

$$Y = ln(\frac{\varepsilon-\sigma/E}{0.002}),\ X = ln(\frac{\sigma}{f_{0.2}}) \tag{5}$$

Equation 3 is transformed into a linear function Y = nX. First, the partial derivative of the sum of squared deviations between the new sample and the linear formula is calculated, and then it is set to 0 to obtain:

$$n_z = (\sum_{i=1}^{m} X_i Y_i)/(\sum_{i=1}^{m} {X_i}^2) \tag{6}$$

However, the direct linear transformation will have a numerical problem [19]. When there is no error between the experimental data and its expression, the fitting error after linear transformation is 0. The above fitting method takes the empirical formula as the expression, and there is a fitting deviation between the experimental data and the empirical formula. When the deviation square sum $R_O(n)$ (n) (Equation 7) of the linear transformation expression is the minimum, the deviation square sum $R_p(n)$ (Equation 8) of the empirical formula is not necessarily the minimum.

$$R_O(n) = \sum_{i=1}^{m}(Y_i - nX_i)^2 = \sum_{i=1}^{m}[ln(\frac{\varepsilon_i-\sigma_i/E}{0.002}) - nln(\frac{\sigma_i}{f_{0.2}})]^2 \tag{7}$$

$$R_p(n) = \sum_{i=1}^{m}(\varepsilon_i - \hat{\varepsilon}_i)^2 = \sum_{i=1}^{m}[\varepsilon_i - (\frac{\sigma_i}{E} + 0.002(\frac{\sigma_i}{f_{0.2}})^n)]^2 \tag{8}$$

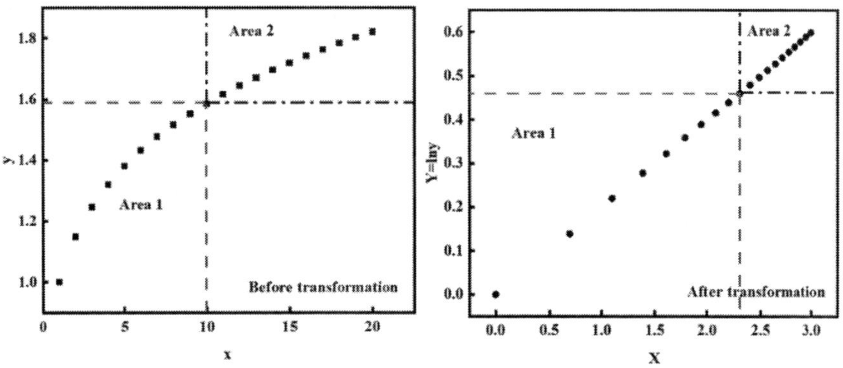

Figure 4. Linear transformation of power function.

Taking the function $y = x^{\frac{1}{5}}$ as an example, the sample (x_i, y_i) is shown in Figure 4. After linear transformation, $Y = \frac{1}{5}X, X = lnx, Y = lny$, new sample $(X_i,\ Y_i)$. It can be seen from the above figure that after linear transformation, Region 1 becomes larger, Region 2 becomes smaller, the original data points gather to the end, and the distribution uniformity changes. Therefore, the original error cannot be evaluated by the transformed error function.

Journal of Physics: Conference Series **2808** (2024) 012062 doi:10.1088/1742-6596/2808/1/012062

3.2.2 Genetic optimization algorithm. The genetic algorithm is a global search optimization algorithm, which simulates the natural evolution process (natural selection, genetic). Starting from the original population, the environmental adaptors were obtained by random selection, crossover, and mutation. The population evolves to the superior area of the search space until the most suitable individual is found, and the optimal solution of the problem is obtained. The calculation steps are as follows:

1. In the strain of [0, 1%], with Equation 1 as the objective function, fitness function such as Equation 9, setting precision m = 3 (decimal point after 3).

$$\varepsilon = sum(abs(\varepsilon_{goal} - \varepsilon_{true})/length(\varepsilon_{true}) \tag{9}$$

Among them, sum ( ) represents the summation function, abs ( ) represents the absolute value function, $length(\varepsilon_{true})$ is the length of the real strain compilation array, ε_{goal} is the fitting strain, and ε_{true} is the real strain.

2. The strain and corresponding stress in the experimental data are taken as individuals, and the individual gene is characterized by a binary system. The chromosome n (coding length) has the following relationship with the accuracy:

$$2^{n-1} \leq (b-a)10^m \leq 2^n - 1 \tag{10}$$

When solving the individual adaptive function value, it is necessary to decode the chromosome. For a chromosome with a length of T (binary string), the value X represented by the decoded chromosome is the following formula:

$$X = a + \frac{b-a}{2^n-1}\sum_{k=1}^{n} b_k 2^{k-1} \tag{11}$$

In the above equation, b and a are the lengths of individual binary encoding, and b_k is the k-th value of the binary string (from left to right).

3. The initial population is generated, and there are 10 populations in this experiment.

4. Selection: Using the roulette wheel selection method, the relative fitness of the individual determines the probability that the individual is selected.

$$P(x_i) = \frac{f(x_i)}{\sum_{j=1}^{N} f(x_i)} \tag{12}$$

$P(x_i)$ is the relative fitness (selection probability) of individual x_i, $f(x_i)$ is the initial fitness of the individual, and $\sum_{j=1}^{N} f(x_i)$ is the original fitness of all populations. The greater the fitness of the individual is, the larger the corresponding fan area on the wheel is, and the greater the probability of selection is.

5. Crossover: Randomly pair two individuals in the population to generate crossover points, and exchange the chromosome genes after the crossover points. The crossover probability is generally between [0.4, 0.99], and its specific value is determined by the population and individual. If the probability is too large, the crossover of generations will be more sufficient, and the original excellent individuals in the population may be destroyed. Therefore, the probability $\partial = 0.4$ is selected for chromosome crossover between individuals to generate new individuals.

6. Variation: Randomly select an individual in the population, and change a gene value of the chromosome with a certain random probability, Step 2 is binary coding of the chromosome, so changing the gene essence is to reverse the binary coding. The mutation probability γ is generally between [0.0001, 0.1], and the probability is affected by population and individual. When the probability is too small, the stability of the solution is high, which easily leads to a local optimum. In this paper, a random mutation operation is performed on random individuals with $\gamma = 0.001$.

7. The new population satisfies the fitness function, and outputs the optimal solution when the fitness function value is minimum, and otherwise returns to Step 4 to continue the iteration. The process is shown in Figure 5:

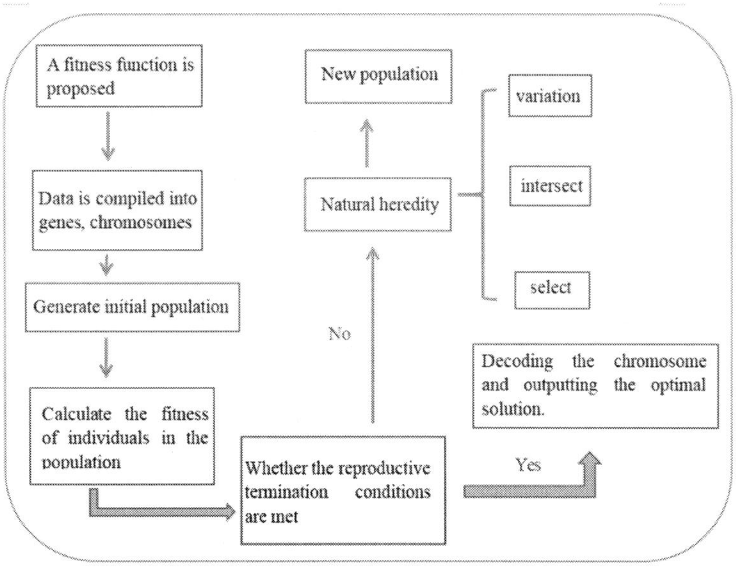

Figure 5. Genetic process.

3.2.3 Comparison of calculation results with traditional methods. According to the steps of the Ga algorithm, the Matlab calculation program is written. By comparison, it is found that the constitutive relationship obtained by the genetic algorithm is consistent with the real stress-strain relationship of the material. Taking the sample A_{1-5} as an example, the hardening index was optimized by two-point method, least square method, and genetic algorithm, $n_L = 40.728$, $n_Z = 8$, $n_g = 20.591$, respectively. The relationship between the constitutive model and the true stress-strain curve is shown in Figure 6. By observing Figure 6a, it is found that the Ga optimization algorithm is closer to the real situation and has a high degree of agreement; the least square method has the lowest degree of agreement; the fitting error of Ga optimization algorithm is close to that of the two-point algorithm, but n_g and n_L are quite different.

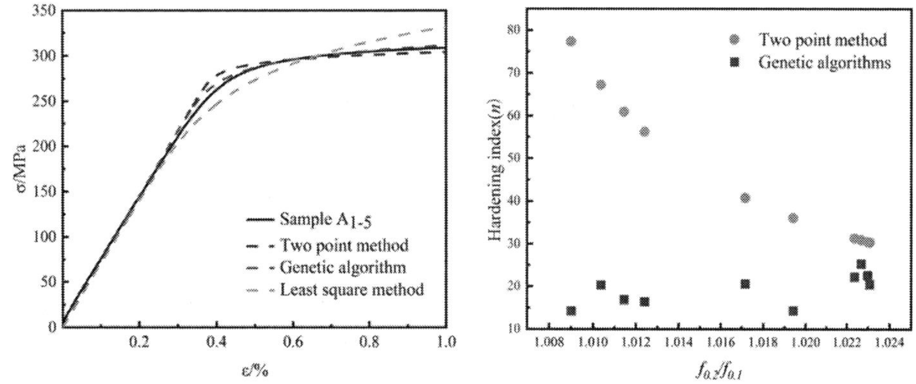

Figure 6. Fitting results of different algorithms and distribution of hardening exponent.

The relationship between the stress ratio $f_{0.2}/f_{0.1}$ and the arithmetic values of n_g and n_L of 10 experimental samples is shown in Figure 6b. It can be seen from Figure 6a that: (1) n_g of all experimental samples is less than n_L; (2) The smaller the value of $f_{0.2}/f_{0.1}$ is, the greater the change of the hardening index of the two-point method is; (3) The variation range of n_g is less than n_L, the former is [14.23～25.258], the latter is [30.381～77.368], and n_g is more stable. Table 3 shows the statistical values of the hardening exponent obtained by the two-point method and genetic

Journal of Physics: Conference Series **2808** (2024) 012062 doi:10.1088/1742-6596/2808/1/012062

optimization algorithm.

Table 3. Hardening index n_L and n_g statistics.

Project	Two-point method		Genetic algorithm	
	n_L	$f_{0.2}/n_L$	n_g	$f_{0.2}/n_g$
average value μ	46.164	7.222	20.316	15.272
standard deviation σ	17.680	2.343	3.516	2.912
variable coefficient σ/μ	0.383	0.324	0.173	0.19
μ-1.645σ	17.08	3.368	14.53	10.481

It can be seen from Table 3 that the coefficients of variation of n_L and n_g are 0.383 and 0.173, and the ga algorithm is more stable in fitting the optimization parameters. The ga algorithm is based on the overall optimization of the constant strain interval of 0 to 1 %, which can better map the mechanical properties of the material in this interval. Steinhsrdt suggests $f_{0.2}/\mathrm{n} = 10$, the value of $f_{0.2}/n_L$ obtained by the genetic algorithm is greater than the recommended value, and the value of $f_{0.2}/n_g$ obtained by two-point method is less than the recommended value [19].

4. Conclusion

(1) The quasi-static tensile test of the 6082-T6 aluminum alloy sample was carried out, and the performance parameters of the material, elastic modulus, Poisson's ratio and specified non-proportional elongation strength were obtained. The elastic modulus $E = 71.54\,GPa$, Poisson's ratio $\nu = 0.31$, specified non-proportional elongation strength $f_{0.2} = 295\,MPa$ and the standard value 264.03 MPa are $+2\%, +3\%,$ and $+ 1.5\%$ respectively. 6082-T6 aluminum alloy plate has better tensile mechanical properties and tensile ductility, but its elastic modulus is small (about 1 / 3 of steel), so it is necessary to pay attention to its stiffness effect in the field of civil engineering.

(2) The strain hardening index n of $Ramber - Osgood$ constitutive model based on the genetic algorithm is about 21, which is more stable than the two-point method and least square method. It is stipulated that the ratio of non-proportional extension strength $f_{0.2}$ to strain hardening index n is greater than the suggested value of Steinhsrdt.

(3) The $Ramber - Osgood$ power hardening constitutive model obtained by the genetic algorithm can be used to describe the constitutive relationship of 6082-T6 aluminum alloy. The constitutive equation is $\varepsilon = \frac{\sigma}{71.54\ \mathrm{GPa}} + 0.002(\frac{\sigma}{264.03\ \mathrm{MPa}})^{20.32}$. Compared with the real stress-strain curve, the constitutive model has high application accuracy and can be used to describe the constitutive relationship in engineering.

Funding: This work was supported by Yunnan Provincial Department of Transportation Science and Technology Plan Project(Yun Jiao Ke Jiao Bian [2021] No. 29)

references

[1] Kato M, Isobe Y, Ohtsuka T, et al. Corrosion Behavior of Aluminum Alloys in Aqueous Solution Containing Ethylene Glycol and NaCl [J]. Zairyo-to-Kankyo, 2009, 40(11): 742-746.

[2] HE Dève, Evans A G. Twin toughening in titanium aluminide [J]. Acta Metallurgica Et Materialia, 1991, 39(6): 1171-1176.

[3] G, Elkabir, and, et al. Microstructural evaluation of a high-energy high-rate p/m processed aluminum alloy [J]. Scripta Metallurgica, 1986, 20(10): 1411-1416.

[4] Cuntz J M, Villatte M. Physico-chemical characterization of aluminium alloy surfaces after sulphochromic pickling prior to bonding [J]. International Journal of Adhesion & Adhesives, 1986, 6(4): 221-224.

[5] Wang Li, Guo Xiaonong, Liu Linlin, et al. Study on hysteretic constitutive model of domestic aluminum alloy materials [J]. Journal of Hunan University: Natural Science Edition, 2018, 45(11): 29-36.

[6] SHEN Zuyan, GUO Xiao-nong. Research on design formula and reliability of aluminum alloy structural members [J]. Progress of building steel structure, 2007, 9 (6): 1-11.
[7] Dou Zhijia, Zhu Xiao, Wei Sun, et al. Ramberg-Osgood constitutive model construction and method research of 2024-O aluminum alloy profile [J]. Nonferrous Metal Processing, 2020, 49(4): 51-54.
[8] Gao Shuai, Ji Yujie.6063 Dynamic Mechanical Constitutive Model of Aluminum Alloy BP Neural Network [J]. Journal of Shenyang Ligong university, 2020,39(4):48-52.
[9] Feng Ruoqiang, Wang Xi, Zhu Jie, et al. Study on the overall stability of assembled single-layer aluminum alloy reticulated shell structure of Beijing New Airport [J]. Journal of Architectural Structure, 2020, 41(4): 11-18.
[10] Feng Zhenyu, Li Henghui, Liu Yi, et al. Comparison of constitutive model and failure model of 7075-T7351 aluminum alloy at medium and low strain rates [J]. Materials Guide, 2020, (12): 12088-12093.
[11] Zhou Guxin, Lang Yujing, Du Xiuzheng, et al. Dynamic mechanical response and J-C constitutive relation of 7A62 aluminum alloy [J]. transactions of nonferrous metals society of China, 2021, 31(1): 21-29.
[12] GB 50429-2007, Code for Structural Design of Aluminum Alloys [S].
[13] GB/T 32498-2016, Tensile Test Method of Metal Matrix Composites at Room Temperature [S].
[14] GB/T 32498-2016, Tensile Test Method of Metal Materials at Room Temperature [S].
[15] Zhang Yu, Wang Binwen, Liu Xiaochuan, et al. Mechanical properties and constitutive relation of 2024-T42 aluminum alloy at low and medium strain rates [J]. Vibration and impact, 2020, 39(2): 249-254.
[16] Zhang Ziqun, Jiang Zhaoliang, Wei Qingyue. 2219 Dynamic mechanical properties and constitutive relation of aluminum alloy [J]. Materials Engineering, 2017, (10): 47-51.
[17] Guo Xiao-Nong, Shen Zuyan, Li Yuanqi, et al. Study on constitutive relation and physical and mechanical properties of domestic aluminum alloy materials for structures [J]. Journal of Building Structure, 2007, (06):110-117.
[18] Demartino A, Landolfo R, Mazzolani F M. The use of the Ramberg-Osgood law for materials of round-housetype [J]. Materials and Structures, 1990, 23(1): 59-67.
[19] Liu Jiang. Nonlinear direct fitting method of power function [J]. Seismological Science of North China, 1984, 1(4): 102-104.

doi:10.1088/1742-6596/2808/1/012063

Influence of pre-tightening force on bearing capacity of composite cylinder

Bo Yang[1*]

[1]Shanghai Institute of Special Equipment Inspection and Technical Research, Shanghai, 200062, China

*Corresponding author's e-mail: yangbo@ssei.cn

Abstract: For investigating the effect of preload on the bearing capacity of the composite hydrogen storage cylinders, the FEM of fiber full-wound aluminum liner cylinder is established by ABAQUS software. Stress calculation was carried out with and without preload. The results show that under working pressure, the application of pre-tightening pressure can cause the reduction of lining stress level and increase fiber layer stress, and the application of pre-tightening pressure can improve the stress distribution of the lining and fiber layer.

1. Overview

Storage and transportation of hydrogen energy and on-board hydrogen storage are the core technologies that affect the development of hydrogen fuel cell vehicles. Hydrogen storage and transportation, as a bridge, connects hydrogen production and application and is a key factor restricting the development scale of the hydrogen energy industry. Up to now, almost all storage equipment and transmission equipment for hydrogen is under pressure, among which a high-pressure hydrogen storage bottle is a common pressure-bearing equipment, which has been widely used in underwater submarines, aerospace, and hydrogen energy fuel vehicles [1-4]. At present, China's hydrogen fuel vehicles mainly use Type III hydrogen storage bottles as energy storage devices, which are all-wound aluminum-lined carbon fiber structures, in which the aluminum lining mainly plays a role in preventing gas leakage and only bears a small portion of the pressure load. As the main pressure-bearing component, the composite material layer stands for most of the container pressure load (more than 75%), and carbon fiber and glass fiber as important composite materials are usually selected [5].

When the load borne by the gas cylinder is within the elastic limit of the lining, the load-carrying level of the gas cylinder is not high, and the tensile characteristics of the fiber layer cannot be fully exerted. When the load is further enhanced, the fiber layer still stays at the elastic deformation stage, while the stress of the metal lining has exceeded its yield limit and plastic deformation occurs at the same time, which enhances gas cylinder bearing capacity. After unloading, the glass fiber can completely recover its elasticity, while the metal lining still has residual deformation, which makes the deformation of the two inconsistent [6-8]. After repeated loading and unloading, the cylinder will be fatigued quickly, so it needs pretreatment before the cylinder is used. For improving the bearing capacity of carbon fiber composite cylinders and ensuring the safety of the cylinder, it is necessary to investigate the pre-tightening force influence on the carrying capacity of composite cylinders.

In this study, the finite element model is established for aluminum alloy lined and fiber-wrapped composite cylinders, and the effect analysis of preload treatment on the load-bearing capacity of the composite cylinder is carried out.

2. Modeling of filament winding composite cylinders

Considering the symmetry of cylinder load and structure, a quarter model of aluminum alloy lined and fiber-wrapped composite cylinder is established. The green part shown in Figure 1 is the carbon winding layer, and the gray part is the inner container with a thickness of 2.7 mm. Aluminum alloy 6061 is

Journal of Physics: Conference Series **2808** (2024) 012063 doi:10.1088/1742-6596/2808/1/012063

chosen as the liner material, and Table 1 lists the material parameters. Material T700/M21 with strong parameters is selected as the winding layer of the composite material, and its properties are shown in Table 2.

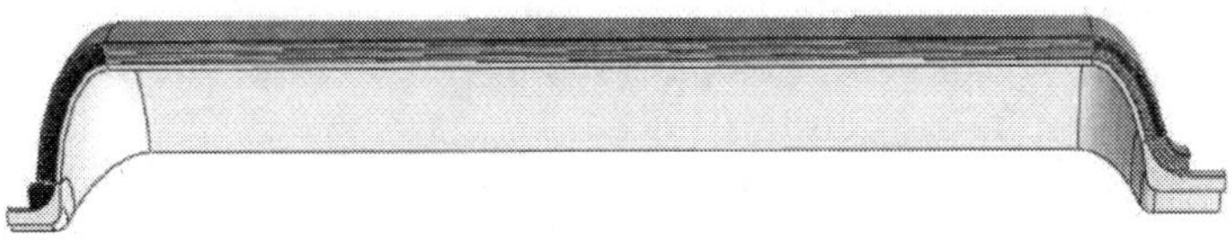

Figure 1. Model diagram of the inner lining and a winding layer of fiber wound composite gas cylinder.

Table 1. Material parameters of cylinder lining.

Material	Young's modulus (GPa)	Poisson's ratio	Yield strength (MPa)	Ultimate strength (MPa)
Aluminum alloy 6061	96	0.324	298	330

Table 2. Mechanical performance parameters of T700/M21.

Parameter	Value	Parameter	Value
E1	130	Xt	2, 080
E2	7.7	Xc	1, 250
E3	7.7	Yt	60
v12	0.3	Y	140
v13	0.3	Sxy	110
v23	0.35	Gft (N/mm)	133
G12	4.8	Gfc (N/mm)	10
G13	4.8	Gmt (N/mm)	0.5
G23	3.8	Gmc (N/mm)	1.6
		Gs (N/mm)	1.6

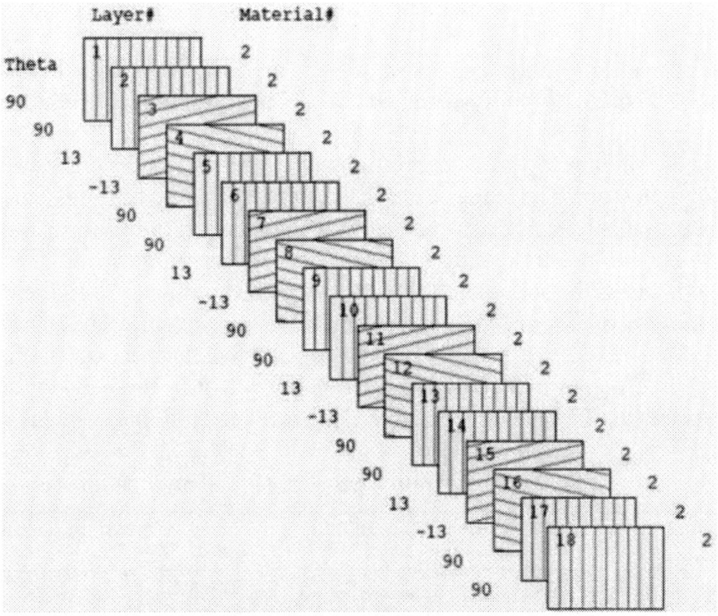

Figure 2. Fiber layer sequence.

Journal of Physics: Conference Series **2808** (2024) 012063 doi:10.1088/1742-6596/2808/1/012063

The winding layer consists of a circular winding layer and a spiral winding layer. According to the calculation, for the single circular-wound layer, the thickness is 0.314 mm, and for the single spiral-wound layer, the thickness is 0.343 mm. The stacking angles, from inside to outside are {[90°] 2, [90°] 2, [±12.8°] 2, [90°] 2, [±12.8°] 2, [90°] 2, and [±12.8°] 2, ...}, as shown in Figure 2.

The boundary conditions of composite cylinders finite element model are determined by the cylinder structure and real working conditions. For the 1/4 finite element model, the 1/4 section of the gas cylinder is subjected to symmetric constraint displacement boundary conditions, the bottle mouth is subjected to axial displacement constraint, and the inner surface of the inner tank is subjected to uniform load. Figure 3 shows the model boundary conditions.

Figure 3. Boundary conditions for composite cylinders.

The fiber-wound layer of the composite cylinder model is set to 22 circumferential-wound layers and 18 spiral-wound layers in ABAQUS. The cylinder part of the cylinder liner adopts the C3D8R three-dimensional solid element, and the fiber winding layer adopts the SC8R continuous shell element. Finally, the aluminum alloy liner is divided into 34, 920 units, and the carbon fiber winding layer is divided into 389, 760 continuous shell units. Figure 4 shows the finite element mesh mode for composite cylinders.

Figure 4. 1/4 finite element mesh model for composite cylinders.

3. Stress analysis of pre-tightening treatment of fiber winding composite cylinders

According to GB/T 35544-2017 "Fully-wrapped carbon fiber reinforced cylinder with an aluminum liner", the cylinder analysis model established by ABAQUS is used to calculate the stress distribution of aluminum gasket and winding layer under pre-tightening pressure, zero pressure after pre-tightening, and nominal working pressure. The material nonlinearity of aluminum bushing, anisotropy of composite material, and geometric nonlinearity of structure are considered in the analysis model. For the sake of illustrating the influence of pre-tightening treatment on load-carrying capacity, two sets of simulation conditions are set up. One is the stress analysis of external load without pre-tightening pressure, and the other is the stress analysis of external load under pre-tightening pressure. The specific application of the load is shown in Table 3. There are three analysis steps in the whole process. The first step is to load the internal pressure to the preset pre-tightening pressure (65 MPa), the second step is to unload the pre-tightening pressure (0 MPa), and then the third step is to load it to the nominal working pressure (55 MPa).

Table 3. Pressurization process of load application.

Pre-tightening pressure (MPa)	After unloading (MPa)	Working pressure (MPa)
65	0	55

The stress distribution of the composite cylinders under pre-tightening pressure, zero pressure after the preload pressure is released, and working pressure are shown separately in Figure 5 to Figure 7.

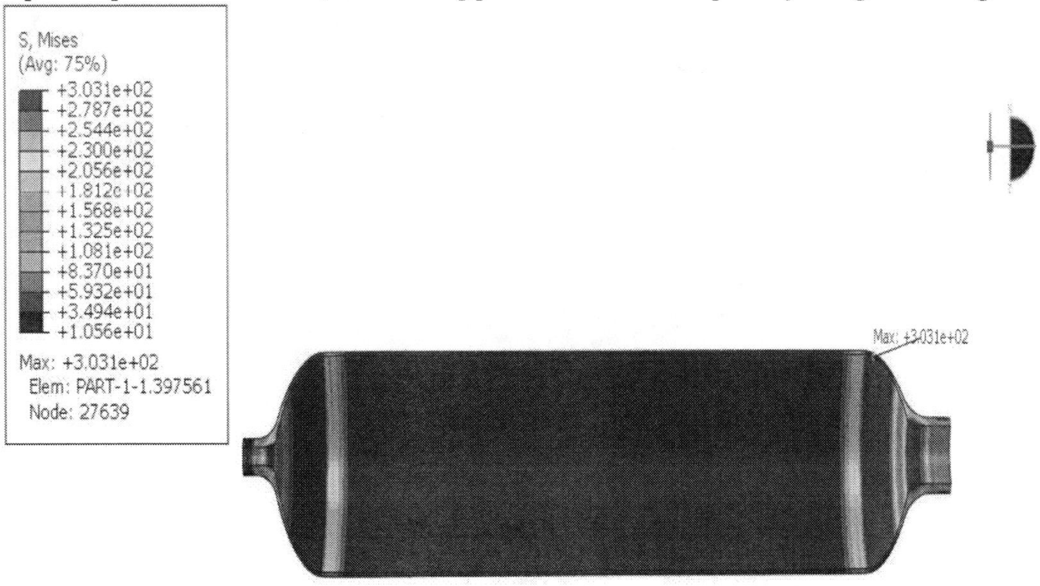

Figure 5. MES (Mises equivalent stress) distribution at 65 MPa.

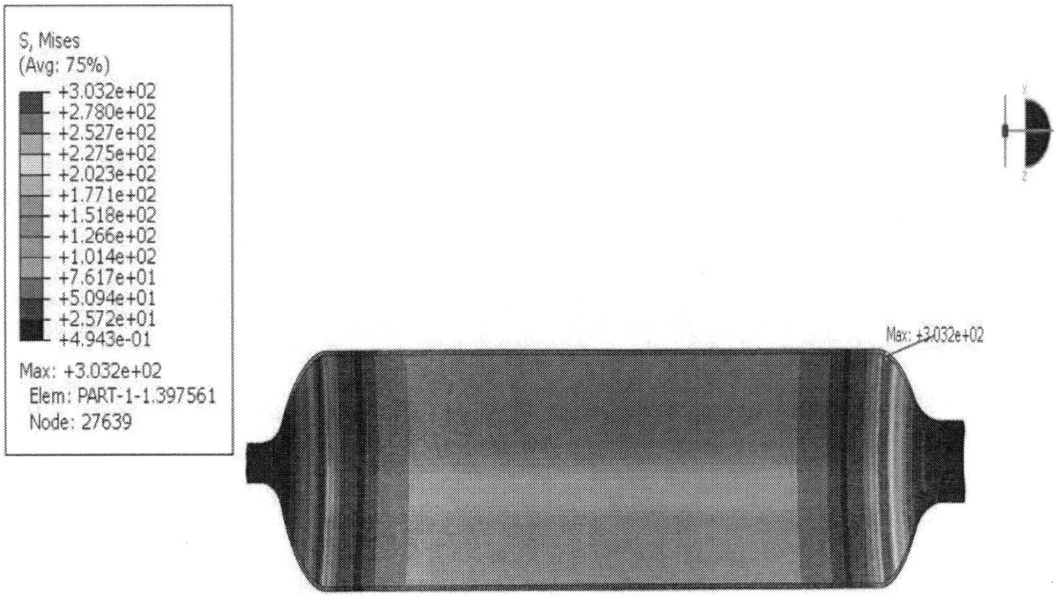

Figure 6. MES distribution at 0 MPa after preloading.

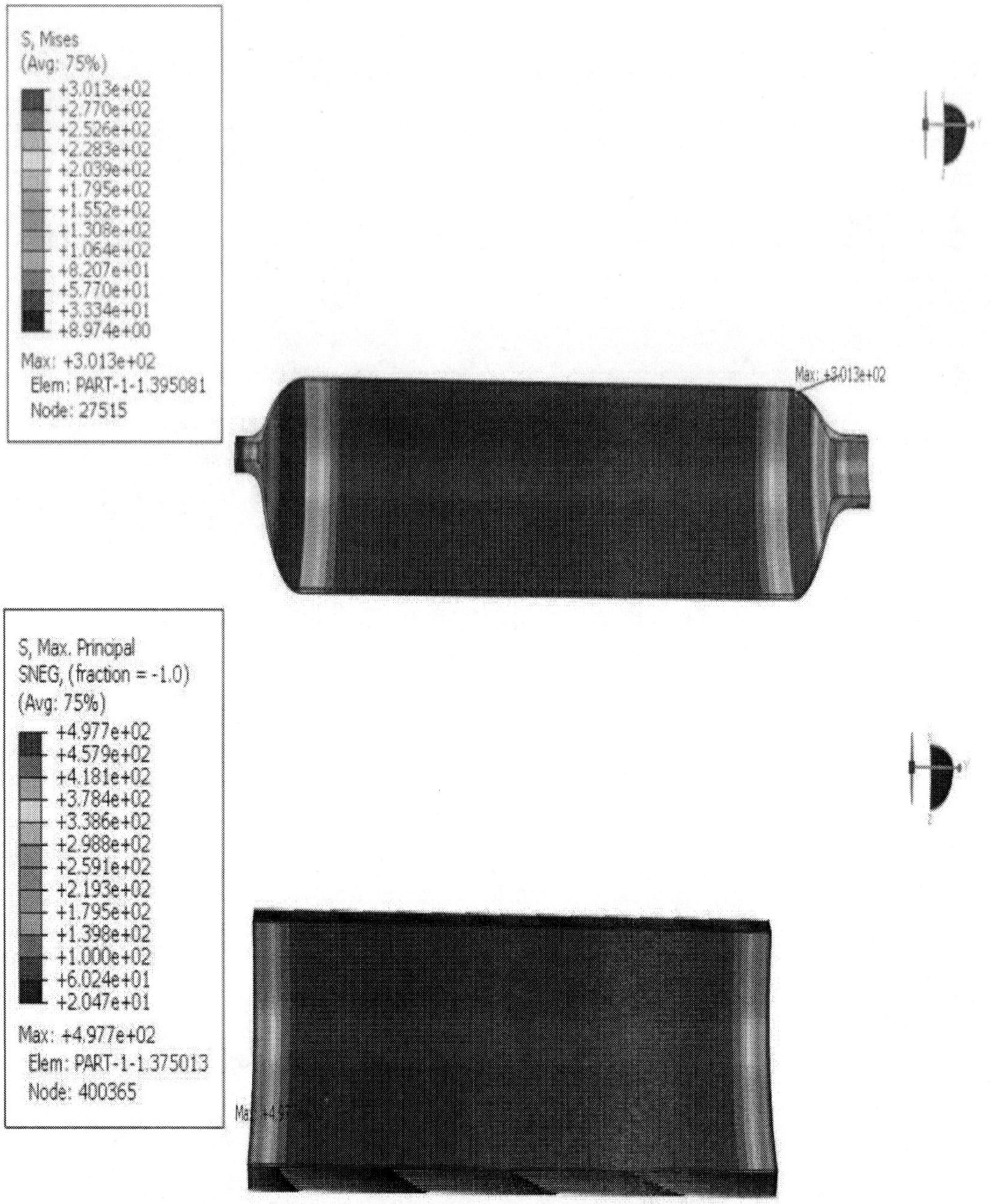

(a) No pre-tightening force

Journal of Physics: Conference Series **2808** (2024) 012063 doi:10.1088/1742-6596/2808/1/012063

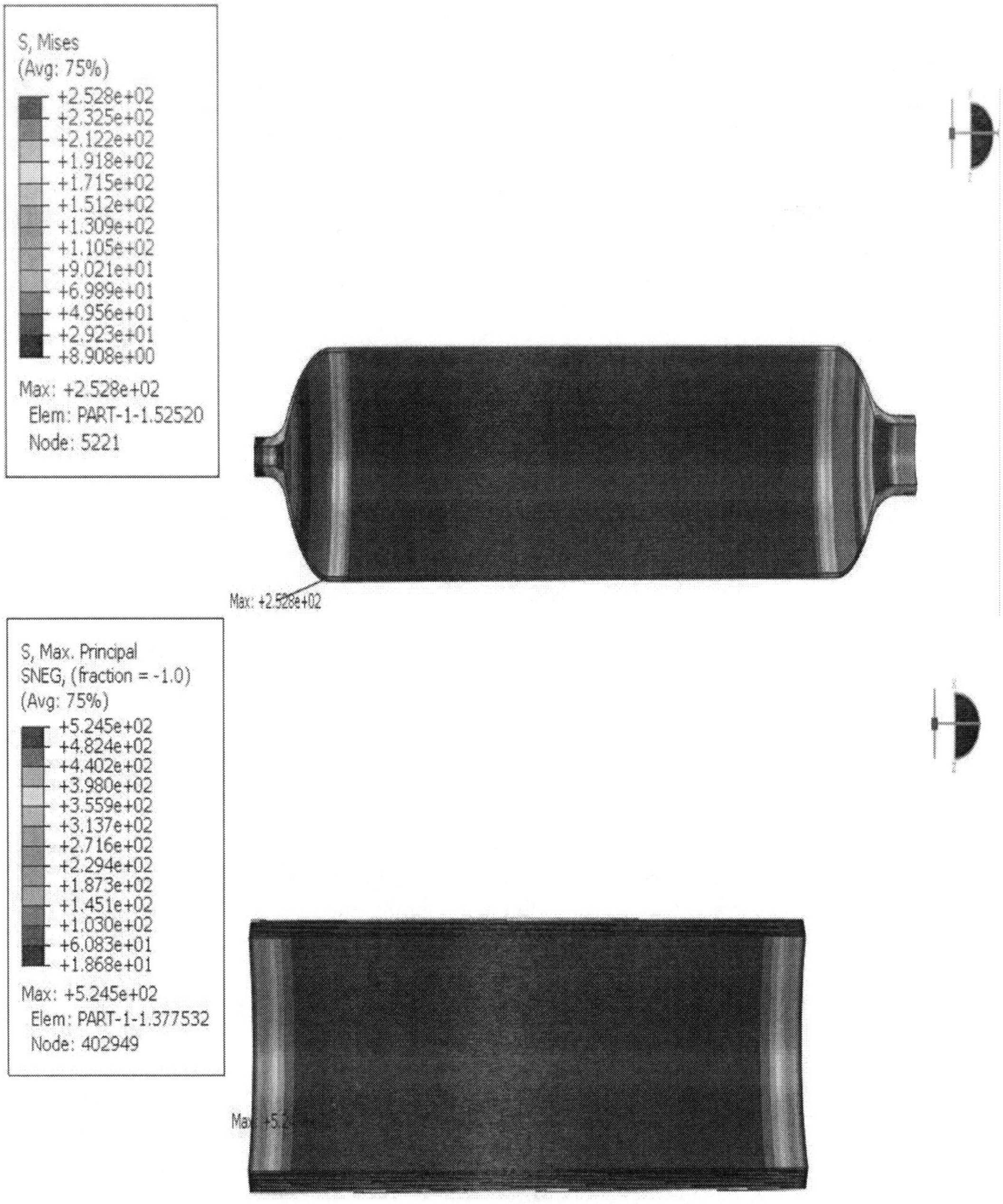

(b) Under preload force

Figure 7. MES (Mises equivalent stress) and FPS (First principal stress) distribution at 55 MPa.

Figure 7 shows a comparison of the MES distribution and the FPS distribution of the carbon fiber layer under working pressure, i.e. 55 MPa, with and without the preload applied to the cylinder. When the working pressure is 55 MPa, applying the pre-tightening force will reduce the MES value of the composite cylinder body, and the stress value will decrease from 301.3 MPa to 252.8 MPa, with a decrease ratio of nearly 16%. At the same time, the maximum FPS of the circumferential carbon fiber layer increases from 497.7 MPa to 524.5 MPa, with an increase ratio of nearly 5.4%. According to the data comparison, the conclusion can be obtained that under working pressure, applying preload can effectively lower the inner liner stress level and enhance the carbon fiber layer stress. The pre-tightening force improves and optimizes the stress distribution of the inner liner and the fiber layer.

4. Conclusion

Due to the structure and material characteristics of the composite cylinder, according to the simulation results, under working pressure, the reason for the improvement of the fiber layer stress distribution of the composite cylinder is that the maximum MISES stress would exceed the yield limit of the lining material under the preload, and the lining material would produce plastic yield under the action of the preload. For aluminum alloy, the Bauschinger effect occurs when the preload pressure is removed, while the residual compressive stress remains and the outer fiber layer generates a certain amount of tensile stress. When the cylinder is used normally under the working pressure, the tension stress generated by the lining and the residual stress generated after pre-tightening are removed. The maximum stress of the lining is weakened and the stress of the fiber layer is increased. In the section direction from the lining to the fiber layer, the stress distribution is homogenized, which improves the bearing capacity of the cylinder.

Through the simulation of a fiber-wound composite cylinder with and without pre-tightening, the simulation results show that under working pressure, the pre-tightening force applied can reduce the stress level of the liner and intensify the fiber layer stress. That is, the application of pre-tightening force can optimize the stress distribution situation of the fiber layer and liner under working pressure.

Acknowledgments

The Research Project of the Shanghai Institute of Special Equipment Inspection and Technical Research (Project No. 2023-07Y) supported this study.

References

[1] BAI Wenyi, LIAO Yingqiang, and LIU Yongqiong. Research progress on optimization of carbon fiber composite winding cylinders [J]. CARBON, 2020 (1): 36-41. DOI: 10.3969/j.issn1001-8948.2020-01-005.

[2] MA Quansheng, WANG Wenyi, and LU Zhaojun. Development and Application Progress of Composite Full Wrap Tank for Hydrogen Storage [J]. Hi-Tech Fiber and Application, 2023,48 (3): 3-19. DOI: 10.3969/j.issn.1007-9815.2023.03.001.

[3] WANG Te, LIU Yongqiong, and LIAO Yingqiang. Application investigation of high-performance carbon fiber for composite pressure cylinder [J]. CARBON, 2015 (4): 23-27. DOI: 10.3969/j.issn1001-8948.2015.04-004.

[4] LI Jian, ZHANG Lixin, LI Ruiyi, YANG Xiao, and ZHANG Ting. High-pressure gaseous hydrogen storage vessels: Current status and prospects [J]. Energy Storage Science and Technology, 2021, 10 (5): 835-1844. DOI: 10.19799/j.cnki.2095-4239.2021.0309.

[5] Geng fagui. Study on cumulative damage performance of fully wound composite cylinders [D]. Liaoning: Dalian University of Technology, 2021. https://d.wanfangdata.com.cn/thesis/ChJUaGVzaXNOZXdTMjAyNDAxMDkSCUQwMjQ5MDI0NBoIYXYybnYzemI%3D.

[6] Wang Xiaoyan. Research on the Load-Carrying Capacity of Composite Vessels with Mental Lining [D]. Heilongjiang: Harbin Institute of Technology, 2007. DOI: 10.7666/d.D273034.

[7] Blanc Vannet P., Papin P., and Weber M. Sample scale testing method to prevent collapse of plastic liners in composite pressure vessels [J]. International Journal of Hydrogen Energy.2019, 44 (17), 8, 682-8, 691. DOI: 10.1016/j.ijhydene.2018.10.031.

[8] Castagnet S. and Grandidier J. C. M. Mechanical Testing of Polymers in Pressurized Hydrogen: Tension, Creep, and Ductile Fracture [J]. Experimental Mechanics, 2012, 52 (3), 229-239. DOI: 10.1007/s11340-011-9484-1.

Journal of Physics: Conference Series **2808** (2024) 012064 doi:10.1088/1742-6596/2808/1/012064

Establishment and prediction of liquid permeability model for metal porous materials

Shengcun Lin[1] and Wenxiu Feng[1*]

[1]School of Automotive Engineering, Liuzhou Vocational & Technical College, Liuzhou, 545005, China

[*]Corresponding author's e-mail: sclin@lzzy.edu.cn

Abstract: The demand for metal porous materials in the field of liquid purification and filtration is increasing, so it is necessary to master the permeability of liquid through porous materials. Currently, the liquid permeability of porous materials is mostly tested through the method of flow pressure difference, and predicting the permeability through the establishment of a model is an innovative topic. Experimental methods determine the average pore size and porosity of metal porous materials by using the mass volume method and the bubble method. Then, a flow pressure difference curve model is established to predict the permeability of other samples. Compared with experiments, the established model accurately predicts the unit pressure drop of flow in porous materials at the equivalent Reynolds number. This work can provide a reference for the study of the permeability of porous materials.

1. Introduction

Porous materials are commonly used in the fields of filtration and permeation [1]. Due to the porous nature of porous materials, pressure fluids with a certain velocity can penetrate porous materials, and the ability of this fluid to penetrate porous materials is called permeability. The permeability of porous materials is mainly affected by two aspects: one is the distribution of the pore structure of the porous material itself, and the other is the characteristics of the fluid itself, such as viscosity [2]. The types of fluids flowing through porous materials can be classified into two categories: gas and liquid. During the development of industry, the demand for metal porous materials in the field of liquid purification and filtration is increasing day by day, so it is essential to test the permeability of liquid [3].

According to the literature review, in addition to Kozeny Carman's empirical formula [4], other scholars have also proposed various liquid permeability models to describe the flow characteristics of liquids in porous materials. For example, an empirical model proposed by Ergun [5] is used to calculate the permeability of fluids in a packed bed. By combining the theoretical models of Blake and Kozeny, and waiting for the Blake-Kozeny formula [6], the influence of the geometric shape and size distribution of particle gaps on permeability was further considered. This article is based on a porous material formed by composite sintering of metal powder and wire mesh. The average pore size and porosity of the material are tested, and the liquid permeability model of the material is obtained through formula derivation to predict its permeability performance. Due to the complex distribution of internal pores in porous materials and the presence of many influencing factors, there is no universal prediction model. Compared with existing models, this paper simplifies the internal complex pores and establishes a mathematical model based on the average pore size of porous materials, which has a reference value.

2. Experimental section

2.1 Pore size testing of porous materials

Different types of porous materials correspond to their applicable pore-size testing methods [7]. This article uses the PSDA-20 pore size analysis tester (Figure 1 (a)) to test the pore size parameters of porous materials. The bubble method [8] is to first immerse the porous material in a liquid, with the aim of wetting all the pores inside the sample. Then the sample is taken out and placed in a sealed container that can be ventilated. The container is placed in a liquid and gas is introduced into the container. When the gas reaches a certain pressure, the liquid inside the porous material will be blown out, and bubbles will be generated during the blowing process. Next, the liquid inside the porous material is blown dry and its flow pressure curve is tested with gas (Figure 1 (b)). The sample soaked in porous material is called a wet sample, and the flow pressure curve tested is called a wet curve. The flow pressure curve measured in the absence of liquid in the internal pores of porous materials is called a dry curve. A semi-dry curve is a straight line that corresponds to half of the gas flow rate of the measured porous material. Through testing, the main parameters for pore size testing of porous materials can be obtained, including maximum pore size, minimum pore size, medium flow pore size, and average pore size.

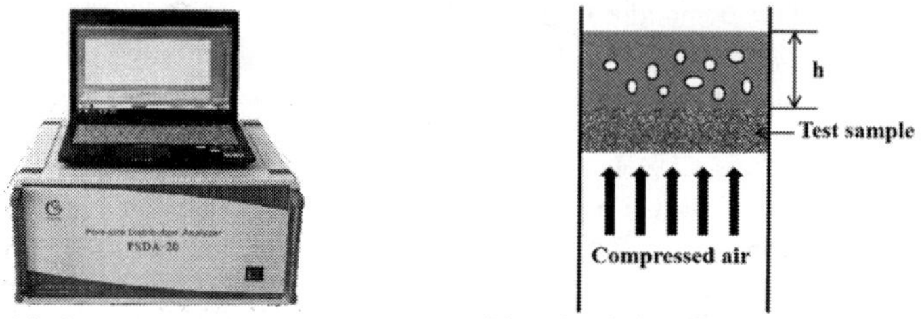

(a) Aperture tester (b) Principle of aperture testing

Figure 1. Bubble method for testing pore size and pore size distribution.

2.2 Porous material permeability testing

The liquid permeability of porous materials is tested by pressurizing the liquid [9, 10], as shown in the schematic diagram of liquid permeability testing in figure 2. Considering the characteristics of pulse water supply in water pumps, the inability to achieve constant pressure during the testing process can lead to poor stability of test data. A water storage tank is introduced between the water pump and the pressure regulating valve, and the water pump is used to store and pressurize the tank to a certain pressure, which can meet the requirement of maintaining stable water pressure during the test process. Therefore, the water storage tank also plays a buffering role during the testing process, ensuring the stability of the test. The water pump draws water from the water source to the water tank and inputs a certain pressure of water into the tank. The water flow in the tank passes through the already set pressure regulating valve and is adjusted by the flow meter to output a certain flow rate of water into the porous material [11]. At this time, the sensor collects the water pressure difference at both ends of the porous material and displays it on the digital meter. The sample is placed inside the nut, and sealing rings are placed at both ends of the sample. The pipes and nuts at both ends are tightly connected to ensure sealing, ensuring that the pressure liquid will not leak out during the test process and obtaining accurate test data.

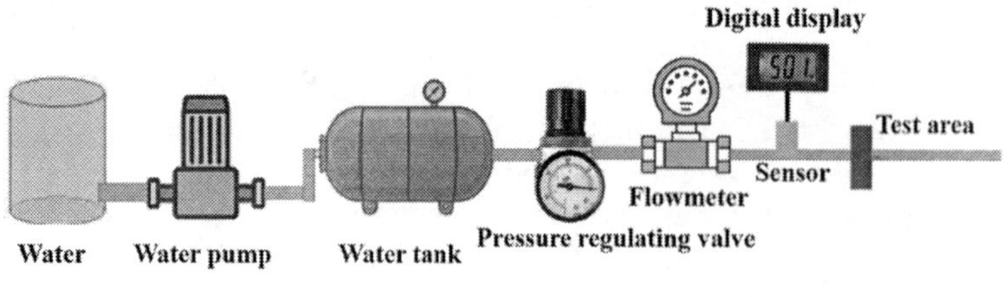

Figure 2. Experimental schematic diagram.

3. Result and prediction

3.1 Aperture test results

The average pore size test results of three samples are shown in Table 1. It is obvious that although the three samples have similar thicknesses, they have different porosity, which is due to different preparation process parameters. The porosity of Sample 1 is 25.6%, corresponding to an average pore size of 6.35 μm. It is 1.68 times that of Sample 3 (with a porosity of 18.2%), which indicates that as the porosity increases, the average pore size inside the porous material also increases. This is because the larger the porosity is, the larger the pores there are, resulting in an increasing average pore size [12, 13].

Table 1. Average pore size test results of three samples.

Sample Number	Porosity (%)	Thickness (mm)	Average Pore Size (μm)
Sample 1	25.6	2.08	6.35
Sample 2	21.2	2.06	4.56
Sample 3	18.2	2.03	3.87

3.2 Liquid permeability test results

Table 2 shows the permeability test results of three samples of hydraulic fluids. It can be seen from the table that as the liquid flow rate increases, the pressure difference formed at both ends of the samples also gradually increases. Comparing the three samples, it is found that under the same flow rate, the smaller the porosity is, the greater the pressure difference forms. This is because samples with lower porosity have a relatively larger number of small pores, resulting in a relatively smaller average pore size [14]. This leads to greater obstacles for fluid flow through the material, resulting in a larger pressure difference. When the liquid flow rate is 0.56 m/s, Sample 3 forms a pressure difference of 588 Kpa, while the maximum pressure of the test liquid is 600 Kpa. Therefore, when the fluid flow rate increases to 0.59 m/s, the digital meter cannot display the numerical value.

Table 2. Test results of hydraulic fluid permeability for three samples.

Fluid Velocity (m/s)	Differential Pressure (Kpa)		
	Sample 1	Sample 2	Sample 3
0.0059	22	19	24
0.00885	30	31	40
0.0118	36	42	58
0.01474	41	58	78
0.01769	50	71	99
0.02064	57	89	122
0.02359	61	108	149
0.02654	72	138	198
0.02949	82	165	223
0.03244	95	198	266
0.03539	110	223	302
0.03833	122	253	341
0.04128	142	286	375
0.04423	160	319	415
0.04718	187	345	455
0.05013	202	378	498
0.05308	220	412	542

0.05603	241	453	588
0.05898	261	512	/

3.3 Permeability formula prediction

Khayargoli et al. [15] found that the relationship between pressure difference, inertia coefficient, porosity, and average powder particle size is:

$$\frac{\Delta p}{h} = a\frac{(1-\varepsilon)^2}{\varepsilon^3 d_s^2}\mu v + b\frac{(1-\varepsilon)}{\varepsilon^3 d_s}\rho v^2 \quad (1)$$

In the formula, Δp is the pressure drop (pa); h is the thickness of the material (m); ε is the porosity of porous materials (in dimensionless units); d_s is the average particle size of the powder particles (m); μ is the dynamic viscosity of the fluid (pa·s); v is the flow velocity of the fluid (m/s); ρ is the density of the fluid (Kg/m^3); a and b are constant.

When a porous material is composed of spherical powder particles of the same size, the average pore size of the material can be calculated as:

$$d_a = \frac{4\varepsilon d_s}{5(1-\varepsilon)} \quad (2)$$

In the formula, d_a is the average pore size of the porous material; ε is the porosity of porous materials; d_s is the particle diameter of the powder. The particle diameter of the 200 mesh stainless steel powder used in this article is 75 μm.

For sintered metal fiber porous materials, the average pore size of the material can be calculated as:

$$d_a = \frac{\varepsilon d_f}{(1-\varepsilon)} \quad (3)$$

In the formula, d_f is the fiber diameter, and the fiber diameter of the 80 mesh stainless steel plain weave wire mesh used in this article is 75 μm.

Through the literature review, there is no formula for calculating the average pore size of porous materials formed by powder and wire mesh composite molding. Considering that the silk screen of this material plays a role in enhancing mechanical functions, the calculation formula for the average pore size is obtained by averaging the two formulas above:

$$d_a = \frac{4\varepsilon d_s}{4.5(1-\varepsilon)} \quad (4)$$

Then, the real porous material is far from the ideal stacking of particles or fibers, and the microporous material in the paper is made by composite rolling and sintering of powder and wire mesh. During the rolling process, the powder and wire mesh undergo significant deformation due to the squeezing force between the rollers, with the wire mesh becoming narrower and wider, and the powder becoming flatter. Based on the average pore size, the particle size of the material can be derived by using the above formula. The formula is:

$$d_s = \frac{1.125(1-\varepsilon)d_a}{\varepsilon} \quad (5)$$

Therefore, the formula for predicting the flow rate pressure difference relationship of the powder wire mesh composite microporous plate prepared in the paper can be derived as follows:

$$\frac{\Delta p}{h} = a * \frac{1}{\varepsilon * 1.125^2 * d_a^2} * \mu * v + b * \frac{1}{\varepsilon^2 * 1.125 * d_a} * \rho * v^2 \quad (6)$$

In the permeability experiment of water working fluid, Curve 1 is extracted and fitted to obtain the following data as shown in figure 3:

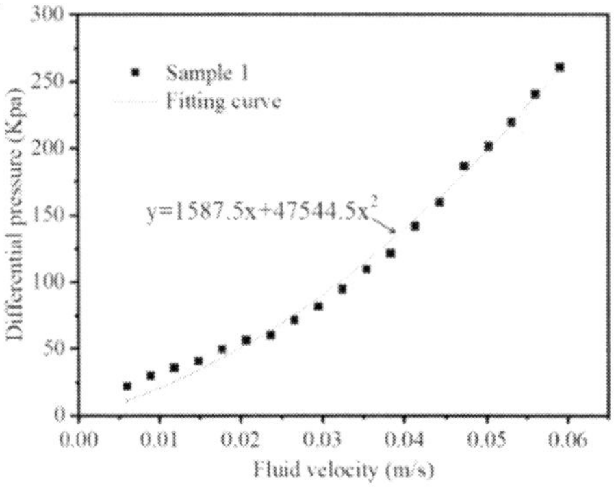

Figure 3. Fitting curve of sample 1.

Therefore, for the No.1 curve of the hydraulic fluid, it can be obtained that:

$$1587.5 = a * \frac{1}{\varepsilon * 1.125^2 * d_a^2} * \mu * h \tag{7}$$

$$47544.4 = b * \frac{1}{\varepsilon^2 * 1.125 * d_a} * \rho * h \tag{8}$$

ε=0.256; d_a=6.3538 μm; μ=1.01E-03 pa·s; ρ=1000 Kg/m^3; h=2.62 mm, it is calculated that:

$$a = 7.847E - 03 \tag{9}$$

$$\mathrm{b} = 8.5E - 03 \tag{10}$$

Thus, the Ergun formula can be concretized to obtain an analytical model for predicting the pressure difference of hydraulic fluids passing through porous materials, which is:

$$\frac{\Delta \mathrm{p}}{h} = 6.262E - 06 * \frac{1}{\varepsilon * d_a^2} * v + 7.556 * \frac{1}{\varepsilon^2 * d_a} * v^2 \tag{11}$$

By substituting the corresponding porosity of the three samples, the average pore size obtained from testing, and the thickness of the permeability sample during actual testing into the analytical model, the pressure difference prediction curves of the three samples under the analytical model can be obtained as shown in the following figure 4:

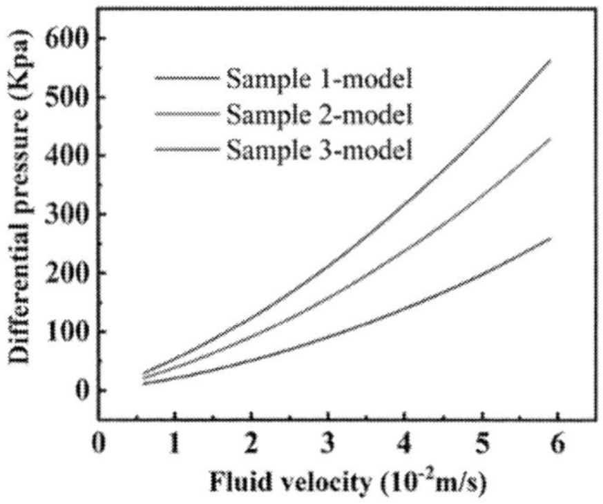

Figure 4. Pressure difference prediction curves for four samples.

3.4 Comparison between prediction model and experimental results

As shown in figure 5, it can be seen that at low equivalent Reynolds numbers, the model accurately predicts the unit pressure drop of flow in porous materials with a deviation of about 10%. As the equivalent Reynolds number increases, the deviation between the calculated value and the experimental value becomes larger and larger, reaching about 28%. This indicates that at high Reynolds numbers, the model can no longer predict the flow characteristics of sintered particle porous materials well. The main reason for this is that as the Reynolds number increases, the wall effect and dispersion effect inside the porous material channel will be enhanced, At this point, the relationship between unit pressure drop and Reynolds number may require the introduction of other composition terms to accurately express it.

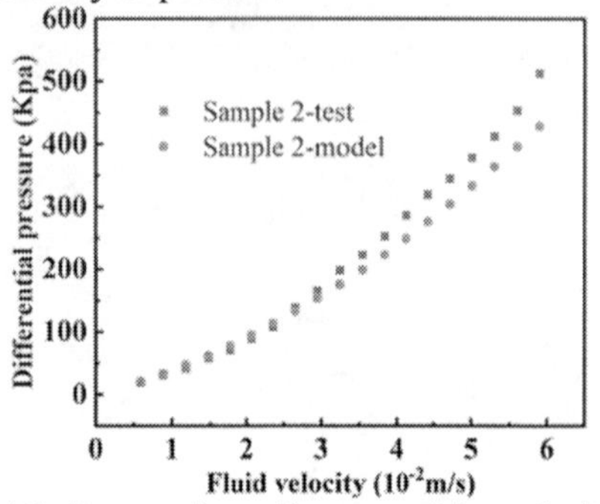

(1) Comparison diagram of Sample 2

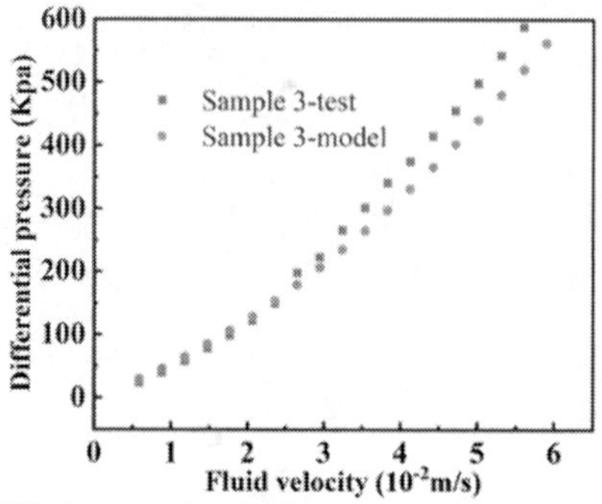

(2) Comparison diagram of Sample 3

Figure 5. Comparison between the prediction model and experimental results of hydraulic fluids.

4. Conclusion

Due to the complex internal pores of porous materials, theoretical prediction of liquid permeability becomes difficult. This article is based on the relationship between pressure difference, inertia coefficient, porosity, and average powder particle size. A theoretical model for flow velocity and pressure difference of porous materials is gradually derived, and the theoretical prediction model is compared with experimental results. It is found that at low equivalent Reynolds numbers, the model accurately predicts the unit pressure drop of flow in granular porous materials, with a deviation of about 10%. This work can provide a reference for the study of the permeability of porous materials.

Acknowledgments

The authors acknowledge financial support from the Basic Ability Improvement Project of Young and Middle-aged Teachers in Higher Education Institutions of Guangxi (2021KY1037) and the Second Batch of High-level Talent Research Start-up Funding Project of Liuzhou Vocational and Technical College in 2023 (2023GCQD05) and General research topics at the school level for Liuzhou Vocational and Technical College in 2023(2023KB02).

References

[1] Li Y., Wang N., Chen W. Application of modified porous materials and pollution treatment in water [J]. Journal of Innovation and Development, 2023.
[2] Krach D., Steeb H. Comparing methods for permeability computation of porous materials and their limitations [J]. PAMM, 2023, 23.
[3] Yao L.X.Z. Permeability of short fiber reinforced powder metallurgy material and its application in porous mold [J]. Powder Metallurgy and Metal Ceramics, 2018, 56(11a12).
[4] Yamazaki R. Kozeny-carman's formula [J]. J. Soc. Powder Technol., Japan, 1974, 11.
[5] Macdonald I.F., El-Sayed M.S., Mow K., et al. Flow through porous media-the Ergun formula revisited [J]. Ind.eng.chem.fundam, 1979, 18(3): 199-208.
[6] Siskovic N., Gregory D.R., Griskey R.G. Viscoelastic behavior of molten polymers in porous media [J]. Aiche Journal, 1971, 17(2): 281-285.
[7] Ruzhen Z. Method for testing pore size of porous materials [J]. Powder Metallurgy Technology, 1992.
[8] Bhatia S.K., Smith J.L. Geotextile characterization and pore-size distribution: part ii. A review of test methods and results [J]. Geosynthetics International, 2015, 3(2): 85-105.
[9] Wu J., Chu X., Zhao G., et al. Test study and application of fracturing stimulation technologies

Journal of Physics: Conference Series **2808** (2024) 012064 doi:10.1088/1742-6596/2808/1/012064

with liquid powder for low-permeability oil reservoir in fy block of daqing oilfield [C]. International Conference on Energy, Environment and Sustainable Development, 2012.
[10] Pan B., Clarkson C.R., Ghanizadeh A., et al. Liquid permeability determination from vacuum imbibition in tight/shale core plugs [J]. Energy & Fuels, 2023(13): 37.
[11] Geng G., Xiao-Ding G. The test system of water permeability of the porous material based on LabVIEW [J]. Journal of Xi'an Polytechnic University, 2012.
[12] Yuan J., Wang Y., Chen X. Study on the evolution of pore structure of anthracite coal under liquid-nitrogen freeze-thaw cycles [J]. ACS Omega, 2022, 7: 4648-4654.
[13] Sim J.G., Kang M., Min K. Effects of porosity gradient and average pore size in the in-plane direction and disposition of perforations in the gas diffusion layer on the performance of proton exchange membrane fuel cells [J]. Journal of Power Sources, 2022.
[14] Yang B., Huang W., Causse P., et al. On the design of test molds based on unidirectional saturated flows to measure transverse permeability in liquid composite molding [J]. Polymer Composites, 2022(4): 43.
[15] Khayargoli P., Loya V., Lefebvre L.P., et al. The impact of microstructure on the permeability of metal foams [J]. 2004.

Influence of high strain rate and catalyst concentration on dynamic impact behavior of vitrimers

Rui Xiang[1] and Le An[1*]

[1]State Key Lab for Strength and Vibration of Mechanical Structures, Department of Engineering Mechanics, Xi'an Jiaotong University, Xi'an, 710049, China

[*]Corresponding author's e-mail: anle2018@mail.xjtu.edu.cn

Abstract. Vitrimers are dynamically cross-linked networks with dynamic covalent bonds, possessing attractive mechanical robustness, self-healing, re-processability, recyclability, and shape memory properties. Therefore, vitrimers hold promise for applications including aerospace and transportation, in which cases the structures will inevitably be subjected to dynamic loading. The objective of this study was to characterize the dynamic impact behavior of vitrimers utilizing the Split Hopkinson Pressure Bar (SHPB) experiments. Five kinds of vitrimers containing different concentrations of catalyst $Zn(acac)_2$ of 0.1, 1, 3, 5, and 10 mol% and two strain rates 2000 s^{-1} and 4000 s^{-1} are considered. These vitrimers exhibit strain rate dependence, whose yield stress and strain softening effect is enhanced when the strain rate is increased from 2000 s^{-1} to 4000 s^{-1}. Catalyst concentration significantly affects the dynamic performance, and the vitrimer with 10 mol% catalyst shows the most pronounced strain softening, the highest yield stress, and the best impact energy absorption properties. The results provide fundamental data and strategies for improving the impact resistance of epoxy for engineering applications.

1. Introduction

Covalent Adaptable Networks (CANs) were investigated by introducing dynamic covalent bonds into a polymer network so that the crosslinked network can experience bond exchange in response to external stimuli like light or heat [1]. Generally, CANs are divided into dissociative and associative CANs based on the different mechanisms of bond exchange. Under the influence of external stimuli, the dynamic covalent bonds within the dissociative CANs will undergo two separate processes of break and reorganization. While in associative CANs (also referred to as "vitrimers"), the two processes happen simultaneously. Thus, vitrimers exhibit stress relaxation and re-processability while maintaining the network integrity [2]. At the same time, vitrimers still perform similarly to conventional thermoset materials with excellent thermal and mechanical performance at service temperatures.

Recently, there have been considerable advances achieved within the vitrimers field, with increasing interest in their applications. Huang et al. synthesized the vitrimer elastomers containing boronic ester bonds, which exhibit mechanical robustness, shape memory, and recyclability [3]. Hu et al. fabricated the epoxy vitrimers featuring remarkable thermal stability, and outstanding self-healing and welding capabilities through dynamic transesterification [4]. Chen et al. demonstrated how to perform digital light processing 3D printing of high-performance vitrimers [5]. Guo et al. introduced a variety of dynamic bonds into PDMS, resulting in a highly stretchable vitrimer whose stretchability is strongly dependent on the loading rate [6]. Therefore, vitrimers are expected to replace traditional thermosets in some applications such as aerospace and automotive, in which cases the structures will inevitably suffer from dynamic loading at different rates. Accordingly, investigating the dynamic response of vitrimers at high strain rates is crucial. Although there have been many studies on various properties and applications of vitrimers under static loading, the dynamic mechanical behavior of vitrimers has rarely been reported.

In 2023, we investigated the dynamic impact properties of hard epoxy/glutaric anhydride vitrimers at impact loads with split Hopkinson pressure bars [7]. Here we aim at the dynamic impact response of soft epoxy/fatty acid vitrimers. We use $Zn(acac)_2$ catalyst at five different concentrations (0.1, 1, 3, 5, and 10 mol%) to synthesize these soft vitrimers. The relation between the catalyst concentration and the bond exchange reaction rate is investigated by stress relaxation experiments. SHPB tests are carried out under strain rates 2000 s^{-1} and 4000 s^{-1}. A detailed discussion on the effects of high strain rate and catalyst concentration on the dynamic impact properties of vitrimers is presented according to the experimental curves of true stress and true strain.

2. Materials and methods

2.1. Raw materials

The epoxy diglycidyl ether of bisphenol A (DGEBA) and the catalyst zinc acetylacetonate ($Zn(acac)_2$) were obtained by Sigma Aldrich. The curing agent was a fatty acids derivatives blend, consisting of approximately 23 wt% dimers and 77 wt% trimers, supplied by Croda.

2.2. Preparation of vitrimer

Firstly, the fatty acids and the catalyst were introduced into a beaker and mixed under 130°C until the particles were fully solubilized. Five different concentrations of catalyst were added: 0.1, 1, 3, 5, and 10 mol% of the catalyst to the COOH groups of the curing agent. Secondly, the mixture is put into a vacuum at a temperature raised from 100°C to 180°C until no air bubbles are generated. Thirdly, DGEBA was introduced into the beaker and mixed at 130°C till homogeneous. The stoichiometry between epoxy and COOH is 1:1. Lastly, the compound was placed under a vacuum to remove air bubbles and then cured at 130°C for 6 hours.

2.3. Test methods

2.3.1. Stress relaxation experiments. Stress relaxation testing was conducted with a rheometer (MCR302, Anton Paar, Austria). The samples were processed as discs with 25 mm diameter, and 1.5 mm thickness. The specimens will first be kept under 180°C for 10 minutes for thermo-equilibrium and then subjected to loading with 1% continuous shear strain. A normal force of 1.5 N was used as a preload force to restrain the specimens.

2.3.2. SHPB dynamic compression experiments. The Split Hopkinson Pressure Bar (SHPB) method can be employed for obtaining the dynamic impact performance of specimens under impact loadings (10^2-10^4 s^{-1}) [7,8]. As shown in Figure 1(a), the test system mainly consists of an air gun, a striker bar, an incident bar, a transmission bar, absorption devices, velocity measuring devices (here, strain gauges), and data acquisition devices. The specimens used for this experiment are cylindrical bars with 5 mm diameter and 5 mm length.

During an experiment, the sample was clamped between the incident and the transmission bar, and the striker bar stroked the incident bar with a specific velocity. Compressive stress wave produced within the incident bar propagated forward, with part of the incident wave propagating through the specimen and the rest being reflected to the incident bar. Strain gauges can record the incident signal ε_i, transmitted signal ε_t, and reflected signal ε_r, as shown in Figure 1(b). According to one-dimensional stress wave theory, specimen strain ε_e, strain rate $\dot{\varepsilon}_e$, and average stress σ_e are given by the following equations [9]:

$$\varepsilon_e(t) = -\frac{2C}{l_0}\int_0^t \varepsilon_r d\tau \quad (1)$$

$$\dot{\varepsilon}_e(t) = -\frac{2C}{l_0}\varepsilon_r \quad (2)$$

$$\sigma_e(t) = \frac{EA_b}{A_0}\varepsilon_t \quad (3)$$

where C is the velocity of the elastic wave within these bars, A_b is the cross-sectional area of these bars, E is the elastic modulus in these bars, l_0 is the original length of the sample, and A_0 is the original area of the cross-section of the sample. Table 1 summarizes the parameters of the SHTB system.

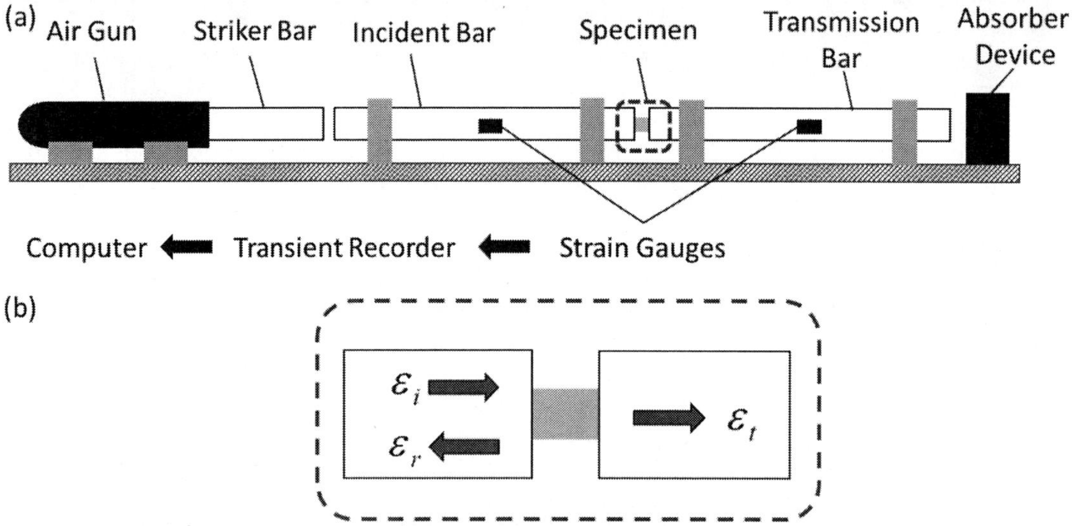

Figure 1. SHPB testing system (a) Illustration of SHPB (b) Illustration of stress waves in the specimen section.

Table 1. The parameters for the SHTB system.

C (m/s)	A_b (mm^2)	E (GPa)	l_0 (mm)	A_0 (mm^2)	Length of incident bar (mm)	Length of transmission bar (mm)
4966	282	113.5	5	19.6	1400	1400

Correlations between true stress σ_T and engineering stress σ_e and true strain ε_T and engineering strain ε_e are shown below:

$$\sigma_T = \sigma_e \cdot (1-\varepsilon_e) \quad (4)$$

$$\varepsilon_T = -\ln(1-\varepsilon_e) \quad (5)$$

3. Results and discussion

3.1. Stress relaxation behavior

The bond exchange reaction rates that occur in the vitrimers were characterized by stress relaxation experiments. Figure 2 exhibits the normalized relaxation modulus of the vitrimers containing 1, 3, 5, and 10 mol% catalyst decrease from 1 to 1/e under 180°C. The higher the concentration of the catalyst, the lower the relaxation time, and the faster the bond exchange reaction rate. However, the bond exchange reactions are so slow that the vitrimer with 0.1 mol% catalysts essentially behaves like an elastomer without much stress relaxation.

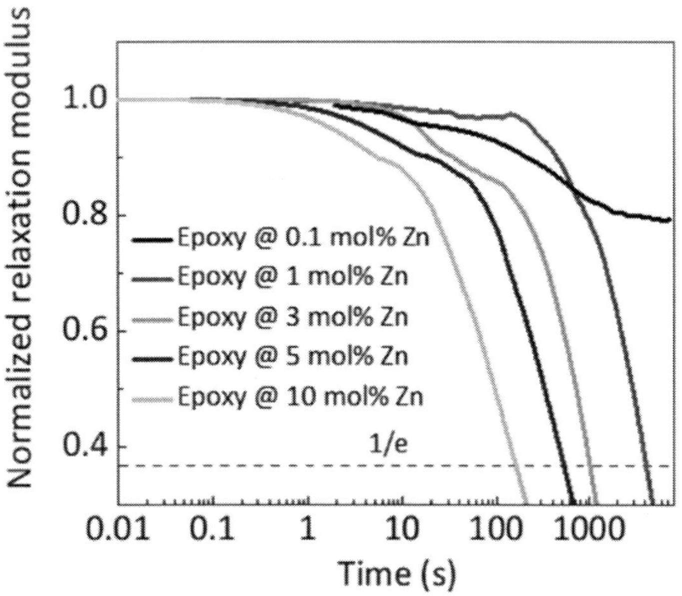

Figure 2. Stress relaxation characteristics at 180°C.

3.2. Dynamic impact response of vitrimers

The true stress-strain histories for vitrimers with five different catalyst concentrations under 2000 s^{-1} and 4000 s^{-1} can be obtained by processing the recorded waves, as shown in Figure 3. For vitrimers containing the same catalyst concentration, the yield stress of the specimen increases as increasing from 2000 s^{-1} to 4000 s^{-1}, showing that vitrimers exhibit a significant strain rate dependence. Moreover, the enhancement of yield stress is greater for vitrimers with higher catalyst content, which indicates that the more catalyst concentration, the greater the sensitivity of the vitrimers to the strain rate.

As can be observed, the high strain rate impact behavior of the vitrimers shows significant non-linear characteristics. For the vitrimers with 1, 3, 5, and 10 mol% catalysts, the curves first undergo a linear elastic deformation, then enter a non-linear elastic deformation phase to reach yield stress, succeeded by strain softening, then a plateau stress phase, and a strain hardening phase until reaching the ultimate stress. The extent of strain softening is more significant at 4000 s^{-1} than at 2000 s^{-1}. What's more, the vitrimer containing 10 mol% catalyst shows the most significant strain-softening effect. However, the curves of the vitrimer with 0.1 mol% catalysts do not have a strain-softening stage.

Figure 4 illustrates the variation of true stress-strain curves for vitrimers with different catalyst concentrations at 2000 s^{-1} and 4000 s^{-1}, separately. This shows catalyst concentration influences the dynamic impact properties of vitrimers significantly. At 2000 s^{-1} or 4000 s^{-1} respectively, the vitrimer with 10 mol% catalyst exhibits the highest stress level, while the vitrimer with 0.1 mol% catalyst exhibits the lowest.

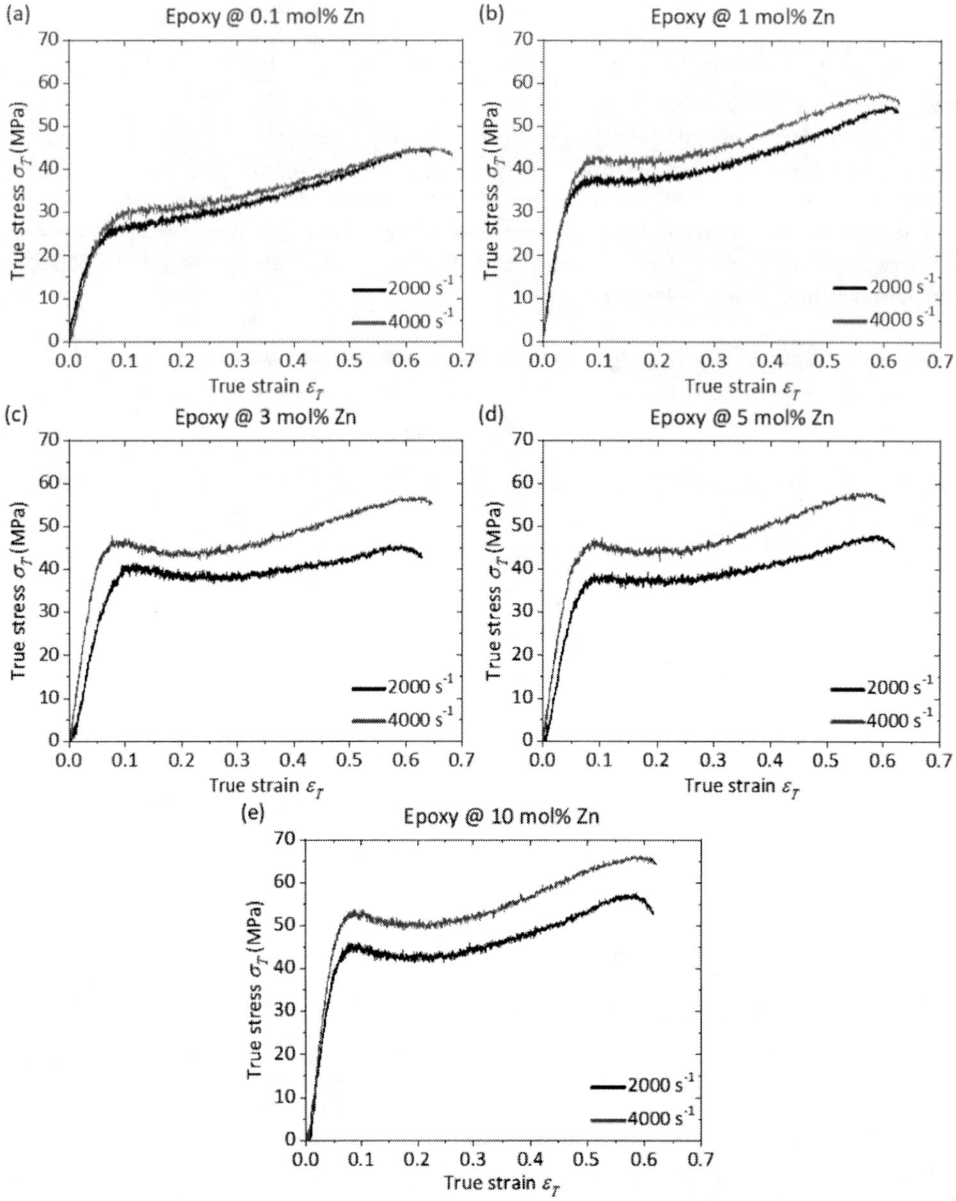

Figure 3. True stress-strain curves for vitrimers with different catalyst concentrations (a) 0.1 mol% (b) 1 mol% (c) 3 mol% (d) 5 mol% (e) 10 mol%.

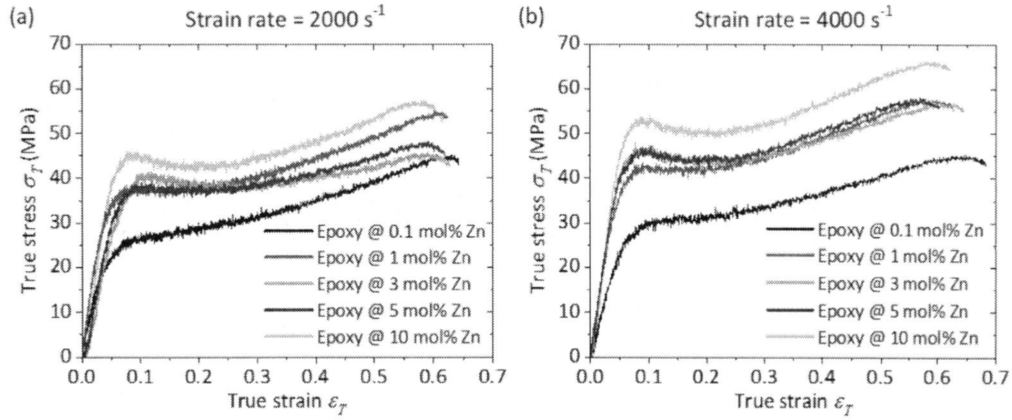

Figure 4. True stress-strain curves for vitrimers under different strain rate (a) 2000 s^{-1} (b) 4000 s^{-1}.

Vitrimers transmit high-frequency mechanical waves at impact loads during SHPB experiments. Research indicates that high-frequency mechanical waves, such as sonication, can increase the fluidity of polymer chains as well as facilitate bond exchange processes [10]. Consequently, bond exchange reactions are conceivable to take place in dynamic impact experiments. This phenomenon can lead to increased energy absorption [11]. To investigate the energy absorption properties of vitrimers under high strain rates, here we calculate the strain energy density by integrating the true stress-strain curve in the strain range 0 ~ 0.55, with the results presented in Figure 5. At 2000 s^{-1}, the energy density absorption rises from 17 MJ/m^3 to 24 MJ/m^3, i.e. by 41%, when the catalyst concentration increases from 0.1 mol% to 10 mol%. Similarly, at 4000 s^{-1}, the increase in absorbed energy density is approximately 56%. Our findings suggest that the vitrimers containing more catalyst may absorb more energy than the vitrimers containing less catalyst at high strain rate loads because of bond exchange, indicating the potential for significant enhancement in the impact performance of thermosets with dynamic polymer networks.

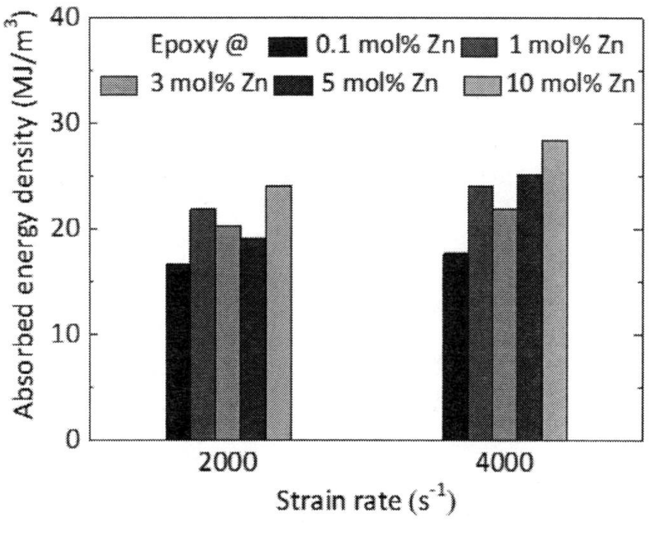

Figure 5. The absorbed energy density of vitrimers with different catalyst concentrations at 2000 s^{-1} and 4000 s^{-1}.

4. Conclusions

Here, we aim to characterize the dynamic impact response of soft epoxy/fatty acid-based vitrimers with different concentrations of catalyst and at different strain rates. Stress relaxation experiments show that the more catalyst concentration, the more rapid the bond exchange reaction rate. The SHPB tests have been carried out on vitrimers with catalyst concentrations of 0.1, 1, 3, 5, and 10 mol% and under high strain rates (2000 s^{-1} and 4000 s^{-1}), respectively. Three conclusions are drawn as follows:

(1) The yield stress for the vitrimers is enhanced when the strain rate increases from 2000 s^{-1} to 4000 s^{-1}, reflecting a strain rate-dependent effect. Moreover, the enhancement effect on yield stress is more pronounced with higher catalyst concentrations, suggesting increased sensitivity of the vitrimers to strain rates.

(2) Catalyst concentration significantly influences the dynamic compressive impact behavior of the vitrimers. The vitrimer with a 10 mol% catalyst exhibits the highest yield stress and the most robust impact energy absorption property.

(3) The strain softening effect is influenced by both strain rate and catalyst concentration. A more pronounced strain softening effect is observed when increasing from 2000 s^{-1} to 4000 s^{-1}. Vitrimer containing 10 mol% of catalyst demonstrates the most significant strain-softening effect.

Acknowledgments

The work received support from the National Natural Science Foundation of China (No. 12002257).

References

[1] Bowman, C., Du Prez, F., *et al.* (2020) Introduction to chemistry for covalent adaptable networks. Polym. Chem., 11(33): 5295-5296.
[2] Montarnal, D., Capelot, M., *et al.* (2011) Silica-Like Malleable Materials from Permanent Organic Networks. Science, 334(6058): 965-968.
[3] Huang, L., Yang, Y., *et al.* (2022) Boronic ester bonds crosslinked vitrimer elastomers with mechanical robustness, shape memory, self-healing, and recyclability properties. Compos. Sci. Technol., 228: 109621.
[4] Hu, Y., Tong, S., *et al.* (2023) Cardanol-based epoxy vitrimer/carbon fiber composites with integrated mechanical, self-healing, re-processable, and welding properties and degradability. Chem. Eng. J., 471: 144633.
[5] Chen, Z., Yang, M., *et al.* (2021) Recyclable thermosetting polymers for digital light processing 3D printing. Mater. Des., 197: 109189.
[6] Guo, H., Han, Y., *et al.* (2020) Universally autonomous self-healing elastomer with high stretchability. Nat. Commun., 11(1): 2037.
[7] An, L., Xiang, R., *et al.* (2023) Dynamic behavior of vitrimers under high–strain rate loadings. Polymer, 282: 126149.
[8] Lv, Y., Wu, H., *et al.* (2023) Experimental and numerical simulation study of fiber-reinforced high strength concrete at high strain rates. J. Build. Eng., 65: 105812.
[9] Miao, Y., Du, W., *et al.* (2022) Characterizing multi-mechanical behaviors for epoxy-like materials under wide strain rate range. Polym. Test., 116: 107804.
[10] Lee, H. H., Kim, Y. W., *et al.* (2019) Fast healing of ionic bonds in tough hydrogels under an acoustic excitation. Extreme Mech. Lett., 33: 100572.
[11] Zhang, H., Wu, Y., *et al.* (2019) Super-stretchable Dynamic Polymer Networks. Adv. Mater., 31(44): 1904029.

Journal of Physics: Conference Series **2808** (2024) 012066 doi:10.1088/1742-6596/2808/1/012066

Numerical study on mechanical responses of AHR and BWD packer under a cyclic loading

Xin Ma [1], Bing Shao [1*], Yongliang Zhang [1] and Xiaomei Wang [2]

[1]School of Mechanical and Automotive Engineering, Qingdao University of Technology, Qingdao, 266525, China
[2]Qingdao Energy and Huarun Gas Co., Ltd., Qingdao, 266580, China

*Corresponding author's e-mail: upcshaobing@163.com

Abstract: For AHR and BWD packers which are commonly used in gas storage, the mechanical behavior and sealing properties are comparatively studied in this paper by the numerical simulations and experimental verification. First of all, through the experimental data fitting, the Yeoh equation is adopted for the constitutive relationship of the hyper-elastic rubber used in the packers. Based on the ABAQUS explicit FEM algorithm, the mechanical behaviors of the packer during the setting process and under working conditions with a cyclic loading are discussed and the stress risky spots of the sealing rubber are investigated. The results obtained in this work could be helpful for the engineering application of the packer in gas storage.

1. Introduction

As a down-hole equipment, the packer is used for sealing the annulus area, isolating the destination formation, and controlling the injected and extracted fluids in oil and gas exploration, development, and storage. Its sealing performance and rated working pressure difference directly affect the oil and gas extraction and regulation efficiency [1, 2]. As the only part of the packer functioning sealing, the mechanical property of the rubber cylinders with limited deformation under the setting pressure and the contact state with the well wall determines the packer's sealing property. However, failures often happen to the rubber cylinders in the down-hole environment [1-3]. Therefore, the focus of investigation on the packer's sealing performance should be put on the rubber cylinder's stress state and deformation.

Two types of packers manufactured by Halliburton, which are AHR and BWD packers, are widely used [4, 5]. In this work, the setting process of AHR and BWD packer and their sealing performance under cyclic loading in gas storage are numerically simulated and discussed.

2. Constitutive equation of rubber

For isotropic materials, like rubber, the strain energy density is decomposed into two components, the strain bias energy and the volume strain energy, which are represented by the two right-hand sides of Eq. (1) [6, 7].

$$W=\sum_{i=1}^{N} C_{i0}\left(\hat{I}_1-3\right)^i+\sum_{k=1}^{N} \frac{1}{D_k}(J-1)^{2k} \tag{1}$$

where N is the order of the selected polynomial; J denotes the volume ratio of rubber deformed or not; $\hat{I}_1$ denotes the first invariant of the Cauchy-Green tensor; C_{i0} and D_k are the input parameters.

Considering the difficulty of converging for a higher-order model on finite element calculation, experimental and numerical simulations were carried out for the material constitutive structure relationship of the nitrile rubber used in the packer. Figure 1 shows the test machine (Figure 1-a), numerical model (Figure 1-b), and comparison of constitutive curves between numerical model and initial test data (Figure 1-c) [8]. It can be seen that the Yeoh model fits optimally with a correlation coefficient of 0.9998.

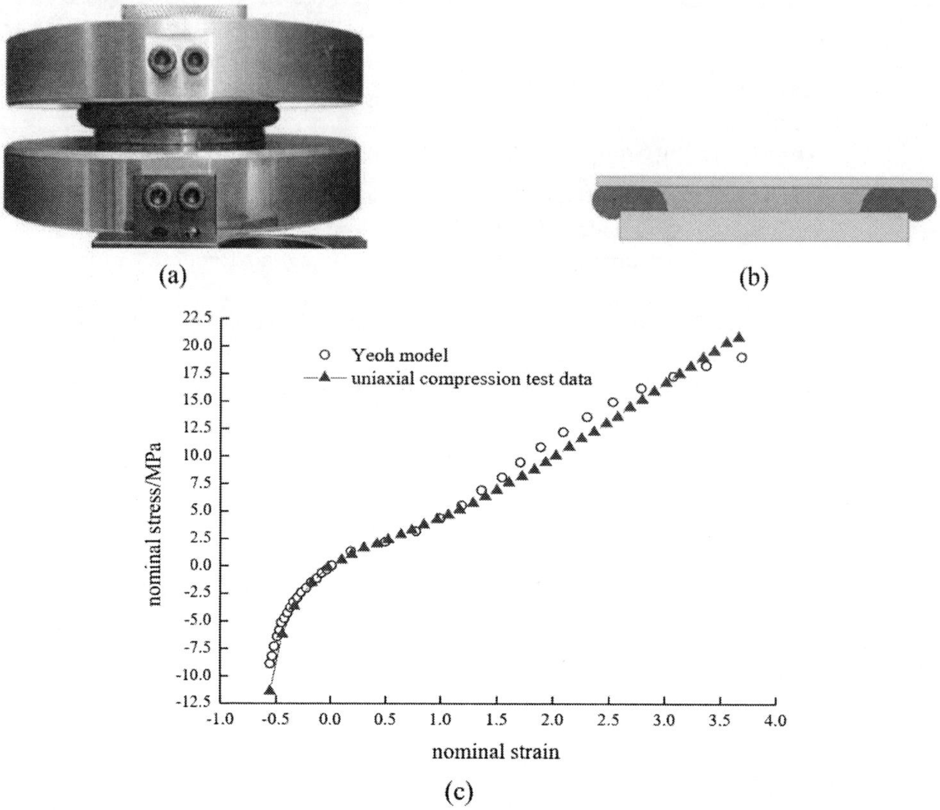

Figure 1. The experiments and simulations on rubber constitutive relationship.

3. AHR/BWD finite element model

The packer is assembled of several parts. Figure 2 shows the overall structure of the Halliburton AHR and BWD packers. This work focuses on the sealing parts, that is, the rubber cylinder in the middle of the packer. The sealing parts mainly include a rubber cylinder, oil pipe, casing, spacer ring, support ring, and other parts. Figure 3 gives the assembly structure of each part. Because the parts involved are all rotating bodies, to save calculation time and reduce calculation volume, the rubber cylinder, central tube, casing, and spacer ring are treated as axisymmetric models. The 1/2 model is used for the AHR model, the total number of elements is 590.2 thousand, the 1/4 model is used for the BWD, and the total number of elements is 772.2 thousand.

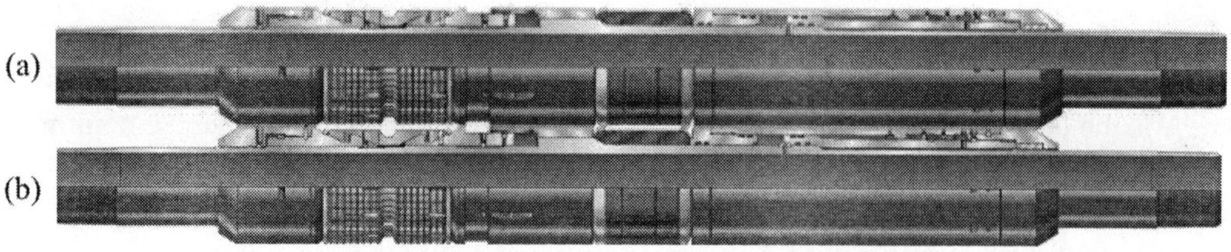

Figure 2. Structure charts of Halliburton AHR (a) and BWD (b) packer.

(a) (b)

Figure 3. Simulations models and assembly diagram for AHR (a) and BWD (b) packer.

In boundary condition settings, the outer surface of the casing is fixed; the displacement of the lateral section of the tubing is axially and normally constrained. Normal constraints are applied to the lateral section of the spacer ring, rubber cylinder, and support ring. In addition, the lower spacer rings of both packers are fully constrained. Table 1 shows the material parameters for different parts.

Table 1. Materials parameters for the packer parts.

Model parts	Materials	Density(kg/m^3)	Young's modulus (MPa)	Poisson's ratio	Yield stress (MPa)
Casing/oil Pipe/spacer/support ring(AHR)	Steel	7800	2.0E5	0.28	600
Support ring(BWD)	Brass	8980	9.0E4	0.343	110
Rubber cylinder(hyper-elastic material)	Rubber	C10: -8.4373; C01: 11.439; C20: -0.1102; C11: 0.496; C02: 2.7311; D1: 0; D2: 0;			

To investigate the effect of alternating pressure changes (i.e. bulging effect) inside the oil pipe on the rubber cylinder's sealing performance, a cyclic load is applied to the inner wall of the oil pipe, and the cyclic amplitude curve is expressed by the Fourier series and the expression is like [8]:

$$\alpha = A_0 + \sum_{n=1}^{N}\left[A_n \cos n\omega(t-t_0) + B_n \sin n\omega(t-t_0)\right] \quad t \geq t_0 \tag{2}$$

$$\alpha = A_0 \quad t \geq t_0$$

where N denotes the quantity of Fourier series terms; ω denotes the circular frequency; t_0 denotes the starting time; A_0 denotes the starting value; A_n and B_n are constants.

4. Analysis of results

Figures 4 and 5 show the contact stress distributions on the axial paths on the inner wall of the casing. It can be seen that the contact stress is highest on the upper rubber cylinder and relatively lowest on the lower one. The overall contact stress distribution on the middle cylinder of AHR is in the shape of a "hump", which means the contact stress at the end is relatively larger and the contact stress in the middle is smaller. Slight protrusion deformation of the two ends of the middle cylinder under the extrusion of the spacer ring makes larger contact stress appear on both ends. For the BWD packer, contact stress on three boundaries among the rubber cylinders is rather high, indicating that the extrusions among the rubber cylinders result in high contact stress.

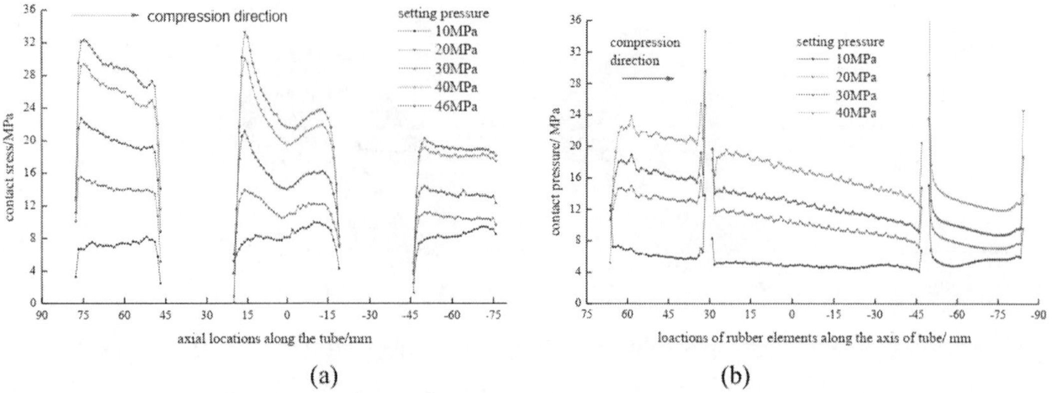

Figure 4. Contact stress of AHR packer (a) and BWD packer (b) rubber with casing.

It can be seen from the Von-Mise stress contours of the two packers in Figure 5, that there always exists a stress concentration in the bulge of the middle rubber cylinder of the AHR packer with the septum ring in the compression direction. The concentrated stress exceeds 300 MPa, which indicates a wear failure very easy to happen. Moreover, the maximum value of the stress on the contacting bulge between the spacer ring and the upper rubber cylinder of the AHR packer is 572.3 MPa under the 46 MPa seating pressure, which is on the verge of yield (yield stress 600 MP) and the safety factor is as low as 1.048. So, it can be seen that this contacting bulge is a key point for the spacer ring and more attention is needed for design, for which a stronger material or an improved shape is suggested.

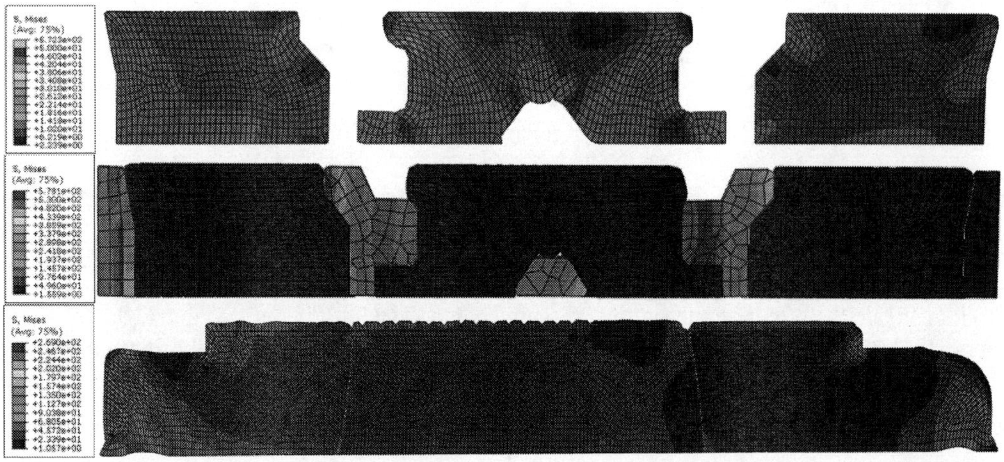

Figure 5. The Mises stress contour of AHR rubber (the upper and middle one) and BWD rubber (the lower one).

For the BWD packer, stress concentration exists on the contact surface of the rubber with the bottom corner of the upper support ring, where the maximum stress value reaches 269 MPa while the stress at the center is only about 30 MPa. Meanwhile, a local bulging rubber cylinder appears accompanied by a tiny separation to the outer wall of the oil pipe, which is extremely detrimental to the sealing performance. Moreover, the upper corner of the support ring also possesses a stress concentration during contact with the rubber cylinder. As the packer remains in service under a harsh environment, the stress concentration points of the rubber are prone to fail first.

5. Conclusion

Finite element simulations are conducted for packers in the setting process and the mechanical behavior and sealing state are investigated. For the AHR packer, there exist large stress concentrations in the bulge of the middle rubber with the septum ring in the compression direction. Also for the BWD packer, stress concentration appears on the rubber cylinder in contact with the bottom corner of the upper support ring, which is extremely unfavorable to the sealing effect. The issues discussed in this paper could be significant for the application or innovative design of packers.

Acknowledgments

This work was supported by the National Natural Science Foundation of China [grant nos. 52374209] and the Shandong Provincial Natural Science Foundation [grant nos. ZR2023QE080].

References

[1] F. Zhang, T. Li, Y. Zhang. Analysis of Free deformation characteristics of packer rubber surface under the initial sealing load. Applied Mathematics and Mechanics, 2019, 40(4): 419-432.
[2] H. Liu, X. Shen, S. Liu, et al. Integrity analysis method of multi-packer string in HTHP gas wells and its application cases. Natural Gas Industry, 2020, 40(7): 83-89.
[3] P. Wang, M. Cai, Z. Liu, et al. Research on key technology of packer rubber barrel for integrated fracturing and completion of gas well. Journal of Petroleum Exploration and Production Technology. 2024. https://doi.org/10.1007/s13202-023-01737-9.
[4] C. Wang. The study of sealing performance and influencing factors of permanent packer. Chongqing University, 2018.
[5] S. Sun, L. Tang. Sealing valuation of underground gas storage in the conditions of multi-period alternating load. Science Technology and Engineering, 2016, 16(6): 23-28
[6] W. Lan, H. Wang, X. Zhang, et al. Sealing properties and structure optimization of packer rubber under high pressure and high temperature. Petroleum Science, 2019, 16: 632-644.
[7] Y. Liu, Z. Lian, J. Chen, et al. Design and experimental research on sealing structure for a retrievable packer. Shock and Vibration, 2020: 1-15.
[8] H. Liu, W. Zhang, Y. Cao, et al. Analysis of the performance improvement effect of combined packing for oil and gas well packers. 2020 IOP Conference Series: Earth and Environmental Science. Sci. 514 022017.

Journal of Physics: Conference Series **2808** (2024) 012067 doi:10.1088/1742-6596/2808/1/012067

Strain-rate dependent dynamic constitutive model of tempered glass

Lei Li [1,*], Bangxian Liu [1] and Erfeng An [2]

[1]Zhengzhou Technical College, Zhengzhou, 450121, China
[2]State Key Laboratory of Explosion Science and Technology, Beijing Institute of Technology, Beijing, 100081, China

*Corresponding author's e-mail: lilei20031010@126.com

Abstract: Tempered glass is essentially different from float glass in dynamic mechanical properties, although it is toughened float glass made by physics and they have the same chemical composition. The dynamic mechanical properties of tempered glass are investigated in this paper by using the modified techniques of the Split Hopkinson Pressure Bar (SHPB), and the curves of stress-strain are obtained in various strain rates. Experimental results show that tempered glass is sensitive to strain rates and its dynamic curves of stress-strain are nonlinear. Further study reveals the phenomenon of "pseudo-plastic yield" is completely different from that of ductile materials, and dynamic Young's modulus and dynamic compressive strength of tempered glass increase with the increasing strain rates. Based on the damage mechanics theory, a kind of constitutive model with damage of tempered glass is given. A strain rate-dependent constitutive model with damage of tempered glass is created by the use of polynomial fitting, according to SHPB experimental results.

1. Introduction

Glass is greatly concerned because of its excellent mechanical properties such as excellent optical properties and high strength. Tempered glass, compared with ordinary glass, has high mechanical strength, impact resistance, and strong features. When crushed, the debris of glass will not hurt people, with anti-theft, anti-fire, and some decorative features, which are widely used. In recent years, with the development of international terrorism, the impact of the glass on the mechanical properties of glass is increasing but mainly concentrated in the transparent armor system and other military aspects. Knight studied with high-speed photography technology Pyrex and small balls of soda-lime glass surface impact fracture process in 1977 (Knight et al., 1977) [1]. LI Lei has studied the strain rate dependent dynamic constitutive equation of float glass under impact, but tempered glass was not involved [2]. The Continuum Damage Mechanics (CDM) model is used to study the impact of the anti-stone of float glass (Sun et al., 2005) [3]. In 2023, dynamic response of toughened laminated glass sheet under impact load was researched by HAO Yunhong [4]. GU Linlin has studied dynamic mechanical properties of cement mortar-granite composites after high temperature by SHPB [5]. A four-point bending method was used to study the tensile strength of borosilicate glass by SHPB and MTS (Nie et al., 2008) [6-9]. In 2024, the research progress of tempered glass at home and abroad are summarized [10] , but the above research rarely involves on the impact properties of tempered glass.

The Split Hopkinson pressure bar (SHPB) is widely used to obtain the stress-strain curve of materials under a high strain rate. The dynamic plastic flow stress of metal materials in the range of 10^2-10^4 s^{-1} was studied. Ideal SHPB experiments on constant strain rate are one sample of one-dimensional stress guaranteeing an important condition for dynamic equilibrium.

As a kind of common building and decoration material, tempered glass is widely used in both residential and public buildings, and its thickness is generally below 15 mm. However, classification methods of architectural glass are also diverse, so we have to be careful if research is not realistic. Based on the application status of existing glass, the main use of SHPB technology is to study the dynamic mechanical properties of ordinary tempered glass in the explosive impact, to provide theoretical guidance for its scientific prevention and control.

2. Experiment

2.1. Materials

In this experiment, glass specimens were provided by Beijing Lopo Hollow Company (steel degree σ_M = 24 MPa). The tempered glass cannot be cut, so the first float glass with a water jet is cut into rounds by grinding to ensure that the specimen surface is without defects. The appearance of the specimen shall make detailed observations prior to the experiment. The process of generating a large crack in the specimen should be removed.

Sample dimensions are as follows: φ48.69 (± 0.50) mm × 14.82 mm. The chemical composition of the sample is shown in table 1, and the mechanical properties of other parameters are as follows: Density ρ=2, 500 kg/m^3, Young's modulus of the material E=70 GPa, and Poisson's ratio v=0.22. For glass and other brittle materials, the failure strain of only a few thousandths of an effective measurement time is very short and does not reach even before the stress of the specimen has been damaged. Uneven stress has a great influence on the dynamic curve material. To extend the rising edge of the incident wave, the dispersion effect is reduced, increasing the effective load time of the sample to ensure a uniform axial stress in the sample. In this paper, a rubber sheet is placed before the incident bar as the pulse shaper with a size of Φ 24 mm× 1.2 mm. To ensure that an accurate test all of the samples will not be polished, the samples to prevent scattering of broke, are radial with double-sided tape. The sample surface is coated with butter impact (or vaseline) to prevent friction effect.

Table 1. Main chemical compositions of float glass.

Composition	SiO_2	Na_2O	CaO	Al_2O_3	K_2O	Other
Percentage (%)	72.50	13.00	9.30	1.50	0.30	3.40

2.2. Laboratory equipment and the theoretical formula

Split Hopkinson pressure bar (SHPB), initially improved by Kolsky, has become an important means of testing materials at high strain rates (10^2-10^4 s^{-1}) under dynamic mechanical properties. According to the uniformity hypothesis, there is a relationship between the incident signal and the transmitted strain wave three groups of the reflected wave signal:

$$\varepsilon_I + \varepsilon_R = \varepsilon_T \tag{1}$$

According to one-dimensional elastic stress wave theory, the stress-strain curve can be obtained with the reflected pulse and transmitted pulse represented among specimens of stress. Strain and strain rate is determined by the following:

$$\sigma(t) = \frac{A}{A_0} E\varepsilon_t(t)$$
$$\varepsilon(t) = -\int_0^t (\dot{\varepsilon}_r(s) ds \quad (2)$$
$$\dot{\varepsilon}(t) = -\frac{2c_0}{L}\varepsilon_r(t)$$

where A_0 is the initial cross-sectional area of the sample, A is the cross-sectional area of the Hopkinson pressure bar, E is Young's modulus of the Hopkinson pressure bar, ε_t *(t)* is the transmission lever in the transmission strain, and ε_r *(t)* is the incident bar reflection strain. SHPB experiment with basic mechanical parameters is as follows: a bar of materials is alloy steel, Density ρ=7, 800 kg/m^3, Young's modulus E=206 GPa, Poisson's ratio ν=0.3, and P-wave velocity of stress wave C_0=5, 065 m/s. The diameters of bullets, incident bar, and transmission rod are 50 mm, with lengths of 800 mm, 2, 000 mm, and 2, 000 mm. We use the optical method to measure the impact velocity and use a strain gauge signal to measure strain. Semiconductor strain gauges and SDY2107 super dynamic strain testing are used, and test signals are collected by Tektronix DP04034 digital storage oscilloscope. Figure 1 shows a typical split Hopkinson pressure bar.

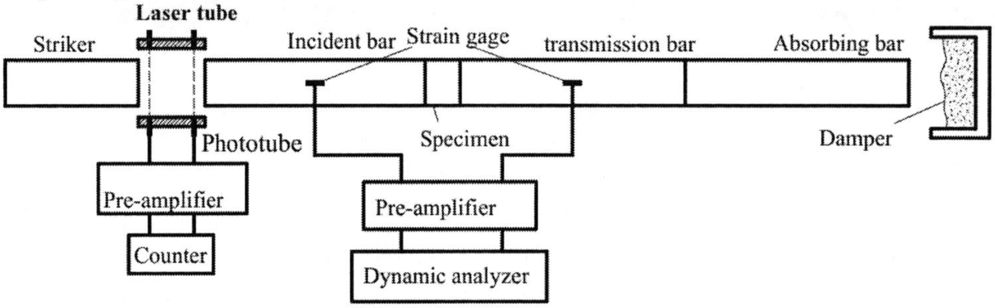

Figure 1. Typical split Hopkinson pressure bar.

3. Results and discussion

3.1. Effectiveness of experimental results

In consideration of the interference of the SHPB test, the test results are often used to test the validity of the test results. According to the one-dimensional elastic stress wave theory, the stress in the specimen should be exactly equal, but due to various interference factors on the test results, the two curves cannot completely overlap. As shown in Figure 2, by comparing the stress time curves of the test specimen at the front- and back-ends of tempered glass, the results show that the two curves coincide well, and the difference is less than 5%. Therefore, the SHPB test results of this paper are valid, and the stress-strain curves can reflect the real quality of the material.

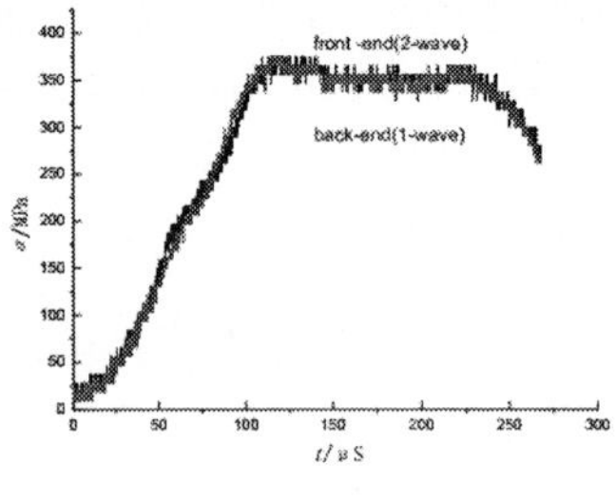

Figure 2. A comparison of axial stresses at the front- and back-ends of tempered glass.

3.2. Results and discussion

To ensure a representative experiment, three samples of each rate are used. Figure 3 is a typical glass of the original waveform. The waveform shows that the use of the pulse shape, the figure of the incident, and the transmitted waveform are trapezoidal, which is different from rectangular waveshaper. Specimen deformation process can determine the stress response to these signals, strain, and strain rate process. Figure 3 shows that the incident is trapezoidal, and the loading speed increases with the increase of the amplitude of the impact. Due to higher transmission wave amplitude and significantly different brittle material (float glass), we can see small reflection amplitudes. The reflected wave has gone through a platform, indicating that the specimen undergoes a constant strain rate, thus reaching the stress uniformly.

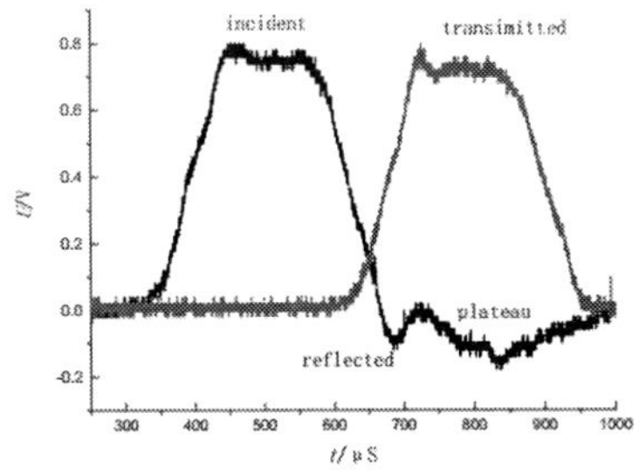

Figure 3. Typical original oscilloscope records of a specimen.

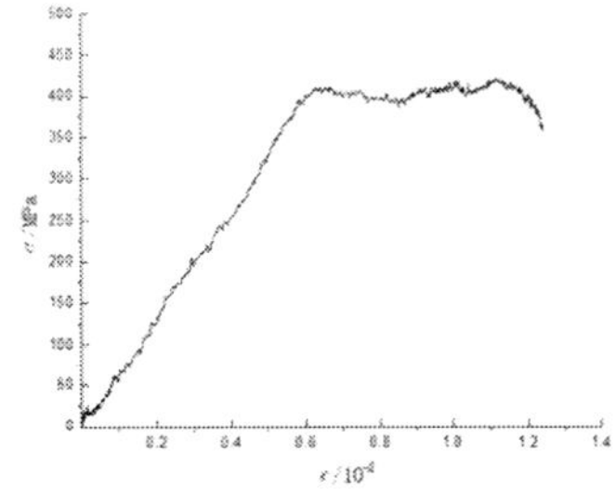

Figure 4. Stress-strain curves of the specimen.

The stress-strain curves of the tempered glass can be obtained by processing the original shock wave of Figure 3. The elastic modulus of E=69.205 GPa can be calculated from Figure 4, which is in agreement with the theoretical value of 70 GPa. From the curves of the specimen in Figure 4, the stress-strain curves show that the stress-strain relationship is elastic, but gradually shows plasticity as the pressure increases. Plastic metals yield very similarly, and we call it "pseudo-plastic yield". The special phenomenon is related to the stress state of the inner complex. After being uniformly heated by the outer surface of the cooling airflow impact and tension, the glass is in a contracted state. However, the internal temperature was still within the range of the softening, so the surface needed to occupy a large space. Such cooling is in a tensile stress state and the surface is in compressive stress. Its stress can be expressed in Figure 5 [11-13].

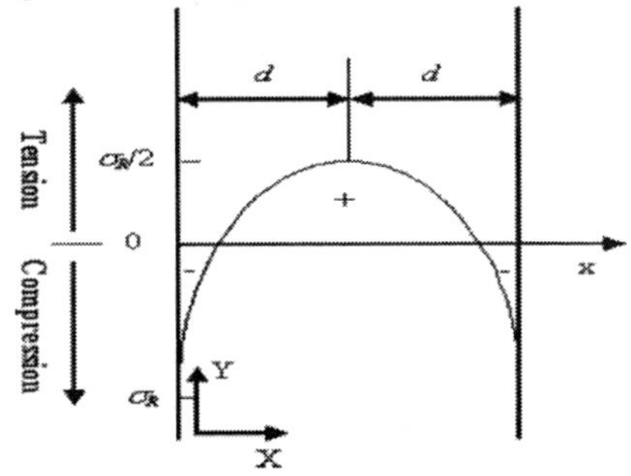

Figure 5. Residual stress profile inside stress-strain curves of the thermally tempered glass.

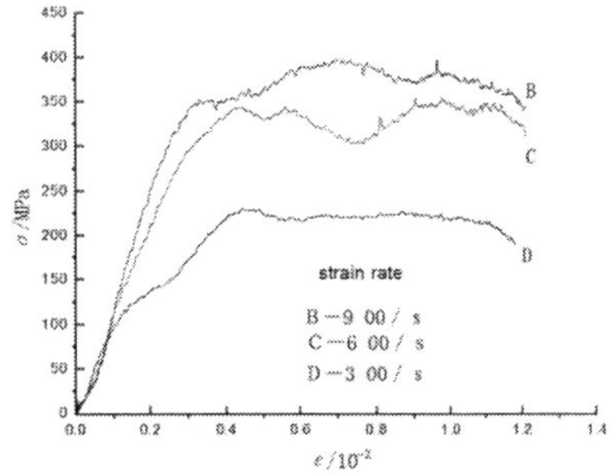

Figure 6. Compressive specimen at various strain rates.

As seen from Figure 5, the maximum compressive stress on the glass surface, the maximum tensile stress in the middle of the thickness of the glass, and the maximum tensile stress is half of the stress of available functions, which are expressed as:

$$\sigma(x,y) = -\sigma_R[1 - 3\times\frac{x}{d} + \frac{3}{2}\times(\frac{x}{d})^2] \tag{3}$$

When the glass surface is under pressure, the internal compressive stress increases the stress parabola stretch, with the balance generating tensile stress and the pressure while increasing tensile stress until rupture. Glass is used in a covalently way with a strong directional characteristic. Compared with the metal material, it hardly occurs dislocation motion. When surface or internal defects arise, it is highly likely to cause stress concentration around it. There are a lot of penny internal micro-cracks under the effect of external loads, due to the presence of microcracks. The mechanical properties of the produced glass have a weakening effect. For the characterization of the effect, the weakening of the material is defined as a certain degree of damage, caused by the material strength reduction, denoted as d. The main mechanism of injury is micro-crack nucleation and growth and its combination of Meso-crack initiation [10-12]. Due to the presence of a large number of randomly distributed microcracks in internal glass, the size varies and these microcracks are activated, resulting in material strength, stiffness deterioration, and eventually cracking damage. In the beginning, injuries, pressure, and tensile stress are very small, lowering the external load performance of the external linear elastic. Glass damage and compressive tensile stress increase, and its internal pressure is greater than the damage caused by the lower material strength of tensile stress increment (where pressure tensile stress increment is recorded as ⊿d). ⊿d-d (⊿d and d are variables) will buffer the external load so that they are in dynamic equilibrium and the external stress-strain curve is expressed as "pseudo-plastic yielding". When the external load overcomes ⊿d-d, tempered glass becomes float glass.

Figure 6 shows that the entire curve can be divided into four stages: linear elastic, non-linear enhancement, pseudo-plastic yield, and extreme damage. When the stress is less than the damage threshold stress σ_{th}, internal material does not damage evolution, all microcracks are only elastically deformed, and extension does not occur. Thus, In the linear elastic stage, the stress-strain curve is nearly a straight line, and the slope is a constant. In the nonlinear strengthening phase, when the stress σ is greater than the damage threshold stress σ_{th} but less than or equal to the maximum critical material stress σ_{cr} (stress is defined as the maximum critical stress when the material begins to happen pseudo-plastic yield σ_{cr}), damage occurs continuously. As the stress increases, more and more stable microcracks expand, and the surface energy barrier has a higher strength bound extension stop. With the increasing expansion of the area of microcracks, microcracks' contribution to effective compliance tensor materials also increased, so they exhibit a non-linear stress-strain relationship. The stress-strain curve showed yield. When $\sigma > \sigma_{cr}$, this is called the "pseudo-plastic yield" stage, which corresponds to a glass of ⊿d-d buffer external load, in the dynamic equilibrium phase. With the constant crack initiation and propagation, D increases and ⊿d-d gradually reduces until it completely overcomes external load, eventually leading to the destruction of specimens. Ultimately, in the fourth stage, it is called the extreme damage stage.

As seen from Figure 6, tempered glass is impacted by the SHPB bar. When the strain rate increased from 300 s^{-1} to 600 s^{-1} to 900 s^{-1}, Young's modulus of glass was 60.51 GPa, 81.43 GPa, and 97.78 GPa, respectively, showing a significant strain rate effect. The compressive strength increased in different degrees.

4. Dynamic constitutive equations

At linear elastic segments and strengthening segments with damage nonlinear stage, the constitutive equation of tempered glass is fitted in the form of Formula (3-3), according to the theory of damage mechanics and an elastic material in the one-dimensional stress conditions with damage constitutive equation. In the "pseudo-plastic yield" segment, the internal complexity stress of tempered glass plays a leading role in the constitutive equations. Its dynamic stress-strain curves and metallic ductile

Journal of Physics: Conference Series **2808** (2024) 012067 doi:10.1088/1742-6596/2808/1/012067

material are very similar, in the "pseudo-plastic yield" segment, with an experience-based Johnson-Cook viscoplastic constitutive model to fit. Based on experimental data and more or less viscoelastic and viscoplastic Johnson-Cook constitutive equation using continuum damage mechanics model and curve fitting methods, it is assumed that there is only limited by microcracks damage evolution of the confining pressure. Other material mechanics parameters remain unchanged, thus confining pressure effect on the material constitutive damage by changing the variables to achieve the establishment of tempered glass with damage dynamic constitutive equations.

According to the damage mechanics, the material is under external loads due to the emergence and development of internal defects and deterioration, and this process is irreversible. Material damage process is this deterioration cumulative process. When accumulated to a certain extent, the material produces macroscopic cracks, resulting in material damage. In the loading process, the material gathered by the micro-cracks and voids caused by the progressive destruction of the material behavior can be used to represent the damage variable D.

When $\sigma \le \sigma_{th}$, the stress-strain relationship is linearly elastic, and the constitutive equation can be expressed as:

$$\sigma = E\varepsilon \tag{4}$$

When $\sigma_{cr} \ge \sigma > \sigma_{th}$, the constitutive equation should contain damage. According to the theory of damage mechanics and properties of the glass material, damage variable D can be expressed as:

$$D = m\varepsilon^n \tag{5}$$

where n is a parameter and m is the amount of strain rate dependent.

Based on the theory of damage mechanics, the elastic material in one-dimensional stress conditions including the damage constitutive equation is:

$$\sigma = E\varepsilon(1-D) \tag{6}$$

where σ and ε are the material of the stress and the strain, E is Young's modulus, and D is the damage variable. We can obtain material containing the constitutive relation of one-dimensional elastic cord injury brittle material by Formulas (5) and (6).

$$\sigma = E\varepsilon(1-m\varepsilon^n) \tag{7}$$

We found that the four polynomials are best (Figure 7) by using the least squares method to fit the experimental curve. Therefore, the dynamic constitutive equation can be expressed as:

$$\begin{aligned} \sigma &= E\varepsilon(1-D) \\ D &= m(C + x\varepsilon + y\varepsilon^2 + \varepsilon^3) \end{aligned} \tag{8}$$

Due to the experimental curve, we selected three representative stress-strain curves, and each set of experimental data was fitted to obtain the stress-strain relationship under different strain rates. Table 2 shows curve fitting parameter values and different strain rates of the correlation index R^2 obtained by fitting, where R^2 is each underlying index close to 1, indicating that curve fitting works well.

Table 2. Fitted values of different strain rates.

$\dot{\varepsilon}$ /s^{-1}	E (GPa)	m	C	x	y	R^2
300	63.34	2.091e-9	3.625	-5.02	3.549	0.99553
600	87.93	-1.864e-8	0.0788	-1.154	2.307	0.99763
900	103.29	-1.89e-8	2.388	-1.6660	0.1557	0.99803

Fitting the data in Table 2, we get:

$$\begin{aligned}
&E(\dot{\varepsilon}) = 637.52\ln(\dot{\varepsilon} - 128.85) + 2966.16 \\
&C(\dot{\varepsilon}) = 13.0266 - 0.0411(\dot{\varepsilon}) + 0.00003(\dot{\varepsilon})^2 \\
&m(\dot{\varepsilon}) = 0.4.3296(E - 80 + 1.71479E - 100\dot{\varepsilon} + 1.1373(E - 130(\dot{\varepsilon})^2 \\
&y(\dot{\varepsilon}) = 3.8817 + 0.00041(\dot{\varepsilon}) - 5.0517(E - 6)(\dot{\varepsilon})^2 \\
&x(\dot{\varepsilon}) = -13.264 + 0.03478\dot{\varepsilon} - 0.00002(\dot{\varepsilon})^2
\end{aligned} \tag{9}$$

The associated index of the fitting curve was 0.98249, 1, 1, 1, and 1, which fits the curve better.

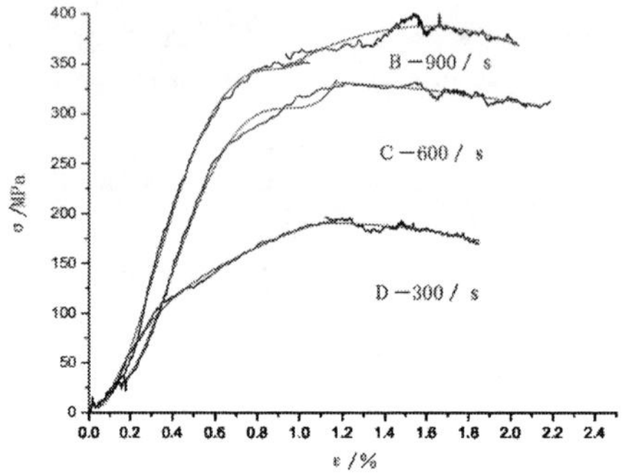

Figure 7. Experimental data and fitted data.

When $\sigma > \sigma_{cr}$, internal stresses in glass play a leading role in the dynamic mechanical properties of materials. Because its stress-strain curve and ductility (metal) are very similar, we use the modified Johnson-Cook model to describe the dynamic stress-strain curve of the segment material. Johnson-Cook constitutive relation is amended as follows:

$$\sigma = (A + B_0\varepsilon + B_1\varepsilon^2)(1 + c\ln(\dot{\varepsilon})^*)(1 - (T^*)^m \tag{10}$$

Three brackets in the formula describe the hardening, strain rate effect, and temperature softening effect of the material, respectively, where σ is the stress, ε is the strain, $(\dot{\varepsilon})^* = \dot{\varepsilon}/\dot{\varepsilon}_0$ is the dimensionless material strain rate, and $\dot{\varepsilon}_0$ is quasi-static strain rate. We take $\dot{\varepsilon}_0 = 10^{-3}s^{-1}$ and $T^* = (T - T_r)/(T_m - T_r)$ for calculation purposes, where T is the initial temperature of the sample, T_r is at room temperature, and T_m is the melting point of the material. Since the split Hopkinson pressure bar test is similar to an adiabatic process, we can obtain $T^* = 0$.

Johnson-Cook constitutive relation simplifies to:

$$\sigma = (A + B_0\varepsilon + B_1\varepsilon^2)(1 + c\ln(\dot{\varepsilon})^*) \tag{11}$$

There are four undetermined coefficients, A, B_0, B_1, C, in the constitutive relation. We need to fit the stress-strain curves under different strain rates. Because of the type described in the second bracket material strain rate effect, it can be used to fit a quadratic polynomial stress-strain curve. Depending on the strain rate, we determine the undetermined coefficients.

Due to the experimental curve, we chose the above three stress-strain curves, and the experimental data were fitted to give different strain rates of stress-strain relationship. After fitting the Johnson-Cook constitutive relation:

$$\sigma = (52.54 + 181.60\varepsilon - 50.54\varepsilon^2)(1 + 0.05472c\ln(\dot{\varepsilon})^*) \tag{12}$$

Based on the above, constitutive relations of tempered glass can be divided into four stages: linear elastic, non-linear enhancement, "pseudo-plastic yield", and extreme destruction, and constitutive relation is the strain rate dependent.

Its constitutive equation is:

$$\begin{aligned}
&\sigma = E\varepsilon \ldots\ldots \sigma \le \sigma_{th} \\
&\sigma = E\varepsilon(1-D) \ldots\ldots \sigma_{cr} \ge \sigma > \sigma_{th} \\
&D = m(y\varepsilon + \varepsilon^2) \\
&E(\dot{\varepsilon}) = 44.58\ln(\dot{\varepsilon} + 135.94) - 224.44 \\
&m(\dot{\varepsilon}) = 0.3634 - 0.00006\dot{\varepsilon} + 3.3748E - 8(\dot{\varepsilon})^2 \\
&y(\dot{\varepsilon}) = 0.04695 - 0.00013\dot{\varepsilon} + 5.85E - 8(\dot{\varepsilon})^2 \\
&\sigma = (52.54 + 181.60\varepsilon 50.54\varepsilon^2)(1 + 0.05472\ln(\dot{\varepsilon})^*) \ldots\ldots \sigma > \sigma_{cr}
\end{aligned} \tag{13}$$

5. Conclusion

The circular rubber sheet as Pulse Shaper SHPB existing device is improved and the end face of stress is compared before and after samples to ensure that the results are credible.

SHPB experiments using an improved method of dynamic mechanical properties of glass were studied to obtain the dynamic stress-strain curve of materials, and finally fit the glass dynamic constitutive equations. The dynamic stress-strain curve of the material can be divided into four stages: linear elastic, non-linear enhancement, "pseudo-plastic yield", and extreme destruction, and the "pseudo-plastic yield" phenomenon is analyzed theoretically. The results showed that tempered glass has the following characteristics: (a) Material dynamic stress-strain curve is a non-linear relationship, and the deformation performance of brittle and ductile material is completely different from the "pseudo-plastic yield"; (b) A higher strain rate within the range dynamic stress of the material is strain rate dependent strain relationships; (c) The initial elastic modulus of the material, failure stress, and failure strain increase with the strain rate increasing; (d) With the application of damage mechanics theory, we give one-dimensional constitutive model with damage tempered glass. According to SHPB results and the use of polynomial fitting to determine model parameters, we obtained tempered glass strain-rate-related damaged constitutive equations.

References

[1] C. G. Knight, M. V. Swain, and M. M. Chaudhry. Impact of small steel spheres on glass surfaces [J]. *Mater Sci.* **12** (1977): 1, 573-1, 586.

[2] LI Lei,AN Erfeng,YANG Jun. Strain rate dependent dynamic con stitutive equation of float glass [J]. Journal of building materials,2011,14(2):202-206(in Chinese).

[3] X. Sun, M. A. Khaleel,R. W. Davies. Modeling of stone-impact resistance of monolithic glass ply using continuum damage mechanics [J]. *Int. J. Damage Mech.* 14 (2005): 165-178.

[4] HAO Yunhong，HU Zhen，GUO Xin，et al. Research on dynamic response of toughened laminated glass sheet under impact load[J]. Chinese journal of applied mechanics，2023，40(4):883-892.

[5]GU Kinlin,YIN Kefei,WANG Zhen,et al. Test study on dynamic mechanical properties of cement mortar-grantie composite layer after high temperature[J].Journal of vibration and shock, 2024,V. 43 N.11,176-184.

[6] Jun Le and Lixin Song. Fracture mechanics analysis of thermally tempered glass plate: fracture induced by an embedded crack [J]. International Journal of Fracture, Vol. 132, N. 4. April 2005: 299-309.

[7] Tokunaga Hitoo and Kaizu Koichi. Impact fracture analysis of thermally tempered glass by the extended distinct element method [M]. Nihon Kikai Gakkai Ronbunshu, A Hen/Transactions of the Japan Society of Mechanical Engineers, Part A, Vol. 71, N 11: 1, 560-1, 566, November 2005 (Language: Japanese).
[8] Jun Le and Lixin Song. Fracture mechanics analysis of thermally tempered glass plate: fracture induced by an embedded crack [J]. International Journal of Fracture (2005) 132: 299-309.
[9] Gao Xuanneng and Wang Shupeng. Deflection of architectural laminated glass under static and explosive loads [J]. Journal of Chinese Ceramics Society, 2008. Vol. 36. No. 10: 1, 477-1483.
[10]WANG Yanhang,YANG Penghui,LI Xianzi， et al. Research Progress on Physical Tempered Glass[J]. Materials Reports, 2024， Vol.38， N.11 :1-6
[11] Lemaitre J. Course of damage mechanics [M]. Beijing: Science Press, 1996.
[12] Nemat Nasser, Mechanics of Mat., 1990, 10, 1~1.
[13 Guan Zhenduo. Physical Properties of Inorganic Materials [M]. Beijing: Tsinghua University Press, 1992.

Study on the effect of permanent magnet guideway surface defect on the mechanical properties of multi-surface high-temperature superconducting permanent magnet levitation system with halbach-type guideway

Jiahao Peng [1,*], Xianfeng Zhao [1] and Yulong Guo [1]

[1]School of Civil Engineering, Lanzhou Jiaotong University, Lanzhou 730070, China

[*]Corresponding author's e-mail: 2429676672@qq.com

Abstract. The purpose of this study is to discuss the effect of defects on the mechanical properties of halbach-type guideway surfaces in high-temperature superconducting permanent magnetic levitation systems with single-surface superconductor distribution versus multi-surface superconductor distribution. The study models the high-temperature superconducting permanent magnetic levitation system based on Maxwell's system of equations and compares it with the experimental data to obtain the variation curve of the levitation force of the levitation system under ideal conditions. Through numerical simulation and data analysis, it is found that the levitation force and guiding force of the high-temperature superconducting permanent magnetic levitation system will regularly decrease with the expansion of the defects, which will lead to the reduction of the system's load-carrying capacity and stability. These findings are of great significance for the design and application of high-temperature superconducting magnetic levitation systems, and provide theoretical support and technical guidance for the practical application of high-temperature superconducting magnetic levitation systems.

1. Introduction

A superconductor is a material that experiences zero resistance at low temperatures. When a superconductor enters the superconducting state at low temperatures, its zero resistance leads to zero internal magnetic field. High-temperature superconductors can be suspended above or below permanent magnets due to their unique anti-magnetism and self-stability under the action of the external magnetic field [1], and the levitation system composed of high-temperature superconductors (HTS) and permanent magnets (PMs) is usually called high-temperature superconducting levitation system (HTSLS), which has a wide range of applications in the fields of transportation, flywheel energy storage, etc. [2]. High-temperature superconducting magnetic levitation trains [3] have the advantages of being more environmentally friendly, faster, and more energy efficient compared with traditional trains and normal-conducting magnetic levitation trains [4-6]. Superconducting maglev trains have great application potential in the field of high-speed train applications, but the key problem is how to improve their carrying capacity [7] and stability ability to resist external disturbances [8].

There are many combinations of permanent magnet guideway (PMG) for high-temperature superconducting maglev trains. Halbach-type permanent magnet guideways are recommended to be preferred as the permanent magnet guideway of choice because the magnetic field strength above the guideway is significantly stronger than that of the other permanent magnet guideways, and because they have a wider range of distribution of the stronger part of the magnetic field [9]. On the basis of Halbach-PMG, Kemal Ozturk et al. used a multi-surface superconductor distribution to fully utilize the magnetic field around the PMG [10].

In addition, the permanent magnet guideway will inevitably be damaged during manufacturing, transportation, and use. Therefore, when the high-temperature superconducting levitation system is in use, when the magnet in the levitation system is damaged, its external magnetic field will also change, which will cause changes in the levitation force in the levitation system and may even lead to changes in the stability of the levitation system [11]. In ordinary magnetic levitation trains, the integrity of the electromagnet needs to be inspected and repaired to ensure the stability of the whole levitation system. In the preparation and utilization process of superconductors, defects may arise due to production procedures and external environmental factors. These defects on the surface of the permanent magnet track can alter the surrounding magnetic field. In high-speed movement, high-temperature superconducting magnetically levitated trains quickly traverse through these magnetic field fluctuations. Consequently, the superconductor can be treated as undergoing high-frequency variations in the magnetic field, leading to high-frequency flux changes and internal AC losses in superconductors. Under the effect of high-frequency magnetic flux change, the superconductor will produce AC loss, high thermal stress, and electromagnetic stress, which will lead to the further increase of the superconductor defects. This behavior causes the attenuation of the levitation force, which will seriously affect the stability of the train [12-13]. Therefore, we also need to consider whether the whole system can maintain safe and stable operation when the permanent magnet is damaged to a certain extent during the actual use of superconducting maglev trains, which will have certain practical significance for the development of high-temperature superconducting maglev trains.

2. The simulation model of HTSLS

2.1 Reasonableness verification of the model

For the control equations of a superconducting levitation system, Maxwell's system of equations is usually used to describe, and a variety of solution methods are derived from it, such as the A method (with the vector magnetic potential A as a variable) and the T method (with the vector potential T as a variable). In recent years, the H-method (with magnetic field strength H as a variable) has received widespread attention due to its fast computational speed and strong convergence. In this paper, the H method is used to establish the control equations of the system, and the basic solution model of the H method is given below. By Maxwell's system of equations:

$$\begin{aligned} J &= \nabla \times H \\ \nabla \times E &= -\mu \frac{\partial H}{\partial t} \\ E &= \rho J \\ B &= \mu H \end{aligned} \tag{1}$$

The electromagnetic control equations for the H method can be obtained by joining the above four equations:

$$\nabla \times (\rho J) = -\mu \frac{\partial H}{\partial t} \tag{2}$$

where ρ is the resistivity, μ is the permeability, B is the flux density, H is the magnetic field strength, E is the electric field strength, and J is the current density.

In this paper, we first model the levitation system of the upper surface distributed superconductor to obtain its levitation force data and select Wan Lei's experimental data[14] as a control. Some of the model parameters are given in Table 1.

Table 1. Model base parameters.

Parameters	Value
HTS Length	60 mm
HTS Thickness	15 mm
PMG Length	150 mm
PMG Thickness	30 mm
Residual Magnetism of PM	0.4 T
Critical Current Density of HTS	10^8 A/mm

Wan used a Stirling chiller to build a low-temperature thermostatic container in the laboratory, on the basis of which a low-temperature test platform was built by combining a temperature controller, a temperature sensor, and a control mainframe. Then a magnetic levitation characteristic test platform was built by combining a high-temperature superconducting test device SCML-1. The experiment puts the superconductor under different cooling temperatures with different field cooling heights and tests a series of levitation force data of the superconducting levitation system in different motion processes.

As shown in Figure 1, the initial position of the two YBCO superconductors is at a height of 40 mm above the NdFeB permanent magnet guideway with zero-field cooling conditions, after which it descends to a height of 10 mm above the guideway at a speed of 1 mm/s. Figure 2 gives the simulation results against the experimental data.

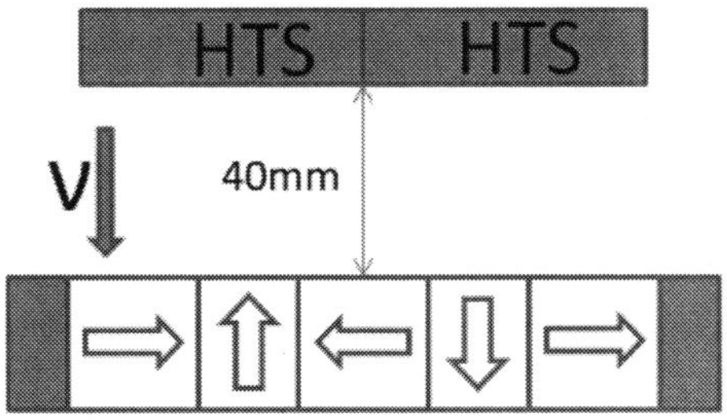

Figure 1. Schematic of the 2HTS model.

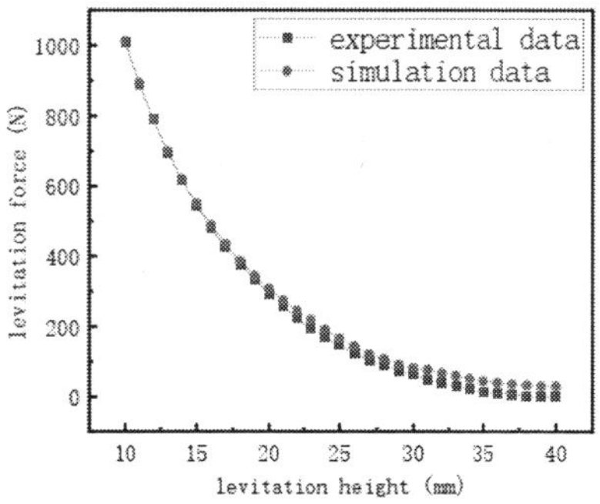

Figure 2. Experimental data and Simulated data.

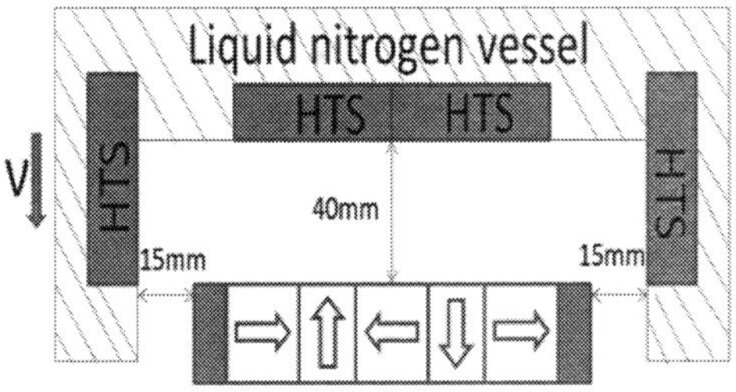

Figure 3. Schematic of the 4HTS model.

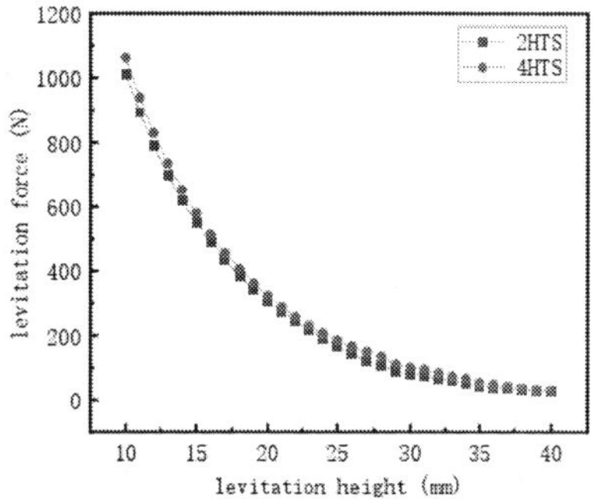

Figure 4. 2HTS data and 4HTS data.

Journal of Physics: Conference Series **2808** (2024) 012068 doi:10.1088/1742-6596/2808/1/012068

From the calculation results, it can be seen that the levitation force of the superconductor increases exponentially in the process of approaching the permanent magnetic guideway, and the simulation results are basically consistent with the experimental data. The levitation system with the multi-surface distribution shown in Figure 3 is further established to analyze the mechanical properties. The results in Figure 4 show that the levitation performance of the system is further improved by the superconductor with multi-surface distribution.

2.2 Effect of defects on the mechanical properties of the levitation system

In order to simulate the defects of the permanent magnet guideway in practical applications, this paper chooses to construct a notch defect in the center of the track surface, and the defect size is set to 1 mm^2, as shown in Figure 5.

Figure 5. Schematic of defects.

Consider the effect of defect depth variation on the maximum levitation force during the motion of the levitation system. Increase the defect depth from 0 to 10 mm.

In the 2HTS levitation system, the maximum levitation force is 1011.9 N when the defect depth is 0, and 999.67 N when the defect depth is 10 mm, and the maximum levitation force decreases by 12.23 N, and the curves of the maximum levitation force with the change of defect depth are given in Figure 6. The maximum levitation force approximates a quadratic polynomial relationship decreasing with increasing defect depth, and the fitting equation is:

$$F_y = 0.04493 D_d^2 - 1.68339 D_d + 1011.978 \tag{3}$$

where F_y is the suspension force and D_d is the defect depth.

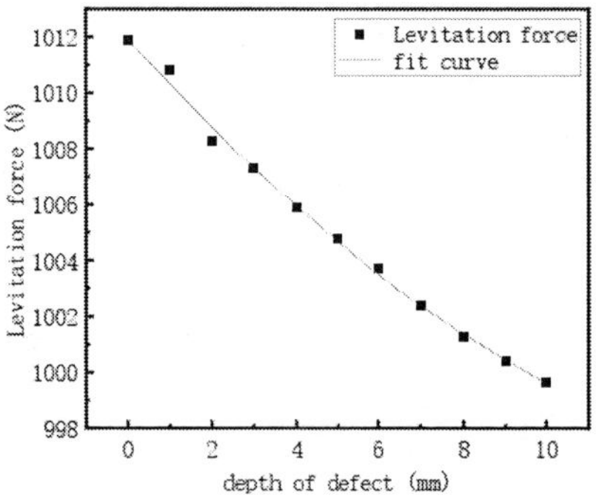

Figure 6. Effect of defect depth on 2HTS levitation force.

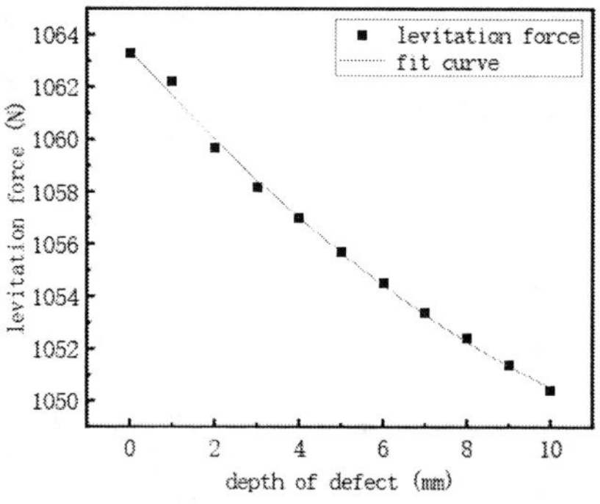

Figure 7. Effect of defect depth on 4HTS levitation force.

In the 4HTS levitation system, the maximum levitation force is 1063.3 N when the defect depth is 0, and 1050.4 N when the defect depth is 10 mm, which is a decrease of 12.9 N in the maximum levitation force, and the curve of the maximum levitation force with the depth of the defect is given in Figure 7. The maximum levitation force approximates a quadratic polynomial relationship decreasing with increasing defect depth, and the fitting equation is:

$$F_y = 0.05233 D_d^2 - 1.81149 D_d + 1063.42587 \quad (4)$$

When the superconductor is at 10 mm above the PMG, the superconductor is then made to produce a lateral displacement of 5 mm in size at that height to obtain the change rule of its guiding force.

In the 2HTS levitation system, when the defect depth is 0, the maximum guiding force is 4.80797 N, and when the defect depth is 10 mm, the maximum guiding force is 4.6979 N, which is decreased by 0.11007 N, and the curve of the maximum guiding force with the defect depth is given in Figure 8. The maximum guiding force decreases approximately linearly with increasing defect depth, and the fitting equation is:

$$F_x = 4.80532 - 0.1054 D_d \quad (5)$$

where F_x is the guiding force and D_d is the defect depth.

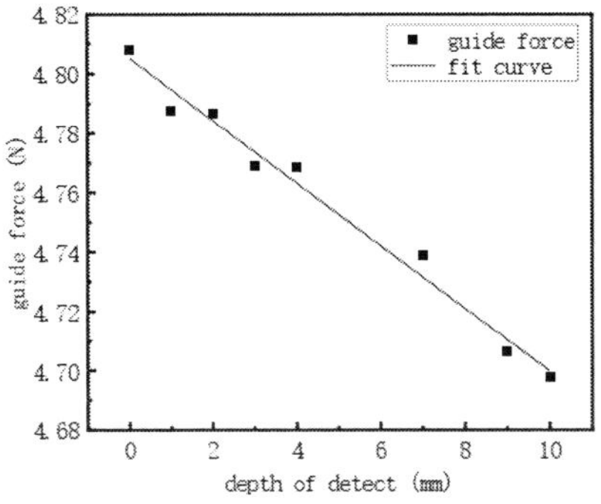

Figure 8. Effect of defect depth on 2HTS guiding force.

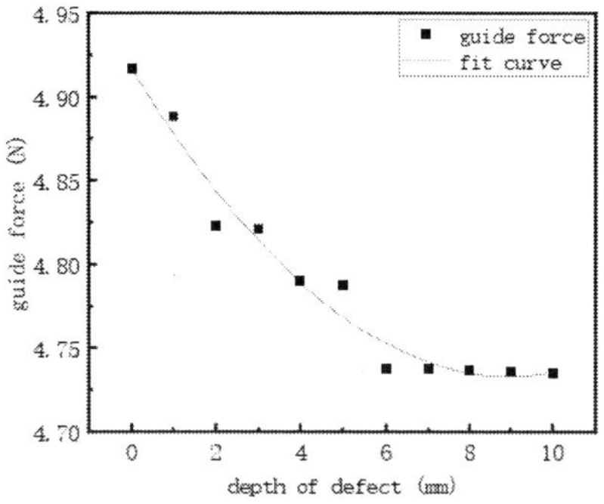

Figure 9. Effect of defect depth on 4HTS guiding force.

In the 4HTS suspension system, the maximum guiding force is 4.91688 N when the defect depth is 0, and the maximum guiding force is 4.73475 N when the defect depth is 10 mm, which decreases by 0.18213 N. The curve of the maximum guiding force with the defect depth is given in Figure 9. The maximum guiding force approximates a quadratic polynomial relationship decreasing with increasing defect depth, and the fitting equation is:

$$F_x = 0.00231 D_d^2 - 0.04126 D_d + 4.91733 \quad (6)$$

The effect of defect width variation on the maximum levitation force during the motion of the levitation system is considered so that the defect width is increased from 0 to 10 mm.

In the 2HTS levitation system, the maximum levitation force is 1011.9 N when the defect width is 0, and 997.57 N when the defect width is 10 mm, and the maximum levitation force decreases by 14.33 N, and the curve of the maximum levitation force with the change of defect width is given in Figure 10. As the defect width increases, the maximum levitation force decreases approximately linearly, and the fitting equation is:

$$F_y = 1012.399 - 1.5449 D_w \quad (7)$$

where D_w is the defect width.

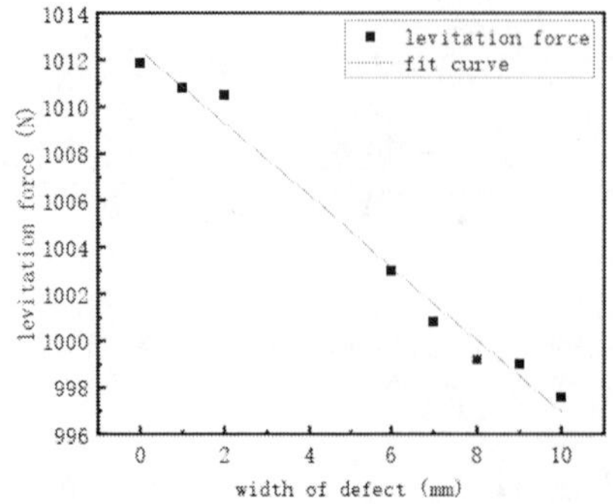

Figure 10. Effect of defect width on 2HTS levitation force.

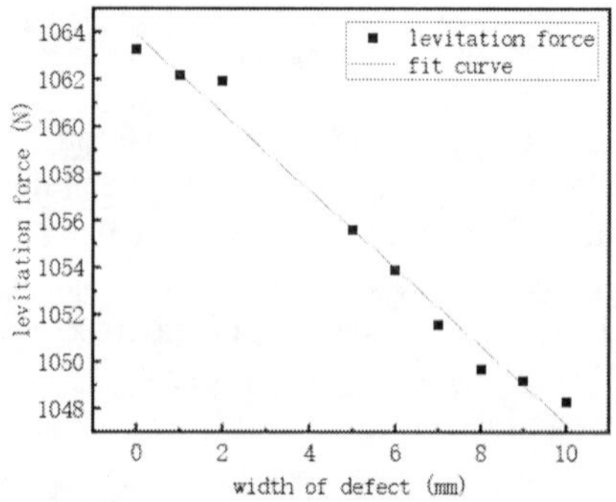

Figure 11. Effect of defect width on 4HTS levitation force.

In the 4HTS levitation system, the maximum levitation force is 1063.3 N when the defect width is 0, and 1048.3 N when the defect width is 10 mm, which is a decrease of 15 N in the maximum levitation force, and the curves of the maximum levitation force with the defect width are given in Figure 11. As the defect width increases, the maximum levitation force decreases approximately linearly, and the fitting equation is:

$$F_y = 1063.885 - 1.6513 D_w \tag{8}$$

In the 2HTS suspension system, the maximum guiding force is 4.80797 N when the defect width is 0, and the maximum guiding force is 4.78136 N when the defect width is 10 mm, which is a decrease of 0.02661 N. The curves for the variation of the maximum guiding force with the defect width are given in Figure 12. As the defect width increases, the maximum guiding force decreases approximately linearly, and the fitting equation is:

$$F_x = 4.80757 - 0.00262 D_w \tag{9}$$

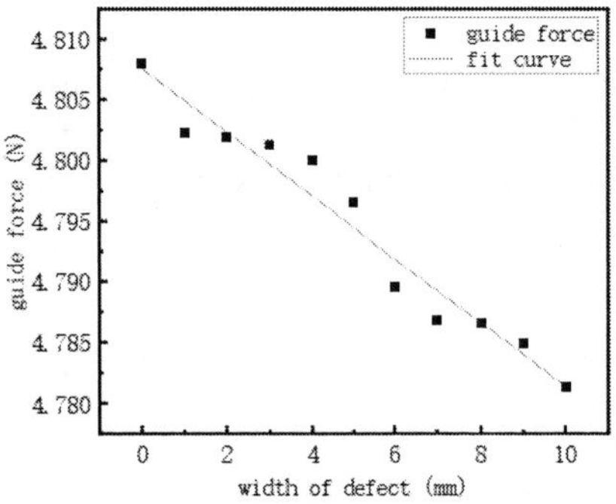

Figure 12. Effect of defect width on 2HTS 2HTS guiding force.

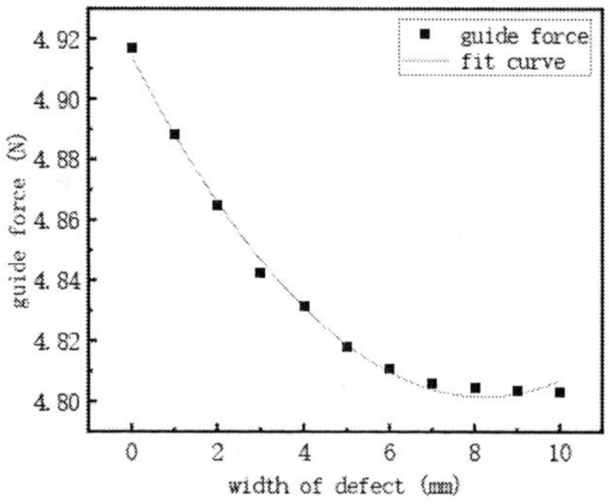

Figure 13. Effect of defect width on 4HTS guiding force.

In the 4HTS suspension system, the maximum guiding force is 4.91688 N when the defect width is 0 and 4.80318 N when the defect width is 10 mm, which is a decrease of 0.1137 N. The curves of the maximum guiding force with the defect width are given in Figure 13. As the defect width increases, the maximum guiding force approximates a quadratic polynomial relationship decreasing, and the fitting equation is:

$$F_x = 0.00167 D_w^2 - 0.02747 D_w + 4.91425 \tag{10}$$

3. Conclusion

Based on Maxwell's system of equations to obtain the control equations of the levitation system, the reliability of the model was determined by comparing it with the experimental data, based on which the surface distribution of superconductors and the influence of defects on the mechanical properties of the magnetic levitation system were further solved, and the following conclusions are obtained.

(1) The maximum levitation force decreases in a quadratic polynomial relationship with the increase of the defect depth, which is 12.23 N in the 2HTS levitation system and 12.9 N in the 4HTS levitation system; the maximum guiding force decreases with the increase of the defect depth, which is 0.11007 N in the 2HTS levitation system and 0.18213 N in the 4HTS levitation system in a linear relationship; the maximum guiding force is 0.11007 N in the 2HTS levitation system and 0.18213 N in the 4HTS levitation system, which is a quadratic polynomial relationship. 0.18213 N in the 4HTS levitation system.

(2) The maximum levitation force decreases linearly with the increase of the defect width by 14.33 N in the 2HTS levitation system and 15 N in the 4HTS levitation system; the maximum guiding force decreases linearly with the increase of the defect width by 0.02661 N in the 2HTS levitation system and quadratically by 0.1137 N in the 4HTS levitation system; the maximum guiding force decreases by 0.11007 N in the 2HTS levitation system and 0.18213 N in the 4HTS levitation system. 0.1137 N.

(3) Adding superconductors on both sides of the track can effectively improve the mechanical properties of the levitation system, but at the same time, the influence of defects on the mechanical properties will also increase.

References

[1] Hellman F, Gyorgy E M, Johnson D W, et al. (1988) Levitation of a magnet over a flat type II superconductor. J. Journal of Applied Physics, 63: 447-450.

[2] Wenxu L, Wenlong L, Jin F. (2020) Research on high-temperature superconducting magnetic levitation. J. Low Temperature and Superconductivity, 48(2): 44-49.

[3] Jianhui Y, Fanpeng M, Yanmeng J, et al. (2022) Study on Static Suspension Characteristics of High Temperature Superconducting Maglev Train. J. Cryogenic, 41(6): 56-61.

[4] Xinyu Y. (2018) Research and application of high-temperature superconducting magnetic levitation in rail transportation. J. Electrotechnical Materials, 4: 27-31.

[5] Jimin X, Fei Z, KIM Y, et al. (2017) Development status and prospect of high-temperature superconducting magnetic levitation bearings. J. China Materials Progress, 36(5): 321-328.

[6] Qianbing C. (2011) Research on key technology of superconducting magnetic levitation flywheel system. D. Graduate School of Chinese Academy of Sciences (Changchun Institute of Optical Precision Machinery and Physics).

[7] Xu Fei, Luo Shihui, Deng Zigang. (2019) Study on Key Technologies and Whole Speed Range Application of Maglev Rail Transport. J. Journal of The China Railway Society, 41(3): 40-49.

[8] Mingliang Z, Xinmeng Y, Liru L, et al. (2023) Vibration Characteristics and Parameter Feasible Rigion of a Type of HTS Maglev[J]. Journal of Vibration and Shock, 42(18): 30-38.

[9] Lulin W, Hua J, et al. (2007) The levitation performance of YBCO above three permanent magnet guideway. J. Cryogenics and Superconductivity, 4: 300-303.

[10] Kemal O, Antonio B, Murat A, et al. (2021) Experimental and Numerical Investigation of Levitation Force Parameters of Novel Multi-surface Halbach HTS-PMG Arrangement for Superconducting Maglev System. J. IEEE Transaction on Applied Superconductivity, 31(7): 3604112.

[11] Donghui J. (2014) Vibration characterization of lumped high-temperature superconductors on permanent magnet tracks. D. Southwest Jiaotong University.

[12] Yufeng Z, Baocai P. (2018) 3D Modeling Effect of Spherical Inclusion on The Magnetostriction of Bulk Superconductors. J. Journal of Low Temperature Physics, 190(3-4): 213-224.

[13] Shaoyan C. (2020) Distribution Characteristics of The Applied Magnetic Field above PMG in

HTS Maglev System. D. Fujian Normal University, 2020.
[14] Wan Lei. (2020) Studies on The Levitation Performance of HTS Maglev Unit With Cryocooler Direct Cooling. D. Southwest Jiaotong University.

Journal of Physics: Conference Series **2808** (2024) 012069 doi:10.1088/1742-6596/2808/1/012069

Lamb wave measurement method for in-plane elastic constants of balanced symmetric laminates based on machine learning

Quan Xie[1], Yaolu Liu[1,2] and Jun Fang[3*]

[1]College of Aerospace Engineering, Chongqing University, Chongqing, 400044, PR China
[2]Chongqing Key Laboratory of Heterogeneous Material Mechanics, Chongqing University, Chongqing, 400044, PR China
[3]Sichuan Provincial Institute for Physical & Chemical Measurement and Nondestructive Testing CO., LTD, Sichuan, 610100, PR China

[*]Corresponding author's e-mail: fangjun0731@sina.com

Abstract. To achieve in-situ measurement of mechanical properties in large-scale composite structures, this paper introduces a Lamb wave measurement based on machine learning to predict the in-plane engineering elastic constants of balanced symmetric laminates. Firstly, we consider that balanced symmetric laminates are equivalent to orthotropic single-layer plates with nine engineering elastic constants. Secondly, by varying these elastic constants and comparing the dispersion curves at different propagation angles, we conclude that, under low frequency-thickness products, the phase velocity of S_0-mode Lamb waves in orthotropic single-layer plates is dependent on four engineering elastic constants: tensile modulus, in-plane shear modulus, and in-plane Poisson's ratio. Subsequently, leveraging this correlation in dispersion curves, a BP neural network model is established using machine learning techniques. Using the neural network model, the goal of predicting engineering elastic constants using phase velocities is achieved. Finally, the effectiveness of this method is verified through theoretical calculations and numerical simulations.

1. Introduction

With the continuous development and innovation of composite material design and manufacturing processes in recent years, fiber-reinforced plastics have gradually replaced traditional metal materials such as aluminium alloys and titanium alloys due to their superior performance advantages, such as high specific strength and specific modulus. They have been widely used in the lightweight process of weight-sensitive fields such as aviation. For complex large-scale plate and shell structures such as fuselages, wings, nacelles, and blades made of fiber-reinforced plastic, the use of in-situ non-destructive testing technology to measure their mechanical properties has become a necessary means of quality control.

Currently, there are two effective non-destructive testing methods for measuring the elastic modulus of composite structures: the vibration method and the ultrasonic method. By using the vibration method, modal parameters like modal damping, natural frequencies, and modal shapes of the specimen are determined through experiments, thereby enabling the assessment of the elastic properties of the composite material. Due to the disadvantages of the vibration method, including bulky equipment, significant errors, complex procedures, and reliance on operator proficiency, the ultrasonic method emerges as a better option. Both bulk waves [1] and guided waves [2] in ultrasound can be used to predict the elastic properties of structural materials. However, compared to bulk waves that require specially prepared samples for convenient detection, guided wave-based non-destructive testing has a multi-modal characteristic, making it more convenient. Usually, among the various modes of guided waves, the velocities or wavelengths of the symmetric S_0 mode and antisymmetric A_0 mode are sufficient to predict elastic constants.

Techniques utilizing guided waves rely on the correlation between the dispersive properties of these waves, including phase velocities, group velocities, and wavelengths, with the elastic constants of composite materials [3]. By establishing a mapping relationship between the elastic constants of composite materials and the propagation speeds of guided waves, various inversion procedures can be employed to calculate the dispersion curves within the composite structure. Cao et al. [4] used an improved particle swarm optimization algorithm to invert the elastic constants and thickness from the zero-order modal Lamb wave, and validated this approach in the isotropic aluminium plate. Wang et al. [5] proposed a method for measuring the macroscopic elastic constants of orthogonally laminated composite plates based on Lamb wave measurements, establishing a mapping relationship between the S_0 group velocity and the tensile modulus. This paper improves the limitation of this method, which can only measure orthogonal laminates, and extends the non-destructive testing of Lamb waves to balanced symmetric laminates. Currently, machine learning has been able to make accurate and rapid predictions in practical complex industrial scenarios, especially in the field of composite materials, where it has gradually become an important tool for analyzing material properties. As the demand for data analysis and processing in ultrasonic non-destructive testing technology continues to increase, machine learning technology is gradually being combined with ultrasonic non-destructive testing technology. Wang et al. [6] proposed a deep learning-based model (ENet) by analysing two adjacent ultrasonic-guided wave signals. This model links the dispersion curves of guided waves with material elastic constants, enabling fully automatic and rapid inversion of the mechanical properties of composite materials. Its accuracy has been experimentally verified.

It can be seen that machine learning has been widely used in material performance analysis, and the speed of material performance prediction leads to other testing methods. Therefore, in order to achieve rapid measurement of the elastic constants of balanced symmetric laminates, this paper adopts a testing method based on the BP neural network. Utilizing the advantages of machine learning, a highly accurate prediction model is trained using a large amount of existing theoretical data. Subsequently, the material elastic constants can be directly obtained through the model and measurement data. Firstly, by analyzing the phase velocities of ultrasonic Lamb waves propagating at different angles in a balanced symmetric laminate, their sensitivity to nine independent elastic constants can be obtained. Then, based on the dispersion curves, the datasets of corresponding characteristic elastic parameters and phase velocities for various angles are constructed. The S_0 phase velocities under different propagation angles are used as inputs, and the corresponding characteristic elastic parameters are used as outputs. BP neural network algorithm is used to establish a prediction model for the elastic constants of balanced symmetric laminates. Subsequent validation and analysis are performed through numerical simulation and theoretical calculation.

2. Materials and Methods

2.1. Characteristics of Lamb wave propagation

During ultrasonic non-destructive testing, when the wavelength of the ultrasonic wave and the thickness of the plate are of the same order of magnitude, the transverse and longitudinal waves of the ultrasonic wave will couple with each other to form Lamb waves that propagate within the plate. The propagation mode of the ultrasonic wave will produce different modal waveforms on the waveform diagram due to differences in frequency and plate thickness. When Lamb waves propagate in plate and shell structures, they can also be divided into symmetric and asymmetric Lamb waves due to different propagation states. The wave speeds of the two are different, and the wave speed of Lamb waves in different modes will increase as the order increases. This is the manifestation of Lamb waves' good dispersion characteristics, which can clearly distinguish the waveforms of different modes in the waveform diagram. In Lamb wave non-destructive testing, pulsed signals are typically used as excitation signals to avoid the superposition of multiple modal Lamb waves, which can hinder dispersion analysis of Lamb wave propagation. Among them, the Lamb wave under the S_0 mode, which has no interference from other higher-order modes at low frequencies and thick thicknesses, and a wave speed higher than the A_0 mode, often appears in the first wave packet position in the waveform diagram, making it very convenient for waveform determination and wave speed analysis. Therefore, the S_0 mode is typically the preferred choice for modal wave speed analysis in Lamb wave measurement methods.

2.2. Balanced symmetric laminates and their corresponding orthotropic single-layer plate

Homogeneous orthotropic single-layer plates, characterized by nine independent elastic constants, serve as equivalents for balanced symmetric laminates. The effective elastic constants $\overline{C}_{ij}$ of The corresponding orthogonally anisotropic single-layer plate for balanced symmetric laminates can be determined as shown in Figure 1, according to the sublaminate theory proposed by Sun [7] and Pagano [8].

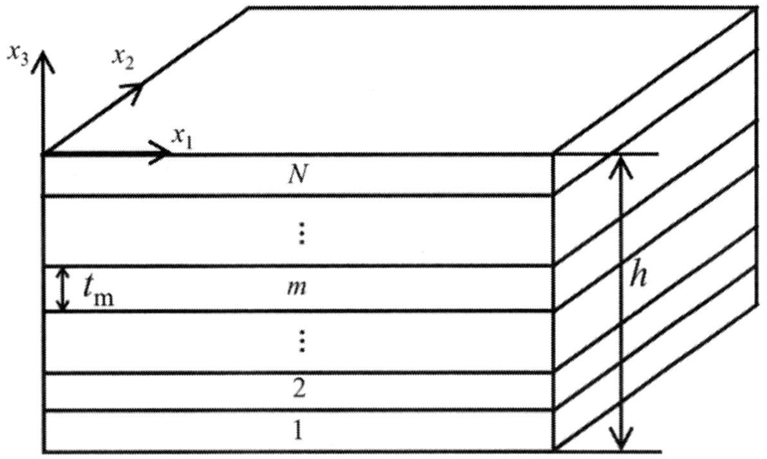

Figure 1. Schematic of the composite plate.

$$\overline{C}_{11} = \sum_{m=1}^{N} v_m C_{11}^{(m)} + \sum_{m=2}^{N} \left(C_{13}^{(m)} - \lambda_{13} \right) v_m \left(C_{13}^{(1)} - C_{13}^{(m)} \right) / C_{33}^{(m)} \tag{1}$$

$$\overline{C}_{12} = \sum_{m=1}^{N} v_m C_{12}^{(m)} + \sum_{m=2}^{N} \left(C_{13}^{(m)} - \lambda_{13} \right) v_m \left(C_{23}^{(1)} - C_{23}^{(m)} \right) / C_{33}^{(m)} \tag{2}$$

$$\overline{C}_{13} = \sum_{m=1}^{N} v_m C_{13}^{(m)} + \sum_{m=2}^{N} \left(C_{33}^{(m)} - \lambda_{33} \right) v_m \left(C_{13}^{(1)} - C_{13}^{(m)} \right) / C_{33}^{(m)} \tag{3}$$

$$\overline{C}_{22} = \sum_{m=1}^{N} v_m C_{22}^{(m)} + \sum_{m=2}^{N} \left(C_{23}^{(m)} - \lambda_{23} \right) v_m \left(C_{23}^{(1)} - C_{23}^{(m)} \right) / C_{33}^{(m)} \tag{4}$$

$$\overline{C}_{23} = \sum_{m=1}^{N} v_m C_{23}^{(m)} + \sum_{m=2}^{N} \left(C_{33}^{(m)} - \lambda_{33} \right) v_m \left(C_{23}^{(1)} - C_{23}^{(m)} \right) / C_{33}^{(m)} \tag{5}$$

$$\overline{C}_{33} = 1 / \left(\sum_{m=1}^{N} v_m / C_{33}^{(m)} \right); \tag{6}$$

$$\overline{C}_{66} = \sum_{m=1}^{N} v_m C_{66}^{(m)} + \sum_{m=2}^{N} \left(C_{36}^{(m)} - \lambda_{36} \right) v_m \left(C_{36}^{(1)} - C_{36}^{(m)} \right) / C_{33}^{(m)} \tag{7}$$

$$\bar{C}_{44} = \left(\sum_{m=1}^{N} v_m C_{44}^{(m)} / \Delta_m \right) / \Delta \tag{8}$$

$$\bar{C}_{55} = \left(\sum_{m=1}^{N} v_m C_{55}^{(m)} / \Delta_m \right) / \Delta \tag{9}$$

where v_m is the volume fraction of the mth lamina, $C_{ij}^{(m)}$ are the elastic constants for the mth lamina in the sublaminate.

$$\lambda_{i3} = \bar{C}_{i3} \quad (i = 1,2,3) \tag{10}$$

$$\Delta = \left(\sum_{m=1}^{N} v_m C_{44}^{(m)} / \Delta_m \right)\left(\sum_{m=1}^{N} v_m C_{55}^{(m)} / \Delta_m \right) - \left(\sum_{m=1}^{N} v_m C_{45}^{(m)} / \Delta_m \right)^2 \tag{11}$$

$$\Delta_m = C_{44}^{(m)} C_{55}^{(m)} - \left(C_{45}^{(m)} \right)^2 \tag{12}$$

$$v_m = t_m / h \tag{13}$$

where h is the thickness of the sublaminate and t_m is the thickness of the mth lamina.

Using the formulas provided above, the stiffness matrix of the corresponding orthotropic single-layer plate for balanced symmetric laminates can be calculated, which gives the elastic moduli of the equivalent single-layer plate. Therefore, the equivalence of the two plates can be verified by comparing their Lamb wave dispersion curves. A glass fiber composite laminate was selected, with the material properties of its single-layer plate shown in Table 1. The density of the composite plate is 2500 kg/m^3, and the ply stacking sequence is $[0/30/-30]_{3s}$. The equivalent stiffness matrix of the corresponding orthotropic single-layer plate is determined using the sublaminate theory.

Table 1. Elastic properties of an orthotropic lamina.

Tensile elastic modulus (GPa)	Shear modulus (GPa)	Poisson ratio
E_1=75	G_{12}=30	v_{12}=0.25
E_2=30	G_{13}=30	v_{13}=0.25
E_3=30	G_{23}=12	v_{23}=0.25

$$C_{[0/30/-30]3s} = \begin{bmatrix} 75.99 & 7.26 & 10.50 & 0 & 0 & 0 \\ 7.26 & 44.71 & 9.64 & 0 & 0 & 0 \\ 10.50 & 9.64 & 33.43 & 0 & 0 & 0 \\ 0 & 0 & 0 & 15.00 & 0 & 0 \\ 0 & 0 & 0 & 0 & 27.00 & 0 \\ 0 & 0 & 0 & 0 & 0 & 26.54 \end{bmatrix} (\text{GPa}) \tag{14}$$

When calculating the dispersion curves of balanced symmetric laminates, the dispersion software requires the input of material properties from Table 1 and the setting of the stacking sequence. Correspondingly, for orthogonally anisotropic single-layer plates, the equivalent stiffness matrices obtained previously need to be inputted. Since Lamb waves propagate in plates along various angles, it is necessary to compare the dispersion curves of the two plates in both the principal and oblique directions to ensure equivalence in different directions. Figure 2 demonstrates a comparative analysis of the dispersion curves at 0° (principal direction) and 45° (oblique direction). It can be observed that, regardless of the direction, the dispersion curves of the two sets of plates align perfectly within the low range of frequency-thickness product (fd) range. This indicates that Lamb waves exhibit identical propagation characteristics in both plates, particularly in the low range of the frequency-thickness product.

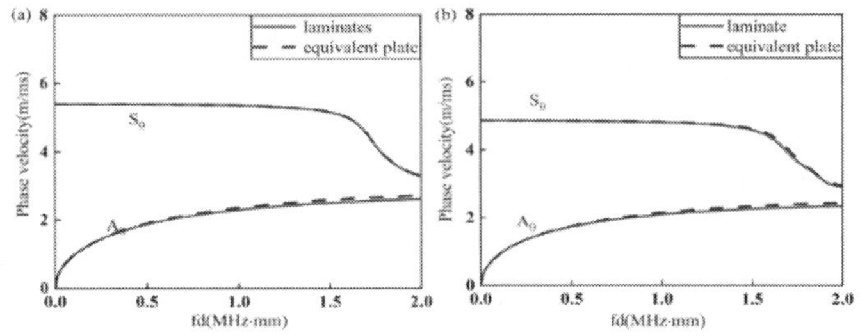

Figure 2. Comparative analysis of Lamb wave dispersion curves at a propagation angle of 0°(a) and 45°(b) for composite laminates versus corresponding orthotropic single-layer plates.

Table 2. Engineering elastic constants of orthotropic single-layer plates.

Tensile elastic modulus (GPa)	Shear modulus (GPa)	Poisson ratio
E_1=75	G_{12}=10	v_{12}=0.1
E_2=75	G_{13}=10	v_{13}=0.33
E_3=10	G_{23}=10	v_{23}=0.33

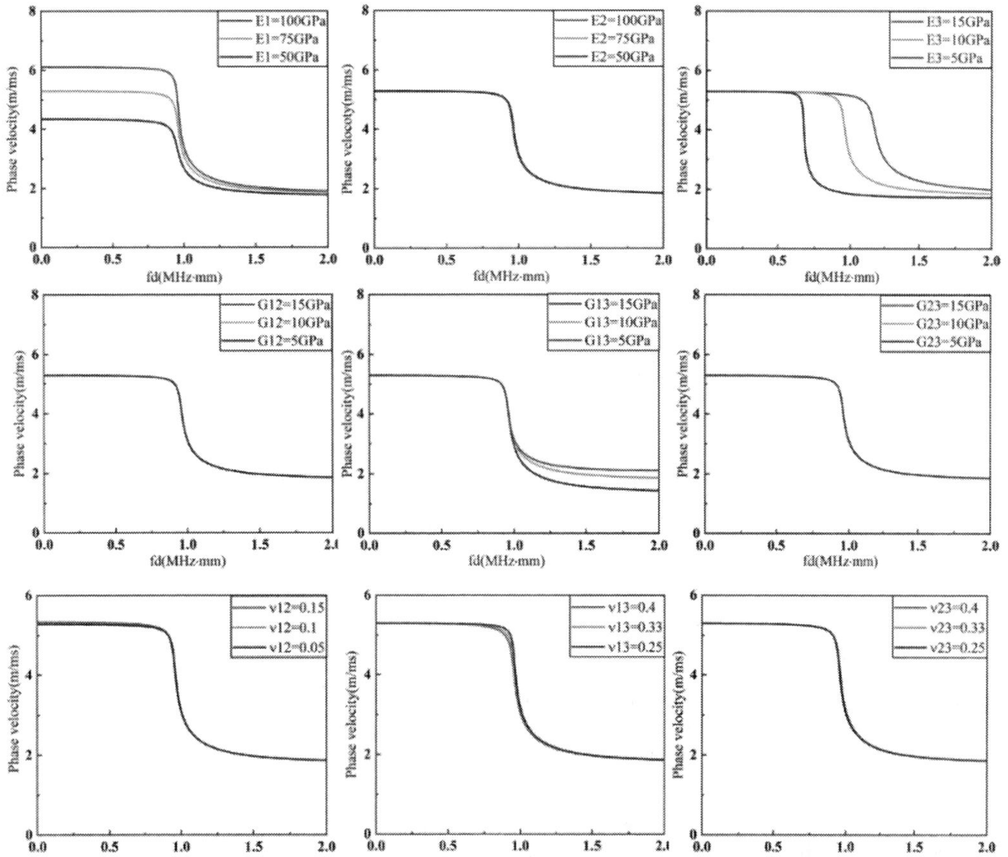

Figure 3. Comparison of S_0 mode dispersion curves for the nine engineering elastic constants of the orthotropic single-layer plate at a propagation angle of 0°.

Journal of Physics: Conference Series **2808** (2024) 012069 doi:10.1088/1742-6596/2808/1/012069

2.3. Sensitivity study

To explore how the nine engineering elastic constants affect the dispersion curves of orthotropic single-layer plates in both principals (0°, 90°) and oblique directions (30°, 45°, 60°), we employed the method of controlled variables, varying the values of each constant while keeping the other eight constants constant, referencing the engineering elastic constants outlined in table 2. Utilizing Origin software for plotting, we compared the S_0 mode dispersion curves for all scenarios under each engineering elastic constant on the same graph. The dispersion curves for the balanced symmetric laminates used in this study were calculated using the dispersion curve calculation software (Dispersion Calculator). Finally, a comparison diagram of dispersion curves for each engineering elastic constant of the S_0 mode at different propagation angles is obtained. Figure 3 illustrates the relationship between the S_0 mode dispersion curves and the nine engineering elastic constants of orthotropic single-layer plates in the principal direction of propagation (0°). Conversely, Figure 4 compares the S_0 mode dispersion curves of the same elastic constants in an oblique propagation direction (45°) within the same orthotropic single-layer plate.

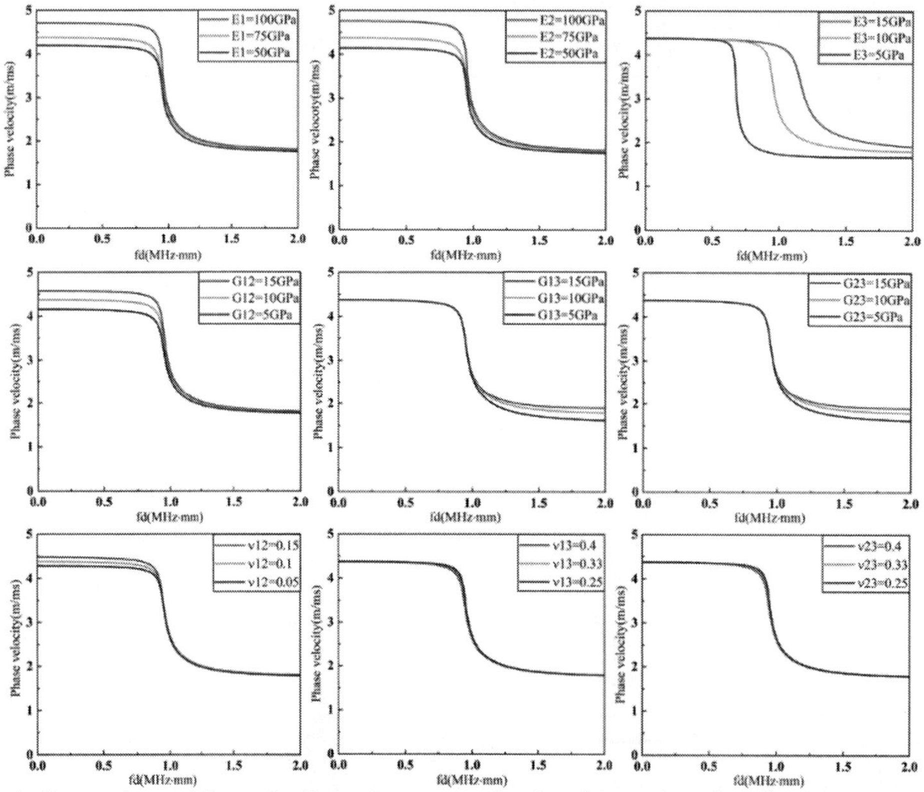

Figure 4. Comparison of S_0 mode dispersion curves for the nine engineering elastic constants of the orthotropic single-layer plate at a propagation angle of 45°.

From the sensitivity comparison chart of the nine engineering elastic constants in the principal direction (0°), it can be observed that when the frequency-thickness products of the orthotropic single-layer plate are in low range, the S_0 mode phase velocity has a significant relationship only with the tensile modulus E_1 in the 0° direction. However, from the sensitivity comparison chart of the nine engineering elastic constants in the oblique direction (45°), it can be seen that the phase velocity of the S_0 mode in the low frequency-thickness products range of the orthotropic single-layer plate is not only related to the tensile moduli E_1 and E_2 but also has a certain correlation with the in-plane Poisson ratio v_{12} and the in-plane shear modulus G_{12}. Therefore, based on the combined analysis of the sensitivity comparison diagrams in both the principal and oblique directions, it can be concluded that the four engineering elastic constants all influence the value of the S_0 mode phase velocity, when the frequency-thickness products of the orthotropic single-layer plate is in low range. Based on this, we can utilize the mapping relationship between E_1, E_2, G_{12}, ν_{12}, and the S_0 mode phase velocity, combined with the concept of machine learning, to establish a corresponding network model for the inversion of these four feature values of the orthotropic single-layer plate. Due to the previously verified equivalence between the balanced symmetric laminates and the corresponding orthotropic single-layer plate, this network model can also be adopted to predict the macroscopic engineering elastic constants of the balanced symmetric laminates.

2.4. Datasets

The datasets employed in this article are exclusively computed using the software Dispersion Calculator. Since only four elastic constants pertain to the phase velocity of the S_0 mode, we solely manipulate these constants while maintaining all other parameters consistent with Table 2. Specifically, the in-plane tensile elastic modulus E_1 and E_2 are varied from 30 GPa to 100 GPa in increments of 2 GPa, yielding a total of 36 distinct values. The in-plane shear elastic modulus G_{12} spans from 5 GPa to 30 GPa with irregular intervals, resulting in 18 values. ν_{12} range between 0.04 and 0.65, and to augment the training samples with smaller in-plane Poisson's ratios, a non-uniform distribution is adopted, leading to 24 values. In aggregate, approximately 560k samples are generated by the Dispersion Calculator for propagation angles of 0°, 30°, 45°, 60°, and 90° to acquire the corresponding dispersion curves.

By extracting the phase velocity of the S_0 mode at a low frequency-thickness product (fd=0.5 MHz*mm), a comprehensive dataset of Lamb wave velocities at five propagation angles is compiled, correlating the four elastic constants for each sample. Arrange all the corresponding nine data in the format of (V_0, V_{30}, V_{45}, V_{60}, V_{90}, E_1, E_2, G_{12}, ν_{12}) per row and save them in a txt file for subsequent machine learning to read the data. The dataset is randomly partitioned into training and testing sets with a 4:1 ratio. The larger training set is employed to refine the parameters of the neural network model, thereby enhancing its predictive accuracy, while the testing set serves to validate the model's performance. To ensure feature uniformity during training and to optimize the training process, input and output variables undergo normalization, which transforms the values into a range between 0 and 1. The normalization formulae for the input and output variables are outlined below:

$$X^* = (X - X_{min}) / (X_{max} - X_{min}) \tag{15}$$

$$Y^* = (Y - Y_{min}) / (Y_{max} - Y_{min}) \tag{16}$$

where X^* and Y^* are the normalized input and output values; X and Y are the unnormalized input and output values; X_{min} and X_{max} are the minimum and maximum values among the unnormalized input values; Y_{min} and Y_{max} are the minimum and maximum values among the unnormalized output values.

2.5. BP neural network

This article employs a modular approach in PyTorch to construct a Backpropagation (BP) neural network. The BP neural network is a machine learning model that takes input feature variables, computes predictions through internal functions, and then optimizes the parameters of those functions based on the differences between the predictions and the actual values corresponding to the input feature variables. The BP neural network structure consists of three layers: an input layer for the input feature variables, an output layer for the predicted results, and a hidden layer that determines the internal parameters of the neural network model. For the neural network used in this training, the input variables are the phase velocities of the S_0 mode at five propagation angles (0°, 30°, 45°, 60°, 90°) obtained from the dataset, while the theoretical values used to correct the predictions are the four engineering elastic constants (E_1, E_2, G_{12}, ν_{12}) from the dataset, as shown in figure 5. To increase the nonlinear fitting capability of the neural network model, the activation function chosen for the BP neural network is torch.nn.Tanh. The sizes of the input and output layers depend on the number of input feature variables and the corresponding number of theoretical values, respectively. The size of the hidden layer is crucial

Journal of Physics: Conference Series **2808** (2024) 012069 doi:10.1088/1742-6596/2808/1/012069

to the BP neural network model as it not only determines the computational efficiency of the model but also directly affects its accuracy. Therefore, it is necessary to use a portion of the data beforehand to determine the size of the hidden layer during the training process. This not only significantly reduces the overall training time but also provides effective initial parameters for the subsequent neural network model.

In addition, the choice of optimization algorithms during the optimization of neural network structure parameters can significantly impact the training efficiency and accuracy of the model. Two widely used optimizers in BP neural networks are Adam and LBFGS. Adam is an adaptive learning optimization method that can automatically adjust the learning rate during training based on the degree of difference in training results, improving the fitting capability of the neural network model. On the other hand, LBFGS is a popular Newton's method optimization technique that stores only a limited number of iterations of information during Newton's method. Sacrificing computational memory for reduced computation time significantly lowers computational complexity and provides rapid convergence. While determining the size of the hidden layer using a subset of the data in the dataset, we can separately discuss the advantages and disadvantages of the two optimization algorithms, providing an effective initial neural network model structure for subsequent training with a large amount of data.

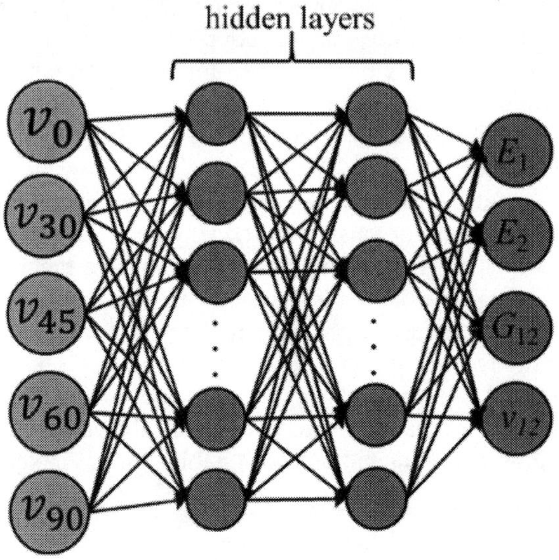

Figure 5. Schematic diagram of BP neural network

3. Results & Discussion

3.1. BP neural network

To train the BP neural network, a normalized training set is used. By varying the number of hidden layers and neurons, the optimal BP neural network structure is selected based on the coefficient of determination (R2), which represents the degree of neural network fitting. At the same time, the mean squared error (MSE) of the training set during the training process should be compared. The predicted values of the neural network should be compared with the corresponding theoretical values in the database, and their relative errors should be calculated. The training results with a relative error of less than 10% are considered accurate predictions, from which the overall accuracy of the model for predicting various engineering elastic constants can be calculated. These values can be used as criteria to evaluate the performance of the neural network model. As seen in Figure 6, the fitting performance using the LBFGS optimizer is generally higher than that of the Adam optimizer. Finally, a neural network structure with four hidden layers, 20 neurons per layer, and the LBFGS optimizer is chosen. The prediction results of the selected network model are shown in Table 3.

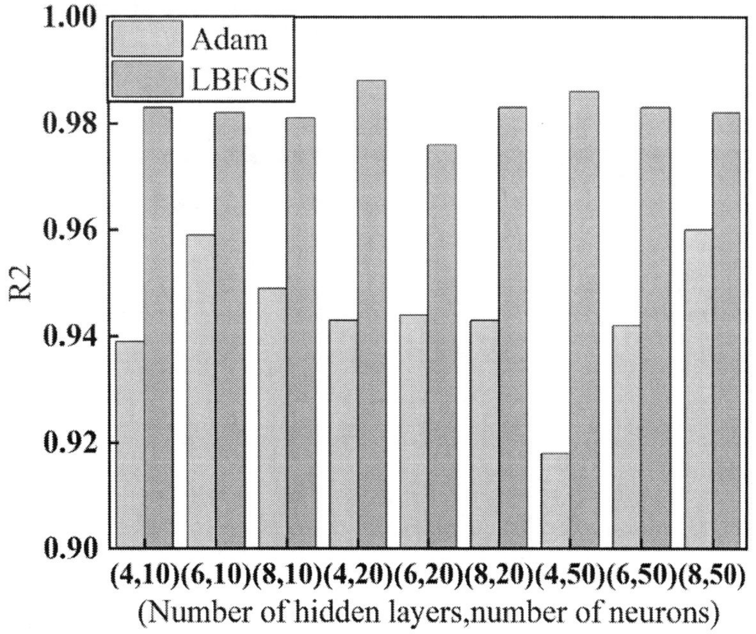

Figure 6. The coefficient of determination for each network structure.

Table 3. Results of the prediction network model.

MSE	R2	Accuracy (relative error<10%)			
		E_1	E_2	G_{12}	v_{12}
0.116	0.992	99.956%	99.956%	99.735%	97.926%

It can be seen that within the acceptable range of 10% relative error, all engineering elastic constants have a high degree of accuracy. Among them, the accuracy of in-place Poisson ratio v_{12} is slightly lower compared to the other three engineering elastic constants, which may be due to some actual values of in-place Poisson ratio v_{12} being smaller, resulting in relatively larger relative errors for some predicted values of in-place Poisson ratio v_{12}. Taking into account the MSE, R2, and accuracy, this neural network model exhibits good performance in inverting engineering elastic constants.

3.2. Numerical simulation

To validate the accuracy of the neural network, a model was constructed using the finite element simulation software (Abaqus) for simulation analysis. The phase velocities of S_0 mode Lamb wave of the orthotropic single-layer plate in various propagation directions were calculated using finite elements and then fed into the previously trained neural network model to assess various errors. The material properties of the orthotropic lamina are presented in Table 1, and the finite element model is shown in Figure 7. The stacking sequence of the finite element model is $[0/30/-30]_{3s}$. The dimensions of the balanced symmetric laminates are 700*700*2.3mm, with a mesh size of 2.5mm and a time step set to 1.0×10^{-8}s. The excitation frequency is 200kHz, and the waveform signal is a 5-cycle sinusoidal windowed signal. The excitation function is defined as:

$$H(t)=\begin{cases}\frac{1}{2}\left(1-\cos\left(2\pi ft/n\right)\right)\sin\left(2\pi ft\right) & (0<t\le n/f)\\ 0 & (t>n/f)\end{cases} \tag{17}$$

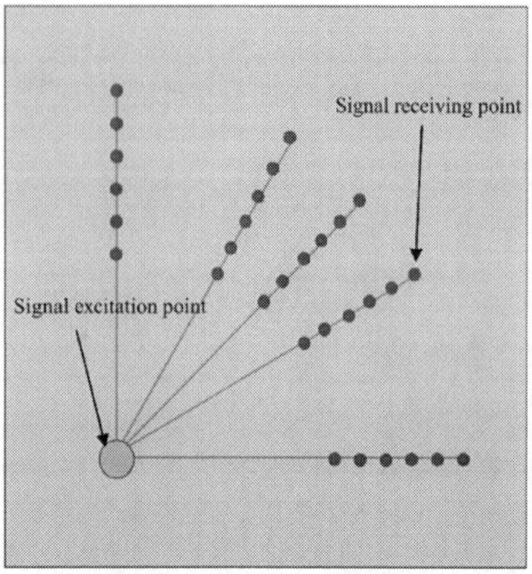

Figure 7. Finite element model.

There are five propagation angles in total for the finite element model, which are 0°, 30°, 45°, 60°, and 90°. Each propagation angle has six signal-receiving points, with a spacing of 30mm between each point. Detect each signal receiving point and output the corresponding in-plane displacement-time curve. Based on the previously known dispersion curve, it can be seen that under low-frequency thickness products, the phase velocity of the S_0 mode is higher than that of the A_0 mode. Therefore, the first wave packet in the displacement-time curve represents the waveform of the S_0 mode. By calculating the time difference of the S_0 waveform at each corresponding point under each propagation angle, the value of the phase velocity of the S_0 mode at that propagation angle can be determined.

The S_0 mode Lamb wave velocities in various directions are calculated and presented in Table 4. Then, these velocities are input into the previously trained neural network model, and the prediction results are shown in Table 5. The theoretical values are the macroscopic elastic constants of balanced symmetric laminates calculated through the stiffness matrix. It can be seen that the errors for the four elastic constants are all within 10%, except for the in-plane Poisson's ratio, which has a relatively large error. This may be due to the small numerical value of Poisson's ratio itself, which can easily lead to larger fluctuations.

Table 4. Phase velocity of S_0 Lamb wave at each propagation angle.

Propagation angle	0°	30°	45°	60°	90°
Phase velocity (m/s)	5266.472	4897.27	4579.514	4284.141	4024.683

Table 5. Material parameters of an orthotropic single-layer plate.

	E_1	E_2	G_{12}	v_{12}
Theoretical Value	72.27	41.68	26.54	0.1009
Prediction Value	74.15	43.98	24.98	0.0925
Relative Error	2.6%	5.52%	5.88%	8.33%

4. Conclusions

First, this article analyses the macroscopic equivalence of balanced symmetric laminates to orthogonally anisotropic plates and investigates the sensitivity of the phase velocities of Lamb waves propagating at various angles in the balanced symmetric laminates to nine elastic constants in the S_0 mode. The results indicate that, within the low frequency-thickness products range, the wave speed of the S_0 mode in orthotropic laminates is correlated with the in-plane tensile modulus E_1 and E_2, the in-plane shear modulus G_{12}, and the in-plane Poisson ratio v_{12}. Based on these findings, a method for predicting the elastic constants of balanced symmetric laminates is explored through a combination of machine learning and ultrasound techniques. This approach enables rapid prediction of four elastic constants of balanced symmetric laminates. Numerical simulations demonstrate that this method exhibits good accuracy within a 10% relative error range, making it suitable for rapid and preliminary predictions of structural elastic constants in industrial applications.

To further enhance the adaptability and accuracy of the prediction model, future work can involve expanding the database by increasing the range of data values and reducing the intervals between them. Additionally, more complex algorithms can be employed to select features and optimize the neural network model, thereby improving its accuracy.

References

[1] Wu, T.T., Liu, Y.H. (1999) On the measurement of anisotropic elastic constants of fiber-reinforced composite plate using ultrasonic bulk wave and laser generated Lamb wave. Ultrasonics, 37:405–12.

[2] Kudela, P., Radzienski, M., Fiborek, P., Wandowski, T. (2020) Elastic constants identification of woven fabric reinforced composites by using guided wave dispersion curves and genetic algorithm. Composite Structures, 249:112569.

[3] Eremin, A.A., Glushkov, E.V., Glushkova, N.V., Lammering, R. (2015) Evaluation of effective elastic properties of layered composite fiber-reinforced plastic plates by piezoelectrically induced guided waves and laser Doppler vibrometry. Composite Structures, 125:449–58.

[4] Cao, H., Guo, S., He, Z., Xie, Y., Zhang, T., Feng, W. (2022) In situ elastic constant determination of unidirectional CFRP composites via backwall reflected multi-mode ultrasonic bulk waves using a linear array probe. Composites Part B: Engineering, 238:109953.

[5] Wang, Y., Liu, Y., Hu, N. (2024) A method for measuring the tensile elastic modulus of cross-ply composite plates based on S0 Lamb waves. Journal of Chongqing University, 2-22:1–11.

[6] Wang, S., Luo, Z., Jing, J., Su, Z., Wu, X., Ni, Z. (2022) Real-time determination of elastic constants of composites via ultrasonic guided waves and deep learning. Measurement, 200:111680.

[7] Sun, C.T., Li S. (1988) Three-Dimensional Effective Elastic Constants for Thick Laminates. Journal of Composite Materials, 22:629-39.

[8] Pagano, N.J. (1994) Exact Moduli of Anisotropic Laminates. In: Broutman, L.J., Krock, R.H., Sendeckyj, G.P. (Eds.), Mechanics of composite materials. Academic Press, Waltham. pp. 32–45.

Journal of Physics: Conference Series **2808** (2024) 012070 doi:10.1088/1742-6596/2808/1/012070

Static and dynamic mechanical behavior and constitutive parameters determination of high nitrogen austenitic stainless steel

Xiaowen Hong[1*], Anshun Shi[1], Yuanyuan Ge[1], Xuefei Fan[1], Chao Qin[1], Guoji Yu[1] and Lujie Wang[1]

[1]No. 52 Institute of China Ordnance Industries, Yantai, 264003, China

[*]Corresponding author's e-mail: hongycxw@163.com

Abstract. In order to obtain the material model parameters for numerical simulation of the ballistic performance of high nitrogen austenitic stainless steel (HNASS) perforated armour, the static and dynamic mechanical properties and material fracture experiments of HNASS have been carried out. Parameters for the Johnson-Cook constitutive and failure model were then adjusted based on the experimental findings. The penetration process of perforated armour against 12.7 mm armour-piercing incendiary shells (API) was simulated by AUTODYN software and the simulation results were compared with the ballistic test results. The findings indicate that using model parameters acquired from mechanical experiments for penetration simulation results in matching failure modes of the perforated armor and penetration depths of body armor with test outcomes, confirming the reliability of the model parameters.

1. Introduction

The steel designed for ballistic protection (e.g. Swedish PRO500 bulletproof steel plate) mainly applies the multi-microalloying method, adding alloying elements such as manganese, chromium, molybdenum, and nickel to the steel ingredient. Due to the performance requirements, the ballistic protective steel is generally martensite to improve its strength and impact resistance. However, the impact toughness of martensitic steel is poor. Therefore, HNASS with good plasticity has become a research hotspot in the field of ballistic protective steel [1-4].

In addition to many excellent properties such as high strength, high toughness, strong corrosion resistance, and low magnetism, HNASS's strength improvement under high strain rate impact is significantly higher than that of martensitic steel, signifying that it has impact hardening characteristics [5]. At present, scholars have studied HNASS from the aspects of dynamic tensile, compression, shear properties, and elastic resistance. Wang et al. [6] conducted a study on the quasi-static and dynamic compression properties of a specific type of HNASS. It demonstrates that the dynamic yield strength and hardness increase as the strain rate rises, suggesting a trend of impact hardening. Chen et al. [7] employed a modified Hopkinson device to perform a punching shear test on high nitrogen steel (HNS), investigating its shear performance under high strain conditions and formulating a dynamic shear damage constitutive relationship. Chen et al. [8] conducted an experimental study on the ballistic performance of HNS with varying thickness and a nitrogen content of 0.56%. The test results demonstrated that high nitrogen steel exhibits a noticeable impact hardening phenomenon and possesses exceptional ballistic performance. Singh et al. [9] conducted an evaluation on the resistance of HNS and Rolled Homogeneous Armor Steel (RHA) against 7.62 mm and 12.7 mm API. The findings revealed that HNS displayed higher sensitivity in terms of dynamic flow stress and strain rate compared to RHA, thereby enhancing its resistance to adiabatic shear localization and impact hardening. Furthermore, they also investigated the penetration performance of a tungsten rod projectile into both HNS and RHA. The analysis of hardness measurements indicated that the volume of material involved in energy dissipation within the HNS target was larger than that of RHA. Additionally, the main failure mechanism observed in both targets was the formation of cracks caused by adiabatic shear bands [10].

The aforementioned studies predominantly concentrate on the mechanical characteristics and ballistic effectiveness of uniform HNS plates. Few were studied on the ballistic performance of HNASS perforated structure, and the mechanical properties of HNASS are lacking. Hence, it is crucial to determine the mechanical properties of HNASS.

In order to obtain the simulation parameters of HNASS perforated armour under the impact of 12.7 mm API, this paper firstly determines a set of Johnson-Cook constitutive and failure parameters suitable for HNASS through a series of quasi-static tensile, compression, notched specimen tensile and dynamic tensile experiments, and then verifies the parameters by ballistic tests, which can provide references for the ballistic protection research of HNASS perforated armour.

2. Johnson-cook constitutive model fitting of HNASS

2.1. *Quasi-static experiment*

The raw material used in the experiment was a 20 mm thick HNASS plate, which was produced by the Iron and Steel Research Institute. Chemical analysis shows that the nitrogen content of the material is as high as 0.88%, and its specific chemical composition is shown in Table 1.

Table 1. Chemical composition of high nitrogen austenitic armour steel.

Elements	N	Mn	C	Ni	Cu	Mo	Cr	W
Content ω (%)	0.88	19.3	0.03	2.01	0.03	0.0001	19.3	0.005

The specimen in the quasi-static compression test is a Φ8 mm × 10 mm smooth cylinder. The size of the quasi-static tensile test specimen shown in Figure 1 refers to the national standard GB/T228.1-2010. The universal material testing machine controlled the quasi-static test. The loading speed was adjusted to 2 mm/min, resulting in a nominal strain rate of 3.3×10^{-3} s^{-1}.

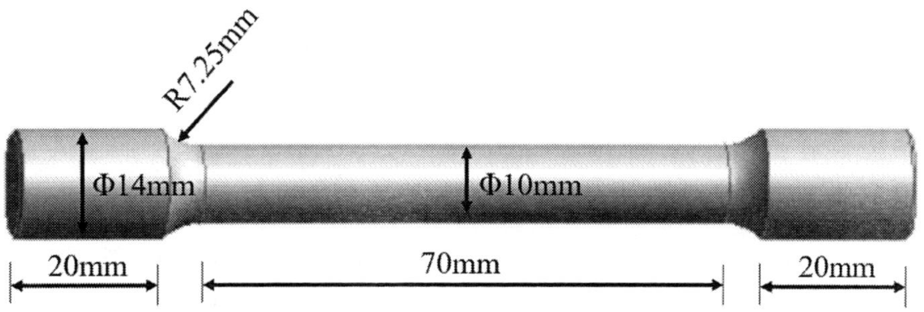

Figure 1. Illustration of quasi-static tensile specimen.

The engineering stress-strain curves of HNASS obtained by quasi-static uniaxial tensile and compression tests are shown in Figure 2.

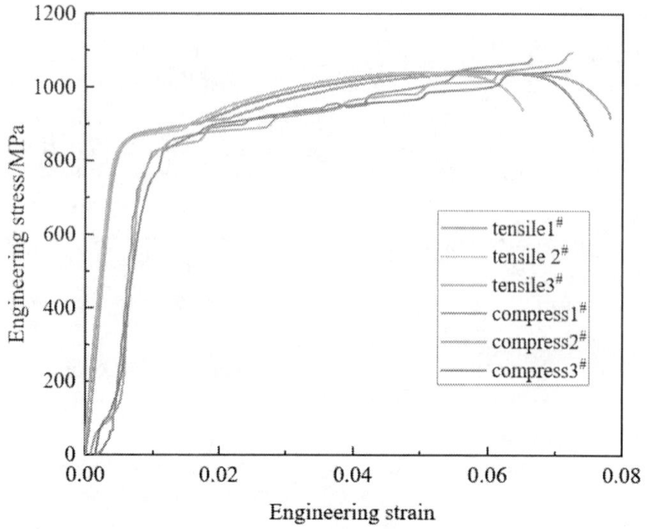

Figure 2. Quasi-static stress-strain curves of HNASS.

The average yield strength of HNS measured from quasi-static tensile tests is 863.12 MPa, while that from quasi-static compression tests is 824 MPa. The tensile yield strength is closely similar to the compressive yield strength. Hence, the tensile and compression constitutive behaviors of HNS can be regarded as identical.

The Johnson-Cook dynamic constitutive model is based on the equivalent stress-strain relationship. The constitutive relation then reads:

$$\sigma_y = \left[A + B\left(\bar{\varepsilon}^p\right)^n\right](1 + C\ln\dot{\varepsilon}^*)\left[1 - (T^*)^m\right] \tag{1}$$

The equivalent strain, denoted as ε_{eq}, is defined in terms of constants A, B, n, C, and m that relate to material properties. The dimensionless equivalent plastic strain rate is determined by $\dot{\varepsilon}^* = \dot{\varepsilon} / \dot{\varepsilon}_0$ that involves the reference strain rate, denoted as $\dot{\varepsilon}_0$. The dimensionless temperature parameter $T^* = (T - T_r)/(T_m - T_r)$, denoted as θ, is a function of the current material temperature (T), the reference temperature (T_r), typically room temperature, and the material's melting temperature (T_m).

The true stress and strain of the material can be obtained by the following two equations:

$$\varepsilon_T = \ln(1+\varepsilon) \tag{2}$$

$$\sigma_T = \sigma(1+\varepsilon) \tag{3}$$

The equivalent stress and strain are as follows:

$$\varepsilon_{eq} = \varepsilon_T - \frac{A}{E} \tag{4}$$

$$\sigma_{eq} = \sigma_T \tag{5}$$

According to Equations (4) and (5), the parameter A in Equation (1) is the quasi-static yield strength of the material, which is 863.12 MPa. The equivalent stress-strain curve (before necking) of each smooth round bar specimen at room temperature was fitted separately (with strain rate under quasi-static tension and T_r at room temperature 293 K), and the average values B and n of the HNASS (i.e. $\dot{\varepsilon}_0$ B=5978.04 MPa and n=1.045) in the J-C constitutive model of HNASS were obtained.

2.2. *SHTB dynamic tensile test*

Figure 3 displays the dynamic tensile test of HNASS conducted using a Φ16 mm SHTB device. The device uses a direct tensile method to load tensile specimens and in the tensile test, maraging steel rods were used to load HNASS specimens. The loading pressure was adjusted to achieve varying tensile conditions at different strain rates. The strain rate in the dynamic tensile test varies between 10^2 and 10^3 s^{-1}. Figure 4 displays the HNASS dynamic tensile test specimens and the corresponding recovered specimens.

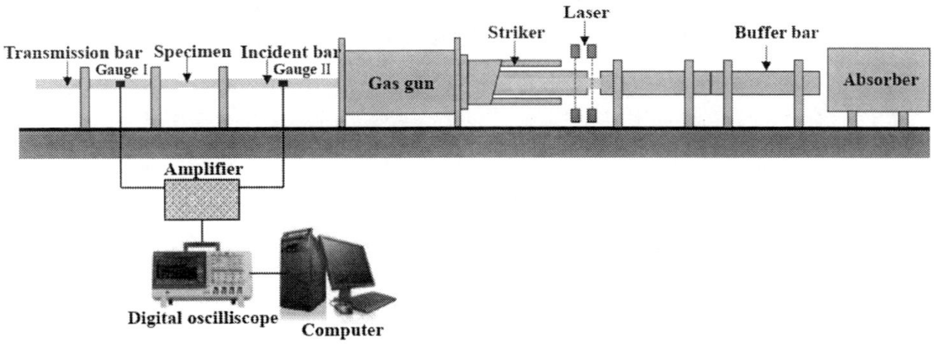

Figure 3. Illustration of the separated hopkinson tensile device.

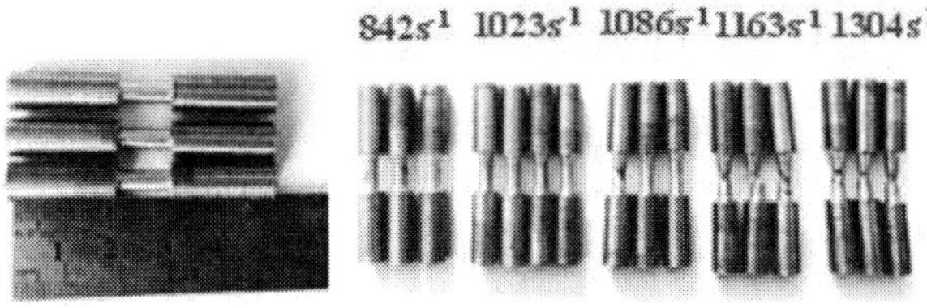

Figure 4. HNASS dynamic tensile test specimens and recycled specimens.

Figure 5 displays the authentic stress-strain curves of HNASS when subjected to dynamic tensile loading at varying strain rates.

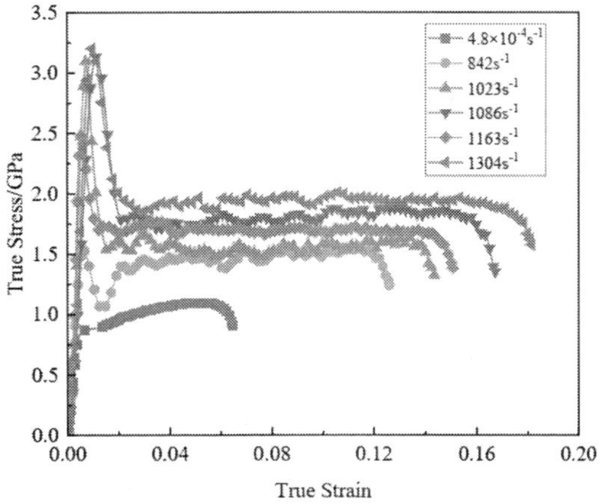

Figure 5. Displays the true stress-strain curves of HNASS under high strain rates.

It could be observed that the HNASS material is very sensitive to the strain rate. The material's yield strength fluctuates between 1000 and 2000 MPa as the strain rate ranges from 10^2 s^{-1} to 10^3 s^{-1}. At a strain rate of approximately 1300 s^{-1}, the material's dynamic yield strength is roughly double its quasi-static tensile yield strength. Due to the absence of clear yield segments in the dynamic tensile stress-strain curves, linear fitting was performed on the elastic and plastic strengthening segments of the engineering stress-strain curve for each specimen. The ordinate value at the intersection of two straight lines was designated as the yield strength of HNASS, and the correlation between the yield strength of HNASS and the strain rate was determined, as depicted in Figure 6.

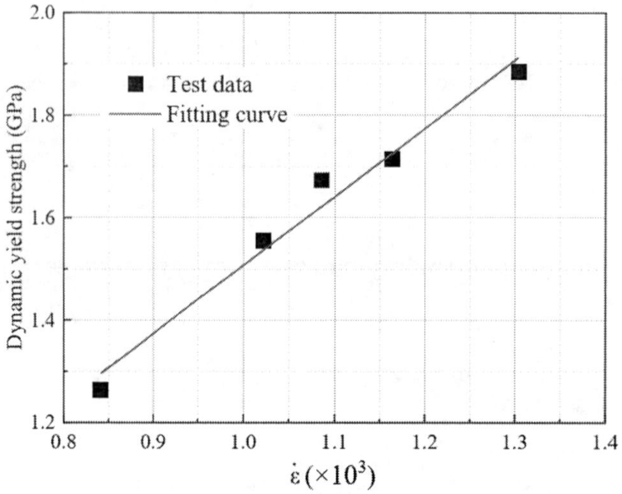

Figure 6. Correlation between dynamic tensile yield strength and strain rate.

In Figure 6, the dynamic tensile yield strength of HNASS shows a linear increase as the strain rate increases, suggesting a significant strain rate strengthening effect. This is consistent with the study results obtained by Wang et al. [6], which show the obvious strain-hardening phenomenon of HNASS in the plastic flow stage. The second term in Equation (1) represents how strain rate affects the material's constitutive relationship. By utilizing the results obtained from static and dynamic tensile testing, and adopting a reference strain rate alongside experimental data collected at various strain rates, the yield stress of HNASS was fitted using the least squares method. $\dot{\varepsilon}_0 = 4.8\times10^{-4}s^{-1}$ $\sigma = A(1 + C\ln\dot{\varepsilon}^*)$. The curve of yield stress of HNASS with dimensionless equivalent plastic strain rate $\ln\dot{\varepsilon}^*$ is shown in Figure 7, where $\dot{\varepsilon}^* = \dot{\varepsilon} / \dot{\varepsilon}_0$. As a result, the value of C in the strain rate influence term is determined.

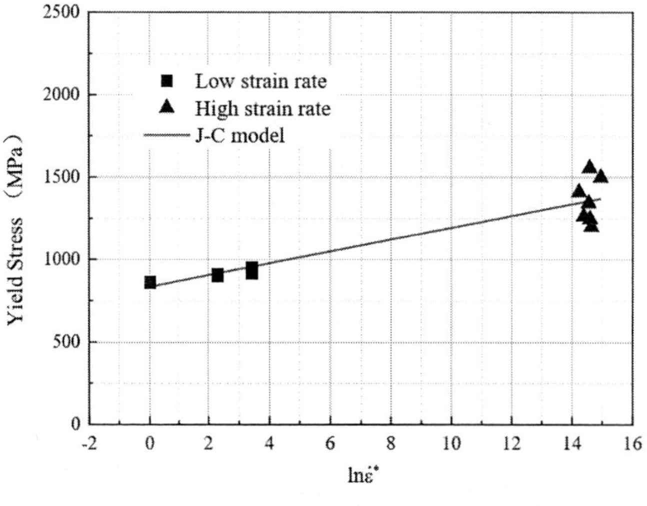

Figure 7. Correlation between the yield stress and the non-dimensional strain rate.

Based on the fitting results shown in Figure 7, a quasi-static yield strength of A= 833.82 MPa was determined, which closely matches the A value obtained from quasi-static tensile experiments. This paper uses the value A obtained from quasi-static tensile experiments, which is 863.12 MPa, and the strain rate strengthening term C = 0.043.

3. Fitting of Johnson-Cook failure model for HNASS

Johnson-Cook considered stress triaxiality, temperature, and strain rate effects and extended the Hancock-Mackenzie fracture model:

$$\varepsilon_f = \left[D_1 + D_2 \exp\left(D_3 \sigma^*\right)\right]\left(1 + D_4 \ln \dot{\varepsilon}^*\right)\left(1 + D_5 T^*\right) \tag{6}$$

where D_1~D_5 are material parameters. The stress triaxiality σ^* is defined as $\sigma^* = \sigma_m/\sigma_{eq}$, where hydrostatic pressure σ_m is obtained by $\sigma_m = (\sigma_{11}+\sigma_{22}+\sigma_{33})/3$ and σ_{eq} is the equivalent stress. $\dot{\varepsilon}^*$ is the dimensionless equivalent plastic strain rate. T^* is a dimensionless temperature parameter. The J-C failure model uses cumulative damage to describe the damage process of materials and the damage evolution of elements is expressed as follows:

$$D = \sum \frac{\Delta \varepsilon_{eq}}{\varepsilon_f} \tag{7}$$

The damage variable, denoted by D, starts at 0. The material fails when D reaches 1. ε_{eq} represents the amount of plastic strain increase in a single time increment. Equation (6) defines the failure strain as $\varepsilon_f = \ln(A_0/A_f)$, where A_0 is the initial minimum cross-sectional area of the specimen and A_f is the area of the fracture surface. The J-C failure model's parameters D_1 to D_5 can be determined by analyzing the material's failure strain under various conditions such as stress triaxiality, strain rate, and temperature. Since no experiments involving temperature effects have been carried out, this paper mainly focuses on the influence of parameters D_1~D_4.

3.1. Influence of stress triaxiality on failure strain

The influence of stress state, i.e. stress triaxiality, on failure strain, was investigated through uniaxial tensile tests on circular bar notched specimens. The specimens had a length (L) of 120 mm and a diameter (d) of 10 mm. The radii R at the notches were 1.5 mm, 3 mm, 6 mm, 9 mm, and 11 mm, consecutively. The notched specimens are shown in Figure 8. By conducting uniaxial tensile tests on notched specimens, a stress triaxiality range greater than 1/3 was achieved. Uniaxial tensile tests were conducted on the notched specimens using a universal material testing machine, with an extensometer gauge length of 50 mm and a tensile speed of 2 mm/min. The average was calculated based on the results of three experiments in each group. The failure strain was determined by measuring the cross-sectional area of the specimens both before and after failure.

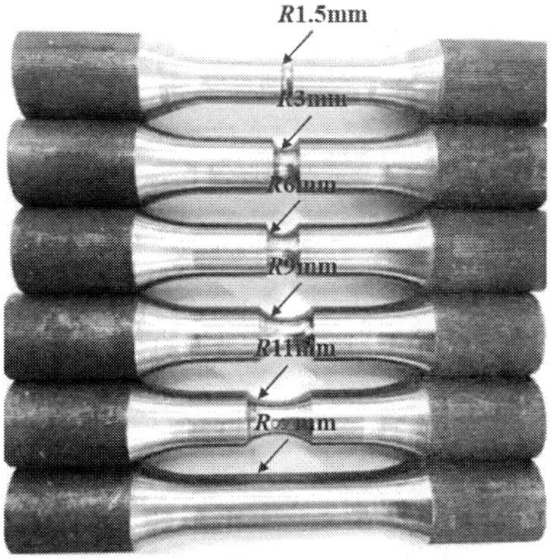

Figure 8. Notch test specimens.

The initial stress triaxiality in the notched specimens in the uniaxial tensile tests can be given according to the Bridgman equation:

$$\sigma_0^* = 1/3 + \ln[1 + a/(2R)] \tag{8}$$

where a (mm) represents the radius of the cross-section at the notch, and R (mm) is the radius of the arc line of the notch.

The first term in Equation (6) is used to fit the failure strain and stress triaxiality, and the parameters D_1~D_3 are determined. Figure 9 illustrates the correlation between HNASS failure strain and stress triaxiality across various stress conditions.

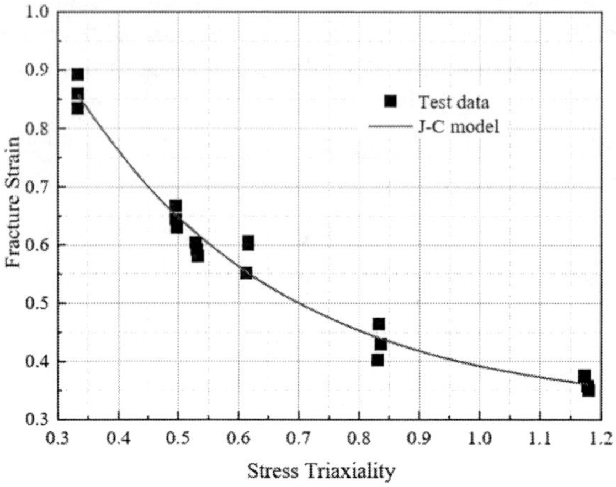

Figure 9. Correlation between the failure strain of HNASS and the stress triaxiality.

It can be seen that the failure strain of HNASS decreases as stress triaxiality increases. By fitting the data with the J-C failure criterion, parameters D_1=0.32, D_2=1.45, and D_3=-2.95 can be obtained.

3.2. *Influence of strain rate on failure strain*

According to the quasi-static tensile test and SHTB test, the relationship between the failure strain and the strain rate of HNASS at low and high strain rates was obtained. The second term in Equation (6) was used to fit the failure strain and dimensionless strain rate to determine the failure parameter D_4. Figure 10 illustrates the correlation between failure strain and dimensionless strain rate. The least squares method was used to fit the J-C failure model, and the strain rate term coefficient D_4=-0.03 was obtained.

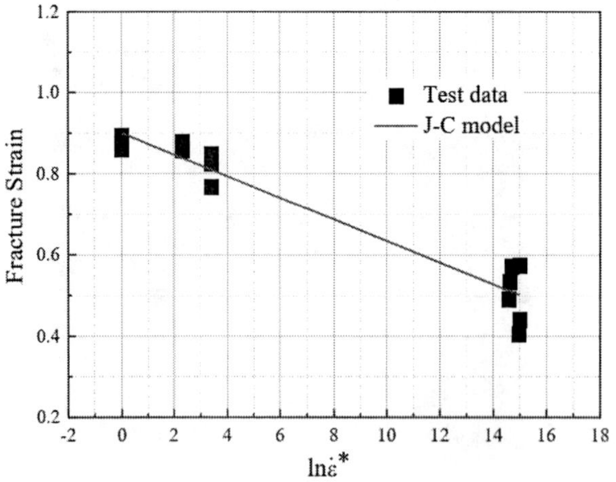

Figure 10. Correlation between failure strain and dimensionless strain rate.

4. Penetration simulation of high nitrogen steel perforated armour against 12.7 mm API

The ballistic tests yielded penetration results for perforated armor made of various materials when subjected to 12.7 mm API. The test arrangement is shown in Figure 11. From left to right, there was a 12.7 mm ballistic gun, a 12.7 mm API, a light curtain velocity-measuring system, a perforated plate, a base plate, and a target frame. The size of the perforated plate was 320 mm × 320 mm × 6 mm.

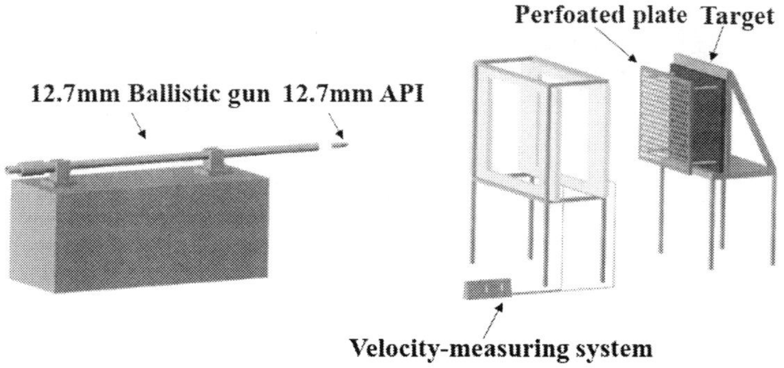

Figure 11. Schematic diagram of ballistic device.

The perforated armor's finite element model was created using TrueGrid software. The size of the perforated armour is 72 mm × 90 mm × 6 mm, the perforation diameter is 10 mm and the hole spacing is 15 mm. We utilized a hexahedral element mesh for modeling, with a mesh size of 0.2 mm. The mesh model was imported into AUTODYN software to establish the projectile and the base armour model, the minimum mesh size of the projectile model is 0.2 mm and the base armour is 0.4 mm. Boundary conditions were set for the perforated armor and the base armor, and the Trajectory contact algorithm was utilized to model the interaction between the projectile and the target using a unit system of millimeters, milligrams, and milliseconds. The FEM model is shown in Figure 12.

Figure 12. FEM model.

The 12.7 mm API model adopted a simplified modeling method, and only a hard-core model was established. The core material was T12A tool steel, and the projectile penetration speed was set to 818 m/s. The kinematic hardening material model was adopted, and the principal stress failure criterion was added to describe the failure in the collision. The parameters of J-C constitutive and failure for the HNASS, determined through mechanical experiments, were applied. The temperature effect has a similar impact on metal materials, so the thermal softening coefficient m is taken as 1. The room temperature T_r is 294 K, the melting temperature T_m is 1520 K, and the material of base armour is high-strength armour steel 6252. The material parameters of the projectile core, HNASS, and 32Mn2Si2MoA armour steel are shown in Table 2.

Table 2. Material parameters.

Parameters	ρ (g/cm^3)	G/GPa	T_m/K	C_p (J/kgK)	A/GPa	B/GPa	C	m
T12A	7.85	80	1800	450	1.976	2.856	0.005	1
HNASS	7.85	76	1800	450	0.863	5.978	0.043	1
32Mn2Si2MoA	7.86	77.5	1793	477	1.455	1.95	0.02	0.58
Parameters	n	$\dot{\varepsilon}_0$ /s^{-1}	D_1	D_2	D_3	D_4	D_5	Erosion
T12A	0.207	0.008	Princ. strain	3.6	Princ. strain	0.1		Failure
HNASS	1.045	0.008	0.32	1.45	-2.95	-0.03	0	Failure
32Mn2Si2MoA	0.52	1.0	0.0001	1.9	-1.717	0.001	4.95	Failure

In order to validate the FEM model, the model parameters obtained from the mechanical experiments were used to simulate the perforated armour anti 12.7 mm API. The typical cases of the simulation results of the perforated armour and the base armour are shown in Figures 13 and 14.

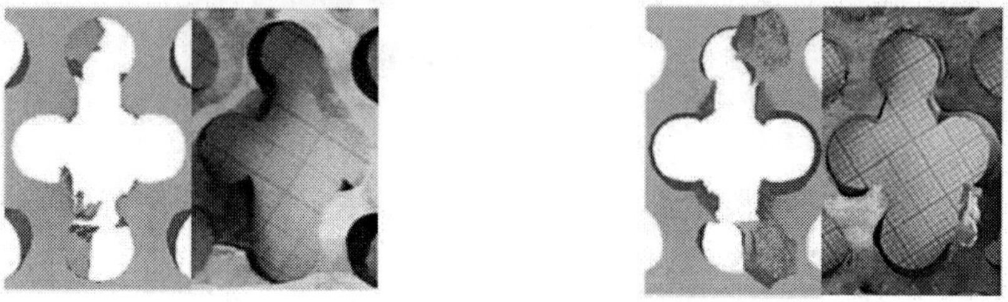

Front face Back face

Figure 13. Penetration damage of high nitrogen steel perforated armour.

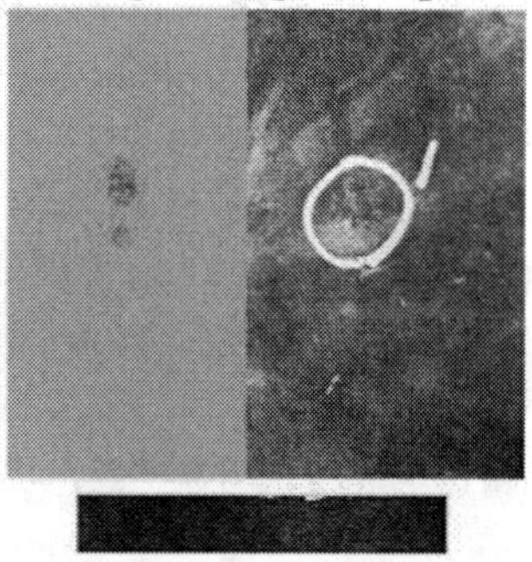

Figure 14. Damage results of the base armour.

Upon comparing the failure mode test results and simulation results of the HNASS perforated plates, it was discovered that the failure modes of the front and back of the perforated plate obtained through simulation closely align with the test results. In Figure 14, the penetration depth of the base armour in one shot was 3.0 mm, and the penetration depth in the simulation was 2.8 mm. The average penetration depth of the base armour in 5 tests was 4 mm, while the average penetration depth in the simulation was 3.7 mm, and the error was within 10%. Thus, it can be inferred that the J-C model parameters obtained in this study are trustworthy.

5. Conclusion

(1) The quasi-static and dynamic mechanical tests and experiments of specimens with different notch sizes were carried out for HNASS material. The J-C constitutive model and failure model parameters for HNASS armor steel material were determined by fitting mechanical experimental data.

(2) The parameters of the J-C constitutive model and failure model were used to simulate the 12.7

mm API penetration process of HNASS perforated armour. The results of numerical simulations show that the front and back failure modes of perforated plates and penetration depth of the base armor are consistent with the ballistic test results, which validate the J-C model parameters proposed in this study.

References

[1] Shi F, Cui W F, Wang LJ, et al. (2006) Advance in the research of high-nitrogen austenitic stainless steels [J]. Shanghai Metals, 28(5): 45.

[2] Zhang P, Chen P F. (2020) Heat treatment processes and resulting microstructure and hardness for a high-nitrogen stainless steel. Heat Treatment, 35(1): 1.

[3] Liu H D, Wang D Z, Wei H D, et al. (2009) Research progress in high nitrogen austenite stainless steel [J]. Special Steel, 30(4): 45.

[4] Li W P, Li J, Wang J X, Gao D H. (2021) Research status on the bulletproof steel. Automotive Digest, 7: 1-6.

[5] Wang Y, Peng X F, Li J, et al. (2022) Research progress on the strengthening mechanism and ballistic performance of high-nitrogen austenitic stainless steels. Iron and Steel, 57(01): 18-38.

[6] Wang Y L, Jia G Z, Zhang T, et al. (2016) Investigation of dynamic mechanical properties and microstructure of 19Cr19Mn2Ni0.88N high nitrogen austenite stainless steel. Proceedings of the 25th National Academic Conference on Structural Engineering (Volume II). No. 52 Institute of China Ordnance Industries, 432-435.

[7] Chen X A, Shao B M, Song S C, et al. (2011) Measurements of shearing property of high nitrogen alloy steel at super high strain rates. Journal of Zhejiang University (Engineering Edition), 45(11): 1948-1952+1979.

[8] Chen W, Liu Y L, Qi Z W, et al. Research on ballistic behavior of high nitrogen austenitic armour steel. Ordnance Material Science and Engineering, 32(6): 51-55.

[9] Singh B B, Sukumar G, Sivakumar K, et al. (2020) A comparative study on the ballistic performance of high nitrogen steel and rolled homogeneous armour steel against 7. 62 mm ball and 12. 7 mm armour piercing projectiles. Materials Today: Proceedings, 41(12): 1188-1194.

[10] Rathod S, Khaire N, Tiwari G. (2022) A comparative study on the ballistic performance of Aramid and Aluminum honeycomb sandwich structures [J]. Composite Structures, 299.

Journal of Physics: Conference Series **2808** (2024) 012071 doi:10.1088/1742-6596/2808/1/012071

Analysis of influencing factors on numerical computation of underwater explosion of slender tapered charge

Houlin Fang[1], Cunxu Liu[1], Qiang Lu[1], Quanshi Guo[1], Shiying Tang[1], Zhiyun Guo[1] and Dezhi Zhang[1, *]

[1]National Key Laboratory of Intense Pulsed Radiation Simulation and Effect, Northwest Institute of Nuclear Technology, Xi'an, Shaanxi, China

[*]Corresponding author's e-mail: zhangdezhi@nint.ac.cn

Abstract: To accurately calculate the shock-wave pressure curve of the underwater explosion of slender tapered charge, this research constructs a two-dimensional axisymmetric numerical model. In this foundation, this research further investigates the influence of the computational domain size, computational domain symmetry, gauge grid size, charge grid size, and biaxial grid uniformity on the shock-wave pressure curve. As a whole, this research reveals the following several findings. To begin with, increasing the size of the computational domain is helpful to weaken the curve oscillation, and so is maintaining the symmetry of the computational domain. Secondly, an excessive ratio of the number of two-dimensional grids, which is manifested by the non-uniformity of biaxial grids, leads to the grid being too slender at the boundary and the oscillation amplitude of the pressure curve increasing obviously. Furthermore, the gauge grid size exerts a notable influence on the peak pressure of the shock wave. Lastly, in the case of a fine grid covering the gauge, the charge grid size generates no obvious influence on the peak pressure.

1. Introduction

In contrast to traditional spherical and cylindrical charges, slender tapered charges, as a special-shaped charge, receive relatively limited research attention. Nonetheless, considerable efforts have been devoted to investigating the linear and cylindrical charges featuring a significant length-diameter ratio. Notably, the foregoing endeavors primarily focus on exploring such aspects as underwater explosion pressure or energy distribution, pressure calculations, stress wave propagation, drilling blasting experiments [1], seismic source studies [2], and charge-making processes [3], etc.

Specifically, by utilizing a two-dimensional axisymmetric Lagrange methodology to compute the underwater explosion pressure distribution of cylindrical charges with different length-diameter ratios, Sternberg [4] found that the shock wave parameters are controlled by the length-diameter ratio, and the shock wave approximates a spherical shape in the case that the length-diameter ratio approaches 1. By comparing the numerical values of cylindrical and spherical charges, Zhang et al. [5] computed the variation of pressure peak value with angle at different distances, deducing the approximate empirical formula of the pressure field of cylindrical charges based on the empirical formula of spherical charges. Based on the multi-material arbitrary-Language-Euler (MM-ALE) numerical model verified by relevant experiments, Huang et al. [6] investigated the influence of the length-diameter ratio of cylindrical charge within the range of 2.0 to 9.2 on the shock wave, thereby analyzing peak pressure, pulse duration, impulse and energy flux of the underwater explosion of slender explosives in detail.

Journal of Physics: Conference Series **2808** (2024) 012071 doi:10.1088/1742-6596/2808/1/012071

Likewise, Ma et al. [7] studied the influence of length-diameter ratio within the range of 1.0 to 8.2 on shock wave and bubble load based on experiments and the ALE method, revealing a remarkable directional enhancement effect as well as the fact that increasing the length-diameter ratio benefits the improvement of the destructive effect of combined loads. Regarding slender tapered charges, William [8] employed tapered charges to produce lower peak pressures and slower pressure plateaus, with numerical computation programs utilized to determine pressure-time histories and pressure nephograms at specified distances. By subdividing the slender tapered charges into several small charge segments based on the superposition principle, Xu et al. [9] realized the pressure superposition of shock waves underwater according to the propagation sequence of detonation wave and plotted the whole shock-wave pressure curve of slender tapered charges, thereby analyzing the distribution patterns of shock-wave pressures at different orientations of the slender tapered charge. In this foundation, Huang et al. [10] put forward the method of shock wave pressure regulation and investigated the energy output law of underwater explosion of slender charges by leveraging independently developed software, indicating that the initiation position and charge shape generated a notable influence on the peak value and duration of shock-wave pressure of underwater explosion. Studies on numerical simulation of slender tapered charge, however, remain limited. Within this context, this paper investigates the influence exerted by computational domain size, computational domain symmetry, gauge grid size, charge grid size, and uniformity of the biaxial grid on the shock-wave pressure curve by constructing a two-dimensional axisymmetric computational model concerning slender tapered charges, with a view to furnishing valuable insights for accurately computing the shock-wave pressure curve of slender tapered charge.

2. Numerical computational model

2.1. Numerical modeling

Given the axisymmetric-body structure adopted by slender tapered charges, this research employs a two-dimensional axisymmetric numerical computation model, as depicted in Figure 1. More precisely, the model features a tapered charge with a small end radius of 15 mm, a large end radius of 20 mm, a charge length of 2,376 mm, and TNT serving as the charge material. In addition, the initiation point is positioned at the center of the small end, with the flow-out boundary set as the boundary condition of the computational domain, and the flow-field pressure gauge is positioned along the charge axis at a distance of 1.5 m from the initiation point. To explore the influence of mesh and computational domain size on the calculation results, this research takes the center of the charge axis as the origin and sets the boundary value of the negative computational domain of the x-axis as $-a$, the boundary value of the positive computational domain of the x-axis as b, and the boundary value of the computational domain of the y-axis as c. Furthermore, the computational domain size can be expressed as $(a + b) \times c$. Notably, the charge length is indicated by l, whereas the distance from the gauge to the end of the charge is indicated by d.

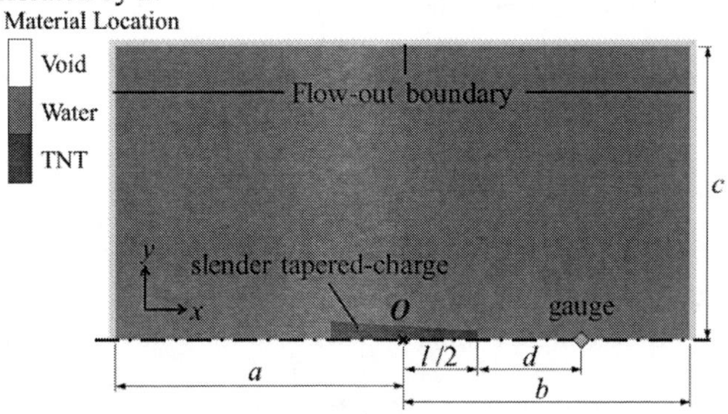

Figure 1. Two-dimensional axisymmetric computational model of slender tapered charge.

Journal of Physics: Conference Series **2808** (2024) 012071 doi:10.1088/1742-6596/2808/1/012071

Through detailed demonstration, Fang et al. [11] validated the computational model of underwater explosion with spherical charges, encompassing the selection of the state formulas of detonation products and water medium, the comparison of calculation results with Cole's empirical formula, and the influence analysis of the grid size. Within this research, the hydrostatic-pressure environment is set by the specific internal energy of the water medium, while the linear viscosity and quadratic viscosity are taken as 0.02 and 1, respectively.

Regarding the selection of computational grid size, this research draws lessons from the methodology proposed by Huang [12]. Specifically, the grid size, denoted as s, is determined to be equal to $C\theta/100$, where C is equal to 1500 m/s, while θ is estimated to be 0.035 ms according to the empirical formula of spherical charges with the same radius of the small end of slender tapered charges. Substituting the aforementioned values into the formula, it is further deduced that s is equal to 0.525 mm. Notably, the value of θ will be larger than the selected value when the peak attenuation speed of shock-wave pressure of slender tapered charges is obviously slowed down. Moreover, taking into account the numerical calculation capacity, the radial grid size of charges is not more than 1 mm. Hence, the calculation results are acceptable.

2.2. Computational condition design

Without a significant increase in the total number of computational grids, this research primarily focuses on the following two strategies to acquire ideal calculation results:

More precisely, the first strategy is to control the size of the computational domain. To determine the complete shock-wave pressure curve of slender tapered charge at the gauge without being affected by the boundary, it is necessary to set the boundary condition of the computational domain as a "flow-out boundary". Furthermore, it is imperative to keep enough distance from the boundary as per the research results of Fang et al. [11]. To this end, this research assumes that the duration of the pressure curve of the slender tapered charge is τ, thus computing the boundary distance requirements of the pressure gauge side:

$$b \geq \frac{1}{2}\left(C\tau + l + 2d\right) \tag{1}$$

where the conclusion that b is greater than or equal to 4.1 m can be drawn by computation when τ is equal to 1.9 ms. Meanwhile, given the absence of a gauge within the opposite area of the pressure gauge, the size can be appropriately reduced. In other terms, the computational domain is asymmetric with respect to both sides of the charge center, provided that a satisfies $\frac{l}{2} < a < b$. The computational domain sizes c and b of the y-axis are basically the same, thereby avoiding the influence of boundary reflection of the y-axis. The second strategy, on the other hand, is to optimize the mesh. Specifically, the grid of the core areas (i.e., the charge center and the gauge) is refined to fulfill the size requirements of the computational domain and tally with the quantitative limitation of grids. Concurrently, some issues related to the mesh of *the* x-axis and y-axis need to be taken into account.

In summary, this research primarily implements the numerical simulation of slender tapered charges from several aspects, encompassing computational domain size, computational domain symmetry, the grid size of the gauge, and the grid size of refined areas. The design conditions are described in Table 1.

2.2.1. Computational domain size. Case 1 and Case 2 differ solely in the computational domain size. However, the same number of grids makes the grid in the y-axis direction in Case 2 finer than that in Case 1. Consequently, to ensure that only the influence factor of computational domain size is analyzed, instead of other influence factors such as grid refinement, this research further designs Case 3, whose grid in the y-axis direction is similar to that in the range of $y \in [0,4]$ of Case 1.

2.2.2. Computational domain symmetry. To reduce the computational domain, refine the core computational domain, and improve the computation accuracy, this research designs Cases 4, 5, and 6.

Journal of Physics: Conference Series **2808** (2024) 012071 doi:10.1088/1742-6596/2808/1/012071

Specifically, the area on the opposite side of the pressure gauge is reduced from 4.1 m to 2.9 m, while the center of the computational domain is biased to one side by 0.6 m, with the required size of the initiation side being at least 2.088 m. In the meantime, the grid size in the *x*-axis direction of the core area is decreased from 3 mm in Case 6 to 2 mm in Cases 4 and 5, and the middle refined area in the *x*-axis direction covers 2.1 m in both directions to cover the gauge. On these grounds, the influence of computational domain symmetry is determined by further comparative analysis.

2.2.3. Grid size of the gauge. Cases 1 to 6 all include the gauge in the core area, with the grid in the *x*-axis direction being refined to 2 mm or 3 mm. To facilitate the comparative analysis of the influence of the grid size of the gauge on the computation results, this research designs Case 7, which can reduce the total number of grids to a certain extent, and the grid refined area fails to cover the monitoring point. In this case, the grid size in the *x*-axis direction near the gauge is approximately 13 mm. Furthermore, this research compares its computation results with those of Case 1 to investigate the influence of the grid size of the gauge.

2.2.4. Charge grid size. The grid size of the charge in the *y*-axis direction is approximately 0.5 mm, whereas the grid size in the *x*-axis direction is 2 to 3 mm. By comparing the selected Case 1 with a 3 mm grid and Case 5 with a 2 mm grid, it is found that there is no significant difference in the ratio of grid number between them.

Table 1. Design of computational conditions.

Case No.	Computational Domain Size $(a+b)\times c$ (m)	Computational Range of *the* *x*-axis $[-a,b]$ (m)	Grid Number of *the x*-axis (Refined Grid Size×Grid Number)	Grid Number of the *y*-axis (Refined Grid Size×Grid Number)	Total Number of Grids ($\times10^4$)	The Ratio of Grid Number (*x*-axis/*y*-axis)
1	10×5	[-5, 5]	2300 (3×2000)	400 (0.5×100)	92	5.75
2	10×4	[-5, 5]	2300 (3×2000)	400 (0.5×100)	92	5.75
3	10×4	[-5, 5]	2300 (3×2000)	339 (0.5×100)	77.97	6.78
4	7×3	[-2.9, 4.1]	3500 (2×3500)	250 (0.5×100)	87.5	14
5	7×3	[-2.9, 4.1]	2335 (2×2100)	346 (0.5×100)	80.79	6.75
6	8.4×3	[-4.2, 4.2]	2200 (3×2000)	350 (0.5×100)	77	6.29
7	9×4.5	[-4.5, 4.5]	1500 (2.5×1000)	500 (0.5×100)	75	3

3. Computation results and analysis

Through the computation of several cases designed above, this research ultimately determines the shock-wave pressure-time history curve for the gauge, conducting a comparative analysis of the computation results according to the grouping of influencing factors.

3.1. Effects of computational domain size

Figure 2 outlines the comparison of the computation results of Cases 1 to 3. The shock-wave pressure curve of slender tapered charge is characterized by the beginning of oscillation of the curve upon the peak of shock-wave pressure. The filter processing, however, can be employed to describe the curve changes, demonstrating a relatively obvious pressure plateau. All curves within Figure 2 are filtered, with a label "(filter)" added to the legend to distinguish the original pressure curves.

The comparison of Cases 1 to 3 indicates that the curves corresponding to the three cases reflect a high coincidence degree and basically the same pressure peak before 1.4 ms. After 1.6 ms, the foregoing three curves exhibit significant differences, which are manifested in diverse degrees of oscillation. Particularly, Case 3 initially presents the largest oscillation, reaching approximately 8 MPa. Based on the computational model of Case 3, increasing the computational domain size (Case 1), coupled with refining the grid (Case 2), can reduce the curve oscillation. Nevertheless, it is difficult to quantitatively determine which of the above two methods exerts a more significant influence.

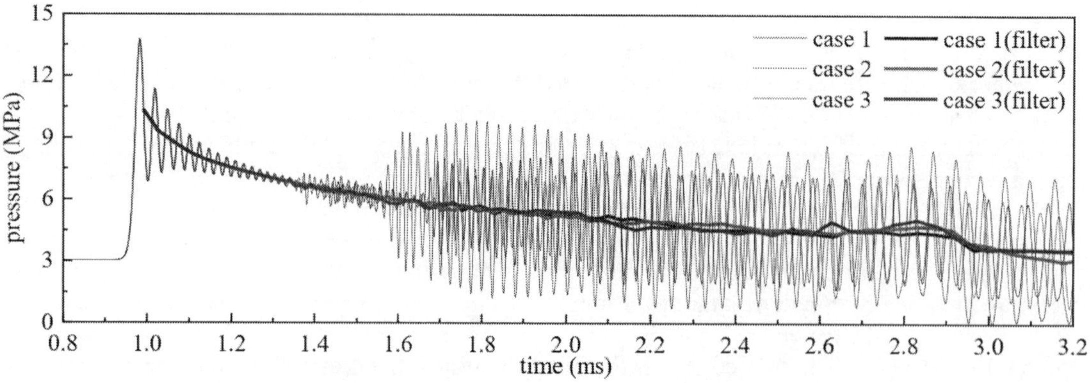

Figure 2. Effects of computational domain on pressure curves.

3.2. Effects of computational domain symmetry

Figure 3 presents the comparison of the computation results of Cases 4 to 6. More specifically, the curves corresponding to the above three cases showcase a high coincidence degree and basically the same pressure peak before 1.2 ms. In addition, these three curves indicate notable differences after 1.5 ms, which are manifested in diverse degrees of oscillation. Of them, Case 4 initially exhibits the largest oscillation, reaching approximately 15 MPa. The reason can be attributed to the non-uniformity of the grid along the *x*-axis and *y*-axis. The curve corresponding to the adjusted Case 5 reflects the obviously weakened oscillation. By further adjusting Cases 4 and 5 with asymmetric computational domain to Case 6 with symmetric computational domain, the curve oscillation is further weakened. In this scenario, the maximum amplitude is reduced to approximately 8 MPa, whereas the large amplitude oscillation time is delayed from 1.5 ms to 1.66 ms.

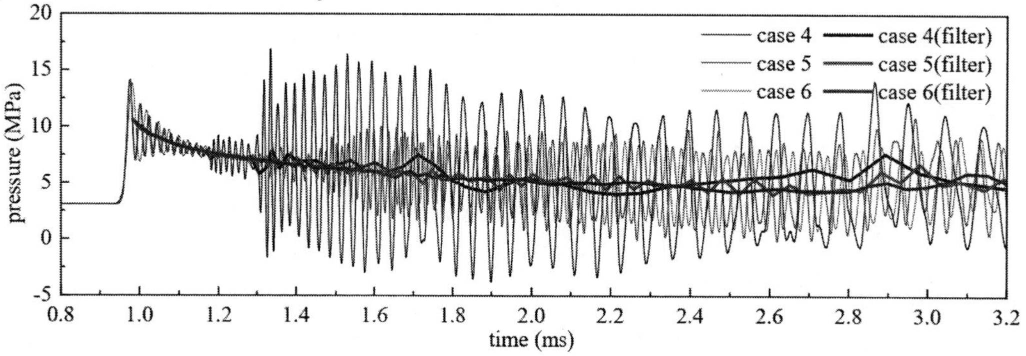

Figure 3. Effects of computational domain symmetry on pressure curves.

3.3. Effects of gauge grid size

Figure 4 illustrates the comparison of the computation results of Cases 1 and 7. The curves corresponding to the two cases show significant differences, which are manifested in the decrease of the peak pressure of Case 7 by about 2 MPa as well as the progressive decrease of the oscillation amplitude of the pressure curve of Case 7 with time. Essentially, the foregoing differences can be attributed to two causes, one of which may be the oversize grid at the gauge that leads to inaccurate positioning within a limited deviation range, whereas the other is the decrease of the computation accuracy of pressure peak caused by the oversize grid at the gauge.

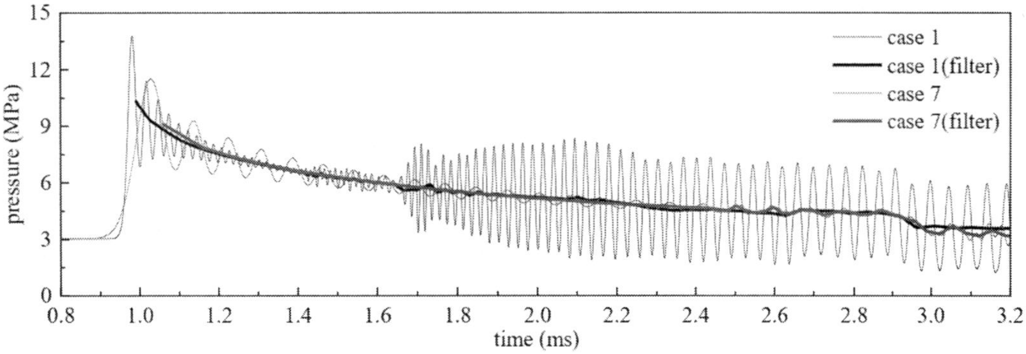

Figure 4. Effects of Gauge Grid Size on Pressure Curves.

3.4. Effects of charge grid size

Figure 5 depicts the comparison of the computation results of Cases 1 and 5. The grid sizes of the curves corresponding to the two cases are 3 mm and 2 mm, respectively. Meanwhile, the pressure peaks of the two are different to some extent, which are 14.3 MPa and 13.8 MPa, respectively, indicating that their influence on the pressure curve is not significant within the selected charge grid size range.

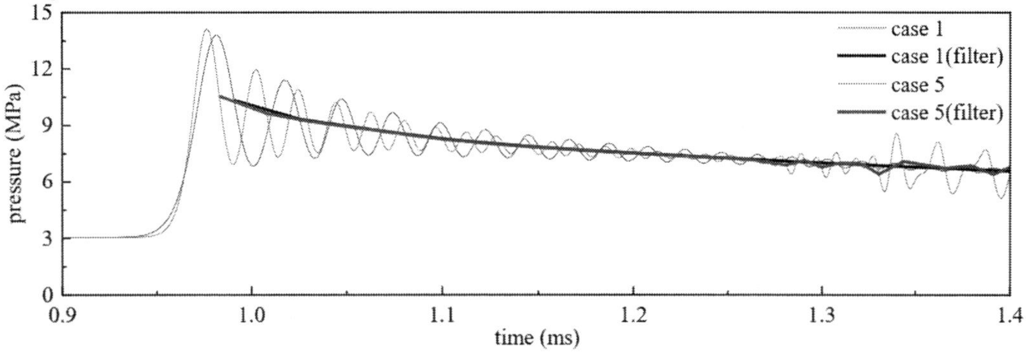

Figure 5. Effects of charge grid size on pressure curves.

Additionally, by synthesizing the characteristics of the pressure curve corresponding to each case, it can be known that the ratio of the number of grids in the *x*-axis and *y*-axis directions exerts a significant influence on the oscillation characteristics of the pressure curve. Particularly, Cases 4 and 7 present the largest ratio difference and the most significant curve oscillation characteristics difference. Based on all the above-mentioned factors, to obtain ideal computation results, it is imperative to ensure the symmetry of the computational domain as well as the coverage of the gauge by the refined grid, intending to avoid the excessive ratio (i.e., the non-uniformity of the biaxial grid) of the number of grids in the *x*-axis and *y*-axis. In this foundation, the computational domain can be enlarged as much as possible as per the computational capacity.

4. Conclusions

This research constructs a two-dimensional axisymmetric numerical model. In this foundation, this research further investigates the influence of the computational domain size, computational domain symmetry, gauge grid size, charge grid size, and biaxial grid uniformity on the shock-wave pressure curve. As a whole, this research reveals the following several findings:

First and foremost, given the long duration of shock-wave pressure, the computational domain size should not be excessively small to eliminate the influence of interface reflection. Concurrently, the computational domain size in a single direction should not be less than 4.2 m. Increasing the computational domain size and maintaining the computational domain symmetry are beneficial to weaken the curve oscillation.

Secondly, the excessively large ratio of the number of grids in the x-axis and y-axis directions (i.e., the non-uniformity of biaxial grids) leads to the excessively slender grids at the boundary as well as the obvious increase of the oscillation amplitude of the pressure curve.

Additionally, the gauge grid size exerts a great influence on the peak pressure of the shock wave. In the scenario where the refined grid of the x-axis covers the gauge, the charge grid size within the range of 2 to 3 mm has no significant effect on the peak pressure.

In closing, the curves with diverse amplitudes of oscillation are relatively consistent except for the area near the peak pressure.

References:

[1] Hu, W. C., Zhang, X. Q., Guo, M. F., *et al.* (2019) Influence of different charging structures on the blasting effect of deep water drilling. Yangtze River, 50: 161-166.
[2] Xu, F., Huang, L. M., Liu, F. L., *et al.* (2022) Optimal charge design method for single-borehole shots. Oil Geophysical Prospecting, 57:1020-1027.
[3] Yue, X. Y., Zhang, H. S., Han, X. L., *et al.* (2021) Process of Improved Hot Mandrel for Large Length-Diameter Ratio Warhead Melting Cast. Chinese Journal of High-Pressure Physics, 35:015302.
[4] Stemberg, H. M. (1995) Underwater detonation of pentolite cylinders. Physics of Fluids, 30: 761-769.
[5] Zhang, C. Y., Guo, R., Liu, R. Z., *et al.* (2017) Equivalent relationship between cylindrical charge and spherical charge for underwater explosion. Torpedo Technology, 25: 66-70.
[6] Huang, C., Liu, M. B., Wang, B., *et al.* (2019) Underwater explosion of slender explosives: Directional effects of shock waves and structure responses. International Journal of Impact Engineering, 130: 266-280.
[7] Ma, T., Wang, J. X., Liu, L. T., *et al.* (2022) Study on directional enhancement effect of underwater explosion for cylindrical explosive with large length-to-diameter ratio and structural response of nearby water-back plate. Ocean Engineering, 266: 112614.
[8] William, E. M. (1992) Simulation of the underwater nuclear explosion and its effects. Monterey, California: Naval Postgraduate School.
[9] Xu, W. Z., Huang, C., Zhang, P., *et al.* (2022) A method for calculating underwater explosion shock wave parameters of slender cone-shaped charges. Explosion and Shock Waves, 42:1-9.
[10] Huang, C., Zhang, P., Zeng, F., *et al.* (2022) A method for adjusting and controlling underwater explosion shock wave. Explosion and Shock Waves, 42: 83201.
[11] Fang, H. L., Zhang, D. Z., Guo, Q. S., *et al.* (2022) Research on the Reflection Characteristics of the Shock Wave Boundary in the Numerical Simulation of Underwater Explosion. In: HEEPS-2022. Xiamen. pp. 2271 012030.
[12] Huang, Z. P. (2006) Explosion and Shock Measuring Technique. National Defense Industry Press.

Journal of Physics: Conference Series **2808** (2024) 012072 doi:10.1088/1742-6596/2808/1/012072

Mechanical properties analysis of bending-torsion coupled blades with different skin area angles

Xiaoyin Jin[1], **Jiang Liu**[1, 2, *]

[1]Shanghai University of Electric Power, Shanghai, 201306, China
[2]Shanghai Engineering Research Center of Energy-Saving in Heat Exchange Systems, Shanghai, 200090, China

*Corresponding author's e-mail: jiangliuxjtu@shiep.edu.cn

Abstract: In this paper, the CFD-FEM method is used to study the mechanical properties of wind turbine bend-twist coupled blades by changing the bias angle of the inner and outer skins under the rated wind speed. The results show that the wind turbine blade's inner and outer skin offset ± 45° respectively, and the maximum stress and maximum strain of the wind turbine blade decreased by 27.7% and 2.1%, indicating that the blade load reduction performance is the best in this case. Changing the skin angle on the blade deformation does not have much effect, and the maximum increase in blade deformation is only 2.5%.

1. Introduction

Wind turbine blades are an important component of wind power generation, and the load on the blades is also increasing. To reduce the weight of wind turbine blades and improve their bearing capacity, we take advantage of the anisotropic characteristics of blade composites to reduce the maximum stress and strain of blades while increasing the bending and torsion deformation of the blades and introduce the concept of bending-torsion coupling by changing the layering thickness, layering angle, layering sequence, and material ratio of the composite laminates inside the wind turbine [1]. As an important part of wind turbine blades, the skin is distributed in the outermost and innermost layers of the wind turbine, and the skin needs to bear the surface force of the wind on the wind turbine blade first [2], so the selection of materials and the study of material ply parameters are very important.

At present, scientific research teams at home and abroad have conducted a lot of research on bendingtorsion coupled blades based on the anisotropy characteristics of composite materials through finite element theory. For the first time, the Sandia team studied the use of the bending-torsional coupling characteristics of the material to reduce the load of the blade, and they proposed to realize the bendingtorsional coupling of the blade by changing the ply angle of the uniaxial cloth at the position of the blade beam cap to achieve the purpose of reducing the load [3]; An Liqiang's research team from North China Electric Power University established a structure model of a 5 MW wind turbine bending-torsion coupling blade, and applied a simple load at the blade tip to study the load reduction mechanism of the bending-torsional coupling blade by changing the material angle of the cap beam area [4], and the results showed that the stress was the lowest when the offset angle was 15° to 20°. Li Chun's research team at the University of Shanghai for Science and Technology added an aerodynamic model based on the structural model of An Liqiang's research team and studied the influence of changing the offset angle of the material in the skin area of the blade under the action of pneumatics to study the mechanical properties of the blade [5], and concluded that the stiffness of the blade is the largest when the material in the skin area is offset to 15°. At present, the scientific research teams at home and abroad only change the same ply angle of the inner and outer sides of the skin area to study, and there are relatively few studies on changing the offset angle of the inner and outer sides of the skin material opposite, so it is significant to study the angle of the inner and outer layer of the wind turbine skin area.

Therefore, in this paper, a bending-torsional coupling blade structure model is established for the NERL5MW blade, and the ply angle of the composite materials with different layers in the inner and outer areas of the skin is changed, and the influence of different offset angles of the materials in the skin area on the mechanical properties of the blade is studied when the offset angles of the inner and outer sides are opposite.

2. Research object

The research model adopts the 5 MW wind turbine blade model of the NREL laboratory, and the wind turbine shell model is established concerning the data [6]. The radial length of the blade is 61.5 m, and the maximum chord length is 4.7 m.

As shown in Table 1, based on establishing the shell model of the wind turbine blade, this paper uses the CFD method to establish the aerodynamic model of the wind turbine, calculate the pressure distribution of the wind turbine blade, establish the wind turbine structure model by layering different materials [6], change the angle of the inner and outer layers of the blade ±θ respectively (as shown in Figure 1), and study the influence of the skin coupling angle on the mechanical properties and stability analysis of the bending-torsion coupling blade based on the FEM.

Table 1. Mechanical parameters of bending and torsion coupled blade materials [6].

Material name	E_x/GPa	E_y/GPa	G_{xy}/GPa	γ	$\rho/kg\cdot m^{-3}$
Fiber-glass	42	14	2.6	0.3	1920
Saertex Fiberglass	13.6	13.6	12	0.5	1780
Triax Fiberglass	28	13.7	7	0.4	1850
Carbon	115	8.4	6	0.3	1220
Foam	0.25	0.25	0.022	0.3	200

Figure 1. Skin-coupled blades.

3. Mathematical models and calculation methods

3.1 Aerodynamic model of wind turbine blades

The aerodynamic model is established, and the outer area, inner area, velocity inlet boundary, and pressure outlet boundary are set respectively.

In this project, the CFD calculation method and the k-ε turbulence model were used [7], the wind speed at the inlet was 11.4 m/s, the rotation speed of the inner basin was 12.1 rpm, and the pressure on the wind turbine blades was calculated by rotating in a counter-clockwise direction.

After calculation, the thrust of the wind turbine is 783 KN and the torque is 3990 KN·m. The theoretical thrust and torque of a 5 MW wind turbine at the rated wind speed are 800 KN and 4150 KN·m[7]. Therefore, the errors between the calculated and theoretical values of thrust and torque at rated wind speed are within 4%, and the accuracy of the aerodynamic model is also verified.

Wind turbine blades are mainly affected by aerodynamic loads, gravity loads, and centripetal loads [8], resulting in stress, strain, and displacement of wind turbine blades. With the decrease in the radial direction, the radial force and centripetal force of the blade need to be increased, and the radial force of the blade increases with the decrease of the radial direction of the blade.

3.2 Wind turbine blade structure model

3.2.1. Laminate Theory. Laminate is a monolithic board formed by stacking multiple unidirectional panels on top of each other according to a given layup angle and layup sequence and then curing. The mechanical properties of laminates are closely related to the material properties, ply angles, and ply sequences of unidirectional laminates, so different laminates can be designed according to specific loads [9].

The establishment of the bending-torsional coupling blade structure model is based on the laminate theory, which combines laminates of different materials and different thicknesses in different areas according to a certain ply sequence to form a new laminate.

3.2.2 Wind turbine blade layup area division and layup principle. The Triax Fiber-glass material used in the skinned area is a combination of four equal-thickness monolayers in the order of [0°/45°/0°/-45°]; the Saertex Fiber-glass material is a combination of four equal-thickness monolayers in the order of [45°/-45°/45°/-45°] [9].

As shown in Figure 2, each cross-section of the blade is divided into five regions, which are the leading-edge region, cap beam region, web region, trailing edge region, and trailing edge reinforcement region. In addition to the region shown in Figure 2, they have two layers of laminate inside and outside of each region. These areas are named as skin areas. The material used for the skin is the Triax Fiberglass material, and the thickness of each layer of the skin is 0.94 mm.

It can be divided into 25 sections along the radius direction [9], from the leaf root region to the aerodynamic region (roughly 0 m - 13.5 m in the radius direction) with a spacing of 1-2 m; from the aerodynamic region to the tip of the leaf (roughly 13.5 m - 61.5 m in the radius direction), with a spacing of 2 – 4 m, roughly more than 120 regions for the design of the ply structure.

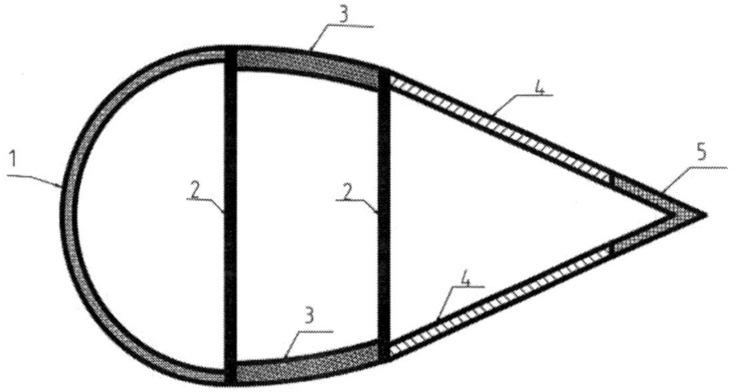

Figure 2. Blade cross-section area.

3.2.3. Structural model validation. To verify the accuracy of the established blade structural model, we will carry out modal calculations for the blade with a skin deflection angle of 0°, calculate the intrinsic frequency of this blade, and compare it with its reference value. Calculating the first three orders of the blade structural intrinsic frequency and the reference value of the intrinsic frequency [10] for comparison, the results are shown in Figure 3, the first three orders of the intrinsic frequency calculated value and the reference value[10] of the error within 10%, this is because of the structural model in the specific crosssection area delineation and design of a certain bias, but these biases have been in the range of controllable, to prove that the structural model has a certain degree of accuracy.

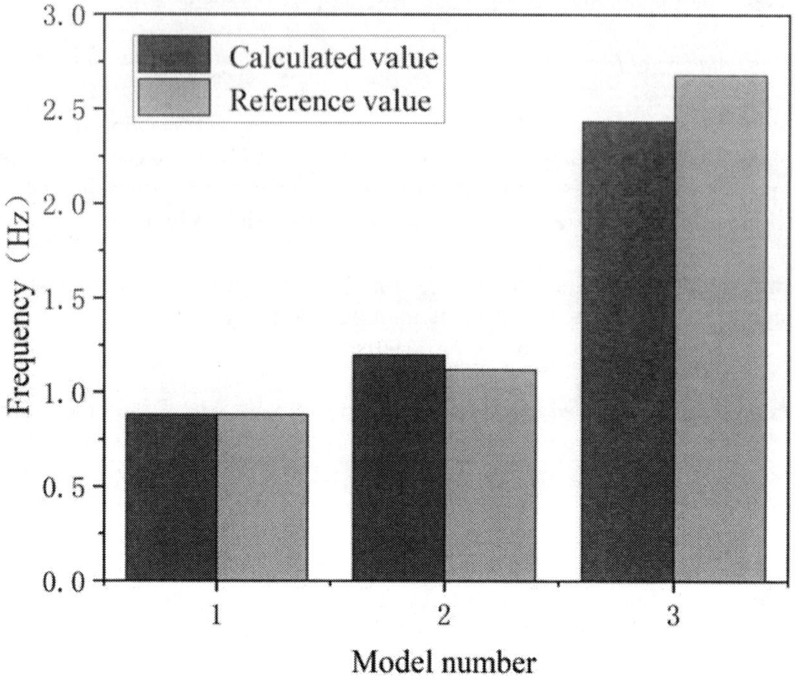

Figure 3. Comparison of calculated and reference values [10] for 0° vibration frequency.

4. Results and analysis

4.1 Modal analysis

Modal analysis refers to the computational and theoretical analysis of the modal parameters. It can be used to determine whether the structural model will resonate or not, as well as to verify the accuracy of the constructed bending-torsion coupled blade [11].

Table 2. Layup materials in each area of the blade section.

Cross-section area	Layup material
Skin	Saertex Fiberglass
Leading edge	Foam+ Triax Fiberglass
Share webs	Foam + Saertex Fiberglass
Cap	Carbon + Triax Fiberglass
Trailing edge	Foam + Triax Fiberglass
Trailing edge enforce	Foam + Fiberglass +Triax Fiberglass

Since the operating speed of the wind turbine blade is 12.1 rpm [12], resonance occurs if the intrinsic frequency is close to 0.606 Hz. As shown in Table 2, the minimum resonance frequency of the blade at θ=0° is about 0.889 Hz, which is far away from the resonance region, and therefore no resonance occurs.

As shown in Figure 4, from the first order to the sixth order, no matter how the deflection angle of the wind turbine skin part material, the vibration pattern is the same as that at 0°; from the first order to the fourth order, with the increase of θ deflection angle, the intrinsic frequency of each order of the blade decreases with the increase of the deflection angle; the material layup angle changes compared to the 0°

in the increase, the sixth order of the blade intrinsic frequency increases the most when θ=45°, this is because the sixth order of the inherent frequency is manifested in the form of torsional deformation. The vibration pattern of inherent frequency is expressed in the form of torsional deformation, and the material has the strongest bending moment and torque resistance and the largest inherent frequency when it is deflected according to θ=45°; among all the blades, the minimum inherent frequency of the first order of the blades is 0.873 Hz, and none of the blades will resonate.

4.2 Analysis of mechanical properties of bending-torsion coupled blade

The deformation, stress, and strain of the wind turbine under the rotational speed of 12.1 r/min, the gravity of the blade itself, and the aerodynamic load of 11.4 m/s rated wind speed are calculated, to optimize the internal and external structural model of the wind turbine blade. In the following, the calculations are carried out by changing the skin layup angle, the inner and outer skins are offset at ±θ, and θ is taken as 0°, 15°, 45°, 60°, and 75° respectively.

4.2.1 Analysis of deformation of bending-torsion coupled blades. The blade waving deformation increases gradually along the radial direction of the blade at any wind speed, and the waving deformation at the tip of the blade is the largest. As shown in Figure 4, the blade tip waving deformation of the wind turbine blade increases with the offset angle θ. When θ=75°, the tip waving deformation is the largest, and the amount of change is 73.6 mm larger than that at 0°, and the amount of deformation is increased by 2.5%, which is due to the decrease of elasticity modulus in the blade waving direction with the increase of the offset angle, and all the maximum deformation of the blade tip of the bias program is less than 5 m, and it will not have collided with the tower in the process of operation, so it is safe. They can avoid collisions during operation, so they are all safe.

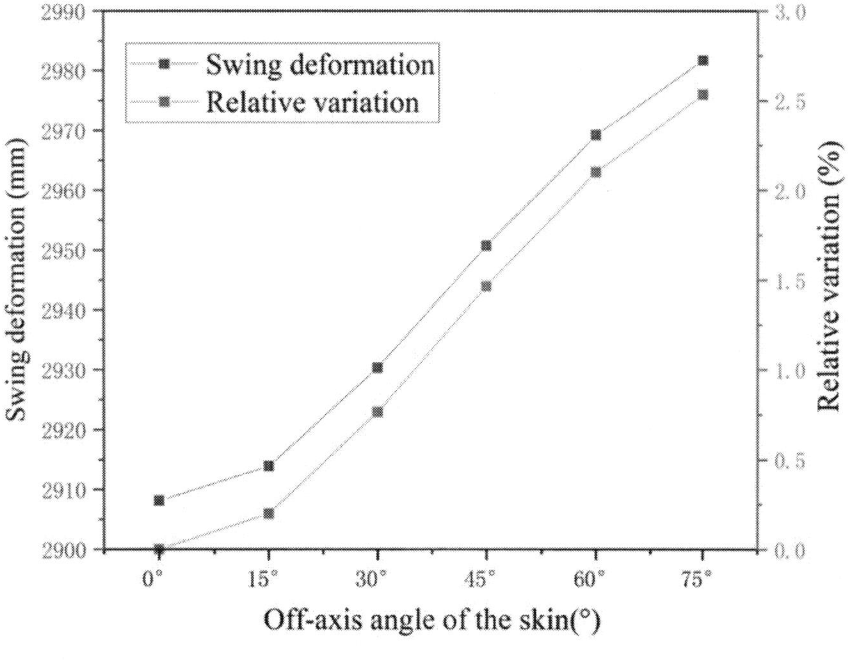

Figure 4. Flap-wise deflection and relative variation of blade tip under different offset angles.

4.2.2 Analysis of mechanical properties of bending-torsion coupled blades. As shown in Table 3, the maximum stress is mainly distributed in the blade cap beam about 40 m distance from the leaf root and 25 m distance; this is because the pressure on the Windward side is larger than the windward side, the cap beam area near the leaf root is more susceptible to greater stress, and the Triax Fiber-glass material is rapidly reduced from the leaf root at 25 m, and Carbon material at the cap beam is increased in the radial modulus of elasticity of the blade is increased at the same time as the axial and tangential Modulus of elasticity of the blade radial modulus of elasticity increases at the same time, axial and tangential modulus of elasticity of the large reduction leads to a large stress, and in the distance from the root of the leaf about 40 m at the cap beam region of the Carbon material is once again a large reduction, and the trailing edge part is no longer set up to strengthen the area of the trailing edge, resulting in this side of the larger stress. As shown in Table 3, the maximum strain is mainly distributed in the windward side of the cap beam at about 40 m from the blade root, which is due to a significant reduction in the modulus of elasticity of the blade compared with that at 25 m from the blade root, resulting in this area being the most likely to become a dangerous area.

Table 3. Maximum stress and strain position of blades.

Blade offset angle	Maximum stress position	Maximum strain position
0°	Windward side, Cap area, 40 m from the root of the leaf	Windward side, Cap area, 40 m from the root of the leaf
15°	Windward side, Cap area, 40 m from the root of the leaf	Windward side, Cap area, 40 m from the root of the leaf
30°	Windward side, Cap area, 40 m from the root of the leaf	Windward side, Cap area, 40 m from the root of the leaf
45°	Windward side, Cap area, 25 m from the root of the leaf	Windward side, Cap area, 40 m from the root of the leaf
60°	Windward side, Cap area, 25 m from the root of the leaf	Windward side, Cap area, 40 m from the root of the leaf
75°	Windward side, Cap area, 25 m from the root of the leaf	Windward side, Cap area, 40 m from the root of the leaf

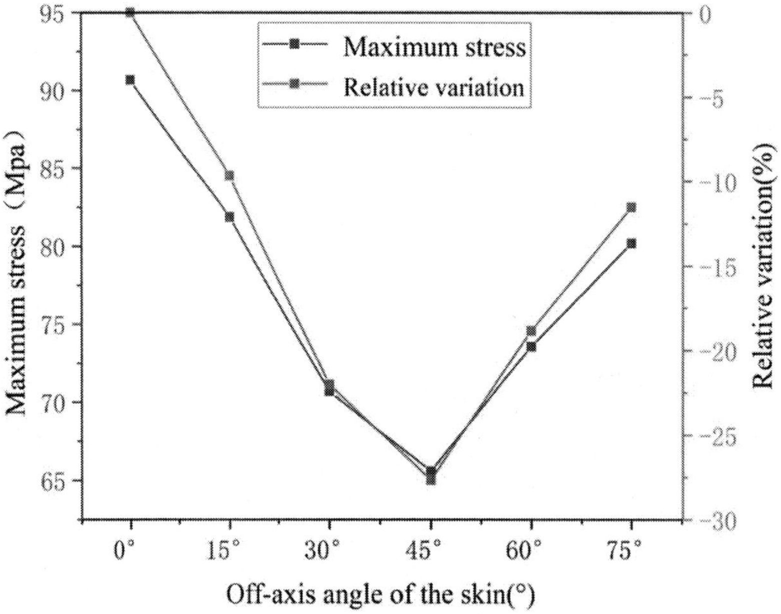

Figure 5. Maximum stress and relative variation at different off-axis angles.

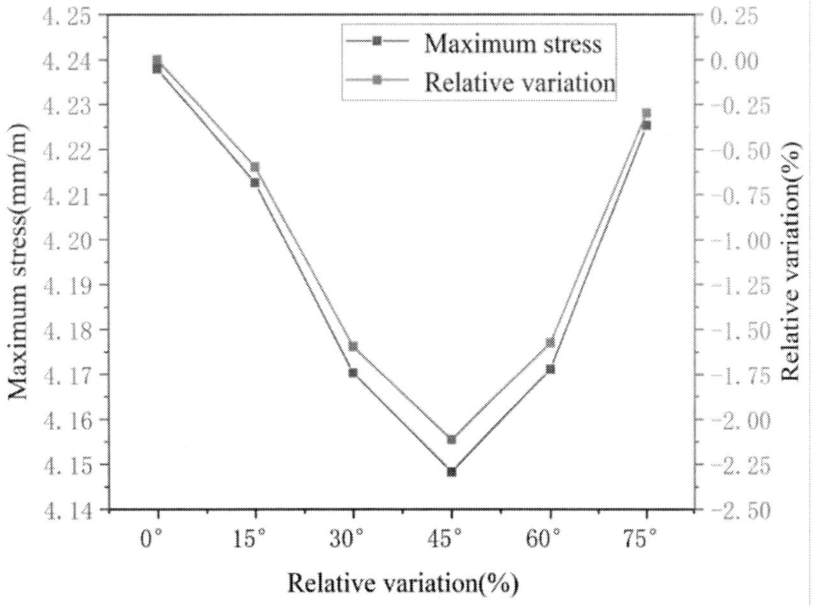

Figure 6. Maximum strain and relative variation at different off-axis angles.

As shown in Figure 5, taking the offset angle 0° as the reference, from 0° to 75°, the blade stress value decreases firstly and then increases with the increase of the angle; when θ=0°, the blade stress value is the largest; when θ=45°, the blade stress value is the smallest, and the change is 25.1 MPa smaller than the maximum, and the change is reduced by 27.7%, which is because the inner and outer skin of the wind turbine is subjected to the shear-directional loading force more often [12]. The increase of blade shear direction elastic modulus is beneficial to reduce the overall blade stress. As shown in Figure 6, the trend of strain value and the trend of stress value is consistent with the change rule; when θ=45°, the blade strain value is the smallest, and the change amount is 2.21 mm/m smaller than the maximum, and the change amount is reduced by 2.11%, which is because the increase of blade shear direction elastic modulus is conducive to the reduction of the overall blade strain.

5. Conclusion

To reduce the maximum stress borne by the wind turbine blade, this paper establishes a bending-twisting coupling structure model for the NERL5MW blade, analyzes the characteristics of the inner and outer skin materials under the bias condition, and obtains the following conclusions:

(1) The skin layup angle has a certain effect on the maximum deformation and maximum strain of the wind turbine bend-twist coupling blade; when the bias angle θ = 75°, the maximum deformation value of the wind turbine blade is the largest, compared with the 0 °, the deformation increases by 2.5%;

(2) Skin laying angle has some effect on the maximum strain of wind turbine bending and torsion coupling blade; when the bias angle θ=45°, the maximum strain value of the blade is the smallest, compared with 0°, the amount of strain is reduced by 2.11%;

(3) The angle of skin layup has a greater influence on the maximum stress of the wind turbine benttwist coupled blade, when the bias angle θ=45°, the maximum stress value of the wind turbine blade is the largest, compared with 0°, the amount of change is reduced by 27.7%;

(4) In summary, the results are optimal when the bias angle of the inner and outer skin of the benttwist coupled blade is ±45° respectively. Subsequent research on the bent-twist coupled blade will optimize the blade structure based on the inner and outer skin bias angle of ±45° respectively.

Acknowledgments
This study was supported by research funds from the Shanghai 2020 Annual Science and Technology Innovation Action Plan: Social Development and Science & Technology Project (No. 20dz1205302).

References

[1] Miao WP, Li C, and Ye Z. (2017) Research on trailing offset control strategy of horizontal axis wind turbine. J. Journal of Solar Energy, 38:23-31.
[2] Wang T G, Li H, and Ye T T. (2015) Structure Design for Wind Turbine Blades. Science Press. Bei Jin.
[3] Zhou X Y, An L Q, and Wang Z Q. (2017) Bend-twist coupling effect of symmetric un-uniform laminate plate beam. J. Axta Materiae Compositae Sinica, 34:1462-1468.
[4] Zhou X Y. Research on Bend-twist Coupling Characteristics of Large Composite Wind Turbine Blade. D. North China Electric Power University.
[5] Sun R, Li C, and Chen W P. (2019) Structural performance analysis of megawatt-class wind turbine blade based on composite layup. J. Journal of Materials Science and Engineering, 37:257-261.
[6] An L Q, Zhou X Y, and Zhao H X. (2013) Modal characterization of 5 MW wind turbine blade. J. Journal of Power Engineering, 33: 890 -894.
[7] Lin X H. (2023). Aeroelastic deformation characteristics and aerodynamic performance analysis of offshore wind turbine blades. D. Shanghai Electric Power of the university.
[8] Jin J T, PAN L J, and PENG C Y. (2013) Finite element analysis method for wind turbine blade crosssection bending stiffness. J. Journal of Solar Energy, 34:196-200.
[9] Zhang L, YAN Yangtian, and Li C. (2021) Analysis of the influence of design parameters of blade layup for a megawatt-class horizontal axis wind turbine. J. Journal of Thermal Power Engineering, 36:133-142.
[10] Miao W P. (2019) Study of fluid-structure interaction response of wind turbine in typhoon environments and wind resistance of the adaptive blade. D. University of Shanghai for Science and Technology.
[11] Liu Y H, WANG Y B, and Li C. (2018) Research on the bending-twisting coupling performance of large wind turbine blades applied to aeroelastic shearing. J. Journal of Power Engineering, 38:1016-1021.
[12] Hu G Y, SUN W L, and Dong P. (2015) Structural mechanical properties analyses for 5MW wind turbine composite blade. J. Renewable EnergyRssource,33:871-875.

Impacts of mandrel parameters on forming quality of tube bending with small bending radius

Tao Wan [1], Jun Fang [1*], Wenxuan Shang [1], Kelu Wang [2] and Shiqiang Lu [2]

[1] School of Intelligent Manufacturing, Jiangxi Science and Technology Normal University, Nanchang 330038, China
[2] School of Materials Science and Engineering, Nanchang Hangkong University, Nanchang 330063, China

[*] Corresponding author's e-mail: 254430930@qq.com

Abstract: To enhance the quality of tube bending with a small bending radius, a three-dimensional finite element (FE) model was developed to simulate numerically controlled (NC) bending of stainless steel tubes made from 21-6-9 alloy. Subsequently, an analysis was conducted to examine the influence of mandrel parameters on the forming quality in terms of the rate of wall thickening, rate of wall thinning, and ovality of cross-section. The findings indicate that increasing the mandrel axial feed or mandrel diameter leads to a slight decrease in wall thickening rate, a noticeable increase in wall thinning rate, and a significant reduction in cross-section ovality. Optimal values for the mandrel axial feed and mandrel diameter are determined to be 0.5 mm-1.5 mm and 5.23 mm-5.43 mm, respectively. These results have important implications for guiding the design of mandrel parameters and enhancing the forming quality of NC bending for stainless steel tubes made from 21-6-9 alloy with a small bending radius.

1. Introduction

Tube bent components are extensively applied to high-end equipment manufacturing clusters such as aerospace, automobile, and shipbuilding due to their exceptional attribute of high strength, lightweight, low loss, and unique hollow structure. Due to its exceptional properties such as superior strength, resistance to corrosion, and high-temperature oxidation, among others, the utilization of 21-6-9 stainless steel tubes in aircraft engine tubing systems has been increasingly gaining attention. However, the bending formability of 21-6-9 stainless steel tubes is challenging due to their high yield ratio and large ratio of yield strength to elastic modulus. This often leads to the occurrence of various defects such as wall thinning, wall thickening, and cross-section ovality, particularly when dealing with a small bending radius ($R/D \leq 2$, R and D denote the bending radius and tube diameter, respectively). To minimize the above defects, the mandrel is usually installed in a tube to provide internal support, which shows that the tube bending forming process heavily relies on the mandrel's pivotal contribution. For the small diameter tube bending forming, the hemispherical mandrel is usually employed, while the mandrel parameters include mandrel axial feed and mandrel diameter. Therefore, a study on the optimum design of mandrel parameters possesses significant theoretical implications and pragmatic utility worth, improving the forming quality of tube bending.

So far, numerous studies have been conducted by both domestic and international researchers to investigate the impacts of mandrel parameters on the forming quality of tube bending. Xu et al. [1] conducted a numerical study on the impact of mandrel axial feed during NC bending with R=3D on

Journal of Physics: Conference Series **2808** (2024) 012073 doi:10.1088/1742-6596/2808/1/012073

various factors including cross-section ovality, wall thickness change, wrinkling, and springback of 0Cr21Ni6Mn9N stainless steel tube, and the suitable mandrel axial feed range was obtained as 1.5-2.0 mm. Li et al. [2] deduced the analytical model of ball-and-socket mandrel parameters including the diameter of the mandrel, axial feed of the mandrel, the thickness of the mandrel ball, the distance between mandrel balls, the radius of mandrel ball as well as the number of mandrel balls, and numerically analyzed the impact of mandrel parameters on the occurrence of wrinkling, crack as well as cross-section ovality of large diameter thin-walled tube during NC bending. Using NC bender, Tian et al. [3] examined the impacts of mandrel extension length and quantity of mandrel balls on alterations in wall thickness for NC bending of large-diameter thin-walled 6061-T4 tubes. A 3D finite element model was developed by Fang et al. [4] to analyze the NC bending process of a stainless steel tube made from 21-6-9 alloy with a bending radius of 3 times the diameter, and the obtained results focused on the influence of different mandrel types and parameters on the quality of cross-sections. Shen et al. [5] numerically studied the impacts of friction factor between mandrel/mandrel ball and tube, the gap between mandrel/mandrel ball and tube, the number of mandrel balls, and mandrel axial feed on spring back angle of a 3A21 rectangular tube during bending with R=40 mm. Jiang et al. [6] conducted a numerical study to analyze the impacts of various mandrel parameters, such as the quantity and thickness of mandrel balls, as well as the mandrel support angle. The focus was on investigating how these factors influence both wall thickness variation and cross-section ovality in ultra-thin-walled tubes during NC bending with an R/D of 2.1. Guo et al. [7] developed a mechanical model with a built-in mandrel of tube-free bending, and the impacts of the position of the mandrel, the number of balls on the mandrel, clearance for mandrel versus tube on stress/strain distribution, cross-section ovality, as well as variation in the thickness distribution of thin-walled stainless steel tubes when subjected to free bending with R=3D. Bai et al. [8] explored the method to determine the shape parameters of hollow mandrels with various geometrical dimensions and applied the ABAQUS software to conduct a mass of cold bending simulation for tubes with various geometrical dimensions and hollow mandrels with various thicknesses and lengths of hollow sections. However, the primary emphasis of the studies above lies in investigating the impacts of mandrel parameters on the quality of tube bending with a conventional bending radius and rarely that with a small bending radius.

To enhance the quality of tube bending, a study was conducted by using the 21-6-9 stainless steel tube measuring D = 6.35 mm with a wall thickness of t = 0.41 mm as the focus. Firstly, a 3D-FE model is developed in ABAQUS software to simulate tube NC bending with a bending radius of 2 times the tube diameter, and subsequently, an investigation is conducted on the impacts of mandrel parameters on the forming quality of tube NC bending. The findings hold immense importance in guiding optimizing mandrel parameters as well as enhancing the forming quality of stainless steel tubes made from 21-6-9 alloy during NC bending.

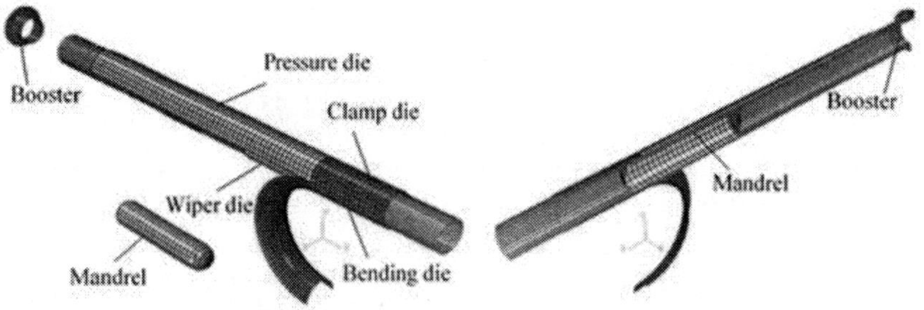

Figure 1. FE model of tube NC bending with a small bending radius.

2. 3D-FE modeling and verification

2.1. 3D-FE model of tube NC bending with a small bending radius

Based on the actual tube NC bending forming process with a small bending radius, the 3D-FE model

Journal of Physics: Conference Series **2808** (2024) 012073 doi:10.1088/1742-6596/2808/1/012073

for NC bending of 21-6-9 stainless steel tube with R= 2D was developed by using ABAQUS software, as shown in Figure 1. The detailed FE modeling process, such as the geometrical model, material model, friction model, and boundary loading conditions can be identified within literary sources [9].

2.2. Development of metrics to evaluate the quality of tube NC bending

Some quality criteria, such as alteration in wall thickness and ovality in cross-section, are applied to estimate the quality of tube NC bending forming. The alteration in wall thickness (wall thickening and thinning) rate I_t and cross-section ovality I_d can be expressed as:

$$I_t = \left|\frac{t - t'}{t}\right| \times 100\% \quad (1)$$

$$I_d = \frac{D - D'}{D} \times 100\% \quad (2)$$

where t' is the minimum or maximum wall thickness of the tube after bending, and D' is a cross-section dimension along the direction of bending radius after bending.

2.3. 3D-FE model verification

To verify the dependability of the 3D-FE model, according to the existing experimental conditions, a series of NC bending experiments were performed on the NC tube bender for the stainless steel tube made from 21-6-9 alloy with R=20 mm and bending angle θ=90°. Table 1 presents a juxtaposition between the outcomes obtained from simulations and those derived from experiments. As observed in Table 1, the relative error is 0.40% for the maximum wall thinning rate, 7.13% for the maximum wall thickening rate, and 7.41% for maximum cross-section ovality. The results demonstrate the reliability of the 3D-FE model and its potential for exploring the impacts of mandrel parameters on forming quality for NC bending of the tube with a small bending radius.

Table 1. Comparison of simulated results and experimental ones.

Forming quality index	Experimental result	Simulated result	Relative error (%)
Maximum wall thinning rate (%)	7.32	7.29	0.40
Maximum wall thickening rate (%)	8.13	8.71	7.13
Maximum cross-section ovality (%)	10.65	9.87	7.41

3. Results and discussion

3.1. Impact of mandrel axial feed on forming quality

Figure 2 shows the effect of mandrel axial feed on the forming quality of the stainless steel tube made from 21-6-9 alloy during NC bending with a small bending radius. The observation depicted in Figure 2 (a) indicates that the maximum wall thickening rate declines from 17.87% to 17.37% with a decrease rate of 2.8% as the mandrel axial feed augments from 0 mm to 2.0 mm. Namely, the mandrel axial feed has no obvious effect on the wall thickening rate. This is because by increasing the mandrel axial feed, the restraint effect of the mandrel within the interior of the tube is enhanced, resulting in a reduction of the wall thickening rate. However, the front of the mandrel gradually separates from the bending section as the mandrel axial feed increases, which causes the wall thickening rate to decline non-obviously. It can be observed in Figure 2 (a) that when the mandrel axial feed is increased from 0 mm to 2.0 mm, there is a corresponding increase in the maximum rate of wall thinning. The maximum wall thinning rate rises from 10.00% to 14.55%, indicating a growth rate of 45.50%. The primary factor contributing to this phenomenon is the direct correlation between the increase of the mandrel axial feed and the resulting increase in friction force exerted on the tube, consequently leading to an amplification of bending deformation force. Namely, the tangential tensile stress augments. As a result, the rate of wall thinning escalates with an increase in the mandrel axial feed.

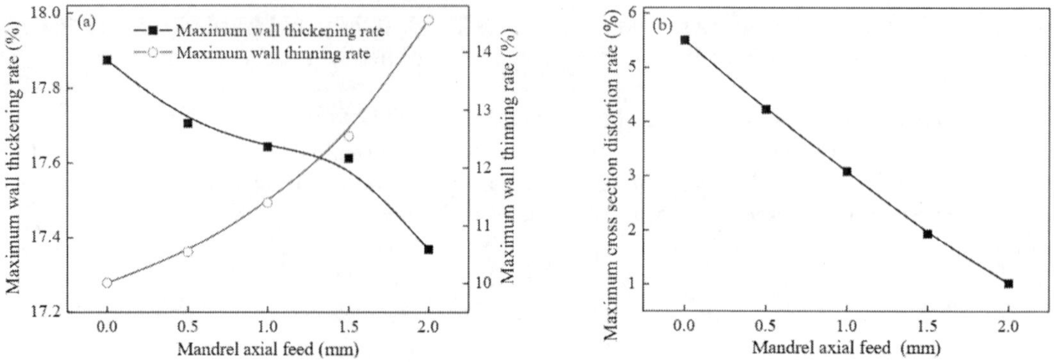

Figure 2. Impact of mandrel axial feed.

It is found from Figure 2 (b) that the maximum cross-section ovality declines from 5.50% to 1.02% with a decrease rate of 81.45% when the mandrel axial feed augments from 0 mm to 2.0 mm. The reasons are as follows: with increasing the mandrel axial feed, the supporting effect of the mandrel enhances, which causes the cross-section ovality to decline. On the other hand, as the mandrel axial feed is raised, there is a corresponding increase in the force exerted on the tube during bending deformation, and then the corresponding tangential tension stress augments, which makes the cross-section ovality augment. However, the supporting effect of the mandrel plays the main role. Hence, as the mandrel axial feed increases, there is a decrease in the maximum cross-section ovality. When the mandrel axial feed is at 0 mm, there is a cross-section ovality of 5.5%, which surpasses the aviation standard limit of 5%. Overlarge mandrel axial feed may cause the bent tube to produce a hump, over-thinning, or even cracking. In the cause of gaining the qualified bent-tube component, the hump and over-thinning should be avoided. Therefore, considering the impact of the mandrel axial feed on wall thickening, wall thinning, and cross-section ovality comprehensively, the mandrel axial feed is optimally maintained within the range of 0.5 mm to 1.5 mm.

3.2. Impact of mandrel diameter on forming quality

Figure 3 illustrates the impact of mandrel diameter on the quality of NC bending for a small bending radius in a 21-6-9 stainless steel tube. It is observed from Figure 3 (a) that the maximum wall thickening rate declines from 18.06% to 17.56%, with a decrease rate of 2.77%, when the mandrel diameter increases from 5.03 mm to 5.43 mm. Namely, the mandrel diameter has little effect on the wall thickening rate. The primary factor is that when the mandrel diameter increases, it enhances the support role provided by the mandrel to the tube, which results in a decline in the wall thickening rate. As also can be found in Figure 3 (a) when the mandrel diameter augments from 5.03 mm to 5.43 mm, the maximum wall thinning rate augments from 9.31% to 11.58%, with an increase rate of 24.38%. The primary factor is that the augmentation of the mandrel diameter increases the friction force exerted by the mandrel on the tube, which results in augmenting the tangential tension stress. Therefore, the wall thinning rate augments.

As depicted in Figure 3 (b), when the mandrel diameter increases from 5.03 mm to 5.43 mm, the maximum cross-section ovality decreases from 6.38% to 2.79%, with a decrease rate of 56.27%. This phenomenon can be attributed to the enhanced support role provided by the mandrel as its diameter increases, leading to a reduction in cross-sectional ovality. When the mandrel diameter is below 5.23 mm, the aviation standard limit of 5% for maximum cross-section ovality is exceeded. Meanwhile, a too-small mandrel diameter will lead to an increase in the risk of wrinkling, especially for tube bending with a small bending radius. In addition, the large mandrel diameter is prone to exacerbate the mandrel wear and tube inside surface graze, which affects the surface quality of the bent tube and the longevity of the mandrel's service life. Therefore, considering the impact of mandrel diameter on the wall thickening rate, wall thinning rate, and ovality in cross-section during NC bending of the stainless steel tube made from 21-6-9 alloy with a small bending radius, the optimal mandrel diameter should be within the range of 5.23 mm to 5.43 mm.

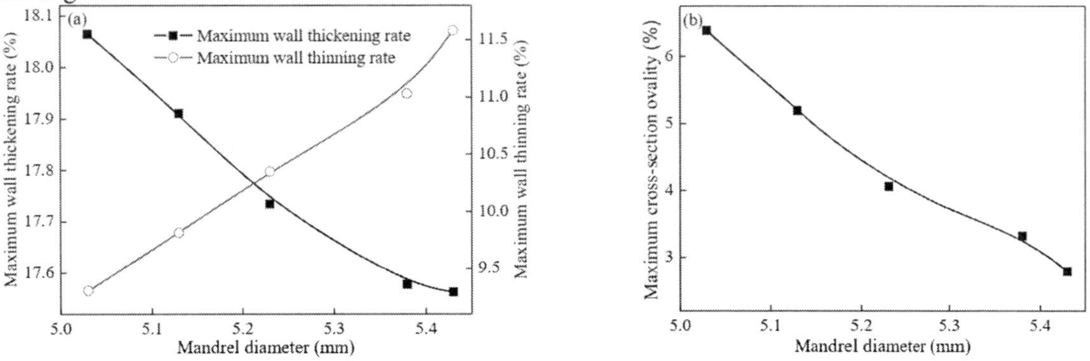

Figure 3. Impact of mandrel diameter.

4. Conclusions

(1) The wall thickening rate decreases as the mandrel axial feed or mandrel diameter increases, but the decrease degree is not large.

(2) With the rise in either the axial feed of the mandrel or its diameter, there is a noticeable escalation in the wall thinning rate, accompanied by a considerable reduction in cross-sectional ovality.

(3) The optimal range for the axial feed of the mandrel is between 0.5 mm and 1.5 mm, while the recommended diameter for the mandrel falls within 5.23 mm to 5.43 mm.

Acknowledgments

This work was supported by Jiangxi Kangte New Energy Technology Co., Ltd. (Grant number: H20230505082339000003).

References

[1] Xu X. M., Lu S. Q., Fang J., et al. (2014). Influence of axial feed of mandrel on NC bending forming quality of 0Cr21Ni6Mn9N stainless steel tube. Forg. Stamp. Technol., 39: 74-77.

[2] Li H., Yang H., Zhan, et al. (2007). Impact of the mandrel on the NC precision bending process for the thin-walled tube. Int. J. Mach. Tool. Manuf., 47: 1, 164-1, 175.

[3] Tian Y. L., Yang H., Li H., et al. (2012). Experimental research on alterations in wall thickness for NC bending of large diameter thin-walled 6061-T4 tubes. Mater. Sci. Technol., 20: 23-29+34.

[4] Fang J., Lu S. Q., Wang K. L., et al. (2013). Investigation on the impact of mandrel usage on the cross-sectional integrity during numerical control bending procedure for small diameter stainless steel 2169 tubes. Adv. Mater. Sci. Eng., 2013: 1-9.

[5] Shen H. W., Liu Y. L., Dong W. Q., et al. (2012). Numerical modeling of the influence of mandrel on spring-back behavior in rotary-draw bending process for rectangular aluminum alloy tubes. Mater. Sci. Technol., 20: 38-43.

[6] Jiang L. F., Wang Y. Q., Lin Y. C., et al. (2020). Impacts of mandrel parameters on forming quality of ultra-thin-walled tube during NC rotary draw bending. J. Xi'an Jiaotong Univ., 54: 144-154.

[7] Guo X. Z., Cheng C., Guo J. X., et al. (2021). Impact of mandrel configuration on forming quality in free bending for the thin-walled tube. Forg. Stamp. Technol., 46: 127-135.

[8] Bai L., Liu J., and Wang Z. A. (2022). Exploration of geometric factors influencing the hollow mandrel in the cold bending procedure for thin-walled tubes. J. Plasticity Eng., 29: 28-35.

[9] Shang W. X., Xiang J. H., Fang J., et al. (2023). Stress/strain analysis for high-strength stainless steel tube during NC rotary draw bending process with small bending radius. J. Plasticity Eng., 30: 204-212.

Journal of Physics: Conference Series **2808** (2024) 012074 doi:10.1088/1742-6596/2808/1/012074

Study on the influence of hole shapes on the fluid and mechanical properties of perforated honeycomb sandwich structures

Pengfei Xiao[1*], Chunping Zhou[1] and Yunfeng Luo[2]

[1]AVIC Research Institute for Special Structures of Aeronautical Composites, Jinan, Shandong Province, 210000, China
[2]Shandong University, Jinan, Shandong Province, 210000, China

[*]Corresponding author's e-mail: 13163339094@163.com

Abstract. Due to the high load and heat transfer capacity, porous honeycomb sandwich structures are widely in engineering. However, in functional requirements, mass and heat transfer should be realized which results and perforated holes being applied in sandwich structures. The shape of the holes plays a crucial role in further improving the load-bearing and flow-resistance performance of the structure. In this study, the influence of hole shape on the flow resistance, permeability, and mechanical properties of perforated honeycomb sandwich structures are systematically studied. Five different hole shapes with the same area are considered, and the flow resistance, permeability, and maximum Mises stress of the perforated honeycomb sandwich structure under the five conditions are calculated. The results show that the hole shape does have a significant influence on the flow resistance, permeability, and mechanical properties of the structure, and especially, it is especially found that the circular structure has the lowest flow resistance and the rectangular structure has the highest load-bearing capacity.

1. Introduction

The active cooling structure is a crucial component in many engineering applications, particularly those involving high-temperature or high-power environments. This structure incorporates advanced cooling mechanisms, such as fans, pumps, or heat sinks, to actively dissipate heat and maintain optimal operating temperatures. By actively circulating a coolant, such as air or liquid, through strategically placed channels or vents, the active cooling structure effectively removes excess heat from critical components, enhancing their performance, reliability, and longevity.

A honeycomb sandwich structure is a special composite structure, usually consisting of two thinner composite face sheets and a honeycomb core material in the middle. Due to its advantages of low density, high strength, and high stiffness, it is widely used in aerospace, communication satellites, automobiles, and other fields [1-3]. The perforated honeycomb sandwich structure is to add holes in the wall of the honeycomb structure. Due to the advantages like low density, high load-bearing, and high heat transfer capabilities of the perforated honeycomb sandwich structure, it has quite potential to be applied in the active cooling heat transfer structure design [4-5]. However, the hole shape of the perforated honeycomb sandwich structures is crucial to further improving the mechanical, fluidic, and thermal performance of the structure [6-7]. Thus, it is of great importance to systematically study the influence of hole shape on the performance such as flow resistance, permeability, maximum stress, and so on for perforated honeycomb sandwich structures [8-9].

In the above context, this paper systematically studies the influence of the hole shape of perforated honeycomb sandwich structures on the fluidic and mechanical properties, such as flow resistance, permeability, and maximum Mises stress. In specific, five different hole shapes with the same area are considered and compared in detail. The results show that the hole shape does play a crucial role in the flow resistance, permeability, and mechanical properties of the structure. At the same time, we found that among these considered structures, the circular structure has the lowest flow resistance and the rectangular structure has the highest load-bearing capacity.

The rest of the paper is organized as follows. In Section 2, how to calculate the flow resistance and permeability of honeycomb sandwich structures is provided. In Section 3, the examples and results are discussed. Finally, a conclusion of this work is given.

2. Flow resistance and permeability of honeycomb sandwich structures

Flow resistance rate [10] is one of the most important parameters of porous materials, which represents the specific flow resistance per unit thickness of isotropic materials, as follows

$$\sigma = \frac{R_S}{d} \tag{1}$$

where σ is the flow resistance rate, with units of $Pa \cdot s / m^2$, while R_s is the specific flow resistance with units of $Pa \cdot s / m$, and d is the sample thickness with units of m. The specific flow resistance R_s is also defined as the ratio of the pressure drop across the sample to the airflow velocity:

$$R_s = \frac{\Delta p}{v} \tag{2}$$

where Δp is the pressure difference across the structure, with units of Pa, v is the airflow velocity with units of m / s.

In addition, the flow resistance R is defined as the ratio of the pressure drop across the sample to the airflow volume velocity as

$$R = \frac{\Delta p}{q_V} = \frac{\Delta p}{Av} \tag{3}$$

In the above equation, q_v is the volume rate, with units of m^3 / s. A A is the cross-sectional area of the sample, with units of m^2. And v υ is the airflow velocity, with units of m / s.

The relationship between flow resistance rate and flow resistance can be obtained by combining Equation (1), Equation (2) and Equation (3), as follows

$$\sigma = \frac{RA}{d} \tag{4}$$

In addition, permeability [11] is also one of the important parameters of porous materials, which can be defined as

$$\kappa = \frac{q_V \mu d}{A \Delta p} \tag{5}$$

In the above equation, q_v is the volumetric air flow rate, with a unit of m^3 / s. A is the cross-sectional area of the sample, μ is the viscosity coefficient of the fluid flowing through, with the unit of $Pa \cdot s$, Δp is the pressure difference on both sides of the sample, and d is the thickness of the sample.

Substituting Equation (3) into Equation (5) yields the relationship between permeability and flow resistance, as follows

$$\kappa = \frac{\mu d}{AR} \tag{6}$$

Similarly, substituting Equation (6) into Equation (4) yields the relationship between permeability

Journal of Physics: Conference Series **2808** (2024) 012074 doi:10.1088/1742-6596/2808/1/012074

and flow resistance rate.

$$\kappa = \frac{\mu}{\sigma} \tag{7}$$

All required relations to calculate the flow resistance and permeability of honeycomb sandwich structures are given above.

3. Results

The dimensions of the honeycomb model are shown in Figure 1. The honeycomb wall panel is perforated. The nitrogen and air mixture flows in from the left and out from the right.

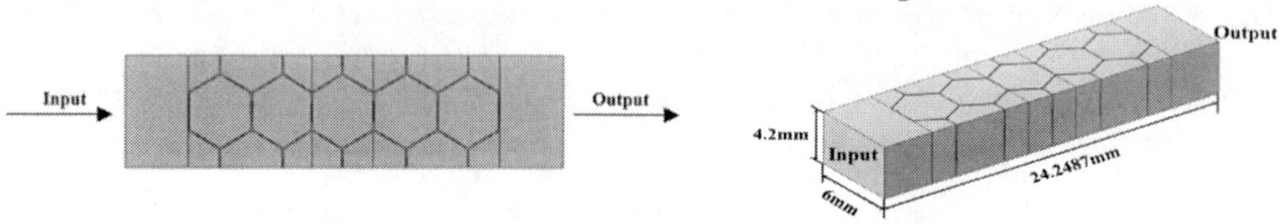

Figure 1. The dimensions of the honeycomb structure.

To study the relationship between flow resistance and load-bearing capacity of honeycombs with different punching patterns, this article selects five honeycomb holes with the same hole area, as shown in Figure 2. The punching area for these five punching patterns is equal to $0.75^2 \times \pi = 0.5625\pi$ mm^2. The honeycomb material is carbon fiber composite material with a density of 160 kg/m^3, Young's modulus of 140 *GPa*, and Poisson's ratio of 0.31 . A load of 20000 N/m^2 is applied to the top plates, while it is fixed in the lower plate.

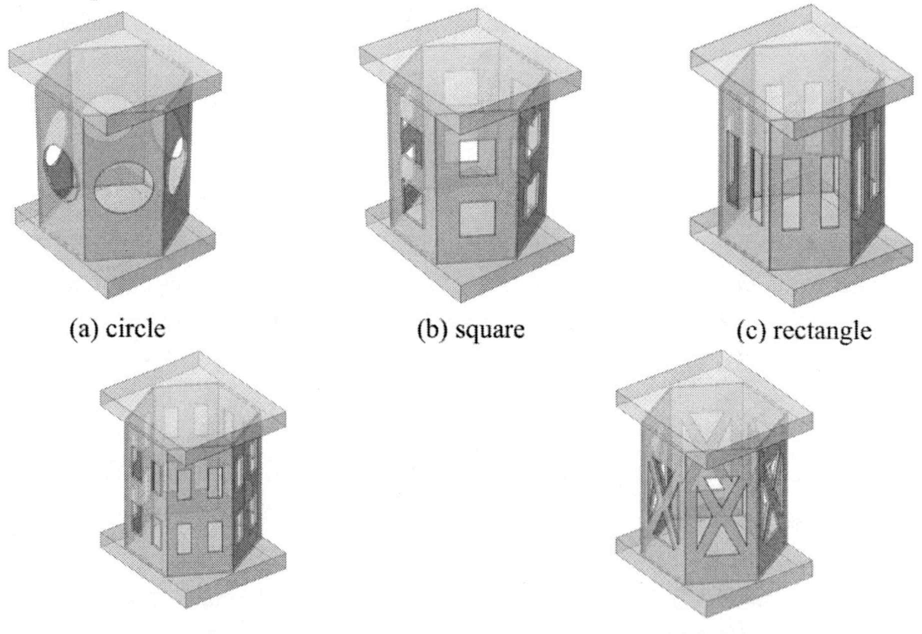

(a) circle (b) square (c) rectangle

(d) cross shape (e) X-shape

Figure 2. Five different punching shapes.

Now, the flow field and stress field can be calculated by the finite element method. The obtained velocity and stress cloud maps are shown in Figure 3 and Figure 4.

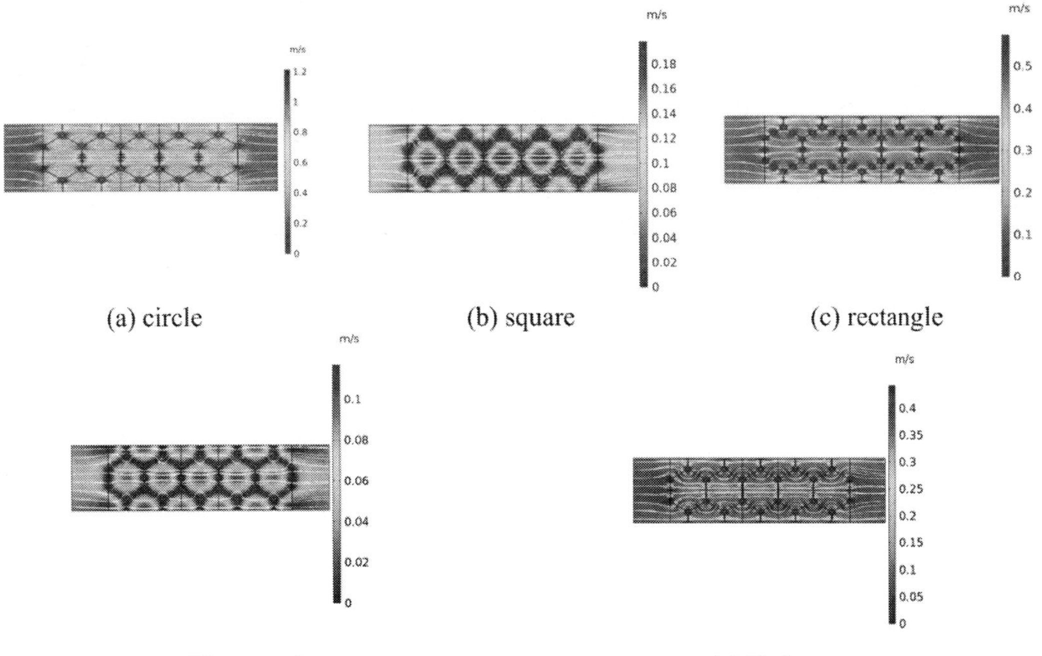

(a) circle (b) square (c) rectangle

(d) cross shape (e) X-shape

Figure 3. Velocity fields for the considered five cases.

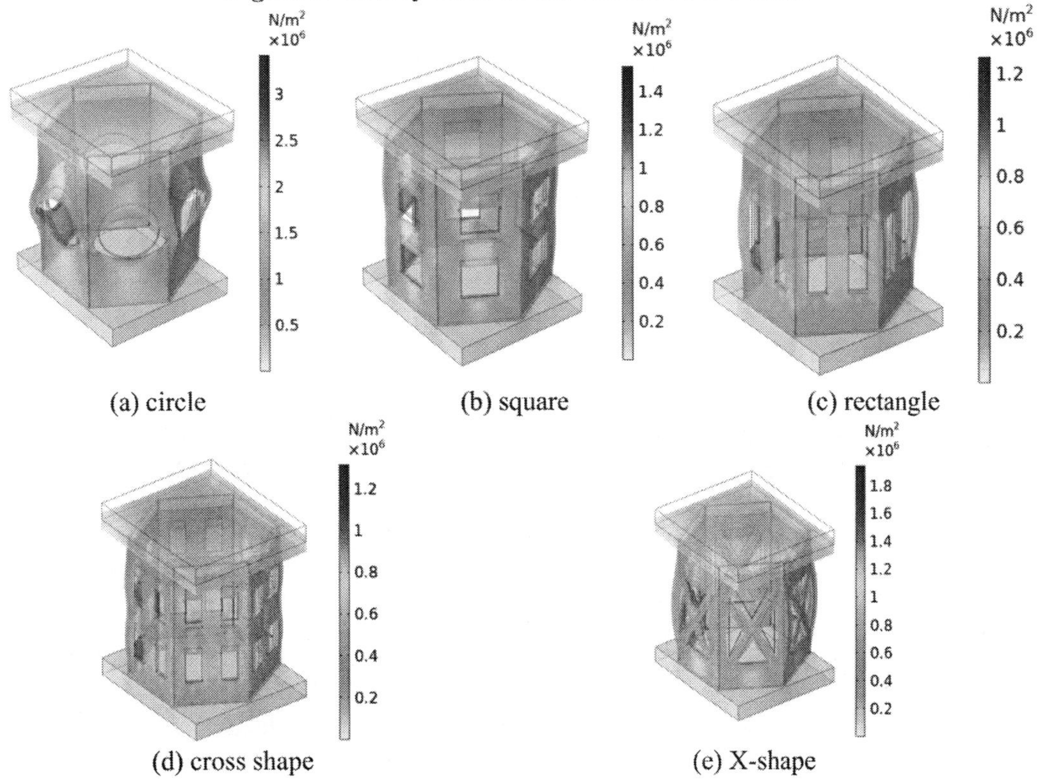

(a) circle (b) square (c) rectangle

(d) cross shape (e) X-shape

Figure 4. Stress fields for the considered five cases.

Now, based on the finite element results, we can calculate the flow resistance, permeability, and maximum stress. The results are shown in Table 1.

Table 1. The flow resistance, permeability, and maximum stress for the considered five cases.

	circle	square	rectangle	cross shape	X-shape
flow resistance ($Pa \cdot s/m^3$)	217.57	353.97	468.86	545.58	589.46
Permeability (m^2)	8.457E-8	5.1982E-8	3.9244E-8	3.3726E-8	3.1215E-8
maximum stress (MPa)	3.5619	1.5333	1.2637	1.3158	1.9448

As seen from the above results, the flow resistances for these five cases considered are 217.57, 353.97, 468.86, 545.58, and 589.46, respectively. Hence, from the angle of flow resistance, the order of these structures from good to bad is the circle structure, square structure, rectangle structure, cross-shape structure, and X-shape structure. Also, the permeabilities are 8.457E-8, 5.1982E-8, 3.9244E-8, 3.3726E-8, and 3.1215E-8, respectively. From the angle of flow permeability, the order of these structures from good to bad is the same in the case of resistance. In addition, the maximum Mises stresses are 3.5619 MPa, 1.5333 MPa, 1.2637 MPa, 1.3158 MPa, and 1.9448 MPa for the five cases above. That is to say, from the angle of load-bearing capacity, the order of these structures from good to bad is the rectangle structure, cross-shaped structure, square structure, X-shape structure, and circle structure, respectively.

Based on the results above, we can see that the circular structure has the lowest flow resistance (217.57) and biggest permeability (8.457E-8), while the rectangle structure has the highest load-bearing capacity (the corresponding maximum stress is the lowest, i.e. 1.2637). Therefore, the rectangle structure can be used as a drilling solution if the spraying power and processing costs allow.

4. Conclusions

This paper systematically studies the influence of hole shape on the flow resistance, permeability, and mechanical properties of the perforated honeycomb sandwich structures. To be more specific, the flow resistance, permeability, and maximum Mises stress for five different perforated honeycomb sandwich structures with different hole shapes are computed and compared. Numerical results show that the hole shape has a significant influence on the flow resistance, permeability, and mechanical properties of the structure, and besides, the circular structure has the lowest flow resistance and the rectangular structure has the highest load-bearing capacity. Thus, based on the above findings, we can conclude that the perforated honeycomb sandwich structure with the circular hole is more suitable for applications in environments with higher temperature requirements because lower flow resistance allows fluids to move faster, thereby carrying away more heat. On the contrary, if the structure has higher requirements for mechanical properties, the rectangular hole is more suitable.

References

[1] Onyibo, E. C.; Safaei, B. Application of finite element analysis to honeycomb sandwich structures: a review. J Reports in Mechanical Engineering 2022, 3 (1), 192-209.

[2] Ramnath, B. V.; Alagarraja, K.; Elanchezhian, C. Review on sandwich composite and their applications. Materials Today: Proceedings 2019, 16, 859-864.

[3] Zhang, Y.; Liu, X.; Zhou, Y.; Shi, Y. Uniaxial compressive performance of an aramid and aluminum honeycomb sandwich structure. Ocean Engineering 2023, 270, 113676.

[4] Mohammadi, H.; Ahmad, Z.; Petrů, M.; Mazlan, S. A.; Johari, M. A. F.; Hatami, H.; Koloor, S. S. R. An insight from nature: honeycomb pattern in advanced structural design for impact energy absorption. Journal of Materials Research and Technology 2023, 22, 2862-2887.

[5] Sobhy, M.; Radwan, A. F. Porosity and size effects on electro-hygrothermal bending of FG sandwich piezoelectric cylindrical shells with porous core via a four-variable shell theory. Case Studies in Thermal Engineering 2023, 45, 102934.

Journal of Physics: Conference Series **2808** (2024) 012074 doi:10.1088/1742-6596/2808/1/012074

[6] Liu, J.; He, X.; Liu, Z.; Cao, X.; Yu, S.; Huang, W. The fluid-structure interaction response of composite auxetic re-entrant honeycomb structures to underwater impact. Thin-Walled Structures 2024, 196, 111465.
[7] Chen, J.; Xie, G.; Yan, H. The effects of ripple amplitude on heat transfer and fluid flow of X-lattice corrugated honeycombs. Science China Technological Sciences 2023, 1-10.
[8] He, M.; Hu, W. A study on composite honeycomb sandwich panel structure. Materials & Design 2008, 29 (3), 709-713.
[9] Sun, G.; Chen, D.; Huo, X.; Zheng, G.; Li, Q. Experimental and numerical studies on indentation and perforation characteristics of honeycomb sandwich panels. Composite Structures 2018, 184, 110-124.
[10] Ren, T.; Zhu, Z.; Wang, J.; Yan, C. Numerical simulation of flow resistance characteristics in a rod bundle channel under rolling conditions. Annals of Nuclear Energy 2023, 181, 109573.
[11] Neumann, M.; Stenzel, O.; Willot, F.; Holzer, L.; Schmidt, V. Quantifying the influence of microstructure on effective conductivity and permeability: Virtual materials testing. International journal of solids and structures 2020, 184, 211-220.

Journal of Physics: Conference Series **2808** (2024) 012075 doi:10.1088/1742-6596/2808/1/012075

Simulation and analysis for residual stresses in anode-supported flat-plate solid oxide fuel cells for ships

Zhaoqiang Li[1*], Lingling Wang[2], Haijun Yang[3], Wendong Ji[4], Yushan Zhang[4] and Baocheng Lu[1]

[1]Non-Commissioned Officer Department, Binzhou Polytechnic, 919 Huanghe 12th Road, Binzhou, Shandong, 256600, China

[2]Mechanical and Electrical Engineering Department, Binzhou Polytechnic, 919 Huanghe 12th Road, Binzhou, Shandong, 256600, China

[3]Training Management Department, Binzhou Polytechnic, 919 Huanghe 12th Road, Binzhou, Shandong, 256600, China

[4]Logistics Management Department, Binzhou Polytechnic, 919 Huanghe 12th Road, Binzhou, Shandong, 256600, China

[*]Corresponding author's e-mail: readerlzq@163.com

Abstract. For investigating the distribution of residual stress in marine SOFC, this paper establishes a two-dimensional axisymmetric model of a SOFC single cell to simulate the temperature change in the cell preparation process, and the results find that: The residual stresses in the cell are mainly concentrated at the bond where the anode support layer and the electrolyte layer combine. At the outermost intersection, there is a maximum stress of 184.43 MPa. Meanwhile, we analyze the interaction of component thickness parameters for residual stresses, and the simulation results show that: The largest first principal stress of the assembly shows a straight-line decreasing trend when the thickness of the anode support layer increases; However, in the same scenario for the electrolyte layer, the maximum stress increases rapidly in the beginning and then increases slowly until it does not increase anymore; There is no effect of the variation in thickness parameters for the cathode and isolation layers on the maximum first principal stress of the module.

1. Introduction

Solid Oxide Fuel Cells (SOFCs) have not only high capacity density but also high energy transformation efficiencies, no pollution to the environment, etc., and they have been considered as the next-generation energy technology with promising applications. Nowadays, there are already a considerable number of ships using SOFC as an electric energy source, and its application in the field of ships will be more and more extensive [1, 2]. In addition, SOFC has the disadvantage of high preparation and operation temperatures. As a ceramic-type battery, high-temperature sintering of battery components is required. Due to the inconsistency in the preparation temperature of each component and the different compositions of each component, the coefficients of thermal expansion are not consistent, resulting in large internal stresses after the cell cooling to 25°C, causing the cell to warp or even crack, which becomes one of the major issues that affect the working performance as well as the service life of the SOFC. Consequently, studying and optimizing the residual stress of SOFC is of great significance for extending the service life of fuel cells. Li *et al.* [3] researched the influence of the interface morphology of the cells on the stress and finally found that the residual stress was smaller when the bonding interface was smooth and gentle. Whereas the normal and shear stresses at the interface varied linearly when the amplitude of the interface between the anode and cathode increased. Severson *et al.* [4] constructed a simulation model of SOFC in order to calculate different residual stresses between anode-supported and electrolyte-supported batteries separately. The study showed that, compared with the anode-supported batteries, electrolyte-supported cells had a higher stress level, and the maximum value occurred at the junction of the anode-support layer and the electrolyte layer. Schleg *et al.* [5] simulated the stress and strain processes in SOFCs by using ABAQUS™ and found that the maximum tensile stresses occurred in the cathode layer, reaching 155 MPa. After 1000 hours of operation at 750°C, the stress was reduced by 20% due to creep, while the constraints of the bipolar plates increased the tensile stress.

2. Theoretical basis

Elastomers deform under the action of forces, and internal forces are generated by uneven internal strains. The stress is the degree of concentration of the distribution of internal force, which can be considered as the value of internal force per unit area on the cross-section. That is, stress and strain are simultaneous, due to a variety of different physical properties of different materials, their stress and strain correspond to different relationships. In elastic mechanics, the elastic constants describing anisotropic materials are generally: the modulus of elasticity E (Young's modulus), Poisson's ratio υ, and shear modulus G. The relationship between the three can be expressed as:

$$G = \frac{E}{2(1+v)} \tag{1}$$

The elasticity matrix of anisotropic materials is generally given by the following equation [6]:

$$D = \frac{E}{(1+v)(1-2v)}\begin{bmatrix} 1-v & v & v & 0 \\ v & 1-v & v & 0 \\ v & v & 1-v & 0 \\ 0 & 0 & 0 & \frac{1-2v}{2} \end{bmatrix} \tag{2}$$

Due to the internal force generated from the temperature variation of SOFC, it can be expressed by coupling the temperature field with the elasticity eigenstructure equation [7]:

$$\varepsilon_{th} = \alpha(T - T_{ref}) \tag{3}$$

3. Research process

3.1 Research object

In this paper, the product of a 10 cm×10 cm anode-supported single cell produced by Ningbo Sofc man Energy Technology Co., Ltd. was taken as an example, the preparation process of the single cell was simulated by Comsol simulation software, and the residual stress was analyzed. The cell size parameters are shown in Table 1.

Table 1. Anode-supported SOFC Dimensional Data.

Structure Name/Material	Thicknesses (μm)	Structure Name/Material	Thicknesses (μm)
Cathode Layer/LSCF+GDC	25	Electrolyte Layer/YSZ	15
Barrier Layer/GDC	3	Anode Layer//NiO-YSZ	400

3.2 The simulation process analysis

For this button cell, the fabrication process was as follows: firstly, the electrolyte YSZ paste was screen-printed on the anode Ni-YSZ billet at 25°C, Then we slowly increased the temperature to 1450°C and held it for 4 hours to produce the Ni-YSZ/YSZ half-cell. Due to the prolonged heat preservation, it can be assumed that the stress inside the half-cell in the process had been fully released (we considered that the internal stress of the battery at this temperature was 0). Subsequently, the cells were slowly cooled down to 25°C. To create a GDC barrier layer, we printed the GDC slurry onto the cooled battery, and then the temperature was slowly increased to 1200°C and held for 5 hours. Then the cells were slowly cooled down to 25°C again, and we printed the LSCF-GDC hybrid cathode layer material onto the isolation layer surface to create the cathode layer. We increased the temperature of cells to 1050°C slowly and maintained this temperature for 2 hours, after then, we slowly cooled down the cells to 25°C again. So the SOFC single-cell preparation process has experienced the temperature change process of 1450°C → 25°C → 1200°C → 25°C → 1050°C → 25°C, and different components experience different processes. So the simulation in this paper mimics the fabrication process of the single cell by first lowering the anode support layer and the electrolyte layer to 25°C at 1450°C (0 stress temperature), and then heating them to 1200°C (0 stress temperature of barrier layer), then reducing to 25°C, then heating to 1050°C (0 stress temperature of cathode layer), and finally reducing to 25°C.

3.3 Establishment of the model
In this paper, the two-dimensional symmetric structure diagram of SOFC single cell was established. As thinner cells, in order to operate easily and observe intuitively, this paper adjusted the scale of the view to x:y=1:30, which is shown in Figure 1. At the same time, physical parameters were added to each component, as shown in Table 2.

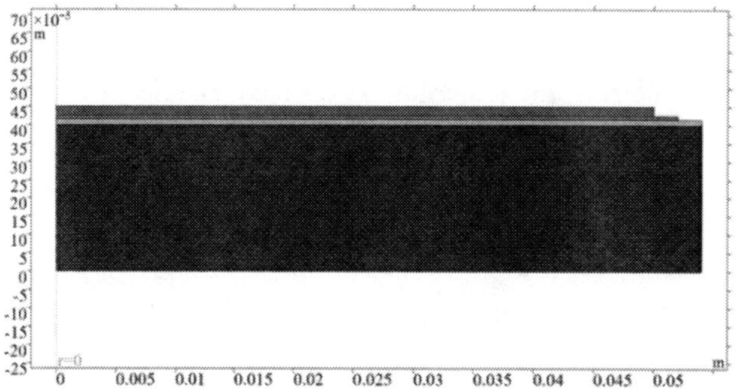

Figure 1. 2D axisymmetric model of anode-supported SOFC single cell.

Table 2. Physical properties parameters of each component [8, 9].

Materials and Properties	Density (kg/m^3)	Young's Modulus (GPa)	Coefficient of Thermal Expansion	Poisson's Ratio	Porosity
NiO-YSZ	6500	105	16.2	0.33	0.4
YSZ	6040	185	10.5	0.313	Dense
GDC	7200	90	12.63	0.32	Dense
LSFC-GDC	6210	10	15.34	0.32	0.4

While the physical field was being set up, we first added Solid Mechanics 1 to establish the thermal expansion of the anode support layer and the electrolyte layer, with 0 Stress temperature set to 1450°C and a target of 1200°C for the temperature change. Then we added Solid Mechanics 2, set the calculation result of Solid Mechanics 1 as pre-stress into the anode layer and electrolyte layer, and then filled the thermal expansion coefficient of the electrolyte layer, the isolation layer, and the anode layer. Meanwhile, a reference temperature of 1200°C and a target temperature of 1050°C were set. The same method was used to add Solid Mechanics 3, set up the pre-stress, and set the target

Journal of Physics: Conference Series **2808** (2024) 012075 doi:10.1088/1742-6596/2808/1/012075

temperature to 25°C. Finally, the model was meshed, solution operations were performed, and the steady state was selected for the study step, with Solid Mechanics 1, 2, and 3 coupled.

Due to the relatively regular structure of this model, the model mainly adopted a mapping method for grid partitioning and locally performed grid refinement. At the same time, a grid refinement coefficient A was introduced, and parameterized scanning was performed on the refinement coefficient A to verify grid independence. The verification results are shown in Figure 2 (a). From the figure, it can be seen that when the grid refinement coefficient increases to 1.5, the maximum first principal stress in the model no longer increases with the increase of grid density. To save computational costs and time, we adopted a grid refinement coefficient A=1.5 as the grid partitioning scheme for this model, corresponding to a number of 8010 grids. The grid partitioning situation is shown in Figure 2 (b).

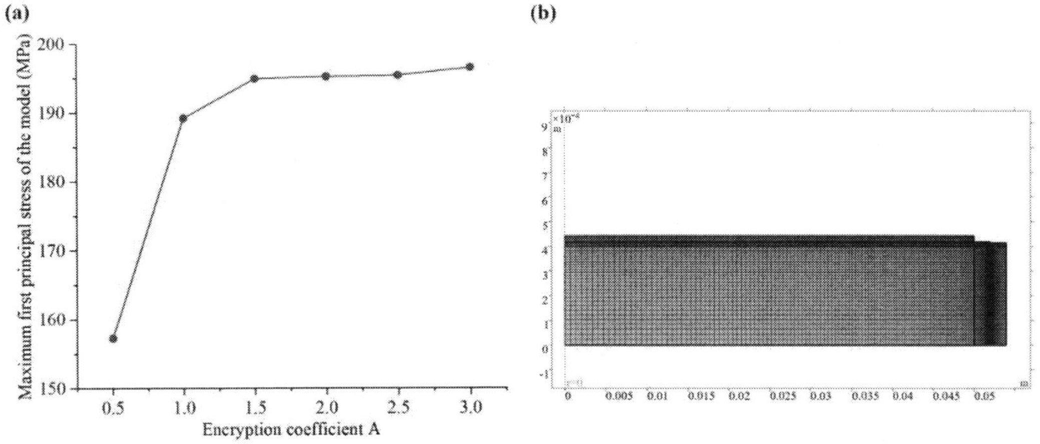

Figure 2. Grid partitioning and independence verification.

4. Results and discussion

4.1 The distribution of remnant stresses

SOFC belongs to the ceramic functional materials, and the first principal stress is generally used to characterize its internal stress. The distribution of remnant stresses for the prepared SOFC single cell is shown in Figure 3 (a). As observed in the figure, the remnant stresses in single cells are primarily concentrated at the boundary between the anode layer and electrolyte layer. Figure 3 (b) indicates the partial enlargement of the single cell, and it is observed that the largest stress is 184.43 MPa, appearing as tensile stress at the outermost junction of the anode layer and the electrolyte layer. The minimum value appears inside the anode layer at -18.63 MPa, which is compressive stress. Since the flexural strength of the single cell is greater than 250 MPa, the thermal stress value is within the tolerable range. Meanwhile, it can be seen that the single cell is offset to the anode side, and the maximum offset occurs at the anode side with a value of 6.73 μm.

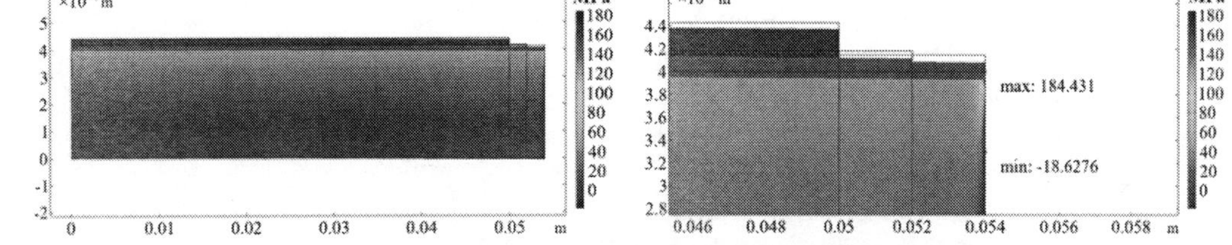

Figure 3. Residual stress distribution in SOFC single-cell.

4.2 The effect of the thickness of components on residual stresses

The residual stress of SOFC single cell could be influenced not only by the fabrication process, but also by the material of the components, the porosity, and the dimensional parameters of each component. To analyze the influence of component thickness on residual stresses, a parametric scan of each component was carried out. The results indicated that the cathode layer as well as the isolation layer depth had no effect on the highest first principal stress of the battery. However, the thickness of the anode support layer and the electrolyte layer influenced the largest first principal stress of the cell, which is illustrated in Figures 4 (a) and (b). In Figure 4 (a), it can be observed that the largest first principal stress of the module shows a straight line downward trend, mainly because the anode layer plays a supportive role for the battery, and the increase in thickness is favorable to the stability of the battery, but compared with the change of the electrolyte layer, the decrease of the maximum first principal stress is small. In Figure 4 (b), As the thickness of the electrolyte layer increases, the largest first principal stress rises rapidly in the beginning, then slowly slows down until it no longer increases. At the same time, the addition of thickness in the electrolyte layer was not conducive to the transport of oxygen ions, so it is important to minimize the electrolyte layer to a thickness that is as low as possible.

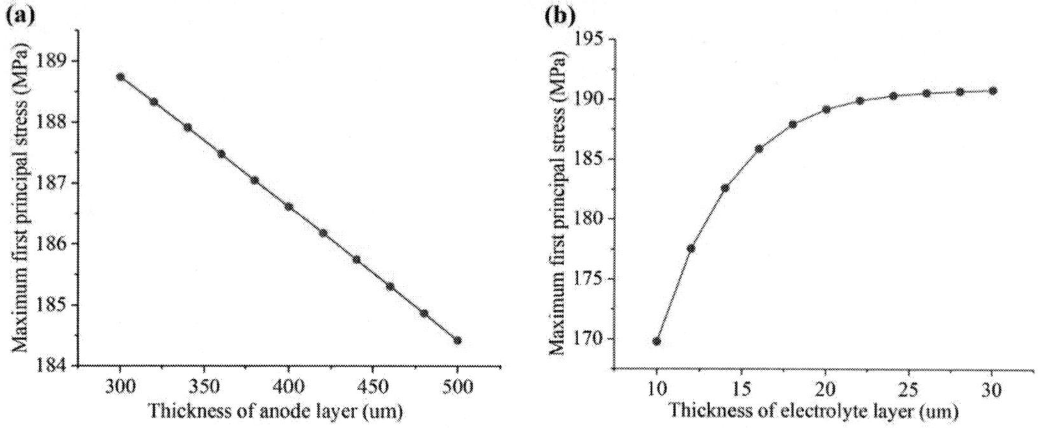

Figure 4. Effect of anode support layer depth dimension on the largest first principal stress.

5. Conclusion

On the basis of the calculation findings of the simulation phantom, we can observe that: in the anode-supported flat-plate SOFC, residual stresses are mainly caused by the inconsistency of the coefficient on thermal expansion in each cell component during the preparation process, which produces tensile or compressive stresses at the interfaces of the cell components when the temperature changes. Furthermore, the dimensions of the components have a large influence on the distribution of residual stresses and can be taken into account to optimize this aspect when designing the cell. In this paper, we only consider the residual stress of the battery, and in the next step, the simulation model can incorporate the operating state of the battery and how moistening affects the thermal stress of the battery to increase the operational stability of the battery and prolong its service life.

References

[1] Wu S, Miao B, Chan S H. Feasibility assessment of a container ship applying ammonia cracker-integrated solid oxide fuel cell technology [J]. International Journal of Hydrogen Energy, 2022, 47(63): 27166-27176.

[2] Baldi F, Moret S, Tammi K, et al. The role of solid oxide fuel cells in future ship energy systems [J]. Energy, 2020, 194: 116811.

[3] Li Q, Li G, Cao G, et al. Effect of interface morphology on the residual stress distribution in solid oxide fuel cell [J]. International Journal of Energy Research, 2020, 44(5): 3497-3509.

[4] Severson H, Assadi M. (2013) Analysis of residual and operational thermal stresses in a planar SOFC. Journal of fuel cell science and technology, 10(6): 061001.
[5] Lugovy M, Slyunyayev V, Brodnikovskyy M, et al. (2020) Residual stress distribution in solid oxide fuel cells: anode-electrolyte and anode-electrolyte-cathode systems. SN Applied Sciences, 2: 1-8.
[6] Osman S, Ahmed K, Nemattalla M, et al. Performance and thermal stresses in functionally graded anode-supported honeycomb solid-oxide fuel cells [J]. International Journal of Hydrogen Energy, 2021, 46(65): 33010-33027.
[7] Sun Y, Hu X, Gao J, et al. Solid oxide electrolysis cell under real fluctuating power supply with a focus on thermal stress analysis [J]. Energy, 2022, 261: 125096.
[8] Jiang, C.; Gu, Y.; Guan, W.; Zheng, J.; Ni, M.; Zhong, Z. 3D thermo-electro-chemo-mechanical coupled modeling of solid oxide fuel cell with double-sided cathodes. Int. J. Hydrog. Energy 2020, 45, 904-915.
[9] Zheng J, Xiao L, Wu M, et al. Numerical Analysis of Thermal Stress for a Stack of Planar Solid Oxide Fuel Cells [J]. Energies, 2022, 15(01): 343.

Journal of Physics: Conference Series **2808** (2024) 012076 doi:10.1088/1742-6596/2808/1/012076

Lightweight design of composite sandwich corrugated structures based on variable density method

Ce Tian[1], Zhimin Tian [1*], Xinwei Cao[1] and Shangwei Dong[2]

[1]Institute of Defense Engineering, AMS, PLA, Beijing, 100850, PR China
[2]Beijing Academy of Emergency Management Science and Technology, Beijing, 101117, PR China

[*]Corresponding author's e-mail: overseatian@sina.com

Abstract. For the lightweight design of composite sandwich corrugated structures, a new topology optimization design scheme is proposed using the variable density method. Multi-working conditions static analysis of the composite sandwich corrugated structure is carried out by finite element software. Then, the topology optimization is carried out by using the minimum flexibility method. Finally, the results of different working conditions are coupled to propose a new composite sandwich corrugated structure. The results indicate that the structure has a weight reduction of 34% and exhibits periodicity and continuity while also meeting the necessary mechanical properties to fulfill the engineering requirements. The optimization of composite sandwich corrugated structures has been achieved, providing a new approach to the lightweight design of composite sandwich structures.

1. Introduction

Composite sandwich corrugated structures are widely used in aerospace [1], mechanical engineering [2], construction engineering [3], and other fields because of their properties such as lightweight and high strength [4,5]. Some data show that for every 10kg of aircraft lowering, it can save 3925kg of oil fuel [6]. For this reason, the research and manufacture of lighter-quality composite sandwich corrugated structures has become the focus of many scholars' research.

However, the traditional composite sandwich structure design is mainly based on empirical optimization design. Since this method has a certain degree of subjectivity and blindness, it is difficult to adapt to the use of different conditions and special environmental needs, so it has the disadvantage of lower cost and efficiency and a limited degree of lightweight enhancement. Topology optimization is a mathematical modeling process used to optimize material distribution and determine optimal force transfer paths in a design area. This is done while considering constraints, loading conditions, and performance indexes. The goal is to reduce material usage and improve structure quality while meeting design requirements. This text describes the use of the variable density topology optimization method to optimize the design of composite sandwich corrugated structures.

2. Static finite element modeling and analysis of composite sandwich structures

The sandwich corrugated structure can be divided into sine, trapezoid, and triangle. This paper mainly optimizes the trapezoid corrugated sandwich structure. Due to the corrugated plate application scenarios, especially in the building assembly structure, aviation machinery skin, and other scenarios, the force is more complex and diverse. In order to find out the universality of optimization of the corrugated structure, the main purpose of this paper is to improve the corrugated structure's compressive and bending resistance. The compressive and flexural finite element model is established according to the common stress characteristics of sandwich structures.

2.1 The finite element model of the composite sandwich corrugated plate was established

Solidworks software is used to model the structure, and the finite element software is imported for calculation. The structure size is shown in Figure 1, where the thickness is 10.5mm, the long side is 100mm, the vertical corrugated side is 90mm, and the eight-node hexahedral grid is used to divide the structure.

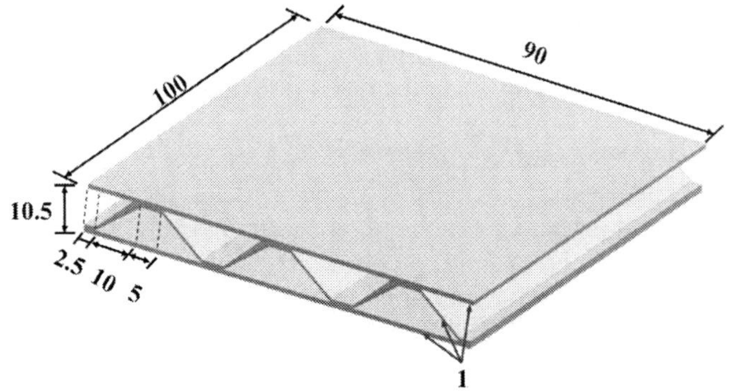

Figure 1. Composite sandwich corrugated plate dimensions (mm).

The optimized structural material is AlSi10Mg, a commonly used metal material for 3D printing, with a density of 2680kg/m^3, elastic modulus of 70.3GPa, and yield strength of 250.6Mpa. The relevant parameters are shown in Table 1. The load is applied in two working conditions. Working Condition 1: the corrugated sandwich plate is under uniaxial compression with fixed constraint at the bottom and the load is 100N. Working Condition 2: The corrugated sandwich plate is subjected to bending surface load, the load width is 10 mm, the length is 95mm, the direction is perpendicular to the corrugated plate ripples, and the simple support strategy is adopted at both ends. The applied load is 100N.

Table 1. AlSi10Mg mechanical parameters.

Densitykg/m^3	Young's modulus /GPa	Yield strength / MPa	Ultimate strength / MPa	Poisson's ratio
2680	70.30	250.61	424.31	0.34

2.2 Static analysis of composite sandwich corrugated plate

According to the typical compression and bending conditions of a corrugated sandwich plate, the deformation and equivalent stress of the specimen are calculated. The calculation results of Working Condition 1 are shown in Figure 2. The maximum displacement is 2.34e^{-5} mm and the maximum equivalent stress is 2.26 MPa. The results of Working Condition 2 are shown in Figure 3, where the maximum displacement is 2.04^{e-3} mm, and the maximum equivalent stress is 30 MPa. The deformation and equivalent stress under a bending load are much greater than that under a uniaxial compression load.

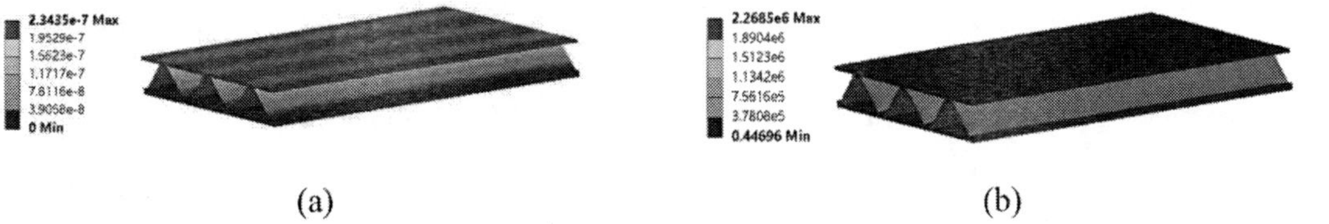

(a) (b)

Figure 2. The calculation results of the composite sandwich corrugated plate under working Condition 1.(a) Displacement cloud map; (b) Equivalent stress cloud map.

Journal of Physics: Conference Series **2808** (2024) 012076 doi:10.1088/1742-6596/2808/1/012076

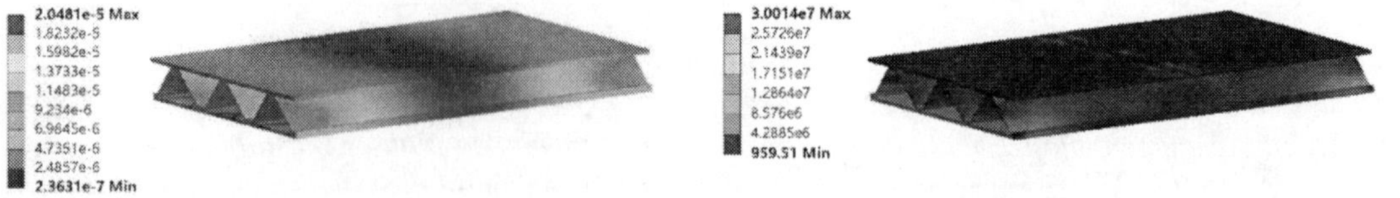

(a) (b)

Figure 3. The calculation results of the composite sandwich corrugated plate under working Condition 2.(a) Displacement cloud map; (b) Equivalent stress cloud map.

3. Variable density method topology optimization mathematical model

3.1 Variable density method interpolation model

The variable density method uses relative density, a type of pseudo-density that reflects the relationship between material density and properties. The method explicitly expresses the correlation between the relative density of the unit and the material properties through the density interpolation model. The relative density ranges from 0 to 1, transforming the topology optimization problem into a continuous variable optimization problem. The density interpolation model penalizes the intermediate densities by introducing a penalty factor, which forces the intermediate densities to approximate the ends, ultimately allowing the optimization to converge to a 0/1 material distribution.

The principle of the density interpolation model is to eliminate the intermediate density by constructing the penalty relation between density and stiffness. Among them, the most representative interpolation model is the Solid Isotropic Microstructures with Penalization (SIMP) method, whose expression is shown in Equation (1).

$$E_i = E(\rho_i) = E_{\min} + \rho_i^p (E_0 - E_{\min}) \quad \rho_i \in [0,1] \tag{1}$$

where: E_i is the elastic modulus of unit i, ρ_i is the relative density of unit i, E_0 is the maximum relative density unit elastic modulus, $E_{\min}$ is the minimum relative density unit elastic modulus, ($E_{\min} < E_i$), and p is the penalty factor.

3.2 Mathematical model of topology optimization

The objective function of topology optimization is the minimum flexibility of the sandwich corrugated structure, and the constraint conditions are external load and volume. The optimization model is shown in Equation (2) :

$$\begin{cases} \text{find: } \boldsymbol{\rho} = \{\rho_1, \rho_2, \rho_3, \cdots, \rho_n\}^T \in \Omega \\ \text{min: } \mathbf{C}(\rho) = \mathbf{F}^T\mathbf{U} = \mathbf{U}^T\mathbf{K}\mathbf{U} \\ \text{s.t: } V = \sum_{i=1}^{n} v_i \rho_i = aV_0 \leqslant V^*, \mathbf{F} = \mathbf{K}\mathbf{U} \\ \quad 0 < \rho_{\min} \leqslant \rho_i \leqslant 1 \end{cases} \tag{2}$$

where: n is the total number of units, Ω is the material relative density optimization domain, $\mathbf{C}$ is the structural flexibility, $\mathbf{F}$ is the load matrix, $\mathbf{U}$ is the node displacement vector matrix, $\mathbf{K}$ is the stiffness matrix, V is the optimized composite corrugated plate volume, v_i is the unit volume of unit i, ρ_i is the relative density of unit i, V_0 is the un-optimized composite corrugated plate volume, a is the volume optimization coefficient, and $\rho_{\min}$ is the lower bound of the relative unit density optimization.

4. Lightweight design of composite sandwich corrugated structure

4.1 Lightweight design requirements
Corrugated sandwich panels are used in practical engineering as mechanical equipment, building facades, aerospace equipment skins, and more. In order to improve the scope of application of the optimized component while ensuring that the topologically optimized component has strong bending pressure, as well as light and high strength characteristics, it should also ensure that its internal structure has continuity and periodicity. This structure should be easy to process and manufacture while meeting the needs of different use conditions, which can be scaled up according to the actual situation and can also be continuously splintered. In addition, the upper and lower panels should have integrity.

4.2 Lightweight design process
Solidworks software is used to model the optimized object, and then the model is imported into Hypermesh software, and topology optimization analysis is carried out with an optistruct finite element analysis solver. Then, the optimization result model is extracted and reconstructed.

4.3 Optimization
The corrugated sandwich structure is meshed using Hypermesh with hexahedral mesh. Since the upper and lower panels of the corrugated sandwich panel should be prepared for integrity, the optimization region is set to be the internal corrugated structure of the composite sandwich panel (see Figure 4). Then, constraints and loads are applied to the corrugated sandwich structure, and three working conditions can be divided according to the different constraints and loads. Working Conditions 1 and 2 are the same as the static load finite element analysis. Working Condition 3 is a coupling working condition, and the loads and constraints of Working Conditions 1 and 2 are applied to the corrugated structure.

Figure 4. Optimized region and non-optimized region.

Working Conditions 1 and 2 are under single load constraints. Two kinds of responses, “compliance index” and “volumef rac”, are set up respectively in the “response” option, indicating that flexibility index and volume fraction are taken as variables for optimization calculation. According to the type of response function, the volume fraction of the optimized retention region is set to 40%.

Working condition 3, with multiple load coupling constraints, “weighted comp” and “volumef rac” responses are adopted, and the volume fraction of the optimized retention area is also 40%. Based on requirements, the loadstep proportional response parameter is set to 1:5.

The objective function of all working conditions is set as minimum flexibility (maximum stiffness), and the penalty function is set as 3 [7]. The SIMP model of the variable density method is used for calculation, and the maximum number of iterations is 80.

4.4 Lightweight result
The final topology optimization results of different working conditions and different parameters are shown in Figure 5, where the red part is retained and the blue part is deleted.

(a) (b) (c)

Figure 5. Topology optimization results under different working conditions. (a) Working Condition 1; (b) Working Condition 2; (c) Working Condition 3.

The result of Working Condition 1 is vertical support, Working Condition 2 is corrugated shape support, Working Condition 3 is coupled condition, and the result is the combination of Working Conditions 1 and 2.

4.5 Composite sandwich corrugated plate structure reconstruction

The topology optimization results of Working Condition 3 lack certain periodicity and regularity, which causes difficulties for actual processing and continuous splicing production, while Working Conditions 1 and 2 show good periodicity and regularity. In addition, in actual engineering, compression or bending is generally only affected, but in order to ensure that the optimized components can be both flexural and compressive at the same time, combined with Working Condition 3, the optimized structural forms of the two working conditions are coupled, and the final results and data are shown in Figure 6.

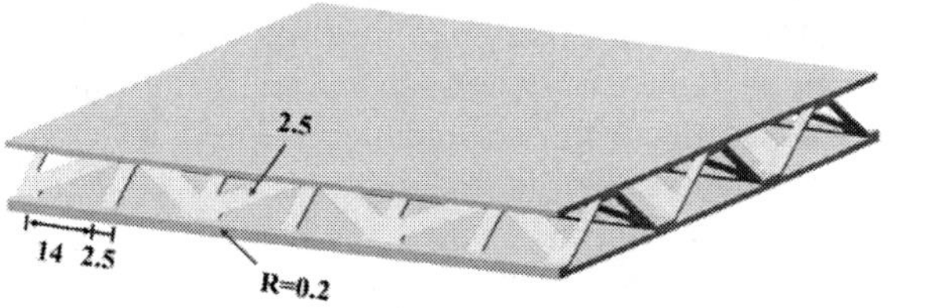

Figure 6. Reconstruct the composite sandwich corrugated plate structure (mm).

5. Finite element analysis of reconstructed composite sandwich corrugated plate

According to the reconstruction results of the composite sandwich structure in the previous section, the mechanical analysis of the reconstructed topological corrugated sandwich structure is carried out by using finite element software. The constraint conditions and loading conditions are the same as those of the non-optimized structural mechanics model, and the displacement and equivalent stress cloud maps of the reconstructed topological corrugated sandwich structure under different working conditions are calculated, as shown in Figures 7-8.

The mass of the non-optimized composite sandwich corrugated plate is 67.68 g, and the mass of the reconstructed composite sandwich corrugated plate is 44.27 g after topology optimization. Under the load constraint of Working Condition 1, the maximum displacement is $6.46e^{-5}$ mm, and the maximum equivalent stress is 4.11 MPa; under the load constraint of Working Condition 2, the maximum displacement is $2.42e^{-3}$ mm, and the maximum equivalent stress is 32.5 MPa. While the change of displacement and equivalent stress is small and does not exceed the yield load, the mass is reduced by 34%.

(a) (b)

Figure 7. The calculation results of the reconstructed structure under Working Condition 1. (a) Displacement cloud map; (b) Equivalent stress cloud map.

Journal of Physics: Conference Series **2808** (2024) 012076 doi:10.1088/1742-6596/2808/1/012076

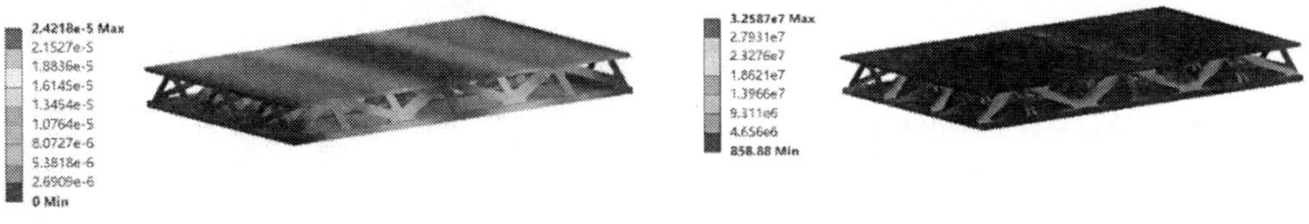

(a) (b)

Figure 8. The calculation results of the reconstructed structure under Working Condition 2. (a) Displacement cloud map; (b) Equivalent stress cloud map.

6. Conclusion

In this paper, the mechanical properties of composite sandwich corrugated structures under different constraints and loads are studied by using the finite element simulation method. Based on the theory of variable density method, Hypermesh software is used to optimize the topology of the structure, and the topology result is reconstructed according to the actual engineering requirements. Finally, the following conclusions were drawn:

(1) The static analysis of the composite sandwich corrugated plate shows that under the same load size, the equivalent stress and deformation of the composite sandwich corrugated plate under bending load are greater than that under uniaxial compression load.

(2) Hypermesh software is used to optimize the topology of the composite sandwich structure. It is found that under uniaxial compression conditions, the corrugated sandwich becomes vertical support, while under bending load conditions, the corrugated sandwich becomes transversely supported in a corrugated shape. Under the coupling of the two loads, the final topology structure is close to the combination of the topology results of the separate action of the two loads.

(3) Combined with the topological optimization results and engineering requirements, the composite sandwich corrugated plate is reconstructed, and the static analysis is carried out. The analysis results show that although the deformation and equivalent stress of the optimized structure increased, it did not exceed the elastic limit, and the structure mass was reduced by 34%, demonstrating an obvious optimization effect.

References

[1] Rejab M, Cantwell W. The mechanical behavior of corrugated-core sandwich panels[J]. Composites Part B: Engineering, 2013, 47: 267-277.

[2] Zhang J, Huang W, Yuan H, et al.Failure behavior of a sandwich beam with GLARE face-sheets and aluminum foam core under three-point bending[J]. Thin-Walled Structures, 2023.

[3] Zu W, Zhimin T. Static loading tests on the composite member of steel plate and foam material [J]. China Civil Engineering Journal, 2006(02): 98-101.

[4] Taghipoor H, Eyvazian A, Musharavati F, et al. Experimental investigation of the three-point bending properties of sandwich beams with polyurethane foam-filled lattice cores [J]. Structures, 2020, 28:424-432

[5] Schaedler T A, Carter W B. Architected cellular materials[J]. Annual Review of Materials Research, 2016, 46: 187-210.

[6] Zhixin ZH, Kun W, Jie W, et al. Dynamic Analysis and Topology Optimization of Aircraft Landing Gear [J]. Machinery Design & Manufacture, 2021, (10): 81-85.

[7] Guofeng ZH, Lei X, Dashuang L, et al. Research on Lightweight Design Method of 3D Printing of Robot Arm Based on Solid Isotropic Microstructures with Penalization [J]. Machinery Design & Manufacture, 2022, (11): 80-84.

Journal of Physics: Conference Series **2808** (2024) 012077 doi:10.1088/1742-6596/2808/1/012077

Research on improving parallel sequence sampling method in ship type optimization

Dongqin Li [1], **Lingfeng Lu** [1,*]

[1]Just School of Shipbuilding and Ocean Engineering, Zhenjiang, Jiangsu, China

*Corresponding author's e-mail: 914102381@qq.com

Abstract: To solve the problem of complex and time-consuming simulation calculations in ship type optimization, the application of approximate model technology based on spatial segmentation sequence sampling is becoming increasingly widespread. Based on the Kriging approximation model and the cross-leaving one error method, a new sampling strategy is proposed during the iterative sampling process. The importance of adding sample subspaces and surrounding subspaces is considered in the design space, and parallel thinking is added to develop an efficient parallel sequence sampling method to improve sample aggregation. The proposed method was applied to optimize ship types, and its effectiveness was verified.

1. Introduction

With the continuous improvement of computer computing power, computer simulation calculation is being applied in more and more fields. This is because computer simulation can replace engineering experiments to a certain extent, saving costs and improving engineering efficiency. Therefore, proxy modeling technology has emerged, establishing a function expression between the response value and the objective function of simulation problems through limited samples. It can explicitly express the relationship between the objective function and design variables of simulation problems and plays an important role in exploring design space and supporting optimization work.

A proxy model must accurately establish the relationship between input and output to replace the actual model. Therefore, when discussing the establishment of a global proxy model, the first thing to consider is the accuracy of the proxy model. After determining the type of proxy model, its accuracy is mainly determined by experimental design methods. The essence of experimental design methods is to ensure that the distribution of samples in the design space can obtain as much model information as possible, thereby constructing an approximate model with high accuracy. At present, experimental design methods are mainly divided into sequential sampling methods and one-time sampling methods. As the name suggests, the one-time sampling method determines the distribution of samples in the design space at once based on the set number of samples. A better one-time sampling method can simultaneously satisfy the filling and projection characteristics of the space. Although the one-time sampling method is relatively simple and efficient, it has two problems: first, it requires determining the number of samples in advance; second is that the proxy model information established without utilizing samples only considers the distribution characteristics of samples in the design space. In our optimization problem of establishing proxy models, the complexity of different design space regions varies. In areas with high nonlinearity, the distribution density of samples needs to be increased, while in areas with low nonlinearity, the distribution density of samples required is relatively low.

The sequence sampling method uses the information of the proxy model established from the basic samples to determine the selection of subsequent samples. In recent years, more and more research results have shown that the accuracy of the proxy model constructed by the sequence sampling method is higher than that of the one-time sampling method. Cross validation error (CV) is a widely used error criterion among the sampling criteria used in sequence sampling methods. Aute et al. [1] proposed a sequence sampling method called SFCVT (space filling cross validation trade off), which considers distance factors while searching for the sample with the highest predicted cross validation error as the new sample. Li et al. [2] proposed a cumulative error sampling criterion based on cross validation error, which selects new sample points by predicting the cumulative error at unknown samples. Viana [3] proposed sampling at the point where Kriging predicted the maximum variance. The above methods have strategies for adding new samples and ultimately improving the accuracy of the proxy model. Surjanovic and Bingham [4] proposed an adaptive sampling method that determines the addition of new samples based on the product of the RBF curvature and distance criteria of the surrogate model. Crombecq et al. [5] proposed a sequence sampling method that estimates the gradient information at the sample location to determine the nonlinear features of the nearby area. The design space is divided into small polygons using the Voronoi diagram method centered on each sample. They determined the target polygon that needs to be added points through gradient and polygon size and then selected new samples in the target polygon according to spatial filling criteria.

From this perspective, sequence sampling methods [6] require optimization algorithms in specific point spaces to find new sample points that meet the "optimal criteria". Inspired by previous research results, this paper proposes a new sampling strategy based on the Kriging approximation model [7] and the cross-leaving one error method during the iterative sampling process. It considers the importance of adding new and surrounding sample subspaces. It incorporates parallel thinking to develop an efficient parallel sequence sampling method to improve sample aggregation, which is applied to ship-type optimization in.

2. Improved parallel sequence sampling method

2.1. Principle of sequence sampling

An effective sequence sampling method [8] should comprehensively consider the strategies of global exploration and local search. Global exploration aims to increase the number of samples in sparsely distributed sample areas, while local search focuses on areas with higher nonlinearity. These two strategies jointly promote the improvement of the accuracy of the proxy model. We adopt a parallel sequence sampling method based on spatial segmentation and use Voronoi diagrams to divide the design space into multiple sub-regions, each containing a sample. The size of sub-regions reflects the sparsity of sample distribution. At the same time, the nonlinearity of the subregion is evaluated by the cross-validation error (CVE) of the samples, and a higher CVE indicates that the sample has a significant impact on model accuracy, reflecting the high nonlinearity of the region. In this way, we select new samples in the target subspace.

2.2. Design space segmentation based on voronoi diagram

This study used Voronoi plots to identify key sampling areas that require additional sample points. Voronoi diagram, also known as the Tyson polygon, is an effective spatial segmentation method proposed by Russian mathematician Voronoi. The efficiency of this method in spatial partitioning makes it widely used in various fields. The core principle is that in an n-dimensional design space, given m sample points P={P1, P2, ..., Pm} (i.e., control points), the design space is divided into m regions C={c1, c2, ..., cm} based on the distance between these points and other points in the space. Each Tyson polygon ci revolves around the control point Pi, ensuring that the distance from any point within ci to Pi is less than the distance to other control points. The advantage of this segmentation method is that it is based on the topological structure between sub-sample points for partitioning and does not rely on the analysis data of sample points. Modifications to any control point can be made within the local range of its Tyson polygon, reducing processing costs.

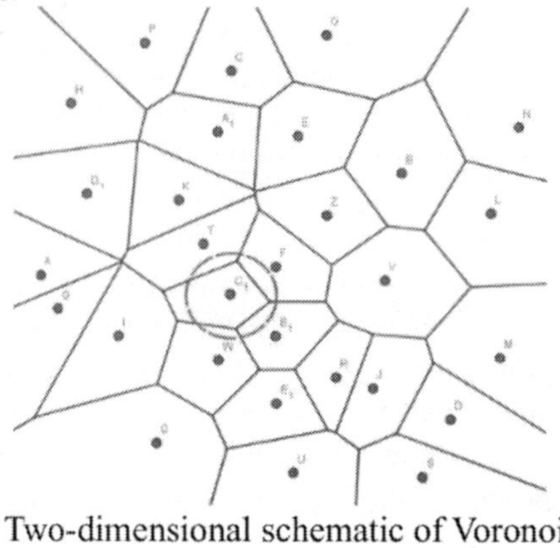

Figure 1. Two-dimensional schematic of Voronoi diagram.

2.3. Positioning target subspace

After establishing the Voronoi plot of the initial sample points, the next step is determining the key sampling areas. Most adaptive sequence sampling methods detect the error of unknown points by constructing an approximate prediction error model. In our method, the prediction error of the control point is taken as the error representing the region, taking advantage of the close relative distance between the points within the Tyson polygon and their control points. The areas with large prediction errors indicate a significant deviation between the approximate model and the true function, so adding more sample points here is necessary. We select the polygon with the highest error as the focus sampling area and add new sample points. This process is repeated, each iteration selecting a key sampling area and adding a sample point. The prediction error of control points is determined by leaving one cross validation (LOOCV). This article proposes an efficient parallel sequence sampling method that optimizes sample distribution by setting a prediction error threshold and simultaneously determining multiple key sampling areas in one iteration. The relative prediction error of sample point Pi is calculated by the Formula (1):

$$e_i = \frac{\left|f(p_i) - f^{-(i)}(pi)\right|}{\left|f(p_i)\right|} \quad (1)$$

$$\begin{aligned} &\max \quad \left\|p - p_{im}\right\| \\ &s.t. h_j(p) \le 0, j = 1, \ldots, m \end{aligned} \quad (2)$$

In the formula, $f(p_i)$ is the actual response calculated by numerical simulation of the sample and $f^{(i)}(p_i)$ is the predicted response value at the sample point, which is predicted by an approximate model composed of the sample point set after removing p_i.

Due to all points falling in the key sampling area being farthest from the control point, adding new samples is simplified as finding the point farthest from the control point in the key sampling area. Because the Tyson polygon is a set of polygons (polyhedra in high-dimensional space), as shown in Figure 1, the adjustment of the new sample can be transformed into solving nonlinear programming problems, and its mathematical formula is Formula (2):

In the formula, P_{im} is the control point of the key sampling area, and $h_j(p) \le 0$ forms the boundary of the key sampling area. The selected points will be used as the next sample point to be evaluated and added to the sample point set to construct a new approximate model.

2.4. Improve the specific process of parallel sequence sampling

(1) We generate initial sampling points and use engineering simulation tools to calculate the response. The LHS method is used to establish the initial samples to meet the basic spatial filling characteristics of the samples.

(2) On the basis of the initial sample, the entire design space is divided into a set of Tyson polygons using the Voronoid diagram. After selecting the type of approximate model, we establish an approximate

model based on the initial sample and use the LOOCV method to determine the important sampling area.

(3) We search for new samples in important sampling areas by solving formulas, simultaneously considering the importance of the surrounding subspace and incorporating parallel strategies to improve sample aggregation: (1) We set a sample error threshold, where the subregions represented by samples exceeding the threshold are defined as more important regions, arranged in order of size as (m1, m2, ... mm); (2) We consider existing computing resources and time costs, incorporate parallel strategies, and freely set the number of new sample points n for each iteration (n<m); (3) We take the sub-regions represented by the top n samples in the error ranking as the newly added sample regions in this iteration, with each new sample point being (n1, n2, ... nn); (4) We select a new sample point n within each key exploration area, and the strategy is as follows: 1. First, we select a point n1 that satisfies the maximum Euclidean distance from the existing sample in the sub-area. Similarly, we select n2. If m1 is adjacent to the m2 subregion, then the n2 point must satisfy Ln1n2>0.5*Lm1m2; 3. Similarly, we select nn. If the sub-region of mn is adjacent to the sub-region of the newly added sample in this iteration, Lnnni>0.5*Lmnmi must be satisfied.

(4) Numerical simulation methods are used to calculate the response values of new samples, and they are added to the sample set to update the original approximate model.

(5) We repeat Steps (2) to (4) until the stop criteria are met. We evaluate the accuracy of the approximate model using the coefficient of determination (R^2), expressed as Formula (3):

$$R^2 = 1 - \frac{\sum_{i=1}^{n}\left[f(p_i) - \hat{f}(p_i)\right]^2}{\sum_{i=1}^{n}\left[f(p_i) - \bar{f}(p_i)\right]^2} \tag{3}$$

In the formula, n is the number of verification points; $f(p_i)$ is the actual value simulated at the i-th validation point; $\hat{f}(p_i)$ is the predicted value calculated by the proxy model for the i-th validation point; $\bar{f}(p_i)$ is the average of the actual values at the verification point. The larger R^2, the more accurate the approximate model created. In general, in complex engineering problems, when it is greater than 0.9, it can be considered that the obtained approximate model meets the accuracy requirements.

The above steps show that the creation of approximate models based on improved sampling methods is an iterative process. This article completes the iterative process using Matlab and Isight platforms to achieve approximate model creation, including initial approximate model generation, approximate model updating, and error analysis.

3. Function examples

This section verifies the effectiveness of the improved parallel sequence sampling method based on spatial segmentation through function examples.

The arithmetic expression of the function is as follows:

$$f(x,y) = 3(1-x^2)e^{-x^2-(y+1)^2} - 10\left(\frac{x}{5} - x^3 - y^5\right)e^{-x^2-y^2} - \frac{1}{3}e^{-(x+1)^2-y^2} \tag{4}$$

The three-dimensional perspective view of the function and the approximate three-dimensional model are shown in Figure 2 and Figure 3. We choose the kriging model as the approximate model type to fit the function. We generate 20 initial sample points using the LHS method and create an initial Voronoid map. We randomly generate 200 sample points as validation points to verify the accuracy of the approximate model. An improved sequence sampling method was used to increase sample points gradually and update the approximate model. Due to the simplicity of approximate fitting of functions compared to practical engineering problems, the iteration steps were appropriately increased to observe the process of increasing samples better step by step. When R^2 >0.95, the iteration stopped.

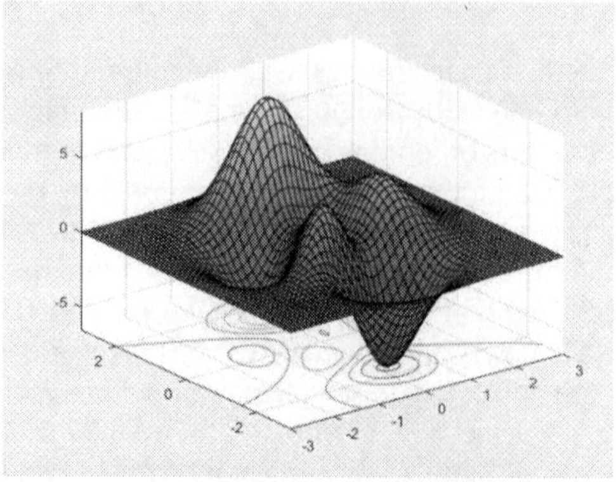

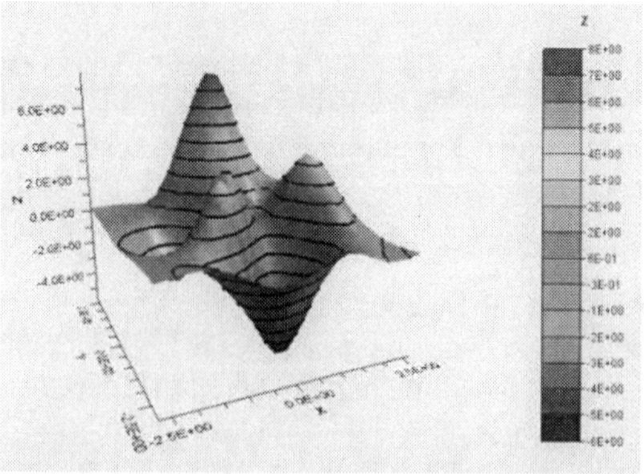

Figure 2. 3D perspective view of functions. **Figure 3.** Function approximation model 3D diagram.

We let N be the number of new sample points added during each iteration, and the improved sequence sampling method uses different N-value parallel strategies to construct a convergence curve for approximate model accuracy, as shown in the figure. The results show that the smaller the N value is, the more iterations it takes to reach convergence and the smaller the overall sample size is. On the contrary, the larger the N value is, the fewer iterations are required to reach convergence, and the larger the overall sample size is. Therefore, when adopting a parallel strategy, the selection of the number of new sample points N for each iteration needs to consider the existing computing resources and computational costs: In the case of sufficient computing resources, the N value can be appropriately increased, and in the case of reducing the number of iterations, the overall number of sample points increases, but the clustering of sample points is improved, and the accuracy of the approximate model is relatively high.

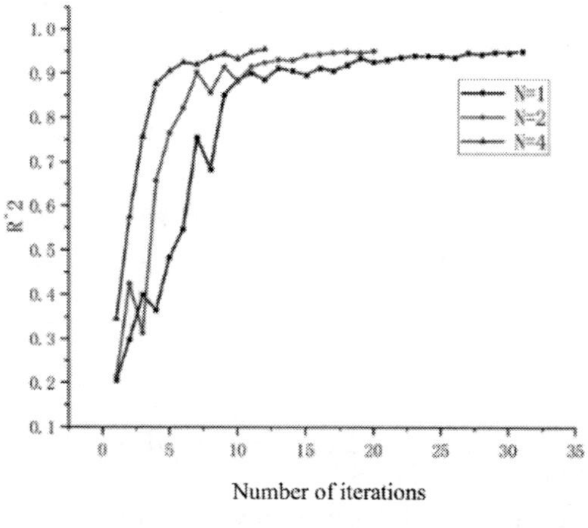

Figure 4. Convergence curves of two-dimensional function examples with different parallel strategies

After adding 4 sample points (a total of 24 sample points), the accuracy reached 0.9, and after adding 12 sample points (a total of 32 sample points), the accuracy reached 0.95. Figure 5 shows the distribution of sample points and their Voronoid plot during sampling. Based on the peak distribution of the function in the figure, it can be seen from the distribution of sample points in Figure 5 that the initial sample points are similar to a uniform distribution. After using an improved sequence sampling method to increase sample points, more sampling points are concentrated in the peak areas with significant changes in function values, and there are relatively few sampling points in flat areas.

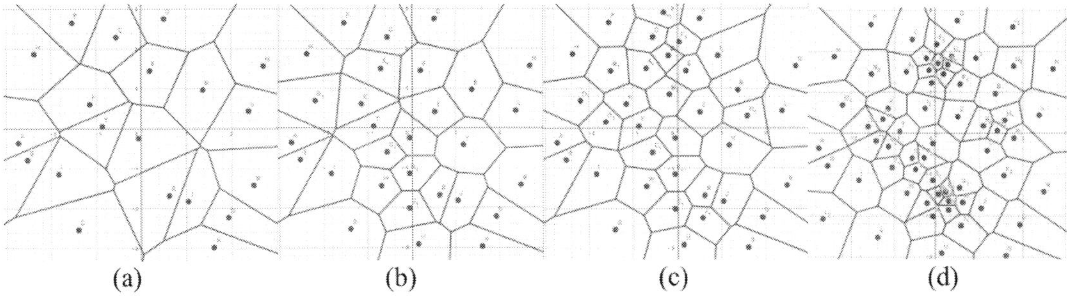

Figure 5. Two-dimensional example (a) 20 initial samples, (b) in the early stage of iteration, (c) in the middle stage of iteration, (d) in the late stage of iteration, sample distribution, and Voronoi plot.

4. Engineering application examples

4.1. Optimization problem

The improved parallel sequence sampling method was applied to construct an approximate model technology in the optimization design of ship types. Based on the static water resistance performance indicators of ships, FFD geometric deformation technology, fine CFD technology, and genetic algorithm NSGA-II were combined to construct a ship-type design optimization framework based on the Isight platform, and the optimal ship type scheme was solved. The three-dimensional model of the ship is shown in Figure 6, and Table 1 shows its main parameters:

Table 1. Main dimensions of real ship.

Lpp/m	B/m	D/m	T/m	Cb
104.12	17.6	9	5.2	0.756

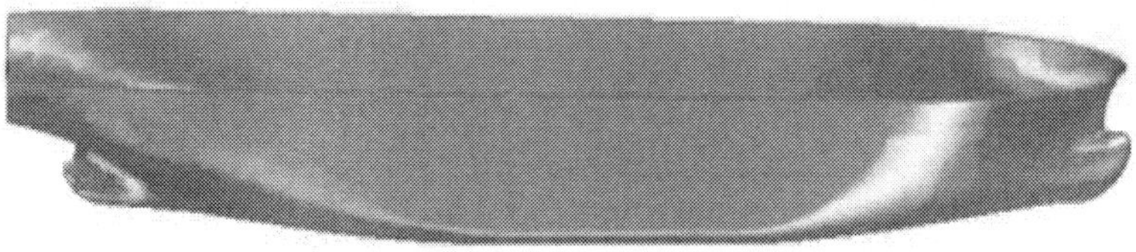

Figure 6. 3D model of ship.

4.2. Optimization results

Based on the number of design variables, we initialize the approximate model and set 50 samples. Therefore, an initial sample set of 50 data was created through SOBOL sampling to maintain the spatial filling characteristics of the samples. The LOOCV method is used for error analysis when applying improved sampling methods to maximize the utilization of existing samples. We remove one sample point from the sample set each time, recalculate the approximation coefficient for each deleted point, and compare the accuracy and approximate output value. Based on 50 samples, an initial approximate model was established, and an improved parallel sequence sampling method was used to add 2 sample points per iteration. After four iterations, a total of 8 new sample points were added. The approximate coefficient R^2 of the constructed model exceeded 0.9, meeting the precision requirements of engineering optimization approximate models.

The output variable of the sample (static water resistance) is implemented using the Computational Fluid Dynamics (CFD) program STAR-CCM. Finally, a ship-type design optimization framework was constructed based on the ISIHT platform, and the genetic algorithm NSGA-II was used to obtain the ship-type scheme with the best overall performance.

Table 2. Comparison of optimization results.

Target	Original	Opti01	Difference/%
Resistance Rt/N	10.9950	10.2201	-7.04
Displacement∇/m^3	7207.7	7115.2	-1.28

Table 2 shows that compared to the original ship, the displacement of the optimized plan has decreased by 1.28%. In Figure 7, the bow and stern lines have become "thinner". The bow line has also undergone significant changes, with a slight increase in the length of the bulbous bow, indicating the direction of optimized design.

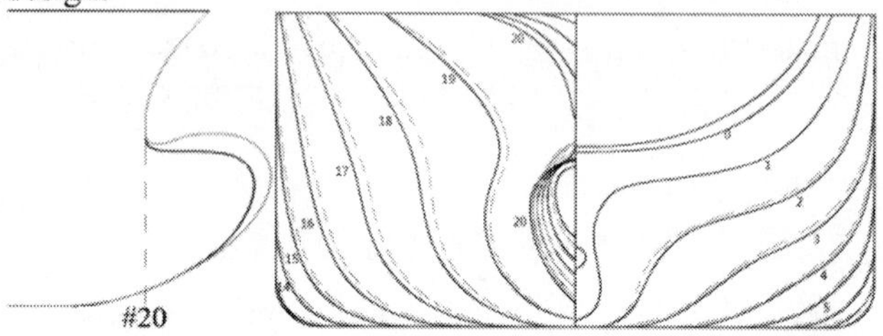

Figure 7. Comparison of the line type of transport ship lines.

The optimization scheme reduced the static water resistance by 7% and successfully improved resistance performance based on little change in drainage volume.

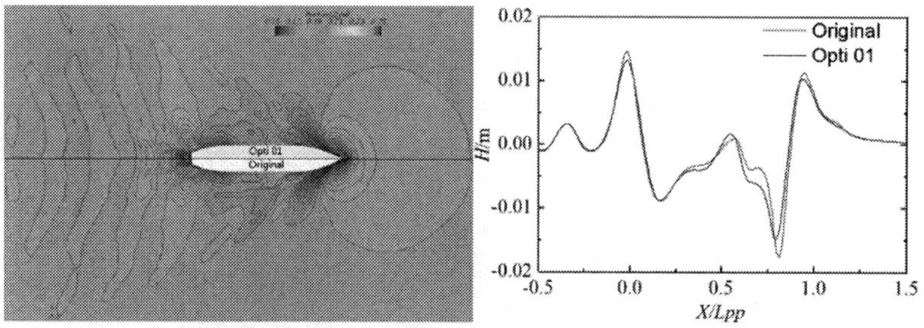

Figure 8. Comparison of waveform and wave cut.

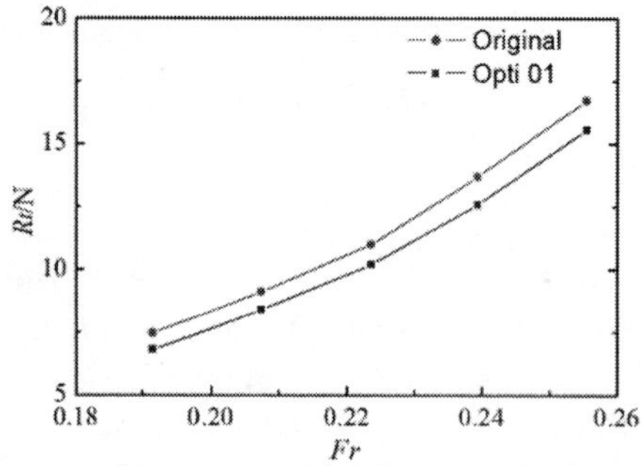

Figure 9. Comparison of static water resistance.

Figure 8 compares free surface wave shapes and longitudinal wave amplitudes at Y=0.12 Lpp between the optimized ship scheme and the original ship in still water. It can be seen from the figure that the wave amplitudes around the optimized ship hull are smaller than those of the original ship. Figure 9 shows the comparison of the static water resistance calculation results between the optimized ship and the original ship at five speeds. It can be seen that the static water resistance performance of the optimized ship is better than that of the original ship.

5. Conclusion

The sequence sampling method proposed in this article iteratively selects new samples step by step throughout the entire design space, ultimately obtaining an approximate model that meets the accuracy requirements. A new sampling strategy is proposed during the iterative sampling process based on the Kriging approximation model and the LOOCV method to balance the sampling efficiency and sample uniformity of adaptive sampling methods. The importance of the surrounding subspace is considered, and parallel thinking is added to develop an efficient parallel sequence sampling method to improve sample aggregation. The proposed method was applied to optimize ship types, and its effectiveness was verified. In practical applications, it has been found that the global approximation model constructed by this method has higher accuracy under the same sample size. The approximate model creation process only requires the initial sample set and the currently created approximate model, which is simple and easy to implement. It is independent of the choice of approximate model type, with strong universality.

References

[1] AUTE V, SALEH K, ABDELAZIZ O, et al. Cross validation based single response adaptive design of experiments for Kriging metamodeling of deterministic computer simulations [J]. Structural and Multidisciplinary Optimization, 2013, 48(3): 581-605.
[2] LI G, AUTE V, AZARM S. An accumulative error based adaptive design of experiments for offline metamodeling [J]. Structural and Multidisciplinary Optimization, 2010, 40: 137-155.
[3] VIANA F A C. SURROGATES Toolbox User's Guide Version 3.0 [EB/OL]. [2021-12-25].
[4] SURJANOVIC S, BINGHAM D. Virtual library of simulation experiments: test functions and datasets [EB/OL]. [2021-12-25].
[5] CROMBECQ K, LAERMANS E, DHAENE T. Efficient space-filling and non-collapsing sequential design strategies for simulation-based modeling. European Journal of Operational Research, 2011, 214(3): 683-696.
[6] SURJANOVIC S, BINGHAM D. Virtual library of simulation experiments: test functions and datasets [EB/OL]. [2021-12-25].
[7] Li Zheng Liang, Wang Tao. A surrogate model sequence sampling method based on k-fold cross validation [J]. Journal of Computational Mechanics, 2022, 39 (2): 244-249.
[8] Hao Jia, et al. Optimization of Sequential Experimental Scheme for Knowledge Driven SSVEP Stimulation of Military Interface [J]. 2021, 49 (7): 113-119.

Structural optimization and drag reduction research of biomimetic hydrofoils inspired by pufferfish

Kairan Wei[1], Lei Zhao[1], Dongpo Zhu[1] and Guizhong Tian[1*]

[1]College of Mechanical Engineering, Jiangsu University of Science and Technology, Zhenjiang, 212100, China

[*]Corresponding author's e-mail: justtgz@163.com

Abstract: The underwater glider will be subjected to fluid resistance during operation, which greatly reduces its endurance and fuel conversion efficiency. Therefore, reducing its surface fluid resistance is an effective way to improve its working capacity and reduce energy consumption. With the development of bionics, people begin to learn various functional surfaces. In this paper, inspired by pufferfish, the tiny spinal protrusions of pufferfish are simplified into microstructure and arranged on NACA0012 airfoil for research. The performance of the bionic microstructure is affected by many factors. The orthogonal test is designed to optimize its structure to evaluate the influence of different heights, bottom radius, arrangement position, and arrangement spacing on the resistance of the hydrofoil. The SST k-ω turbulence model is used for numerical simulation. The bionic hydrofoil has a drag reduction effect of about 6.05 %, which means that the bionic microstructure will be further optimized in the future to be applied to other airfoils.

1. Introduction

With the increasingly severe energy crisis, how to reduce the resistance of ships and underwater vehicles, improve navigation speed, and reduce energy consumption has always been the focus of research. Through natural selection and evolution, organisms adapt to the natural and external environment with the best form and structure, which has many inspirations for the design of artificial structures [1].

Biomimetics is used for human design and manufacturing by imitating the characteristics and functions of natural organisms. In recent years, biomimetic technology has been fully developed. Specifically, many different non-smooth surfaces evolved by organisms have been proven to have good drag-reduction effects. For example, the flexible skin of dolphins reduces drag, the superhydrophobic surface of lotus leaves reduces drag, and the surface structure of shark skin reduces drag. Due to the self-adaptability of the skin, the dolphin can make the water flow stably through itself, delay the turning point of turbulence, and reduce the resistance [2]. The surface of lotus leaves and other plants has the function of self-cleaning, because there are nano-scale micro-protruding structures under the waxy layer, which makes the surface have a large water contact angle and reduces the water resistance [3]. The shield scale structure on the surface of shark skin can suppress and delay the occurrence of turbulence, block the transverse vortex, and reduce the drag [4]

With the deepening of observation, compared with the body surface morphology of complex geometric shapes in the biological world, it is not enough to only use the plate as the object of experimental research. As a common form of curved surface structure, airfoil is widely used in ships, aviation, construction, and other fields. At present, two main strategies are proposed to optimize airfoil performance. One is the active flow control method, which is used to reduce drag and increase lift by adding auxiliary power to the system. Another method is that the passive flow control strategy based on geometric modification can change the lift and drag. They reduce the drag and increase the lift by passively changing the flow, thus beneficially affecting the pressure gradient along the wing [5]. Although the active method is usually better than the passive method, the passive technology is not affected by any type of external energy demand and is easier to implement. For example, Gu et al. analyzed the parametric structure of the mantis shrimp abdomen segment to design the airfoil and achieve drag reduction [6].

In this article, we focus on the hydrofoils and experimentally investigate how the 3D model of the simulated pufferfish body spines arranged on the suction surface passively changes fluid flow. Although previous studies have mostly focused on the effect of simulated pufferfish body spines on flat plate drag reduction, we have demonstrated here that body spines can also reduce the resistance of curved structures. We have studied the mechanisms that lead to this behavior and obtained some interesting insights.

2. Method

The bionic structure of this paper is inspired by the adult pufferfish in the Yangtze River Basin of China. The overall shape is similar to the airfoil, and its skin is covered with a unique spine arrangement, as shown in Figure 1. Approximately 5/8 of the epidermal spinous process structure of pufferfish is within the epidermis, only the tip part is exposed outside the epidermis, and the remaining 3/8 is located within the dermis as a stump structure. These spines can not only protect themselves but also have been proven to have good drag reduction ability [7-8].

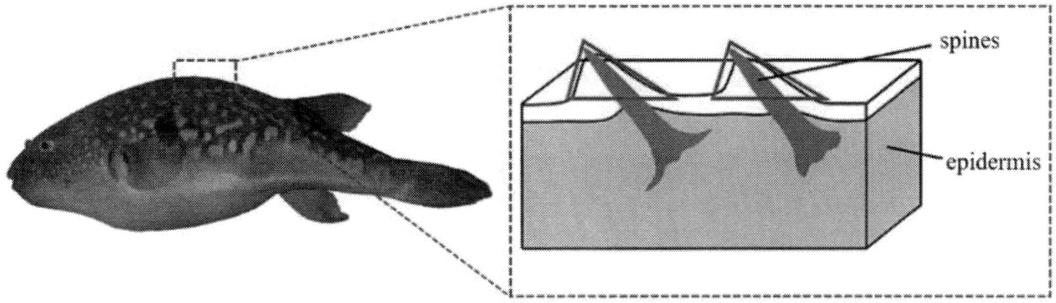

Figure 1. The spinal structure beneath the epidermis.

2.1 Geometry description

As a common structural form in the fields of ships, aviation, and so on, airfoil has been widely used. The NACA airfoil is a series of airfoils developed by the National Aeronautical Advisory Committee (NACA). The NACA0012 selected in this paper is one of the most commonly used typical airfoils. It is an airfoil with a camber of 0 and a relative thickness of 12%. The inspiration for the bionic structure comes from the spinous protrusions on the surface of the pufferfish. Through the morphological characteristics observed by the microscope, it is simplified into a conical component, as shown in Figure 2.

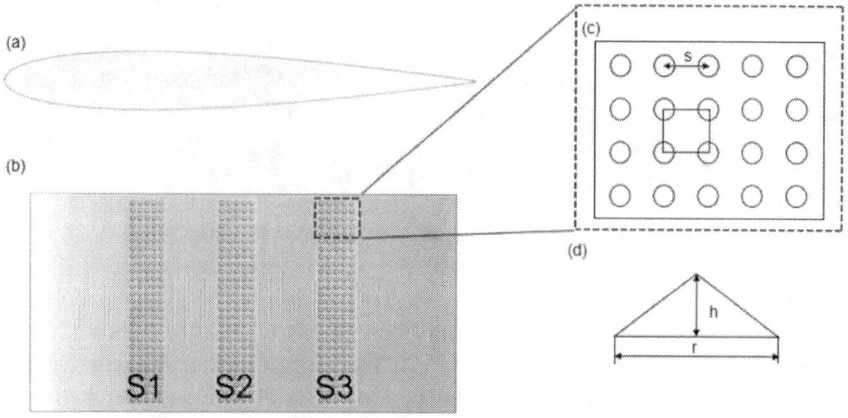

Figure 2. Schematic diagram of bionic hydrofoil.
(a) NACA0012 airfoil (b) bionic hydrofoil top view (c) bionic hydrofoil local top view
(d) bionic structure

2.2 Orthogonal test design

Different from the flat plate, there is a reverse pressure gradient on the NACA0012 with curvature, which means that the influence of body spines on the overall performance will be subject to more factors. Therefore, the orthogonal test is used to analyze the influence degree of multiple factors. In this experiment, four factors are selected, namely: h-body thorn height (A), r-body thorn radius (B), s-adjacent body thorn spacing (C), and body thorn arrangement position (D). Table 1 shows the horizontal values taken for each factor. According to the microscopic bionic structure height parameters observed by the microscope, combined with previous studies, three representative heights of 0.3 mm, 0.4 mm, and 0.5 mm are selected in the overall range. The body thorn radius r is 1 mm, 1.25 mm, and 1.5 mm respectively, and s=3.5 r is obtained according to the average distance of the back thorn of the puffer. The factor D corresponds to the front layout (S1), the middle layout (S2), and the rear layout (S3).

Table 1. Orthogonal test level factor table.

Level	Factors			
	A (h/mm)	B (r/mm)	*C (s*/mm)	D
1	0.3	1	3.5	S1
2	0.4	1.25	4	S2
3	0.5	1.5	4.5	S3

The orthogonal test is an efficient, fast, and economical experimental design method. The orthogonal test can extract representative, uniformly dispersed, and neat test points from the whole experiment for analysis, which can greatly improve work efficiency. The experiment involves four factors and three levels. Without considering the interactive operation of the experiment, the orthogonal table $L_9(3^4)$ is selected. Table 2 Is the corresponding orthogonal test scheme table.

Table 2. Orthogonal Test Design.

NO.	Factors			
	h/mm	r/mm	*s*/mm	Location
1	0.3	1	3.5	S1

2	0.3	1.25	4	S2
3	0.3	2.5	4.5	S3
4	0.4	1	4	S3
5	0.4	1.25	4.5	S1
6	0.4	2.5	3.5	S2
7	0.5	1	4.5	S2
8	0.5	1.25	3.5	S3
9	0.5	2.5	4	S1

2.3 Numerical methodology

2.3.1 Governing equations and numerical methods. This article chooses Ansys Fluent as the solver for CFD. The core and foundation of numerical calculation are to solve the RANS equation for torque conservation and the time-averaged mass conservation equation. The mass continuity equation and Navier Stokes equation are based on incompressible fluids. In various main directions, the mass conservation equation can be expressed as follows:

$$\frac{\partial}{\partial x_i}(\rho u_i) = 0$$

In the equation, ρ is the density of the fluid, and u_i is the component of the velocity in each principal direction (x, y, z). In addition, the momentum conservation equation in each main direction can be written as:

$$\frac{\partial}{\partial t}(\rho u_i) + \frac{\partial}{\partial x_j}(\rho u_i u_j) = -\frac{\partial P}{\partial x_i} + \frac{\partial \tau_{ij}}{\partial x_j} + \rho g_i + F_i$$

where F_i represents the overall force, g_i represents the gravitational acceleration, and τ_{ij} is the tensor of the Reynolds stress and can be expressed by the following equation:

$$\tau_{ij} = \mu\left(\frac{\partial u_i}{\partial x_j} + \frac{\partial u_j}{\partial x_i}\right) - \frac{2}{3}\mu\frac{\partial u_i}{\partial x_i}\delta_{ij}$$

where δ_{ij} is the Kronecker symbol. It can be unified only when i and j are equal, otherwise, it is 0.

In order to better simulate the turbulent boundary layer and other conditions, this paper adopts the k-ω shear stress transport (SST) model. The turbulence model is a two-equation eddy viscosity model proposed by Menter, which ensures the stability and accuracy of the k-ω model near the wall area equation. It also has the independence of free flow away from the wall area [9]. Compared with other turbulence models, especially in the near-wall layer, the capture and simulation of hydrodynamic characteristics are more accurate. At the same time, the model has special independence in boundary layer characteristics and can be effectively used in high Reynolds number environments

2.3.2 Computational domain and boundary conditions. In this paper, the cube around the wing model is used as the three-dimensional calculation domain. As shown in Figure 3, the calculation domain takes the flow velocity direction as the positive direction of the X axis, the height direction of the three-dimensional airfoil as the Z direction, and the direction of the suction surface as the positive direction of the y direction to establish coordinates. The length of the calculation domain is set to 13 times the chord length, in which the distance from the center of the wing to the entrance is three times the chord length, the distance to the exit is 10 times the chord length, the width is set to 8 times the chord length, and the height is set to 8 times the chord length. The computational domain is large enough to avoid the influence of external factors such as countercurrent and induced pressure.

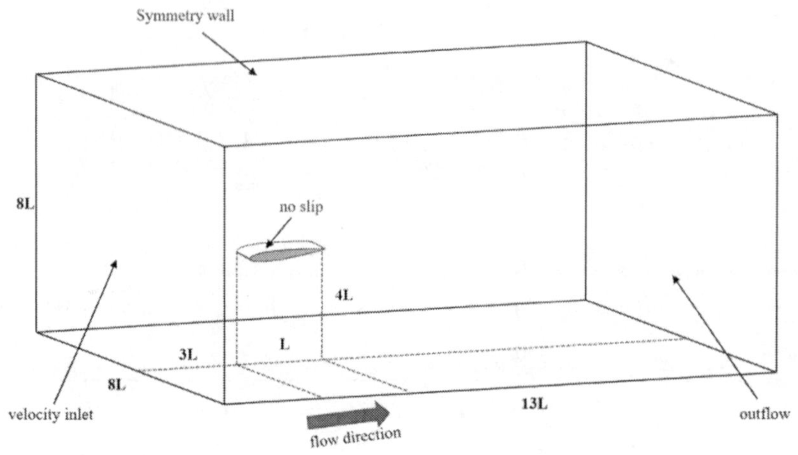

Figure 3. Computational domain and boundary conditions.

The water in the calculation domain is set to 20°C, the density is 998.2 kg/m^3, the dynamic viscosity μ=1.002 g·m^{-1}·s^{-1}, Velocity U=0.5 m/s, and the pressure-velocity coupling adopts the semi-implicit SIMPLE algorithm. In the control equation, all diffusion terms are discretized by the second-order scheme, and the absolute conversion standard of the residual convergence process of the equation is set to 10^{-6} to obtain acceptable accuracy.

2.3.3 Grid independence study. In order to ensure the accuracy of the calculation results, four different resolution grids are generated for the bionic wing, and the corresponding number of grids is 858664, 1173558, 1928079, and 3704309. The specific convergence is shown in Table 3. As the number of grids increases, the total resistance (C_D) and the rate of change decrease continuously. When the number of grids is 3704309, the rate of change of resistance relative to the previous grid is less than 0.5%.

Table 3. Convergence of the Mesh.

Grid	Mesh Resolution	Grid Quality	C_D	ΔC_D
1	Coarse	858664	0.05622	/
2	Medium	1173558	0.05504	2.10%
3	Fine	1928079	0.05426	1.41%
4	Finer	3704309	0.05418	0.15%

Therefore, in order to improve the efficiency of numerical simulation while ensuring accuracy, Grid 3 is finally selected as the unstructured grid division scheme in the research work. The height of the first layer grid is constrained by the value, the growth rate is 1.2, the body spine structure is encrypted, and the overall grid situation is shown in Figure 4.

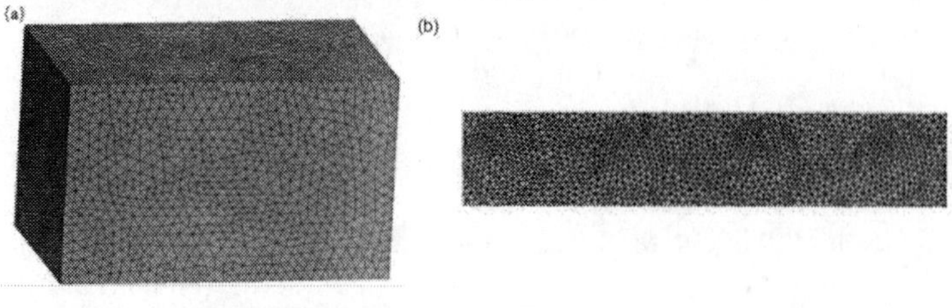

Figure 4. Whole and local grid model of Bionic hydrofoil.

3. Results and discussion

3.1 Analysis of orthogonal test results
Table 4 shows the resistance simulation results in the orthogonal test. Nine groups of test groups have drag reduction effects, of which Groups 4, 5, and 7 have the most obvious effects. The overall drag reduction rate ranges from 4.00% to 6.05%. When the height of the drag reduction element h is 0.4 mm, the radius of the bottom surface r is 1 mm, and the spacing s is 4 mm. The drag reduction effect is the best when arranged in the rear area.

Table 4. Orthogonal experiment simulation results.

NO	Smooth Hydrofoil Resistance	Bionic Hydrofoil Resistance	Drag Reduction Rate
1		0.05426	4.00%
2		0.05418	4.14%
3		0.05371	4.97%
4		0.05310	6.05%
5	0.05652	0.05334	5.63%
6		0.05357	5.22%
7		0.05341	5.50%
8		0.05379	4.83%
9		0.05416	4.18%

3.2 Analysis of orthogonal test results
Table 5 shows the intuitive analysis of the total resistance of the bionic wing. Among the four influencing factors, the height h has the greatest influence on the drag reduction effect, and the range reaches 0.01263. In addition, the extreme difference between the spacing and the arrangement position is 0.00683, which is second. The radius r of the bottom surface has little effect on the drag reduction effect, which is 0.00393. According to the influence of shape factors on the overall drag reduction effect, the primary and secondary order of each factor is A>C>D>B.

Table 5. Visual analysis table of global factor level.

Factors	A	B	C	D
K1	0.04370	0.05183	0.04683	0.04603
K2	0.05633	0.04867	0.04790	0.04953
K3	0.04837	0.04790	0.05367	0.05283
Range	0.01263	0.00393	0.00683	0.00680
Sensitivity	A>C>D>B			

3.3 Influence of bionic structure on flow field characteristics of hydrofoil

3.3.1 Velocity field. To compare the influence of bionic hydrofoil on the velocity field, the fourth group of experiments with better drag reduction effect and smooth hydrofoil is selected for comparison. Firstly, the velocity field of the bionic hydrofoil is analyzed. As shown in Figure 4, when the velocity U is 0.5 m/s, the bionic structure has no great influence on the velocity of the front region of the hydrofoil. However in the tail region, the high-speed flow area in the hydrofoil wake area increases significantly, and the boundary layer becomes thicker, which is equivalent to increasing the thickness of the viscous bottom layer of the boundary layer, resulting in a decrease in friction resistance. Compared with the smooth surface, there is a certain static flow area between the conical microstructures, which increases the thickness of the viscous bottom layer and reduces the velocity gradient of the fluid near the wall, thus reducing the friction resistance.

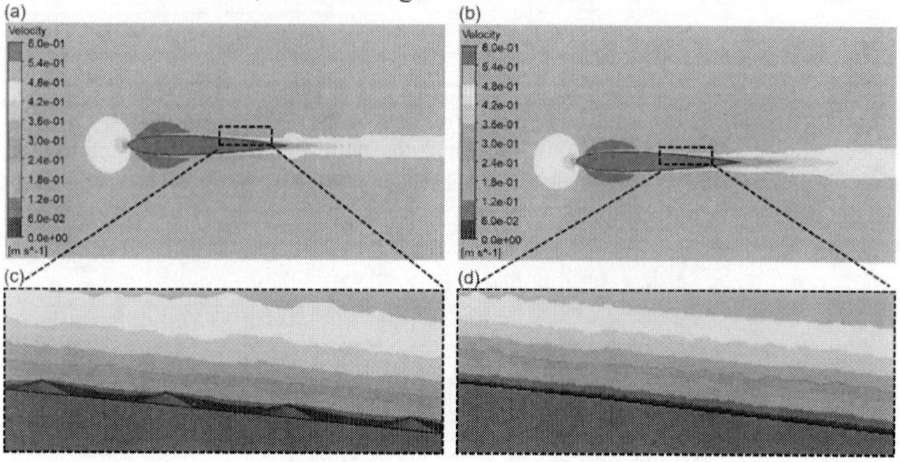

Figure 5. Velocity nephogram (a) bionic hydrofoil (b) smooth hydrofoil (c) bionic microstructure amplification region (d) local amplification region of smooth hydrofoil.

3.3.2 Wall friction coefficient. Figure 6 shows the wall friction coefficient of smooth hydrofoil and bionic hydrofoil, which can reflect the surface shear stress and surface friction. It can be seen from the diagram that the bionic structure has little effect on the wall friction coefficient of the front of the hydrofoil, but after touching the bionic structure, the wall friction coefficient decreases significantly compared with the smooth model. This shows that the bionic structure can reduce the surface friction of the hydrofoil and has a great influence on reducing the overall resistance.

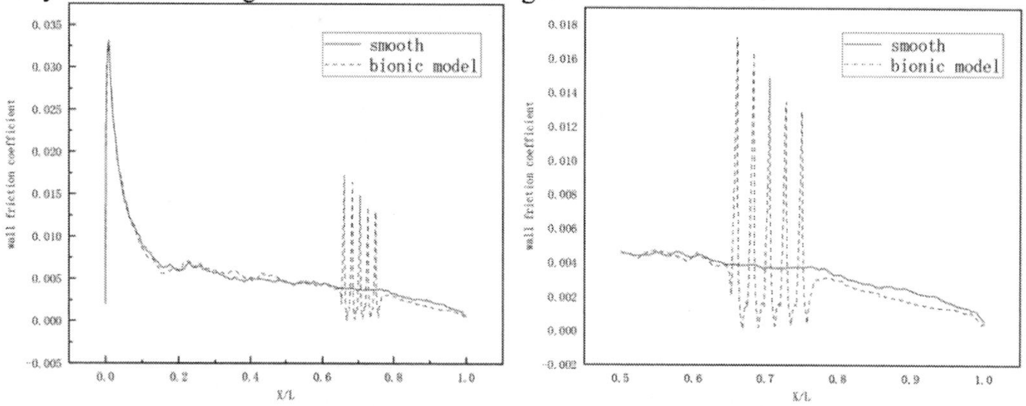

Figure 6. Wall friction resistance distribution diagram and locally enlarged diagram of hydrofoil.

3.3.3 Surface pressure coefficient. Figure 7 shows the distribution of the axial pressure coefficient of the smooth hydrofoil and the bionic hydrofoil and the local enlargement of the bionic structure area. It is not difficult to find that the pressure coefficient curves of the two models in the front are very consistent, but in 0.65 mm < x/L < 0.75 mm, it will suddenly rise after encountering the bionic structure, and reach the minimum at the bottom of the bionic structure. After the bionic structure, the pressure coefficient decreases slowly, but the downward trend is slower than that of the smooth surface. This shows that the bionic structure does not affect the smooth part of the front section, but has a strong influence on the pressure coefficient of the non-smooth surface and the rear section of the hydrofoil.

Journal of Physics: Conference Series **2808** (2024) 012078 doi:10.1088/1742-6596/2808/1/012078

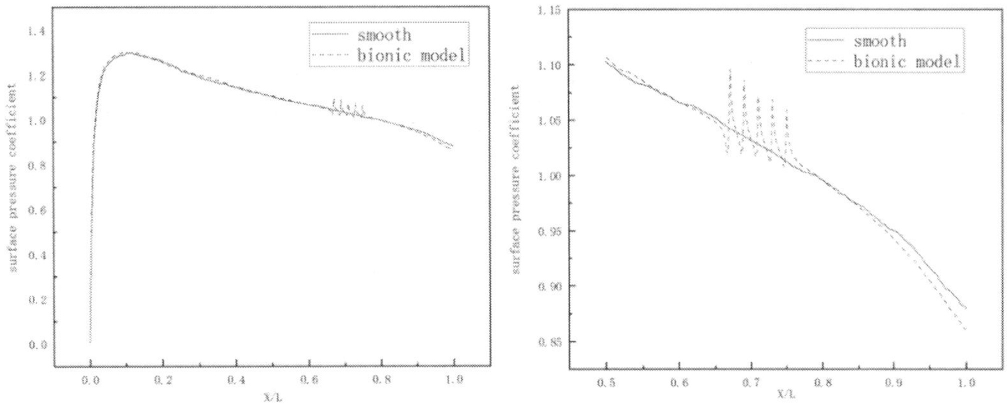

Figure 7. Hydrofoil surface pressure coefficient distribution and local amplification diagram.

4. Conclusions

In this paper, a four-factor three-level orthogonal test is designed to analyze the effects of bionic structure height h (0.3, 0.4, 0.5 mm), bottom radius r (1, 1.25, 1.5 mm), spacing s (3.5, 4 mm), and arrangement position (front, middle, tail) on the hydrofoil. The drag reduction characteristics of the underwater vehicle are analyzed by numerical simulation. The results show that the bionic microstructure can effectively reduce the resistance of the hydrofoil, and the drag reduction rate can reach up to 6.05 %. The order of the factors affecting the drag reduction rate is the height of the bionic structure h, the radius of the bottom surface r, the arrangement position, and the microstructure spacing s. After comparing the velocity field of bionic hydrofoil and smooth hydrofoil, it is found that the velocity between bionic microstructures is reduced, the thickness of the viscous bottom layer is increased, the velocity gradient of fluid near the wall is reduced, and the resistance is reduced. The surface pressure coefficient of the bionic hydrofoil is greatly affected by the bionic microstructure, and it is basically not affected before contact. The bionic microstructure can effectively reduce the wall friction coefficient of the hydrofoil. In addition, the bionic microstructure can effectively reduce the wall friction coefficient behind the hydrofoil, which has a good effect on reducing the overall viscous resistance.

References

[1] Li L, Liu B, Hao H, et al. (2020) Investigation of the drag reduction performance of the bionic flexible coating. Physics of Fluids, 32(8):084103.
[2] Hassanalian M, Abdelmoula H, Mohammadi S, et al. (2019) Aquatic animal colors and skin temperature: Biology's selection for reducing oceanic dolphin's skin friction drag. Journal of Thermal Biology, 84: 292-310.
[3] Wang H, Luo Guihang, et al. (2022) Preparation of a bionic lotus leaf microstructured surface and its drag reduction performance. RSC advances, 12(26):16723-16731.
[4] Jules F M, Jianbo L, Liguo Q, et al. (2023) The optimization of biomimetic sharkskin riblet for the adaptation of drag reduction. Ocean Engineering, 275.
[5] Domel August G, et al. (2018) Shark skin-inspired designs that improve aerodynamic performance. Journal of the Royal Society, Interface 15.139:20170828-20170828.
[6] Gu Y, Xia K, Zhang W, et al. (2020) Airfoil profile surface drag reduction characteristics based on the structure of the mantis shrimp abdominal segment. Archive of Applied Mechanics, (prepublish):1-14.
[7] Zhou H, Liu C, Tian G, et al. (2020) Research on the drag reduction property of puffer (Takifugu flavidus) spinal nonsmooth structure surface. Microscopy Research and Technique, 83(7): 795-803.

Journal of Physics: Conference Series **2808** (2024) 012078 doi:10.1088/1742-6596/2808/1/012078

[8] Zhou H, Zhu Y, Tian G, et al. (2021) Experimental Investigations of the Turbulent Boundary Layer for Biomimetic Surface with Spine-Covered Protrusion Inspired by Pufferfish Skin. Arabian Journal for Science and Engineering, 46(3): 2865-2875.
[9] Aymen M, Hedi K, Salah M A, et al. (2018) Impact of stern design on the hydrodynamic drag of AUV's hull. Indian Journal of Geo-Marine Sciences, 47(1):89-95.

Study on thermo-mechanical coupling model in pantograph-catenary system of rail vehicle

Hangyan Jiang[1], Ning Zhou[1 *], Yao Cheng[1], Haifei Wei[1], Jingke Yan[1] and Weihua Zhang[1]

[1]State Key Laboratory of Rail Transit Vehicle System, Southwest Jiaotong University, Chengdu, 610031, China

[*]Corresponding author's e-mail: zhou_ningbb@sina.com

Abstract: The coupling of the pantograph-catenary system represents a complex multi-disciplinary challenge, intertwining multi-fields and intricate interactions. Based on thermodynamic theory, this paper establishes a thermal-force coupling model for the pantograph carbon slider of high-speed trains, taking into account multiple physical field factors such as heat, dynamics, and electricity. Utilizing a pantograph-catenary contact test bench, we validate a full-scale thermo-mechanical coupling model for the pantograph carbon slider. Thereafter, we assess the coupled temperature and stress outcomes of the carbon slider. The results indicate that dynamic variations in the surface temperature of the carbon slider could lead to the formation of localized high-temperature zones. The phenomena of thermal expansion and contraction, combined with heat accumulation, increase the risk of thermal stress cracks and carbon strip fractures. Longitudinal reverse heat warping deformation occurs in the carbon strip. The airway of the aluminum bracket experiences the highest coupled stress and requires enhanced load-bearing capacity.

1. Introduction

The Pantograph-Catenary (PAC) system is one of the critical subsystems of the high-speed train coupling system [1]. With the rapid development of high-speed railway systems, the operating environment for high-speed trains has become increasingly complex. The thermal effects experienced by the PAC system during operation can lead to reduced equipment performance, shortened lifespan, and even the occurrence of safety accidents [2]. Therefore, studying the thermal-mechanical coupling effects within the PAC system is of great significance for enhancing the system's reliability and efficiency.

Despite significant progress in research on the PAC system in recent years [3], challenges persist due to the system's complexity and the diversity of thermal-mechanical effects. Traditional studies have mainly focused on either thermal or mechanical issues separately, with comprehensive analysis of thermal-mechanical coupling effects being relatively scarce [4-7].

Addressing the shortcomings of existing research, this paper focuses on the pantograph's carbon slider as the research subject. Based on thermodynamic theory, a thermal-mechanical coupling model of the PAC system is established. The model is validated through bench tests, focusing on the carbon slider's thermal-mechanical coupling. Based on this validation, we analyzed the coupled temperature and stress fields of the carbon slider under specific operating conditions. The findings offer a reference for the research on the multiphysical field and fatigue life of the PAC system.

Journal of Physics: Conference Series **2808** (2024) 012079 doi:10.1088/1742-6596/2808/1/012079

2. Thermal analysis theory

2.1. Carbon slider model of pantograph

Because the temperature generated by the thermal load of the carbon slider is propagated to other components of the pantograph, this paper utilizes the pantograph carbon slide model. The relative motion between the pantograph carbon slider and the contact line involves two movements, as shown in Figure 1. One is the relative movement along the direction of the track generated by the running train, and the other is the relative movement along the direction perpendicular to the track caused by the layout of the contact line's stagger. Stagger refers to the distance between the contact line at the positioning point and the center of the pantograph. In straight sections, it is generally staggered on both sides in a direction perpendicular to the track, arranged in a "Z" shape. The main purpose of this is to prevent uneven wear of the pantograph slide and to give the contact suspension good wind stability.

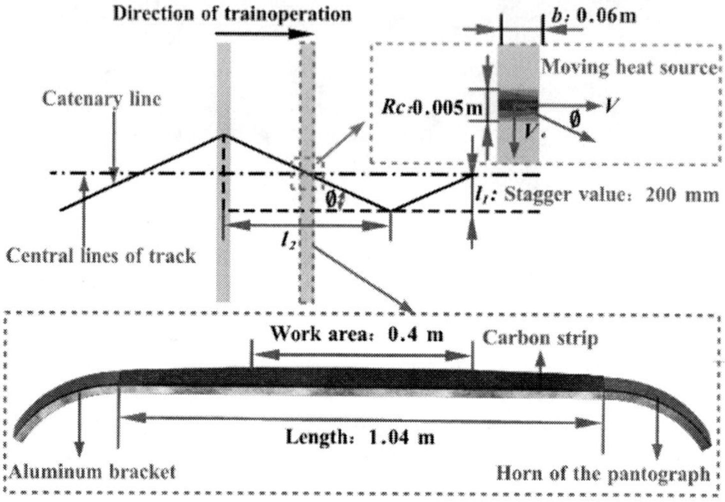

Figure 1. Schematic diagram of carbon slide contact area and carbon slide structure.

$$tan\theta = \frac{2|l_1|}{l_2} = \frac{V_0}{V} \tag{1}$$

$$tan\theta = \frac{2|l_1|}{l_2} = \frac{V_0}{V} \tag{2}$$

$$V_0 = \frac{2|l_1|}{l_2}V \tag{3}$$

where l_1 represents the stagger, l_2 represents the span, V is the train running speed, and V_0 represents the moving speed of the contact area.

The structure of the pantograph carbon slide used by the high-speed train pantograph is shown in Figure 1. The structural parameters and material parameters of this pantograph carbon slide are shown in the table 1.

Table 1. Material parameters of the pantograph carbon slider.

Parameter	Coefficient of expansion (Pa)	Conduction rate (W/m · °C)	Density (kg/m^3)	Elastic modulus (Pa)	Poisson ratio	Specific heat capacity J/(kg · °C)
Carbon slider	6.4 e-6	200	1, 700	12.6e9	0.18	720

Journal of Physics: Conference Series **2808** (2024) 012079 doi:10.1088/1742-6596/2808/1/012079

Aluminum bracket	2.4 e-5	190	2, 800	72e9	0.33	900
Horn	1.35 e-5	150	7, 850	300e9	0.3	502

2.2. Thermal boundary

During the operation of the train, the rise in temperature of the pantograph is the result of the combined action of various heat sources. These heat sources include the Joule heat generated by the contact resistance between the pantograph slider and the overhead contact line under load-carrying conditions and the frictional heat produced by relative sliding. It is assumed that the train is operating without deviation from the line, and arc heat is not considered. We can refer to the calculation formulas in [8] for Joule heat power and frictional heat power.

Considering the construction of the pantograph model, three types of boundary conditions are taken into account: heat flux density boundary conditions, convective heat transfer boundary conditions, and radiative boundary conditions [7], as shown in Figure 2. The solution for the non-steady-state heat conduction of the pantograph carbon slider at each moment is obtained through the ambient temperature and these three types of boundary conditions.

The Joule heat and frictional heat generated by the contact between the pantograph and the overhead line are converted into heat flux density in the form of equivalent heat power for thermo-mechanical coupling analysis. Based on the formulas related to Joule heat power and frictional heat power for various operating conditions found in [8], the heat flux density is then calculated as Equation (4).

$$q = \frac{P}{A} = \frac{P_{\mathrm{J}} + P_{\mathrm{f}}}{A} \tag{4}$$

where q represents the heat flux density, and A is the contact area between the PAC current-carrying friction pairs.

In the heat transfer analysis, the heat flux density is modeled as a time-varying moving heat source. This moving heat source travels at a certain speed over time along a predefined path distributed on the surface of the carbon slider and transfers heat into the interior of the carbon slider.

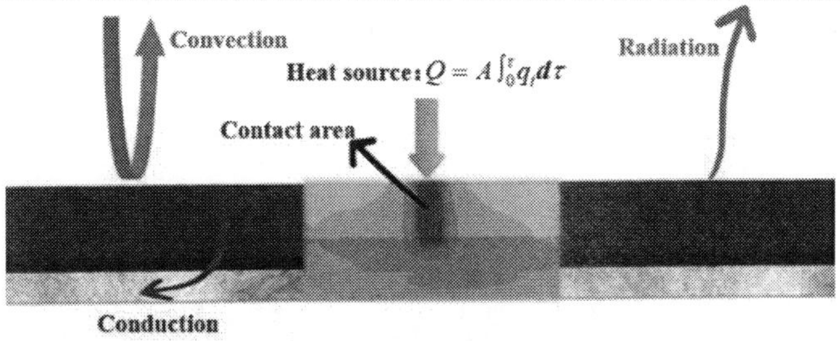

Figure 2. The heat transfer boundary conditions for the pantograph carbon slider.

3. Thermo-mechanical coupling model

The equilibrium equation for the thermo-mechanical coupling analysis of the PAC system is as follows:

$$[\boldsymbol{M}]\{\ddot{u}\} + [\boldsymbol{C}]\{\dot{u}\} + [\boldsymbol{M}]\{u\} = [P(F, T)] \tag{5}$$

$$[\boldsymbol{C}(T)]\{\dot{T}\} + [\boldsymbol{K}(T)]\{T\} = [Q(F,T)] \tag{6}$$

where $[M]$ is the mass matrix, $[C]$ is the damping matrix, $[K]$ is the stiffness matrix, $[P(F, T)]$ is the load vector, $[C(T)]$ is the specific heat matrix, $[K(T)]$ is the heat conduction matrix, $Q(F, T)$ is the heat source vector shown in Equations (5) and (6), u represents the displacement component, T is the temperature, F is the mechanical load, P is the nodal load, and i represents the increment. These equations are solved by using the explicit integration method, as shown in Equation (7).

$$\begin{cases} \dot{u}_{(i+1)}^{N} = \dot{u}_{(i-1)}^{N} + \dfrac{\Delta t_{(i+1)} + \Delta t_{(i)}}{2} \ddot{u}_{(i)}^{N} \\ u_{(i+1)}^{N} = u_{(i)}^{N} + \Delta t_{(i+1)} \dot{u}_{(i+1)}^{N} \end{cases} \tag{7}$$

$$T_{(i+1)}^{N} = T_{(i)}^{N} + \Delta t_{(i+1)} \dot{T}_{(i+1)}^{N} \tag{8}$$

The displacement component uses the central difference integration rule, and the temperature uses the forward difference integration rule. This method incorporates inertia effects, allowing it to simulate transient thermal responses. Therefore, it is suitable for the transient thermal analysis of the carbon slider. In the analysis, the temperature distribution and thermal deformation are computed, which in turn influence the stress analysis. Simultaneously, the stress distribution affects the temperature solution, thereby achieving full coupling.

The analysis and solution process for the pantograph thermal-force coupling model is as follows:

Step 1: We calculate the equivalent thermal power and mechanical load applied on the carbon slider based on the contact force;

Step 2: We set the initial values and boundary conditions for the iterative thermal-force coupling analysis, and conduct the thermal-force coupling analysis on the pantograph carbon slider. Through iterative calculations, we simulate the transfer and distribution of heat and forces within the carbon slider;

Step 3: We obtain the results for temperature distribution and stress distribution.

The solution process diagram is shown in Figure 3.

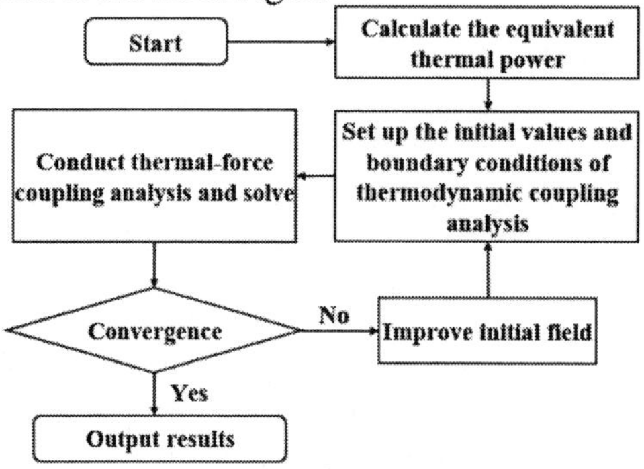

Figure 3. Thermal-force coupling analysis flowchart.

4. Experimental validation

4.1.Experimental introduction

Due to the difficulties in collecting dynamic contact stress distribution during the experiment, the thermal rise of the carbon slider was used to verify the established thermal-force coupling model of the pantograph-catenary system. Through the pantograph-catenary electrical contact test bench, in combination with the FLUKE TiX640 infrared thermal imager, the temperature of the slider contact area was collected. A brand-new carbon slider was monitored in real time under multiple working conditions. Figure 4 (a) displays the test bench used in this experiment. The relative movement of the pantograph to the contact line is primarily facilitated by the contact line turntable, with a diameter of 1, 100 mm. The contact force between the pantograph and the contact line is continuously provided by the normal force motor, with a maximum design capacity of 300 N. The slide base, governed by motor control, can carry out reciprocating vertical movements, thereby simulating the "Z" shaped configuration of the contact network, with a maximum contact wire stagger of 55 mm. Temperature data collection during the experiments employed the FLUKE Tix640 infrared thermal imager shown in Figure 4 (b), which can detect long-wave infrared spectra ranging from 7.5 μm to 14 μm. Simulation calculations were performed under the same working conditions to compute the temperature field of the carbon slider and compared with the experimental data.

Figure 4. (a) Pantograph-catenary electrical contact test bench; (b) Infrared thermal imager.

4.2.Model validation

The experiment compared the temperature rise results of the carbon slider with a contact force of 25 N, a sliding speed of 40 km/h, a pull-out value of 40 mm, and contact currents of 250 A and 300 A, as shown in Figure 5. By taking the envelope of the temperature results from the experiment and the simulation, with a simulation error within 5%, the validity of the established thermal-force coupling model has been verified.

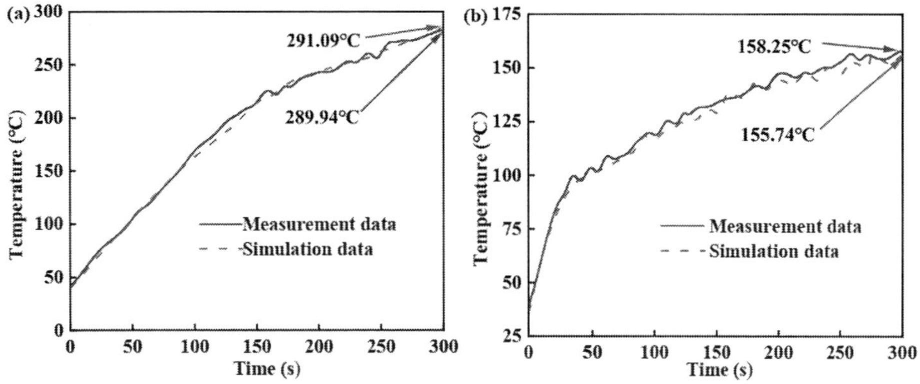

(a) 250 A current; (b) 300 A current.

Figure 5. Comparison of experiment and simulation.

5. Simulation results

5.1.Temperature distribution results

Analyzing the temperature distribution can help identify the heat concentration area of the carbon strip. The contact material in the heat concentration area is prone to damage.

Figure 6 compares the temperature results of various components of the carbon slider. The temperature of each component, ranked from highest to lowest, is as follows: carbon strip, bracket, horn. The heating of the carbon strip is the most complicated because it involves various heat sources and heat transfer modes, and there are complex interactions. The temperature of the aluminum bracket is relatively high because the heat of the carbon strip is transferred to the bracket by heat conduction. In particular, the temperature in the middle area of the bracket near the heat source area of the carbon strip is higher. The temperature of the horn is relatively low because the horn is far from the moving heat source and is mainly affected by the heat conduction of the aluminum bracket and carbon slider. Figure 6(a) shows the temperature rise change of the carbon slider at an ambient temperature of 20°C. The temperature rise of the carbon slider increases exponentially with time. It rises quickly at first, and then the rate of rise decreases, corresponding to a large slope of the curve at the beginning, which gradually decreases afterward.

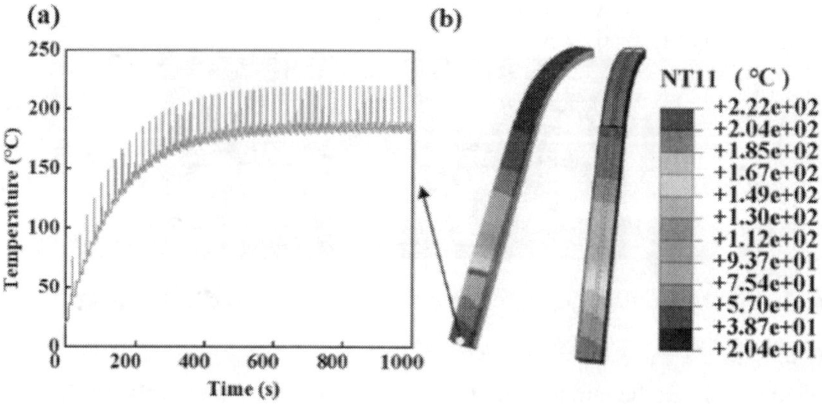

(a) Temperature at the midpoint of the carbon strip; (b) Temperature of the carbon strip and aluminum bracket.

Figure 6. Comparison of carbon slider temperature results.

Figure 7 shows the temperature distribution on the surface of the carbon strip at different times under the action of thermal-mechanical coupling in the carbon slider model. From Figure 7, the temperature distribution on the surface of the carbon strip is uneven, presenting high-temperature and low-temperature areas, with the highest temperature area concentrated in the contact area. This is due to the superimposition of friction heat and current heat in this area, which results in higher temperatures. The surface temperature of the carbon strip dynamically changes with time, and as time progresses, the high-temperature area gradually flows from the contact area to the pull-out edge area. During the conduction and diffusion of heat in the carbon strip, it gradually spreads towards areas away from the heat source. Not only does a heat accumulation area form on the surface of the carbon strip, but heat diffusion also occurs within the carbon strip. Because of heat diffusion, the heat storage areas gradually merge as time progresses, becoming a high-temperature area of the carbon strip. Therefore, during the movement process, heat accumulation areas will form at multiple positions on the surface of the carbon strip. Ultimately, with the diffusion and merging of the heat accumulation areas, the overall temperature of the carbon strip will rise in a short time.

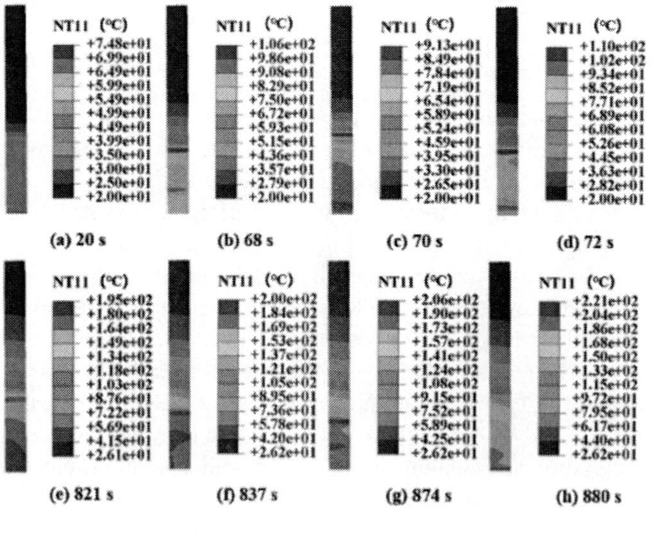

Figure 7. Temporal variation diagram of carbon strip temperature distribution.

5.2.Stress distribution results

Figures 8 (a) and (b) compare the stress results of various components of the carbon slider. The temperature of each component, ranked from highest to lowest, is as follows: bracket, carbon strip, and horn. As shown in Figure 8 (f), the thermal stress in the middle area of the aluminum bracket is significantly higher than the stress at the ends, with the maximum stress being 132 MPa, located in the middle of the airway. This is because the middle area is closer to the heat source and is subject to a greater stress effect. The horn, due to its distance from the heat source, experiences less stress. The maximum stress is 36.1 MPa, located in the area in contact with the slider. The stress on the horn mainly comes from the heat conduction of the carbon strip and the aluminum bracket, as well as the thermal exchange with the environment. Figures 8 (d) and (c) show a stress curve at the midpoint of the carbon strip surface. The stress on the carbon strip grows exponentially over time, first rapidly rising, and then the rate of rise decreases. The steady-state stress of the carbon strip is 19.5 MPa, which is relatively small compared to the 132 MPa stress on the bracket. Figure 8 (e) displays the stress and temperature distribution along the edge of the cross-section of the carbon slider. On the cross-section, the stress of the carbon strip decreases first and then increases from the upper surface to the lower surface along the cross-section. There is a stress discontinuity at the contact point with the aluminum bracket. The temperature gradually decreases from the upper surface to the lower surface along the cross-section, with no obvious discontinuity.

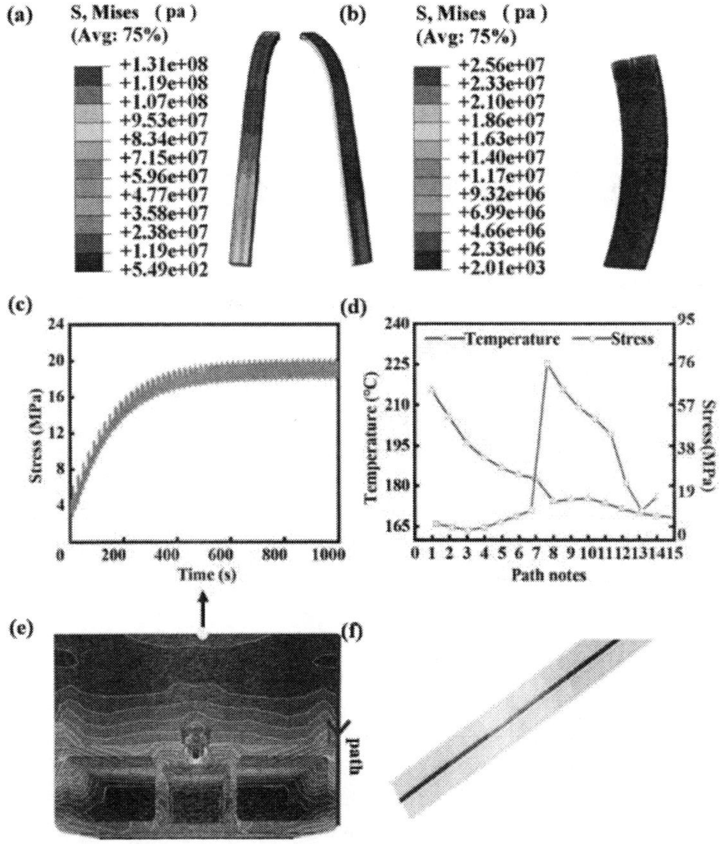

Figure 8. Stress results: (a) Stress of the carbon strip and aluminum bracket; (b) The horn of the carbon slider; (c) Time history of stress at the midpoint of the carbon strip; (d) Stress along the cross-sectional path of the carbon slider; (e) Stress in the middle cross-section of the carbon slider; (f) Stress in the airway of the aluminum bracket.

Figure 9 shows the strain distribution on the surface of the carbon strip when the ambient temperature is 20°C. The surface strain of the carbon strip dynamically changes over time, with the contact area being the main stress concentration area. Over time, the carbon strip undergoes longitudinal anti-thermal warping deformation. This longitudinal anti-thermal warping deformation can cause uneven contact between the carbon strip and the catenary line, leading to an increase in local temperature.

Under high-temperature conditions, the combined effects of thermal stress and mechanical stress can further exacerbate the deformation of the carbon strip and lead to the formation of small cracks and defects. With this cycle repeating over time, the strength and durability of the carbon strip will be greatly reduced.

Over the long term, under high-temperature conditions, stress concentration and deformation can cause the existing cracks to further expand. This can lead to an earlier occurrence of fatigue failure in the carbon strip, thus reducing its fatigue life.

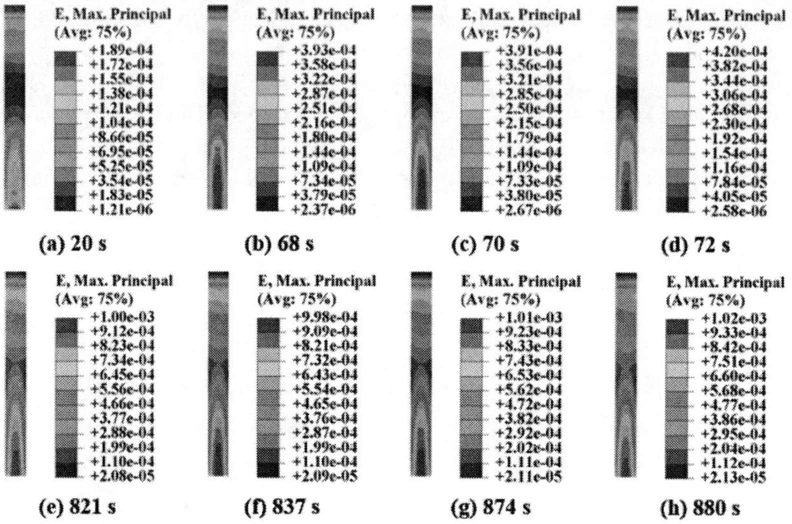

Figure 9. Temporal evolution of strain distribution in the carbon strip.

6. Conclusions

In this study, leveraging thermodynamics theories, a novel thermomechanical coupling model was developed. Comparisons with experimental data have demonstrated that the simulation maintains an error margin of 5%. The principal findings of the research are outlined below:

(1) The operational surface of the carbon slider is prone to localized heat accumulation due to non-uniform heat transfer, leading to the formation of hotspots. These areas of concentrated heat can expand from the center towards the edges of the slider, creating an environment conducive to thermal stress cracking. Continuous uneven heating can further enlarge existing cracks, potentially posing a risk of carbon strip material fracture.

(2) The carbon strip undergoes longitudinal anti-thermal warping deformation. The deformation causes uneven contact and local temperature rise, further intensifying the deformation of the carbon strip. After cycles, the strength and durability of the carbon strip are greatly reduced, and it can also lead to the formation of small cracks and defects on the surface of the carbon strip. By analyzing the special distribution of stress, a basis can be provided for future thermal crack expansion.

(3) Among the carbon slider components, the carbon strip is the component with the highest temperature, so it is necessary to improve the high-temperature mechanical performance of the carbon strip. The high-temperature area is inside the pull-out value in the middle of the carbon slider, where thermal damage is prone to occur. The maximum coupling stress is suffered by the aluminum bracket, with the maximum point being the airway of the aluminum bracket, which needs to be strengthened.

This study provides a new simulation method for the analysis of the thermomechanical coupling response state of the bow-net system and is expected to guide the optimized design of high-speed train carbon sliders. However, the geometric model of this carbon slider does not include the entire bow model of the pantograph. Therefore, in future research, a full-size geometric model of the pantograph will be established, and an improved thermomechanical coupling analysis model considering the actual line impact will be developed to study the thermomechanical coupling characteristics of the carbon slider under complex service environments.

Acknowledgment
This paper is one of the interim achievements of the National Natural Science Foundation of China's Project "Research on the Mechanism and Action Principles of the Influence of Complex Climatic Environments on the Dynamic Current Collection and Service Life of Pantograph-Catenary Systems (52072319)".

References

[1] Zhou N., Cheng Y., Zhang X., Zhi X., and Zhang W. (2023). Wear rate and profile prediction of Cu-impregnated carbon strip for the high-speed pantograph. Wear, 530: 205, 056.
[2] Delcey N. et al. (2020). Analysis of the thermal variations in a moving pantograph strip using an electro-thermal simulation tool and validated by experimental tests. Proceedings of the Institution of Mechanical Engineers, 234 (8): 859-868.
[3] Xiao S., Ye Z., Tong M., et al. (2022). Temperature rise characteristics of rolling pantograph in a high-speed train. High Voltage Technology, 48 (5): 1, 798-1, 807.
[4] Zhi X. et al. (2023). Effect and behaviors of ambient humidity on the wear of metal-impregnated carbon strip in the pantograph-catenary system. Tribology International, 188: 108, 864.
[5] Ma Y., Na Y., Yang L., and Zheng Y. (2021). Computer numerical simulation study of sliding friction temperature of pantograph carbon skateboard. In: 2021 IEEE International Conference on Data Science and Computer Application (ICDSCA). IEEE, pp. 724-727.
[6] El Maakoul A., Remy B., and Degiovanni A. (2019). Modeling of thermal contacts with heat generation: Application to electrothermal problems. International Journal of Heat and Mass Transfer, 140: 293-302.
[7] Liu Z. (2022). Research on the Identification Method of Carbon Strip Wear Amount Based on Temperature Characteristics [D]. Southwest Jiaotong University. DOI: 10.27414/d.cnki.gxnju.2022.002251.
[8] Tao W. (2019). Heat Transfer, Higher Education Press, Beijing.

Journal of Physics: Conference Series **2808** (2024) 012080 doi:10.1088/1742-6596/2808/1/012080

Study on the structure of the cowl based on optimal water dynamic characteristics

Tian Wang[1,] *, Mingxuan Li[1], Jia Tang[1], Yuhan Liu[1] and Wenbo Li[1]

[1]School of Naval Architecture, Ocean and Energy Power Engineering, Wuhan University of Technology, Wuhan, Hubei, 430000, China

*Corresponding author's e-mail: wangtian@whut.edu.cn

Abstract: Fluid mechanics makes us realize the complexity and variability of the fluid force, so the shape of the fluid contact surface will have a great impact on the energy capture power of the generator set. In this paper, a symmetrical airfoil deflector is designed to improve the energy capture power and the conversion efficiency of the current energy unit under the condition of low flow velocity, and the selection of parameters such as inner diameter, chord length, and angle of attack and their growth effects are analyzed by finite element analysis. The three-dimensional finite element model of the underwater fairway is established by using the ANSYS finite element method. Through finite element calculation and analysis, the velocity vector diagram is obtained. The results show that the speed increase efficiency of the aerofoil baffle is 80% higher than that of the traditional one, and the energy capture power is greatly increased.

1. Introduction

The world's fossil resources are in danger of exhaustion. Coastal developed countries such as the United Kingdom, the United States, and Canada attach great importance to the strategic position of ocean energy in the future energy field, issue ocean energy strategic plans, formulate ocean energy development roadmap, guide private capital to invest in the ocean energy field, promote the research and development of ocean energy technology, and promote the development of ocean energy industry. According to the investigation, the theoretical estimated value of ocean current energy in the world is about 108 kW, which has good development value. At present, the main use of ocean current energy is to use the ocean current to drive the turbine for power generation. Only when the flow rate is greater than 1.5 m/s can there be better benefits (the turbine volume is large), which greatly reduces the utilization rate of ocean current energy. It is of great significance for the development and utilization of ocean current energy in the world to design an optimized airfoil fairing structure [5] that optimizes hydrodynamic characteristics. See Figure 1 and Figure 2.

Figure 1. Ocean turbine generator in France.

Figure 2. Ocean current power plant in Japan.

Journal of Physics: Conference Series **2808** (2024) 012080 doi:10.1088/1742-6596/2808/1/012080

2. Study on fairing

The guide hood usually has two kinds of one-way structure and two-way structure. The one-way structure is expanded, and a reversing mechanism is generally required in order to make it able for two-way diversion. The two-way structure is not used, which is generally symmetrical contraction and expansion (that is, venturi tube) type. Luo [1] designed a bidirectional impeller based on BEM theory and airfoil theory. Zhang et al. [2] designed a PDCPD deflector to achieve the lightweight of the deflector and improve its stiffness and strength. Li [3] carried out simulation optimization of the axial fan fairing to further improve the performance of the fairing. The three-dimensional fluid-structure coupling method was used to simulate the underwater monitoring equipment and the underwater monitoring equipment with a guide hood, and the vibration displacement response was obtained [4]. Based on the analysis of the structure and performance of the above fairing, this paper proposes an optimized aerofoil fairing structure, which aims to provide a stable working environment for the inner device of the fairing and optimize the hydrodynamic characteristics.

3. Structure design of airfoil fairing

The structure of the aerofoil deflector is composed of three parts: contraction section, straight section, and expansion section [6]. The contraction section and expansion section are bidirectional symmetrical structures with double cubic curves, and the aerofoil structure with better hydrodynamic characteristics is adopted in its linear line. Among them, the contraction section guides more fluid into the fairing, the linear section plays the role of speed stabilization and forms a stable speed interval in the fairing, and the expansion section generates negative pressure and absorbs more fluid to further increase the fluid speed in the fairing. Figure 3 shows the structure design of the hood.

The airfoil deflector is made of alloy steel, and its anti-corrosion property can adapt to the underwater environment. The neck diameter, chord length, and inner diameter of the airfoil guide hood are determined to be 2.1 m, and 1.5 m, and the optimal angle of attack is determined to be 15° by using the method of estimating the optimal angle of attack of the airfoil guide hood.

Figure 3. Structure of an air duct.

4. Theoretical research and simulation analysis of deflector hood

4.1. Guide hood performance analysis and selection

Compared with a single fan, the diffuser-type guide hood fan studied by Ohya and Karasudani of Kyushu University increases the power by about 2 to 5 times due to the formation of a tail vortex, and more mass flow enters the diffuser-type guide hood. Chen et al. analyzed the structure of 10 curved-type flow guides, and the analysis results showed that the structure of the double-cubic curve xm=0.3 flow guides had the best hydrodynamic performance. Gao et al. obtained the comparison diagram of flow velocity and the comparison of the optimal angle of attack under the same Reynolds number by analyzing different structure fairings through simulation [7].

Figure 4 shows the comparison of water velocities in different hoods. Table 1 shows the optimal angle of attack pairs for the growth rate of the airfoil deflector under the same Reynolds number [8].

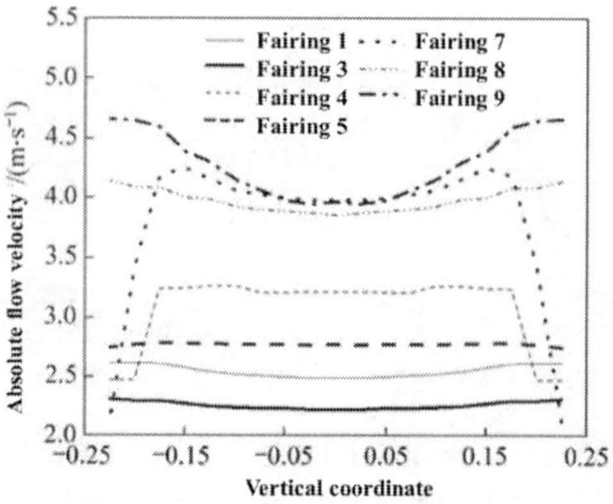

Figure 4. Comparison of water velocities in different hoods.

Table 1. Comparison of the optimal angle of attack at the same Reynolds number.

Chord length/cm	Inlet velocity /(m/s)	Suitable speed range of the impeller When the angle of attack is 13°/(m/s)	Suitable speed range of the impeller When the angle of attack is 14°/(m/s)	Suitable speed range of the impeller When the angle of attack is 15°/(m/s)	Optimal angle of attack to speed up/(°)
30	1	1.3617~1.7391	1.3614~1.7470	1.3342~1.7341	14
45	0.667	0.8846~1.1305	0.8894~1.1366	0.8698~1.1278	14
60	0.5	0.6779~0.8758	0.6798~0.8966	0.6526~0.8536	14

4.2 Simulation analysis of fairing

The expansion section reduces the pressure, forms a vortex to generate negative pressure, absorbs more fluid, and thus increases the flow rate. However, relying only on the growth rate of negative pressure will not produce a stable speed section. It is necessary to add a straight line section, and the contraction section is necessary to facilitate the flow of fluid and play a transitional role.

The effect of using a negative pressure growth rate is better than that of gathering flow growth rate, which indicates that in order to achieve a better growth rate, an expansion section should be added to the deflector to form negative pressure. After the expansion section is added, a vortex forms behind the expansion section to generate negative pressure and attract more fluid, thus increasing the fluid flow rate. In general, the hydrodynamic performance of the airfoil cowl is better.

4.2.1 Optimal parameter selection of airfoil fairing. (1) The selection of the inner diameter depends on the impeller diameter. The speed of the selected stable section suitable for impeller placement decreases with the increase of the inner diameter of the guide hood, that is, the larger the inner diameter is, the lower the suitable working velocity will be.

(2) The larger the string length is, the better the conductivity growth performance will be. However, due to cost constraints, the string length cannot be too large. When the minimum diameter of the airfoil deflector is fixed (inner diameter 20 cm), the chord length of the airfoil is increased by 1~2 times, and the optimal attack Angle of the growth rate is unchanged or reduced by 1°. The speed of the selected stable section suitable for impeller placement increases with the increase of the chord length of the guide hood, that is, the longer the chord length is, the higher the suitable working velocity will be.

(3) The speed of the selected steady velocity section suitable for impeller placement increases first and then decreases as the angle of attack of the airfoil guide hood increases. When the angle of attack is 15°, the maximum speed is obtained, that is, when the angle of attack increases, the suitable working

velocity increases first and then decreases, and the flow rate reaches the highest at 15°. Figure 5 shows the attack Angle and chord length of the hood.

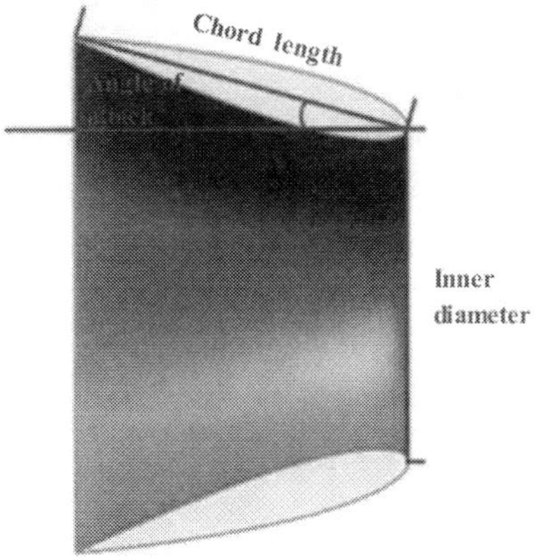

Figure 5. Attack Angle and chord length of the hood.

Combined with the data consulted, the project team selected an airfoil fairing with an inner diameter of 160 mm and an angle of attack of 15° to improve the captured power of the ocean current energy unit, and carried out relevant simulation tests.

4.2.2 Simulation and comparison of the speed increase effect of the hood. The common fairing is a typical three-stage type, including a contraction section, a straight line section, and an expansion section [9]. The Vickers curve of the contraction section and the expansion section is used to simulate and compare the common fairing based on the CFD method. The diameter of the basin is 4 times the maximum diameter of the diversion hood, and the diameter of the flow basin is 5 times and 6 times the maximum diameter of the diversion hood, respectively. A structured grid is adopted, and the near-wall should be encrypted. The entrance of the calculation domain is the velocity entrance (u=2 m/s), the exit is the pressure exit, and the basin wall is the velocity entrance or slip wall (u=2 m/s). The wall of the diversion cover is set as a non-slip wall. After the growth of the fairing, the average velocity is 2.5 m/s, which is 25% higher than that of the inlet velocity. Figure 6 shows the velocity vector diagram of a typical three-stage fairing.

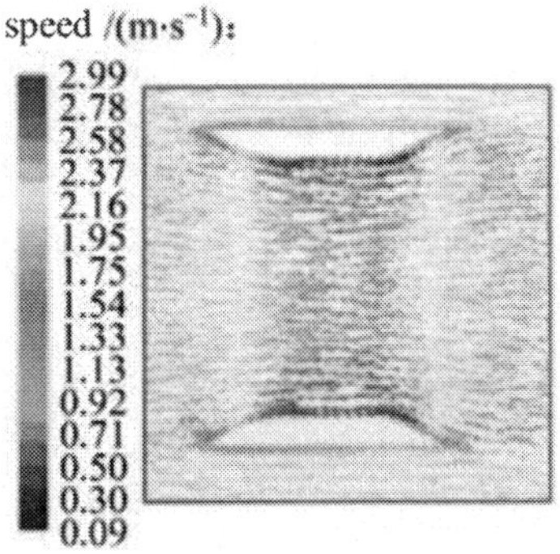

Figure 6. Velocity vector diagram of a typical three-stage fairing.

In the same water flow environment, the speed increase effect of the designed airfoil deflector is simulated and analyzed [10]. Ansys Fluent was used for simulation, and part of the modeling was carried

Journal of Physics: Conference Series **2808** (2024) 012080 doi:10.1088/1742-6596/2808/1/012080

out in SolidWorks based on the CFD method. After mesh division, the mesh number was 643284, and the mesh quality was good. Figure 7 shows Element Quality, and Figure 8 shows Skewness.

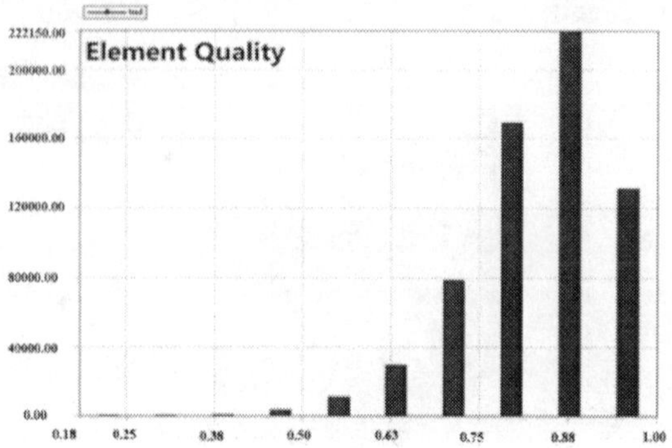

Figure 7. Element quality.

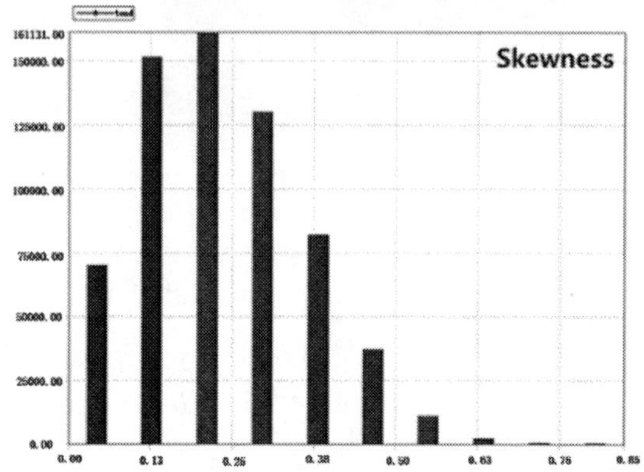

Figure 8. Skewness.

As can be seen from the figure, the expansion section reduces the pressure, forms a vortex, generates negative pressure, absorbs more fluid, and thus increases the flow rate. According to the theoretical calculation, when the inlet flow rate is 2 m/s, the aerofoil fairing designed by the project team increases the growth efficiency by 80% compared with the traditional three-stage type.

5. Conclusion

The symmetrical aerofoil baffle proposed in this paper is compared with the structure of the aerofoil baffle with the optimal xm=0.3 linear curve and the optimal attack Angle of the aerofoil baffle growth rate, which can absorb the ocean current and increase the flow rate by using negative pressure while collecting the flow. Installing the impeller inside the fairing can greatly improve the efficiency of the impeller and make the power generation device work at a lower flow rate. In addition, the guide hood can stabilize the flow, reduce the turbulence of the internal flow field, improve the flow field environment, and enhance the stability of the system operation.

References

[1] Luo R D. Design and hydrodynamic analysis of suspended horizontal axis tidal flow turbine and its fairway [D]. Shanghai Ocean University, 2021. DOI: 10.27314 /, dc nki. Gsscu. 2020.000509.

[2] Zhang Ruijun, ZHANG Jinzhou, LI Jianheng. Research and Development of PDCPD Fairing for a Certain Vehicle [J]. Time Automobile, 2022(03):153-155.

[3] Li Weizhe. Axial flow fan dome and gathering ram simulation optimization [D]. Jiangxi Agricultural University, 2023. DOI: 10.27177 /, dc nki. Gjxnu. 2022.000094.

[4] Dai Shaoling, Guan Guan. Research on three-dimensional fluid-structure coupling of airfoil deflector for underwater monitoring equipment [J]. Marine Standardization Engineer, 2021, 54(06):34-37. DOI:10.14141/j.31-1981.2021.06.007.

[5] Gao Yan-Jing, Gu YA-Jing, Lin Yong-Gang, et al. Hydrodynamic Characteristics of the cowl and its effect on current energy Units [J]. Journal of Central South University (Natural Science Edition), 2021, 52(9): 3164–3173.

[6] Ma Yan. Hydrodynamic Performance Analysis and Optimal Design of Ducted Propeller [D]. Harbin: Harbin Engineering University, 2010.

[7] Cui Baoyu, SI Xiancai, Yuan Peng et al. Numerical study on Effect of solidity on Performance of cowl turbine [J]. Journal of Solar Energy, 2022, 43(02):69-74. DOI:10.19912/j.0254-0096.tynxb.2020-0234.

[8] Sun Ke. Numerical Simulation of Hydrodynamic Performance of Vertical Axis H-shaped Impeller and Guide Hood [D]. Harbin Engineering University, 2008.

[9] Zhang Liang, SUN Ke, LUO Qingjie. Hydrodynamic Design of flow turbine fairway [J]. Journal of Harbin Engineering University, 2007(07):734-737.

[10] Mun G K, Young J J, Jun Y K, et al. A comparative study of purge slot exit shape on the film cooling effectiveness of a gas turbine shroud [J]. International Journal of Thermal Sciences, 2024, 197.

Journal of Physics: Conference Series 2808 (2024) 012081 doi:10.1088/1742-6596/2808/1/012081

Experimental study on the flow capacity of shale fractures with different degrees of sand filling

Ting Chen [1, 2, 3, 4, *], Chunlei Yu [1, 2, 4], Zhigang Sun [1, 2, 4], Min Zhang [1, 2, 4], Qiang Sun [1, 2, 4] and Chenwei Liu [5]

[1]Exploration and Development Research Institute of Sinopec Shengli Oilfield Company, Dongying 257015, China
[2]Shandong Provincial Key Laboratory on Exploration and Development for Unconventional Oil and Gas, Dongying 257015, China
[3]Postdoctoral Research Station of Sinopec Shengli Oilfield Company, Dongying 257002, China
[4]Sinopec Key Laboratory of Shale Oil/Gas Exploration and Production, Dongying 257015, China
[5]China University of Petroleum (East China), Qingdao 266580, China

[*]Corresponding author's e-mail: chenting0828@foxmail.com

Abstract: By conducting tests on the flow capacity of shale fractures with different degrees of sand filling and conducting online CT scans on samples of different lithologies, it is recognized that as the sand concentration increases, the flow conductivity of sand-filled fractures increases and stress sensitivity decreases. Under low effective stress, the large particle size proppant filled sand fractures have higher diversion capacity. As the effective stress increases, the reduction in diversion capacity is significant, and the stress sensitivity is small. The particle-size-filled sand layer is strong. During the process of elastic pressure reduction and failure, the proppant is compacted, the pores of the sand layer are reduced, the crack width is reduced, and the conductivity is reduced.

1. Introduction

As an emerging petroleum resource, shale oil has become a hot and key focus of China's oil and gas exploration and development [1, 2]. The Jiyang shale oil resources in Shengli Oilfield are abundant, and it is an important replacement for the sustainable development of Shengli Oilfield in the next step [3-5]. Currently, elastic development is the main method of shale oil production. Compared with sandstone reservoirs, shale reservoirs have small pore sizes, low permeability, and complex sensitivity [6-10]. As pore pressure decreases, bedding fractures, sand-filled fractures, and pores undergo deformation due to compression, resulting in a decrease in permeability/conductivity, which affects the permeability and development effectiveness of shale [11]. The shale oil reservoir in Jiyang depression has many brittle components, with well-developed interlayered/bedding fractures, which are the main channels for seepage. Related studies have confirmed that fractured low-permeability reservoirs have strong stress sensitivity [12-15]. However, there are currently few experiments on the stress sensitivity of various fractures (bedding fractures and sand-filled fractures) in the Jiyang shale oil reservoir, and the relevant flow patterns still need to be clarified. Therefore, it is necessary to conduct experimental tests on the flow capacity of different types of fractures based on core rock samples taken from the target block and the real geological environment for the Jiyang shale oil reservoir, investigate the characteristics of stress sensitivity of shale bedding fractures and sand-filled fractures, as well as the characteristics of oil-water two-phase flow, clarify the fluid flow patterns within complex fractures in shale, and offer basic experimental findings and theoretical backing for achieving the integration of the construction modulus for shale oil, mathematics, and modeling in the target block.

Zhang et al. [16] conducted indoor diversion capacity experiments that considered closure pressure and proppant to test the effects of proppant fragmentation and embedding on diversion capacity. The results show that the diversion capacity decreases as the closure pressure intensifies. The flow conductivity of ceramic proppants is more stable, and under higher closure pressure, the decrease in flow conductivity is small, and the flow conductivity is maintained the best.

Cooke [17] proposed the influence of embedding this factor on the conductivity of fractures backed by diverse sand concentrations. During volume fracturing, the viscosity of the fracturing fluid directly impacts the distance of proppant transport. Fracturing fluids with low viscosity cannot suspend the proppant, resulting in fast proppant settlement and short transport distance. The fracturing fluid with high viscosity can cause the proppant to travel a longer distance and enter the depths of the fracture. The high sand concentration limits the impact of proppant embedding on the damage to flow conductivity. Therefore, better fracturing effects can be achieved in industrial production by increasing the sand concentration.

Zhu et al. [18] fitted a formula for calculating the flow conductivity of a specified shale reservoir through numerical simulation. The formula considers various factors that affect the flow conductivity of shale reservoirs, including time, pressure, proppant concentration, particle size, type of proppant, etc. The formula was derived with massive data to give people a clearer and deeper understanding of flow conductivity.

Zou et al. [19] conducted indoor diversion capacity experiments to study the attenuation coefficient and diversion capacity of fractures. They concluded that the mode of supporting fracture diversion capacity is related to the loading shut-in pressure. Beneath a low shut-in pressure condition, the diversion mode of fractures with different sand concentrations is the same and is supported by the proppant point. Beneath a high shut-in pressure condition, the insertion and fragmentation of proppant cause a reduction in the width of fractures, accompanied by a change in flow diversion mode. Therefore, increasing the proppant particle size cannot balance the damage to flow diversion capacity caused by proppant insertion. Therefore, the concentration of proppant should be increased. Under high closure pressure, cracks will lose their effectiveness. Therefore, if you want to reduce the impact of proppant embedding, you should increase the laying concentration of proppant. Through comparative experiments, it was found that shale fractures are severely mud-prone, and the use of resin sand support agents can alleviate mud formation.

Wen et al. [20] established a fracture network model and combined it with upgraded conductivity testing instruments to compare the conductivity performance of fracture networks under different forms. The analysis results showed that when the simulated crack width was equivalent to the natural crack width, the influence of vertical cracks parallel to the seepage direction on conductivity was the greatest, followed by water cracks parallel to the seepage direction and vertical cracks perpendicular to the seepage force direction.

Peng et al. [21] improved the API standard diversion chamber and conducted experiments on the permeability of shale reservoir fractures using nitrogen gas. They concluded that under the same proppant application, the diversion capacity increases with the increase of application concentration. The permeability of supported fractures with a certain width is generally higher than that of the reservoir matrix. It is proposed that sand concentration is an important factor in increasing production in shale gas development and extraction.

Warpinski et al. [22] conducted transport experiments and distribution studies of proppants by changing their particle size. They concluded that the final position of proppants in the reservoir is significantly smaller than the distance that slick water can enter the reservoir. It revealed a strong correlation between the proppant particle size and its transport distance, with small particle size

Journal of Physics: Conference Series **2808** (2024) 012081 doi:10.1088/1742-6596/2808/1/012081

proppants having a larger transport distance and large particle size proppants having a smaller transport distance.

Cipolla et al. [23] studied the effect of proppants on the flow conductivity of volume fracturing networks. The experimental results showed that factors such as permeability, proppant type, and paving concentration can affect the production capacity of the network, and it was found that the construction parameters of proppants play an important role in production capacity.

Leonard et al. [24] conducted indoor experiments to study the embedding of proppants and concluded that the embedding of proppants directly affects the seam width. In fracturing and increasing production, the degree of proppant embedding is related to the reservoir hardness and proppant strength. High-strength proppants can be embedded in soft formations to a depth that can reach the particle size of multiple proppants. The seam width will decrease by 10% to 60% due to the insertion of the support agent. Through physical model demonstration and simulation, it was found that the rate of fluid flow decrease in oil wells is linearly related to the cubic width of fractures.

Weaver et al. [25] discovered a positive correlation between the extent of proppant insertion and the intensity of closure pressure, and high closure pressure can significantly reduce the conductivity of fractures. Based on this, Chinese scholars comprehensively study the embedding and diversion capabilities of proppants by combining the characteristics of shale geological reservoirs and proppants through indoor experiments, numerical simulations, and other means.

The flow capacity of sand-filled cracks is influenced by factors such as crack morphology, proppant embedding, fragmentation, and migration. The embedding and fragmentation of proppants are closely related to rock layers, mechanical parameters of proppants, and stress environments. Currently, Jiyang shale oil-filling sand fractures are facing problems with short-term efficiency. It is necessary to conduct targeted testing, explore relevant factors, and guide the design of fracturing technology and development plans. This study is based on physical simulation experiments combined with tests on the flow conductivity, crack morphology, and proppant morphology of sand-filled fractures, revealing the mechanism through which effective stress influences the flow capacity of sand-filled fractures and establishing reasonable sand laying concentration.

2. Experiment

2.1. Experimental principles

Referring to SY/T 6302-2019 "Test Method for Fracturing Proppant Flow Capacity", combined with Darcy's Formula, Darcy's Formula can be converted to obtain the formula for the fracture flow capacity of artificially filled sandstone cores (Figure 1):

$$\mathbf{Q} = \frac{k A \Delta p}{\mu L} \tag{1}$$

$$C_f = k w_f = \frac{Q \mu L}{\Delta p d} \tag{2}$$

In the formula, C_f is the diversion capacity of the sand-filled fracture rock core; k is the permeability of sand-filled joints; w_f is the seam width; μ is the experimental fluid viscosity; Q is the rate of liquid flow; L is the length of the rock core; d is the length of the crack in the cross-section of the rock core; Δp is the pressure difference.

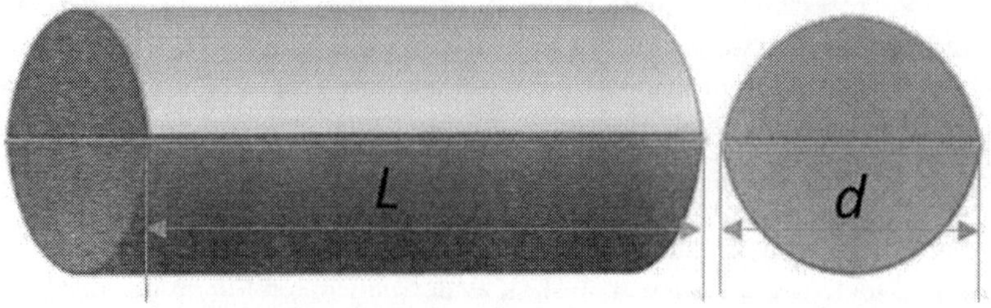

Figure 1. Schematic diagram of sand-filled rock core.

2.2. Experimental materials and equipment

2.2.1. Experimental materials. The kerosene density employed for the experimental process is 0.8 g/cm^3, with a room temperature viscosity of 2.335 mPa·s. The required proppant is quartz sand with particle sizes of 40/70 mesh, 70/120 mesh, and 100/140 mesh. The quartz sand used in the experiment includes sand particles with different particle sizes (Figure 2). The core of this experiment is the actual core sample taken by Shengli oilfield, which is used for sewing and preparing sand-filled core (Figure 3).

Figure 2. Physical image of proppant particles.

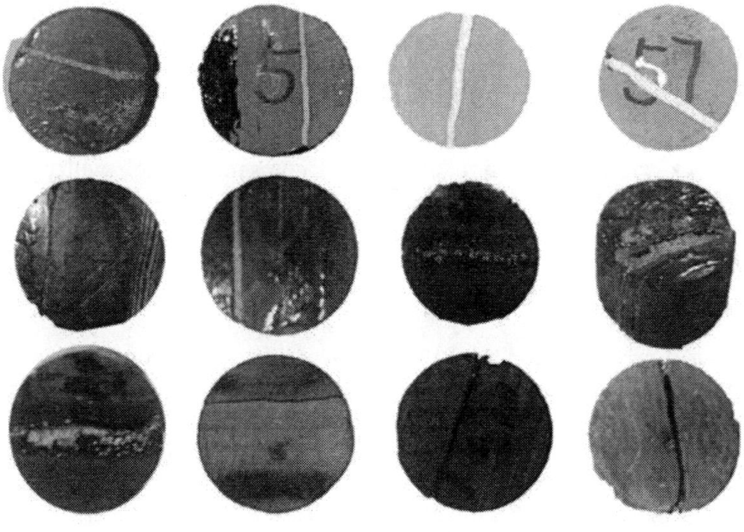

Figure 3. Cores are used to test the diversion capacity of sand-filled fractures.

2.2.2. Experimental equipment. The equipment process, including the ISCO pump (maximum capacity 260 mL and minimum pressure 0.01 MPa), high temperature and high-pressure displacement device, Brazil splitting device core polishing machine, and dynamic particle image analyzer, is shown in Figure 4.

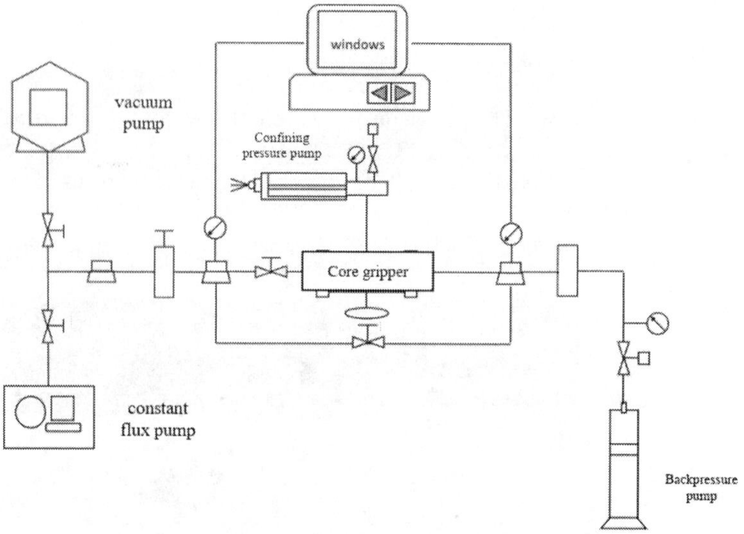

Figure 4. Experimental flowchart.

2.3. *Experimental steps and plan*

2.3.1 Core preparation steps. (1) The top of the splitting machine is aimed at the bedding seams of the quasi-rock core, and the pressure acts on the weak stress surface of the bedding seams locally. Slowly applying pressure opens the rock core and obtains natural rough cracks. (2) The layout chamber of the splitting machine is concave, and the walls on both sides can simulate the process of artificial crack opening under the original formation confining pressure conditions. (3) We measure the length of the structural fracture core and the width of the entrance crack, and calculate the width of the proppant placement crack and the quality of the proppant placement based on the sand concentration and the packing density of the proppant particles. (4) We take out the split rock core, evenly lay the proppant particles on the crack surface, and then fix the two rock cores into a whole with epoxy resin AB adhesive. After the solidification of AB adhesive is completed, a core polishing device is used to polish the semi-finished core to a standard sand-filled rock core of 2.5 cm. (5) After sealing 500 mesh nylon fine mesh at both ends of the core, the core is placed into a gripper to conduct stress sensitivity testing of the sand-filled core under constant confining pressure and variable flow pressure.

2.3.2 Test steps for diversion capacity. (1) We connect the experimental device according to the stress sensitivity testing device diagram of shale artificial fractures and test the airtightness of pipelines and instrument devices. Secondly, the rock sample filled with proppant and fully plastic-sealed material should be placed into the core gripper. (2) Using pressure tracking mode for incremental pressurization of the testing system, the confining pressure gradually increases from a baseline of 0 MPa to a peak of 5 MPa. Then, the pressure gap amidst the confining pressure and flow pressure sustains a constant level of 5 MPa, synchronously increasing the confining pressure and flow pressure until the confining pressure rises to 50 MPa and the flow pressure rises to 25 MPa. This ensures that the artificial fracture permeability does not cause significant damage in the early stage but also avoids damage to the gripper sleeve. (3) During the experiment, by adjusting the needle valve and adjusting the inlet back pressure, the changes in effective stress borne by artificial cracks and proppants were simulated to evaluate the stress sensitivity of artificial cracks under effective stress conditions. (4) We maintain a constant confining pressure, change the fluid pressure, gradually diminish the fluid pressure from a high to a low setting, and obtain the variation law of flow conductivity's change in response to fluid pressure when maintaining a stable confining pressure. (5) We handle the upstream-downstream pressure differential data and calculate the liquid permeability of the rock core.

2.3.3 Experimental plan. The experimental plan and CT scanning experimental plan for the influencing factors of sand filling seam diversion capacity are shown in Tables 1 and 2.

Table 1. Experimental plan for factors influencing the flow capacity of sand filling fractures.

Name	Sanding concentration (kg·m^{-3})	Effective stress (MPa)
40/70 mesh natural sand	0.5, 1, 2, 3	25, 30, 35, 40, 45
70/120 mesh natural sand	0.5, 1, 2, 3	25, 30, 35, 40, 45
100/140 mesh natural sand	0.5, 1, 2, 3	25, 30, 35, 40, 45

Table 2. CT scanning experimental plan.

Serial Number	Fluid pressure (MPa)	Confining pressure (MPa)
1	0	0
2	0	20
3	5	20
4	10	20
5	15	20
6	10	20
7	5	20

3. Experimental results and analysis

3.1. Analysis of experimental results of diversion capacity testing

The influence of proppant sand laying concentration is shown in Figure 5. It can be concluded that the diversion capacity of sand-filled cracks diminishes roughly linearly as the effective stress rises. The main reasons for this are: (1) When pressurizing shale, the spacing between fractures is reduced to a certain extent, causing a reduction in fracture width and compression of the diversion space. At the same time, there are certain grooves on the surface of the cracks, and when the confining pressure is low, these grooves have good connectivity and diversion ability. However, while increasing the confining pressure, these grooves are subjected to compression damage, and the reduction of crack spacing and diversion space results in a decrease in diversion capacity. (2) When the confining pressure of the proppant is small, there are large gaps between adjacent proppant particles, which become good seepage channels and can effectively connect the two ends of the shale. However, the increase in confining pressure causes these spaces to shrink and the support agent to be compacted, some of which may break.

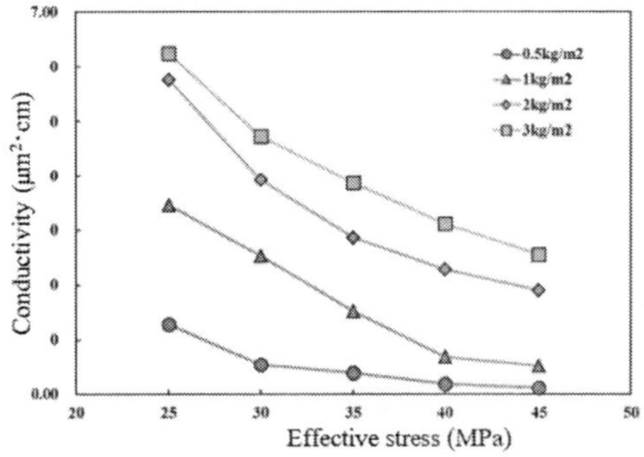

a. 40/70 mesh natural sand

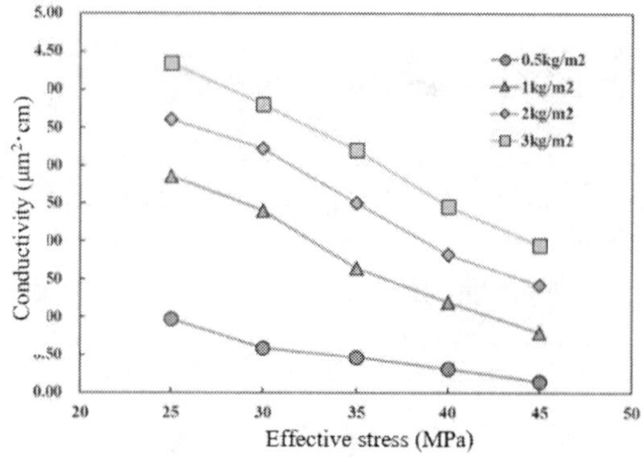

b. 70/120 mesh natural sand

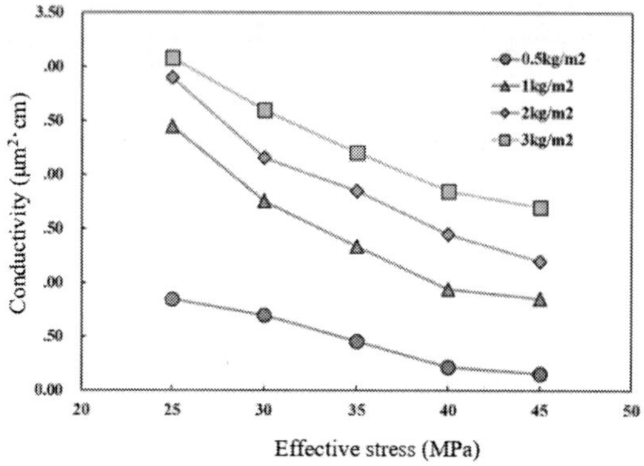

c. 100/140 mesh natural sand

Figure 5. The effect of proppant sand concentration on the flow capacity.

The stress sensitivity of sand-filled rock cores is shown in Table 3. It can be obtained that for the identical type of proppant, the diversion capacity intensifies as the sand concentration rises, and the sensitivity of crack stress is weaker. At a constant sand concentration, an increase in the mesh count of the proppant results in reduced stress sensitivity of the cracks.

Table 3. Stress sensitivity evaluation table for the sand-filled core.

Sanding concentration	40/70 mesh natural sand	70/120 mesh natural sand	100/140 mesh natural sand
0.5	91%	85%	81%
1	85%	72%	65%
2	67%	61%	59%
3	59%	56%	45%

The diversion capacity of quartz sand with varying particle sizes is shown in Figure 6. At a constant sand concentration, the initial flow conductivity is greater when the particle size of the proppant is larger. However, upon an increase in closure stress, the decrease in flow conductivity is greater than that of smaller particles, exhibiting stronger stress sensitivity. The reason is that as the particle size of the proppant increases, the pores created within the cracks also enlarge. In addition, its crushing strength is low, and it is more prone to breakage under closed pressure. The width of the cracks changes more significantly, forming a denser oil flow channel and resulting in a more significant decrease in flow conductivity.

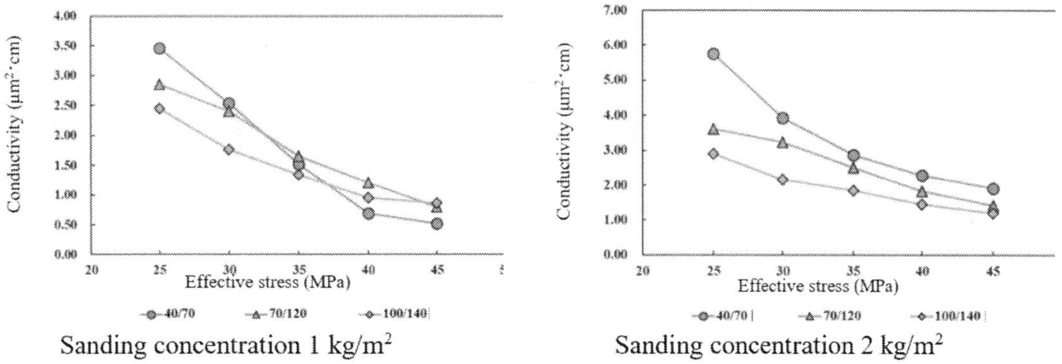

Figure 6. Different sand concentrations and different particle sizes of quartz sand flow conductivity.

3.2. Analysis of experimental results of true triaxial stress online CT

CT scan images of the morphological changes of shale and sandstone interbedded sand fractures are shown in Figures 7 and 8. It can be obtained that (a) for the crack section without filling the proppant (see bottom), as the elastic development flow pressure decreases, the crack gradually closes. (b) As the effective stress increases, microcracks develop around the unfilled portion. (c) As the effect increases, the proppant is compacted and arranged more regularly, resulting in a decrease in porosity and crack width of the sand-filling layer and a decrease in conductivity. (d) There was no evidence of proppant breakage or embedding within the testing pressure range. (e) Sandstone has not developed bedding fractures, and the fracture surface is highly tortuous and irregular, resulting in uneven sand laying. (f) After loading stress, the unsupported (low sand concentration) cracks almost closed, the width of the supporting cracks significantly decreased, and the proppant was compacted. (g) During the fracturing pressure increase process, as the flow pressure increases, the supporting fractures widen, and the contour of unsupported fractures becomes clearer. During the pressure reduction stage, the width of both unsupported and sand-filled fractures decreases.

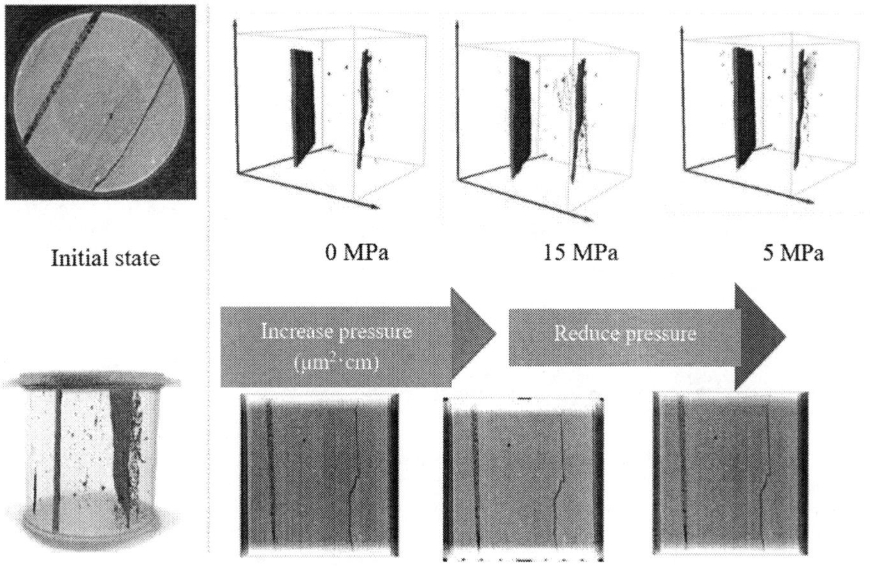

Figure 7. CT scan image of the morphological changes of shale sand 40/70 mesh fractures.

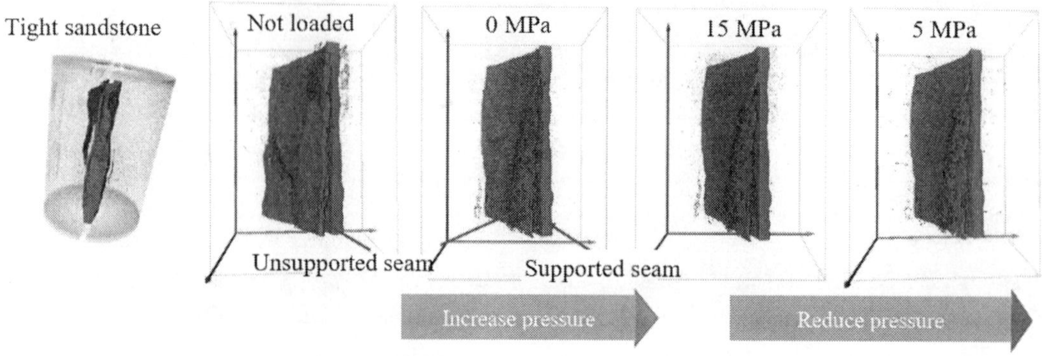

Figure 8. CT scan image of the morphological changes of sandstone sand 70/120 mesh fractures.

The rock core's cross-section and sagittal plane are shown in Figure 9, and the changes in crack width at three points are detected. The results are shown in Figure 10, which can calculate the rate of crack width change during elastic development and pressure reduction. The results are shown in Figure 11. It can be obtained that (a) the seam width gradually increases during the pressure increase process, and the seam width decreases relatively smoothly during the pressure decrease. (b) The larger the crack width is, i.e., the higher the sand concentration is, the less significant the crack width change is (average pressure rate increase: A: 10.67%, B: 6.22%, C: 0.33%). (c) In the unloaded state, the particle accumulation is relatively loose, and after applying stress, the particle arrangement becomes compact, manifested as a significant reduction in crack width. (d) During the elastic mining process, the filling layer is continuously compacted, and the crack width approximately decreases linearly. (e) The stress sensitivity of shale-filled sand fractures is slightly smaller than that of sandstone-filled sand fractures.

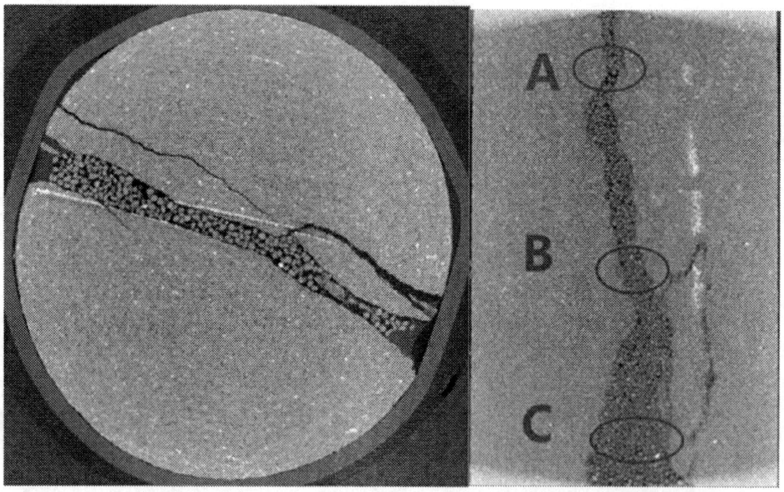

a. Cross section b. Sagittal

Figure 9. Cross section and sagittal plane of crack morphology.

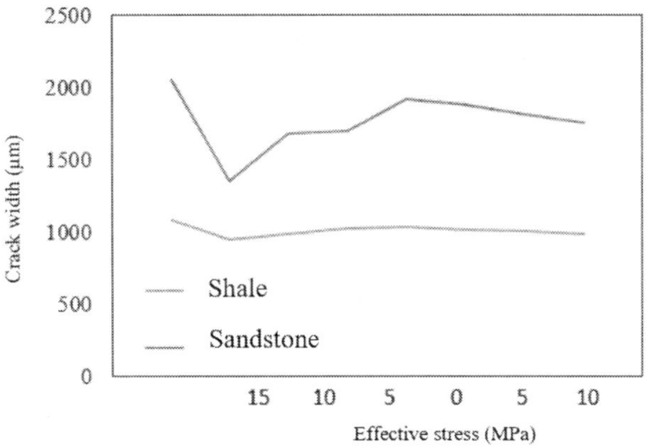

Figure 10. The variation of fracture width with effective stress in different sand-filled rock cores.

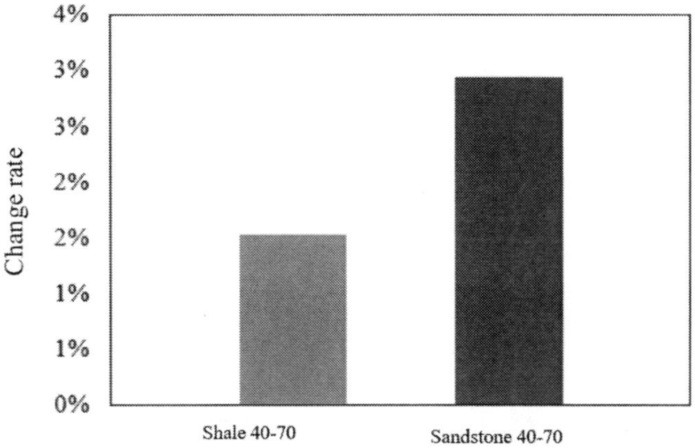

Figure 11. The rate of change in seam width during elastic development for pressure reduction.

4. Conclusion

Based on the results and discussions presented above, the conclusions are obtained as follows:

(1) As the sand concentration increases, the diversion capacity of sand-filled cracks increases, and stress sensitivity weakens.

(2) Under low effective stress, the large particle size proppant has a higher diversion capacity when filling sand fractures. As the effective stress increases, the reduction in diversion capacity is significant, and the stress sensitivity is small. The particle size sand layer is strong.

(3) During the process of elastic pressure reduction and failure, the proppant is compacted, the pores of the sand layer are reduced, the crack width is reduced, and the flow conductivity is reduced. Within the effective stress range of<20 MPa, no fragmentation or embedding phenomenon of 40-70 mesh proppant was observed, and the stress sensitivity of sandstone-filled sand fractures was similar to that of shale.

Acknowledgments

This work received financial backing from the Research project of Sinopec Key Laboratory, "Research on the Flow Law of Elastic Development in Shale Oil Fracturing Production (KLP23024)", and Sinopec Research Project, "Research on integrated technology of shale oil construction module, digital and analog (P22036)".

Journal of Physics: Conference Series **2808** (2024) 012081 doi:10.1088/1742-6596/2808/1/012081

References

[1] Jin Zhijun, Bai Zhenrui, Gao Bo, et al. Has China ushered in the shale oil and gas revolution? [J]. Oil & Gas Geology, 2019, 40(3): 451-458.
[2] Ma Yongsheng, Cai Xunyu, Zhao Peirong. China's shale gas exploration and development: Understanding and practice [J]. Petroleum Exploration and Development, 2018, 45(4): 561-574.
[3] Li Zheng, Wang Xiuhong, Zhu Rifang, et al. Geochemical evaluation of shale oil in lower Es3 and upper Es4 in Jiyang Depression [J]. Xinjiang Petroleum Geology, 2015, 36(5): 510-514.
[4] Liu Yali, Zhang Shun, Liu Huimin, et al. Basic characteristics and exploration prospect of shale in steep slope belt of continental faulted basin: A case study of Jiyang Depression [J]. Journal of China University of Mining & Technology, 2021, 50(6): 1146-1156.
[5] Liu Huimin. Exploration practice and prospect of shale oil in Jiyang Depression [J]. China Petroleum Exploration, 2022, 27(1): 73-87.
[6] Wang Min, Guan Ying, LiI Chuanming, et al. Qualitative description and full-pore-size quantitative evaluation of pores in lacustrine shale reservoir of Shahejie Formation, Jiyang Depression [J]. Oil & Gas Geology, 2018, 39(6): 1 107-1 119.
[7] Shen Yunqi, Jin Zhijun, Su Jianzheng, et al. Characteristics of horizontal and vertical permeability of continental shale oil reservoirs in China: A case from Jiyang Depression in Bohai Bay Basin and Qianjiang Sag in Jianghan Baisn. [J] Oil & Gas Geology, 2022, 43(2): 378-389.
[8] Du Dianfa, Zhang Yaozu, Zhang Lina, et al. Research progress and prospect of seepage mechanism in shale gas reservoirs [J]. Unconventional Oil & Gas, 2021, 8(3): 1-9.
[9] Li Lei, Hao Yongmao, Wang Chengwei, et al. Low-velocity seepage characteristics of single-phase fluid in shale reservoir [J]. Special Oil & Gas Reservoir, 2021, 28(6): 70-75.
[10] Li Jingjing, Sun Guoxiang, Liu Qi, et al. Pore structure and sensitivity of shale reservoir in Lu 1 member of Jimsar Sag [J]. Xinjiang Petroleum Geology, 2021, 42(5): 541-547.
[11] Mei Haiyan, Ma Mingwei, Yu Qian, et al. Productivity and influencing factors of the fractured horizontal well in the shale gas reservoir [J]. Petroleum Geology & Oilfield Development in Daqing, 2019, 38(1): 71-77.
[12] Ma Bingjie, Fan Fei, Sun Zhigang, et al. Experimental study on influencing factors of shale oil flow capacity in laminar shale in Jiyang Depression [J] Petroleum Geology & Oilfield Development in Daqing, 2022, 41(5): 153-159.
[13] Duan Xianggang, Hu Zhiming, Chang Jin, et al. Gas Flow Capacity Experiment in Shale Reservoir [J]. Special Oil & Gas Reservoirs, 2019, 26(3): 143-147.
[14] Gensterblum Y, Ghanizadeh A, Cuss R J, et al. Gas storage capacity and transport in shale gas reservoirs-A review. Part A: Transport processes [J]. Journal of Unconventional Oil and Gas Resources, 2015, 12: 87-122.
[15] Sander R, Pan Z J, Connell L D. Laboratory measurement of low permeability unconventional gas reservoir rocks: A review of experimental methods [J]. Journal of Natural Gas Science and Engineering, 2017, 37: 248-279.
[16] Zhang J, Kamenov A, Zhu D, et al. Propped Fracture Conductivity in Shales [A]. ASME 2013 32nd International Conference on Ocean, Offshore and Arctic Engineering [C]. France: American Society of Mechanical Engineers Press, 2013: 301-302.
[17] Cooke JR C E. Conductivity of fracture proppants in multiple layers [J]. Journal of Petroleum Technology, 1973, 25(9): 101-107.
[18] Zhu Yihui, Jia Changgui, Jiang Tingxue. Calculation of shale supported fracture conductivity under multiple factors [J]. Journal of Oil and Gas Technology, 2014, 36(8): 129-132.
[19] Zou Yushi, Zhang Shicheng, Ma Xinfang. Assessment on the effectiveness of proposed fractures in the fracturing of shale gas reservoirs. Natural Gas Industry, 2012, 32(9): 52-55.
[20] Wen Qingzhi, Li Yang, Hu Lanxiao, et al. Experimental analysis of the flow conductivity of the network-like fractures in shale reservoirs [J]. Journal of Northeast Petroleum University, 2013, 37(6): 55-62.

[21] Peng Kexiang, Li Shaoming, Zhong Chengxu, et al. Experiment Research on Fracture Conductivity of Shale Based on Gas Logging [J]. Contemporary Chemical Industry, 2016, 45(11): 2520-2523.
[22] Warpinski N R, Mayerhofer M J, Vincent M C, et al. Stimulating unconventional reservoirs: maximizing network growth while optimizing fracture conductivity [J]. Journal of Canadian Petroleum Technology, 2009, 48(10): 39-51.
[23] Cipolla C L, Lolon E, Mayerhofer M J, et al. Fracture design considerations in horizontal wells drilled in unconventional gas reservoirs [C]. SPE 119366, 2009.
[24] Malik M A, Hill A D. A New Technique for Laboratory Measurement of Acid Fracture Conductivity [R]. SPE 19733, 1989.
[25] Weaver J D, Nguyen P D, Parker M A, et al. Sustaining fracture conductivity [J]. SPE 94666, 2005.

Journal of Physics: Conference Series **2808** (2024) 012082 doi:10.1088/1742-6596/2808/1/012082

Research on creep mechanical properties of flexible screen functional film layer

Yuhua Ma [1], Weiwei Su [1], Xin Luo [1], Yubin Hu [1], Yiming Yang [1], Geng Zhou [1*] and Yachao Xiao [1]

[1]CSSC Haiwei High-tech Co., Ltd, Zhengzhou, Henan, 450000, China

* Corresponding author's e-mail: zgqz66@163.com

Abstract: The creep behavior of AMOLED flexible screens is a crucial consideration in the design and manufacturing processes. Various factors impact the creep process of AMOLED modules in thermoplastic polymers. However, experimental work typically only focuses on sustained stress over time at room temperature, neglecting other factors affecting creep strain. It is essential to explore the influence of creep on the bending performance of flexible AMOLED modules and analyze the creep curves of each film layer for the bending of flexible screens. Furthermore, the analysis includes the examination of the influence of the thickness and elastic modulus of the protective cover plate and support plate on screen bending at different temperatures, along with further creep analysis of the protective cover plate and the support plate. The mechanical exploration of the protective cover plate and support plate under 6-9 h creep response was analyzed, and the influence of the protective cover plate and support plate on screen bending performance was analyzed. We conduct stress-strain analysis on the protective cover plate at different temperatures (-40~120°C). Before the temperature increases to 80°C, the stress and strain increase with the increase of temperature, and after the temperature reaches 80°C, the stress and strain decrease with the increase of temperature. The temperature of the protective cover plate is 20°C, and the risk of bending failure is relatively low. The creep analysis of the support plate shows a small difference in the fracture strength between the support plate undergoing creep and the fracture strength not undergoing creep. The fracture elongation of the support plate at 20°C is 0.52%, and the fracture strength is 106 MPa. The fracture elongation of the support plate after heating also increases.

1. Introduction

In the 19th century, physicists such as Maxwell, Boltzmann, and Kelvin began studying the viscoelasticity, creep, and recovery of glass, metal, and rubber [1]. After the conclusion of research in the 20th century, additional studies were carried out on synthetic polymers, and the viscoelastic properties of various polymers were examined using different viscoelastic mathematical models [2]. Currently, most experiments rely on the Maxwell and Kelvin-Voigt models, in conjunction with the Boltzmann (Boltzmann superposition principle), to characterize the creep behavior of various materials [3].

Barthel et al. [4] suggested a functional expression to measure the impact of viscoelastic response on the binding energy as a function of contact edge velocity. Also, they proposed a straightforward analytical function expression for creep response.

Murakami et al. [5] elucidated the influence of damage anisotropy on the propagation and fracture frequency of creep cracks. The numerical results were compared with the corresponding results based on isotropic damage theory, and the influence of local fracture criteria was discussed, providing a theoretical basis for creep simulation [6]. However, these works did not pay attention to the service life of AMOLED, especially the creep resistance, which is a very important parameter in AMOLED packaging applications [7]. For example, the protective cover and support plates in AMOLED displays may experience static loads from substrate mismatch during most bending cycles [8-9]. The protective cover plate and support plate layers can maintain long-term loads without excessive deformation [10]. Large deformation in the protective cover plate can cause misalignment of the polarizing film and AMOLED module edges, leading to serious light leakage [11]. This is one of the key issues affecting the bending performance of AMOLED modules [12-13]. This article compares the strain of each film layer of the AMOLED module during creep and explores the mechanical properties of flexible screen bending under different environmental temperatures (-40~120°C) [14-15]. By analyzing the mechanical effects of protective covers and support plates with different stiffness and thickness on the important functional module OLED layer, the optimal bending method and bending temperature for AMOLED flexible screens were obtained. We analyzed the mechanical behavior of the flexible screen after the creep phenomenon at different temperatures. Then we analyzed the creep behavior of the protective cover plate and support plate at the optimal temperature.

2. Composition and theory of AMOLED

2.1. AMOLED bending trajectory

Due to the small bending radius of the flexible AMOLED module, it can easily cause severe deformation of the OCA adhesive layer, leading to detachment due to excessive force. This can damage the film and display area, and the flexible AMOLED screen may fail to meet production standards. To prevent detachment of the OCA layer in flexible AMOLED, it is necessary to keep the entire AMOLED module in a low-stress state to reduce the deformation of the OCA layer. Therefore, optimizing and adjusting the bending motion of the module and having a detailed understanding of the bending stress and deformation process during the machining process are of great help.

In establishing the geometric model, separate modules were used for experiments without adding electronic devices. Two auxiliary steel plates were used to bend the flexible screen. The top is a protective cover plate, the bottom is a support plate for support, and the other functional film layers are in the middle. The various film layers are bonded with OCA optical transparent adhesive. The module has a length of 120 mm and a thickness of 307 mm, and its structure is shown in Figure 1.

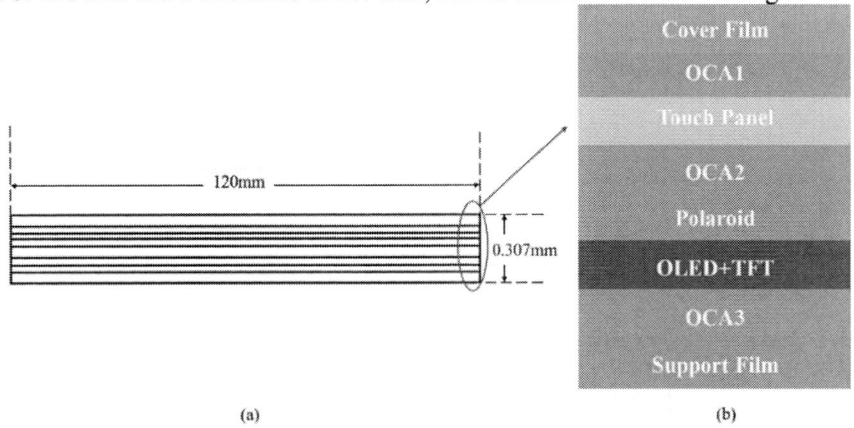

Figure 1. Module structure diagram.

This article simulates the bending motion of a flexible screen, which is a water droplet bending of a relatively mature flexible AMOLED module. It explores the mechanical state of the module after bending is completed. The length of the flexible screen is 120 mm, the width is 71.5 mm, and the thickness is 307 mm μ m. The bending radius is 3 mm. A 180° water droplet bend was made on the AMOLED screen, with auxiliary instruments at both ends of the module to complete the bend, as shown in Figure 2. The steel plate covers many flexible screen parts, making it difficult to see clearly. Figure 2(a) shows the bending diagram of the flexible screen during the experimental process.

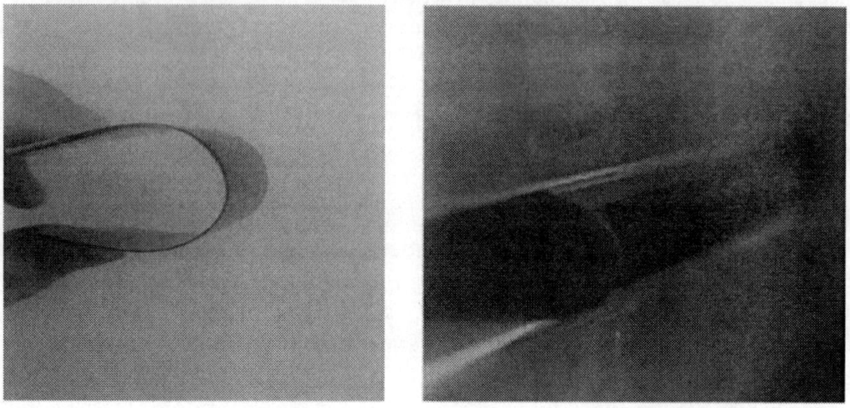

(a) (b)

Figure 2. (a) Schematic diagram of flexible screen water droplet bending; (b) Flexible screen water droplet type experimental bending diagram.

We set two rigid bodies under the module for auxiliary bending and set reference points RP-1 and RP-2 for the two endpoints of steel. The rigid body has two reference points, RP-1 and RP-2. RP-1 is rotating clockwise at 1.57 rad/s, while RP-2 is stationary to support the screen body. When bending, an arc is formed, and the bending part will reduce the stress on the structure. The flexible AMOLED simulation bending motion is shown in Figure 3, and the flexible screen remains bent for 800 seconds to simulate actual use.

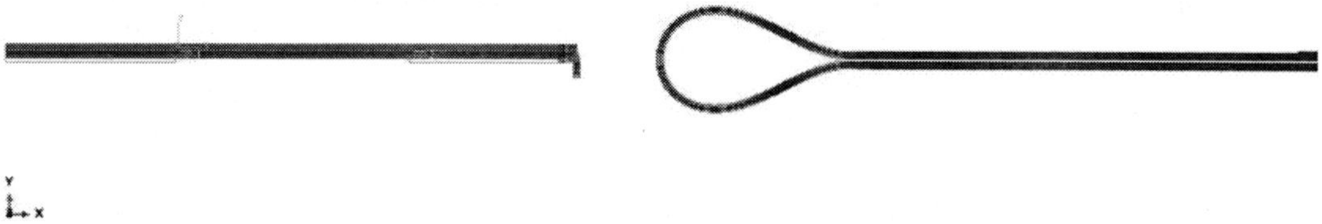

(a) (b)

Figure 3. Simulated bending motion diagram.

2.2. Thermal expansion of film material

Figure 4 shows the strain time curve of the classical creep theory. The phenomenon of stress relief after applying constant stress to the material and maintaining this stress for some time is a creep recovery test.

Journal of Physics: Conference Series **2808** (2024) 012082 doi:10.1088/1742-6596/2808/1/012082

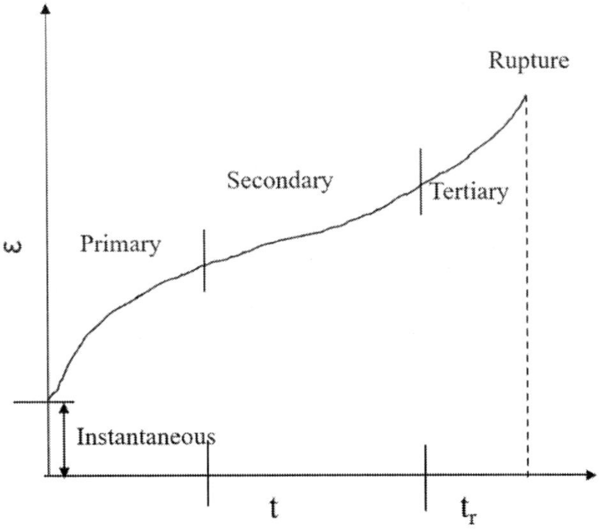

Figure 4. Classic creep stage curve (creep strain changes over time).

The initial stage is the stress-strain curve that immediately increases with the application of stress. Following the instantaneous stage, the focus shifts to the creep strain to observe the pattern of strain increase during the loading cycle, known as a creep test. The reverse elastic strain occurs during the reduction stage, where the material undergoes reduction at a decreasing rate after the load is removed. The recovery process is called delayed elasticity since the recovery phase is within the elastic region. The creep recovery curve begins with elastic recovery, followed by non-elastic recovery (recovery with strain over time), and may eventually result in permanent strain or deformation depending on the material type. Figure 5 illustrates the reaction curve of the creep reduction test.

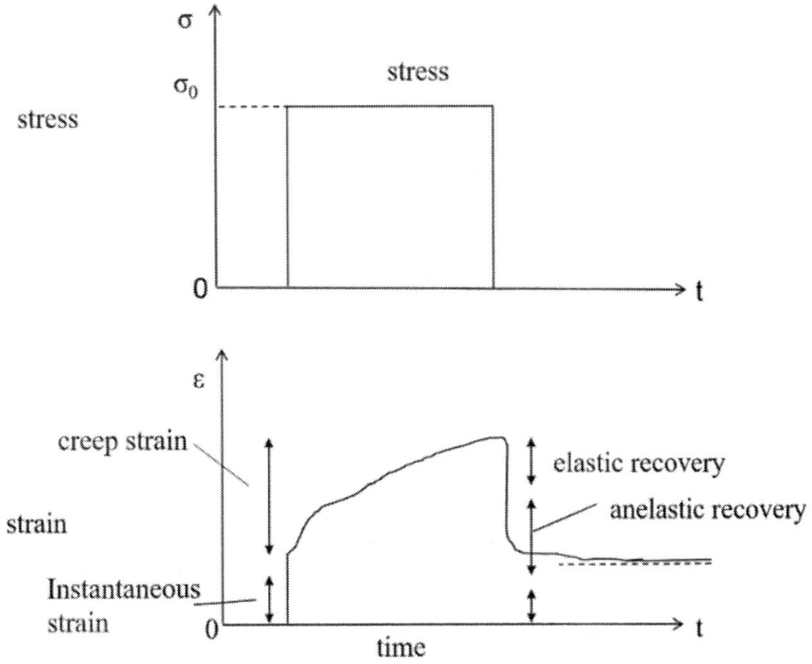

Figure 5. Response to creep recovery test.

The instantaneous elastic strain first appears, then increases with time, and finally recovers as creep strain. The ratio of total strain to constant stress applied is the creep compliance, which is expressed as:

$$J(t)=\frac{\varepsilon(t)}{\sigma_0} \tag{1}$$

where $J(t)$ is the creep flexibility, $\varepsilon(t)$ is the total strain, and σ_0 is the applied constant stress.

3. Creep analysis of AMOLED modules

3.1. Determination of creep coefficient

According to the creep theory of a linear solid model, the instantaneous strain is recorded as the tension of the spring under applied stress. The stress begins to transfer to the damper, and the strain curve gradually increases with time and shows a decreasing trend. After 100 minutes, the elastic spring partially recovered, while the viscous damper did not recover to creep deformation. The WinTest analysis software created a step tensile test with a test definition. We set the "Go to Stress and Hold" command in the definition according to the description of the viscoelastic SLS model.

Figure 6. Electronic universal testing machine.

The function of the recorded creep curve over time is shown in Figure 7. We observe the change of strain rate over time to obtain a creep curve. The specimen fractures are under a load or stress of 20 MPa or above. Therefore, most tests are conducted under a pressure of less than 20 MPa, but a few specimens still have broken. The fracture may be due to the accumulation of material strain during the stretching process reaching the threshold. During the testing process, the same tensile stress was maintained for 22 hours, and then the creep recovery state was achieved by reducing the load by 10 MPa, and the strain rate was recorded as a function of time. Due to the small disturbance of stress during the experiment, the recorded creep curve was not smooth, resulting in many ups and downs.

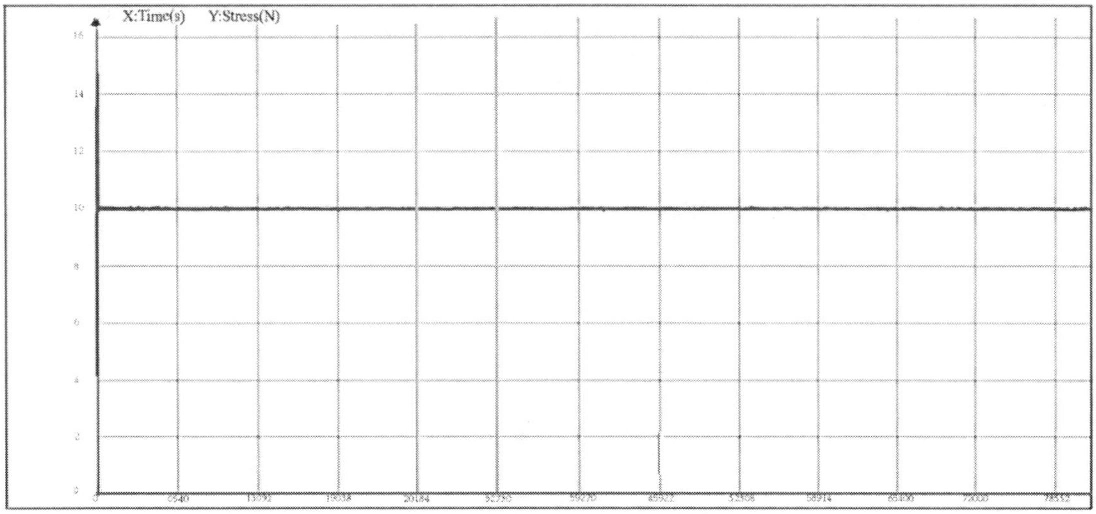

Figure 7. Material creep curve.

3.2. Impact of static bending of the protective cover plate

We analyze the mechanical properties of the protective cover plate under various temperature conditions to investigate the changes during the bending process of the protective cover plate in flexible AMOLED modules. Figure 8 examines the stress-strain curves of the protective cover plate after bending at temperatures of 20°C, 40°C, and 60°C without temperature setting. Figure 9 displays the tensile fracture strength and elongation at the break of the protective cover plate at different temperatures. The stress-strain curve is nonlinear and can be divided into two stages: initial and creep. The figure illustrates the standard linear solid model of the creep curve. Initially linear, the material experiences elastic changes due to molecular orientation. The grid positioning of the protective cover plate is relatively chaotic during the creep stage, and the fiber strength is not high. Figure 9 shows that the fracture strength is 109 MPa, which is significantly higher than the fracture strength at other temperatures without solidification temperature. The elongation at the break of the protective cover plate has slightly increased after temperature treatment. The protective cover plate broke at 20°C with a fracture strength of 98 MPa and at 60°C with a fracture strength of 97 MPa. The fracture strength of the protective cover will fluctuate with the increase in temperature. The ultimate tensile strength of the protective cover plate is highly sensitive to the loading rate. As the creep tensile displacement increases, the fracture strength of the protective cover plate also increases. The bond length and angle of polymer molecules are more likely to undergo instantaneous deformation with increasing temperature. Therefore, different temperatures also significantly improve the longitudinal shrinkage, strength, modulus, etc. of the reinforced protective cover plate. The elastic modulus of the protective cover plate and the elastic modulus before creep remain unchanged with the increase in temperature. Still, the tensile length significantly increases, indicating an improvement in the mechanical properties of the protective cover plate.

Journal of Physics: Conference Series **2808** (2024) 012082 doi:10.1088/1742-6596/2808/1/012082

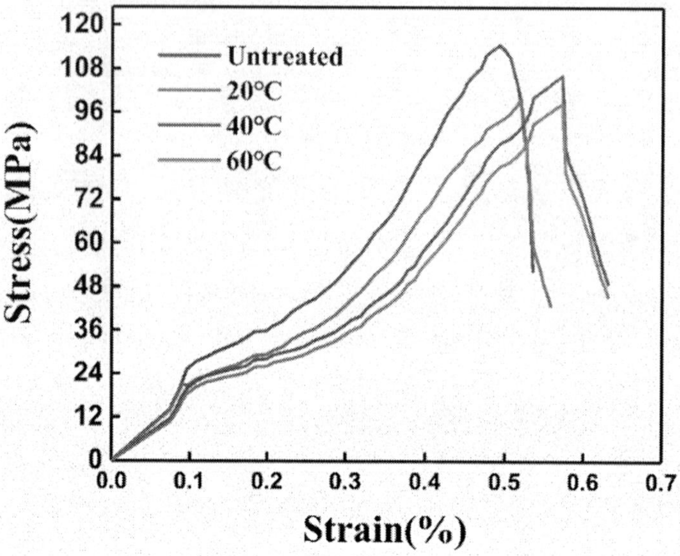

Figure 8. Stress-strain curves of the protective cover plate at different temperatures.

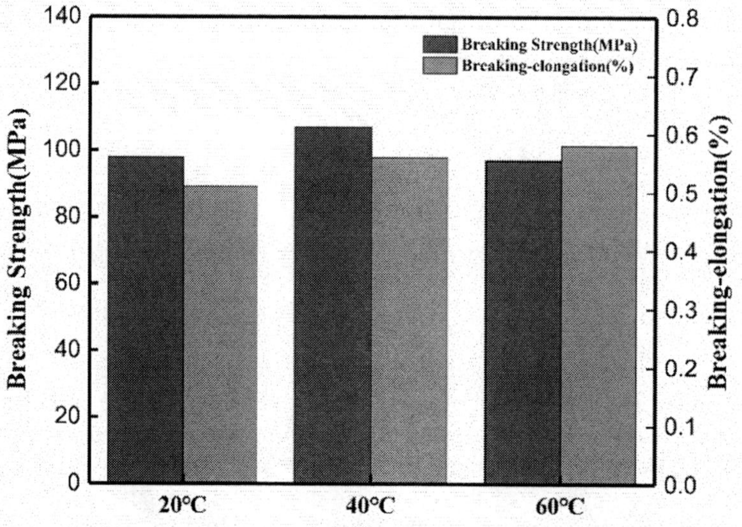

Figure 9. Breaking strength and elongation at the break of the protective cover plate.

3.3. The influence of the stiffness of the support plate on static bending

The support plate is at the bottom of the screen to protect it, with an elastic modulus of E=5.6 GPa. Figure 10 shows the influence of the creep characteristics of the support plate material. Figure 11 shows the predicted creep strain history characteristic curve and compares it with Figures 4-14 to verify the reliability of the support plate creep curve.

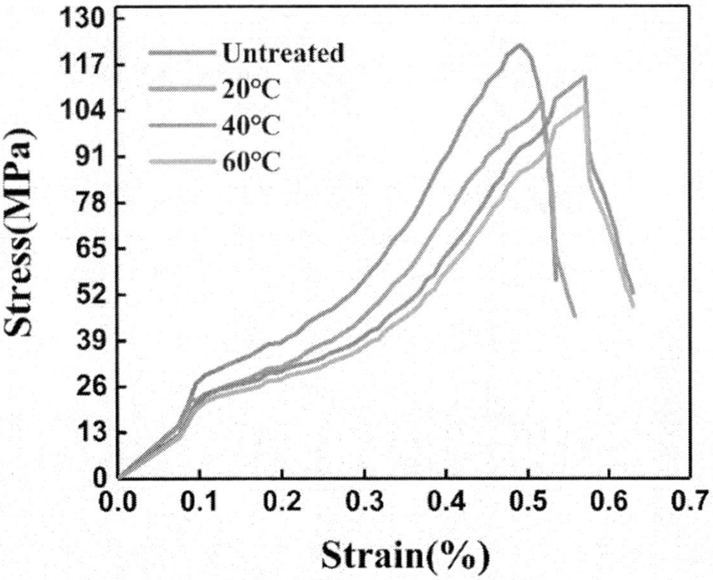

Figure 10. Stress-strain curves of the support plate at different temperatures.

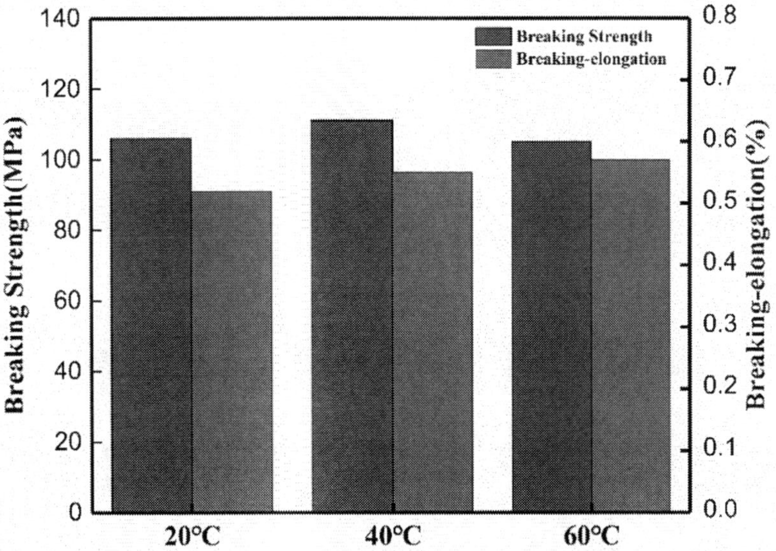

Figure 11. Breaking strength and elongation at break of support plate.

The support plate is a pliable material that is relatively prone to deformation, and its deformation is significant before fracture occurs. In ductile fracture, extensive plastic deformation (necking) occurs before fracture, and ductile fracture is better than brittle fracture. Permissible stress was used to ensure that the working stress of the support plate during bending did not exceed a certain limit. Figure 10 shows the stress-strain curve of the support plate at different temperatures. Figure 11 shows the breaking elongation of the support position.

Figures 10 and 11 show that the stress-strain curve of the support plate is the smoothest at 60°C, and the fracture point is at a strain of 0.57% and a stress of 105 MPa. The fracture point of the support plate at 20°C is at a strain of 0.52% and a stress of 106 MPa. The fracture of the support plate without setting temperature occurs at a strain of 0.49% and a stress of 121 MPa. It can be seen that the stress at the fracture point decreases as the temperature of the support plate increases, and the simulated bending stress of the support plate without temperature is 16 MPa higher than the stress at the fracture point at 60°C. The fracture strength of the support plate after high-temperature treatment significantly decreases, while the support plate without overheating treatment has high fracture strength and the maximum elastic modulus. As the temperature increases, the mesh of the support plate is more prone to instantaneous deformation. After heating, the fracture strength of the support plate is lower than that of the lower temperature support plate. Due to different temperatures, the shrinkage rate of the support plate in the longitudinal direction of stress is different, and there are also significant differences in strength and modulus. As the deformation displacement increases, the modulus of the support plate remains unchanged while the fracture strength significantly increases. The mechanical behavior of the support plate improves after creep occurs.

4. Conclusions

Stress-strain analysis of support plates is at different temperatures (-40~120°C). The stress and strain will increase with the rise in support plate temperature. Lowering the temperature can reduce the force on the support plate and also decrease the strain capacity. The change in support plate thickness significantly affects the stress situation of the OLED layer. The original 3 MPa of the OLED layer has been increased to 76 MPa, and the support plate thickness is 75 microns. The OLED layer experiences the minimum force at point M. The protective cover indicates that the fracture strength fluctuates with increasing temperature. The elongation at the break of the protective cover plate is 0.51%, and the fracture strength at 20°C is 98 MPa. As the temperature rises, the stress-strain curve of the protective cover plate decreases. The creep analysis of the module shows that the flexible AMOLED module is statically bent at 20°C, and the strain capacity of the OLED layer is 0.13%, with the smallest strain capacity in the module. As the creep time increases, the strain rate of each film layer also increases. The strain rate of the OCA layer is 3.35% higher than that of other functional film layers. A relatively uniform distribution of strain rates protects the creep response characteristics of the cover plate and the support plate. Comparing the fracture strength of flexible screens at different temperatures, the fracture strength of the protective cover plate at 20°C is 98 MPa, and the fracture strength of the support plate at 20°C is 106 MPa. The elastic modulus of the protective cover plate and support plate remains unchanged with the increase in temperature, but the elongation at break significantly increases.

References

[1] Cheng A. Chen Y. Jin J. (2019) Study on mechanical behavior and effect of adhesive layers in foldable AMOLED display by finite element analysis. SID Symp. Digest 50: 1060-1063.
[2] Hoonkee, Min, Jaeseob, Lee, Younggu, & Kim, et al. (2010). 2.8-inch wqvga flexible AMOLED using high-performance low-temperature polysilicon TFT on plastic substrates. SID International Symposium: Digest of Technology Papers, 41(2), 706-709.
[3] Yagi, I., Hirai, N., Miyamoto, Y., Noda, M., & Urabe, T. (2012). A flexible full-color AMOLED display driven by outfits. Journal of the Society for Information Display, 16(1), 15-20.
[4] Slanik M L, Nemes J A, Potvin M J, Piedboeuf J C (2000) Time domain finite element simulations of damped multilayered beams using a proxy series representation. Mech. Time-Dependent Mater. 4: 211-230.
[5] Lin Li, Yalong Li, Kun (2018) Reliability and failure mode analysis of foldable AMOLED display module. SID Symp. Dig. Tech. Pap. 49: 899–901.
[6] Kumar R. (1996) Foldable-display systems as a standard platform for multimedia use. IEEE Trans. Consum. Electron. 42: 17-21.
[7] Chiou J Y, Liu Y W, Niu Y F, Liu S F, Kuo C, Huang C Y (2017) Optimization of TFE structure by FTIR analysis and mechanical simulation to achieve excellent encapsulation and high flexibility AMOLED. SID. Symp. Dig. Tech. Papers. 48: 437-440.

[8] Park, S. G., Lee, J. H., Lee, W. K., & Han, M. K. (2012). Investigation of the hysteresis phenomenon of an a-si: h TFT at an elevated temperature for AMOLED displays. Journal of the Society for Information Display, 15(12), 1145-1149.

[9] Lin, Y. C., &Shieh, H. (2004). A novel current memory circuit for AMOLED. Electron devises IEEE Transactions on, 51(6), 1037-1040. [13] Lin, C. L., &Chen, Y. C. (2007). A novel ltps-TFT pixel circuit compensating for TFT threshold-voltage shift and OLED degradation for AMOLED. IEEE Electron Device Letters, 28, 129-131.

[10] Fan, C. L., Lin, Y. S., &Liu, Y. W. (2010). Low-temperature polycrystalline silicon thin film transistor pixel circuits for active matrix organic emitting light-emitting diodes. Ice Transactions on Electronics, 93(5), 712-714.

[11] Choi, B. D., &Byun, C. W. (2010). Data driving methods and circuits for compact and high-image-quality AMOLED mobile displays. IEEE Transactions on Consumer Electronics, 56(2), 1102-1107.

[12] WANG, Xin, ZHANG, Jianfu, ZHANG, & Yanghuan, et al. (2010). Synthesis and thermal expansion of 4j36/zrw_2o_8 composites. Rare Metals, 04(v.29), 35-39.

[13] Gaillard, A., Rogel, R., Crand, S., Mohammed-Brahim, T., & Prat, C. (2012). An averaging pixel structure using microcrystalline-silicon films prepared at high temperature for AMOLED displays. Journal of the Society for Information Display, 15(12), 1137-1143.

[14] Chiou J Y, Liu Y W, Niu Y F, Liu S F, Kuo C, Huang C Y (2017) Optimization of TFE structure by FTIR analysis and mechanical simulation to achieve excellent encapsulation and high flexibility AMOLED. SID. Symp. Dig. Tech. Papers. 48: 437-440.

[15] Lee S M, Kwon J H, Kwon S (2017) A review of flexible OLEDs toward highly durable unusual displays. IEEE Tran. Electron. Devices. 64: 1922-1931.

Journal of Physics: Conference Series **2808** (2024) 012083 doi:10.1088/1742-6596/2808/1/012083

Study on the correlation and calibration of macro and micro parameters of discrete element model for sandstone

Jiangchun Hu [1], Luyang An [1,*], Xuewei Huang [1] and Chenxu Wang [1]

[1]School of Civil & Architecture Engineering, Zhongyuan University of Technology, Zhengzhou, 450007, China

[*]Corresponding author's e-mail: anluyang0416@163.com

Abstract: The effectiveness of numerical simulation of discrete elements for particle flow in sandstone depends on adjusting mesoscopic parameters appropriately. The parallel bonding contact model was used as the constitutive model for particle contacts to explore the relationship between macro and micro parameters and fine-tune the mesoscopic ones. The relationship between macro and micro parameters of sandstone was determined through laboratory tests and numerical simulation comparative analysis. Orthogonal tests, multifactorial analysis of variance (ANOVA), and regression fitting were used for analysis. The results show that each mesoscopic parameter of the parallel bonding contact model has a distinct and significant impact on macroscopic mechanical parameters, with interactive influences between them. The macroscopic cohesive force, normal strength of the parallel bond, and parallel bonding force exhibit non-linear relationships, while the macroscopic internal friction angle and the internal friction angle of the parallel bond have logarithmic relationships. Additionally, there is a linear relationship between the elastic modulus of the parallel bond and its macroscopic counterpart. The numerical simulation results are in close agreement with the laboratory test findings, validating the parameter calibration method.

1. Introduction

In 1979, Cundall and Strack [1] introduced the discrete element method to address discontinuities and inhomogeneities in natural rock. Since then, this method has been widely used in geotechnical engineering [2-4]. The particle flow analysis method, based on discrete elements, represents materials as rigid spherical particles forming aggregates and characterizes the mechanical properties of rock through particle interactions [5]. Understanding the relationship between macro and micro parameters of rock materials is crucial for improving the reliability of numerical simulations and their resemblance to real-world scenarios [6]. While many scholars have studied the relationship between macro and micro parameters of granular flow models, the focus has mainly been on deformation parameters like modulus of elasticity and Poisson's ratio, neglecting strength parameters such as cohesion and angle of internal friction during parameter calibration [7-10]. Therefore, it is essential to consider both deformation and strength parameters for more reliable simulation experiments.

This paper presents a model based on the parallel cohesive contact model and investigates the relationship between macroscopic deformation parameters (cohesion and friction angle), strength parameter (elastic modulus), and micro rock parameters by using multifactor analysis of variance (ANOVA), regression analysis, and other methods. The study provides valuable insights for parameter calibration in numerical simulation tests of sandstone discrete elements.

2. PFC2D parallel bonding model theory and model parameters

2.1. *Parallel bonded contact modelling*

The force-displacement law for forces and moments in the parallel bonded contact model is illustrated in Figure 1 within the PFC2D model. This model accurately represents the interaction between particles by incorporating adhesive and tangential forces at their surfaces. It effectively captures the friction and bonding behaviors of particle contact surfaces, making it ideal for simulating processes such as particle flow, accumulation, and fragmentation. Moreover, the parallel bonded contact model offers computational efficiency in comparison to more intricate contact models. Employing a simplified mechanical model reduces computational complexity and facilitates the quicker completion of large-scale particle simulations.

The parallel bonded contact model comprises two distinct interfacial behaviors. The first interface is an infinitesimal linear elastic interface capable of withstanding forces without tension, similar to the linear model. The second interface, referred to as the parallel bond, is a finite-sized linear elastic interface that can withstand forces and moments and possesses bonding properties. If the maximum stress within the bonded material surpasses its bond strength, the parallel bond will break and the bond will fail, leading to degradation to a linear model [11]. This concept is depicted in Figure 2. When the normal mean stress on the parallel bonds exceeds the tensile strength of the model, the particles will be destroyed in tension. Similarly, if the tangential mean stress on the parallel bonds surpasses their combined action, the rock model will be destroyed in shear.

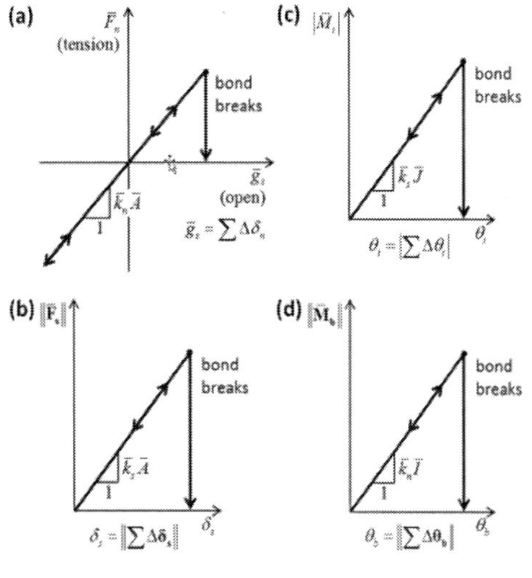

Figure 1. Force-displacement laws for parallel bond and bending moments [11].

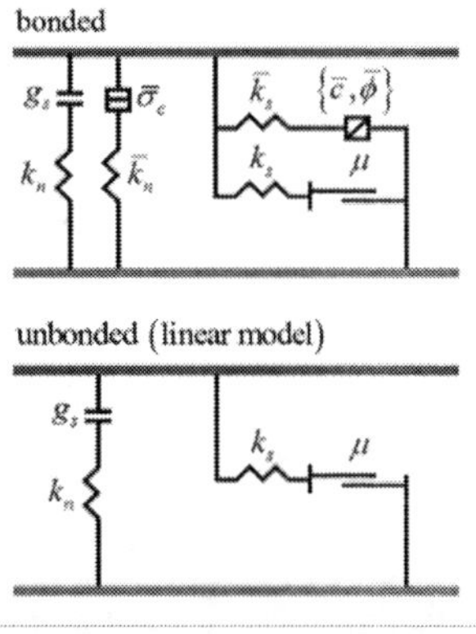

Figure 2. Degradation of parallel bond breaking [11].

2.2. *Main macroscopic and detailed parameters of the PFC2D model*

Xu et al. [8] discovered that the macroscopic strength parameters were influenced by both the finer strength parameters and the deformation parameters. Conversely, the macroscopic deformation parameters were solely affected by the finer deformation parameters. Table 1 illustrates the parallel bonded micro parameters, with the strength parameters including pb_ten, pb_coh, and pb_fa, while the deformation parameters consist of pb_emod and pb_krat.

Table 1. Main macroscopic and micro parameters of the PFC2D parallel bonding model.

Parallel bonding contact detail parameters		Particle size parameters		Macro-parameters	
pb_emod (GPa)	Bond effective modulus	*Rmin* (mm)	Minimum particle size	c (MPa)	Cohesion
pb_krat	Bond normal-to-shear stiffness ratio	*Rmax\Rmin*	Particle size ratio	φ (°)	Friction angle
pb_ten (MPa)	Tensile strength	*ρ*	Particle density	E (GPa)	Effective modulus
pb_coh (MPa)	Cohesion	fric	Friction coefficient		
pb_fa (°)	Friction angle	Krat	Normal-to-shear stiffness ratio		
		emod (MPa)	Effective modulus		

3. Experimental design

3.1. *Rock mechanics experiment*

According to the geological and mineral industry standards of the People's Republic of China DZ/T0276, "Regulation for Testing the Physical and Mechanical Properties of Rock", sandstone specimens measuring 100 mm×100 mm×100 mm were used for conducting variable angle plate shear mechanics tests and uniaxial compression tests. Three sets of experiments were carried out with variable angle plate shear gauges set at angles of 50°, 60°, and 70°, with each group containing three specimens. The average values of the results were calculated. The tests were conducted by using a rock creep instrument, applying a load at a rate of 0.5 MPa/s until specimen failure. The mechanical test of variable angle plate shear showed shear damage along the shear surface, while the uniaxial compression test revealed rock fracture, cracking, and other phenomena. The cohesion (c), angle of internal friction (φ), and modulus of elasticity (E) of the rocks were determined. Three specimens were tested in each group, and the average values of the test results were recorded, as presented in Table 2.

Table 2. Experimental results of rock mechanics.

Number of rock samples	Rock specimen size (mm)	Cohesion (MPa)	Friction angle	Effective modulus (GPa)	Poisson's ratio
12	100×100×100	18.479	43.32°	1.677	0.23

3.2. *PFC2D numerical simulation test*

To simplify the experiment, the parallel bonded contact model made certain assumptions. These include a particle density of 2, 000 kg/m³, a particle radius ratio of 1.66, and a minimum particle radius of 1.25 mm, which is 1/80 of the side length of the model. Furthermore, eight microscopic parameters were chosen [9, 10, 13, 14, 15] for pre-experimentation of efficient computer simulation. For further details on these assumptions and parameter values, we can refer to Table 3 below:

Journal of Physics: Conference Series **2808** (2024) 012083 doi:10.1088/1742-6596/2808/1/012083

(a) Before specimen destruction (b) After specimen destruction

Figure 3. Before and after the destruction of the test specimen.

Table 3. Table of micro parameter levels for the parallel bonding model.

Serial number	Parallel bonded contact model fine view parameters							
	emod (GPa)	krat	pb_emod (MPa)	pb_krat	pb_coh (MPa)	pb_ten (MPa)	pb_fa (°)	fric
1	0.1	0.5	0.1	0.5	0.1	0.1	30	0.3
2	0.2	1.0	0.2	1.0	0.2	0.2	40	0.4
3	0.3	1.5	0.3	1.5	0.3	0.3	50	0.5
4	0.4	2.0	0.4	2.0	0.4	0.4	60	0.6

Numerical simulation experiments were conducted by using an orthogonal experimental design based on Table 3 to determine the macroscopic parameters of elastic modulus, cohesion, and angle of internal friction. The obtained results are presented in Table 4.

Table 4. Orthogonal design of micro parameters of the parallel bonding model and results.

Serial number	Parallel bonded contact model fine view parameters								Macro-parameters		
	emo (GPa)	krat	pb_emod (MPa)	pb_krat	pb_coh (MPa)	pb_ten (MPa)	pb_fa (°)	fric	c (MPa)	tanφ	E (GPa)
1	0.3	1.5	0.4	1	0.3	0.1	40	0.3	0.159	0.785	0.211
2	0.1	2	0.2	2	0.1	0.4	60	0.5	0.146	1.082	0.204
3	0.1	2	0.3	1	0.4	0.2	50	0.3	0.221	1.063	0.392
4	0.2	1.5	0.2	1.5	0.4	0.4	30	0.3	0.481	0.677	0.723
5	0.1	1	0.4	1	0.2	0.4	30	0.6	0.251	0.612	0.392
6	0.3	0.5	0.3	1	0.1	0.3	60	0.6	0.113	0.992	0.193
7	0.3	2	0.4	1.5	0.1	0.2	30	0.4	0.121	0.708	0.199
8	0.1	0.5	0.1	0.5	0.1	0.1	30	0.3	0.112	0.417	0.150
9	0.1	1	0.1	2	0.3	0.2	40	0.4	0.217	0.968	0.375
10	0.3	0.5	0.2	2	0.4	0.1	50	0.4	0.171	0.642	0.173
11	0.4	2	0.1	1.5	0.3	0.3	60	0.3	0.276	0.753	0.447
12	0.2	1.5	0.3	0.5	0.1	0.2	40	0.5	0.131	0.552	0.171
13	0.2	2	0.2	1	0.2	0.3	40	0.4	0.228	0.733	0.391
14	0.1	1.5	0.3	1.5	0.2	0.1	60	0.4	0.137	0.801	0.185
15	0.3	1.5	0.1	2	0.2	0.3	30	0.5	0.303	0.591	0.443
16	0.4	1.5	0.4	2	0.4	0.2	60	0.6	0.216	0.963	0.375
17	0.3	1	0.2	0.5	0.2	0.2	60	0.3	0.176	0.823	0.296
18	0.4	1	0.2	1.5	0.1	0.1	40	0.6	0.127	0.545	0.176
19	0.2	1	0.4	2	0.1	0.3	50	0.3	0.104	1.092	0.216

20	0.1	0.5	0.4	1.5	0.4	0.3	40	0.5	0.366	0.874	0.521
21	0.3	1	0.3	1.5	0.3	0.4	50	0.5	0.300	0.925	0.522
22	0.3	2	0.1	0.5	0.4	0.4	40	0.6	0.409	0.659	0.637
23	0.4	1	0.3	0.5	0.4	0.3	30	0.4	0.400	0.467	0.551
24	0.1	1.5	0.2	0.5	0.3	0.3	50	0.6	0.212	1.137	0.476
25	0.4	1.5	0.1	1	0.1	0.4	50	0.4	0.117	0.898	0.199
26	0.2	0.5	0.4	0.5	0.3	0.4	60	0.4	0.261	1.008	0.483
27	0.4	2	0.4	0.5	0.2	0.1	50	0.5	0.181	0.506	0.221
28	0.2	0.5	0.1	1.5	0.2	0.2	50	0.6	0.211	0.748	0.344
29	0.2	1	0.1	1	0.4	0.1	60	0.5	0.135	0.899	0.208
30	0.4	0.5	0.2	1	0.3	0.2	30	0.5	0.307	0.528	0.398
31	0.4	0.5	0.3	2	0.2	0.4	40	0.3	0.203	1.004	0.454
32	0.2	2	0.3	2	0.3	0.1	30	0.6	0.150	0.683	0.187

4. Analysis of numerical simulation test data of PFC2D

4.1. Multifactor analysis of variance

Table 4 presents the results of a multifactor ANOVA analysis conducted by using SPSS software to explore the relationship between macro and fine parameters. This analysis utilized the F-statistic and sig. value to assess the degree of influence and statistical significance of these parameters. A sig. value below 0.01 indicates a highly significant effect, while a value below 0.05 indicates a significant effect. The specific details regarding the degree of influence and level of significance between macro and micro parameters are outlined in the table.

In Figure 4, the order of the effects of the eight mesoscopic parameters on cohesion (c) is as follows: pb_coh > pb_ten > pb_fa > pb_krat > pb_emod > fric > emod > krat. If the significance (Sig.) of pb_ten and pb_coh is less than 0.01, these two microscopic parameters have a highly significant impact on cohesion (c). Similarly, if the Sig. of pb_fa is less than 0.05, the parallel bonding internal friction angle has a significant impact on cohesion (c). However, the remaining five mesoscopic parameters do not have a significant impact on cohesion (c). Moreover, based on the relative F statistics, it can be observed that the two mesoscopic parameters, parallel bonding normal strength (pb_ten), and parallel bonding force (pb_coh) have a dominant influence on the cohesion force (c).

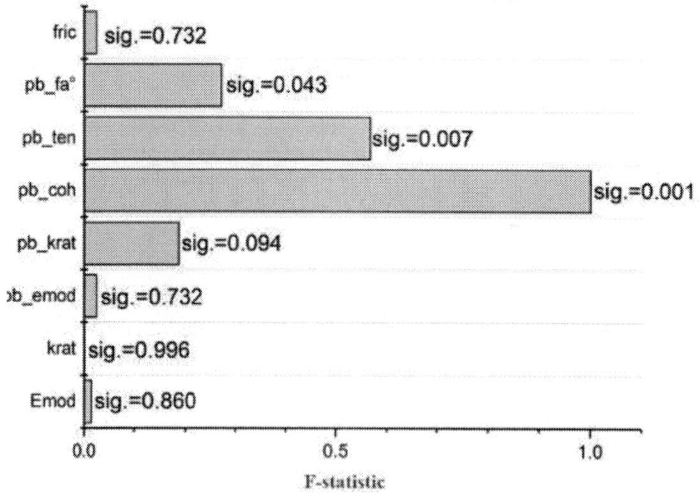

Figure 4. Cohesion-relative F-statistic for multifactorial ANOVAs.

In Figure 5, the parallel friction angle (pb_fa) is found to have a significance level of less than 0.05, suggesting a notable impact on the internal friction angle (φ). Conversely, the other parameters exhibit significance levels exceeding 0.05, indicating that they do not have a statistically significant effect. The figure prominently illustrates that pb_fa plays a dominant role in influencing the internal friction angle.

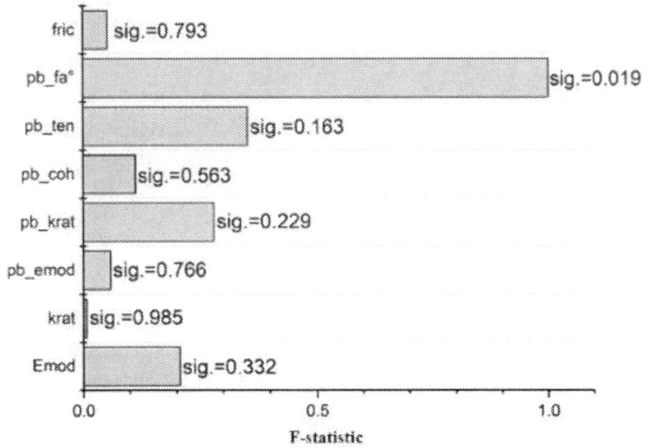

Figure 5. pb_fa-relative F-statistic for multifactorial ANOVAs.

In Figure 6, the order of effects of the eight micro parameters on the elastic modulus E is as follows: pb_emod > emod > pb_krat > pb_ten > pb_fa > krat > pb_coh > fric. The effects of emod, pb_emod, pb_krat, and pb_ten on the elastic modulus are highly significant, with a significance level (Sig.) of less than 0.01. The remaining parameters krat and pb_coh also have significant effects, with a Sig. less than 0.05. When comparing the relative F-statistics of emod, pb_emod, pb_krat, and pb_ten, pb_emod stands out as the most dominant among the micro parameters influencing the modulus of elasticity.

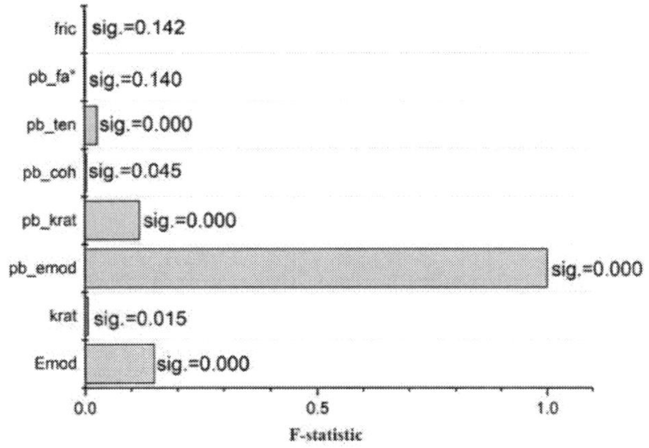

Figure 6. pb_emod-relative F-statistic for multifactorial ANOVAs.

4.2. Macro and micro parametric regression analysis of the parallel bonding model

The study compared the influence of micro parameters on macro-parameters through multiple regression analysis to establish a relational equation.

The primary significant influencing factors of cohesion were found to be pb_ten, pb_coh, and pb_fa. The relative F-statistics revealed that pb_ten and pb_coh had a dominant influence on cohesion, while other factors had a negligible impact. The multiple linear regression analysis of pb_ten and pb_coh produced Equation (1) with a weak fit coefficient of $R^2 = 0.675$.

$$c = 0.560 pb\_coh + 0.424 pb\_ten - 0.29 \tag{1}$$

Based on Equation (1), it is clear that the cohesive force is positively correlated with pb_ten and pb_coh. However, the level of fit is relatively low, implying that the factors affecting cohesion c are

Journal of Physics: Conference Series **2808** (2024) 012083 doi:10.1088/1742-6596/2808/1/012083

complex and may involve a nonlinear relationship. As a result, a multiple nonlinear regression analysis was performed, yielding Equation (2) with a fitting coefficient R^2=0.871, demonstrating a more robust fitting effect.

$$c = 0.168 - 0.228 pb\_coh - 0.364 pb\_ten + 0.084 pb\_coh \times pb\_ten \tag{2}$$

The internal friction angle is primarily influenced by the micro parameter pb_fa. Based on Figure 7, it is evident that the internal friction angle is solely influenced by the micro parameter pb_fa, disregarding other parameters. The average value of tanφ corresponding to pb_fa values of 30°, 40°, 50°, and 60° were determined, as illustrated in Figure 7. The analysis revealed a logarithmic relationship between tanφ and pb_fa, showing that as pb_fa increases, the internal friction angle also increases. A regression analysis produced Equation (3) with a fitting coefficient of R^2=0.969, indicating a strong correlation.

$$\tan\phi = 0.487 \ln(pb\_fa) - 1.053 \tag{3}$$

The modulus of elasticity is influenced by several factors including pb_emod, emod, pb_krat, pb_ten, pb_fa, and krat. Based on the relative F-statistics of pb_emod in Figure 6, it is clear that pb_emod plays a significant role in determining the modulus of elasticity E. Therefore, the quantitative analysis focuses solely on pb_emod without considering other parameters. The mean values of elastic modulus E corresponding to pb_emod values of 0.1, 0.2, 0.3, and 0.4 are used as reference points. In Figure 8, a linear correlation between elastic modulus and pb_emod is observed, with the elastic modulus increasing as pb_emod values increase. The regression analysis of elastic modulus E and pb_emod yields Equation (4) with a fitting coefficient $R^2 = 0.991$, indicating a strong correlation.

$$E = 0.167 \text{pb}\_emod + 0.106 \tag{4}$$

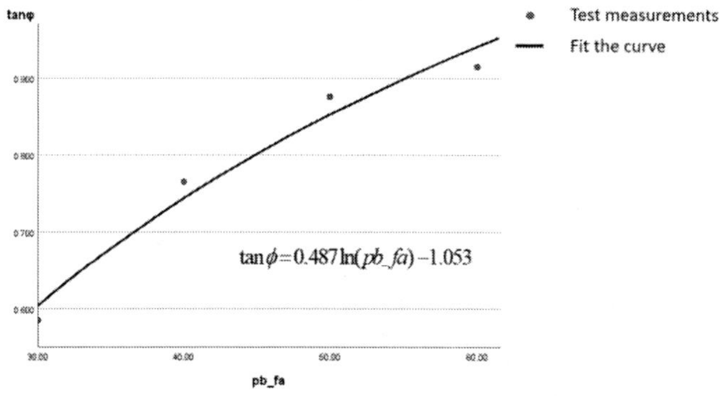

Figure 7. The relationship between pb_fa and tanφ.

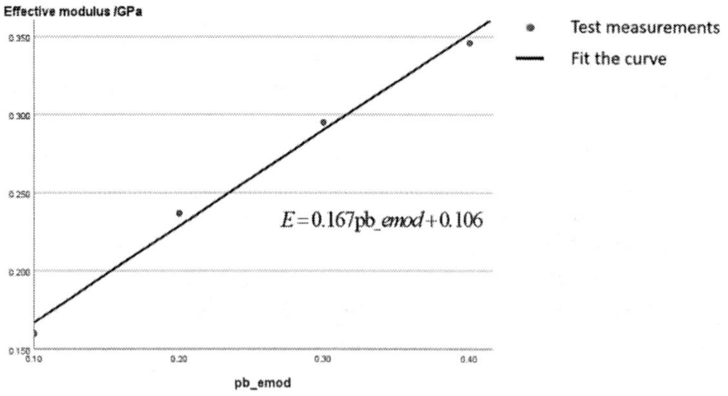

Figure 8. The relationship between pb_emod and E.

Journal of Physics: Conference Series **2808** (2024) 012083 doi:10.1088/1742-6596/2808/1/012083

5. Calibration methods and applications for micro parameters

5.1. Calibration methods for micro parameters

Based on the results of the above analyses, the micro parameters are calibrated by using the trial-and-error method to establish the relationship between the macro and micro parameters. The following outlines the main calibration process:

(1) The determination of the minimum radius of the particles is a crucial step in numerical test simulations. This micro parameter holds great significance and is typically regarded as the primary parameter to be considered.

(2) The angle of internal friction is only affected by the micro parameter pb_fa, and the rest of the parameters are negligible to the angle of internal friction, which can be calibrated according to the logarithmic relationship between the angle of internal friction and pb_fa.

(3) The angle of internal friction is primarily influenced by the micro parameter pb_fa. The remaining parameters have negligible effects on the angle of internal friction. Calibration of the angle of internal friction can be done based on the logarithmic relationship between the angle of internal friction and pb_fa.

The cohesion of a material is significantly influenced by the combination of pb_ten and pb_coh, as noted by Liu et al. [12]. The macro-parameter cohesion is determined by the ratio of the mean to the variance of the micro-parameter parallel bond normal strength (pb_ten) and tangential strength (pb_coh). This ratio plays a critical role in determining the type of damage in the material, with a commonly accepted ratio of 4:1. The values of pb_ten and pb_coh can be calibrated by using Equations (2) and (5), where x_1 represents pb_ten and x_2 represents pb_coh.

$$\frac{x_1 + x_2}{2} = 4 \times \frac{(x_1 + \bar{x})^2 + (x_2 + \bar{x})^2}{2} \tag{5}$$

In addition to the calibrated parameters, there are four micro parameters, emod, krat, pb_krat, and fric, which have not yet been adjusted. While fric has minimal impact on the macro-parameters, emod, krat, and pb_krat significantly affect the modulus of elasticity, albeit to a lesser extent compared to pb_emod's influence on E. For simulation, fric is assumed to be 0.5. It is further assumed that krat=pb_krat=1.5 and emod=pb_emod.

5.2. Experimental verification

The micro parameters necessary for validation were determined by using macro-parameters from laboratory tests. These micro parameters were adjusted and refined through comparison with the laboratory test results. Subsequently, the simulated test results included each micro-parameter and macro-parameter (refer to Table 5).

Table 5. Simulation values of mesoscopic parameters of the parallel bonding model and macroscopic mechanical parameters of rock.

Parallel bonded contact model fine view parameters								Macro-parameters		
emod (GPa)	krat	pb_emod (MPa)	pb_krat	pb_coh (MPa)	pb_ten (MPa)	pb_fa (°)	fric	c (MPa)	tanφ	E (GPa)
2.89	1.5	2.89	1.5	20.941	16.608	60.25	0.5	17.67	39.89	1.531

The error between the simulation results and the laboratory test results, as shown in Tables 2 and 5, is within 10%. This supports the reliability of the conclusions in this paper on using the trial-and-error method to calibrate the fine microscopic parameters of sandstone.

6. Conclusion

The parallel bonding model is used as the intrinsic model for contact in this study. Numerical simulation experiments are conducted to explore the relationship between macro- and micro parameters by using multi-factor ANOVA and regression analysis. The findings of the study are as follows:

(1) Cohesion is influenced by pb_ten and pb_coh together, showing a complex non-linear relationship. The internal friction angle is mainly impacted by pb_fa, following a logarithmic function. The elastic modulus is affected by pb_emod, emod, pb_krat, pb_ten, pb_fa, and krat, with pb_emod having the most significant effect and showing a linear relationship with the elastic modulus.

(2) By examining the macro and micro parameters of the parallel bonding model, the paper proposes a calibration method for these parameters and introduces a trial-and-error approach for calibration. The numerical test results closely match the laboratory results, with an error margin of less than 10%, demonstrating the reliability of the trial-and-error method for calibrating the fine apparent parameters of sandstone.

References

[1] Cundall P. A. and Strack O. D. L. (1980). A discrete numerical model for granular assemblies. J. Géotechnique, 30 (3): 331-336.
[2] Li K., Cheng Q. W., Lin Q. W., et al. (2022). State of the Art on Rock Avalanche Dynamics from Granular Flow Mechanics. J. Earth Sciences, 47 (03): 893-912.
[3] Hu X. J., Bian K., Liu J., et al. (2022). Particle flow code analysis of the effect of a discrete fracture network on rock mechanical properties and acoustic emission characteristics. J. Rock and Soil Mechanics, 43 (S1): 542-552.
[4] Yan L., Lu C. H., Yu S. H., et al. (2022). Test and numerical analysis on combined pressure arch of surrounding rock in adjacent twin tunnels. J. Journal of Beijing Jiaotong University, 46 (03): 103-109.
[5] Deng Y. B., Yang Y. C., Shi D. D., et al. (2017). Refinement and application of variable particle-size methods in 3D discrete element modeling for large-scale problems. J. Journal of Geotechnical Engineering, 39 (01): 62-70.
[6] Abi Erdi, Zheng Y. R., Feng X. T., et al. (2018). Relationship between particle micro and macro mechanical parameters of the parallel-bond model. J. Rock-soil mechanics, 39 (04): 1, 289-1, 301.
[7] Zhou Y., Wu S. C., Jiao J. J., et al. (2011). Research on micromechanical parameters of rock and soil mass based on BP neural network. J. Rock-soil Mechanics, 2011, 32 (12): 3, 821-3, 826.
[8] Xu Z., Wang Z., Wang W., et al. (2022). An integrated parameter calibration method and sensitivity analysis of microparameters on the mechanical behavior of transversely isotropic rocks. J. Computers and Geotechnics, 142: 104, 573.
[9] Chen P. Y., Kong Y., and Yu H. M. (2018). Research on the Calibration Method of Microparameters of a Uniaxial Compression PFC2D Model for Rock. J. Journal of Underground Space and Engineering, 14 (05): 1, 240-1, 249.
[10] Cong Y., Wang Z. Q., Zheng Y., et al. (2015). Experimental study on microscopic parameters of brittle materials based on particle flow theory. J. Journal of Geotechnical Engineering, 37 (06): 1, 031-1, 040.
[11] Itasca Consulting Group LLC. PFC2D user's manual (version 5.0). (2016). Minnesota: Itasca Consulting Group LLC.
[12] Liu C., Chen X. X., Zhang W., et al. (2017). Study on the Calibration Process of Parallel Bonding Meso-structure Parameter in PFC Numerical Simulation. J. Value Engineering, 36 (26): 204-207.
[13] Wang T., Han Y. H., Zhu Y. S., et al. (2018). Method and application of discrete elements of PFC 2D/3D particles. China State Engineering and Construction Press, Beijing.
[14] Shi C., Zhang Q., and Wang N. (2018). Numerical simulation technology and application of particle flow (PFC 5.0). China State Engineering and Construction Press, Beijing.

[15] Yang J. Q. (2018). Study and Application of Parameter Characteristics of Bonded Particle Rock Model. D. Handan: Hebei University of Engineering.

Journal of Physics: Conference Series **2808** (2024) 012084 doi:10.1088/1742-6596/2808/1/012084

Study on the relationship between reflectivity and thickness of radar-absorbing material

Song He[1*], Jun Hu[1], Jinsong Chen[1] and Qiushi Xi[1]

[1]Air Force Early Warning Academy, Wuhan, Hubei, China

[*]Corresponding author's e-mail: hesong_1988@qq.com

Abstract: Radar-absorbing materials (RAM) have been widely used in civilian areas for electromagnetic interference shielding and in the defense sector for radar wave stealth. The thickness of radar-absorbing materials is one of the important parameters affecting their performance. This paper analyzes the relationship between the absorbing performance of radar-absorbing materials and thickness based on transmission line theory. Through theoretical derivation and simulation verification, the concepts of optimal usage thickness and cutoff thickness of materials are derived. The research results have important guiding significance for the application and optimization design of radar-absorbing materials.

1. Introduction

Radar-absorbing materials convert the energy of incident electromagnetic waves into thermal energy or other forms through the loss mechanisms within the material, thereby reducing the reflected electromagnetic waves [1]. This process involves complex electromagnetic theory, including phenomena such as propagation, reflection, refraction, and absorption of electromagnetic waves [2]. The electromagnetic parameters of the material (such as dielectric constant and magnetic permeability) as well as the thickness of the material can have a significant impact on its wave-absorbing performance.

The influence of the thickness of radar-absorbing materials on their wave-absorbing performance is mainly manifested in the propagation and attenuation of electromagnetic waves within the material [3]. Theoretically, the thicker the wave-absorbing material is, the more electromagnetic wave energy it can absorb, and the less energy will be transmitted and reflected, resulting in an increase in reflection loss. However, as the thickness increases, multiple reflections and a certain probability of transmission of electromagnetic waves within the material occur, causing the reflectance to stabilize after a certain thickness, and the reflection loss also reaches a certain upper limit [4]. However, when the thickness is too large, multiple reflections may occur within the material, leading to increased scattering and transmission of electromagnetic waves, thus reducing the wave-absorbing performance. How to utilize the limited space to achieve optimal wave-absorbing performance becomes a key focus in stealth technology research.

2. Theory of reflection loss calculation

After the electromagnetic wave is incident on the material, it undergoes interface reflection, absorption, and interference effects. The calculation of the overall reflection rate after synthesizing multiple effects requires the use of transmission line theory. Figure 1 is a schematic diagram of an electromagnetic wave incident on a wave-absorbing material [5], where θ and θ_m are the incidence angle and refraction angle of the electromagnetic wave, respectively; ε_r and μ_r represent the complex permittivity and permeability of the material. Refractive angle calculation is shown in Formula (1), where ε_r and μ_r respectively represent the complex permittivity and permeability of the material; ε_0 and μ_0 respectively represent the complex permittivity and permeability of air; The real parts of the complex permittivity and permeability, ε_r' and μ_r' are measures of the polarization or magnetization produced by the absorbing material under the electromagnetic field; The imaginary parts of the complex permittivity and permeability, ε_r'' and μ_r'' are measures of the electromagnetic wave loss due to the material under the electromagnetic field. The representation of the material's electromagnetic parameters is shown in Formula (2), and the calculation of the absorbing material's reflection rate is shown in Formula (3).

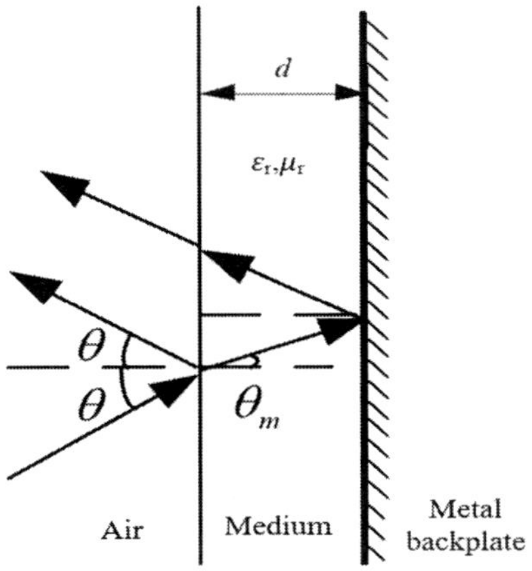

Figure 1. Diagram of the principle of wave absorption in wave-absorbing materials.

$$\frac{\sin\theta}{\sin\theta_m} = \frac{n}{1} = \frac{\sqrt{\varepsilon_r\mu_r}}{\sqrt{\varepsilon_0\mu_0}}(0 \le \theta \le 90) \tag{1}$$

$$\varepsilon_r = \varepsilon_r' - j\varepsilon_r'' \qquad \mu_r = \mu_r' - j\mu_r'' \tag{2}$$

$$R_{TE} = 20\,log_{10}\left(\left|\frac{Z_{in}\cos\theta_m - \eta_0\cos\theta}{Z_{in}\cos\theta_m + \eta_0\cos\theta}\right|\right) \qquad R_{TM} = 20\,log_{10}\left(\left|\frac{Z_{in}\cos\theta - \eta_0\cos\theta_m}{Z_{in}\cos\theta + \eta_0\cos\theta_m}\right|\right) \tag{3}$$

$$Z_{in} = Z_0 j\sqrt{\frac{\mu_r}{\varepsilon_r}}\sqrt{\frac{\mu_0}{\varepsilon_0}}th\left(\frac{2\pi fn\cos\theta_m}{c}d\right) \tag{4}$$

3. Calculation example

The transmission line theory is suitable for the calculation of the total reflection attenuation rate of finite thickness materials, which explains the core idea of impedance matching. The parameters determining the absorption characteristics of radar-absorbing materials include the complex permittivity, complex permeability, material thickness, electromagnetic wave frequency, and angle of incidence, which can be simply expressed as Formula (5):

$$RL = F(\varepsilon_1, \varepsilon_2, \mu_1, \mu_2, f, d, deg) \tag{5}$$

In this paper, pure dielectric absorbing materials are taken as an example. The reflection loss of dielectric materials is closely related to the material thickness and incident wavelength. It is assumed that the electromagnetic parameters are shown in Table 1.

Table 1. Electromagnetic Parameters of Radar-absorbing Materials.

f	ε'	ε''	μ'	μ''
0.5 GHz	3	6	1	0
1.0 GHz	2.5	4	1	0
2.0 GHz	2	3.5	1	0
3.0 GHz	1.5	2	1	0

The reflectivity of the absorbing material with different thicknesses is calculated, as shown in Figure 2.

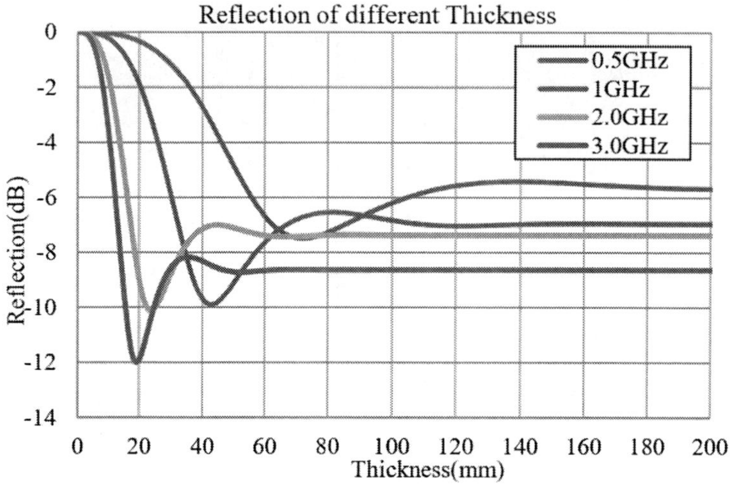

Figure 2. The reflection of different thickness.

From the computational results, it can be deduced that within a certain thickness range, as the thickness increases, the reflectivity gradually decreases, which means the microwave absorption performance is enhanced progressively. When a specific thickness is reached, the optimal reflectivity value can be obtained, referred to as the material's optimal application thickness. Moreover, at different frequency points, there will be a cutoff thickness for the material, meaning that when the material thickness reaches a certain value, even if the thickness continues to increase, the reflectivity remains essentially unchanged. As the frequency increases, both the optimal thickness and cutoff thickness decrease. At the same thickness, as the frequency increases, the reflectivity gradually decreases. With the increase in frequency, both the optimal thickness and cutoff thickness become smaller.

4. Simulation verification

Taking 0.5 GHz as an example, when the material reaches a thickness of 150 mm, the reflectivity remains essentially unchanged, and the corresponding vertical incidence absorbing performance is basically consistent. Therefore, it is possible to simulate the RCS of materials with the same length and width but different thicknesses to compare whether they conform to the absorbing material cutoff thickness theory. This article uses Feko simulation software and the MOM simulation algorithm. The simulation model is 1 m long and 1 m wide. First, the RCS of the metal plate is simulated. The simulation model is shown in Figure 3. Then, material samples with thicknesses of 70 mm, 150 mm, and 300 mm are respectively attached to the metal plate to simulate the RCS. The material sample simulation model is shown in Figure 4.

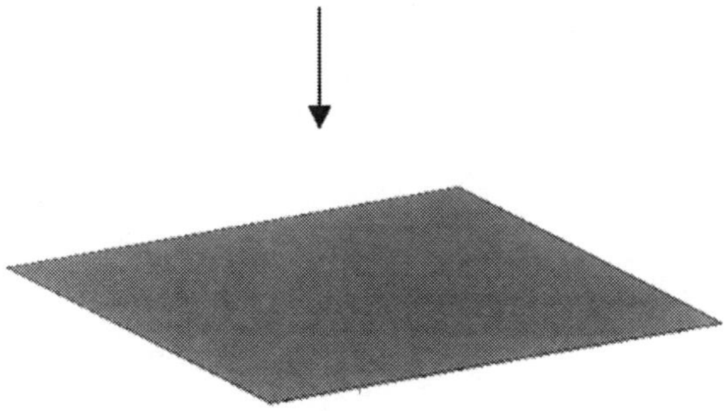

Figure 3. Simulation model of metal plate.

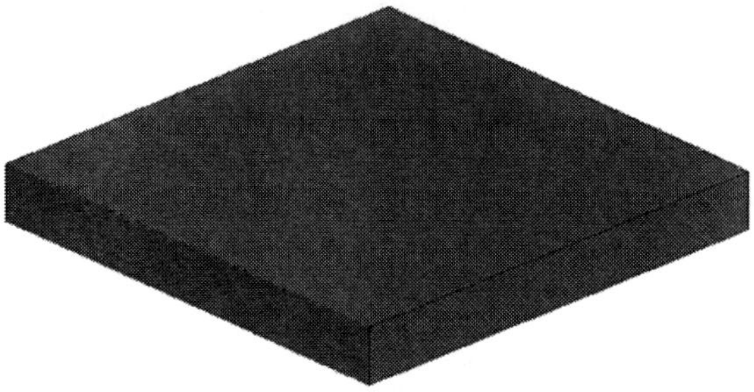

Figure 4. Simulation model of material plate.

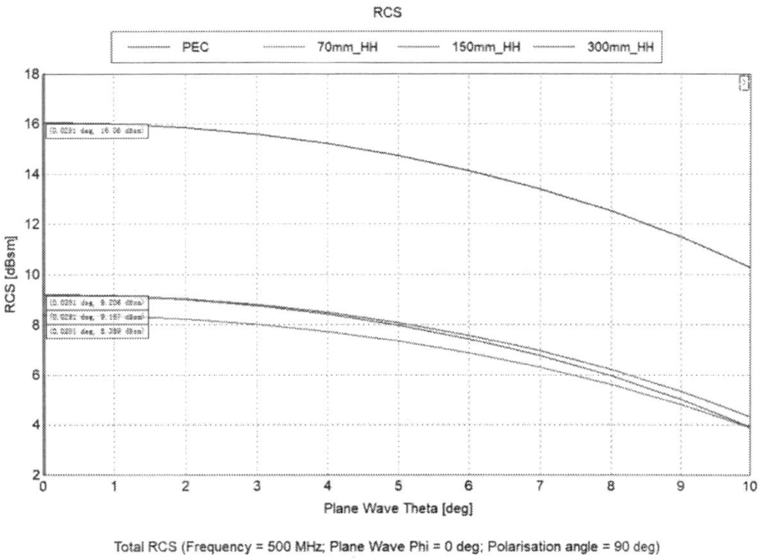

Figure 5. Radar Cross Section (RCS) for flat plates of absorbing materials with different thicknesses.

This simulation mimics the stealth effect brought by the application of different thicknesses of the same material on the metal reflective layer. The simulation results are shown in Figure 5. At 0°, electromagnetic waves are vertically incident. Within the range of 0° to 10°, specular reflection is the main source of scattering for the flat plate. The reflectivity of the material determines the absorbing properties of the flat plate, which is then manifested as the RCS (Radar Cross Section) value of the material plate. According to the simulation results, when the material thickness is 70 mm, the RCS value is the smallest, matching the optimal usage thickness of the material. When the material thickness is 150 mm or 300 mm, the RCS values are basically the same, indicating that once the material reaches a certain cutoff thickness, increasing the material thickness does not result in a significant change in absorbing properties.

5. Conclusion

This article studies the relationship between the reflection rate of wave-absorbing materials and the thickness of the material through theoretical calculations and simulation verification. The research indicates that by optimizing the thickness of the material, the efficiency of the wave-absorbing material can be significantly improved. It is found that the reflection rate of the wave-absorbing material decreases with the increase of the material thickness, but when the thickness exceeds a certain value, the downward trend of the reflection rate becomes slow. This suggests that there is an optimal thickness, and when the material reaches a certain thickness, even if the thickness is increased, it cannot enhance the wave-absorbing performance. Therefore, it is necessary to apply the wave-absorbing structure space reasonably, to ensure the wave-absorbing performance while avoiding unnecessary material waste. In addition, we also find that the electromagnetic parameters of the material are crucial for determining the optimal thickness. These findings have important guiding significance for designing more efficient wave-absorbing materials and have had a positive impact on the development of fields such as stealth technology and electromagnetic compatibility. Future research will focus on exploring the optimal thickness of different types of wave-absorbing materials and how to achieve precise control of these thicknesses in practical applications. At the same time, research will extend to the development of multi-band wave-absorbing materials to meet a wider range of electromagnetic wave absorption needs.

References

[1] Stefan Muhlin, Mohamed Farhat, Carsten Rockstuhl, el al. Cloaking dielectric spherical objects by a shell of metallic nanoparticles [J], Physical Review, 2011(5):195116

[2] Song He, Tong Su, Ling He, Jinsong Chen, Liang Xia. "Inversion of Equivalent Electromagnetic Parameters of Multilayer Absorbing Materials Based on Simulated Annealing Algorithm" [J] Journal of Physics: Conference Series, 2023.

[3] Pamuk G, Kemikliolu U. Radar Absorbing Capability and Fatigue Properties of Unidirectional Carbon Fabric-Reinforced Composites [J]. Fibers and Polymers, 2023. DOI:10.1007/s12221-023-00292-8.

[4] Samukhina Y V, Nikoladze G M, Kulkova T A, et al. Radar-Shielding and Microwave-Absorbing Properties of Composite Materials Based on Shungite [J]. Russian Journal of Physical Chemistry, 2023(2):97.

[5] Suo Qingtao, Xu Baocai, Wang Jianjiang, et al. Current status and progress in the research of low-frequency radar absorbing materials [J]. New Chemical Materials, 2019, 047(004):25-28.

Journal of Physics: Conference Series **2808** (2024) 012085 doi:10.1088/1742-6596/2808/1/012085

Simulation study on the dynamic response of X80 steel pipelines under explosive loads

Yuntian Wang[1], Min Yang[1], Danping Hu[2], Qiujie Wei[1], Ying Zhang[1] and Yuzhu Guo[3]*

[1]School of Urban Construction, Chengdu Polytechnic, Chengdu, 610065, China
[2]LVTC School of Intelligent Construction, Luzhou Vocational and Technical College, Luzhou, 646000, China
[3]School of Civil Engineering, Changsha University of Science & Technology, Changsha, 410114, China

*Corresponding author's email: guoyuzhu5185@163.com

Abstract. This research investigates the dynamic response and failure behavior of X80 steel pipelines under external contact explosion loads, a critical concern for oil and gas pipeline safety due to increased terrorist activities. The dynamic response characteristics are critically analyzed under ten different conditions of explosive charges, providing a detailed evaluation of the damage mechanisms, including denting and perforation. The study finds that wall thickness significantly influences the blast resistance of pipelines, with thicker walls offering better protection against explosive loads. This is substantiated by comparing destruction scenarios and morphologies of perforation holes between X80 and X70 steel pipelines. The paper concludes that increasing wall thickness markedly improves pipeline resilience against explosive destruction, contributing valuable guidelines for designing more robust and safer pipeline infrastructures.

1. Introduction

Pipeline transport is a widely utilized mode of conveyance in the energy sector, with metal thin-walled shell structures (pipelines) representing an important structural form in oil and gas transportation. These structures hold a significant position within China's energy transmission network. In recent years, the frequency of terrorist attacks worldwide has increased, with oil and gas pipelines often being primary targets. The destruction of metal pipelines under intense dynamic loads such as explosions can lead to catastrophic consequences. Therefore, a deep understanding of the dynamic response and failure behavior of metal pipelines under explosive and other intense dynamic loads is significant and practical for the design and evaluation of blast-resistant safety measures in metal pipeline structures.

In response to the failure behaviors of metal pipelines under intense dynamic loads, Zhang et al. [1] conducted experimental studies on the destruction of thin-walled soft steel pipelines by various projectile types, analyzing the characteristics of damage at different impact angles. Shen et al. [2] investigated the damage caused to thick-walled steel pipelines by fragments of varying mass and velocity, establishing a failure criterion for steel pipelines based on the Tresca criterion. Rushton et al. [3] performed internal explosion experiments on 9.5 mm thick seamless steel pipelines and verified the applicability of the Johnson-Cook constitutive model under explosive loads through numerical

Journal of Physics: Conference Series **2808** (2024) 012085 doi:10.1088/1742-6596/2808/1/012085

simulation. Lu et al. [4] studied the damage caused to thin-walled cylindrical shell structures by explosions of acetylene/air mixtures, with results indicating that the ability of the shell model to resist destruction is related to its diameter-to-height ratio. Pan et al. [5] conducted a systematic study on the dynamic response characteristics of cylindrical thin-walled storage tanks under explosive impacts through numerical simulation analysis.

In terrorist attacks targeting oil and gas pipelines, mounting explosives on the exterior of the pipeline is a typical assault method, leading to significant local destruction in metal cylindrical structures under contact explosion loads. Current research is primarily focused on indirect blast impact responses, and direct contact explosion studies are predominantly based on X70 steel [6][7]. There has been little research on the dynamic response and failure behavior of higher-grade X80 steel metal pipelines under external contact explosion conditions. This paper investigates the dynamic response and failure behavior of X80 steel circular pipelines with varying wall thicknesses under contact explosion loads of different explosive masses through experimental studies and numerical simulation methods. The results of this study aim to guide the design and evaluation of blast-resistant safety measures for pipelines.

2. Computational method

2.1. Computational model

The simulation was conducted using ANSYS/AUTODYN, with model dimensions following the actual X80 pipeline sizes used in the West-East Gas Pipeline Project. The wall thicknesses are d = 14.6 and 26.2 mm, and the diameter D = 1016 mm. The internal operating pressure of the pipeline is not considered. The pipeline is placed on the ground, and TNT explosives are used as the blast source, positioned at the center top of the steel pipe. A total of 10 conditions were calculated and analyzed based on the amount of explosive charge, with specific condition parameters as shown in Table 1.

Table 1. Parameters in different condition.

Model No.	TNT/kg	D/mm	d/mm
A1	1.3	1016	14.6
A2	3.3	1016	14.6
A3	5.4	1016	14.6
A4	6.7	1016	14.6
A5	8.7	1016	14.6
B1	1.3	1016	26.2
B2	3.3	1016	26.2
B3	5.4	1016	26.2
B4	6.7	1016	26.2
B5	8.7	1016	26.2

Given the model's symmetry, a 1/4 model was established for the analysis. The ground and pipeline were modeled using the Lagrange method, with the pipeline model for a diameter of 14.6 mm consisting of 180,000 elements and 26.2 mm consisting of 300,000 elements. The air domain surrounding the explosion contact area was modeled using the Euler-3D Material algorithm with a cell size of 10 mm. The boundary of the air domain was set as material outflow, and fluid-structure interaction was modeled using the Euler/Lagrange full contact algorithm. The initiation time for the TNT explosives was set to 0 s. The simulation model is illustrated in Figure 1.

Journal of Physics: Conference Series **2808** (2024) 012085 doi:10.1088/1742-6596/2808/1/012085

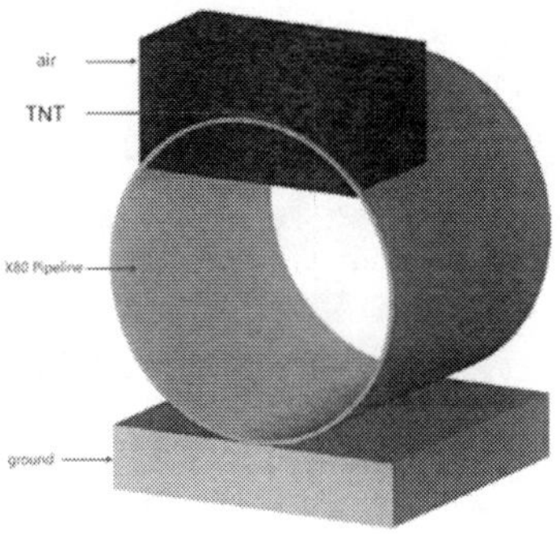

Figure 1. X80 pipeline contact explosion simulation model.

2.2. Model parameters

The bulk behavior is described using the Mie-Grüneisen equation of state, which can accurately depict the dynamic behavior of metals under shock loads. The parameters for the Mie-Grüneisen equation of state are provided in Table 2 [8].

Table 2. Mie-Grüneisen parameters.

Material	ρ_0 (kg/m^3)	$c0$ (m/s)	$S1$	γ
X80 Steel	7890	4569	1.49	2.17

The deviatoric behavior is captured using the Johnson-Cook constitutive model [9]. Ji et al. [6] employed the Johnson-Cook constitutive model to numerically simulate and analyze the direct contact explosion of X70 steel pipelines. The model parameters were determined experimentally, and the parameters for the Johnson-Cook constitutive model were determined as shown in Table 3.

Table 3. Johnson-Cook constitutive model parameters.

Material	A/MPa	B/MPa	n	C	m
X80 Steel	590	448	0.499	0.055	1.00

In the simulation of explosive impact, air can be considered an ideal gas, and its state equation is that of the ideal gas law, expressed as follows [6]:

$$p = (\gamma - 1)\rho_0 e_0 \tag{1}$$

In Formula (1), p represents gas pressure; γ is the adiabatic index of air; ρ_0 is the initial air density; e_0 is the internal energy per unit volume of air.

TNT explosives are used as an equivalent blast source, with the state equation for TNT described by the JWL (Jones-Wilkins-Lee) equation of state. The JWL equation of state is expressed as follows [10]:

$$p = A\left(1 - \frac{\omega}{R_1 V}\right)e^{-R_1 V} + B\left(1 - \frac{\omega}{R_2 V}\right)e^{-R_2 V} + \frac{\omega\psi}{V} \tag{2}$$

In Formula (2), p is the explosive pressure, V is the volume of explosion products per unit volume of TNT, E is the specific internal energy of TNT, and the remaining terms are material constants. The values for each parameter are provided in Table 4 [6].

Table 4. TNT and air material parameters.

ψ_0 (MJ/m3)	ρ_0-TNT (kg/m3)	A (GPa)	B (GPa)	R_1	R_2	ω	γ	ρ_0-air (kg/m^3)	e_0 (MJ/kg)
6.0×10^3	1630	3.737×10^2	3.747	4.15	0.9	0.25	1.4	1.225	206.8

3. Results and analysis

Figures 2 and 3 respectively show the destruction of pipelines with wall thicknesses of 14.6 mm and 26.2 mm under different TNT explosive loads.

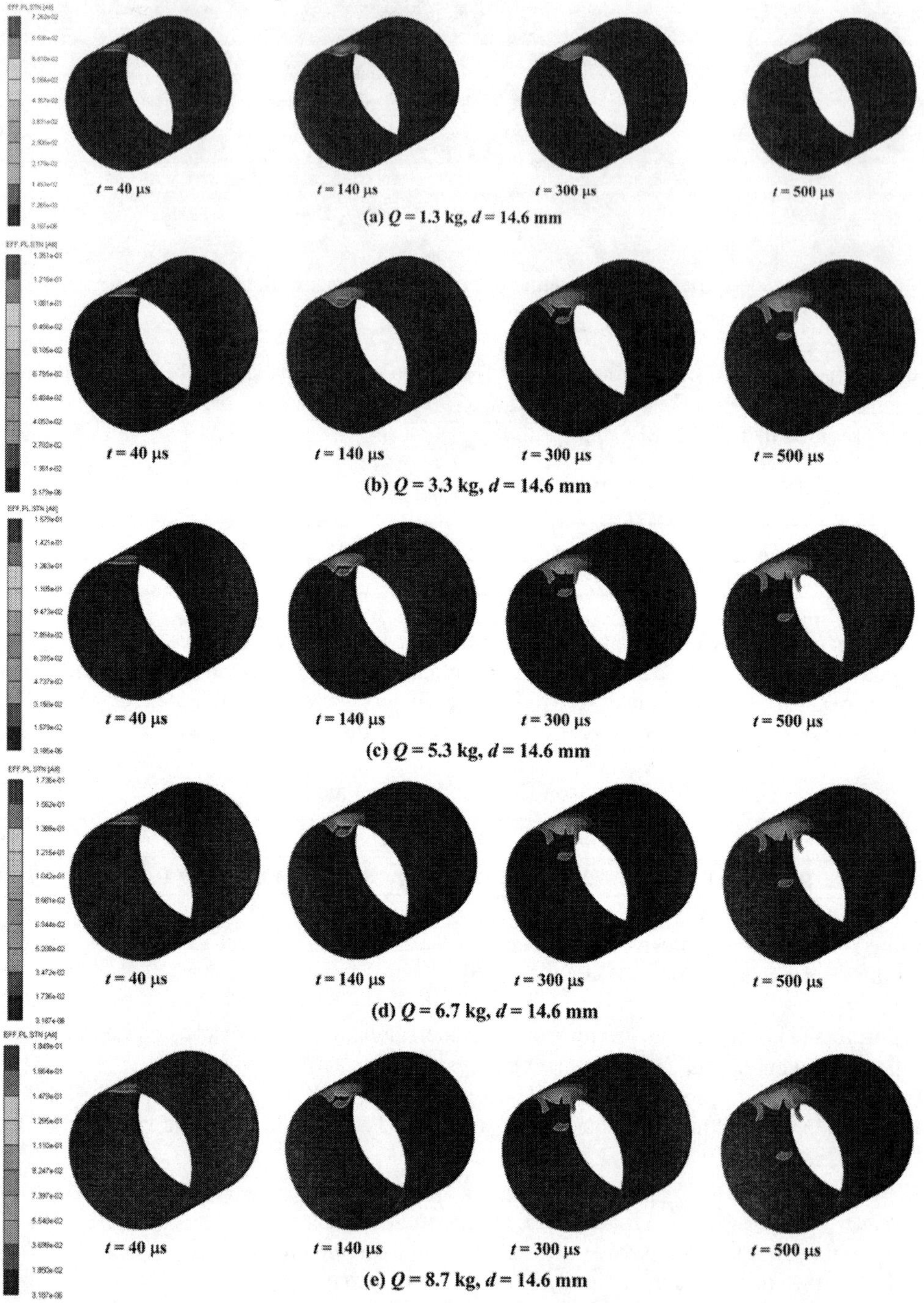

Figure 2. Failure of X80 pipelines with different TNT masses and 14.6 mm wall thickness.

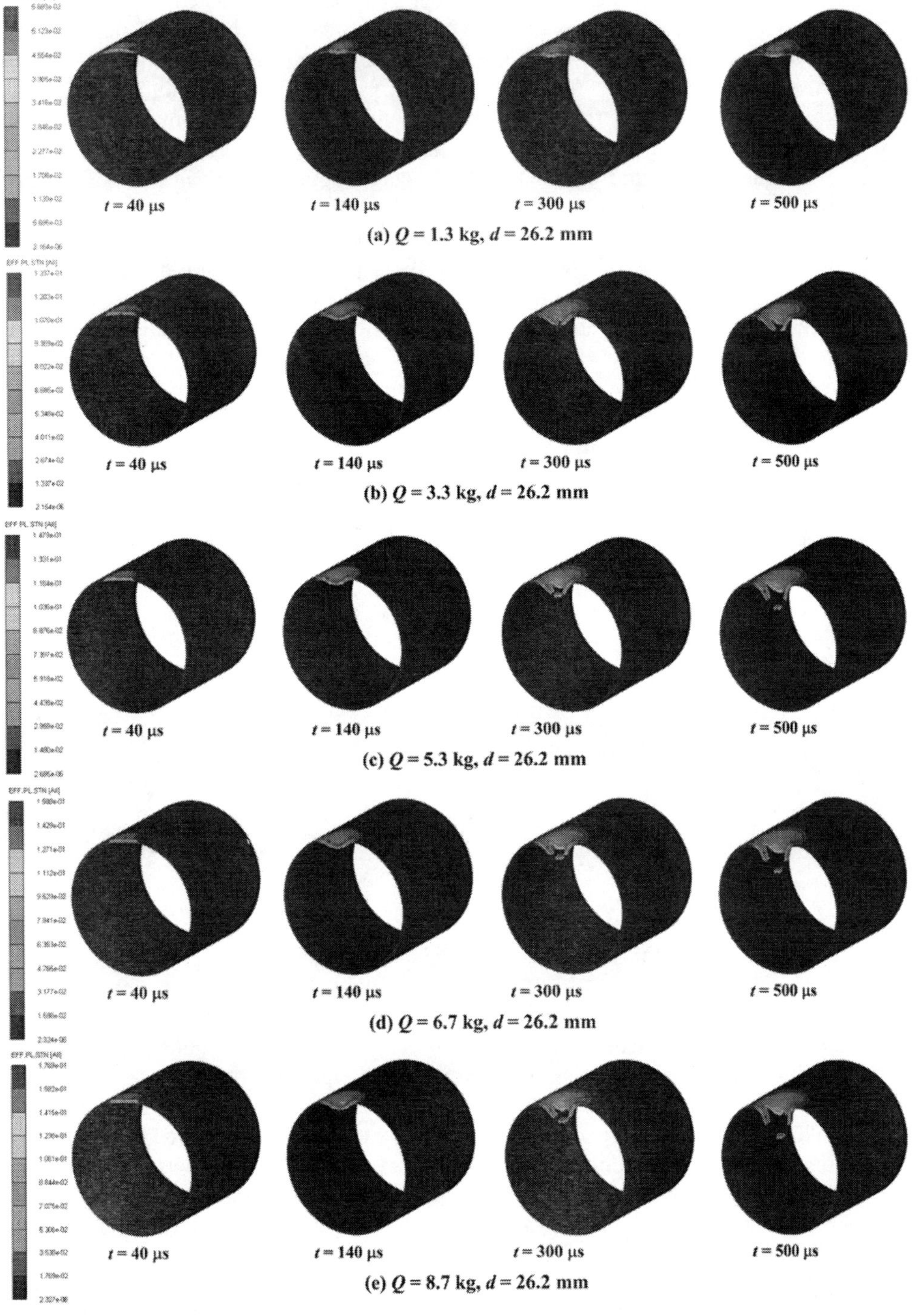

Figure 3. Failure of X80 pipelines with different TNT masses and 26.2 mm wall thickness.

From Figures 2 and 3, it can be observed that under the impact of different masses of TNT explosives, the explosive destruction of X80 pipelines can be divided into two scenarios: one where significant denting occurs without the pipeline being perforated, producing no fragments; and another where the pipeline is perforated and fragments are generated. It is evident that as the mass of TNT explosives increases, the size of the perforation and the extent of damage also increase. Conversely, as the wall thickness of the pipeline increases, the extent of damage is mitigated.

For conditions with perforation, Condition A2 is representative, with a TNT mass of 3.3 kg and a wall thickness of 14.6 mm. As shown in Figure 2(b), after the detonation of the TNT explosives, the impact load causes a dent at the top of the pipeline. The pipeline exhibits significant plastic deformation, and as the deformation expands beyond the material's ultimate fracture strain, the pipeline is perforated, and fragments are generated. The hole expands circumferentially until the circumferential strain equals the crack strain, at which point the expansion stops, reaching the maximum size of destruction [11].

For conditions without perforation, Condition B2 is representative, with a TNT mass of 3.3 kg and a wall thickness of 26.2 mm. As shown in Figure 3(b), after the detonation of the TNT explosives, a hemispherical detonation wave impacts the steel pipe contact surface. The contact surface of the steel pipe experiences significant plastic deformation under the explosive impact load, subsequently causing denting along the direction of the explosion impact. The size of the dent increases with the impact duration. Due to the plastic strain not exceeding the material's ultimate fracture strain and the other parts of the pipeline absorbing and offsetting the energy of the explosion impact wave, only a dent is formed at the top of the pipeline, with no perforation.

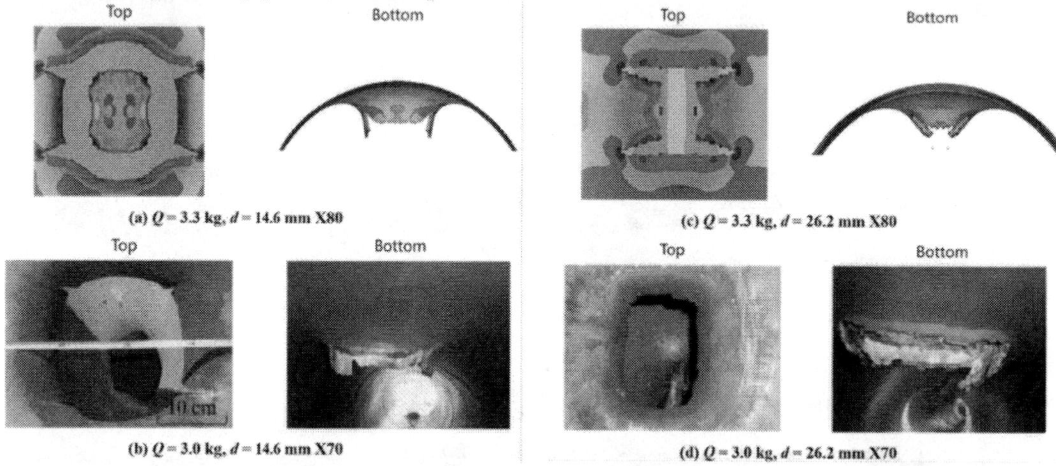

Figure 4. Comparison between the simulation results of X80 pipelines under A2 and B2 conditions and the experimental results of X70 pipelines [6].

The simulation results for Conditions A2 and B2 (3.0 kg TNT, wall thicknesses of 14.6 mm and 26.2 mm) are compared with experimental data of X70 steel pipelines under similar direct contact explosion conditions [6]. The destruction scenarios of both pipeline structures are illustrated in Figure 4. From Figure 4, the morphology of perforation holes caused by TNT contact explosion on X80 steel pipelines is like that on X70 steel pipelines, with the interior of the pipeline showing petal-shaped cracks radiating outward. Under Condition A2, the transverse width of the perforation hole in the 14.6 mm thick X80 pipeline is approximately 17 cm, and for the 14.6 mm thick X70 steel pipe caused by 3.0 kg of TNT explosives, the width of the perforation hole is approximately 18 cm, a difference of 5.6%. Under Condition B2, as shown in Figures 4(c) and (d), the size of the through hole in the 26.2 mm thick X80 pipeline is also smaller than that in the 26.2 mm thick X70 steel pipeline. The comparison results show that under the same wall thickness and TNT explosive mass, the blast resistance of X80 steel pipelines is superior to that of X70 steel pipelines. Additionally, the results validate the accuracy of the analysis model and the bulk and deviatoric constitutive descriptions.

Journal of Physics: Conference Series **2808** (2024) 012085 doi:10.1088/1742-6596/2808/1/012085

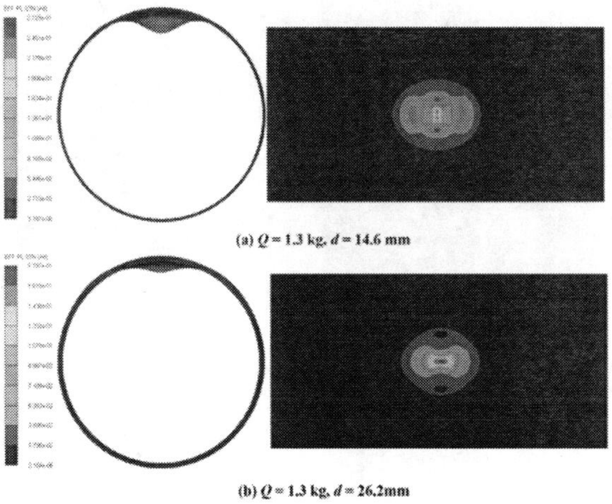

Figure 5. A1 and B1 conditions X80 pipelines depression damage.

In the simulation, there are 4 conditions where the pipelines exhibit significant denting but no perforation, specifically conditions A1 (Q = 1.3 kg, d = 14.6 mm), B1 (Q = 1.3 kg, d = 26.2 mm), B2 (Q = 3.3 kg, d = 26.2 mm), and B3 (Q = 5.4 kg, d = 26.2 mm), with the pipeline destruction scenarios illustrated in Figure 5. From Figure 5, it is apparent that as the wall thickness of the pipeline increases, its ability to resist explosive destruction also improves. The 14.6 mm thick pipeline only avoided perforation with a TNT explosive mass of 1.3 kg but exhibited significant denting with a depth of 248 mm. The 26.2 mm thick pipeline had a dent depth of 164 mm with a TNT explosive mass of 1.3 kg, a reduction of 33% relative to the 14.6 mm pipeline, and neither exhibited perforation. The simulation results indicate that increasing wall thickness significantly enhances the pipeline's resistance to explosive destruction.

Figure 6 presents the time history curve of the vertical velocity of the elements at the center of the dent. From the figure, it is evident that under the impact of the explosive shock wave, the velocity at the center of the dent rapidly increases, reaching a peak at about 0.1 ms. The peak velocity for the 14.6 mm wall thickness is 418 m/s, and for the 26.2 mm wall thickness, the peak velocity is 238 m/s, only 57% of the peak velocity in the 14.6 mm center. Afterward, the velocity rapidly decreases, stabilizing at around 0.5 ms.

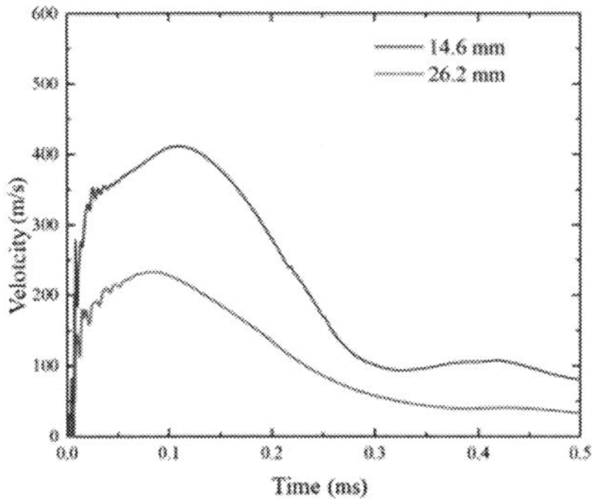

Figure 6. The vertical velocity time history curve of the top of the pipe under A1 and B1 conditions.

Aside from the two conditions, the remaining 8 conditions all exhibit perforation holes and fragments. The morphology of the perforation holes for the 8 conditions is illustrated in Figure 7, the hole area was also measured.

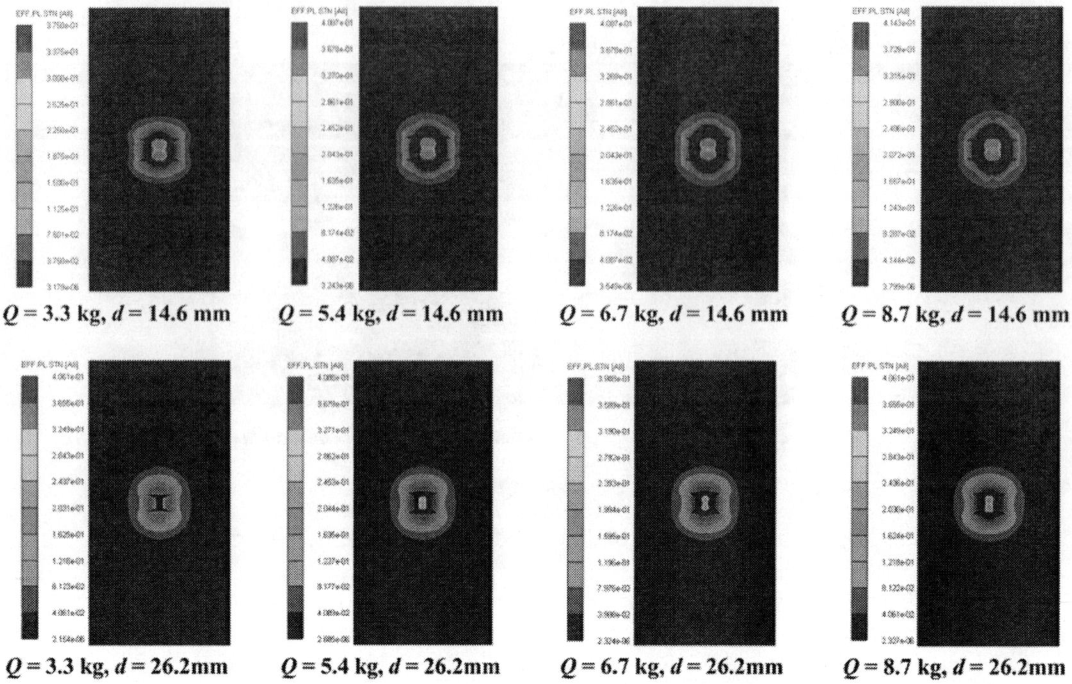

Figure 7. A2-A5 and B2-B5 conditions through the top of the pipe through the hole morphology.

From Figure 7, the morphology of the holes is elliptical, with the size of the holes increasing with the mass of TNT explosives. For the 14.6 mm thick pipeline, when the TNT mass is 8.7 kg, the area of the hole increases by 25% relative to a 3.3 kg charge. Comparing the perforation hole areas for the two wall thicknesses, it can be found that with a TNT mass of 3.3 kg, the perforation hole size in the 26.2 mm thick pipeline is reduced by 34% compared to the 14.6 mm thick pipeline, and at a TNT mass of 8.7 kg, the reduction is 18%. This indicates that increasing the wall thickness of the pipeline significantly enhances its blast resistance, effectively safeguarding the safety of oil and gas pipelines. In the contact explosion experiments on X70 steel pipelines [6], 3.0 kg of TNT explosives caused a through-hole area of 594 cm² in a 14.6 mm thick pipeline. The same wall thickness X80 steel pipeline with a TNT explosive mass of 3.3 kg had a hole area of 570 cm², 4% smaller than that of X70 steel. For a 26.2 mm thick pipeline with TNT at 3.0 kg, the through hole area was 391 cm², and for X80 steel at a TNT mass of 3.3 kg, the hole area was 381 cm², 3% smaller than that of X70 steel. The results show that even with an increased TNT mass of 0.3 kg, the size of the perforation hole formed in the X80 steel pipeline is still smaller than that in the X70 steel pipeline, demonstrating that X80 steel pipelines possess superior blast resistance.

4. Conclusion

The study concludes that the dynamic response of metal pipelines, particularly X80 steel pipelines, to explosive loads, varies significantly with the wall thickness and the amount of explosive used. The results demonstrate that pipelines with greater wall thickness exhibit enhanced resistance to blast-induced damages, evidenced by reduced deformation and smaller breach sizes. This research contributes to the field by providing a comprehensive understanding of the dynamic behavior of pipelines under blast loads, and it offers practical guidelines for designing more resilient pipeline infrastructures against explosive threats. The findings underscore the importance of considering wall thickness in the safety design and evaluation of pipelines in energy transportation networks.

Acknowledgments
The authors gratefully acknowledge the support of the Scientific Research Platform by Chengdu Polytechnic (No. 23KYPT01, No. 23KYPT02), Chengdu Polytechnic Scientific Research Project (No. 23CZYG047), 2022 Luzhou Science and Technology Plan Project (No. 2022-JYJ-130).

References

[1] Zhang T G, Stronge W J. Rupture of thin ductile tubes by the oblique impact of blunt missiles: experiments[J]. International journal of impact engineering, 1998, 21(7): 571-587.
[2] Shen X J, Zhang P X, Sun T, et al. Research on Thick Barrel's Damages Due to Impaction by Fragments of High Explosive Projectile[J]. Acta Armamentarii, 2005, 26(4): 438.
[3] Rushton N, Schleyer G K, Clayton A M, et al. Internal explosive loading of steel pipes[J]. Thin-Walled Structures, 2008, 46(7-9): 870-877.
[4] Lu S Z, Zhang B Y, Wang W, Zhang C W. Experimental Research on Dynamic Response Mechanism of Thin Cylindrical Shell under Blast Loading[J]. Journal of Nanjing University of Science and Technology,2011(05):621-626.
[5] Pan X, Xu J, Jiang J. Simulation analysis of the dynamic response of thin-wall cylindrical tank to shock wave. Journal of Chemical Industry and Engineering, 2008, 59. 796-801.
[6] Ji C, Long Y, Tang X S, Gao Z R, Li Y C. Local Damage Effects of Steel Pipe Subjected to Contact Explosion Loading[J]. Chinese Journal of High-Pressure Physics, 2013, 27(4): 567-574. doi: 10.11858/gywlxb.2013.04.016
[7] Chen L L, Research of Oil and Gas Pipelines under Blast Loads with Numerical Simulation Method[D]. Qingdao: China University of Petroleum (East China), 2017.
[8] Marsh S P. LASL shock Hugoniot data[M]. Berkeley: University of California Press, 1980.
[9] Johnson G R, Cook W H. A constitutive model and data for metals subjected to large strains, high strain rates and high temperatures[C]. Proceedings of the 7th International Symposium on Ballistics, The Hague, The Netherlands, 1983: 541–547.
[10] Kury J W, Hornig H C, Lee E L, et al. Metal acceleration by chemical explosives[C]//fourth symposium (International) on Detonation. US Government Printing Office Washington, DC, 1965: 3-13.
[11] Mu J J, Zhu X, Huang X M, Bai X F. Crevasse computation for a clamped square plate subjected to near-field noncontact underwater explosion [J]. Journal of Vibration and Shock, 2011, 30(1): 37-39.

Journal of Physics: Conference Series **2808** (2024) 012086 doi:10.1088/1742-6596/2808/1/012086

Generalized mixed element method of plane problems in cartesian coordinate and its applications

Yunze Tian [1*] and Guanghui Qing [1]

[1]Faculty of Aerospace Engineering, Civil Aviation University of China, Tianjin, 300300, China

*Corresponding author's e-mail: 1531144608@qq.com

Abstract. In finite element methods, constructing elements in Cartesian coordinates facilitates the introduction of interpolation functions with mechanical properties. This article will start from the generalized mixed variational principle and establish an 8-node mixed finite element method based on the expression of rational finite elements in the Cartesian coordinate system, in which the stress interpolation function satisfies the equilibrium equation of two-dimensional problems. This element provides convenience for the introduction of stress boundary conditions and can solve both displacement and stress variables simultaneously. The numerical analysis of several classic examples shows that the stress interpolation function satisfying the equilibrium equation can improve the accuracy of numerical results for displacement and stress fields. Especially for the sparse mesh model, the accuracy of the shear stress results is significantly better than that of the conventional displacement element.

1. Introduction

In the numerical methods of elasticity problems, the finite element method is the most favored, with displacement elements being the most widely used. The equilibrium element based on the principle of minimum residual energy requires the stress field to satisfy the equilibrium equation within the element and the boundary force inverse condition between the elements, which increases the difficulty of constructing the field function. In 1964, Pian [1] established a hybrid stress element based on the principle of minimum residual energy. However, when calculating the stiffness matrix of such elements, it is necessary to invert the matrix, resulting in slower computational efficiency. Moreover, the stress and displacement field variables of the hybrid stress element are independently selected. Without a systematic method to ensure a reasonable match between the stress and displacement functions, redundant maneuvering deformation patterns may occur. In 1973, Wilson et al. [2] first proposed nonconforming elements, which made the displacement field of the element a complete quadratic equation and improved the properties of the element. In 1984, Pian and Sumihara [3] proposed a rational hybrid stress element based on the H-R variational principle combined with the Lagrange multiplier idea. The stress field of this method does not need to satisfy the equilibrium equation. In recent years, a joint finite element method combining the H-R variational principle and minimum potential energy principle has achieved high computational accuracy [4]. Later, Tong and Yao constructed a 20-node hexahedral symplectic element based on this method and used it to analyze the interlayer stress of composite laminates [5].

In 1967, Herrmann [6] first proposed the Mixed Finite Element Method (MFEM) for analyzing bending problems of plates with different thicknesses and arbitrary shapes. In 1968, Dunham and Pister

[7] established the classical mixed finite element method for plane elastic problems based on the H-R variational principle. Due to its non-positive definite control equation, not using unconventional stabilization techniques during solution may lead to unstable stress and displacement results [8]. Recently, Lederer and Stenberg's further analysis was conducted on the weak symmetry problem in the mixed finite element method [9]. In 1981, Rong [10] proposed the generalized mixed element method, which can avoid the problem of numerical instability in classical mixed finite elements. On the other hand, the proportion of residual energy and potential energy in the system equation can be adjusted by changing the value of the splitting factor, which makes it highly accurate in dealing with certain engineering problems. In recent years, Qing et al. [11] combined the principle of minimum potential energy and the H-R variational principle, considered the superiority of non-conforming displacement elements, and proposed the non-conforming generalized mixed element. Zhao and Qing [12] established a non-conforming generalized mixed finite element based on the enhanced hypothesis strain theory, which improves calculation speed by simplifying integration operations. Yao et al. [13] used this method to analyze the stress concentration problem of perforated plates. He and Qing [14] applied this element to the stress intensity factor analysis at the crack tip, and the calculation accuracy was better than that of the non-conforming displacement element. Wang and Qing [15] applied this element to the analysis of static and dynamic problems in piezoelectric composite laminates, demonstrating its adaptability and reliability in multi-physical field problems. Wang and Qing [16] used this element to analyze the free boundary effect problem of composite laminates with different layup states and compared it with the general displacement element results, obtaining very satisfactory numerical results. Tong and Chen [17] used the non-conforming generalized mixed element method to analyze the vibration problem of reinforced cylindrical shells and obtained more accurate natural frequencies than the non-conforming displacement element method. Yang and Qing [18] applied this method to the stress analysis of variable stiffness laminated panels and obtained good calculation results. He and Qing [19] successfully obtained the stress intensity factor by constructing a singular element based on the non-coordinated generalized mixed variational principle.

In 1996, Zhong and Ji [20] abandoned the coordinate transformation idea of isoparametric elements and directly formulated elements in physical planes, establishing a rational finite element method that better meets the requirements of mechanics for element deformation modes. This method uses displacement analytical solutions of elastic plane problems to construct elements, improving their properties and avoiding volume locking and shear locking. Subsequently, Wang and Zhong [21] constructed a planar 8-node and 20-node rational finite element and verified it through various types of examples. The results showed that the rational finite element has high efficiency and good stability. Mao and Yao [22] applied rational elements to solve problems of spatial anisotropy and demonstrated through examples that the element also has high computational accuracy under distorted grids.

Based on the ideas of rational displacement element and generalized mixed element, this article attempts to construct an 8-node generalized mixed element (RGME8) in the Cartesian coordinate system that satisfies continuity conditions for displacement and equilibrium equations for stress while considering stress and displacement boundary conditions, and conducts case analysis.

2. Basic theory

Assuming that the displacement boundary condition satisfies $\boldsymbol{u}-\bar{\boldsymbol{u}}=0$ in advance and the volumetric force is ignored, the principle of minimum potential energy can be expressed as:

$$\prod_{\mathrm{P}}=\int_{\mathrm{V}}\frac{1}{2}(\nabla\boldsymbol{u})^{\mathrm{T}}\boldsymbol{C}(\nabla\boldsymbol{u})\mathrm{dV}-\int_{S_\sigma}\overline{\boldsymbol{T}}^{\mathrm{T}}\boldsymbol{u}\mathrm{dS} \tag{1}$$

where ∇ represents the differential operator matrix, $\boldsymbol{u}=\begin{bmatrix}u & v & w\end{bmatrix}^{\mathrm{T}}$ is the displacement coordinate, $\boldsymbol{C}$ is the material stiffness matrix, V is the volume of the structure, S is the surface area of the structure, and S_σ is the surface with known stress. For planar problems, $\overline{\boldsymbol{T}}=\begin{bmatrix}\overline{T}_x & \overline{T}_y & \overline{T}_z\end{bmatrix}^{\mathrm{T}}$ is the known stress acting on S_σ.

The H-R variational principle (ignoring volumetric forces) can be expressed as:

$$\prod_{\mathrm{HR}}=\int_{\mathrm{V}}\left[-\frac{1}{2}\boldsymbol{\sigma}^{\mathrm{T}}\boldsymbol{S}\boldsymbol{\sigma}+\boldsymbol{\sigma}^{\mathrm{T}}(\nabla\boldsymbol{u})\right]\mathrm{dV}-\int_{\mathrm{S}_{\sigma}}\overline{\boldsymbol{T}}^{\mathrm{T}}\boldsymbol{u}\mathrm{dS} \tag{2}$$

where $\boldsymbol{\sigma}=\left[\sigma_{xz}\ \ \sigma_{yz}\ \ \sigma_{z}\ \ \sigma_{x}\ \ \sigma_{y}\ \ \sigma_{xy}\right]^{\mathrm{T}}$ is the stress vector and $\boldsymbol{S}=\boldsymbol{C}^{-1}$ is the flexibility matrix of the material.

Combining Equations (1) and (2), the generalized mixed variational principle with parameters can be expressed as:

$$\prod_{\mathrm{GHR}}=-\frac{1}{2}(1-\lambda)\int_{\mathrm{V}}\boldsymbol{\sigma}^{\mathrm{T}}\boldsymbol{C}^{-1}\boldsymbol{\sigma}\mathrm{dV}+(1-\lambda)\int_{\mathrm{V}}\boldsymbol{\sigma}^{\mathrm{T}}(\nabla\boldsymbol{u})\mathrm{dV}+\frac{1}{2}\lambda\int_{\mathrm{V}}(\nabla\boldsymbol{u})^{\mathrm{T}}\boldsymbol{C}(\nabla\boldsymbol{u})\mathrm{dV}-\int_{\mathrm{S}_{\sigma}}\overline{\boldsymbol{T}}^{\mathrm{T}}\boldsymbol{u}\mathrm{dS} \tag{3}$$

where the value range of λ is $0\leq\lambda\leq1$. When $\lambda=0$, Equation (3) is equivalent to the H-R variational principle. When $\lambda=1$, Equation (3) is equivalent to the principle of minimum potential energy. The stress of an elastic plane problem can be represented by the following general solution [23]:

$$\begin{cases}\sigma_{x}=E\sum_{k=1}^{n-1}\sum_{j=1}^{k}j(j+1)a_{m+2}x^{k-j}y^{j-1}\\ \sigma_{y}=E\sum_{k=1}^{n-1}\sum_{j=1}^{k}(k-j+1)(k-j+2)a_{m}x^{k-j}y^{j-1}\\ \tau_{xy}=-E\sum_{k=1}^{n-1}\sum_{j=1}^{k}j(k-j+1)a_{m+1}x^{k-j}y^{j-1}\end{cases} \tag{4}$$

where E is the elastic modulus, n is the number of items, $a_1\sim a_{m+2}$ is any constant, $m=\frac{(n+1)(n+2)}{2}+j$, $k=1\sim n-1$, $j=1\sim k$.

If the first 14 items in Equation (4) are selected, then there are

$$\begin{pmatrix}\boldsymbol{C}_{x}\\ \boldsymbol{C}_{y}\\ \boldsymbol{C}_{xy}\end{pmatrix}=\begin{pmatrix}1 & 0 & 0 & -y & 0 & y & -x & -xy & \frac{x^2}{2} & \frac{y^2}{2} & -\frac{x^3}{3} & x^2y & -xy^2 & y^3\\ 0 & 1 & 0 & 0 & -x & -y & x & \frac{x^2}{2} & \frac{y^2}{2} & xy & -xy^2 & \frac{y^3}{3} & x^3 & -x^2y\\ 0 & 0 & 1 & 0 & 0 & x & y & \frac{y^2}{2} & -xy & -\frac{x^2}{2} & x^2y & -xy^2 & \frac{y^3}{3} & \frac{x^3}{3}\end{pmatrix} \tag{5}$$

where $\boldsymbol{C}_x,\boldsymbol{C}_y$ and $\boldsymbol{C}_{xy}$ represent the vector expressions of the interpolation functions for σ_x,σ_y and τ_{xy}.

As shown in Figure 1, it is assumed that the origin of the Cartesian coordinate system of the element is the centroid of the element.

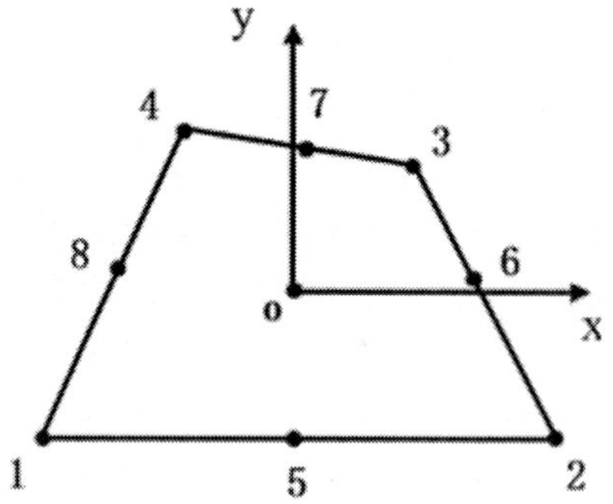

Figure 1. The diagram of the element node.

The node displacement q_{e} and stress p_{e} can be expressed as:

$$\boldsymbol{q}_{\mathrm{e}}=\boldsymbol{T}_{\mathrm{u}}\boldsymbol{a}_{\mathrm{q}} \tag{6}$$

$$\boldsymbol{p}_{\mathrm{e}}=\boldsymbol{T}_{\sigma}\boldsymbol{a}_{\mathrm{p}} \tag{7}$$

where a_q is the generalized displacement field variable of the element, a_p is the generalized stress field variable of the element, T_u and T_σ are the generalized displacement and strain transformation matrices, respectively.

$$T_u = \begin{bmatrix} 1 & x & y & xy & x^2 & y^2 & x^2y & xy^2 & 0 & 0 & 0 & 0 & 0 & 0 & 0 & 0 \\ 0 & 0 & 0 & 0 & 0 & 0 & 0 & 0 & 1 & x & y & xy & x^2 & y^2 & x^2y & xy^2 \end{bmatrix} \tag{8}$$

$$T_\sigma = \begin{pmatrix} C_x & 0 & 0 \\ 0 & C_y & 0 \\ 0 & 0 & C_{xy} \end{pmatrix} \tag{9}$$

By substituting Equations (6) and (7) into Equation (3), there are:

$$\prod_{\text{GHR}} = -\frac{1}{2}(1-\lambda)\, p_e^{\mathrm{T}} \kappa_{pp} p_e + (1-\lambda)\, p_e^{\mathrm{T}} \kappa_{pq} q_e + \frac{1}{2}\lambda q_e^{\mathrm{T}} \kappa_{qq} q_e - q_e^{\mathrm{T}} f \tag{10}$$

where $\kappa_{pp} = \int_V T_\sigma^{-\mathrm{T}} C^{-1} T_\sigma^{-1} \mathrm{d}V$, $\kappa_{pq} = \int_V T_\sigma^{-\mathrm{T}} \nabla T_u^{-1} \mathrm{d}V$, $\kappa_{qq} = \int_V (\nabla T_u)^{-\mathrm{T}} C (\nabla T_u)^{-1} \mathrm{d}V$, and $f = \int_{S_\sigma} T_u^{-\mathrm{T}} \bar{T} \mathrm{d}S$.

Because the elements in the generalized displacement and generalized stress transformation matrices are polynomials concerning for unit integrals, the expression is as follows:

$$\begin{cases} A_e = \iint_e \mathrm{d}x\mathrm{d}y \\ \iint_e x\mathrm{d}x\mathrm{d}y = \iint_e y\mathrm{d}x\mathrm{d}y = 0 \\ J_{mn} = \iint_e x^m y^n \mathrm{d}x\mathrm{d}y \end{cases} \tag{11}$$

where A_e represents the area of element and J_{mn} is the $(m+n)$ ed moment in the Cartesian coordinate system. The integral expression in Equation (11) can be obtained in advance by using exact integration.

For specific elements, a and b in Equation (10) are variational to obtain a mixed element formulation:

$$\begin{bmatrix} -(1-\lambda)\kappa_{pp} & (1-\lambda)\kappa_{pq} \\ (1-\lambda)\kappa_{pq}^{\mathrm{T}} & \lambda\kappa_{qq} \end{bmatrix} \begin{Bmatrix} p_e \\ q_e \end{Bmatrix} = \begin{Bmatrix} 0 \\ f \end{Bmatrix} \tag{12}$$

3. Numerical examples

This chapter uses Mathematica software and Wolfram language for programming and conducts relevant calculation examples and data comparisons. To facilitate the comparison of the computational accuracy of different methods under the same grid, this article refers to the comparison results of relative errors in previous literature as the standard for evaluating accuracy. The calculation equation for error is shown here:

$$\text{Error} = \frac{(\text{Exact solution - Calculated value})}{\text{Exact solution}} \times 100\% \tag{13}$$

3.1. Analysis of the problem of simply supported beam at one end

A rectangular beam with one end simply supported is shown in Figure 2. The geometric parameters of this beam are as follows: t=1, c=1, L=10. The dimensionless material parameters are as follows: E=1000, Poisson's ratio μ=0. 25, p=80. The displacement boundary condition is [24]: when $x_2 = \pm c$, $\sigma_{12} = \sigma_{22}$ =0; when $x_1 = 0$, $\sigma_{11} = 0$, $\sigma_{12} = -60(1-x_2^2)$; when $x_1 = L$, $\sigma_{11} = -120x_1x_2$, $\sigma_{12} = -60(1-x_2^2)$.

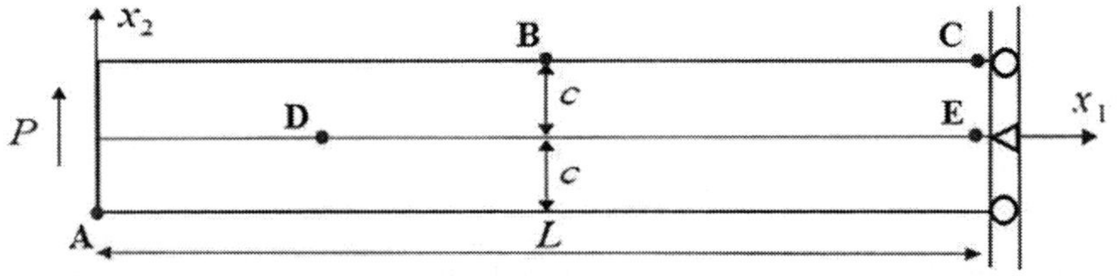

Figure 2. Geometric parameters and loads of a simply supported rectangular beam at one end.

Journal of Physics: Conference Series **2808** (2024) 012086 doi:10.1088/1742-6596/2808/1/012086

This article obtains eight grids, namely Mesh 1-8, through grid refinement. Among them, the density of the first four grids is shown in Tab. 1. The encryption method of Mesh 5-8 is the same as in Table 1.

Table 1. The encryption mode of meshes.

Mesh	Mesh 1	Mesh 2	Mesh 3	Mesh 4
Number of grids	8×2	12×4	16×6	20×8

Figure 3 shows a comparison of the calculation errors between the proposed method (RGME8) and the 8-node displacement element method (Q8), where the horizontal axis represents the grid number, i.e., Mesh 1-8, and the vertical axis represents the calculation error.

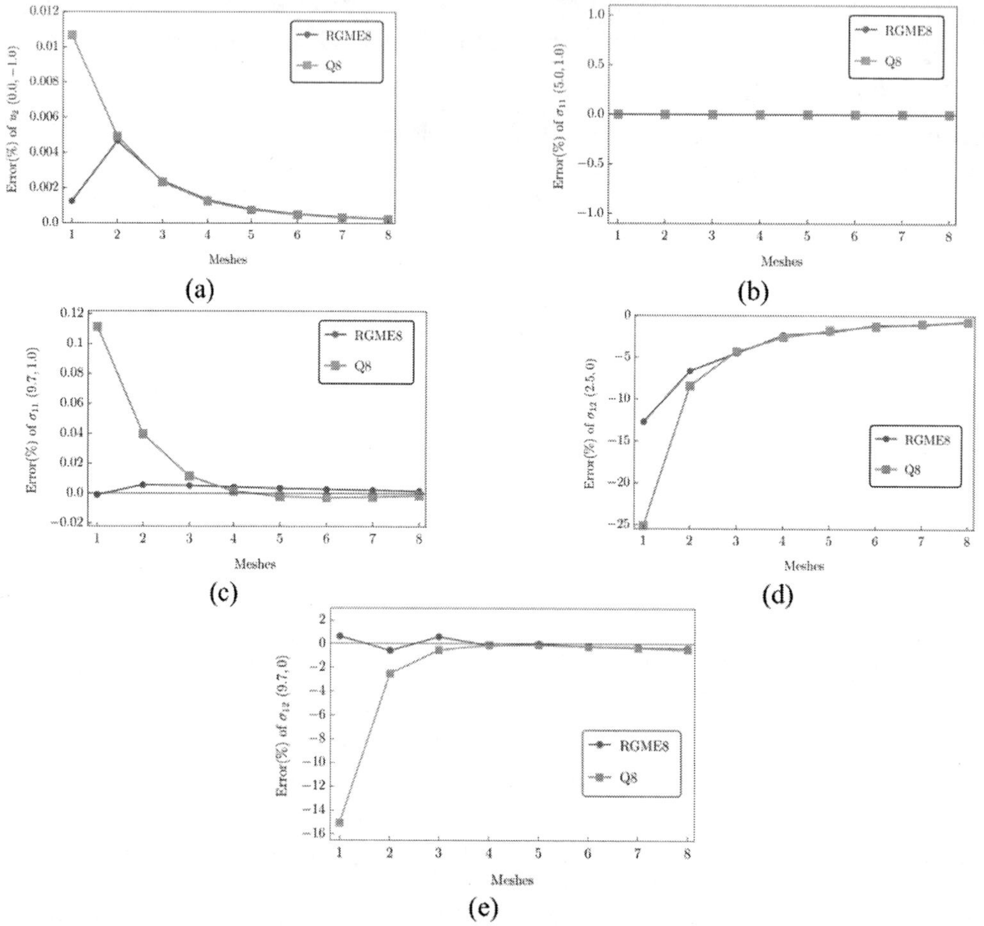

Figure 3. The comparison results of stress in the bending problem of a beam with one end simply supported.

Figure 3 (a) is the comparison results of displacement in y direction at point A (0. 0, -1. 0). Figure 3 (b) the comparison results of σ_{11} at point B (5. 0, 1. 0). Figure 3 (c) is the comparison results of σ_{11} at point C (9. 7, 1. 0). Figure 3 (d) is the comparison σ_{12} results of at point D (2. 5, 0. 0). Figure 3 (e) is the comparison σ_{12} results of at point E (9. 7, 0. 0).

The above five figures show that the RGME8 element is generally superior to the 8-node displacement element (Q8 element) in terms of convergence speed and result accuracy [18]. The

Journal of Physics: Conference Series **2808** (2024) 012086 doi:10.1088/1742-6596/2808/1/012086

introduction of high-order interpolation functions enables both Q8 and RGME8 elements to converge quickly in regions far from the end, such as the normal stress results at point B.

However, the Q8-based model cannot introduce stress boundary conditions and relies on the strain displacement derivative relationship when calculating stress. In the case of sparse grids at the ends of points C and E, the accuracy of normal and shear stress results is significantly lower than that of RGME8, as shown in Figure 3 (d) and (e). The accuracy of shear stress calculation under the sparse grid of RGME8 is also better than Q8 at point D far from the end.

3.2. Analysis of the problem of one-end fixed support beam

To further demonstrate the accuracy advantage of the proposed method in displacement and shear stress calculations, this section selects a rectangular beam with one end fixed as shown in Figure 4. Its geometric parameters are as follows: t=1, c=1, and L=10. The dimensionless material parameters are as follows: E=1000, Poisson's ratio μ=0. 25. The left end of the beam is subjected to a tangential force P=100. The displacement boundary condition is: when $x_1 = L$, u=v=0. The stress boundary conditions are: when $x_2 = \pm c$, $\sigma_{11} = \sigma_{22} = 0$; when $x_1 = 0$, $\sigma_{11} = 0$ and $\sigma_{12} = -75(1-x_2^{\ 2})$.

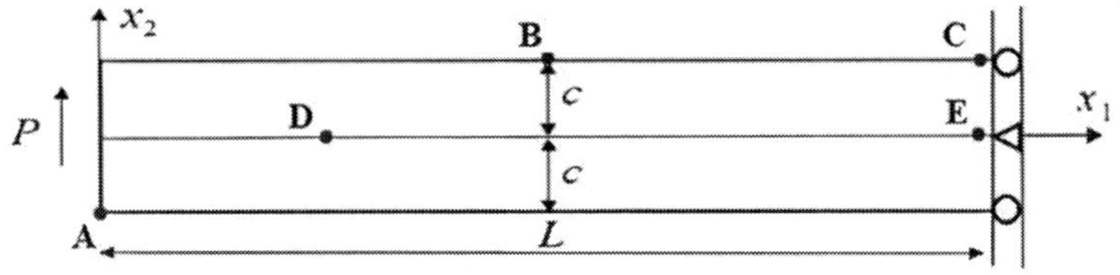

Figure 4. Geometric parameters and loads of rectangular beams with one-end fixed support.

Considering the situation of fixed support at the end, this example adopts a gradient mesh model. Figure 5 shows the grids of Mesh 1 and Mesh 2, and the encryption methods of other models can be derived from this. The latter grid increases by 10 in the x_1 direction and 2 in the x_2 direction compared to the previous one. For example, Mesh 3 is 30×6 and Mesh 4 is 40×8.

(a) Mesh 1 10×2 (b) Mesh 2 20×4

Figure 5. The mesh mode of the fixed beam.

Figure 6 shows a comparison of the displacement errors of the RGME8 method and the 8-node displacement method Q8 in the x_1 and x_2 directions at point A. The position of point A(0.0, -1.0) is shown in Figure 4.

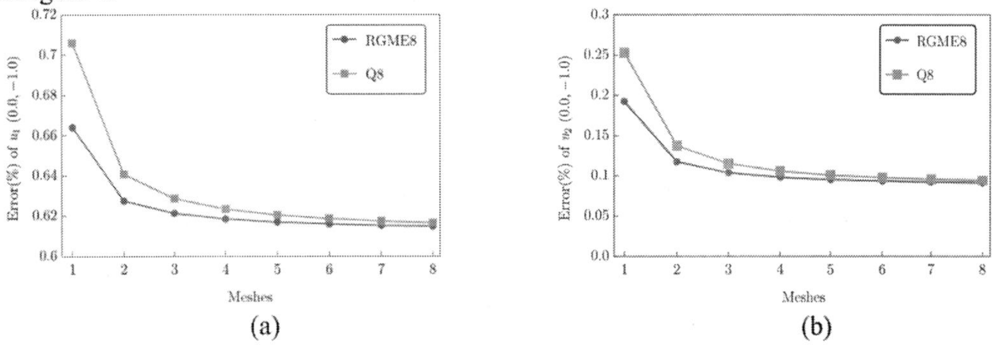

(a) (b)

Figure 6. The comparison results of displacement in the bending problem of a beam with one end fixed.

From the displacement results shown in Figure 6, it can be seen that RGME8 has a higher convergence speed and accuracy than Q8. It is obvious that in the case of sparse grids, the displacement results of RGME8 are close to the exact solution.

Figure 7 shows the comparison of errors between the two methods in calculating the normal stress at point B (7.5, 0.0) and the shear stress at point C (9.0, 0.0).

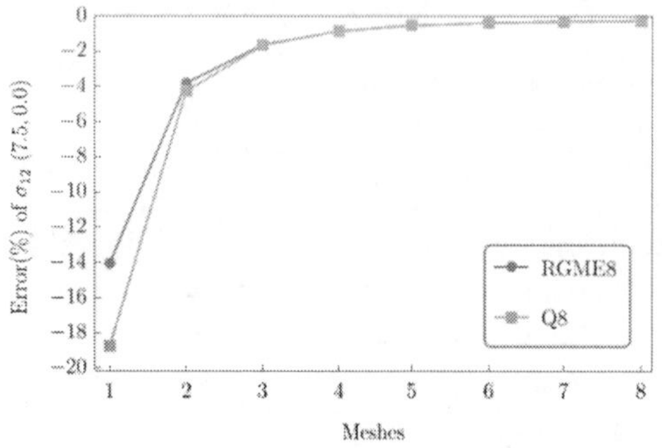

(a)

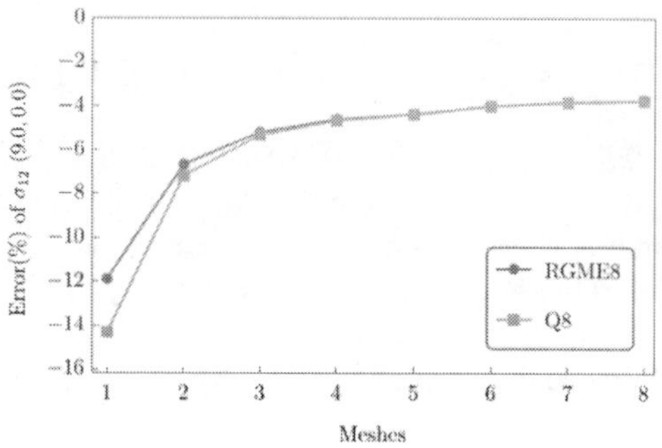

(b)

Figure 7. The comparison results of shear stress σ_{12} in the bending problem of a beam with one end fixed.

From Figure 7(a) and (b), it can be seen that for shear stress, the convergence rates of RGME8 and Q8 are similar. However, for sparse grid models, RGME8 has better shear stress results than Q8.

3.3. Analysis of radial concentrated force at one end of a curved beam

As shown in Figure 8, the inner diameter of the curved beam is a=10, the outer diameter is b=15, one end is fixed, and the other end is subjected to radial force F=120, elastic modulus E=1000, and Poisson's ratio is μ=0. 3. We take the center of curvature as the coordinate origin O.

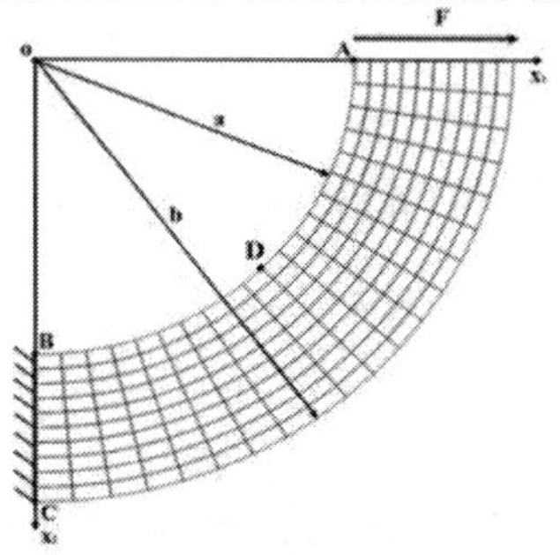

Figure 8. Finite element model of radial concentrated force problem of curved beam.

Table 2 shows the errors and calculation results of the new method (RGME8, 8-node generalized mixed finite element method (CGME8), and 8-node displacement finite element method (Q8)) in the same grid.

Table 2. Numerical results and errors of displacement and stress in curved beams.

Method	U2 (A)	S11 (B)	S11 (C)	S11 (D)
Exact	90.41	2214	-1476	1840.58
RGME8	90.52	2214.96	-1476.43	1839.73
Error	-0.12%	-0.04%	-0.05%	-0.05%

CGME8 [7]	90.53	2206.81	-1484.35	1844.88
Error	-0.13%	0.30%	-0.58%	-0.23%
Q8	90.53	2207.04	-1486.32	1845.2
Error	-0.13%	0.30%	-0.72%	0.25%

The displacement and stress results presented in Table 2 indicate that RGME8 generally outperforms CGME8 and Q8. For example, at the fixed endpoints B and C of the curved beam, the normal stress S_{11} has errors of -0.04% and -0.05% for RGME8 element, 0.30% and -0.58% for CGME8 element, and 0.30% and -0.72% for Q8 element. The error of the RGME8 element in normal stress S11 at point D is -0.05%, while the Q8 element and the CGME8 element are 0.25% and -0.23%, respectively. Obviously, in terms of the displacement result at point A, the situation is also the same.

3.4. Comparison of calculation examples for an infinite plate with central holes

The infinite rectangular plate with a circular hole in the center, as shown in Figure 9, is subjected to a uniformly distributed tensile load $\sigma_0 = 1$ at both ends. The length and width of the board are all *a*; the hole radius is *r*; the material parameter is E=1000, μ= 0.3. Numerical calculations are only performed for the square subregions near the circular hole. Because the subdomain is a square, to reduce the calculation time of the example, a quarter of the subdomain of the board, as shown in Figure 10, can be selected for calculation. We select the center o of the circular hole to establish a coordinate system. To facilitate the representation of stress analysis solutions, we use *o* to establish polar coordinates ρ-θ.

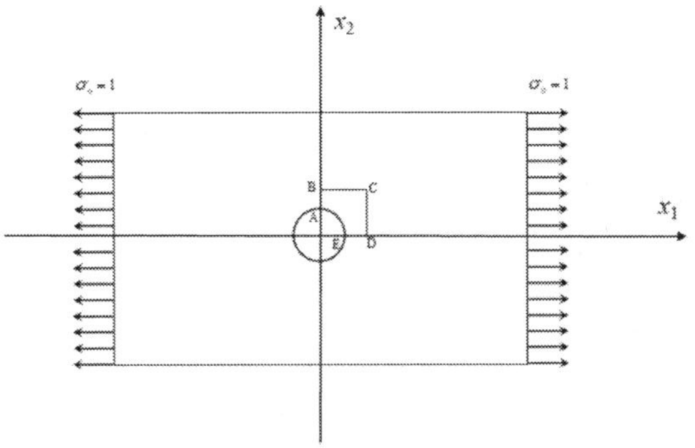

Figure 9. A uniformly stretched infinite plate with a hole.

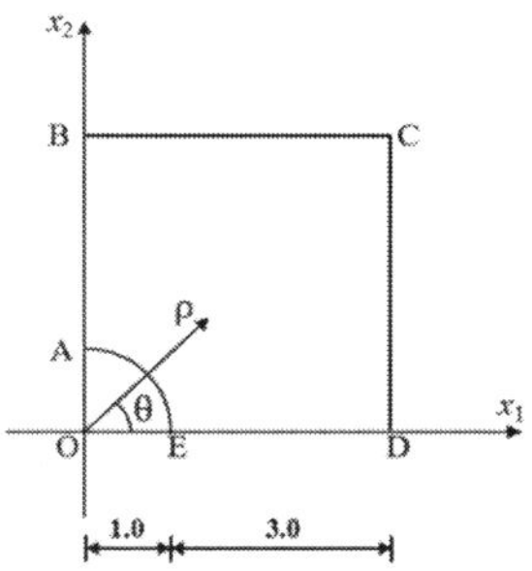

Figure 10. Quarter subregion.

Figure 11 and Figure 12 show the two types of grids used in this example, which are 5 × 5 and 10 × 10, respectively. Considering the issue of containing holes, stress concentration may occur around the circular hole, and further refinement should be carried out around the hole.

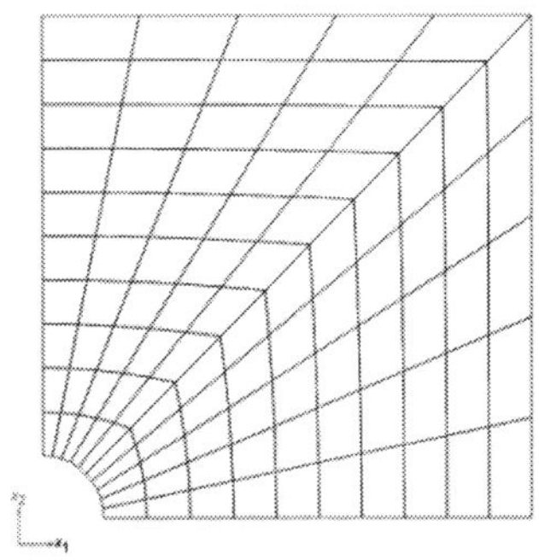

Figure 11. Mesh 1 5×5.

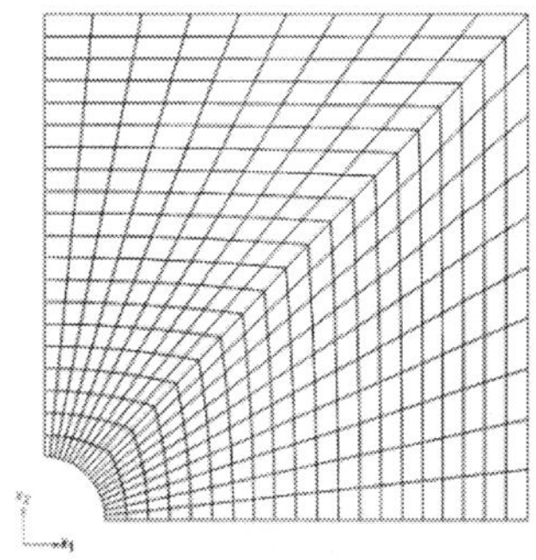

Figue 12. Mesh 2 10×10.

Journal of Physics: Conference Series **2808** (2024) 012086 doi:10.1088/1742-6596/2808/1/012086

Figure 13 shows the stress cloud map of an infinite flat plate with a central hole calculated under Mesh 1. Figure 13(a) is σ_{11} stress nephogram of the infinite slab quarter model with a center hole. Figure 13(b) is σ_{22} stress nephogram of the infinite slab quarter model with a center hole. Figure 13(c) is τ_{12} stress nephogram of the infinite slab quarter model with a center hole. The continuity of stress distribution in the stress cloud diagram can prove the reliability of this method in calculating this example.

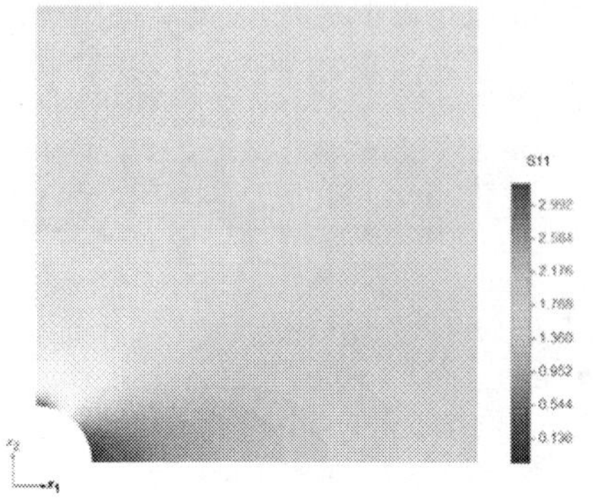

(a)

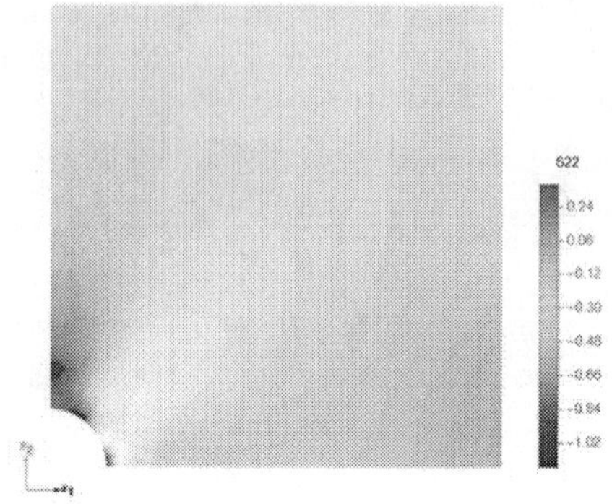

(b)

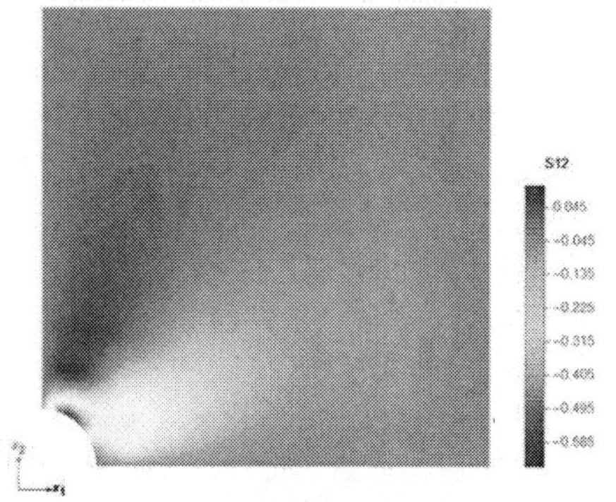

(c)

Figure 13. Stress nephogram of infinite slab quarter model with center hole.

Table 3 presents the calculation results of normal stress and shear stress at point P (ρ =1. 5, θ=45°) for two different grids, Mesh 1 (5 × 5) and Mesh 2 (10 × 10).

Table 3. Results and errors of the stress of infinite slab quarter model with center hole.

Mesh	Method	σ_{11}	σ_{22}	σ_{12}
Mesh 1	RGME8	1.1230	-0.1658	-0.2102
	Error	2.1862%	-11.9514%	5.4005%
	Q8	1.1050	-0.1289	-0.2413
	Error	3.7540%	12.9642%	-8.5959%
Mesh 2	RGME8	1.1392	-0.1563	-0.2228
	Error	0.7752%	-5.5368%	-0.2700%
	Q8	1.1239	-0.1337	-0.2198
	Error	2.1078%	9.7232%	1.0801
Exact		1.1481	-0.1481	-0.2222

Journal of Physics: Conference Series **2808** (2024) 012086 doi:10.1088/1742-6596/2808/1/012086

For the convenience of comparing the accuracy of the calculation results, the exact solutions are listed in the table, and the error relative to the exact solutions is calculated. From the calculation results, it can be seen that the RGME8 element method in this article is significantly closer to the exact solution in stress calculation results. For example, in the calculation of point P, the error of our method in Mesh 1 is only 5.4005%, while the error of the Q8 element is -8.58%; in the mesh of Mesh 2, the error of our method is only 0.2700%, almost close to the exact solution, while the error of the Q8 element is 1.0801%. Under the Mesh 1 grid, the calculation error of normal stress at point P is only 0.7752%, significantly higher than the 2.1078% of the Q8 element. The stress calculation values for other P points are also slightly closer to the exact solution than the Q8 element.

4. Conclusion

This article is based on the general solution of plane static problems and rational finite element theory and uses stress interpolation functions that satisfy equilibrium equations to propose an 8-node generalized mixed element (RGME8) represented in the Cartesian coordinate system on the element domain. The main conclusions are as follows:

(1) The method proposed in this article yields stable convergence and high computational accuracy.

(2) In the case of sparse grids, compared with the displacement and stress results of Q8 and CGME8, RGME8 has high accuracy and stable convergence. The analysis of the bending problem of beams shows that the RGME8 element in this article can improve the accuracy of shear stress.

Acknowledgments

National Natural Science Foundation of China (11502286).

References

[1] Pian T HH. Derivation of element stiffness matrices by assumed stress distributions [J]. AIAA journal, 1964, 2(07): 1333-1336.

[2] Wilson E L, Taylor R L, Doherty W P, et al. Incompatible Displacement Models [J]. numerical & computer methods in structural mechanics, 1973.

[3] Pian T HH. Sumihara K. Rational approach for assumed stress finite elements [J]. Inter J Numer Meth Engng, 1984:1685-1695.

[4] Qing Guang-hui, Wang Bo-ao. Combined finite element method and analyses on its convergence and accuracy [J]. Journal of Civil Aviation University of China, 2019, 37(01):60-64.

[5] Tong Yao, Yao Yu-zhe. 20-Node hexahedron symplectic elements for stress analysis of composite laminates [J]. Applied Mathematics and Mechanics, 2020, 41(05):509-516.

[6] Herrmann L R. Finite-element bending analysis for plates [J]. Journal of the Engineering Mechanics Division, 1967, 93(05): 13-26.

[7] Dunham R S and Pister K S. A Finite Element Application of the Hellinger-Reissner Variational Theorem [C]. Proceedings of the Second Conference on Matrix Methods in Structural Mechanics, 1968.

[8] Li Rui. Three kinds of thermal elastomers based on the generalized mixing method and the symplectic finite element method. [M]. Tianjin: Civil Aviation University of China, 2019.

[9] P. Lederer, R. Stenberg. Analysis of weakly symmetric mixed finite elements for elasticity. Mathematics of Computation. 2022. DOI:10. 1090/mcom/3865.

[10] Rong Tin-yu. Variational principle of splitting modulus in elasticity. Journal of Southwest Jiaotong University, 1981(01): 48-56.

[11] Qing G, Mao J, Liu Y. Generalized mixed finite element method for 3D elasticity problems [J]. Acta Mechanica Sinica, 2018, 34(02): 371-380.

[12] Zhao Zhi-qin, Qing Guang-hui. Improved non-compatible generalized mixed elements and performance analysis [J]. Journal of Applied Mathematics Mechanics, 2019, 40(5): 518-526.

Journal of Physics: Conference Series **2808** (2024) 012086 doi:10.1088/1742-6596/2808/1/012086

[13] Yao Yu-zhe, Qing Guang-hui, Wang Shao-bo. Stress analysis of a plate with a hole based on incompatible generalized mixed element [J]. Journal of Technology Wind, 2020, (20):193-196.

[14] He Yuxuan, Qing Guang-hui. Calculation of SIF based on nonconforming generalized mixed element method. Journal of Structural Engineer, 2021, 37(05):19-24.

[15] Wang Yu-hang, Qing Guang-hui. Analysis of piezoelectric composite laminates based on generalized mixed finite element. Acta Materiae Compositae Sinica, 2022, 39(06):2987-2996.

[16] Wang Xie, Qing Guang-hui. Analysis of laminates free-edge effect based on non-compatible generalized partial mixed elements. Acta Materiae Compositae Sinica, 2022, 39(08):4172-4178.

[17] Tong Yao, Chen Xu-tao. Vibration analysis of stiffened cylindrical shells based on generalized mixed element method [J]. Journal of Structural Engineers, 2022, 38(04):1-8.

[18] Yang Li-zho,Qing Guang-hui. Static analysis of variable stiffness laminates based on mixed finite element [J]. Journal of Engineering Mechanics, 2024, 41(03):19-25.

[19] He Li, Qing Guang-hui. Generalized mixed singular element and stress intensity factor analysis [J]. 1-7. [2024-03-05].

[20] Zhong Wan-xie, Ji Zheng. Rational finite element. Chinese Journal of Computational Mechanics | Chin J Comput Mech, 1996, 13(001):1-8.

[21] Wang Yong-fu, Zhong Wan-xie. A rational hexahedron 8-node finite element. Chinese Journal of Applied Mechanics, 2003, 20(3):131-135.

[22] Mao Ling, Yao Wei-an. Twenty-node rational element for spatial anisotropic elastic problems [J]. Applied Mathematics and Mechanics, 2014, 35(06):589-597.

[23] Jiang Wei. Finite element-relaxed boundary condition method for solving the plane problem of elastic theory. Chinese Journal of Theoretical and Applied Mechanics, 1981(03): 236-247.

[24] Wu Jialong. Elastic mechanics. Beijing: Higher Education Press, 2001.

Journal of Physics: Conference Series **2808** (2024) 012087 doi:10.1088/1742-6596/2808/1/012087

Split-tensile strength and micro-structure analysis of hybrid fibers coal gangue concrete

Jiuyang Li [1], Jingwei Luo [1a], Xinmei Fan [1], Yuepeng Zhu [1], Xiaoyu Wang [1], Li Chen [1], Fan Li [2] and Zicheng Zhang [3]

[1]School of Civil Engineering, Changchun Institute of Technology, Changchun, 130012, China
[2]School of Traffic Engineering, Wuhan Vocational College of Communications, Wuhan, 430065, China
[3]China. Fifth Construction Co., LTD., ChinaConstucion First Group, Shenzhen, 518000, China

*Corresponding author's e-mail: sixiliqing@163.com

Abstract: To make the coal gangue better applied and engineered, through four factors (Coal gangue ceramicist [CGc] Substitution, coal gangue ceramic sand [CGs] Substitution, end-hooked steel fiber [SF], polypropylene fiber [PF]) four-level orthogonal test was conducted to explore the variation of Split-tensile strength of mixed-fiber coal gangue concrete and the strengthening mechanism of hybrid fiber on split-tensile strength of Coal-gangue concrete is analyzed from the microscopic point of view. The results show that the sequence of factors affecting the split-tensile strength is substitution rate of CGs > substitution rate of CGc > incorporation amount of SF > incorporation amount of PF. The optimal admixture combination is: CGc substitution rate of 30%, CGs substitution rate of 25%, SF incorporation rate of 1.00%, and PF incorporation rate of 0.2%. The spatial framework formed by hybrid fibers can enhance the stability of the ITZ to bond inner material tightly, thus enhancing the split-tensile strength of concrete.

1. Introduction

Coal gangue is a by-product of coal mining [1-2]. This by-product accumulates in large quantities as solid waste and is a global problem for all countries. A large accumulation of coal gangue will occupy land resources, also contaminate surrounding water and soil through rainwater [3-4], affecting the physical and mental health of residents. Previous studies have shown that the coal gangue physico-chemical properties are highly similar to those of natural stone sand and can be substituted for aggregate to produce concrete [5-6]. Due to certain defects of coal gangue concrete itself, many scholars choose to elevate the strength of coal gangue concrete by adding fibers [7]. With the deepening of the research, it is found that compared to concrete with only a single fiber, the concrete of hybrid fiber can achieve the effect of 1+1 > 2 due to the different kinds of fiber play a complementary role in common concrete, which can more effectively enhance the mechanical properties of common concrete.

In this paper, the split-tensile strength of steel-polypropylene fibers coal gangue concrete is studied by a four-factor and four-level orthogonal test, using variance analysis and range analysis to reveal the effect of different factors on the split-tensile strength. In addition, we analyze the failure mechanism of steel-polypropylene fiber coal gangue concrete from a microscopic perspective and macro performance. It provides a reference for the resource-utilization of coal gangue in building construction

2. Test overview

2.1 Material

The cement was P·O 42.5 Dinglu brand ordinary Portland cement used in the test. The coarse aggregate is ordinary crushed stone, and the particle size is 5.0~16.0 mm; Natural river sand with a fineness modulus of 2.63, was used for fine aggregate. Coal gangue is through the harmless treatment of ceramsite and ceramic sand, with an apparent density of 2791 kg/m^3, as shown in Figure 1 (a) and (b). The end-hook steel fiber is used, the length is 25 mm, and the tensile strength is 1000~1800 MPa, as shown in Figure 1 (c); The polypropylene fiber used was 50 mm in length, and tensile strength is 720 MPa, as shown in Figure 1 (d).

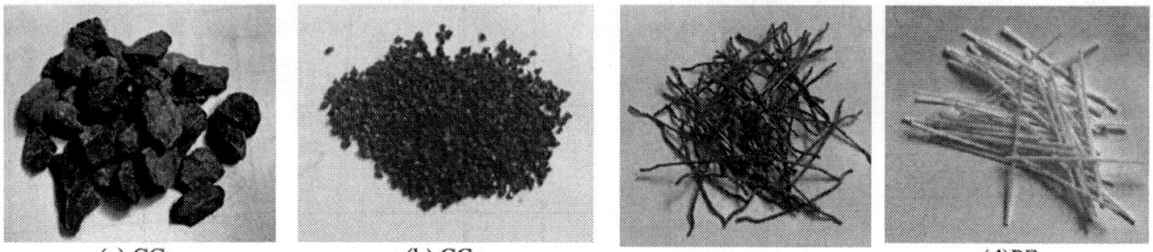

(a) CGc (b) CGs (c) SF (d)PF

Figure 1. Test raw materials.

2.2 Mix proportion design

The C30 concrete base fit calculated in this test is shown in Table 1 according to the "Common Concrete Mix Design Regulations" (JGJ 55-2011). Since the effect of several variables on the split-tensile strength was involved in this experiment, an orthogonal experimental group of L16(44) was designed to analyze it. First of all, the four factors are determined as the replacement rate of CGc (A), the replacement rate of CGs (B), the SF incorporation amount (C), and the PF incorporation amount (D). There are four levels for each factor, and ordinary concrete (PC) is added. Specific factor levels are shown in Table 2.

Table 1. C30 concrete base mix ratio.

Water-cement ratio	Sand rate (%)	Cement (kg·m^{-3})	Stone (kg·m^{-3})	Sand (kg·m^{-3})	Water (kg·m^{-3})
0.6	40	366.67	1087.99	725.33	220

Table 2. Specific test level and factors.

Level	Factor			
	A (%)	B (%)	C (%)	D (%)
1	25	25	0.25	0.1
2	30	30	0.50	0.2
3	35	35	0.75	0.3
4	40	40	1.00	0.4

2.3 Specimen preparation and test method

As fiber is easy to clump, this test put two kinds of fiber separately and dry mix with other materials for 2 mins. After mixing evenly, we add water and wet mix for 3 mins. This method can greatly reduce fiber clump. Three cubes of 100 mm×100 mm×100 mm were set in each group test. We put them into the standard maintenance machine with constant humidity and constant temperature for 28 days, and then take them out for test. Loading tests were carried out on the YAW-2000 pressure test machine and were carried out based on the provisions of Chinese norms (GB/T 50081-2019). Figure 2 shows the loading process of the split tensile test. The splitting and pulling data of the specimen were recorded and collected.

Figure 2. Splitting tensile test loading process.

3. Orthogonal experiment results analysis

The cured specimens were tested, and we recorded the test data in Table 3. Among them, the orthogonal combination with the highest splitting tensile strength is A2B1C2D3, which increased by 57.30% compared with PC.

Table 3. Orthogonal test results.

Orthogonal combination	Splitting tensile strength	Orthogonal combination	Splitting tensile strength
A1B1C1D1	2.37	A3B2C4D3	2.46
A1B2C2D2	2.65	A3B3C1D2	2.42
A1B3C3D3	2.49	A3B4C2D1	1.95
A1B4C4D4	2.36	A4B1C4D2	2.58
A2B1C2D3	2.80	A4B2C3D1	2.46
A2B2C1D4	2.48	A4B3C2D4	1.96
A2B3C4D1	2.59	A4B4C1D3	1.96
A2B4C3D2	2.34	PC	1.78
A3B1C3D4	2.52		

3.1 Range analysis

Range analysis is simple, intuitive, and computant-intensive, and can reflect the results of orthogonal tests by various factors. Through processing the data of orthogonal tests, range analysis results are obtained, as recorded in Table 4.

Table 4 Results of range analysis.

Program	Splitting tensile strength/MPa			
	CGc	CGs	SF	PF
K_1	2.47	2.57	2.31	2.34
K_2	2.55	2.51	2.34	2.50
K_3	2.34	2.37	2.45	2.43
K_4	2.24	2.15	2.50	2.33
R	0.31	0.42	0.19	0.17

Note: K_i is the mean intensity of each factor at the i level

The higher the value of R, the greater the fluctuation of the effect of the factor on the test value. It can be easily seen from Table 4 that the sequence of effecting factors of different factors on split-tensile strength of concrete is as follows: Substitution rate of CGs > Substitution rate of CGc > incorporation amount of SF > incorporation amount of PF.

3.2 Variance analysis

Since the range analysis could not estimate the error of the data, the data were further processed through the analysis of variance. The calculated variance analysis data results are recorded in Table 5.

Table 5. Results of variance analysis.

Factor	Degree of freedom	Mean square	Deviation sum of squares	P	F
CGc	3	0.10	0.31	89.48%	5.17
CGc	3	0.14	0.41	92.62%	6.88
SF	3	0.03	0.10	65.10%	1.63
PF	3	0.02	0.07	56.55%	1.23

As can be seen from Table 5, the sum of deviation sum of squares of CGs is the highest, which has an obvious impact on the split-tensile strength, and its reliability is 92.62%, while the reliability of PF is the lowest, only 56.55%, indicating that PF has a weak impact on the split-tensile strength. The reliability ranking of ANOVA is the same as that of range analysis. It shows that the orthogonal tests are reasonable and reliable.

We plot the data in Table 4 into a more intuitive effect graph, as shown in Figure 3.

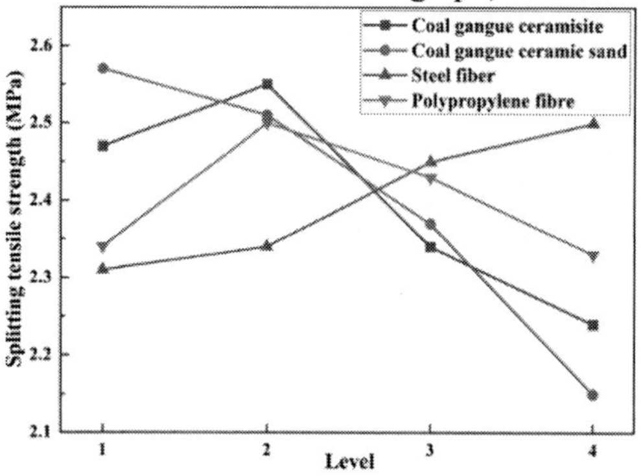

Figure 3 Effect curve.

The split-tensile strength in factor A shows a trend of first rising and then diminishing. As for factor B, the split-tensile strength showed a decreasing trend. The pores in coal-gangue aggregate can augment the water absorption and make the hydration reaction degree near it high. The internal conservation effect is formed, and this improvement is P1. Secondly, coal gangue aggregate can produce a volcanic ash effect, and the calcium silicate gel produced can fill the gaps and make the interior of the coal gangue concrete more compact. And this improvement is denoted as P2. Due to the low strength of coal gangue itself, this intensity reduction effect is denoted by H. When the sum of P1 and P2 is greater than H, the effect of improvement is more than the effect of reduction, and the concrete strength is enhanced. On the contrary, when H is greater than the sum of P1 and P2, the strength of concrete is reduced.

For factor C, the splitting tensile strength showed an increasing trend. For factor D, the splitting tensile strength increased before they were reduced. The main reason is that hybrid fibers with low content overlap with each other in the concrete. When the specimen is broken, more power is required to pull the fibers out at this crack and inhibit the development of cracks. Macroscopically, the split-tensile strength of concrete is augmented. When the PF and SF incorporation rate is too high, the fiber agglomerates together, resulting in a decrease in compactness, thus reducing the enhancement

Journal of Physics: Conference Series **2808** (2024) 012087 doi:10.1088/1742-6596/2808/1/012087

effect. Therefore, when using hybrid fibers, it is necessary to pay attention to the amount of incorporation not being too high.

4. Microstructure analysis

The micromorphology of mixed fiber coal gangue concrete was observed by SEM test, as illustrated in Figure 4.

(a) Steel fiber (b) Polypropylene fiber

Figure 4. Microscopically fiber-cement slurry interface transition zone.

Figure 4 (a) shows the Interfacial transition zone (ITZ) between cement slurry-steel fibers (SF), in which the disoriented distributed steel fibers overlap in cement paste and aggregate. The establishment of the "Hoop-effect" can effectively prevent aggregate from sinking, and its strength is augmented by constraining the internal material of concrete. Furthermore, the ITZ between SF and the aggregates is distributed with a lot of calcium hydroxide (CH) and is strewn with calcium silicate hydrate (C-S-H). In addition, many ettringites (AFt) are also distributed. These hydrated products are easy to bond with the steel fiber and can consume more energy when under load, thereby improving the effect of strengthening and toughening.

The ITZ of polypropylene fiber (PF) and Cement paste is reflected in Figure 4 (b), wherein the surface of polypropylene fiber is smooth and there are fewer hydration reactants attached. Moreover, on account of the certain hydrophobic properties of polypropylene fiber, during the hydration process, there are fewer water molecules near the fibers, which makes the hydration reaction near ITZ less severe and the formation of low-density C-S-H. Therefore, it has little effect on improving the splitting property.

5. Conclusion

(1) The split-tensile strength of SF-PF Coal gangue concrete is stronger than common concrete. Compared to common concrete specimens, the strength value is increased by 57.30%.

(2) According to the calculation results of variance analysis and analysis of range, the order of influencing factors on split-tensile strength is substitution rate of CGs > substitution rate of CGc > incorporation amount of SF > incorporation amount of PF. The optimal admixture combination is: CGc replacement rate of 30%, CGs replacement rate of 25%, SF incorporation rate of 1.00%, and PF incorporation rate of 0.2%.

(3) The spatial skeleton formed by hybrid fibers can enhance the stability of the ITZ. Moreover, the outer surface of the fiber can increase the adsorption area of the cement paste, so that the hydration products can be closely bonded, thus enhancing the split-tensile strength of the concrete.

References

[1] Cong X Y, Lu S, Yao Y, et al. Fabrication and characterization of self-ignition coal gangue autoclaved aerated concrete [J]. Materials & Design, 2016, 97(May):155-162.

[2] LI Jiuyang, Chen Li, Fan Xinmei et al. Research hotspot and frontier analysis of coal gangue concrete based on CiteSpace [J]. Green Technology, 2019,25(08):268-271+275.

[3] Hu Z X, Hu X M, Cheng W M, et al. Performance optimization of one-component polyurethane healing agent for self-healing concrete [J].Construction and Building Materials, 2018, 179:151-159.
[4] Hui Wu, Qingbo Wen, Liming Hu, et al., Feasibility study on the application of coal gangue as landfill liner material, Waste Manag. 63 (2017) 161-171.
[5] M. Zhou, Y. Dou, Y. Zhang, Y. et al., Effects of the variety and content of coal gangue coarse aggregate on the mechanical properties of concrete, Construct. Build. Mater. 220 (2019) 386–395.
[6] Wang Q, Li Z, Zhang Y, et al. Influence of coarse coal gangue aggregates on elastic modulus and drying shrinkage behavior of concrete [J].Journal of Building Engineering,2020:101748.
[7] Li Jiuyang, Chen Li, Hu G. C., et al. Macro and micro analysis of splitting tensile strength of steel fiber coal gangue concrete [J]. Journal of Heilongjiang Institute of Technology (Comprehensive Edition), 2023,23(07):145-150.

Journal of Physics: Conference Series **2808** (2024) 012088 doi:10.1088/1742-6596/2808/1/012088

Study on the effect of polypropylene fiber shape and admixture on the mechanical properties of concrete

Jiuyang Li [1], Xinmei Fan [1,*], Yuepeng Zhu [1], Jingwei Luo [1], Xiaoyu Wang [1], Jinpeng Guo [1], Fan Li [2] and Zicheng Zhang [3]

[1]School of Civil Engineering, Changchun Institute of Technology, Changchun 130012, China;
[2]School of Traffic Engineering, Wuhan Vocational College of Communications, Wuhan 430065, China;
[3]China. Fifth Construction Co., LTD., China Construction First Group, Shenzhen 518000, China

[*]Corresponding author's e-mail: fancy_922@126.com

Abstract: The effect of three different shapes of PPF on the mechanical properties of concrete at different dosages was investigated. Cubic compressive ability test, split tensile ability test, and flexural ability test were conducted on polypropylene fiber concrete to analyze the effect of PPF on compressive ability, tensile compression ratio, and flexural compression ratio. The study shows that as the shape and admixture of polypropylene fibers have an impact on the compressive strength of concrete, monofilament-type polypropylene fibers will have the worst effect on the enhancement of the effect when the choice of a mixture of 0.90% of embossed polypropylene fibers on the role of the compressive ability of the concrete has the best effect. The concrete tensile compression ratio is mainly affected by the mixing amount of polypropylene fibers, and the main factors affect the concrete folding compression ratio of the polypropylene fibers of the shape of the main factors.

1. Introduction

Fiber concrete refers to the addition of a higher modulus of elasticity in ordinary concrete and tensile properties of fiber materials formulated from composite concrete materials, which can play an effective role in improving the effect of the tensile strength of concrete [1,2]. In recent years, common fiber materials include steel fiber, carbon fiber, glass fiber, etc., among which the earliest steel fiber applied in concrete has a huge effect on the tensile and compressive strength of concrete. However, it is easy to wear the mixing equipment in the concrete mixing process, the corrosion resistance is poor [3], and carbon fiber has the same shortcomings as expensive [4]. Glass fiber is a high-strength fiber, but it is easy to break and poor brittleness, which also affects its application range. Polypropylene fiber (PPF) is a synthetic fiber that has the advantages of low density, better ductility, good corrosion resistance, etc. The addition of concrete significantly improves the ductility of concrete and is widely used in building materials [5,6].

This paper takes C50 as the base concrete strength, selects three different shapes of polypropylene fiber (wavy polypropylene fiber, embodied-type polypropylene fiber, mono-filament polypropylene fiber), and blends them into concrete according to different dosages (0, 0.30%, 0.60%, 0.90%, 1.20%). The influence of different shapes of polypropylene fiber on concrete performance under different dosages was studied by analyzing its compressive strength, tensile compression ratio, and folding compression ratio.

2. Test overview

2.1 Test material

The clinker is made of P O42.5R ordinary silicate clinker of Jilin Yatai Cement Co., Ltd.; the coarse aggregate is made of ordinary granite gravel after screening with a grain size of 16~20 mm; fine aggregate is natural river sand has a fineness modulus of 2.61; the water is ordinary tap water; the superplasticizer is naphthalene superplasticizer (CQJ-NX), with a water reduction rate of 18%. The polypropylene fiber (PPF) used in this paper is divided into three types: wavy type, embossed type, and monofilament type, and its appearance is shown in Figure 1. The performance indexes of the three polypropylene fibers are shown in Table 1.

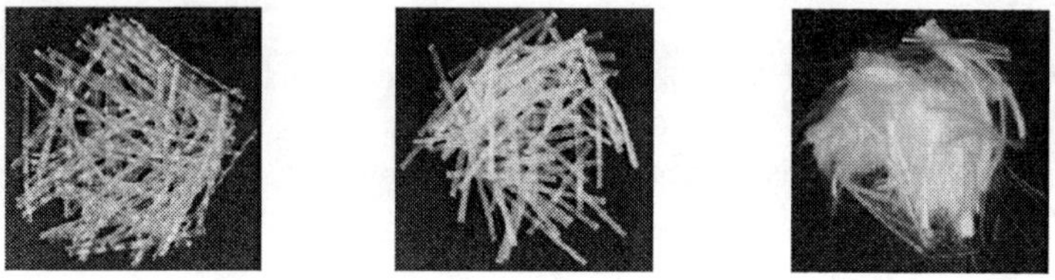

(a) Wavy type (b) Embodied type (c) Mono-filament type

Figure 1. Appearance of polypropylene fiber.

Table 1. Polypropylene fiber performance index.

Fiber shape	Lengths /mm	Caliber /mm	Density/ (g/cm^3)	Tensile strength/ MPa	Modulus of elasticity/ MPa	Elongation /%
Wavy type	30	1.00	0.95	542	5134	25%
Embodied type	25	0.60	0.95	556	5117	25%
Monofilament type	19	0.03	0.91	565	4138	15%

2.2 Experimental mix design

According to norms with C50 concretions as the basis to calculate the material mix ratio, three kinds of polypropylene fibers, namely wavy polypropylene fiber (PPFA), embodied-type polypropylene fiber (PPFB) and monofilament polypropylene fiber (PPFC), are incorporated into ordinary concrete in five small fractions of 0, 0.30%, 0.60%, 0.90% and 1.20%, respectively. The concrete benchmark is shown in Table 2.

Table 2. Concrete mixing ratio.

Water/(kg/m^3)	Cement/(kg/m^3)	Sand/(kg/m^3)	Stone/(kg/m^3)	Water reducer/%
180.4	451	574.8	1193.8	3.38

2.3 Test preparation and test method

The test block was prepared according to norms. Cubic compressive ability, split tensile ability, and flexural ability of three mechanical property tests were conducted, and three specimens were produced for each group. Compressive ability and split tensile ability test blocks of 0.1×0.1×0.1 m are used, and 0.1×0.1×0.4 m size test blocks are used for flexural strength tests.

2.4 Experimental loading

Cube compressive strength test and split tensile strength loading using Changchun Kedar testing machine factory production 2000 kN microcomputer control electric hydraulic servo-hydraulic testing machine, and the bending strength loading capacity uses an MWW-50D bending testing machine. Loading was carried out following the requirements of norms, and the loading process of specimens is shown in Figure 2.

(a) Compressive strength (b) Split tensile strength (c) Flexural strength

Figure 2. Polypropylene fiber concrete testing patches loading.

During the loading process, it was found that the damaged form of the PC was a "waist drum" type, and the surface appeared serious spalling. However, the number of cracks in the PPF test block decreased significantly, the width of cracks decreased, and the test block surface did not appear. It is found that the failure time of PPF blocks is longer than that of PC, and polypropylene fiber can play the role of "bridge" in cement-based materials.

3. Test results and analysis

3.1 Test result

Since the test block used in this experiment is not standard, according to norms, the test results of compressive, splitting, and folding strength should be multiplied by conversion factors of 0.95, 0.85, and 0.85 respectively. The results of the test calculations are shown in Table 3.

Table 3. Mechanical properties of PPF concrete.

Sample	Fiber content /%	Compressive strength/MPa	Splitting tensile strength/MPa	Bending strength/MPa	Tensile compression ratio/$\times10^{-2}$	Folding ratio /$\times10^{-2}$
PC	0	57.75	1.68	3.80	0.029	0.066
PPFA03	0.30%	56.64	1.67	3.49	0.030	0.062
PPFA06	0.60%	58.16	2.83	3.72	0.049	0.064
PPFA09	0.90%	60.24	3.04	3.76	0.050	0.062
PPFA12	1.20%	58.73	2.42	3.82	0.041	0.065
PPFB03	0.30%	56.60	1.96	4.05	0.035	0.072
PPFB06	0.60%	60.09	2.86	3.95	0.048	0.066
PPFB09	0.90%	61.55	1.62	3.76	0.026	0.061
PPFB12	1.20%	60.59	1.60	3.23	0.026	0.053

Journal of Physics: Conference Series **2808** (2024) 012088 doi:10.1088/1742-6596/2808/1/012088

PPFC03	0.30%	54.04	1.63	4.05	0.030	0.075
PPFC06	0.60%	50.65	2.22	4.24	0.044	0.084
PPFC09	0.90%	54.42	2.03	3.74	0.037	0.069
PPFC12	1.20%	55.31	1.69	3.69	0.031	0.067

3.2 Effect of polypropylene fiber on compressive strength of concrete

The link between the measured compressive ability and the change of content of three kinds of PPF reinforced concrete is brought about into a histogram as shown in Figure 3. It can be seen from the figure that for PPFA and PPFB, with the increasing dosage, the compressive strength of PPFA and PPFB shows an increasing trend as a whole, reaching the peak when the dosage is 0.90%, and the compressive strength of PPFA and PPFB rose by 4.30%, 6.57% respectively compared with PC. Later, along with the dosage, the compressive strength began to decrease, but the compressive ability of the concrete cube was still better than the PC strength in the range of the dosage ≤ 1.20%, indicating that the inclusion of PPFA and PPFB in this range could have a favorable influence on the compressive capacity of concrete. After PPFC is incorporated, concrete has a lower compressive capacity than PC. The reason for this is that the shape of PPFC is flocculent, which is not easy to disperse compared with the other two shapes of PPF, and the phenomenon of "agglomeration" occurs when mixing concrete, resulting in voids inside the concrete matrix and the reduction of compactness, this weakens the compressive capacity of the concrete.

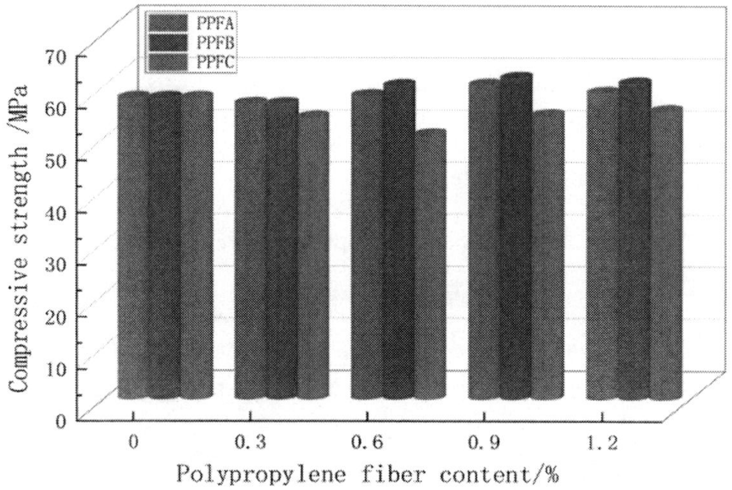

Figure 3. PPF concrete cube compressive strength.

In summary, the amount and shape of PPF will affect the cube compressive ability of concrete, among which mono-filament pPPFis not easy to disperse, which will hurt the compressive ability of concrete. Embossed PPF is good, and the strength increase is the largest when the content is 0.90%.

3.3 Effect of polypropylene fiber on the tensile compression ratio of concrete

The tensile compression ratio of concrete is the ratio of the splitting tensile capacity of concrete to its cubic compressive capacity [7]. Figure 4 shows the effect of PPF mixtures of different shapes on the tensile/compression ratio of concrete.

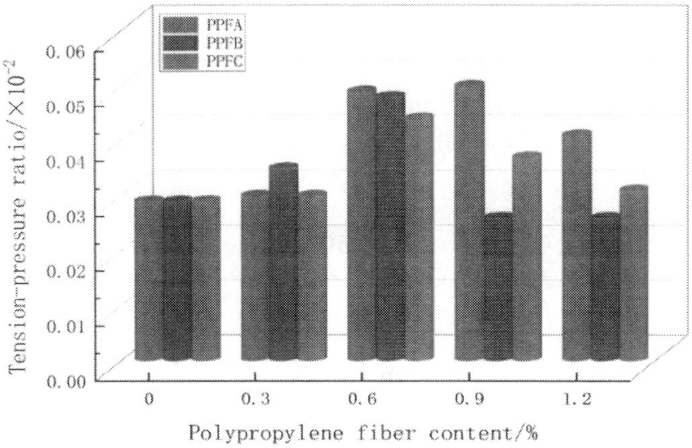

Figure 4. PPF concrete tensile compression ratio.

As can be seen from the figure, with the rise of PPF dosage, its dosage showed a rising and then decreasing trend, the rise is the largest when the content is 0.60%, among which PPFA has the largest increase, which increases by 64% compared with the fiber content of 0.30%. The main reason is that the fiber length of PPFA is longer and the elastic modulus is larger, and the appropriate amount of PPFA can play a good flexible bridging role on the splitting tension of concrete, so the fiber shows better ductility than the other two shapes in the split tensile strength testing. The tensile compression ratio of the three shapes of PPF concrete decreased or slightly increased when 0.60% to 1.20% of the mix was added, indicating that when the dosage exceeds a certain range, the pores in the matrix increase due to the dispersion of the fiber, and the favorable effect of PPF bridging is less than the unfavorable effect of the reduction of concrete compactness, leading to the reduced strength of concrete.

In summary, the content of the three fibers is the main factor affecting fiber concrete. From the point of view of improving the tensile/compression ratio of concrete, the 0.60% wavy polypropylene fiber is the most favorable, with an increase of 67.05% and 63.33%. However, compared crack tensile strength of general concrete, polypropylene fiber-embossed concrete has the highest strength, which is 69.94%.

3.4 Effect of polypropylene fiber on the folding ratio of concrete

The folding ratio of concrete is the ratio of the flexural ability of concrete to the cubic compressive abilities, which is one measure of the toughness of concrete [7]. Figure 5 shows the relationship between the folding ratio of PPF concrete and increasing PPF content.

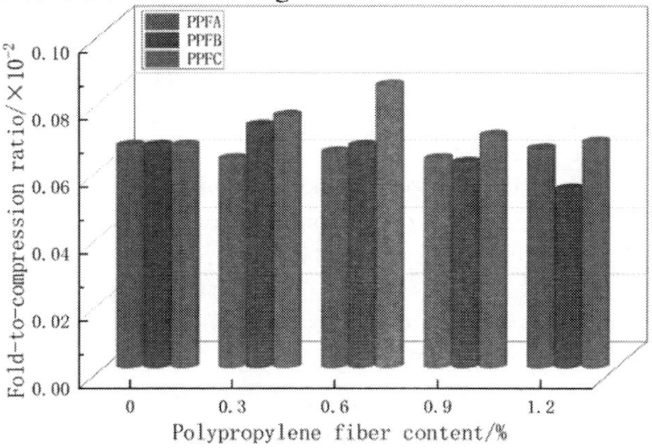

Figure 5. PPF concrete folding ratio.

It can be told from the way they look that the folding ratio of PPF concrete is mainly related to the shape of PPF, and the folding ratio of PPFB and PPFC arises first and then decreases with the rise of PPF contains, while PPFA weakens the folding ratio of concrete when the content is less than 1.20%. This is because the tensile ability of PPFA is smaller than PPFB and PPFC, and it cannot produce a large enough force to contain the concrete at both ends of the fiber during the loading of concrete specimens. Therefore, it was found that the flexural strength of concrete is adversely affected when the content of PPF is less than 1.20%. For concrete mixed with PPFB and PPFC, the changing trend of PPFC is more obvious, because the density of PPFC itself is small, the amount of PPFC per unit volume is large, and its tensile strength is high and toughness is strong.

To sum up, the properties and shape of PPF have a great influence on the concrete's folding ratio, and 0.60% monofilament PPF has the best effect on the folding ratio of concrete.

4. Conclusion

(1) Both the amount and shape of PPF affect the compressive ability of concrete cubes, and 0.90% polypropylene fibers are most favorable for the compressive strength of concrete cubes.

(2) The main factor affecting the tensile/compression ratio of concrete is the dosage of PPF. Within 0~0.60%, the tensile and compression ratios of the three PPF-reinforced concrete gradually increase with the rising of PPF. Combined with fiber shape and cost, 0.60% wavy PPF has the most favorable effect on the split tensile strength of concrete, which is 69.94% bigger than that of ordinary concrete.

(3) The shape of polypropylene fiber has a huge influence on the folding ratio of PPF concrete. Embossed polypropylene fiber and monofilament PPF have an enhanced effect on the folding power of concrete. Among them, the bending power of concrete arises most obviously when the content of monofilament PPF is 0.60%, increased by 13.98% and 7.46% respectively.

References

[1] ZHAO Yongxing, HAO Xiuling. Comprehensive durability analysis of large amounts of polypropylene coarse fiber concrete [J]. Jilin Water Resources, 2023, (07):62-6

[2] Liang N, Geng S, Mao J, et al. Investigation on cracking resistance mechanism of basalt-polypropylene fiber reinforced concrete based on SEM test[J]. Construction and Building Materials, 2024, 411: 134102.

[3] Yan S, Dong Q, Chen X, et al. An experimental and numerical study on the hybrid effect of basalt fiber and polypropylene fiber on the impact toughness of fiber-reinforced concrete [J]. Construction and Building Materials, 2024, 411: 134270.

[4] Xu Y H. Experimental study on tensile and compressive properties of basalt-crude polypropylene fiber dry hardened concrete [D]. Chongqing University,2020

[5] Liang N, Geng S, Mao J, et al. Investigation on cracking resistance mechanism of basalt-polypropylene fiber reinforced concrete based on SEM test [J]. Construction and Building Materials, 2024, 411: 134102.

[6] Yan S, Dong Q, Chen X, et al. An experimental and numerical study on the hybrid effect of basalt fiber and polypropylene fiber on the impact toughness of fiber reinforced concrete[J]. Construction and Building Materials, 2024, 411: 134270.

[7] Tang Jiajun, Pei Changchun. Analysis of tensile compression ratio and folding compression ratio of recycled concrete by orthogonal method [J]. Zhujiang River,2019,40(11):68-71 +77.

Numerical simulation of impactor velocity under dynamic mesh-based double-sided pressurization

Jialiang Zhou[1], Tong Tang[1], Feiyin Li[1], Shaojie Ma[1*], Huifa Shi[1], Saisai Liang[1] and Yingzhong Zhang[1]

[1]Department of Mechanical Engineering, Nanjing University of Science and Technology, Nanjing, Jiangsu, 210094, China

[*]Corresponding author's e-mail: 438729185@qq.com

Abstract: In order to analyze the influence of different initial conditions on the performance of the impactor in a new impact testing device with double-sided pressurization, a two-dimensional numerical model of the impact testing device is established. Numerical calculations are performed on the model by using CFD software, combined with dynamic mesh technology. The influence of the initial pressure on the velocity performance of the impactor is obtained, and numerical validation is conducted with the existing impact testing device. The results show that the impactor under double-sided pressurization can achieve a higher velocity close to the experimental data, confirming the effectiveness and accuracy of this method. The research findings are helpful for predicting the velocity of impactors in impact testing devices under laboratory conditions.

1. Introduction

The impact testing device, unlike propulsion devices powered by gunpowder, is a launch device driven by high-pressure gas. As a commonly used experimental equipment, the impact testing device is frequently employed to test the characteristics of various materials, such as the impact resistance and ductility of automotive materials. It is also commonly applied in physics experiments and scientific research. For instance, by adjusting the air pressure and parameters of the impactor, the motion trajectory, velocity variations, and impact force of the impactor can be studied. In this paper, a computational model for a new type of impact testing device is established, and the accuracy of the computational model is validated through experiments.

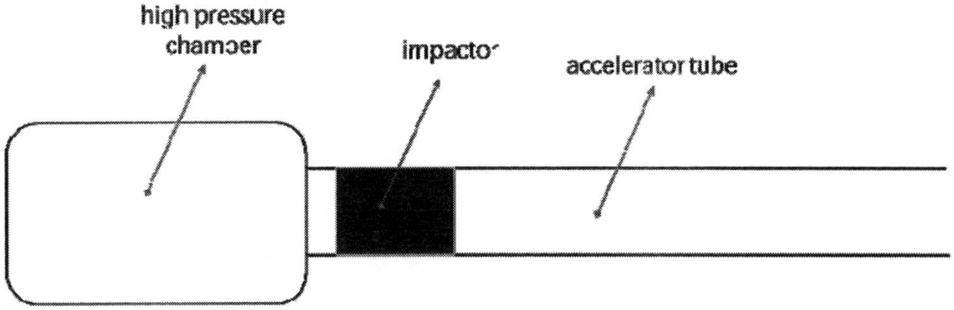

Figure 1. Conventional impact testing device refueling method.

The traditional pressurization method for impact testing devices is shown in Figure 1, where a high-pressure chamber is located at the rear of the impactor. This method has a simple structure but leads to slow pressurization and is unfavorable for the installation of the impactor. In this study, a new type of impact testing device is utilized, employing a low-pressure configuration behind the impactor and high-pressure on both sides. This approach not only shortens the time required for pressurization to the desired level but also facilitates the installation and adjustment of the impactor's shape.

In recent years, with the improvement of computer performance and advances in dynamic mesh simulation technology [1-2], numerical simulation has become a primary means of scientific research. Many scholars both domestically and internationally have utilized numerical modeling methods to conduct related studies [3-5]. Therefore, this paper adopts a numerical simulation approach that couples computational fluid dynamics (CFD) and rigid body dynamics (RBD) models [6]. Based on dynamic mesh technology, the impact testing device's projectile launch process is simulated to investigate the velocity variation of the projectile under double-sided pressurization.

2. Simulation method and computational model

2.1. RBD equations

The impactor is held stationary by a clamp during pressurization of the chamber. Once the chamber reaches the desired pressure, the clamp is released simultaneously with the opening of the control valve. At the moment of launch, the impactor experiences the forces F_G of gravity, support F_T from the barrel, propulsion force F_P from the gas, and viscous friction F_U. Within the barrel, the impactor undergoes single-degree-of-freedom motion. In the numerical simulation process of the impactor's motion, the material characteristics of the impactor are not considered (assuming it to be a rigid body). The motion of the bomb in space is regarded as the movement of the impactor's center of mass. It can be represented by linear velocity $v_c\left(v_x, v_y, v_z\right)$ in an inertial coordinate system $\left(x, y, z\right)$, and both are functions of time.

The translational motion equation of the impactor's center of mass in an inertial coordinate system is given by:

$$\dot{v}_c = \frac{1}{m}\sum F \tag{1}$$

In Equation (1), $\dot{v}_c$ represents the acceleration of the impactor's center of mass in the inertial coordinate system; m represents the mass of the impactor; F represents the external force acting on the impactor's center of mass.

In the simulation calculation, the position of the impactor's center of mass can be determined by integrating Equation (1). In the equation, x_c^n represents the position of the center of mass at the current time step n, then the position of the center of mass at the next time step $n+1$ is given by:

$$x_c^{n+1} = x_c^n + v_c \bullet \Delta t \tag{2}$$

2.2. CFD equations

This paper involves numerical simulation of RBD, thus using the Arbitrary Lagrangian-Eulerian method to describe the N-S equations for the flow field. In the CFD inertial coordinate system, for any control volume V with moving boundaries, the scalar variables (mass ρ, velocity u, energy E), the unsteady conservation-based dynamic mesh flow field calculation equation is given by:

$$\frac{d}{dt}\int_V \rho\Phi dV + \oint \rho\Phi(\vec{u} - \vec{u}_g)\bullet d\vec{A} = (\oint D\nabla\Phi\bullet d\vec{A}) + \int_V S_\Phi dV \tag{3}$$

In Equation (3): $V(t)$ represents the control volume in space whose size and shape change with time; ρ denotes the fluid density; $\partial V(t)$ represents the moving boundary of the control volume; $\vec{u}$ is the fluid velocity vector, while $\vec{u}_g$ represents the mesh motion velocity of the dynamic grid; D represents the dissipation coefficient; S_Φ denotes the source term for the scalar variable.

The turbulence model adopts the widely used $k-\varepsilon$ two-equation model with high computational accuracy.

k Control equation:

$$\frac{\partial k}{\partial t}+u_i\frac{\partial k}{\partial x_i}=P-\varepsilon+\frac{\partial}{\partial x_i}\left(\frac{v_t}{\sigma_k}\frac{\partial k}{\partial x_i}\right) \tag{4}$$

ε Control equation:

$$\frac{\partial \varepsilon}{\partial t}+u_i\frac{\partial \varepsilon}{\partial x_i}=C_{\varepsilon 1}\frac{\varepsilon}{k}P-C_{\varepsilon 2}\frac{\varepsilon^2}{k}+\frac{\partial}{\partial x_i}\left(\frac{v_t}{\sigma_\varepsilon}\frac{\partial \varepsilon}{\partial x_i}\right) \tag{5}$$

In Equations (4) and (5): P represents the term responsible for turbulent kinetic energy production; v_t denotes the viscosity coefficient; $v_t=\frac{\mu_t}{\rho}$ while $\mu_t=C_\mu\frac{\rho k^2}{\varepsilon}$, C_μ, $C_{\varepsilon 1}$, $C_{\varepsilon 2}$, σ_ε, and σ_k are model coefficients.

Additionally, to close the system of equations, there is also an equation of state for the gas:

$$PV=nRT \tag{6}$$

In Equation (6): P represents the gas pressure; V denotes the gas volume; n represents the mass of the gas; R represents the gas constant; T denotes the temperature of the gas.

2.3. Dynamic mesh simulation technology

The dynamic mesh model allows for the movement of boundaries in a specified region, where the mesh automatically adjusts accordingly. The boundary motion can be rigid or elastic. In either case, the positions of the grid nodes within the computational domain must be updated over time, making the dynamic mesh problem inherently unsteady. In this study, layered dynamic mesh surfaces are set on the left and right walls of the impactor. As the impactor moves, the grids adjacent to these two walls are automatically updated.

To ensure the computational accuracy of the aerodynamic forces acting on the rigid body (impactor) within the flow field simulation region, a fluid boundary layer is first generated around the impactor. The fluid boundary layer moves with the impactor to maintain the integrity of the boundary layer near the impactor, and the grid is refined in the vicinity of the impactor. During the calculation process, to prevent the distortion and degradation of the computational grid due to the motion of the rigid body, which significantly affects the simulation accuracy, the present study employs two dynamic mesh techniques: Elastic Smoothing and Local Remeshing [7]. Elastic Smoothing ensures that the grid nodes in the entire simulation domain behave like a spring-connected grid system, achieving a new equilibrium position after each time step update and reducing global grid distortion. Local Remeshing ensures that local grid elements near the rigid body are reconfigured when the distortion or size exceeds a specified threshold, thereby minimizing excessive deformation in the local grid.

2.4. Simulation coupling methods

This paper employs a coupled CFD (Computational Fluid Dynamics) and RBD (Rigid Body Dynamics) equation solver to simulate the impactor launch process [8]. Firstly, using CFD equations, the flow field parameters distributed in the simulation domain at a given time are computed to obtain the aerodynamic forces acting on the impactor. These forces are then passed to the RBD equations to determine the response of the impactor to the aerodynamic forces. Subsequently, the RBD equations

Journal of Physics: Conference Series **2808** (2024) 012089 doi:10.1088/1742-6596/2808/1/012089

calculate the position of the impactor for the next time step, based on which the computational grid is updated through iterative refinement until the dynamic mesh.

2.5. Computational model

The computational model is shown in Figure 2, with the solid regions representing gas and the void regions representing solid materials, as indicated by the labeling on the diagram. The tail section of the impactor serves as a low-pressure propulsion chamber responsible for propelling the impactor into the high-pressure launch area. Both sides are high-pressure chambers that propel the impactor to high speed once it reaches the launch area. After passing through a 4-meter-long acceleration tube, the impactor enters the external air domain.

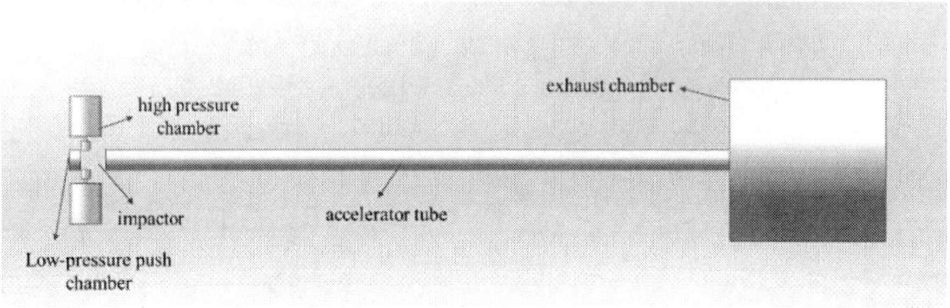

Figure 2. Computational model.

3. Simulation calculation

3.1. Simulation parameter setting

The front and rear surface grids in contact with the air for the test impactor are dynamic grids employing 6 degrees of freedom. The mass of the test impactor is set to 6.384 kg. The air pressure in the propellant chamber is set to 0.5 MPa, while the air pressure in the high-pressure chambers is set to 1.5 MPa. The pressure inside the barrel chamber and the external air domain of the barrel is set to 0 MPa, as shown in Figure 3. All these conditions align with the experimental conditions in the laboratory. After the settings are completed, the initial velocity of all air domains is set to 0, as shown in Figure 4.

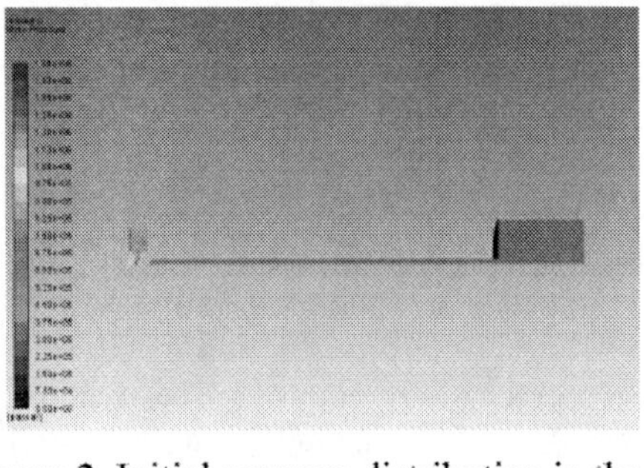

Figure 3. Initial pressure distribution in the air domain.

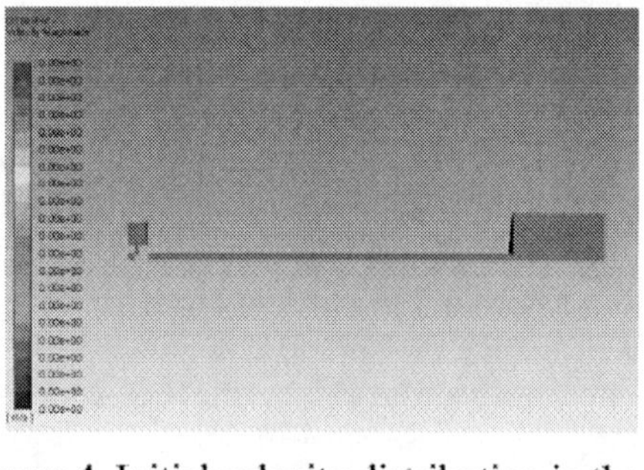

Figure 4. Initial velocity distribution in the air domain.

The solution results file is saved every 20 timesteps with a time step size ranging from 1E-6 to 1E-4, eventually yielding the motion process of the test impactor inside the air cannon.

3.2. Simulation results and analysis

The pressure distribution and velocity distribution in the air domain after the simulation ends are shown in Figure 5 and Figure 6.

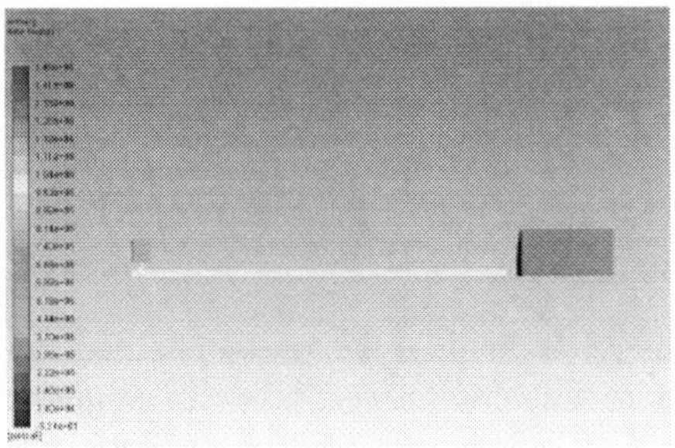

Figure 5. Post-launch air domain pressure distribution.

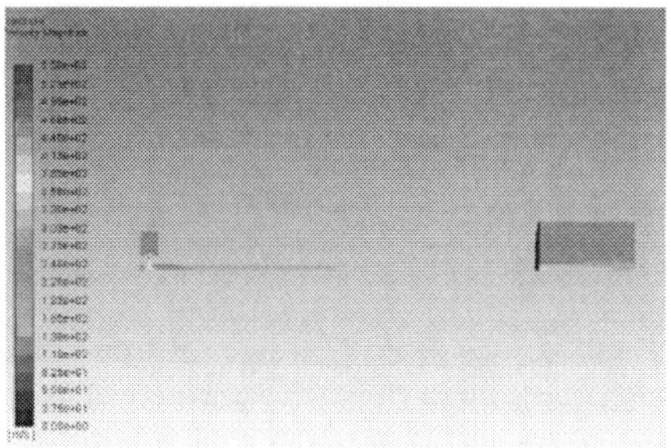

Figure 6. Post-launch air domain velocity distribution.

The data obtained from the simulation is exported in origin and re-graphed to obtain the impactor velocity and displacement versus time as shown in **Figure 7**.

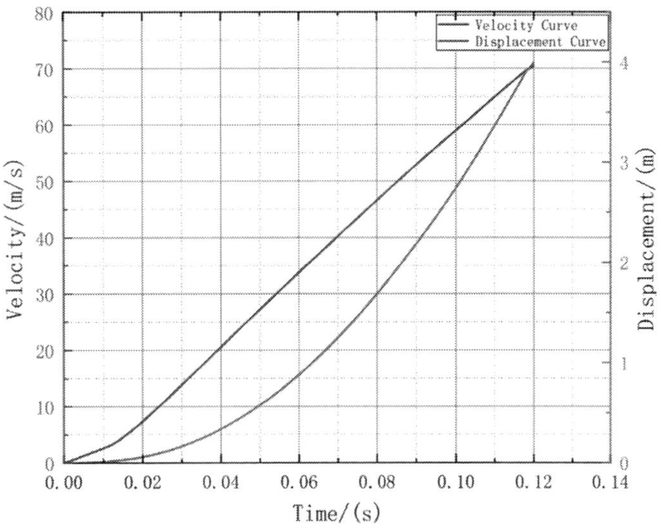

Figure 7. Velocity displacement vs. Time.

According to the simulation results, under the initial conditions where the air pressure in the launching chamber is 0.5 MPa and the air pressure in the high-pressure chamber is 1.5 MPa, the velocity of the impactor at a displacement of 4 m is 70.407 m/s. Keeping other initial conditions unchanged, the results obtained by varying the air pressure in the high-pressure chamber to 0.8 MPa, 1.2 MPa, 1.6 MPa, 1.8 MPa, 1.9 MPa, and 2 MPa are shown in the following Table 1.

Table 1 Simulation Results of the Impactor Velocity Under Different Air Pressures.

Pressure/Mpa	0.8	1.2	1.5	1.6	1.8	1.9	2
Velocity/m/s	34.352	52.163	70.407	72.537	77.993	83.242	93.814

3.3. Result validation and fitting

Experimental tests are conducted by using an air cannon under laboratory conditions, with the same initial conditions as the previous simulation. Different impactor velocities are obtained and compared with the simulation results. The error results are shown in the following Table 2.

Table 2 Comparison of Simulation Results and Experimental Data at Different Air Pressures.

Pressure /Mpa	0.8	1.2	1.5	1.6	1.8	1.9	2
Simulated Impactor Velocity /m/s	34.352	52.163	70.407	72.537	77.993	87.242	93.814
Test Impactor Velocity /m/s	32.476	51.287	68.236	70.137	75.942	84.555	90.9
Inaccuracies /%	5.78	1.71	3.18	3.42	2.7	3.18	3.21

It can be observed that in multiple simulations conducted at different air pressures, the error range is consistently less than 6%, which indicates the accuracy of the simulation results can be guaranteed.

Based on the simulation results, fitting and prediction are performed, resulting in Figure 8 shown below.

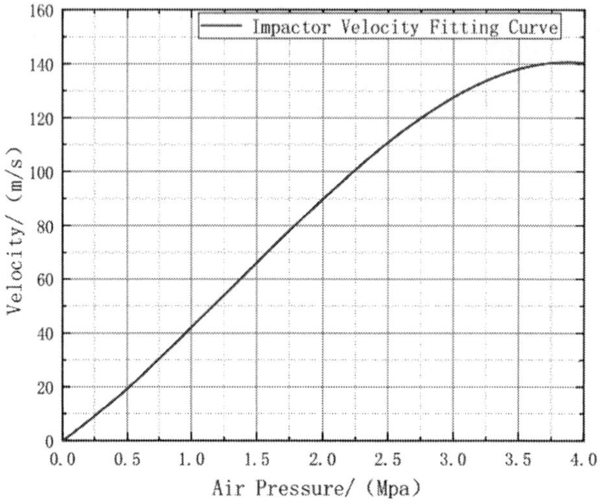

Figure 8. Impactor velocity fitting curve.

According to the literature review, it is known that there is a direct relationship between the velocity of the impactor and the air pressure. However, this relationship is not linear, and an increase in air pressure does not result in an unlimited increase in the impactor velocity. Under certain conditions, such as the limit pressure, mass, and shape of the impactor, and internal principles of the air cannon, factors may limit further increase in impactor velocity. The fitted curve of velocity in the graph shows that the speed change becomes less significant after the increase of air pressure reaches a certain value, which is consistent with the regularity described in the literature.

4. Conclusion

This paper takes a novel air cannon with gas injection on both sides as the computational model. Based on the dynamic mesh technique, the CFD and RBD equations are coupled to simulate and calculate the velocity change process of the impactor after launching. The simulation process and results indicate:

(1) The propulsion method with air intake at both ends is feasible and can achieve a relatively high velocity for the impactor.

(2) By employing the dynamic mesh technique, it is possible to effectively avoid network distortion caused by rigid body motion, which significantly improves the quality of the mesh and computational accuracy.

(3) The simulation results exhibit small discrepancies with actual experimental data, which can provide predictions for the required muzzle velocity before conducting experiments.

(4) Under the propulsion of a high-pressure chamber, the impactor can be approximated as uniformly accelerated linear motion.

References

[1] ZHANG LAIPING, DENG XIAOGANG, ZHANG HANXIN. Reviews of moving grid generation techniques and numerical methods for unsteady flow [J]. Advances in Mechanics, 2010, 40(4): 424-447. (in Chinese)

[2] ZENG ZHENG, WANG GANG, YE ZHENGYIN. Enhanced RBF mesh deformation method and multi-body trajectory simulation [J]. Acta Aerodynamica Sinica, 2015, 33(2): 170-177. (in Chinese)

[3] Gilbert T, Baker C J, Quinn A. Gusts caused by high-speed trains in confined spaces and tunnels [J]. Journal of Wind Engineering & Industrial Aerodynamics, 2013,121(2):39-48.

[4] Minzhang L, Chunguang Z, Huan Z, et al. The environment and energy consumption of a subway tunnel by the influence of piston wind [J]. Applied Energy, 2019,246(5):11-23.

[5] Marta López González, Mónica Galdo Vega, Jesús Manuel Fernández Oro, et al. Numerical modeling of the piston effect in longitudinal ventilation systems for subway tunnels. [J]. Tunnelling and Underground Space Technology, 2014,40(1):22-37.

[6] JIN CHENHUI, WANG GANG, WANG ZEHAN. Numerical simulation of unsteady CFD/RBD in the multibody separation process of cluster munitions [J]. Physics of Gases, 2018, 3(4): 47-63. (in Chinese)

[7] SUI HONGTAO, LI PENGFEI, MA SHIHU, et al. Proficient in CFD dynamic grid engineering simulation and case practice [M]. Beijing: People's Posts & Telecommunications Publishing House, 2013: 15-20. (in Chinese)

[8] ZHU SHIQUAN, CHEN ZHIHUA, ZHANG HUI, et al. Investigations on the influence of control devices on the separation characteristics of a missile from the internal weapons bay [J]. Journal of Mechanical Science and Technology, 2018, 32(5): 2047-2057.

Study on inertial migration behaviours under secondary flow patterns using a fluid-structure interaction approach

Xin Tan [1] **and Ying Lin** [1*]

[1]School of Mechanical Engineering, Shanghai Institute of Technology, Shanghai 201418, China

[*]Corresponding author's e-mail: liny@sit.edu.cn

Abstract: In the present work, the critical aspects of how particles influence the structure of secondary flows are explored. Despite its significance in the manipulation and detection of particles and cells in microfluidic applications, this topic has received limited attention until now. Employing a fluid-structure interaction method, this research examines the two-way coupling between particles and secondary flows at a Reynolds number (Re) of 80, revealing significant insights into the behaviour of particles under such conditions. The investigation explores the intricate relationships between particle transverse migration, rotation, and secondary flow fields, emphasizing how these interactions amplify with both particle size and secondary flow intensity. The findings underscore the critical role of secondary flow vortices in generating drag forces, which in turn influence particle motion. Notably, the study demonstrates that for larger particles (a/H=0.375), the vortex-induced additional force acts repulsively, discouraging their movement along with the secondary flow. Through detailed analysis, this paper elucidates the lateral migration behaviour of particles, their rotational dynamics in response to secondary flows, and the overarching effects of their interactions with secondary flow structures. This research provides insights into the fundamental mechanisms governing particle behaviour in microfluidic environments, offering potential advancements in particle manipulation and detection.

1. Introduction

Inertial migration was initially observed in macroscopic systems [1-2]. It is a pivotal technique in microscopic systems for various applications such as clinical research [3], diagnostics [4], and bioanalysis [5]. This technique offers several advantages, such as cost-effectiveness, portability, and simplicity [6], enabling the enrichment and separation of particles or cells by controlling their lateral migration and ordering [7]. Various microchannel designs, including linearity, cylindrical obstacles, serpentine, or spiral shapes, have been developed and evaluated to achieve these objectives. Particularly, serpentine channels have been highlighted for their efficacy in inertial focusing, primarily due to the secondary flow effects in curved geometries that expedite the stabilization of particle positions [8-10].

Secondary flows arise in curved channels due to velocity discrepancies, with the faster-moving fluid at the centre moving outward and the slower-moving fluid near the walls moving inward to conserve mass. This results in the formation of two counter-rotating vortices perpendicular to the primary flow direction, known as secondary flow [11]. This flow introduces an additional drag force that alters the equilibrium positions of particles within the flow. Recent studies have extensively explored the implications of secondary flows. For instance, Zhang et al. [12] observed through experiments that particles focus into distinct patterns along channel sidewalls and the centre, influenced by the intensity of secondary flow in symmetrical serpentine channels. Cruz et al. [13] achieved sub-micron resolution in particle alignment and separation using numerical simulations of secondary flows. Moreover, numerical simulations by Jiang et al. [14] demonstrated that larger particles (10 μm) are more susceptible to rotation and separation from smaller particles (4 μm) under certain flow intensities, as dictated by the secondary vortex dynamics.

The experimental observation is restricted by the geometrical constraints of curved channels, leading to a reliance on numerical methods as a predictive tool for particle-focusing positions. Among these, the Lattice Boltzmann Method (LBM) [15], Fluid-Structure Interaction (FSI) [16], and Particle Tracing (PT) [17] are notable. Kechagidis et al. [18] used the LBM method for particle migration analysis in microfluidic cross-slot channels. Ying and Lin [19] used the PT method to predict the migration path of particles in serpentine channels. In contrast, the FSI method, as used by Pedrol et al. [20] for capturing particle dynamics in curved channels, offers a comprehensive approach by accurately simulating the interactions between moving particles and the surrounding fluid, demonstrating its superiority in addressing the complexities of secondary flows [20-21].

This paper utilizes the FSI method to investigate the interaction between particles and secondary flow in a serpentine channel with obstacles to induce secondary flow, at a moderate Reynolds number (Re=80). The study first examines how particles affect the structure of the secondary flow field. Subsequently, it analyses the motion of particles of specific sizes (a=8, 15 μm) and blockage ratios (a/H=0.2, 0.375) in response to the induced flow field, providing a detailed exploration of the interplay between particle dynamics and secondary flow phenomena.

2. Method and model

Particles undergo lateral migration due to inertial effects while simultaneously experiencing rotation due to the differential fluid velocity around them in confined channels. Additionally, the flow field is altered by the presence of the particle, specifically due to the obstruction it creates, rather than being distorted by the particle. To study the combined effect of particles and fluid, the FSI module was utilized. FSI focuses on modeling the interaction between moving objects and the surrounding fluid. In this module, the velocity field of the fluid was described by Eulerian, while the solid particle was described by Lagrangian. The flow field and pressure solution for an incompressible flow were solved by the Navier-Stokes equation. The FSI method couples solutions of the flow field and pressure to the solid mechanics equations, linking the forces and displacements on the solid boundary of the particle through the following equation [20-21]:

$$\vec{f} = \vec{n} \bullet [-pI + (\mu(\nabla \vec{u}_{fluid} + (\nabla \vec{u}_{fluid})^T) - \frac{2}{3}\mu(\nabla \bullet \vec{u}_{fluid})I)] \tag{1}$$

where $\vec{n}$ is the normal vector to the particle surface, p denotes the pressure, I stands for the identity matrix, ρ and μ represent the density and viscosity of the fluid, and $\vec{u}_{fluid}$ is the velocity of the fluid-particle interface, respectively.

The simulation model is depicted in Figure 1. The inflow boundary condition was set to simulate stable laminar flow over a specific period, as depicted in Figure 1 (a). The particle was initially positioned at a stable location based on previous studies [14]. Figure 1 (b) presents the mesh configuration used in this simulation. The particle was not subjected to any constraints, allowing it to move freely under the influence of the surrounding flow, thereby becoming part of a dynamic mesh system. An automatic re-meshing feature was utilized to adjust the meshes around the particle, accommodating mesh deformations in the fluid domain caused by the particle's movement. The particle's density was set at 1, 050 kg/m^3, while the density and viscosity of water were specified as 1, 000 kg/m^3 and 0.001 Pa·s, respectively. Mesh independence tests and validation of the FSI method were performed by comparing the simulation's inertial focusing results with those from previous experiments [22]. With a maximum deviation of less than 10%, the FSI method was deemed reliable.

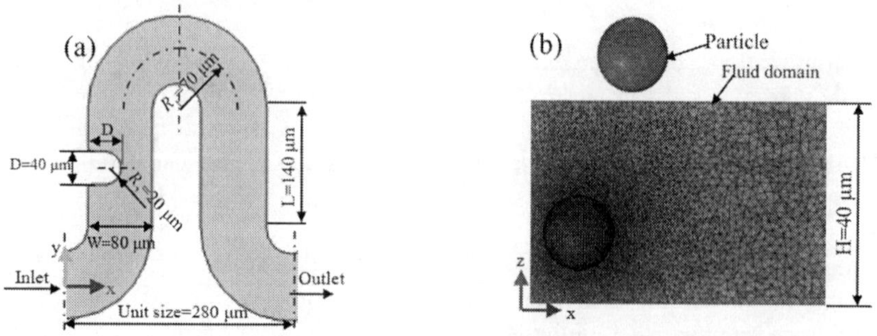

Figure 1. Geometry model: (a) A period unit of the serpentine and the height of the channel is H=40 μm; (b) Meshes of this model.

3. Results and discussion

To investigate the pronounced interaction between particles and secondary flows, a Reynolds number (Re) of 80 was selected to ensure the presence of strong secondary flows within the serpentine channel. Particles were restricted to travel through the second curved channel segment to focus the investigation on the region with significant secondary flows. This decision was made because the intensity of secondary flows in the third segment mirrored the weaker flows found in the first segment. Particle sizes of 8 μm and 15 μm were selected based on clinical diagnostic requirements [23], aiming to elucidate the distinct interactions between secondary flows and particles of different sizes, which are relevant to specific diagnostic applications. The presence of obstacles within the serpentine channel led to variations in the intensity of secondary flows along its length.

3.1. The influence of particles on the secondary flow field

The simulation results, as depicted in Figure 2, illustrate the impact of a freely moving particle driven by fluid on the secondary flow. These results encompass streamlines, vorticity, and particle motion information. Vorticity, representing the curl of velocity, is related to the intensity of secondary flows and the shear rate (velocity gradient) in the flow field.

It is well known that secondary flow consists of two counter-rotating vortices that are symmetrical in cross-section. However, the secondary flow structure is altered under the influence of an 8 μm particle, as depicted in Figure 2 (a). As an 8 μm particle enters the flow field from the left side of the inlet (inner wall of the curved channel), the secondary flow rotates the particle and carries it towards the outer wall. The Cross-Section 1 demonstrates the flattened elliptical vortex within the cross-section. The tendency of the motion is from left to right, with the upper vortex counterclockwise (-) and the lower vortex clockwise (+). The particle rotates clockwise with the vortex and migrates laterally to the right. This figure reveals that the obstruction effect of the particle causes the vortex lines to divert around the particle. The rotation of the particle alters the flow velocity of the secondary flow around it, resulting in four regions with significantly higher vorticity magnitudes than those of the secondary flow. We observe that the vorticity on both sides of the particle is negative, indicating an opposite direction to the vorticity of the secondary flow. Upon closer examination of the two vortices, it becomes evident that the shape of the bottom vortex is not a flat ellipse. In Section 2, the centrifugal force caused by the obstacle produces a weak secondary flow (the larger vorticity is due to increased wall curvature). The vortex is in the middle of the top and bottom walls, and the vortex lines do not form a closed ellipse. The value of the vorticity around the particle is smaller than that of Section 1 but still distinctly differentiates from the secondary flow vorticity. Moreover, the intensity of the secondary flow in the second curved channel is significantly enhanced compared to the first one due to the acceleration of the mainstream by the obstacle in the channel. The secondary flow gradually weakens along the downstream direction as it moves away from the obstacle. Observing the development of the secondary flow within Sections 3 to 5, it is found that vortex centres move from right to left, and the vorticity magnitudes decrease gradually. Within Section 3, the vortex centre of the secondary flow starts at the right wall (inner wall) due to the reversing of the curved channel. At this time, the particle is located above the vortex, rotating counterclockwise to coincide with the secondary flow. The magnitudes of negative vorticity above and below the particles become larger, and the left and right vorticity is opposite to the vorticity of the secondary flow. Subsequently, the particle gradually approaches the vortex centre, as shown in Section 4. Within Section 5, the rotating particle and the vortex centre merge, and irregular vortex streamlines are generated around the particle. Furthermore, in Sections 4 and 5, the bottom vortex centre is slightly leftward than the top one. The above results indicate that the presence of the particle affects the vorticity distribution, shape, and centre position of the secondary flow vortices.

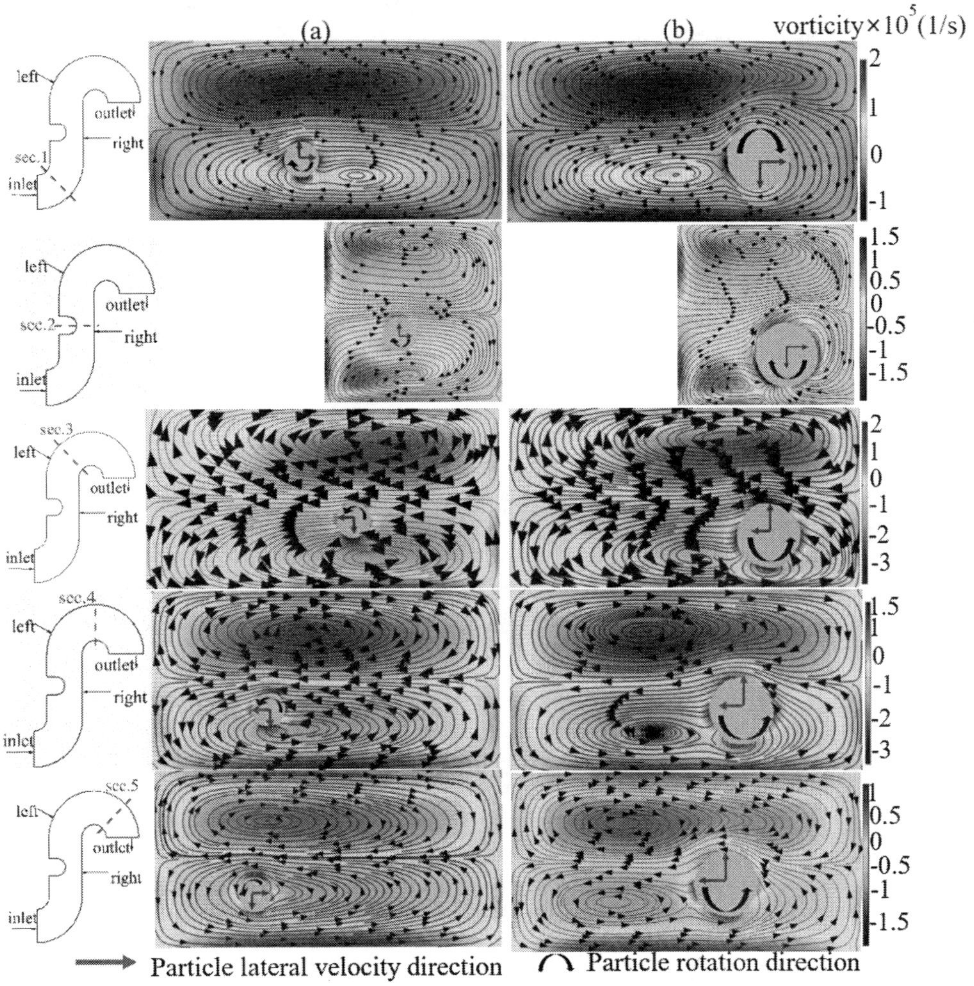

Figure 2. The secondary flow fields under the influence of (a) 8 μm particle and (b) 15 μm particle at *Re*=80 in the channel with obstacle size D=40 μm.

Subsequently, the influence of 15 μm particles on the secondary flow structure, as depicted in Figure 2 (b), was examined and compared with Figure 2 (a). Positioned on the right side (outer wall) of the channel, the 15 μm particle exhibits greater inertia compared to an 8 μm particle. In Section 1, its rotation is slower than that of the 8 μm particle, with smaller vorticity magnitudes around the particle. The pronounced barrier bypassing effect of the 15 μm particle is attributed to its larger size. Moreover, its downstream positioning impedes the rightward movement of the vortex centre, leading to a deviation of the bottom vortex centre to the left compared to Figure 2 (a). Consequently, the rotating particles behave akin to a new vortex, causing the top vortex to be squeezed and concaved by the particle's vortex. In Section 2, the secondary flow compresses into an elliptical vortex in the lower left corner under the repulsive force of the large rotating particle, distinctly altering the flow pattern compared to Figure 2 (a). These phenomena underscore the remarkable influence of the large particle on the secondary flow structure. As the large particle migrates along the right wall into the strong secondary flow channel, it displaces the vortex centre, originally near the right wall (inner wall), to the left, as seen in Section 3. Additionally, there is a tendency to form a tiny ring-like vortex in the region of maximum negative vorticity at the particle's bottom, as illustrated in Section 4. However, this trend diminishes in Section 5 as the secondary flow weakens, indicating a strengthening shearing effect of the particle on the secondary flow with increasing flow intensity. Within Sections 4 and 5, as the vortex centre shifts leftward, the large particle upstream of the secondary flow is pulled leftward by the drag force, no longer repelling the vortex. However, the concave shape of the top vortex persists.

In summary, the behaviour of particles in the secondary flow is dynamic and significantly alters its dynamics. A small particle rotates in the secondary flow and migrates toward the vortex centre, eventually merging with it. On the other hand, a large particle in the secondary flow forms a new larger-sized vortex. At a close range, this new vortex and the secondary flow vortex repel each other. Overall, particles are not merely passive elements in the secondary flow but notably influence its dynamics. Particle migration and rotation modify the velocity distribution of the secondary flow around the particle. The fluid velocity at the particle surface can be expressed as $u = U \pm 0.5\ \omega a$, where U is the particle lateral migration velocity and ω is the particle rotational velocity. To further understand the effect of particles on the structure of the secondary flow, Figure 3 illustrates the velocity profile of the strong secondary flow (in Section 3) past the central cut-off of the particles. The dashed line indicates the velocity profile of the undisturbed secondary flow at the same intercept. In Figure 3, z represents the vertical direction and x is the horizontal direction.

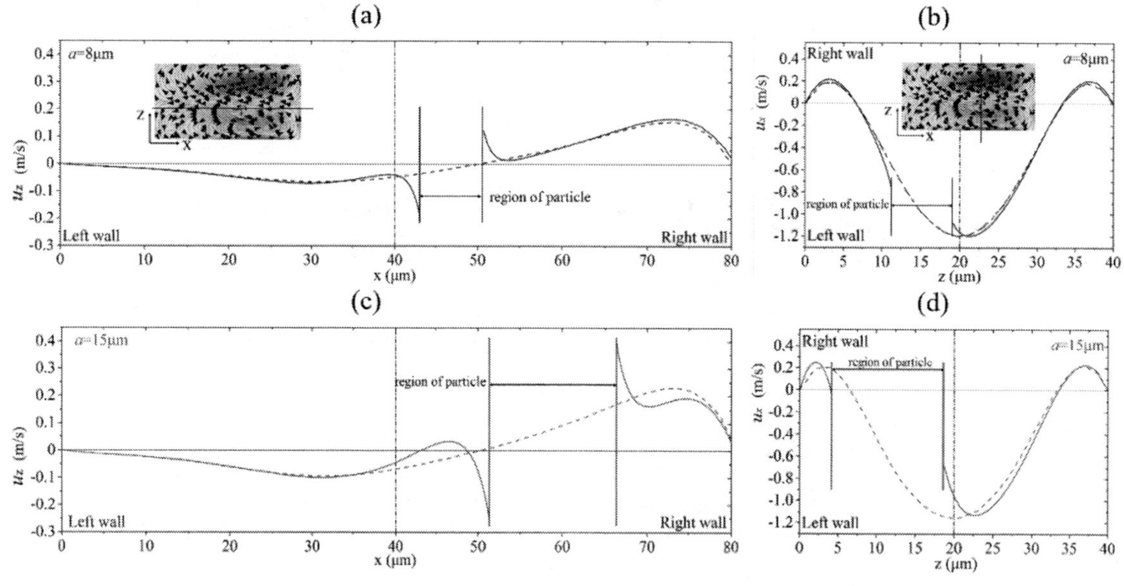

(a) and (c) Velocity in the z direction; (b) and (d) Velocity in the lateral direction.

Figure 3. The velocity profiles of secondary flow in Section 3.

Figure 3 (a) shows the secondary flow velocity profile in the z-direction under the influence of an 8 μm particle. The dashed line indicates that the velocity is zero, and the negative velocity gradient at the vortex centreline is attributed to the leftward secondary flow, which opposes the x-positive direction. However, the counter-clockwise rotation of the particle results in a sharp change in velocity around the particle, opposite to that of the secondary flow, leading to a positive velocity gradient. This change causes the formation of a positive vorticity region to the left and right of the particle.

Figure 3 (b) displays the velocity profile of the secondary flow in the x-direction under the influence of an 8 μm particle. The undisturbed secondary flow exhibits a velocity of zero and a maximum negative velocity gradient (except at the wall) at the vortex centreline. At the channel centre (z=20 μm), the velocity is maximum, and the velocity gradient is zero. The velocity trend around the particle aligns with the original trend of the secondary flow, with all velocity gradients being negative. However, the rotation of the particle leads to a larger velocity gradient, resulting in maximum negative vorticity above and below the particle. Comparing Figures 3 (a) with 3 (c), it is evident that the velocity gradient around the 15 μm particle is larger, and the velocity profile deformation is more pronounced. In Figure 3 (d), not only do the velocity gradient values above and below the particle become larger, but also the maximum velocity position moves upwards. This shift is responsible for the concave shape of the top vortex.

3.2. Exploration of particle lateral migration dynamics

To further elucidate the interaction between particles and secondary flows, this section discusses the lateral migration of particles within modified secondary flows, as illustrated in Figure 4. The figure showcases outcomes in a straightened channel, where movement towards the right wall is deemed positive. Particles experience secondary flow drag forces (F_D) and lift forces (F_L) generated by the shear action of the main flow in channels.

After being released on the left side of the inlet, the 8 μm particle migrates rightward (depicted by the black line), influenced continuously by the rightward drag force F_D and the lift force F_L. Leaving the first curved channel, the particle continues to move to the right due to inertia. Upon entering the obstacle region, the channel narrows to the right, causing the particle to swiftly approach the right wall under the influence of the rightward drag force. As the channel widens to the left in Section 2, a secondary flow is induced to the right by the obstacle. These opposing effects nullify each other, resulting in negligible lateral motion of the particle, which continues migrating along the right wall in the main flow direction. After entering the second curved channel, the direction of the secondary flow reverses, pointing leftward, and the particle accelerates leftward under the influence of the leftward F_D and F_L. Passing through the channel centre in Section 3, the particle persists in its leftward movement, gradually approaching the wall accompanied by an increasing rightward F_L. In Section 4, the F_D and F_L reach equilibrium, leading the particle towards the vortex centre on the left side of the channel. The vortex centre then captures the particle, experiencing negligible lateral position change as depicted in Figure 2 (a) and Section 5. This underscores the significant interaction between secondary flow and particle dynamics.

The general trend of the motion of the 15 μm particle is similar to that of the 8 μm particle, but its position is always within the right side of the channel. Referring to Figure 2 (b), the interaction between the 15 μm particle and the secondary flow is strong, and the vortex exerts a repulsive force (F_V) on the rotating particle. In the first curved channel, the particle is closer to the right wall under the action of the F_D and F_V of the secondary flow, thereby generating a larger leftward wall force F_L to balance it. After leaving the first curved channel, the particle migrates to the right again, driven by inertia and the drag force generated by the narrowed channel. In Section 2, the 15 μm particle is closer to the right wall than the 8 μm particle due to the action of the vortex's repulsive force F_V.

Following a pattern similar to the 8 μm particle, the lateral position of the 15 μm particle remains relatively stable until the second curved channel, where it begins moving leftward due to the reverse secondary flow. However, owing to the repulsive force exerted by the vortex and the greater inertia of the larger particle, the 15 μm particle moves leftward at a slower pace. It fails to cross the vortex centre (Figure 2 (b)). This outcome underscores a direct interaction between the large particle and the ring-like vortex within the secondary flow. A comparison with the trajectory of the 8 μm particle reveals a strengthening interaction between the particle and secondary flow with increasing particle size, ultimately resulting in distinct differences in inertial migration behaviour between the two particle sizes.

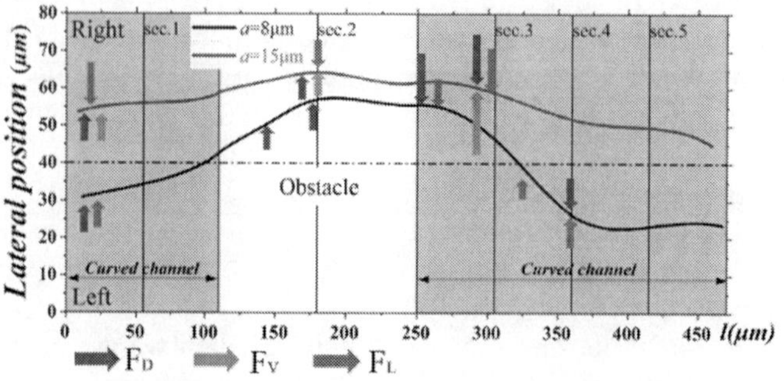

(a) Lateral velocities; (b) Lateral positions.
Figure 4. The lateral migration behaviour of particles.

3.3. Particle's rotation in response to secondary flow

Particles' rotation in response to secondary flow was further explored by examining their rotational velocity. Figure 5 illustrates the rotational velocity (ω) of particles within cross-sectional views, referencing Figure 2, with clockwise rotation considered positive.

Upon entering the secondary flow field, an 8 μm particle initially at rest experiences substantial shear forces from the surrounding flow, leading to a rapid increase in angular velocity. As the intensity of the secondary flow diminishes, the particle's angular velocity also decreases. Between the two curved channels, there is a temporary increase in angular velocity near Section 2 due to the secondary flow induced by the obstacle. However, the overall trend shows a decrease. Within the second curved channel, as the direction of the secondary flow reverses, the particle's angular velocity increases in the negative direction. The rate of rotational change is faster between Sections 3 and 4, after which the particle's rotation gradually slows. Comparing the rotation direction of the particle from Sections 1 to 5, it is observed that the particle's rotation direction aligns with the secondary flow direction. Considering the changes in the vorticity size of the secondary flow depicted in Figure 2, it is evident that the trend of angular velocity variation corresponds with the intensity of the secondary flow in the field. This leads to the conclusion that the stronger the secondary flow is, the greater its impact on the particle's angular velocity is.

The angular velocity of the 15 μm particle significantly differs from that of the 8 μm particle both in the first curved channel and following Section 3. In the weaker secondary flow of the curved channel, the 15 μm particle, being more massive than the 8 μm particle, exhibits a slower increase in angular velocity. Following Section 3, the repulsion between the 15 μm particle and the secondary flow vortex leads to a lower angular velocity compared to the 8 μm particle, which remains at the vortex centre. Overall, particle size has a relatively minor effect on their angular velocity.

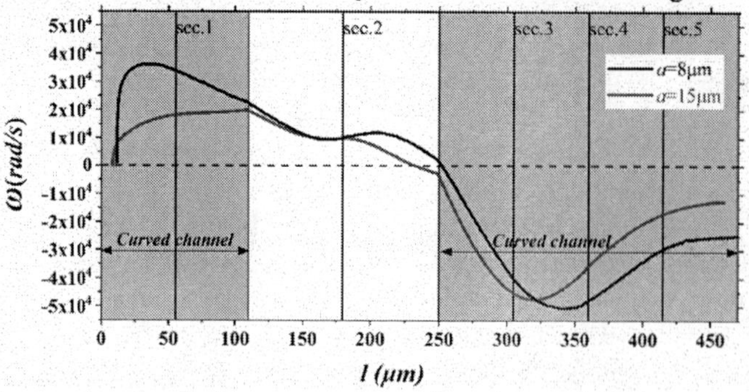

Figure 5. The rotational velocity components of particles within cross-sectional views.

4. Conclusion

In the present work, the intricate interplay between particles and secondary flows is explored by using fluid-structure interaction numerical methods, unveiling the impact of particle-induced obstructions on vortex behaviour and vorticity distribution within secondary flows. Key findings demonstrate that particle size notably influences the structure of secondary flow and their inertial migration. Specifically, larger particles tend to be repelled by vortices, while smaller ones are captured by vortex centres, highlighting a size-dependent separation mechanism within secondary flows. Constraints imposed by computational resources restricted this study to examining single particle dynamics within a partial serpentine model. Future work should expand to full periodic channels to investigate multi-particle interactions and their collective impact on migration dynamics. This advancement could offer deeper insights into particle sorting technologies and improve the efficiency of microfluidic devices in various applications.

References

[1] G. Segre and A. Silberberg (1961). Radial particle displacements in Poiseuille flow of suspensions. Nature, 189 (4760): 209.
[2] Dean W. R. (1928). The streamlined motion of fluid in a curved pipe. Philos. Philos. Mag, 5: 461-512.
[3] Jiang L. Guo K. Chen Y. and Xiang N. (2024). Droplet Microfluidics for Current Cancer Research: From Single-Cell Analysis to 3D Cell Culture. ACS Biomater Sci Eng.
[4] Petchakup C., Chen Y. Y. C., Tay H. M., Ong H. B., Hon P. Y., De P. P., Yeo T. W., Li K. H. H., Vasoo S., and Hou H. W. (2023). Rapid Screening of Urinary Tract Infection Using Microfluidic Inertial-Impedance Cytometry. ACS Sens., 8 (8): 3, 136-3, 145.
[5] Wang L., Abdulla A., Wang A., Warden A. R., Ahmad K. Z., Xin Y., and Ding X. (2022). Sickle-like Inertial Microfluidic System for Online Rare Cell Separation and Tandem Label-Free Quantitative Proteomics (Orcs-Proteomics). Anal. Chem., 94 (15): 6, 026-6, 035.
[6] Torres Castro K., Acuna Umana K., Lesser Rojas L., and Reyes D. R. (2023). Microfluidic Blood Separation: Key Technologies and Critical Figures of Merit. Micromachines (Basel), 14 (11).
[7] Xu Xuefeng, Huang Xiwei, Sun Jingjing, Wang Renjie, Yao Jiangfan, Han Wentao, Wei Maoyu, Chen Jin, Guo Jinhong, Sun Lingling, and Yin Ming. (2021). Recent progress of inertial microfluidic-based cell separation. The Analyst, 146 (23): 7, 070-7, 086.
[8] Amani Ali, Shamloo Amir, Vatani Pouyan, and Ebrahimi Sina (2022). Particles Focusing and Separation by a Novel Inertial Microfluidic Device: Divergent Serpentine Microchannel. Industrial & Engineering Chemistry Research, 61 (38): 14, 324-14, 333.
[9] Shen S., Zhao L., Bai H., Zhang Y., Niu Y., Tian C., and Chan H. (2024). Spiral Large-Dimension Microfluidic Channel for Flow-Rate and Particle-Size-Insensitive Focusing by the Stabilization and Acceleration of Secondary Flow. Anal. Chem., 96 (4): 1, 750-1, 758.
[10] Cha Haotian, Amiri Hoseyn A., Moshafi Sima, Karimi Ali, Nikkhah Ali, Chen Xiangxun, Ta Hang T., Nguyen Nam Trung, and Zhang Jun (2023). Effects of obstacles on inertial focusing and separation in sinusoidal channels: An experimental and numerical study. Chemical Engineering Science, 276 (0009-2509).
[11] Di Carlo D. (2009). Inertial microfluidics. Lab. Chip, 9 (21): 3, 038-3, 046.
[12] Zhang J., Yuan D., Zhao Q., Teo A. J. T., Yan S., Ooi C. H., Li W., and Nguyen N. T. (2019). Fundamentals of Differential Particle Inertial Focusing in Symmetric Sinusoidal Microchannels. Anal. Chem., 91 (6): 4, 077-4, 084.
[13] Cruz Javier, Graells Tiscar, Walldén Mats, and Hjort Klas (2019). Inertial focusing with a sub-micron resolution for separation of bacteria. Lab on a Chip, 19 (7): 1, 257-1, 266.
[14] Jiang Di, Tang Wenlai, Xiang Nan, and Ni Zhonghua (2016). Numerical simulation of particle focusing in a symmetrical serpentine microchannel. RSC Advances, 6 (62): 57, 647-57, 657.
[15] Owen Benjamin, Kechagidis Konstantinos, Bazaz Sajad Razavi, Enjalbert Romain, Essmann Erich, Mallorie Calum, Mirghaderi Fatemehsadat, Schaaf Christian, Thota Krishnaveni,

Vernekar Rohan, Zhou Qi, Warkiani, Majid Ebrahimi, Stark Holger, and Krüger Timm. (2023). Lattice-Boltzmann modeling for inertial particle microfluidics applications – a tutorial review. Advances in Physics: X, 8 (1).

[16] Li Xiang and Lin Ying (2021). Fluid-Structure Interaction Approach to Single Particle in a Square Microchannel. Journal of Physicals: Conference Series, 2097 (1).

[17] Li Xiaohong, Duan Junping, Wang Jiayun, Qu Zeng, Ji Miaomiao, and Zhang Binzhen (2022). Continuous particle separation of microfluidic chip with integrated inertial separation and dielectrophoresis separation. AIP Advances, 12 (3).

[18] Kechagidis K., Owen B., Guillou L., Tse H., Di Carlo D., and Kruger T. (2023). Numerical investigation of the dynamics of a rigid spherical particle in a vortical cross-slot flow at moderate inertia. Microsyst Nanoeng, 9: 100.

[19] Ying Y. and Lin Y. (2019). Inertial Focusing and Separation of Particles in Similar Curved Channels. Sci. Rep., 9 (1): 16, 575.

[20] Pedrol Eric, Massons Jaume, Díaz Francesc, and Aguiló Magdalena (2018). Two-Way Coupling Fluid-Structure Interaction (FSI) Approach to Inertial Focusing Dynamics under Dean Flow Patterns in Asymmetric Serpentines. Fluids, 3 (3): 62.

[21] Mohammadali R. and Bayareh M. (2023). Deformability-based Isolation of Circulating Tumor Cells in Spiral Microchannels. Micromachines (Basel), 14 (11).

[22] Di Carlo, Dino Edd, Jon F., Humphry Katherine J., Stone Howard A., and Toner Mehmet (2009). Particle Segregation and Dynamics in Confined Flows. Physical Review Letters, 102: 094503.

[23] Ebrahimi S., Alishiri M., Pishbin E., Afjoul H., and Shamloo A. (2023). A curved expansion-contraction microfluidic structure for inertial-based separation of circulating tumor cells from blood samples. J. Chromatogr. A, 1705: 464, 200.

Numerical simulation of single abrasive grain grinding of silicon carbide surface

Zhiquan Guo[1*] and Luchao Ma[1]

[1]Shandong University of Science and Technology, Qingdao, China

*Corresponding author's e-mail: guozhiquan02@163.com

Abstract: Silicon carbide (SiC) ceramics are extensively employed in aerospace and mechanical manufacturing owing to their outstanding material properties. However, their inherent high hardness and brittleness pose challenges in product processing. Utilizing the classical Johnson-Holmquist (JH-2) constitutive model in finite element simulation technology can facilitate the analysis of the impact of various grinding parameters on material surface damage. This approach yields valuable insights for selecting optimal processing parameters. The simulation of SiC surface grinding with single diamond abrasive grains demonstrates that grinding force increases with grain penetration depth. The resulting stress elevation promotes the accumulation of surface cracks, exacerbating material surface damage and abrasive wear. Furthermore, the damage to SiC surfaces is affected by grinding speed. A higher speed can lower grinding force, reduce abrasive wear, and enhance SiC surface smoothness. Nonetheless, increasing grinding speed also elevates residual stress in the workpiece due to higher kinetic energy. Hence, controlling the grinding wheel speed within a specific range and minimizing wheel surface penetration depth can effectively enhance SiC surface processing quality and prolong tool lifespan.

1. Introduction

Silicon carbide (SiC) ceramics, known for their characteristic hardness and brittleness, possess outstanding physical, chemical, and mechanical attributes, including high hardness, excellent oxidation resistance, low coefficient of thermal expansion, and high thermal conductivity. These ceramics are extensively utilized across diverse sectors such as the mechanical industry and aerospace [1]. SiC exhibits outstanding chemical stability, posing challenges in achieving effective surface processing by electrochemical machining methods. The utilization of high-hardness abrasive tools like diamond is commonplace for grinding SiC [2]. However, due to the inherent hardness and brittleness of SiC, surface damage, primarily in the form of cracks, is prone to occur during grinding. This can lead to prolonged subsequent grinding and polishing durations, diminishing the service life of components [3]. Consequently, the development of efficient and low-damage grinding technologies for hard and brittle materials, epitomized by SiC ceramics, remains a prominent focus in contemporary advanced manufacturing research.

During the grinding process, variations in the depth of abrasive particles penetrating the workpiece surface lead to three distinct stages: elastic sliding friction, plastic plowing, and brittle damage. As a result of these machining operations, minute levels of plastic deformation and brittle crack propagation are inevitably generated. Consequently, it is imperative to recognize that different grinding depths play a pivotal role in determining the material removal mechanism [4]. In the course of material surface processing, abrasive grains undergo wear due to mechanical action, physical consumption, chemical corrosion, and other factors. When the grinding stress exceeds the strength of the abrasive grain, it results in material breakage, leading to diminished grinding efficiency and deterioration in processing quality [5]. Additionally, varying grinding wheel speeds also influence the extent of surface damage to the material, consequently impacting the surface integrity of the final product.

Currently, the combined approach of experimental characterization and numerical simulation is an efficient and scientific method for studying the grinding of SiC. This method provides strong support for understanding and controlling damage in hard and brittle materials. However, experimental characterization necessitates the preparation of a substantial number of high-quality workpieces and precision diamond grits, rendering it a complex and costly undertaking. With advancements in numerical simulation technology, the simulation of single diamond grit grinding on SiC surfaces can be achieved by constructing a high-precision model integrated with material constraints. This approach can offer valuable insights into deep crack evolution and material damage and aid in the selection of appropriate grinding parameters. For instance, Zhu et al. [6] conducted modeling of single grit grinding on SiC with varying maximum undeformed chip thicknesses. Their study delved into the governing crack initiation and propagation during SiC grinding, while also analyzing the extent of surface damage to SiC workpieces. Notably, their findings revealed an escalation in surface damage of SiC workpieces with increased maximum undeformed chip thickness.

In finite element simulation, diamond abrasive grains are typically treated as rigid bodies, neglecting their deformation and damage during grinding. In practical grinding operations, abrasive grains experience passivation, detachment, and failure due to various mechanisms such as mechanical forces, physical wear, and chemical corrosion, leading to a decline in grinding quality. This study employs the classical Johnson-Holmquist (JH-2) constitutive model and utilizes a robust nonlinear solver within commercial finite element software to conduct finite element numerical simulations on the grinding of SiC workpiece surfaces using single diamond grits. The analysis includes the examination of surface and subsurface damage, as well as abrasive wear on the SiC workpiece under varying grinding rates and depths. The research investigates the impact of different grinding parameters on the grinding process and abrasive wear of SiC workpieces, aiming to provide insights for optimizing parameters in SiC material surface grinding.

2. JH-2 constitutive model

The mechanism of surface damage during single-grit grinding on a workpiece primarily hinges on the abrasive grain penetration depth. Initially, upon contact between the abrasive grain and the SiC surface, a moderate level of elastic extrusion occurs. As the penetration depth increases, the abrasive grains induce plastic plow marks. Subsequently, as the grinding wheel advances, the plow marks on the surface of the workpiece intensify, forming discontinuous fractured craters. Further advancement leads to increased boundary squeezing of the grains, resulting in voids that progress into cracks, ultimately forming continuous craters. Dai et al. [4] categorized the surface scratch morphology into four stages: plastic removal, quasi-plastic removal, quasi-brittle removal, and brittle removal. In practical grinding applications, brittle damage primarily governs the SiC surface grinding process.

The JH-2 constitutive model, developed by Johnson and Holmquist, incorporates the material's strain rate, hydrostatic strain equation of state for stress, and damage evolution. This model is extensively employed to characterize the mechanical response and damage behavior of brittle materials subjected to significant strains, high strain rates, and elevated pressures [7]. According to the JH-2 theory, as the stress level attains the yield state, the material experiences progressive damage accumulation until reaching the point of failure. The normalized strength model associated with this theory is presented as follows:

$$\sigma^* = \sigma_i^* - D\left(\sigma_i^* - \sigma_f^*\right) \quad (1)$$

where the normalized full equivalent stress is denoted as σ^*, and the damage variable D is typically defined as a variable related to the accumulated equivalent plastic strain, $\sum \Delta\bar{\varepsilon}^{\mathrm{p}}$, indicating the fatigue or damage level of the material. It is commonly defined as:

$$D = \frac{\sum \Delta \bar{\varepsilon}^{\mathrm{p}}}{\varepsilon_f^{\mathrm{p}}} \tag{2}$$

The criteria for determining material damage depend on:

$$D = \begin{cases} 0 & \text{undamaged} \\ 0 \sim 1 & \text{progressive loss} \\ 1 & \text{lose efficacy} \end{cases} \tag{3}$$

where $\varepsilon_f^{\mathrm{p}}$ is the fracture plastic strain associated with hydrostatic stress. It can be expressed as:

$$\varepsilon_f^{\mathrm{p}} = D_1 \left(P^* + T^*\right)^{D_2} \tag{4}$$

where D_1 and D_2 are material constants, respectively.

In Equation (1), σ_i^* and σ_f^* are the normalized intact strength and normalized fracture strength, respectively, which are expressed as:

$$\sigma_i^* = A\left(P^* + T^*\right)^N \left(1 + C\ln\dot{\varepsilon}^*\right) \le \left(\sigma_i^*\right)_{\max} \tag{5}$$

$$\sigma_f^* = B\left(P^*\right)^M \left(1 + C\ln\dot{\varepsilon}^*\right) \le \left(\sigma_f^*\right)_{\max} \tag{6}$$

where A, B, C, M, and N are material constants; $\dot{\varepsilon}^* = \dot{\varepsilon} / \dot{\varepsilon}_0$ is the dimensionless strain rate, in which $\dot{\varepsilon}$ and $\dot{\varepsilon}_0$ are the strain rate and reference strain rate, respectively; P^* and T^* denote the normalized pressure and the maximum tensile hydrostatic stress. We can obtain:

$$P^* = P / P_{HEL} \tag{7}$$

$$T^* = T / P_{HEL} \tag{8}$$

where P_{HEL} represents the Hugoniot elastic limit, T denotes the hydrostatic tensile stress, and P represents the hydrostatic compressive stress, which can be expressed as:

$$P = \begin{cases} K_1\mu + K_2\mu^2 + K_3\mu^3 & \mu \ge 0 \\ K_1\mu & \mu < 0 \end{cases} \tag{9}$$

where K_1, K_2, and K_3 are constants; $\mu = \rho / \rho_0 - 1$ denotes the reference density change, in which ρ is the current density and ρ_0 is the reference density, respectively.

When the material undergoes progressive damage, the volume expansion of the brittle material causes an increase in pressure, and the hydrostatic pressure is defined as:

$$P = K_1\mu + K_2\mu^2 + K_3\mu^3 + \Delta P \tag{10}$$

The reduction in strength due to material damage causes a decrease in elastic energy, and the relationship between elasticity and hydrostatic stress can be written as:

$$\left(\Delta P_{n+1}^2 - \Delta P_n^2\right) / K_1 + 2u_{n+1}\left(\Delta P_{n+1} - \Delta P_n\right) = 2\beta\Delta U \tag{11}$$

where the indicator n+1 denotes the current moment, n denotes the previous moment, U is the elastic potential energy, and β is the energy conversion factor, which ranges from 0 to 1.

3. Numerical simulation

3.1 Finite element modeling

The finite element method is highly favored in numerical analysis and extensively used in multiscale mechanical calculations and the prediction of mechanical behavior in engineering structures [8]. Its distinct advantage lies in its ability to address fine-scale mechanical responses under complex internal and external boundary conditions and provide a reasonable explanation of organizational structure interface mechanics. The essence of the finite element method lies in computing discrete displacement increments that meet force balance equations and boundary conditions, determining stress and strain at integration points at each time increment based on discretized material constitutive models, and obtaining numerical solutions to partial differential control equations through iterative numerical methods.

To study crack evolution and abrasive grain damage on SiC surfaces during grinding, this paper develops a two-dimensional finite element model. The model simulates sub-surface damage on SiC workpieces cut by a single diamond abrasive grain, abrasive grain wear, and the evolution of grinding forces. Illustrated in Figure 1, the microelement representing the SiC workpiece in the simulation is a 250 μm×150 μm rectangle with a fixed bottom and constrained displacement in the X and Y directions on both sides. The diamond abrasive grain is a cone with a 50 μm grain size and a top cone half-angle of 45°, featuring a chamfer radius of 5 μm at the tip. A refined mesh distribution is applied to the workpiece's grinding surface and the abrasive grain contact area in the model to mitigate the influence of mesh density on simulation results.

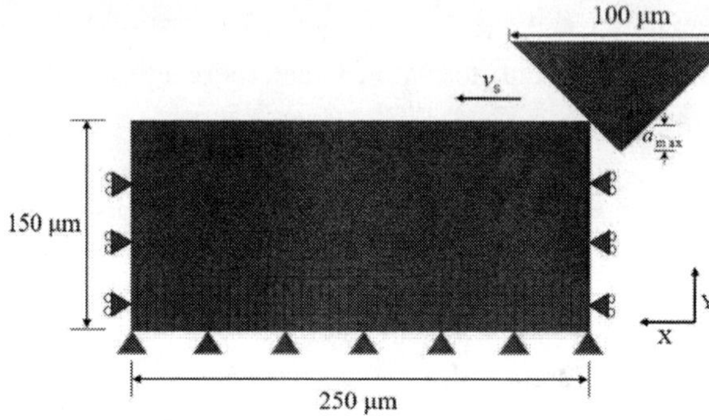

Figure 1. Abrasive grain-workpiece assembly diagram.

Based on the experimental results, the contact between the abrasive grit and the workpiece is considered as hard contact with a normal contact condition. The tangential friction is assumed to be 0.1, and the abrasive grit is oriented parallel to the workpiece (X-direction) with a grinding speed v_s. The grinding depth of the abrasive grit is regulated by varying the depth of penetration a_{max}. Due to the brief contact time between the abrasive grit and the workpiece, thermal expansion effects during grinding are disregarded in this study. Additionally, the workpiece feed speed is significantly slower than the grinding wheel's surface speed, leading to the neglect of its influence in this analysis. Table 1 outlines the material parameters specific to SiC ceramics.

Table 1. Parameters of the JH-2 constitutive model.

Material parameters	Physical significance	Numerical value	Unit (of measure)
A	Material constant	0.96	-
B	Material constant	0.35	-
M	Exponential constant	1.0	-
N	Exponential constant	0.65	-
T	Hydrostatic tension	0.75	GPa
$\dot{\varepsilon}_0$	Reference strain rate	1.0	-
$\left(\sigma_i^*\right)_{max}$	Maximum normalized full strength	12.2	GPa
$\left(\sigma_f^*\right)_{max}$	Maximum normalized fracture strength	1.3	GPa
HEL	Hugoniot pressure	11.7	GPa
P_{HEL}	Hugoniot elastic limit	5.13	GPa
K_1	Hydrostatic stress constant	220	GPa

Journal of Physics: Conference Series **2808** (2024) 012091 doi:10.1088/1742-6596/2808/1/012091

K_2	Hydrostatic stress constant	361	GPa
K_3	Hydrostatic stress constant	0	GPa
β	Energy conversion factor	1.0	-
D_1	Material constant	0.48	-
D_2	Material constant	0.48	-
FS	Failure criterion	0.8	-
$\bar{\varepsilon}^{pl}_{f,\min}$	Minimum plastic strain at fracture	0	-
$\bar{\varepsilon}^{pl}_{f,\max}$	Maximum plastic strain at fracture	1.2	-

3.2 Simulation results

1) Effect of grinding depth on SiC surface damage and diamond grit wear

The grinding depth plays a significant role in influencing the material removal rate, material removal mechanism, and grinding force. Excessive grinding depth has been observed to cause an increase in cracks and plastic chips in SiC ceramics [9]. Dai et al. determined from experimental and simulation data that the tough-brittle transition depth of SiC materials is approximately 0.14 μm [4], indicating that brittle removal mechanisms predominate when the grinding depth exceeds this threshold.

In this section, grinding simulations were conducted to assess the surface damage and diamond grit wear of SiC under brittle removal mechanisms. The simulations were carried out at a constant speed of 30 m/s, with varying grinding depths (5 μm, 7 μm, 8 μm, and 9 μm).

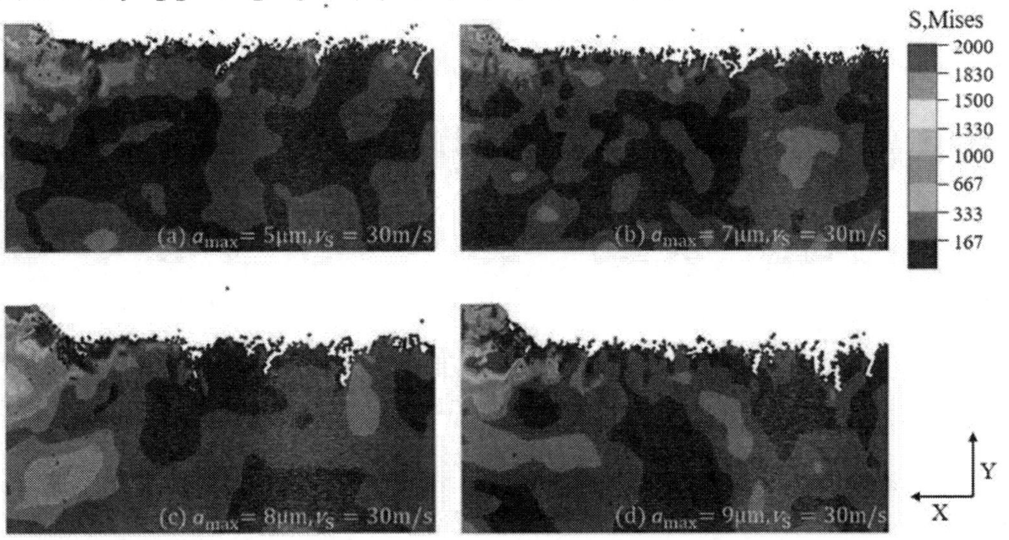

Figure 2. Surface damage of SiC ceramics under various grinding depths.

Figure 2 (a) illustrates that at a grinding depth of 5 μm, stress concentration on the SiC workpiece's surface leads to local unit failure, caused by extrusion between the workpiece surface and the abrasive grain's front knife surface, initiating the sprouting of microcracks. The evolution of cracks is stress-driven, with crack extension aligning with stress flow and material pores coalescing to form fine longitudinal cracks in the stress extension direction. Notably, crack evolution occurs at an angle of approximately 45°, indicating shear stress-driven crack propagation. Progressing to Figure 2 (b), at a 7 μm grinding depth, more surface units fail under extrusion, accumulating on the grits' front blade surface and intensifying friction between the two surfaces. This heightened tangential friction causes stress flow along the X-direction, leading to brittle spalling of some units due to combined tangential stress and longitudinal cracking effects, leaving crushed craters sporadically on the workpiece surface. Subsequently, revealed in Figures 2 (c) and (d), increased grinding depth results in longitudinal crack extension into the subsurface due to tangential stress, resulting in more brittle spalling and surface coverage by crushed craters. The escalation of abrasive penetration depth induces higher surface stress, consequently deepening and widening surface cracks, thus escalating surface/subsurface damage on SiC workpieces, aligning with findings by Gu et al. [5].

In the grinding process, exceeding the material's strength limit initiates damage. The formation of local stress concentration at the abrasive grain workpiece interface during chip contact incurs stress approaching the material strength limit, leading to abrasive grain surface breakage and tool passivation. Additionally, Figure 3 presents a stress damage cloud diagram of the diamond abrasive grains in the post-processing stage.

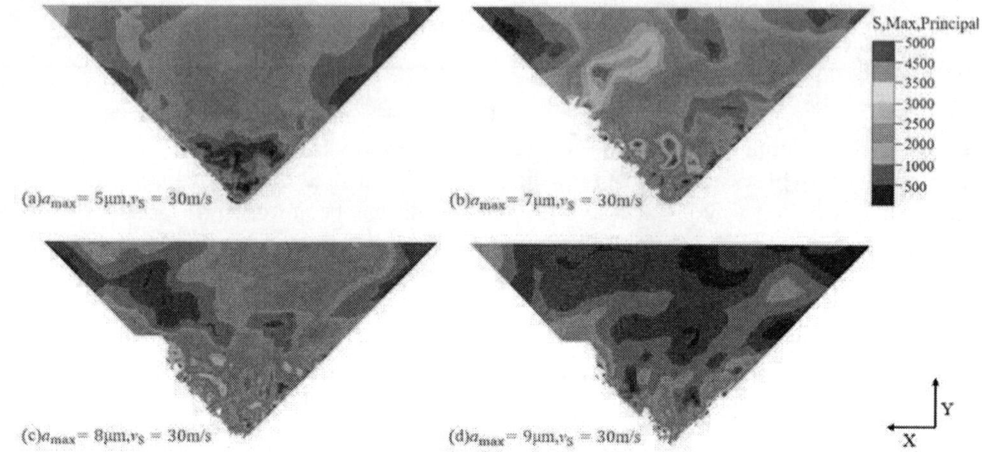

Figure 3. Damage to single diamond grits at different grinding depths.

As demonstrated in Figure 3, at a 5 μm grinding depth, the minimal friction between the workpiece surface and the front face of the abrasive grain results in fewer surface chips. The material removal along the tangential direction by the front face of the abrasive grain causes no surface fragmentation. A comparison of the two cutting surfaces of the abrasive grain reveals wear solely on the front cutting surface without evident tip wear. Upon increasing the depth to 7 μm, the expanded contact surface between the abrasive grain's front cutting surface and the workpiece surface leads to a rise in chip count, escalating abrasive grain damage and generating numerous discontinuous craters on the front cutting surface. Progressing to an 8 μm depth, the craters amalgamate to create a continuous damage plane on the abrasive surface. Subsequent abrasive tip contact with the material surface forms sharp micro-edges, moderately impacting material removal efficiency. Further depth increasing 9 μm induces significant abrasive grit tip abrasion due to friction, contributing to notable damage to the back face of the cutter during continuous cutting. These simulation findings indicate that diamond abrasive grain damage correlates with cutting depth: the deeper the penetration is, the more severe the abrasive grain surface wear is. Based on the simulation outcomes, in the grinding process, crack extension and abrasive wear on the workpiece surface are influenced by stress intensity and material flow direction.

Consequently, grinding force stands as a primary factor governing material damage. Analyzing the dynamic evolution of grinding force during cutting under varying penetration depths can enhance researchers' comprehension of the intricate relationship between material damage and grinding parameters in surface machining processes. This paper depicts the dynamic grinding force during steady-state grinding and plots the force's evolution over time at different penetration depths.

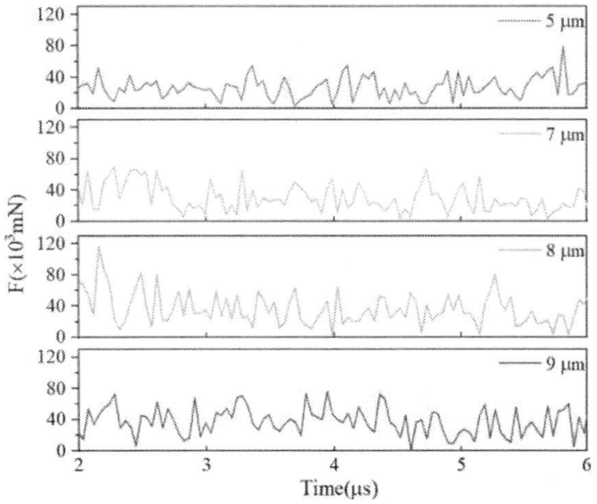

Figure 4. Variation curves of grinding force at different grinding depths.

The analysis of the grinding force image in Figure 4 reveals irregular dynamic fluctuations over time, showcasing a significant correlation between the grinding force and both the surface damage of the workpiece and the extent of abrasive wear. To facilitate a comparison of the grinding force variations across various penetration depths, Figure 5 displays corresponding frequency distribution histograms. These histograms depict typical grinding force values that exhibited an approximately normal distribution throughout the grinding process.

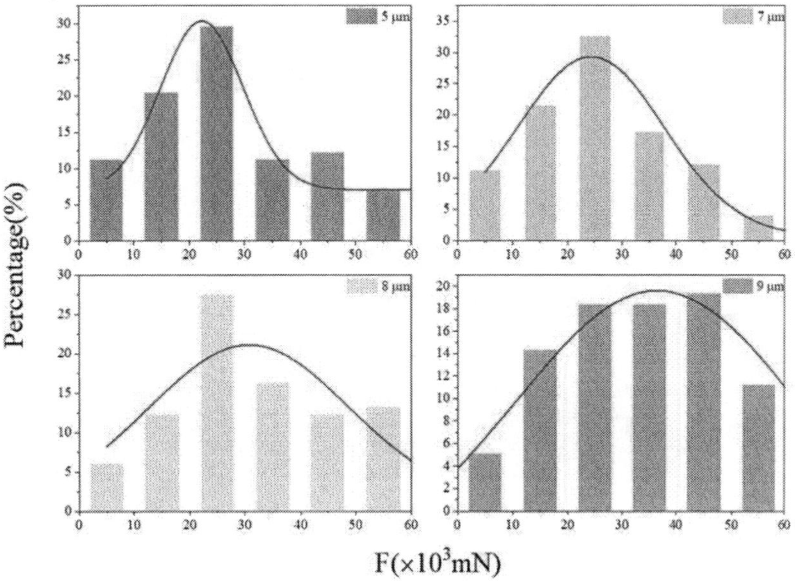

Figure 5. Histogram of the frequency distribution of grinding force at different grinding depths.

The findings depicted in Figure 5 reveal a clear positive correlation between the grinding force and the depth of grinding. Specifically, as the penetration depth of the abrasive grains increases, so does the grinding force. This relationship stems from the escalating grinding energy demand with deeper grinding, which, in turn, necessitates a higher grinding force for material removal. These simulation results are strongly supported by the experimental validation conducted by Ghoreishi et al. [10].

In essence, the heightened penetration depth of diamond grits results in increased surface cracks, residual stresses, workpiece damage, grit wear, and a corresponding elevation in the grinding force

Journal of Physics: Conference Series **2808** (2024) 012091 doi:10.1088/1742-6596/2808/1/012091

essential for material removal, paralleled by a more pronounced grit wear pattern.

1) Effect of grinding speed on SiC surface damage and diamond grit wear

Grinding speed stands out as a highly influential parameter in SiC surface machining, exerting a significant impact on machining efficiency, surface damage, grit wear, and the evolution of the grinding force. Depending on the linear speed of the grinding wheel during processing, grinding operations can be categorized into low-speed, medium-speed, high-speed, and ultra-high-speed grinding. Enhancing the grinding speed within a specific range proves instrumental in boosting material surface processing efficiency and optimizing workpiece surface integrity [11].

Given the hard and brittle nature of SiC ceramics, excessive grinding speeds can lead to material failure during machining, profoundly affecting the surface smoothness of the SiC workpiece. To comprehensively assess the influence of varying grinding speeds on surface damage and abrasive wear, this study employs different grinding speeds (5 m/s, 15 m/s, 30 m/s, and 50 m/s) representing low, medium, high, and ultra-high speeds, respectively, to simulate material damage under diverse machining conditions in SiC surface processing.

Figure 6 illustrates that grinding at a low speed (5 m/s) results in minimal surface stress on the workpiece accompanied by numerous elongated longitudinal cracks, leading to significant material subsurface damage. As the grinding speed transitions to a medium speed (15 m/s), the depth of cracks reduces, with surface cracks aggregating to form pits, and noticeable micro-cracks sprouting in the subsurface. Further increasing the grinding speed to 30 m/s enhances surface damage, where one primary crack dominates on the surface, extending into the material interior perpendicular to the shear stress direction. Ultra-high-speed grinding (50 m/s) demonstrates suppressed material surface damage, with fewer surface cracks and improved surface smoothness. The escalation in grinding speed effectively enhances the surface damage degree of the SiC workpiece, albeit high-speed kinetic energy increases internal material stress. Increased grinding speed facilitates the transformation of narrow and deep longitudinal cracks into wider and shallower tangential cracks, mitigating subsurface damage. This phenomenon can be attributed to the higher dynamic fracture toughness of SiC under high-impact conditions at elevated grinding speeds, inhibiting longitudinal crack propagation and promoting tangential crack formation. For ceramic components, the elimination of tangential cracks through post-grinding polishing highlights the benefit of high-speed grinding in enhancing workpiece surface grinding quality.

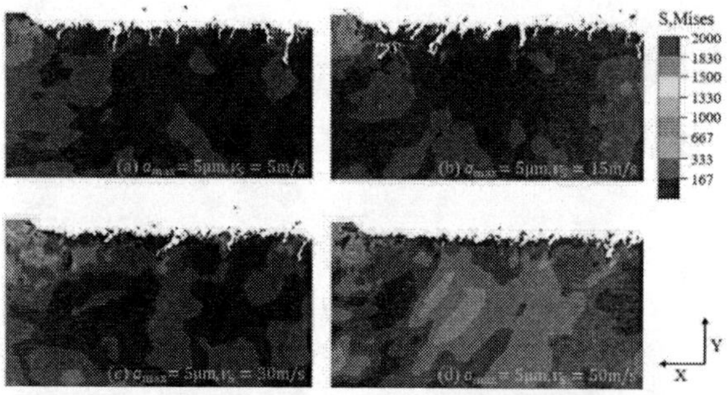

Figure 6. Surface damage of SiC ceramics at different grinding speeds.

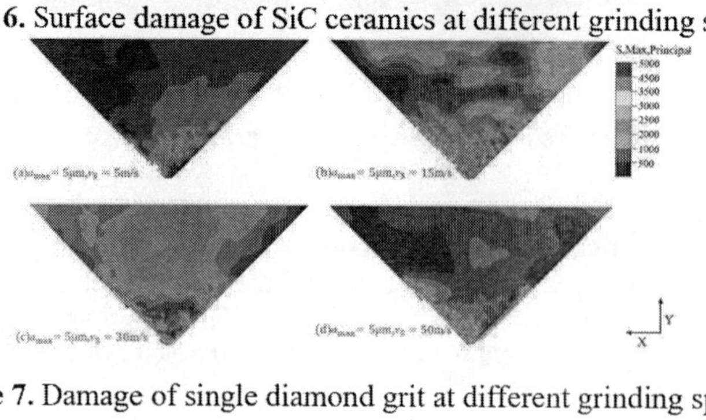

Figure 7. Damage of single diamond grit at different grinding speeds.

As depicted in Figure 7, in low-speed grinding, the shallow grinding depth limits the contact area between the abrasive grit front face and the workpiece, resulting in subtle stress concentration and low abrasive grit surface damage. As the grinding speed escalates, heightened kinetic energy increases localized stress on the abrasive grit front face, facilitating the dissemination of diamond abrasive grit abrasion towards the edges. This process dislodges material surface damage, intensifying the surface wear of the abrasive grains. The acceleration of material removal by the abrasive grains with increasing grinding speed culminates in the formation of a comparatively smooth wear surface on the abrasive grit front face.

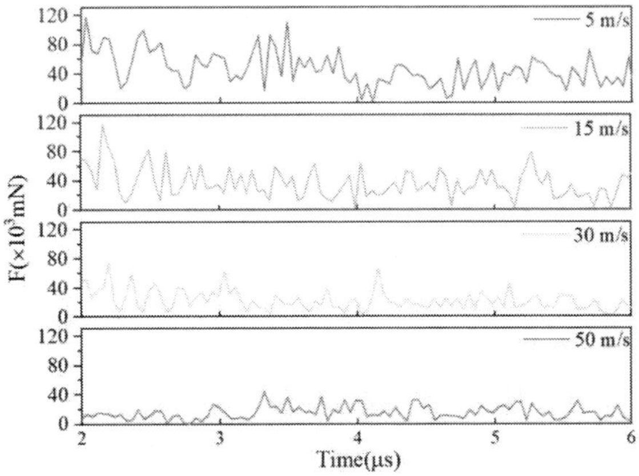

Figure 8. Variation curves of grinding force at different grinding speeds.

Figure 8 displays the evolution of grinding force under various grinding speeds. As the grinding speed increases, there is a notable negative correlation observed in the grinding force, indicating that higher grinding speeds lead to a reduction in the grinding force exerted on the material. This reduction not only diminishes the stress distribution on the workpiece surface but also effectively lowers the residual stress post-machining, extending the product's operational lifespan.

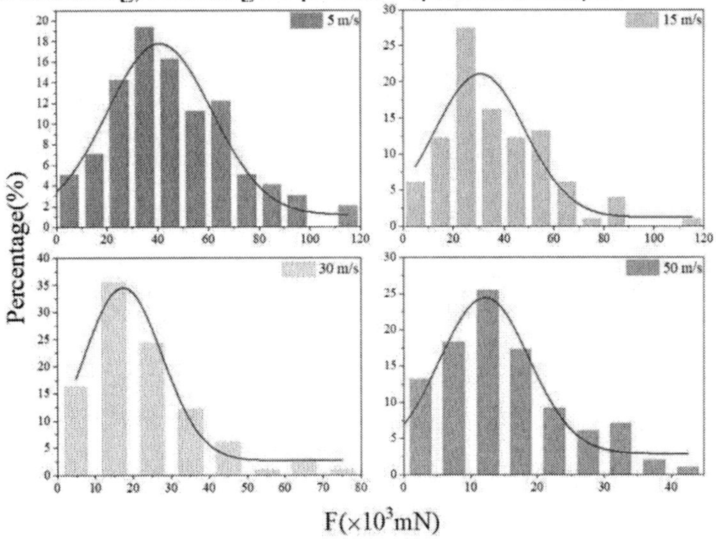

Figure 9. Histogram of the frequency distribution of grinding force at different grinding speeds.

Figure 9 illustrates a reduction in the grinding force on the abrasive grains with an increase in grinding speed. This phenomenon occurs because the higher grinding speed results in more grinding passes per unit of time, leading to a decrease in the grinding force required for each pass.

In essence, elevating the grinding speed within a specific range not only enhances processing efficiency and workpiece surface smoothness but also effectively reduces the grinding force exerted by the abrasive grains. Consequently, this reduction helps diminish residual stress within the workpiece and enhances the product's operational lifespan.

4. Summary

This paper conducts numerical simulations utilizing the JH-2 intrinsic model to analyze the damage to single diamond grits when grinding the SiC surface under varying parameters. The study reveals that an increase in diamond grit penetration depth corresponds to elevated grinding force on the grits, intensifying grit surface damage and ultimately degrading machining quality. Consequently, the formation of microcracks and numerous craters on the material surface leads to heightened workpiece damage. Comparing workpiece surfaces across grinding speeds from low to ultra-high, the data demonstrates enhanced SiC surface integrity with increased grinding depth. Moreover, raising the grinding speed effectively reduces the magnitude of grinding force, and optimizes stress concentration on the abrasive grains' surface, thereby enhancing tool protection. These results emphasize that within a specific range, elevating the grinding speed and selecting suitable grinding depths can significantly enhance SiC workpiece surface integrity, improve processing quality, and prolong tool lifespan.

References

[1] Qinglong A. N., Jie C., Weiwei M., et al. Machining of SiC ceramic matrix composites: a review [J]. Chinese Journal of Aeronautics, 2021, 34 (4): 540-567.
[2] Yan L., Zhang X., Li H., et al. Machinability improvement in three-dimensional (3D) ultrasonic vibration assisted diamond wire sawing of SiC [J]. Ceramics International, 2022, 48 (6), 8, 051-8, 068.
[3] Dai J., Su H., Wang Z., et al. Damage formation mechanisms of sintered silicon carbide during single-diamond grinding [J]. Ceramics International, 2021, 47 (20): 28, 419-28, 428.
[4] Dai J., Su H., Zhou W., et al. Research on the Crack Damage Formation Mechanisms of Polycrystalline Silicon Carbide Ceramics in Grinding Process [J]. Journal of Mechanical Engineering, 2022, 58 (13): 307-320.
[5] Gu Y., Zhu W., Lin J., et al. Subsurface damage in the polishing process of silicon carbide ceramic [J]. Materials, 2018, 11 (4): 506.
[6] Zhu D., Yan S., and Li B. Single-grit modeling and simulation of crack initiation and propagation in SiC grinding using maximum undeformed chip thickness [J]. Computational Materials Science, 2014, 92: 13-21.
[7] Johnson G. R. and Holmquist T. J. Response of boron carbide subjected to large strains, high strain rates, and high pressures [J]. Journal of Applied Physics, 1999, 85 (12): 8, 060-8, 073.
[8] Yazid A., Abdelkader N., and Abdelmadjid H. A state-of-the-art review of the X-FEM for computational fracture mechanics [J]. Applied Mathematical Modelling, 2009, 33 (12): 4, 269-4, 282.
[9] Agarwal S. and Rao P. V. Experimental investigation of surface/subsurface damage formation and material removal mechanisms in SiC grinding [J]. International Journal of Machine Tools and Manufacture, 2008, 48 (6): 698-710.
[10] Ghoreishi R., Roohi A. H., and Ghadikolaei A. D. Analysis of the influence of cutting parameters on surface roughness and cutting forces in high-speed face milling of Al/SiC MMC [J]. Materials Research Express, 2018, 5 (8): 086521
[11] Zhang B., Du Y., Liu H., et al. Experimental study on high-speed milling of SiCf/SiC composites with PCD and CVD diamond tools [J]. Materials, 2021, 14 (13): 3, 470.

Journal of Physics: Conference Series **2808** (2024) 012092 doi:10.1088/1742-6596/2808/1/012092

Analysis of the influencing factors of transmission coefficients based on ZnSe/BeTe heterojunction

Xingyu Li [1 *]

[1]School of Physics, Huazhong University of Science and Technology, Wuhan, Hubei, 430074, China

[*]Corresponding author's e-mail: u202015994@hust.edu.cn

Abstract. As a classic model, the multiple potential barrier structure has been widely studied in the last decades. In this work, we investigate the transmission coefficient of electron tunneling through multiple quantum walls based on ZnSe/BeTe heterojunction. At first, based on the one-dimensional stationary Schrödinger equation and its boundary condition, the equation set is set up. Then, by using the matrix method, the transmission coefficient can be solved by Matlab. The relationship of the transmission coefficient to the energy of the electron, the width of the well, the width of the barrier, the number of barriers, and the potential barrier are simulated. According to the analysis, some phenomena like gap region and miniband can be observed. The number of resonance states in minibands increases when the number of barriers increases while the zones of minibands approximately remain unchanged. This work can be the guide to designing multiple potential barrier nanostructures.

1. Introduction

According to quantum theory, an electron with energy E can pass through a potential barrier whose potential is higher than its energy. This is regarded as the quantum tunneling effect, which cannot be explained by classical physics theory [1, 2]. Particularly, the multiple potential barrier structures have received significant attention worldwide due to their unique properties like miniband structures [3]. For instance, potential barriers can be used as the fundamental building blocks of advanced semiconductors, nanoscale optoelectronic devices, and nano-electronic devices [3, 4]. The advancement of techniques has made it possible to construct an exceedingly excellent class of chalcogenide-based (e.g., BeS, BeSe, BeTe) heterojunctions (e.g., multiple quantum walls) which can be used for optoelectronic devices to meet the rapidly growing energy-related and photonic application needs [5].

Due to the pronounced exciton resonances in optical spectra, the relatively large exciton binding energy, and the large oscillator strength, ZnSe/BeTe heterojunctions have been intensively studied [6]. For the reason that the study of the electron tunneling effect in the multiple barrier structure has theoretical and experimental significance for the design of semiconductor devices, it is meaningful for us to analyze the electron transmission coefficient which is related to the electron tunneling effect [7]. Therefore, based on ZnSe/BeTe heterojunctions with BeTe as the barrier region and ZnSe as the well region, this article aims to apply the matrix method to analyze the electron transmission coefficient tunneling through the heterojunction structure [8].

2. Methodology

The quantum multiple barrier model is detailly discussed in previous studies [8-11]. Multiple barrier structure is shown in Figure 1, where b is the barrier thickness, w is the well thickness, and V_0 is the potential barrier. The unidimensional Schrödinger equation in the x direction can be written as: [9]

$$-\frac{\hbar^2}{2m_j}\frac{d^2\psi_j(x)}{dx^2} + V_j(x)\psi_j(x) = E\psi_j(x) \tag{1}$$

where j is the layer of multiple quantum wall structure, $\hbar$ is the reduced Planck's constant, m_j is the effective mass, $V_j(x)$ is the potential of the barrier, $\psi_j(x)$ is the wave function, and E is the energy of the particles. As for the solution to the equation, since the electron will partially transmitted through the barrier and be partially reflected by the barrier when it propagates to the rectangular potential barrier [8], the solution can be written as:

$$\psi_j(x) = C_{2j}e^{-ik_jx} + C_{2j-1}e^{ik_jx} \tag{2}$$

where the coefficients C_{2j} and C_{2j-1} respectively represent the amplitudes of waves being reflected from and entering the left boundary of the jth barrier. As for k_j, one sets:

$$k_j = \begin{cases} \dfrac{\sqrt{2m_jE}}{\hbar} & (j = 1,3,5,\dots) \\ \dfrac{\sqrt{2m_j(E-V_0)}}{\hbar} & (j = 2,4,6,\dots) \end{cases} \tag{3}$$

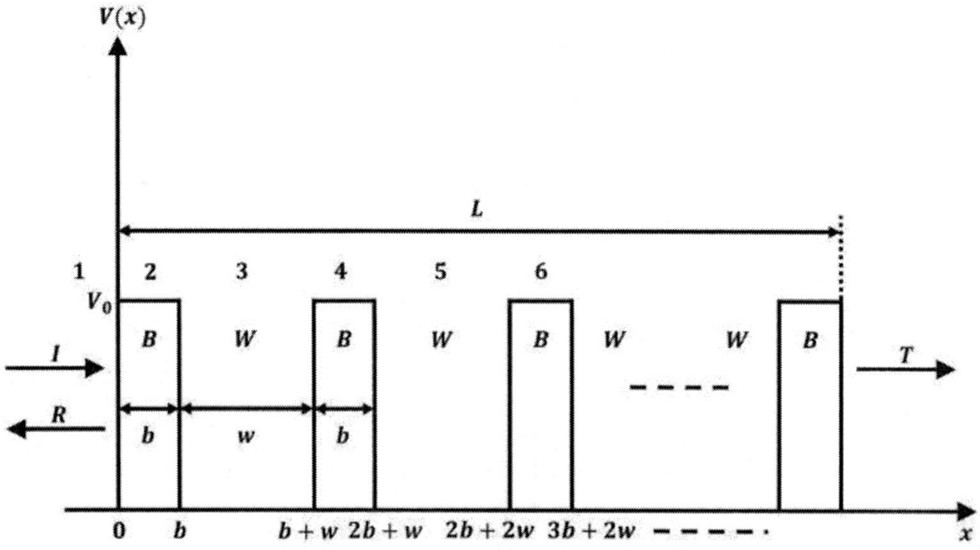

Figure 1. Multiple barrier structure.

According to the boundary condition, when x is equal to the boundary coordinate x_n, the following relations should be satisfied:

$$\begin{cases} \psi_j(x_n) = \psi_{j+1}(x_n) \\ \dfrac{\psi_j'(x_n)}{m_j} = \dfrac{\psi_{j+1}'(x_n)}{m_{j+1}} \end{cases} \quad j,n = 1,2,..,2N \tag{4}$$

In detail, Equation set (4) can be written as:

$$\begin{cases} C_1 e^{ik_1x_1} + C_2 e^{-ik_1x_1} - C_3 e^{ik_2x_1} - C_4 e^{-ik_2x_1} = 0 \\ C_3 e^{ik_2x_2} + C_4 e^{-ik_2x_2} - C_5 e^{ik_3x_2} - C_6 e^{-ik_3x_2} = 0 \\ \vdots \\ C_{4N-1} e^{ik_{2N}x_{2N}} + C_{4N} e^{-ik_{2N}x_{2N}} - C_{4N+1} e^{ik_{2N+1}x_{2N}} = 0 \\ C_1 \frac{k_1}{m_1} e^{ik_1x_1} - C_2 \frac{k_1}{m_1} e^{-ik_1x_1} - C_3 \frac{k_2}{m_2} e^{ik_2x_1} + C_4 \frac{k_2}{m_2} e^{-ik_2x_1} = 0 \\ C_3 \frac{k_2}{m_2} e^{ik_2x_2} - C_4 \frac{k_2}{m_2} e^{-ik_2x_2} - C_5 \frac{k_3}{m_3} e^{ik_3x_2} + C_6 \frac{k_3}{m_3} e^{-ik_3x_2} = 0 \\ \vdots \\ C_{4N-1} \frac{k_N}{m_N} e^{ik_{2N}x_{2N}} - C_{4N} \frac{k_N}{m_N} e^{-ik_{2N}x_{2N}} - C_{4N+1} \frac{k_{N+1}}{m_{N+1}} e^{ik_{2N+1}x_{2N}} = 0 \end{cases} \quad (5)$$

Since the transmission coefficient is the ratio of the amplitudes of wave functions entering and being reflected from the multiple barrier structure [10], the transmission coefficient can be defined as:

$$T = \frac{|C_{4N+1}|^2}{|C_1|^2} \quad (6)$$

By setting [11]:

$$D_2 = \frac{C_2}{C_1}, D_3 = \frac{C_3}{C_1}, \cdots, D_{4N} = \frac{C_{4N}}{C_1}, D_{4N+1} = \frac{C_{4N+1}}{C_1} (\Rightarrow T) \quad (7)$$

Equation set (5) can be rewritten as:

$$\begin{cases} D_2 e^{-ik_1x_1} - D_3 e^{ik_2x_1} - D_4 e^{-ik_2x_1} = -e^{ik_1x_1} \\ D_3 e^{ik_2x_2} + D_4 e^{-ik_2x_2} - D_5 e^{-ik_3x_2} - D_6 e^{-ik_3x_2} = 0 \\ \vdots \\ D_{4N-1} e^{ik_{2N}x_{2N}} + D_{4N} e^{-ik_{2N}x_{2N}} - D_{4N+1} e^{ik_{2N+1}x_{2N}} = 0 \\ -D_2 \frac{k_1}{m_1} e^{-ik_1x_1} - D_3 \frac{k_2}{m_2} e^{ik_2x_1} + D_4 \frac{k_2}{m_2} e^{-ik_2x_1} = -\frac{k_1}{m_1} e^{ik_1x_1} \\ D_3 \frac{k_2}{m_2} e^{ik_2x_2} - D_4 \frac{k_2}{m_2} e^{-ik_2x_2} - D_5 \frac{k_3}{m_3} e^{ik_3x_2} + D_6 \frac{k_3}{m_3} e^{-ik_3x_2} = 0 \\ \vdots \\ D_{4N-1} \frac{k_N}{m_N} e^{ik_{2N}x_{2N}} - D_{4N} \frac{k_N}{m_N} e^{-ik_{2N}x_{2N}} - D_{4N+1} \frac{k_{N+1}}{m_{N+1}} e^{ik_{2N+1}x_{2N}} = 0 \end{cases} \quad (8)$$

Thus, Equation set (8) can be transformed into

$$MD = B \quad (9)$$

where M is a $4N \times 4N$ matrix, while both D and B are $4N \times 1$ matrixes:

$$M = \begin{bmatrix} e^{-ik_1x_1} & -e^{ik_2x_1} & -e^{-ik_2x_1} & \cdots \\ 0 & e^{ik_2x_2} & e^{-ik_2x_2} & \cdots \\ \vdots & \vdots & \vdots & \vdots \\ 0 & 0 & 0 & \cdots \\ 0 & 0 & 0 & \cdots \\ -\frac{k_1}{m_1} e^{-ik_1x_1} & -\frac{k_2}{m_2} e^{ik_2x_1} & \frac{k_2}{m_2} e^{-ik_2x_1} & \cdots \rightarrow \\ 0 & \frac{k_2}{m_2} e^{ik_2x_2} & -\frac{k_2}{m_2} e^{-ik_2x_2} & \cdots \\ \vdots & \vdots & \vdots & \vdots \\ 0 & 0 & 0 & \cdots \\ 0 & 0 & 0 & \cdots \end{bmatrix}$$

$$\rightarrow \begin{matrix} \cdots & 0 & 0 & 0 \\ \cdots & 0 & 0 & 0 \\ \vdots & \vdots & \vdots & \vdots \\ \cdots & -e^{ik_{2N}x_{2N-1}} & -e^{-ik_{2N}x_{2N-1}} & 0 \\ \cdots & e^{ik_{2N}x_{2N}} & e^{-ik_{2N}x_{2N}} & -e^{ik_{2N+1}x_{2N}} \\ \cdots & 0 & 0 & 0 \\ \cdots & 0 & 0 & 0 \\ \vdots & \vdots & \vdots & \vdots \\ \cdots & -\frac{k_{2N}}{m_N}e^{ik_{2N}x_{2N-1}} & \frac{k_{2N}}{m_N}e^{-ik_{2N}x_{2N-1}} & 0 \\ \cdots & \frac{k_{2N}}{m_N}e^{ik_{2N}x_{2N}} & -\frac{k_{2N}}{m_N}e^{-ik_{2N}x_{2N}} & -\frac{k_{2N+1}}{m_{N+1}}e^{ik_{2N+1}x_{2N}} \end{bmatrix} \tag{10}$$

$$D = \begin{bmatrix} D_2 \\ D_3 \\ \vdots \\ \vdots \\ D_{4N} \\ D_{4N+1} \end{bmatrix}, B = \begin{bmatrix} -e^{ik_1x_1} \\ 0 \\ \vdots \\ 0 \\ -\frac{k_1}{m_1}e^{ik_1x_1} \\ 0 \\ \vdots \\ 0 \end{bmatrix} \tag{11}$$

Then, the problem is changed to solving Equation set (9). By using Matlab, it is easy to solve matrix D, and then the transmission coefficient T. To simplify the model, in this report, unless otherwise specified [4, 12], the potential carrier is V_0=2.3 eV, the barrier thickness b is 2.45 nm, the well thickness w is 5.07 nm, the number of barriers N is 2, and the effective mass m_j is depended on:

$$m_j = \begin{cases} m_1 = m_{2N+1} = m_e \\ m_j = 0.17m_e \qquad (j \neq 1, 2N+1) \end{cases} \tag{12}$$

3. Results and discussion

3.1. Relationship between the energy of electron and transmission coefficient

The energy of the electron chosen in this part ranges from 0 eV to 4.0 eV. Figure 2 describes the relationship between the energy of the electron and the transmission coefficient. For the electron whose energy is lower than the potential barrier, the transmission coefficient is 0 except when the energy is close to 1.07 eV, 1.64 eV, and 2.25 eV. When the electron energy is larger than the potential barrier, the transmission coefficient does not seem to fall to 0 when the energy of the electron rises. As shown in Figure 2, the function has several peaks. That is called the resonant tunneling phenomenon which means that an electron tunneling through a barrier structure will resonate when its energy is close to some particular numerical values [13]. This effect can be explained as follows: when the electron touches the barrier, it will reflect and transmit [13]. Once the energy of the electron is so appropriate that the well width is exactly an integer multiple of half probability wavelength, then the phase of waves that transmit barriers are the same [13]. Due to coherent superposition, the amplitude of the last transmitted wave increases significantly. As a result, the resonance tunneling phenomenon appears, which can be observed in suitably built double-barrier structures and multiple-barrier structures [13].

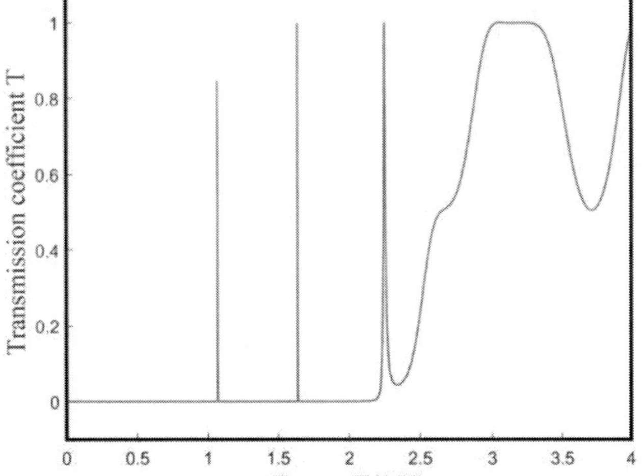

Figure 2. The transmission coefficient as a function of electron energy.

3.2. Relationship between the width of barrier thickness and transmission coefficient.

Here, the barrier thickness is varied from 0 nm to 10 nm. At first, the electron energy is set to be 2.0 eV, which is below the potential barrier. Then, the energy of the electron is set to 3.0 eV, which is higher than the potential barrier. Figure 3 illustrates the relationship between the barrier thickness and transmission coefficient. When the energy of the electron is lower than the the barrier height, the electron transmission coefficient declines rapidly to 0 as barrier thickness increases, which indicates a very small quantum tunneling possibility through the wide barrier. However, as long as E is larger than the potential barrier, the electron has a large probability of going through the barrier. Moreover, the transmission coefficient does not remain at 1 but fluctuates periodically as barrier thickness increases.

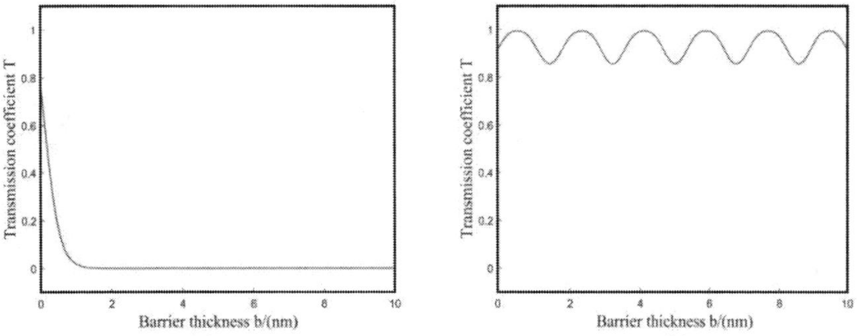

Figure 3. The transmission coefficient as a function of barrier thickness for (a) E = 2.0 eV and (b) E = 3.0 eV.

3.3. Relationship between the width of well thickness and transmission coefficient

Here, the range of the well thickness is from 0 nm to 10.0 nm. At first, the energy of the electron is set to be 2.2 eV, which is below the potential carrier. The energy of the electron is set to be 3.0 eV, which is larger than the potential barrier. Figure 4 describes the relationship between the width of the well thickness and the value of the transmission coefficient. When the energy of the electron is below the potential barrier, the transmission coefficient is 0 with certain exceptions. Therefore, the resonance tunneling phenomenon can be observed when electron energy is lower than the potential barrier.

Meanwhile, when E is larger than the potential barrier height, the transmission coefficient fluctuates periodically as well as thickness rises. However, the transmission coefficient does not remain 1, which can be attributed to the quantum effect.

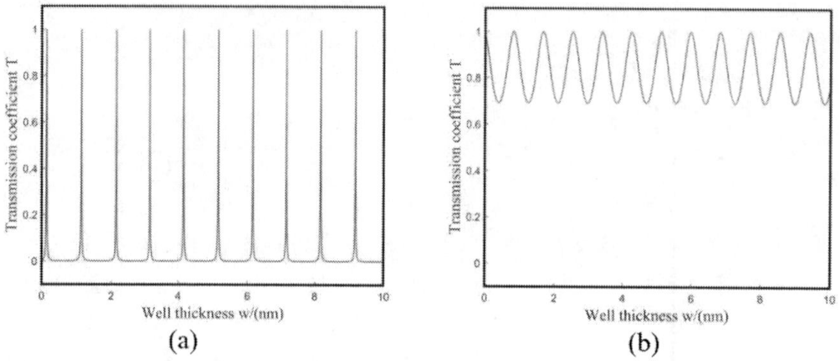

Figure 4. The transmission coefficient as a function of well thickness for (a) E = 2.2 eV and (b) E = 3.0 eV.

3.4. Relationship between the number of barriers and transmission coefficient

Here, the energy of the electron ranges from 1 to 2.5 eV. At first, the number of barriers is chosen to be 10. Then, it is changed to 30. Figure 5 describes the relationship between the number of barriers and the value of the transmission coefficient. It is noticed that there are regions where the transmission coefficient is 0 which means that the electron whose energy is in certain ranges is not able to tunnel through the barrier. That region is regarded as a gap region [14]. Besides, it is also observed that the transmission coefficient exhibits a series of oscillations which indicates the existence of resonant states in the quantum well regions sandwiched between the barriers [15]. These zones where resonance tunneling phenomenon can be observed are called minibands, and there are supposed to be N-1 peaks that achieve perfect transmission (T=1) in each miniband [13, 16]. Furthermore, as the number of barriers increases, the number of peaks will increase while the zones of minibands approximately remain unchanged.

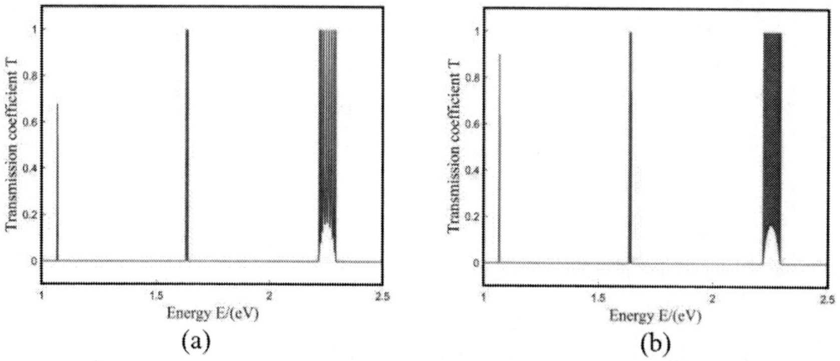

Figure 5. Relationship between the number of barriers to the value of transmission coefficient for (a) N=10 and (b) N=30.

3.5. Relationship between the potential barrier and transmission coefficient

Here, the range of the potential barrier is from 0 eV to 3.0 eV. At first, the energy of the electron chosen is 1.07 eV. Subsequently, the energy of the electron is changed to 1.64 eV. Figure 6 describes the relationship between the potential barrier and the value of the transmission coefficient. When the potential barrier is smaller than the energy of the electron, the transmission coefficient fluctuates without

Journal of Physics: Conference Series **2808** (2024) 012092 doi:10.1088/1742-6596/2808/1/012092

falling to 0. Then, the transmission coefficient decreases significantly to 0 when the potential barrier increases close to the energy of the electron. After that, the transmission coefficient remains 0 as the potential barrier increases except that when the energy of the electron is 2.3 eV, which means that the resonance tunneling phenomenon can be observed when electron energy is lower than the potential barrier. That fact is the same as what has been described in 3.1.

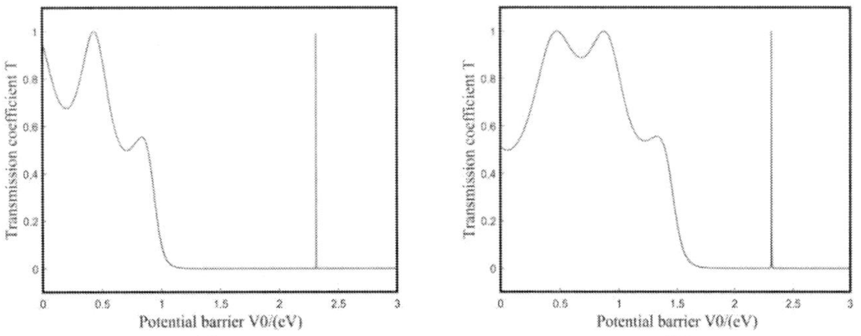

(a) (b)

Figure 6. The transmission coefficient as a function of the potential barrier for

(a) E = 1.07 eV and (b) E = 1.64 eV.

4. Possible application

The resonant tunneling diode, a promising device for terahertz communication technology, is dependent on a double barrier structure, as shown in Figure 7 [17]. While the applied voltage increases, the potential of one barrier declines. Therefore, the tunneling effect can be observed when the applied voltage is suitable. When the applied voltage continues increasing, the rate of the electron going through the barrier decreases, followed by a rise. Figure 7 also shows the current-voltage characteristic of the resonant tunneling diode, which illustrates the process discussed above. When working in the negative differential resistance (NDR) region, the device can interact with a resonant circuit to generate oscillation outputs in the terahertz frequency band [18]. In contrast, when the device works in other regions, it can be used for the detection of terahertz signals [18]. Several reports are giving out the functional relationship between the current of the device and the transmission coefficient based on the double barrier structure, which can explain the experiment results partly [19, 20]. Therefore, according to the applicable requirements, our work can be a reference when analyzing the function of this kind of device.

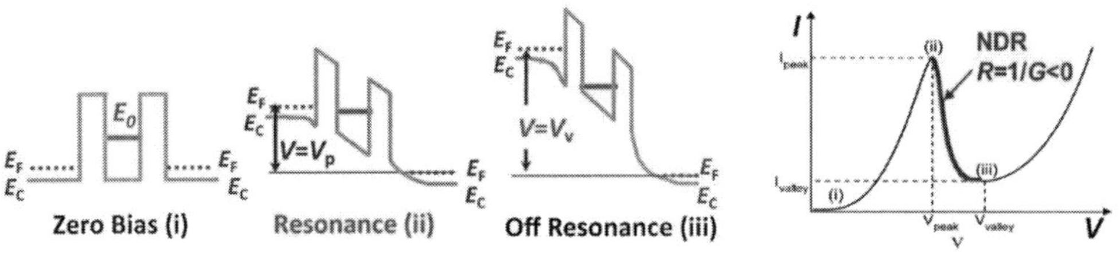

Figure 7. On the left the figure shows the energy diagrams of the RTD under bias and on the right there is the resultant current voltage curve of the RTD. The bias conditions

(i) zero bias (ii) resonance (iii) off-resonance are shown in both parts of the figure [17].

5. Limitations

However, this work also has some limitations. Firstly, the application of the work is limited, as other physical effects, such as interface states and charge transfer, are ignored during the analysis process. In future work, other physical models should be introduced for the analysis of practical ZnSe/BeTe

Journal of Physics: Conference Series **2808** (2024) 012092 doi:10.1088/1742-6596/2808/1/012092

heterojunction especially when the structure is used for particular applications that are not mentioned in this work. Secondly, as Matlab is a software used for numerical simulation, the solution is not completely accurate and the simulation may have problems when choosing the parameter. For instance, the generated figure will be incomplete when 60 is chosen as N. In future work, software, like Wolfram Mathematica, used for symbolic computation will be applied to the simulation to get more precise results. Finally, because the condition is restricted, the simulated result is not verified by this experiment. In the future, the experimental verification of this work will be done.

6. Conclusion

In summary, based on ZnSe/BeTe heterojunction, this work studies the relationship of transmission coefficient to the energy of the electron, the well width, the barrier thickness, the number of barriers, and the potential barrier respectively by the simulation of the computer. Firstly, the equation set is based on one dimensional stationary Schrödinger equation and its boundary condition is changed into matrix form. Secondly, the relationship between the transmission coefficient and those 5 factors above is shown in several graphs respectively. It found that the electron whose energy is close to 1.07 eV, 1.64 eV, and 2.25 eV can tunnel through the barrier whose potential is 2.3 eV, which cannot be explained by classical theory. In terms of the width of the barrier, when the energy of the electron is higher than the potential barrier, the transmission coefficient of the electron will decline significantly to 0 as the width of the barrier rises. As for the width of the well, when the energy of the electron is higher than the potential barrier, the transmission coefficient fluctuates periodically as the width of the well increases. What is worthwhile mentioning is that by increasing the number of barriers, the gap region and the miniband can be observed clearly. Moreover, the transmission coefficient will decrease significantly to 0 when the potential barrier increases close to the energy of the electron. Nevertheless, the effective mass is considered as an invariant parameter when other parameters change, which is also a limitation. Besides, when the value of some parameters (e.g., the number of barriers N, the potential) changes, the practical feasibility is not under consideration. In the future study, as a parameter changes, the related parameters should be calculated and changed. It is a fact that the value of the transmission coefficient is highly related to the efficiency of semiconductor devices. Therefore, before the construction of multiple quantum walls based on ZnSe/BeTe heterojunction, scholars can use a method that is similar to this study for simulation to find suitable parameters that make the transmission coefficient suitable for application. However, the work also has some problems that need to be solved in future work.

References

[1] Tang Y, Harutyunyan H. Optical properties of plasmonic tunneling junctions. The Journal of Chemical Physics, 2023, 158: 6.

[2] Pinotti E, Longhi S. Accelerating Quantum Decay by Multiple Tunneling Barriers. Entropy, 2023, 25(9).

[3] Sabzevar M, Ehsani M H, Splaimani M. Spin-polarised wave-packets transport through one-dimensional nonlinear multiple barriers: Rashba and Dresselhaus Spin-Orbit interactions. Philosophical Magazine, 2023, 103(8): 791-811.

[4] Elkenany E B, Elabsy A M. Comparative investigation between complex energy and transfer matrix techniques for quasi-energy states in heterostructure materials. Journal of Materials Science: Materials in Electronics, 2022, 33(28): 22469-79.

[5] Talwar D N. Impact of stacking sequence on the tight-binding electronic band structures of (BeX)m/(ZnX)m, X = S, Se and Te superlattices. Computational and Theoretical Chemistry, 2022, 1210.

[6] Mund J, Farenbruch A, Yakovlev D R, et al. Optical second- and third-harmonic generation on excitons in ZnSe/BeTe quantum wells. Physical Review B, 2020, 102(12).

[7] Shi P, Li X. Transmission characteristics of one-dimensional periodic quasi-periodic multi-barrier structures. Science Technology and Engineering, 2013, 13(16): 4475-80.

[8] Yao X, Chen L, Liu M, et al. Rational design of Si/TiO_2 heterojunction photocatalysts: Transfer matrix method. Applied Catalysis B: Environmental, 2018, 221: 70-6.
[9] Althib H. Effect of quantum barrier width and quantum resonant tunneling through InGaN/GaN parabolic quantum well-LED structure on LED efficiency. Results in Physics, 2021, 22.
[10] Cirillo E N M, Colangeli M, Rondoni L. Transport in Quantum Multi-barrier Systems as Random Walks on a Lattice. Journal of Statistical Physics, 2019, 176(3): 692-709.
[11] Wang Y, Tang L, Yue Q. MATLAB Analysis and Calculation of Quasi-Bound State in One-dimensional Multi-barrier Structure. Infrared Technology, 2011, 33(2): 85-91.
[12] Yamakawa I, Akanuma Y, Akimoto R, et al. Scanning-tunneling-microscopy observation of heterojunctions with a type-II band alignment in ZnSe/BeTe multiple quantum wells. Applied Physics Letters, 2005, 86(15).
[13] Xu H. Electron Transport in Multiple Barriers. Henan Science, 2012, 30(12): 1710-5.
[14] Mekkaoui M, Jellal A, Bahlouli H. Tunneling of electrons in graphene via double triangular barrier in external fields. Solid State Communications, 2022, 358.
[15] Dakhlaoui H, Belhadj W, Wong B M. Quantum tunneling mechanisms in monolayer graphene modulated by multiple electrostatic barriers. Results in Physics, 2021, 26.
[16] Guzmán E J, Oubram O, Navarro O, et al. Disorder effects on the ballistic transport of gated phosphorene superlattices. Physical Review B, 2023, 107(4).
[17] IRONSIDE C N, FIGUEIREDO J M L, ROMEIRA B, et al. Resonant tunneling diode optoelectronic integrated circuits; proceedings of the Conference on Quantum Sensing and Nanophotonic Devices VII, San Francisco, CA, F Jan 24-28, 2010 [C]. 2010.
[18] Su J, Tan W. Short-Range Terahertz Communication Technology Based on Resonant [J]. Mobile Communications, 2023, 47(05): 32-7.
[19] GIL-CORRALES J A, VINASCO J A, MORA-RAMOS M E, et al. Study of Electronic and Transport Properties in Double-Barrier Resonant Tunneling Systems [J]. Nanomaterials, 2022, 12(10).
[20] Wang D, Su J, Yi H, et al. Analytical Modeling and Band-Structure Engineering of Resonant Tunneling Diode for Biased Terahertz Detector [J]. IEEE Transactions on Electron Devices, 2022, 69(3): 994-8.

Journal of Physics: Conference Series **2808** (2024) 012093 doi:10.1088/1742-6596/2808/1/012093

Establishment and validation of mixed solid degenerate shell elements

Xinlei Jin [1] and Guanghui Qing [1,*]

[1]CAUC, College of Aeronautical Engineering, Tianjin, China

[*]Corresponding author's e-mail: qingluke@126.com

Abstract: Based on the generalized mixed variational principle, a 20-node generalized mixed degenerate shell element is proposed from the 20-node 3D spatial isoparametric element, and the basic assumptions of the plate and shell theory are introduced. Several typical examples are analyzed statically. In the case of the same finite element model, due to the introduction of stress boundary conditions, the result accuracy of the generalized mixed solid degenerate shell element is better than that of the conventional displacement element. The deflections of plate-shell structures with different thickness-to-span ratios are also analyzed, and the generalized mixed solid-degenerate shell element is insensitive to the span-to-thickness ratio. Numerical calculations result in stable convergence, reliability, and high accuracy.

1. Introduction

Shell structures are widely used in engineering practice due to their advantages of material saving, lightweight, high stiffness, and high load-carrying capacity. However, due to the complexity of such engineering structures, it is usually difficult to obtain analytical solutions. Therefore, numerical solution methods such as finite element on such structures are widely used.

Abed-Meraim and Combescure [1] proposed an 8-node strained solid shell unit for analyzing geometric nonlinearities and elastoplastic problems. The cell avoids self-locking by utilizing reduced integrals with not less than two integration points in the thickness direction. Khorasani et al. [2] investigated the buckling problem of honeycomb sandwich panels on elastic foundations and used Navier's method to solve the governing equations. Xing et al. [3-4] proposed the separated variable method for solving the buckling and vibration problems of rectangular plates.

In 1970, Ahmad et al. [5] introduced the concept of degenerate shell elements, which avoids some of the assumptions of 2D shell theory. On the one hand, it is based on the three-dimensional continuum equations, and on the other hand, it only discretizes the midplane of the shell. Obviously, the degenerate shell element is simpler and more general in comparison with other conventional shell elements. In conclusion, the most important feature of degenerate shell elements is that they avoid shell theory based on elastodynamics and are constructed directly from elastic theory in three dimensions. Such units naturally satisfy the coordination condition of C_0 continuity between units. Xu and Ling [6] introduced the corresponding plate-shell assumptions in the spatial isoparametric element to obtain a series of solid degenerate elements suitable for the analysis of thin plate, medium-thick plate-shell, and membrane structures. Huan [7] overcame the self-locking phenomenon for degraded shell elements using the enhanced assumed strain method and the assumed natural strain method and overcame the cell singularity caused by the lack of in-plane rotational stiffness by removing the component of the nodal incremental rotation axis vector in the thickness direction. After Pian [8] successfully established the hybrid stress element from the generalized variational principle in 1964. The idea of establishing multiclass variable finite elements based on various generalized variational principles or modified generalized variational principles has gained the attention of scholars at home and abroad. To overcome the self-locking phenomenon of shell cells, reduced integrals or selected integrals have been widely used in displacement element models. Lee and Pian [9] proposed an 8-node quadrilateral degenerate shell element based on the H-R variational principle by assuming displacements and strains in the cell. This approach is similar to the use of reduced integrals. This method is equivalent to the displacement model using the reduced integration technique and can overcome the self-locking phenomenon in the plate-shell cell.

A deeper study of the generalized hybrid finite element method was carried out by Dang and Rong [10], who applied the method to planar problems with anisotropic materials. Numerical examples show that the generalized hybrid element method well solves the pathological problems occurring in the displacement finite element method, and the displacement and stress results obtained have high accuracy and stability. For the generalized hybrid variational principle, the ratio of potential energy and residual energy can be adjusted by changing the splitting factor to adjust the stiffness of the generalized hybrid model, which meets the needs of elastomer stiffness variability. Qing et al. [11] combined the principle of minimum potential energy with the principle of H-R hybrid variational principle to derive an 8-node uncoordinated hexahedral hybrid element, which was successfully applied to the analysis of the static problems of elastic materials. Wang [12] proposed the non-conforming generalized partial mixed element method and conducted free boundary effect analysis on laminated plate models with different layers. The numerical results of the non-coordinated generalized partial mixed element method are significantly better than those of the displacement finite element method under the same conditions.

According to the basic idea of a solid degenerate shell and the generalized H-R variational principle, this paper attempts to construct a 20-node generalized hybrid solid degenerate shell unit, which is applied to the analysis of plate-shell structures.

2. Methods

2.1. Generalized mixed elementary columns

The parameter-containing generalized H-R variational principle [13] without considering volumetric forces can be expressed as follows:

$$\begin{aligned}\Pi_{\text{GHR}} &= -\frac{1}{2}(1-\lambda)\int_{\text{V}}\boldsymbol{\sigma}^{\text{T}}\boldsymbol{S}\boldsymbol{\sigma}\text{dV} + (1-\lambda)\int_{\text{V}}\boldsymbol{\sigma}^{\text{T}}(\nabla\boldsymbol{u})\text{dV} \\ &+ \frac{1}{2}\lambda\int_{\text{V}}(\nabla\boldsymbol{u})^{\text{T}}\boldsymbol{C}(\nabla\boldsymbol{u})\text{dV} - \int_{\text{S}_{\sigma}}\overline{\boldsymbol{T}}^{\text{T}}\boldsymbol{u}\text{dS}\end{aligned} \tag{1}$$

where $\boldsymbol{\sigma}=\left[\sigma_{xz}\ \ \sigma_{yz}\ \ \sigma_{z}\ \ \sigma_{x}\ \ \sigma_{y}\ \ \sigma_{xy}\right]^{\text{T}}$ is the stress component, $\boldsymbol{u}=\left[u\ \ v\ \ w\right]^{\text{T}}$ is the displacement component, $\boldsymbol{C}$ represents the coefficient matrix for material stiffness, ∇ represents the coefficient matrix for material stiffness, $\overline{\boldsymbol{T}}$ is external load, V is the volume of the structure, S_{σ} is the surface area of the structure, and $\boldsymbol{S}$ is the flexibility matrix of materials. The value range of the parameter λ is $0 \le \lambda \le 1$. When $\lambda = 0$, Eq. (1) is equivalent to the H-R variational principle; when $\lambda = 1$, Eq. (1) is equivalent to the principle of least potential energy.

The coordinate transformation formulas and displacement patterns for the 20-node hexahedral isoparametric unit are respectively.

$$x = \sum_{i=1}^{20} N_i x_i \tag{2}$$

$$u = \sum_{i=1}^{20} N_i u_i \tag{3}$$

The form function of the 20-node space hexahedron isoparametric unit is:

$$N_i = (1+\xi_i\xi)(1+\eta_i\eta)(1+\zeta_i\zeta)/8 \quad (i=1,2,\ldots 8) \tag{4}$$

$$N_i = (1-\xi^2)(1+\eta_i\eta)(1+\zeta_i\zeta)/8 \quad (i=9,10,11,12) \tag{5}$$

$$N_i = (1-\eta^2)(1+\xi_i\xi)(1+\zeta_i\zeta)/4 \quad (i=13,14,15,16) \tag{6}$$

$$N_i = (1-\zeta^2)(1+\xi_i\xi)(1+\eta_i\eta)/4 \quad (i=17,18,19,20) \tag{7}$$

For nodes 9 to 20, ξ_i, η_i, and ζ_i are taken as 0, +1, and -1, respectively.

For a parametric unit such as a 20-node spatial hexahedron, the displacements and stresses are denoted as respectively:

$$\sigma = \boldsymbol{N}_p \boldsymbol{p}_e \tag{8}$$

$$u = \boldsymbol{N}_q \boldsymbol{q}_e \tag{9}$$

where N_p and N_q are the shape function matrices of the displacements and stresses respectively, and p_e and q_e are the displacement vector and stress vector of the unit node respectively.

Substituting Eqs. (8) and (9) into Eq. (1) and performing variational processing, we get:

$$\begin{bmatrix} -(1-\lambda)\boldsymbol{K}_{pp} & (1-\lambda)\boldsymbol{K}_{pq} \\ (1-\lambda)\boldsymbol{K}_{pq}^T & \lambda\boldsymbol{K}_{qq} \end{bmatrix} \begin{Bmatrix} \boldsymbol{p}_e \\ \boldsymbol{q}_e \end{Bmatrix} = \begin{Bmatrix} 0 \\ \boldsymbol{f}_e \end{Bmatrix} \tag{10}$$

where $\boldsymbol{K}_{pp} = \int_V \boldsymbol{N}_p^{\mathrm{T}} \boldsymbol{S} \boldsymbol{N}_p \,\mathrm{d}V$, $\boldsymbol{K}_{pq} = \int_V \boldsymbol{N}_p^{\mathrm{T}} \nabla \boldsymbol{N}_q \,\mathrm{d}V$, $\boldsymbol{K}_{qq} = \int_V \boldsymbol{N}_q^{\mathrm{T}} \nabla^{\mathrm{T}} \boldsymbol{C} \nabla \boldsymbol{N}_q \,\mathrm{d}V$, $\boldsymbol{f}_e = \int_{S_\sigma} \overline{\boldsymbol{T}}^{\mathrm{T}} \boldsymbol{N}_q \,\mathrm{d}S$.

[14] proved that the parameter λ of 0.75 is the most optimal for the generalized hybrid element method, so this optimal value is adopted for all numerical examples in this thesis.

2.2. Basic assumptions for mixed solid degenerate shell elements

In the case of a 20-node solid degenerate shell element, the nodal degrees of freedom are only linear displacements, no angular displacements, and each node has three degrees of freedom. The unit coordinates and displacement interpolation shape functions are the same as those of the 3D solid unit. The 20-node solid degenerate shell element is shown in Figure 1. The node displacement vector of the 20-node 3D isoparametric element is written as:

$$\delta = \left[u_1, v_1, w_1, \ldots, u_5, v_5, w_5, \ldots, u_{20}, v_{20}, w_{20}\right]^{\mathrm{T}} \tag{11}$$

The shape functions, stress vectors, strain vectors, and strain matrices are the same as those of the 20-node 3D isotropic elements.

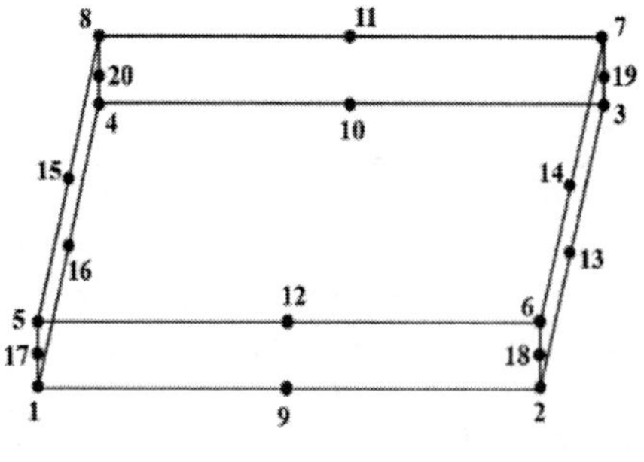

Figure 1. 20-node solid degenerate shell.

According to the plate and shell theory, the stresses satisfy $\sigma_z \ll \sigma_x, \sigma_z \ll \sigma_y$. Therefore, the resulting deformation produced by σ_z is negligible, it can be considered $\sigma_z = 0$ in the physical equations. The elastic stress-strain relationship after introducing this assumption can be simplified as [6]:

$$\begin{Bmatrix}\sigma_x\\ \sigma_y\\ \tau_{yz}\\ \tau_{zx}\\ \tau_{xy}\end{Bmatrix}=\begin{bmatrix}d_1 & d_2 & 0 & 0 & 0\\ d_2 & d_1 & 0 & 0 & 0\\ 0 & 0 & d_3 & 0 & 0\\ 0 & 0 & 0 & d_3 & 0\\ 0 & 0 & 0 & 0 & d_3\end{bmatrix}\begin{Bmatrix}\varepsilon_x\\ \varepsilon_y\\ \gamma_{yz}\\ \gamma_{zx}\\ \gamma_{xy}\end{Bmatrix} \tag{12}$$

$$\varepsilon_z=-\mu(\varepsilon_x+\varepsilon_y) \tag{13}$$

$$d_1=\frac{E}{1-\mu^2},d_2=\mu d_1,d_3=\frac{E}{2(1+\mu)} \tag{14}$$

From Eq. (13), it can be seen that for the strain components that are not independent, integrating Eq. (13) along the thickness direction yields the relative deflection.

$$\Delta w=\int_{-h/2}^{h/2}\varepsilon_z\,\mathrm{d}z=-\mu\int_{-h/2}^{h/2}(\varepsilon_x+\varepsilon_y)\mathrm{d}z \tag{15}$$

After constraining the degrees of freedom, the total number of cell degrees of freedom decreases from 60 to 52. If the assumption that the midplane does not elongate is introduced, then

$$u_{\text{upper}}+u_{\text{lower}}=0 \tag{16}$$

$$v_{\text{upper}}+v_{\text{lower}}=0 \tag{17}$$

where the subscripts "upper" and "lower" represent the nodes corresponding to the upper and lower surfaces respectively. Eqs. (16) and (17) constrain 16 degrees of freedom. Therefore, the cell becomes a 20-node 36-degree-of-freedom shell cell.

3. Results & discussion

For convenience, the 20-node generalized mixed solid degenerate shell element will be abbreviated as GMSE20 and the 20-node displacement element will be abbreviated as H20 in the subsequent sections.

3.1. Four-sided simply supported square plate

A simply supported square plate of length a and a uniformly distributed load q, with a Poisson's ratio μ of 0.3. The square plate dimensions are shown in Figure 2. The center deflection is recorded as:

$$\omega=\alpha\frac{qa^4}{100D} \tag{18}$$

where the bending stiffness of the plate $D=Eh^3/12(1-\mu^2)$, $E=1.00$, $a=1$, $q=1$, h is the thickness of the plate, and α is the center deflection coefficient (dimensionless quantity).

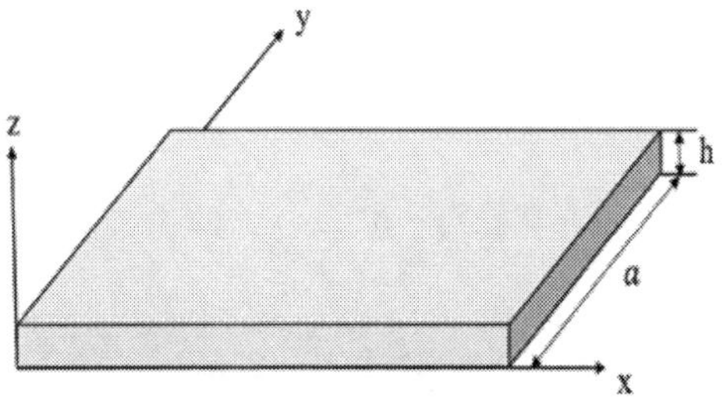

Figure 2. Dimensional of square plate.

The error in the numerical results for all the examples in this paper is calculated as follows.

Journal of Physics: Conference Series **2808** (2024) 012093 doi:10.1088/1742-6596/2808/1/012093

$$Error = \frac{Exact - Element}{Exact} \times 100\% \tag{19}$$

where "Exact" denotes the exact or reference solution and "Element" denotes the solution of the finite element method. The numbers in parentheses in the table are the percent errors of the numerical results.

Table 1. Center deflection coefficient of a four-sided simply supported square plate.

Mesh type (full plate)	Exact	H20	GMSE20
4×4		0.3685(9.33)	0.3774(7.14)
6×6		0.4005(1.44)	0.4021(1.07)
8×8		0.4042(0.54)	0.4048(0.39)
12×12	0.4064	0.4057(0.18)	0.4059(0.13)
16×16		0.4061(0.08)	0.4062(0.06)
24×24		0.4063(0.03)	0.4063(0.02)
32×32		0.4064(0.009)	0.4064(0.004)

Table 1 selects a four-sided simply supported square plate with a thickness-to-span ratio of 0.01 for analysis, with a density range of 4 × 4 to 32 × 32. Under the 4 × 4 grid, the relative calculation error of GMSE20 to the analytical solution is 7.14%, while under the 8 × 8 grid, the relative calculation error of GMSE20 to the analytical solution has rapidly decreased to 0.39%. With the refinement of the grid, the calculation results of both units tend to be analytical solutions. However, analyzing the calculation results of the seven grid types, it can be seen that the calculation accuracy of GMSE20 is better than H20.

Table 2. Center deflection coefficient of four-sided simply supported square plate (16×16).

Thickness-to-span ratio	Exact	H20	GMSE20
0.01	0.4064	0.4061	0.4062
0.1	0.4273	0.4241	0.4249
0.15	0.4536	0.4472	0.4490
0.2	0.4906	0.4813	0.4845
0.25	0.5379	0.5278	0.5329
0.3	0.5956	0.5888	0.5964

As shown by the arithmetic example in Table 2, for both thick and thin plate deflections, the GMSE20 has high accuracy without shear self-locking. When the thickness of the shell unit becomes very thin, its bending stiffness will be much smaller than the shear stiffness. So, when the shell cell undergoes pure bending deformation, its integration point also generates a large false shear stress to resist the deformation of the cell. Under the same mesh (16×16) model, by comparing the computational accuracy of GMSE20 and H20 under different thickness-to-span ratios, it can be seen that the results of GMSE20 are comprehensively better than those of H20. Of course, since GMSE20 can introduce the stress boundary condition, it improves the solution accuracy to a certain extent.

Table 3. Center deflection coefficient of four-sided simply supported square plate.

Mesh type (full plate)	Exact	CT	GMSE20
4×4		0.3512(13.59)	0.3774(7.14)
8×8	0.4064	0.3928(3.35)	0.4048(0.39)
12×12		0.4003(1.51)	0.4059(0.13)

The simply supported square plate at a thickness-to-span ratio of 0.01 is analyzed and compared with the computational results of the displacement unit CT [15], as shown in Table 3. The computational error of the CT unit relative to the analytical solution under the sparse mesh model (4×4) mesh is 13.59%, while the error of GMSE20 is only 7.14%. On the other hand, as the grid is encrypted to 8×8 and 12×12, the resultant errors of GMSE20 have been narrowed down to 0.39% and 0.13%, respectively, whereas those of CT are 3.35% and 1.51%, respectively, and the accuracy of the computational results of CT is still inferior to that of GMSE20.

3.2. Four-sided fixed support square plate

The square plate with side length a is subjected to uniform load q with Poisson's ratio μ of 0.3. We center the deflection equation with the same Eq. (19).

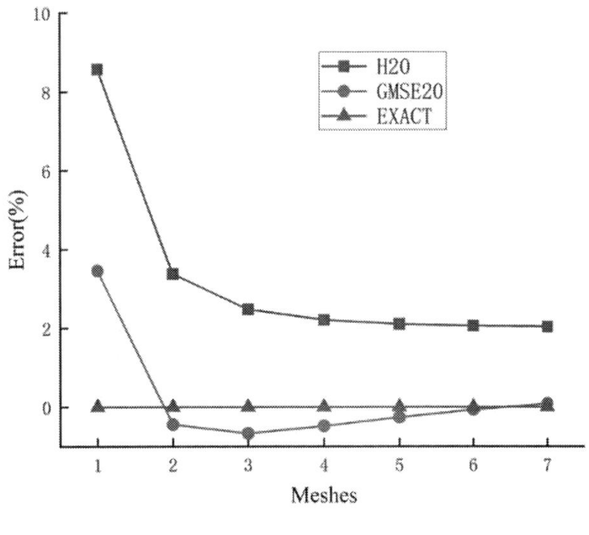

Figure 3. Convergence Diagram of Center Deflection of Four-sided Fixed Supported Square Plate.

Figure 3 shows the analysis of a four-sided fixed supported square plate with a thickness-to-span ratio of 0.1. Mesh 1 to Mesh 7 are 4 × 4, 6 × 6, 8 × 8, 10 × 10, 12 × 12, 14 × 14, and 16 × 16, respectively. When the grid is relatively sparse (4 × 4), the calculation error of GMSE20 relative to the analytical solution is 3.45%, which can meet the accuracy requirements of engineering practice. Gradually densifying the grid, the GMSE 20 element gradually converges to the analytical solution, with errors within 1%. When the grid is 16 × 16, the error is only 0.09%. Compared to H20, when the

Journal of Physics: Conference Series **2808** (2024) 012093 doi:10.1088/1742-6596/2808/1/012093

grid is gradually increased from 8 × 8 to 32 × 32, the error is all above 2%, and the convergence performance is not as good as GMSE20. In theory, one of the main reasons for the good performance of GMSE20 elements may be the introduction of stress boundary conditions.

Table 4. Center deflection coefficient of four-sides fixed supported square plate under different thickness span ratios.

Thickness-to-span ratio	Exact	H20	GMSE20
0.01	0.1265	0.1254(0.85)	0.1270(-0.41)
0.1	0.1499	0.1469(2.04)	0.1498(0.09)
0.15	0.1798	0.1719(4.40)	0.1758(2.23)
0.2	0.2167	0.2075(4.23)	0.2128(1.78)
0.25	0.2675	0.2553(4.56)	0.2625(1.88)
0.3	0.3227	0.3173(1.67)	0.3269(-1.31)

Table 4 shows that under the same grid (16×16) model, the generalized mixed degenerate shell element exhibits good performance in both thin and thick plate conditions. It can be seen that the calculation results of the GMSE20 element are comprehensively better than those of the H20 element under the four-sided fixed support condition. This example indicates that the degenerate shell element of a three-dimensional solid adapts to a wider thickness-to-span ratio.

Table 5. Center Deflection Coefficient of Four-sides Fixed Supported Square Plate.

Mesh type (full plate)	Exact	TCGC-T9	GMSE20
4×4		0.1376(8.21)	0.1447(3.45)
8×8	0.1499	0.1462(2.47)	0.1509(-0.67)
12×12		0.1492(0.47)	0.1498(0.09)

The data in Table 5 were calculated under the condition of a thickness-to-span ratio of 0.1. By analyzing the data in Table 4, it can be concluded that the GMSE20 element in this paper, under the boundary conditions of four fixed supports, has a relatively accurate solution with a calculation error of less than 1% after gradually increasing the mesh to 8 × 8. In general, even if a sparse 4x4 grid is used, with an error of 3.45%, results that meet engineering requirements can still be obtained. For other different grids, compared to the TCGC-T9 element [16], the GMSE20 element has higher accuracy.

3.3. Cantilever plate with free end subjected to lateral force

One end of the cantilever plate is fixed, and the other end is subjected to a vertical downward load. The length and width of the beam gradually reduce the thickness of the cantilever plate to 0.2, 0.1, and 0.01. The ratio of length to thickness of the corresponding structure is 50, 100, and 1000. The material property is $E=10^7$.

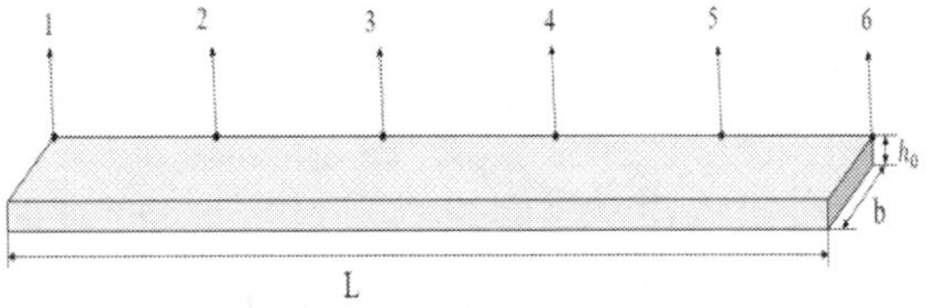

Figure 4. Dimensional of cantilever plate and data sampling points.

We use the generalized mixed degenerate shell element GMSE20 and the reduced integral C3D8R element from ABAQUS in this paper, respectively. The C3D8R element is less prone to shear self-locking under bending loads and has higher accuracy in solving displacement. To prevent the occurrence of zero energy mode when using first-order reduction integration, an hourglass control method is adopted to avoid the hourglass mode. For the convenience of data sampling and analysis, 6 nodes are taken at equal distances from the edge of the cantilever plate, as shown in Figure 4.

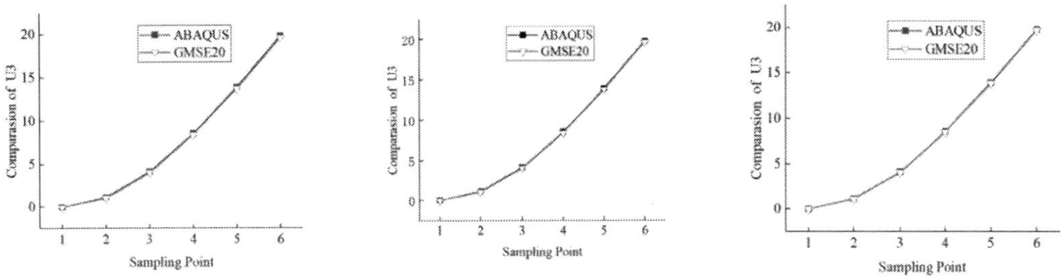

(a) (b) (c)

Figure 5. Compression Deflection Comparison of Cantilever Plate.

As shown in Figures 5, Figure 5 (a)(b)(c) shows the deflection of cantilever plate with thicknesses of 0.01, 0.1, and 0.2, respectively, under different plate thickness conditions, the results of GMSE20 are basically consistent with C3D8R. Prove that the GMSE20 element in this article has no shear self-locking problem.

4. Conclusions

Based on the generalized H-R variational principle, a generalized hybrid solid degenerate shell element (GMSE20) is established for the analysis of three-dimensional plate-shell problems. As a mixed element, it is convenient to introduce the stress boundary conditions of the finite element model. The conclusion is as follows:

1) The three-dimensional solid degenerate shell element column formula is simple and uniform because the interpolation function only requires C0 continuity, the shape function is simple to construct, and the program is convenient to implement;

2) Some numerical examples show that compared with ordinary shell elements, three-dimensional solid degenerated shell elements can adapt to a wider thickness-to-span ratio;

3) The numerical examples show that the generalized mixed solid degenerate shell elements converge stably and the numerical results are reliable. The result precision of the generalized mixed solid degenerated shell element is higher than that of the solid displacement element.

Acknowledgments

National Natural Science Foundation of China (NSFC) (11502286).

Journal of Physics: Conference Series **2808** (2024) 012093 doi:10.1088/1742-6596/2808/1/012093

References

[1] Abed-Meraim F, Combescure A. An improved assumed strain solid–shell element formulation with physical stabilization for geometric non-linear applications and elastic-plastic stability analysis. Int J Numer Methods Eng,2009,80(13):1640-1686.
[2] Khorasani M, Soleimani-Javid Z, Arshid E, et al. Thermo-elastic buckling of honeycomb micro plates integrated with FG-GNPs reinforced Epoxy skins with stretching effect. Composite Structures,2021, 258: 113430.
[3] Yuan Y, Xing Y. An extended separation-of-variable method for eigenbuckling of orthotropic rectangular thin plates. Composite Structures, 2021, 259: 113239.
[4] Wang Z, Xing Y, Sun Q, et al. Highly accurate closed-form solutions for free vibration and eigenbuckling of rectangular nanoplates. Composite Structures, 2019, 210: 822-830.
[5] Ahmad S, Irons B M, Zienkiewicz O C. Analysis of thick and thin shell structures by curved elements. Int J Numer Methods. Eng, 1970(2):419-451.
[6] Xu Xing, Ling Daosheng. Degenerated solid elements [J]. Chinese Journal of Solid Mechanics (Computational Mechanics Special), 2001, 22:1-12.
[7] Wang Huan. Nonlinear Finite element analysis of beam and shell structures based on SiPESC [D]. Dalian University of Technology,2018.
[8] Pian T H H. Derivation of element stiffness matrices by assumed stress distribution. J AIAA, 1964, 2:1333-1336.
[9] Lee S W,Pian T H H.AIAAJ,1978,16:29-34.
[10] Dang Faning, Li Ning, Rong Tingyu. Generalized mixed variational principles and the finite element method [J]. Progress in Natural Science,2002(03):28-32.
[11] Qing G, Mao J, Liu Y. Generalized mixed finite element method for 3D elasticity problems[J]. Acta Mechanica Sinica,2018:1-10.
[12] Wang Xie,Qing Guanghui. Analysis of laminates free-edge effect based on non-compatible generalized partial mixed elements[J]. Acta Materiae Compositae Sinica, 2022, 39(08): 4172-4178.
[13] Qing G, Tian J. Highly accurate symplectic element based on two variational principles[J]. Acta Mechanica Sinica,2017.
[14] Felippa C A. Parametrized multifield variational principles in elasticity: I. Mixed functionals [J]. Communications in Applied Numerical Methods,1989,5(2):79-88.
[15] Fricker A J. An improved three-node triangular element for plate bending.Int J Numer Meth Eng,1985,21:105-114.
[16] Cen Song, Long Zhifei. A new triangular generalized conforming element for thin-thick plates [J]. Engineering mechanics,1998(01):10-22.

Journal of Physics: Conference Series **2808** (2024) 012094 doi:10.1088/1742-6596/2808/1/012094

Mechanical properties and mechanism analysis of modified red layer filler with slag powder

Yuan Gao[1, *], Haoying Niu[2] and Bingbing Zhang[1]

[1]School of Architecture and Construction, Lanzhou Jiaotong University, Lanzhou, Gansu 730070, China

[2]School of Chemistry and Chemical Engineering, Lanzhou Jiaotong University, Lanzhou, Gansu 730070, China

[*]Corresponding author's e-mail: 15530330262@163.com

Abstract. In order to study the mechanical properties of slag micropowder cured red layer filler and its microscopic action mechanism, the mechanical properties of red layer filler with different dosages of slag micropowder were investigated by using test methods such as no lateral limit and straight shear, and the microscopic properties and improvement mechanism of slag micropowder cured red layer filler were analysed by using test methods such as X-ray diffraction, infrared spectroscopy, laser particle size, nuclear magnetic resonance, thermogravimetric analysis, and scanning electron microscopy. The experimental results show that the addition of slag powder can significantly improve the mechanical strength and water stability of the red-layer filler. The microscopic experiments such as X-ray diffraction and infrared spectroscopy proved that the action mechanism of slag micropowder-modified red layer filler is ion exchange and hydration. More importantly, using scanning electron microscopy (SEM), it can be visually observed that the improved red layer filler causes the soil particles to flocculate into sheets or clumps with the increase of hydration gel materials, forming lamellar accumulation or flocculation structures, and the structural integrity is significantly improved.

1. Introduction

The red-bedded mudstone is a clastic rock system formed in arid and hot environments with predominantly terrestrial deposition [1]. As the Gansu red layer is rich in clay minerals, it is easy to weather, collapse in water, and even slime [2], so it is easy to cause highway subgrade diseases [3]. Its water sensitivity and weak expansion have always been the focus of geotechnical engineering. At the same time, China has a high rate of generation of industrial solid waste and a low utilization rate. For example, a group in Gansu Province produces about 1.3 million tons of slag, steel slag, and other industrial waste each year, but only more than 10 tons are applied to traditional building materials. The chemical composition of slag is similar to that of silicate cement clinker, which is mainly silicate and aluminate composed of CaO, MgO, SiO_2, Al_2O_3, etc., with good cementitious activity [4-6]. Especially for the large expansion problem of sulphate saline soil, when slag is used as the curing material, the ions that produce expansion products can be reduced by participating in the reaction, which plays a better role in inhibiting expansion [7]. Therefore, the use of slag for red-layer roadbed improvement is of great significance for road construction in the Gansu region.

Journal of Physics: Conference Series **2808** (2024) 012094 doi:10.1088/1742-6596/2808/1/012094

In past studies, Yang et al. [8-9] explored the feasibility of improving loess with slag micro-powder by configuring different dosages of slag micro-powder loess and utilizing the tests of unconfined compressive strength, shear strength, and wet entrapment in order to explore the effect of slag micro-powder in loess improvement. Zhou et al. [10-11] solidified saline soil with slag micro-powder composite material, and the results showed that with the increase of slag content, the swelling ions involved in the reaction in saline soil became less and less, thus effectively inhibiting the swelling property of saline soil. Zhu et al. [12] showed that slag can effectively inhibit the wet subsidence of loess and improve the compressive strength. Shao et al. [13] found that the increase of slag micronized powder admixture not only significantly reduces the expansion of soft soil, but also improves the mechanical strength of soft soil. Li et al. [14] improved the Yellow River silt by using composite materials such as slag and steel slag, and the experimental results showed that the 7-day unconfined compressive strength of the improved silt with 10% content reached 4.03 MPa. The 7-day immersion strength reduction rate and the 5 times dry-wet cycle strength reduction rate were less than 16%. Li et al. [15] solidified the expansive soil with slag and rice husk ash, and compared with plain soil, the CBR value of the improved soil increased by 7.9 times, and the mechanical strength of the subgrade soil was significantly improved. He et al. [16] solidified the silt with alkali slag and mineral slag, and the 7 d unconfined compressive strength of the modified filler with 30% alkali slag and 8% slag reached 1.23 MPa, which greatly enhanced its deformation resistance and greatly improved the mechanical properties of the silt. Song [17] conducted California Bearing Ratio (CBR) tests on improved soils of steel slag, cement, and lime and found that steel slag-improved soils have the most significant improvement in CBR values and have more potential to be utilized compared to the other two. Zhou et al. [18-19] cured soft soil with slag and fly ash exchange-mixed with cement, respectively, and found that the effect of slag exchange-mixing was most obvious, and the curing ability was even better than cement, which further proved the improvement effect of slag on soil. At present, there are fewer research results on the microstructure of slag-improved mudstone filler, and most of them are for unilateral research on the strength and deformation characteristics of the soil body, etc. It is necessary to further synthesize the research on the macro-mechanical properties of slag-improved red-layer filler as well as the microscopic mechanism of action.

Therefore, through the unconfined compression test and direct shear test, this paper aims to study the change law of compressive and shear strength of red layer filler under different slag content. At the same time, experimental methods such as X-ray diffraction, infrared spectroscopy, thermogravimetric analysis, laser particle size analysis, nuclear magnetic resonance, and scanning electron microscopy were used to reveal the mechanism of slag micropowder's action on the red layer filler in the Gansu area from a microscopic point of view. Based on this, a theoretical basis was provided for the further application of slag micropowder in the curing of the red-layer filler.

2. Experimental material

2.1 Test soil sample

The test soil samples were taken from a work site in Mujiawan, Yongdeng County, Gansu Province, where the naturally weathered red layer fill has low water content and significant wetting disintegration characteristics. The basic physical property test was conducted on the red layer filler, which can be seen in Tables 1 to 3. The red layer filler is a low liquid limit clay with a high content of labile salts, mainly including low-valent cations represented by Na^+, K^+, and more Ca^{2+}, SO_4^{2-} and Mg^{2+}, etc. It is an alkaline medium-sulfuric acid saline soil, whose ore composition consists of ilmenite, kaolinite, quartz, calcite, gypsum, and sodium feldspar, with ilmenite having the highest content of 40%.

Table 1. Indicators of physical properties of red layer fillers.

Liquid limit/%	Plastic limit/%	Plasticity index	Optimal moisture content/%	Maximum dry density/($g \cdot cm^{-3}$)	Free expansion rate/%
34.78	14.82	19.9	14.2	1.966	34.3

Table 2. Results of red layer filler labile salt test.

Sample	Component content/(mmol·kg $^{-1}$)				Total soluble salt /%	PH value
	SO_4^{2-}	Ca^{2+}	Mg^{2+}	Na^+, K^+		
Red layer packing	11.77	31.67	9.34	242.68	0.895	8.74

Table 3. XRD test results of red layer fillers.

Mineral composition	Kaolinite	Illite	Quartz	Calcite	Albite	Gesso
Content /%	11.6	40.0	22.2	13.5	5.7	7.0

2.2 Slag micropowder

XRD diffraction test was carried out on the slag micro powder (Ground Granulated Blast furnace Slag, GGBS), and its mineral composition and content are listed in Table 4. It can be seen that the main components of slag micropowder include calcium oxide $Ca(OH)_2$, silicon dioxide SiO_2, aluminum trioxide Al_2O_3, magnesium oxide MgO, and iron trioxide Fe_2O_3.

Table 4. XRD test results of slag micro powder.

Test material	Chemical composition and content				
	CaO	SiO_2	Al_2O_3	MgO	Fe_2O_3
Slag powder /%	37.5	36.5	18.5	6.8	0.7

2.3 Test program

The slag micro powder mass external admixture of 0%, 5%, 10%, 15%, and 20% and defined as GGBS-0, GGBS-5, GGBS-10, GGBS-15, and GGBS-20 groups of specimens, respectively. The water content of the mixture was uniformly controlled to be 13.8%, the dry density was 1.9 g/cm^3, and the mixture was stirred and smothered for 24 h. Static pressing samples (unconfined samples with a diameter of 50 mm and ring knife samples with a height of 50 mm) were used, and the samples were wrapped in plastic cling film and placed in a curing room with a temperature of 20°C±2°C and humidity of 95% or more for 7 d. According to Highway Geotechnical Test Regulations (JTG3430-2020), unconfined compressive tests and direct shear tests were carried out on the samples, and the axial loads applied in the direct shear tests were 100, 200, 300, and 400 kPa, respectively.

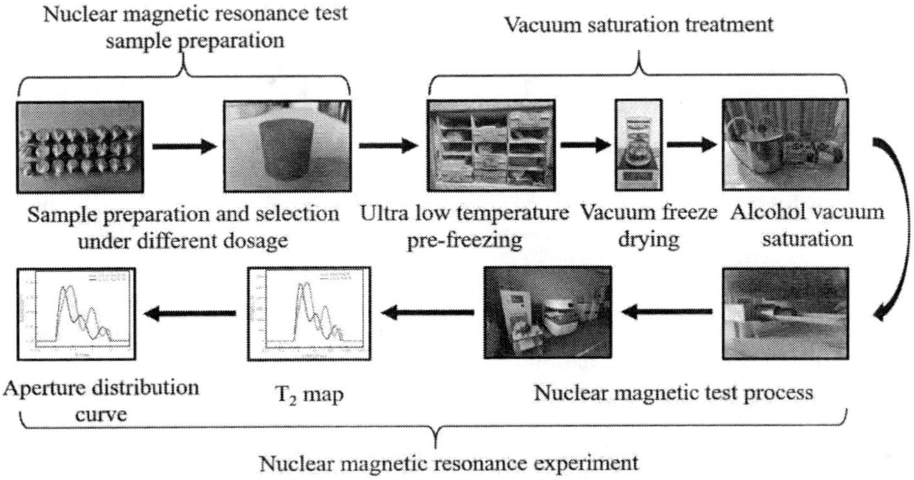

Figure 1. Flow chart of the NMR test.

The nuclear magnetic resonance (NMR) test flow is shown in Figure 1, and the specimens after 7 d of the above standard conditioning were pre-frozen at a constant temperature of -80°C for 12 h at ultra-low temperature. The soil samples were then vacuum dried at -50°C for 24 h, vacuum saturated with 99.7% anhydrous ethanol, and placed into a Neumay MacroMR12-150H low-field nuclear magnetic resonance (NMR) instrument. Based on the pore size distribution curve, the microscopic pore structure of the red layer filler was analyzed under different slag micronutrient dosing.

Similarly, the sample after standard curing was freeze-dried, and a representative cross-section of the sample was selected for SEM test. The test equipment was a German Zeiss (ZEISS GeminiSEM 500) field emission scanning electron microscope (SEM), taking magnification of 3,000 and 10,000 times, to observe and obtain the image of the soil microstructure, and to analyze the microstructure of the soil and the morphology of the hydration products. The SEM test flow is shown in Figure 2.

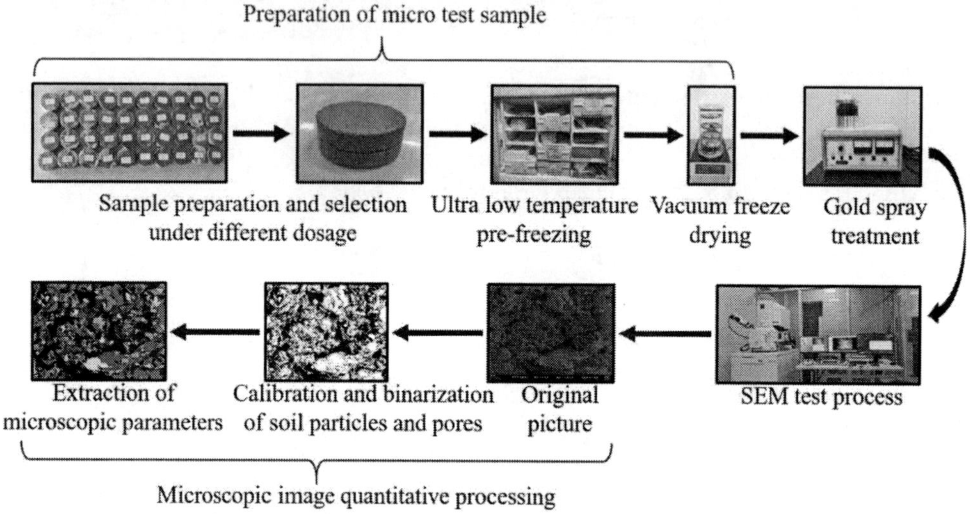

Figure 2. Scanning electron microscope test flow.

3. Analysis of test results

3.1 Unconfined compression test

Figure 3 shows the variation curves of the unconfined compressive strength and rate of strength loss of the modified filler. The unconfined compressive strength and immersed compressive strength of the modified packing gradually increased with the increase of the slag powder content. The unconfined compressive strength increased from 0.8 MPa to 3.1 MPa without immersed water. At the same time, the strength loss rate gradually decreased from 100% to 11.3%, which proved that the addition of slag powder effectively stifles the wet disintegration characteristics of the red layer filler and improves the water stability of the filler.

The reasons are as follows: (1) The calcium ions in the improved soil gradually increased with the increase in the amount of slag powder. After the exchange with sodium ions, the thickness of the double electric layer of clay minerals is reduced, the agglomeration of soil particles is increased, and the soil particles are arranged into a layered stacking structure dominated by surface-to-surface contact, and the structure is improved; (2) The hydration of slag micro powder forms gelling substances such as hydrated calcium silicate (C-S-H). These gels resist moisture intrusion by both filling internal pores and cementing the surrounding soil particles together; (3) The fine slag powder has sharp edges and irregular particles. The cementation between soil particles increases, which is conducive to the further development of the improved compressive strength of the filler. This ultimately led to a significant improvement in the wetting and disintegration characteristics.

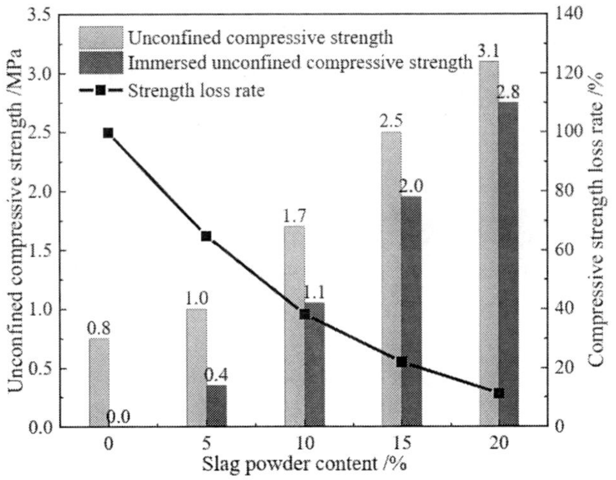

Figure 3. Unconfined compressive strength and rate of strength loss curves of modified red layer fillers.

3.2 Direct shear test

Figure 4 demonstrates the variation curves of the internal friction angle and cohesion of improved soil with different external dosages of slag micronized powder. With the increase of slag micro powder external mixing, the shear strength index of the soil body is gradually increased. The cohesion of the red layer filler is only 116 KPa, and the cohesion of the improved soil increases to 495 KPa when the content of slag powder is 20%. The internal friction Angle of the improved packing gradually increased from 28.09° to 48.66°.

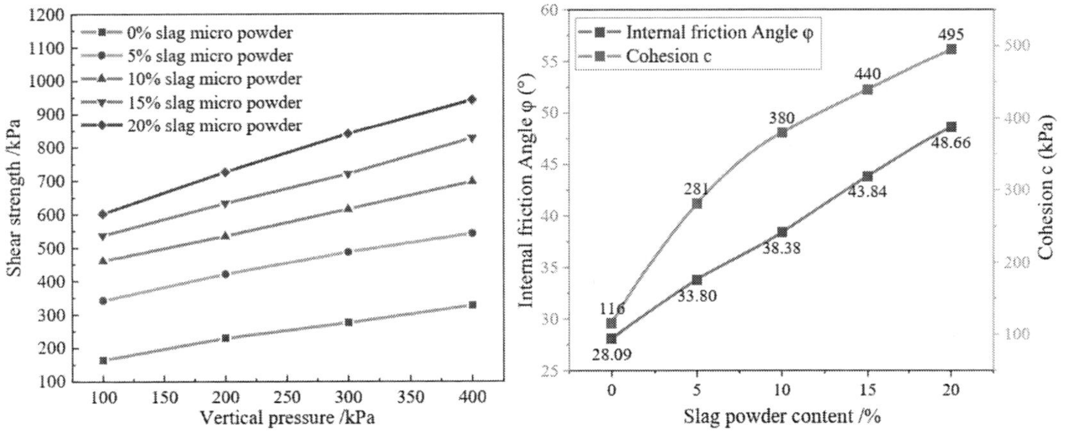

Figure 4. Variation curves of shear strength and c and φ values of modified red layer fillers.

3.3 X-ray diffraction test

In order to further investigate its mechanism of action, its mineral components were analyzed by X-diffraction tests. As shown in Figure 5, due to the fact that in the early stage of hydration, the pH in the liquid phase of the amended soil reaches about 11. Quartz begins to decompose gradually in an alkaline environment, and the [SiO_4] tetrahedron depolymerizes to form $Si0_3^{2-}$. At this time, $Ca(OH)_2$ in the liquid phase is consumed in large quantities, reacts with Ca^{2+} to form silicate particles when it meets water, and generates a large number of amorphous hydrate, which plays a role in reducing soil expansion and strengthening soil. At the same time, the surface of clay particles with Na^+ and K^+, and slag in the water dissolution ionization produces a large number of Ca^{2+} and Al^{3+}, due to the strong

Journal of Physics: Conference Series **2808** (2024) 012094 doi:10.1088/1742-6596/2808/1/012094

attraction of Ca^{2+} and Al^{3+}, will be ion adsorption and exchange with the Na+ and K+ in the clay particles. As a result, the bonding between the crystal cells is enhanced, and flocculation occurs among the soil particles to form large soil aggregates, which reduces the expansibility of montmorillonite and illite in the red layer filler.

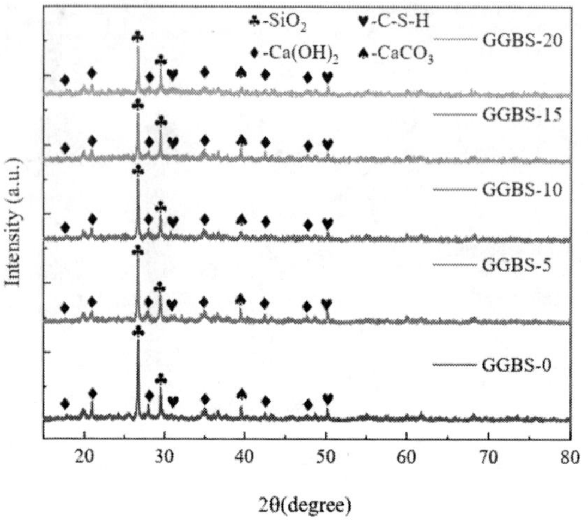

Figure 5. X-diffraction spectra of modified red layer fillers.

3.4 Infrared spectrum test

The information on chemical bonds and functional groups can be analyzed based on the infrared spectrum. In Figure 6, the characteristic peaks at 3420 cm^{-1} and 1640 cm^{-1} correspond to the stretching and bending vibration peaks of the crystal water bond (nH_2O), the characteristic peaks at 730 cm^{-1} correspond to the bending vibration peaks of the Si-OH bond, and the characteristic peaks at 3625 cm^{-1} are the Al^{3+} coordination peaks. The characteristic peaks of the crystal water bond and Si-OH bond were obviously enhanced when slag powder was added to the red layer filler, reflecting the increase of calcium silicate gel produced by the hydration reaction. Meanwhile, the enhanced Al^{3+} coordination peaks of the modified clay indicate the existence of the ion-exchange effect of high-valent Al^{3+} and Ca^{2+} replacing low-valent Na^+ and K^+, which reduces the thickness of the bilayer of the clay particles and thus leads to better agglomeration of the clay particles.

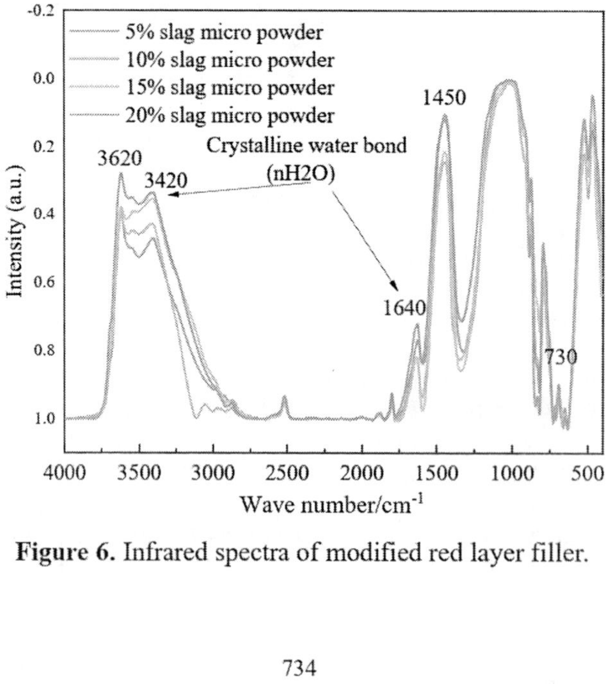

Figure 6. Infrared spectra of modified red layer filler.

3.5 Combined analysis of TG and DTG

As shown in Figure 7, TG and DTG curves under different slag micro-powder content were obtained by using a synchronous thermal analyzer and combined thermogravimetric (TG) and derivative thermogravimetric (DTG) analysis. The peak value of the DTG curve represents the temperature range of the thermal decomposition of hydration products and bonding processes.

As shown in the DTG curves, the modified fillers with different dosages of slag micro powder have more obvious decomposition peaks at 30°C~200°C, 450°C~580°C, 720°C~760°C and 850°C~950°C. At the beginning of heating, the DSC curves all showed large heat absorption peaks accompanied by uniform weight loss (30°C to 200°C). This is mainly due to the loss of initial pore water and the removal of the gelling water of hydrated magnesium silicate M-S-H and the interlayer water on the surface of mineral crystals. Mass loss is caused by dehydroxylation of $Ca(OH)_2$ at 450°C~580°C (peak position 510°C), 720°C~760°C (peak position 742°C) is caused by water decomposition of C-S-H dehydroxy (peak position 742°C), and small endothermic peaks appear at 850°C~950°C. The calcium carbonate produced by $Ca(OH)_2$ carbonization is thermally decomposed into calcium oxide and carbon dioxide at this temperature range (peak position is 876°C). In summary, this is corroborated by the results of the XRD and FT-IR tests.

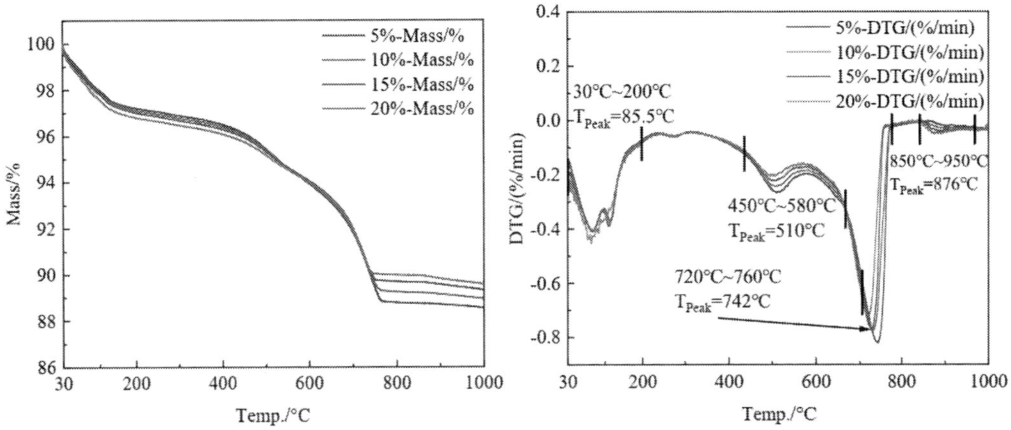

Figure 7. Thermogravimetric (TG) and micro commercial thermogravimetric (DTG) analysis curves of modified fillers.

3.6 NMR and laser particle size analysis

Nuclear magnetic resonance (NMR) testing techniques can more clearly characterize the pore distribution of the soil. (Figure 8(1)) With the increase of slag powder content, firstly, the three peaks in the pore distribution curve shift to the left obviously, and the overall pore size distribution tends to develop towards small pores. Secondly, the peak value of small pores at 1 increased significantly, and the peak value at 2 and 3 decreased slightly. The reason lies in the cementing and filling effect of hydration products, resulting in a gradual decrease in the volume and number of large pores, and part of the large pores are converted into small pores, which leads to an overall increasing trend in the proportion of small pores in the specimen.

The soil particle size gradation was measured by the Mastersizer 3000 laser particle size meter (Figure 8(2)). With the increase of slag content, the particle size gradually decreases in the range of 2 um~50 um, and gradually increases in the range of 100 um~400 um, indicating that the agglomeration between particles is enhanced, and some small particles gradually agglomerate to form agglomerates. On the one hand, it is due to the strong cation exchange, which promotes the flocculation and agglomeration of soil particles. On the other hand, partially hydrated calcium silicate gel is formed due to hydration, which plays the role of filling and cementing. These ion exchange and hydration reactions optimize the particle size distribution of the red layer fillers and enhance the agglomeration between particles, thus showing better mechanical properties.

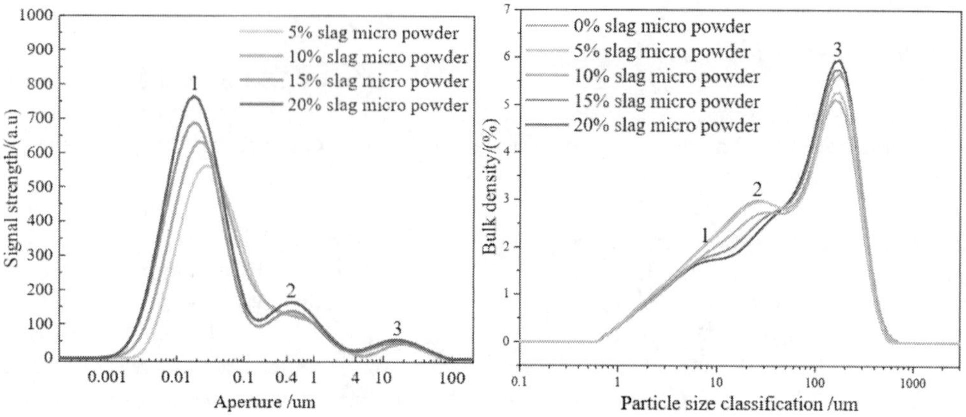

Figure 8. Pore distribution and laser particle size profiles of modified red layer fillers.

3.7 Scanning electron microscope (SEM) tests

3.7.1 Qualitative analysis of microstructure. Figure 9 shows the SEM image of the improved soil with different slag powder content, magnification of 10 k. In plain soil, the soil particles are arranged loosely and piled up disorderly, the cementation is weak, the pore joints are developed, there are many large overhead structure pores, and the point-to-point contact between the particles is mainly, resulting in a poor connection between the particles and the soil structure is relatively loose. With the inclusion of slag powder, a small amount of hydration products were generated, and the C-S-H gels bonded with each other to form sparse flocculent structures, which were distributed on the surface of soil particles, and the surface contact between particles gradually increased. With the further increase of slag powder content, C-S-H gel changes from flocculent structure to rod-like structure, and fills the pores to bond the soil particles to each other, thus further improving the cementation strength of the soil particles. Finally, the C-S-H gel in the modified packing is dense, and the filling and cementation make the modified packing form a dense structure.

As shown in Figure 9(f), calcium silicate hydrate and soil particles are closely cemented together, large pores in the soil gradually disappear, and a large number of small aggregates move closer to each other and aggregate to form larger aggregates, forming dense composite structures. This is because the flocculent cement formed inside the soil further fills the void between the aggregates, wraps and compresses the soil particles, effectively preventing the displacement and deformation of the soil particles, and makes the soil structure more dense, thus improving the mechanical properties and water stability of the improved filler.

(a) 0% Local microscopic map (10 k) (b) 0% Local microscopic map (20 k)

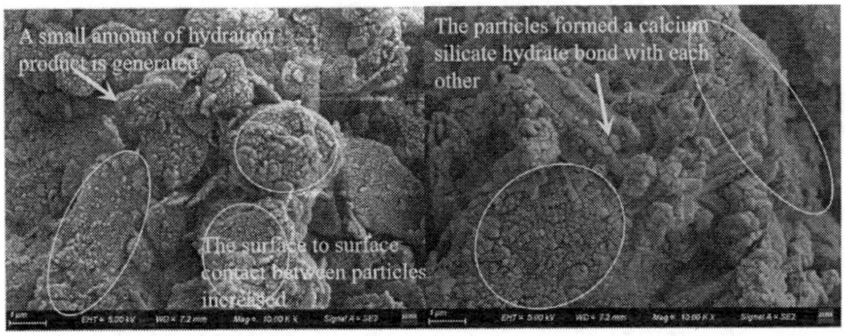

(c) 5% local microscopic map (10 k) (d) 10% local microscopic map (10 k)

(e) Local microscopic map of 15% content (10 k) (f) Local microscopic map of 20% content (10 k)

Figure 9. SEM image of the modified red layer filler.

3.7.2 Quantitative analysis of microstructure. According to the SEM results, 3000 times images were selected for quantitative analysis. PCAS particle analysis software was used to adjust the brightness, binarization, and vectorization of the image, and then the change of soil particle morphology before and after the improvement of the red layer filler was quantified, and the fractal dimension of the particle was calculated. Figure 10 shows the 3 k microscopic diagram, binarization diagram, vectorization diagram, and fractal dimension fitting diagram of the improved soil with different slag fine powder content. Table 5 shows the fractal dimension table of improved soil with different slag powder content.

(a) The content of slag powder is 0%

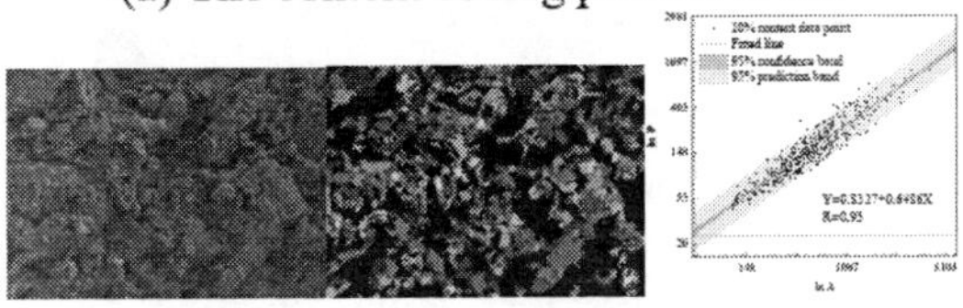

(b) The content of slag powder is 10%

Journal of Physics: Conference Series **2808** (2024) 012094 doi:10.1088/1742-6596/2808/1/012094

(c) The content of slag powder is 20%

Figure 10. Micrographs (3 k times), vector plots and fractal dimension fits of the modified red layer filler.

Table 5. Fractal dimensions of modified red layer fillers.

Dosage /%	0%	5%	10%	15%	20%
Total particle count	676	469	454	349	399
Average particle area (um^2)	1.05	1.10	0.96	1.14	1.15
Average particle circumference (um)	5.99	6.78	6.98	8.68	8.49
The fractal dimension	1.29	1.39	1.53	1.59	1.59

It can be seen from Table 5 that the fractal dimension of slag increases gradually with the increase of the content of slag fine powder. It shows that the complexity of the modified soil structure increases gradually, the roughness of the internal particles is obvious, and the skeleton structure is enhanced. The reason for this is that the increase of hydration gel substances significantly increases the binding force between soils, which causes the soil particles to flocculate into sheets or clumps, forming layered accumulation or flocculation structures, and significantly improving the structural integrity.

The fractal dimensions of unconfined compressive strength and cohesion obtained under different slag fine powder content and the same slag fine powder content were plotted in Figure 11, and linear fitting was performed, with correlation coefficients of 0.93 and 0.97 respectively. It can be seen from Figure 11 that the unconfined compressive strength, cohesion, and fractal dimension have linear relationships, indicating that the change law of fractal dimension of slag micro-powder improved soil can better reflect the change law of its compressive and shear strength indexes. The larger the fractal dimension is, the more obvious the roughness of particles inside the improved filler is and the more complex the soil structure is. Therefore, the mechanical strength index is larger.

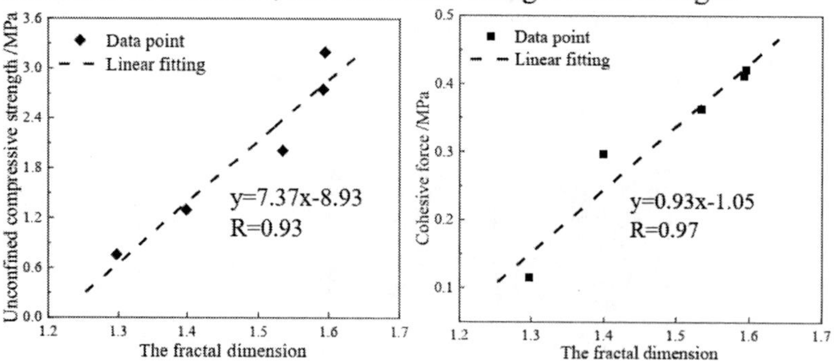

Figure 11. Plot of correlation between fractal dimension and cohesion, angle of internal friction.

The improvement of the microstructure particle morphology is in response to the improvement of the macro-mechanical properties of the improved soil. The improvement of soil properties is more obvious. It can be seen from the above that the qualitative and quantitative analysis of the microstructure is in agreement with the experimental results of the macroscopic mechanical properties.

Journal of Physics: Conference Series 2808 (2024) 012094 doi:10.1088/1742-6596/2808/1/012094

4. Conclusion

Through comparative analysis of the rule of influence of slag fine powder content on the mechanical properties of red layer filler, and using microscopic testing methods to study the mechanism of slag fine powder solidification of red layer filler, the following conclusions were drawn:

1.With the increase of slag powder content, the compressive strength of the improved filler was significantly improved, and its unconfined compressive strength gradually increased from 0.8 MPa to 3.1 MPa, and the increase rate gradually decreased. Similarly, the cohesive force and internal friction Angle of the modified filler increased from 116 KPa and 28.09° to 495 KPa and 48.66°, respectively. This shows that the slag powder has a significant improvement in the mechanical properties of the red layer filler, and the wet disintegration properties are significantly improved.

2.The micro-experiments revealed the action mechanism of the modified filler with slag fine powder: (1) The slag fine powder promoted the flocculation and agglomeration of clay minerals under the action of strong cation exchange. (2) The hydration of slag micro-powder forms gelling substances such as hydrated calcium silicate (C-S-H), which plays the role of filling and cementing, forming a spatial network structure. (3) The slag powder is angular and irregular, which increases the cementation between soil particles, which is conducive to the further development of the improved compressive strength of the filler. In the end, the mechanical strength and the wet disintegration characteristics were obviously improved.

3.Through the in-depth study of the mechanical properties and micro-mechanism of the modified red layer filler with slag micro-powder, the research results of this paper provide a useful reference for the comprehensive utilization of industrial solid waste and the sustainable development of red layer roadbed engineering. At the same time, these research results will help promote the positive interaction between the construction of local transportation infrastructure and environmental protection in China.

References

[1] Xu, Q., Tang, R. Research on red beds and their geologic hazards [J]. Journal of Rock Mechanics and Engineering, 2023, 42(01):28-50.

[2] Yin, H. I., Wen, L. D., Wu, G. Q., et al. Study on disintegration characteristics and law of red layer mudstone under dry and wet cycling conditions [J]. Highway and Transportation Science and Technology, 2023, 40(04):94-102.

[3] Zhou, Y. H. Study on Instability Mechanism and Reinforcement Protection Technology of Red Cut Slope [D]. Southwest Jiaotong University, 2006.

[4] Zhang, L., Guo, L. J., Wei, X. M., et al. Study on filling test of slag-based cementitious materials and microstructure evolution of cementitious body [J]. Nonferrous Metal Engineering, 2023, 13(05):84-93.

[5] Li, D. P., Ping, H. Y., Zhang, K. F., et al. Preparation and hydration mechanism of steel slag-slag-desulfurisation gypsum composite cementitious materials [J]. Science, Technology and Engineering, 2023, 23(06):2558-2566.

[6] Wan, Z. H., Zhang, W. Q., Liu, Z. C., et al. Study on the properties of calcium carbide slag-slag composite cementitious materials [J]. Silicate Bulletin, 2022, 41(05):1704-1714.

[7] Chrysochoou, M., Grubb, D. G., Malasavage, N. E. Assessment of sulfate-induced swell in stabilized dredged material: is ettringite always a problem? [J]. Journal of Geotechnical and Geoenvironmental Engineering, 2012, 138(3):407-414.

[8] Yang, G. H., Li, C., Shang Y. et al. Experimental study on the improvement of pore characteristics of loess by slag micro-powder based on CT scanning technology [J/OL]. Water Resources and Hydropower Technology:1-13.

[9] Yang, G. H., Li, C., Shang, Y., et al. Mechanical properties and shear strength prediction model of slag powder amended loess [J]. Science, Technology and Engineering, 2023, 23(03):1185-1191.

[10] Zhou, Y. X., Yan, P. Y. Autogenous volume stability of solidified saline soils [J]. Journal of Tsinghua University (Natural Science Edition), 2007(12):2089-2094.
[11] Zhou, Y. X., Yan, P. Y. Drying and wetting characteristics of different types of saline soil curing bodies [J]. Journal of Geotechnical Engineering, 2007(11):1653-1658.
[12] Zhu, M. M., Zhu, W. W. Progress of research on the performance and mechanism of slag and industrial waste to improve loess [J]. Disaster Science, 2022, 37(01):129-133.
[13] Shao, L., Li, P. Q., Wang, B. J. Influence of freeze-thaw cycle on the strength of soft soil reinforced with alkali-stimulated blast furnace slag powder [J]. Highway and Transportation Science and Technology, 2022, 39(01):40-47.
[14] Li, X. H., Yang, Y. H., Wei, J. C., et al. Strength and mechanism of slag-steel slag-desulfurisation gypsum-cement stabilised pulverised soil [J]. Journal of Central South University (Natural Science Edition), 2023, 54(06):2382-2390.
[15] Li, L. H., Huang, C., Li, W. T., et al. Study on mechanical and microscopic properties of rice husk ash-slag cured expansive soil [J/OL]. Geotechnical Mechanics, 2023(10):1-13.
[16] He, J., Wang, X. Q., et al. Unconfined compressive strength and microscopic characteristics of alkali slag-slag cured silt [J]. Journal of Applied Basic and Engineering Sciences, 2021, 29(02):376-386.
[17] Song, X. B. Study on Design and Pavement Performance of Steel slag Stabilized soil [D]. Zhengzhou: Zhengzhou University, 2019.
[18] Liang, S, H., Zhou, S. Z., Dai, J., et al. Experimental study of slag and cement curing soft soil in Nansha, Guangzhou [J]. Industrial Construction, 2015, 45(10):116-120.
[19] Zhou, S. Z., Liang, S, H., Dai, J. Experimental study on curing soft soil of Nansha with slag and fly ash [J]. Journal of Water Resources and Construction Engineering, 2015, 13(04):76-79.

Journal of Physics: Conference Series **2808** (2024) 012095 doi:10.1088/1742-6596/2808/1/012095

Stress analysis and calculation of the inner anchor structure of pressure-type anchor

Jian Ma[1], Qi Zhang[2], Hanbo Dan[3], Feng Pan[3], Gang Xu[1], Yong Guo[3] and Qing Lü[1*]

[1]State Grid Jinhua Power Supply Company, Jinhua Zhejiang 321017, China;
[2]College of Civil Engineering and Architecture, Zhejiang University, Hangzhou 310058, China;
[3]China Energy Engineering Group Zhejiang Electric Power Design Institute Co., Ltd, Hangzhou 310012, China

[*]Corresponding author's e-mail lvqing@zju.edu.cn

Abstract: The pressure-type anchor transfers the load to the grouting body via the inner anchor structure at its bottom when bearing the uplift loads. In order to design the inner anchor structure reasonably, the elastic thin plate theory is used to analyze the inner anchor structure under uniformly distributed loads. Based on material properties, the calculation method and the recommended values for the size parameters of the inner anchor structure are given. A finite element model of the inner anchor structure is established with which the rationality of the proposed method and the recommended parametric values are verified by analyzing the stress situation of the internal anchor structure under ultimate load.

1. Introduction

The pressure-type anchor, due to their rational force mode and high pull-out bearing capacity, has been extensively applied in various sectors such as pit supporting, anti-floating of high-rise buildings, transmission line base, etc. [1-3]. The rod body of the pressure-type anchor is a full-length free unbonded rebar or steel strand, through which the upward pulling load is directly transferred to the internal anchor structure at the bottom of the rod by setting up a bearing body, thereby applying compressive stress to the grouting body. As the grouting body is under compression, its high compressive strength and the good bearing capacity of deep rock bodies are fully exploited to obtain a higher pull-out bearing capacity [4-5].

The internal anchor structure is pivotal in channeling external loads to the interior of the anchor rod, which must satisfy the requests of both strength and deformation. The type of internal anchor structure for pressure-type anchor rods is not clearly specified in the official guidelines. The primary methods employed in the field of engineering practice involve the use of prefabricated nuts and welded steel plates [6-7]. The internal anchor structure utilizing nuts usually has a smaller cross-sectional area, leading to potential local compression failure of the grouting body due to stress concentration. Conversely, by using welded steel plates, the size arrangement is flexible, which can lead to uniform stress distribution on the cross-section of the grouting body and substantially lower cost compared to using nuts.

This paper introduces a type of bearing body that couples a welded steel plate with stiffening plates for the design of the internal anchor structure of the pressure-type anchor. The force mode of the internal anchor structure is similar to that of the flange plate [8-10]. By improving the design calculation method of the flange plate in the standard and applying the theory of elastic thin plate to analyze the internal

Journal of Physics: Conference Series **2808** (2024) 012095 doi:10.1088/1742-6596/2808/1/012095

anchor structure under uniformly distributed loads, the design parameters of the internal anchor structure under ultimate loads can be identified. On this basis, the reliability of the method proposed in this paper is further verified through numerical simulation. The relevant achievements can provide a basis for the design of pressure-type anchor rods.

This paper discusses a particular pressure-type anchor, which is employed primarily for the base of a transmission tower, composed of an internal anchor structure made of an anchoring plate and a stiffening plate. The anchor bars pass through the circular holes of the welded steel plate and are reliably connected to the internal anchor structure through welding, as represented in **Figure 1**.

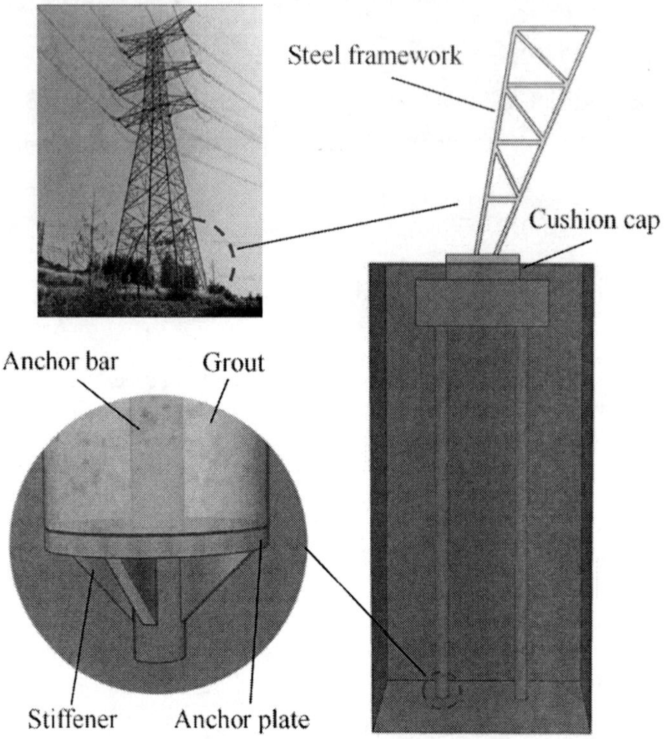

Figure 1. Schematic diagram of internal anchor structure of pressure-type anchor.

When the pressure-type anchor rod is under tensile force, the pull-up load is transmitted from the anchor bar to the internal anchor structure and grouting body, and finally bears the pull-up load through the shear resistance between the grouting body and the drilling surrounding rock. On the one hand, the internal anchor structure must meet the strength requirements, and on the other hand, in order to ensure even compression of the grouting body, the anchor plate should also have high stiffness. Theoretical analysis and numerical simulation are conducted to design parameters such as anchor plate thickness, stiffener plate thickness, and height.

2. Theoretical analysis of internal anchor structure stress

The anchor plate has a circular hole in the center and, due to symmetry, the shape of the force-bearing unit of the anchor plate is fan-shaped. It can be simplified into a calculation model with two sides fixed, one side simply supported, and one side free, as shown in **Figure 2(a)**. The bending moment can be calculated by using the theory of elastic thin plates under uniform load.

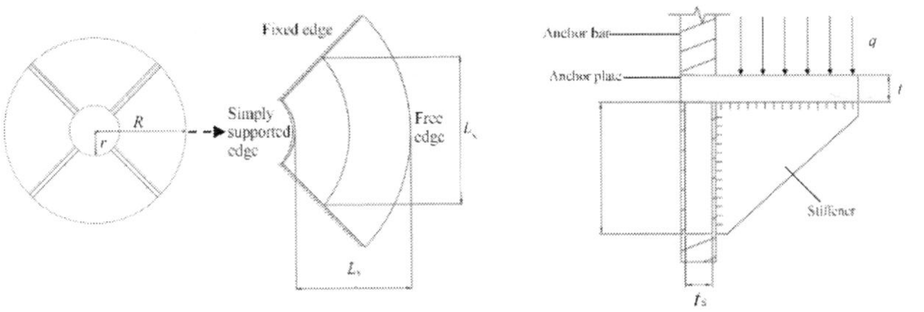

(a) (b)

Figure 2. Calculation and structural schematic diagram.

When the anchor rod is subjected to an upward pull load, the anchor plate will compress the grout body and be subjected to a uniformly distributed load perpendicular to the plate. The design resistance of the anchor rod rebar is used as the design load of the internal anchor structure, indicating that the resultant force of the uniformly distributed load borne by the anchor plate is equal to the design resistance value of the anchor rod. Therefore, the uniformly distributed load on the anchor plate is:

$$q = \frac{T}{\pi(R^2 - r^2)} \tag{1}$$

In the formula: q represents the uniformly distributed load on the anchor plate (MPa), T is the design resistance of the anchor rod (kN); R is the radius of the anchor plate (mm); r is the radius of the rod (mm), as shown in **Figure 2(a)**.

Under uniform load, the anchor plate bends and deforms. According to the theory of elastic thin plates, the maximum bending moment in the plate is:

$$M_{max} = \beta q L_x^2 \tag{2}$$

In the formula: M_{max} represents the maximum bending moment in the anchor plate (N); β is the bending moment coefficient, which can be determined according to **Table 1**; L_x represents the average length of the simply supported edge and the free edge (mm), as shown in **Figure 2(a)**.

The thickness of the anchor plate can be calculated and determined by the following formula:

$$t \geq \sqrt{\frac{5M_{max}}{f_y / \gamma_R}} \tag{3}$$

The simple calculation diagram of the stiffener is shown in **Figure 2(b)**. The main considerations are the pressure shear composite stress at the connection with the anchor rod and the compressive stress at the connection with the anchor plate.

The normal stress at the connection with the anchor rod is calculated by the following formula:

$$\sigma_s = \frac{3\alpha T(R-r)}{4t_s h_s^2} \leq f_c^w \tag{4}$$

In the formula: σ_s is the normal stress at the connection of the stiffener and the anchor rod (MPa); α is the reaction ratio of the stiffener, which can be determined according to **Table 1**; t_s is the thickness of the stiffener (mm); h_s is the height of the stiffener (mm); f_c^w is the design value of the compressive strength of the butt weld (MPa).

Table 1. Calculation coefficient of bending moment of anchor plate and reaction ratio of stiffening plate under uniformly distributed load.

L_y/L_x	β	α
0.35	0.0785	0.67
0.40	0.0834	0.71
0.45	0.0874	0.73

0.50	0.0895	0.74
0.55	0.0900	0.76
0.6	0.0901	0.79
0.65	0.0900	0.8
0.7	0.0897	0.8
0.75	0.0892	0.81
0.8	0.0884	0.82
0.85	0.0872	0.83
0.9	0.086	0.83
0.95	0.0848	0.84
1	0.0843	0.85
1.1	0.084	0.86
1.2	0.0838	0.87
1.3	0.0836	0.88
1.4	0.0835	0.89
1.5	0.0834	0.9
1.75	0.0833	0.91
2.00	0.0833	0.92
>2.0	0.0833	1

The shear stress at the connection with the anchor rod is calculated by the following formula:

$$\tau = \frac{\alpha T}{4t_s(h_s - 2t_s)} \le f_f^w \tag{5}$$

In the formula: τ is the shear stress at the connection of the stiffener and the anchor rod (MPa); f_f^w is the design value of the shear strength of the butt weld (MPa).

The compressive stress at the connection with the anchor plate is calculated by the following formula:

$$\sigma_f = \frac{\alpha T}{4t_s(R - r - 2t_s)} \le f_c^w \tag{6}$$

In the formula: σ_f is the normal stress at the connection of the stiffener and the anchor plate (MPa).

When designing, the material and size parameters of the internal anchor structure need to satisfy inequalities in Formulas (4) to (6) simultaneously.

3. Numerical simulation verification of internal anchor structure stress

Numerical simulation is used for comparative verification to verify the reliability of the theoretical analysis of the internal anchor structure of the anchor-type anchor rod. Taking the commonly used 120 mm and 150 mm anchor rod hole diameters in engineering as examples, the internal anchor structure is welded with Q355 steel. The finite element model of the internal anchor structure is established by using ABAQUS software. The anchor plate and stiffener are treated as line-elastic material models with an elastic modulus of 210 GPa and a Poisson's ratio of 0.3. Two scenarios are selected for comparative verification.

In Scenario 1, the anchor bar is a standard HPB300 steel bar with a diameter of 20 mm, and the anchor plate has a diameter of 100 mm. The model contains a total of 186,740 elements and 638,322 nodes. The inner ring surface of the anchor plate and the anchor bar are bound together, and the inner side surface and the upper side surface of the stiffener are separately bound to the anchor bar and anchor plate, simulating the welded connection between the anchor plate, stiffener, and anchor rod. The upper surface of the anchor bar is fully constrained. A vertically downward uniform load is applied to the anchor plate, with a load size obtained by the calculation of Formula (1), resulting in q=11.3 MPa. The thickness of the anchor plate that satisfies the design conditions is determined to be t=6 mm, with the stiffener thickness t_s=12 mm and height h_s=50 mm, according to the previous theoretical analysis calculations.

The maximum vertical deformation obtained by numerical simulation is only 0.17 mm (**Figure 3(a)**), which meets the requirement. The maximum normal stress at the connection point between the stiffening

Journal of Physics: Conference Series **2808** (2024) 012095 doi:10.1088/1742-6596/2808/1/012095

plate and the anchor rod obtained by numerical simulation is 68.3 MPa, and the maximum shear stress is 54.1 MPa. The maximum normal stress at the connection between the stiffening plate and the anchoring plate is 97.8 MPa. σ_s=71.3 MPa, τ=57.1 MPa, and σ_f=92.7 MPa obtained from Formulas (4) to (6) are very close to the results of numerical simulation, as shown in **Figure 4(a).** Meanwhile, **Figure 3(b)** shows the Mises stress cloud map of the internal anchor structure in Scenario 1. None of the stress in each part of the internal anchor structure has exceeded the strength design value of Q355 steel.

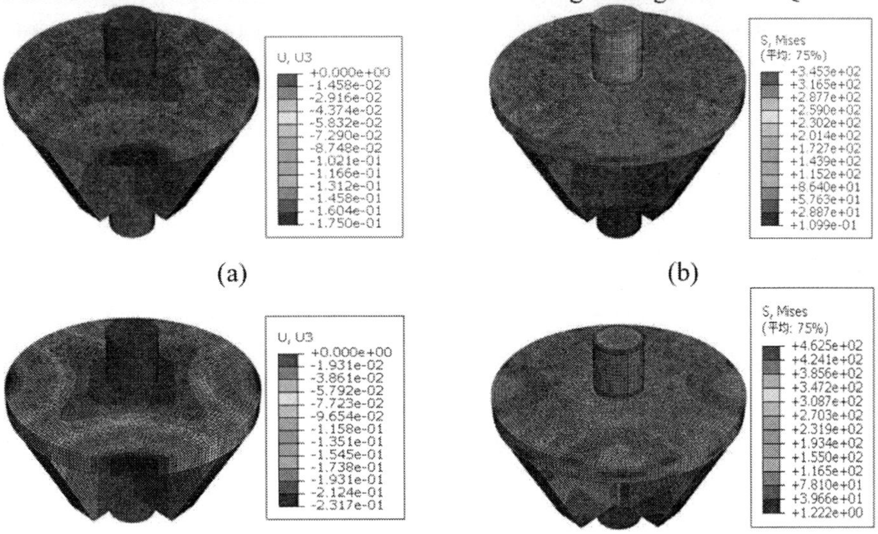

(c) (d)

Figure 3. Numerical simulation nephogram.

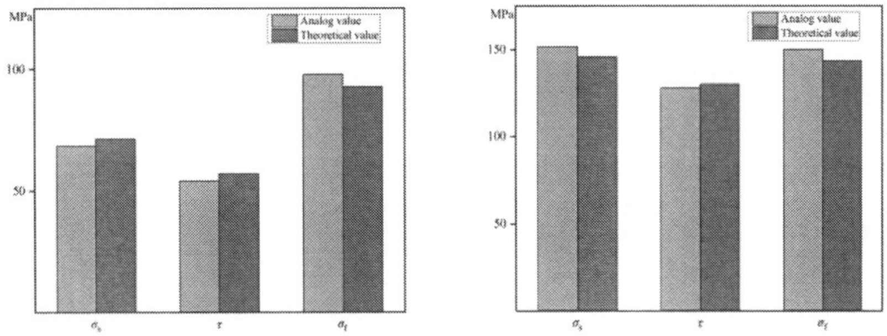

(a) (b)

Figure 4. Comparison chart between analog values and theoretical values.

In Scenario 2, where the anchor bar is an HRB500 steel bar with a diameter of 28 mm and the anchor plate has a diameter of 140 mm, there are a total of 64,460 elements and 76,709 nodes. The size of the uniform load is q=18.1 MPa, which is calculated from Formula (1). According to the theoretical analysis of the last section, the thickness of the anchor plate is determined to be t=10 mm, with the thickness of the stiffener t_s=14 mm, and the height h_s=60 mm. The maximum vertical deformation obtained from numerical simulation is only 0.23 mm (**Figure 3(c)**), which meets the requirement. The maximum normal stress value at the connection between the stiffener and the anchor rod obtained by numerical simulation is 151.7 MPa, the maximum shear stress value is 128.1 MPa, and the maximum normal stress value at the connection between the stiffener and the anchor plate is 150.3 MPa. The σ_s, τ, and σ_f calculated from Formulas (4) to (6) are 145.8 MPa, 130.2 MPa, and 143.5 MPa, respectively, which are very close to the results of the numerical simulation, as shown in **Figure 4(b).** Meanwhile, **Figure 3(d)** is the Mises stress cloud map of the internal anchorage structure in Scenario 2. None of the stresses in all parts of the internal anchor structure has exceeded the strength design value of Q355 steel.

4. Conclusion

The pressure-type anchor rod, due to its rational force distribution and high pull-out resistance, has been widely applied in engineering. The internal anchor structure is key to transferring external loads to the interior of the anchor rod. At present, related specifications have not explicitly provided the style of the internal anchor structure and the calculation method of force. Hence, this paper proposes an internal anchor structure type that welds an anchoring plate and a stiffening plate together. Using the theory of elastic thin plates to conduct a force analysis of the internal anchor structure under a uniform load, the design parameters of the internal anchor structure under a limited load can be determined. Based on this, the reliability of the method proposed in this paper has been verified further by numerical simulation. The results can provide a basis for the design of pressure-type anchor rods.

Acknowledgments

This work received financial support from the State Grid Zhejiang Electric Power Co., Ltd. Technology Project (SGZJJH00JJJS2250128).

References

[1] Tan G. Y., et al. (2021) Study on the structural design of full-length compression post-tensioned prestressed anti-floating anchor. *Building Structure*, 2021, 51(21): 7-10+6. DOI:10.19701/j.jzjg.2021.21.002.

[2] He L. B., Yang M. Z., Miao Y. W. (2022) Analysis on the Principle and Type Selection of Geotechnical Bolt Support in Engineering Construction. *Engineering and Technological Research*, 7(07): 59-62. DOI:10.19537/j.cnki.2096-2789.2022.07.017.

[3] Xue Y. Z., et al. (2022) Application and economic analysis of the rock anchor foundation in UHV transmission line. *Engineering Journal of Wuhan University*. 55(S1):142-147.

[4] Liu Y. J., Fu C. G. (2019) Analysis of Pressure Type Anchor Bolt Stiffness Based on Load Transfer Method and Shear Displacement Method. *Highway Engineering* 44(02): 41-44. DOI:10.19782/j.cnki.1674-0610.2019.02.008.

[5] Lu W., et al. (2024) Analytical method of interfacial stress transfer and bearing capacity of pressure-type anchorage system at the earthen site with crack sealed by grouting. *Chinese Journal of Theoretical and Applied Mechanics*, 56(1): 1-14 DOI: 10.6052/0459-1879-23-340.

[6] Konstantinos D. T., et al.(2016) Analytical Approach of Anchor Rod Stiffness and Steel Base-Plate Calculation Under Tension. *Structures*, 5: 207-218. https://doi.org/10.1016/j.istruc.2015.11.001.

[7] Zhang S., et al. (2023) Study on the Style Design and Anchoring Mechanism of Enlarged Head Anchors. *Sustainability*, 15, 8645. https://doi.org/10.3390/su15118645.

[8] Sawa T., Higurashi N., Akagawa H. (1991) A stress analysis of pipe flange connections. *Journal of Pressure Vessel Technology*, 4: 497-503. https://doi.org/10.1115/1.2928786.

[9] Chen J. L., Ma R. L., He M. J. (1999) The study of flange joint with stiffening rib under tension in tower and guy mast structures. *Structural Engineers*, 4: 16-20. CNKI:SUN:JGGC.0.1999-04-003.

[10] Zhang D. H., Jin X. H., Zeng C. (2011) Analysis of Calculation Formula of Transmission Tower Rigid Flange Joint under BendingMoment. *Electric Power Construction*, (3): 48-50.

AUTHOR INDEX

An, Erfeng 508
An, Ji 215
An, Le 496
An, Luyang 638
Ba, Shuhong 459
Bao, Ronghao 298
Bao, Yuanhao 320
Bi, Linnan 6
Cao, Liangqiang 342
Cao, Tong 1
Cao, Xinwei 578
Cao, Yongxin 99
Chen, Hao 148, 172
Chen, Haojie 23
Chen, Jianfu 396
Chen, Jinsong 648
Chen, Jun 130
Chen, Li 192, 209, 673
Chen, Ling 111
Chen, Peng 245
Chen, Suokun 473
Chen, Ting 616
Chen, Yang 99
Cheng, Huanchao 451
Cheng, Peng 60
Cheng, Wei 179
Cheng, Yao 601
Cui, Yan 252
Cui, Yu 282
Dai, Jingjing 86
Dan, Hanbo 741
Deng, Jiajia 465
Deng, Junyu 451
Deng, Zhian 356
Ding, Feng 1
Ding, Keqin 192
Ding, Miaojiang 438
Ding, Yang 290
Dong, Baoquan 23
Dong, Shangwei 578
Du, Jiahui 230
Duan, Jingjing 342
Duan, Tao 60
Duan, Wenbo 41
Fan, Pingcheng 172
Fan, Xinmei 209, 673, 679
Fan, Xuefei 48, 536
Fang, Houlin 546
Fang, Jun 525, 561
Feng, Guanghua 363
Feng, Wenliang 370
Feng, Wenxiu 489
Fu, Guanggang 314
Gao, Jun 160
Gao, Junlin 142
Gao, Rui 215
Gao, Yan 272
Gao, Yuan 729
Ge, Yuanyuan 536
Gong, Xun 23
Gong, Zihan 148
Guo, Donghui 200
Guo, Dongming 41
Guo, Fengwei 130
Guo, Jinpeng 679
Guo, Kuixiang 282
Guo, Qiang 370
Guo, Quanshi 546
Guo, Yong 741
Guo, Yulong 517
Guo, Yuzhu 653
Guo, Zhiquan 701
Guo, Zhiyun 290, 546
Han, Gaoxiao 148
Han, Hongwei 48
Han, Xiao 179
Han, Yinan 438
Hang, Peng 342
Hao, Qingyun 172
He, An 350
He, Shuaiqi 438
He, Song 648
He, Weidong 111
He, Wenbin 71
Hong, Xiaowen 48, 536
Hu, Danping 653
Hu, Haotian 444
Hu, Jiangchun 638
Hu, Jun 648
Hu, Yubin 628
Hu, Zhifang 223
Hua, Xiaoge 117
Huang, Shuhua 396
Huang, Su 187
Huang, Tao 342
Huang, Xuewei 638

Huang, Yunzhi 432
Ji, Chenli 419
Ji, Jun 314
Ji, Wendong 572
Jia, Xinlei 200
Jiang, Hangyan 601
Jiang, Meijuan 426
Jiang, Ming 282
Jiang, Senhua 78
Jiang, Shuxuan 356
Jiang, Yahong 444
Jiao, Haichao 23
Jin, Fuqiang 426
Jin, Runze 200
Jin, Xiaoyin 553
Jin, Xinlei 720
Kong, Fanfang 314
Li, Dejun 356
Li, Dongqin 584
Li, Fan 209, 673, 679
Li, Feiyin 685
Li, Fule 86
Li, Gang 370
Li, Hang 252
Li, Huanjun 306
Li, Jianping 187
Li, Jin 290
Li, Jiuyang 209, 673, 679
Li, Junliang 71
Li, Lei 508
Li, Lidan 148
Li, Lina 320
Li, Linpeng 459
Li, Mingfu 230
Li, Mingxuan 610
Li, Minxing 298
Li, Sen 272
Li, Wenbo 610
Li, Xian 13
Li, Xingyu 711
Li, Yongning 223
Li, Zhaoqiang 572
Liang, Saisai 685
Liao, Jiaxuan 6
Lin, Baosen 187
Lin, Shengcun 489
Lin, Ying 6, 692
Lin, Ziyue 245
Liu, Bangxian 508
Liu, Changkui 130
Liu, Chenwei 616
Liu, Cunxu 546
Liu, Fanghua 99
Liu, Hongwei 23
Liu, Huan 306
Liu, Jiang 553
Liu, Kai 259
Liu, Kun 71
Liu, Shaoxuan 252
Liu, Shuheng 215
Liu, Tiejun 230
Liu, Yaolu 525
Liu, Yuan 252
Liu, Yuepeng 432
Liu, Yuhan 610
Liu, Yuxin 334
Liu, Ziqi 215
Long, Xin 6
Lu, Baocheng 572
Lu, Lingfeng 584
Lu, Qiang 290, 546
Lü, Qing 741
Lu, Shiqiang 561
Lu, Zhenyang 13
Luo, Jingwei 209, 673, 679
Luo, Jinni 223
Luo, Xin 628
Luo, Yunfeng 566
Lv, Hao 154
Lv, Jianbo 334
Lyu, Ronghao 135
Ma, Jian 741
Ma, Luchao 701
Ma, Shaojie 685
Ma, Xin 503
Ma, Yinliang 419
Ma, Yuhua 628
Mao, Sen 370
Mei, Li 23
Mi, Dewei 60
Miao, You 99
Nie, Jingkai 314
Niu, Haoying 729
Niu, Jiahong 379
Niu, Jinhai 239
Pan, Feng 741
Pan, Futing 142
Peng, Jiahao 517
Qiao, Mengsun 223
Qin, Chao 48, 536
Qin, Hongbin 350
Qin, Jianan 6
Qing, Guanghui 662, 720
Qiu, Bo 1

Qiu, Haomiao....78
Qu, Min....252
Sang, Dengfeng....363
Shang, Wenxuan....561
Shao, Bing....503
Shao, Jiawei....99
Shen, Yu....264
Shi, Anshun....48, 536
Shi, Huifa....685
Shou, Lei....78
Song, Yonglun....13
Su, Shiding....363
Su, Weiwei....628
Su, Zhan....272
Sun, Biao....1
Sun, Hongchang....320
Sun, Long....272
Sun, Qiang....616
Sun, Sujie....48
Sun, Yuanping....166
Sun, Zhigang....616
Tan, Xin....692
Tang, Jia....610
Tang, Jubo....6
Tang, Shiying....546
Tang, Tong....685
Tian, Ce....578
Tian, Guizhong....592
Tian, Xue....166
Tian, Yunjing....223
Tian, Yunze....662
Tian, Zhaoping....396
Tian, Zheng....117
Tian, Zhimin....578
Wan, Tao....561
Wang, Bo....264
Wang, Chaohua....451
Wang, Chenxu....638
Wang, Guowei....379
Wang, Huiming....298
Wang, Jinsong....282
Wang, Ke....230
Wang, Kelu....561
Wang, Lingling....572
Wang, Lujie....536
Wang, Mingyuan....408
Wang, Penghui....432
Wang, Qin....60
Wang, Shun....200
Wang, Tao....111, 142
Wang, Tian....610
Wang, Xiaomei....503
Wang, Xiaoyu....209, 673, 679
Wang, Yajun....408
Wang, Yin....314
Wang, Ying....142
Wang, Yu....314
Wang, Yuntian....653
Wang, Zelong....200
Wang, Ziwen....239
Wei, Haifei....601
Wei, Kairan....592
Wei, Qiujie....653
Wei, Xiongbang....6
Weng, Yu....356
Weng, Zhan....230
Wu, Chao....451
Wu, Gang....356
Wu, Wenjin....130
Wu, Yahu....342
Xi, Qiushi....648
Xia, Kailang....48
Xiang, Rui....496
Xiao, Houjie....350
Xiao, Pengfei....566
Xiao, Yachao....628
Xie, Liming....160, 172
Xie, Quan....525
Xie, Zongming....370
Xing, Weijie....130
Xu, Baosheng....200
Xu, Gang....741
Xu, Lin....465
Xu, Yutian....379
Xue, Dawen....465
Xue, Ji....6
Xue, Xiaojie....396
Xue, Yaping....444
Yan, Jingke....601
Yan, ...342
Yang, Bo....482
Yang, Dedong....314
Yang, Haijun....572
Yang, Jianjun....438
Yang, Min....653
Yang, Ming....298
Yang, Peng....272
Yang, Shuang....356
Yang, Yan....379
Yang, Yiming....628
Yang, Yingying....166
Yi, Xiaolu....426
Yu, Chunlei....616
Yu, Guoji....536

Yu, Xinru 451
Yuan, Binxia 86
Zhai, Yusheng 272
Zhang, Bingbing 729
Zhang, Dezhi 546
Zhang, Geng 1
Zhang, Hao 426
Zhang, Jili 272
Zhang, Jinghui 230
Zhang, Kai 135
Zhang, Liang 396
Zhang, Liangxia 223
Zhang, Liangyong 290
Zhang, Lihui 130
Zhang, Lin 272
Zhang, Ling 179
Zhang, Min 616
Zhang, Qi 741
Zhang, Qiang 130
Zhang, Qinmin 473
Zhang, Ruiliang 272
Zhang, Shoulong 363
Zhang, Suzhou 154
Zhang, Weihua 601
Zhang, Wengui 148
Zhang, Xiaolong 29
Zhang, Ying 653
Zhang, Yingzhong 685
Zhang, Yongliang 503
Zhang, Yushan 572
Zhang, Zengbao 334
Zhang, Zerui 187
Zhang, Zhaoyang 179
Zhang, Zhifeng 272
Zhang, Zhiyong 386
Zhang, Zicheng 209, 673, 679
Zhang, Zuguo 408
Zhao, Bingyao 41
Zhao, Chuangxin 370
Zhao, Gang 282
Zhao, Guilin 215
Zhao, Haitao 23
Zhao, Hui 23
Zhao, Lei 592
Zhao, Lijuan 223
Zhao, Mingxia 93
Zhao, Wenjie 465
Zhao, Xianfeng 517
Zhao, Xin 426
Zhao, Yongfeng 451
Zheng, Chong 314
Zheng, Huifeng 408
Zheng, Tao 200
Zheng, Xiong 282
Zhou, Chunping 566
Zhou, Geng 628
Zhou, Jialiang 685
Zhou, Jinyu 444
Zhou, Ning 200, 601
Zhou, Ping 41
Zhou, Xinmeng 306
Zhou, Zhiliang 200
Zhu, Dongpo 592
Zhu, Xiaoming 334
Zhu, Yongmei 154
Zhu, Yuepeng 209, 673, 679